Coordinate Systems of the World

A comprehensive consolidation of data for the world, this book gives a short precis of each nation, each nation's history, its topography and a chronology of the development of geodetic surveying and coordinate systems for that specific nation. This book is a starting point of information for understanding the world's datums and grids.

Based on the details available for each nation, the reader is given an overall view that can answer questions regarding the sources of spatial information available, their limitations, and the critical things to be aware. The topographic maps compiled over the centuries represent the mixes of technology specifically to that nation. The book provides information and clues regarding existing maps and how those maps and coordinate systems were created.

Coordinate Systems of the World
Datums and Grids

Clifford J. Mugnier

CRC Press
Taylor & Francis Group
Boca Raton London New York

CRC Press is an imprint of the
Taylor & Francis Group, an **informa** business

First edition published 2023
by CRC Press
6000 Broken Sound Parkway NW, Suite 300, Boca Raton, FL 33487-2742

and by CRC Press
4 Park Square, Milton Park, Abingdon, Oxon, OX14 4RN

CRC Press is an imprint of Taylor & Francis Group, LLC

© 2023 Clifford J. Mugnier

Reasonable efforts have been made to publish reliable data and information, but the author and publisher cannot assume responsibility for the validity of all materials or the consequences of their use. The authors and publishers have attempted to trace the copyright holders of all material reproduced in this publication and apologize to copyright holders if permission to publish in this form has not been obtained. If any copyright material has not been acknowledged please write and let us know so we may rectify in any future reprint.

Except as permitted under U.S. Copyright Law, no part of this book may be reprinted, reproduced, transmitted, or utilized in any form by any electronic, mechanical, or other means, now known or hereafter invented, including photocopying, microfilming, and recording, or in any information storage or retrieval system, without written permission from the publishers.

For permission to photocopy or use material electronically from this work, access www.copyright.com or contact the Copyright Clearance Center, Inc. (CCC), 222 Rosewood Drive, Danvers, MA 01923, 978-750-8400. For works that are not available on CCC please contact mpkbookspermissions@tandf.co.uk

Trademark notice: Product or corporate names may be trademarks or registered trademarks and are used only for identification and explanation without intent to infringe.

ISBN: 9781032310343 (hbk)
ISBN: 9781032310404 (pbk)
ISBN: 9781003307785 (ebk)

DOI: 10.1201/9781003307785

Typeset in Times
by codeMantra

This book is dedicated to the love of my life and wife, Miranda Hirezi Mugnier and to my children: Gaston (Gus), André, Jacques, Monique, Theodore (Ted), Philippe, Alyce and to my step children: Sami, Natalie, and Omar, as well as to all nineteen (for now) grandchildren.

Contents

Foreword xvii
Preface.. xix
Acknowledgments xxiii
Author xxvii

A *1*

Islamic State of Afghanistan 3

Republic of Albania 9

Democratic and Popular Republic of Algeria 13

Principality of Andorra 19

The Republic of Angola 23

Antigua and Barbuda 27

The Republic of Argentina 31

Republic of Armenia 35

Aruba and the Netherlands Antillies 39

Commonwealth of Australia 43

Republic of Austria 49

Republic of Azerbaijan 53

B *57*

Commonwealth of the Bahamas 59

State of Bahrain 63

People's Republic of Bangladesh 67

Barbados 71

Republic of Belarus 73

Kingdom of Belgium 77

Belize 81

Republic of Bénin 83

Kingdom of Bhutan 89

Republic of Bolivia 93

Bosnia and Herzegovina 97

Republic of Botswana 101

Federative Republic of Brazil 105

Negara Brunei Darussalam 109

Republic of Bulgaria 113

Burkina Faso 119

State of Burma 123

Republic of Burundi 129

C *133*

Kingdom of Cambodia 135

Republic of Cameroon 139

Canada 145

Republic of Cape Verde 149

Cayman Islands 153

Central African Republic 157

Republic of Chad 161

Republic of Chile 165

The People's Republic of China 169

Republic of Colombia 175

Federal Islamic Republic of the Comores 179

Democratic Republic of the Congo (Kinshasa) 183

Republic of the Congo (Brassaville) 187

Republic of Costa Rica 191

Republic of Croatia 195

Republic of Cuba 199

Republic of Cyprus 203

The Czech Republic 207

D 213

Kingdom of Denmark 215

Republic of Djibouti 219

Commonwealth of Dominica 223

Dominican Republic 225

E 227

The Republic of Ecuador 229

Arab Republic of Egypt 233

Republic of El Salvador 237

Republic of Equatorial Guinea 241

Republic of Estonia 245

Federal Democratic Republic of Ethiopia 249

F 253

The Republic of Fiji Islands 255

Republic of Finland 259

The French Republic 263

Overseas Lands of French Polynesia 267

G 273

Republic of Gabon 275

Republic of The Gambia 279

Georgia 283

Federal Republic of Germany 287

The Republic of Ghana 295

Gibraltar 299

The Hellenic Republic 301

Grenada 305

The Department of Guadeloupe 309

Territory of Guam 313

Republic of Guatemala 317

Republic of Guinea 321

Republic of Guinea-Bissau 325

Co-operative Republic of Guyana 327

H 331

Republic of Haïti 333

The Republic of Honduras 337

Hong Kong 341

The Republic of Hungary 345

I 349

Republic of Iceland 351

Republic of India 355

Republic of Indonesia 361

Islamic Republic of Iran 367

Republic of Iraq	371
Ireland	375
The State of Israel	379
Italian Republic	385
Republic of Côte D'Ivoire	389

J — 393

Jamaica	395
Japan	401
Hashemite Kingdom of Jordan	405

K — 409

Republic of Kazakhstan	411
Republic of Kenya	415
Kiribati	421
The Republic of Korea	427
State of Kuwait	431
Kyrgyz Republic	435

L — 439

Lao Peoples Democratic Republic	441
The Republic of Latvia	445
The Lebanese Republic	451
Kingdom of Lesotho	455
Republic of Liberia	459
Great Socialist People's Libyan Arab Jamahiriya	465
Principality of Liechtenstein	469

Republic of Lithuania 473

Grand Duchy of Luxembourg 477

M 481

Macau Special Administrative Region 483

Republic of Macedonia 487

The Republic of Madagascar 493

Republic of Malawi 497

Malaysia 501

Republic of Maldives 505

Republic of Mali 509

Republic of Malta 513

Republic of the Marshall Islands 517

Department of Martinique 521

Islamic Republic of Mauritania 523

The Republic of Mauritius 527

United Mexican States 531

Federated States of Micronesia 535

Republic of Moldova 539

Principality of Monaco 543

Mongolia 547

Montenegro 551

Montserrat 553

The Kingdom of Morocco 555

The Republic of Moçambique 559

N 563

Republic of Namibia 565

Kingdom of Nepal 569

The Kingdom of the Netherlands 573

New Zealand 577

Republic of Nicaragua 583

Republic of Niger 587

Federal Republic of Nigeria 591

Niue 595

The Kingdom of Norway 599

O 605

Sultanate of Oman 607

P 611

Islamic Republic of Pakistan 613

Republic of Panamá 617

Independent State of Papua New Guinea 621

Republic of Paraguay 625

Republic of Perú 629

The Republic of the Philippines 633

The Republic of Poland 637

Kingdom of Portugal 641

Q 645

State of Qatar 647

R **649**

România 651

Russia Federation 657

Republic of Rwanda 663

S **667**

Sakhalin Island 669

The Islands of Samoa 675

Democratic Republic of São Tomé and Príncipe 679

Kingdom of Saudi Arabia 683

Republic of Senegal 689

Republic of Serbia 693

Republic of Seychelles 697

Republic of Sierra Leone 701

Republic of Singapore 705

Slovak Republic 709

Republic of Slovenia 713

Solomon Islands 717

Federal Republic of Somalia 721

Republic of South Africa 725

The Kingdom of Spain 731

Democratic Socialist Republic of Sri Lanka 735

Federation of Saint Kitts and Nevis 739

Saint Lucia 741

Saint Vincent and the Grenadines 743

Republic of the Sudan 749

Contents

Republic of Suriname 755

Kingdom of Swaziland 759

The Kingdom of Sweden 763

The Swiss Confederation 769

The Syrian Arab Republic 773

T *781*

Republic of China (Taiwan) 783

Republic of Tajikistan 787

United Republic of Tanzania 791

Kingdom of Thailand 795

Democratic Republic of Timor-Leste 799

Togolese Republic 803

Kingdom of Tonga 807

The Republic of Trinidad and Tobago 811

Tunisian Republic 815

Republic of Turkey 819

Turkmenistan 825

Tuvalu 827

U *833*

Republic of Uganda 835

Ukraine 839

United Arab Emirates 843

United Kingdom 847

United States of America 853

Oriental Republic of Uruguay 859

Republic of Uzbekistan 863

V *867*

Republic of Vanuatu 869

The Bolivarian Republic of Venezuela 873

The Socialist Republic of Vietnam 877

Y *881*

Republic of Yemen 883

Z *885*

Republic of Zambia 887

Republic of Zimbabwe 891

Appendix: Yugoslavia 895

Index 897

Foreword

Clifford Mugnier's enjoyment and passion of geodesy and *Coordinate Systems of the World* have resulted in this comprehensive product from around 20 years of close to 200 monthly articles on the grids and datums of every country in the world in the *Photogrammetric Engineering & Remote Sensing* journal.

I first knowingly corresponded with Clifford back in the mid-1990s around the time this "Project" to document the world's geodesy, country by country began.

We met because of our passionate interest in geodesy, initially by correspondence back in the early days of the internet. We did finally meet in person in 2002, when I stayed with him after I traveled to the United States to present a paper at ION in Portland, Oregon.

Having been involved with a number of oil and gas, and mineral exploration companies here in Australia and internationally, I have come across a number of geodetic/coordinate "issues" where details from a number of Clifford's "Coordinate Systems" articles have helped time and again in resolving them, especially where the name of a Grid, Ellipsoid, or Datum, sounds right, but the numerical details are incorrect or vice versa with a similarly named Grid, Ellipsoid, or Datum. This was especially so when the source of the information is dubious, but is all you have.

This is especially so, where, in some countries "Coordinate Systems" information is dealt with secrecy, and only by having access to details in this book have solutions been able to be resolved.

Another item that came up along the way was the different conversion factors used converting from feet to metres. Fortunately, today, there are only three countries still using imperial units: the United States, Liberia, and Myanmar.

Clifford, had an earlier career where he was engaged in research for mapping from the CORONA spy satellite system and worldwide topographic mapping, which has given him an eye and talent for detail and precision.

Given Clifford's background with several different government agencies over the years and military mapping services, he has covered many aspects of mapping, giving him exceptional credentials and experience for this book. And the fact that he has used many of the well-qualified people worldwide to help and verify the details.

I would definitely have this book on my shelves very close to me and used as a constant reference as I have done with Clifford's columns over the last 24 years. I have used a lot of what Clifford has written as reference material for mostly exploration clients all around the world in that period.

Malcolm A.B. Jones, F.S.S.S.I.
Perth, Australia

Preface

Grids and Datums comprise the systems of coordinates used by all large-scale maps for every country of the world. Every country has its own system of Latitude and Longitude, and most of these systems originated some time ago, mostly during the 18th and 19th centuries. Where did they originate? Usually some point important to local citizens was chosen, and observations of the stars were made over a period of time to establish the astronomical Latitude (Φ) of the point. In the Northern Hemisphere, it's pretty easy to get astronomical Latitude as the angle "up" from the horizon to the North Star, Polaris. The Longitude (Λ) of the point, however, must be a function of time ($24^H = 360°$), sometimes the zero point or Prime Meridian was chosen at that local point, and other times one is already established by a nearby country. These starting points represented the origins of national coordinate systems, commonly at local national astronomical observatories. For instance, some classical horizontal datum origins that had their own Prime Meridians include Amersfoort, Netherlands; Bogotá, Colombia; Dehra Dun, India; Tokyo, Japan; Madrid, Spain; Athens, Greece; Quito, Ecuador; Ferro, Canary Islands; Singkawang, Borneo; Potsdam, Germany; and Greenwich, England.

Centuries ago, optical instruments for precisely measuring angles were developed long before the advancement of technologies were developed for precisely measuring distances. In order to compile a map of an area, it is necessary to have some known points positioned about the perimeter of the area with respect to each other such that everything mapped is a product of interpolation. The then available instruments could measure angles relatively easily, but what about distances to provide a consistent scale? During this period, measuring a single distance of a few kilometers involved weeks to months of labor in order for check measurements to verify accuracies of mere centimeters in a few kilometers! Something had to be done to minimize the physical measurement of great distances because of the enormous amount of labor and time required. Willebrord Snell (1580–1626) used the Law of Sines: $a/\sin A = b/\sin B = c/\sin C$, and that allowed the computation of distances of the sides of a triangle with only one known distance and the known interior angles of a triangle. Starting with an origin point (Φ_o, Λ_o), a measured distance to a second point at azimuth (α_o) with reference to true North, one can compute the geodetic coordinates (φ, λ) of the second point, and with measured angles to a third point one can *compute* the distance to that third point and determine the geodetic coordinates (φ, λ) of that third point. This process has been repeated millions of times by nations of the world; each nation using its own starting point or origin. Establishing known points of control through geodetic triangulation has allowed topographers to go about the countryside to develop reliable maps that were (and are) used by governments for equitable systems of taxation and for military defense. This simple concept has been used over the centuries throughout the world. Note that most everyone has used a different starting point, and as a result, everyone has a coordinate system that is consistent only *within* their own country or region. The system of geodetic coordinates (φ, λ) of a nation is called a datum. When developed from an astronomically derived origin point (Φ, Λ), it is called a *classical* datum. Most nations continued their geodetic surveys over the decades, and often previously surveyed points were reobserved with newer instruments and technologies resulting in new mathematical adjustments being performed. The old datums were then renamed, commonly only with a new date. It is common to find several names of datums with the only differences being the years of new adjustments – that usually means that the actual positions have been changed slightly to better fit the new observations. The date does make a difference when it comes to datums.

When various datums do not match each other, as a result maps of one country's border will never match an adjacent country's border – there's always a discontinuity, a gap, or a mismatch. Finding the relationship between two adjacent nations' datums has been the objective of many espionage operations over the decades (and centuries). In some parts of the world, such knowledge remains a military secret. Examples of several countries that still classify this topic as secret are given herein. Since the advent of artificial geodetic satellites like the U.S. NAVSTAR Global Positioning System (GPS), such practical use for military secrecy is entirely a waste of time and continues more out of tradition than actual need. In the mid-1990s, the war in Bosnia demonstrated that precision-guided ordinance can target which window in which floor can be targeted and hit from a launch site from over the horizon. As one U.S. General Officer once let slip, the accuracy worldwide (horizontally and vertically) is ±6 ft (±2 m)! The reality is that although such military secrecy is silly in the 21st century, it still exists. Be aware of this fact in several nations of the world and mind your personal cell phone and/or personal GPS receiver!

Centuries ago, Christiaan Huygens invented the pendulum to regulate a clock. Decades later, based on the time keeping variation of a pendulum-regulated clock when moved north or south some distance, the Englishman Sir Isaac Newton thought that the earth's shape approximated an *oblate* ellipsoid of revolution (flattened at the poles) because of variations in gravity. The Frenchman Cassini thought the opposite in that the shape approximated a *prolate* ellipsoid of revolution (prolonged at the poles) based on measuring north-south distances with triangulation. French expeditions were sent to Sweden and to Ecuador to measure north-south arcs of the meridians, and to the consternation of the French scientists, Newton was proven to be correct! The first ellipsoid of revolution ever published was by the French in the early 1700s. Ever since then, north-south surveyed triangulation arcs have resulted in hundreds of different ellipsoids being computed and published; that is a critical element in the definition of a datum whether it is a classical datum or a modern inertial datum such as WGS84. Why so many ellipsoids? It is because of tiny variations in the earth's gravity field that vary from point to point. Sometimes it's also due to different national standards of physically measuring lengths. The gravity variations are due to the variations in the density of the earth's crust (mud *vs.* limestone *vs.* granite, *etc.*). When a surveyor sets up an optical instrument with a tripod, the surveyor adjusts the instrument and "levels" it using a level bubble with respect to what is apparently "down." However, "down" is not necessarily pointing to the center of the earth; it is more likely pointing "down" somewhere else that has been *deflected* by the local gravity field. Now this deflection doesn't matter when surveying which way water is going to flow. When we are measuring angles to *compute distances* with the Law of Sines, it *does* matter, and is the primary reason why errors accumulate in triangulation chains. Different distances computed in north-south directions (with unknown errors due to variations in the gravity field) result in different ellipsoids being computed for different datums. Gravity has been a major topic of research in geodesy for centuries, and some classical datums even define the tilt of the local gravity field at the datum origin (ξ, η). What completely defines a classical geodetic datum? It's astronomic latitude at the origin (Φ_o), astronomic longitude at the origin (Λ_o), clockwise astronomic azimuth at the origin from true north to a reference point (mire) on the ground (α_o), the semi-major axis of the ellipsoid of revolution (a), the reciprocal of flattening of the ellipsoid ($1/f$) or the semi-minor axis of the ellipsoid of revolution (b), the ellipsoid height of the origin point (h_o), the elevation above mean sea level of the origin point (H_o), and the deflection of the vertical (gravity direction) of the origin point (ξ_o, η_o). When we have the deflection of the vertical *known* at the origin point, then we also have the *geodetic* coordinates of the origin point (φ_o, λ_o). With the exception of a few major continental classical datums, we rarely know all of the *observed* parameters of a classical datum. More frequently than not, many of the theoretical defining parameters are unknown and are just assumed to be zero. Nowadays, modeling the gravity field over a region is said to be modeling the *geoid* to compute elevations from GPS observations. This generally introduces the concept of a map datum, the foundation of all modern large-scale maps. How we use a modern large-scale map involves the manner which the coordinate system of the map is used. The "universal system" of coordinates nowadays is what's possible because of the space age; that is artificial satellite systems such as GPS. For all intent and purposes, the GPS satellites have zero error. None of the existing classical datums match the GPS datum (called WGS84), but through research we have found how to transform among many classical datums and WGS84. Transformation parameters from local native coordinate systems (datums) to WGS84 are known for many local systems and are listed as (ΔX, ΔY, ΔZ). Some relations are intended for more precise applications and more parameters are listed herein.

When beginning to work with a new data set, be it on paper or in digital form the most common coordinate system encountered is a plane Cartesian (X, Y) coordinate system called a "grid." Grid coordinates are commonly favored because distances and directions are easy to work with as ellipsoidal coordinates (φ, λ) are difficult to work with. There used to be a knotty problem that was all the rage in 19th century Europe called "The Principal Problem of Geodesy." Given the (φ, λ), distance, and azimuth (α) to a point, compute the (φ, λ) of the second point for the Direct case, and given the (φ, λ) of Point 1 and the (φ, λ) of Point 2, compute the distance and azimuths between the two points for the inverse case. The problem involves elliptic integrals of the second kind in three dimensions along the surface of the ellipsoid of revolution. Taught to compute the inverse case with a mechanical calculator, the process used to take cartographers 30–45 minutes. Before the invention of mechanical calculators, the manual solution with tables of logarithms and trig tables was even more arduous. In wartime conditions, finding the azimuth and distance to a distant point is a critical item for an artillery officer, and time can cost lives. The practical solution for the soldier is to utilize grid coordinates for calculations rather than (φ, λ). The same goes for cartographers, engineers, surveyors, and tax collectors in peacetime. The relation of the coordinate systems can be expressed simply as: $f(\varphi, \lambda) \rightarrow (X, Y)$, where f is some mathematical function to map a 3D ellipsoidal surface onto a plane (flat) Cartesian surface.

In other words, that's what map projections are for! Cartographers and geodesists have developed many map projections for practical applications in mapping that are directed to "best cases" for different shapes of areas of interest. Sometimes map projections have been used for decades for the simplest of reasons such as being easy to graphically construct in a field tent rather than in a drafting room. Grid coordinates are easier to use than geodetic coordinates in that verbal communications are facilitated, centesimal parts are simpler than sexagesimal parts to work with to plot by hand, and computations on a plane are simple in comparison to a curved surface. For large-scale topographic mapping and geodetic computations, every kind of map projection in the world is listed herein. The mathematical equations are not covered here as they have been covered extensively in the literature and are trivial to find through the internet. Variations and exotic applications are discussed as well as legal ramifications because rights of access, national boundaries, mineral lease boundaries, grid boundaries, nautical boundaries, etc. are extant through mathematical functions. Projection types include the ancient aphylactics such as the polyconic, the polyhedric, the Cassini-Soldner, and the azimuthal equidistant, the conformal and partially conformal such as the conic (multiple types), the Mercator, the Transverse Mercator (multiple types), and the Oblique Mercator (multiple types). Some systems are applicable worldwide such as the ubiquitous UTM and the like, some are only found in one fly-speck island or another. There are over a thousand datums worldwide; there's ten times that for Grid systems.

I started at Louisiana State University in the fall of 1961 with the intention of studying electrical engineering to pursue a career in television stage lighting. After speaking to several upper classmen studying engineering, I changed my mind and decided on a career as a cartographer. After a couple of years of study at LSU I dropped out for a year and found employment as a map draftsman at Offshore Navigation, Inc., a radio-positioning company that served the oil exploration industry and had operations worldwide. I enjoyed the work and noticed that the one group in the company that had the highest respect among the company professionals was the Grid Computations section. It was that which sparked my lifelong interest in coordinate systems and the myriad variations encountered. I completed my undergraduate studies, worked for a while as a civilian cartographer for the Air Force, and soon found myself in the Army Corps of Engineers during the Vietnam War. The Commander of Army Map Service requested that I be assigned under his command, where I spent several years as a Topographic Engineer learning photogrammetry while in the CORONA spy satellite mapping program. After the Army, I worked in several cities in commercial photogrammetric mapping until I entered private practice as a consultant and took part-time employment teaching the mapping sciences at the University of New Orleans. Two decades and seven children later,

I accepted a full-time position with Louisiana State University to teach the mapping sciences, and to continue research in the physical geodesy (gravity) of subsidence in the State of Louisiana and publishing *Grids & Datums*. The columns in *Photogrammetric Engineering and Remote Sensing* were completed for the world, and I figured that my monthly publishing days were pretty much over when the Society suggested that I start updating my old columns! A few years later, another suggestion was that the columns be collected into one volume.

Clifford J. Mugnier
Baton Rouge, Louisiana

Acknowledgments

My first job in mapping was thanks to Mr. Bill Parkhurst, and over the years I have been mentored and helped by many people, among whom were: Prof. John McIntire, Les Schroeder, Bob Schmidt, Allen Cuny, Rudy Lambert, Bill Ose, Al Poppe, Russell Duty, Jerry Hart, Eugene Dells, Louis Duet, Steven R. Baker, Prof. Bill Culp, COL. John Conard, COL. Frank Spacek, COL. John R. Oswalt, Jr., Don Light, Zeno V. Kittrell, Oscar Kluh, Elwin Condon, Betty Russell, Scott Rae, Pauline Smith, Cynthia Buccy, Valerie Namasian, Paul Farrington, Jr., Steve E. Anderson, Benny Munroe, Phil Bizzoco, Ernest Rabb, Jessie Schreiter, Morton Keller, Robert Zurlinden, Carl L. Mistric, Prof. Phil Larimore, Don Eames, Mark Huber, Prof. Michael E. Pittman, Prof. Kenneth L. McManis, John P. Snyder, Kim Tilley, Rae Kelley, Dr. Jim Case, Stan Morain, John W. Hager, J. A. Cavell, Mark Nettles, Otto Churio, Robert Frost, Russell Swan, Leopold Romeijn, Malcolm A.B. Jones, Dr. Muneendra Kumar, Melita Kennedy, Joseph Dracup, Dave Doyle, Jim Stem, Juliana Broadwell, Dave Zilkoski, Gilbert Mitchell, Prof. Javier Urquizo, Roberto Urquizo, Ivan Medina, Prof. Kazuo Kobayashi, Hampton Peele, Prof. George Z. Voyiadjis, Prof. Roy K. Dokka, Jim Brumfield, Randy L. Osborne, Vasily P. Dubinin, Jon Cliburn, Larry Dunaway, Rachel R. Barnett, Erin Verpil, Prof. Gábor Timár, Prof. Ahmed Abdalla, Prof. Irwin Scollar, Prof. Hang Guo, Jim Cain, Prof. Michael Barnes, Craig Rollins, Roger Lott, Jānis Balodis, Prof. Ansis Zariņš, and Dr. Katerina Morosova. Finally, special notice must be given to the Central Intelligence Agency for the Public Domain color maps of all the countries covered herein as well as to the Perry-Casteñeda Map Collection of the University of Texas.

LIST OF SYMBOLS USED

Φ_o = Astronomic latitude of the origin point
Λ_o = Astronomic longitude of the origin point
φ = geodetic latitude of a point
λ = geodetic longitude of a point
h_o = ellipsoid height of the origin point
H_o = elevation of the origin point
α_o = azimuth to another point from the origin point
a = semi-major axis of the ellipsoid of choice
b = semi-minor axis of the ellipsoid of choice
$1/f$ = reciprocal of flattening of the ellipsoid of choice
ξ_o = deflection of gravity in the plane of the meridian of the origin point
η_o = deflection of gravity in the prime vertical of the origin point

REFERENCE ELLIPSOID NAMES AND CONSTANTS USED FOR DATUM TRANSFORMATIONS

Reference Ellipsoid Name	a (meters)[a]	1/f
Airy 1830		
(OSGB 1936)	6377563.396	299.3249646
Modified (Ireland 1965)	299.3249646	299.3249647
Andræ (Old Danish, Old Icelandic)	6377104.43	300
Australian National 1966	6378160	298.25
Bessel 1841		
Ethiopia, Indonesia, Japan, Korea	6377397.155	299.1528128
Namibia	6377483.865	299.1528128
Norway (NGO 1948)	6377492.018	299.1528
Clarke 1858		
Cyprus 1935	6378235.6	294.2606768
Australia (old)	6378293.645	294.26
Clarke 1866		
North America, Central America, Caribbean, and Portuguese Africa	6378206.4	294.9786982
Clarke 1880		
Modified (Arc 1950)	6378249.145	293.465
Cape (South Africa)	6378249.145	293.4663077
Palestine	6378300.782	293.4663077
IGN (France, N. Africa, Syria)	6378249.2	293.4660208
Delambre 1810		
Carte de France (old)	6376985	308.64
Belgium (old)	6376985.228	308.64
Everest 1830		
Indian 1916	6377276.345	300.8017
Brunei, Sabah, Sarawak	6377298.556	300.8017
Indian 1956	6377301.243	300.8017
Pakistan	6377309.613	300.8017
Singapore (Revised Kertau 1948)	6377304.063	300.8017
Malayan RSO	6377295.664	300.8017
Fischer 1960		
South Asia Datum	6378155	298.3
Mercury Datum	6378166	298.3
Fischer 1968 Modified Mercury Datum	6378150	298.3
Geodetic Reference System 1967	6378160	298.2471674
Geodetic Reference System 1980	6378137	298.257222101
Hayford (International 1909, Madrid 1924)	6378188	297
Helmert 1906	6378200	298.3
Hough (Wake-Eniwetok 1906)	6378270	297
Krassovsky 1940	6378245	298.3
Krayenhoff 1827	6376950.4	309.65
NWL-8E (Doppler precise ephemeris)	6378145	298.25
Plessis		
Modified (France map graticules)	6376523	308.64
Reconstituted (*Nord de Guerre*)	6376523.994	308.624807
South American 1969	6378160	298.25

(Continued)

Acknowledgments

Reference Ellipsoid Name	a (meters)[a]	1/f
Struve 1860 (old Spain)	6378298.3	294.73
Svanberg (old Sweden)	6376797	304.2506
Walbeck 1819 (Russia, old Bulgaria)	6376896	302.78
War Office 1926 (McCaw 1924)	6378300	296
World Geodetic System		
1966	6378145	298.25
1972	6378135	298.26
1984	6378137	298.257223563

The number of significant digits used for defining constants may be nationally legislated based on unit conversion criteria and customs. These are normally treated as exact values for coordinate transformation applications.

[a] Some countries in some decades have parameters that reflect non-standard distance measuring lengths. When they exist, notes will be found in discussions of that country.

Author

Clifford J. Mugnier

Born at the French Hospital in New Orleans, Louisiana, he was educated in elementary school in New Orleans and in high school in Balboa, Panama Canal Zone. He attended Louisiana State University in Baton Rouge, LA for 2 years and later transferred to Northwestern State University in Natchitoches, LA. He worked part-time as a Cartographic Draftsman for Offshore Navigation, Inc., in New Orleans. After graduation with a Baccalaureate in Physical Geography and Mathematics, he was employed by the U.S. Air Force Aeronautical Chart and Information Center in St. Louis, Missouri. Trained as a Professional Cartographer, he was issued an Induction Notice for the Armed Forces and enlisted in the U.S. Army. After graduation from Explosives Demolition course and The Engineer School, he was commissioned into the Army as a junior officer. Through an Air Force recommendation, he was assigned to the U.S. Army Map Service where he served 3 years as a Topographic Engineer in production-oriented research in geodesy and photogrammetry for extraterrestrial mapping and for the CORONA satellite program. He was a company commander, and as Captain was the executive secretary of the Army Topographic Scientific Advisory Committee. After completing military service he worked for the Autometric Operation, Raytheon Company in Wayland, Massachusetts as chief of photogrammetric triangulation for 2 years and then moved to Baton Rouge as a manager of the photogrammetry division of Owen and White, Inc., Consulting Engineers. Four years later, he moved to New Orleans and began full-time consulting in geodesy and forensic photogrammetry. He started part-time teaching surveying, geodesy, and photogrammetry at the University of New Orleans for 20 years. In the middle to late 1980s, Mugnier was engaged in geodetic applications for numerous countries in South America as well as precise leveling and relative gravity surveys in metro New Orleans. In the 1990s, he worked in medical applications of high-speed photogrammetry and X-ray photogrammetry for the U.S. Navy Medical Command ejection seat experiments using human and primate test subjects. The late 1990s saw research in using digital cameras for dimensional control in the shipbuilding industry and terrestrial LIDAR for shipbuilding and the chemical process industry. In 2000, he moved to Baton Rouge and started full-time teaching at Louisiana State University. His research continued in the coordinate systems of the world and in the physical geodesy (relative and absolute gravity as well as deflection of the vertical) of subsidence within the State of Louisiana. He is Chief of Geodesy (*Emeritus*) at the LSU Center for GeoInformatics and continues to consult in Forensic Photogrammetry for civil and criminal cases. He has seven children, three stepchildren, and nineteen grandchildren and is an Expert in classical American Precision Target Pistol competition.

Islamic State of Afghanistan 3
Republic of Albania 9
Democratic and Popular Republic of Algeria 13
Principality of Andorra 19
The Republic of Angola 23
Antigua and Barbuda 27
The Republic of Argentina 31
Republic of Armenia 35
Aruba and the Netherlands Antillies 39
Commonwealth of Australia 43
Republic of Austria 49
Republic of Azerbaijan 53

Islamic State of Afghanistan

Afghanistan is comprised mostly of rugged mountains with plains in the north and southwest. Slightly smaller in area than Texas, it borders China (76 km), Iran (936 km), Pakistan (2,430 km), Tajikistan (1,206 km), Turkmenistan (744 km), and Uzbekistan (137 km). The lowest point is Amu Darya (258 m), and the highest point is Nowshak (7,485 m). The Hindu Kush Mountains run southeast to northeast and divide the northern provinces from the rest of the country. The highest peaks are in the northern Vakhan (Wakhan Corridor), and this finger of the country reaches between Tajikistan and Pakistan to connect with China. Thanks to the Library of Congress: "Mountains dominate the landscape, forming a terrigenous skeleton, traversing the center of the country, running generally in a northeast-southwest direction. More than 49 percent of the total land area lies above 2,000 meters. Although geographers differ on the division of these mountains into systems, they agree that the Hindukush [sic] system, the most important, is the westernmost extension of the Pamir Mountains, the Karakorum Mountains, and the Himalayas. The origin of the term Hindukush [sic] (which translates as Hindu Killer) is also a point of contention. Three possibilities have been put forward: that the mountains memorialize the Indian slaves who perished in the mountains while being transported to Central Asian slave markets; that the name is merely a corruption of Hindu Koh, the pre-Islamic name of the mountains that divided Hindu southern Afghanistan from non-Hindu northern Afghanistan; or, that the name is a posited Avestan appellation meaning water mountains."

In early times, Afghanistan formed part of the empires of Persia and of Alexander the Great. The Turkoman dynasty was set up at Ghaznī in the 10th century and was conquered by the Turkic ruler Timur in the 15th century. A steady series of wars and conquering invaders have passed through the country ever since. The Taliban remains a serious challenge for the Afghan government in almost every province. The Taliban still considers itself the rightful government of Afghanistan, and it remains a capable and confident insurgent force fighting for the withdrawal of foreign military forces from Afghanistan, establishment of sharia law, and rewriting of the Afghan constitution (*World Factbook*, 2022).

The first geodetic work in the Afghan region was done for the Northern Trans-Indus Frontier Survey (1852–1869) by the Survey of India, and this was part of the work associated with the "Measure of the Great Arc." In the late 19th century, British authorities in India feared the encroachment of Czarist Russia into Central Asia, Sinkiang, and Tibet. The British obtained a buffer region between Russia and India by extending the Afghan claim to the Wakhan Corridor. Subsequent boundary treaties between Great Britain and Russia were signed in 1873, 1885, and 1895. Treaties were signed between Afghanistan and Russia in 1921, 1932, 1946, 1958, and 1981. All of the surveys performed by the British Survey of India in Afghanistan were based on the Indian principal triangulation that referenced the Everest 1830 ellipsoid where $a=6,377,301.243$ m and $1/f=300.80176$. The datum origin for the subcontinent of India (and most of southern Asia) is considered to be at Kalianpur Hill Station, where $\Phi_o=24° 07' 11.26"$ N and $\Lambda_o=77° 39' 17.57"$ East of Greenwich.

The Office of the Geographer, U.S. Department of State in their International Boundary Study No. 26, Revised in 1983 says: "The Wakhan Corridor River boundary from Eshkashem to Lake Sari-Qul (Victoria) results from Anglo-Russian diplomatic agreements of 1869–73. From Lake Sari-Qul to the Afghanistan-China-U.S.S.R. tripoint, the 218 kilometers of boundary, which follows the watershed of the Vakhanskiy Khrebet Range, was delimited by the Anglo-Russian Pamir Boundary Commission of 1895. The Commission demarcated the boundary at 12 points. The location of the boundary pillars, as noted by the Russian surveyor Zaliessky, was calculated east of the Russian observatory located at Pulkowa [*actually Pulkovo – Ed.*] (13° 19' 38.55" east of the Royal

Greenwich Observatory). The location was recalculated during the Indo-Russian Trigonometrical survey in 1921. Geographic values for the 12 pillars noted on the Wakhan Corridor map are taken from the World Geodetic System (WGS-72) employed by the United States Defense Mapping Agency." The coordinates of the pillars are listed and arithmetic is presented showing a coordinate shift to the Indian 1916 Datum by correcting for both a re-determination of the Madras Observatory longitude as well as a correction for latitude to all points for the entire datum. In *Triangulation in India and Adjacent Countries, Sheet 42.G (Great Pamir)*, Dehra Dun, 1921: "Excluding Trotter's latitude stations which are now impossible to identify, there is one astronomical station of interest in the area. Boundary Pillar No. 1 is described as 'a Conical Stone Pillar, 9 feet high, built at the eastern end of Lake Victoria on a mound rising 10 feet above the level of the lake.' The latitude of this point was observed astronomically by Colonel Wahab (Wauhope) with an 8-inch diameter transit theodolite, the value obtained being $\Phi = 37° 26' 33''$. Zalesky, of the Russian Commission, during the same work, obtained a longitude $\Lambda = 73° 46' 32''$ for the same point by a comparison of local time with that shown by six chronometers brought from Osh, the longitude of which place had been determined telegraphically from Pulkowa [sic] via Tashkent. He considered his probable error not greater than 5 seconds of arc – it is not known whether Struve's longitude value of Pulkowa [sic] ($\Phi = 30° 19' 40.11''$) or the *Nautical Almanac* value $\Phi = 30° 19' 38.55''$ was adopted by the Russians. The interest in these figures lies in the fact that Boundary Pillar No. 1 was fixed trigonometrically by the British Commission; corrected for the latest value of Madras and adjusted to the Indian Triangulation on the Everest spheroid, these values are $\Phi = 37° 26' 27.5''$ and $\Lambda = 73° 46' 30.1''$."

Examination of the listed coordinates of the 12 pillars indicates four different versions of the same points for the Pamir Boundary Commission 1895 (PBC95), the Indian 1916 Datum, the WGS 1972 Datum, and the System 1942 Datum (with an origin at Pulkovo). For instance, "Pillar 1" coordinates for the (PBC95 datum) are $\varphi = 37° 26' 32.2''$ N and $\lambda = 73° 49' 00.6''$ E and, for "Pillar 12," are $\varphi = 37° 21' 25''$ N and $\lambda = 74° 50' 22''$ E. The "other" datum coordinates (for "Pillar 1") show a difference for the Indian Datum 1916 of $\Delta\varphi = -4.7''$ and $\Delta\lambda = -2' 30.3''$, for the WGS 72 Datum of $\Delta\varphi = -8.1''$ and $\Delta\lambda = -2' 32.8''$, and for the System 42 Datum of $\Delta\varphi = -8.4''$ and $\Delta\lambda = -2' 30.3''$. For "Pillar 12," the "other" datum coordinates show a difference for the Indian Datum 1916 of $\Delta\varphi = -4.7''$ and $\Delta\lambda = -2' 30.3''$, for the WGS 72 Datum of $\Delta\varphi = -8.0''$ and $\Delta\lambda = -2' 33.5''$, and for the System 42 Datum of $\Delta\varphi = -8.3''$ and $\Delta\lambda = -2' 31.1''$. Although the differences between the PBC95 and the WGS 72 datums seem plausible at first glance, when we look at how Indian 1916 and System 42 compare with the same points, my conclusion is that the veracity of this "analysis" is somewhere between a "high geodetic crime" and "pulp fiction!" The only thing that is believable is the Pamir Boundary Commission of 1895 coordinates – the remarkably close correspondence of the other three listed datum coordinates have no semblance with reality. Incidentally, the difference between Indian 1916 Datum and WGS72 is about a kilometer, which is more than 10 times that as reported in International Boundary Study No. 26. I have been reading these Office of the Geographer International Boundary Studies for over 30 years, and I must admit that this is the first and only one with which that I have had a problem.

International Boundary Study, Number 89 reports the treaty between Afghanistan and China on 22 November 1963. There are no attempts at geodesy in this report, but an interesting part of the summary observes: "The problem also exists that the geographic coordinates given in the treaty for the initial point of the boundary – 37° 03' North, 74° 36' East – do not conform with the same point in the China-Pakistan agreement. The problem obviously is related directly to the poor quality of mapping in the frontier."

International Boundary Study, Number 6 reports on the treaties between Afghanistan and Iran. The first arbitration, under the supervision of Sir Frederick Goldsmid, occurred in 1872 between Persia and Afghanistan. Between 1888 and 1891, a compromise boundary was laid down as the Hari Rud system and 39 pillars to the south marked the (*arbitration*) award from the Russian tripoint latitude 34° 20' North. In 1896, the Helmand River changed course and the boundary again became a point of conflict. Trig surveys were carried from India and, by 1905, the McMahon Commission placed 90 markers along the boundary from the tripoint on the Kuk-I-Malik Siah to the Kuh Siah.

"The 550-mile boundary is demarcated by 172 pillars, or approximately one pillar for every three miles of boundary."

Thanks to Mr. John W. Hager, now retired from the Defense Mapping Agency (no longer DMA nor NIMA, but the National Geospatial-Intelligence Agency – NGA), there are a number of datums that have been established in Afghanistan. "Bogra Datum – I have no data but note that there is a Bogra Dam on the Helmand at approximately latitude 31° 56′ N, and longitude 64° 44′ E. A guess is that is was a local system used for the construction of the dam."

The oldest of the *local* Afghan datums is the Ishpushta Datum of 1940 where, as Hager states, "Point is Observatory Station at latitude (geodetic) $\varphi = 35°\ 18'\ 53.5''$ N, $\xi = 10.80'' \pm 0.9''$, longitude (geodetic) $\lambda = 68°\ 05'\ 08.53''$ E, $\eta = 7.40'' \pm 2.7''$, Everest. Reference is 'Triangulation in Afghanistan,' published by Survey of India in 1947." I was curious as to where this location is in the country and I noticed that this is about 75 km northeast of the ruins of the Buddhas of Bamian, destroyed by the Taliban regime. The Swiss Office of Topography has done terrestrial photogrammetric restitutions of the originals, and there are a number of fascinating papers in print on the topic of the statues. In 1951, the U.S. Army Map Service (AMS) performed a cartometric analysis of boundary points between Afghanistan and what is now the Republic of Turkmenistan that were based on the Ishpushta Datum of 1940. The geodetic station comparisons between the Afghan Ishpushta Datum minus the Indian Datum of 1916 show a trend of $\Delta\varphi = -6.1''$ and $\Delta\lambda = +32.0''$.

John Hager mentioned something surprising. "Kalianpur Hill station … is the definition of Indian datum except for the ellipsoid. About 1954 AMS produced 1:250,000[-scale] maps in northern Afghanistan on the International Ellipsoid. To differentiate from Indian datum (Everest), the datum name Kalianpur was used. The dividing line is longitude 61° 30′ N, latitude 36° E, thence east to 66°, north to 37°, east to 72°, north to 38° N, thence to longitude 78° E. Note that the horizontal segments are multiples of 1° 30′, and the vertical are 1°, the size of the standard 1:250,000[-scale] map]. When Zone 0 was eliminated, the northern limit of Zone I was redefined, I believe as above."

For the now obsolete India Zone 0, (all of the India Zones were cast on the Lambert Conical Orthomorphic Projection), Everest ellipsoid where $a = 6,974,310.600$ Yards, $e^2 = 0.006637846630200$, Latitude of Origin, $\varphi_o = 39°\ 30'$ N, Central Meridian, $\lambda_o = 68°$ E of Greenwich, Scale Factor at the Parallel of Origin $m_o = 649/650 = 0.998461538$, False Northing, FN = 2,590,000 Yards, and False Easting, FE = 2,355,500 Yards. An example test point for India Zone 0 is: $\varphi = 42°\ 38'\ 51.627''$ N, $\lambda = 61°\ 41'\ 57.291''$ E, X = 1,790,983.28 Yards, and Y = 2,991,605.57 Yards. For India Zone I, Latitude of Origin, $\varphi_o = 32°\ 30'$ N, Central Meridian, $\lambda_o = 68°$ E of Greenwich, Scale Factor at the Parallel of Origin $m_o = 823/824 = 0.998786408$, False Northing, FN = 1,000,000 Yards, and False Easting, FE = 3,000,000 Yards. An example test point for India Zone I is $\varphi = 30°\ 33'\ 49.893''$ N, $\lambda = 62°\ 12'\ 13.613''$ E, X = 2,392,655.35 Yards, and Y = 782,000.02 Yards.

Thanks again to John W. Hager, the Herat North Datum of 1959 origin is $\Phi_o = 34°\ 23'\ 09.08''$ N and $\Lambda_o = 62°\ 10'\ 58.94''$ East of Greenwich, $H_o = 1,111.7$ m and is referenced to the International ellipsoid where $a = 6,378,388$ m and $1/f = 297$. This is the datum that shift parameters are published for by NIMA (now NGA) from local (Herat North 59) to WGS84. To my amazement, Hager said, [This was] "established by the Soviets (Technoexport). There was a joint U.S.S.R. and U.S.A. mapping project at that time, we did the south half of the country, and they did the north half. One part was done by a consortium (I can't remember what they were called) made up of Aero Services Corp. out of Philadelphia and Fairchild Aero Services. They flew Shoran controlled photography at a scale approximately 1:60,000 using B-17s, based on a Shoran measured trilateration net."

Never short for surprises, John Hager transmitted to me a facsimile of a paper translated by John M. Willis of the DMA Aerospace Center in 1990. The surprise paper was entitled *The Local Coordinate System of Kabul (Mestnaya Systema Koordinat Kabula), Geodeziya I Kartografiya*, No. 12, 1988, pp. 21–33. As I have touted for years, the local use of a municipal or county coordinate system must take into account the elevation for the implementation of a simple system for the use of GIS (Geographic Information System) technicians. The author of the Kabul System was Bakhavol' Darvesh, and this individual ingeniously utilized the Australian Map Grid (AMG) with its concomitant ellipsoid for the City of Kabul and surrounding region! The choice of the ellipsoid and scale factor at origin just happened to perfectly fit the elevation and latitude of Kabul. That's what proper cartographic/geodetic design is all about for a project/city coordinate system. Many readers of this journal are professional cartographers and photogrammetrists. We need to design systems that will facilitate a GIS for local governments with the educational level of the average municipal GIS Technician in mind. What's "common sense" to us is not necessarily obvious to the neophyte. Darvesh applied superb design engineering for facile local use. That system has been used since 1984, and I'll bet it's still in use in Kabul. The Republic of Colombia and the states of Minnesota, Wisconsin, *etc.* have designed local city or county coordinate systems that are designed to compensate for heights above the ellipsoid such that Land Surveyors can submit their ground surface measurements and "fit" to the local GIS without mathematical manipulation. As I teach my Louisiana State University students every semester, the "K.I.S.S." principle is preferable to complicated solutions – ("Keep It Simple, Stupid").

The only published datum shift parameters for Afghanistan are those by NGA (*ex-NIMA*) from Herat North 1959 Datum to WGS84: $\Delta a = -251$ m, $\Delta f \times 10^4 = -0.14192702$, $\Delta X = -333$ m, $\Delta Y = -222$ m, and $\Delta Z = +114$ m. The astute reader will notice that there is no accuracy estimate offered, and there is no information offered regarding the number of stations used to compute the shift parameters. This "guess" published by NIMA/NGA is listed as "Non-Satellite Derived Transformation Parameters."

So what is an Infantry soldier to do over there? My youngest son, Philippe, was in the 82nd Airborne Division, 504th Parachute Infantry Regiment over there in 2002–2003. He went on many patrols searching for the "bad guys," and he had the dubious distinction of being a short Louisiana Frenchman; therefore, he had the job of the "Tunnel Rat" for his company. Apart from this awful but necessary duty task, he tells me that his unit was supplied with 1:100,000-scale topographic maps referenced to the WGS84 Datum. Philippe tells me that the maps had the towns and hamlets placed where they should be, and the coordinates on the paper maps appeared to match their personal Global Positioning System (GPS) receivers. The U.S. military has not yet issued personal GPS receivers to all their combat troops, except for the Squad Leaders. As a measure for individual survival, my son tells me that most of the Paratroopers over there purchase their own personal consumer-grade GPS receivers just to record their own treks to retrace if separated. The GPS units are thankfully in stock at the Post Exchanges in Kandahar and elsewhere in country, and they offer a "comfort factor" to the combat soldiers that walk the valleys of Afghanistan looking for the "bad guys." Hopefully, Lieutenant General Clapper of NGA will produce 1:50,000-scale paper topographic maps to support our troops in Iraq better than has been done so far in Afghanistan. Although my youngest son is finished with his term of enlistment in the Army, he's got a lot of buddies that are "rotating" to the Sunni Triangle this month. They personally write to me (from the Task Force rotating out), and the highly computerized NGA needs to better support the "grunt on the ground." There are still bad guys in Afghanistan, and I think that the 1:100,000-scale *paper* maps are not good enough. Air Force brass are obviously not the best judges of what the grunt needs on the ground, in the tunnels, and in the caves. Taking "mapping" out of "NIMA" must not take the map out of the hands of the Infantryman.

Albania
International boundary
Republic boundary
District (rreth) boundary
National capital
District (rreth) center
Railroad
Road
A district has the same name as its center except where noted.
0 20 Kilometers
0 20 Miles
Lambert Conformal Conic Projection, SP 40N/56N
Adriatic Sea
Strait of Otranto
Ionian Sea
Corfu Channel
Lake Scutari
Lake Ohrid
Lake Prespa
ITALY
BOSNIA AND HERZEGOVINA
CROATIA
Montenegro
Serbia
Kosovo
THE FORMER YUGOSLAV REPUBLIC OF MACEDONIA
GREECE
DHIAPONDIOI NISOI
Sazan
Tirana
TIRANË
Shkodër
Durrës
Elbasan
Vlorë
Korçë
Gjirokastër
Sarandë
Berat
Fier
Lushnjë
Kavajë
Krujë
Lezhë
Kukës
Pukë
Peshkopi
Burrel
Bulqizë
Librazhd
Gramsh
Pogradec
Përmet
Tepelenë
Delvinë
Ersekë
Këlcyrë
Çorovodë
Cërrik
Peqin
Rrogozhinë
Kuçovë
Ballsh
Mavrovë
Selenicë
Shijak
Laç
Rrëshen
Koplik
Bajram Curri
Valbonë
Krumë
Maliq
MALËSI E MADHE
TROPOJË
HAS
MIRDITË
DIBRË
KURBIN
MAT
SKRAPAR
DEVOLL
KOLONJË
MALLAKASTËR
Shëngjin
Bar
Ulcinj
Podgorica
Nikšić
Dubrovnik
Kotor
Tivat
Trebinje
Peć
Dečani
Priština
Prizren
Đakovica
Uroševac
Kačanik
Skopje
Tetovo
Gostivar
Kičevo
Debar
Struga
Ohrid
Resen
Bitola
Prilep
Florina
Kastoriá
Grevená
Kónitsa
Ioánnina
Kalabáka
Kérkira
Corfu
Igoumenitsa
Serbia and Montenegro have asserted the formation of a joint, independent state, but this entity has not been formally recognized as a state by the United States.

Republic of Albania

Slightly smaller than Maryland, Albania is bordered by Greece (282 km), Macedonia (151 km), Montenegro (172 km), and Kosovo (112 km). The terrain is mostly mountains and hills with small plains along the coast. The lowest point is the Adriatic Sea (0 m), and the highest point is Maja e Korabit (Golem Korab) (2,764 m) (*World Factbook*, 2022).

"The territories of present-day Albania have been inhabited as far back as 100,000 years ago. It was at the turn of the third millennium *B.C.* that an Indo-European population settled there. As a result of the mixture, a population incorporating the unique cultural and linguistic characteristics of the whole Balkan Peninsula (pelages) was created. Based on this ancient population, the Illyrian people developed through the second millennium and the first century *B.C.* After its fall in the year 30 *B.C.* Illyria came under the control of Roman Empire. With the division of the Roman Empire (395 *A.D.*), Illyria became a part of the Byzantine Empire.

"The country has suffered continuous invasions over the last 1000 years and by the end of the 14th century Albania was occupied by the Ottoman Empire. The subsequent efforts and insurrections for independence eventually brought about the proclamation of the independence of Albania in 1912. After 1912 till the end of the First World War, the country was attacked by neighboring countries. After eleven years of monarchy the country was occupied by Mussolini forces in 1939, marking the end of the monarchy. In 1943 the armies of Hitler occupied the country.

"The resistance against foreign invasion was known as the Anti–Fascist National Liberation front. The Communist party took power in November 1944, when the foreign armies were expelled. Shortly thereafter, a totalitarian regime was established under the communist leader Enver Hoxha. For about 50 years, the regime applied the policy of self-isolation, leaving the country in great economic poverty when it finally emerged from isolation in 1991. The principle of self-reliance applied by the Communist regime prohibited foreign loans, credits and investment. From 1991 until 1997 The Democratic Party led the country. After the unrests of 1997 due to the failure of pyramidal schemes, the Socialist Party with its coalition was in power until 2005.

"Albania joined NATO in April 2009 and in June 2014 became an EU candidate. Albania in April 2017 received a European Commission recommendation to open EU accession negotiations following the passage of historic EU-mandated justice reforms in 2016. Although Albania's economy continues to grow, it has slowed, and the country is still one of the poorest in Europe. A large informal economy and a weak energy and transportation infrastructure remain obstacles" (*World Factbook*, 2022).

The Italian *Instituto Geografico Militare* (IGM) measured a 3,044.2301 m ± 0.04 m baseline at Lecce (in the heel of Italy's boot) in 1872 and observed a connection with the Albanian island of Sazan. "The Triangulation Network that was established by the Military Geographic Institute of Vienna (MGIW) during 1860-1873, in the framework of the construction of the geodetic basis was done for mapping of the Balkans at 1:75000 scale" (*Coordinate Reference Systems Used in Albania to Date, Nikolli, P., & Idrizi, B., FIG Working Week 2011, Morocco, 18–22 May 2011*). The Second Austro-Hungarian Triangulation (II Military Triangulation 1806–1869) used the Vienna University Observatory as the datum origin for regions that included Albania, where $\Phi_o = 48°\ 12'\ 35.50''$ N, $\Lambda_o = 34°\ 02'\ 36.00''$ East of Ferro, the azimuth to Leopoldsberg, $\alpha_o = 163°\ 42'\ 12.27''$ and the Bessel 1841 ellipsoid of revolution where the semi-major axis $(a) = 6,377,397.155$ m and the reciprocal of flattening $(1/f) = 299.1528$. The reference meridian used was Ferro in the Canary Islands, where $17°\ 39'\ 46.02''$ West of Greenwich. The K.u.K. Military Geographic Institute of Vienna observed a 1st Order triangulation net in the Adriatic region that included an Albanian baseline measured just southeast of Lake Scutari at Shkodër in the early 1900s. "(It is only 0.726 m shorter than the Austrian base line measured in 1869.)" (*Mapping of the Countries in Danubian and Adriatic Basins, Glusic, Andrew M., AMS TR No. 25, June 1959, 406 pages.*)

While the Austro–Hungarian lands were mapped with a Polyhedric projection, the Albanian Republic was mapped based on an ellipsoidal Bonne projection (*Cohen, Ruth, Geodetic Memo No. 485, Development of Rinner's Formulas in the Bonne projection and the inverse, Army Map Service, 28 October 1949*), where the projection origin is collocated at the Albanian Datum origin where $\Phi_o = 41°\ 20'\ 12.809''$ N, $\Lambda_o = 19°\ 46'\ 45.285''$ East of Greenwich and thanks to John W. Hager, the azimuth to East Base measured from North, $\alpha_o = 294°\ 38'\ 02.57''$. This datum origin and projection origin is commonly referred to as being on the Tiranë-Durrës Highway in the Laprakë neighborhood of Tiranë. Upon perusal of *Google Earth*™ imagery, it appears that this point is likely centered in the median of a traffic circle/roundabout on that highway. Although some date the Bonne projection origin in 1918 (*op. cit. Nikolli & Idrizi, 2011*), others date the datum origin in 1932 (*Marcussi, A., Lineamenti geoidici della penisola balcanica, Bollettino Geodetica, vol. XXIV, No. 4*). "In 1946, the Yugoslav first order net was tied with the first order net of Albania. Common stations are: 328 Gruda-Griži, 331 Jubani, 332 Taraboš and 245 Cukali" (*op. cit., Glusic, 1959*).

"In the 1955, the specialists of Military Topographic Group of Albania carried out the reconstruction and the densification of the IGM Net in order to grant the request for mapping at 1: 25 000 scale. At the same time, the first-order network was transformed from the IGM System (1934) into the 1942 coordinate system, which was based on the Krassovsky ellipsoid, Gauss-Krüger projection with central meridian $\lambda_o = 21°$(*op. cit., Nikolli & Idrizi, 2011*)." The System 42 Datum of the former Union of Soviet Socialist Republics origin is at Pulkovo Observatory, where $\Phi_o = 59°\ 46'\ 18.55''$ North, $\Lambda_o = 30°\ 19'\ 42.09''$ East of Greenwich, the defining azimuth at the point of origin to Signal A is: $\alpha_o = 317°\ 02'\ 50.62''$ and the ellipsoid of reference is the Krassovsky 1940, where $a = 6,378,245$ m and $1/f = 298.3$. The Russia Belts are a Grid System identical to UTM except that the scale factor at origin is $m_o = 1.0$, and for Albania the System 42 Datum used a central meridian as stated above: $\lambda_o = 21°$.

"In 1960, the Albanian government declared its territorial waters to 'embrace a sea area ten nautical miles in breadth in the direction of the open sea, reckoning from the basic shoreline, which connects the following points on the coast: the mouth of River Bojana – Cape Rodini – Cape Pali – Cape Durres – Cape Lagi – Cape Semani – Sazan Island – Cape Linguetta (*kepi Gjuhëzës*) – in the Corpe (Corfu) Channel territorial waters go the middle of the Channel' (*medium filium acquae – Ed.*) The enumeration of these points in the Adriatic Pilot constituted a new Territorial Sea based on the principle of Straight Baselines." (*International Boundary Study, LIMITS IN THE SEAS, No. 7, Albanian Straight Baselines, Office of the Geographer, U.S. Department of State, February 16, 1970.*)

The New Albanian Net, which was constituted from Triangulation and Leveling, was designed, rebuilt, measured, and calculated by the Military Topographic Institute of Albania (MTI) during 1970–1985. The readjustment is termed the ALB86 System. The Russia Belts TM continued to be used for this readjustment (*op. cit., Nikolli & Idrizi, 2011*). In October 1994, the U.S. Department of Defense occupied **35** existing Albanian triangulation stations and established WGS84 Datum positions in cooperation with MTI personnel. The coordinate system adopted by the Military Topographic Institute of Albania after 1994 is the UTM, Zone 34. This coordinate reference system has transformation parameters for Albania as published by *TR8350.2* where **from** ALB86 **to** WGS84: $\Delta X = +24\ m \pm 3\ m$, $\Delta Y = -130\ m \pm 3\ m$, $\Delta Z = -92\ m \pm 3\ m$.

The National Report of Albania for 2009 listed new seven-parameter Bursa-Wolf transformation results computed by the Department of Geodesy of the Polytechnic University of Tirana **from** ITRF96 (epoch 1998.1) **to** ALB86 as: $\Delta X = +35.758\ m$, $\Delta Y = +11.676\ m$, $\Delta Z = +41.135\ m$, $R_x = +2.2186''$, $R_y = +2.4726''$, $R_z = -3.1233''$, $\delta s = +8.3855$ ppm based on **18** collocated points. (*Gjata, G., et al., EUREF2009 Symposium, Florence, Italy, May 2009.*) The following year, the *Instituto Geografico Militare* of Italy computed a similar transformation based on **90** collocated points in Albania **from** ETRF2000 **to** ALB86 as: $\Delta X=+44.183$ m, $\Delta Y=+0.58$ m, $\Delta Z=+38.489$ m, $R_x=+2.3867''$, $R_y=+2.7072''$, $R_z=-3.5196''$, $\delta s=+8.2703$ ppm. (*Isufi, E., et al., Overview on Albanian Reference Systems, actual situation and the future challenge, BALGEOS, Vienna, January 2010.*) Significant progress has been achieved for a new Albanian Geoid, and the Albanian Association of Geodesy appears to be flourishing.

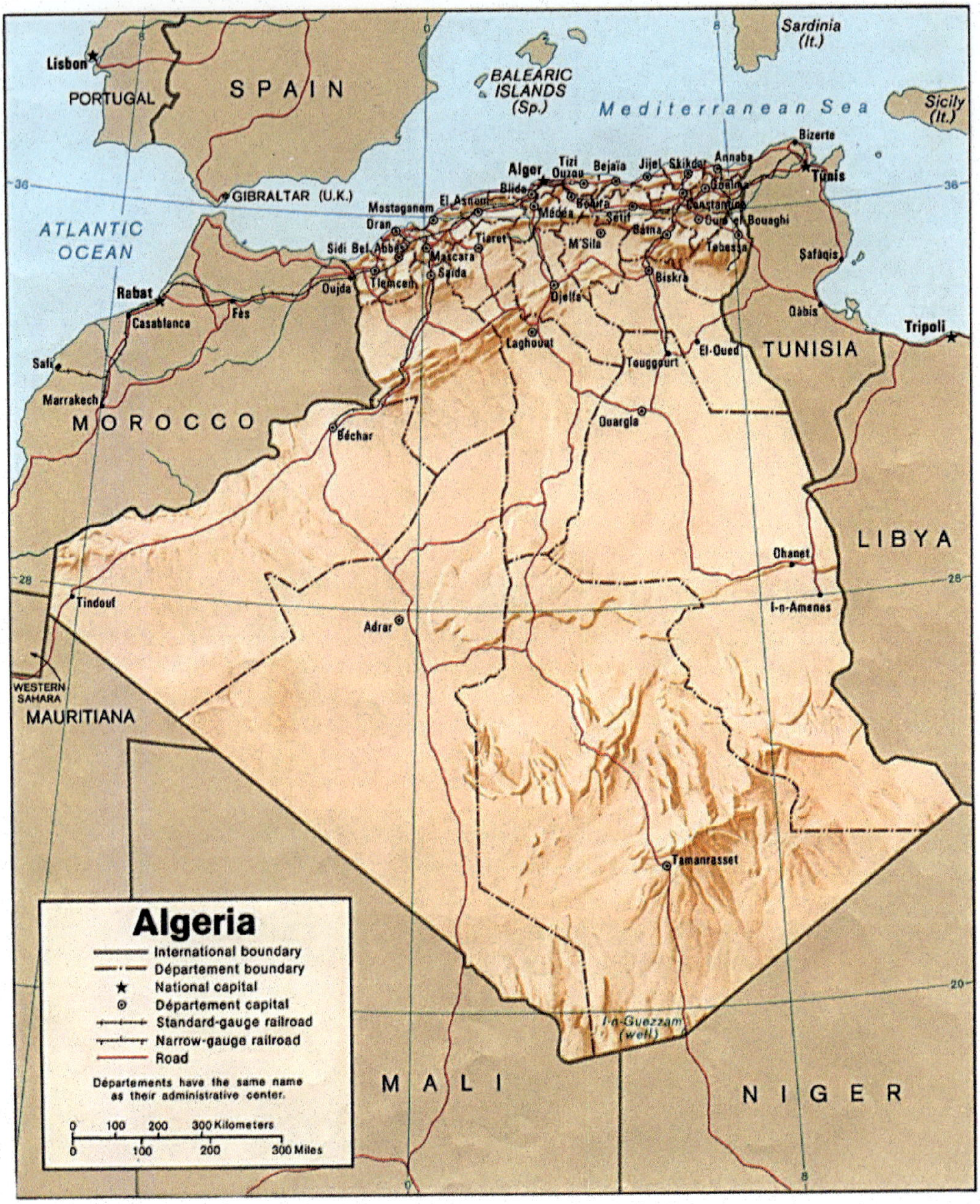

Algeria
International boundary
Département boundary
National capital
Département capital
Standard-gauge railroad
Narrow-gauge railroad
Road
Départements have the same name as their administrative center.
0 100 200 300 Kilometers
0 100 200 300 Miles
SPAIN
PORTUGAL
Lisbon
GIBRALTAR (U.K.)
BALEARIC ISLANDS (Sp.)
Sardinia (It.)
Sicily (It.)
Mediterranean Sea
ATLANTIC OCEAN
MOROCCO
Rabat
Casablanca
Fès
Safi
Marrakech
Oujda
TUNISIA
Tunis
Bizerte
Şafāqis
Qābis
LIBYA
Tripoli
MAURITANIA
WESTERN SAHARA
MALI
NIGER
Alger
Tizi Ouzou
Bejaïa
Jijel
Skikda
Annaba
Blida
Bouira
Guelma
Constantine
Oum el Bouaghi
Mostaganem
El Asnam
Médéa
Sétif
Batna
Tébessa
Oran
Sidi Bel Abbès
Mascara
Tiaret
M'Sila
Tlemcen
Saïda
Biskra
Djelfa
Laghouat
El-Oued
Touggourt
Ouargla
Béchar
Ohanet
I-n-Amenas
Tindouf
Adrar
Tamanrasset
I-n-Guezzam (well)

Democratic and Popular Republic of Algeria

Algeria is the second largest country in Africa after Sudan and was known to the Romans as Numidia. The coastline on the Mediterranean Sea extends for 998 km. Algeria is bounded by Tunisia and Libya on the east, by Niger and Mali on the south, on the west by Mauritania, the former Western Sahara, and Morocco. The highest point is Tahat at 3,003 m, and the lowest point is Chott Melrhir at −40 m. Algeria is mostly high plateau and desert; the Atlas and Saharan Atlas Mountains are in the north along with narrow discontinuous coastal plains.

"Mainly under the rule of the Ottoman Empire until 1705, it was occupied by the French in 1830. Algeria gained independence from France following a referendum of 1 July 1962. Abdelkader Bensalah, became interim head of state on 9 April 2019. Bensalah remained in office beyond the 90-day constitutional limit until Algerians elected former Prime Minister Abdelmadjid Tebboune as the country's new president in December 2019" (*World Factbook*, 2022).

The triangulation of Algeria was carried out by the Dépôt Général de la Guerre from 1854 to 1887. After 1887, the Société Géographique de l'Armée, headquartered in Paris, continued the work. The First-Order net of triangulation consists essentially of two parallel chains and three meridional chains. The parallel chains are the coastal chain from the Moroccan to the Tunisian borders (1860–1868), and the chain Aïn Sefra-Laghouat-Biskra-Gabes (in Tunisia) (1889–1895). There are also two shorter parallel chains: the Guerara tie chain (1909–1910) and the Southern El Oued tie chain (1909–1910). The meridional chains are the Biskra chain (1872–1873, 1899–1902), the Laghouat chain (1886, 1899–1902), and the Saïdia chain (1896–1897). Fill nets of First-Order complementary and Second- and Third-order triangulation were surveyed from 1864, following the pattern of planned mapping. The survey work was executed and adjusted in 94 cartonnés (books of sections), which progressed southward from the coastal areas according to military requirements. These cartonnés were numbered in *chronological* sequence of completion. Note that in the English-language literature of North African Geodesy, one will likely find "carton," the derivative of cartonné.

This original work comprised the Colonne Voirol Datum of 1875, commonly termed Voirol 75. The fundamental point is at the geodetic pillar of the Colonne Voirol Observatory, and the astronomical coordinates are: $\Phi_o = 36° 45' 07.9''$ N ($40^G 8357.8''$), $\Lambda_o = 3° 02' 49.45''$ East of Greenwich ($0^G 7887.3''$ East of Paris). The reference azimuth from south to Melab el Kora is: $\alpha_o = 322° 16' 52.7''$, and the ellipsoid of reference is the Clarke 1880 (IGN), where $a = 6,378,249.2$ m and $1/f = 293.4660208$. The baselines measured for the Algerian triangulation, with dates of execution are: Blida (1854, 1912), Bône (Annaba) (1866, 1885), Oran (1885, 1910), Laghouat (1914), Ouargla (1920), Mercheria (1932), Biskra (1932), and Navarin (1949). The original mapping was cast on the ellipsoidal Bonne projection – the ubiquitous projection *du jour* for the Europeans of the time. The North African (ellipsoidal) Bonne Grid Latitude of Origin: $(\varphi_o) = 35° 06'$ N ($39^G 00$N), the Central Meridian $(\lambda_o) = 2° 20' 13.95''$ East of Greenwich, and some time before WWII, the False Easting and False Northing were changed from 0 to 100 km for each. Interestingly, this old Bonne Grid still influences current mapping in that Grid limits of the Lambert Conic Grids are still defined by Bonne Grid values. The sheet boundaries of the new Lambert Grids are commonly computed by a reversion of the late Prof. Karl Rinner's Bonne power series formulae published in *Zeitschrift für Vermessungswesen* during the 1930s. That reversion allows cartographers to compute the intersection of a constant Bonne Grid value with a chosen arc of the parallel or of the meridian. Those intersections then were used to define the limits with the graticule of the Lambert Conic Grids computed by Mr. John W. Hager of the Defense Mapping Agency (*ex* Army Map Service), in 1974.

Based on original triangulations of the French Army, a local (temporary) Astro station was established in the port city of Oran by Capitaine Faure during 1905–1906. Station Tafaraoui coordinates are: $\Phi_o = 39^G\ 3778.26''$ N, $\Lambda_o = 3^G\ 1532.06''$ East of Paris. The reference azimuth to Tessala is: $\alpha_o = 62°\ 09'\ 57.73''$, and the ellipsoid of reference is the Clarke 1880 (IGN). The observations were later adjusted and used in the 1930 hydrographic survey of that portion of the coast of Algeria and the port of Oran. The Lambert Conic Grid was used by the French Navy for the hydrographic survey.

The reader will notice that I have left off the word "Conformal" when describing the Lambert Conic Grids of Algeria. That is because the original systems that succeeded the ellipsoidal Bonne Grid in 1906 were not fully conformal. There are two original zones: for *Nord Algerie*, the Latitude of Origin $(\varphi_o) = 36°$ North (40^G), the Central Meridian $(\lambda_o) = 2°\ 42'\ (3^G)$ East of Greenwich, and the scale factor at origin $(m_o) = 0.999625544$. For Zone *Algerie Sud*, the Latitude of Origin $(\varphi_o) = 33°\ 18'$ North (37^G), the Central Meridian $(\lambda_o) = 2°\ 42'\ (3^G)$ East of Greenwich also, and the scale factor at origin $(m_o) = 0.999625769$. The False Origin is 500 km for eastings and 300 km for Northings *for both zones*, the same convention as used in the adjacent Kingdom of Morocco. The complete replacement of the Bonne Grid for original topographic mapping in Algeria did not happen until 1942.

During the 19th century, projection table computations were performed by hand, and all formulae were commonly truncated past the cubic term to ignore infinite series terms considered at the time, too small to warrant the extra effort. For instance, the Lambert Conformal Conic projection was used only to the cubic term in the formulae for the tables of the developed meridional distances. This resulted in French Army projection tables that have become part of the arcane lore of computational cartography.

Furthermore, another idiosyncrasy of the French Army formulae is that the Lambert (fully) Conformal Conic normally utilizes one of the principal radii of the ellipsoid called the Radius of Curvature in the Plane of the Meridian (ρ_o). The French Army instead substituted the Length of the Ellipsoid Normal Terminated by the Semi-Minor Axis (υ_o) at the Latitude of Origin (φ_o). Although not strictly conformal, this is the system that was commonly used by the French in all colonies (before WWII) that utilized the Lambert Conic projection (including Syria).

Standard Lambert formulae will not work for Algeria under certain conditions, and the improper use of the fully conformal projection will yield computational errors that can exceed 15 m! As an example, consider a test point where $\varphi = 33°$ N, $\lambda = 3°$ E. For *Nord Algerie* on the French Army Truncated Cubic Lambert Conic Grid, X = 528,064.182 m, Y = –32,764.881 m: for the same test point on the *Nord Algerie* Lambert *fully* Conformal Conic Grid, X = 528,074.691 m, Y = –32,776.731 m. The computational difference between the two formulae at the same test point is $\Delta X = -10.509$ m, $\Delta Y = +11.850$ m, or a total error of 15.839 m! Mathematical elegance is not what matters in a country's coordinate transformations; what matters is computational conformity to local *legal* standards. The certain condition when a fully conformal Lambert Conic will work in Algeria is based on *when* a particular Algerian map was compiled. That is, when the Algerian triangulation was recomputed for the European Datum of 1950, the French dropped usage of the Truncated Cubic version on the old Voirol 75. In summary, for surveys and maps before 1948, one must use the French Army Truncated Cubic Lambert Conic. After 1948, one must use the Lambert fully Conformal Conic. The parameters of the two Lambert zones did not change for the Colonne Voirol Datum of 1875; only the formulae changed. Things soon got more complex.

In 1953–1954, the First-Order coastal parallel chain was re-observed by the French. In 1959, the Institut Géographique National (IGN), Paris, re-adjusted the entire First-Order and First-Order complementary triangulation on European Datum 1950, incorporating the results of all previous surveys and adjustments. The rule of thumb for this datum shift is to increase both latitude and longitude from Colonne Voirol Datum of 1875 to European Datum 1950. The UTM Grid was used for this purpose, as were all datums that were transformed to ED50. Like most countries, the ED50 UTM Grid was reserved for military topographic mapping, and local native systems continued in use. That tradition has resulted in some convoluted transformations being perpetuated in Algeria.

The North Sahara Datum of 1959 was obtained (in 1957–1958) by re-computing the results of the First-Order nets and the First-Order Complimentary nets adjusted to the ED50 but referenced to the Clarke 1880 (modified) ellipsoid where $a = 6,378,249.145$ m and $1/f = 293.465$. The adjustment on the Clarke 1880 (modified) ellipsoid was performed such that it optimized the fit of the shape of the geoid in North Africa, i.e., by reducing to a minimum the sum of the squares of the relative deflections of the vertical in the areas involved. This principle was intended to minimize the mean discrepancies between the geodetic net used in the northern part of Algeria and the astronomic net used primarily in the southern part of Algeria. Some maps were stereo-compiled on the North Sahara Datum of 1959 with the UTM Grid at 1:200,000 scale. However, many maps were not cast on the UTM Grid.

The Lambert North Sahara Auxiliary Grid was directly applied to the geodetic coordinates in accordance with the definition of the *Nord Algerie* Zone with the fully conformal formulae. However, it was never used in any publication or in mapping because of the large discrepancies found between the rectangular coordinates of any given point in the Colonne Voirol Datum of 1875 (Voirol 75) or the North Sahara Datum of 1959. This computational experiment is the reason for the development and subsequent adoption of the Lambert Voirol 60 Grid System. This curious system adds 135 m exactly to the X coordinates and adds 90 m exactly to the Y coordinates of the original *Nord Algerie* Zone parameters. In other words, the Lambert Voirol 60 Grid has a False Easting = 500,135 m and a False Northing = 300,090 m. According to the French Army in June of 1970, "Under these conditions, when we compare the LAMBERT – VOIROL 75 with the LAMBERT VOIROL 60 coordinates, the shift between the two is always less than 50 m in absolute value. This value does not represent a mathematical relation, but rather the result of comparing the two sets of coordinates. It shows up the inaccuracies in the initial VOIROL 75 system. The maps made with the LAMBERT VOIROL 60 rectangular coordinates are all referenced to the geographic coordinates of the NORTH SAHARA geodetic system." The current parlance for this in English is the "Voirol Unified 1960 Grid" on the "North Sahara Datum of 1959." Note that there is no classical origin for this Datum due to the fact that it is derived from ED50.

In 1966, AMS developed a series of conversions on a Carton-by-Carton basis for transforming from Voirol 75 to ED50 with UTM coordinates. As an example of the transformation series for Algeria, the following is for coordinates in UTM Zone 31 whose eastings are greater than 355,000 m: Carton 59: $N = 0.9998873966\ n - 0.10000869984\ e + 691.561$ m, $E = 0.9999391272\ e + 0.0000869984\ n - 416.633$. The stated Root-Mean-Square Error (RMSE) for this Carton is ±0.200 m. The adjacent Carton 60 when used with the appropriate coefficients has a stated RMSE of ±2.759 m!

In recent years, IGN derived a seven-parameter transformation from ED50 to WGS84 for North Africa. The quoted accuracy is ±2 m in X, Y, and Z, and when applying this transformation, the resulting heights are approximately 30 m higher than expected for Algeria. The parameters are: $\Delta X = -130.95$ m, $\Delta Y = -94.49$ m, $\Delta Z = -139.08$ m, $\Delta s = +6.957$ ppm, $R_x = +0.4405''$, $R_y = +0.4565''$, $R_z = -0.2244''$. The U.S. National Imagery and Mapping Agency (NIMA) does not list a three-parameter transformation in TR 8350.2 for transforming from ED50 to WGS84 in Algeria. However, the non-satellite-derived NIMA parameters from Colonne Voirol Datum of 1875 to WGS84 are: $\Delta X = -73$ m, $\Delta Y = -247$ m, $\Delta Z = +227$ m, with no stated accuracy. NIMA states that from Colonne Voirol Unified Datum of 1960 to WGS84 the parameters are: $\Delta X = -123$ m, $\Delta Y = -206$ m, $\Delta Z = +219$ m, and each parameter is stated accurate to ±25 m. NIMA further states that from North Sahara Datum of 1959 to WGS84, the parameters are: $\Delta X = -186$ m, $\Delta Y = -93$ m, $\Delta Z = +310$ m, and each parameter is stated accurate to ±25 m.

Using a 1° by 1° 30′ mesh of ED50 coordinates over Northern Algeria, a set of 54 North Sahara Datum of 1959 and WGS84 coordinates were derived by others using the transformation developed by IGN. I solved for the three-parameter transformation from North Sahara Datum of 1959 to WGS84 using the WGS84 Geoid such that: $\Delta X = -131.798$ m, $\Delta Y = -75.442$ m, $\Delta Z = +329.895$ m. The geodetic residual Root-Mean-Square (RMS) is expressed as meters for $\Delta \varphi = \pm 1.74$ m, for $\Delta \lambda = \pm 1.04$ m, and $\Delta h = \pm 4.52$ m. For comparison, I then solved for North Sahara Datum of 1959 to

WGS84 using the EGM96 Geoid such that: $\Delta X=-156\,m$, $\Delta Y=-77.366\,m$, $\Delta Z=+311.265\,m$. The geodetic residual RMS is expressed as meters for $\Delta\varphi=\pm2.12\,m$, for $\Delta\lambda=\pm2.51\,m$, and $\Delta h=\pm4.35\,m$. In conclusion, since the IGM seven-parameter solution cannot be fully evaluated, the preferred transformation from North Sahara Datum of 1959 to WGS84 Datum in the format given in TR 8350.2 is then: $\Delta X=-159\,m$, $\Delta Y=-77\,m$, $\Delta Z=+311\,m$, $\Delta a=-112.145$, $\Delta f\times 10^4=-0.54750714$.

A significant amount of geodetic research has been performed in recent years. However, the primary areas of research have been devoted to observations of gravity and of crustal motion in the Atlas Mountain range.

"The REGAT ('*REseau Géodésique de l'Atlas*') geodetic network is composed of 53 continuously recording GPS stations distributed in the Algerian Atlas. It spans the whole width of the Algerian coast and reaches 300 km inland, with inter-sites distance of about 100 km. One additional site is located in Tamanrasset in the southernmost part of the country. The network, whose oldest stations started operating in 2007, encompasses the main active tectonic features of the most seismically active segment of the Nubia-Eurasia plate boundary in the Western Mediterranean" (*Heliyon*, Volume 5, Issue 4, April 2019, e01435).

"The comparisons based on different GPS campaigns provide, after fitting by using the four-parameter transformation, an RMS differences±11 cm especially for the north part of the country over distances of 1 to 1000 km and proves that a good fit between the new quasi-geoid and GPS/levelling data has been reached" (*A New Quasi-Geoid Computation from Gravity and GPS Data in Algeria*, S. A. Benahmed Daho, J. D. Fairhead). www.isgeoid.polimi.it/Newton/...2/Benahmed-A%20new%20quasigeoid-revised.pdf

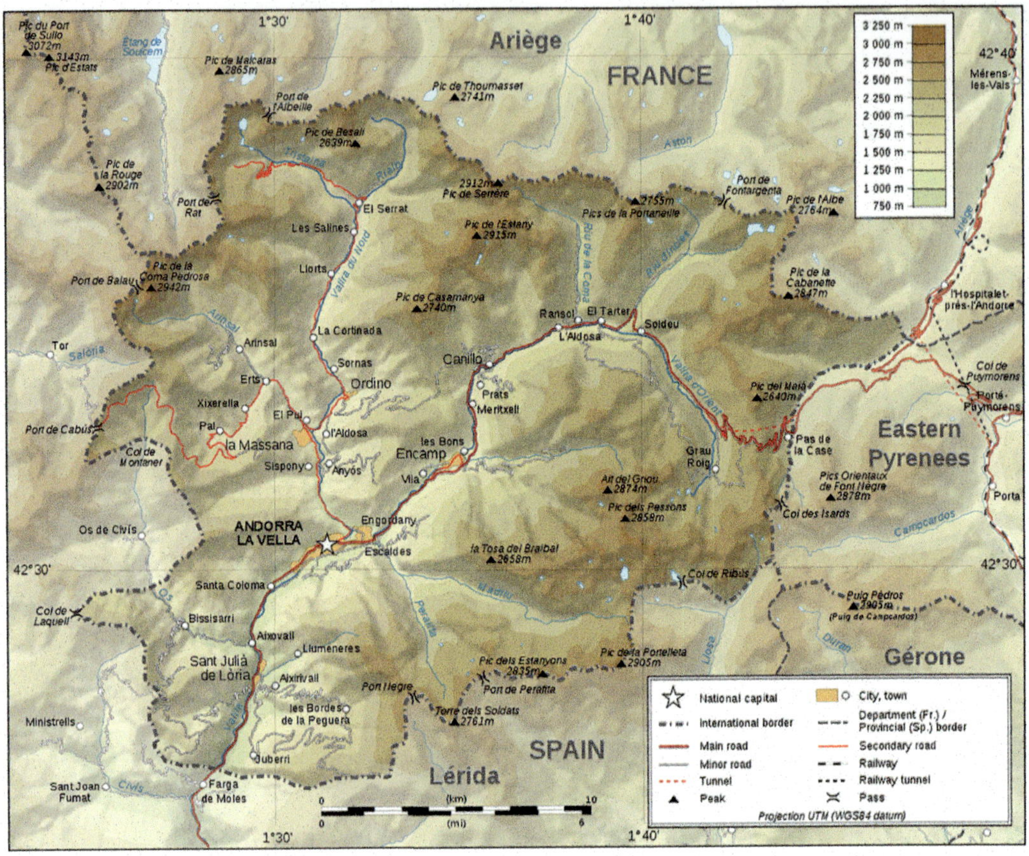

Ariège
FRANCE
Eastern Pyrenees
Gérone
Lérida
SPAIN
ANDORRA LA VELLA
Ordino
la Massana
Encamp
Canillo
Escaldes
Engordany
Sant Julià de Lòria
Pic du Port de Sullo 3072m
3143m Pic d'Estats
Étang de Soulcem
Pic de Malcaras 2865m
Port de l'Albeille
Pic de Besali 2639m
Pic de Thoumasset 2741m
Pic de la Rouge 2902m
Port de Rat
El Serrat
Les Salines
2912m Pic de Serrère
Pic de l'Estany 2915m
2755m Pics de la Portanelle
Port de Fontargenta
Pic de l'Albe 2764m
Mérens-les-Vals
Aston
Pic de la Cabanette 2847m
l'Hospitalet-près-l'Andorre
Port de Balau
Pic de la Coma Pedrosa 2942m
Llorts
Pic de Casamanya 2740m
Ransol
El Tarter
Soldeu
L'Aldosa
La Cortinada
Arinsal
Tor
Sornas
Prats
Meritxell
Pic del Maià 2640m
Col de Puymorens
Porté-Puymorens
Erts
Xixerella
El Puí
Pal
l'Aldosa
Port de Cabús
Col de Montaner
Sispony
Anyós
les Bons
Vila
Pas de la Case
Grau Roig
Pics Orientaux de Font Nègre 2878m
Porta
Col des Isards
Campcardos
Alt del Griu 2874m
Pic dels Pessons 2858m
Os de Civís
la Tosa del Braibal 2658m
Col de Ribus
Puig Pèdros 2905m (Puig de Campcardos)
Santa Coloma
Col de Laquell
Bissisarri
Aixovall
Llumeneres
Pic dels Estanyons 2835m
Port de Perafita
Pic de la Portelleta 2905m
Aixirivall
Port Negre
Torre dels Soldats 2761m
les Bordes de la Peguera
Ministrells
Juberri
Sant Joan Fumat
Farga de Moles
Valira del Nord
Valira d'Orient
Riu de la Coma
Madriu
Perafita
Duran
Ariège
1°30'
1°40'
42°40'
42°30'
3 250 m
3 000 m
2 750 m
2 500 m
2 250 m
2 000 m
1 750 m
1 500 m
1 250 m
1 000 m
750 m
0 (km) 10
0 (mi) 6
National capital
City, town
International border
Department (Fr.) / Provincial (Sp.) border
Main road
Secondary road
Minor road
Railway
Tunnel
Railway tunnel
Peak
Pass
Projection UTM (WGS84 datum)

Principality of Andorra

Bordered by France (55 km) and Spain (63 km); Andorra is 2½ times the size of Washington, DC. The lowest point is Riu Runer (840 m), the highest point is Pic de Coma Pedrosa (2,946 m), and the terrain is rugged mountains dissected by narrow valleys (*World Factbook*, 2015).

"Andorra is the last independent survivor of the March states, a number of buffer states created by Charlemagne to keep the Muslim Moors from advancing into Christian France. Tradition holds that Charlemagne granted a charter to the Andorran people in return for their fighting the Moors. In the 800s, Charlemagne's grandson, Charles the Bald, made Count of Urgell overlord of Andorra. A descendant of the count later gave the lands to the diocese of Urgell, headed by Bishop of Seu d'Urgell. In the 11th century, fearing military action by neighboring lords, the bishop placed himself under the protection of the Lord of Caboet, a Spanish nobleman. Later, the Count of Foix, a French noble, became heir to Lord Caboet through marriage, and a dispute arose between the French Count and the Spanish bishop over Andorra. In 1278, the conflict was resolved by the signing of a pareage, which provided that Andorra's sovereignty be shared between the Count of Foix and the Bishop of Seu d'Urgell of Spain. The pareage, a feudal institution recognizing the principle of equality of rights shared by two rulers, gave the small state its territory and political form. Over the years, the title was passed between French and Spanish rule until, in the reign of the French King Henry IV, an edict in 1607 established the head of the French state and the Bishop of Urgell as co-princes of Andorra. Given its relative isolation, Andorra has existed outside the mainstream of European history, with few ties to countries other than France and Spain. In recent times, however, its thriving tourist industry along with developments in transportation and communications have removed the country from its isolation.

"There has been a redefinition of the qualifications for Andorran citizenship, a major issue in a country where only 35.7% of 78,549 are legal citizens. In 1995, a law to broaden citizenship was passed but citizenship remains hard to acquire, with only Andorran nationals being able to transmit citizenship automatically to their children. Lawful residents in Andorra may obtain citizenship after 25 years of residence. Children of residents may opt for Andorran citizenship after 18 if they have resided virtually all of their lives in Andorra. Mere birth on Andorran soil does not confer citizenship. Dual nationality is not permitted. Non-citizens are allowed to own only a 33% share of a company. Only after they have resided in the country for 20 years, will they be entitled to own 100% of a company. A proposed law to reduce the necessary years from 20 to 10 is pending approval in Parliament. By creating a modern legal framework for the country, the 1993 constitution has allowed Andorra to begin a shift from an economy based largely on tax-free shopping to one based on tourism and international banking and finance. Despite promising new changes, it is likely that Andorra will, at least for the short term, continue to confront a number of difficult issues arising from the large influx of foreign residents and the need to develop modern social and political institutions. In addition to questions of Andorran nationality and immigration policy, other priority issues will include dealing with housing scarcities and speculation in real state, developing the tourist industry, defining its relationship with the European Union, and reforming the investment law to allow up to 100% foreign ownership in activities and sectors considered strategic" (*U.S. Dept. of State Background Note, January 2007*).

The geodetic history of Andorra is intertwined with that of Spain. The recent (65-year) history begins with the end of WWII when the U.S. Army Map Service transformed all of the local datums of Europe and North Africa to the European Datum of 1950 which was referenced to the International ellipsoid. The subsequent transformation of the Spanish Net to ED50 was based on a two-step two-dimensional Helmert transformation. The initial step computed by the Axis (Germans) was still on a Lambert Conformal Conic, but referenced to the Clarke 1880, which was the standard for France. That series of co-located points that started computations in the Pyrenees Mountains was continued into a substantial portion of the Iberian Peninsula. After the war, the U.S. Army Map Service decided that the best way to merge the Madrid Observatory Datum of 1858 into the new ED50 was to use the existing German data as an intermediate step. In computing the new French Datum values of the Spanish triangulation, use was made of the German adjustment of 1938. Values of the New Triangulation of France (NTF) were available for the following 11 Spanish triangulation stations: Forceral, Canigou, Licuses, Rouge, Cabére, Maupas, Anie, Orhi, Baigura, La Rune, and Biarritz. Computations to the Madrid Datum of 1858 were performed by V.U.K.A. 631 (April 1943) and by Kriegskarten und Bermessungsamt, Paris (August 1943).

Since then, the Principality has been exclusively on the European Datum of 1950 where the UTM Grid (Zone 31) has been in use and referenced to the International 1924 ellipsoid, where $a=6,378,388$ m and the reciprocal of flattening $(1/f)=297$. In a similar type of transformation used for the computation of ED50, the Principality has published a new four-parameter (two-dimensional) transform **from** ED50 **to** the European Terrestrial Reference System of 1989 (ETRS89), where $T_x=-129.549$ m, $T_y=-208.185$ m, $\mu=0.0000015504$, and $\alpha=-1.56504''$. This was published effective August 29, 2007, in the Royal Decree Number 1071 of 2007 (*Institut Cartografic de Catalunya*, 2015). Furthermore, a grid-based datum transformation system has been implemented by the government based on the Canadian technique of NTv2, similar (but different) to the U.S. National Geodetic Survey's NADCON software package.

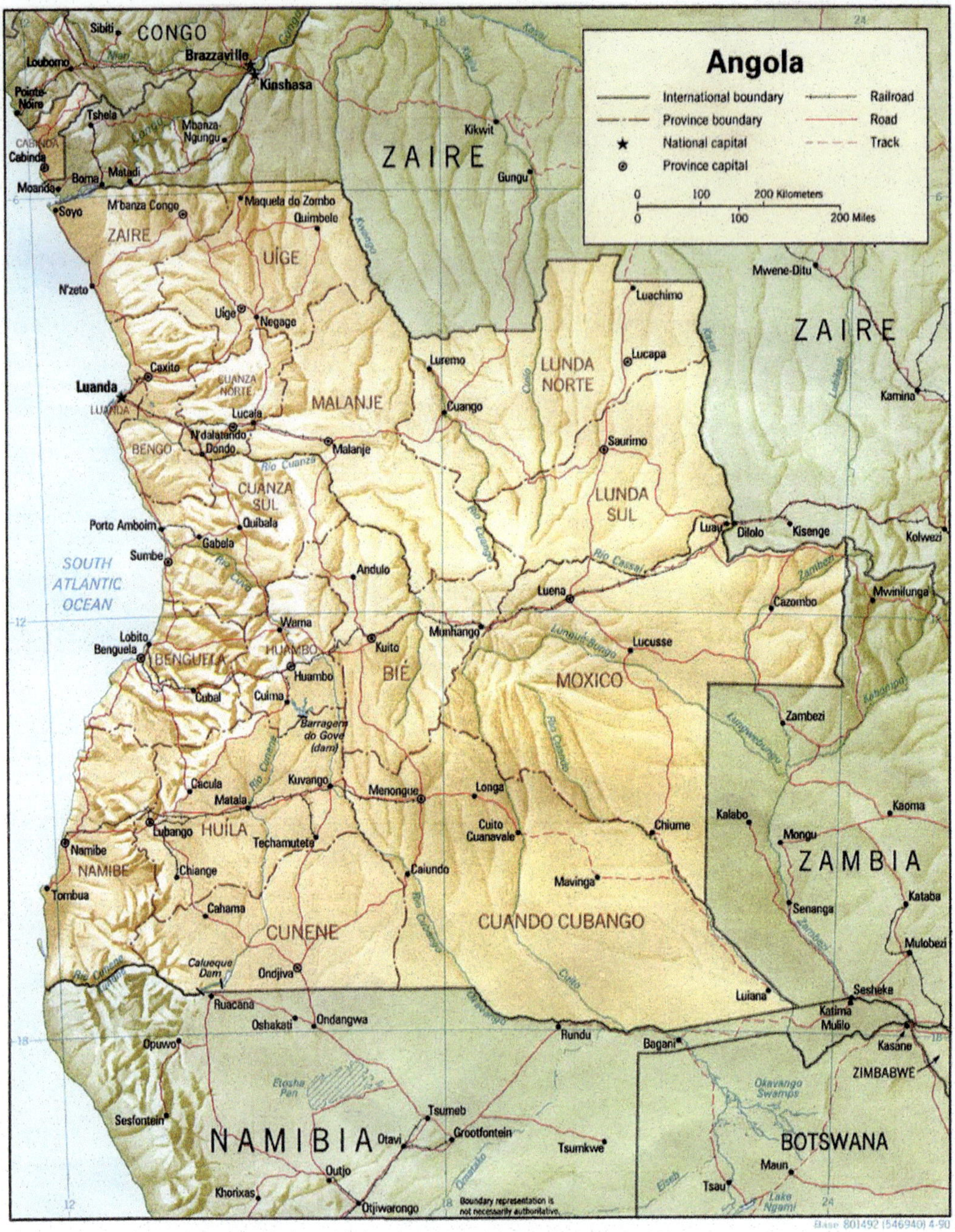

Base 801492 (546940) 4-90

The Republic of Angola

Bordered by the Democratic Republic of the Congo (2,646 km) (of which 225 km is the boundary of discontiguous Cabinda Province), Republic of the Congo (231 km), Namibia (1,427 km), and Zambia (1,065 km). The interior forms part of the Central African Plateau, with elevations that range from 1,220 to 1,830 m (4,000–6,000 ft). The coastal plain is about 1,610 km long (1,000 miles) and varies in width from 48 to 160 km (30–100 miles). The highest point is Mt. Moco in the west at 2,559 m (8,397 ft). The chief rivers include the Congo, Cuanza, and Cuene to the north, and south of the Lunda Divide some flow into the Zambezi River and others flow into the Okavango River.

The original peoples of what is now Angola were probably Khoisan-speaking hunters and gatherers (bushmen). During the first millennium A.D., large-scale migrations of Bantu-speaking people moved into the area and eventually became the dominant ethnolinguistic group of southern Africa. The most important Bantu kingdom in Angola was the Kongo, with its capital at Mbanza Kongo (called *São Salvador do Congo* by the Portuguese). South of the Kongo was the Ndongo kingdom of the Mbundu people. Angola got its name from the title of its king, the **ngola**. In 1483, Portuguese explorers reached Angola, Christianized the ruling family, and engaged in trade and missionary work. By the early 17th century, some 5,000+ slaves were being exported from Luanda annually. Angola received its independence from Portugal in 1975 but has been plagued by civil war and insurrections since independence. In 2010, Joao Lourenco was elected president in August 2017 and became president of the Popular Movement for the Liberation of Angola (MPLA) in September 2018 (*World Factbook*, 2022). A familiar Bantu word in the United States is kwanza, which is Angola's unit of currency.

Angola consists of two geographically separate expanses that include Angola proper and Cabinda. Portuguese authority was not exercised continuously north of the Congo River in the present-day district of Cabinda until a relatively recent date. It was occupied by the Portuguese in 1783, but a French expedition forced them to evacuate the area 11 months later. Portugal laid a definite claim to Cabinda in an additional convention to the Anglo–Portuguese treaty of January 22, 1815. Again, on February 26, 1884, an Anglo–Portuguese treaty acknowledged claims by Portugal that included not only Cabinda and the Congo River inland to Nóqui but the whole Atlantic coast between 5° 12′ and 8° South latitude. This produced a storm of protests in Europe, and Portugal proposed a conference on the Congo that resulted in the Berlin Conference held between November 15, 1884, and February 26, 1885. Consider then, that the borders of Cabinda are in common with Congo (Brazzaville) that was part of the former Congo Français (French Congo), and currently the Republic of the Congo. The southern border is in common with Congo (Kinshasa) once the Belgian Congo later called Congo, then Zaire, and currently the Democratic Republic of Congo. The controlling classical Datum for southern Africa and most surrounding countries of Angola proper is the Cape Datum of 1950 where the point of origin is station Buffelsfontein, where $\Phi_o = -33° 59′ 32.00″$ S, $\Lambda_o = 25° 30′ 44.622″$ East of Greenwich and the azimuth from south to station Zuurberg is: $\alpha_o = 183° 58′ 15″$. The reference ellipsoid for the Cape 50 Datum is the Clarke 1880, where the semi-major axis $(a) = 6,378,249.145$ m, and the reciprocal of flattening $(1/f) = 293.465$. Angola's southern border is with that country once called German Southwest Africa, and Namibia is the only country in the African continent to utilize the Bessel 1841 ellipsoid for its Schwarzeck Datum where for Namibia: the semi-major axis $(a) = 6,377,483.865$ m and the semi-minor axis $(b) = 6,356,165.383$ m. The origin point is Schwarzeck, near Gobabis, where $\Phi_o = -22° 45′ 35.820″$ S and $\Lambda_o = 18° 40′ 34.549″$ East of Greenwich. Our Paul M. Hebert School of Law here at Louisiana State University is helping build a Law Library at the University of Namibia.

In 1891, *ANNALES HYDROGRAPHIQUES* of the French Navy published the telegraphic determination of longitudes for three sites in Angola as determined by Commander Pullen of the Royal English Navy. Those determinations were: São-Paulo de Loanda (Luanda) – at the pavilion slab of the Fort of San Miguel, where $\Phi_o = -08° 48' 24''$ S, $\Lambda_o = 10° 53' 05$ *East of Paris*, Benguela – at the Bureau Télégraphique (Telegraph Office), where $\Phi_o = -12° 34' 43''$ S, $\Lambda_o = 11° 03' 40''$ *East of Paris*, and Moçamedes – at the pavilion slab of Ponta da Noronha, where $\Phi_o = -12° 34' 43''$ S, $\Lambda_o = 11° 03' 40''$ *East of Paris*.

Other than the geodetic work performed mainly by the British and French for surrounding lands, little survey work was done in the first half of the 20th century in Portuguese West Africa (Angola and Cabinda). The Camacupa Datum of 1948 is based on the origin at Campo de Aviação where $\Phi_o = -12° 01' 19.070''$ S, $\Lambda_o = 17° 27' 19.800''$ East of Greenwich, $h_o = 1,508.3$ m. Thanks to Mr. John W. Hager, "This is the principal vertex marked by a concrete monument, constructed on a high part of the Camacupa Air Field, immediately to the north of the runways." (This is) "defined as the 'Datum Point' of the main triangulation network of Angola. A concrete monument with the dimensions 70×60×100 cm. (length E-W × width N-S × height), topped by a white marble slab on which is cut in black: M.G.A. –P.F.-1948; in the center of which is placed the top mark of the base, which is defined as the extreme West of the Geodetic Base of Camacupa. The mark found here is protected by a masonry casing with a metallic cover, easily removed to permit observations over the base when necessary." Remarkably, all datums established by the Portuguese in Angola (and Moçambique) were referenced to the Clarke 1866 ellipsoid (the same as used in the U.S. for the North American Datum of 1927), where $a = 6,378,206.4$ m and $b = 6,356,583.8$ m. The only transformation parameters I have ever been able to scrounge *from* this Datum *to* WGS84 were obtained from Prof. Charles L. Merry at the University of Cape Town, where $\Delta X = -49$ m, $\Delta Y = -301$ m, $\Delta Z = -181$ m; and Prof. Merry estimates the accuracy at ±60 m. According to John W. Hager, "sometime in the 1960s or 1970s, DMA was asked to put Angola, then on Camacupa Clarke 1866, on Camacupa Clarke 1880 and on Arc 50 Datum. The Portuguese provided all the coordinates based on Clarke 1866. They also provided tables to convert from Clarke 1866 to Clarke 1880 assuming that the tangent point of the two ellipsoids was at Camacupa. I think that the tie was on the 12th Parallel South to the Zambian Triangulation. The 6th Parallel south and Bas Congo surveys of Congo occupy common points with the Angola surveys and were adjusted to Arc 50. A comparison of the Angola values showed that Arc 50 in Angola was adequate for mapping purposes. Angola is on the UTM Grid. I did find a local Grid for Luanda and would expect other similar ones. For Luanda, the 1:2,000 scale city map plots directly on top of the UTM Grid of the 1:100,000 map. The 50,000 50,000 intersection is, in UTM N=9,024,000, E=306,000. This then results in a local Grid, TM projection, Clarke 1880 ellipsoid, latitude of origin=0°, longitude of origin=15° E, FN=1,026,000, FE=244,000, scale factor=0.9996. The UTM scale factor at local 50,000 50,000 is 1.00006581. A unity scale factor would be expected for a City Grid, and this is pretty close to unity. The math for the false coordinates is FN=10,000,000−9,012,000+50,000=1,026,000 and FE=500,000−306,000+50,000=244,000."

Thanks to parameters published into the public domain by the European Petroleum Studies Group (EPSG) headed up by Mr. Roger Lott of British Petroleum, there are a number of transformations from the Clarke 1880 version of the Camacupa Datum of 1948. For instance, Camacupa to WGS 72BE: $\Delta X = -37.2$ m, $\Delta Y = -370.6$ m, $\Delta Z = -228.5$ m; this was derived by Geophysical Services, Inc. in 1979. Camacupa to WGS84 used by Conoco for Offshore Block 5: $\Delta X = -42.01$ m, $\Delta Y = -332.21$ m, $\Delta Z = -229.75$ m. Camacupa to WGS84 and used by Topnav at PAL F2, by Elf in blocks 3 and 17 since 1994, and by total in block 2 since 1994: $\Delta X = -50.9$ m, $\Delta Y = -347.6$ m, $\Delta Z = -231$ m. An additional eight versions of parameters are used for the "same" transformation in offshore areas spanning the entire coast of Angola. Another relation published by EPSG is the MHAST Datum (Missão Hidrográfica de Angola), that according to the EPSG is referenced to the International ellipsoid of 1924 where $a = 6,378,388$ m and $(1/f) = 297$. From MHAST to WGS84: $\Delta X = -252.95$ m, $\Delta Y = -4.11$ m, and $\Delta Z = -96.38$ m.

John W. Hager went on to say, "A survey was done across Congo (Kinshasa) connecting Angola proper to Cabinda but the data was destroyed by a fire in Lisbon, so Cabinda is on a local datum. About all the booklet for Cabinda will say is that it is not on Camacupa 1948 Datum. The values of the boundary marks in the northwest of Cabinda are in agreement with those published by (the French) IGN and used by Congo (Brassaville). Several of the U.S. oil companies did work there and came up with a unified scheme but I don't know how valid it was. I think one of the companies was Chevron."

Other datums existing in Angola include Lobito Datum based on the origin point Extremo NE da Base do Lobito, where $\Phi_o = -12°\ 19'\ 00.86''$ S, $\Lambda_o = 13°\ 34'\ 45.67''$ East of Greenwich, Clarke 1866 ellipsoid. Dr. José Carvalho of Maputo, Moçambique states that the Camacupa Datum of 1948 coordinates of the same point are: $\Phi_o = -12°\ 19'\ 01.357''$ S, $\Lambda_o = 13°\ 34'\ 58.375''$ East of Greenwich. Luanda Datum is based on the origin point at Luanda Observatory, where $\Phi_o = -08°\ 48'\ 46.8''$ S, $\Lambda_o = 13°\ 13'\ 21.8''$ East of Greenwich, Clarke 1866 ellipsoid. Moçamedes Datum of 1956 origin point is at the Moçamedes Meteorological Station, where $\Phi_o = -15°\ 11'\ 16.34''$ S, $\Lambda_o = 12°\ 07'\ 34.53''$ East of Greenwich, Clarke 1866 ellipsoid.

Mr. Mal Jones of Perth, Australia, offered the Malongo 1987 Datum that is referenced to the International ellipsoid such that the transformation parameters to WGS84 Datum are: $\Delta X = -254.10$ m, $\Delta Y = -5.36$ m, $\Delta Z = -100.29$ m. The National Surveying and Mapping Organization needs to be provided with legal support and technical documentation. In general, modernization should involve three types of reference nets – Geodetic, Leveling, and Gravimetric (*Sebastian J.F., Kutushev S.B.,* (2015) [Modernization of geodetic net of the Republic of Angola. The base tasks]. *Geodesy and Cartography = Geodezija i kartografija*, 2, pp. 19–24. (In Russian). DOI: 10.22389/0016-7126-2015-896-2-19-24); *however, the article itself was not available (21 January 2019)*. http://geocartography.ru/en/scientific_article/2015_2_19-24

The National Surveying and Mapping Organization, geodetic, leveling and gravimetric reference nets.

In a recent email note from Dr. Rui Fernandes (Universidade da Beira Interior · Department of Computer Science · SEGAL Portugal · Covilhã), he mentioned that he inquired as to what the Angolan government had accomplished in terms of a new datum and coordinate system. Their reply was "We used your past project," and they offered no further details (personal communication, January 2019).

Antigua and Barbuda
National capital
Parish boundary
Railroad
Road
0 5 10 Kilometers
0 5 10 Miles
Lambert Conformal Conic Projection, SP 16°10'N/17°40'N
British Virgin Is. (U.K.)
Anguilla (U.K.)
Puerto Rico (U.S.)
Virgin Is. (U.S.)
SAINT KITTS AND NEVIS
ANTIGUA AND BARBUDA
Montserrat (U.K.)
Guadeloupe (Fr.)
DOMINICA
Caribbean Sea
Martinique (Fr.)
SAINT LUCIA
BARBADOS
SAINT VINCENT AND THE GRENADINES
GRENADA
TRINIDAD AND TOBAGO
VENEZUELA
62°00'
61°40'
NORTH ATLANTIC OCEAN
17°40'
Codrington
Barbuda
Caribbean Sea
17°20'
Antigua
Cedar Grove
Saint John's
ST. GEORGE
ST. JOHN
Parham
ST. PETER
Willikies Village
Bolands
All Saints
ST. PHILIP
ST. MARY
Sweets
Freetown
ST. PAUL
Carlisle
English Harbour Town
The islands of Barbuda and Redonda are dependencies.
17°00'
Redonda
Montserrat (U.K.)
Saint John's Village
Long Ground Village
Plymouth
62°20'

Base 802393 (547694) 5-95

Antigua and Barbuda

The combined area of the three islands, Antigua, Redonda, and Barbuda is 440 km², which is approximately 2.5 times the size of the District of Colombia. The two main islands lie northeast of Montserrat and are comprised mostly of low-lying limestone and coral islands with some higher volcanic areas. The lowest point is the Caribbean Sea, and the highest point is Boggy Peak at 402 m.

Around 2400 *B.C.*, Antigua was settled by the "Siboney" people, an Arawak word meaning stone-people where numerous shell and stone tools have been found throughout the island. The Venezuelan Arawak culture later populated the island from about 35 to 1100 *A.D.* when they were displaced by the Carïbs, the cannibal warriors of the Caribbean. However, the Carïbs did not settle on either Antigua or Barbuda. During his second voyage to the New World Christopher, Columbus landed on the islands in 1493, naming the larger one Santa María de la Antigua in honor of the cathedral in Seville, Spain. However, Antigua had a dearth of fresh water and a number of unfriendly Carïbs so European settlement was not immediate. In 1682, a group of Englishmen from St. Kitts established a successful settlement. Two years later, Sir Christopher Codrington came to the island and initiated sugar cane cultivation, a commercial crop already flourishing in other islands of the Caribbean. Codrington later leased the island of Barbuda from the English Crown in order to raise provisions for his plantations. Barbuda's only town is named after him. West African slaves were brought to the plantations of Antigua by Codrington and others. By the end of the 18th century, Antigua had become an important strategic port as well as a valuable commercial colony. Horatio Nelson arrived in 1784 at the head of the Squadron of the Leeward Island to develop the British naval facilities at English Harbor and to enforce English shipping laws. Nelson spent almost the entire time on board ship, declaring the island to be a "vile place" and a "dreadful hole." Serving under Nelson while in Antigua was William IV, the future king of England. Later during the reign of King William IV, slaves were emancipated in 1834, but in Antigua and Barbuda, they remained economically dependent on the plantations. There was no surplus of farmland, no access to credit, and the economy was built on agriculture rather than manufacturing. Poor labor conditions persisted until just before WWII when a trade union movement was prompted by a member of a royal commission. Antigua became independent in 1967, with Barbuda and the small isle of Redonda as dependencies. Antigua became an associated state of the Commonwealth, and in 1981, it achieved full independent status.

Antigua Island Astro 1943 Datum was observed at the U.S. Navy "astro" point named Bowditch 1943 (station A-14) where thanks to Mr. John W. Hager: $\Phi_o = 17° \ 10' \ 35.633''$ N, $\Lambda_o = 61° \ 47' \ 45.268''$ W, and the orientation is based on the azimuth from James (station A-13) to Pointed Hill (station A-12) as $\alpha_o = 268° \ 33' \ 09.8''$ from north. That "astro" point is located at the northernmost point of land on Antigua Island. The ellipsoid of reference is the Clarke 1880 where $a = 6,378,249.145$ m and $1/f = 293.465$. Mr. Hager went on to mention that, "Also found are (seconds only) latitude = … 36.371″N and longitude = … 45.268″W and values marked U.S.C.&G.S. 1943 of latitude = … 35.506″ and longitude = … 45.380″W. I make the difference on the first at 1.2 meters and on the second at 4.9 meters but have no idea of the significance. Possibly the astro was observed a short distance from the trig station." The British West Indies "BWI Grid" for Antigua and Barbuda is based on the Transverse Mercator where the Central Meridian, $\lambda_o = 62°$ W, the Latitude of Origin $\varphi_o =$ equator, the scale factor at the Latitude of Origin $m_o = 0.9995$, False Easting = 400 km, and False Northing = nil. The formulae are the Gauss-Krüger, but for such a small span of latitude and longitude that includes all three islands; the distinction in this case is irrelevant. The National Imagery and Mapping Agency (NIMA) lists the three-parameter datum shift values (dated 1991) from Antigua Island Astro 1943 Datum (Clarke 1880) to WGS 84 Datum as: $\Delta a = -112.145$ m, $\Delta f \times 10^4 = -0.54750714$, $\Delta X = -270$ m ± 25 m, $\Delta Y = +13$ m ± 25 m, and $\Delta Z = +62$ m ± 25 m, and the solution is based on one station.

DOI: 10.1201/9781003307785-7

The U.S. Army Map Service formed the Inter-American Geodetic Survey (IAGS) in the 1950s and established the headquarters at Fort Clayton, Canal Zone. The associated IAGS School was established at Fort Curundu, Canal Zone, and I remember my father driving me past the facility when I was still in high school. At the time, all I was aware of was the fact that the school trained cartographers in Spanish. By educating local nationals of the nations in the Caribbean region, the IAGS was able to enlist technical workers to help in the classical triangulation and photogrammetric mapping of the Caribbean and Latin America. The IAGS extended the classical triangulation net of North America to south of the border and computed the North American Datum of 1927 (NAD27) throughout Central America and the Caribbean. The NAD27 as realized in the Caribbean Islands that include Antigua and Barbuda is listed by NIMA for the three-parameter datum shift values (dated 1991) from North American Datum of 1927 (Clarke 18676) to WGS 84 Datum as: $\Delta a = -69.4$ m, $\Delta f \times 10^4 = -0.37264639$, $\Delta X = -3$ m ± 3 m, $\Delta Y = +142$ m ± 9 m, and $\Delta Z = +183$ m ± 12 m, and the solution is based on 15 stations in the region.

The U.S. National Geodetic Survey (NGS) performed a high precision geodetic-quality GPS survey of Antigua and Barbuda in 1996 for the position determination of airport runways and appurtenances. Alas, with four points surveyed on Antigua and three points surveyed on Barbuda, no local datum position coordinates were requested (as usual) by NGS from the Surveys Division in the capital of St. John's.

Thanks to Perkins and Parry, *World Mapping Today* 2nd edition, "Antigua was one of the first Caribbean islands to be mapped by the Directorate of Colonial Surveys with publication of a 1:25,000 scale map (DCS 6) in 1946 based on U.S. Army Air Corps air photography with military ground control." A number of different scales have been produced since then, and current large-scale topographic maps of Antigua and Barbuda are available from government sources as well as commercial map sellers. According to Perkins and Parry, a military edition series at 1:50,000 scale utilizes a UTM Grid but apparently (to me), it is on the Antigua Island Astro 1943 Datum because the ellipsoid is reported to be the Clarke 1880.

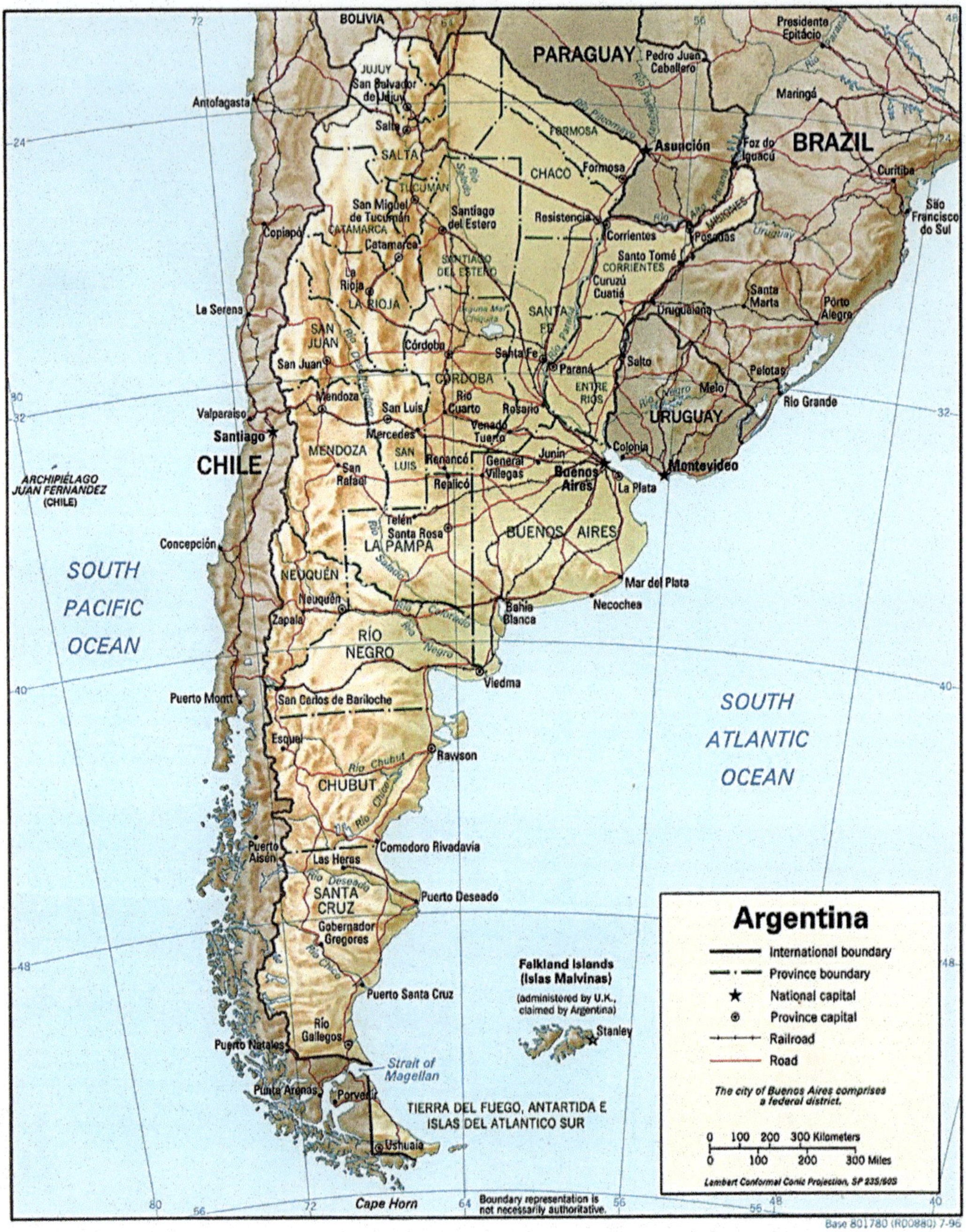

Base 801780 (R00880) 7-96

The Republic of Argentina

Bordered by Bolivia (942 km), Brazil (1263 km), Chile (6691 km), Paraguay (2531 km), and Uruguay (541 km). The northern Chaco and the central Pampas are vast expanses of flat land, which is the home of the Argentine cowboy, the Gaucho. Elevations in Argentina range from −40 m to +6,962 m (+22,841 ft) on Aconcagua, the highest peak in the Western Hemisphere.

The Río de la Plata (Silver River) was discovered by Solís in 1516, and Sebastian Cabot first explored Argentina from 1526 to 1530. Pedro de Mendoza founded the first permanent colony at Buenos Aires (good winds) in 1536. Argentina received its independence from Spain in 1816, and it is the second largest country in South America after Brazil. The years 2003–2015 saw Peronist rule by Nestor and Cristina Fernandez de Kirchner, whose policies isolated Argentina and caused economic stagnation. With the election of Mauricio Macri in November 2015, Argentina began a period of reform and international reintegration (*World Factbook*, 2022).

In 1826, the Topographic Department of the Province of Buenos Aires was founded. A national agency responsible for mapping the entire country was created on the 5th of December 1879 as the *Oficina Topográfica Militar* (Military Topographic Office). By 1901, the Army General Order No. 37 changed the name to the *Instituto Geográfico Militar* (Military Geographic Institute), which is a name that continues to this day. By 1943, the Argentine version of "La Ley" (The Law) was legislated giving the Army's *Instituto Geográfico Militar* (IGM) the national monopoly on large-scale topographic mapping. (See my column on Ecuador that has a short sociological commentary on the common Latin American mapping monopolies – Honduras is an exception.) Argentina's organization or its mapping agency followed the European practice of the time. The early Argentine "Anuarios" (yearbooks) of the 1900s actually detailed the various military topographic organizations of Europe.

In 1887, the old astronomical observatory at Córdoba established the longitude of its meridian circle as: $\Lambda = -68° \ 12' \ 03.3''$ West of Greenwich. The first geodetic-quality astronomical ("Astro") station observed was in 1894 at the geodetic pillar in the Army Barracks in Mendoza, where $\Phi_o = -32° \ 52' \ 54.8''$ South, $\Lambda_o = -68° \ 51' \ 22.8''$ West of Greenwich. At the time, the Argentines were using the Bessel 1841 ellipsoid where the semi-major axis $(a) = 6,377,397.155$ m and the reciprocal of flattening $(1/f) = 299.1528128$. Initial plans for establishing geodetic control in the country were for a perimeter survey of the Atlantic coastline, as well as along the international and provincial borders. This plan was modified in 1912 to consist of $2° \times 2°$ quadrilaterals formed by chains of triangulation. The classical geodetic triangulation network of Argentina reflects strict adherence to this plan as it exists today. Practically everywhere else in the world the chains have been dictated by the topography. That is, triangulation stations commonly are located on the summits of hills and mountains. Since so much of the Argentine country consists of the very flat Chaco and Pampas, the topography had no influence on the shape of the primary triangulation chains. Consequently, since observation towers had to be built anyway, the plan for chains of man-made towers to follow strict graticule lines was a logical system to implement.

The longitude of the IGM circle in Palermo was determined in 1902 which was $\Lambda_o = -58° \ 25' \ 25.05''$ West of Greenwich and was later transferred to Belgrano in 1910, whose value was $\Lambda_o = -64° \ 13' \ 10.8''$ W. This is significant in the development of the Argentine network because of the technology of the time. Longitude transfers were a major technological accomplishment over long distances and were done with telegraph wires. The successes of such longitude (time) transfers document the modern development and settlement of the Americas as well as the world. When we attempt to model the systematic error in these old networks that form the metric base of existing large-scale topographic maps, we need to recognize that the major component of error is time = longitude.

Early Argentine topographic mapping followed the European model of the times, just like the organization of their mapping agency. A common projection of the time was the "Poliédrica" (Polyhedric). As I have pointed out in the past, this projection is mathematically equivalent to the local space rectangular implemented in analytical photogrammetry software. It is an ellipsoidal version of the gnomonic projection, and it is easy to draft a graticule from modest projection tables. The sheets were cut on the graticule and were 2° of latitude by 3° of longitude. That longitudinal spacing was convenient in later years when the IGM changed its basic projection.

Starting in 1894, Astro stations were determined throughout the Republic. By 1919, there was at least one point observed in every province of Argentina except for the northern provinces of Chaco, Formosa, and Catamarca and the southernmost provinces of Santa Cruz and Tierra del Fuego. The majority of these points were located at junctions or planned junctions of railroad tracks. Some of these Astro stations were used as local datums until the national chains of triangulation were able to incorporate the hinterlands into the network. Mendoza 1894 was one example. Others include Paraná 1908, where $\Phi_o = -31° 44' 00.7''$ S, $\Lambda_o = -60° 31' 58.5''$ W, and Tomé 1908, where $\Phi_o = -28° 32' 34.380''$ S, $\Lambda_o = -56° 02' 09.225''$ W. Another old datum that is still occasionally used in connection with oil exploration is the Chos Malal 1914 Datum, where $\Phi_o = -37° 22' 30.3''$ S, $\Lambda_o = -70° 17' 01.8''$ W. All of these old datums were established when the Bessel 1841 was the ellipsoid of reference for Argentina. However, the Chos Malal Datum of 1914 is probably used on the International 1924 ellipsoid, which was later adopted for all mapping in the country since 1926.

In 1926, the IGM adopted a new ellipsoid as well as a new projection for the national topographic series. The Gauss-Krüger Transverse Mercator was selected as the new projection and Grid System. The scale factor at origin ($m_o = 1.0$), the central meridians of the belts (C. M. $= \lambda_o = 72°, 69°, 66°$, etc., West of Greenwich), the False Easting at C. M. $= 500$ km, and the False Northing $= 10,002,288.299$ m. Note that the central meridians of the TM belts are the same interval as the predecessor Poliédrica. The defining parameters of the International ellipsoid (also called the Hayford 1909 and the Madrid 1924) are $a = 6,378,388$ m and $1/f = 297$. By 1926, the entire province of Buenos Aires (and more) was completely triangulated, so the present origin of the Argentine Datum of 1926 was included. The point of origin for the Campo Inchauspe triangulation station is in the town of Pehuajó, where $\Phi_o = -35° 58' 16.56''$ S, $\Lambda_o = -62° 10' 12.03''$ W. Different from most of the world, the proper name of the Argentine classical datum is the same name as the town and the origin point: "Campo Inchauspe Datum."

According to Mr. Rubén C. Rodriguez, by 1954 ten loops formed by chains of double triangles running along even-numbered meridians and parallels were completed and the first datum adjustment of Argentina was performed. All the angle measurements, baselines at chain intersections, and Laplace Azimuths determined at the same intersections and at half distances on meridian chains were included in that adjustment. By 1969, the network had grown to 19 loops, with a few baselines determined with electronic distance meters. There were 5,000 direction observations from 1,000 vertices (stations), and two Argentine geodesists adjusted the network at the U.S. Army Map Service in Washington, DC. The mean error of the least squares adjustment by variation of coordinates was 0.4″. Simultaneously, Dr. Irene Fisher directed the adjustment of the South American Datum of 1969 (SAD69) at Army Map Service (AMS) that included the Argentine data set in all the observations of the entire continent.

As a side note, I was assigned to AMS at the time, and later attended a classified symposium at Cameron Station, Virginia, where Dr. Fisher presented a paper on the SAD 69. I believe I was the only junior officer attending, but there was a veritable constellation of stars with all the generals and admirals there. The flag-rank officers sat in front of the audience, the senior civilian geodesists sat in the middle of the audience, and I sat in the rear with the rest of the peasants. Dr. Fisher walked onto the stage, and the retired Austrian school teacher-turned Senior Geodesist of the Army Map Service and Director of the Department of Defense Gravity Control Library stopped at the edge of the podium. She was about 4′ 10″ tall (1.5 m), Dr. Fisher paused, and turned to the front row of 2-, 3-, and 4-star generals and admirals and she looked at one particular Navy admiral.

She frowned, addressed that giant of a man who was seated, and started scolding the admiral as if he were a child. She told the admiral (who was the commander of the U.S. nuclear submarine fleet) that "Your Captains are not taking the proper observations for their gravity measurements. If your Captains cannot provide the data in proper scientific detail, do not waste my geodesists' time trying to decypher such garbage!" The Admiral cringed, said not a word, but he nodded acknowledgment. The audience was in stunned silence, the rest of the flag-rank officers had tears of silent laughter streaming from their eyes, and the tiny Dr. Irene Fisher turned, stepped onto the platform behind the podium so that she could reach the microphone, and presented her paper on the SAD 69.

The origin point for the South American Datum of 1969 is at station "Chua" in Brazil, where $\Phi_o = -19°\ 45'\ 41.6527''$ S, $\Lambda_o = -48°\ 06'\ 04.0639''$ W, and the azimuth to Uberaga is: $\alpha_o = 91°\ 30'\ 05.42''$. The ellipsoid of reference for the SAD 69 is the "South American Datum of 1969 ellipsoid," where $a = 6,378,160$ m and $1/f = 298.25$. The country of Argentina continued to favor and use the Campo Inchauspe Datum after 1969, and for precise positioning applications in the geophysical industry, the old classical datums prevailed. The Chos Malal Datum mentioned previously is still used for "oil patch" work in the central mountains near Chile, and in Patagonia, the Pampa del Castillo Datum is used for "oil patch" work. The geodetic coordinates of the origin point that bears the same name as the town is: $\phi_o = -45°\ 47'\ 30.2911''$ S, $\lambda_o = -68°\ 05'\ 27.7879''$ W, and $h_o = 732$ m. Of course, both these latter two local classical datums are referenced to the International ellipsoid.

Back in May of 1982, the Deputy Director of IGM informed me that the three-parameter datum shift values from the National Datum (Campo Inchauspe) to WGS 72 were: $\Delta X = +160.69$ m, $\Delta Y = -129.19$ m, $\Delta Z = -84.98$ m; this solution was based on observations at 21 stations. NIMA currently offers a ten-station solution to the WGS 84 Datum as: $\Delta X = -62$ m ± 5 m, $\Delta Y = -1$ m ± 5 m, $\Delta Z = -37$ m ± 5 m. Some years ago, I became privy to an "oil patch" solution for the Pampa del Castillo Datum. From Pampa del Castillo Datum to WGS 84 Datum, the two-point solution reported was: $\Delta X = +27.488$ m, $\Delta Y = +14.003$ m, $\Delta Z = +186.411$ m, but I would strongly recommend truncating the parameters to the closest 25 m! This is a good example of an analyst becoming enraptured with the "power" of the decimal point display and implying that the data is as good as the format statement allowed in the three-parameter solution on the computer. Remember that for trivial single-digit numbers of stations observed for the determination of systematic error from a classical geodetic datum to the WGS 84 Datum, NIMA quotes the accuracy at no better than 25 m. Millimeter level reporting does not equate to millimeter level accuracy. As a test point solution, for geodetic station Lagarto, $\phi = -45°\ 54'\ 36.2683''$ S, $\lambda = -68°\ 29'\ 40.3391''$ W (Campo Inchauspe Datum). The corresponding "oil patch" coordinates are: $\phi = -45°\ 54'\ 40.316''$ S, $\lambda = -68°\ 29'\ 34.389''$ W (Castillo del Pampa Datum), and: $\phi = -45°\ 54'\ 35.4876''$ S, $\lambda = -68°\ 29'\ 44.4146''$ W (WGS 84 Datum). Presumably, the ellipsoid height was constrained to zero. For the geodetic purist, the aforementioned solutions are substantially less than desirable, but they do reflect the common level of quality from some "oil patch" consultants.

The IGM currently publishes the POSGAR positions of its fiducial stations in Argentina that are part of the South American solution of geodetic positions referenced to the WGS 84 system of coordinates. Although in its preliminary stages of adjustment, the current 1999 POSGAR coordinates of station Campo Inchauspe are: $\phi = -35°\ 58'\ 1.9731''$ S, $\lambda = -62°\ 10'\ 14.8175''$ W, $h = 106.697$ m. I consider the coordinate precision quoted by the Argentina *Instituto Geográfico Militar* as significant.

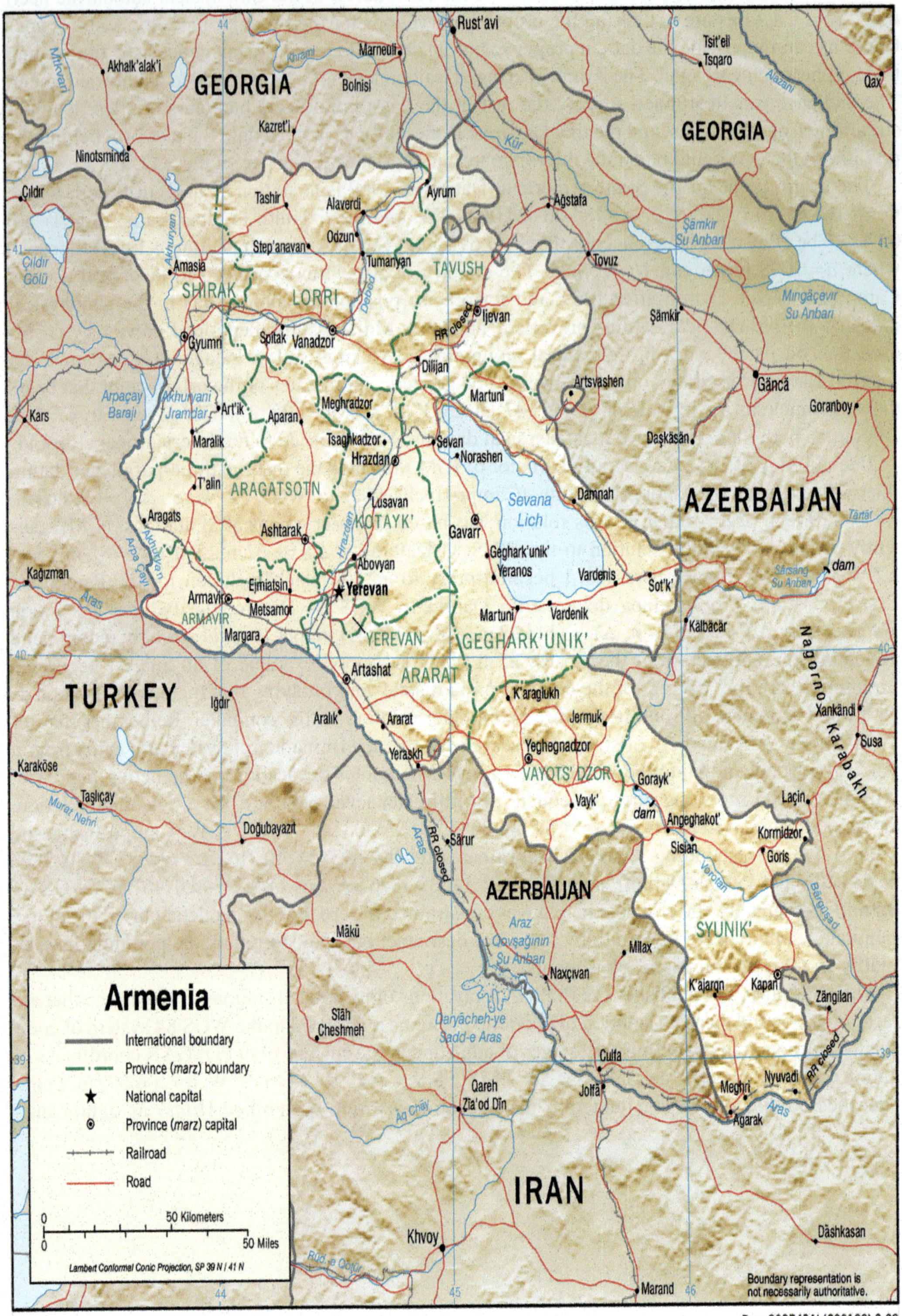

Base 802742AI (C00132) 8-02

Republic of Armenia

Bordered by Azerbaijan (996 km), Georgia (219 km), Iran (44 km), and Turkey (311 km), Armenia is slightly smaller than Maryland. The terrain consists of highland with mountains, little forest land, fast flowing rivers, and good soil in the Aras River valley. The lowest point is the Debed River (400 m), and the highest point is Aragats Lerrnagagat (4,090 m) (*World Factbook*, 2014).

"People first settled what is now Armenia in about 6000 *B.C.* The first major state in the region was the kingdom of Urartu, which appeared around Lake Van in the 13th century *B.C.* and reached its peak in the 9th century *B.C.* Shortly after the fall of Urartu to the Assyrians, the Indo-European-speaking proto-Armenians migrated, probably from the west, onto the Armenian Plateau and mingled with the local people of the Hurrian civilization, which at that time extended into Anatolia (present day Asian Turkey) from its center in Mesopotamia. Greek historians first mentioned the Armenians in the mid-sixth century *B.C.* Ruled for many centuries by the Persians, Armenia became a buffer state between the Greeks and Romans to the west and the Persians and Arabs of the Middle East. It reached its greatest size and influence under King Tigran II, also known as Tigranes or Tigran the Great (r. 95–55 *B.C.*). During his reign, Armenia stretched from the Mediterranean Sea northeast to the Mtkvari River (called the Kura in Azerbaijan) in present-day Georgia. Tigran and his son, Artavazd II, made Armenia a center of Hellenic culture during their reigns. By 30 *B.C.*, Rome conquered the Armenian Empire, and for the next 200 years Armenia often was a pawn of the Romans in campaigns against their Central Asian enemies, the Parthians. However, a new dynasty, the Arsacids, took power in Armenia in *A.D.* 53 under the Parthian king, Tiridates I, who defeated Roman forces in *A.D.* 62. Rome's Emperor Nero then conciliated the Parthians by personally crowning Tiridates king of Armenia. For much of its subsequent history, Armenia was not united under a single sovereign but was usually divided between empires and among local Armenian rulers" (*Library of Congress Country Studies*, 2014).

"The local religious scene in Armenian villages attracted Christian missionaries as early as *A.D.* 40, including the apostles Bartholomew and Thaddeus. According to lore, King Trdat III declared Christianity the state religion in *A.D.* 301. His moment of epiphany came after being cured of madness by St Gregory the Illuminator, who had spent 12 years imprisoned in a snake-infested pit, now located under Khor Virap Monastery. A version preferred by historians suggests that Trdat was striving to create national unity while fending off Zoroastrian Persia and pagan Rome. Whatever the cause, the church has been a pillar of Armenian identity ever since. Another pillar of nationhood arrived in 405 with Mesrop Mashtots' revolutionary Armenian alphabet. His original 36 letters were also designed as a number system. Armenian traders found the script indispensable in business. Meanwhile, medieval scholars translated scientific and medical texts from Greek and Latin. Roman and Persian political influence gave way to new authority when western Armenia fell to Constantinople in 387 and eastern Armenia to the Sassanids in 428. The Arabs arrived around 645 and pressure slowly mounted from Baghdad to convert to Islam. When the Armenians resisted, they were taxed to the point where many left for Roman-ruled territories, joining Armenian communities in a growing diaspora. Better conditions emerged in the 9th century when the Caliph approved the resurrection of an Armenian monarch in King Ashot I, the first head of the Bagratuni dynasty. Ani (now in Turkey) served as capital for a stint. Various invaders including the Seljuk Turks and Mongols took turns plundering and at times ruling and splitting Armenia. By the 17th century, Armenians were scattered across the empires of Ottoman Turkey and Persia, with diaspora colonies from India to Poland. The Armenians rarely lived in a unified empire but stayed in distant mountain provinces where some would thrive while others were depopulated. The seat of the Armenian Church wandered from Echmiadzin to Lake Van and further west for centuries.

"The Russian victory over the Persian Empire, around 1828, brought the territory of the modern-day Armenian republic under Christian rule, and Armenians began immigrating to the region. The Tsarist authorities tried to break the Armenian Church's independence, but conditions were still preferable to those in Ottoman Turkey, where many Armenians still lived. When the latter pushed for more rights, Sultan Abdulhamid II responded in 1896 by massacring between 80,000 and 300,000 Armenians. The European powers had talked often about the 'Armenian Question', considering the Armenians a fellow Christian people living within the Ottoman Empire. During WWI some Ottoman Armenians sided with Russia in the hope of establishing their own nation state. A triumvirate of pashas who had wrested control of the Empire viewed these actions as disloyal, and ordered forced marches of all Armenian subjects into the Syrian deserts. What is less certain – and remains contentious to this day – is whether they also ordered pogroms and issued a decree for Armenians to be exterminated. Armenians today claim that there was a specific order to commit genocide; Turks strenuously deny this. What is inescapable is the fact that between 1915 and 1922 around 1.5 million Ottoman Armenians died.

"The first independent Armenian republic emerged in 1918, after the November 1917 Russian Revolution saw the departure of Russian troops from the battlefront with Ottoman Turkey. It immediately faced a wave of starving refugees, the 1918 influenza epidemic, and wars with surrounding Turkish, Azeri and Georgian forces. It fought off the invading Turks in 1918, and left the final demarcation of the frontier to Woodrow Wilson, the US president. Meanwhile, the Turks regrouped under Mustafa Kemal (later Kemal Ataturk) and overran parts of the Caucasus. Wilson's map eventually arrived without troops or any international support, while Ataturk offered Lenin peace in exchange for half of the new Armenian republic. Beset by many other enemies, Lenin agreed.

"The Armenian government, led by the Dashnaks, a party of Armenian independence fighters, capitulated to the Bolsheviks in 1921. They surrendered in order to preserve the last provinces of ancient Armenia. The Soviet regime hived off Karabakh and Naxçıvan (Nakhchivan) for Azerbaijan. Forced from their homes, hundreds of thousands of survivors regrouped in the French-held regions of Syria and Lebanon, emigrating en masse to North America and France. Remarkably, the Armenians who stayed began to rebuild with what was left, laying out Yerevan starting in the 1920s. Armenia did well in the late Soviet era, with lots of technological industries and research institutes. Armenians voted for independence on 21 September 1991" (*Lonely Planet, 2014*). "In spring 2018, former President of Armenia (2008–18) Serzh Sargsian of the Republican Party of Armenia (RPA) tried to extend his time in power by becoming prime minister, prompting popular protests that became known as the "Velvet Revolution" after Sargsian was forced to resign. The leader of the protests, Civil Contract party chief Nikol Pashinyan, was elected by the National Assembly as the new prime minister on 8 May 2018. Pashinyan's party prevailed in an early legislative election in December 2018, and he was reelected as prime minister" (*World Factbook, 2022*).

By 1912, the majority of Armenia had been covered by 1st-Order classical triangulation surveyed by the Russian Army. The triangulation and 1:42,000 scale topographic mapping was referenced to the prime meridian at Pulkovo Observatory in St. Petersburg, Russia, where $\Phi_o = 59°\ 46'\ 18.55''$ North, $\Lambda_o = 30°\ 19'\ 42.09''$ East of Greenwich. The ellipsoid of reference used at the time was the Walbeck, where $a = 6,376,896$ m and $1/f = 302.78$. Pendulum gravity observations were made near the cities of Dilijan and Gyumri. However, precise leveling lines bypassed Armenia in favor of Georgia and Azerbaijan (*Anuario del Instituto Geografico Militar, Buenos Aires, 1914*). The majority of positioning performed in Armenia during the latter half of the 20th century has been with reference to the traditional System 42 Datum referenced to the Krassovsky 1940 ellipsoid, where $a = 6,378,245.0$ m, and $1/f = 298.3$. With its origin still at Pulkovo Observatory, the defining azimuth at the point of origin to Signal A is: $\alpha_o = 317°\ 02'\ 50.62''$.

On March 11, 2002, the Republic of Armenia passed decision Number 225, declaring WGS 84 as the official Datum of the country. From 2005 to 2008, Geographic Information Systems were carried out in Ashtarak, Tsakhkadzor, Byureghavan, and in the Administrative Districts of Yerevan (Arabkir, Davtashen, Kanaker-Zeytun, Avan, Nor Nork, Achapnyak, Nork-Marash, Malatia-Sebastia, Nubarashen, and Erebuni). By 2010, the Republic planned to implement GPS Continuously Operating Reference Systems throughout the country (*Global Navigation Satellite Systems: Armenian Experience, by H. Vardges, Chisinau Moldova, May 2010*).

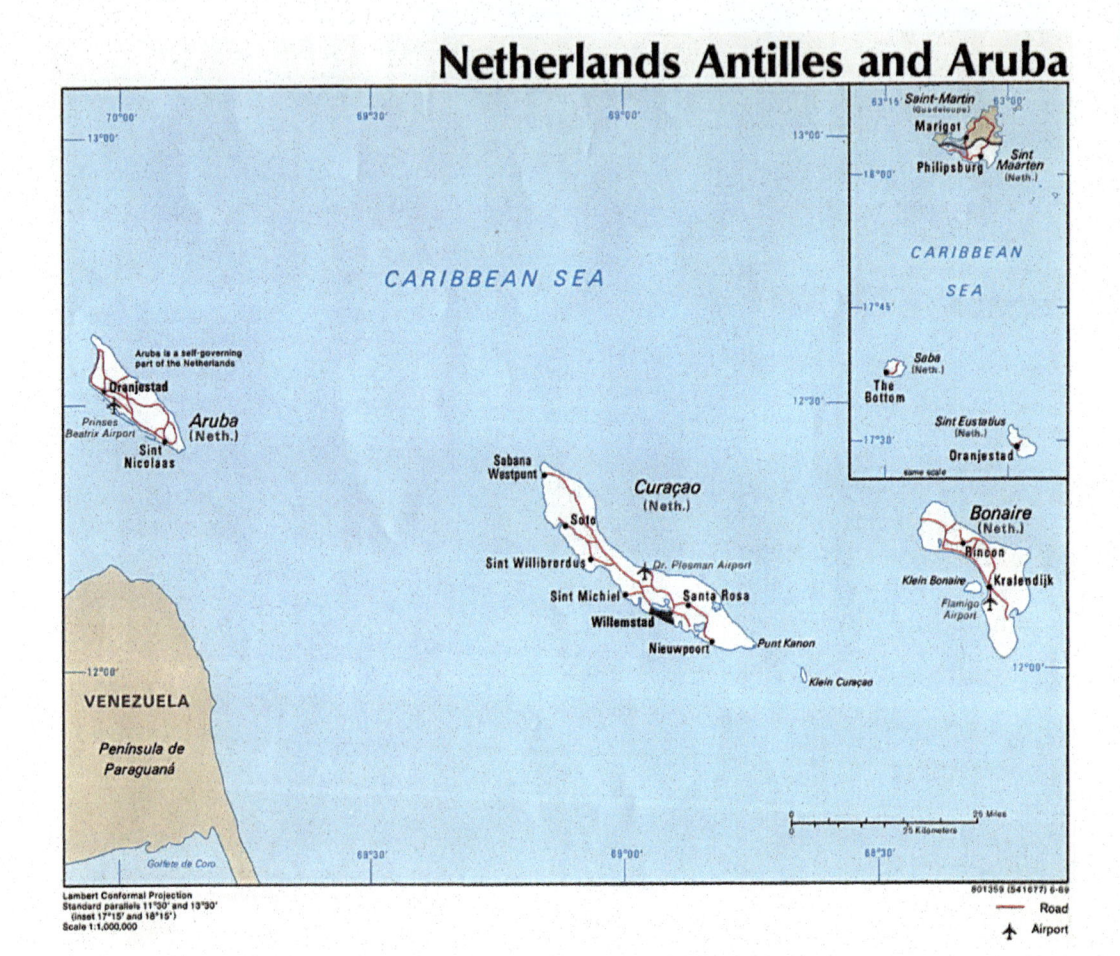

Netherlands Antilles and Aruba

Aruba and the Netherlands Antillies

Aruba has an area of 193 km², and is slightly larger than Washington, DC. The Caribbean island is barren, with the capital in Oranjestad. Its highest point is Ceru Jamanota (188 m) and its lowest point is the Caribbean Sea (0 m).

Originally a single colony of the Kingdom of the Netherlands, these islands consist of two groups: the Windward Islands – Aruba (now independent), Curaçao, Bonaire, and the Leeward Islands – Sint Maarten (shared with French Guadeloupe), Saba, and Sint Eustatius. The Leeward Islands were discovered by Hojeda in 1499 and were occupied by Spain in 1527. Taken by the Dutch in 1634, Curaçao immediately became an important post, trading with Coro, Puerto Cabello, and La Guaira, Venezuela. Once the center of slave trading in the Caribbean, Curaçao was hard hit by the abolition of slavery in 1863. Aruba and Curaçao recovered economically in the early 20th century after oil refineries were built to service the crude oil being produced in Venezuela's Lake Maracaibo. During WWII, Aruba was bombarded by German submarines in 1942. In October 2010, the former Netherlands Antilles was dissolved and the three smallest islands – Bonaire, Sint Eustatius, and Saba – became special municipalities in the Netherlands administrative structure. The larger islands of Sint Maarten and Curacao joined the Netherlands and Aruba as constituent countries forming the Kingdom of the Netherlands.

Part of the Kingdom of the Netherlands, full autonomy in internal affairs was obtained in 1986 upon separation from the Netherlands Antilles; the Dutch government remains responsible for defense and foreign affairs. Offshore banking and oil refining and storage are important to the Aruban economy, but the mainstay is tourism. The Netherlands Antilles has a combined area of 960 km², and the capital is Willemstad on the island of Curaçao.

The original geodetic datums in the Netherlands Antilles were Astro stations established at Willemstad on Curaçao and on Aruba in 1908. Topographic maps were compiled at 1:200,000 scale by the Dienst van het Kadaster from 1911 to 1915. The French observed a position on Sint Maarten in 1949 at Fort de Marigot Astronomic Pillar, where $\Phi_o = 18°\ 04'\ 28.1''$ N, $\Lambda_o = -63°\ 05'\ 14.654''$ West of Greenwich, $h_o = 61.84$ m, and the reference azimuth from the Fort to Mat K.L.M, $\alpha_o = 218°\ 14'\ 07.8''$. The ellipsoid of reference was the International 1909, where $a = 6,378,388$ m and $1/f = 297$.

Geodetic Surveys were conducted by the Dienst van het Kadaster on all the islands comprising the Netherlands West Indies starting in 1951, and each island received a complete classical triangulation network and Transverse Mercator Grid System. Each Grid is on the International ellipsoid, and each Grid has a scale factor at origin equal to unity. About the same time, the U.S. Army Map Service's IAGS started a massive triangulation campaign in the Caribbean and most of Latin America.

The triangulation network of Aruba consists of 24 stations, and all angles were measured with a Wild T-2 (1″ precision) theodolite. The base of the network (Station 2-Station 21) was measured with a standard measuring tape. The origin point for the Aruba Datum of 1951 is at Station No. 8, where $\Phi_o = 12°\ 31'\ 12.360''$ N, $\Lambda_o = -69°\ 59'\ 34.586''$ West of Greenwich. This is also the Grid Origin for the Aruba TM, and the False Easting is 10 km, and the False Northing is 15 km. The IAGS co-located with Jamanota, Station No. 12. From the Aruba Datum of 1951 to the WGS84 Datum implemented as the North American Datum of 1983 by the National Geodetic Survey, $\Delta X = \pm 1$ m, $\Delta Y = +112$ m ± 1 m, and $\Delta Z = -360$ m ± 1 m, and my solution is based on two stations.

The triangulation network of Curaçao consists of 59 stations, and all angles were measured with a Wild T-3 (0.5″ precision) theodolite. The base of the network (Station 8-Station 9) was measured with a Geodimeter Model 2. The origin point for the Curaçao Datum of 1951 is at Station No. 8, where $\Phi_o = 12° 11' 58.145''$ N, $\Lambda_o = -69° 00' 31.791''$ West of Greenwich, and $h_o = 96.66$ m. The IAGS co-located with Taf, Station No. 3 and Chris, Station No. 13. Station No. 8 is also the Grid Origin for the Curaçao TM, and the False Easting is 42,098.45 m and the False Northing is 60,044.53 m. From the Curaçao Datum of 1951 to the WGS84 Datum implemented as the North American Datum of 1983 by the National Geodetic Survey, $\Delta X = -266$ m ± 1 m, $\Delta Y = +109$ m ± 1 m, and $\Delta Z = -361$ m ± 1 m, and my solution is based on three stations.

The triangulation network of Bonaire consists of 35 stations, and all angles were measured with a Wild T-3 (0.5″ precision) theodolite. The base of the network (Station 1-Station 2, 1-3, and 2-3) was measured with a Tellurometer MRA 1. The origin point for the Bonaire Datum of 1951 is at Station Grandi, where $\Phi_o = 12° 10' 46.971''$ N, $\Lambda_o = -68° 15' 06.639''$ West of Greenwich, and $h_o = 98.45$ m. The IAGS co-located with stations: Grandi, Brandaris, and Will. Station Grandi is also the Grid Origin for the Bonaire TM, and the False Easting is 23 km and the False Northing is 20,980.49 m. From the Bonaire Datum of 1951 to the WGS84 Datum implemented as the North American Datum of 1983 by the National Geodetic Survey, the geocentric translations are probably quite similar to those for both Aruba and Curaçao.

The triangulation network of Saba consists of 14 stations, and all angles were measured with a Wild T-3 (0.5″ precision) theodolite. The base of the network (Station 2-Station 4, 4-5, 5-6, 6-7, 7-8, and 8-9) was measured with a Tellurometer MRA 1. The origin point for the Saba Datum of 1951 is at Station Saba No. 1C, where $\Phi_o = 17° 38' 07.606''$ N, $\Lambda_o = -63° 14' 17.187''$ West of Greenwich, and $h_o = 834.20$ m. The IAGS co-located with station Saba. Station Saba is also the Grid Origin for the Saba TM, and the False Easting is 4,714.87 m and the False Northing is 1,967.19 m.

The triangulation network of Sint Eustatius consists of 21 stations, and all angles were measured with a Wild T-3 (0.5″ precision) theodolite. The base of the network (Station 1-Station 1A, 1-1C, 1-5, 1-7, 7-6, 6-5, 11-12, and 12-13) was measured with a Tellurometer MRA 1. The origin point for the Sint Eustatius Datum of 1951 is at Station Quill No. 1A, where $\Phi_o = 17° 28' 33.272''$ N, $\Lambda_o = -62° 57' 37.458''$ West of Greenwich, and $h_o = 600.44$ m. The IAGS co-located with station Quill. Station Quill is also the Grid Origin for the Sint Eustatius TM, and the False Easting is 4,782.15 m and the False Northing is 1,831.37 m.

The triangulation network of Sint Maarten consists of 19 stations, and all angles were measured with a Wild T-3 (0.5″ precision) theodolite. The base of the network (Station 11-Station 14, 14-15, 15-12, 12-9, 9-1, and 1-7) was measured with a Tellurometer MRA 1. The origin point for the Sint Maarten Datum of 1949 is at Station Naked Boy, No. 1, where $\Phi_o = 18° 02' 19.391''$ N, $\Lambda_o = -63° 01' 55.766''$ West of Greenwich, and $h_o = 296.14$ m. The IAGS co-located with Naked Boy, Station No. 1. Naked Boy is also the Grid Origin for the Sint Maarten TM, and the False Easting is 12,598.47 m and the False Northing is 3,999.18 m. From the Sint Maarten Datum of 1949 to the WGS84 Datum implemented as the North American Datum of 1983 by the National Geodetic Survey, $\Delta X = -85$ m ± 1 m, $\Delta Y = +307$ m ± 1 m, and $\Delta Z = +45$ m ± 1 m, and my solution is based on four stations. Based on the Dutch triangulation diagrams I have examined, I would venture to say that the rough geocentric translation parameters for Saba and for Sint Eustatius would be quite similar, if not identical, to that of Sint Maarten.

Since the Netherlands West Indies occupies only part of the island of Sint Maarten, and the other part (Saint Martin) is occupied by the Department of Guadeloupe, I decided to scrounge my research files a bit more. Much to my surprise, I found a number of "DP" points (Dienst Punkt) listed in an old 1971 report of the French Navy. After comparing coordinates of the French with those of the Dutch, I realized that the French Navy occupied four Dutch survey points on St. Martin and came up with their own local French Datum rather than accept the Dutch values! From the Sint Maarten Datum of 1949 to the French Saint Martin Datum of 1951, $\Delta X = +138$ m ± 1 m, $\Delta Y = +246$

m ± 1 m, and ΔZ = −431 m ± 1 m, and my solution is based on four stations. The original Astro observation of the origin point was performed in 1949, therefore the earlier date. Thanks go to Mr. John W. Hager, retired from AMS/DMA/NIMA, to Mr. Dave Doyle, Senior Geodesist, United States National Geodetic Survey, and to Mr. Dwingo E. Puriël, Chief of Cadastral Office (Dienst van het Kadaster, Nederlandse Antillen), Curaçao.

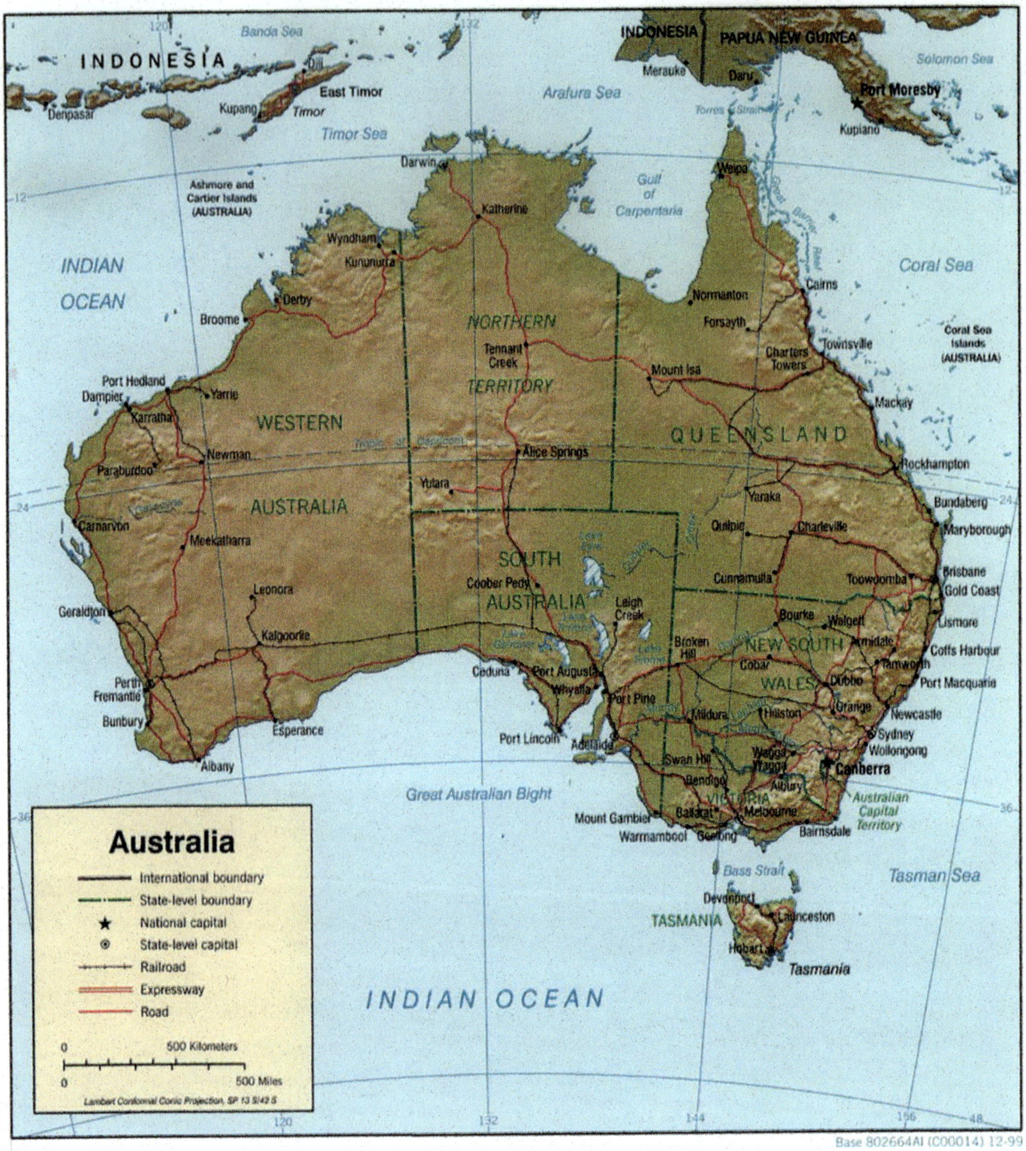

Base 802664AI (C00014) 12-99

Commonwealth of Australia

Located in Oceania, it is a continent between the Indian Ocean and the South Pacific Ocean and slightly smaller than the U.S. contiguous 48 states. It has a coastline of 25,760 km, is mostly a low plateau with deserts; fertile plain in southeast and the highest point is Mount Kosciuszko (2,228 m), the lowest point is Lake Eyre (−15 m), and has a mean elevation of 330 m.

Originally populated by aborigines who probably came from Asia about 40,000 years ago, Australia was first sighted by the Spanish in the early 17th century. In 1606, the Dutch landed on the eastern coast of the Bay of Carpentaria and named it New Holland. The eastern part was claimed by Capt. James Cook in 1770 and named New South Wales. The first English settlement at Port Jackson was mainly populated by convicts and seamen in 1788.

Capt. Matthew Flinders circumnavigated Australia from 1801 to 1803 and exhibited a level of professionalism not previously seen in the hydrographic charting expeditions of others in the British Admiralty such as Vancouver. Flinders received his initial instruction in navigation and chart making as well as tongue-lashings by Capt. Bligh during the successful breadfruit voyage from Tahiti to the Caribbean. It was on that early voyage that Flinders was in charge of the navigation chronometers (*The Admiralty Chart* by RADM G. S. Ritchie, 1995).

Flinders proved the continental unity of New Holland and New South Wales. Named Australia in the 19th century, the entire continent was claimed by the United Kingdom in 1829. The continent of Australia is slightly smaller than the United States; the lowest point is Lake Eyre (−15 m), and the highest point is Mount Kosciusko (2,229 m). The *CIA Factbook* describes the country as mostly low plateau with deserts; fertile plain in southeast and having a generally arid to semiarid climate that is temperate in the south and east and tropical in the north. A further note points out that Australia is the "world's smallest continent but sixth-largest country; population concentrated along the eastern and southeastern coasts; regular, tropical, invigorating, sea breeze known as 'the Doctor' occurs along the west coast in the summer."

The first astronomical "fix" or precise position determination along the coast of southern Australia was by Flinders in 1801 when he wrote: "The latitude of our tents at the head of Port Lincoln, from the mean of four meridian observations of the Sun taken from an artificial horizon was $\Phi = 34° 48' 25''$ S. The longitude from thirty sets of distances of the sun (*sic*) and stars from the moon was $\Lambda = 135° 44' 51''$ E. – (Ritchie, 1995). The enormous size of the country and the fact that this continent is surrounded by water has resulted in many local datums being established in coastal areas and little early geodetic work in the vast interior. Among those lesser datums known to exist on the Clarke 1858 ellipsoid are: Adelaide Observatory, Astro Fixation Western Australia 21, Army OP LA22 Lacrosse Island, Valentine, Australian Pillar, Central Origin 1963, Cookes Pillar Broome, Townsville, Emery Point Lighthouse, Final Sydney 1941, Gladstone Observatory Spot, Old Sydney, Maurice 1962, Melbourne Observatory, Weipa Mission Astro, Mildura Aerodrome, Mt. Rapid Fleurien Peninsula, Plantation Point Jervis Bay, Point Langdon South Base, Groote Eylandt, Port Huon Hospital Bay Observation Spot, New South Wales, New Sydney, and Mount Campbell.

Prior to the Australian Geodetic Datum of 1966, the Clarke 1858 ellipsoid as used in Tasmania was $a = 6,378,293.645$ m and $1/f = 294.26$ and in Australia proper was $a = 6,378,339.78$ m and $1/f = 294.26$. The difference between the two was the Clarke foot $= 0.3047972654$ m versus the British foot of $1926 = 0.30479947$ m. Also, the Everest 1830 ellipsoid was used at least in parts of Western Australia with $a = 6377304.068$ and $1/f = 300.8017$. Of the earlier more important datum origins, there were: Sydney Observatory, where $\Phi_o = 33° 51' 41.10''$ S and $\Lambda_o = 151° 12' 17.85''$ E, Perth Observatory 1899 where $\Phi_o = 31° 57' 09.63''$ S and $\Lambda_o = 115° 50' 26.10''$ E, Darwin Origin Pillar, where $\Phi_o = 12° 28' 08.452''$ S and $\Lambda_o = 130° 50' 19.802''$ E, and Lochmaben Astro Station in Tasmania, where $\Phi_o = 41°\ 38'\ 23.389''$ S and $\Lambda_o = 147°\ 17'\ 49.725''$ E. The astronomic longitudes differed from geodetic longitudes on either the Sydney or Perth origins on an average of 10", which indicated the magnitude of the deflections of the vertical.

During the 1930s, the Australia Belts were devised on the Transverse Mercator projection. Referenced to the Clarke 1858 ellipsoid, and an ersatz military datum, the scale factor was equal to unity; the belts were numbered from 1 to 8 and were 5° wide, starting with a Central Meridian at 116° and continuing east. Each belt had a False Easting at the Central Meridian of 400,000 yards, and the False Northing origin was 800,000 yards at 34° S. A caveat published by the U.S. Lake Survey, New York Office, in 1944 cautioned: "If these false coordinates are used, negative values will result in Tasmania." According to Mr. Mal Jones, "In Tasmania they added 1,000,000 to the False Northings to keep coordinates positive." The Clarke foot was implemented for this grid. A test point was published, where $\varphi = 39°\ 31'\ 12.767''$ S, $\lambda = 143°\ 27'\ 46.321''$ E, X = 631,629.24 yds, Y = 126,892.94 yds.

The least squares adjustment of the Australian geodetic network performed in March 1966 used the Australian Geodetic Datum. This adjustment produced a set of coordinates which, in the form of latitudes and longitudes, was known as the Australian Geodetic Datum 1966 coordinate set (AGD66). The grid coordinates derived from a Universal Transverse Mercator projection of the AGD66 coordinates, used the Australian National Spheroid, and was known as the Australian Map Grid 1966 coordinate set (AMG66), but with opposing mathematical convergence convention to match the northern hemisphere of negative west of the Central Meridian and positive east. New South Wales instituted the Integrated Survey Grid (ISG) where the projection was the Transverse Mercator truncated to the cubic terms since the belts were only 2° wide with a ¼° overlap. The scale factor at origin, $m_o = 0.99994$, the False Easting (FE) = 300 km, and the False Northing (FN) = 5,000 km at the equator. The central meridians $\lambda_o = 141°$, 141°, *etc.* to 153° E.

Thanks to Geoscience Australia, "While much early mapping was based on these origins, some 1:250,000 maps were based only on astronomical observations with an accuracy of the order of 100 metres or more, or by a mixture of astro and conventional surveying. A comparison of coordinates based on different origins of this kind will include differences due to the uncertainty of the astronomical observation as well as the deflections of the vertical and could show differences of several hundreds of metres. For a short period in 1962, geodetic computations were performed on the so-called 'NASA' spheroid with an origin at Maurice as below; but these computations were completely superseded." $a = 6,378,148$ m, $1/f = 298.3$. "From the end of 1962 until April 1965, the computation and adjustment of the Australian Geodetic Survey was done on the '165' spheroid: $a = 6,378,165$ m, $1/f = 298.3$. Prior to April 1963, the 'Maurice' origin used with the NASA spheroid was retained. As a result of these computations, new origin values were determined and from April 1963 to April 1965, computations were made on the 165 spheroid and this new 'Central' origin. Computations still emanated from Maurice whose various coordinates were: 165 Central: S 32° 51′ 13.979″, E 138° 30′ 34.062″, 165 Maurice: S 32° 51′ 13.000″, E 138° 30′ 34.000″, Clarke 1858, Sydney: S 32° 51′ 11.482″, E 138° 30′ 42.29″, and Astronomic: S 32° 51′ 11.341″, E 138° 30′ 25.110″. The Central origin was based on the best mean fit to 155 Laplace stations spread over the whole of Australia except for Cape York and Tasmania. The residual mean deflection was less than 0.1″ in both latitude and longitude whether isostatic topographic corrections were applied to the astronomic values or not. It was therefore considered unlikely that there was a significant artificial component in N with the Central origin. As no observed values of N from geoid surveys existed, it was assumed that N is everywhere zero. *('N' here refers to the separation between the geoid and the ellipsoid – Ed.)*

"In April 1965, it was changed to the spheroid adopted by the International Astronomical Union, and this spheroid was called the Australian National Spheroid: $a = 6,378,160$ m, and $1/f = 298.25$. In May 1965 a complete recomputation of the geodetic surveys of Australia was begun, emanating from the trigonometrical station Grundy, whose coordinates on both the 165, Central datum and the Australian National Spheroid, Central origin were: $\varphi = S\ 25°\ 54'\ 11.078''$, $\lambda = E\ 134°\ 32'\ 46.457$. By December 1965, the total number of Laplace stations in Australia was 533. From these, 275 stations were selected... no corrections for the topography were applied... and it was found that random undulations in the geoid make it impossible to locate a centre for the spheroid with a standard error of less than 0.5 seconds, about 15 metres, even with a very large number of stations.

"The Central origin was therefore retained but is now defined in terms of the Johnston memorial cairn. The Central origin was originally defined in terms of the trigonometrical station Grundy... The spheroid is oriented by defining the minor axis to be parallel to the earth's mean axis of rotation at the start of 1962 and defining the origin of geodetic longitude to be E 149° 00' 18.855" west of the vertical through the photographic zenith tube at Mt. Stromlo. The size, shape, position and orientation of the spheroid are thus completely defined, and together define the Australian Geodetic Datum: Johnston φ_o=S 25° 56' 54.5515", λ_o=E 133° 12' 30.07s71", h=571.2 metres ellipsoid height."

"The Geocentric Datum of Australia 1994 (GDA94) is the new Australian coordinate system, replacing the Australian Geodetic Datum (AGD). GDA is part of a global coordinate reference frame and is directly compatible with the Global Positioning System (GPS). It is the culmination of more than a decade of anticipation and work by the Intergovernmental Committee on Surveying and Mapping (ICSM) and its predecessor, the National Mapping Council (NMC). When the NMC adopted the AGD84 coordinate set in 1984, it *"recognised the need for Australia to eventually adopt a geocentric datum."* This was further recognized in 1988 when ICSM *"recommended the adoption of an appropriate geocentric datum by 1 January 2000.""*

The state of Western Australia has the "Project Grids" that closely correspond to what we use in the United States as State Plane Coordinates. The new Project Grids for the GDA94 Datum as well as for the previous datum for each are used for the following regional areas: Albany GDA94 – λ_o=117° 53' 00", m_o=1.00000440, FE=50km, FN=4,000km, and for Albany AGD84- λ_o=117° 55' 00", m_o=1.000012, FE=50km, FN=4,000km; for Broome GDA94- λ_o=122° 20' 00", m_o=1.00000298, FE=50km, FN=2,200km, and for Broome AGD84- λ_o=122° 20' 00", m_o=1.000003, FE=50km, FN=2,200km; Busselton GDA94- λ_o=115° 26' 00", m_o=0.99999592, FE=50km, FN=3,900km, and for Busselton AGD84- λ_o=115° 26' 00", m_o=1.000007, FE=50km, FN=3,900km; for Carnarvon GDA94- λ_o=113° 40' 00", m_o=0.99999796, FE=50km, FN=2,950km, and for Carnarvon AGD84- λ_o=113° 40' 00", m_o=1.000005, FE=50km, FN=3,050km; for Christmas Island GDA94- λ_o=105° 37' 30", m_o=1.00002514, FE=50km, FN=1,300km, and for Christmas Island WGS84- λ_o=105° 37' 30", m_o=1.000024, FE=50km, FN=1,300; for the Cocos (Keeling) Islands AGD94- λ_o=96° 52' 30", m_o=0.99999387, FE=50km, FN=1,500km, and for the Cocos (Keeling) Islands WGS84- λ_o=96° 52' 30", m_o=1.0, FE=50km, FN=1,400km; for Collie GDA94- λ_o=115° 56' 00", m_o=1.0000190, FE=40km, FN=4,000km; for Esperance GDA94- λ_o=121° 53' 00", m_o=1.00000550, FE=50km, FN=3,950km, and for Esperance AGD84- λ_o=121° 53' 00", m_o=1.000012, FE=50km, FN=3,950km; for Exmouth GDA94- λ_o=114° 04' 00", m_o=1.00000236, FE=50km, FN=2,650km, and for Exmouth AGD84- λ_o=114° 04' 00", m_o=1.000009, FE=50km, FN=2,750km; for Geraldton GDA94- λ_o=114° 35' 00", m_o=1.00000628, FE=50km, FN=3,350km, and for Geraldton AGD84- λ_o=114° 40' 00, m_o=1.000016, FE=50km, FN=3,350km; for Goldfields GDA94- λ_o=121° 30' 00", m_o=1.00004949, FE=60km, FN=3,700km, and for Goldfields AGD84- λ_o=121° 27' 00", m_o=1.000057, FE=60km, FN=4,000km; for Jurien GDA94- λ_o=114° 59' 00", m_o=1.00000314, FE=50km, FN=3,550km, and for Jurien AGD84- λ_o=114° 59' 00", m_o=1.000010, FE=50km, FN=3,550km; for Karratha GDA94- λ_o=116° 56' 00", m_o=0.99999890, FE=50km, FN=2,450km, and for Karratha AGD84- λ_o=116° 56' 00", m_o=1.000004, FE=50km, FN=2,450km; for Kununurra GDA94- λ_o=128° 45' 00, m_o=1.00001650, FE=50km, FN=2,000km, and for Kununurra AGD84- λ_o=128° 45' 00", m_o=1.000014, FE=50km, FN=2,000km; for Lancelin GDA94- λ_o=115° 22' 00", m_o=1.00000157, FE=50km, FN=3,650km, and for Lancelin AGD84- λ_o=115° 22' 00", m_o=1.00000157, FE=50km, FN=3,650km; for Margaret River GDA94- λ_o=115° 10' 00", m_o=1.00000550, FE=50km, FN=3,950, and for Margaret River AGD84- λ_o=115° 06' 00", m_o=1.000014, FE=50km, FN=4,050km; for Perth GDA94- λ_o=115° 49' 00", m_o=0.99999906, FE=50km, FN=3,800km, and for Perth AGD84- λ_o=115° 50' 00", m_o=1.000006, FE=40km, FN=3,800km; for Port Hedland GDA94- λ_o=118° 36'

00″, $m_o = 1.00000135$, FE = 50 km, FN = 2,400 km, and for Port Hedland AGD84 - $\lambda_o = 118°\ 35'\ 00″$, $m_o = 1.000004$, FE = 50 km, FN = 2,400 km.

Datum shifts between the various classical datums and the various scientific reference frames of the GPS satellites are available for cartographic-accuracy transformations. However, for precise geodetic applications, the parameters change monthly because the entire continent is moving to the northeast at about 3 cm/year! For instance, a couple of cartographic transform accuracy parameter sets are given as follows: From Australian Geodetic Datum 1966 (Victoria/New South Wales) to WGS84: $\Delta X = -119.353$ m, $\Delta Y = -48.301$ m, $\Delta Z = +139.484$ m, $R_x = -7.243 \times 10^{-3}$ radians, $R_y = -4.538 \times 10^{-3}$ radians, $R_z = -7.627 \times 10^{-3}$ radians, and $\Delta s = -6.13 \times 10^{-1}$. From Australian Geodetic Datum 1984 to WGS84: $\Delta X = -117.763$ m, $\Delta Y = -51.51$ m, $\Delta Z = +139.061$ m, $R_x = -5.096 \times 10^{-3}$ radians, $R_y = -7.732 \times 10^{-3}$ radians, $R_z = -4.835 \times 10^{-3}$ radians, and $\Delta s = -1.91 \times 10^{-1}$. Note that the aforementioned datum shift parameters are the best available to me. These values have not been verified, nor are accuracy estimates available at this time. Australia is a free and open society. Their geodesy is not a secret and their history, their coordinates, and their datum transformations are an open book – a very large open book, but open. Thanks go to Mr. Malcolm A. B. Jones, "Geodesy Jones" of Perth for the enormous accumulation of Australian historical geodetic documents he has sent to me over the years.

Comprehensive information on upgrades to the Australian Geospatial Reference System can be found at: https://www.icsm.gov.au/upgrades-australian-geospatial-reference-system.

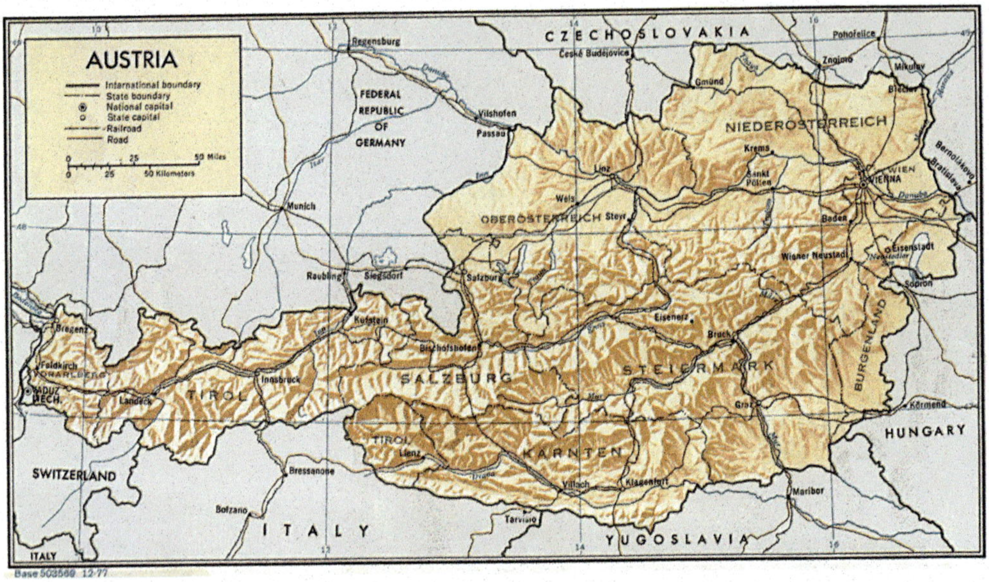

Base 503569 12-77

Republic of Austria

Bordered by Czech Republic (402 km), Germany (801 km), Hungary (321 km), Italy (404 km), Liechtenstein (34 km), Slovakia (105 km), Slovenia (299 km), and Switzerland (158 km). The highest point is Grossglockner (3,798 m), the lowest point is Neusiedler See (115 m), and the mean elevation is (910 m).

On November 1, 996, an area of land popularly known as "*Ostarrichi*" was given by Emperor Otto the Third to the Bishop of Friesing as a gift. In 1156 the *Priviglium Minus* elevated Austria to the status of a Duchy. On ascension of the Habsburg dynasty to power, the lands of Vorarlberg were added with Bohemia and Hungary added as provinces to their holdings. These acquisitions completed under the Habsburg rule were the foundation for the country of Austria today. After the crowning of Maria Theresa as Queen of Bohemia in 1743, her husband Franz was elected Holy Roman Emperor in 1745. As a measure of standing to other royal courts, Queen-Empress Maria Theresa ordered a survey of all of the Hapsburg holdings in 1763. This was the First Topographical Survey (*Josephinishe Aufnahme*) of the Habsburg provinces. The *Liesganig* triangulation and attached supplemental surveys were executed graphically with plane table and alidade. There was no geodetic survey used as a foundation. The associated topographic survey was performed at a scale of 1:28,800 and was based on the Vienna Klafter System where 1 Zoll = 400 Klafters = 758.6 m. Altogether there were about 4,500 sheets surveyed and all of them were kept secret for military purposes. After completion of this first survey in 1787, Emperor Franz I declared Austria an Empire.

The second topographic survey of Austria (*Franziszeische Aufnahme*) was conducted from 1806 to 1869. The Vienna Datum of 1806 was established based on the origin of St. Stephan Turm (St. Stephan's Tower) where $\Phi_o = 48° 12' 31.5277''$ North, $\Lambda_o = 16° 22' 27.3275''$ East of Greenwich. (These longitudes were originally referenced to Ferro, in the Canary Islands which is 17° 39' 46.02'' West of Greenwich.) The defining azimuth of the Datum was from St. Stephan Turm to Leopoldsberg: $\alpha_o = 345° 55' 22.0''$. Coordinates for the mapping were based on 5 Cassini-Soldner Grids. For the provinces of Lower Austria, the Vienna Grid was centered at the datum origin. The Gusterberg (*Kremsmünster*) Grid was for the provinces of *Oberösterreich* and Salzburg, where $\varphi_o = 48° 02' 18.4753'' N$, $\lambda_o = 14° 08' 15.0242''$ East of Greenwich, the *Schöcklberg* Grid (near Graz) was for the Province of *Steiermark*, where $\varphi_o = 47° 11' 54.8745'' N$, $\lambda_o = 15° 27' 59.9472''$ East of Greenwich, the *Krimberg* (*Laibach*) Grid near Ljubljana was for the provinces of Carinthia, Carinola, and Littoral (now Italy), where $\varphi_o = 45° 55' 43.7228'' N$, $\lambda_o = 14° 28' 18.8027''$ East of Greenwich, and the Innsbruck Grid, centered at the tower of the city-parish church (*Pfarrturm*) for the provinces of Tyrol and Vorarlberg, where $\varphi_o = 47° 16' 11.3060'' N$, $\lambda_o = 11° 23' 39.3157''$ East of Greenwich. The Bohnenberger ellipsoid was used from 1810 to 1845, where $a = 6,376,602$ m and $1/f = 324$. The Zach ellipsoid was used from 1845 to 1863, where $a = 6,376,602$ m and $1/f = 324$. From 1847 to 1851, the Walbeck ellipsoid was also used, where $a = 6,376,896$ m and $1/f = 302.78$. Some small-scale derivative mapping was done on the Bonne projection. The main chains of the second military triangulation surveyed in 1806–1829 covered the western part of the Empire (west of the Budapest meridian) and the chain which extended along the Carpathian Mountains to Transylvania. The baselines used were at Wiener-Neustadt (1762) 6,410.90 Klafters, at Wels (1806) 7,904.045 Klafters ±0.360 Klafters, at Raab (1810) 9,429.429 Klafters ±0.010 Klafters, and at Radovec (*Radautz* 1818) 5,199.597 Klafters.

The third topographic survey of the Austro-Hungarian Empire (*Neue Aufnahme*) was conducted from 1869 to 1896 and was mainly based on the Vienna Datum and the Bessel 1841 ellipsoid (actually adopted in 1863), where $a = 6,377,397.15$ m, $1/f = 299.1528$. The Arad, St. Anna Datum of 1840, was initially used in Transylvanian surveys, where $\Phi_o = 46° 18' 47.63''$ North, and $(\Lambda_o) = 39° 06'$

54.19″ East of Ferro (geodetically determined from Vienna). This datum was referenced to the Zach ellipsoid and started in Transylvania and Tyrol. In 1872, the metric system was legally established in the Empire and the scale of 1:25,000 was finally adopted for topographic surveys. The graticule sheet was adopted with the Polyhedric projection in order to eliminate inconsistencies in sheet lines and differences in the size and shapes of sheet lines. The baselines used were at Arad, St. Anna, Hungary (1840) 8,767.578 m, at Partin (Tarnow), Galicia (1849) 5,972.501 m, at Hall, Tyrol (1851) 5,671.317 m, at Weiner Neustadt (1857) 9,484.065 m, at Maribor, Styria (1860) 5,697.405 m, at Josefov (*Josefstadt*) Bohemia (1862) 5,257.266 m, and at Sinj, Dalmatia (1870) 2,475.474 m. The Walbeck ellipsoid was used in computing the chains in Balicia and Bucovina so that ties could be made with the Russian triangulation.

The most common classical datum (prior to the European 1950) found in Austria and still used extensively is the Militärgeographisches Institut (Military Geographic Institute or MGI) Hermannskogel, *Habsburgwarte* 1871 Datum, where $\Phi_o = 48° 16' 15.29'' N$, $\Phi_o = 16° 17' 41.06''$ East of Greenwich and azimuth to Hundsheimer is $\alpha_o = 107° 31' 41.7''$. I developed the seven-parameter datum shift relation between Hermannskogel 1871 and ED 50 for Yugoslavia, but most of the points were not in present Austria. The Austrian government has made transformation parameters available for "AT_MGI to ETRS89 (WGS 84)," where $\Delta X = +577.3$ m, $\Delta Y = +90.1$ m, $\Delta Z = 463.9$ m, $\Delta s = 2.42$ ppm, $R_X = -5.137''$, $R_Y = -1.474''$, $R_Z = -5.297''$. (The seven parameters are stated to be suitable "for applications with an accuracy of about 1.5 m.") "The three-dimensional coordinates of (X, Y, Z) of AT_MGI were derived using ellipsoidal heights, which are computed from leveling heights related to Molo Sartorio (Trieste) and a Geoid related to AT_MGI Datum Hermannskogel and Josefstadt in Bohemia." Note that the sign of the rotations has been changed to conform to the U.S. standard convention as used and published by the NGA and by the NGS. The standard projection for civilian topographic mapping in the republic is the Gauss-Krüger Transverse Mercator where the belts are 3° wide such that for Belt 3, $\lambda_o = 9°$, the False Easting = 3,500 km, for Belt 4, $\lambda_o = 12°$, the False Easting = 4,500 km, *etc.*; there is no False Northing and the scale factor at origin, $m_o = 1.0$.

The military standard 1:50,000 scale topographic series of Austria is on the ED50 Datum, and the parameters published by the NGA from ED50 to WGS84 are: $\Delta a = -251$, $\Delta f \times 10^4 = -0.14192702$, $\Delta X = -86$ m ± 3 m, $\Delta Y = -98$ m ± 8 m, $\Delta Z = -121$ m ± 5 m. The reader is cautioned that the above Austrian-government furnished seven parameters from Hermannskogel 1871 Datum to WGS 84 Datum *may NOT* be truncated to three parameters. Much of this was obtained from AMS Technical Report 25 by Andrew Glusic and from Dr. Suzanne Van Cooten's term paper in a graduate course she took from me years ago.

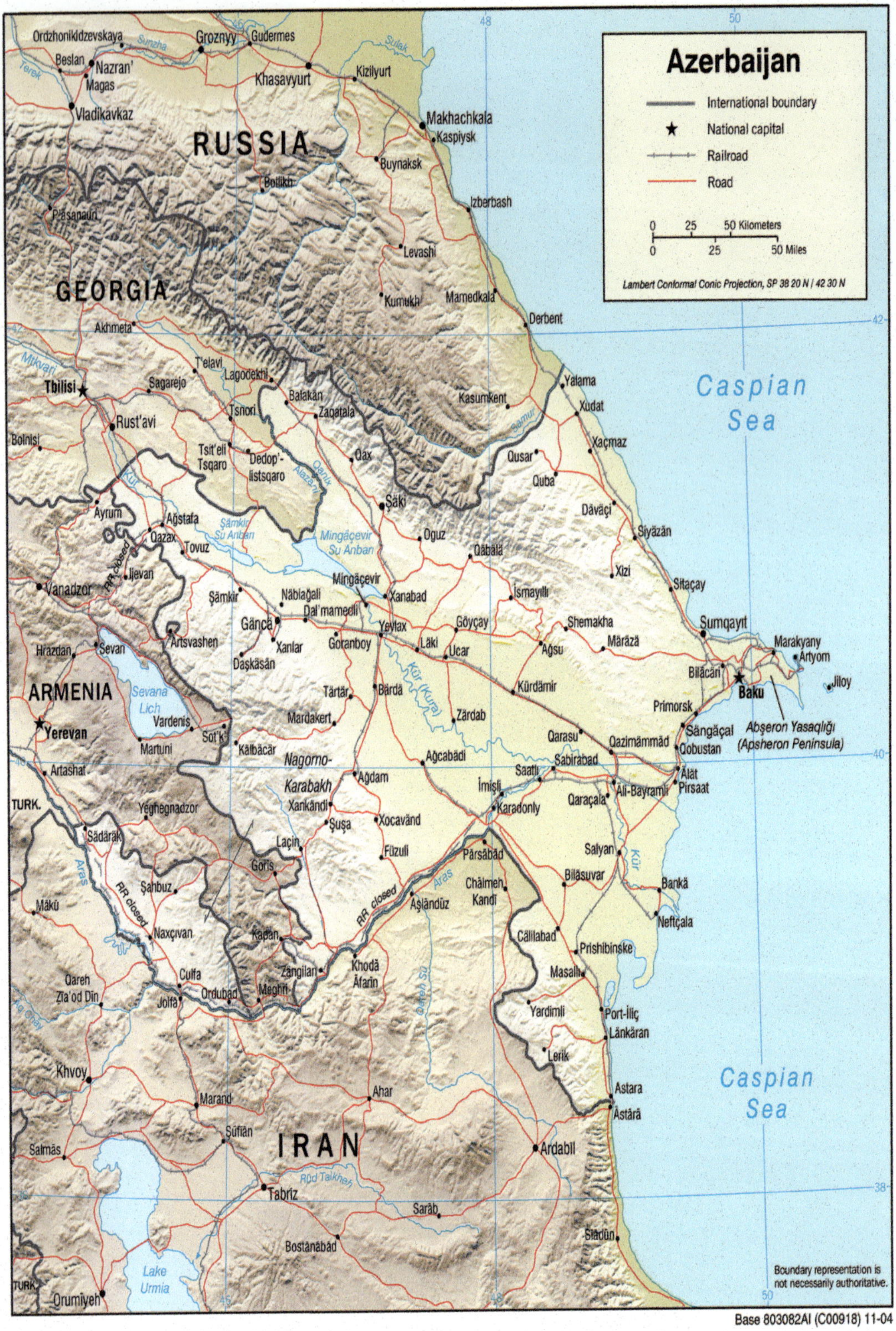

Base 803082AI (C00918) 11-04

Republic of Azerbaijan

Slightly smaller than Maine, Azerbaijan is bordered by Armenia (566 km), Georgia (322 km), Iran (432 km), Russia (284 km), and Turkey (9 km). The lowest point is the Caspian Sea (−28 m), and the highest point is *Bazardüzü Daği* (4,485). "The terrain is large, *Kür-Araz Ovalığı* (Kura-Araks Lowland) (much of it below sea level) with Great Caucasus Mountains to the north, *Qarabağ Yaylası* (Karabakh Upland) in west; Baku lies on *Abşeron Yasaqlığı* (Apsheron Peninsula) that juts into Caspian Sea" (*World Factbook,* 2010 and *NGA Quick Geonames Search,* 2010).

"In the ninth century *B.C.*, the seminomadic Scythians settled in areas of what is now Azerbaijan. A century later, the Medes, who were related ethnically to the Persians, established an empire that included southernmost Azerbaijan. In the sixth century *B.C.*, the Archaemenid Persians, under Cyrus the Great, took over the western part of Azerbaijan when they subdued the Assyrian Empire to the west. In 330 *B.C.*, Alexander the Great absorbed the entire Archaemenid Empire into his holdings, leaving Persian satraps to govern as they advanced eastward. Between the first and third centuries *A.D.*, the Romans conquered the Scythians and Seleucids, who were among the successor groups to the fragmented empire of Alexander. The Romans annexed the region of present-day Azerbaijan and called the area Albania. As Roman control weakened, the Sasanid Dynasty reestablished Persian control. Between the seventh and eleventh centuries, Arabs controlled Azerbaijan, bringing with them the precepts of Islam. In the mid-eleventh century, Turkic-speaking groups, including the Oghuz tribes and their Seljuk Turkish dynasty, ended Arab control by invading Azerbaijan from Central Asia and asserting political domination. The Seljuks brought with them the Turkish language and Turkish customs. By the thirteenth century, the basic characteristics of the Azerbaijani nation had been established. Beginning in the early eighteenth century, Russia slowly asserted political domination over the northern part of Azerbaijan, while Persia retained control of southern Azerbaijan. In the nineteenth century, the division between Russian and Persian Azerbaijan was largely determined by two treaties concluded after wars between the two countries. After the Bolshevik Revolution, a mainly Russian and Armenian grouping of Baku Bolsheviks declared a Marxist republic in Azerbaijan. Azerbaijan was strongly affected by the autonomy that spread to most parts of the Soviet Union under Gorbachev's liberalized regime in the late 1980s. After independence was achieved in 1991, conflict with Armenia became chronic, and political stability eluded Azerbaijan in the early years of the 1990s" (*Library of Congress Country Study,* 1994). "Under the terms of a cease-fire agreement following Azerbaijan's victory in the Second Nagorno-Karabakh War that took place from September-November 2020, Armenia returned to Azerbaijan the remaining territories it had occupied and also the southern part of Nagorno-Karabakh, including the culturally and historically important city that Azerbaijanis call Shusha and Armenians call Shushi. Despite Azerbaijan's territorial gains, peace in the region remains elusive because of unsettled issues concerning the delimitation of borders, the opening of regional transportation and communication links, the status of ethnic enclaves near border regions, and the final status of the Nagorno-Karabakh region. Russian peacekeepers deployed to Nagorno-Karabakh to supervise the cease-fire for a minimum five-year term have not prevented the outbreak of sporadic, low-level military clashes along the Azerbaijan-Armenia border in 2021" (*World Factbook,* 2022).

The earliest geodetic position for Baku *(Baki)*, Azerbaijan is the Khan's Palace Minaret at $\Phi_o = 40°\ 21'\ 57.90''$ N, $\Lambda_o = 49°\ 50'\ 27.57''$ E and is a 1927 Astro position, courtesy of John W. Hager. The likely reference ellipsoid is the Bessel 1841, where $a = 6,377,397.155$ m and the reciprocal of flattening $1/f = 299.1528128$. As was common with all of the satellite countries of the former U.S.S.R., the local datums and coordinate systems were replaced by the unified "System 42" with origin at the Pulkovo Observatory, where $\Phi_o = 59°\ 46'\ 18.55''$ North, $\Lambda_o = 30°\ 19'\ 42.09''$ East of

Greenwich. The defining azimuth at the point of origin to Signal A is: $\alpha_o = 317° \ 02' \ 50.62''$, and the ellipsoid of reference is the Krassovsky 1940, where $a = 6,378,245$ m and $1/f = 298.3$. The "Russia Belts" Grid System is used with the System 42 Datum; identical to UTM except that the scale factor at origin is unity. According to Roger Lott, Chief Surveyor of BP in his document (*Azerbaijan Coordinate Systems, 27 February 1997*), the local Grid designation of the North component is the X axis, and the X′ or Y axis is East. Furthermore, it is referred to as Coordinate System 42 or "CS42." Lott goes on to explain that Zone 8 is used for onshore areas west of 48°E and Zone 9 is used for onshore east of 48°E and the offshore areas of the Caspian Sea. An interesting quirk of this Grid is that the Zone number is used as the digit in the millions place of the False Easting, quite unlike what is used for UTM.

A secondary coordinate system used in Azerbaijan is referred to as the Coordinate System of 1963 (CS63) and is comprised of 3° wide belts rather than the standard military 6° wide Russia Belts. "Onshore Azerbaijan falls within the Transcaucasian block designated A. Official zone nomenclature within block A is unknown at the time of writing. However the zone covering eastern Azerbaijan has easting (Y) values prefixed by 4. If practice is consistent with that used in CS42, this would be designated CS63 zone A-4 with the zone adjacent to the west designated zone A-3" (*ibid., Lott, 1997*). According to Lott, the transformation in this region **from** System 42 Datum **to** WGS84 Datum can be approximated to an accuracy of 0.5 m to 1.0 m as: $\Delta X = -18$ m, $\Delta Y = +125$ m, and $\Delta Z = +83$ m. A test point is provided **from** System 42: $\varphi = 39° \ 59' \ 59.998'' $ N, $\lambda = 49° \ 05' \ 59.991''$ E **to** WGS84: $\varphi = 39° \ 59' \ 59.746''$ N, $\lambda = 49° \ 05' \ 59.967''$ E. A final note is that because of uncertainty in the geoid surface over Azerbaijan, errors in ellipsoid height of ±20 m result from using the above transformation parameters. It is advised that the parameters are not used for three-dimensional transformations.

REFERENCES: A

"Triangulation in Afghanistan," published by Survey of India in 1947.
11 March 2002, the Republic of Armenia passed decision Number 225.
1891, *Annales Hydrographiques*.
29 August 2007 in the Royal Decree number 1071 of 2007 (Institut Cartografic de Catalunya, 2015.
A New Quasi-Geoid Computation from Gravity and GPS Data in Algeria, S. A. Benahmed Daho, J. D. Fairhead. www.isgeoid.polimi.it/Newton/…2/Benahmed-A%20new%20quasigeoid-revised.pdf.
Anuario del Instituto Geografico Militar, Buenos Aires, 1914.
Azerbaijan Coordinate Systems, 27 February 1997, Roger Lott.
Cohen, Ruth, Geodetic Memo No. 485, Development of Rinner's Formulas in the Bonne projection and the inverse, Army Map Service, 28 October 1949.
Coordinate Reference Systems Used in Albania to Date, Nikolli, P., & Idrizi, B., FIG Working Week 2011, Morocco, 18–22 May 2011.
Dr. José Carvalho of Maputo, Moçambique, (Personal Communication – Feb. 2001).
Dr. Rui Fernandes, (Universidade da Beira Interior · Department of Computer Science · SEGAL Portugal · Covilhã), (Personal Communication – Feb. 2001).
Dr. Suzanne Van Cooten, Personal Communication – Nov. 2003.
European Petroleum Studies Group (EPSG) headed up by Mr. Roger Lott of British Petroleum.
Geoscience Australia. https://www.icsm.gov.au/upgrades-australian-geospatial-reference-system.
Gjata, G., et al., EUREF 2009 Symposium, Florence, Italy, May 2009.
Global Navigation Satellite Systems: Armenian Experience, by H. Vardges, Chisinau Moldova, May 2010.
Heliyon, Volume 5, Issue 4, April 2019, e01435.
International Boundary Study, Limits in the Seas, No. 7, Albanian Straight Baselines, Office of the Geographer, U.S. Department of State, February 16, 1970.
International Boundary Study, Number 6.
International Boundary Study, Number 89.
Isufi, E., *et al.*, Overview on Albanian Reference Systems, actual situation and the future challenge, BALGEOS, Vienna, January 2010.
Library of Congress Country Studies, 2014.

References: A

Library of Congress Country Study, 1994.
Lonely Planet, 2014.
Mapping of the Countries in Danubian and Adriatic Basins, Glusic, Andrew M., AMS TR No. 25, June 1959, 406 pages.
Marcussi, A., Lineamenti geoidici della penisola balcanica, Bollettino Geodetica, vol. XXIV, No. 4.
Mr. Dave Doyle, Senior Geodesist, United States National Geodetic Survey, Personal Communication – July 2002).
Mr. Dwingo E. Puriël, Chief of Cadastral Office, (Dienst van het Kadaster, Nederlandse Antillen) Curaçao.
Mr. John W. Hager, (Personal Communication – Feb. 2001).
Mr. John W. Hager, (Personal Communication – Sept. 2014).
Mr. Mal Jones of Perth, Australia, (Personal Communication – Feb. 2001).
Mr. Mal Jones, Personal Communication – Nov. 2003.
Mr. Rubén C. Rodriguez, Personal Communication – Nov. 1999.
National Imagery and Mapping Agency (NIMA) TR 8350.2 (dated 1991).
Pamir Boundary Commission 1895 (PBC95).
Perkins and Parry, *World Mapping Today*, 2nd Edition.
Prof. Charles L. Merry at the University of Cape Town, (Personal Communication – Feb. 2001).
Sebastian J.F., Kutushev S.B., (2015) [Modernization of geodetic net of the Republic of Angola. The base tasks]. Geodesy and Cartography=Geodezija i kartografija, 2, pp. 19–24. (In Russian). DOI: 10.22389/0016-7126-2015-896-2-19-24, however, the article itself was not available (21JAN2019). http://geocartography.ru/en/scientific_article/2015_2_19-24.
The Admiralty Chart by RADM G. S. Ritchie, 1995.
The Local Coordinate System of Kabul, Mestnaya Systema Koordinat Kabula), Geodeziya I Kartografiya, No. 12, 1988, pp. 21–33.
The National Report of Albania for 2009.
The Office of the Geographer, U.S. Department of State in their International Boundary Study No. 26, Revised in 1983.
Triangulation in India and Adjacent Countries, Sheet 42.G (Great Pamir), Dehra Dun, 1921.
U.S. Dept. of State Background Note, January 2007.
World Factbook, 2010 and NGA Quick Geonames Search, 2010.
World Factbook, 2014.
World Factbook, 2015.
World Factbook, 2022.
Zeitschrift für Vermessungswesen during the 1930s.

B

Commonwealth of the Bahamas 59
State of Bahrain 63
People's Republic of Bangladesh 67
Barbados 71
Republic of Belarus 73
Kingdom of Belgium 77
Belize 81
Republic of Bénin 83
Kingdom of Bhutan 89
Republic of Bolivia 93
Bosnia and Herzegovina 97
Republic of Botswana 101
Federative Republic of Brazil 105
Negara Brunei Darussalam 109
Republic of Bulgaria 113
Burkina Faso 119
State of Burma 123
Republic of Burundi 129

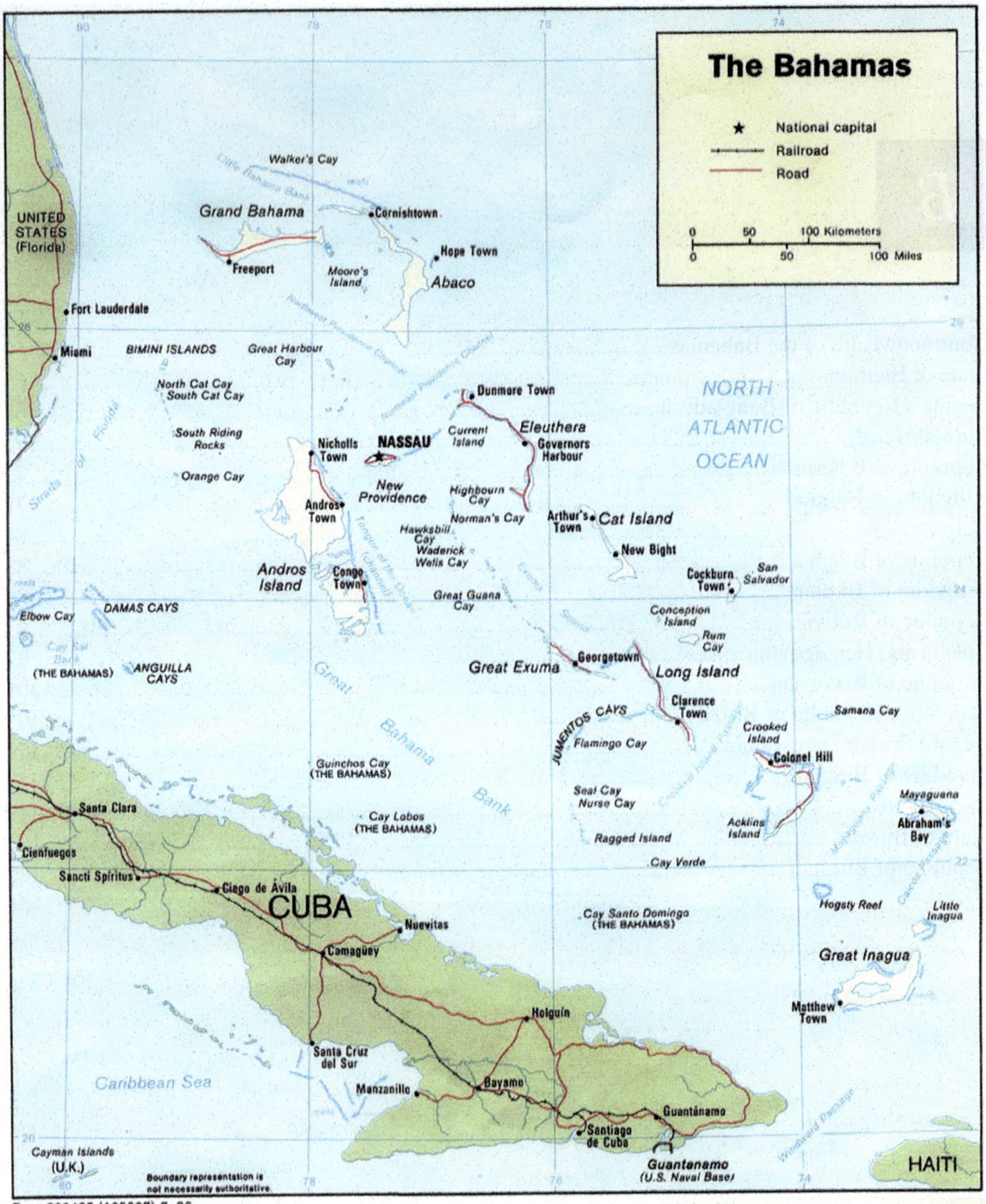

Base 800485 (A05607) 7-86

Commonwealth of the Bahamas

"Comprised of a 700-island and islet archipelago with an additional 2,400 cays and rocks in the North Atlantic Ocean, the total land area is 10,070 km² and it is slightly smaller than Connecticut. With a total coastline of 3,542 km, the terrain of the Bahamas is long, low coral formations with some low rounded hills. The lowest point is the Atlantic Ocean (0 m) and the highest point is Mount Alvernia (63 m) on Cat Island. It extends from eighty kilometers east of Florida to eighty kilometers northeast of Cuba" (*Library of Congress Country Study* 2022).

Inhabited by Lucayan Indians at the time of sighting by Christopher Columbus on October 12, 1492, the islands were assigned to Spain by papal grant. Subsequently occupied only by slave traders and buccaneers, the Bahamas were granted by the British crown to Sir Robert Heath in 1629. The Commonwealth of the Bahamas became independent from the United Kingdom on July 10, 1973. Twenty-two of the main islands are inhabited, 70% of the population of 316,000+ people live on New Providence and 16% live on Grand Bahama.

Prior to World War II, the only surveys performed in the Bahamas were astronomical observations ("Astros") of hazards to navigation and local cadastral-type surveys of some privately held properties. Initial geodetic ties of the islands to the mainland coast of Florida were performed with flare triangulations in the 1960s that were soon followed by BC-4 observations of the PAGEOS satellites. Flare triangulations were performed by simultaneous theodolite observations to parachute flares dropped from airplanes flying at high altitudes in order to make geodetic connections over the horizon. BC-4 observations were performed by photogrammetric triangulations of passive satellite reflections against a background of star fields. Dr. Helmut Schmid (one of the original V-2 scientists) led that geodetic program for the U.S. Coast & Geodetic Survey. The BC-4 program was the intercontinental geodetic program that tied all of the continents into the first worldwide geodetic system. Dr. Schmid was the designer of the BC-4 ballistic camera and was the mentor to Dr. Duane C. Brown, one of the fathers of modern analytical photogrammetry. The Bahamas have been referenced to the North American Datum of 1927 (Clarke 1866 ellipsoid) since the 1960s, where $a=6,378,206.4$ m and $b=6,356,583.8$ m. The datum origin point is Meades Ranch, Kansas (quite a distance away) at: $\Phi_o=39°\ 13'\ 26.686''$ N, $\Lambda_o=-98°\ 32'\ 30.506''$ W, and the reference azimuth to station Waldo is $\alpha_o=75°\ 28'\ 09.64''$ (*PE&RS*, April 2000).

Thanks to John W. Hager, the following positions have been determined in the Bahamas by classical observation techniques. These following geodetic positions (φ, λ) are presumably on the NAD1927 while the astro positions (Φ, Λ) are independent. Elbow Cay Light (ECL), $\varphi=26°\ 32'\ 21.715''$ N, $\lambda=-76°\ 57'\ 10.870''$ W. Also is the Astro Obs. (1940) where $\Phi=26°\ 32'\ 22.500''$ N, $\Lambda=-76°\ 57'\ 15.353''$ W, Astro Obs. to light = 126.79 m, $\alpha=$ S 79° 01' 27" E true. Flamingo Cay Light (FLA), $\varphi=22°\ 52'\ 43.48''$ N, $\lambda=-75°\ 51'\ 38.28''$ W. Great Inagua Island Light, $\varphi=20°\ 55'\ 56.81''$ N, $\lambda=-73°\ 40'\ 37.58''$ W. Great Isaac Light (GIL), $\varphi=26°\ 01'\ 48.30''$ N, $\lambda=-79°\ 05'\ 22.08''$ W. Great Ragged Island Light (GRL), $\varphi=22°\ 11'\ 17.29''$ N, $\lambda=-75°\ 43'\ 16.03''$ W.

Great Stirrup Cay (GSC), $\varphi=25°\ 49'\ 36.41''$ N, $\lambda=-77°\ 53'\ 50.20''$ W. Gun Cay Light (GUN), $\varphi=25°\ 34'\ 30.22''$ N, $\lambda=-79°\ 18'\ 01.18''$ W. Harvey Cay Light (HCL), $\varphi=24°\ 09'\ 16.19''$ N, $\lambda=-76°\ 28'\ 53.95''$ W. Hog Island Light (HIL), $\varphi=25°\ 05'\ 35.3''$ N, $\lambda=-77°\ 21'\ 13.5''$ W. Hogsty Reef Light (HRL), $\varphi=21°\ 41'\ 27.71''$ N, $\lambda=-73°\ 50''\ 56.81''$ W. Hole-in-the-Wall Light (HIW), $\varphi=25°\ 51'\ 32.522''$ N, $\lambda=-77°\ 10'\ 37.938''$ W. Observed astro (1940), $\Phi=29°\ 51'\ 22.320''$ N, $\Lambda=-77°\ 10'\ 37.370''$ W. Corrected in 1945, $\Phi=29°\ 51'\ 21.1155''$ N, $\Lambda=-77°\ 10'\ 36.2901''$ W. Little San Salvador Island (LIT), $\varphi=24°\ 33'\ 53.73''$ N, $\lambda=-75°\ 56'\ 08.00''$ W. Man Island Light (MAN), $\varphi=25°\ 33'$

31.34″ N, λ=−76° 38′ 26.83″ W. Northwest Point Astro (NPA), Φ=22° 27′ 24.42″ N, Λ=−73° 07′ 44.86″ W. Northwest Point Light (NPL), φ=22° 27′ 35.56″ N, λ=−73° 07′ 47.43″ W. Pinder Point Light (PPL), φ=26° 30′ 08.92″ N, λ=−78° 46′ 00.71″ W. Rum Cay Light (RUM), φ=23° 38″ 36.1″ N., λ=−74° 50′ 05.7″ W. Santa Maria Light (SML), φ=23° 40′ 54.54″ N, λ=−75° 20′ 27.60″ W. South Point Light (SPL), φ=22° 50′ 56.48″ N, λ=74° 51′ 14.42″ W. Stocking Island Astro (SIA), Φ=23° 32′ 33.97″ N, Λ=−75° 46′ 10.75″ W. Sweetings Cay Light (SWC), φ=26° 36′ 40.62″ N., λ=−77° 54′ 00.86″ W.

The NGA lists the three-parameter transformation from NAD27 to WGS84 for the Bahamas *excluding San Salvador Island* as ΔX=−4 m±5 m, ΔY=+154 m±3 m, and ΔZ=+178 m±5 m, where the 1987 solution is based on 11 station observations. For San Salvador Island, ΔX=+1 m±25 m, ΔY=+140 m±25 m, and ΔZ=+165 m±25 m, and the 1987 solution is based on 1 station observation. In 1997, the U.S. National Geodetic Survey observed a number of high-order positions in the Bahamas on the NAD83. The only grid ever used in the Bahamas is the UTM.

The U.S. Department of State issued a new paper on Limits in the Seas, No. 128 on the Bahamas Archipelagic and other Maritime Claims and Boundaries on January 31, 2014. "This study analyzes the maritime claims and maritime boundaries of the Commonwealth of The Bahamas, including its archipelagic baseline claim. The Bahamas' Archipelagic Waters and Maritime Jurisdiction (Archipelagic Baselines) Order, 2008 (Annex 1 to this study) took effect on December 8, 2008 and established the coordinates for the archipelagic baselines of The Bahamas. 1 The archipelagic baselines are shown on Map 1 to this study. This Order was made in exercise of the powers conferred by section 3.2 of the Archipelagic Waters and Maritime Jurisdiction Act, 1993 (Act No. 37, Annex 2 to this study). 2 The 1993 Act also established a 12-nautical mile (nm) territorial sea and 200-nm exclusive economic zone (EEZ). The Bahamas ratified the 1982 United Nations Convention on the Law of the Sea (LOS Convention) on July 29, 1983 and consented to be bound by the 1994 Agreement Relating to the Implementation of Part XI of the Convention on July 28, 1995.3" (http://www.state.gov/e/oes/ocns/opa/c16065.htm).

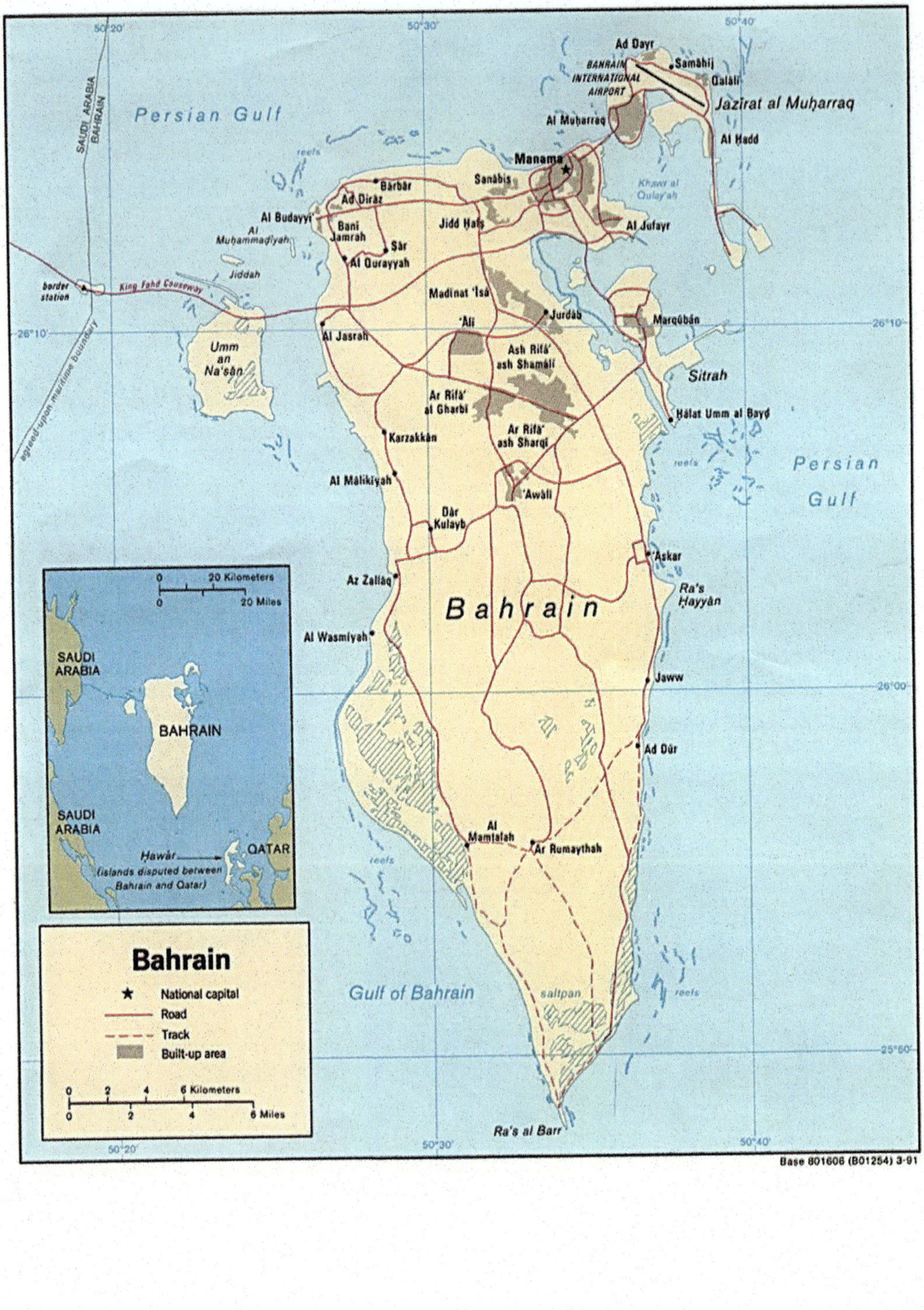

Base 801606 (B01254) 3-91

State of Bahrain

Bahrain is an archipelago in the Persian Gulf and is east of Saudi Arabia. It is 3½ times the size of Washington, DC, and has a total coastline of 161 km. The terrain is mostly low desert plain rising gently to a low central escarpment. The lowest point is the Persian Gulf (0 m), and the highest is Jabal ad Dukhan (122 m).

"Bahrain's history goes back to the roots of human civilization. The main island is thought to have broken away from the Arabian mainland sometime around 6000 B.C. and has almost certainly been inhabited since prehistoric times. The archipelago first emerged into world history sometime around 3000 B.C. as the seat of the Dilmun trading empire. Dilmun, a Bronze Age culture that lasted some 2000 years, benefited from the archipelago's strategic position along the trade routes linking Mesopotamia with the Indus Valley. In the midst of a region rapidly becoming arid, Dilmun's lush spring-fed greenery gave it the image of a holy island in the mythology of Sumeria, one of the world's earliest civilizations, which flourished in what is today southern Iraq. Dilmun had a similar cachet with the Babylonians, whose *Epic of Gilgamesh* mentions the islands as a paradise where heroes enjoy eternal life. Some scholars have suggested that Bahrain may be the site of the biblical Garden of Eden.

"Dilmun eventually declined and was absorbed by the Assyrian and Babylonian empires. The Greeks arrived around 300 B.C., giving the islands the name Tylos. Bahrain remained a Hellenistic culture for some 600 years. After experimenting with Christianity, Zoroastrianism and Manichaeism, in the 7th century many of the island's inhabitants accepted the personal invitation of the prophet Mohammed to convert to Islam.

"After a series of Islamic rulers, Bahrain was conquered by the Portuguese in the early 16th century. The Portuguese used the island as a pearling port and military garrison. In 1602, the Portuguese governor made the fatal mistake of executing the brother of one of the island's wealthiest traders. The trader, Rukn El-Din, proceeded to lead an uprising that soon drove the Europeans from Bahrain. The island then became part of the Persian Empire, but that association was cut short by the arrival of the Al-Khalifa clan, Bahrain's current ruling family.

"In the 1830s, Bahrain signed the first of many treaties with Britain, who offered Bahrain naval protection from Ottoman Turkey in exchange for unfettered access to the Gulf. This arrangement kept the British out of Bahrain's internal affairs until a series of internecine battles prompted the British to install their own choice for emir in 1869" (*Lonely Planet*, 2007).

"Persian ownership was denied by the British in 1928; oil was discovered in 1932, and Bahrain established a Council of State in 1970. It became a member of the United Nations in 1971, and with five other Arab states Bahrain joined the Gulf Cooperation Council in 1981" (*Merriam Webster's Geographical Dictionary*, 3rd Edition).

"On 15 September 2020, Bahrain and the United Arab Emirates signed peace agreements (the Abraham Accords) with Israel – brokered by the US – in Washington DC. Bahrain and the UAE thus became the third and fourth Middle Eastern countries, along with Egypt and Jordan, to recognize Israel" (*World Factbook*, 2022).

Topographic maps of Bahrain have been produced as early as 1825 at a scale of 1 inch to 2 miles. The British Directorate of Military Survey produced maps at a scale of 1:253,440 from original surveys in 1915–1917. Since that time the Directorate has also produced town plans for populated places in Bahrain. A map of Bahrain Island was produced by the Directorate of Military Survey, the British War Office and Air Ministry in 1968 at a scale of 1:63,360 (1 inch to 1 mile) with the UTM grid overprinted. This polychrome map supplied complete coverage of Bahrain and Muharaq Islands. Relief is indicated by contours at 20-foot intervals and was compiled from 1951 to 1953 aerial photography by Hunting Aero-surveys, Ltd., based on control by Bahrain Petroleum Co., Ltd. (BAPCO).

The classical geodetic datum used for Bahrain is the Ain el Abd Datum of 1970 located in the "Arq" Oil Field of Saudi Arabia, where $\Phi_o = 28° 14' 06.171"$ N, $\Lambda_o = 48° 16' 20.906"$ East of Greenwich, thanks to the EPSG Database, and the ellipsoid of reference is the International 1924, where $a = 6,378,388$ m and $1/f = 297$. Thanks to some correspondence I had in 1998 with Mr. Andrew Kopec, the GIS Manager for the Bahrain Centre for Studies and Research, the local grid system used by the Ministry for Housing is a Transverse Mercator where the central meridian, $\lambda_o = 51°$ E, the latitude of origin $\varphi_o =$ Equator, the scale factor at the latitude of origin $m_o = 0.99962$, and False Easting $= 0$, and the False Northing $= 2,000$ km. According to Mr. Kopec, the main product is a 1:1,000 scale topographic map. The same datum and projection is used on marine charts produced by the same ministry at scales of 1:25,000, 1:50,000, and 1:100,000. (Ain el Abd is a brackish spring easily found with Google Earth™.)

The only local transformation parameters I could find were published in the now obsolete TR8350.2, where **from** Ain el Abd 1970 **to** WGS84 are: $\Delta X = -150$ m ± 25 m, $\Delta Y = -250$ m ± 25 m, $\Delta Z = -1$ m, ± 25 m, and the solution is based on only two points.

There are two treaties that have been negotiated among Bahrain, Saudi Arabia, and Iran regarding the division of the mineral resources of the Persian Gulf (continental shelf). Both are variations on the "equidistance principle" of the International Law of the Seas but are acceptable and valid under International Law as all three participating high parties are in agreement. (See *Department of State, Office of the Geographer, "Limits in the Seas, No. 12 and No. 58."*)

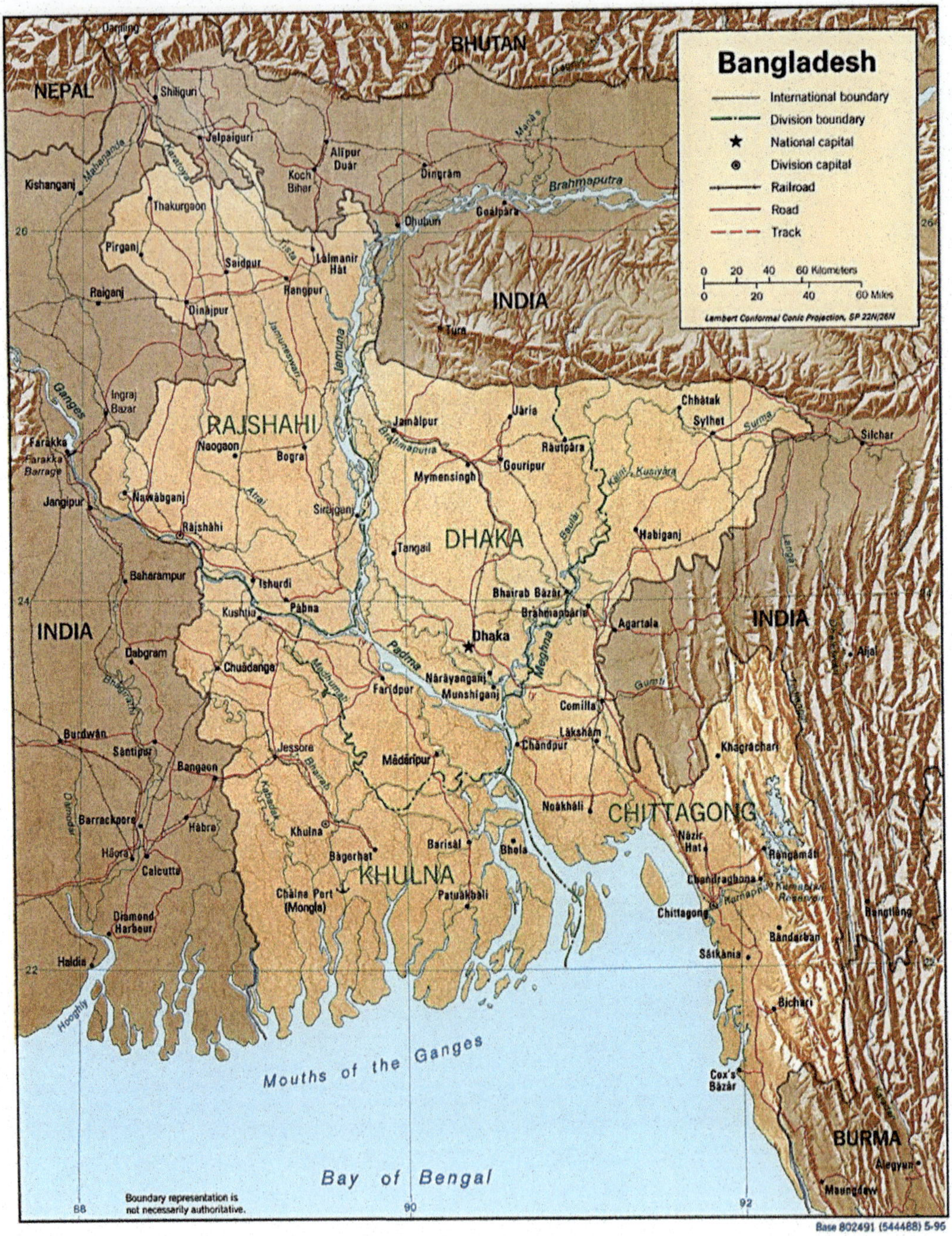

Base 802491 (544488) 5-96

People's Republic of Bangladesh

Bangladesh is slightly smaller than Wisconsin, and is bordered by Burma (193 km), India (4,053 km), and the Bay of Bengal (580 km). The terrain is mostly flat alluvial plain and is hilly in the southeast. The lowest point is the Indian Ocean (0 m), and the highest point is Reng Tlang (957 m) (*World Factbook*, 2022).

"Historians believe that Bengal, the area comprising present-day Bangladesh and the Indian state of West Bengal, was settled in about 1000 B.C. by Dravidian-speaking peoples who were later known as the Bang. The first great indigenous empire to spread over most of present-day India, Pakistan, and Bangladesh was the Mauryan Empire (*circa* 320–180 B.C.), whose most famous ruler was Asoka (*circa* 273–232 B.C.). Although the empire was well administered and politically integrated, little is known of any reciprocal benefits between it and eastern Bengal. The western part of Bengal, however, achieved some importance during the Mauryan period because vessels sailed from its ports to Sri Lanka and Southeast Asia. The Turkish conquest of the subcontinent was a long, drawn-out process covering several centuries. It began in Afghanistan with the military forays of Mahmud of Ghaznī in 1001. By the early thirteenth century, Bengal fell to Turkish armies.

"The Indian subcontinent had had indirect relations with Europe by both overland caravans and maritime routes, dating back to the fifth century *B.C.* The lucrative spice trade with India had been mainly in the hands of Arab merchants. By the fifteenth century, European traders had come to believe that the commissions they had to pay the Arabs were prohibitively high and therefore sent out fleets in search of new trade routes to India. The arrival of the Europeans in the last quarter of the fifteenth century marked a great turning point in the history of the subcontinent. The British East India Company, a private company formed in 1600 during the reign of Akbar and operating under a charter granted by Queen Elizabeth I, established a factory on the Hooghly River in Bengal in 1650 and founded the city of Calcutta in 1690. Siraj ud Daulah, governor of Bengal, unwisely provoked a military confrontation with the British at Plassey in 1757. He was defeated by Robert Clive, an adventurous young official of the British East India Company. Clive's victory was consolidated in 1764 at the Battle of Buxar on the Ganges, where he defeated the Mughal emperor. As a result, the British East India Company was granted the title of ***diwan*** (collector of the revenue) in the areas of Bengal, Bihar, and Orissa, making it the supreme, but not titular, governing power. Henceforth the British would govern Bengal and from there extend their rule to all of India. Pakistan itself had been created on August 15, 1947, largely the result of communal passions pitting Hindus against Muslims. Pakistan was divided into two wings, separated by 1,600 kilometers of Indian Territory, with Islam only a tenuous link between the wings. Of paramount importance to East Pakistanis was the Bangla (before 1971 usually referred to as Bengali) language and culture, a consideration not appreciated by the West Wing of Pakistan until it was too late. Bangladesh, formerly the East Wing of Pakistan, emerged as an independent nation in December 1971" (*Library of Congress Country Studies*).

The Survey of Bangladesh is the national surveying and mapping agency for Bangladesh. The department started its functioning as the "Bengal Survey" on January 1, 1767, under the command of Major James Rennell, the First Surveyor General in undivided India. The Bengal Survey conducted survey and mapping activities until 1947. After the partition of the subcontinent on August 14, 1947, the organization started its new role as the Survey of Pakistan and established a regional office at Dhaka. This regional office was transformed into the Survey of Bangladesh when the country immerged as an independent state in 1971. The original "datum" of the Indian subcontinent is commonly referred to as the Indian Datum of 1916 with origin at Kalianpur Hill, where $\Phi_o = 24°\ 07'\ 11.26''$ N, $\Lambda_o = 77°\ 39'\ 17.57''$ East of Greenwich, and the ellipsoid of reference is the Everest 1830, where $a = 6,377,276.345$ m and $1/f = 300.8017$. The original topographic mapping of Bangladesh by the Survey of India was based on the Lambert Conical Orthomorphic projection, India Zone IIB, where the central meridian, $\lambda_o = 90°$ E, the latitude of origin $\varphi_o = 26°$ N, the scale factor at the latitude of origin $m_o = 0.998786408$, and False Easting = 3,000,000 Indian Yards, and the False Northing = 1,000,000 Indian Yards. (Note that 1 meter = 1.093619 Indian Yards.)

However, the Survey of Bangladesh considers the national datum to be the Bangladesh Datum with its origin at Gulshan (in metropolitan Dhaka), where $\Phi_o = 23°\ 47'\ 49.48502''$ N, $\Lambda_o = 90°\ 25'\ 06.55270''$ East of Greenwich, the ellipsoid height, $h_o = +8.53$ m, and the ellipsoid of reference remains the Everest 1830. The local Grid coordinates of Gulshan on India Zone IIB *as published by the Survey of Bangladesh (SoB)* are Easting = X = 3,044,611.838 Indian Yards, Northing = Y = 733,492.5266 Indian Yards. Furthermore, the WGS84 coordinates of Gulshan are published by *SoB* as: $\varphi = 23°\ 47'\ 52.02714''$ N, $\lambda = 90°\ 24'\ 56.34024''$ E, and ellipsoid height = h = −45.4494 m. The SoB has published a three-parameter transformation **from** the Bangladesh Datum **to** WGS84 Datum but did not offer an accuracy estimate of the transformation parameters. Thanks to a report, *Compatibility of GPS with Local Grid Co-Ordinate System of Bangladesh,* by LTC. Kazi Shafayetul Haque, Director, Survey of Bangladesh (probably published around 1998). Eleven collocated points were published on both datums. I re-computed the transformation myself, and I verified the test points relation to have virtually the same parameters as LTC Haque to within ± 0.43 m. Those published parameters are: $\Delta X = −283.7$ m, $\Delta Y = −735.9$ m, and $\Delta Z = −261.1$ m. Considering the size of the area of Bangladesh, the accuracy of a three-parameter datum shift being within ±0.43 m is really quite excellent! Incidentally, LTC Haque is retired outside of Toronto, Canada, the last time I met with him in 2018.

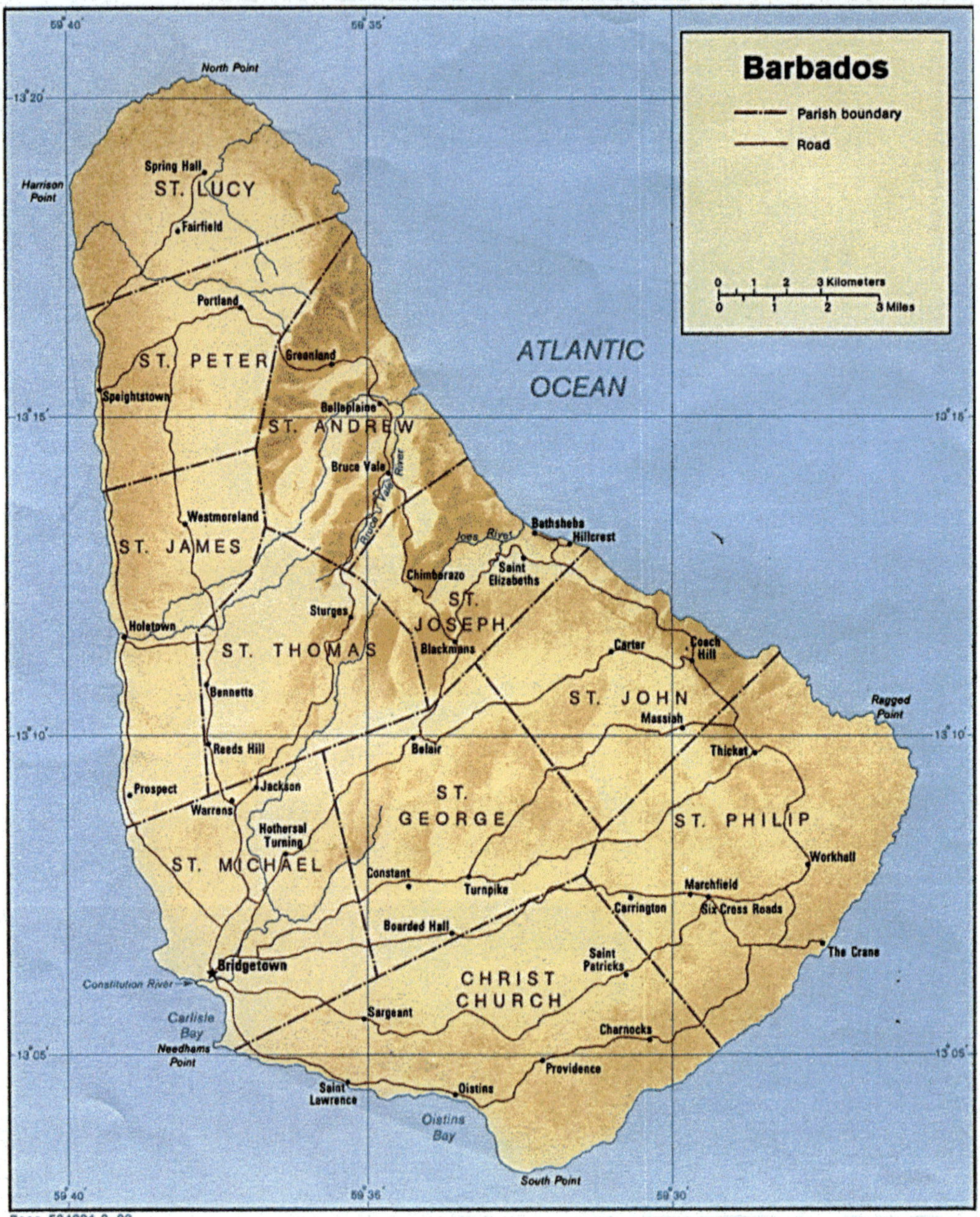

Base 504291 2-80

Barbados

The terrain of Barbados is relatively flat, and it rises gently to a central highland region. The lowest point is the Atlantic Ocean (0 m), and the highest point is Mount Hillaby (336 m). The land area comprises 430 km² and is approximately 2½ times the size of Washington, DC; Barbados is about 170 km due East of St. Vincent.

Probably originally inhabited by Arawaks, Barbados was first visited by Europeans in the 16th century. First claimed by England in the early 17th century, it was first settled under the auspices of William Courteen, *circa* 1625. Slaves were freed in 1838, and it was the seat of government of the Windward Islands from 1833 to 1885, when it was made a separate administration. Barbados achieved independence within the Commonwealth in 1966 (*Merriam-Webster's Geographical Dictionary*, 3rd Edition).

The coordinate systems of Barbados seem to be somewhat of an enigma for a Commonwealth nation. According to a statement found on the Barbados Land Surveyors Association some years ago, they stated that triangulation "was used to establish the major control stations in Barbados in the 1920's" (*sic*). However, the local datum name is H.M.S. Challenger Astro Datum of 1938, and thanks to the EPSG database, the origin at St. Ann's Tower is: $\Phi_o = 13°\ 04'\ 32.53''$ N, $\Lambda_o = 59°\ 36'\ 29.34''$ W, which is on the southern coast just west of the town of St. Lawrence.

The original topographic mapping of British Commonwealth nations was undertaken by the Directorate of Overseas Surveys (DOS), after World War II. Barbados was one of the very first nations to be surveyed and mapped by DOS, and the first series of 1:10,000 scale maps with 20-foot contours was published in 1954–56. Recompiled in 1986, the number of sheets was reduced from 18 to 12. The DOS maps were based on the Challenger Datum of 1938, the ellipsoid of reference is the Clarke 1880 (RGS), where $a = 6,378,249.145$ m, and $1/f = 293.465$. The *original* DOS grid system used for Barbados was the British West Indies, "BWI Transverse Mercator Grid," where the central meridian, $\lambda_o = 62°$ W, the latitude of origin, $\varphi_o =$ equator, the scale factor at the latitude of origin, $m_o = 0.9995$, False Easting = 400 km, and False Northing = nil. The formulae are the Gauss-Krüger, but for such a small span of latitude and longitude the distinction is irrelevant. As is common with the BWI Grid usage, the grid is also used as an "atlas index" numbering system for the popular tourist maps and is not numbered with coordinate values but with an alphanumeric system for facile use to locate tourist interest points. The grid is easy to recover if one is familiar with the standard BWI grid conventions, but the defining parameters are unfortunately obscure to many. (See Grenada.)

The Barbados National Grid (BNG) is also based on a Transverse Mercator projection, referenced to the Challenger Datum of 1938 where the central meridian, $\lambda_o = 59°\ 33'\ 35''$ W, the latitude of origin, $\varphi_o = 13°\ 10'\ 35''$ N, the scale factor at the latitude of origin, $m_o = 0.9999986$, False Easting = 30 km, and False Northing = 75 km.

Some years ago, Mr. Kevin Grootendorst sent some collocated coordinates to me in Barbados. They were represented as one set of coordinates being on the BNG and another on UTM, WGS84. The three-parameter datum transformation of four points well-distributed (a fifth point resulted in being an outlier), about the island *from* Challenger 1938 Datum *to* WGS84 Datum yields: $\Delta X = +60$ m, $\Delta Y = +264$ m, and $\Delta Z = +43$ m, and the accuracy of my solution is about ±4 m for four points. Around that same time, Mr. Leo Romeijn sent another set of transformation parameters to me for exactly the same thing (*from* Challenger 1938 Datum *to* WGS84 Datum), as published by the EPSG parameter set version of "6.tenC," where $\Delta X = +32$ m, $\Delta Y = +301$ m, and $\Delta Z = +419$ m, and the accuracy is stated to be about ±2.5 m, and is "derived at 2 stations (S40 and M1, St. Annes Tower) in 2004." Considering the whopping difference in datum shift parameters between the same two datums on the island of Barbados, I'd say it still remains an enigma …

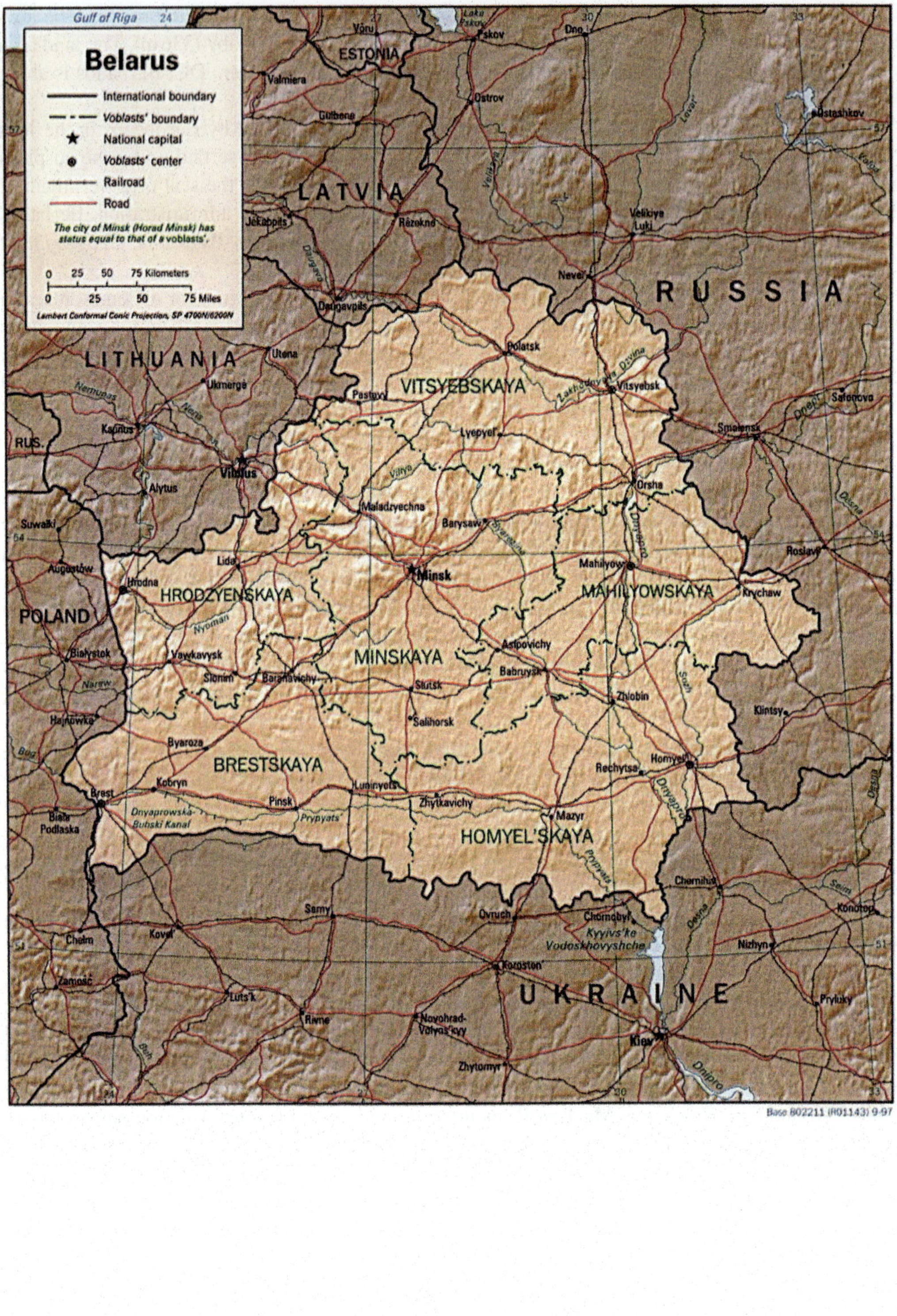

Base 802211 (R01143) 9-97

Republic of Belarus

"Belarus, a generally flat country (the average elevation is 162 meters above sea level) without natural borders, occupies an area of 207,600 square kilometers, or slightly smaller than the state of Kansas. Its neighbors are Russia to the east and northeast, Latvia) to the north, Lithuania to the northwest, Poland to the west, and Ukraine, to the south. Belarus's mostly level terrain is broken up by the Belarusian Range (*Byelaruskaya Hrada*), a swath of elevated territory, composed of individual highlands, that runs diagonally through the country from west-southwest to east-northeast. Its highest point is the 346-meter Mount Dzyarzhynskaya (*Dzerzhinskaya*, in Russian); named for Feliks Dzerzhinskiy, head of Russia's security apparatus under Stalin. Northern Belarus has a picturesque, hilly landscape with many lakes and gently sloping ridges created by glacial debris. In the south, about one-third of the republic's territory around the Prypyats' (*Pripyat'*, in Russian) River is taken up by the low-lying swampy plain of the Belarusian Woodland, or Palyessye (*Poles'ye* in Russian). Belarus's 3,000 streams and 4,000 lakes are major features of the landscape and are used for floating timber, shipping, and power generation. Major rivers are the west-flowing Zakhodnyaya Dzvina (*Zapadnaya Dvina* in Russian) and Nyoman (*Neman* in Russian) rivers, and the south-flowing Dnyapro (*Dnepr* in Russian) with its tributaries, Byarezina (*Berezina* in Russian), Sozh, and Prypyats' rivers. The Prypyats' River has served as a bridge between the Dnyapro flowing to Ukraine and the Vistula in Poland since the period of Kievan Rus'. Lake Narach (*Naroch'*, in Russian), the country's largest lake, covers eighty square kilometers. Nearly one-third of the country is covered with **pushchy** (sing., **pushcha**), large unpopulated tracts of forests. In the north, conifers predominate in forests that also include birch and alder; farther south, other deciduous trees grow. The Belavezhskaya (*Belovezhskaya*, in Russian) Pushcha in the far west is the oldest and most magnificent of the forests; a reservation here shelters animals and birds that became extinct elsewhere long ago. The reservation spills across the border into Poland; both countries jointly administer it" (*Library of Congress Country Studies,* 1995).

"Belarus's origins can be traced from the emergence in the late 9th century *A.D.* of Kievan Rus', the first East Slavic state. After the death of its ruler, Prince Yaroslav the Wise, in 1054, Kievan Rus' split into a number of principalities, each centered on a city. One, Polatsk (*Polotsk*, in Russian), became the nucleus of modern-day Belarus. In 1240, after the Tatar overthrow of Kiev, the dominant principality of Kievan Rus', Belorussia and part of Ukraine came under the control of Lithuania. The resulting state was called the Grand Duchy of Lithuania, Rus', and Samogitia. Because territories inhabited by East Slavs made up about 90 percent of the Grand Duchy, they exerted a great cultural influence on the new state. Official business was conducted in a Slavic language (a predecessor of both Belorussian and Ukrainian) based on Old Church Slavonic, and the law code was based on that of Kievan Rus'. In its early history, the region was known as "Belaya Rus'," "Belorussia," "White Ruthenia," or "White Rus'." (A number of explanations have been proffered for the term "white.") As if this were not confusing enough, the terms "Rus'" and "Russia" have often been confused, sometimes deliberately. The original Rus' was Kievan Rus', which existed for centuries before Muscovy (which would later become Russia) gained significance. Russia later claimed to be the sole successor to Kievan Rus' and often blurred the line between the two. In the Russian language, both "*russkiy*" and "*rossiyskiy*" mean "Russian." Since the late 19th century, national activists have based their attempts to create an independent Belarusian state based on the Belorussian language, which had been kept alive over the centuries mainly by peasants. The stage was set for the emergence of a national consciousness by the industrialization and urbanization of the nineteenth century and by the subsequent publication of literature in the Belorussian language, which was often suppressed by Russian, and later Polish, authorities. It is ironic, then, that the first long-lived Belorussian state entity, the Belorussian Soviet Socialist Republic (Belorussian SSR), was created by outside forces – the *Bolshevik* government in Moscow. And it was those same forces, the communists, whose downfall in 1991 precipitated the existence of an independent Belarus, which has been torn between its desire for independence and a longing for integration with newly independent Russia.

The tsarist Russians performed surveys and topographic mapping of Belarus in the 19th and early 20th centuries, but these works were for military purposes only. They did nothing with respect to individual land ownership registration, and they preferred the sazhen for their unit of measurement. At that time, the Russians preferred the Walbeck 1819 ellipsoid, where $a=6,376,896$ m and the reciprocal of flattening, $1/f=302.78$. Some of these old maps also referred longitudes to Ferro in the Canary Islands; a practice dropped after World War II. The Russian "System 42" Datum is referenced to the Krassovsky 1940 ellipsoid, where $a=6,378,245$ m and $1/f=298.3$. The origin is at Pulkovo Observatory: $\Phi_o=59°\ 46'\ 18.55''$ North, $\Lambda_o=30°\ 19'\ 42.09''$ East of Greenwich, and the defining azimuth at the point of origin to Signal A is: $\alpha_o=317°\ 02'\ 50.62''$. The Grid system used is the Russia Belts which are identical to the UTM Grid except that the scale factor at origin (m_o) at the Central Meridian is equal to unity for each of the 6° belts.

"Since January 01, 2010 *(the)* National Reference Frame – State Geodetic Reference System of 1995 of the Republic of Belarus (SGR95 RB) has been in use. Introduction of the National Reference Coordinate System on the territory of Belarus was caused by the necessity of preservation of the common coordinate space with *(the)* Russian Federation in accordance with the political agreements and demands of the Ministry of Defense.

"The new structure of the National Geodetic Network consists of the: Fiducial Astro-Geodetic Network (FAGS in Minsk). Precise Geodetic Network (Zero order reference network). 1st class Satellite Geodetic Network (First order reference network). National Detail Geodetic Network. Connection between the FAGS station and the coordinate system ITRS/ITRF2005 was exercised with fixed binding to 9 IGS stations. Some details of the campaign are Accuracy (RMS) of the coordinates: ±0.8 mm, ±0.3 mm, ±2.0 mm in North, East and Up components. Network solution includes 4 IGS stations+FAGS stations. Some details of the 1st class Satellite Geodetic Network are Number of points: 846; Accuracy (RMS) of the coordinates: ±2.2 mm, ±1.6 mm, ±4.2 mm in North, East and Up components. *(The)* network solution includes 9 points of the Precise Geodetic Network. 306 points (36%) were compounded with old triangulation points (1st and 2nd classes Astro-geodetic network). *(The)* National Detail Geodetic Network includes 6,268 points" (*National report of Belarus, S. Zabahonski, N. Rudnitskaya, Minsk 2012*).

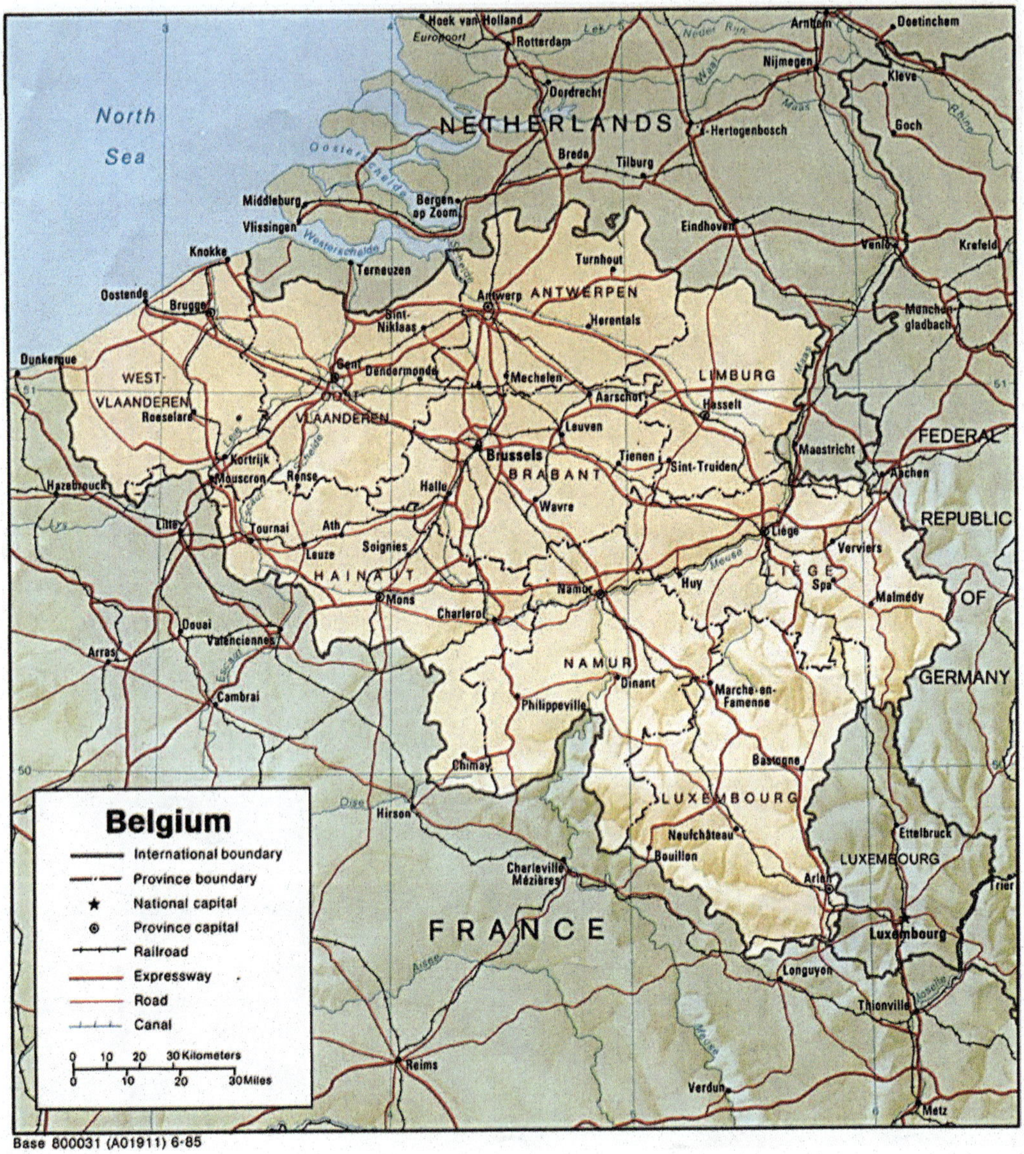

Base 800031 (A01911) 6-85

Kingdom of Belgium

"Belgium is bordered by France (556 km), Germany (133 km), Luxembourg (130 km), and Netherlands (478 km). Its terrain has flat coastal plains in northwest, central rolling hills, and rugged mountains of Ardennes Forest in southeast. Its highest point is Botrange (694 m), and its lowest point is the North Sea (0 m). Its mean elevation is (181 m)" (*World Factbook*, 2022).

The Kingdom of Belgium, birthplace of Mercator, gained its independence from the Netherlands in 1830. Famous Belgian place names have become prominent in many battles, including Waterloo and part of the Battle of the Bulge. Triangulation was originally carried out by Snellius (1617), Cassini (1745–1748), and General Krayenhoff (1802–1811). Although large-scale topographic coverage dates back to 1744, the organized national mapping was carried out by the military *Dépôt Général de la Guerre*, originally founded in 1831. This national mapping organization has gone through a number of name changes until becoming a civilian agency in 1976 with the present name of "*Institut Géographique National de Belgique*" (IGNB).

The classical first-order triangulation of Belgium was carried out between the years 1851 and 1873, when 84 primary stations were established. The *Observatoire Royale d'Uccle* Datum of 1884 was based on the origin of the old Brussels Observatory: $\Phi_o = 51°\ 10'\ 06.895''$ North, $\Lambda_o = 04°\ 22'\ 05.89''$ West of Greenwich, and the ellipsoid of reference was the "Carte de France" (Delambre 1810) where $a = 6376985.228$, and $1/f = 308.64$. The triangulation was dependent on two measured bases at Lommel and Ostend with Beverloo as a base of verification. The ellipsoidal Belgian Bonne Grid was used with this datum from 1884 to 1919 with no False Easting or False Northing and was later referred to as the "Old Convention" Grid. The lack of a false origin is typical for late 19th-century grid systems, and is a common theme found in European and colonial grids of that era. Negative grid coordinates (in the appropriate quadrants) were accepted as a normal way of doing things. Note that the Bonne projection is an authalic (equal area) projection. Many countries in Europe followed the French preference for the Bonne, although it is a miserable projection for large-scale military and civilian engineering applications. The Bonne projection tables were easy to compute on the ellipsoid. That advantage apparently helped to keep it popular, in addition to the prestige afforded by adopting the same system as used by the French.

After the "Great War" (World War I), the Kingdom modified the Grid such that the False Easting = 150 km and the False Northing = 120 km. The "New Convention" Bonne Grid (1919–1950) had a Latitude of Origin $(\varphi_o) = 50°\ 24'$ North, and a Central Meridian $(\lambda_o) = 4°\ 22'\ 05.91''$ West of Greenwich. In post-war years, the Belgians continued to re-observe and re-adjust their triangulation network. In 1924, the International Conference of Geodesy and Geophysics which conferred in Madrid adopted the Hayford ellipsoid as the new standard. The Belgians adopted this new reference ellipsoid for their new Belgian Datum of 1927 with the origin altered to the astronomical values of Lommel Signal at: $\Phi_o = 51°\ 10'\ 08.092''$ North, $\Lambda_o = 04°\ 22'\ 05.91''$ West of Greenwich. The azimuth was from Lommel Signal to Lommel Turm: $\alpha_o = 186°\ 58'\ 47.05''$. During World War II, the connection between the New Triangulation of France (NTF) Datum and the Belgian Datum of 1927 was accomplished in 1940 by considering the geodetic differences in longitude. The longitude difference between the meridians of Brussels and Paris, as determined by a comparison of the two triangulations through Mont Kemmel, gave the geodetic longitude of Brussels as 04° 22′ 12.7″. The observed astronomical difference between Brussels and Greenwich was not used. This discrepancy was one of the major factors that prompted the recomputation of all geodetic control in Western Europe after the war. The Bonne Grid "New Convention" is also referred to as the "Orange Report Net Grid" and was used through World War II until 1950.

After World War II, the Kingdom of Belgium adopted the Lambert Conformal Conic Projection and defined the new grid such that the False Easting = 150 km and the False Northing = 5,400 km at latitude 50° 30′ North. The Belgian Lambert "KM" Grid had a Northern Standard Parallel of $(\varphi_N) = 51°\ 10′$ North, a Southern Standard Parallel of $(\varphi_S) = 49°\ 50′$ North, and a Central Meridian $(\lambda_o) = 4°\ 22′\ 04.71″$ West of Greenwich. Note that the Belgians defined their Lambert Zone with two standard parallels, which is the definition method favored by the French and the Americans. The British prefer to use a Latitude at Origin (φ_o) and a Scale Factor at Origin (m_o). The two methods of zone definition use slightly different formulae to produce the identical projection in most cases. The U.S. Army Map Service (AMS) organized the re-computation of all the classical geodetic datums of Western Europe after the war, and the resultant coordinate system was termed the European Datum of 1950 (ED50). AMS directed the computations by initially computing common points between Belgium and France in 1945. The Belgian stations consisted of Arlon Church, Anlier, Bouillon, Willerzie, Rulles, Bon Secours, Mont Sainte Genevieve, and Mont Kemmel. The French stations were Aumets, Tellancourt, and Sighy-Mont Libert. The computations consisted of a least squares polynomial solution of the coordinates of the respective stations as computed on the conformal plane represented by the UTM Grid, International (Hayford) ellipsoid. The coefficients that were determined in Washington by AMS for these junction points were used to transform the respective interior points to ED50, with UTM used as the unifying connection tool for these two countries. The rest of Western Europe followed with French datum connections being the initial governing factor for the remaining countries. The ED50 remained a NATO military secret for some years. Belgium continued with independent observations and established a new provisional adjustment for the country. The two Vlaanderen provinces, the northern half of Brabant, Antwerpen, and Limburg provinces were essentially held fixed from 1940. The southern provinces of Hainaut and Namur had minimal changes in coordinates, and the region in between received the major adjustment. The Kingdom then published the Belgian 1950 Datum with the "KM" Lambert 50 Grid, all of the Belgian 1927 defining parameters remaining unchanged.

New instruments and re-observations required another adjustment a couple decades later, and the Belgian Datum of 1972 is defined by the origin at the geodetic monument d'Uccle: $\Phi_o = 50°\ 47′\ 57.704″$ North, $\Lambda_o = 04°\ 21′\ 24.983″$ West of Greenwich. The azimuth was from d'Uccle to Kester: $\alpha_o = 262°\ 08′\ 37.95″$. The International ellipsoid continued to be used, and the only change for the new Lambert 72 Grid was that the new central meridian matched the new datum origin $(\lambda_o = 7°)$. Note that the plane coordinates for the Lambert 72 Grid are considerably different from the Lambert 50 Grid. The ellipsoid is defined to be tangent with the geoid at d'Uccle. The vertical datum "*Deuxieme Nivellement General*" (DNG) was established at Oostende by conventional tide gauge observations. Curiously, the DNG is based on Mean Low Water and therefore is −2.33 m with respect to local mean sea level.

The Readjustment of the European Triangulation Network (RETrig) has been involved for decades as an international cooperative effort that started with ED50. The latest fruit of their labors is the ED87 adjustment that is quite comprehensive and remarkably accurate. The Royal Observatory of Belgium is now operating four permanent GPS reference stations, and the accuracies of these observations are now at the sub-centimeter level. The stations are Brussels, Dentergemand, Dourbes, and Waremme.

The Belgian Institut Géographique National has published a PDF document in *French* that addresses the approved transformation parameters **from** a UTM Grid on either the EU50 Datum or on the WGS84 Datum **to** Projection Lambert 72 or **to** Projection Lambert 50. See: https://www.ngi.be/website/fr/. Page 12 of the document has an easy-to-understand flow chart.

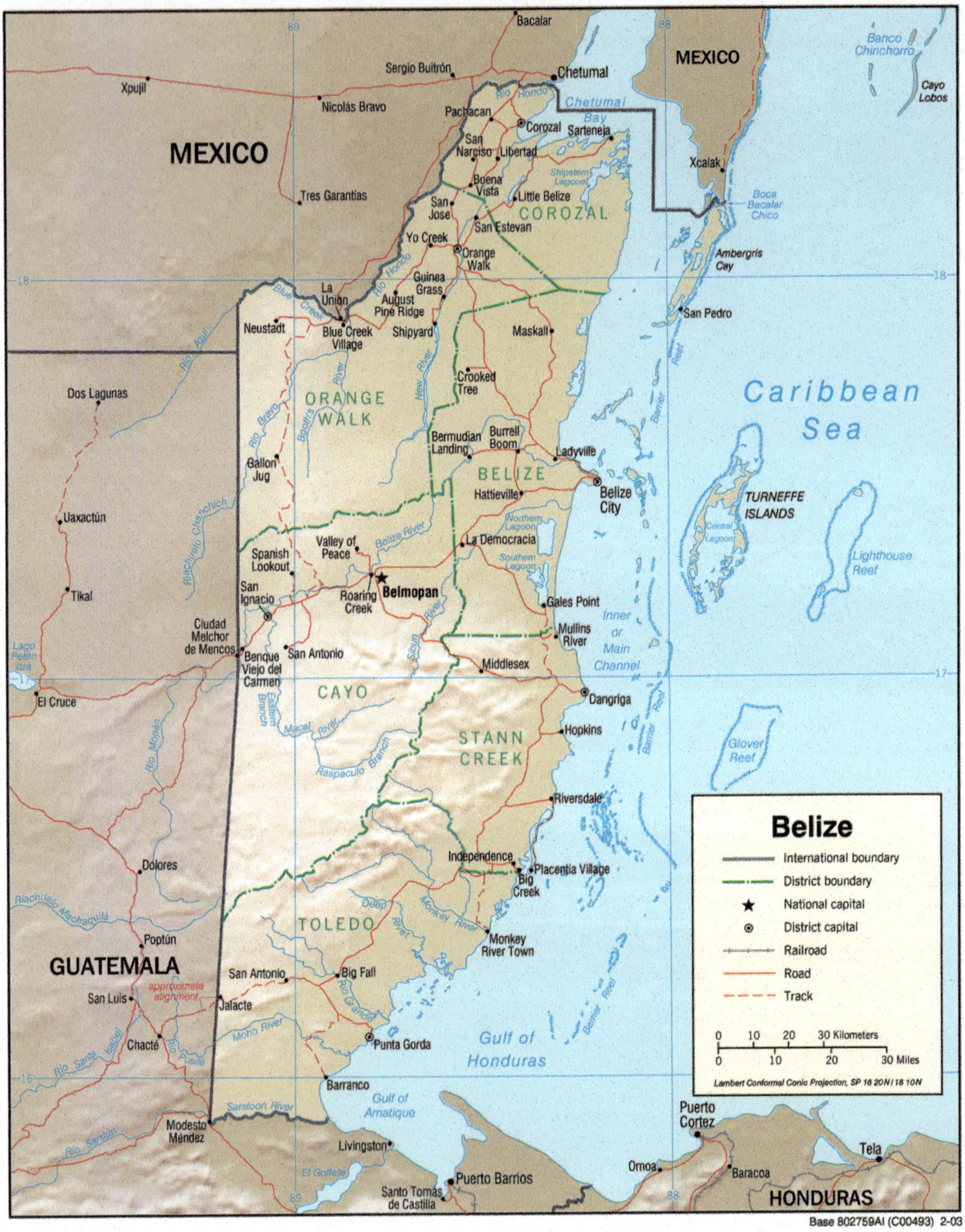

Base 802759AI (C00493) 2-03

Belize

"Slightly smaller than Massachusetts, Belize is bordered by Guatemala (266 km), and Mexico (250 km). With a coastline of 386 km, the lowest point is the Caribbean Sea (0 m), and the highest point is Doyle's Delight (1,160 m). Belize's territorial sea is "12 nautical miles in the north; 3 nautical miles in the south; note – from the mouth of the Sarstoon River to Ranguana Cay, Belize's territorial sea is 3 nautical miles according to Belize's *Maritime Areas Act, 1992*, the purpose of this limitation is to provide a framework for negotiating a definitive agreement on territorial differences with Guatemala" (*World Factbook*, 2009).

The Maya were still in Belize when Christopher Columbus traveled to the Gulf of Honduras during his fourth voyage in 1502. When Cortés passed through the southwestern corner of present-day Belize in 1525, there were settlements of Chol-speaking Manche in that area. "Early in the seventeenth century, on the shores of the Bay of Campeche in southeastern Mexico and on the Yucatán Peninsula, English buccaneers began cutting logwood, which was used in the production of a dye needed by the woolen industry. According to legend, one of these buccaneers, Peter Wallace, called 'Ballis' by the Spanish, settled near and gave his name to the Belize River as early as 1638. English buccaneers began using the tortuous coastline of the area as a base from which to attack Spanish ships. Some of the buccaneers may have been refugees expelled by the Spanish in 1641–42 from settlements on islands off the coasts of Nicaragua and Honduras. Buccaneers stopped plundering Spanish logwood ships and started cutting their own wood in the 1650s and 1660s. Logwood extraction then became the main reason for the English settlement for more than a century. A 1667 treaty, in which the European powers agreed to suppress piracy, encouraged the shift from buccaneering to cutting logwood and led to more permanent settlement. The 1670 Godolphin Treaty between Spain and England confirmed English possession of countries and islands in the Western Hemisphere that England already occupied" (*Library of Congress Country Studies,* 2009). Formerly British Honduras, Belize achieved independence as a parliamentary democracy from the United Kingdom in 1981.

The British Colonial Survey Committee authorized a survey of the colony in 1925. In 1926, work began on a contoured map series of the colony and various town plans, general maps at 16 miles to the inch scale, and two geological maps that were produced by 1928. By the mid-1940s, 6,600 square miles had been topographically mapped, presumably all by planetable and alidade methods. The original datum established for Belize is the Sibun Gorge Datum of 1922 where the astronomical coordinates of the origin are: $\Phi_o = 17° 03' 40.471''$ S, $\Lambda_o = -88° 37' 54.687''$ W, and the ellipsoid of reference is the Clarke 1858, where $a = 6,378,293.645$ m, $1/f = 294.26$. The Colony Coordinates used the datum origin for the Transverse Mercator projection with a scale factor at origin of unity, a False Northing = 445,474.83 ft, a False Easting of 217,259.26 ft, and the unit of measure is where 1 m = 3.28086933 Jamaican feet. Another datum known to exist is called the Jesuit College Flagstaff, probably being the origin for a local hydrographic survey.

With aerial photography flown in 1969 and 1972, a series of 1:50,000 scale topographic maps were produced by the British Directorate of Overseas Surveys (DOS). The coordinate reference system currently used in Belize is the North American Datum of 1927, presumably introduced in the 1950s by the U.S. Army Map Service's Inter-American Geodetic Survey. The commonly used Ocotepeque Datum of 1948 for Honduras, Guatemala, and Nicaragua is not known to have been used in Belize. The available 1:50,000 scale maps of Belize on the NAD27 are over-printed with the UTM Grid. According to *TR 8350.2*, the three-parameter datum shift for Central America including Belize **from** NAD27 **to** WGS84 is: $\Delta X = 0$ m ± 8 m, $\Delta Y = +125$ m ± 3 m, $\Delta Z = +194$ m ± 5 m, and is based on a 19-point solution in 1987.

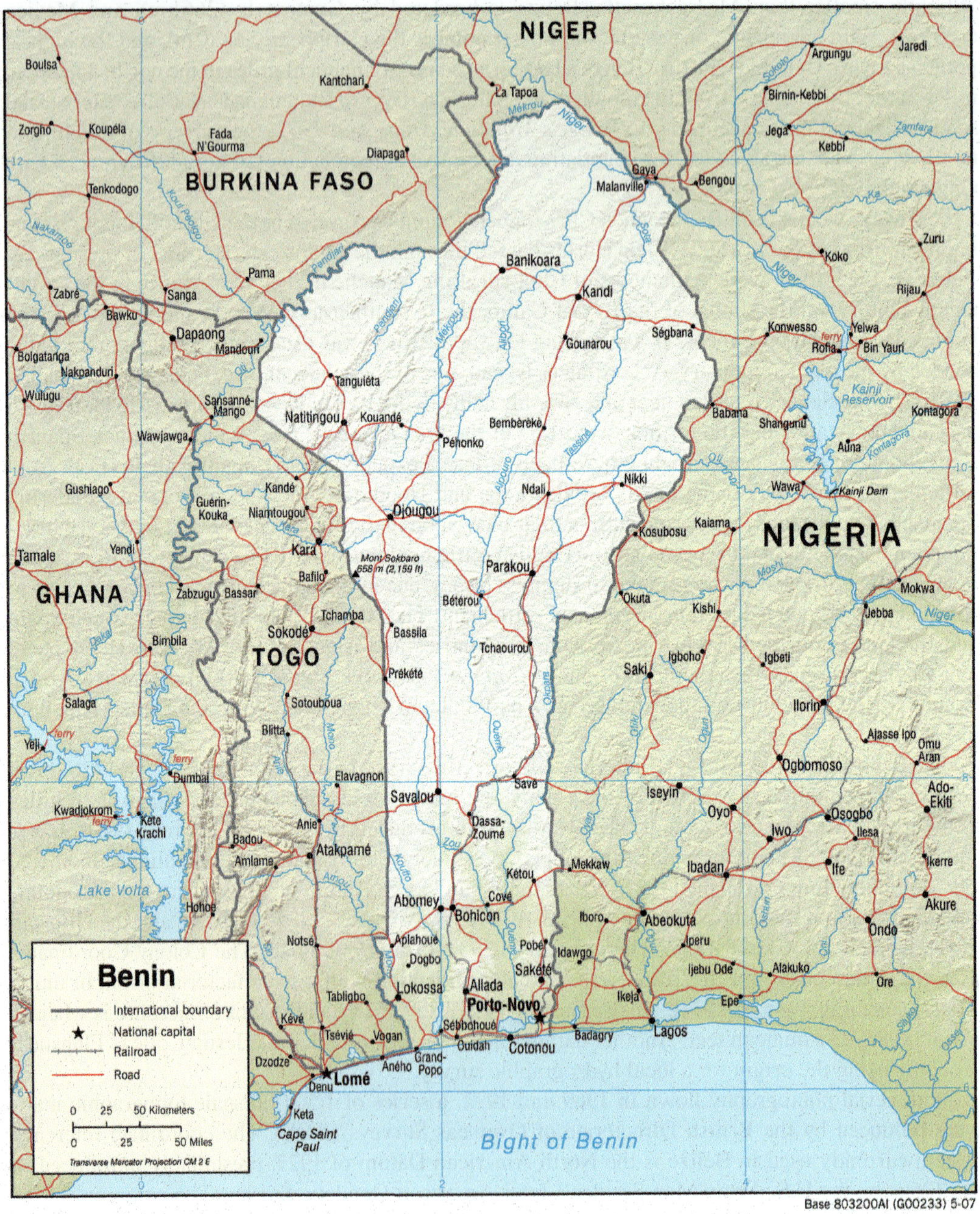

Base 803200AI (G00233) 5-07

Republic of Bénin

Bénin is bordered on the west by Togo (644 km), on the south by the Atlantic Ocean (121 km), on the east by Nigeria (773 km), and on the north by both Niger (266 km) and Burkina Faso (306 km). The lowest point is the Atlantic Ocean, the highest point is Mount Tanekas (641 m), and Bénin is slightly smaller than the state of Pennsylvania. Although the country is mostly flat to undulating plains, there are some hills and low mountains.

The site of the most organized kingdom of West Africa in the 15th century was at Abomey until the Portuguese came to the region in 1485. The kingdom of Great Bénin exerted great influence in the 17th century; the French established a trading presence in Cotonou (1851) with King Gezo. Temporarily suspended, the French re-established former rights at Porto-Novo in 1863, and protection was extended to various other political entities along the coast and in the interior. Dahomey eventually was made an overseas territory of France in 1946, and it became independent in 1960. Bénin is the name Dahomey changed to in 1975.

When the federation of the eight territories constituting former French West Africa came into being in 1904, the *Annexe de l'Institut Géographique National á Dakar* (Senegal) became responsible for the official mapping. At the time, the IGN Annex, Dakar was known as *Service Géographique de l'Afrique Occidentale Française* – SGAOF (Geographic Service of the French West Africa). Topographic mapping of Bénin by SGAOF was largely at the scales of 1:200,000 and 1:500,000. The methods originally used were rapid ground mapping (using planetable and alidade with *graphical* triangulation). The Clarke 1880 was the ellipsoid of reference for these series. In December 1945 the *Cabinet du Directeur, Institut Géographique National* in Paris issued an instruction concerning the systems of projection to be utilized in French West Africa. The instruction detailed that a Gauss (Gauss-Schreiber Transverse Mercator) system of projection was to be used for the group of regular map compilations and related works that included geodesy, topography, photogrammetry, and cartography for a range of scales that included 1:200,000. All of the map series were to be cast on the International Map of the World sheet system based on the graticule. In particular, *Fuseau Dahomey* was defined with a central meridian, $\lambda_o = 0° 30'$ East of Greenwich, a scale factor at origin, $m_o = 0.999$, and both False Easting and False Northing equal to 1,000 km. The limits of the Dahomey Zone grid were defined between 3° West and 4° East. The ellipsoid of reference was defined as the Hayford 1909 (International), where $a = 6,378,388$ m and $1/f = 297$. The detailed measures that were to be taken by the Colonial Inspector Generals in charge included training their staffs to use the new tables of projection.

Within 5 years, French coordinate systems changed to UTM throughout the world with the exception of Madagascar, and for Reunion. In terms of far-reaching developments in grid system usage, this was spectacular! D. R. Cockburn and W. L. Barry of Army Map Service translated the IGN Instruction dated September 20, 1950:

1. The General Directorate has decided to abandon the projection systems now in use in the French Overseas Territories and Departments and to utilize henceforth, in all these territories, a new projection system called the Universal Transverse Mercator (*Mercator Transverse Universelle*), having a unique definition.

   In-so-far as Madagascar is concerned, the use of the Laborde Projection will be continued without change. Similarly for Reunion, the Gauss system, in which the triangulation of the island has been computed, will be retained. With the exception of these two particular cases, the U.T.M. projection will from now on be the only official projection in the French Overseas Territories and Departments.

   Consequently I.G.C. instruction No.1212, dated 12 December 1945, is rescinded.

2. The new projection is a Gauss conformal projection applied to zones of 6° of longitude in width. These zones, identical to those of the 1:1,000,000 International Map of the World, are indicated on the attached index map.

3. For a long time, views have been expressed in the international meetings of geodesists in favor of a universal projection system, which would be adopted by all the countries of the world. Inspector General Tardi proposed himself at the General Assembly of the International Association of Geodesy at Edinburgh (1936), a Gauss projection in 6° zones for the African continent, which is the same as the UTM projection. These views remained the dead issue for a long time. Before 1940, each country was quite satisfied with its own projection system and was reluctant to undertake the enormous task of converting coordinates into a universal system. They were especially reluctant to modify their quad printing plates. However, during the course of the last war, the extension of military operation to vast regions of the globe, the strategic deployments on a great diversity of warfronts entailed the creation of a great number of projection systems (in 1945, over 100 of these systems were in use). As a result, a state of utter chaos ensued, and considerable expense was entailed for the computation of the transformation and the adaptation operations. Consequently, the prospect of a universal projection system aroused much interest in the post war period.

The United States was very much in favor of the project and to facilitate its adoption by the various countries, Gauss projection tables (called a UTM projection) were computed and published. These tables were computed in the sexagesimal angular division system. The American agencies also computed the same tables on a centesimal system.

The *Institut Géographique National*, when asked to adopt the new projection in December of 1949, did not hesitate in agreeing to its use for French Colonial regions with certain exceptions which are explained below. In point of fact, it was entirely possible to adopt this new projection for the major part of the colonial possessions at a very small cost. However, a problem arose for its use in France proper and in North Africa. For France itself, a 6° belt UTM projection leads to very extreme scale changes, *i.e.*, extreme from the point of view of civilian use.

4. Actually, it was not merely in a spirit of international cooperation that the *Institut Géographique National* agreed to the new projection but also because it offers incontestable practical advantages. In December 1949, the situation was as follows:

After long conferences in which various proposals were suggested, we finally adopted the solution proposed by General Laborde for our overseas possessions at the end of 1945. this solution was as follows: A Gauss system (double projection) on the international Ellipsoid with φ_o equaling 0° in French East Africa and French West Africa. For the smaller regions (Guadeloupe, Martinique, Reunion, etc.) the value of φ_o is equal to the mean latitude of the territory, φ_o being the latitude of the central point. This procedure leads to the establishment of separate tables for each value φ_o.

Tables for the conversion of geographic coordinates into rectangular and vice versa (tables which would produce the centimetric precision necessary for geodetic computations) had not been set up at the end of the year 1949 with the exception of tables covering a few small regions. Although this is a very pressing urgency, the Institut Géographique National, due to limit (*sic*) means, has neither the facilities for computing the tables rapidly nor for editing them without detriment to other equally urgent tasks.

Considering on one had the small number of stations to be converted into the new system (for astro points the work involved is insignificant) and considering that the dimensional variations of the sheets already published would be less than the standard size, the Institut Géographique National has agreed to rapidly extend the UTM projection in these territories being aware of the following factors:

That the United States was in a position to immediately deliver to us as many copies as was necessary of the tables computed on the sexagesimal system and contracted to compute the same tables on the grad system; that the United States was able to undertake the conversion of coordinates into the system using data obtained from electronic computing machines.

5. In point of fact, the UTM projection as it has been already adopted (or in the course of being adopted) by a number of countries is not absolutely "Universal." This would have been the case if a uniform ellipsoid had been chosen for all the countries. However, the difficulties entailed in changing ellipsoids is common knowledge and because of this, the basics of the ellipsoids in use for the various continents have been retained. Accordingly, the Clarke 1866 ellipsoid has been kept in use for North America; the International Ellipsoid has been adopted for South America and the Pacific

regions, and the Everest Ellipsoid has been chosen for the East Indies and the adjacent regions. So as to fulfill a request made by the British who have already computed vast geodetic nets on the Clarke 1880 Ellipsoid, the Institut Géographique National has adopted this ellipsoid for the entire African continent. In addition, this ellipsoid was used for French geodetic work previous to 1945.

6. The UTM projection may be defined as having the following intrinsic properties: It is a Gauss conformal projection, a direct projection of the ellipsoid on the plane. Linear values are maintained on the prime meridian of the projection with the exception of a scale-reduction which is defined by the following coefficient: $k_o = 0.9996$.

    The zones have an overall width of 6° in longitude (3° on each side of the central meridian). The zones coincide with those of the 1:1,000,000 International Map. The Greenwich meridian is at the limit of two zones (zone numbers 30 and 31). These basics will suffice to define the projection for any given ellipsoid.

7. The new UTM projection differs from the Gauss projection adopted at the end of 1945 in the sense that it is a direct projection of the ellipsoid on the plane instead of being an indirect projection employing the intermediary of a sphere upon which the ellipsoid is first applied before projecting it on the plane.

    The new projection retains the linear values on the central meridian of each zone to the approximate scale factor. The former projection did not retain linear values on this meridian.

    In toto, the basics of the two projections are, at least within the limits of the proposed narrow zones, absolutely comparable and considered from the view point of practical application it is impossible to give preference to either one or the other. The only advantage of the former projection is that of adapting itself more simply to the extension of latitudinal belts and that this predicament will not arise for overseas geographic services.

8. Covering memo No. 1 in reference to the implementation of the new projection program is to be effective immediately." *Director, Institut Géographique National.*

The instruction quoted above was accompanied with some specific procedures for all of the French colonies, territories, and departments. With respect to French West Africa (and Bénin), IGM explained that AMS agreed to compute the UTM coordinates of all astro points that were observed as control for the 1:200,000 scale topographic maps.

When Army Map Service computed all of the UTM tables for all of the commonly used ellipsoids, the hard-copy output was on a typewriter. As legend tells, the computer operator would occasionally note that the typewriter would stop typing, and the operator would get out of his chair and go to one particular computer cabinet and kick it. Typing would resume, and the operator would go back to his seat as the UTM table continued to print. Those were the days of the minicomputer in the 1960s.

In January and February of 1952, Hydrographic Engineer Bourgoin of the French Navy directed a port survey of Cotonou in support of a new wharf. The survey included depth soundings, current measurements, measurement of the period, direction, and amplitude of the tide, observations of the characteristics of the local sea, soundings in the adjacent lagoon, and granularity determinations of the bottom. A local coordinate system was devised with the origin of $X = 10,000$ m, $Y = 1,000$ m at the church steeple. The Cotonou Lighthouse was calculated to have the coordinates: $X = 10,241.75$ m, $Y = 508.01$ m. This local grid was oriented by tachéomètre (used at the time in Europe for cadastral surveying and distance measurement with subtense bar) to star observation, and scale was provided by a 50-m triangulation baseline. No geodetic coordinates were given for the origin point.

A Franco–German protocol of July 9, 1897 delimited a boundary between German Togo and the French possessions of Dahomey and Soudan (presently Upper Volta). The affirmed convention boundary utilized the lagoon eastward from *Île Bayol* to the Mono River and then follows the river northward to the 7th North parallel, thence various jogs, thalwegs, and meridians to the tripoint with Upper Volta (now Burkina Faso). Straight lines were defined according to French topographic mapping and are therefore cast on the Gauss-Schreiber Transverse Mercator projection. Note that the thalweg is the "thread of the stream" and is not always equidistant between the two riverbanks.

The Bénin–Burkina Faso boundary is demarcated by the Mékrou and Pendjari rivers for about 85% of the distance. Between the rivers, the boundary follows the Chaîne de L'Atacora for about 25 miles to the tripoint with Niger. This boundary (and that with Niger) was determined most recently by a French statute of October 27, 1938, and re-established on September 4, 1947.

The Lagos area of Nigeria was ceded to the United Kingdom by a local monarch in 1861. With the expansion inland of both Bénin and Nigeria, the Anglo–French convention of August 10, 1889, delimited a boundary between the territories from the Gulf of Guinea northward to the parallel of 9° N. The Anglo–French agreement of October 19, 1906, and amendments made by the demarcation protocol of July 20, 1912, determined the final boundary between the two countries. Using numerous beacons (survey monuments), thalwegs of rivers, and straight lines on French 1:200,000 scale topographic maps, the boundary ends at the tripoint beacon with Niger.

In April 1959, the SGAOF name was changed to *Service Géographique, Dakar – SG, Dakar* (Geographic Service, Dakar) and in January 1961, to its present name. Topographic mapping of Bénin by IGN has been largely at the scale of 1:50,000. In the late 1940s and early 1950s, IGN compiled a small amount of topographic mapping at 1:20,000 scale and in the early through mid-1950s a produced a small amount of mapping at 1:100,000 scale. After World War II the French adopted aerial photogrammetry controlled by astronomical points as the means for surveying at the scales of 1:50,000 and 1:200,000. The reader is reminded that when comparing mapping technology of the 1950s to the present, a shirt-pocket consumer grade GPS receiver is about five times more accurate than a classical "astro point."

The closest major classical datum to the Republic of Bénin is the Minna Datum of 1927. The origin is at station L40, which is the north end of Minna Base in the town of Minna, Nigeria, where $\Phi_o = 09°\ 39'\ 08.87"$ N, $\Lambda_o = 06°\ 30'\ 58.76"$ East of Greenwich and the ellipsoid of reference is the Clarke 1880, where $a = 6,378,249.145$ m and $1/f = 293.145$. In 1987, NIMA published the transformation parameters from the Minna Datum of 1927 to WGS84 Datum as: $\Delta a = -112.145$ m, $\Delta f \times 10^4 = -0.54750714$, $\Delta X = -92$ m ± 3 m, $\Delta Y = -93$ m ± 6 m, and $\Delta Z = +122$ m ± 5 m, and this was a mean solution of six stations. In 1990, C. U. Ezeigbo published his solution of 11 stations for a Bursa-Wolf seven-parameter transformation as: $\Delta X = -92.9$ m ± 1.6 m, $\Delta Y = -116.0$ m ± 2.3 m, and $\Delta Z = +116.4$ m ± 2.4 m, $R_x = -0.3" \pm 1.1"$, $R_y = +3.0" \pm 1.7"$, $R_z = +2.2" \pm 1.5"$, and scale $= +1.00002 \pm 0.6 \times 10^{-5}$. This solution is intriguing because one station appeared to be between Lagos and Cotonou, according to the small-scale survey sketch by Prof. Ezeigbo.

Seven CORS stations were established in October 2008 through Millennium Challenge Account, Benin (MCAB) and National Geodetic Service, USA. Trimble NetR5 GNSS systems are being used in all the stations. The stations are being operated by Millennium Challenge Account, Benin (MCAB) and the 1″ GNSS data, and other stations' information is available at NGS website https://www.ngs.noaa.gov/CORS/. Their ITRF00 (1997.0 Epoch) positions were computed in August 2008 using 13–69 days of data. (Source: National Geodetic Service USA website, *AFREF News Letter No. 11, afref@rcmrd.org, May 2010*.)

PROBLEM OF THE USE OF DUAL-FREQUENCY GNSS RECEIVERS IN BENIN, Degbegnon, L., Koumolou, A., *Global Journal of Engineering Science and Research Management*, 5(11): November 2018, ISSN 2349–4506. http://www.gjesrm.com/Issues%20PDF/Archive-2018/November-2018/6.pdf.

According to Degbegnon and Aïzo, the current *local* datum for Bénin is "Datum 58" with its origin at Point 58 southeast of the town of Dosso in the country of Niger and near the border with Nigeria, where $\Phi o = 12°\ 52'\ 44.045"$ N, $\Lambda o = 3°\ 58'\ 37.040"$ E of Greenwich. Thanks to John W. Hager, "Azimuth is 97° 30′ 04.237″ to C. F. L. 1 from north. Elevation = 266.71 meters. Astro observed by IGN in 1968. This was used as a temporary datum pending the adjustment of the 12th Parallel to Adindan. It was for the section of the 12th Parallel in Niger and Upper Volta. Reference is Final Report of the 12th Parallel Survey in the Republic of Niger." Surveyed in 1969 by the French IGN, the ellipsoid of reference is Clarke 1880, where $a = 6,378,249.145$ m and $1/f = 293.465$. This was used as the basis for computation of the 12th parallel traverse conducted in 1966–1970 from Senegal to Chad and connecting to the Adindan triangulation in Sudan. The country is now moving to the WGS84 Datum (realization undefined), and the paper by Degbegnon & Aïzo (below) discusses a Bursa-Wolf transformation between the WGS84 Datum and the local Datum 58.

See also: Degbegnon, L., Aïzo, P., Comparative study and determination of transformation parameters between: the permanent station system, the datum (58) and the Bénin geodetic system, *Research Journal of Recent Sciences*, Vol. 8, No. 4, 1–8, October (2019) ISSN 2277-2502. http://www.isca.in/rjrs/archive/v8/i4/1.ISCA-RJRS-2018-067.pdf.

BHUTAN
Physiography
International boundary
National capital
Elevation point (feet/meters)
Archaeological site
Dzong fortress
Historic/cultural site
National protected area
UNEP-WCMC 2012 WDPA
Scale 1:1,750,000
CHINA
INDIA
GREAT HIMALAYA RANGE
LESSER HIMALAYA
BLACK MOUNTAINS
Western Duars
Eastern Duars
Kangmar
Kurmê
Goinbachacang
Dogên
Gala
Lhozhag
Moinda
Comai
Lhakang
Puma Yumco
In dispute
Kula Kangri
7,538 m
(24,731 ft)
Gangkar Puensum
7,570 m
(24,836 ft)
Jumolhari
7,326 m
(24,035 ft)
Gasa
Jigme Dorji National Park
Wangchuck Centennial Park
Chumbi Valley
Pagrizhen
Xarsingma
Gangtok
THIMPHU
Paro
Drugyel
Haa
Punakha
Wangdue
Phodrang
Trongsa
Jakar
Lhuentse
Senge
Bumdeling
Trashi
Yangtse
Tawang
Sakteng
Trashigang
Mongar
Zhemgang
Black Mountain
Thrumshingla National Park
Royal Manas National Park
Tsimasham
Daga
Damphu
(Tsirang)
Pemagatshel
Dewathang
Samdrup
Jongkhar
Samtse
Gomtu
Gedu
Phuentsholing
Sarpang
Gelephu
Panbang
Phibsoo
Khaling
Rowta
Mainâguri
Dhupgâri
Jalpâiguri
Alîpur
Duâr
Bongaigaon
Kokrajhar
Barpeta
Rangia
Khârupatia
Koch
Bihâr
BANGL.
BANGLADESH
Boundary representation is not necessarily authoritative.
TRANSVERSE MERCATOR PROJECTION; CENTRAL MERIDIAN 90° 27' E
803224AI (G00541) 2-12

Kingdom of Bhutan

Bordered by China (477 km) and India (659 km), the Kingdom is mountainous with some fertile valleys and savanna. The lowest point is Drangeme Chhu (97 m), and the highest point is Gangkar Puensum (7,570 m).

"Although knowledge of prehistoric Bhutan has yet to emerge through archaeological study, stone tools and weapons, remnants of large stone structures, and megaliths that may have been used for boundary markers or rituals provide evidence of civilization as early as 2000 B.C. The absence of Neolithic mythological legends argues against earlier inhabitation. A more certain prehistoric period has been theorized by historians as that of the state of Lhomon (literally, southern darkness) or Monyul (dark land, a reference to the Monpa aboriginal peoples of Bhutan), possibly a part of Tibet that was then beyond the pale of Buddhist teachings. Monyul is thought to have existed between 500 B.C. and A.D. 600. The names Lhomon Tsendenjong (southern Mon sandalwood country) and Lhomon Khashi (southern Mon country of four approaches), found in ancient Bhutanese and Tibetan chronicles, may also have credence and have been used by some Bhutanese scholars when referring to their homeland. Variations of the Sanskrit words Bhota-ant (end of Bhot, an Indian name for Tibet) or Bhu-uttan (meaning highlands) have been suggested by historians as origins of the name Bhutan, which came into common foreign use in the late nineteenth century and is used in Bhutan only in English-language official correspondence. The traditional name of the country since the seventeenth century has been Drukyul – country of the Drokpa, the Dragon People, or the Land of the Thunder Dragon – a reference to the country's dominant Buddhist sect.

"Some scholars believe that during the early historical period the inhabitants were fierce mountain aborigines, the Monpa, who were of neither the Tibetan or Mongol stock that later overran northern Bhutan. The people of Monyul practiced the shamanistic Bon religion, which emphasized worship of nature and the existence of good and evil spirits. During the latter part of this period, historical legends relate that the mighty king of Monyul invaded a southern region known as the Duars, subduing the regions of modern Assam, West Bengal, and Bihar in India.

The introduction of Buddhism occurred in the seventh century A.D., when Tibetan king Srongtsen Gampo (reigned A.D. 627–49), a convert to Buddhism, ordered the construction of two Buddhist temples, at Bumthang in central Bhutan and at Kyichu in the Paro Valley. Buddhism replaced but did not eliminate the Bon religious practices that had also been prevalent in Tibet until the late sixth century. Instead, Buddhism absorbed Bon and its believers. As the country developed in its many fertile valleys, Buddhism matured and became a unifying element. It was Buddhist literature and chronicles that began the recorded history of Bhutan.

"In A.D. 747, a Buddhist saint, Padmasambhava (known in Bhutan as Guru Rimpoche and sometimes referred to as the Second Buddha), came to Bhutan from India at the invitation of one of the numerous local kings. After reportedly subduing eight classes of demons and converting the king, Guru Rimpoche moved on to Tibet. Upon his return from Tibet, he oversaw the construction of new monasteries in the Paro Valley and set up his headquarters in Bumthang. According to tradition, he founded the Nyingmapa sect – also known as the "old sect" or Red Hat sect – of Mahayana Buddhism, which became for a time the dominant religion of Bhutan. Guru Rimpoche plays a great historical and religious role as the national patron saint who revealed the tantras – manuals describing forms of devotion to natural energy – to Bhutan. Following the guru's sojourn, Indian influence played a temporary role until increasing Tibetan migrations brought new cultural and religious contributions.

"There was no central government during this period. Instead, small independent monarchies began to develop by the early ninth century. Each was ruled by a deb (king), some of whom claimed divine origins. The kingdom of Bumthang was the most prominent among these small entities. At the same time, Tibetan Buddhist monks (lam in Dzongkha, Bhutan's official national language) had firmly rooted their religion and culture in Bhutan, and members of joint Tibetan-Mongol military expeditions settled in fertile valleys. By the eleventh century, all of Bhutan was occupied by Tibetan-Mongol military forces" (*Andrea Matles Savada, ed. Bhutan: A Country Study. Washington: GPO for the Library of Congress, 1991*).

"In 1865, Britain and Bhutan signed the Treaty of Sinchulu, under which Bhutan would receive an annual subsidy in exchange for ceding some border land to British India. Under British influence, a monarchy was set up in 1907; three years later, a treaty was signed whereby the British agreed not to interfere in Bhutanese internal affairs, and Bhutan allowed Britain to direct its foreign affairs. This role was assumed by independent India after 1947. Two years later, a formal Indo-Bhutanese accord returned to Bhutan the areas annexed by the British, formalized the annual subsidies the country received, and defined India's responsibilities in defense and foreign relations. In March 2005, King Jigme Singye Wangchuck unveiled the government's draft constitution – which introduced major democratic reforms - and pledged to hold a national referendum for its approval. In December 2006, the King abdicated the throne in favor of his son, Jigme Khesar Namgyel Wangchuck, in order to give him experience as head of state before the democratic transition. In early 2007, India and Bhutan renegotiated their treaty, eliminating the clause that stated that Bhutan would be 'guided by' India in conducting its foreign policy, although Thimphu continues to coordinate closely with New Delhi. Elections for seating the country's first parliament were completed in March 2008; the king ratified the country's first constitution in July 2008. Bhutan experienced a peaceful turnover of power following parliamentary elections in 2013, which routed the incumbent party. In 2018, the incumbent party again lost the parliamentary election. Of the more than 100,000 ethnic Nepali – predominantly Lhotshampa – refugees who fled or were forced out of Bhutan in the 1990s, about 6,500 remain displaced in Nepal" (*World Factbook*, 2022).

"The oldest recorded history suggesting any form of organized land tenure system goes back to the early 17th century when Zhabdrung Ngawang Namgyel instituted the Marthram Chem – a central register of cultivated land, created practically for taxing purposes. Since then, subsequent monarchs gradually re-defined the relationship between man and land, and land tenure evolved to become what it is today – a form of freehold land tenure adapted to the socio-economic environment of the country. The Bhutanese landowner enjoys a very high degree of security of land title and can exercise full rights of alienation barring a few restrictions that are aimed at maintaining an equitable distribution of land. The Bhutanese Land Act of 1979 provides the legal framework" (*An Insight into Surveying, Mapping, and Land Administration, Ministry of Home Affairs, Trashichhodzong, Thimhu, Bhutan, 2nd Edition, 2001*). A systematic cadastral survey was started in 1980 using plane tables and alidades and countrywide coverage was achieved in 1997 in which each parcel is located on a national geodetic frame.

There appear to be six GNSS continuously operating reference stations that provide reference to an International Terrestrial Reference Frame epoch. Located in Phuntsholing, Gelephu, Deothang, Kanglung, Burnthang, and Thimphu, the sites are intended to be used for post-processing of data. Thanks to GeoRepository™, "Bhutan National Geodetic Datum origin is ITRF2000 at epoch 2003.87. Bhutan National Geodetic Datum is a geodetic datum for Geodetic survey. It was defined by information from Department of Survey and Land Records (DSLR), National Land Commission of Bhutan (NLC) via Lantmäteriet Sweden."

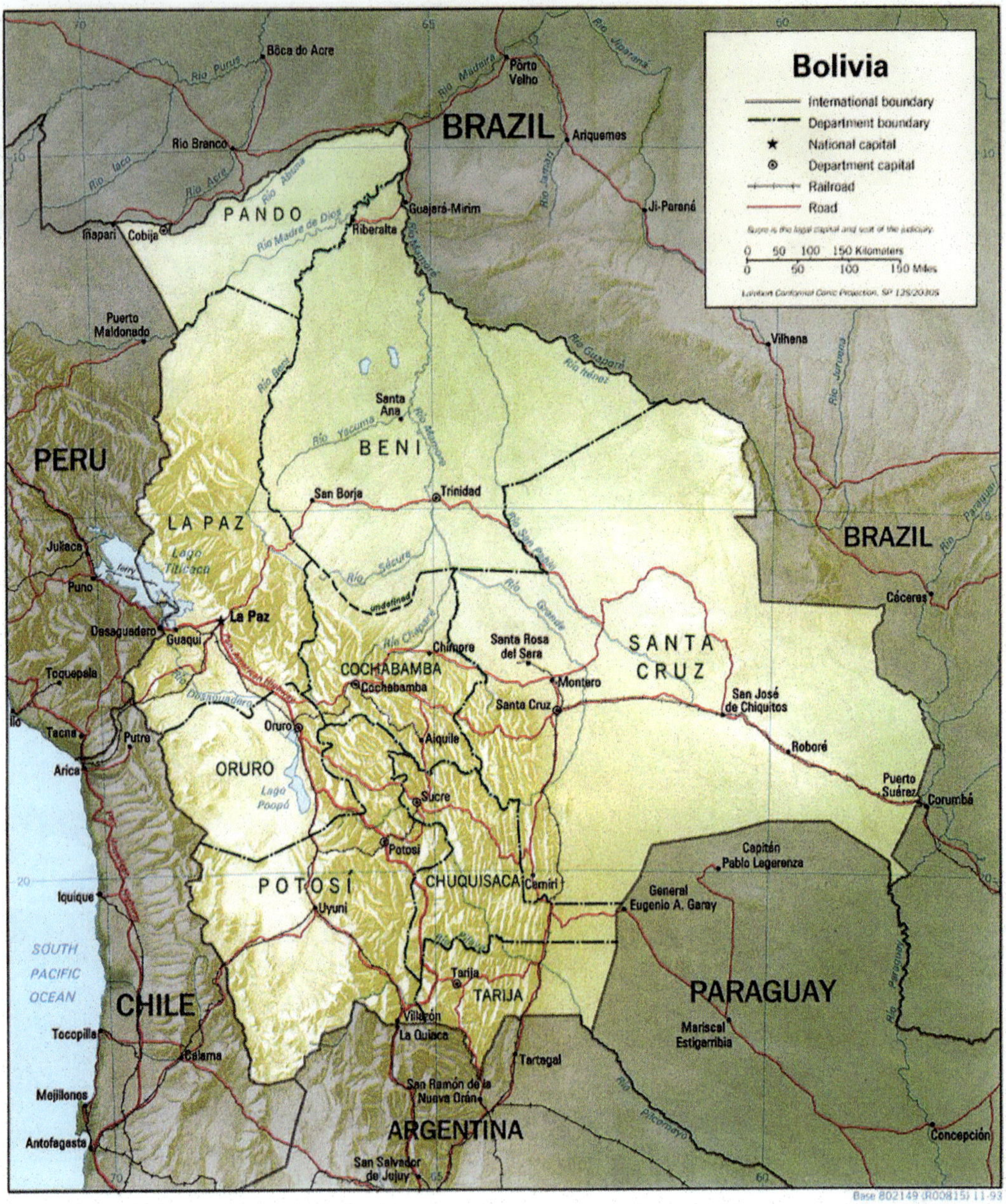

Base 802149 (R00815) 11-93

Republic of Bolivia

Bordered by Argentina (942 km), Brazil (3403 km), Chile (942 km), Paraguay (753 km), and Peru (1,212 km), the eastern part of Bolivia has low, hot, fertile land watered by many rivers. In the central part of the eastern slope of the Andes, there is a high plateau region that includes part of the Gran Chaco. Lake Titicaca is in the western part of the central ranges and the highest peaks include Cerro Illimani at 6,882 m. The lowest point in Bolivia is the Rio Paraguay at 90 m.

The Bolivian highlands were the location of the advanced Tiahuanaco culture, circa 7th–11th centuries. The Aymara Indians followed and were conquered in the 15th century by the Inca Indians. The Incas were subsequently conquered in the 1530s by Hernando Pizarro, the half-brother to Francisco Pizarro who conquered Peru. Bolivia achieved independence from Spain by proclamation of Simón Bolívar on August 6, 1825, when General Antonio José de Sucre invaded Characas. A map of Bolivia that was printed on a 1928 Bolivian postage stamp precipitated a war with Paraguay in the 1930s over the Gran Chaco region! After learning about that postage stamp, I began collecting "map stamps" of the world for many years.

Bolivia has had a cartographical history quite unlike that of any of the other west coast republics of South America. Not only did the period of instrumental surveys begin much late in the Republic of Bolivia, but when in 1921 the American Geographical Society began the compilation of the La Paz 1:1,000,000 sheet; there had never been a general map of the country that could be described as an exhaustive compilation. J. B. Penland, an English explorer and cartographer, opened the era of exploration and discovery in Bolivia with his work on the Altiplano and in the Eastern Cordillera of the Andes that he carried out in 1827 and 1828. His surveys and later work in 1837–1838 were published as charts by the British Admiralty in 1830 and 1839. The most important part of Pentland's work was his determinations of altitudes and astronomical positions, the latter of which were used for many years as the basis for survey work in and for all compiled maps of the areas he covered. His determinations of altitudes in the Cordillera Real not only attracted the attention of other explorers to this hitherto little-known section of the Andes but also gave rise to much controversy. That was because they indicated that the highest crests of the Bolivian section of the Andes were among the highest, if not actually the highest, in the whole Cordillera. Such a position up to that time was believed that the snow-capped volcanoes of Ecuador held that distinction. Meanwhile in 1830, Alcide d'Orbigny, a French anthropologist and geologist, had begun the explorations that resulted in his *Carte Générale de la République de Bolivia* on the scale of approximately 1:1,575,000 published in 1835–1847 in his *Voyage dans l'Amérique Méridionale*.

Thanks to Ms. Sequoia Read of the Defence Geographic and Imagery Intelligence Agency, I received excerpts of a 1935 book entitled, *Die Karten der Cordillera Real und des Talkessels von La Paz (Bolivien)*. The authors were Prof. Dr. Carl Troll, and the famous Photogrammetrist, Dr. Richard Finsterwalder! Finsterwalder detailed his excellent terrestrial Photogrammetric survey of the volcanoes around La Paz.

The "Topographic Desk of the Southeast" was combined with the "Major General Topographic Cabinet of the State" in 1936 to form the Instituto Geográfico Militar (IGM). Initially formed in the city of Cuevo, the institute was relocated to Cochabamba in 1939 and finally moved to La Paz in 1942. By the end of 1948, "La Ley de la Carta" (The Law of the Map) was decreed establishing IGM as the monopoly for mapping the nation. I have discussed this concept common to many Latin American nations as a vehicle to help fund the military because of the low existing tax base. Argentina and Ecuador are no exceptions to this custom.

The southwestern portion of Bolivia is well controlled with classical triangulation arcs, and some high-precision electronic distance meter traverses cover the southeastern portion of the country.

The northern half of Bolivia is relatively devoid of triangulation control. Vertical control does extend to the northern provinces, and 1:50,000 mapping covers the southern half completely and the northern latitudinal half of the northern half.

The Republic of Bolivia is on the Provisional South American Datum of 1956 (PSAD 56) with the origin in the town of La Canoa, Venezuela, where $\Phi_o = 08°\ 34'\ 17.170''$ North, $\Lambda_o = -63°\ 51'\ 34.880''$ West of Greenwich, and the defining azimuth to station Pozo Hondo $(\alpha_o) = 40°\ 22'\ 45.96''$. The ellipsoid of reference is the Hayford 1909 where the semi-major axis, $a = 6,378,388$ m, and the reciprocal of flattening, $1/f = 297$. In 1977, James W. Walker presented and published a paper at the Pan American Institute of Geography and History that included the seven-parameter transformation from PSAD 56 in Bolivia to the WGS 72 Datum, where $\Delta X = +268.20$ m, $\Delta Y = -129.21$ m, $\Delta Z = +408.13$ m, scale $= -1.79024 \times 10^{-5}$, $R_X = -1.549''$, $R_Y = -0.742''$, $R_Z = -0.416''$. However, Walker pointed out that the rotations and scale factor had magnitudes equal to the uncertainty, so the actual parameters had little to do with reality. For that reason, the transformation parameters from the South American Datum of 1969 (SAD69) coordinates for Bolivia to WGS 72 Datum were only offered in terms of three parameters. I have not been able to locate anything in Bolivia that is referenced to SAD69.

All late 20th-century large-scale mapping of Bolivia appears to be referenced to the PSAD 56. The Grid of choice for Bolivia is the Universal Transverse Mercator (UTM). However, a Lambert Conformal Conic does exist in Bolivia that is referenced to the WGS 72 Datum where the Northern Standard Parallel $\varphi_n = -12°$ South, the Southern Standard Parallel $\varphi_s = -20°$ South, the False Northing Latitude of Origin $\varphi_o = -20°$ South, the Central Meridian $\lambda_o = -64°$ West of Greenwich, and the False Easting is 500 km. The ellipsoid of reference is the WGS 72, where $a = 6,378,135$ m and the reciprocal of flattening, $1/f = 298.26$. Other ancient Datums known to exist in Bolivia include the Arequipa Astro from the boundary surveys of the 19th century with Peru, and the Pepito and Poto Astros. Mr. John W. Hager believes that the vertical Datum is referenced to Arica in Chile. Since 2001, Bolivia has entered into a cooperative agreement for the Geocentric Reference System for the Americas, termed "SIRGAS," a Spanish acronym. Extensive geophysical research has been initiated in Bolivia with respect to crustal motion and volcanic activity. The CGPS network in country consists of 43 CGPS station sites, with 11 GPS stations recently installed in response to the Pisagua 8.2 magnitude earthquake on 1 April 2014. As of November 2014, more than 240 GPS stations had been observed with dual-frequency GPS receivers 1,2).

Topography of the Uyuni salt flats determined with kinematic GPS was published in 2008 (3).

Significant gravity surveys were published in 2015 for the Bolivian Central Andean Plateau (4), but there seems to be zero data publicly available on transformation parameters from the PSAD 56 or the SAD69 datums to the current SIRGAS system. The Bolivian military IGM offered no response to my queries.

1. http://www.sirgas.org/fileadmin/docs/Boletines/Bol19/56_Heck_et_al_2014_CrustalMotionBolivia.pdf
2. https://pubs.geoscienceworld.org/gsa/geosphere/article/14/1/65/525536/50-years-of-steady-ground-deformation-in-the
3. https://academic.oup.com/gji/article/172/1/31/2081107
4. http://www.igmbolivia.gob.bo/

Bosnia and Herzegovina

- International boundary
- Inter-Entity Boundary Line (IEBL)
- ★ National capital
- Railroad
- Expressway
- Road

In March 1999, international arbitration made Brčko a neutral district under international supervision.

0 20 40 60 Kilometers
0 20 40 60 Miles
Lambert Conformal Conic Projection, SP 40 N / 56 N

Base 802730AI (R00389) 2-02

Bosnia and Herzegovina

Bordered by Croatia (956 km), Montenegro (242 km), and Serbia (345 km); Bosnia and Herzegovina's terrain is mountains and valleys with the highest point being Maglic (2,386 m), and the lowest point is the Adriatic Sea (0 m). The mean elevation is 500 m (*World Factbook,* 2022).

"The region's ancient inhabitants were Illyrians, followed by the Romans who settled around the mineral springs at Ilidža near Sarajevo in 9 *A.D.* When the Roman Empire was divided in 395 *A.D.,* the Drina River, today the border with Serbia, became the line dividing the Western Roman Empire from Byzantium. The Slavs arrived in the late 6th and early 7th centuries. In 960 the region became independent of Serbia, only to pass through the hands of other conquerors: Croatia, Byzantium, Duklja (modern-day Montenegro) and Hungary. Bosnia's medieval history is a much-debated subject, mainly because different groups have tried to claim authenticity and territorial rights on the basis of their interpretation of the country's religious make-up before the arrival of the Turks. During this period (1180–1463) Bosnia and Herzegovina became one of the most powerful states in the Western Balkans. The most significant event was the expansion of the Bosnian state under Stephen Kotromanić who conquered large parts of the Dalmatian coast and in 1326 annexed the southern province of Herzegovina. The country thus became Bosnia and Herzegovina for the first time. The first Turkish raids came in 1383 and by 1463 Bosnia was a Turkish province *with Sarajevo* as its capital. Herzegovina is named after Herceg (Duke) Stjepan Vukčić, who ruled the southern part of the present republic from his mountain-top castle at Blagaj, near Mostar, until the Turkish conquest in 1482.

"Bosnia and Herzegovina was assimilated into the Ottoman Empire during the 400 years of Turkish rule. Conversion to Islam largely took place during the initial 150 years of Turkish rule and it's generally held that people converted voluntarily. Orthodox and Catholic Christians continued to practice their religions although under certain constraints. As the Ottoman Empire declined elsewhere in the 16th and 17th centuries, the Turks strengthened their hold on Bosnia and Herzegovina as a bulwark against attack. Sparked by the newly born idea of nationhood, the South Slavs rose against their Turkish occupiers in 1875–6. In 1878 Russia inflicted a crushing defeat on Turkey in a war over Bulgaria and at the subsequent Congress of Berlin it was determined that Austria-Hungary would occupy Bosnia and Herzegovina despite the population's wish for autonomy. The Austria-Hungarians pushed Bosnia and Herzegovina into the modern age with industrialization, the development of coal mining and the building of railways and infrastructure. Ivo Andrić's *Bridge over the Drina* succinctly describes these changes in the town of Višegrad. But political unrest was on the rise. Previously, Bosnian Muslims, Catholics and Orthodox Christians had only differentiated themselves from each other in terms of religion. But with the rise of nationalism in the mid-19th century, Bosnia's Catholic and Orthodox population started to identify themselves with neighboring Croatia or Serbia respectively. At the same time, resentment against foreign occupation intensified and young people across the sectarian divide started cooperating with each other and working against the Austria-Hungarians, thus giving birth to the idea of 'Yugoslavism' (land of the southern Slavs). Resentment against occupation intensified in 1908 when Austria annexed Bosnia and Herzegovina outright. The assassination of the Habsburg heir Archduke Franz Ferdinand by a Bosnian Serb, Gavrilo Princip, in Sarajevo on 28 June 1914 led Austria to declare war on Serbia. Russia and France supported Serbia, and Germany backed Austria, and soon the world was at war. These alliances still resonate today, with the Russians and French being seen as pro-Serb, and Austrians and Germans as pro-Croat" (*Lonely Planet,* 2012).

"An original NATO-led international peacekeeping force (IFOR) of 60,000 troops assembled in 1995 was succeeded over time by a smaller, NATO-led Stabilization Force (SFOR). In 2004,

European Union peacekeeping troops (EUFOR) replaced SFOR. Currently, EUFOR deploys around 600 troops in theater in a security assistance and training capacity" (*World Factbook, 2022*).

"The Military Geographic Institute started the survey of the occupied territory (*of Bosnia and Herzegovina – Ed.*) in 1879 with a first order chain which extends between the 35° and 36° meridian through the central part of the province and includes the base line at Sarajevo measured in 1882. In the North the first order chain was attached to these stations of the first order net in Slavonia: 356 Kučerina (of Dublca base extension net); 360 Maksimov Hrast; 362 Kasonja; 363 Lipovica, and in the South to the following stations of the Dalmatian first order chain: 341 Tmor; 343 Rogo; 345 Ostra Glavica; 348 Sniježnica; 350 Orjen. For all these stations the geographic positions referring to the Vienna University (*Datum – Ed.*) can be found in part II of *Positions Rechnungen (Protocol 290B), 1889*. In 1883–1885, the first order net was extended from Sarajevo east and southeast to the Serbian and Montenegro boundary (*Mapping of the Countries in Danubian and Adriatic Basins, Andrew M. Glusic, AMS TR No. 25, June 1959*).

The third topographic survey of the Austro–Hungarian Empire (*Neue Aufnahme*) was conducted from 1869 to 1896 and was mainly based on the Vienna University Datum and the Bessel 1841 ellipsoid (actually adopted in 1863) where $a = 6,377,397.15$ m, the Vienna Datum of 1806 was established based on the origin of St. Stephan Turm (St. Stephan's Tower) where $\Phi_o = 48°\ 12'\ 31.5277''$ North, $\Lambda_o = 16°\ 22'\ 27.3275''$ East of Greenwich. (These longitudes were originally referenced to Ferro, in the Canary Islands which is $17°\ 39'\ 46.02''$ West of Greenwich.) The defining azimuth of the Datum was from St. Stephan Turm to Leopoldsberg: $\alpha_o = 345°\ 55'\ 22.0''$.

"Bosnia and Herzegovina prior to that time had no survey and no reliable maps; therefore, it was planned to start with the cadastral survey as early as 1880. The occupied province still was in a stage of resistance and since survey parties needed the protection of small combat units the K.u.k. Military Geographic Institute was charged with the execution of the cadastral survey; consequently, the cadastral survey was planned and executed in such a manner that its complete utilization in the topographical survey was assured in advance.

"In order to provide the cadastral and topographical survey with a sufficient density of trig points (1 trig point on each 25 square kilometers) the 2nd, 3rd and 4th order triangulation started in 1879, *i.e.*, before the completion of the first order net. This triangulation, usually covering a region which during the next year has to be surveyed by the cadastral survey, started individually from the first order stations belonging to the already mentioned first order chains which surround Bosnia and Herzegovina from the North, West and South. In each region the triangulation was adjusted preliminarily and immediately used in the detail survey.

"For the first and second order stations, which formed the so called 'main net,' the geographic positions of the Vienna System were computed from different starting points in different regions. These geographic positions are the basis for the "uniform graticule system" in Bosnia and Herzegovina. Triangulation of Bosnia and Herzegovina comprise: Base line Dubica, Base line Sarajevo, 30 First order stations, 136 Second order stations, 2,094 Third and Fourth order trig points, totaling 2,260 trig points; of them 54 churches, mosques and monuments.

"The cadastral survey of Bosnia and Herzegovina is unique for the entire Monarchy since it is in position completely identical with the topographical survey. The cadastral sections are cut in the "uniform graticule system." The polyhedric (*identical to the photogrammetric Local Space Rectangular coordinate system – Ed.*) sheet of $15' \times 30'$ at 1:75,000 scale is divided into 16 cadastral sections at 1:12,500 scale, 64 cadastral sections at 1:6,250 scale and 256 sections at 1:3,125 scale which were used only in the survey of closed settlements. In order to meet the cadastral requirement, the spheroidal Soldner (*Cassini-Soldner – Ed.*) coordinates with the origin at center of each 1:75,000 sheet for all lower order stations of Military triangulation were computed. The cadastral survey was completed in 1885. Reduced by pantograph to 1:25,000 scale, it served as base for the topographical survey.

"In 1944, the Germans transformed the Soldner coordinates with the origin at the center of each 1:75,000 sheet, into Yugoslavian Reduced Gauss-Krüger coordinates (*where the scale factor at* *origin ($m_o = 0.9999$), the central meridians of the belts (C.M. = λ_o = 15°, 18°, 21° East of Greenwich) and the False Easting at C.M. = 500 kilometers – Ed.).* It should be remembered that the sheets were positioned on geographic positions of Vienna University System computed from preliminary adjusted regional nets with different starting points used in computation; meanwhile the Yugoslav triangulation represents a homogeneous, final adjusted net oriented on Hermannskogel *(1871 – Ed.)* datum. Linear conformal transformation was made by the sheets where, because of lack of a sufficient number of identical trig points, the transformation elements were derived from the differences between the Soldner and Yugo Reduced Gauss-Krüger coordinates of 2 sheet corners and the central point of each sheet. In order to furnish the Yugo Reduced G.K. coordinates needed in the comparison of the geographic positions of sheet corners and centers in Vienna University system were transferred into the system of the Yugoslav Triangulation by the application of corrections derived by Dr. Ledersteger for the purpose of the recasting of sheets. These corrected geographic coordinates were then converted into reduced Gauss-Krüger coordinates.

"The comparisons of the reduced Gauss-Krüger coordinates of the Military triangulation of Bosnia and Herzegovina, obtained from this transformation, with the Yugo Red. Gauss-Krüger coordinates show at identical stations disparities up to 18 meters. The average values of disparities on each sheet were taken as blanket corrections which have to be applied to the transformed coordinates of stations belonging to the corresponding sheet. By application of these corrections the transformed coordinates should be in sympathy with Yugoslav Triangulation in limit of ±5 meters and would meet only artillery and cartographic requirements.

"The corrections run as follows: dN from +7.00 m to −6.51 m, dE from +9.95 m to −13.70 m.

"Since the plane table sections are positioned upon geographic positions of the Vienna University system computed from preliminary adjusted regional nets the sheet corners are not in sympathy with the adjusted K. un k. M.T. first order net but have displacements within limits of the above-mentioned corrections" *(op.cit., A.M. Glusic, 1959).*

In 2007, funding was received to implement a number of EUREF permanent stations in Bosnia and Herzegovina. Thirty-four permanent stations were installed (17 in the Federation of BiH and 17 in Republika Srpska). The Geodetic Society of Bosnia–Herzegovina became a Member Association of FIG in 2010 and invited a member of the FIG Council to attend the Congress 2011, held on September 28–30, 2011 in Hidza near Sarajevo. Reconstruction of the cadaster began in 2007 and is reported to be making good progress.

"The Federal Administration for Geodetic and Real Property Affairs of FBiH performs administrative and other technical tasks that are under the authority of the Federation of Bosnia and Herzegovina, relating to: survey, real estate cadastre establishment and updating, utility cadastre, with the exception of the tasks for which the authority was devolved by the law to the cantons and municipalities, mapping of the territory of the Federation of Bosnia and Herzegovina, geodetic and cartographic tasks of relevance to defense, keeping technical archives of the originals of plans and maps of basic geodetic works and other data resulting from geodetic works, land consolidation, special purpose land survey, real property records, making property reprivatization proposals, inspection supervision of works related to survey, real estate cadastre, land cadastre, and utility cadastre.

"The geodetic profession in the Federation of Bosnia and Herzegovina is decentralized with its headquarters in Sarajevo, in contrast to the Republic Administration of the Republika Srpska that is headquartered in Banja Luka. Both Administrations cooperate on legislative and professional basis and are partners in various projects and programs" (http://www.fgu.com.ba/en/about-us-558.html, 2022).

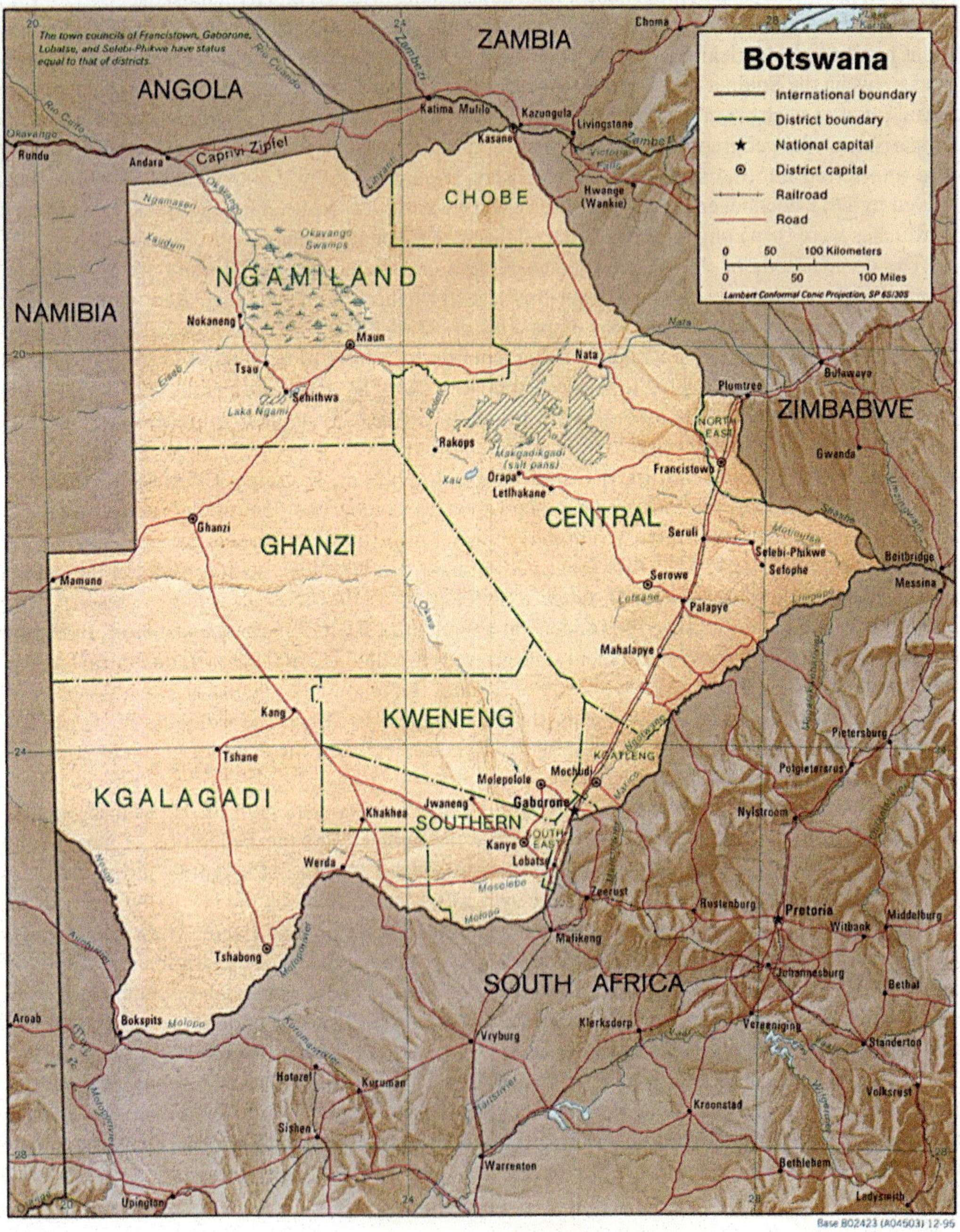

Base 802423 (A04503) 12-95

Republic of Botswana

The republic is landlocked, it is slightly smaller than Texas, and it is bordered by Namibia (1,360 km), South Africa (1,840 km), and Zimbabwe (813 km). The terrain of Botswana is predominately flat to gently rolling tableland with the Kalahari Desert in the southwest. The lowest point is the confluence of the Limpopo and Shashe Rivers (513 m), and the highest point is Tsodilo Hill (1,489 m).

The Boskop people inhabited southern Africa for thousands of years during the Middle Stone Age, and their descendants include the present-day San (Bushmen) people of Botswana. Most of the present population (95%) is known as the "Batswana." The region was occupied by the British at the instigation of the colonial administrator, Sir Cecil Rhodes in 1884. It was organized as a British Protectorate the following year and was divided into the British Bechuanaland and the Bechuanaland Protectorate. Botswana gained independence in 1966, it is a member of the British Commonwealth, and the discovery of diamonds and copper has allowed it to become one of the strongest countries in Africa.

"More than five decades of uninterrupted civilian leadership, progressive social policies, and significant capital investment have created one of the most stable economies in Africa. The ruling Botswana Democratic Party has won every national election since independence; President Mokgweetsi Eric MASISI assumed the presidency in April 2018 following the retirement of former President Ian KHAMA due to constitutional term limits. MASISI won his first election as president in October 2019, and he is Botswana's fifth president since independence. Mineral extraction, principally diamond mining, dominates economic activity, though tourism is a growing sector due to the country's conservation practices and extensive nature preserves. Botswana has one of the world's highest rates of HIV/AIDS infection, but also one of Africa's most progressive and comprehensive programs for dealing with the disease" (*World Factbook*, 2022).

In *GIM International* (September 1994), Mr. B. B. H. Morebodi, Director of Surveys and Lands, reported: "The history of cadastre in Botswana can be traced back to 1890 when the Foreign Jurisdiction Act in Great Britain established and regulated the power of Queen Victoria's Government in the then territory of British Bechuanaland. Subsequently various acts defined the boundaries of the Crown Lands in Botswana, the territories occupied by tribes and land vested in the British south Africa Company. The earliest registration of deeds was effected at Vryburg in South Africa but was moved to Mafeking in 1908 following the establishment of the Office of Registrar of Deeds for the Bechuanaland Protectorate in 1907. From 1908 to independence in 1966 the Registrar of Deeds Office remained in Mafeking.

"Throughout its history the cadastre has always been numerical. Surveying was under the control of the Surveyor-General in Cape Town until the 1959 Land Survey Act established the Office of director of surveys and Lands in Botswana. In 1960 an act was passed establishing the Office of Registrar of Deeds and regulating the transfer of land in Botswana. The Dept. of Surveys and Lands was physically moved from Mafeking to Gaborone in 1969.

"There is no cadastral mapping in Botswana as such. The registration of title to property is affected by a system of registration of deeds. To be capable of registration a deed must be supported by a survey diagram (deed plan) approved by the Director of Surveys. Data from numerical cadastral surveys is compiled into a number of plans called "Compilations" which are regularly updated to show the current land parcel situation. This is not however equivalent to a registry index map.

"The Department of Surveys and Mapping exercises statutory control over all surveying for cadastral purposes and national mapping but not over surveying for engineering purposes. Surveys for the cadastre re-effected in two ways:

a. New township surveys in which a large number of new stands are surveyed simultaneously. This is generally done by the Department, but sometimes registered surveyors are given this work on contract.
b. Surveys of single plots and mutations, in this case most of the work is done by private surveyors

The Land Survey Act of 1959 provides that all data generated for cadastral purposes whether by government surveyors or private surveyors must be subjected to an examination approval process by the Department. Only approved diagram and general plans are accepted by the Registrar of Deeds for registration." The Survey Division of the Department utilizes GPS equipment.

"The Mapping Division is responsible for the National Mapping Program and work is carried out in-house as well as through contract services. It provides topographic maps at small, medium, and large scales in hardcopy as well as digital formats. The first maps by the British Directorate of Colonial Surveys were published in 1955 at a scale of 1:125,000. By the time the series was discontinued in 1966, a total of 104 sheets had been printed in two colors with black used for detail and blue for hydrographic features. Contours were not compiled but form lines were used for prominent hills. The British Directorate of Overseas Surveys published the first series at a scale of 1:50,000 in 1967. These were conventional line and symbol topographic maps with a contour interval of 50 feet and covered the more densely populated eastern part of the country. The sheets were printed in 5 colors. The sheets are cast on the Gauss-Krüger Transverse Mercator projection. The astronomic coordinates of the initial point of the Cape Datum near Port Elizabeth are for Buffelsfontein where $\Phi_o = 33°\ 59'\ 32.000''$ S and $\Lambda_o = 25°\ 30'\ 44.622''$ E. In 1944, D. R. Hendrikz of the South African Trigonometrical Survey wrote, "For the computation of the geographical coordinates of the stations of the Geodetic Survey, Sir David Gill adopted the numerical values of the semi-major and semi-minor axes of Clarke's 1880 figure or $a = 20,926,202$ ft and $b = 20,854,895$ ft. At that time this result was the most recent determination of the figure of the Earth. But, because the baselines were reduced to S.A.G. (*South African Geodetic – Ed.*) feet, the computations were really carried out on a 'Modified Clarke 1880 Spheroid' defined by $a = 6,378,249.145\ 326$ int. metres and $b = 6,356,514.966\ 721$ int. metres. It may be remarked, in passing, that this value of the flattening for this spheroid is $1/f = 293.466\ 307\ 656$ which differs slightly from the value 293.465 given by Clarke himself.' There are 5 belts for the Botswana TM where the Central Meridians are: 21°E, 23°E, 25°E, 27°E, and 29°E. The scale factor is unity, and my information shows no False Easting or False Northing for any of the belts."

In April 1990, Professor Merry and J. Rens of the University of Cape Town published a paper in *Survey Review* that described their solution for datum shift parameters in southern Africa that included Botswana. Although they achieved a seven-parameter solution for nine points only in Botswana, they found that a combined solution of 28 points in Botswana, Lesotho, the eastern half of South Africa, and Swaziland yielded a simpler solution. Their recommended three transformation parameters for the region that includes the Republic of Botswana from the Cape Datum to the WGS 84 Datum is $\Delta X = -136.0$ m± 0.4 m, $\Delta Y = -105.5$ m± 0.4 m, $\Delta Z = -291.1$ m± 0.4 m. The Hartebeesthoek 94 Datum is now the official coordinate system of the Republic of South Africa and presumably also may someday be of the Republic of Botswana.

In regard to the Botswana National Geodetic Reference System 2002 (BNGRF 2002), "A new Geodetic Reference System WGS84 was established in 2002 as a necessary step to ensure not only that the cadastral survey, and mapping were based on it, but also that for Global and regional environment and other purposes, seamless topographic and cadastral mapping could be attained. The Geodetic System whilst established on already monumented old Trigonometrical stations was carried out systematically across the country to allow for a strong network that would allow minimal deviations and progressive densification. Suffice it to say that whilst the cadastral survey is still being compiled in both WGS84 and the old datum system, the conversion to WGS84 has been fully provided for. An integrated system will thus ensure dexterity in information enhancement through over laying of variable data sets, especially as the high accuracy of the system, gained through the use of GPS and modern sophisticated software, far surpasses that of the past. The Land Survey Act requires that each parcel of land for cadastral survey in both urban and rural areas be tied to the Geodetic Network. The resultant erection of reference marks and Geodetic Stations in and around all settlements and other areas across the country, has also acted in favour of increased surveying for registration of title especially in rural areas" (*Botswana – Department of Surveys and Mapping (DSM) Cadastral Information System*, Bryson B. H. Morebodi, Promoting Land Administration and Good Governance 5th FIG Regional Conference Accra, Ghana, March 8–11, 2006).

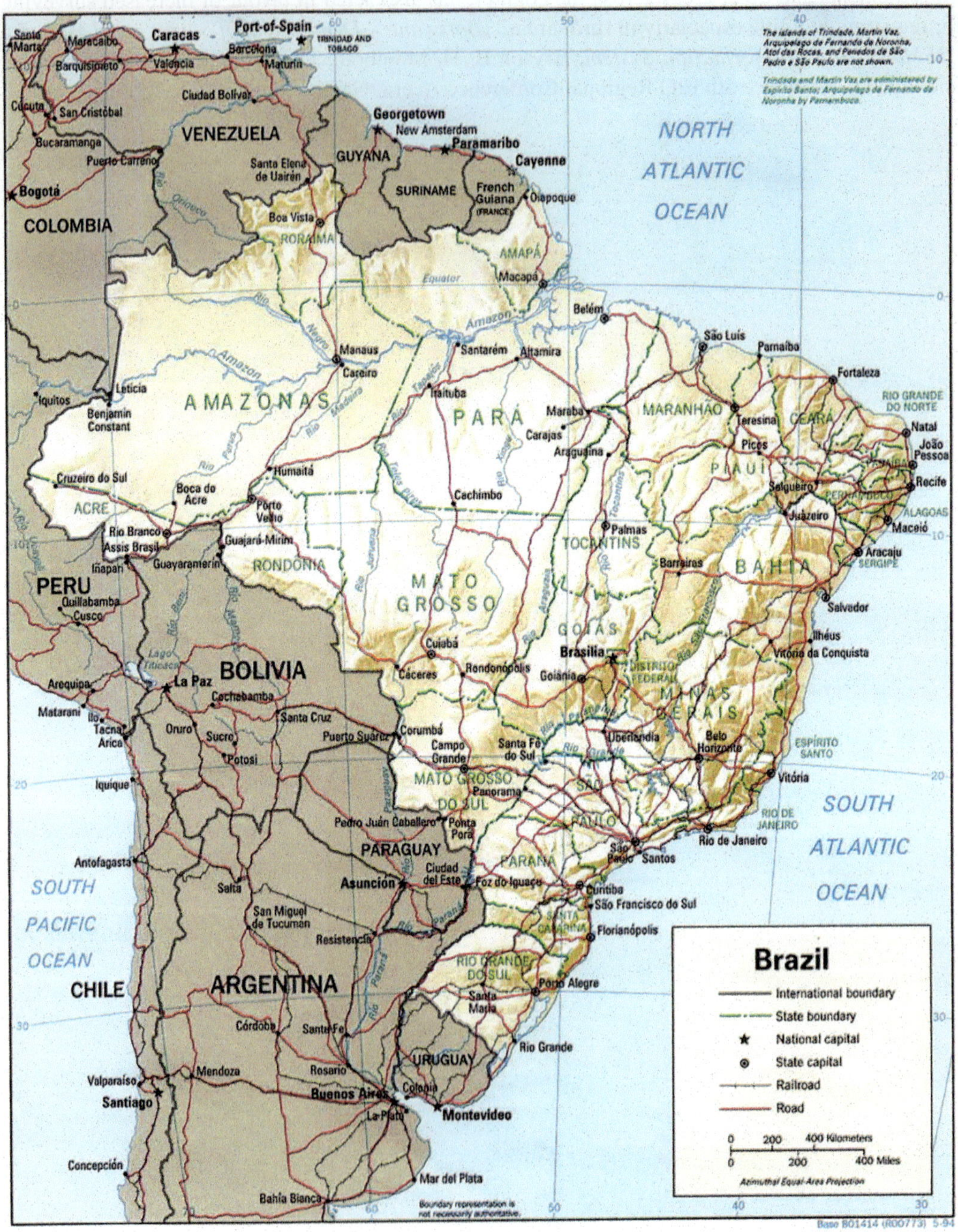

Base 801414 (R00773) 5-94

Federative Republic of Brazil

Brazil is bordered by Argentina 1,261 km, Bolivia 3,423 km, Colombia 1,644 km, French Guiana 730 km, Guyana 1,606 km, Paraguay 1,365 km, Peru 2,995, Suriname 593 km, Uruguay 1,068 km, and Venezuela 2,200 km. With a coastline of 7,491 km, the lowest point in Brazil is the Atlantic Ocean (0 m) and the highest point is Pico da Neblina (3,014 m). Brazil is slightly smaller in area than the United States.

The northern coast of Brazil was explored by Vicente Pinzón, a Spaniard in 1500, although theoretically already allotted to Portugal by the 1494 Treaty of Tordesillas. Brazil was not formally claimed by discovery until the Portuguese navigator Pedro Álvars Cabral accidentally touched there also in 1500. "The native Indians spoke languages that scholars have classified into four families: the *Gê* speakers, originally spread along the coast and into the central plateau and scrub lands; the *Tupí* speakers, who displaced the *Gê* on the coast and hence were the first met by the Portuguese; the *Carïb* speakers in the north and in Amazônia, who were related distantly to the people who gave their name to the Caribbean; the *Arawak* speakers in Amazônia, whose linguistic relatives ranged up through Central America to Florida; and, according to sociologist Donald Sawyer, the *Nambicuara* in northwestern Mato Grosso. Brazil's history prior to becoming an independent country in 1822 and a republic in 1889 is intertwined mainly with that of Portugal. Unlike the other viceroyalties of Latin America, which divided into twenty countries upon attaining independence, the Viceroyalty of Brazil became a single nation, with a single language transcending all diversities and regionalisms. Brazil is the only Portuguese-speaking Latin American country, and its Luso-Brazilian culture differs in subtle ways from the Hispanic heritage of most of its neighbors" (*Library of Congress Country Studies, 2008*).

Three years after independence from Portugal, the *Comissão do Império do Brazil* (Imperial Commission of Brazil) was formed in 1825 – the first official mapping agency of the country and also of South America. Two state agencies appeared before the end of the century: the *Comissão Geológica do Estado do São Paulo* (Geological Commission of the State of São Paulo), in 1866, and the *Comissão Geográfica e Geológica* (Geographical and Geological Commission), the official mapping agency for the state of Minas Gerais in 1892. In the late 19th century, the *Instituto Geográfico e Geológico do Estado São Paulo* produced the first 1:100,000 scale base map of the nation. In 1909, the *Diretoria do Serviço Geográfico do Exército* (Directorship of the Army Geographic Service) produced some 1:10,000 scale sheets, and in 1913 or 1914, terrestrial photogrammetry was first used in Brazil. In 1917, the *Serviço Geográfico Militar* (Military Geographic Service) was founded. In 1922, the first map compiled using aerial photogrammetric procedures was published of the Distrito Federal (*Topographic Mapping of the Americas, Australia, and New Zealand, M.A. Larsgaard, 1984*).

The *Conselho Nacional de Geografia* (National Council of Geography) is responsible to the *Instituto Brasileiro de Geografia e Estatística – IBGE* (Brazilian Institute of Geography and Statistics). Founded in 1937, this civilian agency is the primary national source of public information and data on geodesy and mapping.

With such an enormous country and long history, there have been extensive surveys and mapping projects with concomitant grids and datums established in Brazil. According to **TM 5–248**, *Foreign Maps*, "In 1925 the Brazilian Government adopted the Hayford International ellipsoid of 1909. ($a = 6,378,388$ m, $1/f = 297$. – *Ed.*) Prior to 1925 the Clarke spheroid of 1866 had been used." ($a = 6,378,206.4$ m, $b = 6,356,583.8$ m. – *Ed.*) In 1952, a cooperative agreement was signed between the Brazilian Army and the U.S. Army Map Service and the Inter-American Geodetic Survey (IAGS). The following paragraph is a compilation of all known Astronomical positions in Brazil, mostly published by the various Anuários of the *Diretoria do Serviço Geográfico do Exército*. Some of these may indeed represent small datums established for local city uses but are not generally considered major datums within the country.

The "Abrolhos Lighthouse" is approximately: $\Phi_o=-17°\ 57'\ 32''$S, $\Lambda_o=-38°\ 38'\ 42''$W, and is likely the origin for a local hydrographic survey. Located on Santa Bárbara Island and part of an archipelago, it is a popular tourist destination for viewing whales. The "Barreira do Monte 1941" origin is: $\Phi_o=-08°\ 00'\ 31.1''\pm0.2''$ S, $\Lambda_o=-34°\ 51'\ 15.2''\pm0.6''$ W and is in the State of Pernambuco. The "Base Este do Forte de Coimbra 1945" origin is $\Phi_o=-19°\ 55'\ 17.00''$ S, $\Lambda_o=-57°\ 47'\ 33.75''$ W and is located on the Paraguay River in the State of Mato Grosso. The site of many historical battles dating from the 18th century, it maintains an artillery garrison to this day. The "B of Base Luis Correia" origin in the State of Piauí is approximately $\Phi_o=-02°\ 53'\ 01''$ S, $\Lambda_o=-41°\ 39'\ 58''$ W. The "Centro da Base Mato Grosso 1937" origin is: $\Phi_o=-20°\ 27'\ 01.7''$ S, $\Lambda_o=-54°\ 38'\ 58.3''$ W, and $H_o=578.38$ m. The "Extremo B2 da Base da Forteleza 1944" origin is: $\Phi_o=-03°\ 42'\ 31.4''$ S, $\Lambda_o=-38°\ 34'\ 36.6''$ W and is located in the State of Ceará, a popular coastal destination of Brazilian tourists. The "Extremo S.E. da Base 1939" at point B_2 origin is: $\Phi_o=-21°\ 42'\ 11.7''$ S, $\Lambda_o=-43°\ 24'\ 55.4''$ W, $H_o=68.50$ m, and is in the State of Minas Gerais. The islands of Fernando de Noronha are located 540 km off the Recife coast; the archipelago is made up of 19 islands, and only two of them are inhabited; the largest island has an area of 16.2 km^2. There are a few versions about the discovering of the island; the most accepted version is that Italian navigator Américo Vespúcio came across Noronha in 1503, when at service of the Portuguese King, he commanded a fleet which explored the Brazilian coast. In past centuries, Fernando de Noronha was invaded by the Dutch, the French, and the British, all wanting to establish a base in a strategic spot. In the 19th century, the island became a penal colony; during this period, the forests were cut down to prevent prisoners from using the trees to build boats and escape. More recently during the Brazilian Military Period, Noronha was used to isolate political prisoners. In 1942, Noronha became a Federal Territory. After the Federal Constitution of 1988, Noronha was turned into a District of the State of Pernambuco. The origin is approximately $\Phi_o=-03°\ 54'\ 00''$ S, $\Lambda_o=-32°\ 25'\ 00''$ W. The State capital of Ceará is Fortaleza, and the origin of "Fortaleza 1942" is at "Porta da Base," where $\Phi_o=-03°\ 45'\ 46.7''$ S, $\Lambda_o=-38°\ 31'\ 23.4''$ W. The "Grupo Escolar de Piquete 1926" origin point is located in the municipality of Guaratinguetá, São Paulo, where $\Phi_o=-22°\ 36'\ 58.10''$ S, $\Lambda_o=-45°\ 10'\ 44.22''$ W, $H_o=657.31$ m. "Itararé" is also located in São Paulo State and its approximate coordinates are: $\Phi_o=-24°\ 06'\ 45''$ S, $\Lambda_o=-49°\ 19'\ 44''$ W, $H_o=740$ m. "Leste N.E. Base" has similar coordinates as "Itararé" in that: $\Phi_o=-24°\ 06'\ 55''$ S, $\Lambda_o=-49°\ 17'\ 55''$ W. The "Pico do Jaraguá" origin is located on the highest mountain in the city of São Paulo, where $\Phi_o=-23°\ 27'\ 08.054''$ S, $\Lambda_o=-46°\ 45'\ 51.201''$ W. The "Extremo Nordeste da Base de Leste 1904" origin has its latitude listed as: $\Phi_o=-30°\ 20'\ 52.00''$ S and is located in Pôrto Alegre, the capital of the State of Rio Grande do Sul. The "Macuco 1946" origin in Minas Gerais State is: $\Phi_o=-22°\ 27'\ 03''$ S, $\Lambda_o=-44°\ 26'\ 37''$ W. $H_o=449.04$ m. "Observatório da Comissão 1906" at the point listed as Observatório da Comissão da Carta Geral do Brasil *Pillar in West Room* is: $\Phi_o=-30°\ 02'\ 14.885''$ S$\pm0.074''$, $\Lambda_o=-51°\ 13'\ 12.660''$ W$\pm0.36''$, and at the point listed as *Mark on the Terrace* is: $\Phi_o=-30°\ 02'\ 14.762''$ S $\pm0.07''$, $\Lambda_o=-51°\ 13'\ 12.660''$ W$\pm0.36''$ and has an azimuth to Cavalhada as: $\alpha_o=38°\ 40'\ 16.35''\pm0.35''$, both being located in Pôrto Alegre. The "Observatório Nacional do Rio De Janeiro 1890" origin is: $\Phi_o=-22°\ 53'\ 42.15''$ S, $\Lambda_o=-43°\ 13'\ 22.5''$ W, $H_o=35.18$ m. "Pilar Astronômico 1926" origin north of Sorocaba, São Paulo, is: $\Phi_o=-23°\ 25'\ 34''$ S, $\Lambda_o=-47°\ 35'\ 44''$ W. "Pilar Astronômico de Ponta Grossa 1948" origin is listed as: $\Phi_o=-25°\ 11'\ 09''$ S, $\Lambda_o=-50°\ 08'\ 37''$ W, but to my surprise I noticed the point is actually shown in *TM 5-248, Foreign Maps* on page 226! The scaled coordinates from that exemplar Brazilian map are: $\varphi_o=-25°\ 11'\ 31.8''$ S, $\lambda_o=-50°\ 08'\ 47.3''$ W, and the point is labeled "Datum," with $H_o=794$ m. "Pilar do Forte 1937" origin is: $\Phi_o=-22°\ 22'\ 39''$ S, $\Lambda_o=-41°\ 46'\ 14''$ W. In the city of Salvador, Bahia, the "Pronto Socorro 1937" origin is: $\Phi_o=-12°\ 59'\ 31.43''$ S, $\Lambda_o=-38°\ 31'\ 34.95''$ W, and the azimuth from north to Hospicio is: $\alpha_o=67°\ 52'\ 47.20''$. Located in the vicinity of Santa Vitória do Palmar, Rio Grande do Sul, "Sul S.O. Base 1912" (Extremo S.O.) origin is $\Phi_o=-33°\ 29'\ 25.85''$ S$\pm0.10''$, $\Lambda_o=-53°\ 19'\ 41.12''$ W$\pm0.33''$.

In the State of Santa Catarina, "Três Barras Extremo N.E. da Base 1952" at the Pilar N.E. da Base de Três Barras origin is: $\Phi_o = -26°\ 07'\ 46.0''\ S \pm 0.5''$, $\Lambda_o = -50°\ 18'\ 28.80''\ W \pm 0.53''$, $H_o = 775.627$ m. In Rio Grande do Sul, the "Val de Serra 1947" origin is: $\Phi_o = -29°\ 28'\ 14.30''\ S \pm 0.03''$, $\Lambda_o = -53°\ 42'\ 03.27''\ W \pm 0.24''$. Thanks to John W. Hager for all of these Astro stations!

The major geodetic datums found in Brazil include: Aratu Datum, Córrego Allegre Datum of 1949, Provisional South American Datum of 1956, South American Datum of 1969, SIRGAS, and WGS84. The Córrego Allegre Datum of 1949 *original* origin coordinates are:

$\Phi_o = -19°\ 50'\ 15.14''\ S$, $\Lambda_o = -48°\ 57'\ 42.75''\ W$, $H_o = 683.81$ m, and azimuth to Chapada das Areias: $\alpha_o = 128°\ 21'\ 48.96''$. The origin coordinates of Córrego Allegre *were later revised* to be: $\Phi_o = -19°\ 50'\ 14.91''\ S$, $\Lambda_o = -48°\ 57'\ 41.98''\ W$, the defining azimuth was unchanged, and the components of the deflection of the vertical, $\xi = \eta = 0$, and the geoid height was constrained to zero (N=0 m). As stated previously, the ellipsoid of reference was the International Hayford 1924. According to *TR 8350.2*, the three-parameter transformation values from Córrego Allegre Datum of 1949 to WGS84 are: $\Delta X = -206$ m ± 5 m, $\Delta Y = +172$ m ± 3 m, $\Delta Z = -6$ m ± 5 m, and this is based on a 17-station solution. Two major grid systems were associated with the Córrego Allegre Datum of 1949, both being based on the Gauss-Conform Transverse Mercator projection. The 3° belt system started with Belt I at $\lambda = -28°\ 30'\ W$ and continued east to Belt XVI at $\lambda = -73°\ 30'\ W$. No False Easting or False Northings were used and the Scale Factor at the Central Meridian was equal to unity. The 6° belt system started with Belt I at $\lambda = -30°W$ and continued east to Belt VIII at $\lambda = -72°W$. The False Easting for each belt was 500 km at the Central Meridian, the False Northing was 5,000 km at the Equator, and the Scale Factor at the Central Meridian was 0.999333333. Note that Northing is X, and Easting is Y. On certain small scale maps Brazil uses the Rectangular Polyconic projection, primarily for their 1:1,000,000 International Map of the World Series. I have a copy of those projection tables in Portuguese given to me by my first cousin, Professor Daryl Domning of Howard University when he was doing research on Amazon River Manatees in Manaus during the 1970s. Also included are tables for the Lambert Conformal Conic projection with standard parallels at 0° 40′N and 3° 20′N, and the Polar Stereographic with standard parallels at 76°S and 80°S.

The Provisional South American Datum of 1956 (PSAD56) origin is at La Canoa, Anzoátegui Province, Venezuela, where $\Phi_o = +08°\ 34'\ 17.170''\ N$, $\Lambda_o = -63°\ 51'\ 34.880''\ W$, and the defining azimuth to station Pozo Hondo: $\alpha_o = 40°\ 22'\ 45.96''$. The PSAD56 ellipsoid of reference is also the International 1924. Note, however, that *TR 8350.2* does not list any datum transformation parameters for PSAD56 to WGS84 in the country of Brazil, although it presumably has been used in the northern reaches of the Amazon Basin.

The astronomic origin coordinates of the South American Datum of 1969 (SAD69) at station Chuá are: $\Phi_o = -19°\ 45'\ 41.34''\ S \pm 0.05''$, $\Lambda_o = -48°\ 06'\ 07.80''\ W'' \pm 0.08''$, and the astronomic azimuth to Uberaba: $\alpha_o = 271°\ 30'\ 05.42''$. The corresponding adjusted geodetic coordinates of Chuá are: $\varphi_o = -19°\ 45'\ 41.6527''\ S$, $\lambda_o = -48°\ 06'\ 04.0639''\ W$, $H_o = 763.28$ m, and the geodetic azimuth to Uberaba: $\alpha_o = 271°\ 30'\ 04.05''$. The components of the deflection of the vertical, $\xi = +0.31''$, $\eta = -3.52''$, and the geoid height was constrained to zero (N=0 m). The SAD69 ellipsoid of reference is the International 1967, where $a = 6,378,160$ m and $1/f = 298.25$. According to *Resolução da Presidência da República nº 23, de 21/02/1989*, the following official Brazilian National transformation parameters are from SAD69 to WGS84: $\Delta X = -66.87$ m ± 0.43 m, $\Delta Y = +4.37$ m ± 0.44 m, $\Delta Z = -38.52$ m ± 0.40 m. By altering the characterization of the Brazilian Geodetic System and according to *Resolução da Presidência da República nº 24, de 25/02/2005*, the following official Brazilian National transformation parameters are from SAD69 to SIRGAS2000: $\Delta X = -67.35$ m, $\Delta Y = +3.88$ m, $\Delta Z = -38.22$ m. Note that the GRS80 ellipsoid parameters are $a = 6,378,137$ m and $1/f = 298.257222101$.

The Aratu Datum to WGS84 *mean* three-parameter transformation is: $\Delta X = -158$ m, $\Delta Y = +315$ m, $\Delta Z = -148$ m; however, there are numerous variants used in the offshore basins by Petrobras as well as by various oil company producers, according to the current *EPSG Database, version 6.18*. Brazil now uses the UTM Grid on their current map sheets.

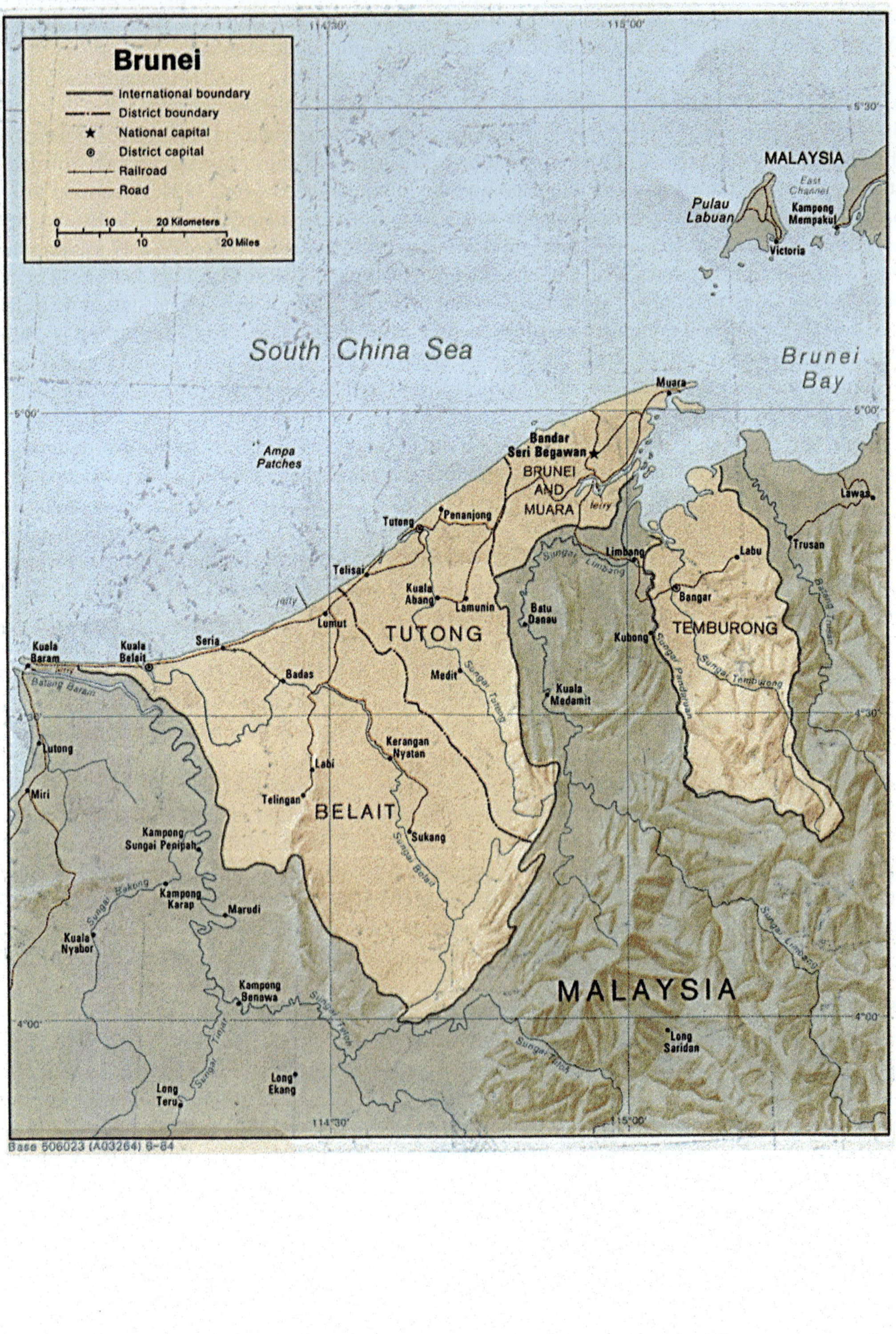

Base 506023 (A03264) 6-84

Negara Brunei Darussalam

Bordered by Malaysia (381 km), Brunei is slightly smaller than Delaware. The lowest point is the South China Sea (0 m), and the highest point is Bukit Pagon (1,850 m) which is on the border with Sarawak, Malaysia.

East Borneo has been settled from *circa* 1st century B.C. by Hindu Pallavas from southeast India. Chinese and Arabic records indicate that this ancient trading kingdom existed at the mouth of the Brunei River as early as the 7th or 8th century A.D. This early kingdom was apparently conquered by the Sumatran Hindu Empire of Srivijaya in the early 9th century, which later controlled northern Borneo and the Philippines. It was subjugated briefly by the Java-based Majapahit Empire but soon regained its independence and once again rose to prominence. In the early 15th century, with the decline of the Majapahit kingdom and widespread conversion to Islam, Brunei became an independent sultanate. It was a powerful state from the 16th to the 19th century, ruling over the northern part of Borneo and adjacent island chains. But Brunei fell into decay and lost Sarawak in 1841, becoming a British protectorate in 1888 and a British dependency in 1905. Set up as the Union of Malaya in 1946, Brunei later formed the Federation of Malaya which became independent in 1957. Malaysia was formed in 1963 when the former British colonies of Singapore and the East Malaysian states of Sabah and Sarawak on the northern coast of Borneo joined the Federation (*Background Notes, U.S. Dept. of State,* 2011).

The Borneo Triangulation network in Sabah and Sarawak consists of the Borneo West Coast Triangulation of Brunei and Sabah (1930–1942). "In 1934 it was decided to cooperate with the Survey of Brunei and to establish the second Sarawak base line near Marudi, some 330 miles from the Kuching baseline, and to rely on a future connection to the Jesselton (North Borneo) base to provide a check on the work northeast of Marudi. The Marudi base line was measured jointly by Officers of the Sarawak and Brunei Surveys. The base net was observed and extended southwestwards towards Sibu by Sarawak, while Brunei carried it northeastwards to the North Borneo Border and established stations in the Fifth Division of Sarawak. *The Triangulation of Brunei*, published by the Surveyor General, Federated Malay States and Straits Settlements, explains how this section of the chain is based on astronomical observations at Labuan *(Island – Ed)*, and is computed in terms of the Sarawak Eastern Meridian of Reference" (*The Primary Triangulation of Sarawak*, W. Harnack, Empire Survey Review, no. 42, pp. 206–214). The Eastern System is in terms of the Marudi Base (probable error ± 1/1,000,000), the latitude and longitude observed at Labuan and the azimuth observed at Timbali (Labuan). The azimuth observed at Miri agrees within 0.04″ (*op. cit., Harnack*). In 1947, readjustment of this triangulation was undertaken by the Directorate of Overseas Surveys (DOS), United Kingdom, to establish a local geodetic reference system known as Borneo Triangulation 1948 and was established with the origin at Bukit Timbalai, Labuan Island, where $\Phi_o = 05° 17′ 03.55″$ North, $\Lambda_o = 115° 10′ 56.41″$ East of Greenwich. The reference ellipsoid used for the BT 1948 is the Modified Everest, where $a = 6,377,298.556$ m and $1/f = 300.8017$. The BT68 results from the readjustment of the primary control of East Malaysia (Sabah, Sarawak plus Brunei) made by DOS, and the old Borneo West Coast Triangulation of Brunei and Sabah (1930–1942), the Borneo East Coast Triangulation of Sarawak and extension of the West Coast Triangulation in Sabah (1955–1960), and some new points surveyed between 1961 and 1968. The Hotine Rectified Skew Orthomorphic (RSO) defining parameters for BT 1948 and for BT68 consist of: Conversion Factor (1 chain = 20.11676512 m, *from Sears, Jolly, & Johnson, 1927*), projection origin $\varphi_o = 4° 00′$ N, $\lambda_o = 115° 00′$ E, Scale Factor at Origin (m_o) = 0.99984, basic or initial line of projection passes through the Skew Origin at an azimuth of $(\gamma_o) = Sin^1(0.8)$ or $53° 19′ 56.9537″$, False Easting = False Northing = zero. The new GDM2000 RSO for Borneo and the Geodetic Datum Brunei Darussalam

2009 (GDBD2009) for Brunei retains the exact same parameters as for BT 1948 and BT 68 *except* for the basic or initial line of projection that passes through the Skew Origin at an azimuth of $(\gamma_o) = -Sin^{-1}(-0.8)$ or 53° 18′ 56.91582″, and the fact that the new ellipsoid of revolution is the GRS 80, where $a = 6,378,137$ m and $1/f = 298.2572221$. According to *TR 8350.2*, the transformation parameters **from** "Timbalai 1948 Datum" **to** WGS 84 Datum (Sarawak, Sabah and Brunei) are: $\Delta X = -679$ m ± 10 m, $\Delta Y = +669$ m ± 10 m, $\Delta Z = -48$ m ± 12 m, and the solution is based on eight satellite stations. According to the *EPSG Database, version 7.6*, **from** BT48 **to** GDBD2009, the Coordinate Frame Rotation Transformation is: $\Delta X = -689.594$ m, $\Delta Y = +623.840$ m, $\Delta Z = -65.936$ m, $R_x = +0.02331″$, $R_y = -1.17094″$, $R_z = +0.80054″$, $\delta s = 5.88536$ ppm. The Position Vector Transformation used by Sarawak Shell Oil Co. is: $\Delta X = -533.4$ m, $\Delta Y = +669.2$ m, $\Delta Z = -52.5$ m, $R_x = 0.0″$, $R_y = 0.0″$, $R_z = +4.28″$, $\delta s = 9.4$ ppm (*op. cit., EPSG, v. 7.6*).

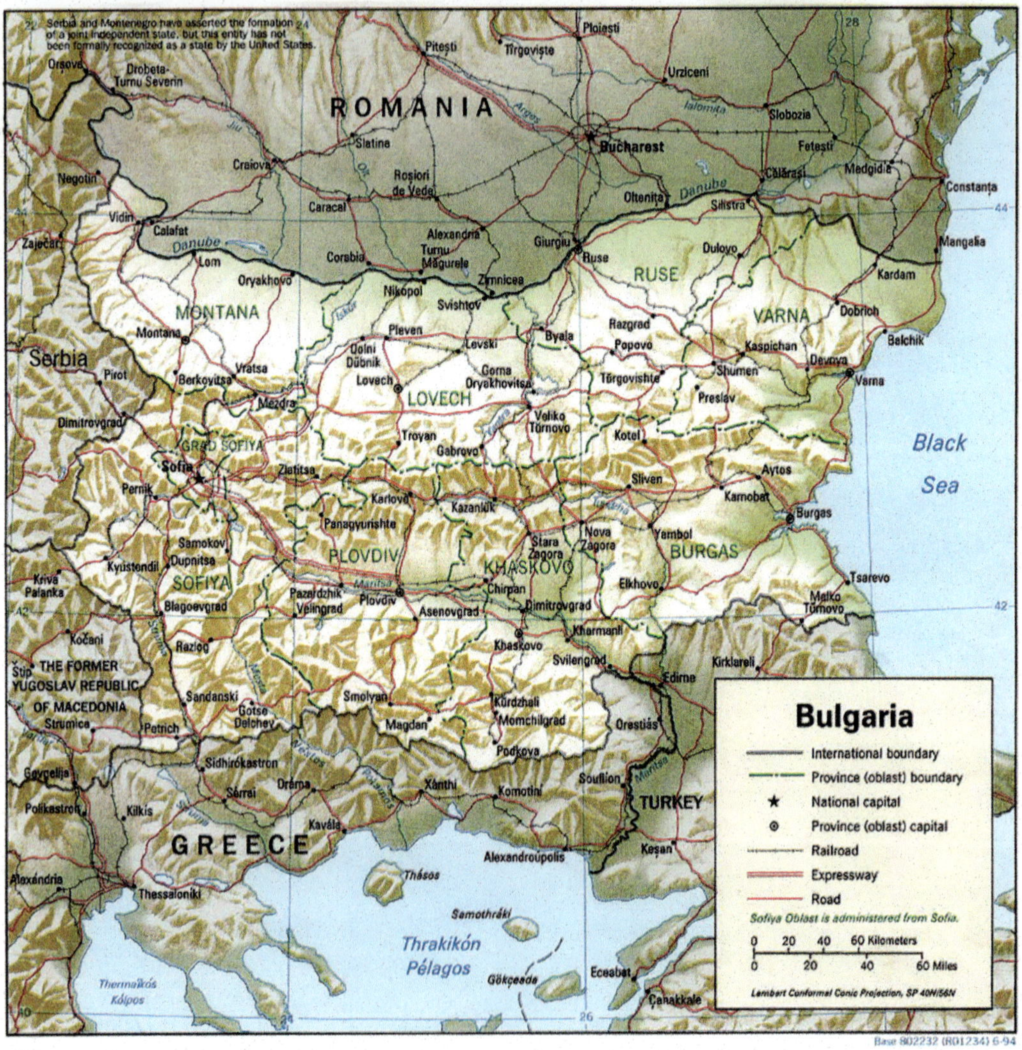

Base 802232 (R01234) 6-94

Republic of Bulgaria

Bulgaria is bordered by Greece (494 km), Macedonia (148 km), Rômania (608 km), Serbia (318 km), and Turkey (240 km). All of its borders have been either intervisibly monumented for nearly a century or comprise riparian boundaries according to the *Rule of the Thalweg*. The terrain of Bulgaria is mostly mountainous with lowlands in the north (Danube Valley) and in the southeast. The lowest point is the Black Sea, the highest point is Musala (2,925 m) and the area of Bulgaria is slightly larger than Tennessee.

Bulgaria was invaded by the Bulgars, a Ural-Altaic people who lived between the Don River and the Caucasus Mountains in the 6th century *A.D.* In the 7th century they settled in Bessarabia, crossed the Danube River, became Slavicized and they became the first Slavic power in the Balkans. Bulgaria was part of the Ottoman Empire from the Turkish conquest (1340–1396) until September 22, 1908. Invaded by the USSR in 1944, its latest constitution was adopted on July 12, 1991. The country joined NATO in 2004 and the EU in 2007.

The first known map of Bulgaria, "Map of the Danube's Downstream" was published in Rome by A. Zaferi in 1560. Johan van der Brugen published a travel map of Bulgaria in 1737, Priest Constantin's map was published in Vienna by D. Davidovich in 1819 at a scale of 1:350,000, and Hristo G. Danov produced a 1:1,000,000 map of European Turkey in 1863. The earliest large-scale geodetic surveys of Bulgaria were in 1877 at the start of the Russo–Turkish War. This triangulation was based initially on astronomic fixes from the military campaigns in Bulgaria of 1828–1829. An instrumental survey was undertaken at that time following the main roads and rivers, with land in between field sketched. Between 1828 and 1833 the Russian troops surveyed the greater part of Eastern Rumelia, North, and West Bulgaria at 1 and 2 Verst scales (1:42,000 and 1:84,000). The projection used was the Russian favorite at the time, the Müffling or Polyhedric. The Turkish authorities agreed to allow Russian military surveyors to reconnoiter between 1867 and 1869 in order to ascertain suitable locations for the subsequent triangulation! 31 new fixes were determined astronomically, and five itineraries were carried out which formed the basis of a new triangulation chain.

The triangulation carried out during the Russo–Turkish war of 1877–1879 by Russian Military Topographers is known as the Lebedev Net after the Colonel in charge of geodetic observations. The field observations were carried out by three main groups, Col. Lebedev in the west, Col. Jarnefeldt in the center and in overall charge of the topographers and plane tablers, and Maj. Zhdanov in the east. Staff Captain Schmidt carried out the triangulation of the Dobrudzha. Because of military exigencies, the plane table survey was carried out concurrently with the triangulation. As a result of this, the coordinates of many of the triangulation stations in the central part of Bulgaria had not been calculated by the time the plane table mapping was performed. The plane table sheets therefore had to be aligned along the local magnetic meridian, and this gave rise to important irregularities when the mapping was later published. The specifications of the Datum established by the Lebedev triangulation are known in Bulgaria as the "Russian Triangulation." The origin was at the minaret of the main mosque in Kyustendja (now Constanța, Rômania), where $\Phi_o = 44°\ 10'\ 31''$ North, $\Lambda_o = 28°\ 39'\ 30.55''$ East of Greenwich. Note that this longitude is a ***correction*** from that published for Rômania, thanks to Dr. Momchil Minchev of the Bulgarian Geoinformation Company. The reference azimuth from East baseline Pyramid to West baseline Pyramid on the Kyustendja Base $\alpha_o = 305°\ 15'\ 01.7''$, and the ellipsoid used was the Walbeck 1819, where $a = 6{,}376{,}896$ m (2,988,853 sazhens), and $1/f = 302.78$.

The Lebedev (Russian) Triangulation measured six baselines at Kujustendja (Constanța) and Turnu Măgurele in Rômania, and at Vidin, Kyustendil, Plovdiv, and Burgas in Bulgaria. There were 52 astronomic observations performed, of which 47 stations were used to determine deflections of the vertical. In the central part of Sofia there were three original Russian Triangulation points, one in a watchtower and two minarets that were Turkish modifications to existing orthodox churches. Although the watchtower along with the minarets were torn down after the national liberation from the Ottoman Empire, their original locations were later recovered by Prof. Vladimir Hristow in 1930 and transferred to a new astronomic tower. The Russian military surveyors did not perform any differential leveling in Bulgaria, but heights were observed via barometric leveling and trigonometric leveling. Curiously, no ties were made to geodetic leveling lines in Russia but were referred to the sea levels of the Black, Marmora, and Aegean Seas using ten marks. On the Black Sea, the marks were at Kujustendja, Shablya, Balchik, Varna, and Burgas. On the Sea of Marmora, the marks were at Kuchuk Kainarji (Kaynardzha), Ereğli, and Terkidağ (Rodosto), all in present Turkey. On the Aegean Sea, the marks were at Dedé Agach (Alexandroúpolis – in present Greece), and reportedly on the island near the city of "Kadykioy" (which doesn't make sense since Kadiköy is a suburb of İstanbul). The assumption was that the levels of all three seas were identical. Lebedev's wartime survey was carried out under very difficult conditions, and it represents a remarkable achievement. During the years 1877–1879, 180 topographers spent only 3,500 man-days and 100,000 rubles in surveying 120,000 km² and in reconnoitering another 1,400 km². A total of 1,274 points were trigonometrically fixed and heighted. Unfortunately, the Russian Triangulation has been lost mainly due to a penchant for monumenting points by burying glass bottles in the ground!

The first Bulgarian surveying institution was the Military Topographic Service, established by royal decree in 1891. Later renamed the State Geographic Institute in 1920 under the Ministry of War, its mission was to establish a new national geodetic network. Prior to World War I, the Military Topographic Service relied on reambulation of old Russian mapping to provide their own map series. Difficulty arose after the upheavals of the 2nd Balkan War in 1913 and World War I when many of the old Russian trig beacons (wooden towers) were destroyed and recovery became impossible. Parts of Thrace and Macedonia were added to Bulgaria, and these areas had never been surveyed by the Russians. Col. Volkoff, director of the Geographic Institute, started a new topographic survey of Bulgaria. The years 1921–1925 were devoted to a through reconnaissance and the erection of triangulation signals. During that period, 76 primary stations (40–69 km apart) and 230 secondary stations (15–25 km apart) were established with Hildebrandt and Bamberg (broken elbow) precise theodolites. The framework was completed by establishing 5,000 points of lower order spaced about 4–6 km apart in mountainous regions and about 3–4 km apart in the lowlands. The principal and secondary triangulations were completed between 1925 and 1929. By 1935, the coordinates of 3,820 points covering almost 7,500 km² had been adjusted and completed. The scale and azimuth of the triangulation was controlled from four bases, measured with Invar wire during 1928–1929 at Ruse, Lom-Palanka, Sofia, and Yambol. The coordinates were calculated by the determination of nine primary station latitudes (Laplace stations), and the longitude of Sofia Observatory was obtained by telegraph from Pottsdam, Germany under the direction of Prof. Hristow. The difference observed and corrected was $1^h\ 33^m\ 19^s.87$ corresponding to a longitude for the observatory of $23°\ 19'\ 58.05''$ East of Greenwich. The observatory was geodetically linked to a station of the Sofia base extension net at the triangulation pillar of Cherni-Vrŭkh ("Black Peak," the highest point of Vitosha Mountain, 2290 m, south of Sofia). This was to give a starting value for the adjustment which was carried out by the Benoit method of least squares compensation. Astronomic azimuths were measured in 1930 and 1931 to strengthen the adjustment. Prof. Hristow's determination of the origin of the Bulgarian Datum of 1930 ("System 1930") at Cherni-Vrŭkh is: $\Phi_o = 42°\ 33'\ 54.5526''$ N, $\Lambda_o = 23°\ 16'\ 51.9603''$ East of Greenwich, and the ellipsoid of reference is the International 1924, where $a = 6,378,388$ m, and $1/f = 297$. The orientation was defined from Cherni-Vrŭkh to Mescit Karmek as $\alpha_o = 309°\ 55'\ 21.752''$. This was the point that Prof. Hristow connected with the ancient Russian Triangulation points demolished in Sofia that were the old watchtower and the two minarets. Plane coordinates were computed on the Bulgarian Gauss-Krüger Transverse Mercator devised by Professor Vladimir Hristow where the central meridians were 21°, 24°, and 27° East of Greenwich (zones 7, 8, and 9), the scale factor at origin was 0.9999 on the central meridian, and the False Easting was 500 km.

The False Northing was zero at 4,540,198.36 m corresponding to 41° N on the International 1924 ellipsoid until 1942 when the Equator was adopted as the origin of the ordinate axis. For this Grid, the Northing coordinates are labeled "X," and the Easting coordinates are labeled "Y." Furthermore, the False Easting for zone 8, Y = 8,500 km, and for zone 9, Y = 9,500 km.

A new framework of precise leveling commenced at the same time as the triangulation. Completed in 1929, the network consisted of 18 closed loops with a total length of 6,445 km. Leveling was run along railroads and roads as well as to all trig points within 4 km of the main route. The levels were referenced to mean sea level at Varna where tide gauge observations were performed continuously between 1928 and 1931. (Note that a full Metonic cycle is 18.67 years.) After 1938, leveling ties with Rômania, Greece, and Yugoslavia yielded vertical Datum discrepancies. With Yugoslavia, the mean of three connections showed the Varna Datum to be 0.60 m higher than that of the Trieste Datum on the Adriatic Sea. With Greece, the mean of five connections showed the Varna Datum to be 0.15 m higher than that of the Kavalla Datum on the Aegean Sea which in turn was 0.24 m higher than the Thessaloniki Datum, also on the Aegean. With Rômania, the mean of three connections showed the Varna Datum to be 0.35 m higher than that of the Constanţa Datum, also on the Black Sea.

In 1947, after World War II the State Geographic Institute was closed and its functions were assumed by the Military Topographic Service of the Bulgarian Army. In 1951 the General Board of Geodesy and Cartography (GUGKK) was established, and in 1954 the National Survey was formed. The Geodetic "System 1950" was adopted in 1950 and is the "System 1930" Datum recomputed on the Krassovsky 1940 ellipsoid, where $a = 6,378,245$ m, and $1/f = 298.3$. The result of the new computations was that a shift of approximately 3.5″ in position occurred, but the original azimuth at Cherni-Vrŭkh was constrained. The Grid system changed such that the False Eastings were equal to 500 km at the central meridians, and the scale factor at the central meridians was changed to unity. The military version of the Gauss-Krüger Transverse Mercator used 6° belts, while the civilian version remained the same as used for the "System 1930" which was 3° belts. Note that the Gauss-Krüger Transverse Mercator projection formulae used by the Soviets for Russia and all of the Warsaw Pact nations was the one developed by Prof. Hristow!

In the 1960s, it was decided to use the "System 50" only for military applications and to introduce a set of plane coordinates for civilian use. The new system consists of four Lambert Conformal Conic projections on the Krassovsky 1940 ellipsoid; the four overlapping zones are K3, K7, K5, and K9. Each zone is defined with a latitude of origin and a scale factor at origin, and each zone has a different initial azimuth of the central meridian in order to rotate (and obfuscate) the grid. The parameters remain secret to this day, and this practice (including zone rotation) is typical of many of the former Soviet Bloc countries. The plane coordinates are used as "local" grids, and the life span of the secrets is dwindling fast as civilians introduce high-precision GPS receivers into the country. Outside of Bulgaria, it is easy to purchase the formerly secret Russian Military Topographic maps at 1:50,000 scale of Bulgaria. These materials are secret only within the Bulgarian borders!

The Russian "System 42" Datum is referenced to the Krassovsky 1940 ellipsoid. The origin is at Pulkovo Observatory: $\Phi_o = 59° 46' 18.55''$ North, $\Lambda_o = 30° 19' 42.09''$ East of Greenwich, and the defining azimuth at the point of origin to Signal A is: $\alpha_o = 317° 02' 50.62''$. This was used in Bulgaria only for military applications.

The Geodetic "System 1942–83" Datum is actually a misnomer created by a novice in Bulgarian government service. Although "System 42" does exist for Bulgaria, *viz.* the 1:50,000 Military Topo series, the "System 1942–83" was introduced by the Army in 1983 as the "Unified Astro-Geodetic Network of Central and Eastern Europe." This is a special-purpose military Datum and is still classified a secret.

According to Dr. Valentin Kotzev, Director of the Central Laboratory of Geodesy, in June of 2001 the Bulgarian Government issued a decree for the adoption of a new geodetic system. The "Bulgarian Geodetic System 2000" is based on the fundamental parameters of the Geodetic Reference System of 1980. The European Terrestrial Reference Frame 1989 (ETRF-89) is introduced as the national coordinate system. The ETRF-89 was extended in Bulgaria during campaigns in 1992 and 1994. The new height system is defined as part of the Unified European Leveling Network. The horizontal coordinates are going to be computed on a single Lambert Conformal Conic projection defined with two standard parallels and a single central meridian yet to be defined. The cadastral system of Bulgaria is prescribed on the new grid. I am told that the new system is controversial. I have personally noticed that controversy is usually created when two or more geodesists get together.

Professor Vladimir Hristow (1902–1979) was the son of Kiril Hristov, one of the greatest Bulgarian poets and writers of the 20th century. Prof. Hristow was one of the greatest geodesists of the 20th century, and he received his Ph.D. from Leipzig, Germany in 1925. Some of his accomplishments have just been chronicled in this column, but he left a record of hundreds of papers published worldwide that speaks for itself.

Thanks to Mr. John W. Hager, Dr. Valentin Kotzev, Dr. Momchil Minchev, and the Defence Geographic and Imagery Intelligence Agency of the United Kingdom.

NATIONAL PERMANENT GNSS (GLOBAL NAVIGATION SATELLITE SYSTEMS) NETWORK

It consists of more than 20 permanent GPS/GNSS stations whose data are archived, proceeded, and analyzed in the center for procession and analysis. The procession and analysis of the measurements permits to carry out a monitoring of the current crustal motions in Bulgaria and Balkan Peninsula and, along with the seismic information, to evaluate the seismic risk.

CENTER FOR PROCESSION AND ANALYSIS OF GPS/GNSS MEASUREMENTS

The Center was established in 2002–2003 in connection with the obligations of the Department of Geodesy in the building and maintaining of the (new) State GPS network of the Republic of Bulgaria. Equipped with high-technology computer techniques and possessing modern software for processing and analysis of GPS/GNSS measurements as: *GAMIT/GLOBK*, elaborated in the Massachusetts Technological Institute; *Bernese 5.2*, of the Astronomical Institute of the University of Bern, software packages *QOCA*, FONDA, and DYNAPG.

In the Center are archived, proceeded, and analyzed the measurements from the permanent GPS/GNSS stations on the territory of Bulgaria, Balkan Peninsula, and Europe. The principal Hemus-NET permanent stations are the stations of the National GPS/GNSS Permanent Network of the National Institute of Geophysics, Geodesy and Geography (NIGGG) including the stations maintained together with NAVITEK Network; also are noted permanent stations of the NAVITEK Network with EUREF permanent stations.

The first permanent stations of the National Permanent Network were settled in the mid-2007 in the framework of the Project, financed in the Science for Peace NATO program, "Monitoring of the deformation of the Earth's crust in the Central-Western Bulgaria and Northern Greece with the Global positioning GPS – *Hemus-NET*.

The main permanent network Hemus-NET consists of eight GPS/GNSS stations – six on Bulgarian territory and two on the territory of Greece:

- *Dragoman (DRAN)*, meteorological station Dragoman;
- *Kustendil(KUST)*, meteorological station Kyustendil;
- *Sandanski (SAND)*, meteorological station Sandanski;
- *Sofia (SOFA)*, roof of the Agency for Geodesy, Cartography and Cadastre, Sofia;
- *Pazardjik (PAZA)*, meteorological station Pazardzhik;
- *Yundola (YUND)*, roof of the research base of the University of Forestry, Sofia;
- Kato Nevrokopi, Greece (NEVR);
- Lemnos island, Greece (LEMN).

The permanent GPS/GNSS stations are stabilized on the *reinforced concrete posts*.

Data from the permanent GPS/GNSS stations of the main permanent network Hemus-NET have free access at:

- Bulgaria – at anonymous ftp server ftp://195.96.249.3;
- Greece – via web server http://egelados.gein.noa.gr/services/GPS/GPS_DATA/.

During the time from 2007 up to now, NIGGG has installed a considerable number of new permanent stations, united in a National Permanent GPS/GNSS Network – Hemus-NET.

The settled stations of the National network at the moment are:

- Plana (PLA1), National Geodetic Observatory in Plana;
- Panagyurishte (PANG), National Magnetic Observatory in Panagyurishte;
- Valandovo (VALA), Republic of Macedonia;
- Vidin (VIDI);
- Oryahovo (ORIA);
- Pleven (PLEV);
- Troyan (TROY), Military-geographic service of Bulgarian Army;
- Rakovski (RAKO);
- Kardzhali (KARJ), Station of the National Seismological Network of NIGGG;
- Harmanli (HARM);
- Elhovo (ELHO);
- Rozhen (ROZE), National Astronomical Observatory in Rozhen;
- Varna (VARN), Astronomical Observatory in Varna;
- Provadia (PROV), Station of the National Seismological Network of NIGGG;
- Varna (VTAG), permanent GPS/GNSS station in the mareographic station in Varna of the National Mareographic Network.
- In the near future will be installed permanent GPS/GNSS stations in Tran and Pernik in the framework of the Project "Monitoring and Information System for Current Crustal Motions and the Seismic Risk via the National Gnss, Seismic and Accelerometric Networks," sponsored by Scientific Research Fund, by agreement ДФНИ ИО1/4–27. 11 2012.

Additionally will be installed a receiver in the mareographic station in Burgas from the National Mareographic Network.

With the intended three new stations in Tran and Pernik and the mareographic station in Burgas the total number of permanent GPS/GNSS stations from the National Permanent GPS/GNSS Network – Hemus-NET is 26.

The permanent GPS/GNSS stations in Vidin, Oryahovo, Pleven, Kardzhali, Harmanli, Elhovo, and Rakovski are installed and operated jointly with the permanent network *NAVITEK*, possession of SB Group. The data from all permanent GPS/GNSS stations of the network NAVITEK are given to NIGGG, according to a binding contract, and are used for the research purposes of the National Institute:

http://niggg.bas.bg/wp-content/uploads/2013/09/gnss_en.html

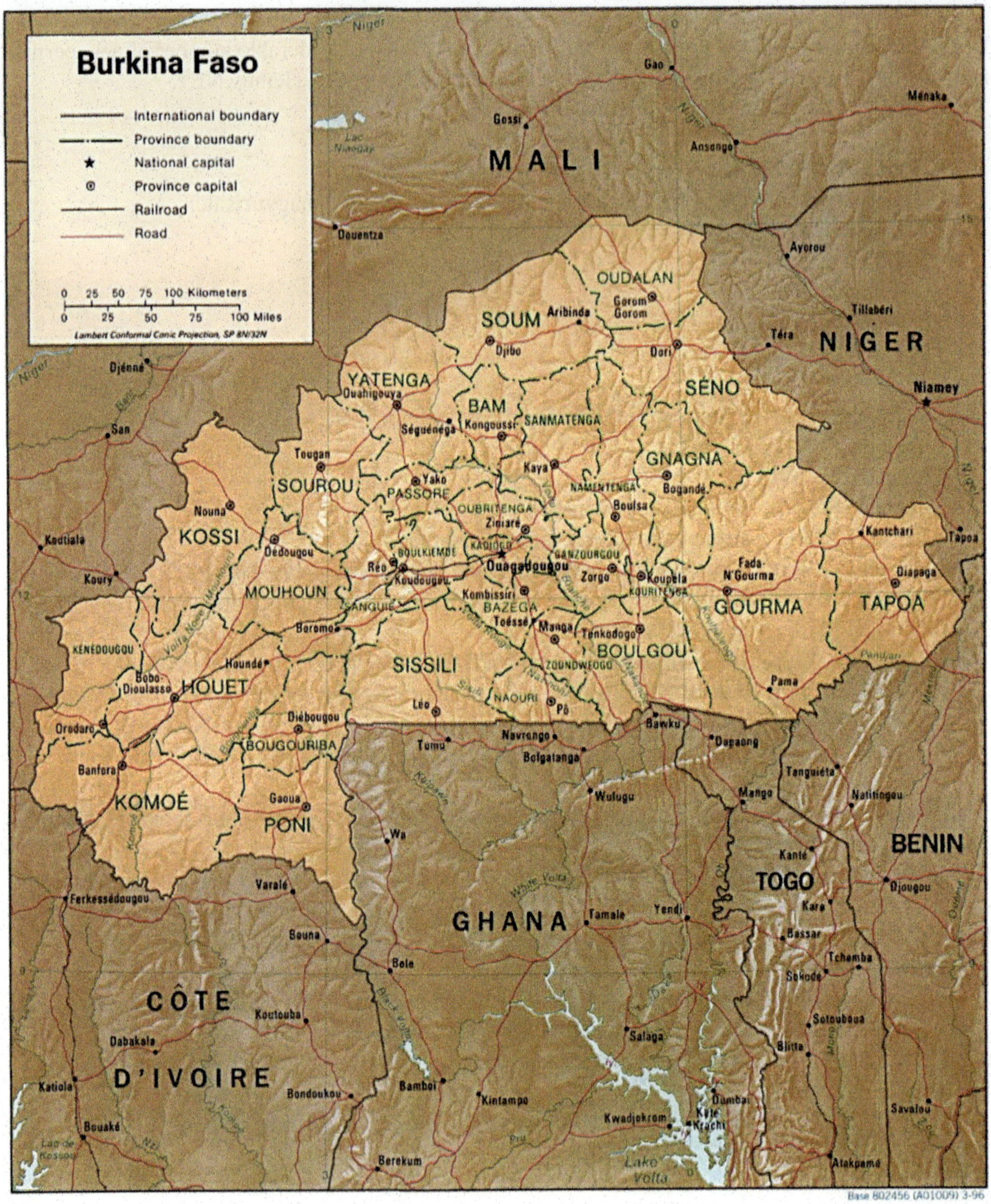

Base 802456 (A01009) 3-96

Burkina Faso

Landlocked Burkina Faso is bordered by Benin (306 km), Ghana (548 km), Côte d'Ivoire (584 km), Mali (1,000 km), Niger (628), and Togo (126 km). Slightly larger than Colorado, the lowest point is the Mouhoun (Black Volta) River (200 m), and the highest point is Tena Kourou (749 m). The terrain is mostly flat to dissected, undulating plains with hills in the west and southeast (*World Factbook*). The major cities are the capital Ouagadougou (Ouaga), and Bobo-Dioulasso.

During the early part of the 15th century, the Mossi horsemen of Ghana invaded the area and established a long-lived empire. The Mossi maintained a cavalry that successfully defended its West African territory from hostile Muslim neighbors, and as a result the former Upper Volta is not predominately Muslim. The French came to the area in the late 19th century, established a protectorate (1895–1897), and later partitioned former areas of the empire to Mali, Niger, and Côte d'Ivoire. Local natives of Upper Volta were "blackbirded" to work on the French plantations of Côte d'Ivoire in the south. For approximately 60 years, the French favored the Côte d'Ivoire at the expense of Upper Volta. Independence from France came in 1960, and a military coup in 1966 was the first of two decades of coups that culminated in Captain Thomas Sankara taking control and changing the name of the country to Burkina Faso which means "country of the incorruptibles" (*Lonely Planet*). "In November 2015, Roch Marc Christian Kabore was elected president; he was reelected in November 2020."

"Terrorist groups – including groups affiliated with Al-Qa'ida and the Islamic State – began attacks in the country in 2016 and conducted attacks in the capital in 2016, 2017, and 2018. By late 2021, insecurity in Burkina Faso had displaced 1.4 million people and led to significant jumps in humanitarian needs and food insecurity. In addition to terrorism, the country faces a myriad of problems including high population growth, recurring drought, pervasive and perennial food insecurity, and limited natural resources. It is one of the world's poorest countries" (*World Factbook*, 2022).

The French *Institut Géographique National* (IGN) had the initial mapping and geodetic responsibility and shared that with the *Annexe de l'Institut Géographique National á Dakar* when the federation of the eight territories constituting French West Africa came into being in 1904. At the time, the IGN Annex, Dakar was known as the *Service Géographique de l'Afrique Occidentale Française* – SGAOF. The IGN later established a network of 200 "Astro" stations (astronomic positions) after 1950 that initially served as the basic control of French West Africa at the scale of 1:100,000 for Burkina Faso. The reference ellipsoid was the Clarke 1880 (IGN), where $a = 6,378,249.2$ m, $b = 6,356,515.0$ m, and computed $1/f = 293.4660208$, the projection and Grid is the Universal Transverse Mercator (UTM). These compilations were also used to make maps at smaller scales. Earlier mapping of portions of Burkina Faso were compiled by SGAOF from 1923 at a scale of 1:200,000, entitled *Carte de l'Afrique de l'Ouest au 200.000^e*. Pre-1952 sheets were compiled from ground surveys or trimetrogon photography; the remaining sheets were compiled from aerial photography and astronomic control. Final sheets were checked and contoured in the field. Relief was indicated by contours at 40- or 50-meter intervals with some supplementary relief portrayed by escarpment, cliff, rock outcrop, sand, and sand dune symbols. Post-1952 sheets were cast on the UTM Grid, Zones 30 and 31 (*U.S. Army TM-5-248, Foreign Maps*). In 1950, the SGAOF performed the classical triangulation of the capital city, Ouagadougou.

In 1958, the French IGN and the SGAOF established 54 stations throughout the country by classical triangulation. In 1960, IGN and the U.S. Army Map Service established 46 stations along the 12th parallel. This triangulation throughout Africa was along the 12th parallel and it started at the origin of the Blue Nile Datum of 1958 in Egypt, at station Adindan, where $\Phi_o = 22° \ 10' \ 07.1098'' \ N$

and $\Lambda_o=31°\ 29'\ 21.6079''$ East of Greenwich, deflection of the vertical: $\xi=+2.38''$, $\eta=-2.51''$. As I stated in my column on Ethiopia, "Adindan" is the name of the origin point and it is *not* the name of the datum; an almost universal mistake found in reference works including the DMA/NIMA/NGA *TR 8350.2*.

In 1979, a Doppler satellite survey was undertaken throughout Africa and included the observation of 16 stations in Burkina Faso. The survey was defined on the WGS72 Datum and final results were computed on the Clarke 1880 ellipsoid for collocated points of the 1958 network of stations in Ouagadougou, apparently on the Blue Nile Datum of 1958. In 1997, 55 points were observed with GPS receivers by the Institut Géographique du Burkina (IGB) in cooperation with the government of Switzerland. Since 1998, this first-order network of GPS observations has been densified with 217 additional points in the southwestern part of Burkina Faso. Thanks go to Alain Bagre in his report to the FIG for the details of the geodetic history of his country.

The NGA lists transformation parameters for the Blue Nile Datum of 1958 (Adindan) to WGS84 Datum based on a *single point* as: $\Delta a=-112.145\,m$, $\Delta f\times 10^4=-0.54750714$, $\Delta X=-118\,m\pm 25\,m$, $\Delta Y=-14\,m\pm 25\,m$, and $\Delta Z=+218\,m\pm 25\,m$.

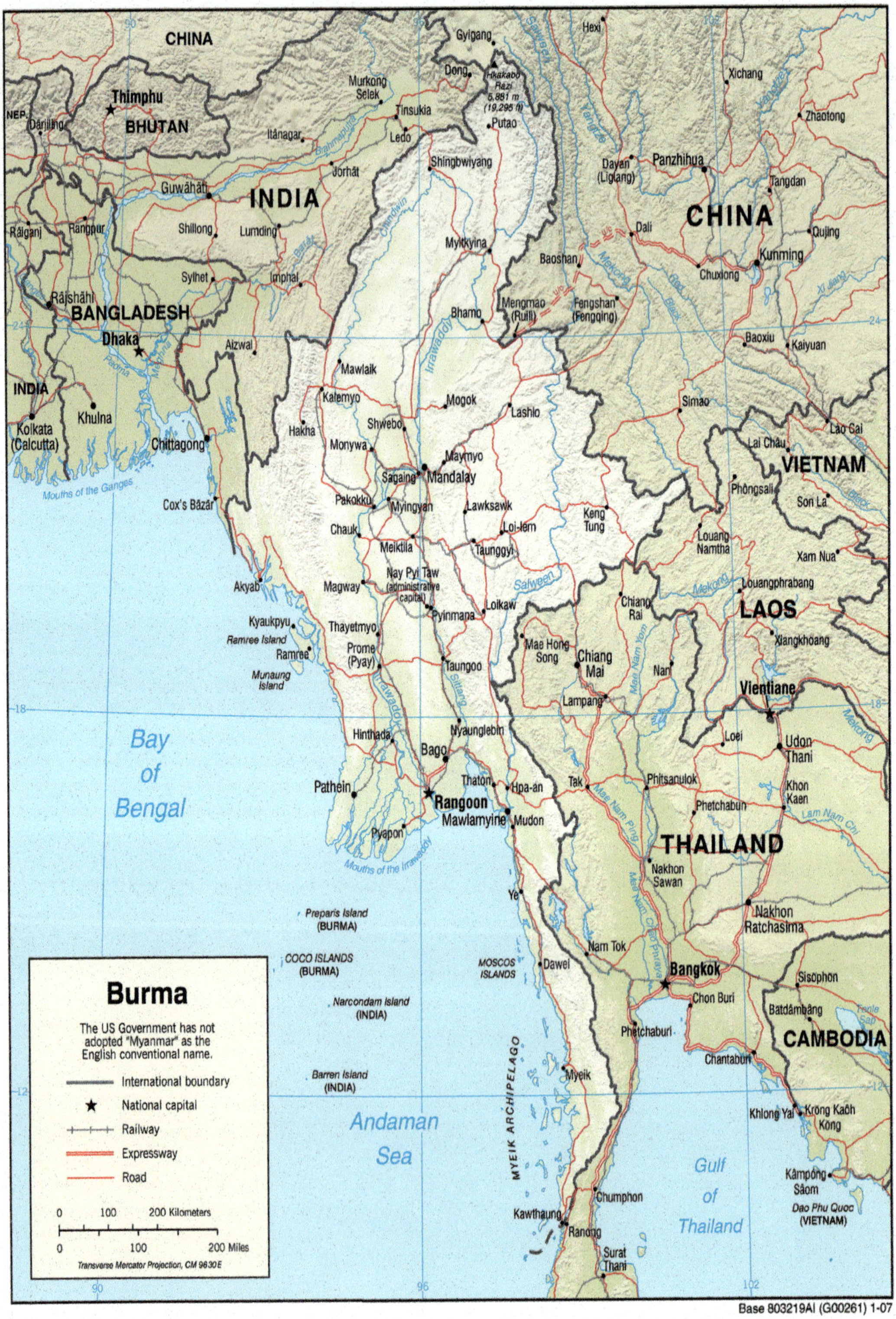

Base 803219AI (G00261) 1-07

State of Burma

Burma is bordered by: Bangladesh (193 km), China (2,185 km), India (1,463 km), Laos (235 km), and Thailand (1,800 km). Slightly smaller than Texas, Burma is comprised of central lowlands ringed by steep, rugged highlands; the lowest point is the Andaman Sea (0 m), and the highest point is Hkakabo Raze (5,881 m) (*World Factbook* and *NGA GeoNames Server*, 2013).

"Tracing Burmese conflicts back to the 9th century, the Himalayan Bamar people, who comprise two-thirds of the population, were at war with the Tibetan Plateau's Mon people. The fight went on for so long that by the time the Bamar came out on top, the two cultures had effectively merged. The 11th-century Bamar king Anawrahta converted the land to Theravada Buddhism and inaugurated what many consider to be its golden age. He used his war spoils to build the first temples at Bagan (Pagan). Stupa after stupa sprouted under successive kings, but the vast money and effort poured into their construction weakened the kingdom. Kublai Khan and his Mongol hordes swept through Bagan in 1287, hastening Burma's decline into the dark ages. There's not much known about the centuries that followed. History picks up again with the arrival of the Europeans – first the Portuguese, in the 16th century, and then the British, who had already colonized India and were looking for more territory in the East. In three moves (1824, 1852, and 1885), the British took over all of Burma. The Burmese king and queen were exiled to India and their grand palace at Mandalay was looted and used as a barracks to quarter British and Indian troops. The colonial era wrought great changes in Burma's demographics and infrastructure. Large numbers of Indians were brought in to work as civil servants, and Chinese were encouraged to immigrate and stimulate trade. The British built railways and ports, and many British companies grew wealthy trading in teak and rice. Many Burmese were unhappy with the colonial status quo. A nationalist movement developed, and there were demonstrations, often led, in true Burmese fashion, by Buddhist monks. During WWII, the Japanese, linked with the Burmese Independence Army (BIA), drove the British out of Burma and declared it an independent country. But the Japanese were able to maintain Burmese political support for only a short time before their harsh and arrogant conduct alienated the Burmese people. Towards the end of the war, the Burmese switched sides and fought with the Allies to drive out the Japanese. Independence followed in 1948" (*Lonely Planet*, 2013). As of 2022, Burma is under authoritarian military rule.

"The credit of the first surveys of the Brahmaputra in Assam, in 1794, and that of the Irrawady River in Burma go to Thomas Wood. The mission also collected interesting information about people, tribes and general geography of Assam and Burma, about which nothing whatever had been known before" (*Survey of India through the Ages*, LTGEN. S.M. Chadha, 1989–90).

According to Guy Bomford, "At the end of 1930 I.M. Cadell was in charge of 15 Party (Triangulation), working in eastern Burma, with a detachment (M.N.A. Hashmie) in the extreme south of Burma. No 17 Party (levelling), with H.P.D. Morton in charge, was working as usual with three detachments on high precision levelling and five on lower order work.

"The programme of Cadell's own detachment was to observe at seven stations to close a primary triangulation circuit, including a connection with the Siamese (now Thai) system, and then to measure a baseline at Kenton (21°15′ N, 99°30′ E), with its extension to the adjacent primary. From central Burma a motor road went as far as the Salween, and in fair weather at the right time of the year, motors could reach Kengtung. Elsewhere there were plenty of tracks fit for mules and ponies, but these animals had to come from Yunnan. The base level of the country is below 1000 feet in the Salween and Mekong valleys, but elsewhere hill tops are between 5000 and 8000 feet.

"By mid-December, working with a 5½inch Wild theodolite, Cadell had completed the Siamese connection and had closed the primary circuit except for one station, when he began to get ill with pneumonia. He managed to reach his last station on the 18 December, three days march from the main road, but died there ten days later. Cadell was very tough. He travelled light and took little notice of feeling ill, but it is hard to see what he or anyone else could have done about it.

"News of Cadell's death reached Dehra Dun on the 31 December, and on the 4 January I left to tidy things up, and to complete the programme if possible. From Dehra Dun it took nine days to reach Salween, where mules were waiting for me, and I got to Cadell's last station on 17 January. The long delay had naturally disarranged the helio squads and it was another 12 days before all were in place and the observations completed. After that measurement of the baseline and its extension went smoothly.

"*Kengtung baseline.* This was the first baseline measured in India since 1882. At Kengtung there is a flat open plain of rice-fields about ten miles by five with hills all round. An ideal site for a baseline. Cadell had reconnoitered the site the previous year, and an assistant L.R. Howard had spent November and December 1930 clearing the line. building stations and laying out pegs for the tripods every 24 metres. For the base measurement we used six invar wires, two simultaneously for the south to north measure and two others for north to south. The other two were used as sub-standards against which the working wires were compared every morning. I had standardized all the wires in our Dehra Dun observatory in August 1930, and they were of course re-standardized on our return in April 1931. For their coefficients of expansion, we had planned to rely on their Sèvres (1908) certificates, but as soon as we started work it was apparent that these certificates were wrong, as was shown by the difference between the two working wires between early morning (13°C) and the afternoon (33°C). To get a provisional figure we set up tripods over ten 24 m bays and made repeated measures with all the working wires in the morning and in the afternoon.

"The squad of about 35 survey khalasis* and a recorder had had some preliminary practice at Dehra Dun, and after the first two days everything went very smoothly. An assistant P.K. Chowdhury and I read the wires, and Howard with 15 khalasis set the tripods over the pegs and recorded their heights above the pegs. The heights of the pegs were obtained by leveling along the line before and after the measurement. To start with we only managed 60 bays a day, but later increased it to 120. When we had finished, we re-measured the first two days' work, where the discrepancies between fore and back were greater than elsewhere.

"*Khalasis are members of the most junior grade in the Survey Department, and include helio men, chain men, instrument carriers *etc*. In the 1930s their pay was between £1 and £2 per month according to skill and length of service. A recorder or computer was a member of the Lower Subordinate service, which also included plane-tablers, traversers and draughtsmen, the backbone of the department, whose pay was between £3 and £10 per month. Assistant, as used in this article, implies members of either the Upper Subordinate service or of the Class II Gazetted service.

"*Triangulation: Southern Burma.* While the work described above was proceeding, Hashmie's detachment was working on a connection between the long Burma Coast triangulation series and the Siamese triangulation in the Kra Isthmus (latitude 10°N), observing at eight stations with a 12-inch micrometer theodolite. They had severe trouble with rain, cloud and thick jungle, and one new station which was placed on what appeared to be a good hill with a clear view to the east had to be re-sited after observations had been made to it from the west. This practically involved a fresh start. Worse still, an outbreak of beriberi resulted in the death of six khalasis. Fortunately, an improvement in the weather in January and February made it possible to get the connection completed.

"*Base measurement.* The programme of the triangulation party in 193-233 was to measure three baselines and their extensions in Burma, at Mergui, Amherst, and Kalemyo (23°20′ N, 94°E) with I.H.R. Wilson in charge, in preparation for which I standardized the six invar wires. The result was disconcerting. Since the last measurement in 1931, one 24-metre wire had decreased by 0.77 mm, one had increased by 2.69 mm, and the other four had changed by between 0.10 and 0.28 mm, all these figures being larger than they ought to be. Clearly some extraordinary misfortune had occurred, but when or how could not be discovered. However, the wires had been re-standardized, the field party was ready to depart, and there was nothing to be done but to carry on and hope for the best.

"Wilson duly measured the three baselines with daily comparisons between the working wires and the two kept as field standards, which showed that there was at least reasonable stability. Re-standardization at Dehra Dun in April 1933 showed changes of at most 0.24 mm (1 in 100 000) in the four working wires, and of 0.35 and 0.42 in the field standards. These figures are larger than one would have liked, but in the circumstances they might have been worse. The six wires were then re-measured in May 1933 to give the current rate of change of each wire. Given the lengths of the wires in September 1932, April 1933 and May 1933, and their relative lengths at the times of the measurement of each of the three baselines, I deduced a probable figure for the length of the mean wire at each baseline, which had an apparent probable error of 1 ppm or less, and I felt confident that the error did not exceed 1 in 300,000. In view of the likely error in the base extensions, an error of 3 ppm in the baseline itself is no disaster. But we had learned that damage to a wire may make it unreliable for at least the next year.

"*Geoid section.* For the field season 193-233 I planned to observe a geoid section across Burma, to start the long section from Siam to Persia referred to earlier. I used a 60° Jobin prismatic astrolabe, but since the astrolabe has no impersonal device, I also took with me a portable bent Transit with a so-called impersonal micrometer, which I used in addition to the astrolabe at every fourth station. My personal equation with the Transit was determined by four nights observations at Dehra Dun both before and after the field season. Anticipating a little, the field value of (Transit minus Astrolabe) varied between +0.08″ and +0.14″, mean +0.12″. So in fact I got substantially the same result as I would have if I had got along without the Transit. I regretted not having calibrated myself on the astrolabe instead of on the Transit.

"For my triangulated (geodetic) position I had chosen a line through country which had been fairly recently surveyed topographically, and there were plenty of recognizable topo triangulation stations and intersected points, generally hill tops. During the preceding months I had selected promising sites from the maps about one day's march apart, say 12 miles in hilly country, and had obtained the coordinates of points likely to be useful. So at each station I observed angles (with a small Wild) to six or eight recognizable points, and generally observed a Polaris azimuth at Sunset. In a few cases I traversed from a single nearby point or set up a short base and found a distance.

"The routine was that we marched one day and pitched camp at or close to a place where I could get a triangulated fix, which I at once did. The next morning, I computed the fix semi-graphically to ensure that all was well with it and set up the astro instruments in time to get the Rugby 16:30 (Burma time) signal. After sunset I observed for two hours with the astrolabe (30 stars), and finally got the 00:30 Rugby signal. I carried three chronometers which I compared with each other at wireless times, and during the star observations. While waiting for the 00:30 signal I marked up the chronograph sheets and worked out the comparisons between the three clocks, as a precaution against unsuspected trouble there. We moved on again the next day.

"Every fourth station where the Transit was to be used, required eight-time stars and two azimuth stars with the bent Transit, about two hours work after completing the astrolabe. Apart from the trouble of setting up the Transit and getting it adjusted in level and azimuth, it set me back with reading the chronograph sheets and the clock comparisons, some of which had to wait until the next station. It was in fact a considerable nuisance.

"I had with me a Survey of India recorder, 20 khalasis, 15 coolies from the Wa States (north of Kengtung), about 40 ponies and mules from Junnan with a dozen Chinese drivers, and two interpreters. Burma does not produce ponies, mules or coolies, and I got them from Syed Ismail of Kengtung, who had done the same for Cadell in previous years. Syed Ismail was a very useful man, with much influence in Kengtung. He was a Chinese Moslem, of which there is a colony in Yunnan.

"We started work at Monywa, on the Chindwin. The coolies and ponies arrived only 12 hours late, having travelled the last 200 miles by rail, preceded by 200 miles on foot from Kengtung, and about the same again from their own countries, Syed Ismail's work. We left Monywa on 2 November and worked through Gangaw and Haka (22°40′ N, 93°20′ E) in the Chin Hills up to the Indian frontier. We then marched back to Monywa and worked eastwards through Ava, Taungyyi and Mongpan (20°20′ N, 98°20′ E) to the Burma-Siam Indo China (now Laos) trijunction, south of Kengtung, 44 stations in all, finishing on 24 February. I then observed longitude for two nights at Kengtung to complete a Laplace station there" (*Geodetic Surveys in India 1930–35, Guy Bomford, Survey Review, No. 200, April 1981, pp. 65–78*).

"It is interesting to note that the Burma Survey Party manned by Survey of India personnel, continued to be under the technical control of the Surveyor General of India till the end of the Second World War" (*op. cit., Chadha, 1989–90*).

The datum used in Burma is the Indian Datum of 1916 with origin at Kalianpur Hill (1880), where $\Phi_o = 24°\ 07'\ 11.26''$ N, $\Lambda_o = 77°\ 39'\ 17.57''$ East of Greenwich, and the ellipsoid of reference is the Everest 1830, where $a = 6,377,276.345$ m $= 6,974,310.6$ Indian Yards, and $1/f = 300.8017$. Nowadays, some refer to the Indian Datum of 1975 which is a misnomer; it's still Indian Datum of 1916, but Burma is part of a regional adjustment performed by the U.S. Army Map Service/Defense Mapping Agency Hydro/Topo Center in 1975. The still-currently used Grid is the India Zone IIIB (Lambert Conical Orthomorphic/Lambert Conformal Conic), where the central meridian, $\lambda_o = 100°$ E, the latitude of origin $\varphi_o = 19°$ N, the scale factor at the latitude of origin $m_o = 0.998786408$, and False Easting $= 3,000,000$ Indian Yards $= 2,743,185.69$ m, and the False Northing $= 1,000,000$ Indian Yards $= 914,395.23$ m. Note that India Zone IIIB as defined by the Survey of India in the early 20th century is a secant conic projection (2 standard parallels) but defined with (φ_o, m_o) as found on page 248 of the ***Manual of Photogrammetry***, 6th edition, 2013. According to *TR8350.2*, the three-parameter transformation in most of Thailand **from** Indian Datum 1975 **to** WGS84 Datum is: $\Delta X = +210$ m ± 3 m, $\Delta Y = +814$ m ± 2 m, and $\Delta Z = +289$ m ± 3 m, based on 62 collocated points in 1997. My guess is that these shift parameters are ballpark for Burma.

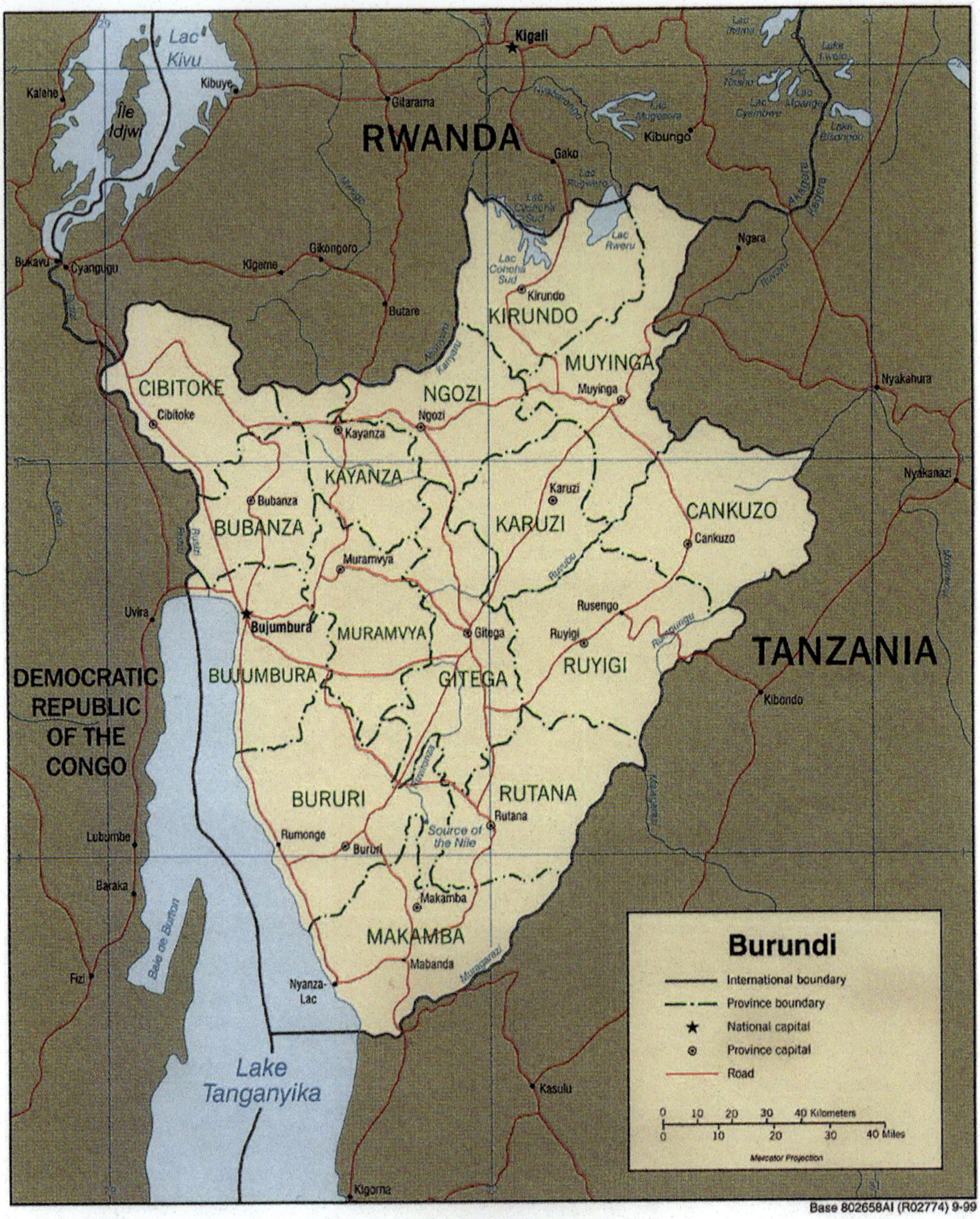

Base 802658AI (R02774) 9-99

Republic of Burundi

Slightly smaller in area than the state of Maryland, Burundi is bordered by the Democratic Republic of the Congo (233 km), Rwanda (290 km), and Tanzania (451 km). Its terrain is hilly and mountainous, dropping to a plateau in east, with some plains. The highest point is an unnamed elevation on Mukike Range (2,685 m) and the lowest point is Lake Tanganyika (772 m). The mean elevation of Burundi is 1,504 m (*World Factbook* 2022).

"In the 16th century, Burundi was a kingdom characterized by a hierarchical political authority and tributary economic exchange. A king (*mwani*) headed a princely aristocracy (*ganwa*) that owned most of the land and required a tribute, or tax, from local farmers and herders. In the mid-18th century, this Tutsi royalty consolidated authority over land, production, and distribution with the development of the *ubugabire* – a patron-client relationship in which the populace received royal protection in exchange for tribute and land tenure. Although European explorers and missionaries made brief visits to the area as early as 1856, it was not until 1899 that Burundi came under German East African administration. In 1916 Belgian troops occupied the area. In 1923, the League of Nations mandated to Belgium the territory of Ruanda-Urundi, encompassing modern-day Rwanda and Burundi. The Belgians administered the territory through indirect rule, building on the Tutsi-dominated aristocratic hierarchy. Following World War II, Ruanda-Urundi became a United Nations Trust Territory under Belgian administrative authority. After 1948, Belgium permitted the emergence of competing political parties. Full independence was achieved on July 1, 1962. In the context of weak democratic institutions at independence, Tutsi King Mwambutsa IV established a constitutional monarchy comprising equal numbers of Hutus and Tutsis. In 1972, an aborted Hutu rebellion triggered the flight of hundreds of thousands of Burundians. Civil Burundi's civil war officially ended in 2006 under a South Africa-brokered cease-fire agreement with the last of Burundi's rebel groups. Today the government is focused on rebuilding its infrastructure and reestablishing external relations with its regional neighbors" (*Dept. of State Background Notes*, 2009).

"The Burundi-Tanzania boundary has a length of 451 km, of which 24 km are in Lake Tanganyika. Tripoints with the Congo (Léopoldville) and Rwanda are located on the median line of Lake Tanganyika and at the confluence of the thalwegs of the Mwibu and Kagera, respectively. East of Lake Tanganyika, the boundary is demarcated by streams and pillars.

"According to the terms of the British mandate for East Africa, the boundary between Tanganyika and Urundi was delimited as follows: Thence the course of the Kagera downstream to meet the western boundary of Bugufi; Thence this boundary to its junction with the eastern boundary of Urundi; Thence the eastern and southern boundary of Urundi to Lake Tanganyika.

"An Anglo-Belgian protocol signed at Kigoma on 5 August 1924, afforded the precise alignment and demarcation of the Tanganyika-Urundi boundary which included Bugufi in the British mandate. The protocol of 1924 is the basis for the present Burundi-Tanzania boundary. In 1934 an Anglo-Belgian agreement was reached on the water rights of the respective mandates of the two states along the boundary" (*International Boundary Study, Burundi–Tanzania Boundary, No. 70 – 16 May 1966, Department of State*).

"Extending from west to east between the Rusizi and Kagera rivers, the Burundi-Rwanda boundary has a length of 290 km. Most of the boundary follows streams, of which the Ruwa *(Luhwa)*, Kanyaru *(Akanyaru)*, and Kagera are the principal rivers. It also passes through Lac Cohoha *(Lac Cyohoha-Sud)* and Lac Rweru *(Lac Rugwero)* and utilizes a number of straight-line segments between streams. There are no known boundary pillars demarcating the boundary. Ordinance No. 21/258 issued by the Vice-Govenor General of Ruanda-Urundi on 4 August 1949 officially delimited the boundary between the two residencies of the Belgian Trusteeship. Both Ruanda and Urundi became independent on 1 July 1962 as the Republic of Rwanda and the Kingeom of Burundi, respectively.

"The tripoint with Congo (Léopoldville) is located at the confluence of the Ruwa and Rusizi, presumably at the junction of the thalwegs of the two rivers. The Rusizi was established by Belgium and Germany as the boundary between their respective territories by a convention signed on 11 August 1910. Fourteen years later a protocol between Belgium and the United Kingdom established the common Boundaries of Tanganyika and Ruanda-Urundi at the point where the Present Tanzania tripoint is located at the confluence of the thalwegs of the Mwibu and Kagera. Prior to WW I, the tripoint of Ruanda and Urundi with the remainder of German East Africa was apparently at the confluence of the Ruvuvu and Kagera. Following WW I, the Kisaka or Kissaga (now known as Gisaka) district in northeast Ruanda was included in the Tanganyika mandate for a short time as the possible location of an all-British route for a railroad to Uganda. Likewise, the Bugufi area between the Ruvuvu and Kagera and westward to the present Burundi-Tanzania boundary, which apparently had been previously administered by Urundi, was included in Tanganyika" (*International Boundary Study, Burundi–Ruanda Boundary, No. 72 – 1 June 1966, Department of State*).

Burundi has three Grid systems found in the country: the Burundi Gauss-Krüger Transverse Mercator, where the central meridian $(\lambda_o) = 30°$ E, scale factor at origin $(m_o) = 1.0$, False Easting = 500 km, False Northing = 1,000 km, the Congo Oriental TM Fuseau 30°, where $\lambda_o = 30°$ E, $m_o = 1.0$, False Easting = 220 km, False Northing = 565 km, and the East Africa TM Belt G, where $\lambda_o = 32° \ 30'$ E, $m_o = 0.9995$, False Easting = 400 km, False Northing = 4,500 km. The ellipsoid of reference is the Clarke 1880, where $a = 6,378,249.145$ m and $1/f = 293.465$. Note that the False Northings are rather large presumably because the entire country is in the Southern Hemisphere.

The published relation between Arc 1950 Datum and WGS84 Datum in Burundi by the NGA is as follows: $\Delta X = -153$ m ± 20 m, $\Delta Y = -5$ m ± 20 m, $\Delta Z = -292$ m ± 20 m, and this solution was based on three points in 1991. Interestingly, "Analysis of TR8350.2 contour charts suggest Arc 1960" (*EPSG v. 7.1*). However, the ellipsoid of reference for both Arc 1950 and for Arc 1960 Datums is the Clarke 1880.

REFERENCES: B

1835–1847 Voyage dans l'Amérique Méridionale, Alcide d'Orbigny.
1935, Die Karten der Cordillera Real und des Talkessels von La Paz (Bolivien), Prof. Dr. Carl Troll, Dr. Richard Finsterwalder.
1998, Mr. Andrew Kopec, Bahrain Centre for Studies and Research.
An Insight into Surveying, Mapping, and Land Administration, Ministry of Home Affairs, Trashichhodzong, Thimhu, Bhutan, 2nd Edition, 2001.
Andrea Matles Savada, ed. *Bhutan: A Country Study*. Washington: GPO for the Library of Congress, 1991.
April 1990, C. Merry and J. Rens, Survey Review.
Background Notes, U.S. Dept. of State, 2011.
Bryson B. H. Morebodi, Promoting Land Administration and Good Governance 5th FIG Regional Conference Accra, Ghana, March 8–11, 2006, Botswana – Department of Surveys and Mapping (DSM) Cadastral Information System.
Comparative study and determination of transformation parameters between: the permanent station system, the datum (58) and the Bénin geodetic System, Degbegnon, L., Aïzo, P., *Research Journal of Recent Sciences*, Vol. 8(4), 1–8, October (2019) ISSN 2277=2502. http://www.isca.in/rjrs/archive/v8/i4/1.ISCA-RJRS-2018-067.pdf
Compatibility of GPS with Local Grid Co-Ordinate System of Bangladesh, LTC. Kazi Shafayetul Haque, Director, Survey of Bangladesh, 1998.
D. R. Cockburn and W. L. Barry of Army Map Service translated the IGN Instruction dated 20 September 1950.
Defence Geographic and Imagery Intelligence Agency of the United Kingdom, Personal Communication – Dec. 2001.
Degbegnon, L., Koumolou, A., Problem of the Use of Dual-Frequency GNSS Receivers in Benin, *Global Journal of Engineering Science and Research Management*, 5(11): November 2018, ISSN 2349-4506. http://www.gjesrm.com/Issues%20PDF/Archive-2018/November-2018/6.pdf.

References: B

Department of State, Office of the Geographer, "Limits in the Seas, No. 12 and No. 58."
Department of Survey and Land Records (DSLR), National Land Commission of Bhutan (NLC) via Lantmäteriet Sweden.
Dept. of State Background Notes, 2009.
DMA/NIMA/NGA TR 8350.2.
Dr. Momchil Minchev of the Bulgarian Geoinformation Company, Personal Communication – Dec 2001.
Dr. Valentin Kotzev, Director of the Central Laboratory of Geodesy, Personal Communication – Dec 2001.
EPSG Database, version 6.18.
EPSG Database, version 7.6.
GIM International, September 1994, Mr. B. B. H. Morebodi.
Guy Bomford, Geodetic Surveys in India 1930–35, Survey Review, No. 200, April 1981, pp. 65–78.
http://niggg.bas.bg/wp-content/uploads/2013/09/gnss_en.html.
http://www.fgu.com.ba/en/about-us-558.html, 2022.
http://www.ngi.be/Common/articles/G/naviguer_avec_GPS.pdf.
http://www.state.gov/e/oes/ocns/opa/c16065.htm.
International Boundary Study, Burundi–Ruanda Boundary, No. 72 – 1 June 1966, Department of State.
International Boundary Study, Burundi–Tanzania Boundary, No. 70 – 16 May 1966, Department of State.
John W. Hager, Personal Communication – July 2004.
Library of Congress Country Studies, 1995.
Library of Congress Country Studies, 2008.
Library of Congress Country Studies, 2009.
Library of Congress Country Study, 2022.
Lonely Planet, 2004.
Lonely Planet, 2012.
Lonely Planet, 2013.
Mapping of the Countries in Danubian and Adriatic Basins, Andrew M. Glusic, AMS TR No. 25, June 1959.
Merriam-Webster's Geographical Dictionary, 3rd Edition.
Mr. Kevin Grootendorst, Personal Communication – Oct. 1998
Ms. Sequoia Read, Personal Communication – Feb. 2013.
National Geodetic Service USA, website, AFREF News Letter No.11, afref@rcmrd.org, May 2010.
National report of Belarus, S. Zabahonski, N. Rudnitskaya, Minsk 2012.
Professor Daryl Domning of Howard University, Personal Communication – 1985.
Resolução da Presidência da República nº 24, de 25/02/2005.
Sears, Jolly, & Johnson, 1927.
Survey of India through the Ages, LTGEN. S.M. Chadha, 1989–90.
The Triangulation of Brunei, published by the Surveyor General, Federated Malay States and Straits Settlements.
TM 5-248, Foreign Maps.
Topographic Mapping of the Americas, Australia, and New Zealand, M.A. Larsgaard, 1984.
TR 8350.2.
U.S. Army TM-5–248, Foreign Maps, 1963.
World Factbook, 2009.
World Factbook, 2022.
World Factbook and NGA GeoNames Server, 2013.

C

Kingdom of Cambodia 135
Republic of Cameroon 139
Canada 145
Republic of Cape Verde 149
Cayman Islands 153
Central African Republic 157
Republic of Chad 161
Republic of Chile 165
The People's Republic of China 169
Republic of Colombia 175
Federal Islamic Republic of the Comores 179
Democratic Republic of the Congo (Kinshasa) 183
Republic of the Congo (Brassaville) 187
Republic of Costa Rica 191
Republic of Croatia 195
Republic of Cuba 199
Republic of Cyprus 203
The Czech Republic 207

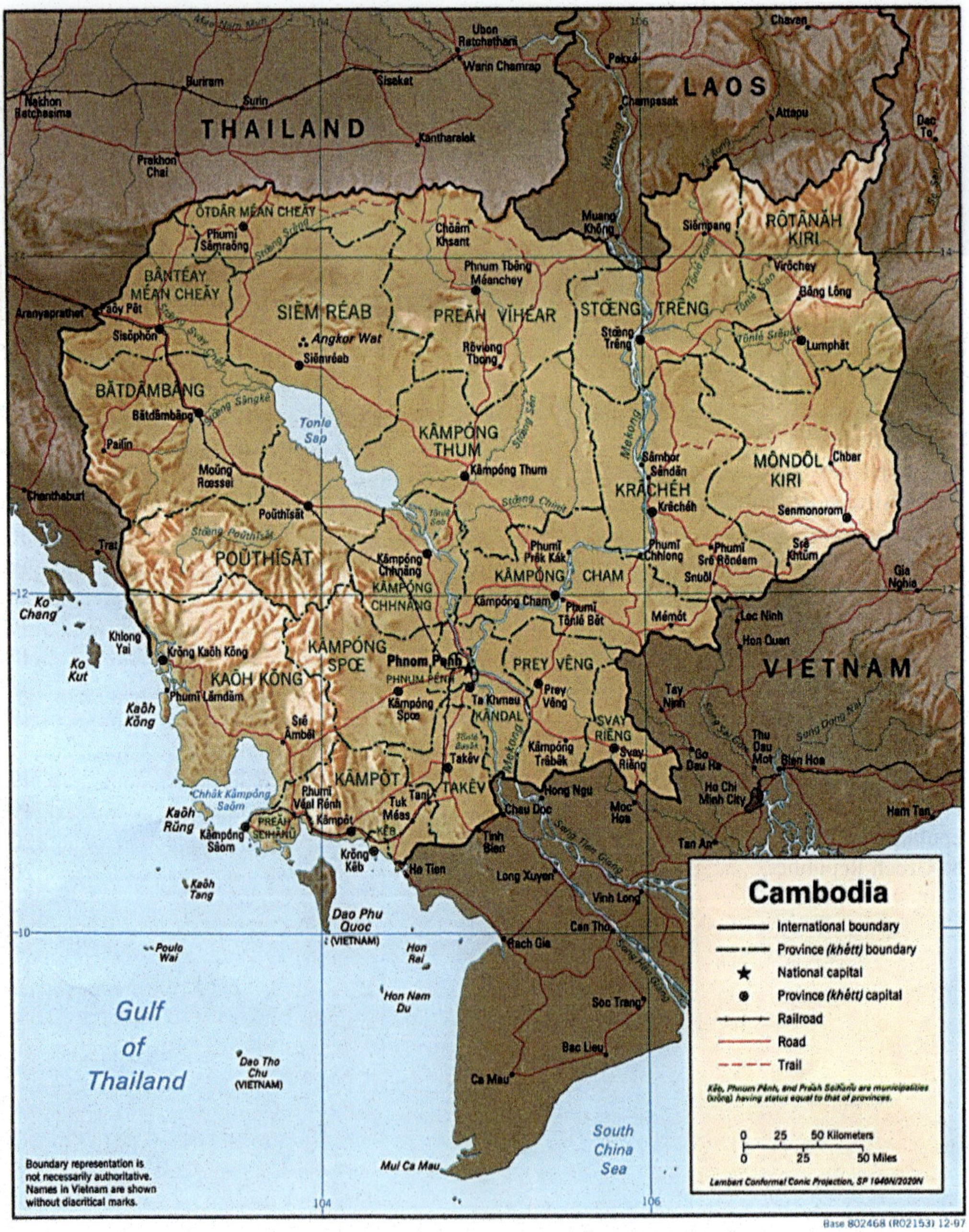

Base 802468 (R02153) 12-97

Kingdom of Cambodia

Cambodia is bordered by Laos (555 km), Thailand (817 km), and Vietnam (1,158 km), the terrain is mostly low, flat plains, and mountains in southwest and north. The highest point is Phnum Aoral (1,810 m), the lowest point is the Gulf of Thailand (0 m), and the mean elevation is 126 m.

Inhabited for millennia, the area's history was unrecorded until the Chinese chronicles of the last 2,000 years. "Historians surmise that by the first century *A.D.*, a small number of Khmer (or Cambodian) states already existed on the fringes of the earliest recorded state in the region, the empire of Funan. Centered in the Mekong Delta of present-day Vietnam, Funan derived its power from commerce. With its port of Oc Eo on the Gulf of Thailand, Funan was well-placed to control maritime traffic between India and China. According to Chinese annals, Funan was a highly developed and prosperous state with an extensive canal system for transportation and irrigation, a fleet of naval vessels, a capital city with brick buildings, and a writing system based on Sanskrit. The inhabitants, whose adherence to Indian cultural institutions apparently coexisted with Mahayana Buddhism, were organized into a highly stratified society. When the small Khmer states to the northwest of the Mekong Delta emerged into recorded history, it was to make war upon the declining empire of Funan. Between *A.D.* 550–650, these Khmer states overran their adversary, which fell apart, losing its tributary states on the Kra Isthmus and along the Gulf of Thailand" (Library of Congress Country Studies).

"Most Cambodians consider themselves to be Khmers, descendants of the Angkor Empire that extended over much of Southeast Asia and reached its zenith between the 10th and 13th centuries. Attacks by the Thai and Cham (from present-day Vietnam) weakened the empire, ushering in a long period of decline. The king placed the country under French protection in 1863 and it became part of French Indochina in 1887. Following Japanese occupation in World War II, Cambodia gained full independence from France in 1953. In April 1975, after a five-year struggle, Communist Khmer Rouge forces captured Phnom Penh and evacuated all cities and towns. At least 1.5 million Cambodians died from execution, forced hardships, or starvation during the Khmer Rouge regime under Pol Pot. Local elections were held in Cambodia in April 2007, and there was little in the way of pre-election violence that preceded prior elections. Cambodia has strong and growing economic and political ties with its large neighbor to the north, China. More than 40% of foreign investment in the country in 2019 came from China, and Beijing has provided over $15 billion in financial assistance since the 1990s. The CPP also partly sees Chinese support as a counterbalance to Thailand and Vietnam and to international criticism of the CPP's human rights and antidemocratic record" (*World Factbook,* 2022).

Most original mapping of Cambodia was accomplished during the period 1886–1954 when the French controlled the area, then called Indochina. The French mapping agencies were considered "official." The first mapping of Indochina of any importance was performed by the *Bureau Topographique* which was set up as a section of the General Staff after the French assumed control of the area in 1886. This organization started triangulation work in Indochina in 1902 and compiled topographic maps as well as boundary and river maps. The 1:500,000 scale series was completed in 1899 and was the most important of these maps. The major mapping organizations were the *Service Géographique de l'Indochine – SGI* (Geographic Service of Indochina); the *Institut Géographique National* – IGN (National Geographic Institute), in Paris; and the *Service Cartographique des Forces Terrestres d'Extrême Orient – F.T.E.O.* (Cartographic Service of the Far East Land Forces). The SGI was established as the topographic mapping agency in Indochina in 1899 and disbanded in April 1955. The *Service Cartographique des F.T.E.O.* was set up in 1949 as a bureau under the General Staff of the French Army in Indochina to serve as the military mapping agency of the country. These mapping organizations, principally *SGI*, have produced regular surveys, based on first-, second-, and third-order triangulation, for 55% of Indochina. The remainder of the area was covered by reconnaissance surveys. Cambodia achieved its full independence in 1954, and in 1955 the *Service Géographique de Forces Armées Royales Khmères (FARK) – SGK* (Geographic Services of the Royal Cambodian Forces) was organized as the mapping agency for the Royal Cambodian Government. On *SGI* sheets geographic coordinates are in grads or degrees or both. Longitude values in grads are referred to Paris, and those in degrees are referred to Greenwich. Both Degrees and grads are given on the *SGI* 1:25,000 scale and 1:100,000 scale series. The *SGI* 1:25,000 scale, 1:100,000 scale, and 1:400,000 scale series and some sheets of the 1:50,000 scale series carry the ellipsoidal French Indochina Bonne grid, a relic of the 19th century. Some of the later editions of all series by *SGI* and all reprints by the U.S. Army Map Service carry the UTM grid (*Foreign Maps, TM 5–248*).

"In 1993–1994, monumented GPS control points were established in and around Phnom Penh city for a mapping project by IGN-France International. The horizontal datum connection for this survey was made from Taiwan with the coordinates being in GRS 1980 datum (*sic*). Extensive documentation on the observations and adjustments are available. A primary network of 9 stations was established surrounding Phnom Penh using Ashtech dual-frequency GPS receivers. Coordinate accuracies of less than 0.01 meter at 95% confidence level were achieved. A further 85 lower order stations were established within the primary network to provide control for aerial photography and terrestrial network traverses.

In 1994 the Government of Finland granted funds to undertake the mapping of the Mekong River in Cambodia. As part of this project a GPS network was established to provide accurate horizontal and vertical control for the mapping. The project established monumented GPS points along the Mekong and Bassac South of Phnom Penh and along the Mekong to Kampong Cham and along the Tonle Sap and the highway to Kampong Chhnang. The network was then extended through to the Vietnam border and then on to the South China Sea. A horizontal datum connection was made from South Laos. It was the intention that this network be based on the Indian 1975 datum, however lack of any network at the southern end of the traverse made it difficult to undertake any adjustment. It is believed that there will not be any practical problems in the use of the 'Mekong datum' in the relative proximity of the network along the river.

The GPS stations used for the transformation parameter computations were distributed within a narrow longitudinal area in the vicinity of the Mekong River and within this area it is believed that the transformation parameters will be correct. With the deterioration of the geodetic system in Cambodia, the European Commission agreed to provide assistance, to re-establish a geodetic network. In 1997 survey work commenced on establishing a geodetic network, with an aim of having at least one primary network point in each province. This network was based on the WGS84 datum. Following the establishment of this primary network, consisting of 22 stations, a number of secondary networks were established in Pailin, Siemreap and Battambang. In the first two provinces the control was funded through other aid projects but in Battambang the provincial government provided the funding. In 1998, through a program funded partly by the Australian Government, GPS observations were undertaken to strengthen the 1997 network. It was part of the Asia Pacific Regional Geodetic Project 1998, Geodetic Observation Campaign (APRGP98). The Australian Land Information Group (AUSLIG), a group within the Australian Federal Government, was the coordinating agency on behalf of the Permanent Committee on GIS Infrastructure for Asia and the Pacific Region (PCGIAP).

The observations were conducted between 19–29 November 1998. Geodetic agencies in the region contributed data from continuous tracking GPS, from GPS observations on key national geodetic stations and other high precision geodetic techniques such as Side Looking Radar (SLR) and Very Long Base Interferometry (VLBI).

The project was designed to coincide with a worldwide campaign. The objects of the campaign were to: *(i)* strengthen the precise regional geodetic network; *(ii)* assist in establishment of a regional datum; *(iii)* assist in monitoring the tectonic motion of the network sites; and *(iv)* to assist with the development of transformation parameters for national datums.

In Cambodia, a bracing network was observed using Phnom Penh as a focus with observations also being made at stations in Sihanoukville, Svay Rieng, Stung Treng and Siemreap. All GPS observation data was transferred to AUSLIG, in Australia, for archiving and processing. In all there was data for 80 sites in 16 nations. Processing of the data was completed in mid 1999 and the results were presented at a workshop in Vietnam in July 1999. The results of the adjustment of the GPS observations were included in the reports presented during the workshop. The adjustment information and the final coordinates for the 5 points observed in Cambodia as part of the AGRPG98 program provide an excellent basis from which to undertake future geodetic activity in Cambodia. *(However, the actual coordinates of any collocated points have not been publicly released. – Ed.)* From the reports and data, it would appear that the network that has been established provides a sound base from which to establish future control networks and from which to base GPS observations for the project required for controlling aerial photography and terrestrial traversing" (*Chris Lunnay, Director, Land Equity International, Australia*).

"There is a plan to complete (within 2008–2010), the 1st order GPS network (about 100 points) to the existing GPS network that has been established since 1997. There were 31 points of 1st order and 92 points of 2nd order that have been installed in Kampong Chhnang, Pursat and its adjacent areas. The study on improving parameters for converting from Indian 1960 datum to WGS84 is also started" (*Chharom Chin, Senior GIS/RS Officer, Geography Dept., Ministry of Land Management, Urban Planning and Construction, Cambodia, 22–24 February 2007*).

The legendary "origin" of the Indian Datum as defined in 1900 and labeled as Indian 1916 origin at Kalianpur Hill Station: $\Phi_o = 24°\ 07'\ 11.26''$ N, $\Lambda_o = 77°\ 39'\ 17.57''$ East of Greenwich, the initial azimuth to Surantal from south is: $\alpha_o = 190°\ 27'\ 05.10''$. The ellipsoid of reference is the Everest 1830, where $a = 6{,}377{,}276.345$ m and $1/f = 300.8017$. John W. Hager tells me that "in 1954, the triangulation of Thailand was adjusted to Indian 1916 based on 10 stations on the Burma border. In 1960, the triangulation of Cambodia and Vietnam was adjusted holding fixed two Cambodian stations connected to the Thailand adjustment of stations from the Cambodian–Vietnam adjustment. Various regional "adjustments" in the 20th century have been performed on the Indian Datum of 1916 that have resulted in a garbled mishmash of discordant coordinate systems that have little to no continuity across national borders much less even with countries such as Cambodia. Some of these "datum adjustments" present daunting challenges to the mapping professional charged with the task of "making sense of the situation."

A close friend of mine is Mr. Malcolm A. B. Jones of Perth, Australia. Mal, or "Geodesy Jones" as he is commonly known, sent to me his best guess as to a set of transformation parameters from "Indian Cambodia" to "WGS84," based on his extensive personal geodetic holdings: $\Delta X = +225$ m, $\Delta Y = +854$ m, $\Delta Z = +301.9$ m, $R_x = R_y = R_z = 0$, $\delta s = 0.38$ ppm. An example "Test Point" is Phnom Anlong Svay (1512) where for "Indian Cambodia" datum, $\varphi = 14°\ 22'\ 32.1733''$ N, $\lambda = 103°\ 44'\ 36.8591''$ E, h = 0.0 m. For WGS84: $\varphi = 14°\ 22'\ 38.2868''$ N, $\lambda = 103°\ 44'\ 23.6058''$ E, h = −25.022 m. Test points are always nice to have as a reassuring tidbit of data for one's transformation software … however, I was provided a data set of classical triangulation data points (a "Trig List"), for Cambodia sometime in the past and *lo and behold*, a point therein is listed as "PN ANLONG SAI" for "Indian 54," where $\varphi = 14°\ 22'\ 32.3966''$ N, $\lambda = 103°\ 44'\ 36.9339''$ E, and h = +457.10 m – **that's over seven meters difference!** Cambodia has a Database of Land Mine Fields, and when the United States Army Corps of Engineers (in 1968), *graciously* provided me with the Officer Candidate School Engineer Demolition training to become a "Sapper," I was inculcated with the knowledge that a 64th of an inch between one's fingers (when on one's hands and knees) can make the difference in a mine field between life and death! *Seven meters uncertainty just is* **not** *going to fit into one's "comfort zone" in Cambodia.* The Kingdom of Cambodia certainly has the history to be paranoid about invaders but maintaining secrecy about datum transformations is not going to provide encouragement for foreign workers to come in to provide assistance in decommissioning land mine fields! My advice is to stay on the paved roads until the government releases their transformation parameters for Cambodia!

Cameroon
International boundary
Province boundary
National capital
Province capital
Railroad
Road
0 50 100 Kilometers
0 50 100 Miles
Mercator Projection
NIGER
NIGER
Lake Chad
Nguru
Komadugu Yobe
Makari
Fotokol
N'Djamena
Kousseri
Potiskum
Maiduguri
Bama
Kari
Waza
Mora
EXTRÊME-
NORD
Mokolo
Maroua
Mubi
Bourrah
Yagoua
Bongor
Bauchi
Gombe
Guider
Kaélé
CHAD
Jos
Fianga
Figuil
Léré
Chari
NIGERIA
Garoua
Yola
Logone
Laï
Lagdo Reservoir
Pala
Jalingo
NORD
Doba
Moundou
Poli
Tchollíré
Lafia
Benue
Bénoué
Mbé
Makurdi
Wukari
Beli
Vina
Touboro
Baïbokoum
Tignère
Ngaoundéré
Banyo
ADAMAOUA
NORD-
OUEST
Nkambe
Wum
Meiganga
Ngaoundal
Tibati
Kumbo
Bankim
Mbam
Mbakaou Reservoir
Ouham
Bea
Bozoum
Ogoja
Ikom
Mbengwi
Bamenda
Bouar
Baboua
Garoua Boulaï
Mamfe
Foumban
SUD-
OUEST
Mbouda
Bafoussam
Dschang
OUEST
Yoko
CENTRAL
AFRICAN
REPUBLIC
Befang
Bangangté
CENTRE
Lom
Calabar
Mundemba
Nkongsamba
Sanaga
Bélabo
Kadei
Kumba
Bafia
Nanga Eboko
Minta
Bertoua
Batouri
Lobaye
Yabassi
Monatélé
Berbérati
Buea
LITTORAL
Douala
Nanga
Gamboula
Nola
Malabo
Edéa
Yaoundé
Ndélélé
Abong Mbang
Luba
Isla de Bioko
EQUATORIAL
GUINEA
Eséka
Mfou
Akonolinga
EST
Yokadouma
Mbalmayo
Lomié
Ngoulémakong
Kribi
Ebolowa
Sangmélima
Bight of
Biafra
SUD
Djoum
Dja
Campo
Ambam
Moloundou
REPUBLIC OF
THE CONGO
Ebebiyin
Bitam
Minvoul
Souanké
Ngoko
Bata
Ouesso
Santo
Antonio
SAO TOME AND PRINCIPE
Ilha do
Principé
Mongomo
Oyem
GABON
EQUATORIAL
GUINEA
Sangha
Base 802576 (R02413) 8-98

Republic of Cameroon

Cameroon is bordered by the Central African Republic (797 km), Chad (1,094 km), Republic of the Congo (523 km), Equatorial Guinea (189 km), Gabon (298 km), and Nigeria (1,690 km). Slightly larger than California, the coastline is 402 km along the Bight of Biafra by the Atlantic Ocean. The highest point is Fako (4,095 m), the lowest point is the Atlantic Ocean, and the terrain is diverse with a coastal plain in the southwest, a dissected plateau in the center, mountains in the west, and plains in the north.

Bantu speakers were among the first groups to settle Cameroon, followed by the Muslim Fulani in the 18th and 19th centuries. Treaties with tribal chiefs in 1884 brought the area under German domination. After World War I, in 1919 the League of Nations gave the former German colony of Kamerun to a French mandate of over 80% of the area, and to the British a mandate over 20% of the area that was adjacent to Nigeria. After World War II, the country came under a UN trusteeship in 1946, self-government was granted. France set up Cameroon as an autonomous state in 1957, and the next year its legislative assembly voted for independence. Cameroon became an independent republic on January 1, 1960. The official name of the country became the Republic of Cameroon in 1983. "The country has generally enjoyed stability, which has enabled the development of agriculture, roads, and railways, as well as a petroleum industry. Despite slow movement toward democratic reform, political power remains firmly in the hands of President Paul Biya" (*World Factbook*, 2022).

"Official topographic mapping of former French Equatorial Africa (which included Cameroon) was carried out by the French *Institut Géographique National*, Paris – IGN; *Institut Géographique National, Paris (Annexe de Brazzaville)*, prior to 1959 known as the *Service Géographique de l'Afrique Équatoriale Française et du Cameroun* – SGAÉF, and the *Institut Géographique National, Paris (Annexe de Yaoundé)*, formerly a field unit under the SGAÉF. After 1959–1960 the IGN (*Annexe de Yaoundé*) was known as the *Service Géographique à Yaoundé*" (*TM-5-248 Foreign Maps 1963*).

The Clarke 1880 was the ellipsoid of reference for these series. In December 1945 the *Cabinet du Directeur, Institut Géographique National* in Paris issued an instruction concerning the systems of projection to be utilized in French West Africa. The instruction detailed that a Gauss (Gauss-Schreiber Transverse Mercator) system of projection was to be used for the group of regular map compilations and related works that included geodesy, topography, photogrammetry, and cartography for a range of scales that included 1:200,000. All of the map series were to be cast on the International Map of the World sheet system based on the graticule.

Within 5 years, French coordinate systems changed to UTM throughout the world with the exception of Madagascar, and for Reunion. In terms of far-reaching developments in grid system usage, this was spectacular! D. R. Cockburn and W. L. Barry of Army Map Service translated the IGN Instruction dated 20 September 1950:

1. The General Directorate has decided to abandon the projection systems now in use in the French Overseas Territories and Departments and to utilize henceforth, in all these territories, a new projection system called the Universal Transverse Mercator (*Mercator Transverse Universelle*), having a unique definition.

   "In so far as Madagascar is concerned, the use of the Laborde Projection will be continued without change. Similarly for Reunion, the Gauss system, in which the triangulation of the island has been computed, will be retained. With the exception of these two particular cases, the U.T.M. projection will from now on be the only official projection in the French

Overseas Territories and Departments. Consequently I.G.C. instruction No. 1212, dated 12 December 1945, is rescinded.

2. The new projection is a Gauss conformal projection applied to zones of 6° of longitude in width. These zones, identical to those of the 1:1,000,000 International Map of the World, are indicated on the attached index map.

3. For a long time, views have been expressed in the international meetings of geodesists in favor of a universal projection system, which would be adopted by all the countries of the world. Inspector General Tardi proposed himself at the General Assembly of the International Association of Geodesy at Edinburgh (1936), a Gauss projection in 6° zones for the African continent, which is the same as the UTM projection. These views remained the dead issue for a long time. Before 1940, each country was quite satisfied with its own projection system and was reluctant to undertake the enormous task of converting coordinates into a universal system. They were especially reluctant to modify their quad printing plates. However, during the course of the last war, the extension of military operation to vast regions of the globe, the strategic deployments on a great diversity of warfronts entailed the creation of a great number of projection systems (in 1945, over 100 of these systems were in use). As a result, a state of utter chaos ensued, and considerable expense was entailed for the computation of the transformation and the adaptation operations. Consequently, the prospect of a universal projection system aroused much interest in the post war period.

"The United States was very much in favor of the project and to facilitate its adoption by the various countries, Gauss projection tables (called a UTM projection) were computed and published. These tables were computed in the sexagesimal angular division system. The American agencies also computed the same tables on a centesimal system.

"The *Institut Géographique National*, when asked to adopt the new projection in December of 1949, did not hesitate in agreeing to its use for French Colonial regions with certain exceptions, which are explained below. In point of fact, it was entirely possible to adopt this new projection for the major part of the colonial possessions at a very small cost. However, a problem arose for its use in France proper and in North Africa. For France itself, a 6° belt UTM projection leads to very extreme scale changes, *i.e.*, extreme from the point of view of civilian use.

4. Actually, it was not merely in a spirit of international cooperation that the *Institut Géographique National* agreed to the new projection but also because it offers incontestable practical advantages. In December 1949, the situation was as follows:

"After long conferences in which various proposals were suggested, we finally adopted the solution proposed by General Laborde for our overseas possessions at the end of 1945. This solution was as follows: A Gauss system (double projection) on the international Ellipsoid with φ_o equaling 0° in French East Africa and French West Africa. For the smaller regions (Guadeloupe, Martinique, Reunion, *etc.*) the value of φ_o is equal to the mean latitude of the territory, φ_o being the latitude of the central point. This procedure leads to the establishment of separate tables for each value φ_o.

"Tables for the conversion of geographic coordinates into rectangular and vice versa (tables which would produce the centimetric precision necessary for geodetic computations) had not been set up at the end of the year 1949 with the exception of tables covering a few small regions. Although this is a very pressing urgency, the *Institut Géographique National*, due to limit (*sic*) means, has neither the facilities for computing the tables rapidly nor for editing them without detriment to other equally urgent tasks.

"Considering on one hand the small number of stations to be converted into the new system (for astro points the work involved is insignificant) and considering that the dimensional variations of the sheets already published would be less than the standard size, the Institut Géographique National has agreed to rapidly extend the UTM projection in these territories being aware of the following factors: That the United States was in a position to immediately deliver to us as many copies as was necessary of the tables computed on the sexagesimal system and contracted to compute the same tables on the grad system; that the United States was able to undertake the conversion of coordinates into the system using data obtained from electronic computing machines.

5. In point of fact, the UTM projection as it has been already adopted (or in the course of being adopted) by a number of countries is not absolutely "Universal." This would have been the case if a uniform ellipsoid had been chosen for all the countries. However, the difficulties entailed in changing ellipsoids are common knowledge and because of this, the basics of the ellipsoids in use for the various continents have been retained. Accordingly, the Clarke 1866 ellipsoid has been kept in use for North America; the International Ellipsoid has been adopted for South America and the Pacific regions, and the Everest Ellipsoid has been chosen for the East Indies and the adjacent regions. So as to fulfill a request made by the British who have already computed vast geodetic nets on the Clarke 1880 Ellipsoid, the Institut Géographique National has adopted this ellipsoid for the entire African continent. In addition, this ellipsoid was used for French geodetic work previous to 1945.

6. The UTM projection may be defined as having the following intrinsic properties: It is a Gauss conformal projection, a direct projection of the ellipsoid on the plane. Linear values are maintained on the prime meridian of the projection with the exception of a scale-reduction which is defined by the following coefficient: $m_o = 0.9996$.

"The zones have an overall width of 6° in longitude (3° on each side of the central meridian). The zones coincide with those of the 1:1,000,000 International Map. The Greenwich meridian is at the limit of two zones (zone numbers 30 and 31). These basics will suffice to define the projection for any given ellipsoid.

7. The new UTM projection differs from the Gauss projection adopted at the end of 1945 in the sense that it is a direct projection of the ellipsoid on the plane instead of being an indirect projection employing the intermediary of a sphere upon which the ellipsoid is first applied before projecting it on the plane."

Thanks to Mr. John W. Hager, "The French published trig lists for Cameroon that consisted mainly of astro points. They seemed to use any source they could, and the accuracy varies. Without having access to those records, I cannot make any definite statements. Some astros were probably even established by and for the U.S. Air Force in the mid 1940s using the 60° astrolabe. This was in conjunction with a project to provide control for the World Aeronautical Chart (W.A.C.). The area formerly British Cameroons was surveyed as part of the Nigerian first order network and possibly the second order network. It was originally computed on Minna Datum *(of 1927 – Ed. The origin is at station L40, which is the north end of Minna Base in the town of Minna, Nigeria where: $\Phi_o = 09° 39' 08.87 N$, $\Lambda_o = 06° 30' 58.76'$ East of Greenwich and the ellipsoid of reference is the Clarke 1880 where $a = 6,378, 249.145 m$ and $1/f = 293.145$).* This is included in the French trig lists. This area was reunified with Cameroon in 1961. I would suspect that many of the surveys published in *Annales Hydrographique* were included in those trig lists. The only horizontal datum I list is Douala at Douala Météo *(I. G. Dufour, 1947–48)*, $\Phi_o = 4° 01' 11.1'' N$, $\Lambda_o = 9° 42' 31.4'' E$, $\alpha_o = 2.1437$ grads from I to IV, Clarke 1880. Reference is Cameroun, Ville De Douala, Triangulation Principale, Triangulation Secondaire, *11 Nov. 1965*. This was part of a Doppler survey. Also found in Annales Hydrographiques, *4th e Series, Vol. I, p. 159 ff.* Also in the same reference is a grid; projection Gauss Laborde, International (**yes**) ellipsoid, φ_o = equator, $\lambda o = 10° 30' E$, $m_o = 0.999$, FN=FE=1,000,000 meters."

"The new projection retains the linear values on the central meridian of each zone to the approximate scale factor. The former projection did not retain linear values on this meridian. In *toto*, the basics of the two projections are, at least within the limits of the proposed narrow zones, absolutely comparable and considered from the viewpoint of practical application it is impossible to give preference to either one or the other. The only advantage of the former projection is that of adapting itself more simply to the extension of latitudinal belts and that this predicament will not arise for overseas geographic services.

8. Covering memo No. 1 in reference to the implementation of the new projection program is to be effective immediately." *Director, Institut Géographique National.* The instruction quoted above was accompanied with some specific procedures for all of the French colonies, territories, and departments. With respect to French West Africa (and Cameroon), IGM explained that AMS agreed to compute the UTM coordinates of all astro points that were observed as control for the 1:200,000 scale topographic maps.

"Curiously, the European Petroleum Survey Group, EPSG (now OGP) database lists a different set of defining parameters for the "Douala 1948 Datum fundamental point: South pillar of Douala base; $\Phi_o = 4°\ 00'\ 40.64''$ N, $\Lambda_o = 9°\ 42'\ 30.41''$ E. Superseded by Manoca 1962 datum (code 6193), International Ellipsoid. Derived at Manoca tower assuming the pyramid on the tower and the centre of the tower reservoir are co-located. This assumption carries a few meters uncertainty.

"Furthermore, another local datum used for topographic mapping reported by the EPSG is: "Garoua where the fundamental point is IGN astronomical station and benchmark no. 16 at Tongo: $\Phi_o = 8°\ 55'\ 08.74''$ N, $\Lambda_o = 13°\ 30'\ 43.19''$ E. (of Greenwich)."

"Finally, the EPSG lists: "Manoca 1962 Datum where the fundamental point is the reservoir centre at the Manoca tower ("tube Suel"), $\Phi_o = 3°\ 51'\ 49.896''$ N, $\Lambda_o = 9°\ 36'\ 49.347''$ E. Used for topographic mapping, and it is referenced to the Clarke 1880 ellipsoid (IGN). The intent of the Bukavu *(Zaire – Ed.)* 1953 conference was to adopt the Clarke 1880 (RGS) ellipsoid (code 7012) but in practice this datum has used the IGN version. Derived via WGS 72BE. Can be implemented as a single position vector transformation with parameter values of $\Delta X = -56.7\,m$, $\Delta Y = -171.8\,m$, $\Delta Z = -38.7\,m$, $R_x = R_y = 0''$, $R_z = 0.814''$, $\delta s = -0.38\,ppm$."

Thanks again go to Mr. John W. Hager for his help with this enigma.

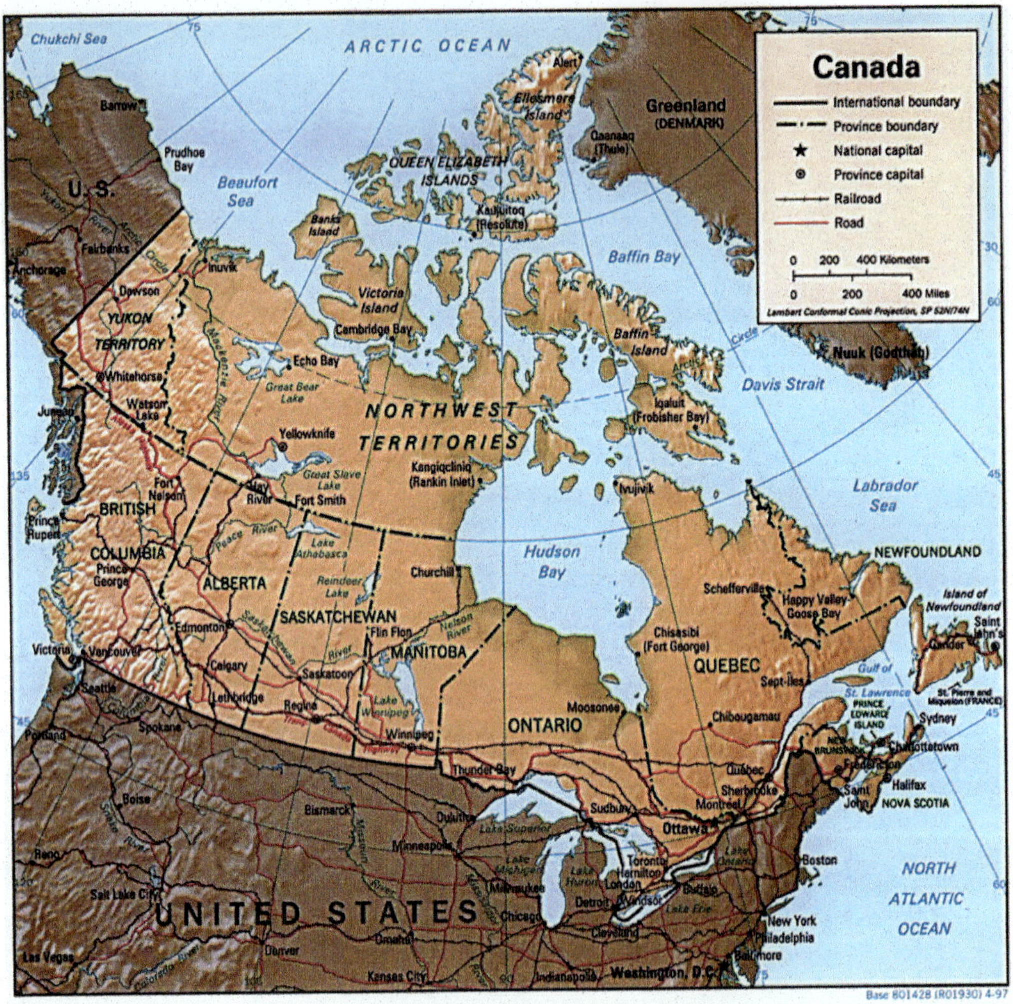

Base 801428 (R01930) 4-97

Canada

Canada is in Northern North America, bordering the North Atlantic Ocean on the east, North Pacific Ocean on the west, and the Arctic Ocean on the north, north of the conterminous United States. Bordering the United States (8891 km, includes 2475 km with Alaska), Canada is the world's largest country that borders only one country. The Canadian Arctic Archipelago – consisting of 36,563 islands, several of them some of the world's largest – contributes to Canada easily having the longest coastline in the world. The highest point is Mount Logan (5,959 m), the lowest point is the Atlantic/Pacific/Arctic Oceans (0 m), and the mean elevation is (487 m) (*World Factbook*, 2022).

The Dominion of Canada was established in 1867 by the union of New Brunswick and Nova Scotia with the Provinces of Canada. The Statute of Westminster of 1931 removed the United Kingdom. Parliament authority over the dominions and the constitution was patriated to Canada in 1982. "Economically and technologically, the nation has developed in parallel with the US, its neighbor to the south across the world's longest international border. Canada faces the political challenges of meeting public demands for quality improvements in health care, education, social services, and economic competitiveness, as well as responding to the particular concerns of predominantly francophone Quebec. Canada also aims to develop its diverse energy resources while maintaining its commitment to the environment" (*World Factbook*, 2022).

The original geodetic surveys of Canada were computed on the Bessel 1841 ellipsoid. In 1880, when the U.S. Coast & Geodetic Survey chose to adopt the ellipsoid that had been published by Colonel A. R. Clarke of the *British Royal Engineers* in 1866 for North America, it was only logical for Canada to adopt the same ellipsoid. The original first-order geodetic networks included only a few networks in Canada. With a few second-order networks, they were adjusted between 1927 and 1932 to produce the original system of coordinates known as NAD27. Southern Quebec, Ontario, and the southern boundary of the western provinces were included. Subsequently, first-order arcs were extended northwards from stations in the original figures and were adjusted while holding the original stations fixed.

This practice was the same as used in the United States and Mexico with respect to new observations made after the original datum adjustment. The North American Datum of 1927 was a spectacular achievement in geodesy *at the time*. In certain cases, local datums were later computed for optimum accuracy at the expense of overall continuity. Some of the major readjustments included: the Prince Edward Island free adjustment of 1968, the 1971 Newfoundland adjustment, the 1972 Nova Scotia readjustment, the Southern Ontario adjustment of 1974, the Northwestern Ontario adjustment of 1974, and the Southern Quebec adjustment of 1978.

With the advent of the Transit satellite system, the Maritime Scientific Adjustment (MSA) was developed in 1971. The Canadian Council on Surveying and Mapping adopted a resolution (June 23, 1975) that the Geodetic Survey of Canada (now Geomatics Canada) should make an adjustment in 1977 of the primary horizontal control net of Canada on the NAD1927. Only the Maritime Provinces are still on the Average Terrestrial System (ATS 77), and they have not yet moved to NAD83 or WGS84.

Early 1-inch maps of Canada (1-inch equals 1 statute mile) by the Geological Survey of Canada as of 1844 were based on the polyconic projection until the Army replaced it with the 1932 Lambert Conformal Conics that included the Maritime Zones and the Eastern Zones. Although the U.S. military was using the World Polyconic Grid system in U.S. Survey yards (3600/3937), the Lambert projection sheets of the Canadian Army were gridded in International yards (1 inch=2.54 cm). On the other hand, Topographical Surveys, part of the Department of the Interior, published one sheet in 1926 on the polyconic projection, and from 1927 on used the Deville-Peters 1924 Transverse Mercator (Gauss) projection with 8° zones for the National Topographic Series (NTS).

There was a lively discussion (and correspondence) for years between the two mapping organizations that centered on the relative merits of aesthetics versus convenience for artillery fire control. On sheets of the NTS that were of use to the military, the Army added a 1,000-yard rectangular grid overprint based on the Army Lambert systems. That was called the "British Grid," a system introduced in England in 1919 and adopted by many Commonwealth countries. Later changed to the "Modified British Grid," it was also overprinted on many existing Army 1-inch maps cast on the polyconic projection. The "Modified British Grid" did *not* have the failing that numerical references would repeat themselves every 10,000 yards, but would continue differently until 100,000 yards, a decided safety advantage to avoid calling "friendly" artillery fire onto your own positions. The Canadian Lambert projection grids continued in use until the 1-inch series was converted by the Army Survey Establishment to 1:50,000 scale maps in 1948 with the UTM grid system and the associated Military Grid Reference System (MGRS).

In 1945, a special Lambert Conformal Conic projection was established for Prince Edward Island and later changed to an oblique double stereographic in 1959 along with a separate stereographic zone for New Brunswick. Sometime after World War II, probably also in 1959, five 3° wide Gauss-Krüger Transverse Mercator belts were established for eastern Canada but seem to have been actually used only for a couple belts covering Nova Scotia. These two belts (zones 4 and 5) were used with the millions digit of the False Easting being equal to the zone number. Although this convention is not uncommon throughout the world, this is the only such usage known with respect to False Eastings for North America.

The grid systems established in 1959 for the Maritime Provinces were later documented (1970s) in a series of publications from the University of New Brunswick which are among the most elegant ever written for the computational use by geodetic surveyors and mappers. These documents are genuine "how to" books and not just math proofs. Those readers that are interested in the topics of this column are strongly urged to obtain copies of this series by Thompson, Krakiwsky, *et al.*

As the Provinces move towards the adoption of NAD83, a change in grid systems is inevitable. Current plans for New Brunswick are to retain the Oblique Double Stereographic, but the False Easting will be changed to 3500 Km and the False Northing will be changed to 4,500 km. All other constants will remain the same, excepting the ellipsoid. The Province of Quebec will use the "Modified Transverse Mercator" (MTM) with eight zones of 3° (2–10 from east to west), for example, zone 7 has a central meridian of 70° 30′ W. The scale factor at origin is 0.9999 and the False Easting is 304800 m for all zones. All other provinces seem to be content with the NAD83 UTM (for now). Surprisingly, not a single Province seems interested (or is aware of) in changing the ellipsoid parameters to simplify the sea-level reduction for GIS/survey applications as done by Minnesota, Wisconsin, and Colombia. The Canadian national government and some provinces have gotten commercial in their support (for a price) of transforming to NAD83. The software to convert Latitude & Longitude between the NAD27 and the NAD83 is available for a couple hundred dollars, the geoid software is available for several hundred dollars more. The Canadian government also offers a GPS correction service for a price.

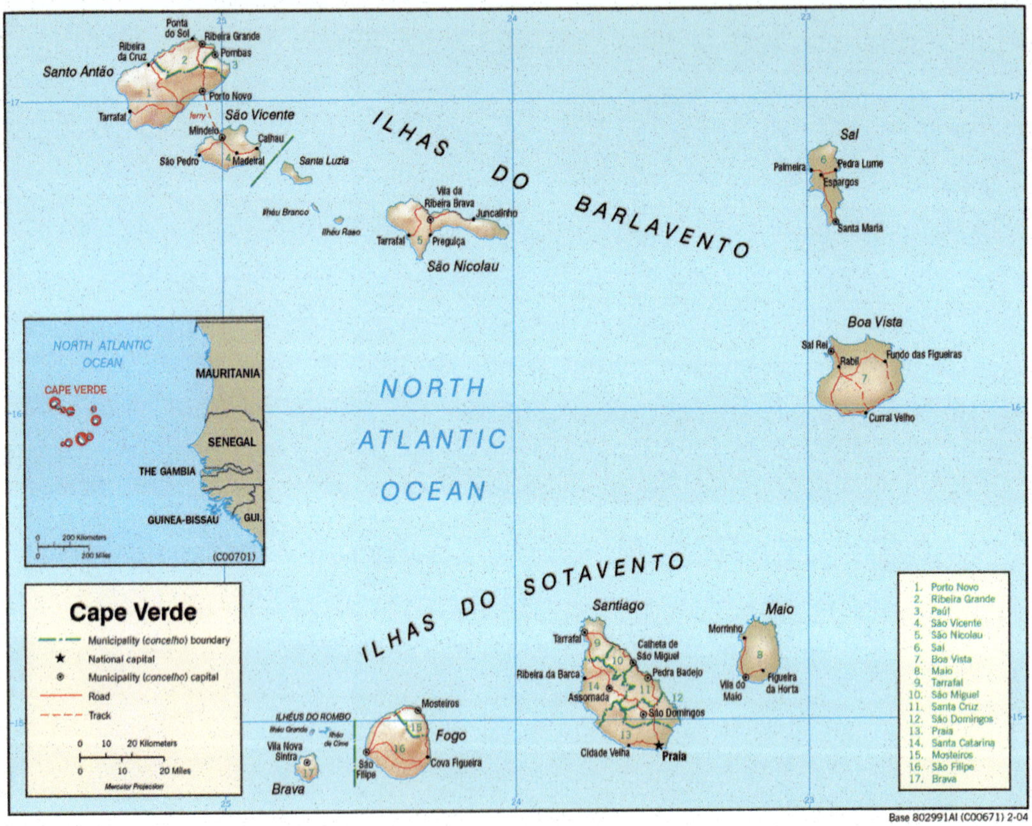

Base 802991AI (C00671) 2-04

Republic of Cape Verde

Cape Verde is located in Western Africa and is a group of islands in the North Atlantic Ocean, west of Senegal. The coastline is 965 km, and the terrain is steep, rugged, rocky, and volcanic; the lowest point is the Atlantic Ocean (0 m), and the highest point is Mt. Fogo (2,829 m), a volcano on Fogo Island (*World Factbook*, 2022).

"In 1462, Portuguese settlers arrived at Santiago and founded Ribeira Grande (now Cidade Velha) – the first permanent European settlement city in the tropics. In the 16th century, the archipelago prospered from the transatlantic slave trade. Pirates occasionally attacked the Portuguese settlements. Sir Francis Drake sacked Ribeira Grande in 1585. After a French attack in 1712, the city declined in importance relative to Praia, which became the capital in 1770. With the decline in the slave trade, Cape Verde's early prosperity slowly vanished. However, the islands' position astride mid-Atlantic shipping lanes made Cape Verde an ideal location for re-supplying ships. Because of its excellent harbor, Mindelo (on the island of São Vicente) became an important commercial center during the 19th century. Portugal changed Cape Verde's status from a colony to an overseas province in 1951 in an attempt to blunt growing nationalism. On June 30, 1975, Cape Verdeans elected a National Assembly, which received the instruments of independence from Portugal on July 5, 1975" (*U.S. Dept. of State Background Note, 24 February 2010*). "Among the nine inhabited islands, population distribution is variable. Islands in the east are very dry and are home to the country's growing tourism industry. The more western islands receive more precipitation and support larger populations, but agriculture and livestock grazing have damaged their soil fertility and vegetation. For centuries, the country's overall population size has fluctuated significantly, as recurring periods of famine and epidemics have caused high death tolls and emigration" (*World Factbook*, 2022).

Astro Stations are reported to include Boa Entrada (village near Praia), Guido do Cavaleiro (mountain on São Antão), Monte Curral (in Espargos, on Sal), Monte Topona (660 m), and Monte Bissau (on São Nicolau). During World War II, the United States Lake Survey, a branch of Army Map Service computed a set of projection tables for the Cape Verde Islands Zone based on the Lambert Conical Orthomorphic projection in 1943. The chosen ellipsoid of reference was the Clarke 1880, where $a = 6,378,249.145$ m, and $1/f = 293.465$ "*as given in the Royal Geographical Society Technical Series No. 4.*" The secant Lambert zone had the following defining parameters: Latitude of Origin, $\varphi_o = 15°$ N, Central Meridian, $\lambda_o = 25°$ West of Greenwich, Scale Factor at the Parallel of Origin, $m_o = 0.999365678$, and False Northing, FN = False Easting, FE = 300 km. The limits of the zone were: North: Parallel of 18° N; East: South along 100,000 m Easting line of South Sahara Zone to parallel of 16° N, thence East along this parallel to 19° W, thence South along this meridian to 13° N; South: Parallel of 13° N; West: Meridian of 27° W. A test point provided on São Vicente Island is: $\varphi = 16°\ 43'\ 21.332''$ N, $\lambda = 25°\ 00'\ 00.000''$ N, X = 820,238.68 m, Y = 496,243.76 m. The choice of the Clarke 1880 ellipsoid for this ersatz datum is probably based on the generalization that most West African colonies back then used that ellipsoid. However, we now know that the Portuguese were famous for actually using the Clarke 1866 ellipsoid for all of their crown colonies in the 19th and 20th centuries.

"The new republican regime of Portugal sought to counter the negative image of Portugal abroad which concerned the lack of its promotion of civilization and progress in their colonies. The Secretary-General of the Portuguese Geographic Missions (*Missão Geográficas*), Ernesto de Vasconcelos adopted on 11 December 1911 a 'series of steps' aimed at reversing that situation. Priority was given to the establishment of a geodetic triangulation network program in Angola, Moçambique, Cape Verde, and São Tome and Principe" (*In the 125 years of IICT Portuguese Science in the Tropics: African Project to the Emptying of Policies under the Third Republic,*

Manuel Lobato, IICT, Lisbon 2007). The Portuguese Geographic Missions (*Missão Geográficas*) performed geodetic surveys and mapped using plane tables and alidades in Cape Verde from 1918 to 1921 and from 1926 to 1932. The maps were published at scales of 1:30,000, 1: 50,000, and 1:75,000 (*AS Missãoes Geodésicas na Comissão de Cartografía (1883–1936), Paula Cristina Santos, IICT, Lisbon, 2007*).

A photogrammetric survey was started by the *Serviço Cartográfico do Exército* (Portuguese Army Map Service) for 1:25,000 scale mapping and is now completed. This map series is referenced to the International ellipsoid, where $a=6,378,288$ m and $1/f=297$. The only grid shown is the UTM, Zone 26N.

On January 31, 2014, The U.S. Department of State published *Limits in the Seas No. 129, Cabo Verde Archipelagic and other Maritime Claims and Boundaries*.

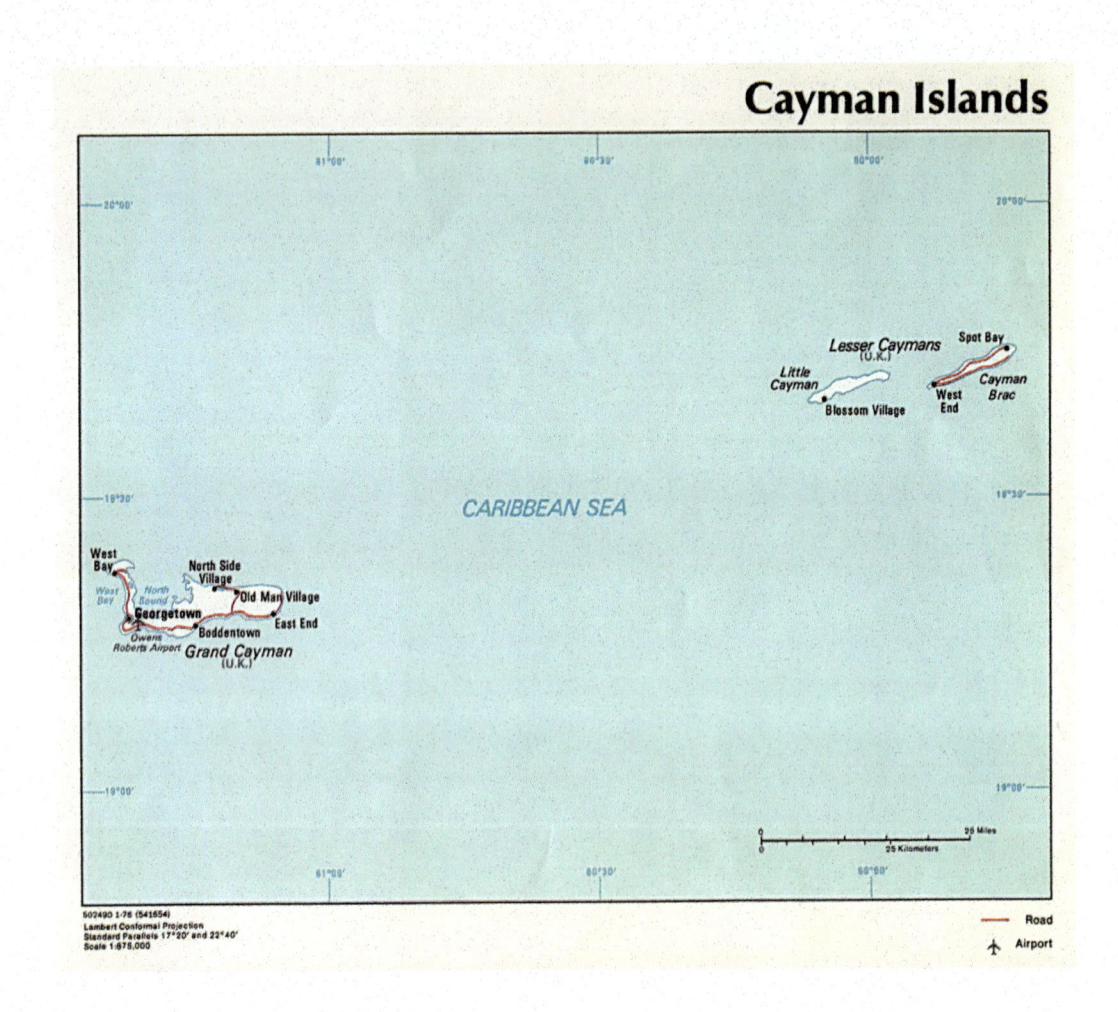

Cayman Islands

Cayman Islands

Located in the Caribbean Sea, the three-island group (Grand Cayman, Cayman Brac, and Little Cayman) is (240 km) south of Cuba and (268 km) northwest of Jamaica. The Cayman Islands are 1.5 times the size of Washington, DC and the terrain is low-lying limestone base surrounded by coral reefs. The highest point is 1 km SW of the Bluff on Cayman Brac (50 m), and the lowest point is the Caribbean Sea (0 m).

The islands of Grand Cayman, Cayman Brac, and Little Cayman comprise this dependency of the United Kingdom. Columbus visited the islands in 1503, but the Spanish never settled there. The Cayman Islands (including Jamaica) were ceded to Britain in 1670 under the Treaty of Madrid. Later, the islands were colonized by settlers from Jamaica and remained a parish of Jamaica. In 1962, when Jamaica became independent, the Cayman Islands became a "direct" British Dependent Territory. The territory has transformed itself into a significant offshore financial center.

Property "Platt Books" (*sic*) were the first recorded land grants of land ownership on Grand Cayman (five parcels) and are dated 1730 and 1740. The Royal Navy performed its original hydrographic survey of Grand Cayman Island in 1773. In 1835, Captain Owen, R.N. determined the longitude difference between Fort George (Grand Cayman) and Morro Light, Havana Cuba. Captain Owen's survey produced the first Admiralty Chart of all three islands. Finlay's North Atlantic Directory of 1895 differed with that longitude by 49 seconds, and the discrepancy continued unresolved for 60 years. The difference was reflected between the Admiralty Chart No. 462 and the U.S. Army Map Service (AMS) 1:50,000 topographic maps of the islands. The AMS maps were based on the U.S. Naval Hydrographic Office Chart No. 0043 (20th) edition of 1933. H.M.S. Vidal, a hydrographic survey ship initiated a new survey in September 1954. The following month, using a geodetic astrolabe, Captain Owen's "Observation Spot" was validated to within 20 feet. The coordinates of the first Cayman Island geodetic datum, "Fort George Observation Spot 1835 Datum" are based on the Origin: $\Phi_o = 19° \ 17' \ 48.01''$ North, $\Lambda_o = 81° \ 23' \ 05.83''$ West of Greenwich. The original datum was the basis of the Admiralty Chart No. 462 and was centered on the ruins of the old Fort George in Georgetown. Furthermore, a new "Observation Spot" was obtained by that astrolabe determination: $\Phi_o = 19° \ 17' \ 45.02''$ North, $\Lambda_o = 81° \ 23' \ 06.83''$ West of Greenwich and was the basis of the new edition of Admiralty Chart No. 462 for the plan of Georgetown Anchorage.

The current reference system for the island of Grand Cayman was originally set by the Inter-American Geodetic Survey (IAGS). It was lost for a number of years after a road was built over it but was recovered in 1995. GC 1 1959 Datum is based on the Origin: $\Phi_o = 19° \ 17' \ 54.43''$ North, $\Lambda_o = 81° \ 22' \ 37.17''$ West of Greenwich, with the defining orientation based on the azimuth from GC 34 to GC 10 being $\alpha_o = 334° \ 13' \ 25.97''$. The datum is referenced to the Clarke 1866 ellipsoid. The scale was established by Tellurometer™, a brand of electronic distance meter (EDM) invented and built in South Africa that uses microwaves rather than the currently more popular type of gizmo that uses Infra-Red (IR) light. The datum was established for the subsequent photogrammetric mapping performed by the British Directorate of Colonial Surveys (DCS), later renamed to the Directorate of Overseas Surveys (DOS). The original DCS survey consisted of 51 classical triangulation stations subsequently densified with DI-10 (IR EDM) traverse for 357 additional points. The establishment of elevations was accomplished by a series of both single-run loops and double-run spurs for a total of 147 stations. The fundamental benchmark on Grand Cayman Island is called "Vidal Spot," established from the observations of 3 months. There is some question on the validity of this benchmark elevation with respect to local mean sea level. (The full Metonic Cycle of 18.6 years is used at permanent tide gages for the determination of local mean sea level.) In 1977, the 512 Specialist Team of the Royal Engineers "on secondment" to DMA/HTC performed Doppler Satellite observations but results referenced to the NWL9D Datum were never sent to the Government of the Cayman Islands (accuracy was ±1.5 m). A subsequent survey by the U.S. National Geodetic Survey determined the WGS 84 coordinates of a number of existing stations on Grand Cayman Island in 1996–1997.

The current reference system for the islands of Little Cayman and Cayman Brac, "LC 5 Astro Owen Island DOS 1961 Datum," is based on the Origin: $\Phi_o = 19°$ 39′ 46.324″ North, $\Lambda_o = 80°$ 03′ 47.91″ West of Greenwich. The defining orientation is based on **two** azimuths, which is very unusual, but it is logical since there is one azimuth per island. The azimuth on Cayman Brac from CB 2 to CB 1 is $\alpha_o = 236°$ 37′ 36.33″, and the azimuth on Little Cayman from LC 5 to LC 6 is $\alpha_o = 256°$ 19′ 02.03″. This datum is referenced to the Clarke 1866 ellipsoid also. The height Datum was based on the fundamental benchmark for Little Cayman, which was 3.266 m. The height datum for Cayman Brac was based on BM 13, which was 3.720 m. The original DOS survey consisted of 27 points, and there was a closed traverse around each island. A braced connection between the two islands was performed with angles and Tellurometer distances. Subsequently, 300 additional points were added with DI-10 distances.

During the Fall of 1997, with the gracious help of the Government Office of the Chief Surveyor of the Cayman Islands, I acquired the "complete" data set of their classical datums. I provided my graduate class with these data, which consisted of those points common to the two classical datums and to the WGS84 Datum. The results of that homework assignment developed a three-parameter solution for both datums, and a seven-parameter solution for the Grand Cayman Datum. The three-parameter transformation from the "GC 1 1959 Datum" to the WGS 84 Datum is: $\Delta X = +67.757$ m, $\Delta Y = +106.114$ m, $\Delta Z = +138.813$ m. The highest residual of six common points did not exceed 0.90 m. On the other hand, if we use a seven-parameter transformation, the accuracy is improved tenfold. Those parameters are: $\Delta X = -0.75$ m, $\Delta Y = +7.93$ m, $\Delta Z = +153.87$ m, Scale $= -13.63 \times 10^{-6}$, $R_x = -0.52$ arc seconds, $R_y = -6.49$ arc seconds, and $R_z = 1.46$ arc seconds. Only two points in common were observed by the U.S. National Geodetic Survey between WGS 84 and the local datum for the Little Cayman and Cayman Brac islands. Therefore, only a three-parameter transformation is possible: $\Delta X = +44.423$ m, $\Delta Y = +108.983$ m, $\Delta Z = +151.666$ m. The highest residual for this two-point fit was 0.34 m, and this relation should be considered quite approximate. The British West Indies (BWI) Grid is not used in the Cayman Islands. The Cayman Islands use the standard UTM Grid for both Datums on all three islands, and all islands are in Zone 17 (Central Meridian $= 93°$ West).

The coordinate systems of the Cayman Islands have been completely updated. The new datum is the Cayman Islands Geodetic Datum of 2011 (CIDG11). The ellipsoid by default is the GRS80 because the ITRF05(2011) is the reference system. With respect to the old datums, the following transformation from Grand Cayman Geodetic Datum of 1959 (GCGD59) to CIDG11 is: $\Delta X = -179.483$ m, $\Delta Y = -69.379$ m, $\Delta Z = -27.584$ m., $R_x = +7.862″$, $R_y = -8.163″$, $R_z = -6.0427″$, $\delta s = -13.925$ ppm; the ESRI/Trimble Coordinate Frame Rotation convention being utilized. Transformation accuracy of ±1 foot is claimed by the government. With respect to Cayman Brac and Little Cayman, the following transformation from Sister Islands Geodetic Datum of 1961 (SIGD61) to CIDG11 is: $\Delta X = +8.853$ m., $\Delta Y = -52.644$ m., $\Delta Z = +180.304$ m., $R_x = +0.393″$, $R_y = +2.323″$, $R_z = -2.96″$, $\delta s = -24.081$ ppm; the ESRI/Trimble Coordinate Frame Rotation convention being utilized. Transformation accuracy of ±1 foot is claimed by the government.

The new plane coordinate system for all of the Cayman Islands is the Cayman Islands National Grid (CING11) now using the Lambert Conformal Conic projection, where Latitude of False Origin is: 19° 20′ N, Longitude of False Origin is: 80°34′ W, Latitude of 1st Standard Parallel is 19° 20′ N, Latitude of 2nd Standard Parallel is 19°42′ N, Easting at False Origin is 2,950,000 feet, Northing at False Origin is 1,900,000 feet, and the International Foot convention is used, where 1 meter = 0.3048 feet. These parameters are defined by the National Government as published as EPSG Code 6391.

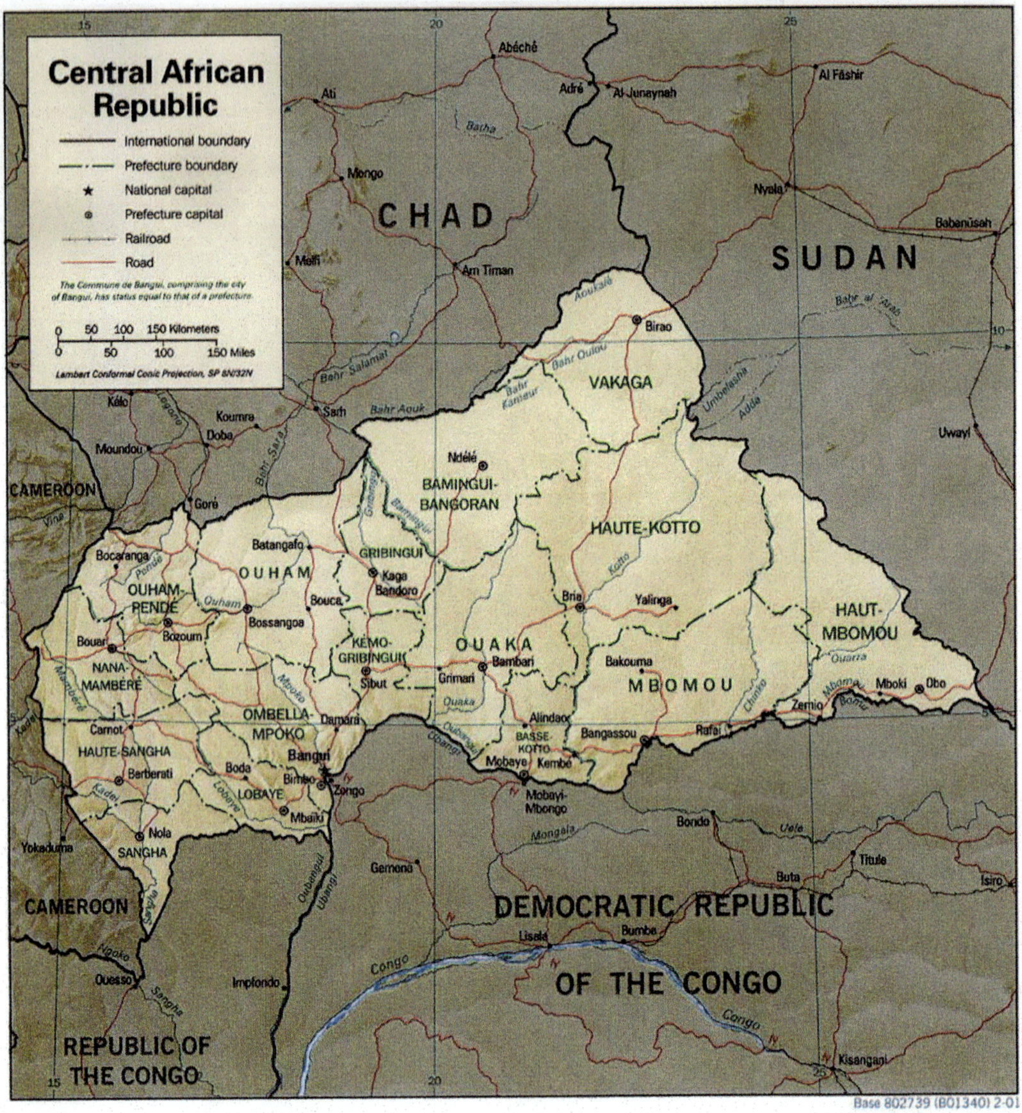

Base 802739 (B01340) 2-01

Central African Republic

Landlocked and almost the precise center of Africa, the C.A.R. is bordered by Cameroon (797 km), Chad (1,197 km), Zaire (1,577 km), the Republic of the Congo (Brazzaville) (467 km), and Sudan (1,165 km). The lowest point is the Oubangui River (335 m), the highest point is Mont Ngaoui (1,420 m) *(NGA GeoNames Search)*, and the Central African Republic is slightly smaller than Texas.

"The Central African Republic (C.A.R.) appears to have been settled from at least the 7th century on by overlapping empires, including the Kanem-Bornou, Ouaddai, Baguirmi, and Dafour groups based in Lake Chad and the Upper Nile. Later, various sultanates claimed present-day C.A.R., using the entire Oubangui region as a slave reservoir, from which slaves were traded north across the Sahara and to West Africa for export by European traders. Population migration in the 18th and 19th centuries brought new migrants into the area, including the Zande, Banda, and M'Baka-Mandjia. In 1875 the Egyptian sultan Rabah governed Upper-Oubangui, which included present-day C.A.R." *(U.S. Department of State Background Notes, 2012).*

"French interests in Central Africa date from early coastal trade and missionary work in Gabon. In 1839 Captain L. E. Bouet-Willaumez completed the first of a series of treaties with local chieftains for coastal rights, so that by 1862 French authority extended along most of the littoral of Gabon. Libreville was founded by the French in 1849, and shortly thereafter, a number of French explorers such as Compiegne, Marche, and Brazza penetrated into the interior. During expeditions in 1874 and 1883, Pierre Savorgnan de Brazza explored the territory of the upper Ogooue and parts of the Congo drainage area. The Berlin Conference of 1885 recognized French claims to the lands explored by DeBrazza, which included the territory along the right bank of the Congo. In 1886 Gabon was administered by a lieutenant governor under the authority of Commissioner DeBrazza. Later the same year, De Brazza was appointed commissioner general over the colonies of Gabon and French Congo (*Congo Française*). A French decree of 1888 united Gabon and French Congo into a single administrative entity until 1903, and during most of this period the combined units were known as the French Congo. Northwestward of French Congo, the territory of Ubangi-Shari (*Oubangi-Chari*) was formally established by the French in 1894. Starting from Libreville in 1897, a mission headed by Captain Marchand crossed unmapped lands to reach Fashoda (*Kodok*) and to establish French claims in the area of the upper Nile. Marchand reached his objective the following year, only to discover British forces already in possession of the area. Before Marchand reached Fashoda, however, the French had made plans to occupy the Chari-Baguirmi region south of Lake Chad, and in 1899–1900 three expeditions were organized with the aim of joining France's possessions in central, west, and north Africa. An expedition marched southward from Algeria, a second moved eastward from the Niger area, and a third travelled northward from the French Congo all meeting on April 21, 1900, at Kousseri (Fort Foureau) south of Lake Chad. The campaign was successful in linking together France's African possessions and in expanding the French Congo northward to include Chad as a military territory.

"In 1903 French Congo consisted of the colonies of Gabon and Middle Congo (*Moyen Congo*) the territory of Ubangi-Shari, and the military territory of Chad. Two years later the territory of Ubangi-Shari and the military territory of Chad were merged into a single territory. The colony of Ubangi-Shari – Chad was formed in 1906 with Chad under a regional commander at Fort-Lamy subordinate to Ubangi-Shari. The commissioner general of French Congo was raised to the status of a governor generalship in 1908; and by a decree of January 15, 1910, the name of French Equatorial Africa was given to a federation of the three colonies (Gabon, Middle Congo, and Ubangi-Shari - Chad), each of which had its own lieutenant governor. In 1914 Chad was detached from the colony of Ubangi-Shari and made a separate territory; full colonial status was conferred on Chad in 1920. In accordance with the constitution of the Fourth French Republic, adopted October 26, 1946, both Chad and Ubangi-Shari became overseas territories within the French Union. As provided in the constitution of the French Republic in 1958, a French Community was established superseding the French Union. Chad and Ubangi-Shari elected to become autonomous members of the French Community and adopted the names of the Republic of Chad and the Central African Republic, respectively. On June 4, 1960, the constitution of the Fifth Republic was amended so that states might become independent and sovereign republics without forfeiting membership in the Community. After agreements of transfer of power and cooperation were signed in Paris on July 12, 1960, Chad proclaimed its independence on August 11, 1960, and the Central African Republic became independent two days later" (*International Boundary Study No. 83 – July 15, 1968, U.S. Department of State*). "Elections organized by a transitional government in early 2016 installed independent candidate Faustin-Archange Touadera as president; he was reelected in December 2020. A peace agreement signed in February 2019 between the government and the main armed factions has had little effect, and armed groups remain in control of large swaths of the country's territory" (*World Factbook*, 2022).

On 12 December 1945, the French *Institut Géographique National* issued an instruction (I.G.C. Instruction No. 1212) concerning the *systèmes de projection* for French Equatorial Africa (*Afrique Equatorial Française*). The Clarke 1880 (IGN) was the ellipsoid of reference for these series, where $a = 6,378,249.200$ and $1/f = 293.4660213$. The instruction detailed that a Gauss (Gauss-Schreiber Transverse Mercator) system of projection was to be used for the group of regular map compilations and related works that included geodesy, topography, photogrammetry, and cartography for a range of scales that included 1:200,000. All of the map series were to be cast on the International Map of the World sheet system based on the graticule. For the C.A.R., two zones were devised: *Fuseau A.E.F. Centre* where the Central Meridian, $\lambda_o = 17° \, 40' \, E$, Scale Factor at Origin, $m_o = 0.9999$, False Easting = False Northing = 1,000,000 m.; and *Fuseau A.E.F. Est* where the Central Meridian, $\lambda_o = 24° \, 30' \, E$, Scale Factor at Origin, $m_o = 0.9999$, and False Easting = False Northing = 1,000,000 m.

Within 5 years, French coordinate systems changed to UTM throughout the world with the exception of Madagascar and for Reunion. In terms of far-reaching developments in grid system usage, this was spectacular! D. R. Cockburn and W. L. Barry of Army Map Service translated the IGN Instruction dated September 20, 1950 (*see Republic of Cameroon*).

Apparently, there is no information on the current or past datums for the Central African Republic. This situation is likely similar to many other former French colonies in Africa such that mapping was and is controlled by sparsely spaced astronomical observations, or "Astro Stations."

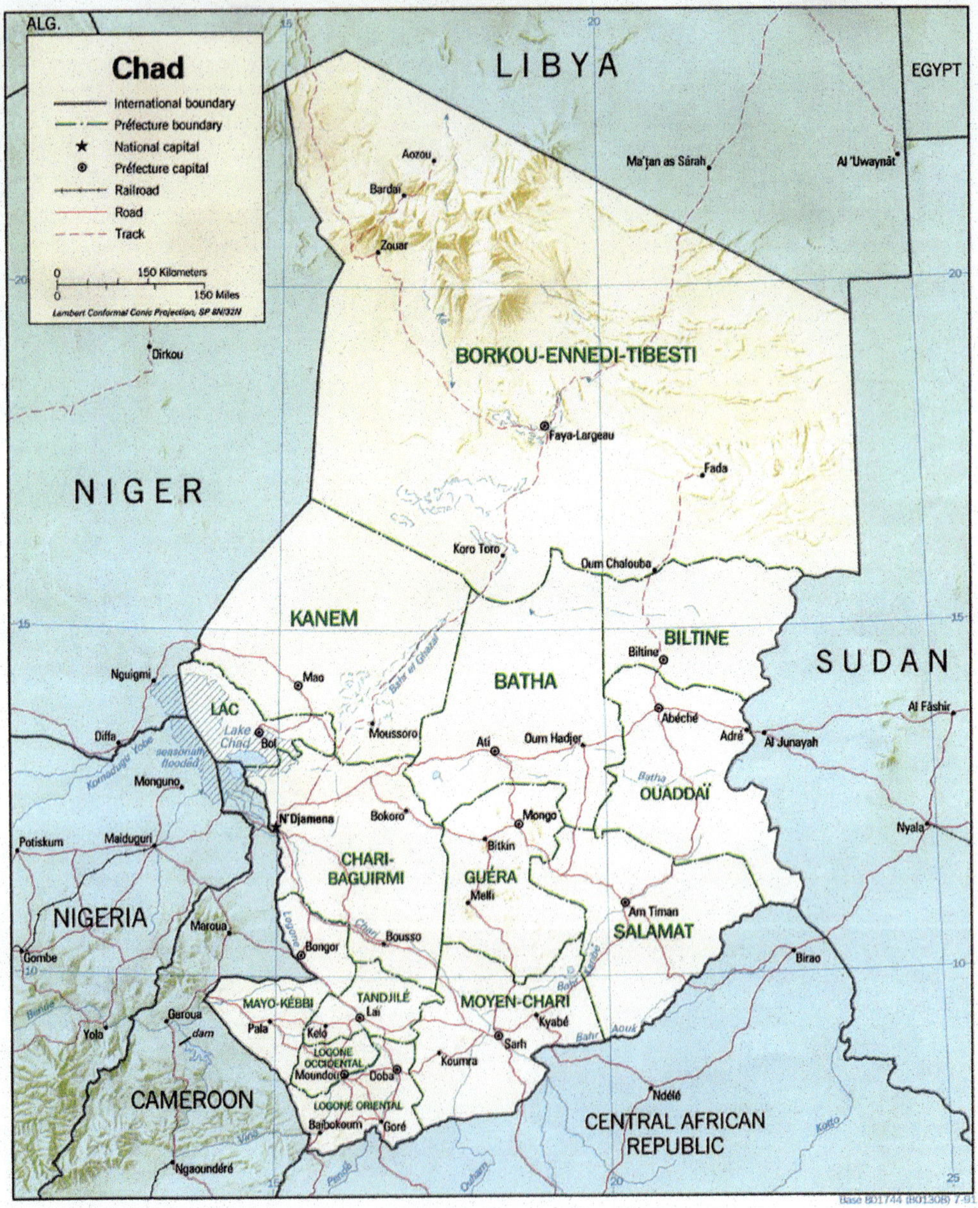

Base 801744 (B01308) 7-91

Republic of Chad

Slightly more than three times the size of California, Chad is bordered by Cameroon (1,094 km), Central African Republic (1,197 km), Libya (1,055 km), Niger (1,175 km), Nigeria (87 km), and Sudan (1,360 km). The terrain has broad arid plains in the center of the country, desert in the north, mountains in the northwest, and lowlands in the south. The lowest point is Djourab (160 m), and the highest point is Emi Koussi (3,415 m) (*World Factbook*, 2014).

"The territory now known as Chad possesses some of the richest archaeological sites in Africa. During the seventh millennium *B.C.*, the northern half of Chad was part of a broad expanse of land, stretching from the Indus River in the east to the Atlantic Ocean in the west, in which ecological conditions favored early human settlement. Rock art of the 'Round Head' style, found in the Ennedi region, has been dated to before the seventh millennium *B.C.* and, because of the tools with which the rocks were carved and the scenes they depict, may represent the oldest evidence in the Sahara of Neolithic industries. Many of the pottery-making and Neolithic activities in Ennedi date back further than any of those of the Nile Valley to the east.

"In the prehistoric period, Chad was much wetter than it is today, as evidenced by large game animals depicted in rock paintings in the Tibesti and Borkou regions. Recent linguistic research suggests that all of Africa's languages south of the Sahara Desert (except Khoisan) originated in prehistoric times in a narrow band between Lake Chad and the Nile Valley. The origins of Chad's peoples, however, remain unclear. Several of the proven archaeological sites have been only partially studied, and other sites of great potential have yet to be mapped.

"Toward the end of the first millennium *A.D.*, the formation of states began across central Chad in the *Sahelian* zone between the desert and the savanna. For almost the next 1,000 years, these states, their relations with each other, and their effects on the peoples who lived in 'stateless' societies along their peripheries dominated Chad's political history. Recent research suggests that indigenous Africans founded most of these states, not migrating Arabic-speaking groups, as was believed previously. Nonetheless, immigrants, Arabic-speaking or otherwise, played a significant role, along with Islam, in the formation and early evolution of these states.

"Most states began as kingdoms, in which the king was considered divine and endowed with temporal and spiritual powers. All states were militaristic (or they did not survive long), but none was able to expand far into southern Chad, where forests and the tsetse fly complicated the use of cavalry. Control over the trans-Saharan trade routes that passed through the region formed the economic basis of these kingdoms. Although many states rose and fell, the most important and durable of the empires were Kanem-Borno, Bagirmi, and Wadai, according to most written sources (mainly court chronicles and writings of Arab traders and travelers).

"European interest in Africa generally grew during the nineteenth century. By 1887 France, motivated by the search for wealth, had driven inland from its settlements on central Africa's west coast to claim the territory of Ubangi-Chari (present-day Central African Republic). It claimed this area as a zone of French influence, and within two years it occupied part of what is now southern Chad. In the early 1890s, French military expeditions sent to Chad encountered the forces of Rabih Fadlallah, who had been conducting slave raids (*razzias*) in southern Chad throughout the 1890s and had sacked the settlements of Kanem-Borno, Bagirmi, and Wadai. After years of indecisive engagements, French forces finally defeated Rabih Fadlallah at the Battle of Kousséri in 1900" (*Library of Congress Country Studies,* 1988).

"Chad, part of France's African holdings until 1960, endured three decades of civil warfare, as well as invasions by Libya, before a semblance of peace was finally restored in 1990. The government eventually drafted a democratic constitution and held flawed presidential elections in 1996 and 2001. In 1998, a rebellion broke out in northern Chad, which has sporadically flared up despite several peace agreements between the government and the insurgents. In 2005, new rebel groups emerged in western Sudan and made probing attacks into eastern Chad despite signing peace agreements in December 2006 and October 2007. In June 2005, President Idriss Deby held a referendum successfully removing constitutional term limits and won another controversial election in 2006. Sporadic rebel campaigns continued throughout 2006 and 2007. The capital experienced a significant insurrection in early 2008, but has had no significant rebel threats since then, in part due to Chad's 2010 rapprochement with Sudan, which previously used Chadian rebels as proxies. Deby in 2011 was reelected to his fourth term in an election that international observers described as proceeding without incident. Power remains in the hands of an ethnic minority. In January 2014, Chad began a two-year rotation on the UN Security Council" (*World Factbook*, 2014).

"Trig lists were compiled by the French Equatorial Africa (*Afrique Équatoriale Française*, AEF) but these were of the astros that the French used to position their 1:200,000 series. The 12th Parallel Traverse ran through the country from near N'Djamena to the Sudanese triangulation near Geneina in Sudan. Temporary coordinates would have been computed on the Sudan datum (*Blue Nile Datum with origin at Adindan – Ed.*). Upon completion of the traverse from Dakar, the whole would have been computed on a common datum, I don't remember which but probably Adindan. Go to Wikipedia for N'Djamena, scroll to Gallery. What does the temporary radio station look like to you? Possibly an early satellite tracking station. Observatory in the background. Brand new military tires on the trailer axels. Back to AEF trig lists. One station was located at Ft. Lamy in a 'Tata.' Check that in your French dictionary and you probably won't find anything reasonable. It is a fortified Sudanese encampment (and also the central fort) where a chieftain lives with his family, his herd, *etc.*" (*Personal Communication, John W. Hager, 23 May 2014*).

In regard to Mr. Hager's comments, that's a BC-4 satellite station. The 12-inch focal length camera case is on the ground next to the little observatory. (I have one of those BC-4 cameras myself.) Since the BC-4 system had to be synchronized into a network, that's what the electronics hut was for. They worked at 600 cycles/second. While I was stationed at Army Map Service & TOPOCOM, I went over to NOAA in Rockville, MD to see the system they had worked up for the program. It was designed by Dr. Helmut Schmid, a German V-2 Rocket Scientist that had originally designed the BC-4 cameras for his missile tracking work at Aberdeen Proving Grounds. (Dr. Duane C. Brown was his protégé.) I was introduced to Dr. Schmid who was so impressed by a Corps of Engineers Captain that he looked me up and down, grunted, and walked away!

The Adindan Datum origin is in the *Blue Nile* Province of Sudan, and sometimes it is called the Blue Nile Datum. Referenced to the Clarke 1880 ellipsoid, where $a=6,378,249.145\,m$, $1/f=293.465$; the origin is at: $\Phi_o=22°\ 10'\ 07.1098''$ N, $\Lambda_o=31°\ 29'\ 21.6079''$ E. According to *TR8350.2*, the transformation parameters **from** Adindan Datum **to** WGS84 for Sudan are: $\Delta X=-161\,m\pm25\,m$, $\Delta Y=-14\,m\pm25\,m$, $\Delta Z=+205\,m\pm25\,m$. Based on observations at 14 stations in Sudan, there are no parameters published specifically for Chad. Being a former French Colony, the UTM Grid System is likely the only *operational* plane coordinate system ever used for Chad.

CHILE

Physiography

Republic of Chile

According to the *World Factbook*, Chile is slightly smaller than twice the size of Montana, it is bordered by Argentina (5,150 km), Bolivia (861 km), Peru (160 km), and on the west by the South Pacific Ocean (6,435 km). The lowest point is the Pacific Ocean (0 m), and the highest point is Cerro Aconcagua (6,962 m). Chile has low coastal mountains with a fertile central valley and the rugged Andes in the east; the climate is temperate with desert in the north and is cool and damp in the south. Chile also claims Easter Island (*Isla de Pascua*) and *Isla Sala y Gomez*.

Originally inhabited by the Araucanians, the northern part of Chile was conquered by the Incas during the 15th century *A.D.* The Spanish Conquistador, Diego de Almagro invaded in 1536–1537, and Pedro de Valdivia began the settlement of Santiago in 1541 under the viceroyalty of Peru. The independence of Chile was assured by the victory of the soldier and statesman Jose de San Martín at Maipo in 1818, thanks to *Webster's New Geographical Dictionary*.

In 1881, the Army General Staff was reorganized, and the Army Geographic Service *(Servicio del Ejercito Geográfico)* was created. Surveys and 1:25,000 scale planetable mapping commenced in 1891 with the establishment of the *Oficina Geográfica*. A variety of ellipsoids have been employed in the Republic of Chile, the original one used by the Army was the Bessel 1941, where $a = 6,377,397.155$ m and $1/f = 299.1528128$. The northern part of Chile that includes Iquique, Tocopilla, and Antofagasta comprises the *Region Salitrera* or Saltpeter Region where nitrates have been historically mined. Originally won by conquest from Bolivia and Peru, the classical triangulation in that northern region was computed on the Clarke 1866 ellipsoid, where $a = 6,378,206.4$ m and $b = 6,356,583.8$ m. The fundamental origin point was at the now destroyed astronomical observatory of Cerro Santa Lucía. The Catastro Salitrero Transverse Mercator (Gauss-Krüger) grid had a Central Meridian $(\lambda_o) = 69°\ 38'\ 46.52''$ West of Greenwich, a Scale Factor at Origin $(m_o) =$ unity, a False Northing Latitude of Origin $(\varphi_{FN}) = 20°\ 37'\ 15.05''$ South, a False Easting = nil, and a False Northing = 1,420,472.60 m. (The grid origin was not at the old observatory of Cerro Santa Lucía but specifically at "Pilar de Pintado," the 1st order triangulation station of the net.) The azimuth from Pintados to Noria was: $\alpha_o = 323°\ 58'\ 15.18''$ reckoned from the North. According to a memo to the U.S. Army Map Service and the Inter-American Geodetic Survey in 1952 from COL. Rodolfo Concha Muñoz, "the observations were made in Iquique and a geodetic connection was made between Iquique and Pintados-Astronómico." The chain of triangulation figures span latitudes from 18° S to 26° S, and curiously were published on the Normal Mercator projection, according to an *Instituto Geografico Militar* report to the International Union of Geodesy and Geophysics in September of 1936. Chile south of 26° S was published on the Gauss-Krüger Transverse Mercator projection and was referenced to the International 1924 ellipsoid, where $a = 6,378,388$ m and $1/f = 297$. The old triangulation of Chile was based on an origin point at the (now destroyed) old astronomical observatory of Santiago at Quinta Normal. The IGM Gauss-Krüger Transverse Mercator projection reported to have been used from 1858 to 1924 on the Bessel 1841 ellipsoid had a Central Meridian $(\lambda_o) = 70°\ 41'\ 16.894''$ West of Greenwich, a Scale Factor at Origin $(m_o) =$ unity, a False Northing Latitude of Origin $(\varphi_{FN}) = 33°\ 26'\ 42.453''$ South, and a False Easting and False Northing = nil. This grid origin I believe corresponds to Quinta Normal. A revised fundamental point of origin at the Reppsol transit telescope was later constructed a few meters away from the old origin point.

Using a grid system adopted in 1924, the Chilean IGM TM Belts 1–3 referenced to the International 1924 ellipsoid had for Belt 1: $\lambda_o = 69°$ W, False Easting (FE) = 1,500 km, Belt 2: $\lambda_o = 72°$ W, (FE) = 2,500 km, and Belt 3: Belt 2: $\lambda_o = 75°$ W, (FE) = 3,500 km.

LTC. Mario Ugarte Olea, Technical Deputy Director of IGM, sent a detailed letter to me on July 9, 1982, in which he detailed operational military mapping of his nation. For 1:50,000 scale mapping, the Provisional South American Datum of 1956 with origin at La Canoa, Venezuela, was used for Chilean mapping between the latitudes of 17° 30′ S and 43° 30′ S. Note that the origin point at La Canoa, Anzoátegui Province, Venezuela is where $\Phi_o = 08°\ 34'\ 17.170''$ North, $\Lambda_o = -63°\ 51'\ 34.880''$ West of Greenwich, and the defining azimuth to station Pozo Hondo $(\alpha_o) = 40°\ 22'\ 45.96''$. Note that the PSAD56 position of "Pilar de Pintado" (Cerro Santa Lucía as listed above) is: $\varphi = 20°\ 37'\ 33.092''$ South, $\lambda = 69°\ 39'\ 21.807''$ West of Greenwich, thanks to John W. Hager in a note to me on July 1, 1982.

From the latitudes of 43° 30′ S and 56° 00′ S, the South American Datum of 1969 was used where the origin at station Chua is $\Phi_o = 19°\ 45'\ 41.6527''$ South, $\Lambda_o = 48°\ 06'\ 04.0639''$ West of Greenwich. Other scales of mapping compiled by IGM were reported to be referenced to the UTM grid and the coverage was based on the same basic latitudinal divisions for classical geodetic datums.

The SIRGAS project is a continental effort to convert all local South American datums to the WGS84 Datum. Originally conceived and spearheaded by Dr. Muneendra Kumar, now retired as the Chief Geodesist of DMA/NIMA/NGA, this system of observations and adjustments will eventually bring the entire continent of South America into a unified coordinate reference system. Chile is an active participant in this international project, and transformation parameters will likely be publicly available within the next few years.

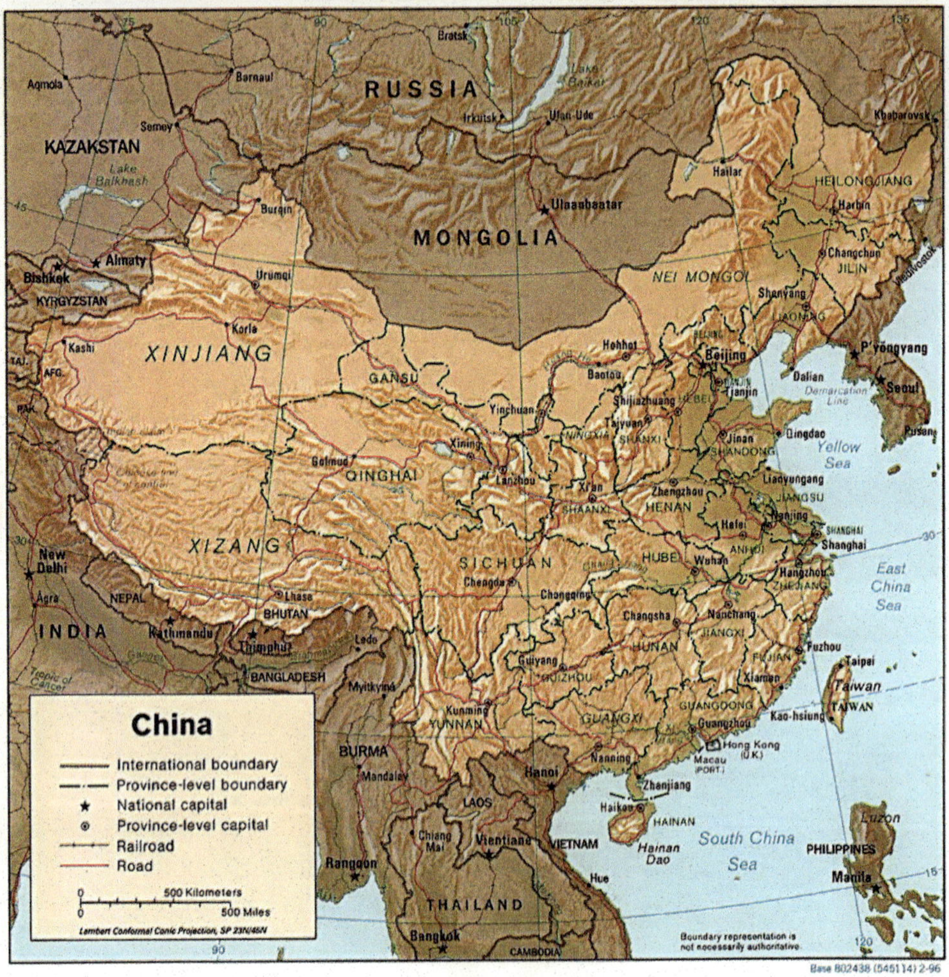

Base 802438 (545114) 2-96

The People's Republic of China

China is bordered by Afghanistan (91 km), Bhutan (477 km), Burma (2,129 km), India (2,659 km), Kazakhstan (1,765 km), North Korea (1,352 km), Kyrgyzstan (1,063 km), Laos (475 km), Mongolia (4,630 km), Nepal (1,389 km), Pakistan (438 km), Russia (northeast) (4,133 km) and Russia (northwest) (46 km), Tajikistan (477 km), and Vietnam (1,297 km). The terrain is mostly mountains, high plateaus, deserts in west, plains, deltas, and hills in east. The highest point is Mount Everest (8,849 m), the lowest point is *Turpan Pendi* (Turfan Depression) (–154 m), and the mean elevation is (1,840 m).

The Chinese civilization spread originally from the Yellow River valley where it probably existed about 3000 B.C. The first valid historical evidence is of the Chou Dynasty (1122–255 B.C.). For most of its 3,500 years of history, China led the world in agriculture, crafts, and science. It fell behind in the 19th century when the Industrial Revolution gave the West superiority in military and economic affairs.

China has 23 provinces, 5 autonomous regions, and 4 municipalities for its administrative divisions. Note that China considers Taiwan as its 23rd province, and it has recently acquired two other special administrative regions: the British Colony of Hong Kong, and the Portuguese Colony of Macau.

The father of scientific cartography was Pei Hsiu (224–271 A.D.). In 1707–1717, Emperor K'anghsi commissioned a group of Jesuits to carry out a survey of the Chinese Empire. In 1902, the Manchu government established the Military Survey Institute and a 1:1,000,000 map series was planned. By 1911, the revolution changed the political horizons, and no noticeable progress was made until 1927 when the first modern series of geodetic surveys were performed by the military until the Sino–Japanese war of 1937–1945. In 1928, the Central Bureau of Land Survey (CBLS) under the Army General Staff was formed, and First-Order triangulation began in 1929 in Chekiang province. In 1930, the CBLS organized the first training class in photogrammetry in Nanking. In 1931, large-scale triangulation projects were also started in Anhwei, Hunan, Hupeh, Kiangsi, and Kiangsu provinces. That same year CBLS started flying aerial photography with one airplane, and by 1935 it was using seven airplanes with cameras using 135 mm and 210 mm focal length lenses.

The Japanese established the Manchurian Principal System (Datum) of 1933 at the origin point, Huan-his-ling (*Shinkyo*), where $\Phi_o = 43°\ 49'\ 36.62''$ North, $\Lambda_o = 125°\ 18'\ 15.42''$ East of Greenwich, and the defining azimuth to station Ta-hei-shan is $\alpha_o = 204°\ 46'\ 54.497''$. Of course, according to Japanese tradition established with the Tokyo Datums of 1892 and 1918, the Bessel ellipsoid of 1841 was referenced where the semi-major axis $(a) = 6,377,397.155$ m and the reciprocal of flattening $(1/f) = 299.1528128$. *(See Korea – Ed.)*. Back in June of 1980, Mr. Frank Kuwamura of the Defense Mapping Agency wrote to me in which he offered that: "If we denote the Tokyo Datum of 1918 with a 'T' subscript, and the Manchurian System of 1933 with a 'M' subscript, then $\phi_T = \phi_M + \phi'$ and $\lambda_T = \lambda_M + \lambda'$, where $\phi' = -8.8386'' + 0.00430\ \Delta\phi \cdot 10^{-4} - 0.43463\ \Delta\lambda \cdot 10^{-4} + 0.00001\ \Delta\phi^2 \cdot 10^{-8} + 0.00021\ \Delta\phi\Delta\lambda \cdot 10^{-8} + 0.01041\ \Delta\phi^2 \cdot 10^{-8} + 0.0017\ \Delta\phi^3 \cdot 10^{-12}$, and $\lambda' = 17.8824'' + 0.82927\ \Delta\phi \cdot 10^{-4} - 0.40987\ \Delta\lambda \cdot 10^{-4} + 0.03866\ \Delta\phi^2 \cdot 10^{-8} - 0.03971\ \Delta\phi\Delta\lambda \cdot 10^{-8} - 0.01008\ \Delta\phi^2 \cdot 10^{-8} + 0.002454\ \Delta\phi^3 \cdot 10^{-12} - 0.00186\ \Delta\phi^2\Delta\lambda \cdot 10^{-12} - 0.000978\ \Delta\phi\Delta\lambda^2 \cdot 10^{-12} + 0.0016\ \Delta\phi^3 \cdot 10^{-12}$. Furthermore, $\Delta\phi = \phi_M - \phi_o$ and $\Delta\lambda = \lambda_M - \lambda_o$ where the units are in arc seconds and ϕ_o and λ_o represent the coordinates of the Manchurian System of 1933 origin point (listed above)."

The CBLS established the Nanking Datum of 1935, where $\Phi_o = 32°\ 04'\ 19.7445''$ North, $\Lambda_o = 118°\ 50'\ 18.5354''$ East of Greenwich. A Gauss-Krüger Transverse Mercator Grid is defined at the Datum origin. The scale factor at origin $(m_o = 1.0)$; the False Easting and the False Northing = zero. The ellipsoid of reference is the International (also called the Hayford 1909 and the Madrid 1924) where the semi-major axis $(a) = 6,378,388$ m and $1/f = 297$.

In addition to the CBLS surveys and maps, there were myriad local surveying and mapping activities pursued throughout the 20th century in the People's Republic of China. For instance, the Chihli River Commission used a Grid for 1:50,000 maps where the Central Meridian is 116° 25′ 24″ East of Greenwich, the Central Parallel is 37° 20′, the False Easting and False Northing=zero. What projection? Probably a Lambert Conformal Conic because of a paper published by J. T. Fang of the National Geological Survey of China. In 1949, J. T. Fang published a series of papers in *Empire Survey Review* concerning the Lambert Conformal Projection as applied to China. "It has been decided by the Central Land Survey of China to adopt the Lambert conformal projection as the basis for the co-ordinate system, and, in order to meet the requirements of geodetic work, the whole country is subdivided into eleven zones bounded by parallels including a spacing of 3½ degrees in latitude-difference. To each of these zones is applied a Lambert projection, properly chosen so as to fit it best. The two standard parallels of the projection are situated at one-seventh of the latitude-difference of the zone from the top and bottom. Thus, the spacing between the standard parallels is 2½ degrees. This gives a maximum value of the scale factor of less than one part in four thousand, thus reducing the distortions of any kind to a reasonable amount." Fang later went on to explicitly list some of the parameters of the "Fifth Zone" where, "The standard parallels of this zone are at latitudes 34° 10′ N and 36° 40′ N. Thus, φ_o=35° 25′ 11.84746″ as referenced to the International ellipsoid (also called the Hayford 1909 and the Madrid 1924) where a=6,378,388 meters, $1/f$=297." Example computations are given by Fang for Fourth and Fifth Zone transformations.

In the late 1970s, I had the bright idea of going into business as a consulting cartographer/geodesist. I moved my family back home to New Orleans and started pursuing the "oil patch" clientele. Lo and behold, I received a telephone call from Houston about the South China Sea. The People's Republic of China tendered bids for the exploration and development of hydrocarbon resources (oil and gas) in the South China Sea and was looking for companies to perform geophysical exploration of its outer continental shelf. A U.S.-owned company (identity to remain anonymous) "cooked up" a specification for a projection and Grid System for geophysical exploration in the South China Sea. The ellipsoid (and presumably the Datum) was the World Geodetic System of 1972. However, the projection was chosen as the Lambert Conformal Conic with two standard parallels and a latitude of origin equal to the arithmetic mean of the standard parallels. That sort of thing will work on a sphere, but on an ellipsoid (WGS72), it is a mathematical impossibility!

I worked up the two different Grids based on the two exclusive presumptions: hold to the two standard parallels and let the latitude of origin "float," or hold to the latitude of origin and let the two standard parallels "float." I termed those two mathematical possibilities as "PRC South China Sea I" and "PRC South China Sea II." To this day, I get phone calls to the effect, "Hey Cliff, ever hear about PRC South China Sea _XX_?" It never ceases to amaze me that they actually find (and produce) oil out there …

There are some traditional Grid Systems associated with China. "China Belts I and II" are Gauss-Krüger Transverse Mercator Grids referenced to the Clarke 1880 ellipsoid (ersatz World War II systems), where the Central Meridians are at 119° (Belt I) and 113° (Belt II). Scale factor at origin=0.9994, the False Northing=−2,210,000 m and the False Easting=400 km. These specifications are part of the "British Grids," and although richly romantic in history; they are lacking in *provenance*.

Current Grid Systems attributed to the People's Republic of China find their roots in the Russian (USSR) origins of assistance. For instance, the Russia Belts for China are identical with the UTM specification with the exceptions or variations with respect to the scale factor at origin is unity rather than 0.9996, and the ellipsoid of reference is the Krassovsky 1940, where the semi-major axis (a)=6,378,245 m and $(1/f)$=298.3. A variation on this is known as the 3° Belts, and the location of the Central Meridians are simply a (half) scalar of the 6° belts.

"Surveying & mapping datum is a key infrastructure of national economy, social development, state security and information construction, an important foundation for determining natural geographic elements and the geometric configuration and spatial & temporal distribution of earth surface artificial facilities, and the initial numerical data for various surveying tasks and the basis reference for accurate demonstration of the geographic space distribution in the real world by making use of map. China's surveying & mapping datum consists of geodetic datum, vertical datum, sounding datum and gravity datum.

"Since its founding in 1949, with a view of meeting the requirements of economic construction, the People's Republic of China has set up China Geodetic Coordinate System (Beijing Coordinate System, CGCS 1954), Huanghai Vertical Datum 1956 and Gravity Fundamental Network 1957, which represent the first-generation datum reference system for surveying and mapping of China.

"Since 1980s, China has strengthened upgrading and reconstruction of the first-generation datum reference system for surveying and mapping and gradually built a national horizontal control network consisting of 48,000 points: namely, China Geodetic Coordinate System 1980 (Xi'an Coordinate System *where the ellipsoid of reference is the IUGG 1975 and $a = 6,378,140 m$ and $1/f = 298.256$, $\Phi_o = 34° 32' 27'' N$, $\Lambda_o = 108° 55' 25'' E$, [Shiisi Village, Yongle Township, Jingang County, Shaanxi Province] – Ed.*), which has a significantly-improved precision than CGCS 1954; it has completed the National Vertical Datum System 1985 comprised of the Class-A national leveling network with 100 loops and a total distance of 93,000 kilometers and the Class-B national leveling network with a total distance of 136,000 kilometers. Compared with the Huanghai Vertical Datum System 1956, National Vertical Datum System 1985 features higher density, higher precision, and more rational structure; moreover, China has completed its National Gravity Fundamental Network 1985, which is comprised of 6 gravimetric datum points, 46 basic gravimetric points, and 163 first-class gravimetric points. As the second-generation datum reference system for surveying and mapping of China, Gravity Fundamental Network 1985 features higher density and precision compared with Gravity Fundamental Network 1957.

"Since 1990s, China has further accelerated the modernization of the datum reference system for surveying and mapping. In 1997, China completed its high-precision national GPS Network A and Network B, realizing the nationwide coverage with three-dimensional geocentric coordinates, 2 orders of magnitude higher than the national horizontal control network 1980. In 2003, China built the national GPS geodetic control network 2000 that is comprised of 2,500 points. In 2004, China built the national geodetic control network 2000, which consists of approximately 50,000 points and enjoys much higher positioning precision than ever. In 2003, China built the National Fundamental Gravity Network 2000, which is a new-generation national gravity datum comprised of 19 datum points and 119 fundamental points. Compared with National Fundamental Gravity Network 1985, National Fundamental Gravity Network 2000 features higher precision and more rational density and distribution of points. In 2001, China set up its first decimeter-level quasi-geoid model (CQG2000), on which the sea level elevation can be determined rapidly by making use of satellite space positioning technologies and geoid model, thereby providing a supplementary approach to traditional leveling height survey within a certain range of precision.

"So far, China has basically set up a relatively complete datum reference system for surveying and mapping, including national GPS geodetic control network 2000 comprised of over 2,500 points and national geodetic control network 2000 comprised of approximately 50,000 points, as well as the national fundamental gravity network comprised of 19 datum points and 119 fundamental points, marking the start of the construction of the modern Chinese datum reference system for surveying and mapping.

"Since 2006, China has further accelerated optimization and upgrading of the existing datum reference system for surveying and mapping, focused on replacing the traditional reference ellipsoid-centric and two-dimensional coordinate system with the geocentric and three-dimensional coordinate system, intensified the construction of the reference station for satellite positioning continuous operation and connected it to the international coordinate reference framework and maintained the coordinate framework through the reference station and the satellite geodetic control network; meanwhile, China has attached greater importance to acceleration of the geoid refinement, improvement of the resolution and precision of geoids as well as gradual replacement of traditional national high-level leveling; in addition, China has strived to increase the quantity of ground absolute gravimetric points, focused on speeding up the development of measurement of airborne gravity and satellite gravity and enriching gravimetric data with different scales.

"On July 1, 2008, the high-precision, geocentric, three-dimensional, and dynamic national geodetic coordinate system 2000, which is China's new-generation geodetic coordinate system, was officially put into operation. Now, the state-level fundamental surveying & mapping results have been transferred to the National Geodetic Coordinate System 2000, and the existing fundamental surveying & mapping results have been or have been basically shifted to the Geodetic Coordinate System 2000 in Zhejiang, Gansu, Jiangxi, Fujian, Guangdong, Shandong, Henan, Beijing, and Shanghai. The construction of satellite positioning continuous operation reference stations has been unfolded step by step. The satellite navigation positioning continuous operation reference station network with centimeter-level real-time positioning precision has been completed or is being built in 24 provinces (autonomous regions, municipalities), and over 1,200 continuous operation reference stations have been newly built. The work related to quasi-geoid refinement has been carried out in 26 provinces (autonomous regions, municipalities), of which 25 have completed the work and 23 have reached the centimeter-level precision. Based on the quasi-geoid refinement, GPS C network construction and Class-C leveling survey have been carried out in most provinces, municipalities, and autonomous regions.

"In June 2012, the project of infrastructure construction of the national modern datum reference system for surveying and mapping (phase I), which is a major special surveying & mapping project of China during the 12th "Five-year Plan" period, was officially kicked off. The project is expected to, in the next 4 years, build 150 reference stations, reform 60 GNSS positioning continuous operation reference stations and make direct use of 150 reference stations, thus forming a national satellite positioning continuous operation datum network consisting of 360 reference stations; build 2,500 satellite geodetic control points, make direct use of 2000 points, thus forming a 4500-point national satellite geodetic control network, which constitutes the new-generation geodetic datum framework together with the national satellite positioning continuous operation datum network; build and rebuild 27,400 height control points, lay 110 leveling bedrock points, deploy the national Class-A leveling network with a length of 122 thousand kilometers and form the national modern vertical datum framework; deploy 50 national gravimetric datum points, improve the national gravity datum infrastructure; build 1 national surveying & mapping datum data center and form the national modern surveying & mapping datum management service system. Through the implementation of infrastructure construction of the national modern datum reference system for surveying and mapping, China plans to complete a high-precision, geocentric, dynamic, practical and uniform national datum reference system for surveying and mapping in 2015.

"At the end of 2012, Beidou (COMPASS) navigation satellite system covered the most areas in the Asia-Pacific region and was officially put in commercial operation. And construction and maintenance of the modern datum reference system for surveying and mapping that is based on GNSS (GPS, GLONASS, GALILEO and COMPASS) have started in some areas in China." (Approved by: National Administration of Surveying, Mapping and Geoinformation of China 28 Lianhuachi West Road, Haidian District, Beijing, 100830, China, 2013-12-20.)

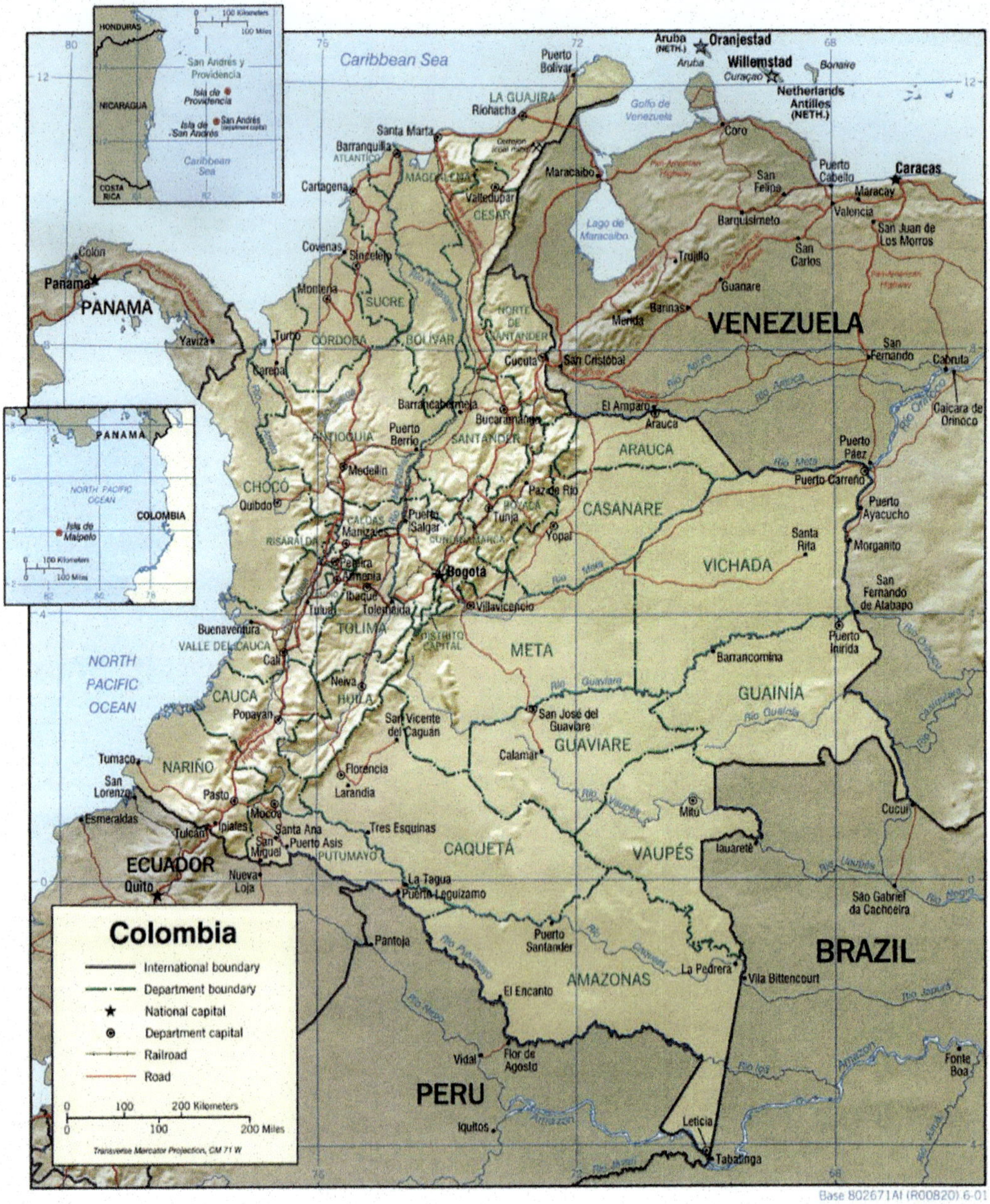

Base 802671AI (R00820) 6-01

Republic of Colombia

Colombia is bordered by Brazil (1,790 km), Ecuador (708 km), Panama (339 km), Peru (1,494 km), and Venezuela (2,341 km). Colombia has flat coastal lowlands, central highlands, high Andes Mountains, and eastern lowland plains (Llanos). Its highest point is Pico Cristobal Colon (5,730 m), its lowest point is the Pacific Ocean (0 m), and its mean elevation is (593 m).

The City of Santa Fé de Bogotá was founded in 1538, the region *Gran Colombia* (modern Panamá, Colombia, Venezuela, and Ecuador) achieved independence from Spain in 1819 and reorganized as a republic in 1886. "A decades-long conflict between government forces, paramilitaries, and antigovernment insurgent groups heavily funded by the drug trade, principally the Revolutionary Armed Forces of Colombia (FARC), escalated during the 1990s. More than 31,000 former United Self Defense Forces of Colombia (AUC) paramilitaries demobilized by the end of 2006, and the AUC as a formal organization ceased to operate. In the wake of the paramilitary demobilization, illegal armed groups arose, whose members include some former paramilitaries. After four years of formal peace negotiations, the Colombian Government signed a final peace accord with the FARC in November 2016, which was subsequently ratified by the Colombian Congress. The accord calls for members of the FARC to demobilize, disarm, and reincorporate into society and politics. The accord also committed the Colombian Government to create three new institutions to form a 'comprehensive system for truth, justice, reparation, and non-repetition,' to include a truth commission, a special unit to coordinate the search for those who disappeared during the conflict, and a 'Special Jurisdiction for Peace' to administer justice for conflict-related crimes. The Colombian Government has stepped up efforts to expand its presence into every one of its administrative departments. Despite decades of internal conflict and drug-related security challenges, Colombia maintains relatively strong democratic institutions characterized by peaceful, transparent elections and the protection of civil liberties" (*World Factbook*, 2022).

The *Instituto Geográfico Militar* was founded in 1935 and later renamed in 1950 as the *Instituto Geográfico "Agustín Codazzi" – IGAC*, the national mapping agency of Colombia that is responsible for all civilian and military mapping. The cartographer, Agustín Codazzi, produced early 19th-century maps of *Gran Colombia*. In fact, these famous maps have even appeared on Venezuelan postage stamps. Furthermore, IGAC provides training services in the mapping sciences for many Latin American countries, a service previously provided solely by the U.S. Inter-American Geodetic Survey (IAGS).

The first classical geodetic datum established in Colombia is the Bogotá Datum of 1941. The defining parameters are referenced to the International ellipsoid (also called the Hayford 1909 and the Madrid 1924), where $a = 6,378,388$ m and $1/f = 297$, with an origin of $\Phi_o = 4°\ 35'\ 56.57''$ N, $\Lambda_o = 74°\ 04'\ 51.30''$ W, and $h_o = 2,633.6$ m, corresponding to the National Astronomic Observatory in Bogotá as of 1935. The azimuth to station Suba, $\alpha_o = 359°\ 30'\ 10.0''$, and the initial defining adjustment of the datum included 5 invar baselines, 133 geodetic stations, 12 astro stations, and "the ellipsoid normal is coincident with the direction of the plumb line."

In 1946, the country was divided into four tangent meridian belts, each belt 3° wide as measured from the observatory. For each belt, the False Eastings and False Northings are equal to 1,000 km each at the intersections of the central meridians (λ_o) and the latitude of the observatory. The belts are lettered as "B" (Bogotá), "Ec" (East central or Este central), "E" (East or Este), and "O" (West or Oeste). There may be a new belt "OO" (Western West or Oeste Occidente) to cover *Isla del Malpelo* in the Pacific Ocean and *Isla de San Andres* in the Caribbean Sea. The projection system used in Colombia by definition is the Gauss Conform Transverse Mercator, a particular truncation of the Gauss Schreiber which was used in the United States for the North American Datum of 1927.

The boundary treaty between Colombia and Panamá was ratified and mutually exchanged on January 31, 1925. There is no indication of the geodetic datum used to provide the geodetic coordinates listed to a tenth of an arc second. Fortunately, the 14 boundary points are physically marked and are presumably recoverable.

The boundary history between Colombia and Brazil is considerably older, dating back to 1494. The latest treaty with Brazil was signed in 1928, but the coordinates of the boundary markers were not published until 1946, presumably on the Bogotá Datum of 1941.

The land boundary with Venezuela is apparently stable; it is straddled largely by a continuous chain of quadrilaterals. However, the territorial limits of the sea seem to be in dispute, particularly with respect to fishing rights.

Other datums that exist in Colombia (at least theoretically) include the Provisional South American Datum of 1956 (PSAD 56) and the South American Datum of 1969 (SAD 69). Both PSAD 56 and SAD 69 seem to be largely ignored by Colombia although both datums are extended into other countries through Colombia and are extensively used elsewhere. The majority network of Colombia, which had consisted of 21 fixed stations and 679 adjustable stations, was adjusted to the SAD 69 in 1972. However, a free adjustment was also done in 1972 to the total network, which consisted of 1 fixed station and 950 adjustable stations. This free adjustment appears to be the system currently in use under the original name of the Bogotá Datum of 1941. The World Geodetic System of 1984 datum is used extensively in Colombia, albeit just a tool to extend the classical datum.

In 1980, IGAC established a special "city grid" projection for the City of Santa Marta, the oldest city in Colombia (1525). This special system was similar in philosophy to the systems developed for counties in Wisconsin and Minnesota for the convenience of the local surveyors and GIS applications. Specifically, this "city grid" was established at a particular elevation that corresponded to the average elevation of the city. Local surveyors could then ignore the reduction to sea level correction when transferring measured field distances to the city maps. The system was still based on the Gauss Conform Transverse Mercator, but the False Eastings and False Northings were modified by truncating to only five digits from the observatory at Bogotá.

The reference ellipsoid (International) was modified to have its semi-major axis increased by 29 m, corresponding to the average elevation of the city above mean sea level determined at Buenaventura. Curiously, the system was reversed back to the standard national coordinate system in 1994, while other cities were published with their own special plane coordinate systems such as Villavicencio 1994 (450 m), Leticia 1994 (100 m), and Armenia 1995 (1,470 m).

Although the Colombian definition uses the Gauss Conform Transverse Mercator for these systems, the Local Space Rectangular sometimes used in analytical photogrammetric calculations could just as easily be used with identical results. We commonly use such systems for analytical rectification projects if the project area is free of relief such as a plateau or a flood plain.

Note that these new city systems were orthophoto projects and not just rectified photos. Grids used in Colombia without special modifications to the ellipsoid include the Ciudad de Bogotá (City) Grid and the Ciudad de Medellín (City) Grid. The Grid System used for the Ciudad de Cartagena is the standard "B" belt of the National System.

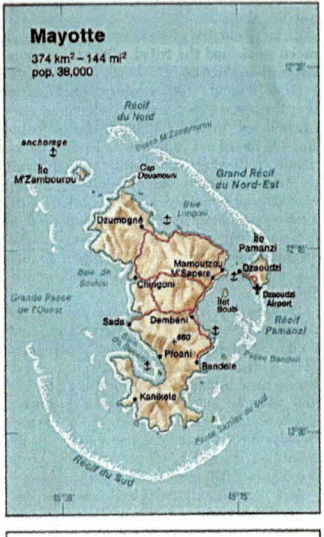

Grande Comore
1,148 km² – 443 mi²
pop. 149,000

Anjouan
424 km² – 164 mi²
pop. 95,000

Mayotte
374 km² – 144 mi²
pop. 38,000

Moheli
290 km² – 112 mi²
pop. 12,000

★ National capital
elevations in meters
Scale 1:700,000

All-weather road
Seasonal road

0   5   10   15 Miles
0   5   10   15 Kilometers

Federal Islamic Republic of the Comores

The Comores are in southern Africa, a group of islands at the northern mouth of the Mozambique Channel, about two-thirds of the way between northern Madagascar and northern Mozambique and slightly more than 12 times the size of Washington, DC. Comprised of volcanic islands, the interiors vary from steep mountains to low hills where the highest point is Karthala (2,360 m) and the lowest point is the Indian Ocean (0 m) (*World Factbook*, 2022).

According to the Library of Congress Country Studies, the Comoros is an archipelago comprised of Grand Comore (Njazidja), Anjouan (Nzwani), and Mohéli (Mwali). Mahoré (Mayotte) is administered by France but is claimed by the Comoros. The islands are situated in the western Indian Ocean, about midway between the island of Madagascar and the coast of East Africa at the northern end of the Mozambique Channel. The archipelago has served in past centuries as a steppingstone between the African continent and Madagascar, as a southern outpost for Arab traders operating along the East African coast, and as a center of Islamic culture. The name "Comoros" is derived from the Arabic *kamar* or *kumr*, meaning "moon," although this name was first applied by Arab geographers to Madagascar. In the 19th century, Comoros was absorbed into the French overseas empire, but it unilaterally proclaimed independence from France on July 6, 1975. The island republic has since had 19 coups or coup attempts since their independence. "In closely contested elections in 2016, former President Azali Assoumani won a second term, when the rotating presidency returned to Grande Comore. A new July 2018 constitution removed the presidential term limits and the requirement for the presidency to rotate between the three main islands. In August 2018, President Azali formed a new government and subsequently ran and was elected president in March 2019" (*World Factbook*, 2022).

Little is known of the first inhabitants of the archipelago, although a 6th-century settlement has been uncovered on Nzwani by archaeologists. Historians speculated that Indonesian immigrants used the islands as stepping-stones on the way to Madagascar prior to *A.D.* 1000. Because the Comoros lay at the juncture of African, Malayo–Indonesian, and Arab spheres of influence, the present population reflects a blend of these elements in its physical characteristics, language, culture, social structure, and religion. Local legend cites the first settlement of the archipelago by two families from Arabia after the death of Solomon. Legend also tells of a Persian king, Hussein ibn Ali, who established a settlement on Comoros around the beginning of the 11th century. Bantu peoples apparently moved to Comoros before the 14th century, principally from the coast of what is now southern Mozambique; on the island of Nzwani they apparently encountered an earlier group of inhabitants, a Malayo–Indonesian people. A number of chieftains bearing African titles established settlements on Njazidja and Nzwani, and by the 15th century they probably had contact with Arab merchants and traders who brought the Islamic faith to the islands. A watershed in the history of the islands was the arrival of the Shirazi Arabs in the 15th and 16th centuries.

The first Europeans to visit the islands were the Portuguese, who landed on Njazidja around 1505. The islands first appear on a European map in 1527, by Portuguese cartographer Diogo Roberos. Dutch 16th-century accounts describe the Comoros sultanates as prosperous trade centers with the African coast and Madagascar. Intense competition for this trade, and, increasingly, for European commerce, resulted in constant warfare among the sultanates, a situation that persisted until the French occupation. The sultans of Njazidja only occasionally recognized the supremacy of one of their number as *tibe*, or supreme ruler. By the early 17th century, slaves had become Comoros' most important export commodity, although the market for the islands' other products also continued to expand, mainly in response to the growing European presence in the region. To meet this increased demand, the sultans began using slave labor themselves following common practice along the East African coast. Beginning in 1785, the Sakalava of the west coast of Madagascar began slaving raids on Comoros. They captured thousands of inhabitants and carried them off in outrigger canoes to be sold in French-occupied Madagascar, Mauritius, or Réunion to work on the sugar plantations, many of which French investors owned. The island of Mahoré, closest of the group to Madagascar, was virtually de-populated. Comoran pleas for aid from the French and the other European powers went unanswered, and the raids ceased only after the Sakalava kingdoms were conquered by the Merina of Madagascar's central highlands. After the Merina conquest, groups of Sakalava and Betsimisaraka peoples left Madagascar and settled on Mahoré and Mwali. Prosperity was restored as Comoran traders again became involved in transporting slaves from the East African coast to Réunion and Madagascar. Dhows carrying slaves brought in huge profits for their investors. On Comoros, it was estimated in 1865 that as much as 40% of the population consisted of slaves! For the elite, owning a large number of slaves to perform fieldwork and household service was a mark of status. On the eve of the French occupation, Comoran society consisted of three classes: the elite of the Shirazi sultans and their families, a middle class of free persons or commoners, and a slave class consisting of those who had been brought from the African coast or their descendants. There is a long history of inter-island hostility, reflected in a Comoran saying: "Grand Comore rules, Anjouan works and Mohéli sleeps."

Datum origin points, thanks to John W. Hager are: "On Île Mohéli is Bangoma (east base) at $\Phi_o = 12°\ 16'\ 55.1''$ S, $\Lambda_o = 43°\ 45'\ 03.9''$ E, $\alpha_o = 247°\ 02'\ 29.544''$ to Ditsoni from north, International (ellipsoid), $H_o = 2.56$ meters. On Île d'Anjouan is Chanda at $\Phi_o = 12°\ 11'\ 06.6''$ S, $\Lambda_o = 44°\ 27'\ 24.6''$ E, $\alpha_o = 144°\ 25'\ 31.84''$ to Dziani from north, International (ellipsoid), $H_o = 823.8$ meters. On Grande Comoro is M'Tsaoueni (north base) at $\Phi_o = 11°\ 28'\ 32.2''$ S, $\Lambda_o = 43°\ 15'\ 42.15''$ E, $\alpha_o = 175°\ 31'\ 53.886''$ to Hahaia from north, International (ellipsoid), $H_o = 5.47$ m. John W. Hager states: "All of these are from *Archipel des Comores, Répertoire des Coordonnées et Altitudes des Pointes Géodésiques*, Paris, 1974. My intuition is that these were published by I.G.N. and probably taken from *Annales Hydrographiques*." Thanks to Mark Nettles, the station for the south base is station Domoni and the length of the baseline is 2746.387 m. The IGM Brigade determined 29 fundamental points (first order triangulation), 5 second-order points, and 20 detail points (third order triangulation). All of these islands are on the UTM Grid, thanks to the policy of the French Institut Géographique National since 1950. Although France administers the Island of Mayotte, the Comores still claims the island. The three-parameter datum shift published by I.G.N. from the Combani 1950 Datum to WGS84 is: $\Delta X = -382$ m, $\Delta Y = -59$ m, $\Delta Z = -251$ m. Mayotte is also on the UTM Grid.

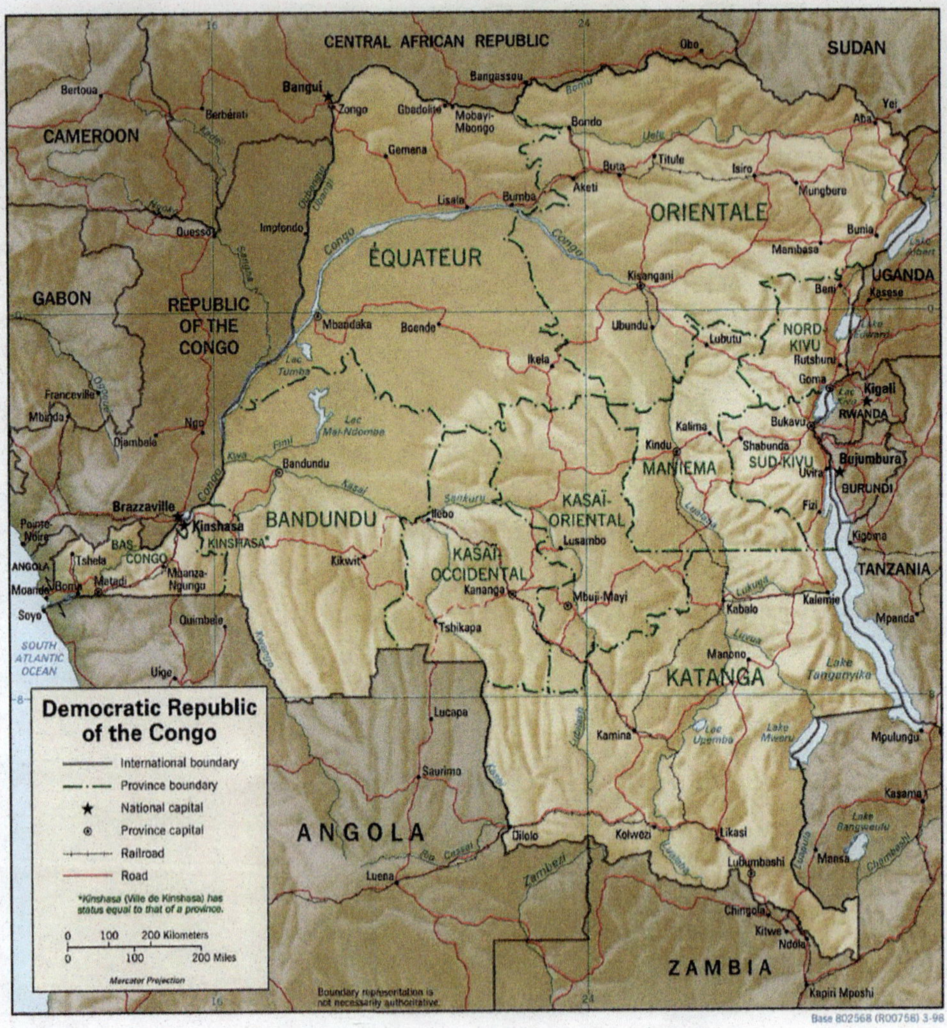
Democratic Republic of the Congo
International boundary
Province boundary
National capital
Province capital
Railroad
Road
*Kinshasa (Ville de Kinshasa) has status equal to that of a province.
0 100 200 Kilometers
0 100 200 Miles
Mercator Projection
Boundary representation is not necessarily authoritative.
CENTRAL AFRICAN REPUBLIC
SUDAN
CAMEROON
GABON
REPUBLIC OF THE CONGO
UGANDA
RWANDA
BURUNDI
TANZANIA
ANGOLA
ZAMBIA
SOUTH ATLANTIC OCEAN
ÉQUATEUR
ORIENTALE
NORD-KIVU
SUD-KIVU
MANIEMA
KASAÏ-ORIENTAL
KASAÏ-OCCIDENTAL
BANDUNDU
BAS-CONGO
KINSHASA*
KATANGA
Bangui
Zongo
Gbadolite
Mobayi-Mbongo
Bangassou
Obo
Yei
Aba
Bondo
Buta
Titule
Isiro
Aketi
Mungbere
Bumba
Lisala
Gemena
Impfondo
Ouesso
Bertoua
Berbérati
Bunia
Mambasa
Kisangani
Beni
Kasese
Mbandaka
Boende
Ubundu
Lubutu
Butembo
Rutshuru
Goma
Kigali
Bukavu
Ikela
Kalima
Kindu
Shabunda
Uvira
Bujumbura
Fizi
Kigoma
Franceville
Mbinda
Djambala
Ngo
Bandundu
Lac Mai-Ndombe
Lac Tumba
Brazzaville
Kinshasa
Pointe-Noire
Tshela
Boma
Matadi
Mbanza-Ngungu
Moanda
Soyo
Quimbele
Uige
Kikwit
Ilebo
Lusambo
Kananga
Mbuji-Mayi
Tshikapa
Kabalo
Kalemie
Mpanda
Manono
Lake Tanganyika
Kamina
Lac Upemba
Lake Mweru
Mpulungu
Kasama
Lake Bangweulu
Mansa
Lucapa
Saurimo
Dilolo
Kolwezi
Likasi
Lubumbashi
Luena
Chingola
Kitwe
Ndola
Kapiri Mposhi
Congo
Ubangi
Uele
Kasai
Sankuru
Lualaba
Zambezi
Lake Albert
Lake Edward
Lac Kivu
Base 802568 (R00758) 3-98

Democratic Republic of the Congo (Kinshasa)

The Democratic Republic of the Congo (Kinshasa), formerly called Zaire and prior to that, it was called the Belgian Congo; lies on the equator and has borders with the Republic of Congo (Brazzaville) (2,410 km), the Central African Republic (1,577 km), Sudan (628 km), Uganda (765 km), Rwanda (217 km), Burundi (233 km), Tanzania (459 km), Zambia (1,930 km), and Angola (2,511 km), and it has a small coastline on the Atlantic. The central region has an equatorial climate with high temperatures and heavy rainfall, with different climatic cycles in the northern and the southern regions. Comprising an area of slightly less than one-fourth of the United States, the country has a 37 km coastline with a 12 nm territorial sea. The terrain is a vast central basin on a low-lying plateau with mountains in the east. The lowest point is the Atlantic Ocean (0 m), and the highest point is Pic Marguerite on Mont Ngaliema or Mount Stanley (5,110 m).

French is the official language and Christianity is the majority religion. Archaeological evidence of past societies in the Congo are scanty due to the rain forest and tropical climate covering the northern half of the country and much of the Congo River Basin. Equatorial Africa has been inhabited since at least the Middle Stone Age. Late Stone Age cultures flourished in the southern savanna from approximately 10,000 B.C. and remained until the arrival of Bantu-speaking peoples during the first millennium B.C.

From the Office of the Geographer, U.S. Department of State in International Boundary Study No. 127, "Initially France established claims in the Congo basin through penetration of the territory from bases in Gabon and by treaties with local rulers. In a series of expeditions between 1875 and 1882, Pierre Savorgnan de Brazza, an officer of the French navy, explored much of the territory between the Ogooué and Congo rivers including the Niari valley. In 1880 de Brazza negotiated a treaty with the ruler of the Teke kingdom, which secured part of the north bank of the Congo for France, but because of quiet anchorage, he constructed a station on Kintamo Island near the south bank.

"During this period with an expedition from East Africa, the explorer Henry M. Stanley descended the Congo River to its mouth in 1877. King Leopold II of Belgium later retained his services to establish stations and to make treaties with the people of the Congo basin. In 1881 de Brazza met Stanley who was advancing eastward thorough the cataract area of the lower Congo River. Stanley arrived in the vicinity of Stanley Pool (Pool Malebo) on July 27 of the same year and founded Léopoldville on the south bank of the Congo at the site of present-day Kinshasa. Shortly thereafter, the French post on Kintamo Island was moved to the north bank of the Congo, where it became known as Brazzaville. …"

"In the meantime, King Leopold had shown great interest in the development of Africa. He invited explorers, geographers, and philanthropists of various states to a conference on central Africa at Brussels on September 12, 1876. An African International Association was organized at the conference with headquarters at Brussels. It was agreed that branches of the association in various states would be known as national committees, and King Leopold headed the Belgian National Committee.

"Following the historic trip of Stanley down the Congo in 1877, King Leopold shifted his primary interests in tropical Africa from the east to the west coast. Therefore, in 1878 another committee was organized under the title of the Committee for Upper Congo Studies, which later was known as the International Association of the Congo. The association was in effect a development company with King Leopold being the principal stockholder, and Stanley was commissioned by the

King for service under the International Association of the Congo. Between 1879 and 1882, Stanley established stations and made treaties with numerous African chiefs, many of which were in the upper part of the Congo basin.

"Early in 1884 several states recognized the Association as a governing power on the Congo River. In an exchange of notes between France and the Association of the Congo, April-May 1884, France was accorded the right of preemption of preferential right to the region of the Congo and in the valley of Niadi-Quillou (Niara-Kouilou) should the Association of the Congo dispose of its territorial possessions.

"The Berlin Conference of 1884–1885 recognized King Leopold as the sovereign head of state for the International Association of the Congo. On July 1, 1885, the name of the entity was changed to the Congo Free State, which was retained until it became a Belgian colony in 1908. A treaty for the cession of the Congo Free State to Belgium was signed at Brussels on January 9, 1895, and submitted to the Belgian Chamber of Deputies for approval shortly thereafter; however, it was withdrawn without any formal action being taken. An arrangement made between Belgium and France relative to the French right of preemption of a Belgian colony of the Congo was signed at Paris on February 5, 1895, but it was not ratified in consequence of the withdrawal of the treaty of cession. A second treaty for the cession of the Congo Free State to Belgium was signed on November 28, 1907 and approved by a Belgian Law of October 18, 1908. The treaty of cession was followed by an arrangement between Belgium and France governing the French preferential right to the Belgian Congo on December 23, 1908."

The first geodetic surveys in the Congo were part of the 1911–1914 boundary survey between Northern Rhodesia (now Zambia) and the southern Congolese province of Katanga. The fundamental (origin) point of the Katanga Triangulation is Station "A" of the Tshinsenda baseline (in Zambia) [Chain I], where $\Phi_o = 12° 20' 31.508''$ S and $\Lambda_o = 28° 01' 02.465''$ E. The altitude of the point was 1,331.31 m, as determined by trigonometric leveling from the 30th Arc triangulation performed in 1912. The ellipsoid of reference used by the Belgians for the computation of the triangulation in Katanga Province was the Clarke 1866, where $a = 6,378,206.4$ m and $b = 6,356,583.8$ m. They also referenced the Clarke 1866 Tables as published by the U.S. Coast & Geodetic Survey. In 1954, *Comité Spécial du Katanga, Les Travaux Géodésiques du Service Géographique et Géologique* was published. The Tshinsenda Baseline [Chain I] was measured in 1912 with a length of 4,152.9912 m with the final value being adjusted with the 1923 base at Nyanza, both surveyed by the Katanga-Rhodesia Boundary Commission. The deflection of the vertical was constrained to zero at point "A." Subsequent geodetic survey connections to the Katanga triangulation by the Arc 1950 Datum provided a couple of common points. I computed the transformation from the Katanga Datum of 1912 to the Arc 1950 Datum as: $\Delta X = +44$ m, $\Delta Y = +46$ m, $\Delta Z = +34$ m, and I'd guess that for southern Katanga these parameters are good to ±25 m. The projection adopted for the general map of Katanga was the Lambert Conical Orthomorphic with two standard parallels at $\varphi_N = 6° 30'$ S and $\varphi_S = 11° 30'$ S and a central meridian, $\lambda_o = 26°$ E. However, the reason why such emphasis was placed on the province to begin with was the presence of large deposits of copper ore. With mining property values being high, a cadastral coordinate system was implemented at the same time such that a Gauss-Krüger Transverse Mercator grid was defined with a central meridian, $\lambda_o = 26°$ E and a False Northing Latitude of Origin = 9° S and the Scale Factor at Origin = 0.9998.

The classical triangulation of Katanga required a number of baselines to be measured in order to provide a uniform reference scale to the datum. Those baselines in addition to the 1912 Tshinsenda distance included: Kitanga (1922) [Chain II], 3,695.0250 m, Mutene (1922) [Chain III], 1,554.9333 m, Nyanza (1923) [Chain XI], 4,881.8892 m, Kilambo (1929) [Chain VI], 6,601.2811 m, Pweto (1929) [Chain IX], 5,018.0550 m, Bululwe (1923) [Chain VIII], 10,516.9679 m, Gandajika (1947) [Chain XIII], 12,955.3016 m, and Kita Mulambo (1951) [Chain XIV], 9,187.7147 m.

An interesting *"faux pas"* (goof) in geodetic lore is the "Gan Datum" of the Congo. The Gandajika baseline was noticed by someone in the DMA and was confused with the Maldives Island of "Gan." As a result, the actual datum transformation for the island of Gan was incorrectly attributed to the

Katanga region of Congo (Kinshasa). That has since been rectified. There is no "Gan Datum" in the Congo. I wonder what is left of the true Indian Ocean Gan Datum after the catastrophic tsunami of December 26, 2004!

The published relation between Arc 1950 Datum and WGS84 Datum in the Congo (Kinshasa) by the NGA is as follows: $\Delta X = -169 \text{ m} \pm 25 \text{ m}$, $\Delta Y = -19 \text{ m} \pm 25 \text{ m}$, $\Delta Z = -278 \text{ m} \pm 25 \text{ m}$, and this solution was based on two points in 1991. The current grid system used for the People's Republic of the Congo (Kinshasa) is the UTM.

Thanks to Ms. Melita Kennedy of ESRI for prodding me for answers on the legendary "Gandajika Datum," to Mr. John W. Hager for the answers on the history of the same legend, and to Mr. Mal Jones of Perth, Australia, for the source material on the triangulation of Katanga. I am informed that a GPS Survey of the Congo (Kinshasa) is currently in the planning stages by private concerns.

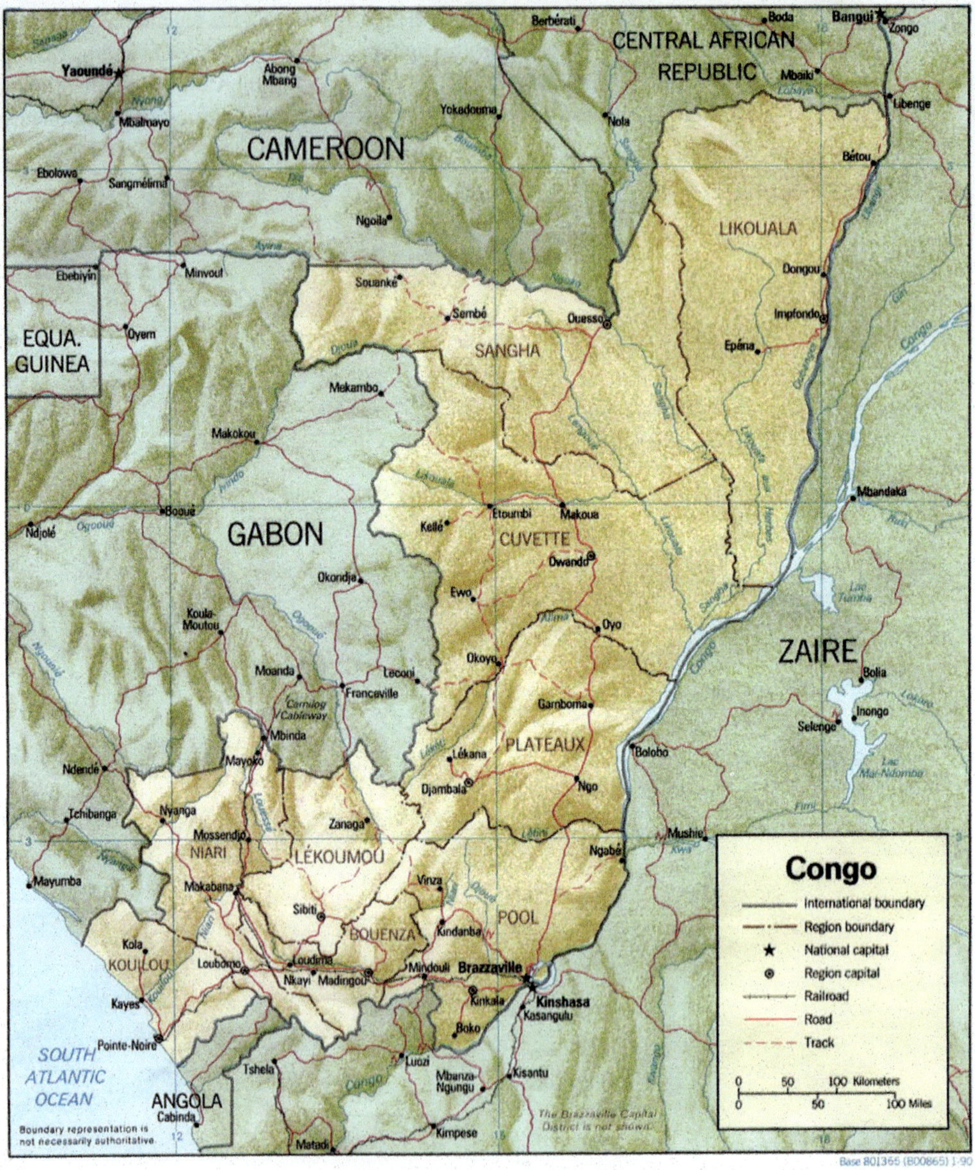

Base 801366 (B00865) 1-90

Republic of the Congo (Brassaville)

Bordered by Angola (231 km), Cameroon (494 km), Central African Republic (487 km), the Democratic Republic of the Congo (1,775 km), and Gabon (2,567 km). The terrain is coastal plain, southern basin, central plateau, and northern basin. Its highest point is Mont Nabeba (1,020 m) and its lowest point is the Atlantic Ocean (0 m). Its mean elevation is (430 m) (*World Factbook*, 2022).

"First inhabited by Pygmies, Congo was later settled by Bantu groups that also occupied parts of present-day Angola, Gabon, and Democratic Republic of the Congo, forming the basis for ethnic affinities and rivalries among those states. Several Bantu kingdoms – notably those of the Kongo, the Loango, and the Teke – built trade links leading into the Congo River basin. The first European contacts came in the late 15th century, and commercial relationships were quickly established with the kingdoms – trading for slaves captured in the interior. The coastal area was a major source for the transatlantic slave trade, and when that commerce ended in the early 19th century, the power of the Bantu kingdoms eroded. The area came under French sovereignty in the 1880s. Pierre Savorgnon de Brazza, a French empire builder, competed with agents of Belgian King Leopold's International Congo Association (later Zaire) for control of the Congo River basin. Between 1882 and 1891, treaties were secured with all the main local rulers on the river's right bank, placing their lands under French protection. In 1908, France organized French Equatorial Africa (AEF), comprising its colonies of Middle Congo (modern Congo), Gabon, Chad, and Oubangui-Chari (modern Central African Republic). Brazzaville was selected as the federal capital" (*U.S. Department of State Background Note, 2010*).

"The Angola–Congo (Brazzaville) boundary is delimited by the Franco–Portuguese convention of May 12, 1886, between Ponta Chamba and boundary pillar D. Ponta Chamba is situated near the Atlantic Ocean at the confluence of the Rio Loema and Rio Lubinda2 and boundary pillar D is located inland at the end of the median line between these two rivers. The remainder of the boundary is delimited to the tripoint with Congo (Kinshasa) at or near the confluence of the Bidihimba and Rio Chiloango by the Franco–Portuguese arrangement of January 23, 1901, which interprets the convention of May 12, 1886, in this sector. The boundary is demarcated clockwise by pillars A through J, including additional intervening pillars. Boundary pillar A is located on Ponta Chamba and J is located on a hill about 0.5 mile southwest of the Congo (Kinshasa) tripoint" (*International Boundary Study No. 105 – October 15, 1970*).

"Cameroon and Congo (Brazzaville) have a common boundary of about 325 miles. It follows a drainage divide for approximately 21 miles, a parallel of latitude for 85 miles, and rivers for 219 miles. Tripoints with both Gabon and the Central African Republic are situated on the thalwegs of rivers. The Gabon tripoint is located in the Ayina at 2° 10′ 20″ N., and the Central African Republic tripoint is located in the Sangha at about 2° 13′ 20″ N" (*International Boundary Study No. 110 – May 14, 1971*).

"The Central African Republic–Congo boundary extends between the Sangha and Ubangi rivers and is about 290 miles long. For much of this distance, it follows the drainage divide between the Lobaye and the Ibenga. The boundary also follows a long straight-line sector adjacent to the Cameroon tripoint and passes along the Gouga river adjacent to the Zaire tripoint. The line is not demarcated by pillars" (*International Boundary Study No. 145 – July 17, 1974*).

"The Congo – Zaire [*Republic of the Congo (Brazzaville) – Democratic Republic of the Congo (Kinshasa) – Ed.*] boundary is approximately 1,010 miles long. From the Angola tripoint to the Congo River, it follows the Shiloango, the Congo – Niari drainage divide, straight-line segments, and various other rivers for a distance of about 220 miles. The remainder of the boundary consists of the Congo River for 500 miles and the Ubangi for 290 miles to the Central African Republic tripoint. Except in Stanley Pool, the exact alignment of the boundary in the Congo River sector is indefinite" (*International Boundary Study No. 127 – September 8, 1972*).

The earliest published geodetic data of the Congo (Brazzaville) is that by Captain Dion of the French *l'Institut Géographique National* in which the defining datum origin at the astronomical station *Pointe-Noire* in 1948 was observed as: $\Phi_o = 04°\ 47'\ 00.1''$ S and $\Lambda_o = 11°\ 51'\ 01.55''$ E, the reference azimuth to the Lighthouse *(Phare)* at *Pointe Noire* is: $\alpha_o = 242°\ 45'\ 04''$, and the ellipsoid of reference is the Clarke 1880, where $a = 6,378,249.145$ m and $1/f = 293.465$. A subsequent survey in 1959 by the French Navy was done in collaboration with the French Oil Company, *Société des Pétroles de l'Afrique Équatoriale Française* or *S.P.A.É.F.* By this time, geodetic surveys performed by the French were commonly computed on the UTM grid as it was in this case, using Zone 32, South. Wild-Heerbrugg T-3 theodolites and AGA Geodimeter electronic distance meters were used in the coastal geodetic survey in order to establish control for TORAN equipment, which was a French version of the standard type of hyperbolic lattice radio-positioning equipment used for over the horizon offshore seismic surveys (*Annales Hydrographiques, Mission Géodésique au Moyen Congo, Avril-Septembre 1959*). TORAN was quite similar to DECCA, LORAN-A, and Raydist in that the Surveyor/Navigator had to manually keep track of the hyperbolic lane count by using a pencil to number the saw-teeth output of a plotter! Skywaves were an interesting annoyance that required no small amount of ingenuity to prevent losing one's lane count. Later developments to come up with an "automatic" lane count, as with the Argo system, yielded remarkable blunders when the skipper would change course.... I entered the arcane world of offshore radio positioning surveying and mapping during the early 1960s.

With the war over in Congo (Brazzaville), the local economy is recovering and exploration and production of hydrocarbons as well as other mining ventures are opening up much of the undeveloped interior. At present though, there is but a single observed point published by NGA *(TR8350.2)* for the transformation **from** the Point Noir Datum of 1948 **to** the WGS 84 Datum: $\Delta X = -148$ m ± 25 m, $\Delta Y = +51$ m ± 25 m, and $\Delta Z = -291$ m ± 25 m. Note that there likely is little to no classical control to be found in the undeveloped interior, anyway.

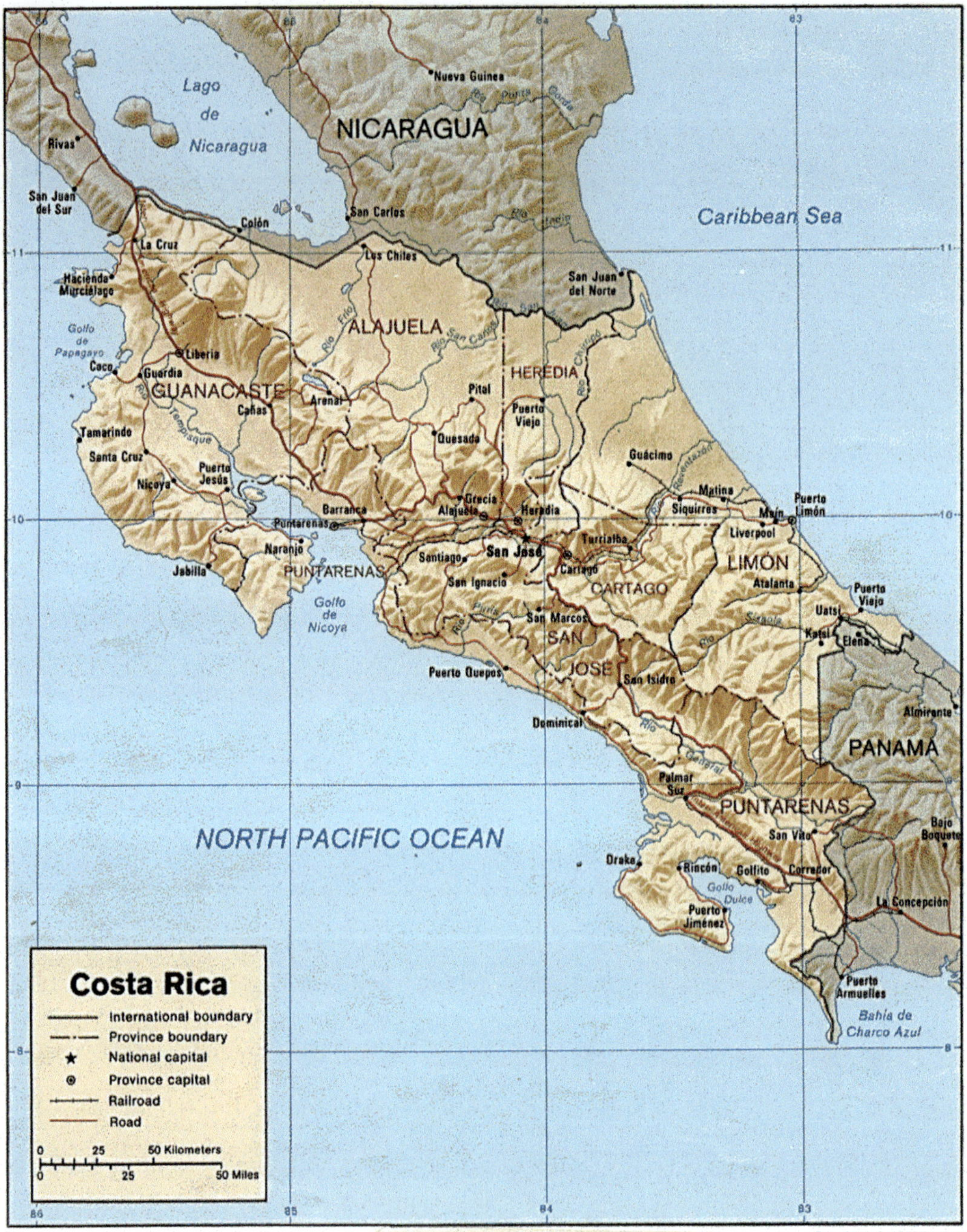

Base 801015 (A05363) 7-87

Republic of Costa Rica

Slightly smaller than West Virginia, Costa Rica is bordered by Nicaragua (309 km) and by Panamá (330 km), and has a coastline on both the Caribbean Sea and the North Pacific Ocean totaling 1,290 km. The climate is tropical with a dry season from December to April when it can rain about every other day and the rainy season when it can rain at least once a day – just like Panamá. Costa Rican terrain is mainly coastal plains separated by rugged mountains comprised of the *Cordillera de Talamanca*. The lowest points are sea level on the Pacific Ocean (lower than the Caribbean), and the highest points include *Cerro Irazú* (3,432 m), *Cerro Barú* (3,475 m), and *Cerro Chirripó* (3,819 m).

Inhabited by several indigenous tribes when reached by Christopher Columbus on his last voyage of 1502, the region became a Spanish province in 1540, although a permanent settlement was not established until the 1560s. "Christopher Columbus who stayed for 17 days in 1502 was so impressed by the gold decorations worn by the friendly locals that he promptly dubbed the country Costa Rica, 'the rich coast.' The 18th century saw the establishment of settlements such as Heredia, San José and Alajuela. It was not until the introduction of coffee in 1808, however, that the country registered on the radars of the 19th-century white-shoe brigade and frontier entrepreneurs looking to make a killing. Coffee brought wealth, a class structure, a more outward-looking perspective and, most importantly, independence. Costa Rica achieved independence from the Spanish Crown in 1821. A bizarre turn of events in 1856 provided one of the first important landmarks in the nation's history and served to unify the people. During the term of coffee-grower-turned-president Juan Rafael Mora, a period remembered for the country's economic and cultural growth, Costa Rica was invaded by the US military adventurer William Walker and his army of recently captured Nicaraguan slaves. Mora organized an army of 9,000 civilians that against all odds, succeeded in forcing Walker & Co. to flee. The ensuing years of the 19th century saw power struggles among members of the coffee-growing elite and the institution of the first democratic elections, which have since been a hallmark of Costa Rican politics" (*Lonely Planet*, 2007).

Cocos Island, called Coco by Costa Rica, is 4.8 km wide with a circumference of 20.8 km, has two good harbors in Chatham and Wafer Bays, and is 480 km southwest of Costa Rica and 880 km west southwest of Panamá City. It is formerly a whaling base and is now in good tuna fishing waters and is one of the strategic points of defense for the Panama Canal. An old legend says that treasure was buried there by the corsair Benito Bonito in 1818–1819, and more by one of his lieutenants in 1826. Another story has it that wealthy Spaniards, including the treasurer of the Cathedral, fleeing from Lima, Peru in 1821 with gold and jewels on the *Mary Dier* were murdered in their bunks by the crew under an English Captain Thompson and that 12 boatloads of treasure worth about $30 million were afterwards buried on Cocos.

In 1888, the *Instituto Fisico Geográfico* (Physical Geography Institute) was established under the direction of the Swiss Professor Henri Pittier, but the project lasted only until 1912. The Ocotepeque Datum of 1935 was established in Honduras at Base Norte, where $\Phi_o = 14°\ 26'\ 13.73''$ North ($\pm 0.07''$), $\Lambda_o = 89°\ 11'\ 39.67''$ West of Greenwich ($\pm 0.045''$), $h_o = 806.99$ m, the defining astronomic azimuth to Base Sur is: $\alpha_o = 235°\ 54'\ 20.37''$ ($\pm 0.28''$), and the ellipsoid of reference is the Clarke 1866, where $a = 6,378,206.4$ m and $1/f = 294.9786982$. The corresponding geodetic coordinates are: $\varphi_o = 14°\ 26'\ 20.168''$ North, $\lambda_o = 89°\ 11'\ 33.964''$ West, and $H_o = 823.40$ m above mean sea level. The defining geodetic azimuth to Base Sur is: $\alpha_o = 235°\ 54'\ 21.790''$. The differences between these two sets of coordinates are due to the local gravimetric deflection of the vertical as determined by Sidney H. Birdseye of the U.S. Coast & Geodetic Survey (USC&GS).

Nothing much happened (geodetically) anywhere in Central America during World War II. After the war, the U.S. Army Map Service (AMS) established the Inter-American Geodetic Survey (IAGS) headquartered in the U.S. Canal Zone of Panamá at the picturesque Fort Curundu with most classrooms at Fort Clayton, as I remember. Co-operative agreements negotiated with most countries in Latin America included the Republic of Costa Rica. However, *"La Ley Numero 59"* (The Law, Number 59) was promulgated to establish the *Instituto Geográfico de Costa Rica* in 1944. In 1946, the *IGCR* signed an agreement with IAGS. The previous geodetic boundary work of the USC&GS (that established the Ocotepeque Datum of 1935) was integrated into the new IAGS observations along with the chains of quadrilaterals observed in Guatemala, Honduras, and Nicaragua (with the IAGS). With connections to the classical triangulation in Mexico, the North American Datum of 1927 (NAD 27) was eventually introduced to the Republic of Costa Rica. Of particular convenience was the fact that the Ocotepeque Datum of 1935 was referenced to the same ellipsoid! (That was courtesy of the USC&GS.) The IAGS observations in Costa Rica were supervised by Bill Parkhurst, my first mentor that I described for some of his later work in the Sudan and he trained Captain Frank Spacek, an officer in the U.S. Army Corps of Engineers (attached to the IAGS), in Costa Rica, and later Colonel Spacek was my Commanding Officer at Army Map Service in the 1970s when I was stationed there as a Captain.

The IAGS method of computing chains of quadrilaterals was the "Army way" of doing things. That way was different from the USC&GS because AMS was concerned with unifying the mishmash of Datums in post-war Europe. The planning and design for that spectacular computational chore in Europe was ultimately accomplished with a conformal projection and Grid. The complexity of existing systems in the Americas was much simpler, but the Topographic Engineers of AMS (and IAGS) were trained according to the "Army way." That way consisted of computing classical triangulation on a conformal Grid. Central American countries, with the exception of Belize, are greater in east-west extent than in north-south extent. Therefore, the Lambert Conformal Conic projection was used as the basis of all IAGS-developed Grid systems in Central America for triangulation computations and the published "Trig Lists" of coordinates.

The Costa Rica Lambert Conformal Grid (1946–present) on the Ocotepeque Datum of 1935 covers two secant zones, *Norte* and *Sud* (North and South Zones). Both zones use the same Central Meridian $(\lambda_o) = 84°\ 20'\ 00''$ West of Greenwich, a False Easting of 500 km, and a scale factor at origin $(m_o) = 0.99995696$. Zone Norte has a Latitude of Origin $(\varphi_o) = 10°\ 28'$ North, and the False Easting = 271,820.522 m. Zone Sud has a Latitude of Origin $(\varphi_o) = 9°\ 00'$ North, and the False Easting = 327,987.436 m. Of course, these parameters will develop two *different* pairs of standard parallels for each zone because of their *different* respective latitudes of origin.

These *secant* Lambert Zones are defined according to the "British Method" which provides a Latitude of Origin (φ_o), and a Scale Factor at Origin (m_o). The common way to do it in the United States is the "American or French Definition" which provides for a pole-ward standard parallel and an equator-ward standard parallel for a *secant* zone. Much commercial GIS software as well as much "geodetic transformation" software mistakenly term this "British Definition" as a "tangent conic with one standard parallel." Patently false, this silly misclassification shows that the mathematical foundation of a package's programming staff is apparently ignorant of cartographic/geodetic transformations and that they are presumably only capable of copying formulae from a book. For those that can work with a bit of algebra, the math published in Chapter 3 of the *Manual of Photogrammetry*, 5th and 6th editions give sufficient detail for a competent mathematician to equate the two Lambert "definitions" as mathematically equal. (*I did provide sufficient clues.*) One secant Lambert zone defined with a latitude at origin and a scale factor at origin less than one *is the same* as defining the zone with two standard parallels! A Lambert conformal conic zone is tangent if and only if (*iff*), the scale factor at origin is equal to unity! (That's "one" for those from Rio Linda.)

With regard to relating the old stuff to the new stuff, from Ocotepeque Datum of 1935 to WGS 72 Datum: $\Delta X = -193.798$ m, $\Delta Y = -37.807$ m, $\Delta Z = +84.843$ m. Furthermore, from Ocotepeque Datum of 1935 to the NAD1927: $\Delta X = +205.435$ m, $\Delta Y = -29.099$ m, $\Delta Z = -292.202$ m. In Costa Rica, the fit of Ocotepeque Datum 1935 to WGS 72 is better than ±3 m; the fit of NAD 27 to WGS 72 is better than ±6 m. On the other hand, *La Registro Nacional de la Republica de Costa Rica* offers the following parameters from Ocotepeque 1935 Datum to WGS84 Datum as: $\Delta X = -66.66$ m ± 0.27 m, $\Delta Y = -0.13$ m ± 0.27 m, $\Delta Z = +216.80$ m ± 0.27 m, $\delta s = -5.7649$ ppm ± 0.2.5657 ppm, $R_x = +1.5657''$ ±1.0782″, $R_y = +0.5242''$ ±0.5318″, $R_z = +6.9718''$ 0.8313″. No information is offered by the Costa Rican government regarding the sense of the rotations nor are example transformations given on the official web site. Beware of the sense of the rotations! A cryptic note is presented on the *Proyección CRTM90-WGS84* which is a "modified" UTM with a central meridian of 84°. Apparently, someone in Costa Rica thought a Transverse Mercator was a good idea (**not!**) for a predominately East-West country. So much for the excellent advice and design of the IAGS ... a modified UTM is now only a mouse click away! The future arguments and disagreements between GIS Technicians and Licensed Land Surveyors will go on *ad nauseum* because the field-observed distances won't match the "correct" GIS-determined distances on the "modified" UTM-style *Proyección CRTM90-WGS84*! Cartographic ignorance is on the march – maybe the local Academics can convince the government regarding the error of their ways. Military-style scale factors (0.9996) are good for military uses, but civilian-style scale factors (0.99995696) are much better for civilian uses, *especially* for laypeople.

Croatia

- International boundary
- County (Županija) boundary
- ★ National capital
- ◉ County (Županija) center
- Railroad
- Expressway
- Road

0   50 Kilometers
0   50 Miles

Lambert Conformal Conic Projection, SP 40N/56N

County (Županija)

1. Bjelovarsko-Bilogorska
2. Brodsko-Posavska
3. Dubrovačko-Neretvanska
4. Grad Zagreb (city)
5. Istarska
6. Karlovačka
7. Koprivničko-Križevačka
8. Krapinsko-Zagorska
9. Ličko-Senjska
10. Međimurska
11. Osječko-Baranjska
12. Požeško-Slavonska
13. Primorsko-Goranska
14. Šibensko-Kninska
15. Sisačko-Moslavačka
16. Splitsko-Dalmatinska
17. Varaždinska
18. Virovitičko-Podravska
19. Vukovarsko-Srijemska
20. Zadarska
21. Zagrebačka

Base 802844AI (C00039) 11-01

Republic of Croatia

Slightly smaller than West Virginia, Croatia is bordered by Bosnia and Herzegovina (932 km), Hungary (329 km), Serbia (241 km), Montenegro (25 km), and Slovenia (455 km). The lowest point is the Adriatic Sea (0 m), the highest point is Dinara (1,831 m), the terrain is geographically diverse with flat plains along the Hungarian border, and low mountains and highlands near the Adriatic coastline and islands (*World Factbook*, 2012).

"Excavations in Krapina have revealed that the area has been inhabited since the Paleolithic Age. The initial Roman province of Illyricum was gradually enlarged during a series of wars that brought much of the Dalmatian coast within their control. By 11 *B.C.*, Rome conquered much of the interior, which was inhabited by the Pannonian tribe, extending the empire's reach to the middle and lower Danube. The realm was reorganized into Dalmatia (the former Illyricum), and Upper and Lower Pannonia, which covered much of the interior of modern Croatia" (*Lonely Planet*, 2012).

"The Croats are believed to be a Slavic people who migrated from Ukraine and settled in present-day Croatia during the 6th century. After a period of self-rule and the establishment of an independent kingdom, Croatians agreed to the *Pacta Conventa* in 1091, submitting themselves to Hungarian authority. By the mid-1400s, concerns over Ottoman expansion led the Croatian Assembly to invite the Habsburgs, under Archduke Ferdinand, to assume control over Croatia. Habsburg rule proved successful in thwarting the Ottomans, and by the 18th century, much of Croatia was free of Turkish control. The Austrian monarchy also acquired control over Dalmatia at the close of the Napoleonic wars following centuries of rule by the Venetian Republic. In 1868, Croatia gained domestic autonomy under Hungarian authority. Following World War I and the demise of the Austro-Hungarian Empire, Croatia joined the Kingdom of Serbs, Croats, and Slovenes (the Kingdom of Serbs, Croats, and Slovenes became Yugoslavia in 1929). During World War II, German and Italian troops invaded and occupied Yugoslavia and set up a puppet, Fascist regime to rule a nominally independent Croatian state. This regime, under the hardline nationalist Croatian Ustasha party, was responsible for the deaths of large numbers of ethnic Serbs, Jews, Roma, and other civilians in a network of concentration camps. It was eventually defeated by the Partisans, led by Josip Broz Tito, in what was essentially a civil war as well as a struggle against the Axis occupiers. The pro-Yugoslav Partisans included many ethnic groups, including a large number of Croatians, and were supplied in large part by the United States and the United Kingdom. Yugoslavia changed its name once again after World War II. The new state became the Federal Socialist Republic of Yugoslavia and united Croatia and several other republics together under the communist leadership of Marshal Tito. After the death of Tito and with the fall of communism throughout Eastern Europe, the Yugoslav federation began to unravel. Croatia held its first multi-party elections since World War II in 1990. Long-time Croatian nationalist Franjo Tudjman was elected President, and 1 year later, Croatia declared independence from Yugoslavia. Conflict between Serbs and Croats in Croatia escalated, and 1 month after Croatia declared independence, the Yugoslav Army intervened, and war erupted. The United Nations mediated a cease-fire in January 1992, but hostilities resumed the next year when Croatia fought to regain one-third of the territory lost the previous year. A second cease-fire was enacted in May 1993, followed by a joint declaration the next January between Croatia and Yugoslavia. However, in September 1993, the Croatian Army led an offensive against the Serb-held self-styled 'Republic of Krajina.' A third cease-fire was called in March 1994, but it, too, was broken in May and August 1995, after which Croatian forces regained large portions of the Krajina, prompting an exodus of Serbs from this area. In November 1995, Croatia agreed to peacefully reintegrate Eastern Slavonia, Baranja, and Western Sirmium under terms of the Erdut Agreement, and the Croatian government re-established political and legal authority over those territories in January 1998. In December

1995, Croatia signed the Dayton peace agreement, committing itself to a permanent cease-fire and the return of all refugees" (*U.S. Dept. of State Background Notes, 2012*). The country joined NATO in April 2009 and the EU in July 2013.

The coordinate systems that have been used for Yugoslavian lands have been quite diverse through history. Geographic coordinates have been based on the Prime Meridians of Paris, France; Ferro, Canary Islands; Greenwich, England; and also, some temporary usage of the Vienna University Observatory, where $\Lambda = 16°\ 22'\ 49.98''$ East of Greenwich (later offset to Paris).

The names of the classical horizontal datums found in Croatia include Hermannskogel 1871, K.u.K. VGI Vienna University System 1892, System 42, and European 1950. The Hermannskogel 1871 datum used the Bessel 1841 ellipsoid, where $a = 6,377,397.155$ m and $1/f = 299.1528128$; the Vienna University System 1892 used the now obsolete Zach 1812 ellipsoid, where $a = 6,376,385$ m and $1/f = 310$; the System 42 used the Krassovsky 1940 ellipsoid, where $a = 6,378,245$ m and $1/f = 298.3$; and the European 1950 Datum used the International ellipsoid, where $a = 6,378,388$ m and $1/f = 297$. The "Parisian" system of mapping (based on the Prime Meridian of Paris, where the offset from Greenwich is accepted as $\lambda = 2°\ 20'\ 13.95''$ E) was cast on the polyhedric projection from 1878 to 1959. The mapping equations for the ellipsoidal polyhedric projection are given in Chapter 3 of the *Manual of Photogrammetry* in the 5th edition and the 6th edition in which it is identical to the Local Space Rectangular (LSR) when $\alpha = h = 0$.

The Hermannskogel 1871 Datum origin is at: $\Phi_o = 48°\ 16'\ 15.29''$ N, $\Lambda_o = 33°\ 57'\ 41.06''$ East of Ferro, where Ferro $= 17°\ 39'\ 46.02''$ East of Greenwich and azimuth to Hundsheimer is $\alpha_o = 107°\ 31'\ 41.7''$. The most common grid found on that datum is the Yugoslavia Reduced Gauss-Krüger Transverse Mercator. The scale factor at origin ($m_o = 0.9999$), the central meridian of the belt that covers Croatia is $\lambda_o = 18°$ East of Greenwich, and the False Easting $= 500$ km. The Ministry of Finance used the non-reduced version only between 1938 and 1940, where $m_o = 1.0$.

The K.u.K. VGI Vienna University System 1892 origin is at: $\Phi_o = 48°\ 12'\ 35.50''$ N, $\Lambda_o = 16°\ 22'\ 49.98''$ East of Greenwich. The K.u.K. VGI Vienna University System 1892 Datum established two baselines in Croatia at Sinj and at Dubica (*Andrew M. Glusic, Mapping of the Countries in Danubian and Adriatic Basins, Army Map Service Technical Report No. 25, June 1959*).

The System 42 Datum origin is at Pulkovo Observatory, where $\Phi_o = 59°\ 46'\ 18.55''$ N, $\Lambda_o = 30°\ 19'\ 42.09''$ East of Greenwich. The defining azimuth at the point of origin to Signal A is: $\alpha_o = 317°\ 02'\ 50.62''$.

The European 1950 Datum origin is at Helmertturm, Pottsdam (Germany) where $\Phi_o = 52°\ 22'\ 53.9540''$ N, $\Lambda_o = 13°\ 04'\ 01.1527''$ East of Greenwich. The defining azimuth to station Golmberg is: $\alpha_o = 154°\ 47'\ 32.19''$,

About 70 years ago, the Army Map Service transformed Hermannskogel 1871 Datum to the European Datum 1950. However, large data sets (including cadastral) still survive on that old datum. The author examined the relation between the two datums and computed new transformations. Twenty-two points were used that are common to both datums throughout the former Yugoslavia and a simple three-parameter shift analysis yielded the following: $\Delta X = +770.417$ m, $\Delta Y = -108.432$ m, $\Delta Z = +600.450$ m. The accuracy of this transformation when expressed in terms of actual geodetic coordinates is latitude change $(\Delta \phi) = \pm 3.74$ m, longitude change $(\Delta \lambda) = \pm 4.54$ m, and ellipsoid height change $(\Delta h) = \pm 12.70$ m. On the other hand, a seven-parameter shift analysis yielded the following: $\Delta X = +758.53$ m, $\Delta Y = +259.52$ m, $\Delta Z = +542.18$ m, Scale $= -6.0 \times 10^{-6}$, Z-rotation $(\omega) = +11.29''$, Y-rotation $(\psi) = +2.06''$, and X-rotation $(\xi) = -5.66''$. The accuracy of this transformation when expressed in terms of actual geodetic coordinates is Latitude change $(\Delta \phi) = \pm 1.07$ m, Longitude change $(\Delta \lambda) = \pm 1.44$ m, and Ellipsoid Height change $(\Delta h) = \pm 0.64$ m. For example, station Vel Gradiste has the following EU50 coordinates: $45°\ 09'\ 17.3501''$ N, $18°\ 42'\ 44.9479''$ E, 0.0 m. and the following Hermannskogel 1871 coordinates: $45°\ 09'\ 14.4675''$ N, $18°\ 43'\ 00.7696''$ E, 0.0 m. The Yugoslavian Reduced Grid coordinates are Northing $(X) = 5,001,303.81$ m, Easting $(Y) = 556,359.65$ m.

The current geodetic situation in Croatia has significantly changed. The Republic of Croatia adopted the Decree on establishing a new official geodetic datum and map projection in the Official Gazettes 110/04 and 117/04. The decree defined the horizontal datum of the Republic of Croatia for the European Terrestrial Reference System for the epoch 1989.0 (ETRS89). The materialization of this reference system is represented by 78 geodetic points defined in the ETRS89 system and is named the Croatian Terrestrial Reference System for epoch 1996.55 or abbreviated HTRS96. The vertical datum is determined by the geoid surface being mean sea level for epoch 1971.5 at five tide gauges along the Adriatic coast at Dubrovnik, Split, Bakar, Rovinj, and Kopar. This system is known as HVRS71. The Croatian gravimetric system is defined at six absolute gravimetric points and 36 points of a 1st order relative gravity network and is known as HGRS03. The map projections used in Croatia include the Gauss-Krüger Transverse Mercator projection with a central meridian at $\lambda_o = 16°\ 30'$ E, and a scale factor at origin of $m_o = 0.9999$ for cadastral and detailed topographic applications. For general state cartographic purposes, the Lambert Conformal Conic projection is defined with standard parallels of 43° 05′ N and 45° 55′ N. Since Croatia joined NATO in 2013, for military applications the UTM grid has been adopted. All projections are referenced to the HTRS96, and by definition of the ETRS89 System, the ellipsoid of reference is the GRS80, where a = 6,378,137 m and $1/f = 298.257223563$ (*M. Bosiljevac and Ž. Bačič, The Implementation of New Official Geodetic Datum and Map Projections in the Republic of Croatia, FIG Congress 2010, Sydney, Australia 11–16 April 2010*).

A unique transformation model has been established by the University of Zagreb based on more than 1,800 points between Hermannskogel 1871 and HTRS96. While I used 22 points in and around Croatia to achieve a fit of slightly over ±1 meter, their seven-parameter transformation achieved a fit of ±76.5 cm and their (complex) HTRS96/HDKS transformation achieved a fit of ±8.5 cm! This sort of phenomenal transformation accuracy will likely dispel any doubt in recovering cadastral property boundaries from long ago. The software appears available for public use within the Republic.

Cuba
Line of separation
National capital
Province boundary
Province capital
Railroad
Expressway
Road
Lambert Conformal Conic Projection, SP 20N/22N
Gulf of Mexico
Straits of Florida
NORTH ATLANTIC OCEAN
Caribbean Sea
Yucatan Channel
Windward Passage
U.S.
FLORIDA KEYS
Key West
Marathon
Key Largo
Homestead
Nassau
New Providence
Eleuthera
Andros Island
EXUMA CAYS
Cat Island
San Salvador
Rum Island
Great Exuma
Long Island
Crooked Island
Acklins Island
Great Inagua
THE BAHAMAS
JUMENTOS CAYS
RAGGED ISLAND RANGE (THE BAHAMAS)
Cay Sal Bank (THE BAHAMAS)
Guinchos Cay (THE BAHAMAS)
Cay Lobos (THE BAHAMAS)
Cay Santo Domingo (THE BAHAMAS)
ARCHIPIÉLAGO DE SABANA
ARCHIPIÉLAGO DE CAMAGÜEY
ARCHIPIÉLAGO DE LOS JARDINAS DE LA REINA
Havana
CIUDAD DE LA HABANA
LA HABANA
Matanzas
Cárdenas
Mariel
Artemisa
Minas de Matahambre
San Cristóbal
Arroyos de Mantua
PINAR DEL RÍO
Pinar del Río
La Fe
Guane
Golfo de Batabanó
Surgidero de Batabanó
Güines
MATANZAS
VILLA CLARA
CIENFUEGOS
Cienfuegos
Santa Clara
Placetas
Sancti Spíritus
SANCTI SPÍRITUS
CIEGO DE ÁVILA
Ciego de Ávila
Morón
Florida
Nuevitas
Camagüey
CAMAGÜEY
Puerto Padre
Las Tunas
LAS TUNAS
Holguín
HOLGUÍN
Banes
Moa
Baracoa
Mayarí
GUANTÁNAMO
Guantánamo
U.S. Naval Base Guantanamo Bay
Santa Cruz del Sur
Golfo de Guacanayabo
Manzanillo
Bayamo
GRANMA
Pilón
SANTIAGO DE CUBA
Santiago de Cuba
Nueva Gerona
ISLA DE LA JUVENTUD (special municipality)
Isla de la Juventud
Cayo Largo
Little Cayman
Cayman Brac
Grand Cayman
George Town
Cayman Islands (U.K.)
Montego Bay
Ocho Rios
JAMAICA
Port Antonio
Kingston
Savanna la Mar
Navassa Island (U.S.)
HAITI

Republic of Cuba

The island of Cuba is slightly smaller than Pennsylvania and is the largest country in the Caribbean. With a coastline of 3,735 km, the terrain is mostly flat with rugged hills and mountains in the southeast; the lowest point is the Caribbean Sea (0 m), and the highest point is Pico Turquino (1,970 m) (*personal communication, Dr. Ernesto Rodriguez Roche 2010, Havana Cuba*). The only land boundary is with the U.S. Naval Base at Guantanamo Bay (29 km) (*World Factbook*, 2010).

"It's thought that humans first cruised from South America to Cuba around 3500 B.C. Primarily fishers and hunter-gatherers, these original inhabitants were later joined by the agriculturalist Taino, a branch of the Arawak Indians. Christopher Columbus sighted Cuba on 27 October 1492, and by 1514, Diego Velázquez de Cuéllar had conquered the island for the Spanish crown and founded seven settlements. When captured Taino Chief and resistance fighter Hatuey was condemned to die at the stake, he refused baptism, saying that he never wanted to see another Spaniard again, not even in heaven! Cattle ranching quickly became the mainstay of the Cuban economy. Large estates were established on the island under the encomienda system, enslaving the Indians under the pretext of offering instruction in Christianity. By 1542, when the system was abolished, only around 5000 Indians (of an estimated 100,000 half a century before) survived. Undaunted, the Spanish imported African slaves as replacements. Unlike in the North American slave trade, Cuba's African slaves retained their tribal groupings, and certain aspects of their culture endure" (*Super CubaTravel*, 2010). "Father Bartolomé de Las Casas was in Cuba 1511–1512, and it was probably his observations there of the abuse and decimation of the native Caribs (who had only a short time before dispossessed the Arawaks) that launched him on his career as '*Apostle of the Indians*'" (*Boundaries, Possessions, and Conflicts*, 1938). Thanks to *InfoPlease, 2010*, "In the early 1800s, Cuba's sugarcane industry boomed, requiring massive numbers of black slaves. A simmering independence movement turned into open warfare from 1867 to 1878. Slavery was abolished in 1886. In 1895, the poet José Marti led the struggle that finally ended Spanish rule, thanks largely to U.S. intervention in 1898 after the sinking of the battleship *Maine* in Havana harbor." The Spanish–American War had begun, and Teddy Roosevelt prevailed at the Battle of San Juan Hill, including my grandfather, Gaston Mandeville Mugnier of the *Louisiana Regulars* although I think he was stuck in Florida tending Teddy's horses. In 1959, Fidel Castro completed a successful revolution and deposed the Batista regime to form the closed communist society extant to this day.

The first map known of Cuba was produced by Juan de la Cosa in 1500. In the 17th century, an improved map of Cuba was compiled by Gerhardus Mercator. In the middle 1800s, the Spanish Army produced a topographic map of the island called *"Mapa de Vivas."* The U.S. Army Map Service Inter-American Geodetic Survey (IAGS) established an office in Cuba in 1947. The *Instituto Cartográfico Nacional (ICN)* was designated as the national agency to collaborate with the IAGS in the cartographic plan (*MAPPLAN*). In the first 8–10 years of cooperative work the operations consisted of establishing a geodetic network covering the entire territory as well as the installation of 10 tide gauges in selected locations. Aerial photography was flown in 1956, and 100% Cuban coverage of topographic maps at 1:50,000 scale was completed in 1960 in cooperation with the IAGS (*IAGS Report [in Spanish] by Army Map Service, ca. 1960*).

"In 1912 Washington Hydrographic Office executed astrogeodetic determinations on 23 stations located in the whole contour of the Cuban coast. Two decades later, in the frame of the program for navigation aids, the US Navy Hydrographic Office carried out oceanographic works as well as third and fourth order triangulation of the Cuban coast and keys, in order to obtain the astrogeodetic coordinates of great number of sites" (*Roche & García, FIG Congress 2010, Sydney, Australia*). SHORAN surveys between the Florida Keys and Cuba were executed in 1950, allowing the IAGS to extend the North American Datum of 1927 (NAD 27) into Cuba. The NAD 27 origin is at station Meades Ranch in Kansas, where $\Phi_o=39°\ 13'\ 26.686''$ N, $\Lambda_o=98°\ 32'\ 30.506''$ W, the reference azimuth to station Waldo is $\alpha_o=75°\ 28'\ 09.64''$, and the ellipsoid of reference is the Clarke 1866, where $a=6,378,206.4$ m and $b=6,356,583.8$ m. The IAGS established two secant zones of plane coordinate systems for Cuba based on the Lambert Conformal Conic projection, *Norte,* and *Sud* (North and South Zones). The Cuba Norte Central Meridian $(\lambda_o)=81°$ West of Greenwich, a False Easting of 500 km, and a scale factor at origin $(m_o)=0.999936020$. Zone Norte has a Latitude of Origin $(\varphi_o)=22°\ 21'$ North, and the False Easting $=280,296.016$ m. Zone Sud Central Meridian $(\lambda_o)=76°\ 50'$ West of Greenwich, a False Easting of 500 km, and a scale factor at origin $(m_o)=0.999948480$. Zone Sud has a Latitude of Origin $(\varphi_o)=20°\ 43'$ North, and the False Easting $=229,126.939$ m. These *secan*t Lambert Zones are defined according to the "British Method" which provides a Latitude of Origin (φ_o), and a Scale Factor at Origin (m_o).

"Between 1951 and 1953 aerial-electronic missions were repeated, applying the High Precision Range Precision Range (HIRAN) trilateration method. The selected stations were linked to the fundamental triangulation network. After the campaign was over and a common adjustment by the parametric method performed, all the stations were referred to the NAD 27. It was considered that a relative precision of 1:113,000 was achieved. Up to 1958, the basic network was constituted by 87 stations, while another 181 stations belonged to other types of networks.

"The astrogeodetic network modernization was carried out on 1970–1973. The new network covered the whole Cuban territory, ..., being conformed *(sic – Ed.)* by 237 first order triangulation stations, 15 lineal bases, and 28 Laplace stations. After the densification task was finished, 490 and 1,903 of second and third *(order – Ed.)* stations respectively were established. Between 1981 and 1985 the basic horizontal fourth order network was developed in prioritized rural areas, settling down 3,500 station *(sic – Ed.)* by mean of the triangulation or traverse methods, depending on the terrain topography. From 1985 to 1990 fourth order and first category densification in urban areas was carried out, with about 28,806 points. Between 1989 and March 1990, a national Doppler campaign was carried out. Works were developed in close collaboration with the Soviet Geodetic Services ... Doppler network consisted of fourteen stations; six of them are located on key territories. The Doppler network adjustment took place in the WGS72 *(Datum – Ed.)* by the short arc approach" *(Roche & García, Ibid.).* The short arc approach was pioneered by the late Dr. Duane C. Brown, Honorary Member of the ASPRS – and he is generally regarded as the Father of Analytical Photogrammetry.

A GPS campaign was carried out in 1998, and some additional GPS densification was performed in the vicinity of Havana in 2001. Although transformation parameters were developed between the modernized astrogeodetic system (NAD 27) and WGS 84, efforts are planned to establish a cutting-edge system of reference stations similar to the National Continuously Operating Reference System (CORS) of the United States. The seven-parameter transformation is reported to be a Coordinate Frame Rotation **from** NAD27 **to** WGS 84: $\Delta X=+2.478$ m, $\Delta Y=+149.752$ m, $\Delta Z=+197.726$ m, $R_x=-0.526356$, $R_y=-0.497970$, $R_z=+0.500831$, $\delta s=+0.6852386$, the rotations are in arc seconds and the scale factor is in parts per million *(ibid., Dr. Ernesto Rodriguez Roche).* The accuracy is reputed to be good to a meter. A test point offered by Mr. Pablo Velazco Villares from the "PROJ mailing list" is **from** NAD 27: $\varphi=20°\ 22'\ 16.1900''$ N, $\lambda=76°\ 38'\ 28.0824''$ W **to** WGS 84: $\varphi=20°\ 22'\ 17.8949''$ N, $\lambda=76°\ 38'\ 27.1732''$ W.

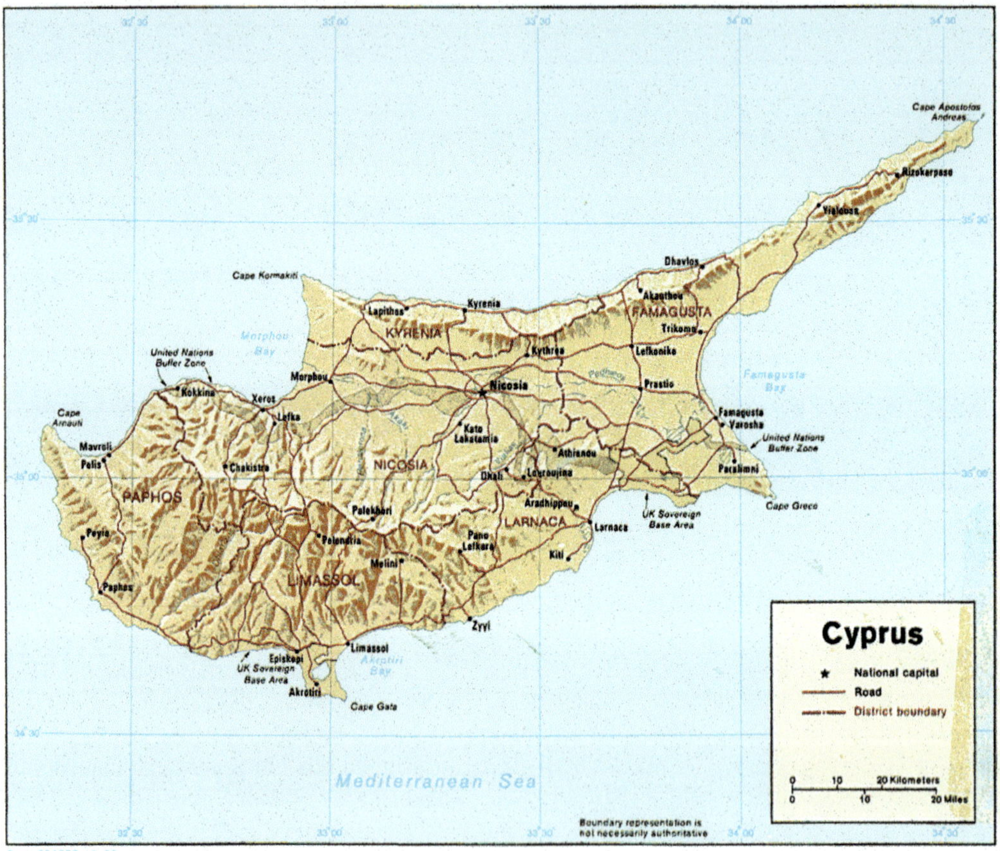

Base 504556 10-80

Republic of Cyprus

Cyprus has an area of 9,250 km² and is about 0.6 times the size of Connecticut. The island comprises two almost parallel mountain ranges going from east to west and an almost flat plain extending between the ranges from Morpho to Famagusta through Nicosia, the capital. The coastline is 648 km, the lowest point is the Mediterranean Sea (0 m), and the highest point is Olympus (1,952 m).

The island of Cyprus has evidence of human inhabitation dating back to 6,000 B.C. Influenced by Minoan and Mycenaean cultures, the island was colonized by the ancient Greeks and Phoenicians. Richard the Lionhearted captured Cyprus during the Third Crusade in 1191 A.D. The government was administered by Great Britain from 1878 to 1914 and became a British Crown Colony in 1925. Cyprus became an independent republic in 1960. Unrest and armed strife between Turkish and Greek inhabitants prompted establishment of a UN Peacekeeping force in 1964. Turkish troops invaded in 1974 and established a separate Turkish Federated State of Cyprus, which was proclaimed in 1975. (Thanks to *Webster's Geographical Dictionary*.) "The entire island entered the EU on 1 May 2004, although the EU acquis - the body of common rights and obligations – applies only to the areas under the internationally recognized government and is suspended in the 'TRNC' – *(Turkish Republic of Northern Cyprus – Ed)*. However, individual Turkish Cypriots able to document their eligibility for Republic of Cyprus citizenship legally enjoy the same rights accorded to other citizens of EU states" (*World Factbook*, 2022).

Cyprus has appeared on maps since about 1200 A.D.; the first Italian map of Cyprus was produced in 1478 (Bartolomeo Zamberti dalli Sonetti) and reproduced in 1490. Subsequent versions appeared in 1528 (Genedetto Bordone), 1538 (Matheo Pagano), and 1585–1595 (G. F. Camocio). The first hydrographic survey work in Cyprus was performed in 1849 by Lord Lieutenant John T. Browne of the Royal Navy and by Captain Thomas Graves of the H.M.S. *Volage*. During that same period, several benchmarks were established on public buildings in Limassol and Cape Gata, all referenced to chart datum. Records also show that real property conveyances were recorded in a cadastre.

Lieutenant H.H. Kitchener, later known as Lord Kitchener of Khartoum, based the first known accurate map of Cyprus on a classical triangulation survey from 1878 to 1883. Kitchener, assisted by a number of Royal Engineers started the survey in September 1878. In February 1883, he reported to the Chief Secretary of Cyprus that the triangulation and survey had been completed and that all the original observations and coordinates had been sent to Stanford's Geographical Establishment in London for publication. The planimetric map (scale: 1 inch = 1 mile) was published in 15 sheets, copies of which are still maintained in the offices of the Department of Lands and Surveys in Nicosia.

In July 1880, a law was passed providing for a fiscal survey. This commenced in 1883, following the completion of the planimetric survey. Kitchener's successor was Grant who was Director of Surveys in 1884 and was responsible for a resurvey of Cyprus at a scale of 4 inches to 1 mile. The survey was only performed with a magnetic compass and the basic triangulation was used as control. Therefore, the survey was insufficiently reliable to form a basis for a land registration system and was even unsuitable for its intended fiscal purpose.

A new survey was started in 1904 for the purpose of a cadastre in the Famagusta district. The survey commenced in the southeastern portion of Famagusta and was based on Kitchener's triangulation. The planetable survey was purely graphical, and no field notes were recorded. Subsequently called the "unsound survey," it continued for 7 years over the Famagusta plains. Nevertheless, it has formed the basis for a general registration of land titles and has been in use until the present!

In 1909 a law was enacted that required compulsory registration of titles to all real property in Cyprus as well as forming the basis for a uniform assessment for taxation. The existing planetable compilation of the Famagusta plains was adequate, but it was not suitable for extension into the hilly areas. The system became unusable when the survey reached the hills of the northern range, which extends into the Karpas peninsula. A subsequent minor triangulation of the Karpas peninsula was undertaken in September 1911.

Captain Lyons, D.Sc., F.R.S., Director-General of the Survey of Egypt was invited to Cyprus in November 1911 for a couple of weeks to consult on the need for further geodetic control of the island. Lyons' report of 1912 is still the basis of the design of the cadastral survey used to the present day. He recommended that the basic scale of 1:2,500 was the best for rural areas and the urban areas should be mapped at a scale of 1:250 or larger. He recommended that a new and more accurate triangulation be undertaken to replace the Kitchener triangulation (1878–1882) as well as because many of the original monuments had become lost or destroyed.

The uniform re-triangulation of the island commenced in 1913 and lasted 18 months, including computations. On completion of the primary triangulation, it was broken down into third-order and fourth-order surveys for detail mapping. After Famagusta and Karpas, the survey of Kyrenia was undertaken with the same specifications. However, it soon became evident that the legislated deadline for a comprehensive island-wide mapping could not be accomplished in 10 years with the methods currently being employed. Therefore, a tacheometric planetable survey at a scale of 1:5,000 was initiated for the remainder of the rural areas of Cyprus. By 1929, 20 years after the law had been enacted, all real property on the island was surveyed by one method or another and that comprised the "cadastral survey."

For the cadastral survey, the island was divided into 59 rectangular sheets, each comprised of 8×12 square mile areas. Each of these sheets was further divided into 64 parts called cadastral maps, with each map covering an area of $1 \times 1\frac{1}{2}$ square miles. The survey of Cyprus was cast on the Cassini-Soldner projection where the latitude of origin $(\varphi_o) = 35° 00' 00''$ N and the central meridian $(\lambda_o) = 33° 20' 00''$ E, referenced to the Clarke 1858 ellipsoid where $a = 6378235.6$ m, and $1/f = 294.2606768$. In 1928, Brigadier Winterbotham recommended that all cadastral maps be reduced graphically and mechanically in scale to produce the 2 inches to 1 mile scale series of maps as well as the ¼-inch scale Cyprus Administration map and the ½-inch scale Cyprus Motor map.

In 1931, the British Naval ship *HMS Ormonde* made astronomical observations at several triangulation stations in Cyprus. The origin point for Famagusta Bastion Datum at Cavalier Flagstaff is, where $\Phi_o = 35° 07' 02.044''$ N, $\Lambda_o = 33° 57' 17.053''$ E, $\alpha_o = 228° 22' 22''$ to station Acheritan, and is referenced to the favorite of the British Admiralty at the time, the Clarke 1858 ellipsoid. In similar fashion, the origin point for Kantara Datum on the Karpas Peninsula was observed, where $\Phi_o = 35° 20' 43.05''$ N, $\Lambda_o = 33° 59' 59.228''$ E, and of course the Clarke 1858 was the reference ellipsoid. When the reduced astro positions were compared to the published triangulation positions of the collocated points, it was noted that Cyprus had been located about 1 minute of arc too far to the east and about 30 seconds of arc too far south, or about 6,000 feet too far east and 3,000 feet too far south. The Director of Surveys and the British Admiralty's Hydrographic Department agreed to adopt the HMS *Ormonde*'s values for longitude and to leave the latitude unchanged. In order to avoid any shifting of the sheet lines of the cadastral survey, it was decided to subtract 1 minute of arc from the longitude of each triangulation point as well as from the point of origin of the Grid system. The new 1935 basic point of origin for the island's Cassini-Soldner Grid system was modified to be such that: the latitude of origin $(\varphi_o) = 35° 00' 00''$ N and the central meridian $(\lambda_o) = 33° 19' 00''$ E, and it was still referenced to the Famagusta Bastion Datum of 1935, commonly referred to as the Cyprus 1935 Datum. *HMS Ormonde's* calculations were proven correct when the British Army connected Cyprus trigonometrically to Syria in 1944 and to Turkey in 1954, the latter with the European Datum of 1950.

During World War II, the Director of Military Surveys of the British Army undertook the preparation of a 1:50,000 scale topographic map of Cyprus in color based on the 2 inches to 1 mile series, revised in the field by planetable methods and verified by aerial photography. The altitudes were obtained from aneroid altimeter traverses, which were in turn connected to the trigonometric control already heighted. This provided vertical control for photogrammetric plotting by Multiplex *(the same type of instrument that I used when I learned how to "push a dot" – Ed.)* and produced contours at a 100-foot interval with dubious accuracies. The map series of the island was divided into 16 sheets, and was cast on the Cassini projection, but was referenced to the Clarke 1880 ellipsoid, where $a = 6,378,249.145$ m and $1/f = 293.465$. The map series established by the British Army General Staff, Geographical Section (GSGS) and published by the U.S. Army Map Service, U.S. Lake Survey in 1944 is where the "Cyprus Grid" is referenced to the Clarke 1880 ellipsoid, the latitude of origin $(\varphi_o) = 35°\ 00'\ 00''$ N and the central meridian $(\lambda_o) = 33°\ 19'\ 00''$ E, the same as the Cyprus Datum of 1935, but with a False Easting = 200 km, and a False Northing = 150 km. The limits of the GSGS Cyprus Grid were for the north, the parallel of 34° 45′ N eastwards to 33° 45′ N and thence a loxodrome to 36° 10′ N, 35° E; the East limit was the meridian of 35° E; the South limit was the parallel of 34° N; and the West limit was the meridian of 31° 50′ E. The color of the Grid was purple. A "Cyprus Grid" test point provided by the Lake Survey is: $\varphi = 34°\ 30'\ 17.366''$ N, $\lambda = 33°\ 54'\ 39.271''$ E, X = 254,573.63 m, Y = 95,229.58 m.

Although Cyprus was triangulated by Kitchener in 1878, re-triangulated in 1913 on Lyons' recommendations, and trilaterated in 1962 with Tellurometer instruments, the Island of Cyprus had never been covered by an adequate network of elevation benchmarks. The need for a First Order leveling net was realized when topographic mapping was started in 1964 for development projects. The First Order leveling started in 1964 and completed in 1966 provided for consistent contouring for the island as a whole.

The existing geodetic network has proven to suffer from both observational and computational errors. The Department of Lands and Surveys decided to establish a new Primary Geodetic Network for the basis for both the Resurvey program and the Land Information System. After a number of studies and tests, it was decided to adopt the GRS80 ellipsoid, and a local datum. A Local Transverse Mercator (LTM) projection system was also chosen. The Cyprus LTM has the same Central Meridian (33° E) as the zone 36 of the UTM System, but a better scale factor (0.9999) was used. With these elements, the maximum scale error in any place of the country is 1:7500. Using GPS technology, a new primary and secondary network has already been established. The primary network consists of 40 points and the secondary network consists of 254 points. The Tertiary network will consist of points, 200–500 m apart.

To add to the bewildering array of ellipsoids, grids, and datums comes the AMS/NATO convention of casting all 1:50,000 scale topographic map series on the European Datum of 1950 as mentioned above when the British Royal Engineers connected Cyprus to Turkey (and earlier to Syria). For the military (only) topographic maps of Cyprus, the published transformation parameters *from* ED50 *to* WGS84 are: $\Delta X = -104$ m ± 15 m, $\Delta Y = -101$ m ± 15 m, $\Delta Z = -140$ m ± 15 m, and this solution was based on 4 station observations. Someday, perhaps the Cypriot government will publicly release the transformation parameters among their local coordinate systems and an ITRF epoch.

The majority of the historical details were garnered from a UN Report prepared by A. Christofi of the Department of Lands and Surveys, Cyprus in 1970. Thanks go to Mr. John W. Hager for the coordinates of the various datum origins and helpful insight regarding Admiralty Office customs. "Geodesy/Hydrography/Photogrammetry is responsible for maintaining the Geodetic Network of Cyprus as well as developing specialised services and mapping products. The Branch deals with maintaining and developing a DTM/DSM for Cyprus and extensively supports the maintenance of orthophotos, satellite imagery, aerial photography, and LIDAR data. Recent data have been captured during 2014 which will be used extensively throughout the Department in order to support its daily operations, as well as being provided though the web. The Department also acts as the official Hydrographic organisation for Cyprus and deals extensively with Nautical and Aero Nautical Mapping" (*Republic of Cyprus Department of Lands and Surveys 2022*).

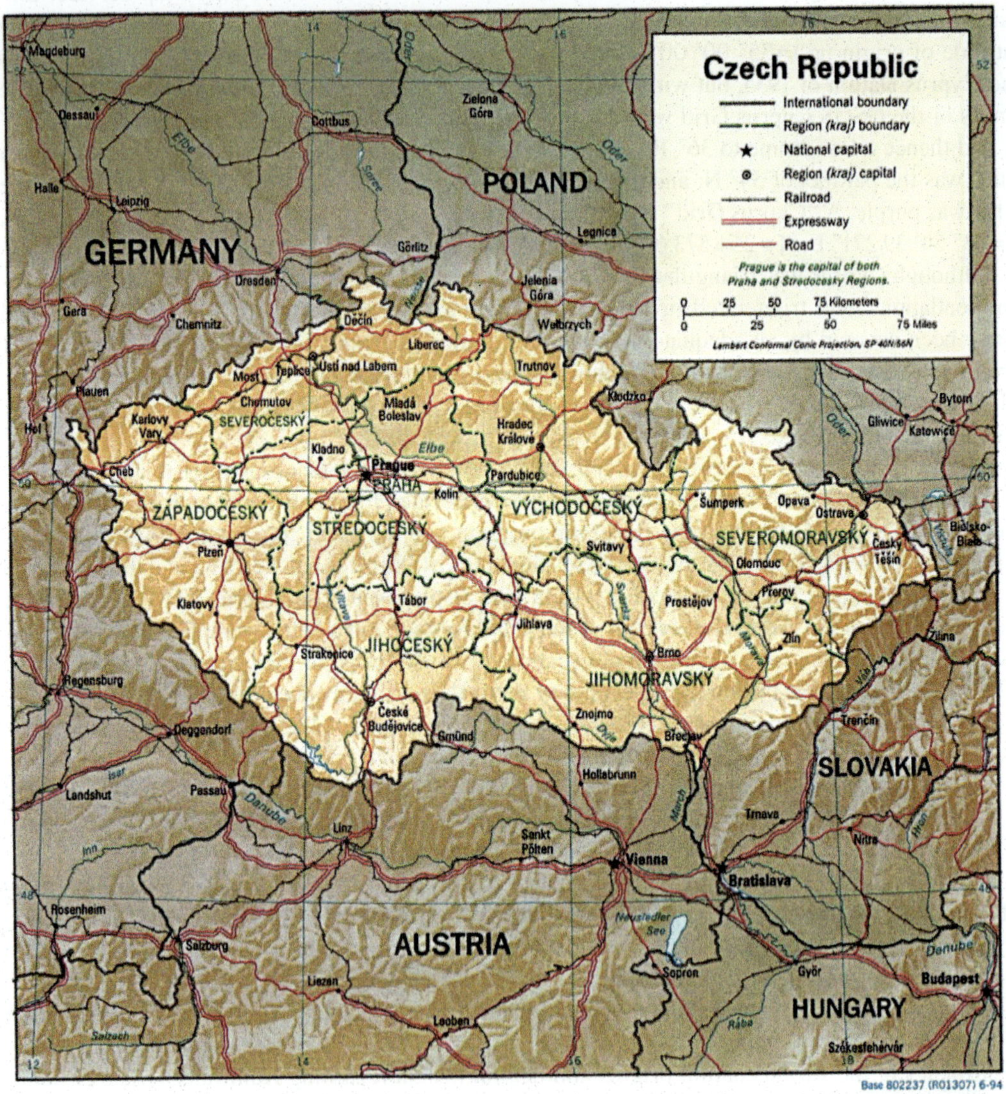

Base 802237 (R01307) 6-94

The Czech Republic

Czechia is bordered by Austria (402 km), Germany (704 km), Poland (699 km), and Slovakia (241 km). The Czech Republic is north of the Danube Valley and is largely mountainous. Bohemia, in the west, consists of rolling plains, hills, and plateaus surrounded by low mountains; Moravia, in the east, consists of very hilly country. The lowest point of the republic is along the Elbe River (115 m), and the highest point is Snezka (1,602 m) in the Krkonose Mountains.

Once part of the Holy Roman Empire, the first Czechoslovakian Republic (1918–1939) was formed by Czechs and Slovaks from territories that were formerly part of the Austro–Hungarian Empire. Those provinces were Bohemia, Moravia, Silesia, and the northern part of Hungary settled by Slovaks and Ruthemians (Slovakia and Carpatho-Ukraine). Hitler supported the ethnic Germans living in the Sudeten region as an excuse to the German annexation, which reduced it to the Protectorate of Bohemia and Moravia (1939–1945). The remaining portions of the republic were broken up among neighboring countries during World War II. After the war, it emerged as the Second Czechoslovakian Republic, but without the Carpatho-Ukraine. The USSR gained control, and Czechoslovakia became part of the Warsaw Pact in 1955. Consequently, in discussing the surveying and mapping of this country, a differentiation is made with respect to the activities of four separate epochs, *i.e.*, the First Czechoslovakian Republic, the Protectorate, the Second Czechoslovakian Republic, and the Czech Republic. On January 1, 1993, the country peacefully split into its two ethnic components, the Czech Republic and Slovakia. "The Czech Republic joined NATO in 1999 and the European Union in 2004. The country formally added the short-form name Czechia in 2016, while also continuing to use the full form name, the Czech Republic" (*World Factbook*, 2022).

The original triangulation of the region by first-order methods was by the III K. und k. military triangulation of the Austro-Hungarian Empire. The cadastral grids employed by the Hapsburgs were the Böhmen Soldner (Cassini-Soldner) with a $\varphi_o = 48°\ 02'\ 20.5''$ N, $\lambda_o = 14°\ 08'\ 24.15''$ East of Greenwich, and the Mähren Soldner with a $\varphi_o = 48°\ 12'\ 32.75''$ N, $\lambda_o = 16°\ 22'\ 36.58''$ East of Greenwich. No false origins were used according to the European convention of the time. However, the entire region was not covered by 1918 as published in the *Ergebnisse der Triangulierungen* (Triangulation Results). The responsibility for the survey activities in the First Republic was divided among the following agencies: *Triangulacni Kancelar Ministerstva Financi* (Triangulation Office of the Ministry of Finance); *Katastralni Mérrické Urady* (Office of Cadaster); *Nivelacni Urad Ministerstva Verejnych Praci* (Leveling Office of the Ministry of Public Works); and *Vojensky Zemepisny Ústav* (Military Geographic Institute). The duties of the civilian agencies had some overlap among themselves as well as with the military, but the records of all agencies could be utilized for military purposes. The territory was covered by the old cadastral triangulations with origins at Gusterberg, St. Stephan Tower (Vienna), Gellerthegy, Pschow, and coordinates referring to Vienna University Datum, St. Anna Datum, and Hermannskogel Datum. A new first-order net was started in 1920 and was completed in 1957. The basic cadastral trigonometric net was connected with the first-order nets of Austria, Germany, Poland, and Romania (through the Carpatho-Ukraine and Slovakia).

Between 1918 and 1932, the Military Geographic Institute (MGI) applied the Lambert conformal conic projection for triangulation computations and mapping. This was based on the Hermannskogel Datum of 1871 referenced to the Bessel 1841 ellipsoid of revolution where the semi-major axis $(a) = 6,377,397.155$ m and the reciprocal of flattening $(1/f) = 299.1528128$. The Hermannskogel 1871 Datum has its origin with $\Phi_o = 48°\ 16'\ 15.29''$ N, $\Lambda_o = 33°\ 57'\ 41.06''$ *Est de l'Ile de Fer* (East of Ferro Island in the Canaries), where Ferro $= 17°\ 39'\ 46.02''$ East of Greenwich and azimuth to station

Hundsheimer is $\alpha_o = 107°\ 31'\ 41.7''$. The secant Lambert Grid had the standard parallels of $\varphi_N = 50°\ 15'$ N and $\varphi_S = 48°\ 30'$ N, a Central Meridian $(\lambda_o) = 35°\ 45'$ East of Ferro, a False Easting = 1,000 km, and a False Northing = 500 km. This point corresponds to the center of the southern sheet line of the 1:75,000 sheet titled "4260 Vsetin." According to Prof. Veverka, "This so called 'provisional military mapping' was only episode and only 3% of the state territory were mapped. Author of projection was Captain Benes, and this projection is practically Lambert because map sheets from this mapping are not available."

Josef Krøvák (*commonly spelled Krovák or Krovak*) prepared the Conformal Oblique Conic Projection of Czechoslovakia in 1922 for the preparation of cadastral (tax) maps and topographic maps of medium scales for the civil geodetic service of Czechoslovakia. The "starting meridian" was termed Ferro where the MGI usage differed from the civilian definition listed above. The MGI used the relation of Ferro = $17°\ 39'\ 45.90''$ East of Greenwich. The Krøvák Projection is a double projection in that the oblique conic is projected from the Gaussian Sphere, where the radius = 6,380,703.6105 m. The Gaussian Sphere was "invented" by Carl Friederich Gauss and is also commonly known as the "conformal sphere." It is simply the geometric mean of the ellipsoidal normal (at a point) terminated by the semi-minor axis and the radius of the ellipsoid (at the same point) in the plane of the meridian. To be succinct, it's $[\eta\rho]^{-1/2}$ evaluated in this case at $\varphi = 49°\ 30'$ N. For the Mapping Scientists and Photogrammetrists that do not live in South Louisiana, this is the same formula used in the commonly used radius of the earth for the "sea-level correction" in establishing ground control. For the Czech Republic, they used the Bessel 1841 ellipsoid at that latitude. Whenever one sees the term "double projection," the generating sphere is usually the Gaussian Sphere. The Krøvák oblique cone has a pole centered at $\varphi = 59°\ 42'\ 42.6969''$ N, $\lambda = 42°\ 30'$ East of Ferro (southwest of Helsinki, Finland). The spherical cartographic coordinates are transformed into the rectangular plane coordinates of the uniform cadastral system. For this purpose, the reduced (0.9999) Gaussian Sphere is projected on the surface of an oblique cone touching the sphere around the central cartographic parallel, having a cartographic latitude of $78°\ 30'$ N, with the vertex in the extended axis connecting the center of the sphere at the rotation angle of $30°\ 17'\ 17.3031''$. This was still in use as of 2000 and known as System – Jednotne Trigonometricke Sítì Katastralni or S-JTSK (System of the Unified Czech/Slovak Trigonometrical Cadastral Net). The Czechs state (Prof. Ing. Bohuslav Veverka, Dr Sc., Prague, November 1997) that the "scale, location and orientation of the S-JTSK on the surface of the Bessel's ellipsoid was derived from the results of the historical Austro/ Hungarian military surveys in the years 1862–98. There are 42 identical points on the Czech territory used for transformation computations. Astronomical orientation was measured only on the Hermannskogel trigonometrical point in Austria, scale factor was derived from the basis of the geodetic length in Josefov." Professor Veverka published programs written in the Pascal and Visual Basic languages that perform the direct and inverse transformations with the Krovak Projection. Note that the X-axis normally coincides with the meridian $42°\ 30'$ East of Ferro increasing South, and the orthogonal Y-axis is increasing West. A test point provided lists: $\varphi = 48°\ 07'\ 46.2973''$, $\lambda = 35°\ 42'\ 35.2147''$, Y = 504,691.675 m, X = 1,289,068.724 m. The Krøvák Projection was officially adopted by the Czech military in 1932.

In 1939, the Germans found that during their occupation of the "Protectorate of Bohemia and Moravia," only 5% of the Protectorate's territory was covered by the new topographic survey. The *Landesvermessungsamt Böhmen und Mähren* (Land Survey of Bohemia and Moravia) was formed. The subsequent triangulation was incorporated into the final *Reichsdreiecksnetz* (Triangulation Net of the Empire), with the datum origin being at Pottsdam, where $\Phi_o = 52°\ 22'\ 53.9540''$ North, $\Lambda_o = 13°\ 04'\ 01.1527''$ East of Greenwich. The defining azimuth to station Golmberg is: $\alpha_o = 154°\ 47'\ 32.19''$, and the ellipsoid of reference is the Bessel 1841. For the purpose of incorporation, the *Reichsdreiecksnetz* was extended over the territory of the Protectorate, and 36 first-order stations were reobserved with two new base lines measured at Podébradý and Kromériÿ. The Grid system used was the German Army Grid or *Deutsche Herres Gitter* (DHG) which had the exact same parameters of the UTM Grid except that the scale factor at origin was unity. The DHG is exactly the same as the USSR's Grid (Russia Belts) except for the ellipsoid.

After World War II, the agencies responsible for geodetic, topographic, and cartographic activities in the Second Czechoslovakian Republic were in a stage of re-organization up to the end of 1953. During the years 1953–1954, those agencies were subsequently organized according to the pattern established in the USSR. The *Ustredni Správa Geodesie a Kartografie* – USGK (Central Administration of Geodesy and Cartography) was established. The *Zakladni Trigonometrica Sit* – ZTS (Basic Trigonometric Net), included the first-order net of the Protectorate and the first-order net established in 1949–1953 in Slovakia. The adjustment of the net was carried out by the method of Pranis-Praniévitch on the Krassovsky 1940 ellipsoid where $a=6,378,245$ m, and $1/f=298.3$. The Datum is defined as "System 42" where the origin is at Pulkovo Observatory: $\Phi_o=59°\ 46'\ 18.55''$ North, $\Lambda_o=30°\ 19'\ 42.09''$ East of Greenwich. The defining azimuth at the point of origin to Signal A is: $\alpha_o=317°\ 02'\ 50.62''$. The "Russia Belts" Grid System is used with the System 42 Datum; identical to UTM except that the scale factor at origin is unity.

A civilian version used since 1952 is a modification of the Russia Belts system in that the False Northing at origin was $\varphi=49°\ 30'$ North, the False Northing$=200$ km, the False Easting$=500$ km, and the scale factor at origin $(m_o)=0.99992001$. Everything else remained the same as the standard Gauss-Krüger Transverse Mercator Grid. Boundary treaties with adjacent countries refer to ancient datums and grids that include the old double stereographic projections of the 19th and early 20th centuries.

As far as the civic sphere it seems that S-JTSK (Krøvák Projection) is alive and well in the Czech Republic for the 21st century. This system is officially used in Czech State Cadastre (maps 1:1,000, 1:2,000, 1:5,000 and 1:2,880). In S-JTSK are also produced Basic State Maps (1:10,000, 1:20,000, 1:50,000, 1:100,000, 1:200,000) where digital form for whole state territory is in development. Civilian mapping (cadastral and basic) is directed by Zememericky urad (Surveying Office) in Prague.

"In the Czech GIS society system WGS 84 plays an important and increasing role, especially in the field of photogrammetry and remote sensing. Applied research in the department is in the first place directed to the present needs of the branch of the Czech Office of Surveying, Mapping and Cadastre in the field of geodetic control networks, their update and integration within the modern European geodetic control. Recently the project of incorporation of the Czech and Slovak Astrogeodetic Network, together with the primary networks of Austria, Hungary, and Germany, into the European terrestrial continental reference system ED87 has been accomplished. The department participated in the extension of the new European reference frame EUREF to the Czech and Slovak Republics, in the establishment of the Czech zero-order GPS reference network and in its densification to the average density 1 station/400 km^2 as well as in the direct connection of the Czech GPS reference network with reference networks of Germany and Austria. In the establishment of GPS reference networks, the department takes part by both the observations and the processing by the scientific software package. The most topical task is the establishment and monitoring of the Czech National Geodynamical Network by GPS technique. The network consists of 32 stations at which GPS observations have been periodically repeated (twice a year) since the spring 1995. These observations are supplemented by levelling and gravity measurements. The objective of this project is a stability assessment of the horizontal control network and its integration with the vertical control network of the Czech Republic. The department takes an active part in the process of incorporation of the Czech national geodetic control into the continental European geodetic control by participation in the international projects EUREF and EUVN (its realization is now in its initial stage). With the help of the national GPS reference networks the present user terrestrial reference frame (S-JTSK/95) has been improved and is now passing through large scale disputations. For the cadastral offices the local networks, densified by themselves by GPS techniques, are tied to the national GPS reference network. A reference quasi geoid model was computed for the territory of the Czech Republic with an accuracy of about 5 cm which makes it possible to determine sea level heights from GPS observations at any place of the state territory. The stability of the reference frame is continuously monitored by evaluation of observations of a cluster of European IGS stations.

For the calibration and processing of the GPS technology, the calibration network was established at Geodetic Observatory Pecny and Skalka. In the field of the geodetic control networks the department collaborates with the Land Survey Office and with the Department of Advanced Geodesy of the Faculty of Civil Engineering of the Czech University of Technology in Prague."

Thanks to Prof. Ing. Bohuslav Veverka, Dr. Sc. for editing many fine points.

REFERENCES: C

A. Christofi, UN Report, Department of Lands and Surveys, Cyprus 1970.
Andrew M. Glusic, Mapping of the Countries in Danubian and Adriatic Basins, Army Map Service Technical Report No. 25, June 1959.
Annales Hydrographiques, 4th Series, Vol. I, p. 159 ff.
Annales Hydrographiques, Mission Géodésique au Moyen Congo, Avril-Septembre 1959.
Boundaries, Possessions, and Conflicts, 1938.
Cabinet du Directeur, Institut Géographique National, 1945.
Cameroun, Ville De Douala, Triangulation Principale, Triangulation Secondaire, 11 Nov. 1965.
Chharom Chin, Senior GIS/RS Officer, Geography Dept., Ministry of Land Management, Urban Planning and Construction, Cambodia, 22–24 February 2007.
Chris Lunnay, Director, Land Equity International, Australia, Personal Communication, (Mar. 2008).
COL. Rodolfo Concha Muñoz, U.S. Army Map Service and Inter-American Geodetic Survey, 1952.
Comité Spécial du Katanga, Les Travaux Géodésiques du Service Géographique et Géologique, 1954.
Dr. Ernesto Rodriguez Roche, Personal Communication (2010, Havana, Cuba).
EPSG, 2007.
Foreign Maps, TM 5–248.
Frank Kuwamura, Defense Mapping Agency, Personal Communication, June 1980.
http://www.state.gov/e/oes/ocns/opa/c16065.htm.
IAGS Report (in Spanish) by Army Map Service, ca. 1960.
Institut Géographique National (I.G.C. Instruction No.1212).
International Boundary Study No. 105 – October 15, 1970.
International Boundary Study No. 110 – May 14, 1971.
International Boundary Study No. 127 – September 8, 1972.
International Boundary Study No. 145 – July 17, 1974.
International Boundary Study No. 83 – July 15, 1968, U.S. Department of State.
J. T. Fang, Empire Survey Review, 1949.
John W. Hagar, Personal Communication (May 2005).
John W. Hager, Personal Communication (1 July 1982).
John W. Hager, Personal Communication, (23 May 2014).
John W. Hager, Personal Communication (Apr. 2007).
John W. Hager, Personal Communication (Apr. 2007).
John W. Hager, Personal Communication (Aug 2004).
La Registro Nacional de la Republica de Costa Rica.
Library of Congress Country Studies, 1988.
Library of Congress Country Studies, Library of Congress Country Studies, 2008.
Limits in the Seas No. 129, Cabo Verde Archipelagic and other Maritime Claims and Boundaries.
Lonely Planet, 2012.
M. Bosiljevac and Ž. Bačič, The Implementation of New Official Geodetic Datum and Map Projections in the Republic of Croatia, FIG Congress 2010, Sydney, Australia 11–16 April 2010.
Malcolm A. B. Jones, Personal Communication (Apr. 2008).
Malcom A. B. Jones, Personal Communication (May 2005).
Manual of Photogrammetry, 5th & 6th editions, Chapter 3.
Manual of Photogrammetry, Chapter 3, 5th and 6th editions.
Melita Kennedy, Personal Communication (May 2005).
Mercator Transverse Universelle in French Territories, Translated by D. R. Cockburn and W. L. Barry, Army Map Service, 20 September 1950.
National Administration of Surveying, Mapping and Geoinformation of China 28 Lianhuachi West Road, Haidian District, Beijing, 100830, China, 2013-12-20.

References: C

Pablo Velazco Villares, "PROJ mailing list," 2010.
Paul M. Dibb, Lands and Survey Dept., Cayman Islands Government, Personal Communication (Oct. 1997).
Paula Cristina Santos, AS Missãoes Geodésicas na Comissão de Cartografía (1883–1936), IICT, Lisbon, 2007.
Prof. Ing. Bohuslav Veverka, Dr. Sc., Personal Communication (Dec. 1999).
Republic of Cyprus Department of Lands and Surveys 2022.
Roche & García, FIG Congress 2010, Sidney, Australia.
Royal Geographical Society Technical Series No. 4.
TM-5-248 Foreign Maps 1963.
U.S. Department of State Background Note, 2010.
U.S. Department of State Background Notes, 2010.
U.S. Dept. of State Background Note, 24 February 2010.
U.S. Dept. of State Background Notes, 2012.
Webster's Geographical Dictionary.
Webster's New Geographical Dictionary.
World Factbook, 2007.
World Factbook, 2010.
World Factbook, 2012.
World Factbook, 2014.
World Factbook, 2022.

D

Kingdom of Denmark 215
Republic of Djibouti 219
Commonwealth of Dominica 223
Dominican Republic 225

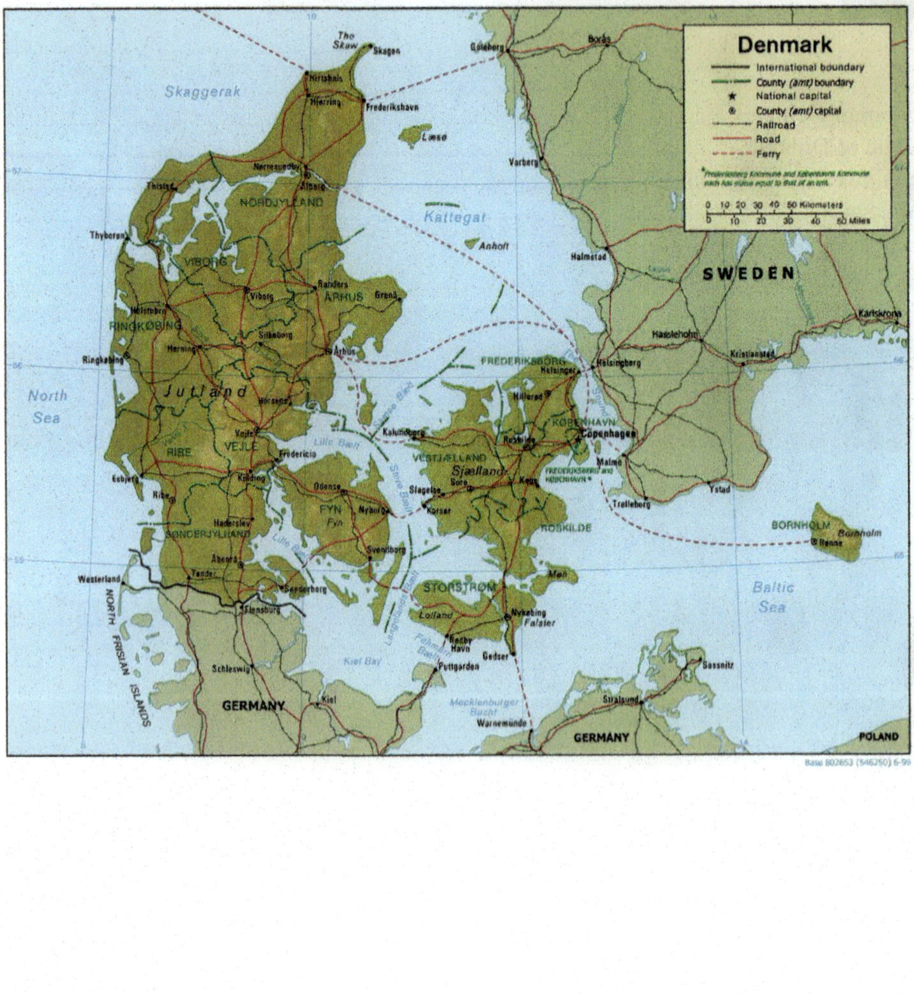

Base 802853 (546250) 6-99

Kingdom of Denmark

Denmark is in northern Europe, bordering the Baltic Sea and the North Sea, on a peninsula (Jutland) that is north of Germany (140 km), as well as a number of islands that include Sjælland, Fyn, Falster, Lolland, Langeland, and Bornholm. Also, parts of Denmark are self-governing Greenland and the Færoe Islands. The kingdom is slightly less than twice the size of Massachusetts; the lowest point is Lammerfjord (−7 m) and the highest is Ejer Bavenjoj (173 m). The coastline is 7,314 km, and the kingdom controls the Danish Straits (Skagerrak and Kattegat), linking the Baltic and North Seas.

According to *Webster's New Geographical Dictionary,* Danes, a Scandinavian branch of Teutons *circa* 6th century A.D., settled the area. Denmark participated in raids on England, France, and the Low Countries in the 8th to 10th centuries. The Danes converted to Christianity in the 10th and 11th centuries, and the United Danish kingdom included Schleswig, Southern Sweden, England, and intermittently Norway. "During the Napoleonic Wars Britain attacked Copenhagen twice, inflicting heavy damage on the Danish fleet in 1801 and leaving much of Copenhagen ablaze in 1807. The Swedes then took advantage of a weakened Denmark, successfully demanding that Denmark cede Norway to them. The 19th century might have started off lean, dismal and dominated by a small Frenchman with a big ego, but by the 1830s Denmark had awakened to a cultural revolution in the arts, philosophy, and literature. A democratic movement in Denmark led to the adoption of a constitution on 5 June 1849, which in turn led to the formation of a Danish constitutional monarchy. Germany took control of Schleswig in southern Jutland, after its inhabitants, people of both Danish and German heritage, revolted against the new constitution," thanks to Lonely Planet. "It joined NATO in 1949 and the EEC (now the EU) in 1973. However, the country has opted out of certain elements of the EU's Maastricht Treaty, including the European Economic and Monetary Union, European defense cooperation, and issues concerning certain justice and home affairs" (*World Factbook*, 2022).

The first triangulation and plane table mapping *(Målebordsblade)* of Denmark was performed by the Royal Danish Society of Sciences and Letters *(Videnskabernes Selskab),* from 1762 to 1821. The *Målebordsblade* of the Færoe Islands was performed from 1791 to 1795 and for many years was the basis of hydrographic charts of the surrounding waters. The General Staff was established in 1808, and the Quartermaster General started a new *Målebordsblade* in 1809 at a scale of 1:20,000 for a planned final compilation scale of 1:60,000. In 1830, this plan changed the scale to 1:80,000. From 1842 to 1887, the country was plane tabled at a scale of 1:20,000 and final sheets were published at that same scale beginning in 1864. A second survey of the country was started in 1887, but coverage was only partial. Based on the old existing *Målebordsblade* of the Færoe Islands, the waters surrounding the islands were hydrographically surveyed from 1899 to 1903 and in 1908. From 1924 to 1937 that part of Slesvig that was returned to Denmark was surveyed. The *Geodætisk Institut* was established in 1928 by merging the Danish Geodetic Survey (*Den Danske Gradmåling*) and the Topographic Division of the General Staff (*Generalstabens Topografiske Afdeling*) *and* was responsible for the topographic mapping of the kingdom. In the early 1950s the final publication scales were changed from 1:20,000 and 1:40,000 to 1:25,000 and 1:50,000. The third topographic mapping of Denmark was commenced in 1965 with modern photogrammetric techniques. On September 8, 1987, the Danish Cadastral Department (*Matrikeldirektoratet*), the Hydrographic Department (*Søkortarkivet*), and the *Geodætisk Institut* were transferred to the Ministry of Housing and renamed *Kort-og Matrikelstyrelsen* (KMS).

Over 20 years ago, John W. Hager, now retired of Army Map Service, later DMA/NIMA/NGA, answered my question regarding the coordinates of the datum origin point for the Old Danish Datum of 1878 at station *283 Agri Bavenhöe*, where $\Phi_o = 56°\ 13'\ 48.217''$N, $\Lambda_o = 2°\ 02'\ 22.629''$

West of Copenhagen, or $\Lambda_o = 10°\ 32'\ 17.271''$ East of Greenwich. The azimuth of reference was from *No. 283 Agri Bavenhöe* to station *No. 282 Lysnet* where $\alpha_o = 294°\ 31'\ 14.17''$, clockwise from north. His reference was from the German Army's *Planheft Dänemark – Planheft Übersichten der Kartengitter Europa, Nordafrika und Vorderasien (Berlin – 1942)*.

From that same reference, the earlier Danish triangulation 1817–1870 was reported to have the basic reference at the "Nicolai Tower" or Round Tower *(Runde Tårn)* of the Old Copenhagen Observatory, where $\Lambda_o = 12°\ 34'\ 40.35''$ East of Greenwich. That "basic reference" is likely what the ephemeris was based on with regard to transits of the local meridian at the old observatory's meridian circle. The ellipsoid of reference was the Danish (Andræ) ellipsoid, where $a = 6,377,104.43$ m and $1/f = 300$. The later triangulation of 1926–1933 was referenced to the Hayford 1924 (International) ellipsoid, where $a = 6,377,388$ m, and $1/f = 297$. The datum origin point was actually at station *Punkt Nicolai*, where $\Phi_o = 55°\ 40'\ 42.937''$ N, $\Lambda_o = -0°\ 00'\ 19.908''$ West of *Old* Copenhagen Observatory and the azimuth of reference was from *Punkt Nicolai* to *Store Möllehöi*, $\alpha_o = 14°\ 15'\ 25.18''$ clockwise from south, and the baseline was at *Amager 1838* with a length of 2,701.0732 m, ±4.5 mm. Later determinations of the length of this baseline were *Amager* 1911: 2,701.1242 m, ±1.4 mm and *Amager* 1934: 2,701.137 m, ±2 mm. Note that both of these last two measurements were performed with invar apparatus. With today's GPS technology as a basis of comparison, it's rather humbling to read what they used to accomplish *(in the old days)*, with such simple tools and profound determination and perseverance. A recent note from John W. Hager states, "In the 1817–1870-time frame, longitude may have been measured positive to the west. This does make some sense where Copenhagen is the prime meridian than most of the country, Bornholm excluded, would have positive longitudes."

An object of my curiosity for decades regarding Danish enigmas (hopefully all solved herein) was the references to the "Buchwald Projection," mentioned in the *Planheft* (Top Secret Nazi books on Grids and Datums) as well as other equally obscure sources, as far as the English language is concerned. During the 1990s, I used to lecture at the Stennis Space Center in southern Mississippi when I was still a member of the Faculty of Civil Engineering at the University of New Orleans. My students were comprised mostly of employees of U.S. Naval Oceanographic Office (NAVO) and the Naval Research Laboratory (NRL). I spent a lot of time rummaging through the Maury Oceanographic Library, which is part of NAVO next door to the local branch of the NRL. I found the 1924, Number 1 issue of *"Bulletin Géodésique"* that contained an obituary for Frants Andreas Buchwaldt (1874–1923), pages 104–106. Apparently, Captain Buchwaldt of the Danish Navy was as revered a scientist as Captain Maury was of the U.S. Navy. The Danish System of 1934 that used the Buchwaldt projection for Denmark was a Lambert Conformal (tangent) Conic projection, where $\varphi_o = 56°\ 13'\ 48.217''$ N, $\lambda_o = 10°\ 32'\ 17.271''$ East of Greenwich, the scale factor at origin by definition, $(m_o) = 1.0$, and False Easting = False Northing = 200 km.

Thanks to an undated manuscript (*circa* 1950s) by the late Jacob A. Wolkeau of Army Map Service (AMS), a Circular dated May 31, 1935, was issued by the Minister of the Interior for the German Reich and Prussia, entitled *Amalgamation of Land Survey Offices*. "The circular implements the technical proposals of the 1921 Survey Committee in regard to the unification of the triangulation and altametric system of the several states and the adoption of the Prussian Geographical Coordinate system and datum throughout the Reich. ... The progress so far achieved is confined in the main to North Germany ... while in Mark Silesia the main triangulation has been completed and connection affected with the new Danish Triangulation." An analysis of the observations of classical triangulation in Europe later published by AMS included the "System 34" of Denmark that was observed from 1926 to 1933. Of the 64 triangles in the adjustment of observations with Hildebrand theodolites, the average error of figure closure was 0.43", and the maximum error was 1.823".

A Danish Military Tangent Lambert projection was known to exist from 1926 to ? (unknown) and was referenced to the Danish (Andræ) ellipsoid. The Latitude of Origin, $\varphi_o = 55°$ N, the Central Meridian, $\lambda_o = 10°\ 22'\ 40.35''$ East of Greenwich, the Scale Factor at Origin by definition, $(m_o) = 1.0$, and False Easting = False Northing = zero. A "British Grid" that covered Denmark was the *Northern*

European Zone III (1925–1948). It was also a Lambert Conformal Conic projection; but was used in the secant case such that: the Latitude of Origin, $\varphi_o = 57°\ 30'$ N, the Central Meridian, $\lambda_o = 20°$ East of Greenwich, the Scale Factor at Origin, $(m_o) = 0.99904$, False Easting = 900,000 m, and False Northing = 543,365.71 m. The ellipsoid of reference for this ersatz datum was the Bessel 1841, where $a = 6,377,397.155$ m and $1/f = 299.1528$.

In January of 1988, I received a note from John W. Hager regarding the Færöes: "Enigma for 1987. A 1:20,000 scale series of maps for the Faeroes *(Second edition, Anden Udgave 1941 – Ed.)* – no grid, no graticule. How do you position them? Finally found a book in Danish that gave the clues. Sheet size is 5/4 mil west to east by 1 mil south to north. Is mil in Danish the same as mil in English? Book also says 9.42 km by 7.53 km, so a mil is a mil. Datum point is Tørshavn, $\Phi_o = 62°\ 00'\ 49.1''$ North, $\Lambda_o = -06°\ 45'\ 22.5''$ West of Greenwich. Lambert Conformal Conic projection, Latitude of Origin, $\varphi_o = 62°$ North, Longitude is 3 minutes west of the datum, or $\lambda_o = -06°\ 48'\ 22.5''$ West of Greenwich, False Northing and False Easting each equal to zero, Scale Factor $(m_o) = 1$. Now the kicker, y-axis positive north and x-axis positive west. The origin of the sheet lines goes back to an older series *(Målebordsblade – First edition, Første Udgave 1901 – Ed.)* that was 1 mil north-south by 5/6 mil east-west. The grid for the sheet lines thus starts at 0 m north and –3,140 m in easting. All this on the Danish ellipsoid, $a = 6,377,104.43$ m, and $1/f = 300$." (Note that the Tørshavn Datum has a date associated with an astronomical observation in 1896.)

In May 1998, Anna B.O. Jensen and Finn Bo Madsen published *A New Three-Dimensional Reference Network in Denmark*. The new system is called REFerencenet DanmarK or REFDK and is based on 94 stations that include the larger islands. The transformation parameters, modified to the standard rotations used in the United States, from European Datum 1950 to REFDK are: $\Delta X = -81.070$ m, $\Delta Y = -89.360$ m, $\Delta Z = -115.753$ m, $R_x = +0.48488''$, $R_y = +0.02436''$, $R_z = +0.41321''$, and $\Delta s = 1.0 - 0.540645 \times 10^{-6}$. The average error of transformation is estimated at ±20 cm. A test point provided by the *Bundesamt für Kartographie und Geodäsie* is as follows: **FROM** Denmark ED50: $\varphi = 56°\ 51'\ 00.0''$ North, $\Lambda_o = +08°\ 26'\ 24.72''$ East **TO** ETRS89 (REFDK): $\varphi = 56°\ 50'\ 57.82''$ North, $\Lambda_o = +08°\ 26'\ 19.88''$ East.

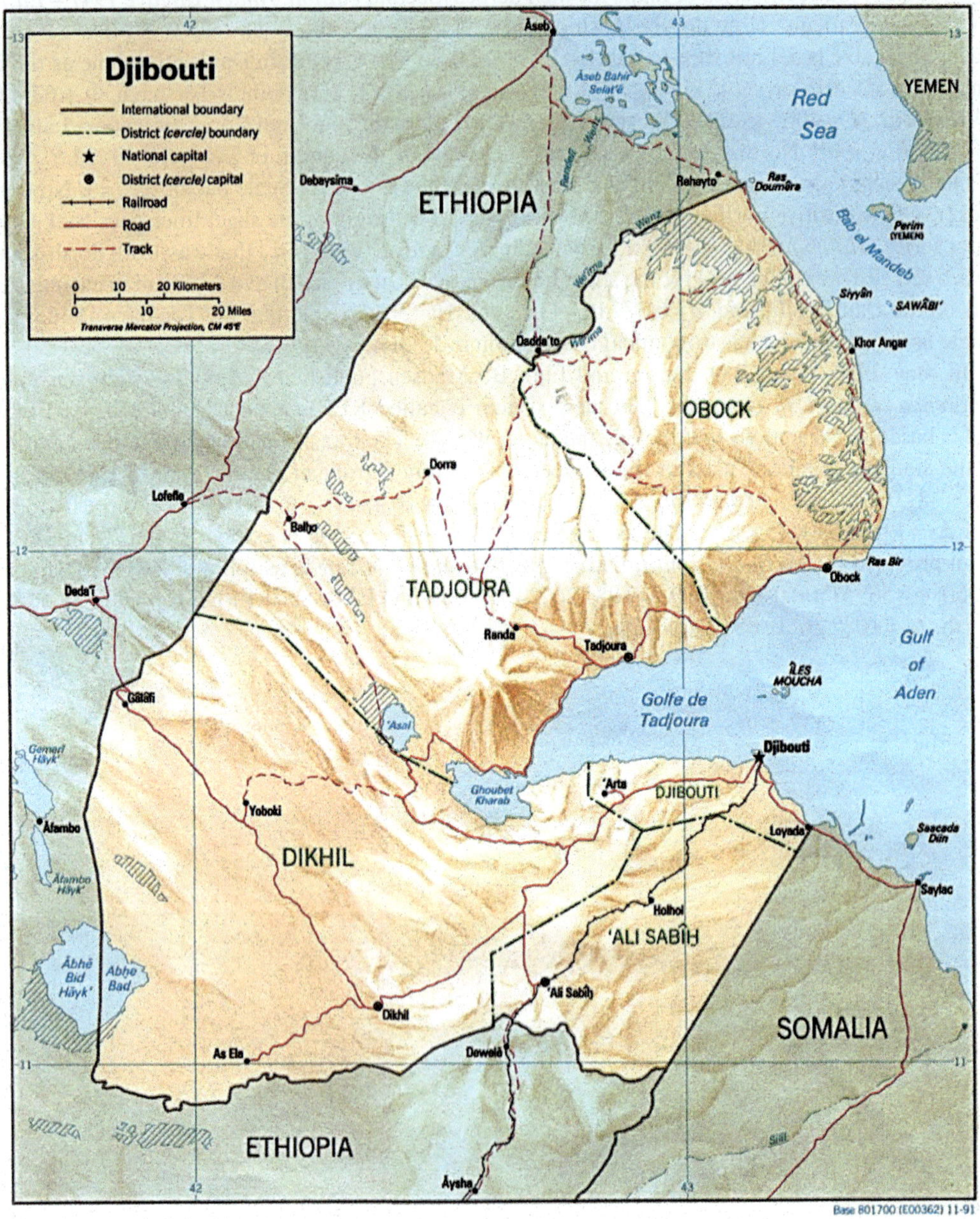

Base 801700 (E00362) 11-91

Republic of Djibouti

Djibouti is slightly smaller than the state of Massachusetts, and is bordered by Eritrea (125 Km), Ethiopia (342 km), and Somalia (61 km). The coastline along the Red Sea is 314 km, the lowest point is Lac Asal at −155 m, and the highest point is Moussa Ali (2,021 m) (*World Factbook*, 2022).

Thanks to *Lonely Planet*, "Despite the inhospitable climate, Djibouti's arid plains have been populated since the Paleolithic era, fought over by Afar and Somali nomadic herds people. Islam spread its prayer rugs from around 825 A.D. in a region that was then used as grazing lands by several tribes, including the Afars from eastern Ethiopia and the Issas from Somalia. Arab traders controlled the region until the 16th century, but the Afar sultans of Obock and Tadjoura were in charge by the time the French arrived in 1862. The French were seeking to counterbalance the British presence in Aden on the other side of the Bab al-Mandab Strait and, after negotiating with the sultans for the right to settle, they bought the place for 10,000 thalers."

"In 1884 and 1885, France expanded its territorial holdings both along the coast and inland of the Gulf of Tadjoura by signing treaties of protection with the Sultans of Tadjoura and Gobad and various chiefs of the Issas Somalis. In 1884, following the completion of Anglo-Somali treaties replacing the earlier trade accords, the protectorate of British Somaliland was established. France and the United Kingdom reached an agreement, February 2–9, 1888, that delimited a boundary between their respective holdings on the Gulf of Aden:

"The protectorates exercised, or to be exercised by France and Great Britain shall be separated by a straight line starting from a point of the coast situated opposite the wells of Hadou [at Loyada], and leading through the said wells to Abassouen; from Abassouen the line shall follow the caravan road as far as Bia-Kabouba, and from this latter point it shall follow the caravan route from Zeyla {Zeila} to Harrar [Hārer] passing by Gildessa [Jaldēsa]. It is expressly agreed that the use of the wells of Hadou shall be common to both parties.

"The Government of Her Britannic Majesty recognizes the Protectorate of France over the coasts of the Gulf of Tajourra [*Golfe de Tadjoura*], including the group of the Mushah Islands [*Îles Musha*] and the Islet Bab [*Île Bab*], situated in the Gulf, as well as over the inhabitants, tribes, and fractions of tribes situated to the west of the line above mentioned.

"The Government of the French Republic recognizes the Protectorate of Great Britain over the coast to the east of the above lines as far as *Bender Ziadah* [Bender Ziada], as well as over the inhabitants, tribes, and fractions of tribes situated to the east of the same line."

The Anglo–French agreement of 1888 determines the alignment of the present Djibouti–Somalia boundary. In 1892 the town of Djibouti was made the capital of the French territory, which became known in 1896 as French Somaliland (*Côte Français des Somalis*). Following WWII, the colony of French Somaliland was made an overseas territory of the French Union; in 1958 it became an overseas territory of the French Community. On July 5, 1967, French Somaliland was renamed the French Territory of the Afars and Issas (*Territoire Français de Afars et Issas*). The French Territory of the Afars and Issas became independent on June 27, 1977, as the Republic of Djibouti." (*International Boundary Study No. 87 [Rev.] Djibouti – Somalia Boundary May 18, 1979, Office of the Geographer, Bureau of Intelligence and Research.*) The boundary with Ethiopia is considerably more complex and is treated fully in *International Boundary Study No. 154*.

The first geodetic and hydrographic expeditions of the French Navy to the coast of French Somaliland were in 1864, 1881, and 1888 to the ports of Obock, Tadjura, and the anchorage of Khor Ambadu. In 1889, a survey was conducted of the port of Djibouti by *Lieutenant de Vaisseau* Cacqueray and *Commandant le Météore*. The *first* Djiboutian geodetic network was originally established by *Lieutenant de Vaisseau* M.R. de Carfort of the French Navy and *Commandant la*

Canonnière l'Etoile during the 1890–1891 triangulation of the littoral (coastal) cape colony, *d'*Obock, north of the city of Djibouti and on the northern coast of the Gulf of Tadjura. The origin of the local coordinate system and datum, Cape Obock 1890, was at the Obock Signal Post (*Mât de Signaux*), where $\Phi_o = 11° 57' 18''$ N, $\Lambda_o = 40° 56' 58.9''$ East of Paris, the azimuth to the Ras-Bir Lighthouse, $\alpha_o = 74° 08' 46''$, the baseline distance was 8,118.08 meters, and the ellipsoid of reference was the Germain, where $a = 6,378,284$ m and $1/f = 294$. The entire triangulation spanned the distance from Djibouti to the Isle of Périm (*Annales Hydrographiques 2e Série, Tome Treizième, Année 1891*).

The *second* Djiboutian Geodetic Survey was conducted by the famous French Hydrographer, M. A. Gougenheim, from December 5, 1927, to April 3, 1928. Gougenheim noted that Carfort's 1891 determination of the lighthouse at Périm was: $\Phi = 12° 39' 15.44''$ N, $\Lambda = 43° 25' 53.38''$ East of Greenwich, but that the British Navy actually adopted: $\Phi = 12° 39' 13''$ N, $\Lambda = 43° 25' 53''$ E. Gougenheim said that the two positions are satisfactory, especially since M. de Carfort admitted that the triangulation of Obock (including Djibouti) to Périm had a relative closure of 1/2,000 with ±2" for (*la difference de latitude*) Périm-Djibouti and ±0.5" for (*la difference de longitude*). Gougenheim inferred that the slight difference was likely due to the deviation of the vertical (*la deviation de la verticale*). The great Ayabelle Lighthouse (*Phare d'Ayabelé*) was therefore determined (in 1928) to have the coordinates: $\Phi = 11° 33' 10.36''$ N, $\Lambda = 43° 01' 23.02''$ East of Greenwich as based on published hydrographic charts of the region.

The British Admiralty Office executed a survey from Aden to *l'Île Périm* in 1876 and in 1904. A point in Aden was observed in 1876–1877 by Captain W.J. Heaviside with a 10-pound theodolite for the determination of latitude, and the longitude of the point was observed in 1882 by Dr. Gill of the Cape Town Observatory. In 1901–1904, the British Survey of India performed a small triangulation survey in the Gulf of Tadjura that included the port of Djibouti, *Mt. Désiré*, and the *Îles Moucha*. The ellipsoid used was not indicated in their report. A position later adopted by the British Admiralty Office for the lighthouse at *Périm* was: $\Phi = 12° 39' 13.05''$ N, $\Lambda = 43° 25' 52.82''$ East of Greenwich.

In 1933–1934, an Italian mission (*bâtiments hydrographes Ammiraglio Magnaghi et Ostia*) executed a complete triangulation survey from Assab to *l'Île Dumeira* and to the Italian-French frontier border. *La Triangulation d'Assab* established two astro stations: Signal Humarrasuh, where $\Phi = 12° 44' 24.38''$ N, $\Lambda = 43° 01' 06.78''$ E, and Signal Gaabla, of particular importance, where $\Phi_o = 12° 42' 43.33''$ N, $\Lambda_o = 43° 08' 07.36''$ E, a baseline was also observed there of 1,032.15 m and with an azimuth from the South End to the North End of $\alpha_o = 336° 39' 21.4''$ observed with a Wild theodolite and the ellipsoid of reference was the Bessel 1841, where $a = 6,377,397.15$ m, $1/f = 299.1528$. The Italian observations of the geodetic distance from Gaabla-Humarrasuh (1933–1934) was 13,060.72 m, with an azimuth of $\alpha = 283° 45' 52.9''$, and the French observations of the geodetic distance (1934–1935) was 13,060.53 m, with an azimuth of $\alpha = 283° 45' 57.6''$. These differences were attributed to differences in the observations of the deflection of the vertical because of the ellipsoids and geoids adopted.

Note that nowadays when we perform a block aerotriangulation of an area in the United States with GPS RTK for camera station control, it's pretty easy to shrug off the magnitude of the areas and the distances that we map. Look at the distances that were physically measured on the ground with invar tapes and the corresponding checks! That is less than a foot in 8 miles as determined by two different European militaries three-quarters of a century ago!

On December 7, 1934, the beginning date of survey, the French Naval officers of the Frigate Fleet Bougainville were as follows: Frigate Captain Mazen, Commandant; Corvette Captain Robain, second in command; *Lieutenant de Vaisseau* Lassave, *Lieutenant de Vaisseau* Guyot (in charge of hydrography); *Enseignes de vaisseau de 1re classe* Salmon (in charge of triangulation); Burser Allain-Dupré; and *Enseignes de vaisseau de 2e classe* Parfond. Ensign Salmon reported that he used a Wild universal theodolite #13 and had 6 male helpers furnished by the local militia. Using a S.O.M. (*Société d'Optique et de Mecanique*) prismatic astrolabe and a Bréguet chronometer controlled by radio signals from *la Croix d'Hins Bordeaux*, he determined the position of the Ayabelle Lighthouse as: $\Phi = 11°\ 33'\ 10.21''$ N, $\Lambda = 43°\ 07'\ 23.83''$ East of Greenwich. With that as an origin, he initially established a local coordinate system with the Musha Lighthouse as: X=+9,729.76, Y=+19,595.54, Maskali Lighthouse as: X=+2,891.62, Y=17,626.72, and Ras Duan (*Dallaï*) benchmark as: X=–3,591.45, Y=+33,382.60. For calculating his chain of triangles from the French–Italian frontier border to *l'Îles Musha*, Ensign Salmon established a secant Lambert Conformal Conic projection (just south of *l'Îles Musha*), where Latitude of Origin (φ_o) = 11° 42′ N, Central Meridian (λ_o) = 43° 12′ East of Greenwich, Scale Factor at Origin (m_o) = 0.99987634, "corresponding to the two standard parallels that are a distance of one *grade* (1^G) apart on either side of the parallel of origin" (where $400^G = 360°$). The ellipsoid of reference chosen was the International (Madrid 1924), where $a = 6,378,388$, $1/f = 297$. The ensign was particularly proud of his choice of origin for his secant Lambert as it was prominently displayed on the Trig List diagram, a rarity in such diagrams for the Annals of the French Navy. (*Reconnaissance Hydrographique de La Côte Française des Somalis au Nord d'Obock, Annales Hydrographiques, 3ᵉ Série, Tome Quatorzième, Annés 1935–1936, Paris 1937.*)

According to DMA/NIMA/NGA *TR 8350.2*, the three-parameter datum shift from "Ayabelle Lighthouse" datum on the **Clarke 1880** ellipsoid to WGS 84 Datum is: $\Delta X = -79$ m, $\Delta Y = -129$ m, and $\Delta Z = +145$ m, each component is listed with an uncertainty of ±25 m and this relation was determined with **one** collocated point as of 1991. NGA needs to loosen up and release a better collection of shift relations to the myriad of native datums in the world.

I once thought that the "local" coordinate system for Ayabelle Lighthouse was the azimuthal equidistant projection. I'm now pretty sure that was wrong – since I have actually read the French papers in my files (regarding the Lambert Conformal Conic), but prior to that – it probably resulted in a lot of other folks taking that as "gospel." I do have a blurb regarding the "Ayabelle Datum," an associated (probably wrong) grid system and shift parameters to WGS84. It is similar to the NGA published shift values, so they are plausible: $\Delta X = -65.958$ m, $\Delta Y = -120.429$ m, and $\Delta Z = +148.056$ m. The decimal points to the millimeters suggest that somebody actually made some observations – but that's just a guess. Beware of "published relations" in Djibouti. Extra anonymously supplied decimal points do not automatically relate to reality. The French *Institut Géographique National* (IGN) initiated topographic mapping of Djibouti in the 1940s and completed its 1:100,000 scale in the 1960s. In September 1950, IGN reported that "triangulation was in progress," and that the computation was being performed on the Clarke 1880 ellipsoid and they were using the "Gauss-Laborde System" for the triangulation computations (*meaning a local Transverse Mercator rather than the Laborde oblique – Ed.*). The ellipsoid of reference is still the Clarke 1880, and the grid on the native sheets is the UTM.

Dominica Passage
Caribbean Sea
Cape Capuchin
Vieille Case
Calibishie
Portsmouth
Glanvillia
Prince Rupert Bay
SAINT JOHN
SAINT ANDREW
Wesley
Toulaman
Marigot
Caribbean Sea
SAINT PETER
Morne Diablatins 1447 meters
Salibia
Colihaut
Morne Raquette
SAINT JOSEPH
Castle Bruce
Salisbury
Layou
Saint Joseph
SAINT DAVID
Pont Cassé
SAINT PAUL
Rosalie
Massacre
Laudat
La Plaine
Roseau
Roseau
SAINT GEORGE
SAINT PATRICK
Pointe Michel
SAINT LUKE
Soufrière
Berekua
Grand Bay
Scotts Head
SAINT MARK
Martinique Passage
British Virgin Is (U.K.)
Anguilla (U.K.)
Virgin Is (U.S.)
Puerto Rico (U.S.)
ANTIGUA AND BARBUDA
SAINT KITTS AND NEVIS
Guadeloupe (Fr.)
DOMINICA
Caribbean Sea
Martinique (Fr.)
SAINT LUCIA
BARBADOS
SAINT VINCENT AND THE GRENADINES
GRENADA
TRINIDAD AND TOBAGO
VENEZUELA
Dominica
National capital
Parish boundary
Road
0 3 6 Kilometers
0 3 6 Miles
Base 801516 (546539) 4-90

Commonwealth of Dominica

Dominica is bordered by Guadeloupe to the north and by Martinique to the south. Slightly more than four times the size of Washington, DC, the lowest point is the Caribbean Sea (0 m), and the highest point is *Morne Diablotins* (1,447 m) (*Google Earth, 2012*). The Commonwealth of Dominica passed legislation on August 25, 1981, declaring its claim to its territorial sea of 12 nautical miles, its Exclusive Economic Zone (EEZ) of 200 nautical miles and enabling legislation for straight baselines. (*See Shore and Sea Boundaries, Vols. 1 & 2, 1962, Aaron Shalowitz, Vol. 3, 2000, anon. – Ed.*) Maritime boundaries with France (Guadeloupe and Martinique) were agreed on December 23, 1988.

"Dominica's first inhabitants, the Ortoroids, arrived from South America around 3100 B.C., and lasted on the island until around 400 B.C. Next came the Arawaks, who settled in about 400 A.D. By 1400, the Kalinago or 'Carïbs,' moved aggressively up the Caribbean from South America, eliminating the Arawak from the region, including Dominica. When Columbus ushered in the era of colonization to Dominica in 1493, the same fate that befell the Arawaks would threaten the Carïbs. Ignoring the Kalinago name of 'Waitukubuli,' Columbus renamed the island Dominica as he first made landfall on a Sunday. The Carïbs successfully resisted efforts of Spanish colonization, but the British and French followed from the 1600s on, battling each other, and the Carïbs, to claim the Island. Through the many battles and ravaged by disease, the Carïbs gradually lost control of the island, fleeing back to South America. However, today approximately 2,000 Carïbs remain on the island, most living in the Carïb Territory in northeast Dominica. You many note that many of village names in and around Dominica are a mix of Carïb, French and English, reflecting the power struggles of the last 500 years. On November 3rd, 1978, the island was finally granted its independence from Britain" (*Discover Dominica Authority ©, 2012*).

The first geodetic survey of the island was performed after WWII. The island local datum origin is in *Pointe Michel (Google Earth, 2012)* at Position M.12, where $\Phi_o = 15° \ 15' \ 25.74''$ N, $\Lambda_o = 61° \ 23' \ 10.85''$ W, $\alpha_o = 346° \ 55' \ 22.8''$ from north to Morne Daniel (*probably in Roseau – Ed.*), the ellipsoid of reference is the Clarke 1880, where $a = 6,378,249.145$ m and $1/f = 293.465$. The British West Indies "BWI Grid" for Dominica is based on the Gauss-Krüger Transverse Mercator where the Central Meridian, $\lambda_o = 62°$ W, the Latitude of Origin $\varphi_o =$ equator, the scale factor at the Latitude of Origin $m_o = 0.9995$, False Easting = 400 km, and False Northing = nil. The reference source for this is (*Island of Dominica, British West Indies, Directorate of Colonial Surveys, Bushy Park, 1947*), *courtesy of Mr. John W. Hager, personal communication, 15 Nov. 2009*. The first edition of topographic maps for Dominica was produced by the Directorate of Colonial Surveys in three 1:25,000 scale sheets in 1961 from aerial photography flown in 1956. In 1996, the U.S. National Geodetic Survey (NGS) visited the island and performed GPS observations at the airport, but all observations were made only at newly monumented positions by NGS. No local control on the Dominica 1945 Datum was collocated by NGS. There is no currently published information to transform Dominica 1945 Datum to any other datum.

"The Seismic Research Centre (formerly the Seismic Research Unit) of The University of the West Indies grew out of a Colonial Development and Welfare (CDW) project. Since 2007, in collaboration with a number of other regional and international agencies, the SRC established a number of Continuously Operating Reference Stations (CORS) in several islands in the Eastern Caribbean (including Grenada, St. Vincent, Dominica, Saint Lucia, Antigua and St. Kitts). These permanently installed instruments record data continuously which are streamed to SRC in Trinidad via the Internet and VSAT."

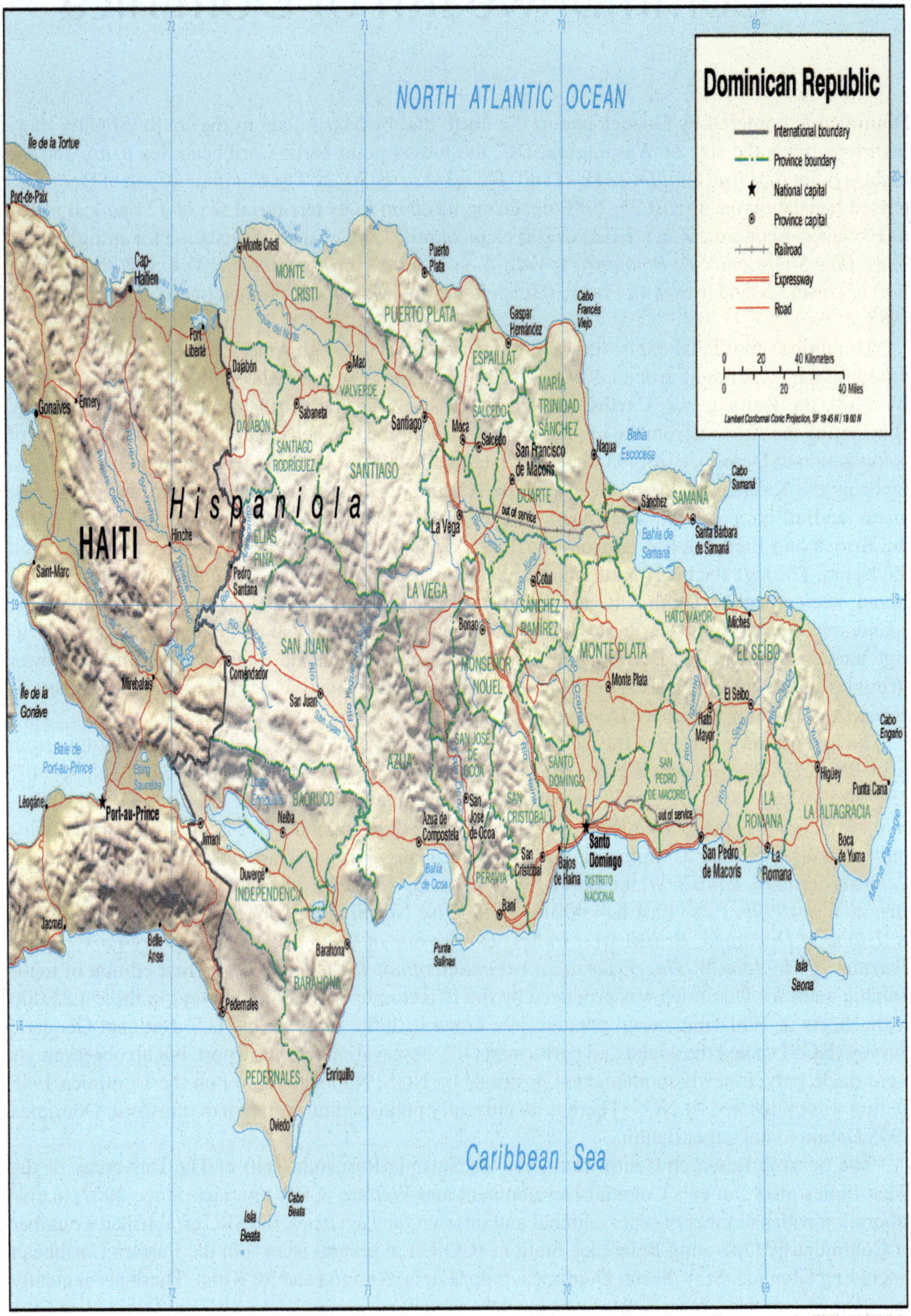

Base 803020AI (C00722) 11-04

Dominican Republic

The area of the Dominican Republic is slightly more than twice the size of New Hampshire. The coastline is 1,288 km long and its only land boundary is with Haïti (275 km). The terrain is comprised of rugged highlands and mountains interspersed with fertile valleys. The lowest point is *Lago Enriquillo* (−46 m), and the highest point is *Pico Duarte* (3,175 m).

"Visited by Christopher Columbus in 1492, the island of Hispaniola became the center of Spanish rule in the West Indies. The native Indians were expatriated by the Spaniards and were then replaced by African slaves. The western part of the island was occupied by buccaneers in the 17th century and was ceded to France in 1697 (St. Domingue, now Haïti), while the eastern part remained under Spanish rule (Santo Domingo). Eventually the entire island became to be ruled by Haitian President Jean-Pierre Boyer from 1822 until the eastern two-thirds revolted and formed the Dominican Republic in 1844. The latest constitution was adopted in 1966. "Following the two-term presidency of Danilo Medina Sanchez (2012–2020), Luis Rodolfo Abinader Corona was elected president in July 2020" (*World Factbook*, 2022).

The first map of Santo Domingo, Hispaniola was compiled by Columbus in 1492. Around 1500 Juan de la Cosa mapped the entire island. Hydrographic surveys were conducted by the French Hydrographer Francés Amedee-Francais Frezier from 1719 to 1725. The first medium-scale map was by Sir Robert Schomburgh in 1858; the second was prepared by General Casimiro N. de Noya in 1906 and was revised in 1921. The Second Brigade of the U.S. Marine Corps mapped the country in 1922 based on prior surveys performed by the U.S. Geological Survey from 1919 to 1921. The third comprehensive map of the Dominican Republic was compiled by Ing. Casimiro Gómez, and it was published by the Rand McNally Company in 1938. In 1940 the *Instituto Geográfico y Geológico de la Universidad de Santo Domingo* was formed. Later reorganized as the *Instituto Cartografico Universitario*, a joint agreement was signed in 1946 with the Latin American branch of the U.S. Army Map Service, the Inter-American Geodetic Survey (IAGS). Although the initial astro station was observed at Samaná Fort, the IAGS carried the North American Datum of 1927 into the island of Hispaniola through a classical triangulation chain that eventually spanned the entire West Indies. The Samaná Fort Datum is likely the same as the North American Datum of 1927, and the reference ellipsoid is the Clarke 1866, where $a=6,378,206.4$ m and $1/f=294.9786982$.

When the IAGS established First Order triangulation stations in the nations of the Caribbean and Latin America, they also developed plane coordinate systems as a service to the cadastral surveyors of each nation. The system devised for the Dominican Republic was based on the same rationale that the U.S. Coast & Geodetic Survey used for the State Plane Coordinate Systems of the United States. The rule of thumb was for regions that are predominately north-south in extent; the Transverse Mercator projection was used. For regions that are predominately east-west in extent, as is the Dominican Republic, the Lambert Conformal Conic projection was used. Furthermore, to facilitate computations by cadastral surveyors, the maximum scale factor for a zone was designed to be better than 1 part in 10,000. The excellent system developed by the IAGS for the Dominican Republic has a maximum scale factor of 1 part in 11,238 for the entire country! The Lambert Conformal Conic for the Dominican Republic is defined by the following parameters: Central Meridian, $\lambda_o=71° 30'$ West, Latitude of Origin, $\varphi_o=18° 49'$ North, Scale Factor at Origin, $m_o=0.999911020$, False Northing = 277,063.657 m, False Easting = 500 km. The NGA *TR 8350.2* document gives the three-parameter datum shift for the Caribbean from NAD27 to WGS84 as: $\Delta X=-3$ m ± 3 m, $\Delta Y=+142$ m ± 9 m, and $\Delta Z=+183$ m ± 12 m, and is based on 15 stations as of 1991.

In 1996, the U.S. National Geodetic Survey performed a GPS survey of the Dominican Republic. One of the points published by NGS was of an existing mark at the Naval Academy in Santo

Domingo Province, where φ=18° 28′ 02.92622″, λ=69° 52′ 32.11417″ West of Greenwich, and h=−16.537m, and these coordinates are on the North American Datum of 1983. Sometime since then, the Dominican Republic has published a document, *"Coordenadas de la Red de Estaciones Permanentes"* ("Coordinates of the Network of Permanent Stations.") The document details the coordinates of 262 points with one duplicate labeled as NAD27 and ITRF 2000 along with a seven-parameter transformation shown on a separate page. I was quite impressed with such a detailed document until I compared coordinates of a common point between the NGS publication and the Dominican publication. The NGS coordinates for the "Naval Academy" on the NAD83 **are identical** to the Dominican coordinates for "Academia" that are labeled "NAD27." The latitudes and longitudes for the points incorrectly listed as NAD27 are shown in degrees-minutes-seconds. The latitudes and longitudes for the points listed as ITRF are shown in decimal degrees but in fact are "packed" degrees-minutes-seconds! I tried transformation solutions between eight of the points listed with extended precision to see if I could make sense of what column represented what datum and what ellipsoid. Nothing matched the published seven parameters of the Dominican government! Regardless of which set of coordinates I used for either the GRS80 ellipsoid or the Clarke 1866 ellipsoid, three-parameter solutions and seven-parameter solutions never matched (closer than 100+ meters) what is published, whatever it is. Until some clarification is issued by the Dominican government, users are cautioned against using the published Trig List and Datum Shift parameters. I recommend using the NGA values; at least the user will be within a few feet of the truth.

REFERENCES: D

A New Three-Dimensional Reference Network in Denmark, Anna B.O. Jensen and Finn Bo Madsen, May 1998.
Annales Hydrographiques 2e Série, Tome Treizième, Année 1891.
Bulletin Géodésique, pp. 104–106, Number 1, 1924.
Coordenadas de la Red de Estaciones Permanentes, date unknown.
Discover Dominica Authority© 2012.
Google Earth, 2012.
International Boundary Study No. 154.
International Boundary Study No. 87 (Rev.) Djibouti – Somalia Boundary May 18, 1979, Office of the Geographer, Bureau of Intelligence and Research.
Island of Dominica, British West Indies, Directorate of Colonial Surveys, Bushy Park, 1947.
John W. Hager, Personal Communication (January 1988).
John W. Hager, Personal Communication (November 15, 2009).
Lonely Planet, 2006.
Lonely Planet, 2008.
Målebordsblade - First edition, Første Udgave 1901.
Planheft Dänemark – Planheft Übersichten der Kartengitter Europa, Nordafrika und Vorderasien (Berlin – 1942).
Reconnaissance Hydrographique de La Côte Française des Somalis au Nord d'Obock, Annales Hydrographiques, 3e Série, Tome Quatorzième, Annés 1935–1936, Paris 1937.
Shore and Sea Boundaries, Vols. 1 & 2, 1962, Aaron Shalowitz, Vol. 3, 2000, anon.
Webster's New Geographical Dictionary.
World Factbook, 2022.

E

The Republic of Ecuador 229
Arab Republic of Egypt 233
Republic of El Salvador 237
Republic of Equatorial Guinea 241
Republic of Estonia 245
Federal Democratic Republic of Ethiopia 249

DOI: 10.1201/9781003307785-57

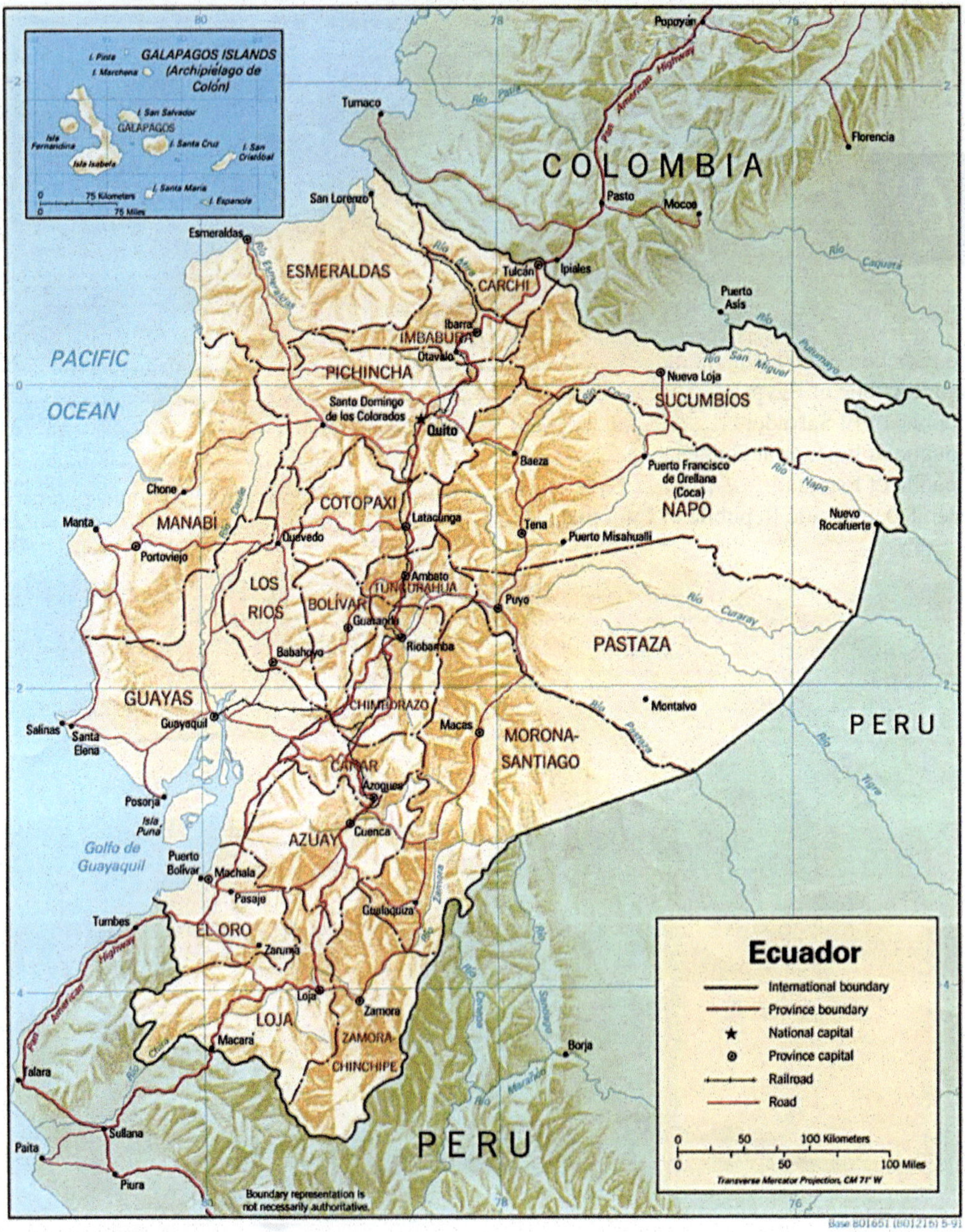

Base 801651 (801216) 5-91

The Republic of Ecuador

Ecuador is bordered by Colombia (708 km) and Peru (1,529 km) with a coastline of 2,237 km. Its terrain is coastal plain (*costa*), inter-Andean central highlands (*sierra*), and flat to rolling eastern jungle (*oriente*). The highest point is Chimborazo (6,267 m), the lowest point is the Pacific Ocean (0 m), and the mean elevation is 1,117 m.

Sir Isaac Newton said that the earth is flattened at the poles (oblate ellipsoid). Monsieur Jacques Cassini said that the earth is prolonged at the poles (prolate ellipsoid). (*See also Sweden.*) The French mission to Quito to determine the length of a degree of latitude (and to help settle the argument) started in 1735. The survey was under the direction of Charles-Marie de la Condamine with Louis Godin, and Pierre Bouguer of France. The Spanish representatives included Capt. Jorge Juan y Santacilla, Captain Antonio de Ulloa (later Governor of Louisiana), and Don Pedro Vicente Maldonado y Sotomayor, a local gentleman of Riobamba. It took 10 years to triangulate the 200 km meridional arc from Tarqui to Cotchesqui and the French expedition proved that the Englishman's theory was correct! The survey is known as the French Mission Datum which found that $a = 6,397,300$ m and $1/f = 216.8$. Another result of the observations was the "toise of Peru," a standard of linear measurement subsequently used in France. Simón Bolívar later liberated South America from the Spanish Crown, and Ecuador was established as a republic in 1830. No further geodetic surveys were undertaken for almost 200 years after the French Mission Datum was completed, and it was never later used for mapping.

On June 30, 1927, the Technical Commission for the "Survey of the National Topographic Map" was created to coordinate the various existing systems of geographic and plane rectangular coordinates being employed for official engineering applications. There seems to be no surviving record of those earlier systems. On April 11, 1928, the *Servicio Geografico Militar* (SGM), "Military Geographic Service," was tasked with the actual job of implementing that survey. The origin of the Ecuador Datum of 1928 was at the Astronomical Observatory of Quito, where $\Phi_o = 00° 12' 47.313''$ South, $\Lambda_o = 78° 30' 10.331''$ West of Greenwich, and $h_o = 2,908$ m. The International ellipsoid (also called the Hayford 1909 and the Madrid 1924), where $a = 6,378,388$ m and $1/f = 297$, was used. The defining azimuth from the datum origin has been lost, but that old observatory is still there in a downtown park. The Ecuadorian Army started their 1:20,000 map series (20 m contour interval) with the ellipsoidal Flamsteed projection, and the sheets were cast on the graticule without a grid overprint. (The Flamsteed is a sinusoidal projection tangent at the equator.) The sheet line intervals were at integer minutes of longitude (east and west) from the meridian of the Quito Observatory and at integer minutes of latitude (north and south) from the equator. The initial mapping on the Ecuador Datum of 1928 started south of Quito in Riobamba, Maldonado's hometown. *(Beware of the rotisserie chicken & French Fries there! – Ed.)*

In January 1930, SGM changed the scale of their series to 1:25,000 and the contour interval to 25 m. This series continued until 1947 when a number of momentous changes occurred in the history of the surveying and mapping activities of all Latin America. According to Ing. Vincente E. Avila, the Ecuadorian SGM substituted their name to *Instituto Geografico Militar* (IGM), "Military Geographic Institute," about the same time they attended the Pan American Institute of Geography and History (PAIGH) meeting in Mexico City. At that special meeting, COL. Floyd Hough of the U.S. Army Map Service presented a proposal to change the military mapping systems of the western world to the Universal Transverse Mercator (UTM) Grid. Furthermore, in 1947, the Inter-American Geodetic Survey (IAGS) was established by Army Map Service (AMS) and was headquartered at Corozal, Panama Canal Zone. In the 1950s, several of my classmates at Balboa High

School had parents that worked for IAGS. Hough later reiterated that proposal in 1952 at the Ciudad Trujillo PAIGH meeting, and IAGS published it in 1956.

The new "standard" topographic mapping format adopted by the Ecuadorian IGM after that 1947 meeting was at a scale of 1:50,000 with 40 m contours. That new series was still cast on the graticule, but it was based on a sheet interval of 10 minutes of latitude and 15 minutes of longitude. The IAGS assisted IGM in establishing its classical triangulation net and by 1951, Ecuador was the first country in South America to have a completely modern geodetic network. A new datum for all of South America was envisaged at the time. The Provisional South American Datum of 1956 (PSAD56) was established with an origin at *La Canoa*, Venezuela, as a joint venture among the Republic of Venezuela, IAGS, and other countries. (I once interviewed the Venezuelan "Father" of PSAD 56, Dr. Romero in Caracas.) The chains along the Andes Mountains were recomputed on the PSAD 56 where the origin at *La Canoa* (1951) is: $\Phi_o = 08°\ 34'\ 17.170''$ North, $\Lambda_o = 63°\ 51'\ 34.880''$ West of Greenwich, and $h_o = 178.870$ m. The azimuth to station *Pozo Hondo* (1951) is: $\alpha_o = 40°\ 22'\ 45.96''$. The old Datums initially included in the IAGS re-computations were: Old Trinidad 1910, *Loma Quintana* 1911 (Venezuela), *Bogotá* 1941 (Colombia), and *Quito* 1928. The International ellipsoid is the reference surface used for the PSAD 56. The Ecuadorian IGM continued its mapping on the International ellipsoid, it recast existing sheets to the PSAD 56, and it has incorporated the UTM Grid since then for all military mapping. The South American Datum of 1969 has never been used for mapping in Ecuador.

Most mapping of Ecuador has been flown and compiled by IGM because of *"la Ley de la Cartografía Nacional de 1978,"* "the 1978 Law of National Mapping." Essentially, this formed a near-perfect monopoly for the benefit of the Army so that most original mapping must be done by IGM. This law has had the result of allowing the establishment of a large well-equipped national agency that is intended to meet the needs of both the military and local government, while helping to support other activities of the military. This sort of mapping arrangement is the rule, rather than the exception for much of Latin America. Military security for mapping is also a major concern for a country that has had combat as recently as the 1990s over boundaries.

In the early 1980s, the City of Guayaquil contracted with IGM for a cadastral mapping project. The compilation scale was 1:1,000, and 1982 photo control was by aerotriangulation from traverse densification of classical Second Order (IGM/IAGS) triangulation. Like most countries in South America, the Ecuadorian cartographers have agreed with the advice of AMS and have shunned the use of the UTM Grid for civilian cadastral mapping. The initial grid established by IGM was a Gauss-Krüger Transverse Mercator. The scale factor at origin, $m_o = 1.0$, the central meridian of the belt, $\lambda_o = 79°\ 53'\ 05.8232''$ West of Greenwich, the False Easting at $\lambda_o = 624$ km, and the False Northing $= 10,000,051.000$ m. However, the population of Guayaquil had doubled in the following 10 years, and the city experienced substantial environmental difficulties because of that explosion of people.

In the early 1990s, the Republic of Ecuador contracted with the University of New Orleans (UNO) for an environmental assessment and subsequent remediation design for the sanitary sewer system of Guayaquil. UNO is an Environmental Protection Agency "Center of Excellence." It is the contact point in the U.S. for foreign governments seeking technical help with urban waste management technology. UNO went into a partnership with a local university in the city, the *Escuela Politecnica del Litoral (ESPOL)*. The analysis and design needed a GIS to maximize efficiency and minimize costs, so UNO and *ESPOL* performed a differential GPS survey of the City of Guayaquil. We occupied a number of existing IGM triangulation stations as well as new photo-identifiable points based on recent IGM aerial photography. Co-located fiducial points were observed according to DMA/NIMA specifications, and NIMA performed the subsequent fiducial point adjustment to the precise ephemeris, thanks to Dr. Muneendra Kumar.

The UNO/IGM solution from PSAD 56 to WGS 84 for a 7-parameter Molodensky model (using the PSAD 56 origin at La Canoa *with Northern latitude*) yielded: $\Delta X = -263.91$ m, $\Delta Y = -25.05$ m, $\Delta Z = -285.81$ m, scale $= -3.61 \times 10^{-6}$, $R_z = -36.88''$, $R_y = -3.42''$, $R_x = +3.54''$. A computational test point, for instance, is station "*Panoramico*" where the PSAD 56 coordinates are: $\varphi = 02°\ 11'\ 43.9093''$ South, $\lambda = 79°\ 52'\ 45.4601''$ West, and $h = 68.614$ m. The "Panoramico" WGS 84 coordinates are: $\varphi = 02°\ 11'\ 55.8406''$ South, $\lambda = 79°\ 52'\ 53.4010''$ West, and $h = 68.530$ m. I prefer to use the Molodensky model to the Bursa-Wolf model in cases like this when the local area is at a great distance from the datum origin. Note that the net computational results (of shift transformations) are identical to the Bursa-Wolf model, but the shift parameters in this case are less correlated and thus they appear more meaningful with the Molodensky model.

A new grid was devised for the City of Guayaquil based on the Normal Mercator projection, WGS 84 Datum. The origin is at the Rotunda of *Simón Bolívar* in downtown Guayaquil. The scale factor at origin, $m_o = 0.999272829$, the central meridian, $\lambda_o = 79°\ 52'\ 45.16''$ West of Greenwich, the False Easting at $\lambda_o = 500$ km, and the False Northing $= 2,242,320.510$ m at $N = 02°\ 11'\ 33.09''$ South. The ellipsoid of reference is the WGS84, where $a = 6,378,137$ m and $1/f = 298.257223563$. Because most of our UNO graduate students on the project were visiting professors from the *ESPOL* University in Guayaquil, the name was dubbed the "*ESPOL La Rotunda Grid*." The city has recently contracted with IGM for a new cadastral mapping system and GIS. They are continuing with the new datum and grid. IGM is currently participating in the SIRGAS project to establish a single, unified datum for all of South America using GPS techniques with technical assistance from NIMA/National Geospatial-Intelligence Agency (NGA).

Ecuador has been adjusted to the new South American Datum called "SIRGAS95," Epoch 1995. The original classical network was re-observed with GPS collocation at 135 stations, and the republic now has 35 Continuously Operating Reference Stations (CORS), managed by the IGM. Transformation parameters from PDAD56 to SIRGAS95 in Ecuador are not publicly available at the IGM website at present.

Egypt
International boundary
National capital
Railroad
Expressway
Road
0 50 100 150 Kilometers
0 50 100 150 Miles
Lambert Conformal Conic Projection, SP 24N/29N
Mediterranean Sea
Red Sea
LIBYA
SUDAN
ISRAEL
JORDAN
SAUDI ARABIA
SYRIA
LEBANON
CYPRUS
Nicosia
Sinai
Cairo
Alexandria
Port Said
Suez
Suez Canal
Nile
Lake Nasser
Lake Nubia
Aswan High Dam
Aswān
Luxor
Asyūṭ
Sīwah
Wādī Ḥalfā
Abū Sunbul
Gulf of Suez
Gulf of Aqaba
Amman
Beirut
Damascus
GAZA STRIP
GOLAN HEIGHTS*
Administrative Boundary
*The Golan Heights is Israeli-occupied Syria; the West Bank and Gaza Strip are Israeli occupied with current status subject to the Israeli-Palestinian Interim Agreement--permanent status to be determined through further negotiation.
Boundary representation is not necessarily authoritative.
Base 802525 (R00622) 5-97

Arab Republic of Egypt

Egypt is bordered by the Gaza Strip (11 km), Israel (266 km), Libya (1,115 km), and Sudan (1,273 km); its coastline on the Mediterranean Sea and on the Red Sea totals 2,420 km. Slightly more than three times the area of New Mexico, the terrain is comprised mainly of a vast desert plateau interrupted by the Nile River Valley and its delta. The lowest point is the Qattara Depression (−133 m), and the highest point is Mount Catherine (2,629 m).

The roots of Egyptian civilization go back more than 6,000 years to the beginning of settled life along the banks of the Nile River. "The regularity and richness of the annual Nile River flood, coupled with semi-isolation provided by deserts to the east and west, allowed for the development of one of the world's great civilizations. A unified kingdom arose circa 3200 B.C., and a series of dynasties ruled in Egypt for the next three millennia. The last native dynasty fell to the Persians in 341 B.C., who in turn were replaced by the Greeks, Romans, and Byzantines. It was the Arabs who introduced Islam and the Arabic language in the 7th century and who ruled for the next six centuries. A local military caste, the Mamluks, took control about 1250 and continued to govern after the conquest of Egypt by the Ottoman Turks in 1517. *(The Mamluks were such fierce and respected fighters that the U.S. Marine Corps adopted their ivory-hilted sword design for Officer's ceremonial dress – Ed.)* Following the completion of the Suez Canal in 1869, Egypt became an important world transportation hub, but also fell heavily into debt. Ostensibly to protect its investments, Britain seized control of Egypt's government in 1882, but nominal allegiance to the Ottoman Empire continued until 1914. Partially independent from the UK in 1922, Egypt acquired full sovereignty with the overthrow of the British-backed monarchy in 1952. The completion of the Aswân High Dam in 1971 and the resultant Lake Nasser have altered the time-honored place of the Nile River in the agriculture and ecology of Egypt.

"In January 2014, voters approved a new constitution by referendum and in May 2014 elected former defense minister Abdelfattah Elsisi president. Egypt elected a new legislature in December 2015, its first House of Representatives since 2012. Elsisi was reelected to a second four-year term in March 2018. In April 2019, Egypt approved via national referendum a set of constitutional amendments extending Elsisi's term in office through 2024 and possibly through 2030 if reelected for a third term" (*World Factbook* 2022).

The first serious mapping of Egypt on modern lines was undertaken in 1798 by Napoleon for his Survey of Egypt. A Topographic Section was formed which consisted of four officers, an astronomer, and four "intelligent soldiers." Bases were measured at Alexandria and Cairo by the *"Service Topographique de l'Armee d'Egypte,"* and topographic maps were compiled with a 10 km grid with an origin at the great pyramids of the North in Memphis, *"Le centre de la projection correspond à l'axe de la grande pyramide du Nord à Memphis."* Referenced to the Plessis ellipsoid, where $a = 6,375,738.7$ m, and $1/f = 334.29$, the projection was the ellipsoidal Bonne, the "standard" projection for Europe at the time. The maps were later published of Egypt and the Palestine in 1808 (*see The History of Lebanon – Ed.*).

In 1874, a number of expeditions were led by British scientists to various European colonies in Africa and the Indian Ocean in order to simultaneously observe the transit of Venus for the purpose of precisely determining differences in longitude. Locations included Mauritius, Rodrigues, Réunion, St. Paul, and Egypt. Helwân Observatory situated on Az Zahra Hill in the Al Moqattam Hills, Qalyûbîya of Cairo was utilized for the observations, and the station was termed "F_1," where $\Phi_o = 30°\ 01'\ 42.8591''$ N, $\Lambda_o = 31°\ 16'\ 33.6''$ East of Greenwich, the initial LaPlace azimuth being measured from Station O_1 (Helwân) to Station B_1 (Saccara), $\alpha_o = 72°\ 42'\ 01.20''$ from South, and $H_o = 204.3$ m, based on mean sea level at Alexandria. This is considered the origin of the

"Old Egypt Datum of 1907." A point of much-heated contention among Geodetic Surveyors has been the "rumor" that the Clarke 1866 ellipsoid was once used in Egypt.

It just so happens that M. Sheppard, Director General of the Survey of Egypt reported (*in French*) to the Secretary General of the Geodesy Section of the International Union of Geodesy and Geophysics that the initial geodetic work performed in Egypt was computed on the Clarke 1866 ellipsoid, where "*a, demi-grande axe equatorial* = $6.378.206^m$ (*sic*), $\alpha = 1/295,0$. (*sic*)." Sheppard went on to say that all cultivated lands in the Nile Valley that were based on 2nd and 3rd order triangulations (for cadastral applications) initially used this ellipsoid, but that a later controlling chain of triangulation spanning the length of the Nile Valley was computed with the later adopted Helmert (1906) ellipsoid, where $a = 6,378,200$ m and $1/f = 1/298.3$. Everything was later re-calculated on the Helmert ellipsoid and also on the International 1924 ellipsoid, where $a = 6,378,388$ m and $1/f = 297$ (*Bulletin Géodésique, No. 8*, 1925).

"Although a great deal of survey work of various kinds had been done in Egypt from time immemorial up to 1907, and this, as regards the Framework, culminated in a certain amount of modern triangulation of secondary order executed between the years 1898 and 1907, it was not until the latter year that a triangulation of the first order was undertaken. The secondary triangulation was necessary to control the modern Cadastral Survey, but the urgency with which the latter had to be completed made it impracticable to undertake work of full first-order accuracy at the beginning. Accordingly, although the existing Survey of Egypt, as founded by the late Sir Henry Lyons, dates from 1898, the geodetic survey proper may be said not to have started until 1907" (*Geodesy in Egypt*, Empire Survey Review, No. 60).

The British established a number of grid systems in 1929 for the Old Egypt Datum of 1907, each Gauss-Krüger Transverse Mercator Belt being designated with a different color: Purple Belt – Central Meridian $(\lambda_o) = 27°$ E, Latitude of Origin $(\varphi_o) = 30°$ N, Scale Factor at Origin $(m_o) = 1.0$, False Easting = 700 km, False Northing = 200 km, and South of the False Origin add 1,000 km to the northings; Red Belt – Central Meridian $(\lambda_o) = 31°$ E, Latitude of Origin $(\varphi_o) = 30°$ N, Scale Factor at Origin $(m_o) = 1.0$, False Easting = 615 km, False Northing = 810 km and South of the False Origin add 1,000 km to the northings; Green Belt – Central Meridian $(\lambda_o) = 35°$ E, Latitude of Origin $(\varphi_o) = 30°$ N, Scale Factor at Origin $(m_o) = 1.0$, False Easting = 300 km, False Northing = 100 km and South of the False Origin add 1,000 km to the northings.

In 1930, after a re-adjustment of the classical network, the New Egypt Datum of 1930 was published, also referenced to the Helmert 1909 ellipsoid. The Purple, Red, and Green Belts were retained with no modifications. The common abbreviation for the new datum is "EG30." This remains the current classical system used in Egypt for civilian mapping purposes.

After WWII, the U.S. Army Map Service (AMS) undertook the re-computation of all classical datums that covered lands involved in the European Theater of War, including all of North Africa. That new unified system was the European Datum of 1950 (ED50) and was referenced to the International 1924 ellipsoid. The longitude of F_1, Venus Station at Helwân, was changed to $\Lambda_o = 31° 16' 37.05''$ East of Greenwich as a result of observations for the deflection of the vertical, but this change was only with respect to its re-computation on the International 1924 ellipsoid. All coordinates on ED50 for Egypt were published on the UTM Grid and generally are considered useful only for military mapping purposes, not for civilian use. A number of datum shift algorithms on the complex plane were developed by AMS to convert directly from the Purple, Red, and Green Belts to the UTM ED50 Grids. Although classified as Secret at the time, they were declassified in 1956.

TR8350.2 lists a couple of datum shifts from classical to the WGS84 Datum. **From** Old Egyptian 1907 **to** WGS84: $\Delta X = -130$ m ± 3 m, $\Delta Y = +110$ m ± 6 m, and $\Delta Z = -13$ m ± 8 m, and is based on 14 stations used in the derivation. This compares favorably with my own computations but using fewer stations. **From** ED50 **to** WGS84, $\Delta X = -130$ m ± 6 m, $\Delta Y = -117$ m ± 8 m, and $\Delta Z = -151$ m ± 8 m, and is *also* based on 14 stations used in the derivation.

Using the full Molodensky 7-parameter datum shift model (including the datum origin coordinates) with 23 points, I derived the following **from** Egypt 1930 **to** WGS84: $\Delta X = -137.5$

m ± 0.5 m, $\Delta Y = +105.0$ m ± 0.4 m, and $\Delta Z = -18.1$ m ± 0.4 m, $\delta s = +4.38 \times 10^{-6} \pm 1$, $Rx = -5.0'' \pm 0.70''$, $R_y = +1.59'' \pm 0.48''$, $R_z = +1.51'' \pm 0.26''$. A solution with 19 differentially weighted points and 5 check points for a much larger area of Egypt yielded the following 7-parameter Bursa-Wolfe model **from** Egypt 1930 **to** WGS84: $\Delta X = -88.832$ m ± 0.02 m, $\Delta Y = +186.714$ m ± 0.03 m, and $\Delta Z = +151.82$ m ± 0.01 m, $\delta s = -6.413 \times 10^{-6} \pm 1.84$, $R_x = -1.305'' \pm 2.21''$, $R_y = +11.216'' \pm 1.57''$, $R_z = -6.413'' \pm 1.84''$ (Shaker, Prof. A.A., et al., *Remove-Restore Technique for Improving the Datum Transformation Process*, FIG Working Week 2007, Hong Kong, 13–17 May 2007).

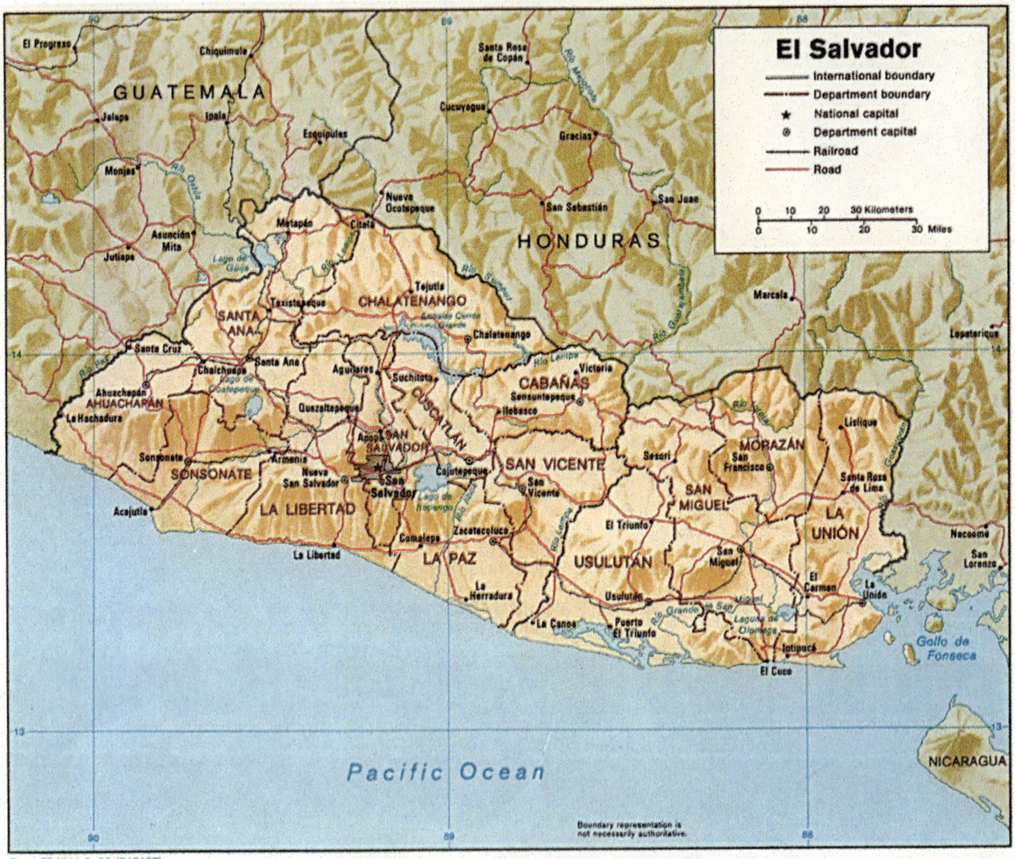

Base 504511 8-80 (545437)

Republic of El Salvador

El Salvador (The Savior) is the smallest Spanish-speaking nation in the Western Hemisphere. It is located on the western side of the Central American isthmus. The country is slightly smaller than Massachusetts, and its land boundaries are with Guatemala (203 km) and Honduras (342 km). El Salvador's coastline is 307 km on the Pacific Ocean and its terrain is mostly mountainous with a narrow coastal belt and central plateau. The lowest point is the Pacific Ocean (0 m), and the highest point is Cerro El Pital (2,730 m). El Salvador is known as the "Land of Volcanoes," and is one of the most seismologically active regions on earth, situated atop three large tectonic plates.

When the Spanish Conquistadores first entered into Central America in the early 16th century, Indians of the Pipil tribe occupied the area now known as El Salvador. The Pipil were a subgroup of a nomadic people known as the Nahua, who had migrated into Central America around 3,000 B.C. They eventually fell under the Maya Empire, which dominated Central America until about the 9th century A.D. According to the Library of Congress *Country Studies*, "Pipil culture did not reach the advanced level achieved by the Maya; it has been compared, albeit on a smaller scale, to that of the Aztecs in Mexico. The Pipil nation, believed to have been founded in the 11th century, was organized into two major federated states subdivided into smaller principalities."

The Spaniards were initially defeated in 1524 when they attempted to enter the area, and it took until 1528 with two more expeditions to finally subdue the Pipil nation. The fierce warrior Atlacatl is revered to this day to the exclusion of Alvarado who finally overcame the natives. "In this sense, the Salvadoran ambivalence toward the conquest bears a resemblance to the prevailing opinion in Mexico, where Cortes is more reviled than celebrated."

Topographic Mapping of the Americas, Australia, and New Zealand states, "To 1930, the only detailed accurate surveying done in El Salvador – a country that became independent in 1821 – was the surveying done for the Intercontinental Railroad Commission and a few surveys related to possible road routes." However, Nicaragua and Honduras had been squabbling regarding their border for over 30 years, and an arbitration agreement was signed in 1930 and soon thereafter surveyed by the U.S. Coast & Geodetic Survey. El Salvador, Guatemala, and Honduras signed a protocol on March 26, 1936 accepting *Cerro Monte Cristo* as the tripoint of the boundaries of the three states. It was during this era that the first geodetic surveys were performed on the border between El Salvador and Guatemala from 1937 to 1940.

The *Dirección General de Cartografía* was established in 1946 to continue the geodetic surveys and to implement a topographic mapping of the country. At the end of the following year 1947, the IAGS (part of the U.S. Army Map Service) signed a cooperative agreement with El Salvador. By 1958, all classical geodetic triangulation had been completed in the country, and topographic mapping was well underway. The national mapping agency is now known as the *Instituto Geográfico Nacional "Ing. Pablo Arnoldo Guzmán."*

The original horizontal datum of El Salvador is the Ocotepeque Datum of 1935 which was established by the U.S.C.&G.S. at *Base Norte* (in Honduras), where $\varphi_o = 14° 26' 20.168"$ North, $\lambda_o = 89° 11' 33.964"$ West of Greenwich, and $H_o = 806.99$ m above mean sea level. The defining geodetic azimuth to Base Sur is: $\alpha_o = 235° 54' 21.790"$, and the ellipsoid of reference is the Clarke 1866, where $a = 6,378,206.4$ m and $1/f = 294.9786982$. The corresponding astronomic observations at that mountainous location are: $\Phi_o = 14° 26' 13.73"$ North ($\pm 0.07"$), $\Lambda_o = 89° 11' 39.67"$ West ($\pm 0.045"$), and the defining astronomic azimuth to Base Sur is: $\alpha_o = 235° 54' 20.37"$ ($\pm 0.28"$). The difference between these two sets of coordinates is due to the local gravimetric deflection of the vertical. The grid system used in El Salvador is based on the Lambert Conformal Conic projection where the Central Meridian, $\lambda_o = 89° 00' 00"$ West, the Latitude of Origin, $\varphi_o = 13° 47' 00"$ North, the Scale Factor at Origin,

m_o=0.999967040, the False Easting=500,000 m, and the False Northing=295,809.184 m. Sometime after that, the National Geospatial-Intelligence Agency (NGA) computed all the classical triangulation of Central America on the North American Datum of 1927. The 1987 published datum shift parameters for Central America based on 19 stations (*TR 8350.2*) from NAD27 to WGS84 are: ΔX=0 m±8 m, ΔY=+125 m±3 m, ΔZ=+194 m±5 m.

After the disaster when Hurricane Mitch hit Central America in 1998, the U.S. National Geodetic Survey established a number of GPS Continuously Operating Reference Stations in the region. They observed a number of stations in El Salvador in order to establish and publish a High Accuracy Reference Network on the North American Datum of 1983. The National Geodetic Survey has posted all of the data, including various other sources of geodetic coordinate data in El Salvador and the region on their website. The documentation of their geodetic observations is a superb resource for the researcher, and it is freely available. Thanks, NGS.

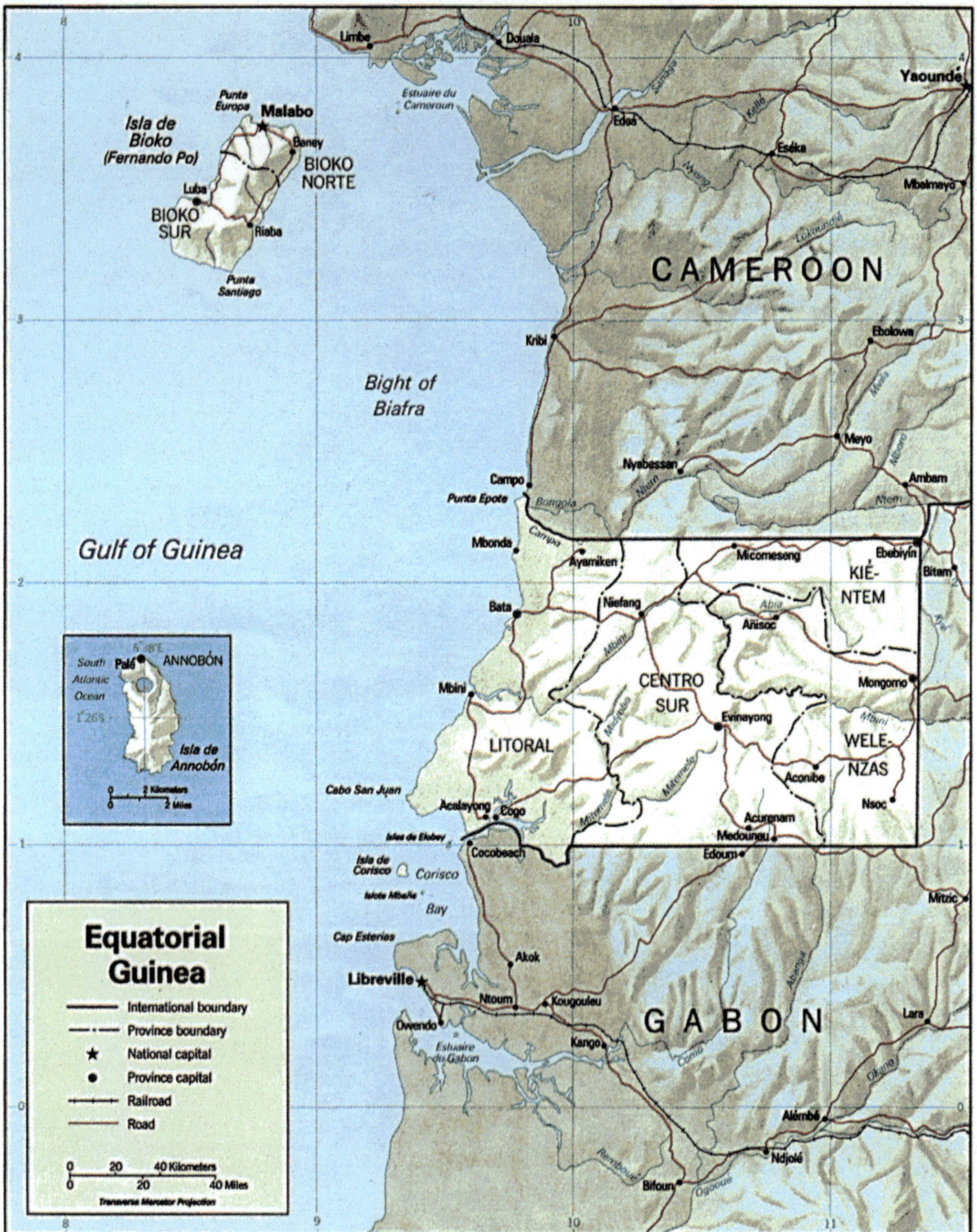

Base 801513 (E00309) 1-92

Republic of Equatorial Guinea

Bioko Island lies about 40 km from Cameroon. Annobón Island lies about 595 km southwest of Bioko Island. The larger continental region of Rio Muni lies between Cameroon and Gabon on the mainland; Equatorial Guinea includes the islands of Corisco, Elobey Grande, Elobey Chico, and adjacent islets. Bioko Island, called Fernando Póo until the 1970s, is the largest island in the Gulf of Guinea – 2,017 km². It is shaped like a boot, with two large volcanic formations separated by a valley that bisects the island at its narrowest point. The 195 km coastline is steep and rugged in the south but lower and more accessible in the north with excellent harbors at Malabo and Luba, and several scenic beaches between those towns. On the continent, Rio Muni covers 26,003 km² (10,040 sq. mi.). The coastal plain gives way to a succession of valleys separated by low hills and spurs of the Crystal Mountains. The Rio Benito (*Mbini*), which divides Rio Muni in half, is not navigable except for a 20 km stretch at its estuary. Temperatures and humidity in Rio Muni are slightly lower than on Bioko Island. Annobón Island, named for its discovery on New Year's Day 1472, is a small volcanic island covering 18 km². The coastline is abrupt except in the north; the principal volcanic cone contains a small lake (*U.S. Department of State Background Note*, March 2009).

The first inhabitants of the region that is now Equatorial Guinea are thought to have been Pygmies, of whom only isolated pockets remain in northern Rio Muni. Bantu migrations between the 17th and 19th centuries brought the coastal tribes and later the Fang. Elements of the latter may have generated the Bubi, who immigrated to Bioko from Cameroon and Rio Muni in several waves and succeeded former Neolithic populations. The Annobón population, native to Angola, was introduced by the Portuguese via São Tomé. The Portuguese explorer, Fernando Po (*Fernão do Póo*), seeking a route to India, is credited with having discovered the island of Bioko in 1471. The Portuguese retained control until 1778 when the island, adjacent islets, and commercial rights to the mainland between the Niger and Ogoue Rivers were ceded to Spain in exchange for territory in South America. From 1827 to 1843, Britain established a base on the island to combat the slave trade. The Treaty of Paris settled conflicting claims to the mainland in 1900, and the mainland territories were united administratively under Spanish rule. The majority of the Equatoguinean people are of Bantu origin. The largest tribe, the Fang, is indigenous to the mainland, but substantial migration to Bioko Island has resulted in Fang dominance over the earlier Bantu inhabitants.

Initial *published* geodetic surveying in only the southern region was done solely by the French Navy. The Spanish Navy apparently had done some hydrographic surveys prior to 1954 *as reported below* by the French. Lafargue of the French Navy established the Gabon River Datum in 1914 at Cape Esterias as: $\Phi_o = +0°\ 36'\ 48.58''$ North, $\Lambda_o = +9°\ 19'\ 19.02''$ East of Greenwich. The defining azimuth was from Banda to Tranchée: $\alpha_o = 158°\ 59'\ 56.0''$. The usual Hatt Azimuthal Equidistant Grid had the same origin as the datum. The triangulation stations monumented by Lafargue for this survey were pyramids constructed of cement, and the south side was imprinted "M.H.A.E.F. 1921."

According to Mr. John W. Hager, "In 1951, William Mussetter of the U.S. Coast & Geodetic Survey was on his way to observe astronomical stations (astros) in the Sudan on the 30th Meridian and stopped in Gabon to observe the astro at Cap Esterias. On 9–13 September 1951, Mussetter observed the position as: $\Phi_o = +0°\ 36'\ 48.65''$ North, $\Lambda_o = +9°\ 19'\ 19.06''$ East. References are: [1] *Annales Hydrographiques*, 4e série, Tome Neuvième, Année 1958, "*Triangulation De La Baie De Corsisco (1954–1955),*" p. 75, and [2] the same, Tome Dixième (1959–1960) p. 82. A further reference is *Repertoire des Coordonnees des Points Astronomiques et Géodésiques*; Feuille 1/200,000 Libraville NA-32-IV, Paris 10–58. This is the trig list compiled by IGN."

In 1955, Mannevy of the French Navy reoccupied an astro station at Cape Esteiras for the triangulation of the Bay of Corisco, North of Libreville. The Cadastral Service of Libreville measured a 3 km baseline in Libreville in concert with the French Hydrographic Mission. The Cape Esteiras Datum of 1955 origin is identical to the Gabon River Datum of 1914. The Department of Public Works of Gabon (*Travaux Publics du Gabon*) assisted Mannevy by building towers for the triangulation of the Bay of Corisco. The triangulation was performed with WILD T3 theodolites using eight sets of angles. The design of the towers and targets were patterned after those used in Madagascar. The first computations on the Universal Transverse Mercator Grid in Gabon were computed by Mannevy with the Clarke 1880 ellipsoid, *Fuseau 32* (zone 32), where Central Meridian $(\lambda_o) = +9°$ East of Greenwich. The Cape Esteiras 1955 Datum is used for Libreville and points north in Gabon and Equatorial Guinea. The Equatoguinean connection of these French Navy Geodetic and Hydrographic surveys was actually performed in November of 1954 by *M. le Capitaine de Frégate Guyot*, assisted by M. Dars. Entry to the Spanish Isle of Corisco was authorized by diplomatic relations established between the French vessel *Beautemps-Beaupré* and the *Station Commandant Canovas de Castillo*. The Spanish authorities also offered to aid in the local administration of the survey to furnish one important document produced by their Spanish Geographic Service (recent chart at 1:100,000 scale) which showed large divergences from the French Navy chart. *Les autorités espagnoles offraient également l'aide de l'administration locale et fournissaient une importante documentation provenant du Service Géographique (carte au 1/100 000 récente) qui montrait de larges divergences avec la carte marine française.* The triangulation station established on the south shore of Corisco Island was Station *Gobé*, but curiously, the French did not publish its coordinates. The approximate scaled coordinates are: $\varphi = +0° 53'$ N, $\lambda = +9° 20'$ E (*Levé de la Baie de Mondah [6 Novembre – 3 Décembre 1954] Mission Hydrographique de la Côte Ouest d'Afrique, Annales Hydrographique, pp. 60–62*).

Datums known (*or I thought so – Ed.*) to exist in Equatorial Guinea include: Annobón Island Datum; Biao, Bioko Island Datum; Kogo, Rio Muni Datum; Rio Benito, Rio Muni Datum; Gabon 1951; and M'Poraloko Datum, the latter two primarily found in Gabon.

The following details are according to John W. Hager: "For Annobón at *Punto Astronómico Palé* (P. A.) $\Phi_o = 1° 24' 04.5''$ S, $\Lambda_o = 5° 37' 50.1''$ E, GRS 80. This I got from *Mapa De La Republica De Guinea Ecuatorial, Isla De Annobon (Ambó), 1:10,000*. I believe that this was one of the Spanish maps with a coordinate list printed on the sheet. For Kogo or Cogo at $\Phi_o = 1° 04' 48.50''$ N, $\Lambda_o = 9° 41' 39.45''$ E. This was from *Guinea Continental Española; Reseñas de los Puntos Astronómicos*, a collection of astros with descriptions that AMS obtained from the Spanish. I never saw this used as a datum. Also, with the same name is $\Phi_o = 1° 04' 53''$ S, $\Lambda_o = 9° 41' 51''$ E. This was a position obtained 25 Feb. 1952 by observing a total solar eclipse. Rio Benito at $\Phi_o = 1° 35' 06.50''$ S, $\Lambda_o = 9° 36' 58.50$ E. This was from the same source as Kogo, and I never saw used as a datum."

The latest edition of TR*8350.2* by the NGA lists the three-parameter shift from M'Poraloko (*sic*) Datum to WGS 84 Datum as: $\Delta X = -74$ m, $\Delta Y = -130$ m, $\Delta Z = +42$ m. Note that only one point was used to determine the published shift, and the accuracy is stated at ±25 m for each component.

John W. Hager commented: "*Phare du Cap Lopez* at $\Phi_o = 0° 37' 54.2''$ S, $\Lambda_o = 8° 42' 13.2''$ E, Clarke 1880. (*Phare is French for lighthouse – Ed.*) This datum for the *Société des Pétroles d'Afrique Équatoriale Française (SPAEF)* is M'Poraloko, and *Phare du Cap Lopez* is connected to it by the main *SPAEF* triangulation north from Port Gentil."

The only grid system known to exist is the UTM, other than the small Hydrographic system described above. All offshore mineral rights boundary treaties with other countries have referenced only the WGS 84 Datum. Significant hydrocarbons have been found in offshore areas, so geodetic surveys and detailed topographic mapping may accompany major economic growth in the republic.

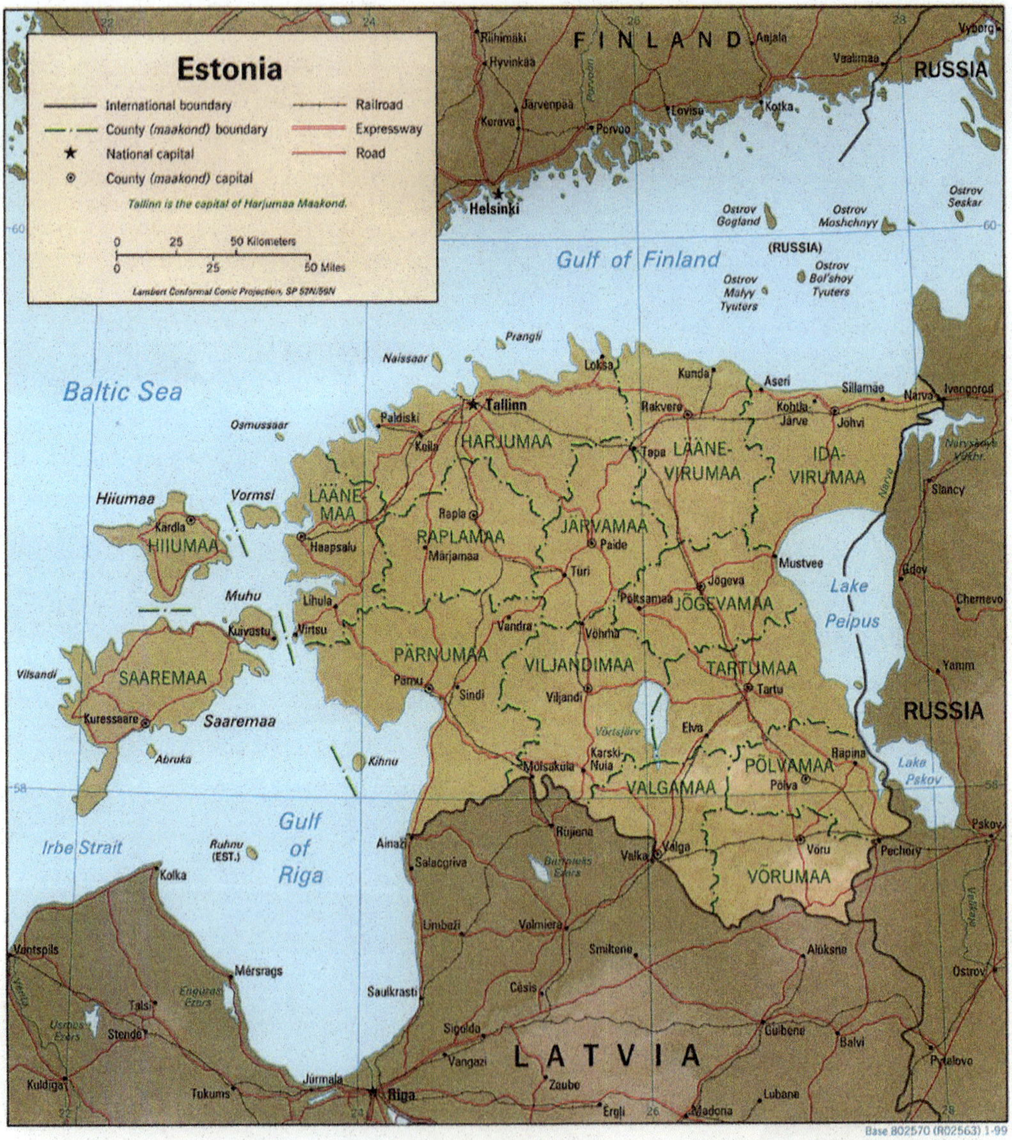

Base 802570 (R02563) 1-99

Republic of Estonia

Estonia is slightly smaller than New Hampshire and Vermont, combined. Bordered on the north by the Gulf of Finland, on the east by Russia (324 km), on the south by Latvia (333 km), and on the west by the Gulf of Riga and the Baltic Sea, the total coastline is 3,794 km. The republic is comprised mostly of marsh and lowlands; the lowest point is the Baltic Sea (0 m), and the highest point is *Suur Munamagi* (318 m).

The original Estonians were conquered by the Danes who founded Revel in 1219. Revel is the former name of Tallinn, the current capital of Estonia. Remains of the ancient Revel fort settlement survive to this day, a testament to the construction skills of the craftsmen under the rule of King Valdemar II. Developed as a trading port and member of the Hanseatic League in the 13th century, it was sold to the Teutonic Knights in 1346 and on dissolution of the order it passed to Sweden in 1546. Later taken by Russia in 1710 and except for the period of independence from 1918 to WWII, it remained a Russian annexation until its present independence in 1991 (*Merriam Webster's Geographical Dictionary*, 3rd Edition).

The first period of geodetic surveying and mapping in Estonia was from 1845 to 1920. The principal mapping agency for this period was the *Korpus Voyennykh Topografov – KVT* (Corps of Military Topographers) under the Russian Imperial General Staff. Survey activities in Estonia began with Struve's triangulation in 1811/1816–1819 and by Tenner's first-order network of 1820–1832 published by Czarist Russia in 1843 and 1847. Tenner later supplemented his primary net with lower order stations. The Tenner chains were originally computed on the Walbeck 1819 ellipsoid where the semi-major axis $(a) = 6,376,895$ m, and $1/f = 302.7821565$; they were later recomputed on the Bessel 1841 ellipsoid, where $a = 6,377,397.155$ m, and $1/f = 299.1528$. The Triangulation of the Coast of the Baltic Sea Datum (1829–1838) was based on the origin point at the Observatory of Tallinn, where $\Lambda_o = 24° 47' 32.55''$ East of Greenwich, and the ellipsoid of reference was on the Walbeck 1819. The fundamental point of both the Triangulation of Finland and St. Petersburg Territory Datum (1891–1903) and of the Baltic Sea Triangulation Datum (1910–1915) is the Pulkovo Observatory (1913 position), where $\Phi_o = 59° 46' 18.54''$ N, $\Lambda_o = 30° 19' 38.55''$ E, $\alpha_o = 200° 38' 35.0''$ Signal A to Kabosi. However, both of these old datums are referenced to the Bessel 1841 ellipsoid, as was the Pulkovo Datum of 1913. Dr. Gábor Timár, Raivo Aunap, and Gábor Molnár published a paper, *Datum Transformation Parameters Between the Historical and Modern Estonian Geodetic Networks* (ISSN 1406-6092) in 2004 that provided the following parameters: **from** Triangulation of the Coast of the Baltic Sea Datum **to** WGS84: $\Delta X = +822$ m, $\Delta Y = +380$ m, $\Delta Z = +649$ m; **from** Triangulation of Finland and St. Petersburg Territory Datum **to** System 42 Datum (the current datum in Russia with origin at Pulkovo Observatory where the new 1942 origin parameters of Pulkovo Observatory are: $\Phi_o = 59° 46' 18.55''$ N, $\Lambda_o = 30° 19' 42.09''$ East of Greenwich, and the reference azimuth to *Bugrõ*: $\alpha_o = 120° 06' 42.305''$, and is referenced to the Krassovsky 1940 ellipsoid, where $a = 6,378,245$ m, and $1/f = 298.3$): $\Delta X = +389$ m, $\Delta Y = +228$ m, $\Delta Z = +664$ m; **from** Baltic Sea Triangulation Datum **to** System 42 Datum $\Delta X = +361$ m, $\Delta Y = +275$ m, $\Delta Z = +664$ m; from System 42 Datum (in Estonia) to WGS84 Datum: $\Delta X = +22$ m, $\Delta Y = -128$ m, $\Delta Z = -87$ m; **from** Triangulation of Finland and St. Petersburg Territory Datum **to** WGS84 Datum: $\Delta X = +411$ m, $\Delta Y = +100$ m, $\Delta Z = +577$ m; and **from** Baltic Sea Triangulation Datum **to** WGS84 Datum: $\Delta X = +383$ m, $\Delta Y = +147$ m, $\Delta Z = +577$ m.

During these inter-war years, the Russians were also actively re-computing their survey information in the Baltic States. Prior to 1932, the Russian horizontal control (used by Russia) of the Baltic States was always referenced to Dorpat Observatory at Tarbu in Estonia. In 1932 the Russians set up Pulkovo Observatory 1932 as their horizontal datum and origin reference to the Bessel 1841 ellipsoid, and later revised this to Pulkovo 1942, now properly termed "System 42" (datum) referenced to the Krassovsky 1940 ellipsoid. According to Dr. Gábor Timár, et al., Ottomar Douglas established a new geodetic adjustment known as the Estonian 1937 System Datum with the fundamental point at *Varesmäe*, where $\Phi_o = 59°\ 18'\ 34.465''$ N, $\Lambda_o = 26°\ 33'\ 41.441''$ East of Greenwich, and the reference azimuth to *Aseri*: $\alpha_o = 39°\ 54'\ 06.256''$. **From** Estonian 1937 System Datum **to** WGS 84 Datum: $\Delta X = +373$ m, $\Delta Y = +149$ m, $\Delta Z = +585$ m, and the average horizontal transformation accuracy is estimated to be about 1 m with maximum error less than 2 m.

With regard to grid systems, the main coordinate system used for geodetic and cartographic activities in Estonia before the Soviet occupation was developed by Fritz Oja and is termed the Historical Lambert System or *Ajalooline Lamberti Süsteem*, where for the Estonian North Zone (*Põhja-Eesti*), the Central Meridian, $\lambda_o = 25°$ East, the Latitude of Origin, $\varphi_o = 59°\ 06'$ North, the Scale Factor at Origin, $m_o = 0.999975$, the False Easting (**Yo**) = 200,000 m, and the False Northing (**Xo**) = 200,000 m. (*Ed. – note the transposition of "X" and "Y" component identifications – the Estonians are consistent with this convention for all grid systems used in their nation, **including UTM**)*! For the Estonian South Zone (*Lõuna-Eesti*), the Central Meridian, $\lambda_o = 25°$ East, the Latitude of Origin, $\varphi_o = 58°\ 06'$ North, the Scale Factor at Origin, $m_o = 0.999975$, the False Easting (**Yo**) = 200,000 m, and the False Northing (**Xo**) = 88,634.86 m. The ellipsoid of reference was the Bessel 1841. Apparently, this was used in connection with all of the Estonian Datums that were referenced to the Bessel 1841 ellipsoid.

The "*O-series maps*" were introduced by the Soviet military (O-34 and O-35), after April 7, 1946. Referenced to the System 42 Datum, the grid system was the "Russia Belts" on the Gauss-Krüger Transverse Mercator for Zone 34 ($\lambda_o = 21°$ East, FE = 4,500,000 m) and Zone 35 ($\lambda_o = 27°$ East, FE = 5,500,000 m), all zones having a False Northing of zero at the equator, and a scale factor at origin (m_o) = 1.0.

The "*C-series maps*" were introduced by the Soviets for civil use in Estonia in 1963 and were deliberately mantled in the typical Soviet penchant for obfuscation for the sake of obfuscation. This "custom" by the KGB/FSB is termed "*maskirovka*." Also referenced to the System 42 Datum, the grid system was a modification of the "Russia Belts" on the Gauss-Krüger Transverse Mercator where the interval spacing (zone widths) of the central meridians were at 3° intervals such that ($\lambda_o = 21°\ 57'$ E, $24°\ 57'$ E, $27°\ 57'$ E), FE = 250,000 m, **all zones having a False Northing of zero NOT at the equator, but at $\varphi = 00°\ 06'$ North** (*maskirovka*) and a scale factor at origin (m_o) = 1.0.

In keeping with the Soviet penchant for obfuscation or *maskirovka*, Soviet legislation about construction activities for every town in Estonia had a local coordinate system based on a local geodetic network. Most of these goofy systems appear to continue to be a mystery to the local inhabitants as to how the local geodetic network was connected to the state geodetic system, or System 42 Datum. An example offered by the Estonian government for one of these "Local Urban Systems" (*Linnade Kohalikud Süsteemid*), "designed" for the capital of Tallinn is as follows: Gauss-Krüger Transverse Mercator (*Faussi Mercatori Põiksilindriline*), $\lambda_o = 24°$ East, FE (**Y$_o$**) = 24,000 m, False Northing (**X$_o$**) = 6,536.000 m at the equator, and a scale factor at origin (m_o) = 1.0. (*See also the goofy coordinate system for Sakhalin Island – Ed.*)

"TM Baltic '93" is designed to give a common reference and mapping frame for Estonia, Latvia, and Lithuania as follows: Gauss-Krüger Transverse Mercator, $\lambda_o = 24°$ East, FE (**Y$_o$**) = 500,000 m, False Northing (**X$_o$**) = 0 m at the equator, and a scale factor at origin (m_o) = 0.9996, euphemistically referred to as a "modified UTM."

Finally, the main official coordinate system (grid system) currently in Estonia is based on the GRS80 ellipsoid, on EUREF-89, and the Lambert Conformal Conic projection. The origin of coordinate parameters "was chosen to match coordinates" with "TM Baltic '93," where $\lambda_o = 24°$ East, the Latitude of Origin, $\varphi_o = 57°\ 31'\ 03.19415''$ N, the southern Standard Parallel, $\varphi_S = 58°\ 00'$ N, the Northern Standard Parallel, $\varphi_N = 59°\ 20'$ N, the False Easting (**Yo**) = 500,000 m, and the False Northing (**Xo**) = 6,375,000 m.

"ESTPOS is the new Estonian GNSS-RTK permanent stations network consisting of continuously operating reference stations. All surveying and mapping is based on geodetic control data.

These data are used in a wide range of application, *e.g.,* geoid calculation, subsidence observations, geodynamic research, meteorology, *etc.*

"The height system is the European Vertical Reference System (EVRS). EVRS is a kinematic reference system which is defined by the Normaal Amsterdams Peil (NAP). A solid Earth-based zero-tide system is used, and heights are expressed as geopotential numbers. To calculate normal heights from geopotential numbers, the normal gravity values of the reference ellipsoid GRS80 are used. The EVRS heights are abbreviated as EH2000. The calculation of EH2000 heights is based on the EVFR2007 solution. The Estonian geoid model is EST-GEOID2017. The EST-GEOID2017 model is used for transformations between EUREF-EST97 ellipsoidal heights and EH2000 normal heights. The gravity system EG2000 is realized through a set of gravity values of the first order gravity network points. Gravity values of the first order gravity network points at the epoch 2000 are based on the absolute gravity measurements performed according to the standards of IAGBN" (https://maaamet.ee/en/spatial-data-and-maps/geodesy).

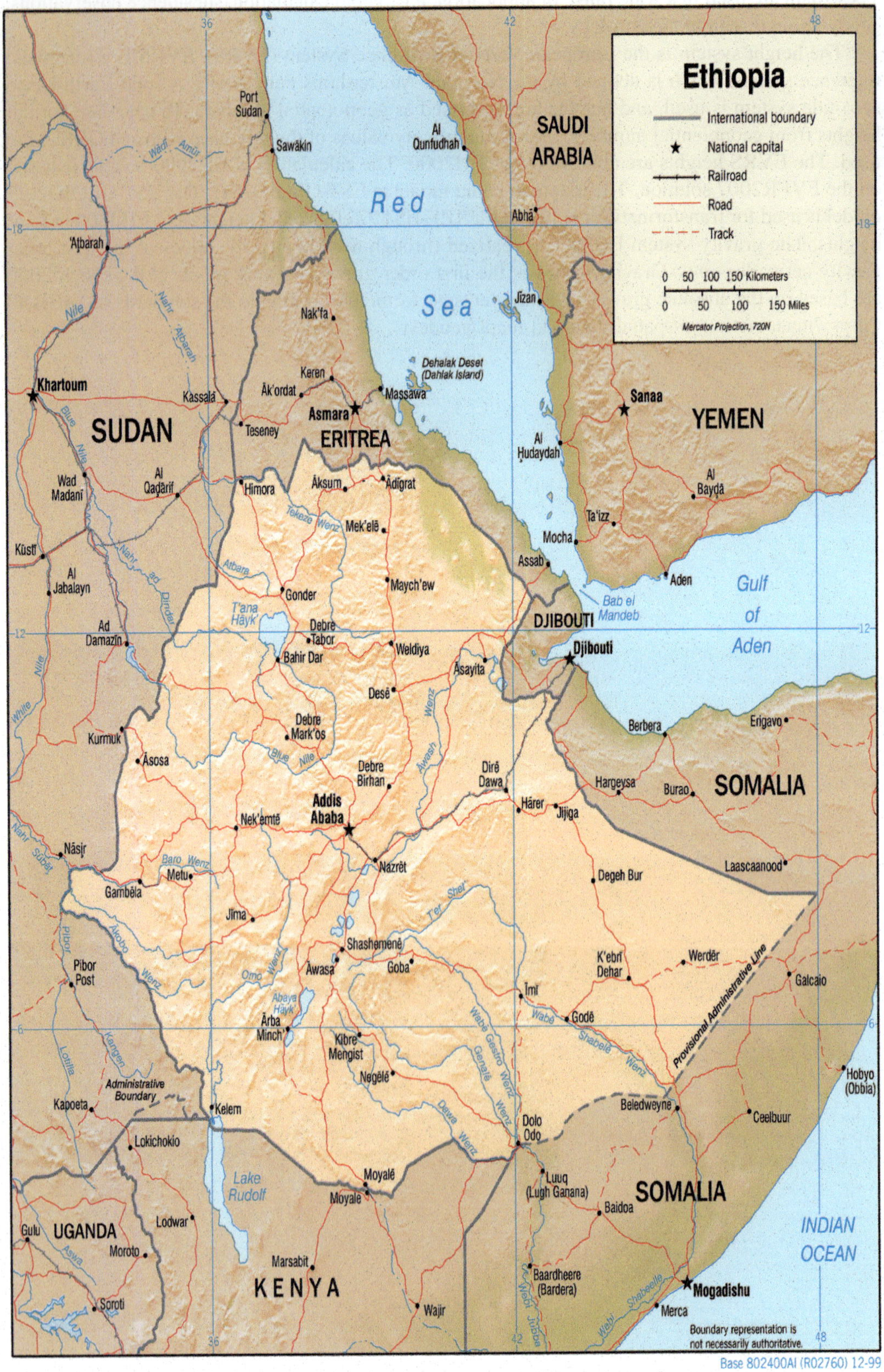

Base 802400AI (R02760) 12-99

Federal Democratic Republic of Ethiopia

Ethiopia is an ancient country occasionally called Abyssinia in Northeast Africa. It is bounded on the north by Eritrea (912 km), on the east by Djibouti (337 km) and Somalia (1,626 km), on the south by Somalia and by Kenya (830 km), and on the west by Sudan (1,606 km). Ethiopia has an area slightly less than twice the size of Texas, it is landlocked, and is mainly high plateau with a central mountain range divided by the Great Rift Valley. The lowest point is *Denakil* (−125 m), and the highest point is *Ras Dashen Terara* (4,620 m).

Modern Ethiopia began with the reign of Tewodros II and the conquest of other chiefs in 1855. Later made Italian in 1882 and claimed as an Italian Protectorate by the Treaty of Ucciali in 1889; the coastal region was made a separate Italian Colony in 1890 and named Eritrea. Territorial integrity of Ethiopia was recognized by Great Britain, France, and Italy in 1906. Invaded by the Italians in 1935, organized with Eritrea and Italian Somaliland as the Empire of Italian East Africa, the region was liberated by the British in 1941. The new constitution was adopted in 1987. Ethiopia's entire coastline along the Red Sea was lost with the *de jure* independence of Eritrea on April 27, 1993.

Very little mapping was done of Ethiopia in the 19th century. One of the earliest investigations of the region resulted in a "*Report of the German Expedition to East Africa, 1861 and 1862*," published in 1864 by Munzinger. The British explorer, Sir Samuel Baker wrote of the region in 1867 in "*The Nile Tributaries of Abyssinia.*" The Italians made some ground surveys in the region of Eritrea from 1888 to 1891, and these surveys formed the basis of the old *Carta della Colonia Eritrea* (Map of the Colony of Eritrea) at a scale of 1:50,000 which was published from 1909 to 1938. Another series was also published during the same time by the Italians at a scale of 1:100,000. New surveys of 1935–1938 resulted in one sheet being compiled by a Santoni stereoplotter with 50 m contours. Smaller scale series were derivative compilations at the time.

The Ethiopian Datum of 1936 was established by the Italians at the West End of Metahara Base (10,083.560 m), where $\Phi_o = 8°\ 53'\ 22.53'' \pm 0.18''$ N, $\Lambda_o = 39°\ 54'\ 24.99'' \pm 0.63''$ East of Greenwich, the reference azimuth to Monte Fantalli was: $\alpha_o = 13°\ 05'\ 21.97'' \pm 0.43''$ and the presumed ellipsoid of reference was the International 1924, where $a = 6,378,188$ m and $1/f = 297$. The check base for the chain commencing at Metahara was Giggiga base which was 12,962.620 m in length. Halfway north from Metahara base to the now Eritrean port city of Massawa was the North End of Alomalà base (18,211.982 m), where $\Phi_o = 12°\ 24'\ 56.56'' \pm 0.13''$ N, $\Lambda_o = 39°\ 33'\ 30.42'' \pm 0.30''$ East of Greenwich, and the reference azimuth to the South End of Alomalà base was: $\alpha_o = 180°\ 00'\ 03.88'' \pm 0.18''$. The old Italian 1:50,000 scale series mentioned above is based on the old datum origin located in the (now) Eritrean capital city of Asmara. The coordinates of that origin elude me, and I suspect that the observations may have been made by the Italian navy rather than by the *Istituto Geografico Miliare* of *Firenze* (Florence). In any case, there was no grid printed on any of the above series, even though the cartographic and geodetic work was done by the Italian military!

The Blue Nile River Basin Investigation Project was funded by the United States, and the geodetic work was performed by the U.S. Coast & Geodetic Survey. The origin of the geodetic work was in southern Egypt near Abu Simbel, south of Lake Nasser, at station Adindan, where $\Phi_o = 22°\ 10'\ 07.1098''$ N, $\Lambda_o = 31°\ 29'\ 21.6079''$ East of Greenwich, deflection of the vertical: $\xi = +2.38''$, $\eta = -2.51''$, and the ellipsoid of reference was the Clarke 1880 (modified), where $a = 6,378,249.145$ m and $1/f = 293.465$. The Blue Nile Datum of 1958 appears to be the established classical datum of Ethiopia and much of North Africa. Adindan is the name of the origin, it is *not* the name of the datum; a most common mistake found in many "reference works." The Ethiopian Transverse Mercator grid is based on a central meridian where $\lambda_o = 37°\ 30'$ E, scale factor at origin where $m_o = 0.9995$, False Easting = 450 km, and False Northing = 5,000 km. (*1957–1961 Ethiopia Geodetic Survey, Blue Nile River Basin*, U.S. Department of Commerce, U.S. Coast & Geodetic Survey, 563 pages.)

Ethiopia and Kenya signed a boundary treaty on June 9, 1970, and the field surveys for the demarcation of the border were performed by British surveyors. The datum used for that survey was the Arc 1960 Datum, referenced to the Clarke 1880 (modified) ellipsoid. The grid system used at the time was the East Africa Transverse Mercator Belts H, J, and K where the central meridians are: $\lambda_o = 37°\ 30'$ (H), $42°\ 30'$ (J), $47°\ 30'$ (K), the scale factors at origin are all where $m_o = 0.9995$, all False Eastings = 400 km, and False Northings = 4,500 km.

An International Boundary Commission has been formed by the United Nations to establish and demarcate a boundary between Ethiopia and Eritrea. The boundary has been researched and established, but the demarcation remains to be performed at the present time. The 125-page document published by the United Nations in April 2002 makes for some fascinating reading. The datum of record of the Commission is the WGS84 and will be used for the demarcation survey someday.

There are two sets of parameters published by NGA for transforming from the Blue Nile Datum of 1958 to the WGS84 Datum: the mean solution for Sudan and Ethiopia is based on a 22-station solution, where $\Delta X = -166\ m \pm 5\ m$, $\Delta Y = -15\ m \pm 5\ m$, $\Delta Z = +204\ m \pm 5\ m$. The solution for Ethiopia is based on an 8-station solution where, $\Delta X = -165\ m \pm 3\ m$, $\Delta Y = -11\ m \pm 3\ m$, $\Delta Z = +206\ m \pm 3\ m$.

A 2019 thesis covers a detailed analysis of datum transformations in the northeast area of Ethiopia, but has no new data points, just more elaborate math models:

Determination of Parameters for Datum Transformation between WGS 84 and ADINDAN-Ethiopia, Hassen, A. M. http://etd.aau.edu.et/bitstream/handle/123456789/23487/Abubeker%20Mohammed.pdf?sequence=1&isAllowed=y

As referenced above, the United Nations has now published some boundary monuments in a report: *Eritrea-Ethiopia Claims Commission – Statement by the Eritrea-Ethiopia Boundary Commission with Annex* (*List of Boundary Points and Coordinates*), November 27, 2006. https//legal.un.org/riaa/cases/vol XXVI/771–799.pdf.

REFERENCES: E

1957–1961 Ethiopia Geodetic Survey, Blue Nile River Basin, U.S. Department of Commerce, U.S. Coast & Geodetic Survey, 563 pages.
Annales Hydrographiques, 4e série, Tome Neuvième, Année 1958
Bulletin Géodésique, no. 8, 1925.
Dr. Gábor Timár, Raivo Aunap, and Gábor Molnár, Datum Transformation Parameters between the Historical and Modern Estonian Geodetic Networks (ISSN 1406-6092).
Equatorial Guinea, U.S. Department of State Background Note, March 2009
Eritrea-Ethiopia Claims Commission – Statement by the Eritrea-Ethiopia Boundary Commission with Annex (List of Boundary Points and Coordinates) 27 November 2006. https//legal.un.org/riaa/cases/vol XXVI/771–799.pdf.
Geodesy in Egypt, *Empire Survey Review*, No. 60.
Guinea Continental Española; Reseñas de los Puntos Astronómicos.
Hassen, A. M., Determination of Parameters for Datum Transformation between WGS 84 and ADINDAN-Ethiopia. http://etd.aau.edu.et/bitstream/handle/123456789/23487/Abubeker%20Mohammed.pdf?sequence=1&isAllowed=y.
https://maaamet.ee/en/spatial-data-and-maps/geodesy.
John W. Hager, Personal Communication, Aug 2009.
la Ley de la Cartografia Nacional de 1978.
Levé de la Baie de Mondah (6 Novembre – 3 Décembre 1954) Mission Hydrographique de la Côte Ouest d'Afrique, *Annales Hydrographique*, pp. 60–62.
Library of Congress Country Studies, El Salvador.

Merriam Webster's Geographical Dictionary, 3rd Edition.
Munzinger, Report of the German Expedition to East Africa, 1861 and 1862, published in 1864
Repertoire des Coordonnees des Points Astronomiques et Géodésiques; Feuille 1/200,000 Libraville NA-32-IV, Paris 10–58.
Shaker, Prof. A.A., *et al.*, Remove-Restore Technique for Improving the Datum Transformation Process, FIG Working Week 2007, Hong Kong, 13–17 May 2007.
Sir Samuel Baker, "The Nile Tributaries of Abyssinia," 1867.
Topographic Mapping of the Americas, Australia, and New Zealand.
TR 8350.2
Triangulation De La Baie De Corsisco (1954–1955).
World Factbook, 2022.

F

The Republic of Fiji Islands 255
Republic of Finland 259
The French Republic 263
Overseas Lands of French Polynesia 267

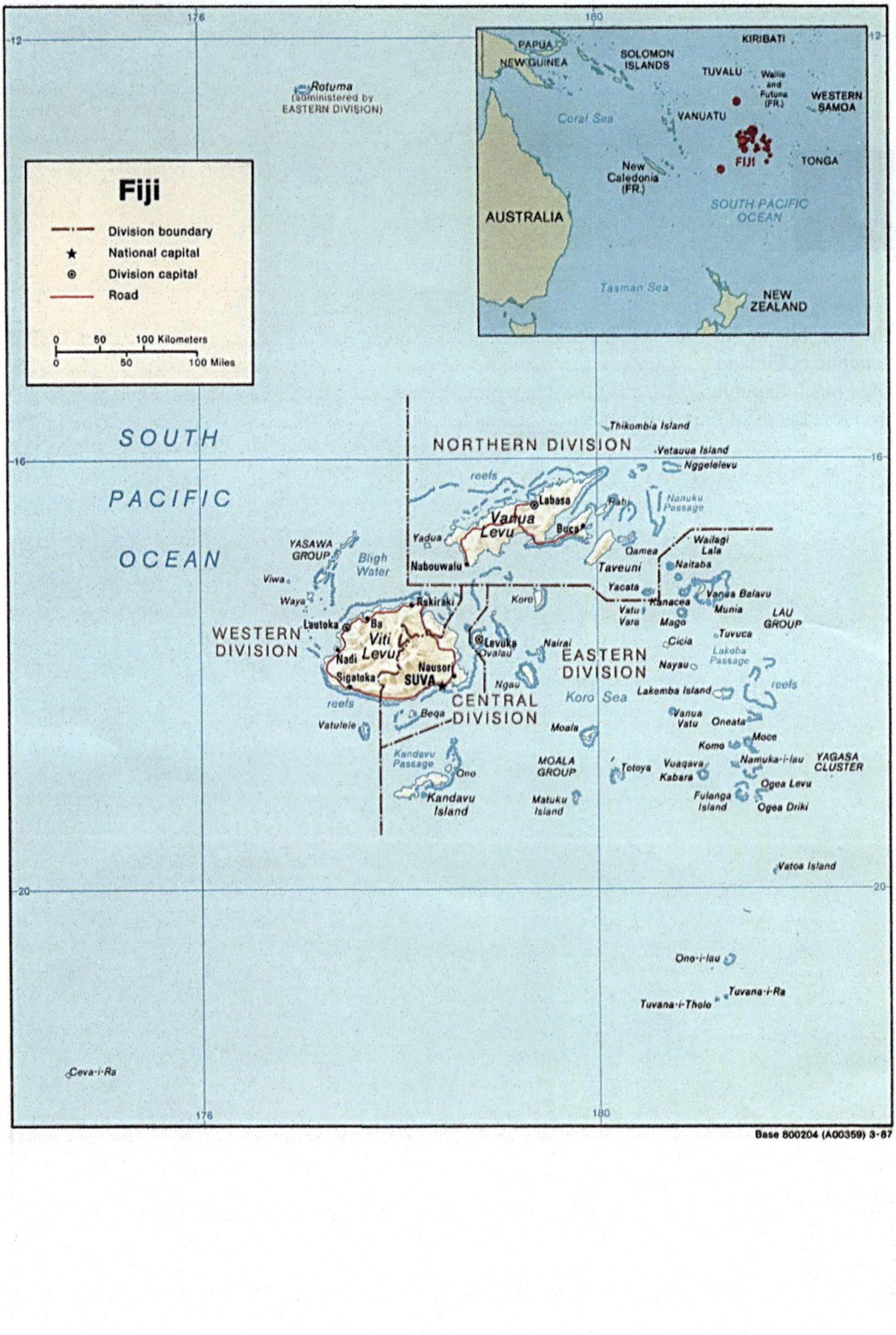

Base 800204 (A00359) 3-87

The Republic of Fiji Islands

Fiji consists of over 300 islands and islets in the South Pacific Ocean of which approximately 110 islands are inhabited. The three largest islands are: Viti Levu, Vanua Levu, and Tavenui. The capital is Suva (on Viti Levu) in the Central District. With a population of almost 813,000 people, the republic gained independence from the UK on October 10, 1970. Fiji has a total coastline of over 1,100 miles; its lowest elevation point is the Pacific Ocean, and its highest point is Tomanivi at 1,324 m (4,344 ft).

Endowed with forest, mineral (gold and silver), and fish resources; Fiji is one of the most developed of the Pacific Island economies. The main sources of income are derived from sugar processing and from the tourist industry. Chief islands include Viti Levu, Vanua Levu, Taveuni, Kandavu, Koro, Ngau, and Oval au. Discovered by Tasman in 1643, visited by Captain Cook in 1774, used by escaped convicts from Australia from 1804, surveyed by Commander Charles Wilkes of the U.S. in 1840; Fiji was offered to Great Britain in 1858 and annexed by the UK in 1874.

European, American, and Australian settlers introduced the concept of land boundaries with monuments, and surveys of properties derived from native people started as early as the 1840s. Angular measurements were recorded in degrees and minutes while linear measurements were recorded in fathoms or chains and links. (The recognized Fijian unit of "fathom" was the distance between the tips of the fingers of a man's outstretched arms!) By 1873, surveys were recognized only if performed by licensed Land Surveyors. The Deed of Cession of 1874 began the Fijian status as a British Crown Colony. Four different categories of land were established: first was the land which could be proved to have become the *bona fide* property of Europeans or other foreigners. Secondly established were the lands which were in the actual use or occupation of some chief or tribe at the date of signing of the deed. Thirdly established lands were required for the probable future support and maintenance of some chief of tribe. Fourthly were all lands not contained in these three previous categories and which were therefore the property of the Crown. Implementing proper surveys according to these categories proved difficult. The Chief Secretary of the Commissioner of Lands traveled to Sidney to engage surveyors for this purpose in 1874. He was authorized to offer an annual salary of £200 for an initial probationary period of 3 months and £300 per annum thereafter. Four surveyors were engaged; one declined before sailing because he was getting married, and another took one look at Levuka, got back aboard the ship, and returned to Sidney. Government surveyors were discouraged from hiring boats for transportation and were told to walk instead to keep expenditures to a minimum. Penny-pinching for survey operations continued for a quarter-century or so until a comprehensive geodetic survey was realized to be the only solution to a critical need for unified positional control.

On October 26, 1908, Captain Ley, an officer of the Corps of Royal Engineers arrived in Fiji with Mr. Travis Rimmer, his civilian assistant. Shortly afterwards, Ley was succeeded by Captain G. T. McCaw, R.E. Thanks to Mr. Russell Fox of the Ordnance Survey of Great Britain for the following "micro biography" on McCaw: 1870 – born in Lurgan, County Armagh, Ireland; 1893 graduated in arts and science from Trinity College, Dublin; 1893–1903 employed by Irish Land Valuation Dept.; 1903–1906 Geodetic Survey of Rhodesia; 1906–1909 with the 30th Arc Survey in Uganda; 1910–1917 Officer in Charge for Trig Survey of Fiji; 1917–1918 Geodetic advisor at British Army General HQ, France; 1919–1936 Tech Assistant at Geographical Section, General Staff (GSGS), British War Office, 1931 first Editor of the *Empire Survey Review*; G.T. McCaw OBE, CMG died in 1942. In later years his protégé, Brigadier Martin Hotine was head of the Directorate of Overseas Surveys. Hotine addressed McCaw as "Uncle" in his correspondence! *Also, the "War Office 1926" ellipsoid is also termed "McCaw 1924."*

The technical instructions initially issued for the Survey of Fiji were "to cover Viti Levu with a system of triangles with sides from 10 to 20 miles long; to measure a primary (baseline) and a check base with probable errors not exceeding 1/50,000 ...." The site of the primary base was approximately 3 miles long and oriented NNE–SSW running along the tongue of land between the Navua River and the coast. McCaw and Rimmer measured the baseline in 1910 with a probable error of ±0.013 feet, i.e., an accuracy of 1/150,000. For the check base at Lautoka, a line 3 miles 5 furlongs in length was selected from a point near the boundary of the Vitogo Estate. The final check was ±0.067 feet. Fieldwork on Viti Levu was completed, and the party returned to England in March 1912. The Viti Levu Datum of 1912 was referenced to the Clarke 1880 ellipsoid, where $a = 20,926,202$ feet (317,063.6667 chains) and $1/f = 293.4663077$. I think it is no coincidence that the ellipsoid parameters and units used for Fiji are exactly the same as used in British Africa since McCaw had already surveyed in Rhodesia and Uganda. The Latitude of Origin was obtained astronomically from station Mona Vatu, where $\Phi_0 = 17°\ 53'\ 28.285''$ South ±0.75″, and the Longitude origin was obtained astronomically at station Suva Longitude pillar (concrete pier in the rear of the offices of the Pacific Cable Board and within a few feet of the high water mark) with telegraphic cable signals from both Vancouver, Canada, and from Moreton Bay, Queensland, Australia, where $\Lambda_0 = 178°\ 25'\ 35.835''$ East of Greenwich ±0.835″. The defining azimuth from north was determined at Mona Vatu to Tikituru to be $\alpha_0 = 205°\ 17'\ 21.17''$ from observations at Rasusuva and from Nath Ova. His main instrument was a 6-inch diameter theodolite with an alidade (telescope) level with a par value of $\pi = 5''$. "It was quickly noticed that the bubble used to 'jam' in the vial ...." He obtained another vial and observed that "a second alidade level was obtained from the makers and proved not a whit better than the first ..."! He figured a way around it, and all his work served as a model for decades – his own reports of his work in Fiji were used as textbooks on how to do it throughout the British Empire. McCaw chose the Cassini-Soldner projection for the basis of the Viti Levu Grid where the Latitude of Origin $(\varphi_0) = 18°\ 00'$ South, and the Central Meridian $(\lambda_0) = 178°\ 00'$ East (per specifications of the Secretary of State). The Scale Factor at Origin (m_0) by definition is unity on the Cassini-Soldner, the False Northing = 7,040 chains (88 miles), and the False Easting = 5,440 chains (68 miles). Note that 1 m = 0.049709782 chains for Fiji. NGA Technical Report 8350.2, dated January 3, 2000, lists the three-parameter shift from Viti Levu Datum of 1916 to WGS 84 Datum as: $\Delta X = +51\ m$, $\Delta Y = +391\ m$, $\Delta Z = -36\ m$. These shift parameters were originally published in 1987 and have an uncertainty of ±25 m in each component. This relation is the result of a single station occupation.

The Trig Survey party returned to Fiji in September 1912 to extend the triangulation of Viti Levu across to Vanua Levu and Taveuni. The baseline for this part of the work was measured from Nasekula toward the sea at Tua Tua and the final measurement over the 3½ miles was accepted at ±0.015 feet. McCaw reported that the Vanua Levu Datum of 1915 was referenced to the Clarke 1880 ellipsoid also, and the Latitude origin was obtained astronomically from station Numuiloa where $\Phi_0 = 16°\ 23'\ 38.36''$ South ±0.529″, and the Longitude origin at station Suva pillar on Viti Levu was adopted. The defining azimuth from north observed at Nasenla (East Base) to Vatia is $\alpha_0 = 123°\ 57'\ 28.75''$. The grid for the Vanua Levu Datum of 1915 is where the Latitude of Origin $(\varphi_0) = 16°\ 15'$ South, and the Central Meridian $(\lambda_0) = 179°\ 20'$ East, the Scale Factor at Origin $(m_0) = 1.0$.

McCaw was concerned about the orientation of these islands with respect to the systematic distortions inherent in the Cassini-Soldner projection. "If an origin be chosen" (as per the Grid Origin above), "and lines be drawn SW and SE from this origin, they will be found to traverse centrally the principal land surfaces. Thus the SW axis, passing near the main stations of Mbulembulewa, Nathau, and Navotuvotu, crosses midway the fine Ndreketi plain and Mbua Province in the extreme west. The SE axis passes up the Mbuthaisau Valley, crosses about midway the Natewa Embayment, passes through the Thakaundrove Peninsula, and leaving the latter near the main station of Navivvia, crosses Taveuni at about one sixth of its length from its centre of figure. The important Lambasa plain lies so close to the origin that the error of projection in this neighborhood (sic) will be very small." He developed a correction factor for this datum that resulted in the most unique projection for a grid in the world; the Hyperbolic Cassini-Soldner where the abscissae (Eastings) are reduced by the factor $1/_3$ X^3O sin $1''$ and where the factor $O = \frac{1}{2} \nu\rho$ sin $1''$. In my column on the Czech Republic, the reader may recall that the Radius of the Gaussian Sphere $= [\nu\rho]^{-1/2}$. The False Northing $= 16,628.885$ chains, and the False Easting $= 12,513.318$ chains because McCaw took the Hyperbolic Cassini-Soldner coordinates of the "natural" origin in the Viti Levu system of the meridian 179° 20′ and the parallel 16° 15′. These manipulations, as I understand them, were to minimize the distortions in the land areas of the main islands while maintaining a consistent-appearing grid coordinate system for Viti Levu, Vanua Levu, Tanvenui, *etc*. The field party left Fiji in 1915 and the final report was published in 1917 after meeting the geodetic demands of WWI. An example point in the McCaw report lists the Vanua Levu Datum of 1915 coordinates of "Ndana" (Station 21) as: $\phi = -16°$ 50′ 29.2435″ S, $\lambda = 179°$ 59′ 39.6115″ E. Thanks to Ms. Litea Biukoto of the South Pacific Applied Geoscience Commission (SOPAC) for a copy of that report. A report printed in 1985 lists the Hyperbolic Cassini-Soldner coordinates of "Dana" as: $X = 1,601,528.90$, $Y = 1,336,966.00$. Thanks to Mr. Mal Jones of Perth, Australia, for a copy of that report.

Mr. Russell Fox informs me, "In 1956 the New Zealand Surveyor General's Office readjusted McCaw's work on one datum (Fiji 1956) using the International Spheroid and the UTM Grid. That was used on military mapping from circa 1960 onwards. DOS (Directorate of Overseas Surveys) did Tellurometer (electronic distance meter or EDM) traversing in the 1960's to strengthen and densify the trig network."

In 1978, two senior personnel from the New Zealand Department of Lands and Survey undertook a review of survey and mapping activities in Fiji. As a result, the Australian Army Survey Corps (RASVY) established a number of Doppler (TRANSIT satellite) stations for mapping control and determination of the exclusive economic zone. In 1984 field work began by the New Zealanders for further Doppler surveys and for the re-computation of McCaw's Datums of Fiji. First-order astronomic latitudes longitudes and azimuths were observed at a total of eight stations. In general, these coincided with either the original McCaw stations or with the Doppler stations. New first-order EDM traverses were undertaken around the periphery of Viti Levu, through Vanua Levu, and across the 100 km wide strait between the two islands. By late 1985 all the observations required for the primary network had been completed. According to Hannah and Maseyk (*Survey Review* 30, 231 – January 1989), "In the first instance the complete set of Ley-McCaw triangulation data as given by the historic records was accepted in total. The only exception to this being the deletion of five stations which had been positioned weakly by either resection or intersection." Note this is the same general philosophy as adopted by the U.S. National Geodetic Survey when they did the North American Datum in the middle 1980s. The result of the New Zealand observations and adjustment was the Fiji Geodetic Datum of 1986. A new unified Cartesian system was also devised; it is called the Fiji Map Grid (FMG) and is based on the Gauss-Krüger Transverse Mercator projection, where Latitude of Origin $(\varphi_o) = 17°$ 00′ South, and the Central Meridian $(\lambda_o) = 178°$ 45′ East. The Scale Factor at Origin $(m_o) = 0.99985$, the False Northing $= 4,000$ km, and the False Easting $= 2,000$ km. The ellipsoid of reference is the WGS 72, where $a = 6,378,135$ m and $1/f = 298.26$. Mr. Franck Martin of SOPAC lists transformation parameters on their Internet site as: "from WGS84 to Fiji Geodetic Datum" as determined by the Forestry Department of Fiji as: $\Delta X = +35.173$ m, $\Delta Y = -136.571$ m, $\Delta Z = +36.964$ m, Scale $= +1.537 \times 10^6$, $R_x = -1.37$ arc seconds, $R_y = +0.842$ arc seconds, and $R_z = +4.718$ arc seconds. I have not verified these transformation parameters with respect to the WGS84 Datum, but in 1985 the Fiji Department of Lands and Survey listed the FMG coordinates of "Dana" as $X = 2,132,200.63$ m, $Y = 4,016,984.99$ m. Thanks to Mr. John Hagar for his generous help and counsel.

Finland
International boundary
Province (lääni) boundary
National capital
Province (lääni) capital
Railroad
Road
Helsinki is the capital of Uusimaa Lääni.
0 50 100 Kilometers
0 50 100 Miles
Lambert Conformal Conic Projection, SP 47N/62N
NORWAY
SWEDEN
RUSSIA
ESTONIA
Barents Sea
Gulf of Bothnia
Gulf of Finland
Baltic Sea
Lake Ladoga
Arctic Circle
LAPPI
OULU
VAASA
KESKI-SUOMI
KUOPIO
POHJOIS-KARJALA
MIKKELI
TURKU JA PORI
HÄME
KYMI
UUSIMAA
AHVENANMAA
ALAND ISLANDS
Helsinki
Tromsø
Alta
Lakselv
Kirkenes
Nikel
Pechenga
Severomorsk
Murmansk
Inarijärvi
Ivalo
Muonio
Kolari
Kiruna
Malmberget
Bodø
Jokkmokk
Mo
Mosjøen
Kandalaksha
Kandalakshskaya Zaliv
Kelloselkä
Kuoloyarvi
Kemijärvi
Rovaniemi
Tornio
Kemi
Taivalkoski
Luleå
Storuman
Oulu
Raahe
Skellefteå
Kostomuksha
Yushkozero
Kem'
Oulujärvi
Kajaani
Ylivieska
Umeå
Kokkola
Pietarsaari
Östersund
Örnsköldsvik
Iisalmi
Lieksa
Pielinen
Vaasa
Kuopio
Ilomantsi
Ånge
Härnösand
Seinäjoki
Joensuu
Sundsvall
Äänekoski
Kaskinen
Virrat
Jyväskylä
Pieksämäki
Varkaus
Parkano
Hudiksvall
Jämsä
Savonlinna
Sortavala
Näsijärvi
Päijänne
Pori
Tampere
Mikkeli
Rauma
Hämeenlinna
Imatra
Saimaa
Lahti
Kouvola
Lappeenranta
Gävle
Uusikaupunki
Vaalimaa
Borlänge
Forssa
Hyvinkää
Kotka
Vyborg
Turku
Salo
Porvoo
Maarianhamina
St. Petersburg
Västerås
Uppsala
Hanko

Republic of Finland

Finland is located in northern Europe and has a 1,126 km coastline on the Baltic Sea, the Gulf of Bothnia, and the Gulf of Finland, excluding islands and coastal indentations. Bordered by Norway (729 km), Sweden (586 km), and Russia (1,313 km), Finland is slightly smaller than Montana. The lowest point is the Baltic Sea (0 m), and the highest point is *Haltiatunturi* (1,328 m). Note that the southwestern border of Finland with Sweden is the Tornio River where the French Royal Academy of Sciences performed one of their triangulations in the 18th century in a futile attempt (by *Pierre-Louis Moreau de Maupertuis*) to prove that Sir Isaac Newton was incorrect in his theory on the shape of the Earth being an oblate ellipsoid.

According to the Library of Congress *Country Study of Finland*, "present-day Finland became habitable in about 8,000 B.C., following the northward retreat of the glaciers, and at about that time Neolithic peoples migrated into the country. According to the legends found in the Finnish folk epic, the *Kalevala*, those early inhabitants included the people of the mythical land *Pohjola*, against whom the *Kalevala* people – identified with the Finns – struggled ... Both the traditional and modern theories agree that in referring to this prehistoric age one should not speak of a Finnish people, but rather of Finnic tribes that established themselves in present-day southern Finland, gradually expanded along the coast and inland, and eventually merged with one another, absorbing the indigenous population. Among those tribes were the *Suomalaiset*, who inhabited southwestern Finland and from whom was derived *Suomi*, the Finnish word for Finland. The *Tavastians*, another Finnic tribe, lived inland in southern Finland; the *Karelians* lived farther east in the area of the present-day *Karelian Isthmus* and *Lake Ladoga*. On the southern coast of the Gulf of Finland were the Estonians, who spoke a Finno-Ugric language closely related to Finnish. North of the Finns were the Lapps (or *Sami*), who also spoke a Finno-Ugric language, but who resisted assimilation with the Finns."

Thanks to the *History of the Finnish Geodetic Institute*, "In 1748, a special surveying committee was established in Finland, and as its first observator (*sic – Ed.*) was named Jakob Gadolin. He measured in 1748–1750 and 1752–1753 the *Turku-Åland* triangle chain, and from there on across the *Åland Sea* to the Swedish side at *Grisslehamn*. In 1754, Johan Justander continued the work of Gadolin, who had gone to *Turku* University to become Professor of Physics, by extending the triangulation Eastward from *Turku*, along the coast of the Gulf of Finland, arriving in *Helsinki* in 1774.

"The next important triangulation in our country was performed 1801–1803, when the Swedish Jöns Svanberg repeated Maupertuis' measurements along the original triangle chain, at the same time extending them to both North and South. The interest of Svanberg was not however mapping but a check on the flattening ratio of the Earth.

"Also, the *Russian-Scandinavian* grade measurement, planned by one of the great names in astronomical history, Wilhelm Struve, was done to check the dimensions of the Earth. This measurement started at the Danube Delta, ran through Bielo-Russia and Estonia to Finland and from here onward into the Norwegian fiords and the Arctic Sea. It entered Finland from Gogland island (*Suursaari*) and, passing west of *Loviisa* to the area of *Jyväskylä* and from there on through *Kajaani, Oulu, Torno* and *Muonio* to *Hammerfest*. The measurements were started in 1816, and the angle observations in the part located in Finland were done in 1830–1845.

"Of the triangulations which took place during the last century *(19th – Ed)*, also the *Baltic triangulation* must be mentioned. It was done by the Hydrographic Department of the Russian Naval Ministry in 1828–1838. This measurement ran along the coast of the Gulf of Finland from the Eastern border to *Åland* and from there to Sweden, where it connected to the points *Söderarm*,

Arholm and *Grisslehamn* determined by Swedish geodesists. All in all, 398 triangulation points were measured on the Finnish territory.

"Neither of these measurements mentioned above, the Russian-Scandinavian and the Baltic triangulation, was very useful to Finnish mapping. In both measurements the triangulation point set was poorly marked in the terrain, as a result of which points were later difficult to find, and a large part of them soon vanished completely. Also, the measurement precision was questionable."

From my readings of the common practices of the Czarist Topographic Corps throughout Russia in the 19th and early 20th centuries, they commonly erected triangulation targets of up to 30 m high made of scantlings (split wooden rails). Their permanent "monuments" in addition to the scantling towers were buried glass vodka bottles! Locals would soon scavenge the towers for firewood and thus most of the enormous geodetic efforts were lost forever ...

"Besides the geodetic measurements summed up above, also original geodetic research was done in Finland during the last century *(19th – Ed.)*, especially by the *Turku* astronomer H.J. Walbeck. In 1819, at the age of 25 only, Walbeck published the work *De forma et magnitudine telluris ex dimensis arcubus meridiani definiendis* (Determining the figure and size of the Earth by means of meridian grade measurements). In it he, as one of the first to do so, applied the least squares procedure known from mathematical statistics to the determination of the figure and dimensions of the Earth. He derived the dimensions of the Earth from five different grade measurements, obtaining $a = 6,376,896$ m and $1/f = 1:302.8$, values used in Russian geodetic measurements for almost a century. Through his publication, Walbeck got acquainted with the famous director of the *Tartto* Observatory, Wilhelm Struve, and planned on Struve's request the course through Southern Finland of the triangulation chain belonging to the Russian-Scandinavian grade measurement." *(Note that in this context, a "grade" is 1/400th of 360° – Ed.)*

Thanks to Marko Ollikainen in *The Finnish Coordinate Reference Systems*, "When the Finish Geodetic Institute (FGI) started carrying out triangulation in 1919 in Finland, the Hayford ellipsoid ... was chosen as the reference ellipsoid. The dimensions of the ellipsoid, also known as the International ellipsoid 1924, are as follows: $a = 6,378,388.0$ m, $1/f = 297.0$." Many early map sheets produced by the Government Printing Office (*Maanmittaushallituksen*), referenced the prime meridian at Helsinki which is +24° 57' 16.5" East of Greenwich (*AMS Geodetic Memo 915.0263, 1947*).

In 1922, the Gauss-Krüger Transverse Mercator projection was selected for mapping purposes. The majority of the *mapping* was referenced to Helsinki Observatory and was known as the Helsinki System, or VVJ (*Vanha Valtion Järjestelmä*), where the origin was at: $\Phi_o = 60°\ 11'\ 02.33''$ N and $\Lambda_o = 24°\ 57'\ 08.94''$ E. The VVJ Grid System is based on 3° Belts, the scale factor at origin is equal to unity (1.0), and the Central Meridian of the various zones with False Eastings are as follows: Zone 0 = 18°E, FE = 500 km, Zone 1 = 21°E, FE = 1,500 km, Zone 2 = 24°E, FE = 2,500 km, Zone 3 = 27°E, FE = 3,500 km, Zone 4 = 30°E, FE = 4,500 km, Zone 5 = 33°E, FE = 5,500 km.

When the Finnish Geodetic Institute published the results of its first triangulation chain from Åland to Helsinki in 1924, the origin chosen was at *Hjortö*, in Korppoo. The *Hjortö* System soon proved unsuitable and was abandoned in 1927. In 1931, a new system was published partly based on gravimetric measurements and was referred to as the "second adjusted system." The fourth system used by the Finnish Geodetic Institute was the all-European ED-50, or European Datum 1950; the starting coordinates of which were defined such that they produced a least squares solution for the deflection of the vertical for approximately 100 points. The results of the adjustment of the primary triangulation of Finland published in 1967 were given on the ED-50 System.

The ED-50 was recognized as a good improvement to the Helsinki System of 1931. It was used as a model for a subsequent adjustment known as the KKJ 1970 System (*Kartastokoordinaattijärjestelmä*). However, it differs from the ED-50 in that it is a three-parameter transformation from ED-50 to match the VVJ system as closely as possible through two translations and one rotation. The ellipsoid of reference remains unchanged from the VVJ and the ED-50. The difference between VVJ and KKJ coordinates is 2 m on average, and the maximum difference is approximately 10 m

(*Eino Uikkanen's Homepage*). The KKJ Grid system is the same as the VVJ Grid System, except that there is also a single KKJ Grid for the entire country that corresponds to the parameters listed for Zone 3. The University of *Jyäaskylä* has prepared a system of transformation parameters consisting of translations and polynomial coefficients for transforming back and forth between the 48 VVJ municipal coordinate systems and the KKJ national coordinate system. Access to these parameters is through the website of the National Land Survey of Finland.

The seven-parameter transformation from KKJ to the European Terrestrial Reference System of 1989 (ETRS89) is: $\Delta X = -96.062$ m, $\Delta Y = -82.428$ m, $\Delta Z = -121.754$ m, $R_x = -4.801''$, $R_y = -0.345''$, $R_z = +1.376''$, and $\Delta s = +1.496 \times 10^{-6}$. The average error of transformation is estimated at ± 0.8 m and the maximum error at ± 2 m, based on a least squares solution of 90 collocated points. The hierarchy of rotations *appears* to match the standard U.S. convention rather than the left-handed preference of many European publications. The reader is cautioned to find a test point for verification of this guess! Thanks go also to Dr. Roy Ladner of the U.S. Naval Research Laboratory who submitted a research paper on mapping and geodesy of Finland for a graduate credit course at the University of New Orleans back in 1999.

France
International boundary
National capital
Railroad
Expressway
Road
0 50 100 Kilometers
0 50 100 Miles
Lambert Conformal Conic Projection SP 36N /48 30 N
UNITED KINGDOM
NETHERLANDS
BELGIUM
LUX.
GERMANY
SWITZERLAND
LIECH.
AUS.
ITALY
MONACO
ANDORRA
SPAIN
North Sea
English Channel
Strait of Dover
CHANNEL ISLANDS
Bay of Biscay
Gulf of Lion
Ligurian Sea
Mediterranean Sea
Corsica
Birmingham
Ipswich
Fishguard
Cardiff
Bristol
London
Reading
Dover
Southampton
Brighton
Portsmouth
Weymouth
Plymouth
Penzance
Amsterdam
The Hague
Utrecht
Rotterdam
Oostende
Dunkerque
Calais
Gent
Antwerp
Brussels
Liège
Lille
Valenciennes
Charleroi
Luxembourg
Alderney
Guernsey (U.K.)
Sark
Jersey (U.K.)
Cherbourg
Le Havre
Amiens
Rouen
Caen
Paris
Reims
Châlons-sur-Marne
Metz
Saarbrücken
Karlsruhe
Mannheim
Stuttgart
Nancy
Strasbourg
Ulm
Troyes
Brest
Lorient
Rennes
Le Mans
Orléans
Mulhouse
Freiburg
Bodensee
Belle-Île
Angers
Tours
Dijon
Besançon
Basel
Zürich
Vaduz
Saint-Nazaire
Nantes
Bourges
Bern
Île d'Yeu
Lausanne
Lake Geneva
Poitiers
Île de Ré
La Rochelle
Île d'Oléron
Geneva
Limoges
Clermont-Ferrand
Lyon
Chambery
Milan
Bergamo
Saint-Étienne
Grenoble
Turin
Piacenza
Bordeaux
Valence
Alessandria
Genoa
Dordogne
Avignon
Nîmes
Nice
Cannes
Santander
San Sebastián
Bayonne
Pau
Toulouse
Montpellier
Marseille
Toulon
Bastia
Bilbao
Vitoria
Pamplona
Perpignan
Burgos
Andorra la Vella
Port-Bou
Girona
Ajaccio
Zaragoza
Río Duero
Río Ebro
Bonifacio
Seine
Loire
Rhône
Garonne
Danube

The French Republic

France is bordered by Andorra (55 km), Belgium (556 km), Germany (418 km), Italy (476 km), Luxembourg (69 km), Monaco (6 km), Spain (646 km), and Switzerland (525 km). Located in Western Europe, bordering the Bay of Biscay and English Channel, between Belgium and Spain, southeast of the UK; bordering the Mediterranean Sea, between Italy and Spain; slightly more than four times the size of Georgia; slightly less than the size of Texas. The terrain of France is mostly flat plains or gently rolling hills in north and west; the remainder is mountainous, especially Pyrenees in south, and Alps in east. The highest point is Mont Blanc (4,810 m), the lowest point is the Rhone River delta (−2 m), and the mean elevation is (375 m).

Originally known in the south as *Gallia Narbonensis*, a province of the Roman Empire from 121 B.C.; the north and central parts known as Gaul were conquered by Caesar from 58 to 51 B.C. In 1789, the French Revolution overthrew the Royal government, and tumultuous times continued for over a hundred years that included Napoleon Bonaparte and his First Empire. The northern part was ravaged by fighting in WWI (1914–1918) and was conquered and controlled in WWII (1940–1944) by the Germans. The Fourth Republic was established in 1945, the Fifth Republic in 1958. President Charles de Gaulle resigned in 1969 following defeat of a referendum on constitutional reforms.

The *Dépôt de la Guerre* was established in 1688 and was responsible for all military surveys. When the English Sir Isaac Newton suggested that the shape of the Earth is an oblate ellipsoid, France countered that Cassini's triangulation showed that the Earth is a prolate ellipsoid. France sent expeditions to Stockholm and Quito to only prove the Englishman's theory correct! (See Ecuador and Sweden.) The first published ellipsoid was by Bouguer and Maupertuis in 1738 because of their historic expeditions. During the Napoleonic Wars, Cassini was the first topographer to utilize a grid overprint on a topographic map for the "coordinated" control of artillery fire. In apparent deference to the great German cartographer, Cassini chose the Bonne equal-area projection. That projection choice also for France's first major topographic map series of 1818–1887 influenced the rest of the world for over a century with the *Carte de l'État-major au* 1:80,000. The Old Triangulation of France Datum of 1818 was referenced to the DeLambre ellipsoid where the semi-major axis $(a) = 6,376,985$ m and the reciprocal of flattening $(1/f) = 308.64$. The *Carte de l'État-major* Latitude of Origin is $\varphi_o = 45°\ 10'\ 00''$, and the Central Meridian is $\lambda_o = 2°\ 20'\ 13.95''$ East of Greenwich in today's convention – but actually zero at the time. The French considered Paris as the prime meridian for the world, and many other countries agreed. The French Navy began publishing the *Annals Hydrographiques* in 1844 that contained latitude and longitude coordinates of stations observed astronomically worldwide, as well as details on local datums that had been established. That first *tome* has an amusing passage regarding the "friendly natives" encountered in an expedition to Tahiti – not all was geodesy and hydrographic surveying!

During the second half of the 19th century, France was the epicenter of mathematical cartography in the world. Tissot and Germain published their monumental works on projections, and the *Service Géographique de l'Armée* was founded in 1887. The New Triangulation of France commenced in 1887 (NTF 1887), and the small-scale *Dépôt de la Guerre* (*Carte de l'État-major*) map series was continued until 1915 on the Bonne projection. Germain developed his Projection of Minimum Deformation, and it was chosen for the new large-scale topographic map series based on the NTF 1887. The ellipsoid of reference was the modified DeLambre 1810 or Plessis Reconstituted where the semi-major axis $(a) = 6,376,523.994$ m and the reciprocal of flattening $(1/f) = 308.624807$.

During this time, the French were developing a philosophy of categorizing map projections, and this penchant when combined with another factor created some curious developments. Since logarithms were the universal tool for hand computations, great algebraic effort was made to simplify equations whenever possible. With numerical techniques being practically restricted to expressions in the form of infinite series, equations were always truncated to only yield the necessary computational precision (for the geographic area of interest). Extra digits were too expensive in terms of labor to waste on niceties. The tables of equations for the different categories of map projections showed terms only to the third power, the cubic. When Germain's Projection of Minimum Deformation was truncated to the cubic, it became identical to the Lambert Conformal Conic when truncated to the cubic. Since the late Heinrich Lambert was senior to Germain, the former's name was given to the new projection adopted for the new large-scale map series. The French tables were computed with the Lambert Conformal Conic projection of the developed meridional distance formulae truncated to the cubic term. This resulted in a not strictly conformal projection, but it was deemed "close enough" at the time. This convention for the "French Army Truncated Cubic Conic" was also applied to practically all of the French Colonies for decades to include Syria, the Levant or Palestine, Morocco, Algeria, Tunisia, and French Indochina (Lao, Cambodia, and Vietnam).

When the officers of the French Army fled France after the Kaiser invaded their country, they carried many of their surveys and map manuscripts to London. The United States commissioned some mathematicians and geodesists with the Coast Survey into the Corps of Engineers and sent them to London to assist the Royal Engineers and the French. Upon arrival, they noticed the scarcity of the Tables of Projection for the northern war zone (*Nord de Guerre*) and sent one copy back to Washington for tabular extension and duplication. In Washington, it was noticed that the documented formula was truncated at the cubic term. They apparently decided to develop tables that not only had greater latitudinal extent, but they also decided to use more terms for the developed meridional distances. Shortly after a group of mathematicians was assigned to perform the task, others at the Coast Survey decided that it was a nifty idea to use a conformal projection for a basis of survey computations. Another group was assigned to perform the same task but instead of using the Plessis ellipsoid for the *Nord de Guerre* Zone of France, they used the Clarke 1866 ellipsoid for the United States. Computations for both tables were completed at the same time, and both manuscripts were sent to the Superintendent of Documents at the same time. When the printing office sent the crates of tables back to the Coast Survey, the wrong crate was shipped to London. After some consternation, the proper crate arrived only to cause further consternation. The fully conformal tables for the Lambert Conformal Conic would not cast a graticule to match the existing *Nord de Guerre Zone* sheets based on the French Army Truncated Cubic Conic. The new tables were discarded, the Royal Engineers and the French Army Engineers "made do" with what they already had. **Moral: Use the same projection formulas as the originating country uses. It is not "correct" unless it matches native work.** The French Kilometric Quadrillage (AEF) of 1918 was based on the Fully Conformal Lambert Conic where the Latitude of Origin was $\varphi_o = 49° 30' 00''$, and the Central Meridian was $\lambda_o = 7° 44' 13.95''$ East of Greenwich, the Scale Factor at Origin $(m_o) = 0.999509082$, and *both* the False Eastings and False Northings were 500 km. The ellipsoid of reference was the Clarke 1866, and this grid was never used. On the other hand, the French *Nord de Guerre Zone* (1914–1948) *was* used, and it was based on the French Army Truncated Cubic Conic where the Latitude of Origin was $\varphi_o = 49° 30' 00''$, the Central Meridian was $\lambda_o = 7° 44' 13.95''$ East of Greenwich, the Scale Factor at Origin $(m_o) = 0.999509082$, and the False Easting was 500 km and the and False Northing was 300 km. The ellipsoid of reference was the Plessis Reconstituted.

After WWI, the French developed four Lambert zones for the country. From 1920 to 1948, Zone I (*Nord*) parameters were where the Latitude of Origin was $\varphi_o = 49° 30' 00''$, the Central Meridian was $\lambda_o = 2° 20' 13.95''$ East of Greenwich, the Scale Factor at Origin $(m_o) = 0.999877340$, and the False Easting was 600 km, and the False Northing was 200 km. The ellipsoid of reference was the Clarke 1880 where the semi-major axis $(a) = 6,378,249.2$ m and the reciprocal of flattening $(1/f) = 293.4660208$. In France, for Zone II (*Center*), the Latitude of Origin was $\varphi_o = 46° 48' 00''$, the Central Meridian was $\lambda_o = 2° 20' 13.95''$ East of Greenwich, the Scale Factor at Origin $(m_o) = 0.999877419$, and the False Easting was 600 km and the False Northing was 200 km. In France, for Zone III (*Sud*), the Latitude of Origin was $\varphi_o = 44° 06' 00''$, the Central Meridian was

$\lambda_o = 2° 20' 13.95''$ East of Greenwich, the Scale Factor at Origin $(m_o) = 0.999877501$, and the False Easting was 600 km, and the False Northing was 200 km. In France, for Zone IV (*Corse*) Corsica, the Latitude of Origin was $\varphi_o = 42° 09' 54''$, the Central Meridian was $\lambda_o = 2° 20' 13.95''$ East of Greenwich, the Scale Factor at Origin $(m_o) = 0.999940004$, and the False Easting was 600 km, and the False Northing was 200 km. All four of these zones were based on the French Army Truncated Cubic Conic projection.

In 1940, the mapping agency was renamed the *Institute Géographique National (IGN)*. The French government declared, "*C'est pourquoi il fut decide en 1948 de subsituer des formulas rigoureuses aux développements limités, en conservant les mêmes limites de zones et les mêmes modules d'homothétie.*" In other words, France went to the rigorous fully conformal formulae in 1948 for the Lambert Conic. Since then, only Algeria has deemed to do the same in the early 1960s according to Mr. Roger Lott, the (*retired*) Chief Surveyor of British Petroleum. The other old French Colonies, although now independent, still use the French Army Truncated Cubic. Some old colonies still consider Paris as the origin of longitudes. One caution for the U.S. practitioner: the French do not use degrees-minutes-seconds for angular measurement; they use Grads where $400^G = 360°$.

After 1948, the new French Lambert zones (I–IV) retained the same parameters as listed above for the republic. Only the formulae changed. During that same era, the U.S. Army Map Service directed the re-computation of all the triangulations of Europe and the Mediterranean to the European Datum of 1950. Everything was sequentially tied to France, and the unifying tool was the UTM Grid. France is covered by UTM zones 30–32 referenced to the EU50 Datum and the International ellipsoid, where $a = 6,378,388$ m and $(1/f) = 297$. IGM currently publishes their general national three-parameter transformation from NTF to WGS84 as: $\Delta X = +168$ m, $\Delta Y = +60$ m, $\Delta Y = -320$ m.

The IGN.FR website is only offered in French, but Google Translator seems to do a good job of translating into English for this particular case (*likely edited by IGN*). The country has complete dense coverage of public GPS CORS sites, and IGN offers a free computational web page that accepts the user's choice of up to four National CORS sites to process with their own observations in a manner somewhat akin to the U.S. National Geodetic Survey's "OPUS" system. The IGN website is well-designed and covers practically anything a potential user could ask for regarding France proper as well as for all its Overseas Territories.

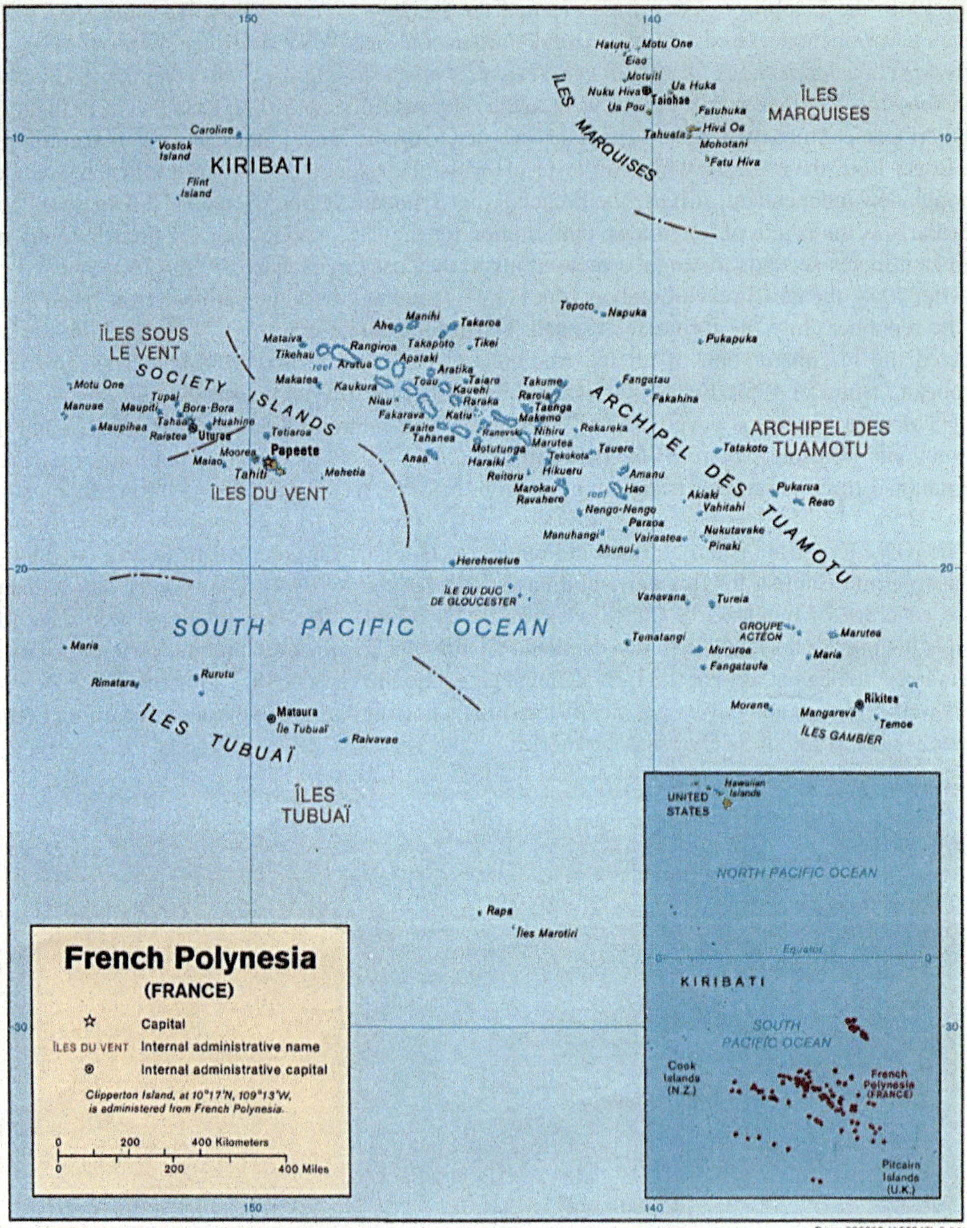

Base 800203 (A00342) 5-89

Overseas Lands of French Polynesia

French Polynesia consists of five archipelagoes: *Archipel Des Tuamotu (Îles Australes), Îles Gambie, Îles Marquises, Îles Tubuai,* and the Society Islands. Slightly less than one-third the size of Connecticut, the lowest point is the Pacific Ocean (0 m), and the highest point is *Mon Orohena* (2,241 m).

"Archaeological evidence suggests that the Marquesas Islands may have been settled about 200 B.C. from western Polynesia. In subsequent dispersions, Polynesians from the Marquesas migrated to the Hawaiian Island about 300 A.D. and reached the Society Islands by about the 9th century. Large chieftainships were formed on Tahiti, Bora-Bora, and Raiatea. Teriaroa, north of Tahiti, was a royal retreat, and Taputapuatea, on Raiatea, was the most sacred shrine in the islands. European contact with the islands of French Polynesia was gradual. The Portuguese navigator Ferdinand Magellan sighted Pukapuka Atoll in the Tuamotu group in 1521. The southern Marquesas Islands were reached in 1595. The Dutch explorer Jacob Roggeveen in 1722 discovered Makatea, Bora-Bora, and Maupiti. Capt. Samuel Wallis in 1767 reached Tahiti, Moorea, and Maiao Iti" (*Encyclopedia Britannica.com*, 2014).

"With his ships *La Boudeuse* and *L'Etoile,* Louis-Antoine de Bougainville arrived in Tahiti in April 1768, less than a year after Wallis. At this time Wallis was still homeward bound, so Bougainville was completely unaware that he was not the first European to set eyes on the island. His visit only lasted nine days, but Bougainville was a more cultured, considered man than Wallis and had no unfriendly clashes with the Tahitians. Bougainville explained that the Tahitians 'pressed us to choose a woman and come on shore with her; and their gestures, which were not ambiguous, denoted in what manner we should form an acquaintance with her.' Bougainville's reports of Venus-like women with 'the celestial form of that goddess', and of the people's uninhibited attitude towards matters sexual, swept through Paris like wildfire" (https://www.britannica.com/place/French-Polynesia). In early 19th-century volumes of the French Navy's *Annals Hydrographique*, further descriptions can be found of these storied customs of Tahiti!

"The history of the Society Island group is virtually that of Tahiti, which was made a French protectorate in 1842 and a colony in 1880. French missionaries went to the Gambier group in 1834, and in 1844 a French protectorate was proclaimed, followed by annexation in 1881. The Tubuai Islands were also evangelized from Tahiti, and as late as 1888 Rimatara and Rurutu sought British protection, which was refused. They were placed under the French protectorate in 1889 and annexed in 1900. The Tuamotus were part of the kingdom of the Pomare family of Tahiti, which came originally from Fakarava Atoll. These islands were claimed as dependencies of Tahiti within the protectorate by France in 1847 and became part of the colony in 1880. In the Marquesas, Nuku Hiva was annexed to the United States in 1813 by Capt. David Porter of the frigate *Essex*, but the annexation was never ratified. French occupation of the group followed the landing of forces from a French warship, requested by the chief of Tahuata (near Hiva Oa). Soon after there was a quarrel with the French; in 1842 the chiefs ceded sovereignty to France. The islands were administered as the French Colony of Oceania. French Polynesia was made an overseas territory of France in 1946" (*op. cit., Britannica*, 2014).

There seems to be no geodetic work performed by the French Navy as directly reported in *Annales Hydrographiques* or by the *Institut Géographique National* during the 18th and 19th centuries with the single exception below regarding Tahiti. Reports begin to appear during the 1930s that consist of geodetic work performed in concert with hydrographic surveys, and primarily represent astrolabe observations of island datum origins. (Wild T-3 theodolites were also used for triangulation, but apparently without the astrolabe attachment for astronomical use.) These positions sometimes have baselines listed and sometimes also list orientation azimuths, but most are just summarized as single positions corrected for deviation of the vertical (*Après Correction des Déviations de la Verticale*). Without exception, for French Polynesian classical positioning the ellipsoid of reference is the International 1924, where $a = 6,378,388$ m, and $1/f = 297$.

Sur La Déviation de la Verticale À Tahiti, M.A. Gougenheim, 1939, *Annales Hydrographiques*, Annales 1940–1945, Paris 1946: Punaauia (South End of Base) No. 1: $\Phi_o = 17°\ 38'\ 08.0''$ S, $\Lambda_o = 151°\ 56'\ 40.5''$ E, and $\alpha_o = 178°\ 04'\ 38.4''$ Punaauia to Tataa. For this single published position on Tahiti *only*, this was from the analysis of triangulation from 1844, and those computations were based on the *Dépôt de la Guerre* ellipsoid where $1/f = 305$.

For New Caledonia (*Nouvelle-Calédonie*) triangulation performed in 1935, Nouméa South Base origin point for a local hydrographic survey: $\phi_o = 22°\ 16'\ 35.0''$ S, $\lambda_o = 166°\ 26'\ 03.5''$ E. At Pointe Paagoumène, astrolabe observations were: $\Phi_o = 20°\ 29'\ 18''$ S, $\Lambda_o = 164°\ 10'\ 57''$ E, and at *Sommet Tano*: $\Phi_o = 20°\ 29'\ 01.1''$ S, $\Lambda_o = 164°\ 24'\ 27.5''$ E, $H_o = 461.83$ m (*by M.A. Canuel, Enseigne de Vaisseau, Annales Hydrographiques, Années 1938–1939, Paris 1946*).

For the Marquises, Haavei Bay (*Ua-Huka* or *Île de Huka*) Astro: $\Phi_o = 08°\ 56'\ 00.8''$ S, $\Lambda_o = 139°\ 35'\ 57.7''$ W, and $\alpha_o = 270°\ 59'\ 22''$ (Solar), with a baseline = 255.08 m. For *Île Hiva Oa*, $\Phi_o = 09°\ 48'\ 27.5''$ S, $\Lambda_o = 139°\ 02'\ 02.7''$ W with a baseline = 336.31 m (*Mission Hydrographique en Polynésie Française, 1956–1960, Annales Hydrographique, 1960*).

For the Isle of Fangataufa origin, $\Phi_o = 22°\ 16'\ 38.8''$ S, $\Lambda_o = 138°\ 45'\ 41.7''$ W with a baseline = 1,381.85 m, ±5 cm. Eleven first-order points (9 monumented) were established on the island, and 21 second-order points were established, all with a Wild T-3 theodolite. On the Isle of Hao, the astro point was: $\Phi_o = 18°\ 06'\ 29.5''$ S, $\Lambda_o = 140°\ 54'\ 25.5''$ W, surveyed in 1958 (*Annales Hydrographique, 1969*).

For the archipelago *Îles Australes*, the Tubuai Island astro station is where: $\Phi_o = 23°\ 20'\ 40.2''$ S, $\Lambda_o = 149°\ 29'\ 08.6''$ W, and $\alpha_o = 40°\ 14'\ 16''$ from *Mataura Terme Sud* to *Terme Nord*. Lieutenant de vaisseau (Naval Lieutenant) *Vallaux* performed the observations with 16 sets of a Wild T-3 theodolite in February 1969. On Tuamoto Atoll, the astro station is where: $\Phi_o = 14°\ 28'\ 36.9''$ S, $\Lambda_o = 145°\ 02'\ 11.8''$ W, and $\alpha_o = 52°\ 42'\ 32.2''$ from *Terme Sud* to *Terme Nord* on a baseline of 5,213.36 m, observed in April 1969. On Tureia, the astro station is where: $\Phi_o = 20°\ 46'\ 20.5''$ S, $\Lambda_o = 138°\ 34'\ 10.0''$ W, and was observed in June 1969 (*Annales Hydrographique, No. 739, 1974*).

For the Isle of Pitiahe, the origin is where: $\phi_o = 16°\ 28'\ 28.942''$ S, $\lambda_o = 152°\ 14'\ 55.059''$ W, $h_o = 19.0$ m, observed at *Terme Sud* by M. J.P. Thouin, *Ingénieur en Chef de l'Armement (Hydrographie)* in February 1983. This observation is referenced to the WGS 72 Datum and ellipsoid where $a = 6,378,135$ m, and $1/f = 298.26$ (*Mission Océanographique du Pacifique 1 Oct 1982 – 3 Avril 1984, Annales Hydrographique*).

In 1954, a summary of practically all astro observations performed in French Polynesia was published in *Annales Hydrographique*. The listing included actual observational results along with deviations of the vertical and the following subsequent adopted positions (Tables F.1–F.5).

A subsequent report on the Marquises offered some more recent transformation details. In particular, for Manihi Atoll, the 1963 cadastre was oriented in 1988 by the *Mission Océanographique du Pacifique* **from** (*MOP 88*) **to** the WGS72 datum as: $\Delta X = +133.9$ m, $\Delta Y = +183.7$ m, $\Delta Z = +162.4$ m. Furthermore, a number of transformation parameters were offered **from** *IGN 72* Datum (referenced to the International 1924 ellipsoid) **to** WGS 84 datum where: $a = 6,378,137$ m and $1/f = 298.25723563$, where:

TABLE F.1
Archipel des Tuamotu et Îles Gambier

Ahunui	19° 36′ 30.5″ S	140° 25′ 06.6″ W
Akiaki	18° 33′ 19.2″ S	139° 13′ 02.2″ W
Anuanu Raro	20° 24′ 42.4″ S	143° 32′ 12.8″ W
Ankuanu Ruga	20° 36′ 15.1″ S	143° 17′ 30.5″ W
Fagataufa	22° 16′ 38.8″ S	138° 45′ 41.7″ W
Gambier (*Rikitea*)	22° 07′ 02.3″ S	134° 58′ 09.8″ W
Hercheretue (*Otetou*)	19° 51′ 50.7″ S	144° 59′ 57.5″ W
Makatea (*Temao*)	15° 49′ 31.4″ S	148° 16′ 40.9″ W
Manuhagi	19° 10′ 53.7″ S	141° 14′ 56.1″ W
Maria	22° 01′ 10.5″ S	136° 12′ 10.4″ W
Marutea	21° 28′ 48.9″ S	135° 38′ 17.4″ W
Maturei Vavao	21° 27′ 17.5″ S	136° 24′ 36.8″ W
Morane	20° 36′ 15.1″ S	143° 17′ 30.5″ W
Mururoa	21° 49′ 59.0″ S	138° 47′ 04.0″ W
Nukutavake (*Tavava*)	19° 16′ 37.9″ S	138° 46′ 24.2″ W
Nukutipipi	20° 41′ 23.9″ S	143° 03′ 09.6″ W
Paraoa	19° 07′ 19.3″ S	140° 43′ 09.4″ W
Pinaki	19° 23′ 24.2″ S	138° 40′ 56.7″ W
Pukaroa (*Marautagaroa*)	18° 16′ 14.4″ S	137° 04′ 02.3″ W
Reao (*Tapuarava*)	18° 27′ 54.5″ S	136° 27′ 46.3″ W
Takakoto (*Tumukuro*)	17° 20′ 39.8″ S	138° 27′ 09.6″ W
Tematagi	21° 37′ 32.7″ S	140° 37′ 35.0″ W
Temoe	23° 18′ 59.5″ S	134° 28′ 56.5″ W
Tureia (*Fakamaru*)	20° 46′ 12.9″ S	138° 33′ 54.5″ W
Vahaga	21° 19′ 18.2″ S	136° 39′ 38.1″ W
Vahitahi (*Temanufaara*)	18° 46′ 32.2″ S	138° 51′ 22.2″ W
Vairaatea	19° 19′ 43.9″ S	139° 13′ 04.2″ W
Vanavana	20° 46′ 11.4″ S	139° 08′ 25.9″ W

TABLE F.2
Archipel de la Société

Bellingshausen Atoll	15° 48′ 32.7″ S	154° 32′ 02.3″ W
Bora-Bora	16° 30′ 25.8″ S	151° 44′ 57.5″ W
Huahine	16° 42′ 47.9″ S	151° 02′ 05.3″ W
Maiao (*Tubuai – Manu*)	17° 38′ 17.7″ S	150° 38′ 15.1″ W
Maupiti	16° 26′ 53.1″ S	151° 14′ 51.8″ W
Meetia	17° 52′ 41.1″ S	148° 03′ 40.2″ W
Moorea (*Papetoai*)	17° 29′ 31.5″ S	149° 52′ 16.1″ W
Mopelia Atoll	16° 47′ 01.9″ S	153° 58′ 26.1″ W
Raitea-Tahaa	16° 33′ 21.0″ S	151° 29′ 06.2″ W
Scilly Atoll	16° 30′ 36.5″ S	154° 38′ 47.1″ W
Tahiti (*Punaauia-IGN*)	17° 38′ 05.0″ S	149° 37′ 00.0″ W
Tahiti Astro	17° 32′ 26.0″ S	149° 34′ 17.9″ W
Tahiti (*Pointe Venus*)	17° 29′ 41.8″ S	149° 29′ 37.2″ W
Tetiaroa Atoll	17° 01′ 47.7″ S	149° 33′ 43.6″ W
Tupai Atoll (*Motu-iti*)	16° 14′ 09.0″ S	151° 50′ 20.5″ W

TABLE F.3
Îles Australes

Maria Atoll	21° 47' 42.7" S	154° 41' 56.3" W
Raevavae (*Rairua*)	23° 52' 04.8" S	147° 41' 17.2" W
Rapa (*Ahurei*)	27° 37' 15.4" S	144° 20' 04.7" W
Rimatara (*Anapoto*)	22° 38' 38.8" S	152° 49' 03.4" W
Rurutu (*Moerai*)	22° 27' 13.9" S	151° 20' 04.6" W
Tubuai (Mataura)	23° 20' 40.2" S	149° 29' 08.6" W

TABLE F.4
Îles Marquises

Eiao (*Baie Vaitua*)	07° 59' 38.7" S	140° 42' 42.2" W
Fatu Hiva (*Hanavave*)	10° 27' 53.3" S	138° 39' 46.6" W
Hiva Oa	09° 48' 27.5" S	139° 02' 02.7" W
Nuku Hiva (*Taiohae*)	08° 54' 56.2" S	140° 05' 35.1" W
Ua Pou (*Hakahetau*)	09° 21' 33.6" S	140° 06' 16.1" W
Ua Uka (Vaipae)	08° 56' 17.8" S	139° 34' 18.1" W

TABLE F.5
Supplementary Observations (Uncorrected for Deflections)

Mopelia	16° 46' 20.0" S	153° 56' 46.8" W
Mururoa	21° 50' 22.3" S	138° 56' 44.1" W
Nukutavake (*Tavava*)	19° 16' 28.4" S	138° 46' 19.9" W
Pukarua	18° 16' 08.8" S	137° 04' 10.4" W
Reao (*Tapuarva*)	18° 27' 47.4" S	136° 27' 57.8" W
Takakoto (*Tumukuru*)	17° 20' 56.8" S	138° 27' 01.6" W
Tureia (*Fakamaru*)	20° 45' 57.7" S	138° 33' 56.8" W
Tureia (*Sud du Lagon*)	20° 50' 47.5" S	138° 31' 24.0" W
Vahitahi (*Temanufaara*)	18° 46' 29.9" S	138° 51' 28.0" W
Vairaatea	19° 19' 30.0" S	139° 13' 03.0" W

Source: Mission Géodésique des Tuamotu, (Juin 1950–Septembre 1953) par M. François Vallaux, Lieutenant de Vaisseau, Annales Hydrographiques, 1954, pp. 90–116.

	ΔX	ΔY	ΔZ
Nuku Hiva*	+133 m*	+191 m*	+183 m*
Hiva Oa	+332 m	−11.5 m	+60 m
Tahuata	+332 m	−11.5 m	+60 m
Ua Pou	+132 m	+186 m	+192 m
Ua-Huka	+84 m	+274 m	+64.5 m
Fatu Iva	+60 m	+284.5 m	+115.5 m
Eiao	+332 m	−64 m	+251 m

Source: Mission Océanographique du Pacifique, Échelon de Polnésie, Dec 1982–Jul 1990, Annales Hydrographiques, Vol. 19, No. 766, 1993.

* In regard to the Nuku Hiva transformation parameters, the French *Institut Géographique National* website more recently listed different values **from** local **to** WGS 84 as:

	ΔX	ΔY	ΔZ
Nuku Hiva (IGN72)	+84 m	+274 m	−251 m

Furthermore, *IGN* also offered the following transformation parameters **from** local **to** WGS 84:

Nouvelle-Calédonie (Lifou)	+336 m	+223 m	−231 m
Nouvelle-Calédonie (MHNC72)	−13 m	−348 m	+292 m

Thanks to the librarians at the Maury Oceanographic Library that suffered my rummaging through their stacks in the 1990s while I was lecturing to the U.S. Naval Oceanographic Office and to the Naval Research Laboratories at Stennis Space Center in Mississippi.

G

Republic of Gabon 275
Republic of The Gambia 279
Georgia 283
Federal Republic of Germany 287
The Republic of Ghana 295
Gibraltar 299
The Hellenic Republic 301
Grenada 305
The Department of Guadeloupe 309
Territory of Guam 313
Republic of Guatemala 317
Republic of Guinea 321
Republic of Guinea-Bissau 325
Co-operative Republic of Guyana 327

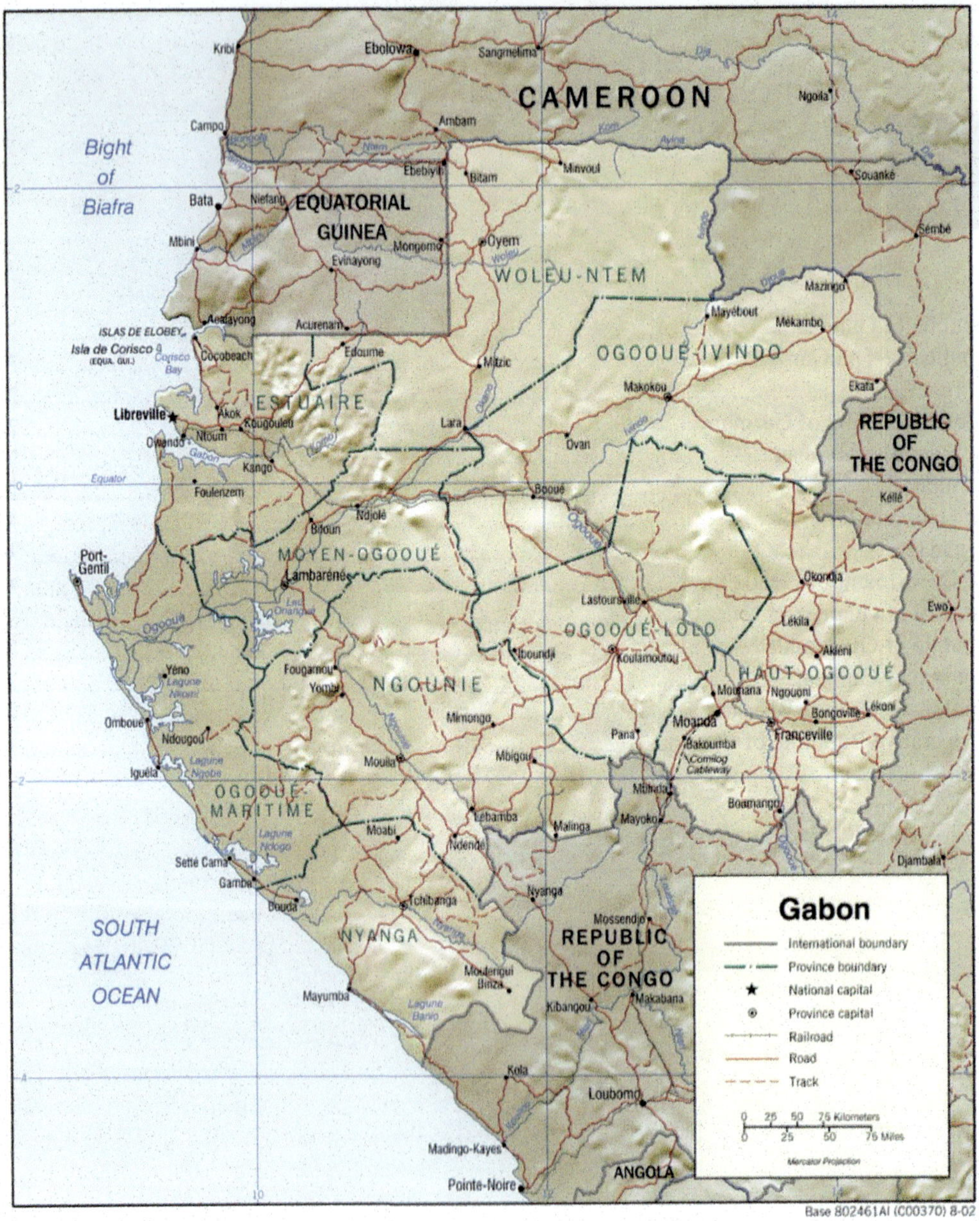

Base 802461AI (C00370) 8-02

Republic of Gabon

The Republic of Gabon is in west central Africa, bordering the Atlantic Ocean at the Equator, and is bordered by Cameroon (349 km), Republic of the Congo (2,567 km), and Equatorial Guinea (345 km). The terrain is a narrow coastal plain; hilly interior; savanna in east and south. The highest point is *Mont Bengoue* (1,050 m), the lowest point is the Atlantic Ocean (0 m), and the mean elevation is (377 m).

The Portuguese settled on the Island of Sao Tomé and discovered the estuary of the Como in 1473. The country is named after the Portuguese word "*gabão*," a coat with sleeve and hood bearing a resemblance to the shape of that estuary. While the French were establishing trading posts in the 18th century, the trade in black slaves began to flourish in Lambaréné and Cape Lopez. At the beginning of the 19th century, the French chased slave traders away and gained the trust of local chiefs. The capital, Libreville, was created in 1849, and Gabon became a French colony in 1883. From 1910 to 1958, it was part of French Equatorial Africa or "*Afrique Equatorial Française*" (l'AEF). Gabon became independent in 1960.

In March 1886, Le Pord of the French Navy used a sextant and two chronometers to determine the position of Libreville as: $\Phi_o = 0°\ 23'\ 15''$ North, $\Lambda_o = +7°\ 06'\ 30''$ East of Paris. In September of 1890, Serres of the French Navy determined the differences in longitude between Libreville and Kotonou with the aid of a telegraph. Audoin of the French Navy established the first classical horizontal datum in Gabon in July of 1911 at Akosso: $\Phi_o = 0°\ 42'\ 45.9''$ South, $\Lambda_o = 8°\ 46'\ 56''$ East of Greenwich. The defining azimuth was determined from Akosso to station signal Alugubuna as: $\alpha_o = 341°\ 47'\ 04.1''$. Akosso 1911 Datum is referenced to the Clarke 1880 ellipsoid, where $a = 6,378,249.145$ m and $1/f = 293.465$. This datum origin also defined the point of origin for the first Grid used in Gabon, which was based on the Hatt Azimuthal Equidistant Projection.

Hatt was the Hydrographer of the French Navy in the late 19th century, and his projection became the standard "local" projection for individual hydrographic surveys. This projection became synonymous with "*Systémé Rectangulaire Usuel*" (Usual Rectangular System) for French hydrographic surveys. Note that until the late 1980s, the Hatt Azimuthal Equidistant projection (and Grid) was also used for the military topographic series of Greece. Of particular note for the Akosso Grid is that one of the points located in the original triangulation was "*Phare du Cap Lopez*" (Cape Lopez Lighthouse), where $X = -10,450.02$ m, and $Y = +10,809.11$ m. In subsequent triangulations and adjustments, this point would gain particular prominence in Gabon.

The following year, Audoin established the Owendo Datum of 1912, where $\Phi_o = +0°\ 17'\ 43.9''$ North, $\Lambda_o = +9°\ 29'\ 35.55''$ East of Greenwich and the usual Grid was the Hatt at the Datum origin. In 1914, Lafargue of the French Navy established the Gabon River Datum at Cape Esterias as: $\Phi_o = +0°\ 36'\ 48.58''$ North, $\Lambda_o = +9°\ 19'\ 19.02''$ East of Greenwich. This point would also gain prominence in the history of the classical Datums of Gabon. Lafargue established another Datum at the North base end (invar baseline) of Pointe Banda in 1921, in the village of Sainte-Marie, where $\Phi_o = -3°\ 50'\ 03.4''$ South, $\Lambda_o = +11°\ 00'\ 46.6''$ East of Greenwich. The defining azimuth was from Banda to Tranchée: $\alpha_o = 158°\ 59'\ 56.0''$. The usual Hatt Grid had the same origin as the Datum. The triangulation stations monumented by Lafargue for this survey were pyramids constructed of cement, and the south side was imprinted "M.H.A.E.F. 1921."

There was no topographic survey of the AEF until after World War II. During the war, the Gabon Belt Transverse Mercator ("*Fuseau Gabon*") was used such that the ellipsoid of reference was the Clarke 1880. The unit of measurement was the meter, the Central Meridian, $(\lambda_o) = 12°$ West of Greenwich, and the Latitude of Origin, (φ_o), by definition was the equator. The Scale Factor at Origin $(m_o) = 0.99931$, the False Northing $= 1,500$ km, and the False Easting $= 800$ km. The particular math model for this zone was presumably the Gauss-Krüger, identified by the French as *"Projection de Gauss."*

A geodetic astronomical station in the town of Mporaloko, east of Port Gentil, was set by the *"Société des Pétroles d'Afrique Équatoriale"* (SPAFE), the Equatorial French Africa Petroleum "Society" (corporation). This position was transferred to the lighthouse at Cape Lopez such that the Cape Lopez Datum of 1951 is synonymous with the Mporaloko Datum of 1951, where $\Phi_o = 0°\ 37'\ 54.2''$ South, $\Lambda_o = 8°\ 42'\ 13.2''$ East of Greenwich. The Mporaloko 1951 Datum is currently used for most of Gabon that is south of Libreville.

In 1954, Sauzay of the French Navy recovered two of the monuments set by Lafargue in 1921. These two points were Table and Babar. The coordinates adopted for Table were: $\Phi_o = 3°\ 49'\ 13.9''$ South, $\Lambda_o = 10°\ 00'\ 55.0''$ East of Greenwich. Sauzay observed a solar azimuth from Table to Babar such that: $\alpha_o = 348°\ 09'\ 05.5''$. The baseline distance adopted was a mean of the new observations with WILD™ invar tapes and with those observed by Lafargue in 1921. The Hatt Grid used for this Datum had the same origin with a False Easting and False Northing of 10 km.

In 1955, Mannevy of the French Navy reoccupied an astro station at Esteiras for the triangulation of the Bay of Corisco, North of Libreville. The Cadastral Service of Libreville measured a 3-km baseline in Libreville in concert with the French Hydrographic Mission. The Cape Esteiras Datum of 1955 origin is: $\Phi_o = 0°\ 36'\ 48.58''$ North, $\Lambda_o = 9°\ 19'\ 19.02''$ East of Greenwich. The Department of Public Works of Gabon (*Travaux Publics du Gabon*) assisted Mannevy by building towers for the triangulation of the Bay of Corisco. The triangulation was performed with WILD T3 theodolites using eight sets of angles. The design of the towers and targets were patterned after those used in Madagascar. The first computations on the Universal Transverse Mercator Grid in Gabon were computed by Mannevy with the Clarke 1880 ellipsoid, Fuseau 32 (zone 32), where Central Meridian $(\lambda_o) = +9°$ East of Greenwich. The Cape Esteiras 1955 Datum is used for Libreville and points north in Gabon.

The latest edition of "TR8350.2" by the National Imagery and Mapping Agency (NGA) published in July of 1997 lists the three-parameter shift from M'PORALOKO (*sic*) Datum to WGS 84 Datum as: $\Delta X = -74$ m, $\Delta Y = -130$ m, $\Delta Z = +42$ m. Note that only one point was used to determine the published shift, and the accuracy is stated at ±25 m for each component. Such a level of accuracy is acceptable for military artillery purposes or shirt-pocket Global Positioning System receivers when one is working with maps at a scale of 1:25,000 or 1:50,000. But an uncertainty of ±25 m is useless for the precision necessary for seismic geophysical exploration or municipal mapping.

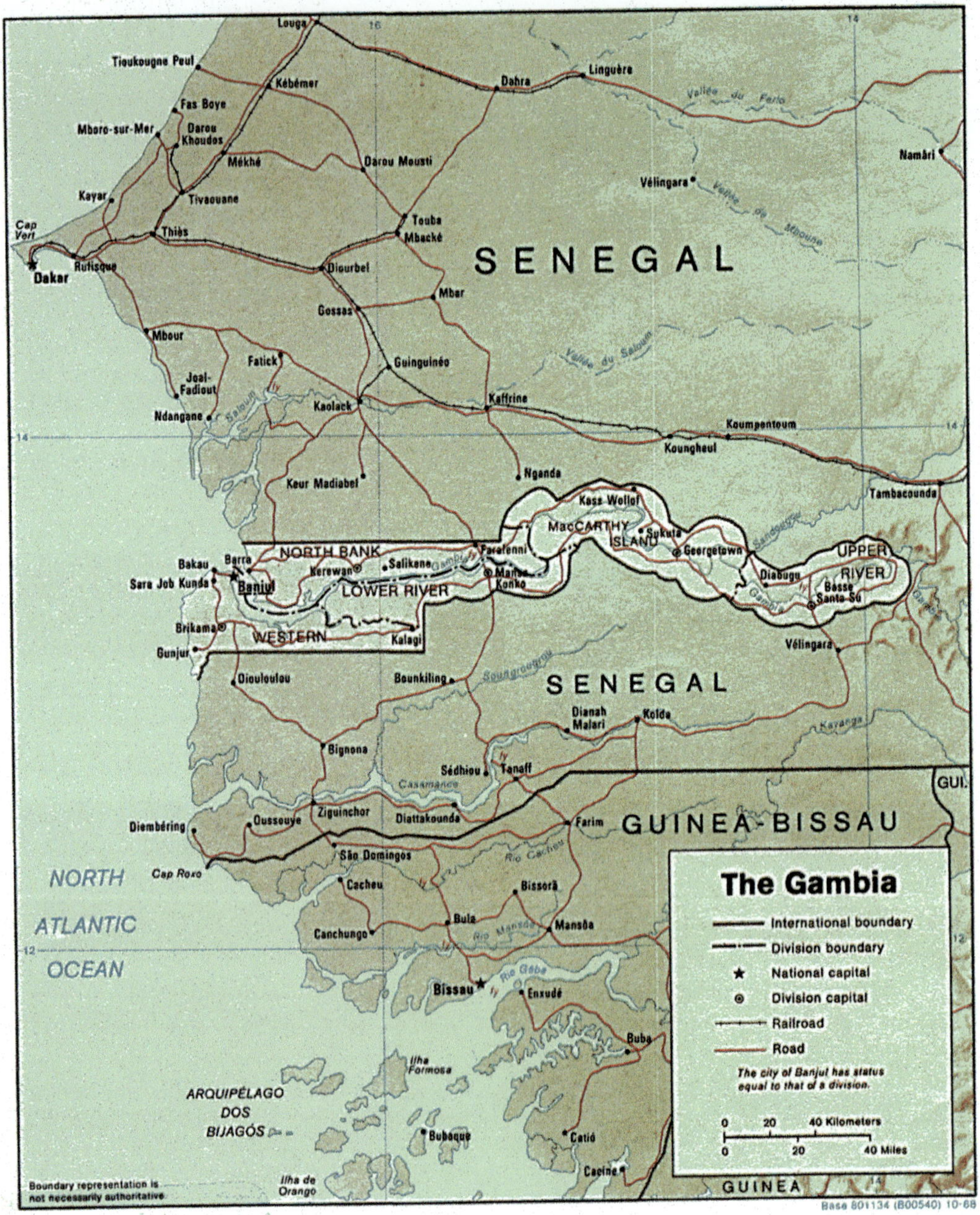

Base 801134 (B00540) 10-88

Republic of The Gambia

Slightly less than twice the size of Delaware, The Gambia is surrounded by Senegal (740 km) and is comprised of the flood plain of the Gambia River flanked by some low hills. The lowest point is the Atlantic Ocean (0 m), and the highest point is an unnamed elevation point (53 m). The Gambia has a coastline of 80 km, and the territorial sea it claims is 12 nautical miles; it is the smallest country in Africa (*World Factbook*, 2012).

"The Gambia was once part of the Mali and Kaabu Empires. The first written accounts of the region come from records of Arab traders in the 9th and 10th centuries A.D. Arab traders established the trans-Saharan trade route for slaves, gold, and ivory. In the 15th century, the Portuguese took over this trade using maritime routes. At that time, The Gambia was part of the Kingdom of Mali. In 1588, the claimant to the Portuguese throne, Antonio Prior of Crato, sold exclusive trade rights on the Gambia River to English merchants; this grant was confirmed by letters patent from Queen Elizabeth I. In 1618, King James I granted a charter to a British company for trade with The Gambia and the Gold Coast. During the late 17th century and throughout the 18th, England and France struggled continuously for political and commercial supremacy in the regions of the Senegal and Gambia Rivers. The 1783 Treaty of Versailles gave Great Britain possession of The Gambia, but the French retained a tiny enclave at Albreda on the north bank of the river, which was ceded to the United Kingdom in 1857. As many as 3 million slaves may have been taken from the region during the 3 centuries that the transatlantic slave trade operated. It is not known how many slaves were taken by Arab traders prior to and simultaneous with the transatlantic slave trade. Most of those taken were sold to Europeans by other Africans; some were prisoners of intertribal wars; some were sold because of unpaid debts, while others were kidnapped. Slaves were initially sent to Europe to work as servants until the market for labor expanded in the West Indies and North America in the 18th century. In 1807, slave trading was abolished throughout the British Empire, and the British tried unsuccessfully to end the slave traffic in The Gambia. They established the military post of Bathurst (now Banjul) in 1816. In the ensuing years, Banjul was at times under the jurisdiction of the British governor general in Sierra Leone. In 1888, The Gambia became a separate colonial entity. An 1889 agreement with France established the present boundaries, and The Gambia became a British Crown Colony, divided for administrative purposes into the colony (city of Banjul and the surrounding area) and the protectorate (remainder of the territory). The Gambia received its own executive and legislative councils in 1901 and gradually progressed toward self-government. A 1906 ordinance abolished slavery. During World War II, Gambian troops fought with the Allies in Burma. Banjul served as an air stop for the U.S. Army Air Corps and a port of call for Allied naval convoys. U.S. President Franklin D. Roosevelt stopped overnight in Banjul en route to and from the Casablanca Conference in 1943, marking the first visit to the African Continent by an American president while in office. After World War II, the pace of constitutional reform quickened. Following general elections in 1962, full internal self-government was granted in 1963. The Gambia achieved independence on February 18, 1965, as a constitutional monarchy within the British Commonwealth. Shortly thereafter, the government proposed conversion from a monarchy to a republic with an elected president replacing the British monarch as chief of state. The proposal failed to receive the two-thirds majority required to amend the constitution, but the results won widespread attention abroad as testimony to The Gambia's observance of secret balloting, honest elections, and civil rights and liberties. On April 24, 1970, The Gambia became a republic following a referendum" (*Background Notes, 11/15/11*). "In 1994, Yahya Jammeh led a military coup overthrowing the president and banning political activity. Jammeh won every presidential election until 2016. In December 2016, after 22 years of authoritarian rule, President Jammeh lost to

Adama Barrow during free and fair elections. Due to The Gambia's poor human rights record under Jammeh, international development partners had substantially reduced aid to the country. These channels have now reopened under the administration of President Barrow. Since the 2016 election, The Gambia and the US have enjoyed improved relations. US assistance to the country has supported military education and training programs, capacity building, and democracy-strengthening activities" (*World Factbook*, 2022).

"The Gambia Datum of 1941 (code GAI) located at East Base 'O' where: $\Phi_o = 13°\ 27'\ 20.035''$ N, $\Lambda_o = 16°\ 34'\ 22.350''$ W, azimuth $\alpha_o = 34°\ 55'\ 45''$ to 'A' Barra from north, ellipsoid of reference is both Clarke 1858 and Clarke 1880 (Trig List is Clarke 1880). The Gambia Trig Lists 1, 2, and 3 were published by the Directorate of Military Survey, 1953. Source is Triangulation Data, West Africa, River Gambia, 'H.M.S/ Challenger,' 1941 (Di302A) with corrections from E6325 Press 31M, 1941. The station is a brass disc let into cement projecting a few inches above surface situated about 5 yards inshore of Highline at the N.W. corner of Bathurst. It is 358° & 550 ft. from the light beacon.

"There is also a code DOC for the datum D.O.S. 415. I have no data. My guess on the ellipsoid is that the British Admiralty used Clarke 1858 *(where: $a = 6,378,235.6\ m$ and $1/f = 294.26$ – Ed.)*. There are no UTM tables for 1858 so they used 1880 for the trig list. Clarke 1880 was also the IGGU ellipsoid for Africa. I never bothered to see what the difference is but it most likely is insignificant. Grid. The Gambia Grid is Cassini-Soldner, Clarke 1858. Unit of measurement is the foot. Origin is the datum point, East Base. Scale Factor is unity. False northing and false easting are 0. The UK published tables; the limits are 13° N to 14° N. You have probably figured out the extent of the country so you can see that the very limited extent of the tables is more than adequate" (*Personal communication, John W. Hager, 8 November 2012*).

According to *Annals Hydrographique*, a 1965 geodetic survey in the vicinity of Senegal, *Travaux Géodésiques au Sénégal (Avril 1965–Juin 1965)* consisted of a series of triangulation points that included two points on the exterior boundaries of The Gambia: Djinnack (to the north), where $X = 333,289.86\ m$, $Y = 1,504,202.99\ m$, and Fanjara (to the south), where $X = 316,602.53\ m$, and $Y = 1,490,293.33\ m$, and the coordinates are on UTM Zone 28, Clarke 1880 where $a = 6,378,249.145\ m$ and $1/f = 293.465$. The French Geodesists used a WILD T-3 theodolite, Tellurometer EDM, and used Bilby towers (USC&GS standard).

According to Mr. Russell Fox, ex-International Library Manager of the Ordnance Survey, Southampton, UK, "We believe that the datum shift from HMS Challenger Astro to WGS84 is approximately: delta Latitude +1".385 North, delta Longitude +5".025 East, *i.e.*, WGS84 positions are roughly 150 m ENE of HMS Challenger Astro Datum positions. I doubt whether the data exists for the computation of reliable three-parameter transformation parameters for Gambia. What is needed are quality GPS observations at a handful of existing trig stations" (November 17, 1998).

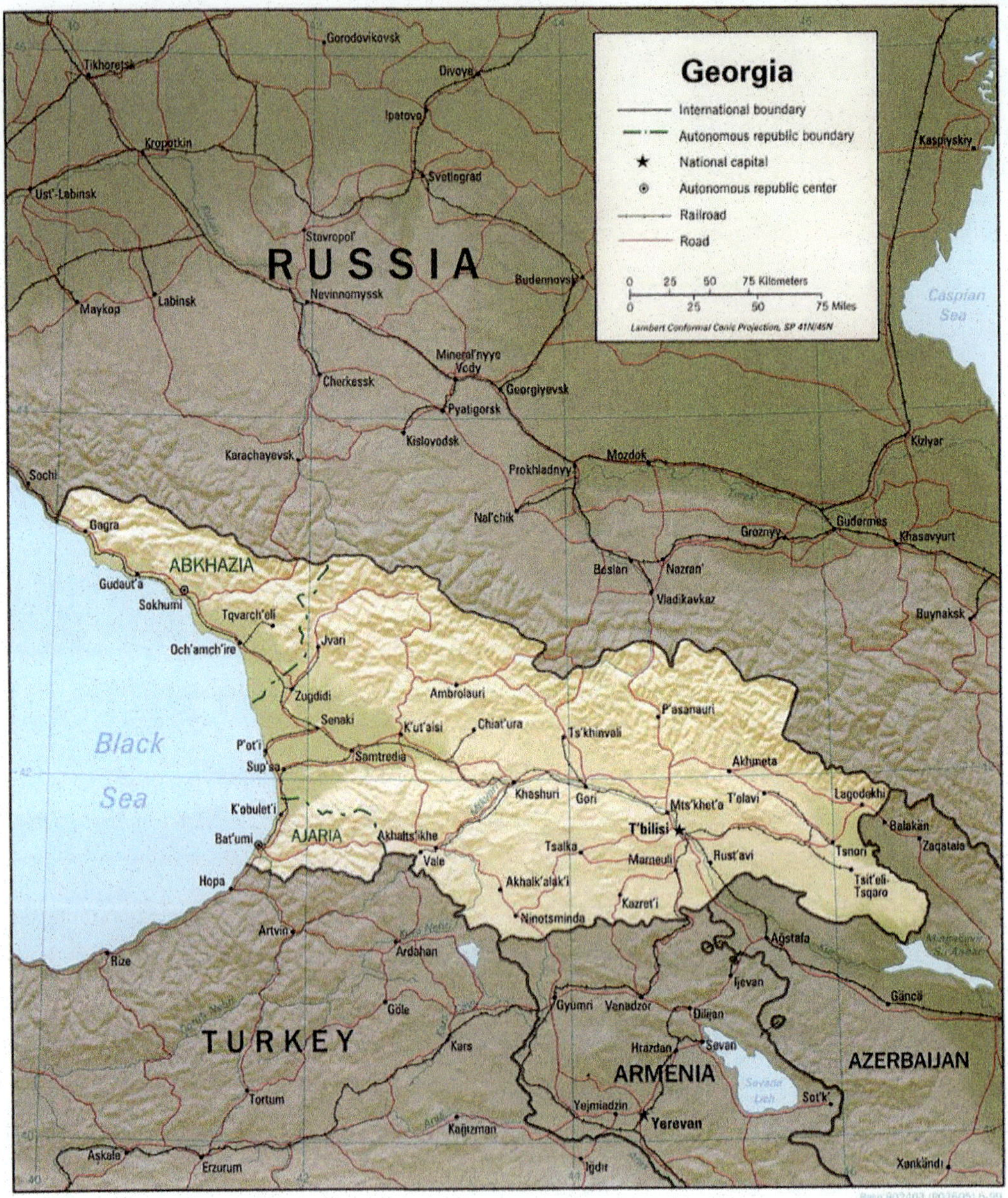

Base 802403 (R02605) 5-99

Georgia

Bordered by Armenia (164 km), Azerbaijan (322 km), Russia (723 km), and Turkey (252 km) Georgia is slightly smaller than South Carolina. The lowest point is the Black Sea (0 m), and the highest point is *MT'a Shkhara* (5,201 m). Thanks to the *CIA Factbook 2012*, the terrain of Georgia is "largely mountainous with Great Caucasus Mountains in the north and Lesser Caucasus Mountains in the south; Kakheti's Dablobi (*Kolkhida* Lowland) opens to the Black Sea in the west; Mtkvari River Basin in the east; good soils in river valley flood plains, foothills of *Kolkhida* Lowland."

"The Georgians know themselves as *Kartvelebi*, tracing their origins to Noah's great-great-grandson Kartlos. In classical times the two principal kingdoms were Colchis in the west (legendary home of the Golden Fleece and site of Greek colonies) and Kartli (also known as *Iveria* or *Iberia*) in the east and south, including some areas in modern Turkey and Armenia" (*Lonely Planet*, 2012).

"Georgia's recorded history dates back more than 2,500 years. Georgian, a South Caucasian (or 'Kartvelian') language, unrelated to any outside the immediate region, is one of the oldest living languages in the world and has its own distinct alphabet. Located in the picturesque Mtkvari River valley, Georgia's capital, Tbilisi, is more than 1,550 years old. In the early 4th century, Georgia became the second nation in the world to officially adopt Christianity. "The independent Republic of Georgia was established on May 26, 1918, in the wake of the Russian Revolution. Georgia became a Soviet Socialist Republic the following year. During the Soviet period, Georgia was one of the wealthiest and most privileged republics, and its Black Sea coastline was a popular holiday destination for the Soviet elite. On April 9, 1991, shortly before the collapse of the Soviet Union, the Supreme Council of the Republic of Georgia declared independence from the U.S.S.R." (*Background Note, U.S. Department of State*, 2012).

The first Georgian to compile an overview of the territory was Vakhushti Batonishvili who made 27 water-colored maps contained in an atlas of Europe and Georgia. The first geodetic surveys were commenced in 1818 by General Staff officers of the Tsarist Russia which included five astronomical positions, including the city of Gori.

The "First Tbilisi astronomical latitude and longitude were determined in 1828–1829 at the concrete gabion fixed in the garden of Transcaucasian Commander-in-Chief. In 1847 those data were assigned to Avlabari Astronomical Observatory and the relevant amendments were made thereto. In 1861 the first astronomical azimuth was also assessed in Georgia (*from the middle of the mark of the concrete gabion of Tbilisi Physical Observatory – along the center of the gabion arranged on the territory of Village Teleti*)" (Mrs. Ekaterine Meskhidze, Chief of International Relations, National Agency of Public Registry, Tbilisi, Georgia – personal communication 6 April 2012).

"In 1847 the Jäderin Invar apparatus was used to measure a 9.1586 km baseline on the right bank of the Mtkvari River for a chain of figures between Tbilisi, Georgia and Ganja, Azerbaijan *(approx. 173 km – Ed.)*. This chain was computed on the Walbeck 1819 ellipsoid *(a = 6,376,896 m, 1/f = 302.78 – Ed.)* from 1847 to 1853 referenced to the Ferro meridian *(where Ferro = 17° 39' 46.02" West of Greenwich – Ed.)*. By 1853 a network of 1st, 2nd, and 3rd order triangulation was adjusted as in place within Georgia. The vertical datum was established in the port city of Poti at the Black Sea tide gauge staff which was 15.2 cm below the tide gauge staff in the port city of Batumi.

"The specified elevation system of the Black Sea was active till 1946. From 1946 in the former Soviet Union, including the territory of Georgia, there was introduced a new Baltic elevation system from the zero of Kronstadt pole. In 1973–1977 rebalancing of the whole level net of the Soviet Union was carried out and after that it was named 1977 Baltic Elevation System which has been officially functioning on the territory of Georgia till now.

"During the 1920s, the Corps of Topographical Engineers of Tzarist Russia for the first time in Transcaucasia began creating geodetic-control-networks by triangulation method, which subsequently resulted in different-scale verst and semi-verst topographical plans" (*op. cit., Eka Meskhidze, 2012*).

In 1924, the metric system of measurement was introduced into Georgia by the Soviet Union, and network triangulation calculations were carried out in a unified system based on an origin at the Pulkovo meridian of the Pulkovo Observatory, where $\Phi_o=59°\ 46'\ 18.55''$ North, $\Lambda_o=30°\ 19'\ 42.09''$ East of Greenwich. The defining azimuth at the point of origin to Signal A is: $\alpha_o=317°\ 02'\ 50.62''$. At the time, the reference ellipsoid was the Bessel 1841, where $a=6,377,397.155$ m and $1/f=299.1528128$. However, also at that time the cadastral land survey of Georgia was still using the centesimal system of angles, or the Grad, so the Gauss-Krüger Transverse Mercator projection was introduced, also based on the Bessel 1842 ellipsoid. In 1932, the Soviet Union introduced a new adjustment called System 32 still referenced to the Bessel ellipsoid.

"In 1939 based on one of the field expeditions of Ukrainian Southern Aerogeodetic Enterprise, which operated in Tbilisi by that time, Transcaucasia Aerogeodetic Enterprise was created first and afterwards it was reorganized into Georgian Aerogeodetic Enterprise and finally starting from the 60s it was formed as 4th Closed Aerogeodetic Enterprise of Tbilisi which was subordinate to the State Committee of Geodesy and Cartography at the Council of Ministers of the former USSR. This authority was also in charge of Tbilisi Cartographic Factory, Transcaucasia Territorial Inspectorate of State Geodetic Supervision and Tbilisi Topographic Technical School founded in 1934 which was the only specialized (specializing in topography- geodesy and cartography) technical school in the whole Transcaucasia. The abovementioned organizations played a major role in development of geodesy and cartography as an independent field in Georgia as well as in skilled staff training. As a result of implementation of topographic-geodetic, cartographic, and photographic operations having been performed on the territory of Georgia for numerous decades, throughout Georgia there were created state geodetic horizontal-vertical control networks and different-scale state topographic maps. Training of engineering-technical staff employed in the field was carried out basically at Tbilisi Topographic Technical School, Georgian Polytechnic Institute and Tbilisi State University" (*op. cit., Eka Meskhidze, 2012*).

In 1942, the system was changed again to the "System 42" referenced to the Krassovsky 1940 ellipsoid, where $a=6,378,245$ m and $1/f=298.3$, and the origin remained the same at the Pulkovo Observatory. The System 42 remained the datum in Georgia until 1999. "And finally, now in Georgia a new World Coordinate System is operating, which was approved by Decree of the President of Georgia, No. 206 dated April 30, 1999, on WGS-84 Ellipsoid, in the Universal *(Transverse – Ed.)* Mercator projection".

For oil exploration purposes, Georgia still utilizes the Coordinate System 42 (CS42), a 6° wide belt system nearly identical to the UTM except for a central scale factor of unity (1.0) and the Coordinate System 63 (CS63) which is a 3° Transverse Mercator. CS42 zone 8 is used east of 42°E longitude. The Black Sea coast falls in zone 7 where the Central Meridian is 39°E and a False Easting of 7,500,000 m is used, and zone 8 where the Central Meridian is 45°E an a False Easting of 8,500,000 m is used. **From** System 42 Datum **to** WGS84 Datum in the area can be approximated as: $\Delta X=+18$ m, $\Delta Y=-125$ m, $\Delta Z=-83$ m, with an accuracy of 0.5–1 m (*Roger Lott, 27 February 1997*).

"In 1998 the first Georgian framework law was made "On Geodetic and Cartographic Activity", the purpose of which was to adjust legal relations between state authorities and natural and legal entities engaged in the sphere of geodetic and cartographic activity. Thus, according to the basic law of the field and the provision approved by Order of the Chief of LEPL National Agency of Public Registry, the main activity of Geodesy and Cartography Service constitutes – "State regulation in the sphere of topography, geodesy, cartography, land cadastre and GPS survey activities and practical implementation of state administration policy" (*op. cit., Eka Meskhidze, 2012*).

Georgia now has a network of 13 GPS Continuously Operating Reference Stations. This consists of seven class A installations throughout the republic, and six class B installations around Tbilisi. Georgia has become a member of EUPOS in 2011 and is connected to the EUREF Permanent Tracking Network.

My son, Ted a USMC Major spent a year in Afghanistan in combat operations with a Georgian battalion as a liaison Intelligence Officer.

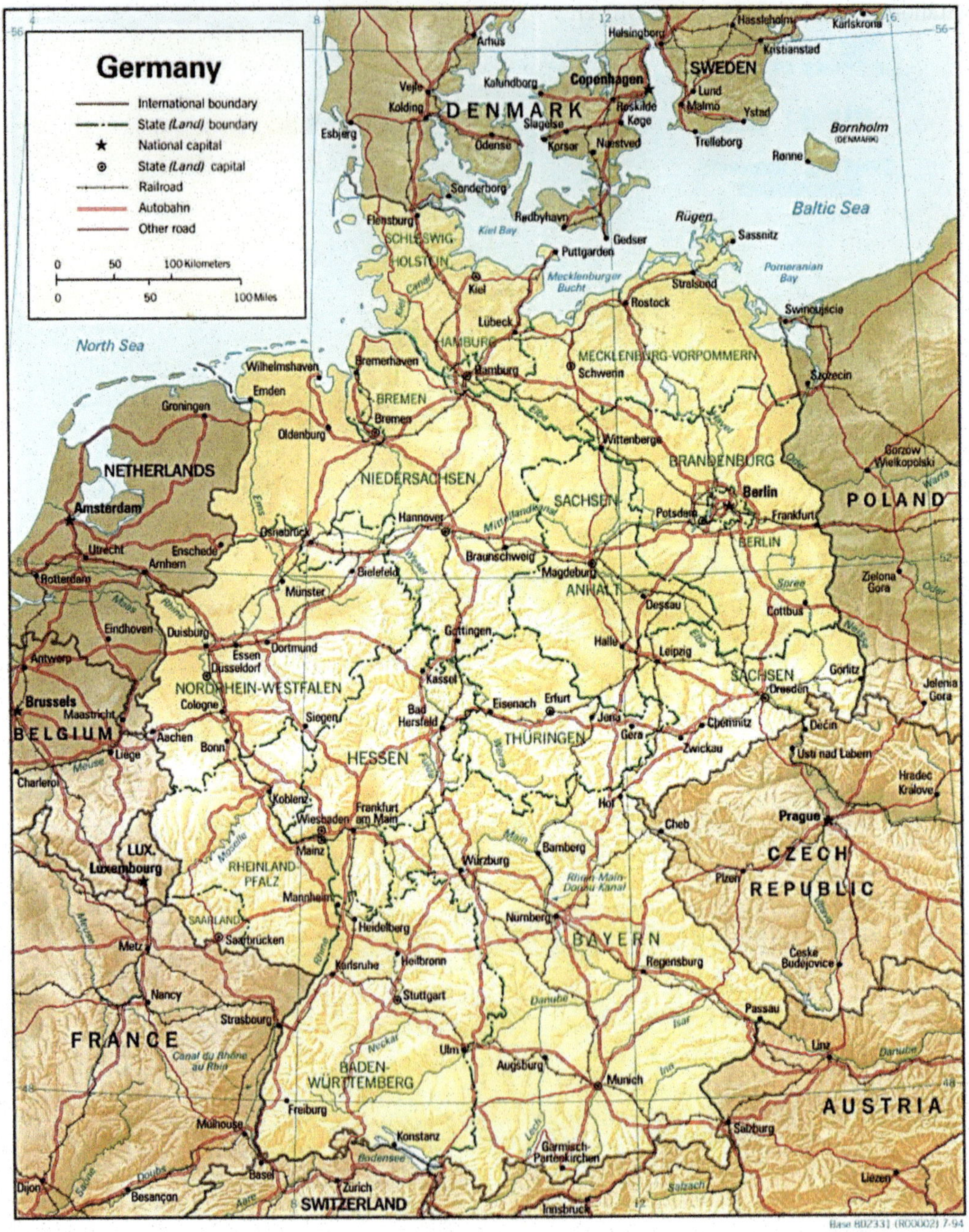

Base 802331 (R00002) 7-94

Federal Republic of Germany

Germany is bordered by Austria (801 km), Belgium (133 km), Czech Republic (704 km), Denmark (140 km), France (418 km), Luxembourg (128 km), the Netherlands (575 km), Poland (467 km), and Switzerland (348 km). The terrain has lowlands in the north, uplands in the center, and Bavarian Alps in the south. The lowest point in Germany is *Neuwndorf bei Wilster* (−3.54 m), and the highest point is *Zugspitze* (2,963 m) (*World Factbook*, 2015).

"The Germanic tribes, which probably originated from a mixture of peoples along the Baltic Sea coast, inhabited the northern part of the European continent by about 500 B.C. By 100 B.C., they had advanced into the central and southern areas of present-day Germany. At that time, there were three major tribal groups: the eastern Germanic peoples lived along the Oder and Vistula rivers; the northern Germanic peoples inhabited the southern part of present-day Scandinavia; and the western Germanic peoples inhabited the extreme south of Jutland and the area between the North Sea and the Elbe, Rhine, and Main rivers. The Rhine provided a temporary boundary between Germanic and Roman territory after the defeat of the Suevian tribe by Julius Caesar about 70 B.C. The threatening presence of warlike tribes beyond the Rhine prompted the Romans to pursue a campaign of expansion into Germanic territory. However, the defeat of the provincial governor Varus by Arminius at the Battle of the Teutoburg Forest in *A.D.* 9 halted Roman expansion; Arminius had learned the enemy's strategies during his military training in the Roman armies. This battle brought about the liberation of the greater part of Germany from Roman domination. The Rhine River was once again the boundary line until the Romans reoccupied territory on its eastern bank and built the Limes, a fortification 300 kilometers long, in the first century *A.D.*

"The second through the sixth centuries was a period of change and destruction in which eastern and western Germanic tribes left their native lands and settled in newly acquired territories. This period of Germanic history, which later supplied material for heroic epics, included the downfall of the Roman Empire and resulted in a considerable expansion of habitable area for the Germanic peoples. However, with the exception of those kingdoms established by Franks and Anglo-Saxons, Germanic kingdoms founded in such other parts of Europe as Italy and Spain were of relatively short duration because they were assimilated by the native populations. The conquest of Roman Gaul by Frankish tribes in the late fifth century became a milestone of European history; it was the Franks who were to become the founders of a civilized German state.

"One of the largest Germanic tribes, the Franks, came to control the territory that was to become France and much of what is now western Germany and Italy. In *A.D.* 800 their ruler, Charlemagne, was crowned in Rome by the pope as emperor of all of this territory. Because of its vastness, Charlemagne's empire split into three kingdoms within two generations, the inhabitants of the West Frankish Kingdom speaking an early form of French and those in the East Frankish Kingdom speaking an early form of German. The tribes of the eastern kingdom – Franconians, Saxons, Bavarians, Swabians, and several others – were ruled by descendants of Charlemagne until 911, when they elected a Franconian, Conrad I, to be their king. Some historians regard Conrad's election as the beginning of what can properly be considered German history.

"German kings soon added the Middle Kingdom to their realm and adjudged themselves rulers of what would later be called the Holy Roman Empire. In 962 Otto I became the first of the German kings crowned emperor in Rome. By the middle of the next century, the German lands ruled by the emperors were the richest and most politically powerful part of Europe. German princes stopped the westward advances of the Magyar tribe, and Germans began moving eastward to begin a long process of colonization. During the next few centuries, however, the great expense of the wars to maintain the empire against its enemies, chiefly other German princes and the wealthy and powerful papacy and its allies, depleted Germany's wealth and slowed its development. Unlike France or England, where a central royal power was slowly established over regional princes, Germany remained divided into a multitude of smaller entities often warring with one another or in combinations against the emperors. None of the local princes, or any of the emperors, were strong enough to control Germany for a sustained period.

"Germany's so-called particularism, that is, the existence within it of many states of various sizes and kinds, such as principalities, electorates, ecclesiastical territories, and free cities, became characteristic by the early Middle Ages and persisted until 1871, when the country was finally united. This disunity was exacerbated by the Protestant Reformation of the sixteenth century, which ended Germany's religious unity by converting many Germans to Lutheranism and Calvinism. For several centuries, adherents to these two varieties of Protestantism viewed each other with as much hostility and suspicion as they did Roman Catholics. For their part, Catholics frequently resorted to force to defend themselves against Protestants or to convert them. As a result, Germans were divided not only by territory but also by religion.

"The terrible destruction of the Thirty Years' War of 1618–48, a war partially religious in nature, reduced German particularism, as did the reforms enacted during the age of enlightened absolutism (1648–1789) and later the growth of nationalism and industrialism in the nineteenth century. In 1815 the Congress of Vienna stipulated that the several hundred states existing in Germany before the French Revolution be replaced with thirty-eight states, some of them quite small. In subsequent decades, the two largest of these states, Austria, and Prussia, vied for primacy in a Germany that was gradually unifying under a variety of social and economic pressures. The politician responsible for German unification was Otto von Bismarck, whose brilliant diplomacy and ruthless practice of statecraft secured Prussian hegemony in a united Germany in 1871. The new state, proclaimed the German Empire, did not include Austria and its extensive empire of many non-German territories and peoples.

"Imperial Germany prospered. Its economy grew rapidly, and by the turn of the century it rivaled Britain's in size. Although the empire's constitution did not provide for a political system in which the government was responsible to parliament, political parties were founded that represented the main social groups. Roman Catholic and socialist parties contended with conservative and progressive parties and with a conservative monarchy to determine how Germany should be governed.

"After Bismarck's dismissal in 1890 by the young emperor Wilhelm II, Germany stepped up its competition with other European states for colonies and for what it considered its proper place among the great states. An aggressive program of military expansion instilled fear of Germany in its neighbors. Several decades of military and colonial competition and a number of diplomatic crises made for a tense international atmosphere by 1914. In the early summer of that year, Germany's rulers acted on the belief that their country's survival depended on a successful war against Russia and France. German strategists felt that a war against these countries had to be waged by 1916 if it were to be won because after that year Russian and French military reforms would be complete, making German victory doubtful. This logic led Germany to get drawn into a war between its ally Austria-Hungary and Russia. Within weeks, a complicated system of alliances escalated that regional conflict into World War I, which ended with Germany's defeat in November 1918.

"The Weimar Republic, established at war's end, was the first attempt to institute parliamentary democracy in Germany. The republic never enjoyed the wholehearted support of many Germans, however, and from the start it was under savage attack from elements of the left and, more important, from the right. Moreover, it was burdened during its fifteen-year existence with serious economic problems. During the second half of the 1920s, when foreign loans fed German prosperity, parliamentary politics functioned better, yet many of the established elites remained hostile to it. With the onset of the Great Depression, parliamentary politics became impossible, and the government ruled by decree. Economic crisis favored extremist politicians, and Adolf Hitler's National Socialist German Workers' Party became the strongest party after the summer elections of 1932. In January 1933, the republic's elected president, Paul von Hindenburg, the World War I army commander, named a government headed by Hitler.

"Within a few months, Hitler accomplished the 'legal revolution' that removed his opponents. By 1935 his regime had transformed Germany into a totalitarian state. Hitler achieved notable economic and diplomatic successes during the first five years of his rule. However, in September 1939 he made a fatal gamble by invading Poland and starting World War II. The eventual defeat of Hitler's Third Reich in 1945 occurred only after the loss of tens of millions of lives, many from military causes, many from sickness and starvation, and many from what has come to be called the Holocaust. Germany was united on October 3, 1990" (*Library of Congress Country Study,* 2015).

The following is taken verbatim from **The Western European Datums by Jacob A. Wolkeau of Army Map Service (unpublished manuscript, circa 1950)**. "The survey and mapping position of pre-war Germany presented even greater complexities than were to be found in France. Practically each German State had its own Survey Department with its own triangulation and mapping system. The triangulations were based on different origins computed on different figures of the earth, and, in general, unconnected across the frontiers of the States. The maps were reproduced on a variety of projections and with no common system of symbols or methods of showing relief, and moreover, at the frontiers, sheets did not include detail in adjoining States. The absence of control by one central survey and mapping authority was further emphasized by the lack of uniformity in the precision of the various State triangulations and the varied rates of progress in map reproduction and map revision.

"The triangulation of the former *Pruβischen Landesaufnahme (Prussian Land Survey – Ed.)*, extending over the whole of Prussia and almost all the small North German States, and the separate triangulations in Mecklenberg, Saxony, Bavaria, Wurtemberg, Hessen and Baden, dates from the first half of last century *(19th century – Ed.)*. The history of the Prussian measures is contained in twenty-four volumes entitled *Abrisse Koordinaten und Hohen,* which give all the information necessary for a practical trigonometrical knowledge of Prussia and the Rhine Provinces.

"The particulars of the other States of the German Confederation are not generally available, but they are probably now only of historical interest, in view of proposals which have been made for the coordination of all German surveys.

"In 1921 a special Survey Committee was appointed to draw up plans for the improvement of the survey and mapping position, and in 1923 this Committee made various recommendations to secure a unified triangulation net and a single coordinate system for the topographical maps of the Reich. The complete union and adjustment of the triangulations of the various States into a single network free of distortion and residual errors desirable as it may be from a scientific point of view, was realized to be impracticable, as it would imply a complete re-computation of most of the nets, a project which was financially impossible. The Survey Departments of the various States, which were still being maintained as separate entities, were therefore enjoined to employ a simplified solution of the problem, in which only the specially disturbing errors of the older parts of the network were to be eradicated and the bulk of the existing network preserved intact. In this process the chains and networks of the *Pruβischen Landesaufnahme* lying West of the 13th Meridian, and the West and east Prussian networks were to be left undisturbed and joined by the modern chain between Berlin and Schubin bases, which was to be freshly adjusted. The remainder of the triangulations of the Reich were to be joined to the above system and the Prussian Geographical Coordinate system and datum adopted for the whole of the Reich.

"The origin of the national survey of Prussia completed in 1859 was the old Observatory at Enckeplatz, Berlin. The accepted values for the latitude and longitude were: $\Phi_o = 52° 30' 16.68''$ N, $\Lambda_o = 13° 23' 55.23''$ (East of Greenwich). The longitudes of the Prussian Survey were referred to the Island of Ferro in the Canaries and the Meridian of Ferro was conventionally fixed as 20° West of Paris. The new fundamental azimuth obtained from the adjustment of the Berlin-Schubin chains the line Potsdam-Golmberg, whose value is $\alpha_o = 154° 47' 32.41''$. The Bessel Spheroid used by the *Pruβischen Landesaufnahme* is adopted for he proposed unified triangulation system. The data of this spheroid, which fits the Geoid particularly well in Mid-Europe are $a = 6,377,397.15500$ metres, $b = 6,356,078.96325$ metres, $f = 1/299.1528128$.

"At the end of the Great War no less than about fifty rectangular co-ordinate systems were in general use throughout Germany. For the purpose of the standardization of the topographical maps, the Committee therefore recommended the adoption of a single co-ordinate system for the whole of Germany plotted on the Gauss transverse conformal projection directly from the spheroid on to the plane by Krüger's formulae *(Gauss-Krüger Transverse Mercator – Ed.)*.

"By the circular of the 31st of May 1935, the uniform production, printing and publishing of the following maps for the whole territory of the Reich is centralized in the Reich Survey office. The uniform production of the 1/50,000 map is an innovation and as a temporary measure some of the sheets of the 1/100,000 of Bavaria and Wurtemburg will still be produced by local authorities on behalf of the Reich Survey Office."

This system was known as the DHG or *Deutches Herres Gitter* (German Army Grid) and has identical specifications for the U.S. Army's UTM with just a few differences. The DHG was referenced only to the Bessel 1841 ellipsoid, the scale factor at origin is unity (1.0) and each 6° zone is numbered Eastward from Greenwich, England, and the 500 km False Easting is preceded by the zone number in the millions unit place.

"The *Prußischen Landestriangulation* Datum origin is at Rauenberg where: $\Phi_o = 52°\ 27'\ 12.0210''$ N, $\Lambda_o = 31°\ 02'\ 04.9280''$ (East of Ferro), and the azimuth from Rauenberg to Berlin-Marienkirche, $\alpha_o = 19°\ 46'\ 04.87''$ *(Bulletin Géodésique, Juillet-août-septembre 1936, page 232)*."

Original German triangulation stations of the *"Reichsdreiecknets"* and the *"Landesdreiecknetze"* adjusted and unified into a provisional *Reichsfestpunktfeld* with the origin Potsdam, *Helmertturm* on Bessel spheroid, where $\Phi_o = 52°\ 22'\ 54.81''$ N, $\Lambda_o = 13°\ 04'\ 01.725''$ East of Greenwich and the azimuth to Golmberg, whose value is $\alpha_o = 154°\ 47'\ 33.61''$ *(Geodetic Analysis of the M745 Series 1:50,000 Scale Maps of Germany, Eugene De Bor, Army Map Service Geodetic Memorandum No. 1584, April 1965)."*

The *Berliner Sternwarte* was the origin for the *Prußischen Landesaufnahme* Soldner from 1822 to 1876. In 1879, the *Prußischen Landesaufnahme* Soldner utilized a new origin at the *Helmertturm* in Pottsdam. However, the *Prußischen Landestriangulation* used the *Berliner Sternwarte* from 1832 to 1923. The Old Bavarian Datum of 1801 was the reference origin for the *Bayern (München)* Soldner from 1802 to 1873 as well as the *Rheinpfalz (Mannheim)* Soldner from 1819 to 1911. Note that *Mannheim Sternwarte* is easy to find with Google Earth at: $\phi = 49°\ 29'\ 11''$ N, $\lambda = 08°\ 27'\ 35''$ E. Furthermore, the Old Bavarian Datum of 1801 was also used for the *Wurttemburg (Tubingen)* Soldner from 1792 to 1841. Afterwards, the new datum used was the *Helmertturm* in Pottsdam from 1841 to 1923. The *Berliner Sternwarte* datum of 1832 was used for the *Saxonia (Großenhain)* Soldner from 1862 to 1923.

The Old Bavarian datum of 1801 was used for the *Baden (Mannheim)* Topographic Soldner from 1824 to 1923, the *Baden (Mannheim)* Forest Soldner from 1846 to 1923, and the *Baden (Mannheim)* Cadaster Soldner from 1841 to 1923. The *Pottsdam Helmertturm* datum was used for the *Oldenburg (Bremerle)* Soldner from 1835 to 1923, and the *Göttingen* datum of 1820 was used for both the *Hannover (Göttingen)* Soldner from 1866 to 1923 as well as the *Kurfurstentum Hessen (Kassel)* Soldner from 1821 to 1899. *The Pottsdam Helmertturm* datum was used for the *Hessen-Darmstadt* Soldner from 1808 to 1923 and for the *Herzogtum Nassau (Schaumburg)* Soldner from 1862 to 1923. The *Pottsdam Helmertturm* datum was used for the *Hessen-Darmstadt* Soldner from 1808 to 1923 as well as the *Herzogtum Nassau (Schaumburg)* from 1863 to 1923, the *Elsaß–Lothringen (Sausheim)* 1889–1923, the *Elsaß–Lothringen (Delmen)* 1889–1923, the Hamburg Soldner 1887–1923, the *Schwarzburg-Sondershausen* Soldner and the *Schwarzburg-Sondershausen Prußischen Landestriangulation* 1851–1923, and the *Thüringen (Seeberg)* Soldner from 1848 to 1923, and the *Thüringen (Seeberg) Prußischen Landestriangulation* 1880–1923 (Dipl.-Ing. Heinz W. Kloos, personal communication, 1981–1982). For all of these Grids, there were multiple periods in which the first usages were local Cartesian systems with no geodetic position, later updated to a geographic position at the Grid origin but computed on an equivalent sphere, and finally updated to ellipsoidal positions for final adjustments and calculations. Note that most of these systems fell into disuse

Federal Republic of Germany

in 1923 when everything began to be coordinated into a Reich-wide unified triangulation net as discussed above, but the original Cadastral coordinates may be of interest to real property owners.

The 40 Prussian Cassini-Soldner origin points used for the Prussian Land Survey and Cadaster established as of 1879 are: 1. *Kucklinsberg* – ϕ=54° 27′ 36.8055″ N, λ=39° 37′ 18.3505″ E; 2. *Paulinen* – ϕ=54° 17′ 21.1583″ N, λ=38° 23′ 59.3555″ E; 3. *Markushof I* – ϕ=54° 03′ 31.728″ N, λ=37° 02′ 24.369″ E; 4. *Turmberg* – ϕ=54° 13′ 31.8753″ N, λ=35° 47′ 32.4975″ E; 5. *Kauernick I* – ϕ=54° 13′ 31.8753″ N, λ=35° 47′ 32.4975″ E; 6. *Thorn (Rathausturm)* – ϕ=53° 00′ 42.5371″ N, λ=36° 16′ 26.1154″ E; 7. *Heinrichsthal* – ϕ=53° 42′ 46.4118″ N, λ=35° 09′ 48.3641″ E; 8. *Gollenberg (Denkmal)* – ϕ=54° 12′ 30.8584″ N, λ=33° 53′ 46.4441″ E; 9. *Gnesen (südl. Domturm)* – ϕ=52° 32′ 17.5346″ N, λ=35° 15′ 40.2180″ E; 10. *Josefsberg* – ϕ=51° 59′ 15.6770″ N, λ=35° 52′ 01.5980″ E; 11. *Schroda (kath. Kirche)* – ϕ=52° 13′ 52.9454″ N, λ=34° 56′ 40.6334″ E; 12. *Pschow (östl. Kirchturm)* – ϕ=53° 00′ 42.5371″ N, λ=36° 16′ 26.1154″ E; 13. *Rummelsberg (Aussichtsturm)* – ϕ=50° 42′ 12.6833″ N, λ=34° 46′ 44.4210″ E; 14. *Gröditzberg I* – ϕ=51° 10′ 41.4963″ N, λ=33° 25′ 40.5751″ E; 15. *Kaltenborn* – ϕ=51° 55′ 44.5335″ N, λ=32° 19′ 43.6659″ E; 16. *Bahn I* – ϕ=53° 06′ 06.6450″ N, λ=32° 22′ 05.2034″ E; 17. *Greifswald (Nikolaikirche)* – ϕ=54° 05′ 49.1594″ N, λ=31° 02′ 43.7053″ E; 18. *Müggelsberg* – ϕ=52° 25′ 07.1338″ N, λ=31° 17′ 37.9332″ E; 19. *Götzer Berg* – ϕ=52° 26′ 14.1346″ N, λ=30° 23′ 43.7870″ E; 20. *Torgau (Stadtkirche)* – ϕ=51° 33′ 40.9038″ N, λ=30° 40′ 27.3695″ E; 21. *Burkersroda (Kirchturm)* – ϕ=51° 10′ 35.6276″ N, λ=29° 18′ 29.0172″ E; 22. *Inselsberg* – ϕ=50° 51′ 08.5674″ N, λ=28° 08′ 03.9542″ E; 23. *Magdeburg (nördl. Domturm)* – ϕ=52° 07′ 34.5112″ N, λ=29° 18′ 07.8117″ E; 24. *Ostenfeld I* – ϕ=54° 28′ 12.6744″ N, λ=26° 54′ 02.7995″ E; 25. *Rathkrügen* – ϕ=53° 49′ 06.2171″ N, λ=27° 42′ 31.9268″ E; 26. *Bungsberg (Aussichtsturm)* – ϕ=54° 12′ 39.9835″ N, λ=28° 23′ 34.9115″ E; 27. *Celle (Stadtkirche)* – ϕ=52° 37′ 32.6709″ N, λ=27° 44′ 54.8477″ E; 28. *Kaltenborn* – ϕ=51° 47′ 47.2820″ N, λ=27° 56′ 28.1079″ E; 29. *Silberberg* – ϕ=53° 43′ 52.4446″ N, λ=26° 43′ 27.8973″ E; 30. *Windberg* – ϕ=52° 52′ 51.1814″ N, λ=25° 11′ 50.2361″ E; 31. *Hermannsdenkmal* – ϕ=51° 54′ 46.8583″ N, λ=26° 30′ 25.8667″ E; 32. *Münster (Überwasserkirche)* – ϕ=51° 57′ 55.7151″ N, λ=25° 17′ 24.0598″ E; 33. *Bochum (Peter-Paul Kirche)* – ϕ=51° 29′ 01.4472″ N, λ=24° 53′ 16.1696″ E; 34. *Homert* – ϕ=51° 15′ 53.2853″ N, λ=25° 46′ 24.7338″ E; 35. *Kassel (Martinskirche, südlicher Hauptturm)* – ϕ=51° 19′ 06.4736″ N, λ=27° 10′ 07.3106″ E; 36. *Schaumburg (Schloß)* – ϕ=50° 20′ 25.7627″ N, λ=25° 38′ 41.0936″ E; 37. *Fleckert* – ϕ=50° 11′ 15.4516″ N, λ=25° 16′ 21.3944″ E; 38. *Köln (Dom. Dachreiter)* – ϕ=50° 56′ 33.2607″ N, λ=24° 37′ 32.3136″ E; 39. *Langschoß* – ϕ=50° 40′ 02.5936″ N, λ=23° 57′ 21.6853″ E; 40. *Rissenthal* – ϕ=49° 28′ 40.8655″ N, λ=24° 25′ 31.1162″ E *(Planheft GROSSDEUTSCHES REICH, Nur für Dienstgebrauch!, Reichsministerium des Innern, Juli 1944)*.

In regard to "Koordinaten-System Names" (Datums), for *Alpen and Donau-Reichsgaue*, **Hermannskogel** – ϕ=48° 16′ 15.29″ N, λ=33° 57′ 41.06″ E; *St. Stephan*, Wien (Vienna – Ed.) – ϕ=48° 12′ 31.5277″ N, λ=34° 02′ 27.3275″ E; *Gusterberg bei Kremsmünster* – ϕ=48° 02′ 18.4753″ N, λ=31° 48′ 15.0242″ E; *Schöcklberg bei Graz* – ϕ=47° 11′ 54.8745″ N, λ= 33° 07′ 59.9472″ E; *Krimberg bei Laibach* – ϕ=45° 55′ 43.7228″ N, λ=32° 08′ 18.8027″ E; **Innsbruck**, südl. Pfarrturm – ϕ=47° 16′ 11.3060″ N, λ=29° 03′ 39.3157″ E. For *Galizien*, **Löwrnburg** in Lemberg (Kopiec Unij) – ϕ=49° 50′ 55.2459″ N, λ=41° 42′ 29.5684″ E. For *Bayern*, **München**, Frauenkirche, nördl. Turm – ϕ=48° 08′ 20.00″ N, λ=29° 14′ 15.00″ E. For *Pfalz*, **Mannheimer Sternwarte** – ϕ=49° 29′ 13.7″ N, λ=26° 07′ 23.4″ E. For *Württemberg*, **Tübingen**, Sternwarte – ϕ=48° 31′ 12.4″ N, λ=26° 42′ 51.0″ E (also see *150 Jahre Württembergische Landesvermessung 1818–1968, Landesvermessungsamt Baden-Württemberg, Stuttgart 1968* – Ed.). For *Sachsen*, **Großenhain** – ϕ=51° 18′ 20.050″ N, λ=00° 00′ 00.000″ (00° 09′ 33.180″ measured from Berlin). For *Baden*, **Mannheim**, Sternwarte – ϕ=49° 29′ 11.00″ N, λ=26° 07′ 35.80″ E. For *Schwerin*, Schloßturm **Mecklenburg** – ϕ=53° 37′ 29.6900″ N, λ=01° 58′ 29.2″ (westl. Berlin), 29° 05′ 12.05″ (ö Ferro). For *Oldenburg*, **Oldenburg**, Schloßturm – ϕ=53° 08′ 22.447″ N, λ=25° 52′ 51.334″ E. For *Hannover*, **Göttingen**, Sternwarte – ϕ=51° 31′ 47.85″ N, λ=27° 36′ 28.30″ E. For *Hessen-Darmstadt*, **Darmstadt**, Knopfmitte des Stadtkirchturms – ϕ=49° 52′ 20.27″ N, λ=26° 19′

25.5768″ E. For *Herzogtum Nassau*, **Schaumburg** – ϕ=50° 20′ 23.63″ N, λ=25° 38′ 29.61″ E, (see also *Die Landesvermessung des Herzogthums Nassau, Wiesbaden 1868*). For *Kurfürstentum Hessen*, **Kassel** *Martinsturm* – ϕ=51° 19′ 06.509″ N, λ=27° 09′ 56.956″ E. For *Elsaß-Lothringen*, **Sausheim** – ϕ=47° 47′ 29.8430″ N, λ=25° 03′ 17.9059″ E and **Delmen** – ϕ=48° 54′ 47.2184″ N, λ=06° 21′ 41.6101″ East of Greenwich. For *Hamburg*, **Hamburg**, *Michaeliskirche* – ϕ=53° 32′ 55.7″ N, λ=9° 58′ 41.75″ ö Greenwich. For Schwarzburg-Sondershausen, **Possen**, *Turm* – ϕ=51° 20′ 23.86″ N, λ=28° 31′ 30.45″ E (Triangulation von Thüringen 18∞1–1855), – ϕ=51° 20′ 24.6378″ N, λ=28° 31′ 34.0489″ E (Preuß. Landestriangulation). For *Thüringen*, **Seeberg**, *Sternwarte* – ϕ=50° 56′ 05.2″ N, λ=28° 23′ 43.5″ E (**Planheft GROSSDEUTSCHES REICH, 1944** ibid.), (*Handbuch der Vermessungskunde, Dr. Wilhelm Jordan, Stuttgart, 1897*), (*Die Trigonometrischen und Polygonmetrischen Rechnugen in der Feldmeszkunst, F. G. Gausz, Stuttgart, 1922*).

"To compute Ferro values to Greenwich the German Geodetic Survey used the value Ferro-Greenwich=17° 39′ 59.411‴" (*Heinz W. Kloos – personal communication, 26 July 1982*).

According to *TR 8350.2*, the *mean* solution for the transformation **from** European Datum 1950 (ED50) **to** WGS84 is: $\Delta X = -87$ m± 3 m, $\Delta Y = -98$ m± 8 m, $\Delta Z = -121$ m± 5 m. Because of the myriad coordinate systems in the Federal Republic of Germany listed above, the error envelope listed with the transformation parameters for WGS84 is probably realistic. Localized transformations can offer better accuracies when only individual networks or chains are involved, but transformation accuracies are still dependent on the accuracies of determination of the original positions. Some of the positions given above were determined over 200 years ago.

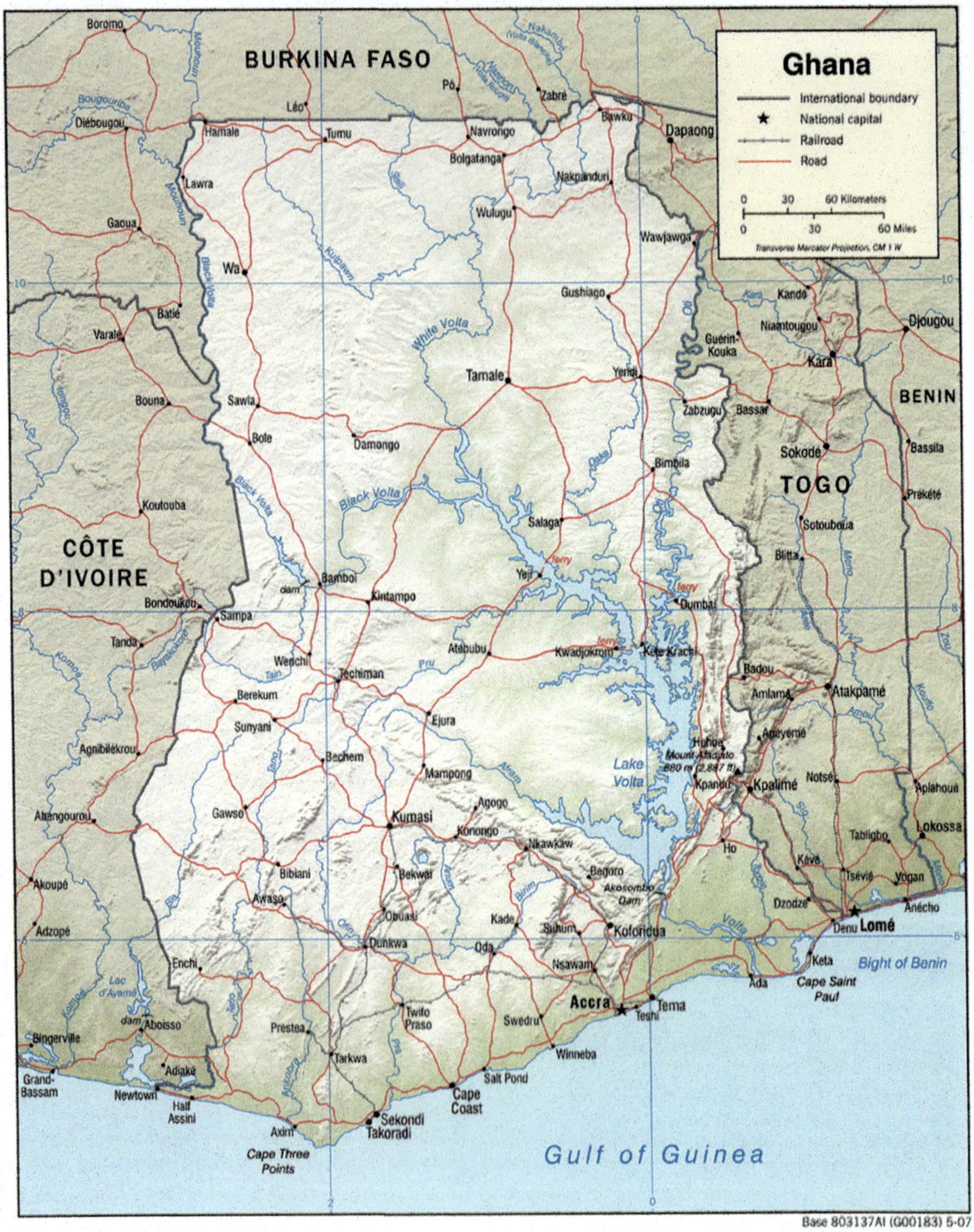

Base 803137AI (G00183) 5-07

The Republic of Ghana

The Republic of Ghana lies on the western coast of tropical Africa. Ghana extends for a maximum of 672 km from north to south between latitudes 4.5° N and 11° N, and for 536 km east to west between longitudes 3° W and 1° E. It is bordered to the west by *Côte d'Ivoire* (Ivory Coast) (668 km), to the north by Burkina Faso (548 km), to the east by Togo (877 km), and to the south by the Gulf of Guinea and the Atlantic Ocean (539 km).

Based on archaeological evidence, it has been established that Ghana was inhabited by humans 300,000 years ago. By 2,000 *B.C.*, domesticated animals such as cattle and guinea fowl were being raised. Modern Ghana takes its name from the ancient Kingdom of Ghana, some 800 km to the north of the present-day capitol of Accra, which flourished up to the 11th century *A.D.* In 1471, the Portuguese arrived at the "*De Costa da el Mina de Ouro*" (the Coast of Gold Mines). In 1486, slaves from Benin were introduced to the port of Elmina. This was the beginning of the slave trade on the Gold Coast, and eventually the more profitable slaves replaced the gold trade. The British Crown Colony of the Gold Coast received its independence from England in 1957.

On May 21, 1929, Capt. J. Calder Wood, M.C. of the Gold Coast Survey Department wrote: "In June 1904, observations for latitude were taken by Capt. F. G. Guggisberg, R.E. (now Sir F. G. Guggisberg, K.C.M.G., D.S.O., Governor of the Gold Coast Colony), from a pillar in the compound of the house of the Secretary for Native Affairs in Accra. Fifteen pairs of stars were observed with a zenith telescope, giving a final probable error of 0.360." *(Zenith telescopes were used to observe the deflection of the vertical – the local tilt of the gravity field – Ed.)* This point was subsequently connected by traverse to the Gold Coast Survey Beacon No. 547 in Accra. The longitude of Accra was determined by the exchange of telegraphic signals with Cape Town in November and December 1904, and the resulting longitude of G.C.S. 547 obtained."

"The pillar G.C.S. 547 was connected to the pillar at Leigon, eight miles from Accra, by means of triangulation, and the resulting values of the Leigon pillar have been adopted as the basic latitude and longitude for the Colony. Subsequent determinations of latitude during the last two years at points throughout the country which have been accurately connected with Leigon by triangulation tend to indicate that the latitude and longitude observations taken at Accra are seriously influenced by local attraction. The conditions obtaining near the Coast do not appear to be reproduced further inland, and it may happen in course of time, when more data are available, that a new basic latitude and longitude will be adopted from inland observations which will give a better datum point for the county as a whole. Sufficient comparison between the astronomical and trigonometrical or geodetic values have not yet been obtained to warrant an immediate change."

The astronomical values observed for (G.C.S. 121) Leigon were: $\Phi_o = 5° 38' 54.39''$ North, and $\Lambda_o = 0° 11' 52.65''$ West of Greenwich. The corresponding geodetic coordinates of the same Accra Datum of 1929 origin point that were computed from the triangulation were: $\varphi_o = 5° 38' 52.270''$ North, and $\lambda_o = 0° 11' 46.080''$ West of Greenwich. In 1928, as other observations were being taken at Leigon, it was decided to observe an azimuth there to test an intermediate side of the triangulation. The azimuth observed from Leigon to station (G.C.S. 113) Asofa was: $\alpha_o = 264° 48' 48.78''$. The ellipsoid of reference is the War Office 1926 (McCaw 1924), where $a = 6,378,300$ m (20,926,201.2257 feet) and $1/f = 296$. Note that this conversion from feet to meters is specific to Ghana, and the elevation of Leigon (H_o) = 147.46 m. As of late 1996, the transformation parameters from Accra Datum of 1929 to WGS 84 Datum were based on three co-located points, where $\Delta X = -199$ m, $\Delta Y = +32$ m, $\Delta Z = +322$ m.

Capt. Wood continued, "For the purposes of cadastral work geographical coordinates are inconvenient, and in their place plane rectangular coordinates on the transverse mercator (*sic*) projection have been adopted. The whole Colony has been placed on the same origin, the central meridian being the meridian 1° W and the origin of the X coordinates 4° 40′ N. 900,000 is added to all Y coordinates, in order to avoid negative coordinates, and the maximum scale error has been reduced in the customary manner by reducing the scale of the projection by 1/4000, so that the scale error nowhere exceeds this value except on the extreme edges of the Colony."

For instance, "Leigon (G.C.S. 121). District Accra. Locality west of road junction at 8th mile Acra-Dodowa road. Approach through low bush from junction of Aburi and Dodowa roads. Bare hill with good all-round view. Concrete pillar 14″ high by 9½″ by 7″ with iron pipe as centre mark. Beacon double quadripod of squared timber, 17′ high, centered over old mark."

The geodetic coordinates of Leigon equate to Colony coordinates of $X = 356,084.33$ ft, $Y = 1,192,117.91$ ft. As I always point out; if your double- or quadruple-precision software does not exactly match the local transformation results, your software is wrong for that country. Colony coordinates are truncated Gauss-Schreiber Transverse Mercator transformations that are similar to those used in the United States for the NAD 1927 State Plane Coordinate Systems. They are not computed the same way as the Gauss-Krüger Transverse Mercator transformations normally done with other Grid systems such as the UTM and the NAD 1983 State Plane Coordinate Systems.

According to J. Clendinning in *Empire Survey Review*, January 1934: "The *Akuse-Obuasi*, *Obuaasi-Nsuta* and *Akuse-Apam* chains were computed in terms of geographical coordinates which were later converted into rectangular coordinates on the projection system; but all recent work, both triangulation and traverse, has been computed directly on the projection system, the necessary corrections to distances and bearings as measured on the ground being applied to enable this to be done." In that same paper, Clendinning also commented on the surroundings of the Gold Coast in that "The denseness of the forest and the topographical features of the country naturally determine the nature of the survey." With that terrain in mind for Ghana, the *Textbook of Topographical Surveying*, Her Majesty's Stationery Office, London, 1965, pp. 221, 222 offered the following description.

> 311. The rope and sound traverse
>
> In West Africa, the minor bush tracks maintain their general direction for considerable distances but wind about just sufficiently to prevent a sight being taken down the track. A system of survey which has been used in these conditions is as follows. The bearings of the traverse legs are observed from a compass station to sound. This sound may be either a call or a whistle from a forward chainman. The length of the legs is measured with a rope. This rope is made of a length allowing for the wind in the path. Thus, it is normally found that a rope 310 ft. long pulled tight along the path is equivalent to 300 ft. measured direct along the leg of the traverse. The actual allowance must be found by experience of local conditions. The length of the rope should be checked and adjusted before and after work and in the middle of the work.
>
> Using this system minor bush tracks can be traversed rapidly, and it is possible to work through trackless bush doing only a minimum amount of cutting. When undergrowth is thick or thorny a shorter rope (100 ft.) may be easier to manipulate. The accuracy of this work is greater than would be expected. There appears to be no systematic error in the observations of bearing and the errors consequently tend to cancel out during the traverse. For work at a scale of 1/125,000, traverses should not be carried for more than five miles by this method. Heights are observed by aneroid barometers. On the Gold Coast, an average of five miles per day can be maintained, and distances of 15 miles have been achieved on single days. If, however, work is continued into the afternoon, as is necessary for such an output, it is necessary to re-run later that part of the aneroid heights which was observed in the period when the diurnal wave is unreliable.

In the *Canadian Cartographic Journal*, a letter to the editor told how they were cutting costs by eliminating the forward chainman. "They tied the rope to the tail of a pig and sent it down the path. When the rope became taut, the tail of the pig was pulled, and it squealed so they could observe a bearing on that sound. They were working on further improving the method by breeding a special pig that would, in addition, monument the spot when its tail was pulled."

In 1977, the control network was readjusted including extra observations and an improved computation technique. As part of the readjustment the datum was changed. The new Datum was Leigon Datum of 1977, in which the old geographical values for (GCS 121) Leigon were held fixed at the old values and the War Office ellipsoid was replaced by the Clarke 1880 (modified) ellipsoid, where $a = 6,378,249.145$ m and $1/f = 293.465$. The False Easting was changed to 274,319.736 m. In 1991, the National Imagery and Mapping Agency (NIMA) published the three-parameter datum shift values from Leigon Datum of 1978 to WGS 84 based on eight co-located points, where $\Delta X = -130$ m ± 2 m, $\Delta Y = +29$ m ± 3 m, $\Delta Z = +364$ m ± 2 m. Out of curiosity, I decided to try an eight-point solution for a seven-parameter Bursa-Wolf shift from Leigon Datum of 1977 to WGS 84 Datum. The resultant was $\Delta X = -110.90$ m, $\Delta Y = -9.13$ m, $\Delta Z = +69.46$ m, Scale $= +3.11 \times 10^6$, $R_x = -1.30$ arc seconds, $R_y = +9.53$ arc seconds, and $R_z = +0.32$ arc seconds. The increase in goodness of fit was about ten-fold over the three-parameter solution. An example data point is station (G.C.S. T20/24) "Wa" on the Leigon Datum of 1977, where $\varphi = 10° 02' 47.7770''$ North, and $\lambda = 2° 28' 17.3746''$ West, and an MSL elevation of $H = 359.7$ m with coordinates of "Wa" on the WGS 84 Datum, where $\varphi = 10° 02' 56.3328''$ North and $\lambda = 2° 28' 16.6377''$ West, and an ellipsoid height of $h = 385.1$ m. Of course, on the WGS 84 Datum the coordinates are expressed on the UTM Grid, and not on Colony Coordinates.

Ghana has installed a new GPS network of Continuously Operating Reference Stations (Poku-Gyamfi, Y., & Schueler, T., *Renewal of Ghana's Geodetic Reference Network*, LNEC, Lisbon, 2008). Numerous publications have been devoted to an exploration of various math models for modeling the transformation parameters from Accra Datum of 1929 and Leigon Datum of 1977 to WGS84 Datum: Poku-Gyamfi, Y., & Hein, G., *Framework for the Establishment of a Nationwide Network of Global Navigation Satellite System*, 5th FIG Conference, Ghana, 2006; Dzidefo, A., *Determination of Transformation Parameters between the World Geodetic System 1984 and the Ghana Geodetic Network*. Master's Thesis, Department of Civil and Geomatic Engineering, Kwame Nkrumah University of Science and Technology, Ghana, 2011; and Ziggah, Y.Y., *et al.*, *Novel Approach to improve Geocentric Translation Model Performance Using Artificial Neural Network Technology*, Boletin de Ciências Geodésicas – Online Version, ISSN 1982-2170, 2017. Work and analyses appear to be primarily in the "Golden Triangle Region" of Ghana.

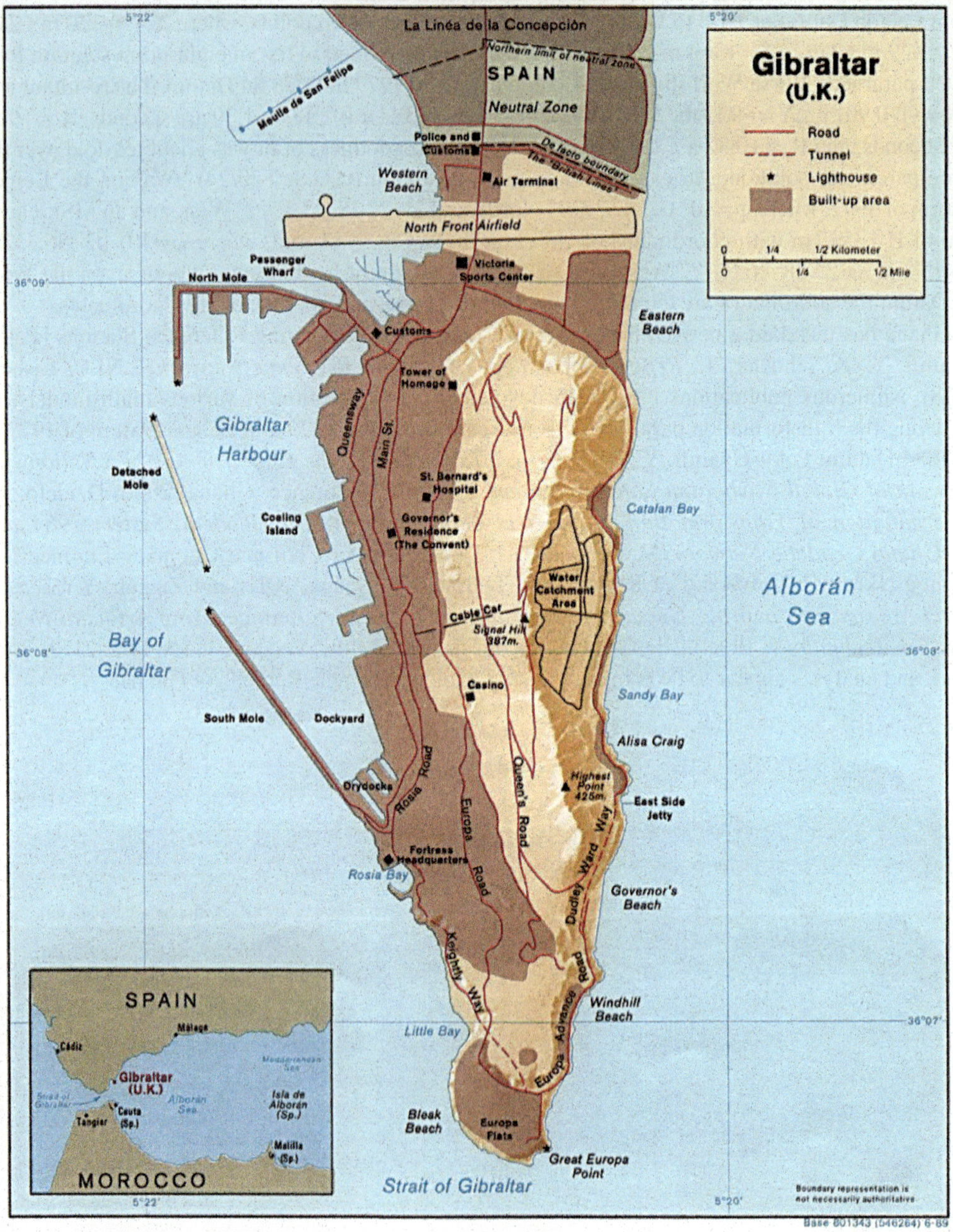

Base 801343 (546264) 6-89

Gibraltar

Slightly less than one-half the area of Rhode Island, Gibraltar is bordered by Spain (1.2 km). The lowest point is the Mediterranean Sea (0 m), and the highest point is the Rock of Gibraltar (426 m).

Captured and fortified by *Ṭāriq ibn Ziyād* in 711 A.D., Moorish invader of Spain, later taken by the Spanish in 1462. It was then captured by the British in 1704 during the War of Spanish Succession (*Merriam-Webster's Geographical Dictionary*, 3rd Edition, 1997). "Strategically important, Gibraltar was reluctantly ceded to Great Britain by Spain in the 1713 Treaty of Utrecht; the British garrison was formally declared a colony in 1830. In a referendum held in 1967, Gibraltarians voted overwhelmingly to remain a British dependency. The subsequent granting of autonomy in 1969 by the UK led to Spain closing the border and severing all communication links. A series of talks were held by the UK and Spain between 1997 and 2002 on establishing temporary joint sovereignty over Gibraltar. In response to these talks, the Gibraltar Government called a referendum in late 2002 in which the majority of citizens voted overwhelmingly against any sharing of sovereignty with Spain. Since the referendum, tripartite talks on other issues have been held with Spain, the UK, and Gibraltar, and in September 2006 a three-way agreement was signed. Spain agreed to remove restrictions on air movements, to speed up customs procedures, to implement international telephone dialing, and to allow mobile roaming agreements. Britain agreed to pay increased pensions to Spaniards who had been employed in Gibraltar before the border closed. Spain will be allowed to open a cultural institute from which the Spanish flag will fly. A new non-colonial constitution came into effect in 2007, but the UK retains responsibility for defense, foreign relations, internal security, and financial stability. Spain and the UK continue to spar over the territory. Throughout 2009, a dispute over Gibraltar's claim to territorial waters extending out three miles gave rise to periodic non-violent maritime confrontations between Spanish and UK naval patrols and in 2013, the British reported a record number of entries by Spanish vessels into waters claimed by Gibraltar following a dispute over Gibraltar's creation of an artificial reef in those waters. Spain renewed its demands for an eventual return of Gibraltar to Spanish control after the UK's June 2016 vote to leave the EU, but London has dismissed any connection between the vote and its continued sovereignty over Gibraltar" (*World Factbook*, 2022).

In regard to its economy, "Self-sufficient Gibraltar benefits from an extensive shipping trade, offshore banking, and its position as an international conference center. The British military presence has been sharply reduced and now contributes about 7% to the local economy, compared with 60% in 1984. The financial sector, tourism (almost 5 million visitors in 1998), shipping services fees, and duties on consumer goods also generate revenue. The financial sector, the shipping sector, and tourism each contribute 25%–30% of GDP. Telecommunications accounts for another 10%. In recent years, Gibraltar has seen major structural change from a public to a private sector economy, but changes in government spending still have a major impact on the level of employment" (*World Factbook*, 2009).

According to the *Bundesamt für Kartographie und Geodäsie* (Federal Office for Cartography and Geodesy), *Außenstelle* (Branch) *Leipzig,* Information and Service System for European Coordinate Reference Systems, *Karl-Rothe-Straße* 10–14, D-04105 Leipzig, the transformation *from* European Datum 1950 *to* WGS 84 Datum is: $\Delta X = -116.8$ m ± 1 m, $\Delta Y = -106.4$ m ± 1 m, $\Delta Z = -154.4$ m ± 1 m. Of course, the ellipsoid of reference for the European Datum of 1950 is the International 1924, where $a = 6,378,388$ m, $1/f = 297$. The *BKG* comments: "Although there is a large scale and orientation error, the small area of Gibraltar makes a 3 parameter shift acceptable at 1 m accuracy. Otherwise very large shifts and rotations will result." The only grid extant is the UTM.

Greece
International boundary
National capital
Railroad
Road
0 50 100 Kilometers
0 50 100 Miles
Lambert Conformal Conic Projection, SP 12N/38N
Base 802487 (A01424) 5-96
ALBANIA
THE FORMER YUGOSLAV REPUBLIC OF MACEDONIA
BULGARIA
TURKEY
Black Sea
Sea of Marmara
Aegean Sea
Ionian Sea
Sea of Crete
Mediterranean Sea
NORTHERN SPORADES
CYCLADES
DODECANESE
Shkodër
Tiranë
Durrës
Vlorë
Berat
Pogradec
Gjirokastër
Skopje
Prilep
Bitola
Blagoevgrad
Sandanski
Plovdiv
Dimitrovgrad
Khaskovo
Kŭrdzhali
Pedkova
Edirne
Kırklareli
Lüleburgaz
İstanbul
Üsküdar
İzmit
Bosporus
Dardanelles
Eceabat
Çanakkale
Bandırma
Bursa
Balıkesir
Manisa
İzmir
Denizli
Bodrum
Marmaris
Keşan
Serrai
Dráma
Kavála
Xánthi
Komotiní
Alexandroúpolis
Thásos
Samothráki
Límnos
Édhessa
Flórina
Kastoría
Véroia
Kozáni
Kateríni
Thessaloníki
Grevená
Corfu
Kérkira
Igoumenítsa
Ioánnina
Kalabáka
Tríkala
Lárisa
Vólos
Kardhítsa
Árta
Préveza
Levkás
Lamía
Karpenísion
Agrínion
Mesolóngion
Ámfissa
Khalkís
Évvoia
Skíros
Thívai
Mégara
Athens
Piraiévs
Argostólion
Kefallinía
Zákinthos
Pátrai
Pírgos
Corinth
Tripolis
Návplion
Kalámai
Sparta
Yíthion
Kíthira
Mitilíni
Lésvos
Khíos
Sámos
Ikaría
Ándros
Tínos
Síros
Ermoúpolis
Náxos
Thíra
Kos
Ródhos
Rhodes
Kárpathos
Meyísti
Khaniá
Kastéllion
Réthimnon
Iráklion
Crete
Sitía

The Hellenic Republic

Greece is slightly larger than the state of Alabama, and it is bordered by Albania (282 km), Bulgaria (494 km), Turkey (206 km), and Macedonia (246 km). The lowest point in Greece is the Mediterranean Sea, and the highest point is Mount Olympus (2,917 m).

The region of present-day Greece was occupied in the Paleolithic period, and Indo-European invasions began about 2000 B.C. Ancient Greece was never unified, but the city-states of Athens and Sparta dominated while other cities shifted alliances over the centuries. Alexander the Great conquered most of the Mediterranean region and spread Greek culture throughout the known world. However, Greece was conquered by Rome in 146 B.C., and by 1456 A.D. Greece was completely under the Ottoman Turk Empire. Greece won its independence from Turkey in the war of 1821–1829 and celebrates its Independence Day on 25 March (1821). The former Kingdom of Greece is now a parliamentary republic; the monarchy was rejected by referendum on 08 December 1974. Beginning in 2010, Greece entered three bailout agreements – with the European Commission, the European Central Bank (ECB), the IMF, and the third in 2015 with the European Stability Mechanism (ESM) – worth in total about $300 billion. The Greek government formally exited the third bailout in August 2018.

In 1889, the Greek Army Geographical Service was formed, and classical triangulation commenced immediately. The agency name was later changed to the Hellenic Military Geographical Service (HMGS). The initial starting point for the triangulation was the Old Athens Observatory, where $\Phi_o = 37° 58' 20.1''$ North, $\Lambda_o = 23° 42' 58.5''$ East of Greenwich, and was referenced to the Bessel 1841 ellipsoid where the semi-major axis $a = 6,377,397.155$ m and the reciprocal of flattening $1/f = 299.1528128$. The *Yeografikí Ipiresía Stratoú* map series at 1:20,000 scale was produced from 1926 through 1947, and had the Greek Military Grid shown on some sheets. The series covered the northern border and scattered strategic areas throughout Greece. The Greek Military Grid was based on the Hatt Azimuthal Equidistant projection, a system originally presented on the sphere by Guillaumme Postel.

Hatt was the Hydrographer of the French Navy, and later taught at a university in Paris. Apparently, he made quite an impression on a Greek student because the Hatt projection, used by the French Navy for local grids of hydrographic surveys, became the national Grid system of Greece in the 20th century. The Hatt projection is quite similar to the Azimuthal Equidistant projections used in Yemen, Guam, and Micronesia; the differences are based on the mathematics used to compute the ellipsoidal geodesic for the direct and inverse cases. Gougenheim, another French Hydrographic Engineer, later published a number of treatises on the geodesic that were later picked up by Paul D. Thomas who published a treatise for the U.S. Navy on the same subject in the 1970s. Thomas presented extensive computational proofs of Gougenheim's work that established the standard for "hand and mechanical calculator" solutions of the geodesic for global applications of the United States Navy. In the 1980s, Thomas' work inspired my research partner at the University of New Orleans, the late Dr. Michael E. Pittman, to publish his original solution of the "Principal Problem of Geodesy" in the *Surveying and Mapping* journal of the ACSM. The other projection variants mentioned above for Yemen and Guam used the ellipsoidal geodesic solutions developed by Puissant and by Andoyer-Lambert. The Azimuthal Equidistant for Micronesia was developed by the late John P. Snyder using the "Clarke Long Line Formula" originally developed by Colonel A. R. Clarke of the British Royal Engineers. Hatt's projection was an enormous influence on European cartography worldwide for many decades, and I am often amused to see contemporary software packages list unknown projections and grid systems as *"Systémé Rectangulaire Usuel"* with no further information. That "Usual Rectangular System" found worldwide is the Hatt projection!

Nevertheless, the only country that adopted the Hatt Azimuthal Equidistant projection as the national grid was Greece. In 1980, Brigadier General Dimitri Zervas, Commander of the Hellenic Military Geographical Institute, sent a treatise to me entitled *Η ΑΖΙΜΟΥΘΙΑΚΗ ΙΣΑΠΕΧΟΥΣΑ ΠΡΟΒΟΛΗ ΤΟΥ ΗΑΤΤ, 1963* (HATT AZIMUTHAL EQUIDISTANT PROJECTION). The 21-page tome was written entirely in Greek by John Bandecas, but General Zervas mercifully penciled-in English translations of paragraph headings so that I could understand the mathematics presented. That was my one and only experience with Greek Geodesy in which the only thing I could comprehend was the Greek symbols for standard geodetic terms in the math!

Coordinates later published for the Datum Origin were $\Phi_o=37°\ 58'\ 18.680''$ North and $\Lambda_o=23°\ 42'\ 58.815''$ East of Greenwich, but curiously the National Topographic Series published by the Greek Military used the *Athens Observatory as their national prime meridian*. The map series at various scales were based on integer minute differences from the Athens meridian. The basic series were based on 30- by 30-minute blocks with latitudes of $36°\ 30'$ N to $42°\ 00'$ N and longitudes of $4°\ 30'$ W of Athens to $3°\ 30'$ E of Athens ($23°\ 42'\ 58.815''$ East of Greenwich).

The "Revised Military Grid" used in some national applications after World War II was based on the Lambert Conformal Conic projection. Using the same central meridian as the Athens Observatory, the three **tangent** zones had latitudes of origin of $35°$, $38°$, and $41°$ with False Eastings of 1,500, 2,500, and 3,500 km, respectively, and all three zones had False Northings of 500 km. The "Old Military Grid" used from 1931 to 1941 had the same parameters except that there were no false origins.

From 1925 to 1946, there were two "British Grids" used by the Allied Forces. The Mediterranean Zone was a *secant* Lambert Conical Orthomorphic where the central meridian was $29°$ East of Greenwich, and the latitude of origin was $39°\ 30'$ N, the scale factor at origin was 0.99906, the False Easting was 900 km, and the False Northing was 600 km. The Crete Zone was a *tangent* Lambert Conical Orthomorphic where the central meridian was Athens ($24°\ 59'\ 40''$ East of Greenwich), the latitude of origin was $35°$ N, the scale factor at origin was 1.0 (by definition of a *tangent* zone), the False Easting was 200 km, and the False Northing was 100 km.

There is a new reference system used in Greece nowadays. It is called the Greek Geodetic Reference System of 1987 (GGRS87) where at origin Dionysos $\Phi_o=38°\ 04'\ 33.8107''$ North, $\Lambda_o=23°\ 55'\ 51.0095''$ East of Greenwich, $h_o=481.743$ m (http://crs.bkg.bund.de/crseu/crs/eu-description.php?crs_id=Y0dSX0dHU1M4NyslMkYrR1JfVE0=) and the new Greek Grid is based on the Transverse Mercator projection (presumably Gauss-Krüger), where $\phi_o=0°$, $\lambda_o=24°$ E, the False Easting = 500 km, and the scale factor at origin $(m_o)=0.9996$. Generally, I have serious doubts concerning any "new" grid system that uses some non-standard variant of the UTM Grid, but I understand that this particular one was devised by Professor Veis of the Technical University of Athens. If Professor Veis approved of this new Grid, then there certainly must be a valid technical reason for the curious parameters chosen. Thanks for the above parameters go to Mr. Yannis Yanniris, a photogrammetrist in Athens.

The National Imagery and Mapping Agency has published datum shift parameters from the European Datum of 1950 in Greece to the WGS84 Datum, where $\Delta X=-84$ m, $\Delta Y=-95$ m, and $\Delta Z=-130$ m; however, this solution is based on only two points and the accuracy of the components is stated to be ±25 m. The European Petroleum Studies Group has published shift parameters from GGRS87 to WGS84 as being $\Delta X=-199.87$ m, $\Delta Y=+74.79$ m, and $\Delta Z=+246.62$ m. The EPSG published no accuracy estimates for their parameters, so ***caveat emptor***.

The Hellenic Positioning System (HEPOS) is based on 98 GPS Continuously Operating Reference System (CORS) sites operated by the commercial organization KTIMATOLOGIO, S.A. The Hellenic Terrestrial Reference System 2007 (HTRS07), based on the ETRS89/ETRF05 reference frame underlies HEPOS, and an evaluation of 2,430 geodetic benchmarks revealed that the accuracy level is better than 10 cm in terms of the rms value for the total transformation error (Katsampalos, K., Kotsakis, C., Gianniou, M., *Hellenic terrestrial reference system 2007 (HYTS07): a regional realization of ETRS89 over Greece in support of HEPOS,* Bollettino di Geodesia e Scienze Affini, Instituto Geografico Militare, Vol. LXIX, No. 2–3, May/December 2010).

A recent publication addressed two-dimensional cartometric transformations of old Greek topographic maps to the HTRS87 Datum as discussed in my original *Grids and Datums* column.

(Moschopouos, G., Demirtzoglou, N., Mouratidis, A., Perperidou, G., Ampatzidis, D., Transforming the old map series to the modern geodetic reference system, *Coordinates*, Vol. XVI, No. 7, July 2020.)

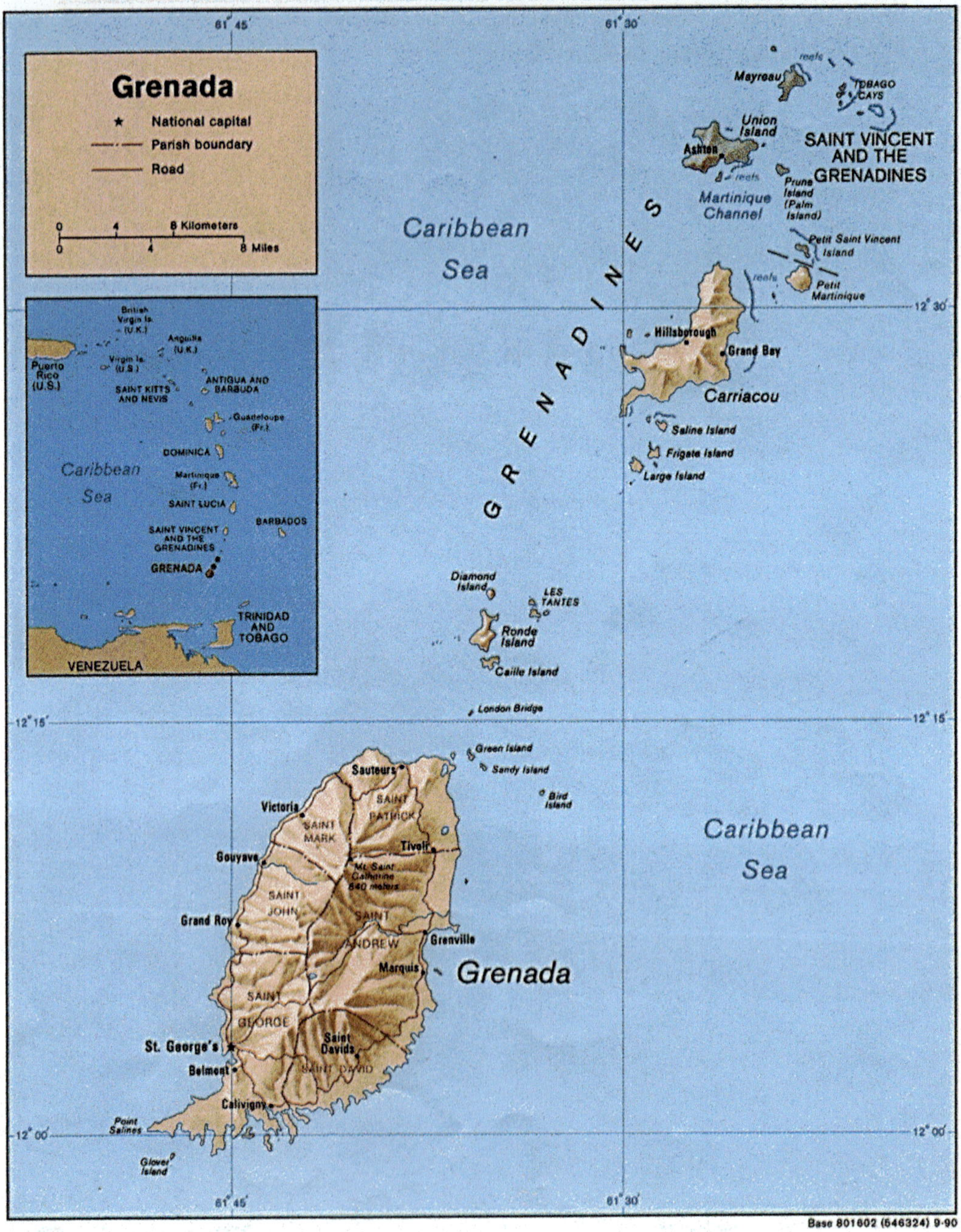
Grenada
National capital
Parish boundary
Road
0 4 8 Kilometers
0 4 8 Miles
Caribbean Sea
GRENADINES
SAINT VINCENT AND THE GRENADINES
Mayreau
TOBAGO CAYS
Union Island
Ashton
Martinique Channel
Prune Island (Palm Island)
Petit Saint Vincent Island
Petit Martinique
Hillsborough
Grand Bay
Carriacou
Saline Island
Frigate Island
Large Island
Diamond Island
LES TANTES
Ronde Island
Caille Island
London Bridge
Green Island
Sandy Island
Bird Island
Sauteurs
Victoria
SAINT PATRICK
SAINT MARK
Tivoli
Gouyave
Mt. Saint Catherine 840 meters
SAINT JOHN
Grand Roy
SAINT ANDREW
Grenville
Marquis
Grenada
SAINT GEORGE
St. George's
Saint Davids
SAINT DAVID
Belmont
Calivigny
Point Salines
Glover Island
61° 45'
61° 30'
12° 30'
12° 15'
12° 00'
Puerto Rico (U.S.)
British Virgin Is. (U.K.)
Anguilla (U.K.)
Virgin Is. (U.S.)
SAINT KITTS AND NEVIS
ANTIGUA AND BARBUDA
Guadeloupe (Fr.)
DOMINICA
Martinique (Fr.)
SAINT LUCIA
SAINT VINCENT AND THE GRENADINES
BARBADOS
GRENADA
TRINIDAD AND TOBAGO
VENEZUELA
Base 801602 (546324) 9-90

Grenada

Dubbed the "Spice Island" because of its impressive production of nutmeg, mace, cinnamon, ginger, and cloves, Grenada has a rugged mountainous interior of rainforests and waterfalls and an indented coastline with protected bays and secluded beaches. Grenada is comprised of the islands of Grenada, Carriacou, and Petit Martinique. Located just north of Trinidad and Tobago, and just south of St. Vincent, the area of Grenada (340 km²), is twice the size of Washington, DC. With a coastline of 121 km, the terrain is volcanic in origin with central mountains. The lowest point is the Caribbean Sea, and the highest point is Mount Saint Catherine (840 m).

Discovered by Christopher Columbus in 1498 on his third voyage, the island was not settled until 1609 when the English attempted to establish tobacco plantations. Native Carïb Indians made numerous raids on the English settlers and convinced them to abandon the island. In 1650, the governor of Martinique purchased Grenada from the Carïbs, and re-settled the island with about 200 French citizens. After a year of subsequent raids by the Carïbs, a contingent of French soldiers was sent to Grenada to secure the island. The Carïbs were routed at Sauteurs Bay, but rather than surrender the entire Carïb population leaped to their deaths from the island cliffs. Thanks to *Lonely Planet* (2004), "The French then set about establishing plantations of indigo, tobacco, coffee, cocoa and sugar, which were worked by African slaves. Grenada remained under French control until captured by the British in 1762. Over the next two decades it teetered between the two colonial powers until it was ceded to the Brits in 1783. It remained under British rule until independence, though animosity lingered between the British colonialists and the minority French settlers, with violence erupting periodically. In 1877, Grenada became a Crown Colony. In 1967, Grenada became an associate state within the British Commonwealth. Grenada and the neighboring Grenadine Islands of Carriacou and Petit Martinique adopted a constitution in 1973 and became an independent nation in 1974."

"In 1979, a leftist New Jewel Movement seized power under Maurice Bishop ushering in the Grenada Revolution. On 19 October 1983, factions within the revolutionary government overthrew and killed Bishop and members of his party. Six days later the island was invaded by US forces and those of six other Caribbean nations, which quickly captured the ringleaders and their hundreds of Cuban advisers. The rule of law was restored, and democratic elections were reinstituted the following year and have continued since then" (*World Factbook*, 2022).

The British Directorate of Colonial Surveys (DCS) flew the first aerial photography in 1951 of Grenada. The original geodetic surveys of the island were performed by DCS in 1953, and the origin point is the astronomical station GS 8, Santa Maria (at the Santa Maria Hotel yard), where $\Phi_o = 12° \, 02' \, 36.56''$ N and $\Lambda_o = 61° \, 45' \, 12.495''$ West of Greenwich. The defining azimuth to G5 North Extension is $\alpha_o = 207° \, 30' \, 46.55''$ East of North, and scale is defined by the length from G1 West Base (Grand Anse Rum Distillery Hill) to G2 East Base (SE of the Grand Anse Rum Distillery chimneys) of 1,991.394 m. The height of Santa Maria (H_o) = 160.24 ft, determined by leveling from the Colony benchmark at St. Georges Harbor which is 3.17 ft above mean sea level. The ellipsoid of reference is the Clarke 1880, where $a = 6,378,249.145$ m, $1/f = 293.465$. The grid system used for Grenada is the BWI Transverse Mercator Grid where the central meridian, $\lambda_o = 62°$ W, the latitude of origin φ_o = equator, the scale factor at the latitude of origin $m_o = 0.9995$, False Easting = 400 km, and False Northing = nil. The formulae are the Gauss-Krüger, but for such a small span of latitude and longitude that includes all three islands; the distinction in this case is irrelevant. As is common with the BWI Grid usage, the grid is used as an "atlas index" numbering system for the popular tourist maps and is not numbered with coordinate values but with an alphanumeric system for facile use to locate tourist interest points. The grid is easy to recover if one is familiar with the standard BWI grid conventions, ***but the defining parameters were unfortunately obscure to many:***

In Grenada, four Navy A-7 Corsair aircraft strafed a U.S. Army command post, inflicting 17 American casualties (Doton, *Acquisition Quarterly Review*, 1996). That tragedy highlighted the Services' failure to establish a common positional picture. Each Service brought its own maps and map systems to the fight. The ground forces were unable to accurately describe a point on the ground to the supporting pilots. Air, ground, and sea Services planned and operated using separate maps referenced to three distinctly different coordinate systems. Accustomed to large-scale maps depicting terrain in familiar grids, Army units deploying from Fort Bragg used maps constructed by the Army's 100th Engineer Company (Cartographic), from a tourist map with an arbitrary grid overlay. Despite pictures of palm trees in the margins, the map was excellent. Constructed by British military engineers, the base map included highly accurate survey data replete with topographic contours. The American Army engineers merely added black grid lines for ground troops to use as a grid reference system. While this worked well for the Army, coordinates from the gridded overlay were useless to any combatant without a copy of the modified tourist map. Some historians link the strafing of the U.S. Army command post to this lack of a common positional picture. "Ground units experienced difficulty in orienting themselves and in directing supporting gunfire and airstrikes. [This] inadvertent airstrike...has been blamed partly on this chart confusion problem" (Rivard, *DTIC* 1985). The failure to create a common reference for planning highlighted the Services' utter lack of attention to planning the joint fight. The "tourist map" debacle merited considerable media attention, providing further grist for 1986 Goldwater-Nichols Act proponents.

(Gruetzmacher, Holtery, and Putney, *Joint Forces Staff College Joint and Combined Staff Officer School*, #02-02, 2002)

A GPS survey by the U.S. National Geodetic Survey (NGS) occupied the station GS 15, Fort Frederick in 1996. I computed a single-point datum shift relation from Grenada 1953 Datum to WGS 84 Datum as: $\Delta X = +72$ m, $\Delta Y = +213$ m, and $\Delta Z = +93$ m. Thanks to Dennis McCleary of NGA for validation that the Santa Maria "astro" position was the same as the geodetic position I received from Dave Doyle of NGS.

Dominica, Guadeloupe, and Martinique

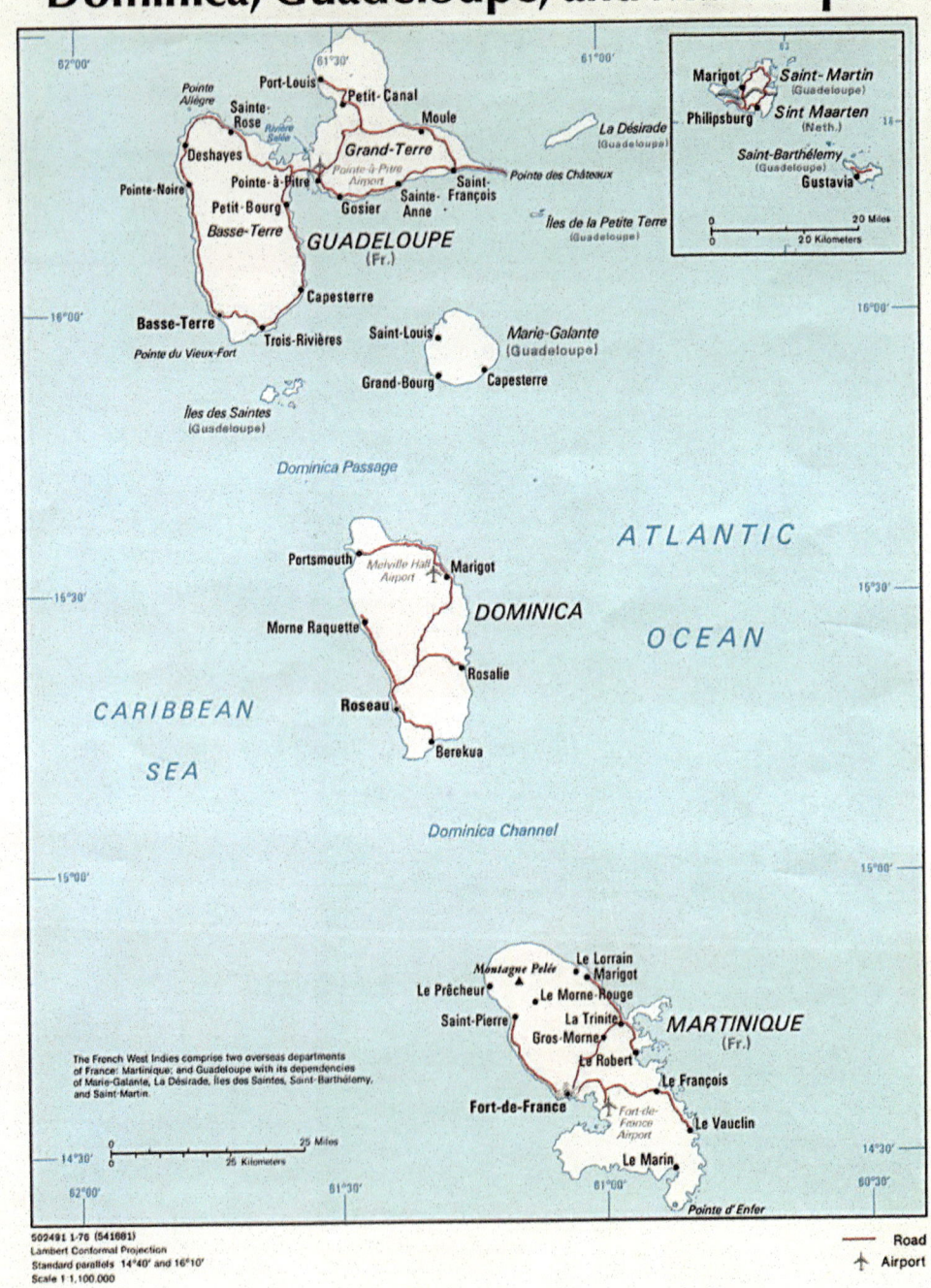

The Department of Guadeloupe

Guadeloupe is an archipelago of inhabited islands near the southern part of the Leeward Islands. This includes *Basse-Terre* (or Guadeloupe proper), *Grande-Terre, Marie-Galante, La Désirade, Îles des Saintes,* and *Saint Barthélemy*. Part of *Saint Martin* is included, but it is 300 km from *Basse-Terre* and is located in the northern end of the Leeward Island Archipelago north of 18° N. (The other part of *Saint Martin* is *Sint Maarten*, which has been Dutch since 1648.) The highest peak is on *Basse-Terre*, which is the active volcano of *Soufrière* at 1,467 m (4,813 ft).

The ancient fishermen of the Arawak Indian tribe were the apparent original inhabitants of the islands of Guadeloupe several hundred years before Christ. Around the 9th century *A.D.*, the cannibal warriors of the Caraïbes tribe ate the last of the Arawaks! Columbus landed on the island of Karukera (Island of Beautiful Waters) on November 3, 1493, and renamed it Guadeloupe. The Spaniards had no interest in settling the island, and in 1635 the Compagnie des Îles d'Amérique introduced volunteer farmers from the French provinces of Normandie, the Bretagne, and the Charente. Working conditions were not conducive to further European labor, so the slave trade with Africa was encouraged. The Compagnie sold Guadeloupe to Charles Houël, who started the major economic growth of the island with plantations of sugar, coffee, and cocoa. During the 18th century, Guadeloupe was a favorite base for buccaneers and pirates. On February 4, 1794, the Convention in Paris voted for the abolition of slavery and sent Victor Hughes to Guadeloupe to control the implementation. Many estate owners departed via the guillotine. On March 19, 1946, Guadeloupe was changed from a colony to a French Overseas Department.

The first topographic surveys of the interior of Guadeloupe were performed by the Royal French Army Topographic Engineers from 1763 to 1768. The first hydrographic charting expeditions to Guadeloupe were run by Ploix in 1867 and by Caspari in 1869, both of the French Navy, and they utilized the old topographic surveys to fill in details of the interior. Subsequent surveys performed in the service of the government as well as private surveys were based mainly on the work of Ploix and Caspari. Over time these dependent surveys were haphazard in planning and execution and were not unified. Different maps compiled from these various survey sources resulted in serious deficiencies in reliability. The paucity of reliable cartographic works proved to be unfavorable for the continued economic development of the island.

In 1946, Principal Hydrographic Engineer Roger Grousson led the *Mission Géodésique de la Guadeloupe* (Geodetic Mission of Guadeloupe) to establish a new coordinate system for the colony. A prismatic astrolabe was used to astronomically establish an origin point for the Guadeloupe Datum of 1947 at the pillar of station Sainte-Anne-1, where $\Phi_o = 16°\ 15'\ 17.3''$ North, $\Lambda_o = 61°\ 25'\ 33.4''$ West of Greenwich. The defining azimuth was determined from *Sainte-Anne-1* to *Petite-Terre* Lighthouse (35 km away) as: $\alpha_o = 104°\ 44'\ 32.6''$. Guadeloupe Datum of 1947 is referenced to the International 1924 ellipsoid, where $a = 6,378,288$ m and $1/f = 297$. This datum origin also defined the point of origin for the first Grid used in Guadeloupe, which was based on the Gauss-Schreiber Transverse Mercator Projection. The Guadeloupe Grid is defined at the Datum origin such that the scale factor at origin (m_o) is equal to unity, False Easting = 57,927.49 m, and False Northing = 9,616,949 m. (This equates *approximately* to a Central Meridian of −61° 30′ West and a False Easting of 50 km.)

The geodetic system used in Guadeloupe is the IGN system established in 1951–1952. This system constituted a synthesis of the surveys performed by: Grousson in 1946–1948 (First-Order triangulation), by Maillard in 1948–1948 (First-Order and Second-Order triangulation), and by

Bouchilloux in 1951–1952 (Second-Order triangulation). The first aerial photography was flown by the IGN in 1946. In all probability, the Guadeloupe Grid was the tool used for the computations of these surveys. Things started changing in the geodesy of the Caribbean during these post-war years, because the U.S. Army Map Service (AMS) founded the Inter-American Geodetic Survey (IAGS) in 1946. The geodesy of Europe was changing at this same time also, and the Universal Transverse Mercator (UTM) Grid was concocted by the Army. AMS started the computational adjustment of the European Datum of 1950 with the UTM Grid as the unifying tool. By the late 1940s, AMS was ardently pushing for the member countries of the Pan American Institute of Geography and History to adopt the UTM Grid. France was in agreement with the "universal" adoption of the UTM Grid for all of its colonies in principle, but delayed adoption of the UTM in those areas that were "in the pipeline" already with active triangulation. Therefore, only the Central African French colonies immediately adopted the UTM.

By 1957, Guadeloupe was using the UTM Grid. IGN flew new aerial photography of *St. Martin* in 1968–1969, and Hydrographic surveys were performed in 1971–1972, 1979, and 1984 in all the islands of Guadeloupe. However, "in 1979 an inventory of triangulation monuments listed in the IGN Trig List revealed that the majority of points had been disturbed by vandals. In particular, those destroyed stations included the fundamental point of the Datum origin (*Sainte-Anne-1*), and the two First-Order end-points of the baseline."

In April and May of 1984, a geodetic survey party verified the following four points: *Cape Marie Galante* I (First-Order IGN), *Sainte Anne II* (Second-Order IGN), *Le Chameau* ("the camel" – Third-Order IGN), and *Phare de la Petite Terre* (lighthouse – Second-Order IGN). As it turned out, *Sainte Anne II* was one of the reference marks set by IAGS when they observed a triangulation survey in 1951–1952 (RM). Since Bouchilloux of the IGM was there at the same time, my guess is that most of the field work performed was by IAGS and he was the IGM representative supervisor for France. The standard operating procedure for IAGS was to set several reference marks at important triangulation monuments. That policy paid off in Guadeloupe – the French Navy recovered IAGS RM1 and IAGS RM3. They decided that RM3 was to be the "reset mark" *Sainte Anne II*, where $\varphi = 16° 15' 51.804 N$,

$\lambda = 61° 23' 49.850 W$, $H = 86.96 m$. Oh, and note that *Petite Terre* lighthouse was also the original defining azimuth mark for the Datum. This emphasis on the recovery of existing trig points indicates that the original Guadeloupe Datum of 1947 was continuing to be perpetuated. By 1984, the U.S. Navy Transit doppler satellite system had been in common use for almost 6 years with the unclassified broadcast ephemeris. The NWL 10 D Datum is referenced to the WGS 72 ellipsoid, where $a = 6,378,135 m$ and $1/f = 298.26$.

The published results of the 1984 geodetic survey were listed in Guadeloupe Datum of 1947 UTM coordinates (Zone 20) and in NWL 10 D Datum coordinates. I computed the three-parameter shift values, where $\Delta X = -462.452 m$, $\Delta Y = -55.272 m$, $\Delta Z = -289.216 m$. An example test point computation can be used to verify these shift parameters where for *Sainte Anne II* (IAGS RM2), the NWL 10 D Datum coordinates are $\varphi = 16° 15' 42.730 N$, $\lambda = 61° 24' 04.700 W$. I would consider the accuracy of this transformation to be reliable to within ±10 m on *Grand-Terre* and *Marie-Galante* Islands and within ±20 m on the *Îles des Saintes*.

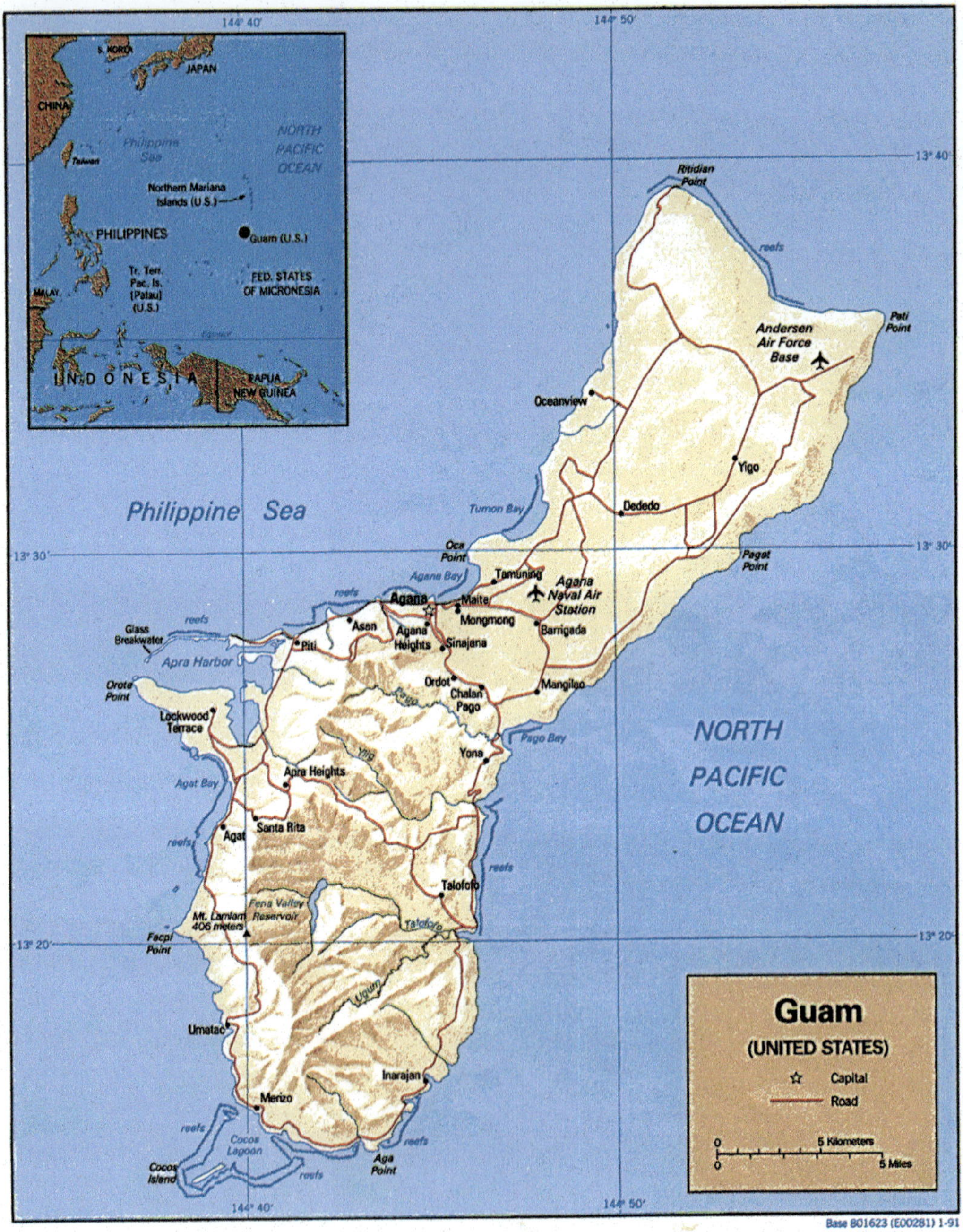

Base 801623 (E00281) 1-91

Territory of Guam

The island is about three times the size of Washington, DC, the lowest point is the North Pacific Ocean (0 m), and the highest point is Mount Lamlam (406 m). Guam is the largest and southernmost island in the Mariana Islands archipelago (*World Factbook*, 2009).

The original inhabitants of Guam are believed to have been of Indo-Malayan descent originating from Southeast Asia as early as 2000 B.C., and having linguistic and cultural similarities to Malaysia, Indonesia, and the Philippines. The Chamorro flourished as an advanced fishing, horticultural, and hunting society. They were expert seamen and skilled craftsmen familiar with intricate weaving and detailed pottery making who built unique houses and canoes suited to this region of the world. Ceded to the United States by Spain in 1898, Guam is located about three-quarters of the way from Hawaii to the Philippines. "Guam became a hub for whalers and traders in the western Pacific in the early 1800s. During the 1898 Spanish-American War, the US Navy occupied Guam and set up a military administration. The US Navy opposed local control of government despite repeated petitions by Chamorro. Japan invaded Guam in 1941 and instituted a repressive regime. During the US recapture of Guam in 1944, the island's two largest villages were destroyed. After World War II, political pressure from local Chamorro leaders led to Guam being established as an unincorporated organized territory in 1950 with US citizenship granted to all Chamorro. In a referendum in 1982, more than 75% of voters chose closer relations with the US over independence, although no change in status was made because of disagreements on the future right of Chamorro self-determination. The US military holds about 29% of Guam's land and stations several thousand troops on the island. The installations are some of the most strategically important U.S. bases in the Pacific; they also constitute the island's most important source of income and economic stability" (*World Factbook*, 2022).

About 40 years ago, a friend of mine, Mr. Don Eames now retired from the Survey Section of the U.S. Army Corps of Engineers, New Orleans District informed me that the Guamanian phrase for "hello, how are you" is pronounced, "hafa tata manoa." We greeted each other with that phrase for decades.

In 1904, the U.S. Coast and Geodetic Survey (USC&GS) performed telegraphic longitude determinations between Guam and both Manila and the Island of Midway (*USC&GS Annual Report 1904, Appendix Telegraphic Longitudes, pp. 257–312*). The first geodetic survey of the island was by Butler of the U.S. Army Corps of Engineers from 1911 to 1913 in which station "Togcha (USE or U.S. Corps of Engineers)" (formerly "Lee No. 7") is where: $\Phi_o = 13° 22' 38.490'' $ N, $\Lambda_o = 144° 45' 51.560''$ East of Greenwich, and the ellipsoid of reference is the Clarke 1866, where $a = 6,378,206.4$ m, $b = 6,356,583.8$ m, or $1/f = 294.9786982$. Additional triangulation network surveys were performed in 1945 and in 1949.

Upon the official request of the Territory of Guam, a completely new triangulation scheme was planned by the USC&GS, and in 1963 the observations for the survey were completed. The 1963 work superseded all previous surveys, although many of the old, marked stations were incorporated in the new scheme. The main scheme was accomplished under the specifications of first-order class II, whereas the subsidiary work was accomplished under specifications for second-order classes I and II. The datum for the work officially designated as the Guam 1963 Datum was essentially astronomic. The geographic position of station "Togcha (USE)," formerly "Lee No. 7," as determined by the Butler Survey was used as the initial geographic position on which all other positions were based. The triangulation scheme was orientated by astronomic azimuths taken over the line "Togcha (USE)" – "Macajna (GG or Guam Geodetic)" in both directions. The mean of these two astronomic determinations, after accounting for the $\Delta\alpha$, was adopted as the azimuth of this line. *(The "Δα" commonly refers to the convergence angle as computed on a Transverse Mercator projection – Ed.)* Length control for the main scheme consisted of 12 lines, measured with the Tellurometer *(A microwave electronic distance meter – Ed.)*, distributed as follows: four lines in the northeast region of the island; five lines in the central area; and three lines in the southern mountainous region. All geographic positions were converted to plane rectangular coordinates based on a Modified Azimuthal Equidistant projection centered on station "Agana Monument 1945," where $\varphi_o=13°\ 28'\ 20.87887''$ N, $\lambda_o=144°\ 44'\ 55.50254''$ E, and False Easting = False Northing = 50 km. The ellipsoid of reference remained the Clarke 1866 *(State Plane Coordinates by Automatic Data Processing, Charles N. Claire, USC&GS Pub. No. 62-4, 1968, pp. 35–39)*. There were 29 main scheme stations; 87 supplemental stations; 30 intersection stations; and 13 traverse stations, totaling 159 stations.

By 1993, the existing Guam 1963 Datum no longer met the needs of Guam in spatially locating the cadastre or supporting Guam's Land Information System. Much of the network had been destroyed over time and there had been no extension of the network into new areas of development since 1963. Consequently, the quality of cadastral surveys had been seriously affected with resultant gaps and overlaps. Aware of the recent developments in geodetic surveying and the Government of Guam Department of Land Management's (DLM) lack of experience, the South Australian Department of Environment & Natural Resources was contacted to provide the technical and administrative support for the DLM to establish the 1993 Guam Geodetic Network (1993 GGN). A total of 2,629 survey marks comprises the 1993 GGN, of which 2,203 new marks have been established. The network comprises 28 primary stations; 216 secondary stations; and 2,385 tertiary stations.

Guam Public Law 23–31 established the North American Datum of 1983 (NAD83) as the coordinate reference system for Guam, replacing the Guam 1963 Datum. The reference ellipsoid is the GRS 1980, where $a=6,378,137$ m and $1/f=298.257\ 222\ 101\ ...$ The law also established the Guam Map Grid, replacing the 1963 map grid. The defining parameters of the Transverse Mercator projection are where: Central Meridian, $\lambda_o=144°\ 45'$ E, Latitude of False Origin, $\varphi_{FN}=13°\ 30'$ N, False Easting = 100 km, False Northing = 200 km, and the scale factor at origin is equal to unity. It is reported that the maximum scale factor on the island is 1.000006 and that occurs at Point Pati on the northeast coast near Anderson AFB. The transformation parameters **from** Guam 1963 Datum **to** NAD83 are: $\Delta X=+201.686$ m, $\Delta Y=+66.350$ m, $\Delta Z=+457.288$ m, scale $=+2.327\times10^{-6}$, $R_X=+2.925''$, $R_Y=+6.188''$, $R_Z=-14.106''$. "Note that rotations are positive anti-clockwise about the axes of 1963 GGTN coordinate system when viewing the origin from the positive axes" (*The 1993 Guam Geodetic Network, Andrew Dyson, South Australian Dept. of Environmental and Natural Resources, November 1995*).

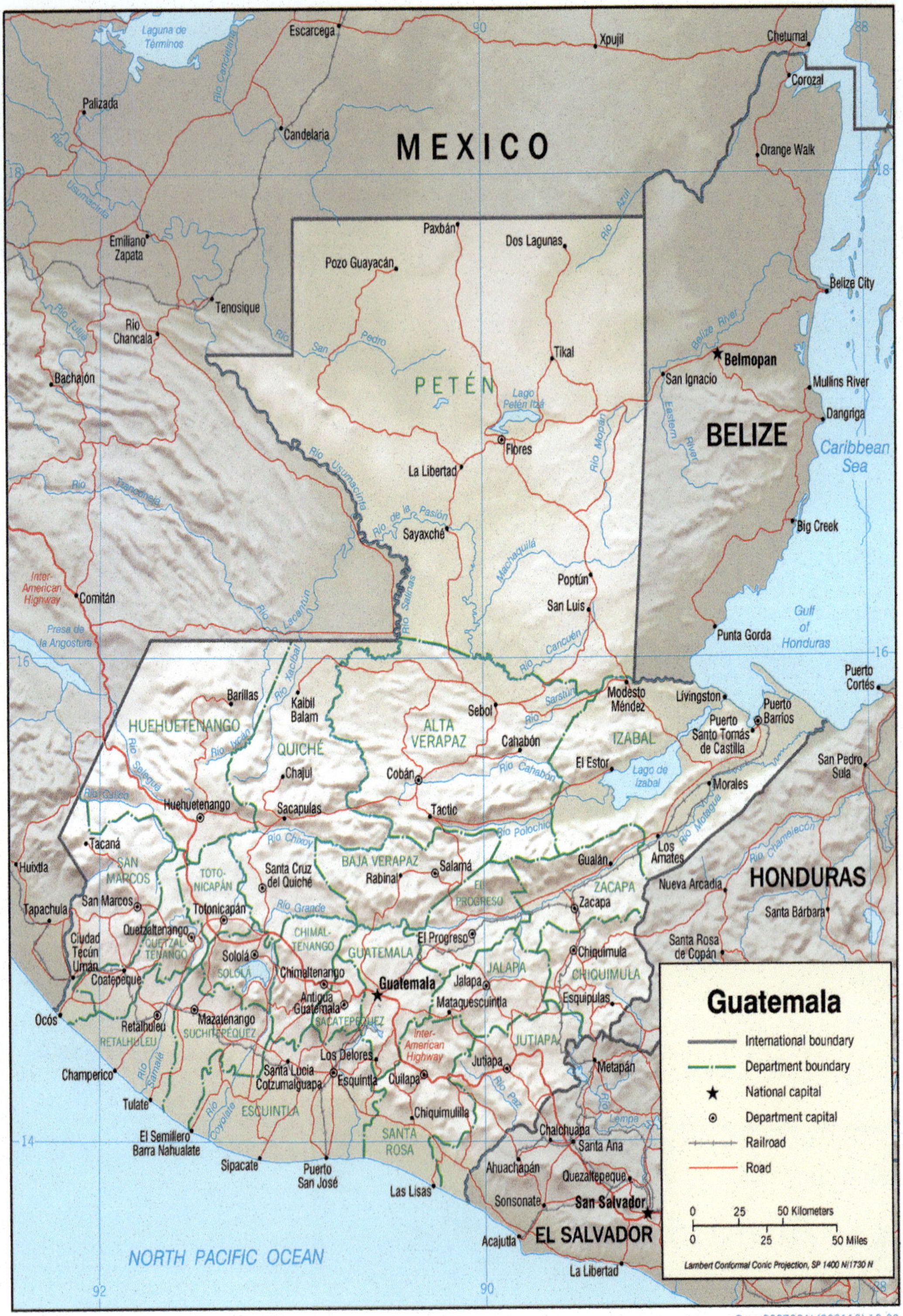

Base 802723AI (C00113) 12-00

Republic of Guatemala

Guatemala is bordered by Belize (266 km), El Salvador (203 km), Honduras (256 km), Mexico (962 km), the North Pacific Ocean and the Gulf of Honduras – Caribbean Sea/Atlantic Ocean totaling 400 km. The lowest point is the Pacific Ocean (0 m), and the highest point is Volcán Tajumulco (4,211 m). My youngest daughter, Alyce Marie was born in Guatemala.

Guatemalan history dates to proof of human settlers as far back as 10,000 B.C., with possible human activity as far back as 18,000 B.C. Predominantly hunters and gatherers, these original inhabitants only began to establish permanent settlements from 2000 B.C. to 250 A.D. From 250 A.D. to 900 A.D., the first nation states and cities were formed with the large-scale architecture that characterized the ceremonial ancient Maya. From 900 to 1500 A.D., the important early Maya cities flourished during this time, including *El Mirador*. It was not only the most populated pre-Colombian city in the Americas, but also the region's first politically organized state. During that latter period, *Teotihuacan* found near present-day Mexico City replaced *El Mirador* as the prime Mesoamerican power. The greatest Maya cities from this era were *Tikal*, *Palenque* (Mexico) and *Copán* (Honduras). By then, the Maya had already established the most advanced form of writing in the New World, as well as they developed their mathematic and artistic genius. By 1500 A.D., the ancient Mayan civilizations begin their mysterious fall and as the Aztecs begin their domination of Mesoamerica, the coming of the Spanish conquistadors changed everything (*Destination360*, 2008).

"When Pedro de Alvarado came to conquer Guatemala for the king of Spain in 1523, he found the faded remnants of the Mayan civilization and an assortment of warring tribes. The remaining highland kingdoms of the *Quiché* and *Cakchiquel* Maya were soon crushed by Alvarado's armies. Their lands were carved up into large estates and their people were ruthlessly exploited by the new landowners. The subsequent arrivals of Dominican, Franciscan, and Augustinian friars could not halt this exploitation, and their religious imperialism caused valuable traces of Mayan culture to be destroyed. Independence from Spain came in 1821, bringing new prosperity to those of Spanish blood (creoles) and even worse conditions for those of Mayan descent. The Spanish Crown's few liberal safeguards were abandoned. Huge tracts of Mayan land were stolen to cultivate tobacco and sugar cane, and the Maya were further enslaved to work that land. Since independence, the country's politics have been colored by continued rivalry between the forces of the left and right – neither of which has ever made it a priority to improve the position of the Maya" (*Lonely Planet*, 2008).

The country was originally mapped in 1527 by Fernando Colón, the son of Christopher Columbus, and was known as "*Mapa Español Oficial*" (Official Spanish Map). The name Guatemala first appeared in a 1529 map by Rivero. The country was mapped in outline form by 1600, and this cartographic convention continued into the 20th century. By 1930, planimetric sketch maps were virtually the only maps available of Guatemala with the exception of a 1923 map prepared by Ing. Claudio Urrutia who employed various scientific instruments that were not identified. In 1934, the Geographic Branch of the U.S. Army issued a 1:250,000 scale map of the country. The Guatemala-Honduras boundary extends between the Caribbean Sea and the tripoint ($\varphi = 14°\ 25'\ 20.06''$ N, $\lambda = 89°\ 21'\ 28.46''$ W), with El Salvador on *Cerro Monte Cristo* (mountain). That tripoint in Honduras is in the *Departamento de Ocotepeque* (state or province). Boundary disputes between Honduras and Guatemala began shortly after the dissolution of the Federation of Central America in 1843. In accordance with the terms of the Treaty of Arbitration that was signed in Washington in 1930, the disputed line was submitted to the Chief Justice of the United States for a decision on the delimitation of the boundary. Because available topographic data were inadequate for the boundary work in many of the disputed areas, the Tribunal directed the making of an aerial photogrammetric survey.

Chief Justice Hughes appointed Sidney H. Birdseye of the U.S. Coast & Geodetic Survey (USC&GS) as Chief of the boundary demarcation commission after Mr. Birdseye completed the photogrammetric mapping of the area. Birdseye's commission erected 1,028 pillars and completed its work between 1933 and 1936. The award of the Special Boundary Tribunal was based on the principle of *uti possidetis* as of 1821. Thus, the award referred to the territory under the administrative control of Guatemala and Honduras at the time of their independence from Spain. Furthermore, El Salvador, Guatemala, and Honduras signed a protocol on March 26, 1936, accepting *Cerro Monte Cristo* as the tripoint of the boundaries of the three states. A fascinating aspect of the Guatemala-Honduras boundary is that part of the boundary is "established on the right banks of the Tinto and Motagua rivers at mean high water mark, and in the event of changes in these streams in the course of time, whether due to accretion, erosion or avulsion, the boundary shall follow the mean high-water mark upon the **actual right banks of both rivers**" (emphasis added – Ed.). Boundaries commonly *do not* change with avulsions. Apparently, this was intended to avoid future squabbles. On the other hand, the April 9, 1938, Guatemala-El Salvador boundary treaty contained Article II in which: "No change in the bed of frontier rivers, whether due to natural causes such as alluvium deposits, landslides, freshets, *etc*. or to artificial causes such as the construction of public works, the deepening of channels for water-supply, *etc*., shall affect the frontier as determined at the time of demarcation, which shall continue to be the international boundary even though a stream may have completely abandoned its original bed." This boundary commission completed the construction of 530 monuments and pillars in 1940.

The Ocotepeque Datum of 1935 was established at Base Norte, where $\varphi_o = 14° 26' 20.168''$ North, $\lambda_o = 89° 11' 33.964''$ West of Greenwich, and $H_o = 806.99$ m above mean sea level. The defining geodetic azimuth to Base Sur is: $\alpha_o = 358° 54' 21.790''$ (*Memoria de la Dirección General de Cartografía, Guatemala, Sept. 1957*), and the ellipsoid of reference is the Clarke 1866, where $a = 6,378,206.4$ m and $1/f = 294.9786982$. The corresponding astronomic observations at that mountainous location are: $\Phi_o = 14° 26' 13.73''$ North (±0.07″), $\Lambda_o = 89° 11' 39.67''$ West (±0.045″), and the defining astronomic azimuth to Base Sur is: $\alpha_o = 358° 54' 20.37''$ (±0.28″) (*Informe Detallado de la Comisión Técnica de Demarcación de la Frontera entre Guatemala y Honduras*). The difference between these two sets of coordinates is due to the local gravimetric deflection of the vertical. There was no grid system associated with this Datum, although that's not surprising since Mr. Birdseye was with the USC&GS. Their custom was to compute their chains of quadrilaterals in geodetic coordinates.

Various surveying and mapping commissions were created in Guatemala for the boundary surveys of the 1930s. In 1945, the *Departamento de Mapas y Cartografía* (Dept. of Maps and Cartography) was formed and then in 1964 was renamed the *Instituto Geográfico Nacional de Guatemala* (National Geographic Institute of Guatemala). On November 7, 1946, the Inter-American Geodetic Survey (IAGS) of the U.S. Army Map Service established the *Guahonels Project* which included Guatemala, Honduras and El Salvador. In April of 1950, the IAGS reorganized the project in Guatemala. The IAGS established two Lambert Conformal Conic zones for the Republic of Guatemala: Guatemala North Lambert Zone – both zones use the same Central Meridian $(\lambda_o) = 90° 20' 00''$ West of Greenwich and False Easting of 500 km. Zone Norte has a Latitude of Origin $(\varphi_o) = 16° 49'$ North, the False Northing = 292,209.579 m, and the scale factor at origin $(m_o) = 0.99992226$. Zone Sud has a Latitude of Origin $(\varphi_o) = 14° 54'$ North, the False Northing = 325,992.681 m, and the scale factor at origin $(m_o) = 0.99989906$. For those facing the conundrum of the British definition of secant Lambert Conical Orthomorphic zones, I am told that the latest (*free*) version of NGA's GeoTrans software will indeed accommodate such defining parameters even though many high-priced GIS packages will not. (*Hint, hint – Ed.*)

After Hurricane Mitch devastated a large area of Guatemala, the U.S. National Geodetic Survey established three Continuously Operating Reference Stations in the country to assist in disaster recovery. Those stations have been helpful in establishing the SIRGAS reference system for Latin America. No transformation parameters have been published from either the Ocotepeque Datum of 1935 or the North American Datum of 1927 to the WGS84 Datum.

Republic of Guatemala

I have been informed that Guatemala may have adopted the Transverse Mercator projection – no idea on the number of zones. No information is available from the Guatemalan government, but since SwedeSurvey has recently provided assistance for a cadastral mapping program and since the Swedes now have a fondness for TM systems, ... I suppose they might have something to do with that possible adoption. The old 1:50,000 scale series of Guatemalan topographic maps are based on the two Lambert zones with the UTM Grid overprint. Someday the government may release such transformation parameter information to the general public and join the rest of the free world! Until then, I suppose Guatemala will remain a closed society, at least from a cartographic/geodetic point of view.

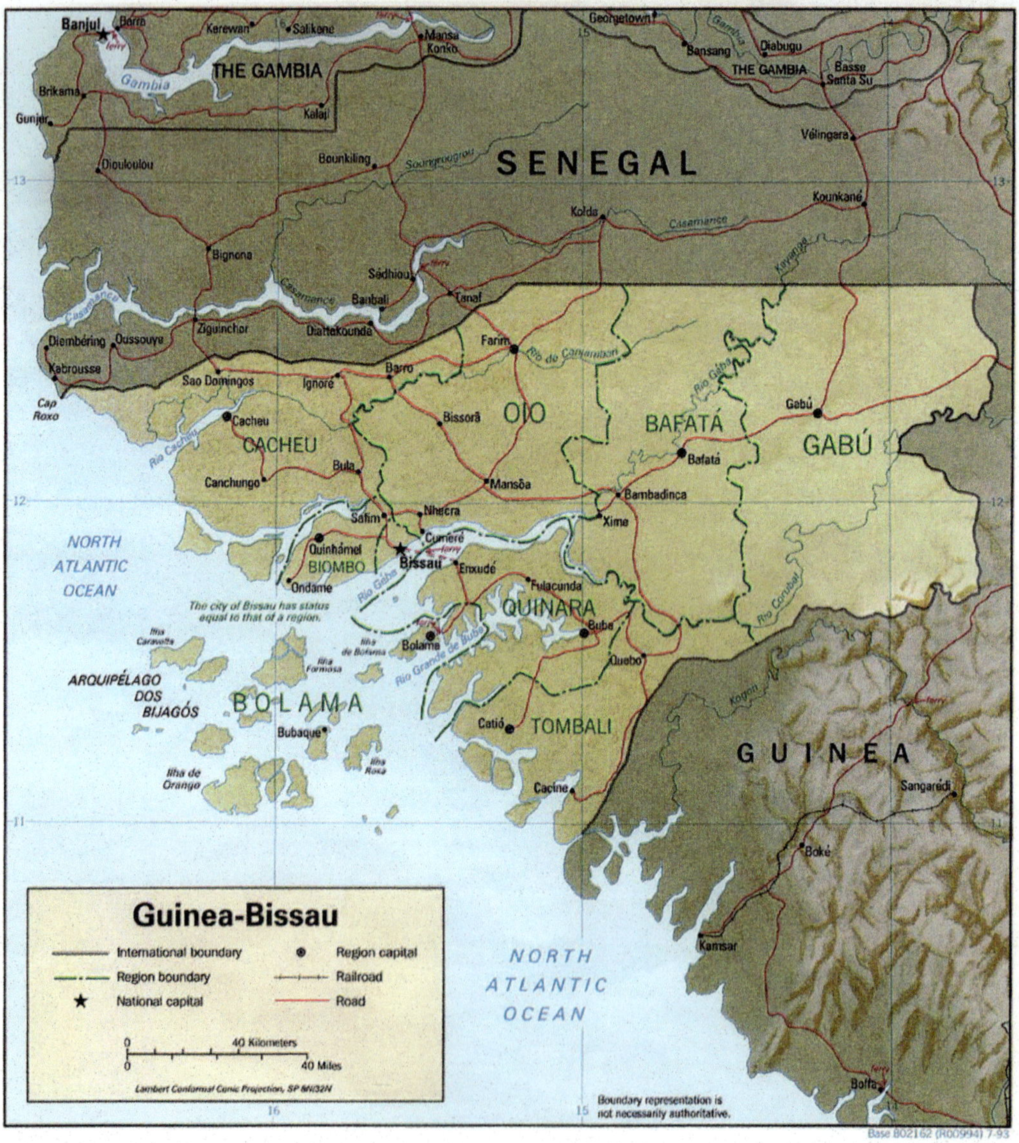
Guinea-Bissau
International boundary
Region boundary
National capital
Region capital
Railroad
Road
0 40 Kilometers
0 40 Miles
Lambert Conformal Conic Projection, SP 8N/32N
Boundary representation is not necessarily authoritative.
SENEGAL
THE GAMBIA
GUINEA
NORTH ATLANTIC OCEAN
CACHEU
OIO
BAFATÁ
GABÚ
BIOMBO
QUINARA
BOLAMA
TOMBALI
ARQUIPÉLAGO DOS BIJAGÓS
The city of Bissau has status equal to that of a region.
Banjul
Barra
Kerewan
Salikene
Mansa Konko
Brikama
Gunjur
Kalaji
Georgetown
Sansang
Diabugu
Basse Santa Su
Vélingara
Diouloulou
Bounkiling
Soungrougrou
Kolda
Kounkané
Casamance
Bignona
Sédhiou
Tanaf
Bambali
Ziguinchor
Diattakounda
Diembéring
Oussouye
Kabrousse
Cap Roxo
Sao Domingos
Ignoré
Barro
Farim
Rio de Canjambari
Rio Geba
Gabú
Cacheu
Bissorã
Rio Cacheu
Bafatá
Bula
Canchungo
Mansôa
Bambadinca
Nhacra
Safim
Xime
Quinhámel
Cumeré
Bissau
Enxudé
Ondame
Rio Geba
Fulacunda
Rio Corubal
Buba
Bolama
Ilha de Bolama
Ilha Caravela
Ilha Formosa
Rio Grande de Buba
Quebo
Bubaque
Catió
Ilha de Orango
Ilha Roxa
Cacine
Kogon
Sangarédi
Boké
Kamsar
Boffa
ferry
Base 802162 (R00994) 7-93

Republic of Guinea

The seacoast is marshy and is about 274 km long, the interior rises to hilly and plateau regions. The highest point is Mount Nimba (1,752 m), near the tripoint with *Cote d'Ivoire* and Liberia. Going clockwise from the North Atlantic Ocean to the west, Guinea shares borders with Guinea-Bissau (421 km), Senegal (363 km), Mali (1,62 km), *Cote d'Ivoire* (816 km), Liberia (590 km), and Sierra Leone (794 km).

The original inhabitants of Guinea were forced out of the area around 900 A.D., and numerous kingdoms were subsequently established. By the mid-1400s, the Portuguese visited the area, and a slave trade was established. The area was under active trade with the British, French, and Portuguese in the 17th century; and the coastal region was declared a French Protectorate in 1849. Administered at various times by Senegal and the *Rivieres du Sud*, the territory of French Guinea was made a colony in 1893. The Federation of French West Africa, which included Guinea as a member, was established in 1895. Its status was changed to that of an overseas territory in 1946, and on October 2, 1958, Guinea became the first state of former French West Africa to become independent. Guinea includes the Los Islands, an island group west of the capitol city of Conakry. "On 5 September 2021, COL. Mamady Doumbouya led a military coup by a National Committee of Rally and Development that arrested President Conde, suspended the constitution, and dissolved the government. Doumbouya is declared President on 17 September 2021. A charter of transition, issued in late September, made Doumbouya transitional president for an undefined period, and on 1 October 2021, he was formally sworn in" (*World Factbook* 2022).

The Guinean Maritime boundary is defined in large part by a single, unique (in the world) Straight Baseline. By Decree of the President of the Republic in 1964; the limits of the territorial waters are fixed "to the north, by parallel of altitude 10° 56′ 42.55″ North, and to the south, by parallel of altitude 9° 03′ 18″ North, along a distance of 130 sea miles seaward, reckoning from a straight line passing by the south-west of Sene Island of the Tristao group, and to the south, by the south-west foreland (cape) of Tamara Island, at low tide." The boundary between Guinea and Guinea-Bissau was established through a Franco–Portuguese convention of May 12, 1886. The demarcation of the 384 km boundary with straight lines between 58 markers and along thalwegs of rivers was completed and approved by 1906. In 1915 an *arrêté* (decision) by the Governor General of French West Africa promulgated a French decree establishing a 328 km boundary between French Guinea and Senegal. Early in 1934 an *arrêté* promulgated a decree of the previous December changing the French Guinea – Senegal boundary in the sector between the head of the Tanague River and the junction of the Bitari and Koïla Kabé. A Franco–Liberian convention of December 8, 1892, delimited a boundary between the possessions of France (*Cote d'Ivoire* and French Guinea) and Liberia inland from the mouth of the Cavalla River to the tripoint with Sierra Leone. That 560 km boundary with French Guinea was redrawn on September 18, 1907. Further surveys and commissions settled the matter with several *procès verbaux* (verbal proceedings) finally in 1926. The Guinea–Sierra Leone boundary has a length of approximately 648 km. Established by Anglo–French Convention of June 28, 1882, a boundary was delimited from the Atlantic Ocean inland along the drainage divide of the Great Scarcies and Mélikhouré (Rivers) to an indefinite point in the interior. Later determined by field surveys, the last agreement fixing the boundary was signed on September 4, 1913. In places, the boundary measurements are described to the closest half-meter. Reading between the lines, I'd guess that the boundary commission surveyors had people literally looking over their shoulders during that process!

When the federation of the eight territories constituting French West Africa came into being in 1904, the *Annexe de l'Institut Géographique National á Dakar* had the local responsibility for topographic mapping. Also known as the *Service Géographique de l'Afrique Occidentale Française – SGAOF* (Geographic Service of the French West Africa), topographic mapping of Guinea has been largely at the scales of 1:200,000 and 1:500,000. This agency has performed a small amount of mapping at the scales of 1:20,000, 1:50,000, and 1:100,000. Topographic mapping of Guinea was in the past largely the result of rapid ground surveys. After World War II, the French adopted aerial photography controlled by astronomical points ("Astro" stations), as the means of surveying and mapping at 1:50,000 and 1:200,000. These compilations were also used for derivative mapping at smaller scales. There is complete coverage of the country at 1:500,000 scale, and at 1:200,000 scale. The latter consists of sheets mainly based on ground surveys. A small portion of Guinea has 1:50,000 topographic sheets compiled, mostly by the French IGN in the coastal west, and by a cooperative agreement with the Japanese (JICA) for some sheets around Kankan and Kérouané-Macenta.

The oldest coordinate system in Guinea that I have been able to locate (with help) is the Conakry Datum of 1905, where $\Phi_o = 9° \ 30' \ 58.997 N$, $\Lambda_o = 13° \ 42' \ 47.483''$ West of Greenwich, $\xi_o = -4.50''$, $\eta_o = -0.02''$, and the ellipsoid of reference is the Clarke 1880 (IGN), where $a = 6,378,249.2$ m and $1/f = 293.4660208$. Thanks go to both Mr. John W. Hager, retired from NGA and to Mr. Russell Fox of the Ordnance Survey of the United Kingdom. The origin point is at the Public Works Building in Conakry, and John W. Hager went on to say: "Reingold cites *Les Manuels Coloneaux*, '*Cartographie Coloniale*,' Paris, 1935 and '*Catalogue de Positions Géographiques*,' Paris, 1923 give the position to the nearest second. *Annales Hydrographiques*, 4e Série, Tome 1, Année 1950, Paris 1951, p. 155 gives the above but is listed as 3rd order. A position for the Railway Astro Pillar is given as latitude = 9° 30' 54.5" N and longitude = 13° 42' 47.1" W, a difference in position of 138.6 meters. I would assume that the astro pillar was not permanently marked."

Some minor hydrographic surveys were performed by the French Navy, and these were based on local Astro stations that served as origins of Grids computed on the Hatt Azimuthal Equidistant projection. The Tabola River survey had its origin at Cabrion Base North End (1936), where $\Phi_o = 9° \ 56' \ 08.1'' N$, $\Lambda_o = 13° \ 54' \ 42.4''$ West of Greenwich. The defining azimuth to Cabrion Base South End was $\alpha_o = 123° \ 34' \ 00''$, and the baseline length was 1,017.537 m. I was wondering why the French performed a survey in such a tiny locale that did not even appear on the standard CIA map of the country. I examined my *Carte Générale of Guinée* and noticed that there is a road to there through the town of Koba that winds north up into the hills. Apparently, something valuable was being trucked out of those hills to the port of Taboriya. Hager found one at "Binari 1949 (code BIN) at the I.G.N. Astro, latitude = 10° 30' 26.2" N, longitude = 14° 38' 45.0" W (1) or ... 41.03" (2) or... 40.0" (3), Clarke 1880. Position (1) is from *Annales Hydrographiques*, 4e série, Tome Sixième, Année 1955, p. 247. Position (2) is from *Annales Hydrographiques*, 4e série, Tome Dixième, Années 1959–1960, p. 65, Paris 1961. Position (3) is a footnote to (2) and refers to the 1954 survey by M. Sauzey."

There have been some other rather curious coordinate systems devised for Guinea during the 20th century. Prior to and during World War II, there were a number of military Grids that were collectively termed the "British Grids." These were all documented and computed into projection tables by the U.S. Army. One published by the U.S. Army Corps of Engineers Lake Survey in 1943 was the Guinea Zone based on the Lambert Conical Orthomorphic Projection Tables. The defining parameters are Latitude of Origin, $(\varphi_o) = 7°$ N, Central Meridian $(\lambda_o) = 0°$ (Greenwich), Scale Factor at the Parallel of Origin $(m_o) = 0.99932$, False Northing, FN = 500 km, and False Easting, FE = 1,800 km. The wording for the projection is characteristically British, as is the method of presenting the defining parameters, and the projection is definitely the fully conformal version rather than the French Army version of the time. Furthermore, the parameters given for the Clarke 1880 ellipsoid were the British version, where $a = 6,378,249.145$ m and $1/f = 293.465$. This Grid continued in use by the U.S. Army Map Service for a couple decades after World War II.

Immediately after World War II, the French *Institut Géographique National* devised a number of Grids for *l'A.O.F.* usage as of December 12, 1945. The region of French Guinea was to be covered by two *fuseau*, or zones: "*Fuseau Sénégal*" with $\lambda_o = 13° \ 30'$ West of Greenwich, and "*Fuseau Cote d'Ivoire*" with $\lambda_o = 6° \ 30'$ West of Greenwich. The scale factor at origin, $(m_o) = 0.999$, and the ellipsoid of reference was to be the International (Hayford 1909), where $a = 6,378,388$ m and $1/f = 297$. Since there was not a great deal of existing mapping in French West Africa at the time, most Datums were established by Astro shots and few classical chains of quadrilaterals that had been surveyed. The introduction of a new ellipsoid was therefore not of major geodetic importance to existing cartographic work. French Navy Hydrographic surveys of the late 1940s in Guinea were cast on the *Fuseau Sénégal* Grid. When I was in college, I once read a science fiction novel about a disgruntled cartographer on a lonely expedition to a new planet. He chose *risqué* names for his gazetteer, and that fact went undiscovered for many years. While perusing the report of the French Navy hydrographic survey of the mouth of the Saloum River (*Mission Hydrographique de la Côte Ouest d'Afrique, 11 Mai 1950 – 18 Mai 1952*), guess what I found? Yep, an American vulgarism and an American gangster's name for triangulation stations!

The U.S. Army Map Service concocted the Universal Transverse Mercator (UTM) Grid System for worldwide use in 1948. France had been trying to gain an international consensus for some sort of similar system, and quickly adopted the UTM for most of its colonies. As of September 30, 1950, all new surveying and mapping of French Guinea was done on the UTM Grid. That situation remains to this day. The only information available on a Datum shift from local to WGS84 for the entire country of Guinea is the entry in NIMA's TR8358.2 for "Dabola Datum," where $\Delta a = -112,145$, $\Delta f \times 10^4 = -0.54750714$, $\Delta X = -83$ m ± 15 m, $\Delta Y = +37$ m ± 15 m, and $\Delta Z = +124$ m ± 15 m. This four-point solution is published by NIMA as of 1991. Since there is only a 1:200,000 scale map published of Dabola, and there are no 1:50,000 scale topographic maps nearby, I am unable to find a plausible reason for the choice of this transformation name or location other than it is more or less in the center of the country. I have found no other evidence of such a Datum in existence.

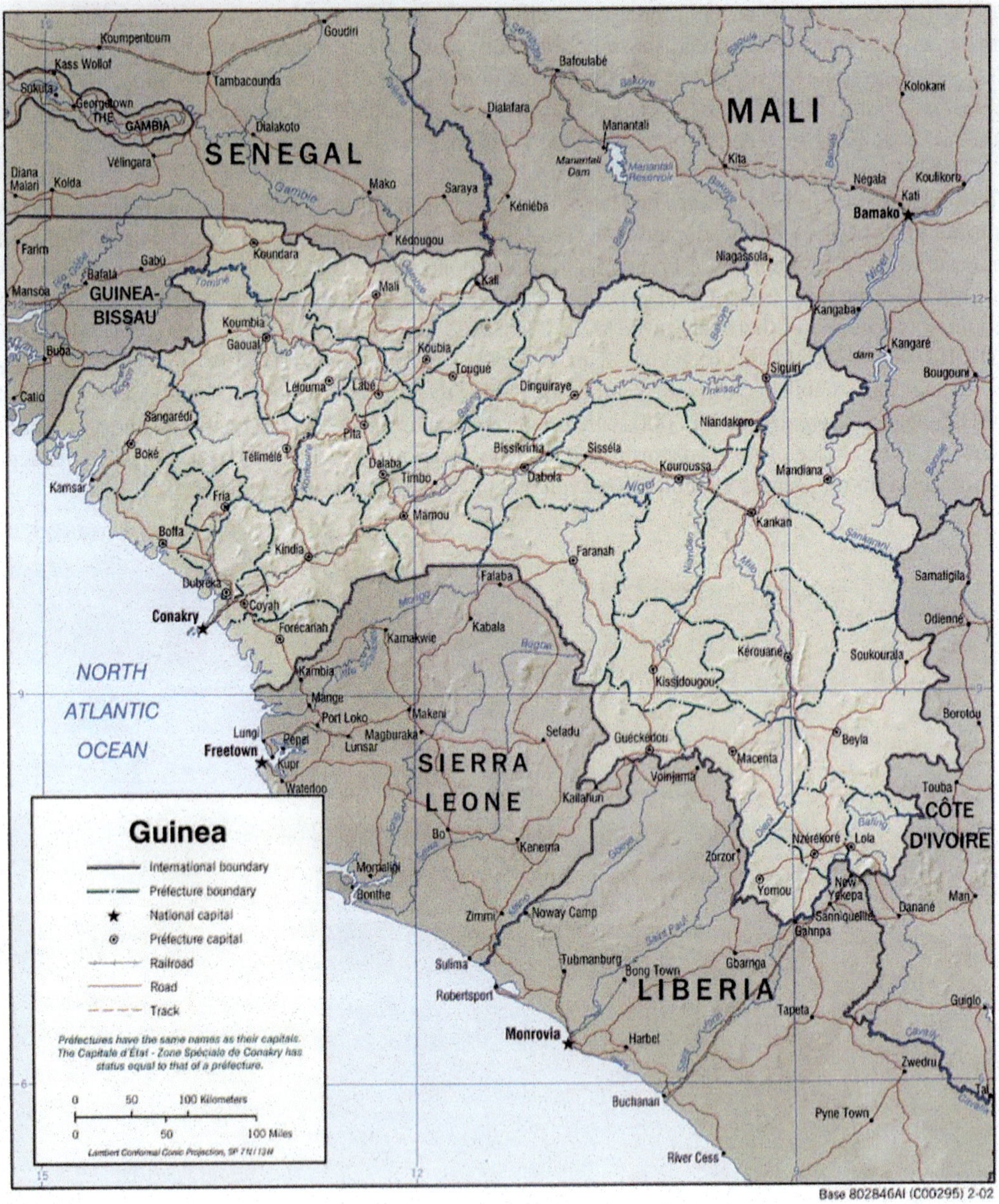

Base 802846AI (C00295) 2-02

Republic of Guinea-Bissau

"Bordered by Guinea, (386 km), Senegal (338 km), and the Atlantic Ocean (350 km), the lowest point is the Atlantic Ocean (0 m), and the highest point is an unnamed elevation in the eastern part of the country (300 m). Slightly less than three times the size of Connecticut, the terrain is mostly low coastal plain rising to savanna in the east" (*World Factbook*, 2010).

"More than 1,000 years ago the coast of Guinea-Bissau was occupied by agriculturists using iron implements. They grew irrigated and dry rice and were also the major suppliers of marine salt to the western Sudan. At about the same time, the area came under the influence of the Mali Empire and became a tributary kingdom known as Kaabu. After 1546 Kaabu was virtually autonomous; vestiges of it lasted until 1867" (*Britannica Concise Encyclopedia*, 2010).

"The rivers of Guinea and the islands of Cape Verde were among the first areas in Africa explored by the Portuguese in the 15th century. Portugal claimed Portuguese Guinea in 1446, but few trading posts were established before 1600. In 1630, a 'captaincy-general' of Portuguese Guinea was established to administer the territory. With the cooperation of some local tribes, the Portuguese entered the slave trade and exported large numbers of Africans to the Western Hemisphere via the Cape Verde Islands. Cacheu became one of the major slave centers, and a small fort still stands in the town. The slave trade declined in the 19th century, and Bissau, originally founded as a military and slave-trading center in 1765, grew to become the major commercial center. Portuguese conquest and consolidation of the interior did not begin until the latter half of the 19th century. Portugal lost part of Guinea to French West Africa, including the center of earlier Portuguese commercial interest, the Casamance River region. A dispute with Great Britain over the island of Bolama was settled in Portugal's favor with the involvement of U.S. President Ulysses S. Grant. Before World War I, Portuguese forces, with some assistance from the Muslim population, subdued animist tribes and eventually established the territory's borders. The interior of Portuguese Guinea was brought under control after more than 30 years of fighting; final subjugation of the Bijagos Islands did not occur until 1936. The administrative capital was moved from Bolama to Bissau in 1941, and in 1952, by constitutional amendment, the colony of Portuguese Guinea became an overseas province of Portugal" (*U.S. Department of State Background Notes,* 2010). "Since independence from Portugal in 1974, Guinea-Bissau has experienced considerable political and military upheaval. In 2014, Jose Mario Vaz won a free and fair election. In June 2019, Vaz became the first president in Guinea-Bissau's history to complete a full presidential term. After winning the 2019 presidential elections, Umaro Sissoco Embalo was sworn in as president" (*World Factbook*, 2022).

In 1912, a geodetic/hydrographic mission (*Missão Geo-Hidrográfica*) was undertaken. Geodetic survey work was also undertaken between 1948 and 1955 as control for a 1:50,000 scale topographic map series, now complete. The results were also published as Rio Geba Hydrographic Charting 1956 (*Engenharia Geográfica nos Séculos XIX e XX, João Matos*). The coordinate system used was the UTM on the Hayford (International 1909) ellipsoid, where $a = 6,378,388$ m and $1/f = 297$. The local datum is based on Bissau Base Northwest Pillar. Thanks to John W. Hager, "*Extrême NW de la base géodésique de Bissau* where $\Phi_o = 11°\ 51'\ 27.60''$ N±0.08″, $\Lambda_o = 15°\ 37'\ 06.77''$ W±0.10″, $\alpha_o = 304°\ 11'\ 16.0''$ ±0.34″ to SE Base End **from south**, International ellipsoid. Reference is: *Rapport sur Les Travaux Géodésiques Exécutés Dans Les Provinces Portugaises D'Outre-Mer, Junta Das Missões Geográficas E De Investigações Do Ultramar, Toronto, September 1957.*"

TR 8350.2 lists the transformation from Bissau Datum to WGS 84 as: $\Delta X = -173$ m, $\Delta Y = +253$ m, $\Delta Z = +27$ m, ±25 m in each of the three components, and is based on collocation at two points as of 1991.

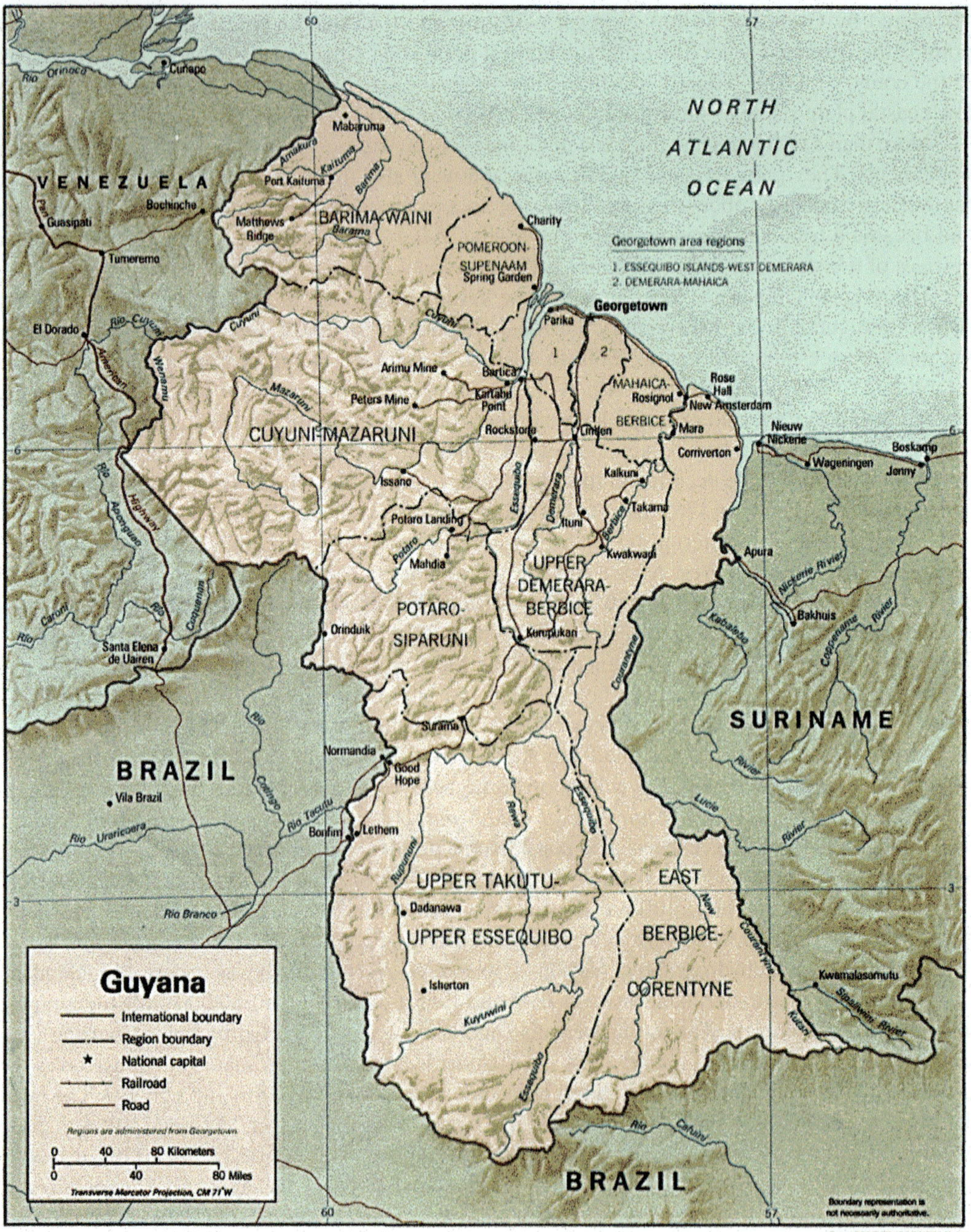

Base 801767 (R00008) 7-91

Co-operative Republic of Guyana

Slightly smaller than Idaho, Guyana borders Brazil (1,119 km), Suriname (600 km), and Venezuela (743 km). Its coastline on the Atlantic Ocean is 459 km long; it claims a territorial sea of 12 nautical miles, and it claims mineral rights to 200 nautical miles or to the outer edge of the continental margin. Guyana terrain is mostly rolling highlands with a low, swampy coastal plain and a savanna in the south. The lowest point is the Atlantic Ocean, and the highest point is Mount Roraima (2,835 m) in the Pakaraima Range along the Venezuela–Brazil border.

The region known as Guiana or Guyana is on the northeastern coast of South America. It is comprised of the former British Guiana, now Guyana, the former Netherlands Guiana, now Suriname, and French Guiana. Originally inhabited by the Surinen Indians, the coast was probably sighted by Christopher Columbus in 1498 and by Ojeda and Vespucci in 1499. Vicente Yáñez de Pinzón in 1500 was the first to sail close along the shore. He entered some of the rivers, and the Oyapock River was at first called the Yáñez Pinzón River by the Dutch, and later called the Vincent Pinzón River by the French. Initially, both Spanish and Portuguese mariners avoided the coast between the Oyapock and Orinoco rivers. In 1597–1598, a Dutch expedition examined the river mouths from the Amazon to the Orinoco. This started a series of colonies settled initially by the Dutch, then by the French and the English that all attempted to simultaneously fight the climate, the jungle, and the Indians. The country was eventually ceded to Great Britain by the Dutch in 1814, and it became the British Guiana Crown Colony in 1831. Guyana achieved independence in 1966, it became a republic in 1970, and it adopted a new constitution in 1980. "Early elections held in May 2015 resulted in the first change in governing party and the replacement of President Donald Ramotar by current President David Granger. After a December 2018 no-confidence vote against the Granger government, national elections were be held before the scheduled spring 2020 date" (*World Factbook*, 2020).

According to Mr. Russell Fox, formerly of the Ordnance Survey's International Library, "In the first half of the 20th century the Lands and Mines Department had observed some high-order astro fixes and traverses in the coastal zone and along the main river valleys leading to the mining areas in the interior of the northern part of British Guyana. There were also a few railway traverses by other entities, such as the Bauxite Company. The only triangulation appears to have been the British Guyana-Brazil boundary survey of the 1930s. The absence of a national triangulation network resulted from logistical, topographical, and economic limitations. In the early 1950s the Directorate of Colonial Surveys (DOS) observed some astro fixes, assessed the existing Lands and Mines Department and Bauxite Company work and agreed with them on an adjustment strategy to produce a unified set of coordinates from the disparate (albeit high quality) traverses and astro fixes. The method was to accept certain fixes and traverses and to adjust weaker ones to fit; there was not a single datum station. The system adopted was the British Guyana Grid on the Transverse Mercator projection referenced to the International ellipsoid where $a=6,378,388$ and $1/f=297$. Latitude of Origin, φ_o=equator, Central Meridian, $\lambda_o=59°$ West of Greenwich, False Easting=900,000 feet, False Northing=nil, and the Central Scale Factor, $m_o=0.99975$. The U.S. Inter American Geodetic Survey observed a HIRAN trilateration between Venezuela and Brazil in the late 1960. DOS decided to use that as the basis of a new national network for Guyana and, in 1971, computed an adjustment based on the 1970 Aerodist values of HIRAN stations Atkinson and Rose. DOS called the datum Provisional South American 1956. The International ellipsoid was retained but the UTM Grid replaced the British Guyana Grid. That system was used on all subsequent DOS mapping.

Note that the 1978 readjustment of the Aerodist trilateration by Matti Jaakkola was not used by DOS. The Royal Engineers and DOS did a lot of Doppler work, DOS to control the aerial photography for its 1:50,000 and large-scale coastal zone map series."

According to Mr. John W. Hager, now retired from NGA, the DOS astro fix at Georgetown Lighthouse was published as a mean of determinations observed in 1926 and 1951, where $\Phi_o = 6° 49' 31.12"$ N and $\Lambda_o = 58° 09' 52.76"$ West of Greenwich. The International ellipsoid was used at the time, as published in the *Guyana Trig List*, 3rd Edition, 1967. Hager went on to say, "Five stations were established in Guyana as part of the HIRAN net from the Caribbean to Brazil. Only station Eagle was held fixed in the Terra Surveys Limited of Canada network established about 1968. The position of Eagle is on the Provisional South American Datum of 1956 where, for Aerodist Station Eagle, $\varphi = 5° 13' 23.6660"$ N and $\lambda = 59° 06' 10.0549"$ W."

The origin of the Provisional South American Datum of 1956 (PSAD56) is at La Canoa, Anzoátegui Province, Venezuela, where $\Phi_o = 08° 34'17.170"$ N, $\Lambda_o = 63° 51' 34.880"$ W, and the defining azimuth to station Pozo Hondo $\alpha_o = 40° 22' 45.96"$. La Canoa is about 657 km from Georgetown Lighthouse, and the Zanderij Datum origin in Suriname is about 361 km from Georgetown Lighthouse.

Of all of the borders of Guyana, only the Brazil–Guyana boundary appears to be currently stable and uncontested. Reading the history of the region is a veritable jumble of squabbles among the European powers over centuries that were reflected in the whole of the Guianas by the British, Dutch, French, Venezuelan, and Portuguese. Of late, there has even been some gunboat diplomacy between Guyana and Suriname over the mineral (oil and gas) resources of the offshore continental shelf. Some of the diplomatic position papers are available over the internet and make for some fascinating reading about the history of the border negotiations. In like fashion, I was involved in the offshore boundary research for a new diplomatic treaty between Venezuela and Trinidad and Tobago that extended the boundary line beyond the Gulf of Paria.

The latest transformation parameters available from NIMA regarding the shift from the PSAD56 to the WGS84 Datum in Guyana are $\Delta X = -298$ m ± 6 m, $\Delta Y = +159$ m ± 14 m, $\Delta Z = -369$ m ± 6 m, and this is based on a solution of nine points in Guyana. Early in 1997, the U.S. National Geodetic Survey published NAD83 coordinates of several local marks in Guyana that were observed by NGS personnel. Being consistent in the NGS "policy" that I consider to be ill conceived and wasteful of U.S. taxpayer funds, no local datum coordinates were researched by NGS personnel nor published in the NGS data sheets. Considering the expenses of a geodetic expedition, consistently not having "sufficient funds" to research a local survey office nor occupy original datum origin points is a specious argument. Let's try harder, NGS!

Guyana has established eight GPS CORS sites throughout the republic, and during the first phase of the Guyanan government's infrastructure development project, a geodetic survey was conducted in 2019. The localized Provisional South American Datum of 1956 (PSAD56) was bypassed and a new "grid system" (*sic*) was established on the WGS84 ellipsoid (Phil Wright, *A grid for Guyana*, 2020).

REFERENCES: G

Annales Hydrographiques, 4e Série, Tome 1, Année 1950, Paris 1951, p. 155.
Annales Hydrographiques, 4e série, Tome Dixième, Années 1959–1960, p. 65, Paris 1961.
Britannica Concise Encyclopedia, 2010.
Brunei Background Notes, 11/15/1.
Bulletin Géodésique, Juillet-août-septembre 1936, page 232.
Bundesamt für Kartographie und Geodäsie (Federal Office for Cartography and Geodesy), Außenstelle (Branch) Leipzig, Information and Service System for European Coordinate Reference Systems, Karl-Rothe-Straße 10–14, D-04105 Leipzig.
Capt. J. Calder Wood, M.C., Gold Coast Survey Department, May 21 1929.
Catalogue de Positions Géographiques, Paris, 1923.
CIA Factbook, 2012.

Destination360, 2008.
Dipl.-Ing. Heinz W. Kloos, Personal Communications 1981–1982.
Doton, Acquisition Quarterly Review, 1996.
Dzidefo, A., Determination of Transformation Parameters between the World Geodetic System 1984 and the Ghana Geodetic Network. Master's Thesis, Department of Civil and Geomatic Engineering, Kwame Nkrumah University of Science and Technology, Ghana, 2011.
Engenharia Geográfica nos Séculos XIX e XX, João Matos.
Eugene De Bor, Geodetic Analysis of the M745 Series 1:50,000 Scale Maps of Germany, Army Map Service Geodetic Memorandum No. 1584, April 1965.
Gambia Trig Lists 1, 2, and 3, Directorate of Military Survey, 1953.
General Dimitri Zervas, Commander of the Hellenic Military Geographical Institute, Η ΑΖΙΜΟΥΘΙΑΚΗ ΙΣΑΠΕΧΟΥΣΑ ΠΡΟΒΟΛΗ ΤΟΥ ΗΑΤΤ, 1963.
Georgia, Background Note, U.S. Department of State, 2012.
Germany, *World Factbook*, 2015.
Gruetzmacher, Holtery, and Putney, Joint Forces Staff College Joint and Combined Staff Officer School, #02-02, 2002.
Guyana Trig List, 3rd Edition, 1967
Heinz W. Kloos – Personal Communication, 26 July 1982.
http://crs.bkg.bund.de/crseu/crs/eu-description.php?crs_id=Y0dSX0dHUlM4NyslMkYrR1JfVE0=.
Phil Wright, A grid for Guyana, Land, London: Copyright Royal Institution of Chartered Surveyors, Jan/Feb 2020 (Jan/Feb): 6–8.
Informe Detallado de la Comisión Técnica de Demarcación de la Frontera entre Guatemala y Honduras.
J. Clendinning, Empire Survey Review, January 1934.
John W. Hager, Personal Communication, 08 November 2012.
Katsampalos, K., Kotsakis, C., Gianniou, M., Hellenic terrestrial reference system 2007 (HYTS07): a regional realization of ETRS89 over Greece in support of HEPOS, Bollettino di Geodesia e Scienze Affini, Instituto Geografico Militare, Vol. LXIX, N.2–3, May/December 2010.
Library of Congress Country Study, 2015.
Lonely Planet, 2004.
Lonely Planet, 2008.
Lonely Planet, 2012.
Memoria de la Dirección General de Cartografía, Guatemala, Sept. 1957.
Merriam-Webster's Geographical Dictionary, 3rd Edition, 1997.
Mission Hydrographique de la Côte Ouest d'Afrique, 11 Mai 1950–18 Mai 1952.
Moschopouos, G., Demirtzoglou, N., Mouratidis, A., Perperidou, G., Ampatzidis, D., Transforming the old map series to the modern geodetic reference system, *Coordinates*, Vol. XVI, No. 7, July 2020.
Mr. Russell Fox, ex-International Library Manager of the Ordnance Survey, Southampton, UK, (Personal Communication, 17 November 1998).
Mrs. Ekaterine Meskhidze, Chief of International Relations, National Agency of Public Registry, Tbilisi, Georgia – Personal Communication 6 April 2012.
Planheft GROSSDEUTSCHES REICH, Nur für Dienstgebrauch!, Reichsministerium des Innern, Juli 1944.
Poku-Gyamfi, Y., & Hein, G., Framework for the Establishment of a Nationwide Network of Global Navigation Satellite System, *5th FIG Conference*, Ghana, 2006.
Poku-Gyamfi, Y., & Schueler, T., *Renewal of Ghana's Geodetic Reference Network*, LNEC, Lisbon, 2008.
Principal Hydrographic Engineer Roger Grousson, Mission Géodésique de la Guadeloupe, 1946
Rapport sur Les Travaux Géodésiques Exécutés Dans Les Provinces Portugaises D'Outre-Mer, Junta Das Missões Geográficas E De Investigações Do Ultramar, Toronto, September 1957.
Reingold, Les Manuels Coloneaux, "Cartographie Coloniale," Paris, 1935.
Rivard, DTIC 1985.
Roger Lott, Personal Communication, 27 February 1997.
Russell Fox, formerly of the Ordnance Survey's International Library, Personal Communication March 2003.
State Plane Coordinates by Automatic Data Processing, Charles N. Claire, USC&GS Pub. No. 62-4, 1968, pp. 35–39.
Textbook of Topographical Surveying, Her Majesty's Stationery Office, London, 1965, pp. 221, 222.
The 1993 Guam Geodetic Network, Andrew Dyson, South Australian Dept. of Environmental and Natural Resources, November 1995.
The Western European Datums by Jacob A. Wolkeau of Army Map Service (unpublished manuscript, circa 1950).
TR8350.2, NGA, July 1997.

Travaux Géodésiques au Sénégal (Avril 1965 – Juin 1965), Annals Hydrographique, 1965.
Triangulation Data, West Africa, River Gambia, "H.M.S/ Challenger," 1941.
USC&GS Annual Report 1904, Appendix Telegraphic Longitudes, pp. 257–312.
World Factbook, 2009.
World Factbook, 2010.
World Factbook, 2012.
World Factbook, 2020.
World Factbook, 2022.
Ziggah, Y.Y., *et al.*, Novel approach to improve geocentric translation model performance using artificial neural network technology, *Boletim de Ciências Geodésicas* – Online Version, ISSN 1982-2170, 2017.

H

Republic of Haïti 333
The Republic of Honduras 337
Hong Kong 341
The Republic of Hungary 345

Haiti
International boundary
Department boundary
National capital
Department capital
Road
0 10 20 30 40 Kilometers
0 10 20 30 40 Miles
Mercator Projection
CUBA
Cayo Babo
Windward
Passage
Île de la
Tortue
NORTH ATLANTIC OCEAN
Port-de-Paix
NORD-OUEST
Môle
Saint-Nicolas
Baie de
Henne
Le Borgne
Cap-Haïtien
Limbé
NORD
Grande Rivière
du Nord
Fort-Liberté
NORD-EST
Ennery
Gonaïves
Saint-
Raphaël
Monte Christi
Pepillo
Salcedo
Dajabón
Golfe
de la
Gonâve
ARTIBONITE
Hispaniola
Lafond
Saint-Marc
Hinche
Verrettes
CENTRE
Montrouis
Canal de
Saint-Marc
Caribbean
Sea
Île de
la Gonâve
La Cayenne
Anse à
Galets
Mirebalais
Comendador
San
Juan
DOMINICAN
REPUBLIC
OUEST
Baie de
Port-au-Prince
Mannevile
Port-au-Prince
Croix de
Bouquets
Léogâne
Pétionville
Kenscoff
Jimaní
Lago
Enriquillo
Duvergé
Jérémie
Grande
Cayemite
Canal du Sud
Dame-
Marie
Pestel
Petit Trou
de Nippes
Anse d'Hainault
GRAND'ANSE
Miragoâne
Petite Goâve
Trouin
Les Anglais
SUD
Aquin
SUD-EST
Jacmel
Marigot
Belle-Anse
Les Cayes
Côtes-de-
Fer
Bainet
Port-Salut
Île à Vache
Pedernales
Enriquillo
Caribbean Sea
74
73
72
20
19
18

Republic of Haïti

Slightly smaller than Maryland, Haïti borders the Dominican Republic (360 km). The lowest point is the Caribbean Sea (0 m), and the highest point is *Chaîne de la Selle* (2,680 m) (*spelling courtesy of the NGA Geonames search engine, 2010*). "Haïti occupies the mountainous portion of the island of Hispaniola. Its land area includes numerous small islands as well as four large islands: *Île de la Gonâve* to the west, *Île de la Tortue* off the north coast, and *Île à Vache* and *Île Grande Cayemite*, situated, respectively, south, and north of the southern peninsula. Five mountain ranges dominate Haïti's landscape and divide the country into three regions – northern, central, and southern. The northern region has the country's largest coastal plain, the *Plaine du Nord*, which covers an area of 2,000 square kilometers. The north's major mountain range, the *Massif du Nord*, buttresses this plain. The central region consists of the Central Plateau, which covers an area of more than 2,500 square kilometers, as well as two smaller plains and three mountain ranges. The *Guayamouc River* splits the Central Plateau and provides some of the country's most fertile soil. Haïti's southern region contains a series of small coastal plains as well as the mountains of the *Massif de la Selle*" (*Library of Congress Country Profile, 2006*).

"Before the arrival of Europeans, Arawak (also known as Taino) and Carïb Indians inhabited the island of Hispaniola. Although researchers debate the total pre-Columbian population (estimates range from 60,000 to 600,000), the detrimental impact of colonization is well documented. Disease and brutal labor practices nearly annihilated the Indian population within 50 years of Columbus's arrival. Spain ceded the western third of the island of Hispaniola to France in 1697. French authorities quelled the island's buccaneer activity and focused on agricultural growth. Soon, French adventurers began to settle the colony, turning the French portion of the island, renamed Saint-Domingue, into a coffee- and sugar-producing juggernaut. By the 1780s, nearly 40 percent of all the sugar imported by Britain and France and 60 percent of the world's coffee came from the small colony. For a brief time, Saint-Domingue annually produced more exportable wealth than all of continental North America. As the indigenous population dwindled, African slave labor became vital to Saint-Domingue's economic development. Slaves arrived by the tens of thousands as coffee and sugar production boomed. Under French colonial rule, nearly 800,000 slaves arrived from Africa, accounting for a third of the entire Atlantic slave trade" (*ibid., Library of Congress*).

"After a prolonged struggle, Haïti became the first black republic to declare independence in 1804. The poorest country in the Western Hemisphere, Haïti has been plagued by political violence for most of its history. A massive magnitude 7.0 earthquake struck Haïti in January 2010 with an epicenter about 15 km southwest of the capital, *Port-au-Prince*. On 4 October 2016, Hurricane Matthew made landfall in Haiti, resulting in over 500 deaths and causing extensive damage to crops, houses, livestock, and infrastructure. Currently the poorest country in the Western Hemisphere, Haiti continues to experience bouts of political instability" (*World Factbook, 2022*).

"Haïti issued a decree in 1972 that altered its claimed territorial sea and contiguous zone. "The basic system utilized in the establishment of the limits of the Haïtian territorial sea is obscure. While the law calls for measurement of the territorial sea from the low-water baseline of the coast, this system has not been utilized. Furthermore, the turning points were plotted on a very small-scale map and not from a reasonably scaled nautical chart. As a result, the problem of interpretation is compounded by positional difficulties. The Haïtian Government described the system as utilizing *droites paralleles* from the most seaward points of the Haïtian coast. The government has basically drawn a 'system of straight baselines' in a unique manner. The Convention on the Territorial Sea and the Contiguous Zone, for example, states that a system of straight baselines may be utilized for deeply indented coasts or coasts fringed with islands. The Haïtian coastline, in places contains islands. These, however, have not been used as the basepoints with the exception of *Tortuga*. Gonave represents an indentation of the coast, but the scale of the resulting system dwarfs the physical features upon which the system has been developed. The breadth of the territorial sea (as plotted) measures not 12 nautical miles as decreed but from less than 12 to more than 40 nautical miles" (*International Boundary Study, Series A, LIMITS IN THE SEAS No. 51, Straight Baselines: Haïti, Office of the Geographer, U.S. Dept. of State*, 1973).

The first-known geodetic surveys of the Republic of Haïti were performed by the U.S. Army Map Service Inter-American Geodetic Survey (IAGS) in 1946. The triangulation connected all of the islands in the Caribbean to the North American Datum of 1927 of which the origin is at Meades Ranch, Kansas, where $\Phi_o = 39°\ 13'\ 26.686''$ N, $\Lambda_o = 98°\ 32'\ 30.506''$ W, the reference azimuth to station Waldo is $\alpha_o = 75°\ 28'\ 09.64''$, and the ellipsoid of reference is the Clarke 1866, where $a = 6{,}378{,}206.4$ m and $b = 6{,}356{,}583.8$ m. The IAGS established a secant plane coordinate system for Haïti based on the Lambert Conformal Conic projection, the Central Meridian $(\lambda_o) = 71°\ 30'$ West of Greenwich, a False Easting of 500 km, and a scale factor at origin $(m_o) = 0.999911020$, a Latitude of Origin $(\varphi_o) = 18°\ 49'$ North, and the False Easting $= 277{,}063.657$ m.

In 1996, the U.S. National Geodetic Survey performed a brief GPS geodetic survey of *Port-au-Prince* in support of airport surveys and occupied an old 1955 IAGS point named "FORT NATIONAL," where $\varphi = 18°\ 32'\ 51.15343''$ N, $\lambda = 72°\ 19'\ 51.64606''$ W, $H = 45.433$ m, $h = 71.353$ m. The original 1955 IAGS coordinates of that *same* point on the NAD 27 datum are: $\varphi = 18°\ 32'\ 48.8236$ N, $\lambda = 72°\ 19'\ 53.0451''$ W, $H = 45.433$ m. Thanks to John W. Hager, two "astro" stations have been observed in Haïti, *Cap Dame Marie* Astro (code CPM) at $\Phi = 18°\ 36'\ 47''$ N, $\Lambda = 74°\ 25'\ 53''$ W, and Fort Islet Lighthouse (code FIH) at $\Phi = 18°\ 33'\ 31.33''$ N, $\Lambda = 72°\ 20'\ 59.03''$ W, both presumably referenced to the Clarke 1866 ellipsoid. The datum shift parameters published by NGA for that area (*but specifically **not** for Haïti*), of the Caribbean in which the IAGS established control are **from** NAD1927 **to** WGS84: $\Delta X = 3$ m ± 3 m, $\Delta Y = +142$ m ± 9 m, $\Delta Z = +183$ m ± 12 m. Curiously, the shift parameters that work out for the single Haïtian point, "FORT NATIONAL" **from** NAD1927 **to** WGS84 are: $\Delta X = -13$ m, $\Delta Y = -95$ m, $\Delta Z = -197$ m. Of course, no transformation accuracy estimate is possible for that single point in the middle of *Port-au-Prince*, even though both datum positions are known to such high precision.

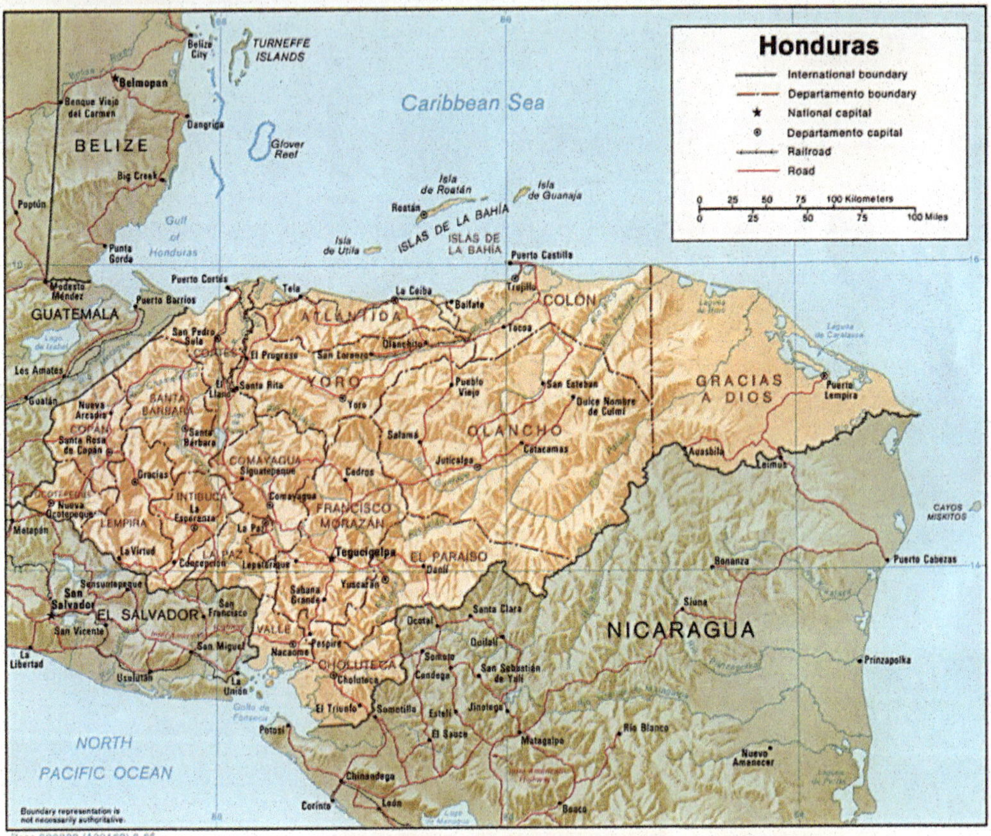

Base 800322 (A00162) 3-85

The Republic of Honduras

Slightly larger than Tennessee, Honduras borders Guatemala (244 km), El Salvador (391 km), and Nicaragua (940 km). The terrain is mostly mountains in interior, narrow coastal plains; the highest point is *Cerro Las Minas* (2,870 m), the lowest point is the Caribbean Sea (0 m), and the mean elevation is (684 m).

Maya settlement at Copán in western Honduras is evident from 1000 B.C. Columbus first set foot on the American mainland at Trujillo in northern Honduras in 1502 and named the country after the deep water off the Caribbean coast. ("*Hondo*" means "depths" in Spanish.) Comayagua was established as the capital in the cool highlands of central Honduras in 1537 and remained the political and religious center of the country until *Tegucigalpa* (Lempira Indian language for "silver mountain") became the capital in 1880. Honduras gained independence from Spain in 1821. It was briefly part of independent Mexico, but then declared independence as a separate nation in 1838.

"After two and a half decades of mostly military rule, a freely elected civilian government came to power in 1982. During the 1980s, Honduras proved a haven for anti-Sandinista contras fighting the Marxist Nicaraguan Government and an ally to Salvadoran Government forces fighting leftist guerrillas. The country was devastated by Hurricane Mitch in 1998, which killed about 5,600 people and caused approximately $2 billion in damage. Since then, the economy has slowly rebounded" (*World Factbook*, 2022).

The northern coast is on the Caribbean Sea and the country's only exit to the Pacific Ocean to the south is through the Gulf of Fonseca, a condominium water body shared with El Salvador and Nicaragua. The Nicaragua–Honduras border crosses the Central American isthmus at its widest part and is 917 km (573 miles) long. Continuing disputes between the two nations led to the submission of the problem to the King of Spain for arbitration in 1906. The matter was temporarily resolved, but no surveying or mapping of the border was performed at that time.

The oldest known map, called "*Mapa Español Oficial*" (Official Spanish Map), was compiled in 1527 by Fernando Colón, the son of Admiral Christopher Columbus. Prominent features were mapped on the northern coast, but little detail of the country's interior was shown. The first complete map of Honduras was by Professor Jesus Aguilar Paz, a pharmacist and cartographer. In 1915, he undertook the labor to produce an adequate map for use in the country's schools without official support. The map was compiled at a scale of 1:500,000 and is extraordinarily exact when the method of compilation (no geodetic control) is taken into consideration. The professor published his map in 1933.

The Guatemala–Honduras boundary is 256 km (160 miles) long and extends between the Caribbean Sea and the tripoint with El Salvador on *Cerro Monte Cristo* (mountain). That tripoint in Honduras is in the *Departamento de Ocotepeque* (state or province). Boundary disputes between Honduras and Guatemala began shortly after the dissolution of the Federation of Central America in 1843. In accordance with the terms of the Treaty of Arbitration that was signed in Washington in 1930, the disputed line was submitted to the Chief Justice of the United States for a decision on the delimitation of the boundary. Because available topographic data were inadequate for the boundary work in many of the disputed areas, the Tribunal directed the making of an aerial photogrammetric survey. Chief Justice Hughes appointed Sidney H. Birdseye of the U.S. Coast & Geodetic Survey (USC&GS) as Chief of the boundary demarcation commission after Mr. Birdseye completed the photogrammetric mapping of the area. Birdseye's commission erected 1,028 pillars and completed its work between 1933 and 1936. The award of the Special Boundary Tribunal was based on the principle of *uti possidetis* as of 1821. Thus, the award referred to the territory under the administrative control of Guatemala and Honduras at the time of their independence from Spain.

Furthermore, El Salvador, Guatemala, and Honduras signed a protocol on March 26, 1936, accepting Cerro Monte Cristo as the tripoint of the boundaries of the three states. A fascinating aspect of the Guatemala–Honduras boundary is that part of the boundary is "established on the right banks of the Tinto and Motagua rivers at mean high water mark, and in the event of changes in these streams in the course of time, whether due to accretion, erosion or avulsion, the boundary shall follow the mean high-water mark upon the *actual right banks of both rivers*" (emphasis added). Boundaries commonly *do not* change with avulsions. Apparently, this was intended to avoid future squabbles.

The Ocotepeque Datum of 1935 was established at Base Norte, where $\varphi_o = 14° 26' 20.168''$ North, $\lambda_o = 89° 11' 33.964''$ West of Greenwich, and $H_o = 806.99$ m above mean sea level. The defining geodetic azimuth to Base Sur is: $\alpha_o = 235° 54' 21.790''$, and the ellipsoid of reference is the Clarke 1866, where $a = 6,378,206.4$ m and $1/f = 294.9786982$. The corresponding astronomic observations at that mountainous location are: $\Phi_o = 14° 26' 13.73''$ North ($\pm 0.07''$), $\Lambda_o = 89° 11' 39.67''$ West ($\pm 0.045''$), and the defining astronomic azimuth to Base Sur is: $\alpha_o = 235° 54' 20.37''$ ($\pm 0.28''$). The difference between these two sets of coordinates is due to the local gravimetric deflection of the vertical. There was no Grid system associated with this datum, although that's not surprising since Mr. Birdseye was with the USC&GS. Their custom was to compute their chains of quadrilaterals in geodetic coordinates.

Nothing much happened (geodetically) anywhere in Central America during World War II. After the war, the U.S. Army Map Service (AMS) established the IAGS headquartered in the U.S. Canal Zone of Panamá. Co-operative agreements negotiated with most countries in Latin America included the Republic of Honduras. In November of 1946, the *Comisión Geográfica Nacional (CGN)* was established under the Secretary of War to satisfy the conditions of agreement with the IAGS. The previous geodetic boundary work of the USC&GS was integrated into the new IAGS observations along with the chains of quadrilaterals observed in Guatemala (with the IAGS). With connections to the classical triangulation in Mexico, the North American Datum of 1927 (NAD 27) was eventually introduced to the Republic of Honduras. Of particular convenience was the fact that the Ocotepeque Datum of 1935 was referenced to the same ellipsoid! (That was courtesy of the USC&GS.)

The IAGS method of computing chains of quadrilaterals was the "Army way" of doing things. That way was different from the USC&GS because AMS was concerned with unifying the mishmash of datums in post-war Europe. The planning and design for that spectacular computational chore in Europe was ultimately accomplished with a conformal projection and Grid. The complexity of existing systems in the Americas was much simpler, but the Topographic Engineers of AMS were trained according to the "Army way." That way consisted of computing classical triangulation on a conformal Grid. Central American countries, with the exception of Belize, are greater in east-west extent than in north-south extent. Therefore, the Lambert Conformal Conic projection was used as the basis of all IAGS-developed Grid systems in Central America for triangulation computations and the published "Trig Lists" of coordinates.

The Honduras Lambert Conformal Grid (1946–present) on the Ocotepeque Datum of 1935 covers two secant zones, Norte and Sud. Both zones use the same Central Meridian (λ_o) = 86° 10' 00" West of Greenwich and False Easting of 500 km. Zone Norte has a Latitude of Origin (φ_o) = 15° 30' North, the False Easting = 296,917.439 m, and the scale factor at origin (m_o) = 0.999932730. Zone Sud has a Latitude of Origin (φ_o) = 13° 47' North, the False Easting = 296,215.903 m, and the scale factor at origin (m_o) = 0.999951400.

In 1960, the Nicaragua–Honduras border dispute was finally settled by the International Court of Justice (ICJ). The determination was that the 1906 Award of the King of Spain should be carried out based on the line of *uti possidetis*. In 1969, El Salvador and Honduras went to war over another border dispute. This war involved six contested "*bolsones*" (pockets) of land encompassing a total area of 436.9 km^2, two islands (*Meanguera* and *El Tigre*) in the Gulf of Fonseca, and the right of passage for Honduras to the Pacific Ocean from its southern coast. A peace treaty was signed in 1980, an arbitration agreement was signed in 1986, and the entire matter was resolved in 1993 after an ICJ ruling the previous year.

Military mapping of Honduras at 1:50,000 scale (100% coverage) and smaller is on the NAD 27 and uses the UTM Grid exclusively. An estimate of the transformation parameters among the active datums in Honduras can be gleaned from those pertinent to nearby Costa Rica. For example, (in Costa Rica) from Ocotepeque 1935 to WGS 72: $\Delta X = -193.798$ m, $\Delta Y = -37.807$ m, $\Delta Z = +84.843$ m. Furthermore, (in Costa Rica) from Ocotepeque 1935 to NAD1927: $\Delta X = +205.435$ m, $\Delta Y = -29.099$ m, $\Delta Z = -292.202$ m.

In Costa Rica, the fit of Ocotepeque Datum 1935 to WGS 72 is better than ±3 m; the fit of NAD 27 to WGS 72 is better than ±6 m. Considering the quality of classical geodetic work done by the IAGS, the accuracy of these identical transformation parameters *applied in Honduras* would probably be less than double these values.

A primary GPS network of Continuously Operating Reference Stations (CORS) was established in 2000. Honduras is now (as of 2004) part of the SIRGAS network of Latin American countries, and it has developed Second-Order and Third-Order networks for the entire republic. A high-order network was observed and published in 2011 for the capital city of Tegucigalpa through the cooperation of the Spanish University of Alcala, the *Universidad Nacional Autonoma de Honduras*, the *Instituto de la Propiedad*, and the *Direccion General de Catastro y Geografia (Red Geodesica Activa de Honduras Y Su Enlace con SIRGAS, 2004) (ESTABLECIMIENTO DE UNA RED GEODESICA EN TEGUCIGALPA (HONDURAS) MEDIANTE TECNOLOGIAS GPS Y ENLACE CON LAS REDES DE REFERENCIA OFICIAL DE CENTROAMERICA, 2010).*

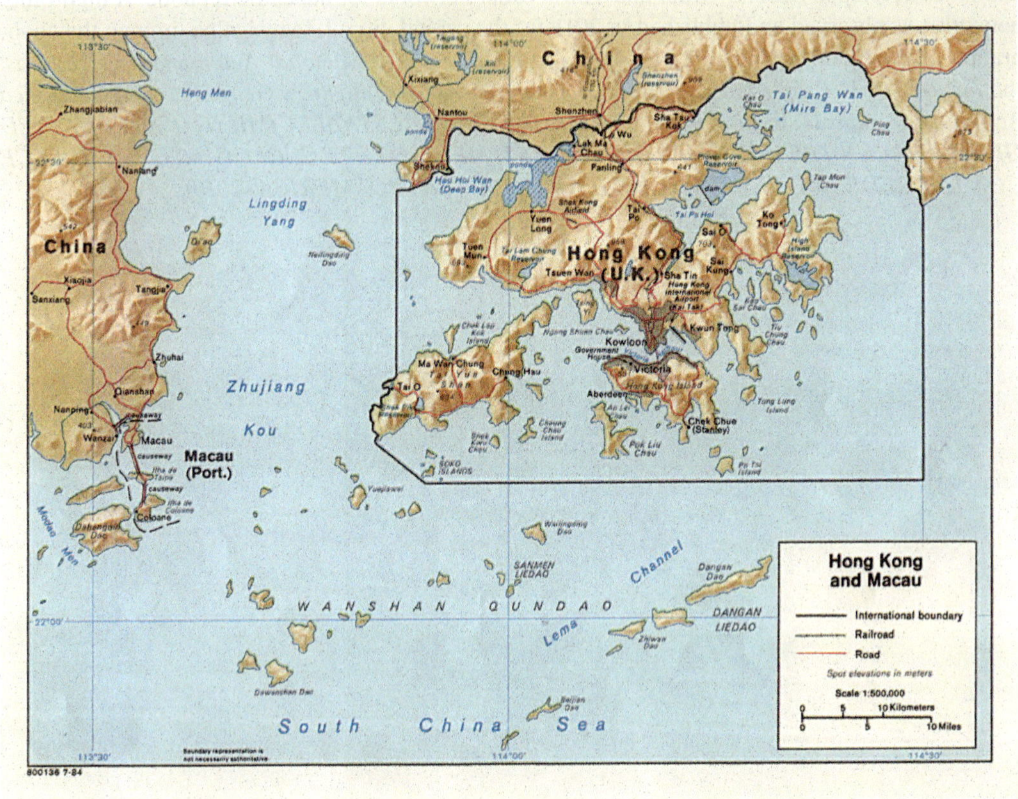
Hong Kong
and Macau
International boundary
Railroad
Road
Spot elevations in meters
Scale 1:500,000
0 5 10 Kilometers
0 5 10 Miles
China
China
Hong Kong
(U.K.)
Macau
(Port.)
Heng Men
Lingding
Yang
Zhujiang
Kou
Zhangjiabian
Nanlang
Xiaojia
Sanxiang
Tangjia
Zhuhai
Qianshan
Nanping
Wanzai
Macau
causeway
causeway
causeway
Ilha de
Taipa
Ilha de
Coloane
Coloane
Modao
Men
Xixiang
Nantou
ponds
Shekou
Hau Hoi Wan
(Deep Bay)
Shenzhen
Shenzhen
(reservoir)
Sha Tau
Kok
Lo Wu
Lak Ma
Chau
Fanling
Tai Pang Wan
(Mirs Bay)
Ping
Chau
Tap Mun
Chau
Plover Cove
Reservoir
dam
Tai Po
Tai Po Hoi
Yuen
Long
Tuen
Mun
Sai O
Ko
Tong
Sai
Kung
High
Island
Reservoir
Tsuen Wan
Sha Tin
Hong Kong
International
Airport
(Kai Tak)
Kwun Tong
Kowloon
Government
House
Victoria
Hong Kong Island
Aberdeen
Chek Chue
(Stanley)
Ma Wan Chung
Chung Hau
Tai O
Tai Yue
Shan
934
Chek Lap
Kok
Island
Cheung
Chau
Island
Pok Liu
Chau
Tung Lung
Island
Po Toi
Island
SOKO
ISLANDS
Yuejiawei
Wailingding
Dao
SANMEN
LIEDAO
Channel
Lema
Dangan
Dao
DANGAN
LIEDAO
Zhiwan
Dao
W A N S H A N   Q U N D A O
Dawanshan Dao
Beijian
Dao
South China Sea
113°30'
114°00'
114°30'
22°30'
22°00'
Boundary representation is
not necessarily authoritative
800136 7-84

Hong Kong

Hong Kong is six times the size of Washington, DC, and is located in Eastern Asia, bordering the South China Sea and China. The terrain is hilly to mountainous with steep slopes; lowlands in north; the highest point is Tai Mo Shan (958 m), and the lowest point is the South China Sea (0 m).

The British occupied the Island in 1839 and the Treaty of Nanking in 1842 ceded Hong Kong to Great Britain. The Convention of 1860 added Stonecutter's Island and in 1898 the New Territories were leased to Great Britain. The 99-year lease expired on July 1, 1997, and Hong Kong was restored to the People's Republic of China as a separate administrative region. "Since the turnover, Hong Kong has continued to enjoy success as an international financial center. However, dissatisfaction with the Hong Kong Government and growing Chinese political influence has been a central issue and led to considerable civil unrest. In June 2020, the Chinese Government passed a security law for Hong Kong that would criminalize acts such as those interpreted as secession, subversion, terrorism, and collusion with foreign or external forces. Critics said the law effectively curtailed protests and freedom of speech and was widely viewed as reducing Hong Kong's autonomy, while Beijing said it would return stability. The law was met with widespread international condemnation and criticism that it effectively ended the 'one country, two systems' guiding principle of Hong Kong's Basic Law. Since its passing, authorities have used the law to detain pro-democracy activists and politicians, oust opposition lawmakers, and raid media offices. In March 2021, Beijing reduced the number of directly elected seats in Hong Kong's legislature, furthering its efforts to curtail political opposition and protests" (*World Factbook*, 2022).

Triangulation stations first appeared on a map of Hong Kong produced in 1845 by Lt. Collinson of the Royal Engineers. Additional maps were produced in 1899/1900 and 1903/1904 with triangulation stations shown, but survey records no longer exist for those triangulations. The 2nd Colonial Survey Section of the Royal Engineers (Survey of India) did basic triangulation in 1924; others adjusted the observations in 1928–30, and again in 1946. This was adopted collectively as the main triangulation of Hong Kong.

Early planimetric plane table mapping was at 1:600 scale in Hong Kong and Kowloon. The Hong Kong New Territories Datum of 1924 (HKNT24) is defined on the Clarke 1880 ellipsoid, where $a = 6,378,249.145$ m, $1/f = 293.4650$, and 1 meter = 3.280869330 Hong Kong feet. Trig "Zero" was 38.4 feet due south of the transit circle of the Kowloon Royal Observatory on Victoria Peak such that the datum origin was: $\Phi_o = 22° 18' 12.82'' $ N, $\Lambda_o = 114° 10' 18.75''$ E, and azimuth from "Tai Mo Shan" (Trig 67.2) to "Au Tau" (Trig 94) = $292° 59' 46.5''$. The Cassini-Soldner Grid system was used with its origin the same as the HKNT24 Datum and with coordinates of 5.18 Ft. North and 0.38 Ft. East. Such strange-looking coordinates for a Grid origin are common for the 19th and early 20th centuries, and large areas of Hong Kong actually had negative coordinates!

In 1963, a re-triangulation was carried out because the HKNT24 network could not meet the accuracy requirements for large-scale mapping and cadastral surveys. The Hong Kong Datum of 1963 (HK63) was defined on the *older* Clarke 1858 ellipsoid, where $a = 6,378,235.6$ m and $1/f = 294.2606768$. HK63 Datum, again with its origin at Victoria Peak, was used to develop a new Cassini-Soldner Grid with the false origin southwest of Lantau Island so that the coordinates in the Hong Kong New Territories were all positive values. The False Easting was 120,000 ft. and the False Northing was 50,000 ft. Subsequent mapping was performed at 1:1,200 scale. As mentioned previously, the Cassini-Soldner was one of the old (but popular) aphylactic projections. The reader should note that an aphylactic projection is not conformal, it is not equal-area, and it is not azimuthal. It was easy to construct with very simple tools and with modest projection tables, and it was popular for British colonial and "expedition" mapping. In *TR 8350.2*, the transformation from HK63 to WGS84 is $\Delta X = -156$ m ± 25 m, $\Delta Y = -271$ m ± 25 m, $\Delta Z = -189$ m ± 25 m.

In 1963, the standard computational tool of the geodetic surveyor (and photogrammetrist) was an electric-powered mechanical calculator and was equipped with the very fancy square-root keys (the cost was 4–6 months' worth of paychecks). A rotary "pepper mill" Curta calculator (made in Liechtenstein and costing 2–3 weeks' worth of paychecks) would do just fine in a tent. (Note that the IBM 1620 electronic computer was the ultimate prize of a private geodetic survey corporation. Sometimes those computers had as much as 8K of ferrite-core memory! The equivalent British machine used in Hong Kong was the I.C.L. 2970 computer.)

Anyway, in the 1960s, the aphylactic projections were still commonly used cartographic projections including the polyconic used in the United States, but they were a nightmare for the control surveyor. When I used to do control surveys with a T-2 theodolite and an electronic distance measuring (EDM) instrument or an invar tape, a significant correction for systematic error was (and still is) for the "scale factor" – the difference between true (geodetic) distance and map (grid) distance. Although such computations are straightforward with a conformal projection, **with an aphylactic projection the scale factor varies as a function of the azimuth of the line being measured!** Things were just dandy for the cartographer and the photogrammetrist with such grids, but the field surveyor was perpetually immersed in exasperating daily calculations because of the HK63 Grid.

Cadastral survey computations were pure drudgery in Hong Kong back in the 1960s on a *day-to-day* basis! In line with the metrication policy of the 1970s, the British Imperial (Cassini-Soldner) Grid was converted to metric units of measure in 1975–77 with the Grid origin further shifted 3,550 m to the west such that False Easting = 33,016 m and False Northing = 15,240 m. Some 3,000 sheets were also converted to 1:1,000 scale. A Photogrammetric Unit was formed in the Lands Department in 1976.

With the introduction of EDM instruments to Hong Kong in the late 1970s, the distances between hilltop triangulation control points were resurveyed in 1978–79 to improve the consistency and accuracy of the control network. In this re-survey and adjustment, a new geodetic datum called Hong Kong 1980 (HK80) was adopted. The definition of the new datum was now referenced to old Trig 2 on Partridge Hill, where $\Phi_o = 22° 18' 43.68''$ N, $\Lambda_o = 114° 10' 42.80''$ E, and the azimuth was re-referenced from old Trig 67 (now lost) to Trig 94 turned as $\alpha_o = 292° 52' 58.4''$. The Hayford (International) 1909 ellipsoid was adopted, where $a = 6,378,188$ m and $1/f = 298$. The *conformal* transverse Mercator projection (Gauss-Krüger) was used for the new rectangular grid system and was known as the Hong Kong 1980 Grid. The new False Easting = 836,694.05 m and the new False Northing = 819,069.80 m were referenced to the new datum origin. The scale factor at origin was unity ($m_o = 1.0$), the same as the implicit definition of any Cassini-Soldner Grid.

Since the miserable aphylactic grid was gone forever, the Cadastral Land Surveyor's life was lifted from "sweat shop" conditions. The re-observation of the control network of Hong Kong had been accomplished by laser Geodimeter® model 600 positioning of survey points by trilateration rather than by microwave Tellurometer® or by Electrotape® observations because of nearby intentional "radio jamming." A total of 162 lines had been measured, and the standard deviation of the combined network residuals was 0.45 parts per million.

In 1990, the Survey and Mapping Office, Lands Department, started to apply the GPS technique for fixing positions of survey control points. A territory-wide observation on a network of 15 stations (12 of which are at existing trig stations) was carried out by the No. 512 Specialist Team, Royal Engineers of the UK Military Survey using GPS (NAVSTAR) and Doppler (TRANSIT) satellite techniques. The network was adjusted with high accuracy results. This survey provided a rigid link between the local HK80 Geodetic Datum and the global WGS84 Datum. Currently, the WGS84 Datum is used with a UTM Grid for tourist and hiking maps. The cadastral and engineering surveys continue to use the HK80 Datum for day-to-day use, although aided by differential GPS.

In Hong Kong, all heights and levels on land refer to the Principal Datum which was formerly known as the Ordnance Datum. The HKPD was originally determined by observation of the tides from 1887 to 1888. A later observation of the full metonic cycle (18.6 years) was performed by the Royal Observatory from 1965 to 1983 at North Point, Victoria Harbor. "Mean Sea Level" is approximately 1.23 m above HKPD. The original monument was "Rifleman's Bolt," a copper bolt fixed in the Hong Kong Naval Dockyard by personnel from H.M. Surveying Vessel "Rifleman" in 1866. It is now preserved for its historical value at the eastern wall of Blake Block in H.M.S. Tamar.

The Chart Datum, formerly known as the Admiralty Datum, is approximately the level of Lowest Astronomical Tide and is adopted as the zero point for Tide Tables since 1917. The Chart Datum is approximately 1.38 m below HKPD. The Royal Engineers determined WGS84 ellipsoid heights in 1991. In general, the WGS84 heights are 2.4 m higher in the west and 0.4 m in the east of Hong Kong. The accuracy of this geoid separation is estimated to be better than ±0.15 m.

Since the turnover, the new people in charge have censored all of the above details in keeping with a closed society.

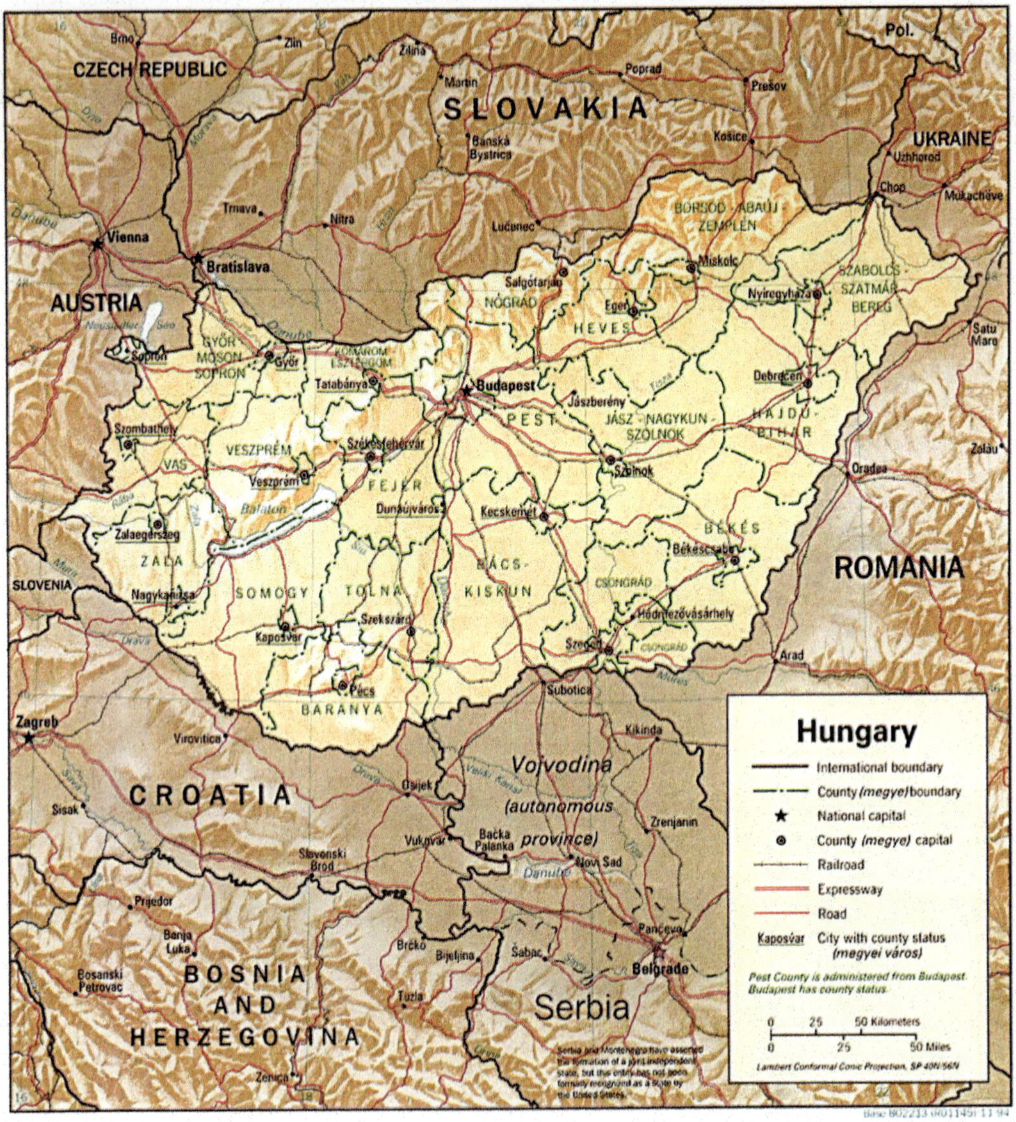

Base 802213 (R01145) 11-94

The Republic of Hungary

Hungary is bordered by Austria (321 km), Croatia (348 km), Romania (424 km), Serbia (164 km), Slovakia (627 km), Slovenia (94 km), and Ukraine (128 km). The terrain is mostly flat to rolling plains; hills and low mountains on the Slovakian border; the highest point is Kekes (1,014 m), the lowest point is the Tisza River (78 m), and the mean elevation is (143 m).

Vaik "came" to the Magyar Duchy in the year 997 *A.D.* through conquest. He applied for and received the title of Apostolic King of Hungary from Pope Sylvester II and was crowned in Budapest in 1000 *A.D.* under the Christian name of Stephan. He died in 1037, and was later canonized as Saint Stephan, becoming the Patron Saint of Hungary. A substantial amount of Hungarian folklore is based on St. Stephan. "The kingdom eventually became part of the polyglot Austro-Hungarian Empire, which collapsed during World War I. The country fell under communist rule following World War II. In 1956, a revolt and an announced withdrawal from the Warsaw Pact were met with a massive military intervention by Moscow. Under the leadership of Janos Kadar in 1968, Hungary began liberalizing its economy, introducing so-called 'Goulash Communism.' Hungary held its first multiparty elections in 1990 and initiated a free market economy. It joined NATO in 1999 and the EU five years later" (*World Factbook*).

Private topographic and cartographic activities in the Austro–Hungarian Empire started in the middle of the 16th century. In 1763, Queen-Empress Maria Theresia ordered the survey and topographic mapping of all the Provinces of Hapsburg. There have been five separate and distinct topographic surveys of Hungary.

The first topographic survey of Hungary was performed from 1763 to 1787 and was termed the "*Josephinishe Aufnahme.*" The Liesganig triangulation and attached supplemental surveys were executed graphically with plane table and alidade. There was no geodetic survey used as a foundation. The associated topographic survey was performed at a scale of 1:28,800 and was based on the Vienna Klafter System where 1 Zoll = 400 Klafters = 758.6 m. Altogether there were about 4,500 sheets surveyed and all of them were kept secret for military purposes.

The second topographic survey of Northern Hungary (*Franziszeische Aufnahme*) was conducted from 1810 to 1866. The Vienna Datum of 1806 was established based on the origin of St. Stephan Turm (St. Stephan's Tower), where $\Phi_o = 48°\ 12'\ 34.0''$ North, $\Lambda_o = 34°\ 02'\ 15.0''$ East of Ferro. (Ferro is in the Canary Islands which is $17°\ 39'\ 46.02''$ West of Greenwich.) The defining azimuth of the datum was from St. Stephan Turm to Leopoldsberg: $\alpha_o = 345°\ 55'\ 22.0''$. Coordinates for the mapping were based on a Cassini-Soldner Grid centered at the datum origin. The Bohnenberger ellipsoid was used from 1810 to 1845, where $a = 6,376,602$ m and $1/f = 324$. The Zach ellipsoid was used from 1845 to 1863, where $a = 6,376,602$ m and $1/f = 324$. From 1847 to 1851 the Walbeck ellipsoid was also used, where $a = 6,376,896$ m and $1/f = 302.78$.

An Austrian Cadastral triangulation was established in 1817 and the origin established for Hungary was in Budapest at the Gellérthegy Observatory ("*hegy*" is Hungarian for hill) for another Cassini-Soldner Grid. The Gellérthegy Grid origin used from 1817 to 1904 is: Latitude of Origin $(\varphi_o) = 47°\ 29'\ 15.97''$ North, with a Central Meridian $(\lambda_o) = 36°\ 42'\ 51.57''$ East of Ferro. The second military triangulation surveyed in 1806–1829 consisted of the main chains (of quadrilaterals) that covered the western part of the monarchy (west of the meridian of Budapest) and the chain that extended along the Carpathian Mountains to Transylvania. In 1863, the Bessel ellipsoid of 1841 was prescribed for use in the triangulation instructions.

The Third Topographic Survey of Hungary (*Neue Aufnahme*) was conducted from 1869 to 1896 and was based on the Arad, St. Anna Datum of 1840, where the origin was: $\Phi_o = 46°\ 18'\ 47.63''$ North, $(\Lambda_o) = 39°\ 06'\ 54.19''$ East of Ferro (geodetically determined from Vienna). This datum was referenced to the Zach ellipsoid and was used for the Third Topographic Survey of Hungary. The defining azimuth to Kurtics was determined astronomically, but the angular value was not published and is now lost. The metric system was legally established in 1872, and the 1:25,000 mapping scale was introduced along with the polyeder (polyhedric) projection to eliminate inconsistencies in map sheet lines. Remember that in other countries I have pointed out that the polyhedric projection is mathematically equivalent to the local space rectangular (LSR) coordinate system that is commonly used in computational photogrammetry.

In 1874, the Budapest Stereographic Projection was defined at Gellérthegy where the Latitude of Origin $(\varphi_o)=47°\ 29'\ 14.93''$ North and the Central Meridian $(\lambda_o)=36°\ 42'\ 51.69''$ East of Ferro. The orientation of this system was defined as the azimuth to Széchényihegy, $\alpha_o=100°\ 47'\ 14.34''$ which was in sympathy with the azimuth to Nagyszal, $\alpha_o=191°\ 28'\ 52.19''$ derived from the Vienna University Observatory. (In the derivation carried out in 1861–63, the Walbeck ellipsoid was used.) The regions of Central and Southern Hungary were topographically surveyed from 1881 to 1884. Northern Hungary was surveyed from 1875 to 1878 and Western Hungary was surveyed from 1878 to 1880, the latter two by the use of cadastral planimetry.

The fourth topographic survey of Hungary was carried out in essentially the same manner as the third survey. All plane table sheets of the cadastral survey were reduced with a pantograph and were published on the polyhedric projection at 1:75,000 scale. From 1896 to 1898, three trig points were required per plane table sheet. From 1898 to 1903, 10 trig points per sheet were required, and that was increased to 20 trig points thereafter. Tacheometry was introduced for this survey, and mapped distances could be estimated only if less than 100 m from the instrument. In 1905, photogrammetry replaced the plane table with the stereocomparator (sort of an un-digitized analytical plotter).

In 1907, Dr. A. Fasching derived the position of the eastern tower of the former observatory at Gellérthegy where the Latitude of Origin $(\varphi_o)=47°\ 29'\ 09.6380''$ North and with a Central Meridian $(\lambda_o)=36°\ 42'\ 53.5733''$ East of Ferro and $\alpha_o=100°\ 47'\ 07.90$ to the Laplace station Széchényihegy. In 1908, a system of three cylindrical projections was introduced, all with the Central Meridian of Budapest. The oblique cylinders touch the Gaussian sphere along the great circles perpendicular to the meridian at the following origins: $48°\ 42'\ 56.3180''$ N, $47°\ 08'\ 46.7267''$ N, and $45°\ 34'\ 36.5869''$ N. For the orientation, the azimuth Gellérthegy- Széchényihegy was used, hence the common X-axis of the three cylindrical projections form an angle of 6.44 arc seconds with the Budapest Stereographic Grid of 1874. The stereographic projection was used only for the computation of the First and Second-Order nets – the cylindrical projection was used for the actual mapping.

Also in 1908, the invention of the stereoautograph by Captain von Orel of the Military Geographic Institute of Vienna allowed the compilation, including contours, to be done completely mechanically (as opposed to *in-situ* field work). Note that this phenomenal breakthrough in mapping with photogrammetry used terrestrial photographs! The Zeiss 1911 Stereoautograph, the next wonder of photogrammetry, was used to compile the last sheets of the never completed fourth topographic survey of the Austro–Hungarian Empire. According to Andrew Glusic of Army Map Service (from 50+ years ago):

> Warfare and the Map. The Armies have been using the maps for more than two centuries. The Napoleonic Wars gave a special impulse to the use of geographic maps in warfare; consequently, in the European Armies mapping services were created, of which many are known as Military Geographic Institutes. It was the military who through the XIX century in Europe as well as in the colonies was responsible for the largest part of the geodetic and topographic surveys. In these surveys the military aspects dominated; particularly at their outsets the scientific purposes were not considered, and many times also technical requirements were ignored. There was solely one goal: – to produce a military map. This military map should include all such information of the area concerned which the military leaders need for the planning and execution of movement, combat, accommodation, and supply. The enormous technical progress in the last century largely influenced the application of strategical and tactical principles in warfare; therefore, the nature of warfare changed and consequently the requirements for the

military maps. In order to avoid the lack of adequate maps in any future war, the nature of the warfare together with the corresponding changes which would affect the standards of the mapping have to be considered in advance within limits of possibility and also proper measures should be undertaken at the time. The Austro-Hungarian military authorities passed up the proper time for such considerations; hence the single tactical map of the Monarch – 1:75,000 special map – trailed far behind the requirements imposed by the changes of warfare in World War I.

A new topographic survey of Hungary was started in 1927. The oblique stereographic projection was used for the "Budapest System" with the origin at the base point of the East Tower of the Astronomical Observatory of Budapest at Gellérthegy. (The observatory was torn down and replaced by a stone fortress on Gellért hill. The old point was then later found to be on the rampart of that fortress, and a National Report to the IUGG portrayed a photograph of the point marked by a gazebo-like canopy!) The origin of the coordinate system was defined, where Latitude of Origin $(\varphi_o) = 47° 29' 09.6380''$ North, with a Central Meridian $(\lambda_o) = 36° 42' 53.5733''$ East of Ferro (Ferro = $17° 39' 46.02''$ West of Greenwich as derived from astronomic observations in 1907). The defining azimuth was from Gellérthegy to Nagyszal, where $\alpha_o = 191° 28' 52.19''$ as retained from the 1874 datum values from Vienna University. The ellipsoid of reference was the Bessel 1841 where: the semi-major axis $(a) = 6,377,397.155$ m and the reciprocal of flattening $(1/f) = 299.1528128$. The False Easting and False Northing were each 500 km.

The northern part of Transylvania, occupied by the Hungarian Army in World War II, was mapped with a system defined as the "Marosvásárhely Stereographic System" with a datum origin point at the cadastral triangulation station Kesztej where: Latitude of Origin $(\varphi_o) = 46° 33' 09.12''$ North, and with a Central Meridian $(\lambda_o) = 42° 03' 20.955''$ East of Ferro. The defining azimuth from Kesztej to Tiglamor was: $\alpha_o = 146° 57' 41.052''$. The False Easting and False Northing were each 600 km, and this local Datum and Grid was referenced on the Bessel 1841 ellipsoid.

The Hungarian surveying and mapping agencies went through a series of reorganizations after World War II through 1952–54 when the country began to follow the pattern established by the USSR. The Gauss-Krüger Transverse Mercator projection was adopted in 1957 with the Grid parameters of a scale factor at origin equal to unity, and false origin 500 km West and 5,000 km North. Since 1954, the Krassovsky 1940 ellipsoid was used where: the semi-major axis $(a) = 6,378,245$ m and the reciprocal of flattening $(1/f) = 298.3$. Prior to 1957, the Central Meridians (λ_o) for military mapping were 18° and 21°, and after 1957, 15° and 21° were used. For cadastral mapping since 1957, the Central Meridians of 17°, 19°, 21°, and 23° were used. The Hungarian Datum of 1957 with origin coordinates at Erdóhegy and associated parameters were kept secret for military purposes.

The new national Hungarian Datum of 1972 (HD 72), also known as the "Unified National Horizontal Network of 1972" (EOVA Datum of 1972), is defined with origin coordinates at Szólóhegy, where $\Phi_o = 47° 17' 30.44''$ North, $\Lambda_o = 19° 36' 10.18''$ East of Greenwich. The defining azimuth is from Szólóhegy to Erdóhegy: $\alpha_o = 209° 55' 27.79''$. The corresponding geodetic parameters of this origin are: $\varphi_o = 47° 17' 32.6156''$ North, $\lambda_o = 19° 36' 09.9865''$ E. The defining geodetic azimuth is: $\alpha_o = 209° 55' 26.64''$ and the ellipsoid of reference is the Geodetic Reference System (GRS) 1980. For the origin, h_o = ellipsoid height, H_o = height above the Baltic Sea, and N_o = geoid undulation. Therefore, $h_o = 235.80$ m $= H_o + N_o = 229.24$ m $+ 6.56$ m. The published transformation parameters from HD-72 to WGS 84 are: $\Delta X = -5.3$ m, $\Delta Y = +157.77$ m, $\Delta Z = +31.6$ m, $k = -2.11$ ppm, $R_z = -1.11''$, $R_y = -0.50''$, $R_x = -0.97''$. Remember the Hungarian transverse cylindrical grids of 1908 mentioned several paragraphs ago? Well, the new system is based on a new definition of that old Hungarian favorite.

The *Egységes Orzágos Vethleti rendszer* (Uniform National Projection system) "EOV Grid" is a conformal double transverse cylindrical projection. The Grid is defined at: $\varphi_o = 47° 08' 39.8174''$ North, Central Meridian, $\lambda_o = 91° 02' 54.8584''$ E. The false origin is 200 km east (+X), 650 km north (+Y), and the scale factor (m_o) is $= 0.99993$. The national fundamental benchmark is at Nadap, where $H_o = +173.1638$ m above the Baltic Sea and $H_o = +173.8388$ m above the Adriatic Sea.

The Institute of Geodesy Cartography and Remote Sensing (*FÖMI*) according to the Hungarian Government's Decree No. 1312/2016 (13 June) by legal succession – with integration into the Government Office for the Capital, Budapest – will be dissolved. *FÖMI* will continue its professional activities within the organization framework of the Government Office for the Capital, Budapest from January 1, 2017.

"The GNSS reference system is the 3-dimensional geocentric Cartesian system ETRS89 (realized in Hungary by the OGPSH). For practical surveying and mapping works, the local HG72 system has to be used. The transformation between the two independent systems is possible based on common points. The accuracy of the transformation cannot exceed the accuracy of the geodetic networks (or sub-networks). There are therefore many possible transformations within the limits. However, it is reasonable to apply the transformation in a standardized way. To this end, the EHT software was developed and released by the FÖMI SGO in 2002. Its method is based on a local transformation using OGPSH points and provides the best available accuracy. The program automatically selects the suitable nearby reference points from its data base. It works for the whole territory of the country. The new version released in 2008 is already capable of transforming the coordinates in both directions. The EHT is available free of charge from the gnssnet.hu web site" (Institute of Geodesy, Cartography & Remote Sensing, 2016).

REFERENCES: H

Establecimiento De Una Red Geodesica En Tegucigalpa (Honduras) Mediante Tecnologias Gps Y Enlace Con Las Redes De Referencia Oficial De Centroamerica, 2010.
Institute of Geodesy, Cartography & Remote Sensing, 2016.
International Boundary Study, Series A, Limits in the Seas No. 51, Straight Baselines: Haïti, Office of the Geographer, U.S. Dept. of State, 1973.
Library of Congress Country Profile, 2006.
NGA Geonames Search Engine, 2010.
Red Geodesica Activa de Honduras Y Su Enlace con SIRGAS, 2004.
World Factbook, 1999.
World Factbook, 2022.

I

Republic of Iceland 351
Republic of India 355
Republic of Indonesia 361
Islamic Republic of Iran 367
Republic of Iraq 371
Ireland 375
The State of Israel 379
Italian Republic 385
Republic of Côte D'Ivoire 389

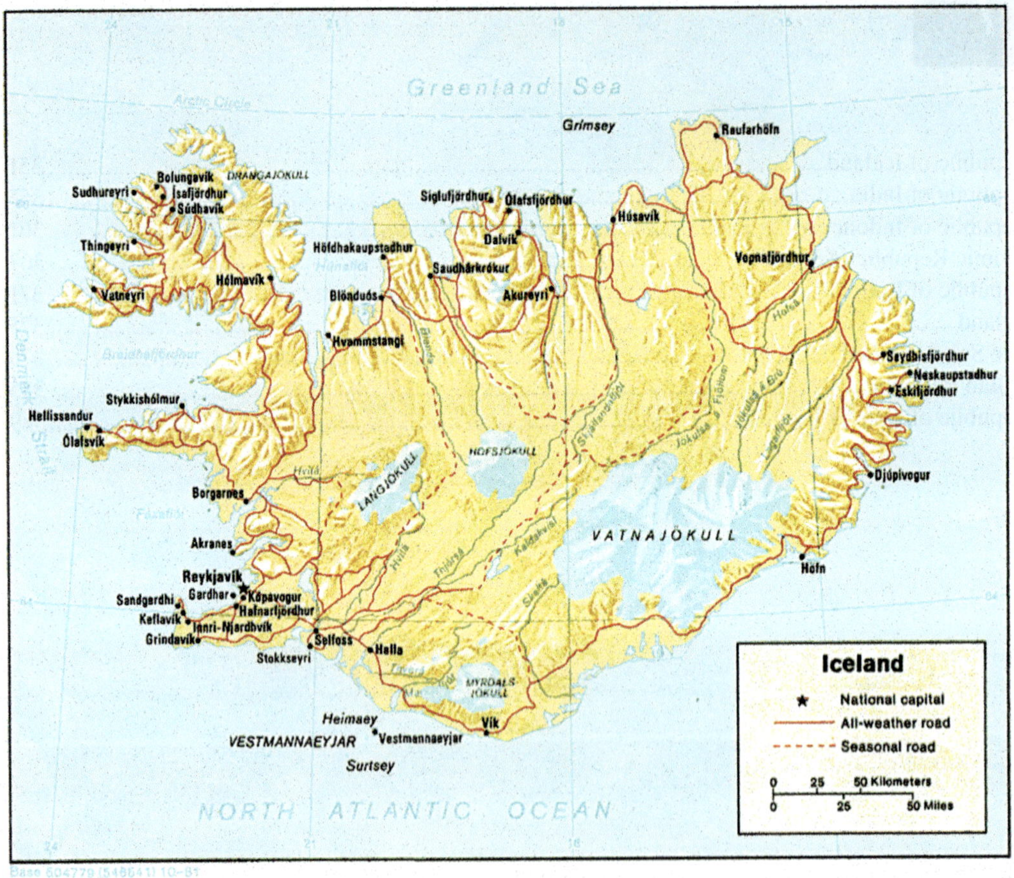
Greenland Sea
Arctic Circle
Grimsey
Raufarhöfn
Bolungavík
DRANGAJÖKULL
Sudhureyri
Ísafjördhur
Súdhavík
Siglufjördhur
Ólafsfjördhur
Húsavík
Thingeyri
Dalvík
Höfdhakaupstadhur
Saudhárkrókur
Vopnafjördhur
Hólmavík
Vatneyri
Blönduós
Akureyri
Hvammstangi
Seydhisfjördhur
Neskaupstadhur
Eskifjördhur
Stykkishólmur
Hellissandur
Ólafsvík
HOFSJÖKULL
LANGJÖKULL
Djúpivogur
Borgarnes
VATNAJÖKULL
Akranes
Höfn
Reykjavík
Gardhar
Kópavogur
Sandgardhi
Hafnarfjördhur
Keflavík
Innri-Njardhvík
Grindavík
Selfoss
Hella
Stokkseyri
MYRDALS-JÖKULL
Heimaey
Vík
VESTMANNAEYJAR
Vestmannaeyjar
Surtsey
NORTH ATLANTIC OCEAN
Iceland
National capital
All-weather road
Seasonal road
0 25 50 Kilometers
0 25 50 Miles
Base 504779 (546641) 10-81

Republic of Iceland

The land area of Iceland is slightly smaller than Kentucky. The terrain is mostly plateau interspersed with mountain peaks and ice fields. The coast is deeply indented by bays and fiords. With a coastline of 4,988 km, the highest point is *Hvannadalshnúkur* (2,119 m).

Iceland's first inhabitants were Irish monks, who regarded the island as a hermitage until the early 9th century. "*Lyoveldio Island*" was permanently settled mainly by Norwegians and by some British Isles Vikings in 874 A.D. The National Assembly or "*Althing*" was the world's first parliamentary system in 930. Christianity was adopted (under threat of sword!) in 999, and Iceland united with Norway in 1262. Leif Eiriksson was born in Iceland, and he sailed from Greenland to become the first European to reach North America (Vinland the Good) in 1000. In 1380 Iceland united with Denmark, and by Act of Union in 1918, became an independent kingdom in personal union with Denmark. Iceland became a constitutional republic and independent from Denmark on June 17, 1944.

"The second half of the 20th century saw substantial economic growth driven primarily by the fishing industry. The economy diversified greatly after the country joined the European Economic Area in 1994, but Iceland was especially hard hit by the global financial crisis in the years following 2008. The economy is now on an upward trajectory, fueled primarily by a tourism and construction boom. Literacy, longevity, and social cohesion are first rate by world standards" (*World Factbook*, 2022).

The maritime claim is 200 nautical miles or to the edge of the continental margin, and the territorial sea claim is 12 nautical miles. Iceland's maritime claims use the "straight baseline system," which are ellipsoidal loxodromes (rhumb lines) that connect 31 points on the coastline perimeter.

According to the National Land Survey of Iceland or *Landmælingar Íslands* (LMÍ), "It is believed that *Guðbrandur Þorláksson*, the bishop at Hólar, was the first Icelander to be involved in mapmaking. Guðbrandur lived from 1541 to 1627 and measured the global position of Hólar with amazing precision. A map named after him was published in 1590. Björn Gunnlaugsson, a teacher at *Bessastaðaskóli School*, made a map of Iceland in 1844. The map was named after him and used for the next 100 years!"

The Reykjavík Datum of 1900 was established by the Danish Army General Staff and published in the *Geodætisk Instituts Publikationer VII, Island Kortlægning* where $\Phi_o = 64°\ 08'\ 31.88''$ N and $\Lambda_o = 34°\ 30'\ 31.5''$ West of *Københavns* (Copenhagen) or $\Lambda_o = 21°\ 55'\ 51.15''$ West of Greenwich. The ellipsoid of reference was the Danish (Andræ) 1876 where $a = 6,377,019.25666$ m and $1/f = 300$. The classical triangulation was initially used for plane table mapping at the scale of 1:50,000, but the publication scale was changed to 1:100,000 7 years later. Thirty years thereafter, the first oblique aerial photography was flown in Iceland. The original grid system used for the Reykjavík Datum of 1900 was the Islandic Conformal Conic where the latitude of origin, $\varphi_o = 65°\ 00'$ N, the central meridian, $\lambda_o = 19°\ 01'\ 19.65''$ West of Greenwich, and the scale factor at origin, $m_o = 1.0$ (tangent conic). There was no false origin used with the grid associated with the Reykjavík Datum of 1900. However, since the conformal conic projection used in Denmark at the time was termed the "Buchwaldt projection," that term might have been also used in Iceland. (Colonel Frants Andræs Buchwaldt (1874–1923) was the director of the Geodetic Service of Denmark. I have noticed that one reference published by *LMÍ* includes a Danish paper authored by another Buchwaldt in 1976.) The datum shift parameters (in standard military "three-parameter Molodensky" form) from the Reykjavík Datum of 1900 to the WGS 84 Datum are: $\Delta a = 1118$ m, $\Delta f = -0.0000195$, $\Delta X = -636$ m, $\Delta Y = +21$ m, $\Delta Z = -934$ m. An official notice states, "All maps on a scale of 1:100 000 published by *Landmælingar Íslands* are currently provided in Reykjavík 1900 datum. In the available map scales Reykjavík 1900 differs significantly from WGS84. ... In case of the 1:50 000 map scale the available maps are provided in two different series first published before and after 1955. For those published after 1955 the horizontal datum is Hjörsey 55 as indicated in the map legend. For those published before this threshold the phrase *Horizontal Datum is based on the Astronomic Station of Reykjavík; 21°55′51.15″ West of Greenwich, 64°08′31.88″ N* indicates the Reykjavík 1900 datum. For a variety of technical reasons, it is impossible to obtain transformation parameters with high accuracy and nationwide validity. However, for navigational purposes (*e.g.*, hiking) a set of transformation parameters has been derived from graphical comparisons. Although they proofed (*sic*) to be useful for orientation with the *LMÍ* maps they do NOT provide geodetic accuracy! In average, from WGS84 towards Reykjavík 1900, a point requires a horizontal shift of approx. 200 m westwards resp. 25 m southwards (± 25 m)."

The same document that originally listed the Reykjavík Datum of 1900 included another, the Akureyri Datum of 1900. Thanks to John W. Hager, "For Akureyri I have latitude (Φ_o)=65° 40′ 15.2″ N±0.2″ or 15.8″ N±0.1″ and longitude (Λ_o)=18° 05′ 12.6″ W±09.0″. ... I was in Akureyri earlier this year and tried, using my GPS, to locate the position but was not able to do so. Time was very limited. Suspect that the point is in the area of the botanical gardens and the hospital." Although this apparent "astro" station represents an obsolete local datum, the U.S. Army Map Service (AMS) noticed that several 1:100,000 scale maps in the region of the town of Akureyri "did not agree with the control values by several seconds *(of arc – Ed.)*." This notation was dated 20 December 1946 by William W. Baird, AMS.

When Denmark was occupied during World War II, Iceland petitioned for independence. That was granted in 1944 as mentioned above, and the United Kingdom and the United States subsequently moved in because of wartime concern for the island's vulnerability. An interim datum was computed apparently for cartometric purposes and is locally termed the Reykjavík 1945 datum referenced to the Hayford 1909 (International 1924) ellipsoid. In 1955, a new classical triangulation and geodetic survey was initiated by Denmark and the United States. The following year, the *LMÍ* was founded. The new survey established the Hjörsey Datum of 1955 where Φ_o=64° 31′ 29.26″ N and Λ_o=22° 22′ 05.84″ West of Greenwich and the ellipsoid of reference is the Hayford 1909 (International 1924) where a=6,378,388 m and $1/f$=297. The grid system devised by AMS for the new datum was the Icelandic Gauss-Krüger Transverse Mercator with four belts (1–4) where the central meridians, λ_o=15° W, 18° W, 21° W, and 24° W, the scale factor at origin, m_o=1.0, and the False Easting of each belt=500 km. However, it appears that *LMÍ* ignored the Transverse Mercator devised by AMS and instead utilized another Lambert Conformal Conic zone where the latitude of origin, φ_o=65° 00′ N, the central meridian, λ_o=18° West of Greenwich, and the scale factor at origin, m_o=1.0 (*another* tangent conic), and the False Easting=False Northing=500 km. The datum shift parameters published by NGA in *TR8350.2* from the Hjörsey Datum of 1955 to the WGS 84 Datum are: Δa=−251 m, Δf=−0.14192702, ΔX=−73 m±3 m, ΔY=+46 m±3 m, ΔZ=−86 m±6 m, and this solution is based on six points. A seven-parameter Bursa-Wolfe transformation published by *LMÍ* from the Reykjavík Datum of 1900 to the Hjörsey Datum of 1955 *(rotations changed to the U.S. Standard – Ed.)* is where ΔX=+629.020 m, ΔY=−214.701 m, ΔZ=+1,028.364 m, R_x=−4.154″, R_y=+0.269″, R_z=+2.279″, Δs=−3.729.

A geodetic surveying campaign was carried out by "Icelandic and German agencies for the purpose of establishing a new horizontal geodetic datum in Iceland. The work culminated in a GPS-campaign named ISNET93 during 3–13 August 1993. The associated new geodetic datum is named ISN93. It will replace the Hjörsey-1955 datum established by terrestrial observations in 1955–56." – *GPS-mælingar í grunnstöðvaneti 1993*. A seven-parameter Bursa-Wolfe transformation published by LMÍ from the Reykjavík Datum of 1900 to the ISN93 *(rotations changed to the U.S. Standard – Ed.)* is where ΔX=+556.020 m, ΔY=−168.701 m, ΔZ=+942.364 m, R_x=−4.154″, R_y=+0.269″, R_z=+2.279″, Δs=−3.729, and the ellipsoid of reference is the GRS 1980 where a=6,378,137 m, and $1/f$=298.257222101. *LMÍ* does offer a free interactive coordinate transformation service (*cocodati*) via the Internet. The new grid system adopted is the *secant* Lambert Conformal Conic projection with standard parallels 65° 45′ N and 64° 15′ N, and central meridian 19° W. False Eastings=False Northings=500km at 65° N and 19° W. The GPS network consists of 119 stations, of which 63 are pillars and the remainder are benchmarks in bedrock. Thanks to Mr. Gunnar Þorbergsson for his historical accounts of the Icelandic datum relations recorded in exquisite detail.

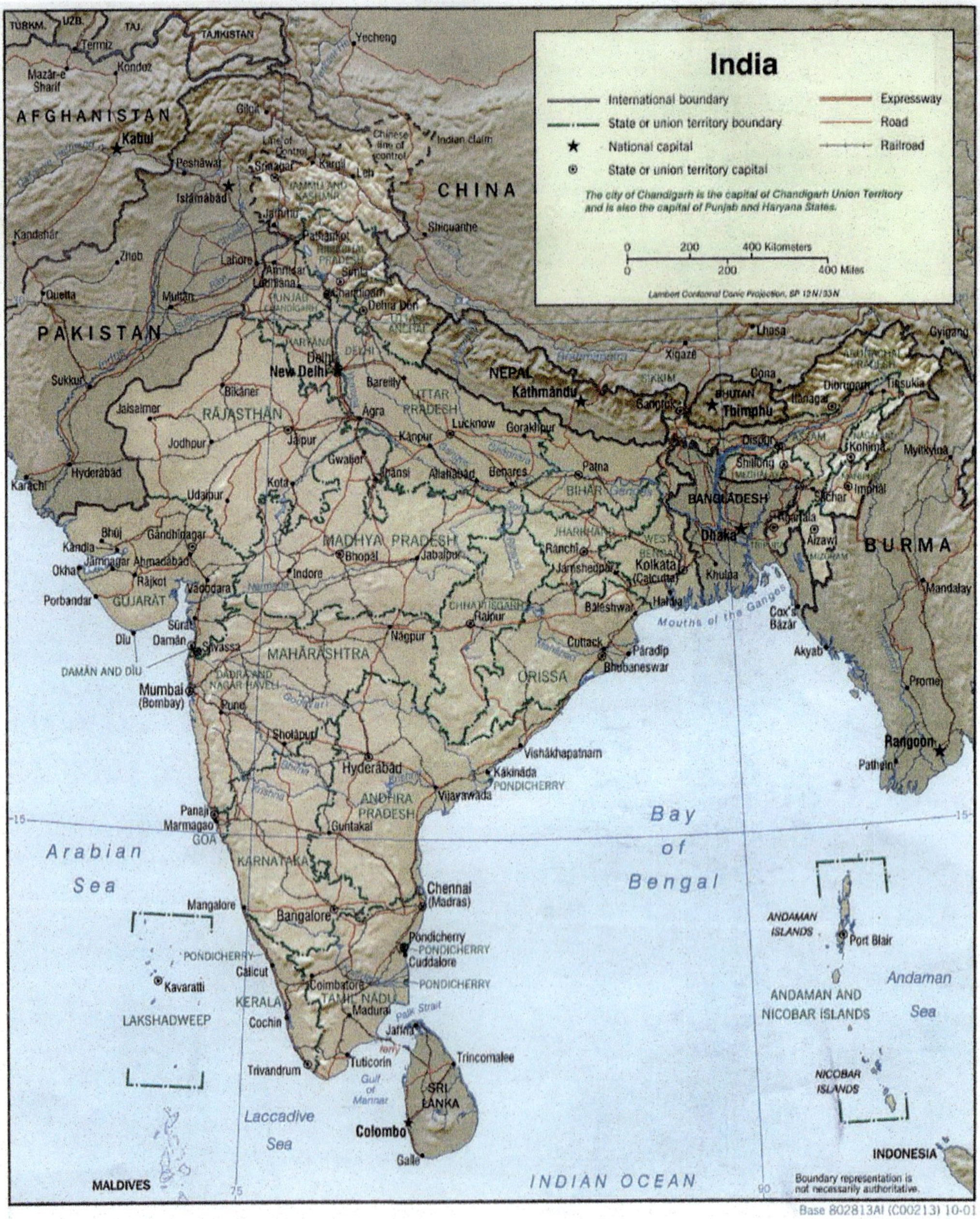

Base 802813AI (C00213) 10-01

Republic of India

Slightly more than one-third the size of the United States, India is bordered by Bangladesh (4,142 km), Bhutan (659 km), Burma (1,468 km), China (2,659 km), Nepal (1,770 km), and Pakistan (3,190 km). The terrain is upland plain (Deccan Plateau) in south, flat to rolling plain along the Ganges, deserts in west, and Himalayas in north. The highest point is *Kanchenjunga* (8,586 m), and the lowest point is the Indian Ocean (0 m).

"The Indus Valley civilization, one of the world's oldest, flourished during the 3rd and 2nd millennia *B.C.* and extended into northwestern India. Aryan tribes from the northwest infiltrated the Indian subcontinent about 1500 *B.C.*; their merger with the earlier Dravidian inhabitants created the classical Indian culture. The Maurya Empire of the 4th and 3rd centuries *B.C.* – which reached its zenith under Ashoka – united much of South Asia. The Golden Age ushered in by the Gupta dynasty (4th to 6th centuries *A.D.*) saw a flowering of Indian science, art, and culture. Islam spread across the subcontinent over a period of 700 years. In the 10th and 11th centuries, Turks and Afghans invaded India and established the Delhi Sultanate. In the early 16th century, the Emperor Babur established the Mughal Dynasty, which ruled India for more than three centuries. European explorers began establishing footholds in India during the 16th century.

"By the 19th century, Great Britain had become the dominant political power on the subcontinent and India was seen as the 'Jewel in the Crown' of the British Empire. The British Indian Army played a vital role in both World Wars. Years of nonviolent resistance to British rule, led by Mohandas Gandhi and Jawaharlal Nehru, eventually resulted in Indian independence in 1947. Large-scale communal violence took place before and after the subcontinent partition into two separate states – India and Pakistan. The neighboring countries have fought three wars since independence, the last of which was in 1971 and resulted in East Pakistan becoming the separate nation of Bangladesh. India's nuclear weapons tests in 1998 emboldened Pakistan to conduct its own tests that same year. In November 2008, terrorists originating from Pakistan conducted a series of coordinated attacks in Mumbai, India's financial capital. India's economic growth following the launch of economic reforms in 1991, a massive youthful population, and a strategic geographic location have contributed to India's emergence as a regional and global power. However, India still faces pressing problems such as environmental degradation, extensive poverty, and widespread corruption, and its restrictive business climate is dampening economic growth expectations" (*World Factbook*, 2022).

"The art, culture and kingdoms of India could not have spread through centuries and countries without knowledge of its geography. In the Vedic literature of over 5000 years ago, the knowledge of land was presented in a graphical form which described the extent and shape of territories. The *Brahmand Purana* of 500 *B.C.* to 700 *A.D.* gives evidence of the art of modern map-making. The art of surveying and techniques of mensuration of areas are described in *Sulva Sutra* (science of mensuration) and in the *Arth Shastra* of *Chanakya* written in the 3rd century *B.C.* The golden age of Indian Renaissance in the 5th century saw the towering genius Arya Bhat who wrote *Surya Siddhant* and calculated the earth's circumference to be 25,080 miles – less than 200 miles off modern measurements of the equator. Chinese and Arab travelers and many adventurers also contributed to Indian geography. Sher Shah Suri and Todar Mai's revenue maps, based on regular land survey systems, were well known in the medieval period, and continued to be in practice during the mid-eighteenth century. Even today, the six huge instruments in masonry built by Raja Sawai Jai Singh in the heart of New Delhi in 1724 attract tourists from all over the world. These were designed and built by him to study the movements of the sun, moon, and planets. Such instruments were also built in Jaipur to measure, among other things, time, and eclipses. Another observatory was built by him in Ujjain in 1723 to forecast eclipses and movements of the sun as well as indicate the correct time. According to records, Rajaraja I of Tanjore (985–1011 A.D.) carried out careful surveys of the lands and cultivation. This shows that there must have been many other surveys of which no clear records have been preserved. However, information is available of the surveys instituted by Akbar during the 16th century; measurements being made by a hempen rope which was replaced by a *'jarib'* of bamboos joined by iron rings. Settlement operations included the measurement and classification of lands, and fixation of rates. Systematic surveys commenced in the 18th century.

"The Survey of India traces its birth to the appointment of Major James Rennell as Surveyor General of Bengal, by Lord Robert Clive and his council, on the first of January 1767. He placed all available surveyors under Major Rennell's orders, amongst them being the Frenchman Claude Martin, who later became famous as the founder of the *La Martiniere Schools*. By 1773, Rennell completed surveys of the possessions before relinquishing the post of Surveyor General in 1777. Rennell surveyed Bengal and Bihar, an area of over 1500 sq. miles, producing a continuous and uniform set of maps. The surveys, however, were far from complete or accurate in detail but were sufficient to meet the needs of the time. Rennell continued his interest in England, and his first Map of Hindustan reached India in 1783. The early history of surveys in India followed the East India Company's expanding areas of influence and conquest. The next Surveyor General, Thomas Call, like many others who followed him, undertook the task of compiling an atlas embracing the whole of India. On the initiative of John Tringle, who surveyed routes with great enthusiasm, a military *'Corps of Guides'* was established. This *Corps* also contributed largely to the surveys of the Madras Presidency for the next 30 years. It was in 1787 that Michael Topping, a marine officer, broke away from the eternal method of Perambulator Traverse and ran a 300-mile line of triangles along the coast from Madras to Palk Strait. It was he who built a permanent astronomical observatory in Madras in 1793 and founded the first surveying school in 1794. In 1796 and 1810, the Presidencies of Bombay and Madras got their own Surveyors General with the appointment of Lt Gen. Charles Reynolds and Col. Colin Mackenzie as the respective Surveyors General. It was on the first of May 1815 that the Directors, finding it wasteful to maintain three separate and independent Surveyors General, appointed Mackenzie as the Surveyor General of India. The credit the first surveys of the Brahmaputra in Assam in 1794, and that of the Irrawaddy River in Burma go to Thomas Wood. The mission also collected interesting information about people, tribes and general geography of Assam and Burma, about which nothing whatever had been known before. India was one of the earliest countries to establish a regular government survey organization and to commence systematic surveys – a few years before even the Ordnance Survey of UK.

"It was very fortunate that a man of the genius and resolution of Lambton was in the subcontinent to lay the foundation of the 'Great Trignometrical Survey of India' a few years before similar projects were undertaken by France and England. In November 1799, he put forward his proposal for a Mathematical and Geographical Survey that should extend right across the Peninsula from sea to sea, controlled by astronomical observations carried out on scientific principles, capable of extension in any direction and to any distance. He started his work from Madras where, in early 1802, he measured the famous base line at Saint Thomas' Mount as a start for his triangulation, north and south through Carnatic India and across the Peninsula, with his famous 36-inch great theodolite. He completed a meridional arc from Cuddalore to Madras observing latitude at both ends and obtaining a value for the length of a degree that was essential for his scientific work. By 1815, he had nearly covered the whole Peninsula south of the river *Kistna* (Krishna) with a network of triangulations braced by main cross belts. To him goes the distinction of measuring the longest geodetic arc closest to the equator, from Cape Comorin to the 18° parallel.

"In 1806, a subaltern came to India at the tender age of sixteen. He was none other than Lieutenant George Everest. He joined Lambton in 1818. Lambton died at work on 20 January 1823 at Hinganghat at the age of 70. General Walker recognizing his work wrote in 1870, 'of all Col. Lambton's contributions to geodesy, the most important are his measurements of meridional arcs, the results of which have been employed up to the present time in combination with those of other parts of the globe, in all investigations of the figure of the earth.' Lambton's mantle fell on the worthy shoulders of George Everest. Everest felt the need for basing the surveys on a rigid reference framework. This raised the problem of finding a suitable reference spheroid to fit the shape of the earth's gravity equi-potential surface for India and the adjacent countries. Everest realized that the Indian subcontinent was too large for basing surveys on an osculating sphere, let alone a tangent or secant plane. Everest, therefore, started his control work from Kalianpur in Madhya Pradesh, more or less in the centre of India. Here he made astronomical observations and treated the astronomical latitude, longitude, and the plumbline at that place as error-free. With Kalianpur as the center, he conceived covering the length and breadth of India by a gridiron of triangular chains, as opposed to the network of triangles conceived by Lambton. He brought to surveying greater accuracy and rigorous observational procedures besides devising and refining the instruments. He introduced the observation of astronomical azimuths from pairs of circumpolar stars, ray traces for long lines, *etc.* His redesigned 36-inch great theodolite is famous today. He replaced the chain with Colby's baseline apparatus and 10-foot compensation bars, with which he measured various bases. He completed the great meridional arc from Cape Comorin to Banog in the first Himalaya near Mussoorie, a length of 2400 km. Everest made the government agree to the revision of Lambton's work, based on more accurate instruments and the procedures as laid down by him. Later, in 1830, he was appointed as the Surveyor General of India but, much against the wishes of the then government, he continued to devote much time to the Great Meridional Arc. This was completed by him in 1841 and he utilized the last 2 years of his service in its computations and adjustments. The work and norms laid down by Everest have stood the test of time. The Everest spheroid, evolved by him in the year 1830, is not only still being used by India but also by Pakistan, Nepal, Burma, Sri Lanka, Bangladesh, Bhutan and other south-east Asian countries.

"We can only grasp the significance of his monumental work if we can visualize India of the early nineteenth century – without communications and full of jungles, wild animals, robbers and disease. The average length of a side of the triangulation was about 31 miles, the maximum being about 62 miles. One cannot imagine how such long-distance observations were planned, laid down on the ground, line of sight cleared of all trees and sometimes even houses, and how big rivers and swamps were crossed. Everest, devoted to his work, did all this despite his partial paralysis and bad health. Based on his conceptualization, the gridiron network today covers the entire country and forms a solid foundation for accurate surveys and mapping for defense, development and efficient administration. It was with the help of this gridiron network that the highest peak of the world was observed and discovered in 1852 and its height declared as 29,002 ft. – *i.e.* about 8840 meters. After fresh observations and computations, the Survey of India declared its height in 1954 as 8848 meters. In 1975, the Chinese put a metallic beacon on Everest and observed it from 9 stations. They also carried out sufficient astronomical and gravimetrical measurements, the coefficient of refraction was reliably determined, and the final result of the determination was declared as 8848.13 ± 0.35 meters. Sir Thomas Holdick concluded in the *Standard* of January 24, 1905, that 'It was officers of the Survey of India who placed his name just near the stars, then that of any other lover of eternal glory of the mountains and let it stay in witness to the faithful work not of one man but scores of men.' Everest was the first from amongst the eight Surveyors General of India to be knighted" (*Survey of India through the Ages, by Lt. General S. M. Chadha, delivered at the Royal Geographical Society on 8 November 1990, on the eve of Sir George Everest Bicentenary celebration by the Surveyor General of India*).

"The bulk of the geodetic triangulation in India was carried out with large theodolites between 1802 and 1882. Its simultaneous adjustment involved a decade's labor. In the 20th century, very little was done in the way of geodetic triangulation – only a few outlying series in Baluchistan and Burma having been observed. The presence of Military Survey Companies in the different theaters during WWII and in the period immediately following it enabled important gaps between the triangulation of India and its neighboring countries like Iraq, Iran, Siam and Malaya to be filled. A continuous chain of triangulation now exists from Syria to Malaya" (*Geodetic Work in India – War and Post-War, B.L. Gulatee, Empire Survey Review, No. 77, Vol. X, 1950*).

The origin of the *Indian Adjustment of 1916* is at station Kalianpur (Strong Base) where $\Phi_o = 24°$ 07' 11.26" N, $\Lambda_o = 77° 39' 17.57"$ East of Greenwich, $\alpha_o = 190° 27' 05.10"$ from Kalianpur to Surantal, and the ellipsoid of reference is the Everest 1830 *(India)* where $a = 6,377,301.243$ m, $1/f = 300.8017$ *(UK Military Survey, 1982)*. The adjustment depended on baselines evenly distributed throughout India. The use of LaPlace stations had not yet been adopted by the Survey of India; consequently, errors in azimuth and position were introduced. These errors are particularly evident in the triangulation series of southern India. The 1880 adjustment has, however, remained the basis of all Indian triangulation and mapping. Therefore, there is no such thing as an "Indian Datum"; it is only an adjustment! *(Personal communication, James M. Nettles, 21 June 1997)*.

"In 1924 the "Minute Mesh" was introduced. This is a reference system consisting of meridians and parallels at one-minute intervals: descriptive references are given by a convenient system of lettering, and all survey computations are done in spherical terms in the usual way" *(Geodetic Report, Survey of India, CPT. G. Bomford, R.E., 1930)*. "As the result of a decision arrived at the artillery survey conference held at Akora on 12th January 1926, two forms and a set of tables were prepared for the conversion of the spherical co-ordinates to rectangular, and *vice versa*, on Lambert's conical orthomorphic projection. This projection is also known as Lambert's second projection with two standard parallels" *(Geodetic Report, Survey of India, CPT. G. Bomford, R.E., 1928)*.

Seven separate India Zones were created in 1926 by the Survey of India, all seven having the same scale factor at origin $(m_o) = 823/824 = 0.998786408$, the same False Easting $= 3,000,000$ Indian Yards, and the same False Northing $= 1,000,000$ Indian Yards. The following parameters differ: Zone I has $\varphi_o = 32° 30'$ N & $\lambda_o = 68°$ E; Zone IIA has $\varphi_o = 26°$ N & $\lambda_o = 74°$ E; Zone IIB has $\varphi_o = 26°$ N & $\lambda_o = 90°$ E; Zone IIIA has $\varphi_o = 19°$ N & $\lambda_o = 80°$ E, Zone IIIB has $\varphi_o = 19°$ N & $\lambda_o = 100°$ E, Zone IVA has $\varphi_o = 12°$ N & $\lambda_o = 80°$ E, and Zone IVB has $\varphi_o = 12$ N & $\lambda_o = 104°$ E. As a hint for the readers that need the actual two standard parallels for each of the above India Zones expressed with the British Method of defining Lambert parameters; for India Zone I, the equivalent standard parallels are 35° 18' 50.3486" N and 29° 39' 18.7703" N. The requisite equations to solve for the other India Zones are in Chapter 3 of the *Manual of Photogrammetry*, editions 5 and 6.

Because there is no unified datum in existence for the sub-continent of India, there is a significant difference in transformation parameters from Indian 1916 to WGS84 from region to region. The Survey of India is slowly releasing geodetic and cartographic data to the general public after maintaining a significant degree of secrecy for centuries. Curiously, the India Lambert Zones are for restricted military use, and civilian applications in GIS for India seem to prefer the U.S. military's Universal Transverse Mercator Grid System.

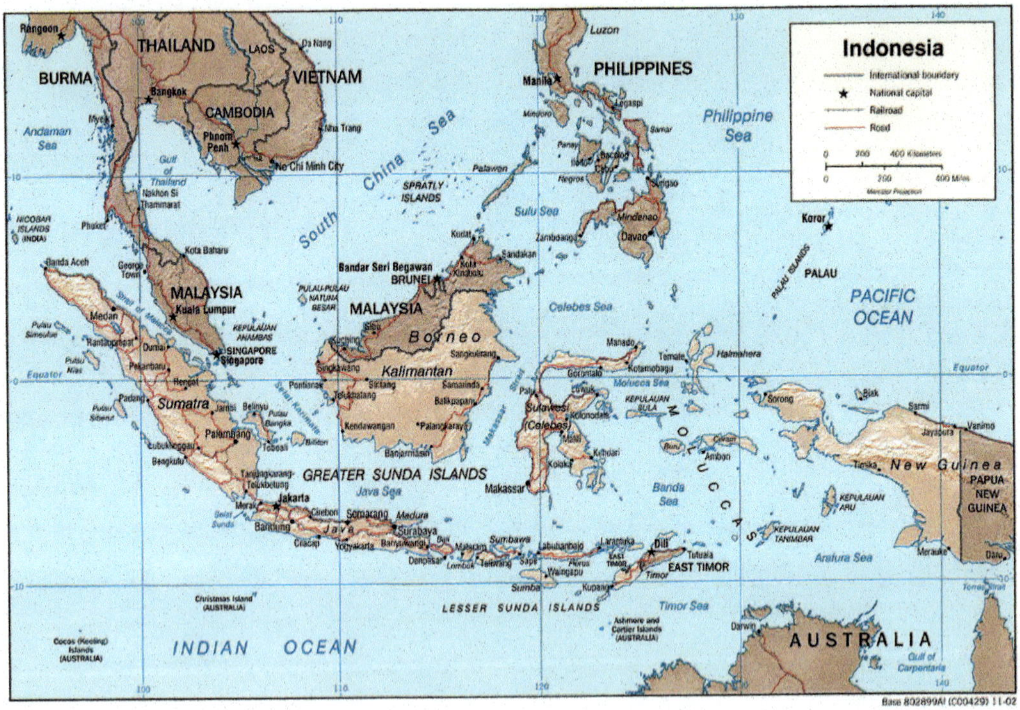

Base 802899AI (C00429) 11-02

Republic of Indonesia

Slightly less than three times the size of Texas, Indonesia is bordered by Timor-Leste (228 km), Malaysia (1,782 km), and Papua New Guinea (820 km). The lowest point is the Indian Ocean (0 m), the highest point is *Puncak Jaya* (5,030 m), and the region is tectonically unstable with some 400 volcanoes, of which 100 are active (*World Factbook*, 2009).

"Beginning in the 1890s, paleontologists discovered fossil remains of creatures on the island of Java that, while probably not the direct ancestors of modern humans, were closely related to them. These Javan hominids, known by scientists as **Homo erectus**, lived 500,000 years ago and some possibly as long as 1.7 million years ago. Evidence of probable descendants of the Trinil *erectus*, known as **Homo soloensis** or Solo Man, was found at Ngandong, also in Central Java; these descendants are thought to have evolved between 500,000 and 100,000 years ago. Assemblages of stone tools have not clearly been tied to **Homo soloensis**, but there is evidence that these early **Homo sapiens** had a rudimentary social organization (small hunting and gathering bands) and used simple tools around 40,000 years ago. Although Indonesia is extremely diverse ethnically (more than 300 distinct ethnic groups are recognized), most Indonesians are linguistically – and culturally – part of a larger Indo-Malaysian world encompassing present-day Malaysia, Brunei, the Philippines, and other parts of insular and mainland Asia" (*Library of Congress Country Study*, 2009). The Dutch began to colonize Indonesia in the early 17th century; Japan occupied the islands from 1942 to 1945. Indonesia declared its independence after Japan's surrender, but it required four years of intermittent negotiations, recurring hostilities, and UN mediation before the Netherlands agreed to transfer sovereignty in 1949. Indonesia's first free parliamentary election after decades of repressive rule took place in 1999. Indonesia is now the world's third-largest democracy, the world's largest archipelagic state, and home to the world's largest Muslim population. "In 2005, Indonesia reached a historic peace agreement with armed separatists in Aceh. Indonesia continues to face low intensity armed resistance in Papua by the separatist Free Papua Movement" (*World Factbook*, 2022).

"In 1850 the Geographical Service was founded as a part of the Navy Department, and Mr. De Lange was sent from Holland in order to determine the geographical positions of various stations in the Archipelago by astronomical observations; de Lange was the first who made some elementary triangulations in the Netherlands East Indies. In 1857 Dr. Oudemans, afterwards professor in astronomy at the University of Utrecht, came to India *(Indonesia – Ed.)* in order to execute astronomical observations for surveying purposes, but soon he convinced the government of the necessity of a regular triangulation of Java, not only for the topographical survey but also for scientific purposes. The primary triangulation of Java, started in 1862, was finished in 1880 and soon after that the Geographical Service was dissolved. The whole system has 114 points, situated on the highest mountains, and three bases *(Simplak [3,887.696 m.], Tangsil [5,040.73 m.], and Logantoeng [4,175.874 m.] – Ed.)*. The coordinates of the Java triangulation-system are founded on a latitude-and-an azimuth determination at Genoek; for the computation of the longitudes the meridian of Batavia (former time-signal station) was taken as zero. *(Note that the origin of Goenoeng Genoek 1873 is where $\Phi_o = 06° 07' 39.520'' S$, $\Lambda_o = +106° 48' 27.790''$ East of Greenwich, deflections of the vertical where $\xi = -0.20''$, $\eta = +11.0''$, and the ellipsoid of reference is the Bessel 1841 where $a = 6,377,397.155 m$, $1/f = 299.1528128$ – Ed.)*

"In 1883 a beginning was made with the survey of West Sumatra, and a Triangulation Brigade was founded as a part of the military Topographical Service. The primary triangulation in Sumatra consists of 118 points, its angles were measured with 27 cm micrometer-theodolites from Pistor and Matins, Wegener and Wanschaff. It consists of three sections, the triangulation of West Sumatra (1883–1896), that of South Sumatra (1895–1909) and that of the Residency of Sumatra's Eastcoast

(1907–1916). The linear dimensions of Sumatra's Westcoast system are computed from the base line at Padang, measured in 1881 with steel tapes; those of South Sumatra are deduced from the Java-system, while in the triangulation-system of Sumatra's East coast a base line was measured at Sampoen in 1910 ([6,666.476 m.] – Ed.). This base was the first one measured with invar wires, since then 5 other bases were measured in this way in the Netherlands East Indies. The position of the West Sumatra-system on Bessel's ellipsoid was until now deduced from a latitude- and azimuth determination at Padang, the longitudes being calculated relative to the meridian of the western end of the base line ([3,986.149 m.] – Ed.). (Note that the origin of Padang BP-A 1884 is where $\Phi_o = 00°\ 56'\ 38.414''$ S, $\Lambda_o = +100°\ 22'\ 08.804''$ East of Greenwich, $H_o = 3.190$ m, and deflections of the vertical are where $\xi = -6.0''$, $\eta = null$, as computed in the Java system – Ed.)

"The primary triangulation in Celebes (Sulawesi – Ed.) was begun in 1911. The angles of the primary system were measured as in Sumatra; three bases (Djeneponto 1911 [10,476.168 m.], Tondano 1915 [7,205.647 m.], Korodolo 1920 [6,567.518 m.]) were measured with the Jäderin apparatus (used with 25 m. brass & invar wires – Ed.). Its place on Bessel's ellipsoid was fixed by a latitude-and an azimuth-determination at Montjong Lowé P1, (Note that the origin of Montjong Lowé P1 1911 is where $\Phi_o = 05°\ 08'\ 41.42''$ S, $\Lambda_o = +119°\ 24'\ 14.94''$ East of Greenwich – Ed.) and a telegraphic longitude-determination at Makassar (Ujungpadang – Ed.) in 1891. (Note: The French published the 1891 astronomic coordinates of Makassar as: $\Phi_o = 05°\ 08'\ 09.8''$ S, $\Lambda_o = +119°\ 24'03.4''$ East of Greenwich (Annales Hydrographiques, 2e Série, Tome Quatorziéme, année 1892, Paris, pp. 351–355. – Ed.)

"The primary triangulation in the Small Soenda Archipelago was extended in the period 1912–1918 eastward over Bali and Lombok. The primary triangulation only contains 11 points. The primary triangulation in Bangka consists of an isolated system of 14 points. At Bakem a base line was measured (1926) with the Jäderin apparatus and invar wires ([5,221.763 m.] – Ed.), while longitude, latitude and azimuth were determined (1926) at Bukit Rimpah. (Note that the origin of Bukit Rimpah 1926 is where $\Phi_o = 02°\ 00'\ 40.16''$ S, $\Lambda_o = +105°\ 51'\ 39.76''$ East of Greenwich – Ed.) A beginning with this triangulation was made in 1917, but from 1921–1926 the work was temporarily stopped.

"Triangulation computations were carried out in the projection of Mercator and from the coordinates in this projection the geographic coordinates and those in the polyhedral projection are derived. In the flat and swampy parts of Sumatra and Borneo (Kalimantan – Ed.), covered with primeval forests, a triangulation is practically impossible and would anyhow be too expensive. So, the necessary fixed points for the surveying, which are chosen near the rivers, are determined astronomically." (Excerpted from: Geodetic Survey in the Netherlands East Indies, Schepers, J.H.G., and Schulte, Capt. F.C. A., Topographische Dienst, 1931, 17 pages.) There are two classical Indonesian datum origins on Borneo: Gunung Serindung Ep. A 1962 where $\Phi_o = +01°\ 06'\ 10.60''$ N, $\Lambda_o = +105°\ 00'\ 59.82''$ East of Greenwich, and Gunung Segara P5 1933 where $\Phi_o = 00°\ 32'\ 12.82''$ S, $\Lambda_o = +117°\ 08'\ 48.47''$ East of Greenwich, both referenced to the Bessel 1841 ellipsoid. For Timor, Koepang (Kupang) Datum is known to exist.

A major geodetic network commonly used in the 1980s "that stretches from North Sumatra through to Java and the Lesser Sunda Islands is the Genuk Datum with origin at Gunung Genuk P-520 where: $\Phi_o = 06°\ 26'\ 53.350''$ S, $\Lambda_o = +110°\ 55'\ 05.836''$ East of Greenwich, $H_o = 716.700$ m. In an attempt to provide a means of connection between individual datums, and to provide a means of readjustment for geodetic data into a single datum the Indonesian Government adopted the parameters of the Geodetic Reference System of 1967 (GRS-67) as that of the Indonesian National Spheroid (INS). These parameters were referenced to the Datum Point at Padang in West Sumatra and the geodetic and Cartesian coordinates, obtained from satellite observations at the Datum Point, were used to compute Cartesian coordinates for a spheroid having the approximate GRS-67 parameters. The two sets of coordinates were differenced to provide a set of transformation constants between WGS72 satellite datums and the adopted geodetic datum ID-74. The Indonesian National Spheroid (INS or ID-74) is defined where $a = 6,378,160$ m, and $1/f = 298.247$" (A Report on *the State of Geodesy in Java by L. J. Quilty, Jakarta, 1995).* The Indonesian Datum 1974 coordinates of the origin "Padang BP-A 1884" are: $\Phi_o = -00°\ 56'\ 37.980''$ S, $\Lambda_o = +100°\ 22'\ 08.467''$ East of Greenwich, $h_o = H_o = 7.500\,m$ (*Datum Geodesi Nasional 1995 (DGN-95) Yang Geosentrik, Subarya, C., Matindas, R.W., Bakosurtanal, 27 pages, 1995).*

"The *Topografische Dienst* has employed several grids and projections in mapping the Netherlands Indies. On Sumatra they employed a Mercator Grid the specifications of which are believed to be as follows: Origin=Equator, meridian of Padang, Scale Factor=none, False coordinates of origin=none. The *Topografische Dienst* also may have employed another Mercator Projection originating at the Equator and the Meridian of South Sumatra, 3° 15' West of Batavia. The *Topografische Dienst* used these projections to compute the positions of some of their third-order points. Generally, the Mercator coordinates were transformed to another projection, the Polyhedric. This projection is really a series of projections each covering a 20×20-minute quadrilateral and having as its origin the center of the quadrilateral. Each quadrilateral corresponds to a sheet of the 1/100,000 map series of the Netherland Indies. The projection used to cover one sheet is considered to be a Lambert Conformal, but in practice, over so small an area, it is indistinguishable from a Polyconic, Transverse Mercator, or any other projection having the same origin. The origin specifications of one such projection are an odd multiple of 10 minutes from the Equator; and odd multiple of 10 minutes from the meridian of Padang or that of South Sumatra (3° 15' West of Batavia).

The *Topografische Dienst* designed numerous Bonne Grids for the various residencies of Java. These have been abandoned presumably because they were not of the conformal type and hence were awkward for survey work. All positions in Java are now expressed in terms of the Polyhedric Projection, as described for Sumatra. The origin is the intersection of the Equator with the meridian of Batavia.

In Borneo, in addition to the Polyhedric coordinates, a system of Mercator coordinates, similar to that employed on Sumatra was used. The origin, however, is the intersection of the Equator with the meridian of 7° East of Batavia.

The *Topografische Dienst* used a Mercator Grid in Celebes similar to that described for Sumatra, except that the origin is at the intersection of the Equator with the meridian of 15° East of Batavia. The longitudes of the sheet lines on Celebes likewise are based on the 15th meridian east of Batavia. In the Lesser Sundas, a Mercator Grid whose central meridian was 8° 15' West of Batavia was used. In Halmahera and Amboina, a Mercator Projection with the central meridian 22° East of Batavia was employed. The Prime Meridians used by the *Topografische Dienst* are as follows, all with reference to Greenwich: Batavia, +106° 48' 27.79" East; Padang, +100° 22' 01.42" East; South Sumatra, +103° 33' 27.79" East; Singkawang, +108° 59' 41.00" East; Middle Celebes, +121° 48' 27.79" East.

The Netherlands East Indies Southern Zone covers Java and the Lesser Sunda Islands. Its specifications are: Projection: Lambert Conformal Conic; ellipsoid: Bessel 1841; Unit: Meter; Origin: 8° South, 110° East; False Coordinates of Origin: 550,000 meters East, 400,000 meters North; Scale Factor: 0.9997; Limits: North: From the junction of the zero Easting grid line of the zone and the loxodrome shoes end points are 7° South –104° 30' East and 5° South –107° East; Northeast to end point at 5° South, 107° East; thence east on the parallel of 5° South to 117° 30' East; thence south along this meridian to 7° South thence east along this parallel to 137° East. East: Meridian of 137° East. South: West to 125° East on the parallel of 11° South; thence south along this meridian to 12° South; thence west along this parallel to 120° East; thence north along this meridian to 11° South; thence west along this parallel to the Zero Easting grid line of the zone. West: Zero-meter Easting grid line of the zone. Early in 1942, DSvy *(UK Military Directorate of Survey – Ed.)* published some maps on Java on which this grid appeared with an increase of 3,000,000 meters to the Eastings and 1,000,000 meters to the Northings.

The remainder of the East Indies, except for a small section of New Guinea south of 7° South, is covered by the Netherlands East Indies Equatorial Zone whose specifications are: Projection: Mercator; ellipsoid: Bessel 1841; Unit: Meter; Origin: Equator, 110° East; False Coordinates of Origin: 3,900,000 meters East, 900,000 meters North; Scale Factor: 0.997; Limits: North: Along the parallel of 7° North to the meridian of 98° 40′ East; thence south along this meridian to 4° 40′ North; thence loxodrome to point 103° 50′ East –0° 30′ North; thence east along the parallel of 0° 30′ North to 105° East; thence north along this meridian to 7° North; thence east along this parallel to 119° 30′ East; thence south along thins meridian to 5° North; thence east along this parallel to 165° East. East: Meridian of 165° East. South: Along the parallel of 5° South to the meridian of 153° 30′ East thence south along this meridian to 7° South; thence west along this parallel to 117° 30′ East; thence north along this meridian to 5° South; thence west along this parallel to 107° East; thence loxodrome to point 104° 30′ East –7° South; thence along parallel of 7° South to 94° East. West: Meridian of 94° East" (*Notes on East Indies Maps, Theater Area T, Army Map Service, pp. 149–157, March 1945*).

According to *TR8350.2*, **from** Djakarta (Batavia) or Genuk Datum **to** WGS84: $\Delta X=-377$ m±3 m, $\Delta Y=+681$ m±3 m, $\Delta Z=-50$ m±3 m. However, according to the *EPSG v.6.18*, "Note: The area of use cited for this transformation (Sumatra) is not consistent with the area of use (Java) for the Batavia (Genuk) coordinate reference system. Derived at 5 stations." An *EPSG v.6.18* alternative is **from** Djakarta (Batavia) or Genuk Datum **to** WGS84: $\Delta X=-378.873$ m, $\Delta Y=+676.002$ m, $\Delta Z=-46.255$ m, reportedly "used by ARCO offshore NW Java area." Another *EPSG v.6.18* alternative is **from** Djakarta (Batavia) or Genuk Datum **to** WGS84: $\Delta X=-377.7$ m, $\Delta Y=+675.1$ m, $\Delta Z=-52.2$ m, reportedly "used by PT Komaritim for Nippon Steel during East Java Gas Pipeline construction." Also, according to *TR8350.2* for Banga and Belitung Islands, **from** Bukit Rimpah Datum **to** WGS84: $\Delta X=-384$ m, $\Delta Y=+664$ m, $\Delta Z=-48$ m; however, no accuracy estimate is available. According to Bakosurtanal, **from** ID74 Datum **to** DGN95 Datum: $\Delta X=-1.977$ m±1.300 m, $\Delta Y=-13.06$ m±1.139 m, $\Delta Z=-9.993$ m±3.584 m, $R_x=-0.364''\pm0.109''$, $R_y=-0.254''\pm0.060''$, $R_z=-0.689''\pm0.042''$, $\delta s=-1.037\times10^{-6}\pm0.177\times10^{-6}$, and the model is defined as a *Coordinate Frame Rotation*. According to *TR8350.2*, **from** ID74 Datum **to** WGS 84: $\Delta X=-24$ m±25 m, $\Delta Y=-15$ m ±25 m, $\Delta Z=+5$ m±25 m. According to the *EPSG v.6.18*, for the area of south west Sulawesi, **from** Makassar Datum **to** WGS 84: $\Delta X=-587.8$ m, $\Delta Y=+519.75$ m, $\Delta Z=+145.76$ m. According to *TR8350.2*, from Gunung Segara Datum **to** WGS 84: $\Delta X=-403$ m, $\Delta Y=+684$ m, $\Delta Z=+41$ m. According to the *EPSG v.6.18*, **from** Gunung Segara Datum **to** WGS 84: $\Delta X=-387.06$ m, $\Delta Y=+636.53$ m, $\Delta Z=+46.29$ m, as originally obtained from Shell Oil Co. Furthermore, for northeast Kalimantan **from** Gunung Segara Datum **to** WGS 84: $\Delta X=-403.4$ m, $\Delta Y=+681.12$ m, $\Delta Z=+46.56$ m, also from Shell. For east Kalimantan–Mahakam delta area, **from** Gunung Segara Datum **to** WGS 84: $\Delta X=-404.78$ m, $\Delta Y=+685.68$ m, $\Delta Z=+45.47$ m, from Total Indonesia. These variations of transformation values from one datum to another are normal because of the variability of systematic errors encountered with classical geodetic datums. The large number of such published values for the Republic of Indonesia is due to the historical level of interest for the exploration and production of natural resources.

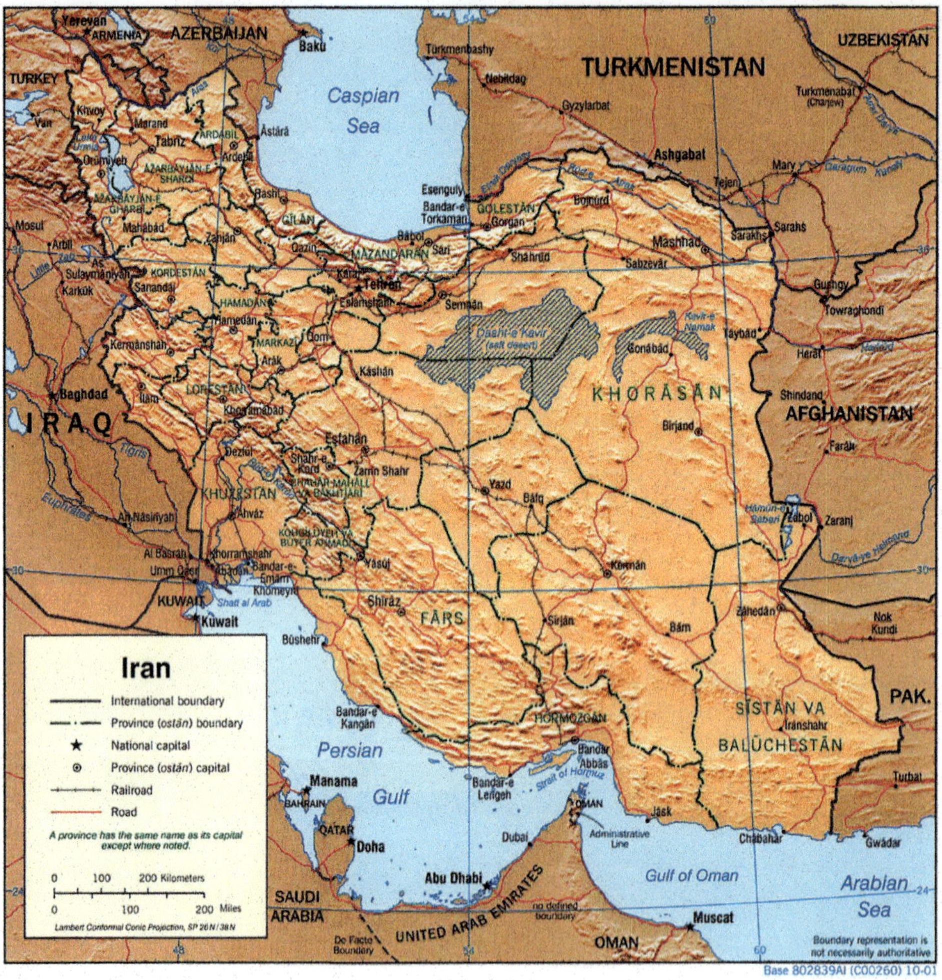

Base 802839AI (C00260) 10-01

Islamic Republic of Iran

Slightly smaller than Alaska, Iran is bordered by Afghanistan (936 km), Armenia (35 km), Azerbaijan (611 km), Iraq (1,458 km), Pakistan (909 km), Turkey (499 km), and Turkmenistan (992 km). The lowest point is the Caspian Sea (–28 m), and the highest point is *Kūh-e Damāvand* (5,671 m) (*World Factbook* and *NGA GeoNames Server*, 2013).

"Archaeologists suggest that during Neolithic times small numbers of hunters lived in caves in the Zagros and Alborz Mountains and in the southeast of the country. Iran's first organized settlements were established in Elam, the lowland region in what is now Khuzestan province, as far back as the middle of the 3rd millennium *B.C.* Elam was close enough to Mesopotamia and the great Sumerian civilization to feel its influence, and records suggest the two were regular opponents on the battlefield. The Elamites established their capital at Shush and derived their strength through a remarkably enlightened federal system of government that allowed the various states to exchange the natural resources unique to each region. The Elamites' system of inheritance and power distribution was also quite sophisticated for the time, ensuring power was shared by and passed through various family lines. By the 12th century *B.C.*, the Elamites are thought to have controlled most of what is now western Iran, the Tigris Valley, and the coast of the Persian Gulf. They even managed to defeat the Assyrians, carrying off in triumph the famous stone inscribed with the Code of Hammurabi. Having rapidly built a mighty military force, Cyrus the Great (as he came to be known) ended the Median Empire in 550 *B.C.* when he defeated his own grandfather – the hated King Astyages – in battle at Pasargadae. The Achaemenids introduced the world's first postal service, and it was said the network of relay horses could deliver mail to the furthest corner of the empire within 15 days. The end of the First Persian Empire finally came at the hands of Alexander the Great, king of Macedonia. The Parthians had settled the area between the Caspian and Aral Seas many centuries before. Under their great King Mithridates (171–138 *B.C.*), they swallowed most of Persia and then everywhere between the Euphrates in the west and Afghanistan in the east, more or less re-creating the old Achaemenid Empire. A crucial chapter in Persian history started when the Arabs defeated the Sassanians at Qadisirya in *A.D.* 637, following up with a victory at Nehavand near Hamadan that effectively ended Sassanian rule. By the time of Mohammed's death in 632 the Arabs were firm adherents of Islam. The Persians found plenty to like in Islamic culture and religion, and happily forsook Zoroaster for the teachings of Mohammed without much need of persuasion.

"During WWI both Britain and Russia occupied parts of Iran while the Turks ravaged the partly Christian northwest. Inspired by the new regime in Russia, Gilan (the west Caspian area) broke away in 1920 to form a Soviet republic under Kuchuk Khan. The weak Qajar shah seemed unable to respond, so Britain backed charismatic army officer Reza Khan, who swiftly retook Gilan before ousting Shah Ahmad. Reza Shah, as he became known, set himself an enormous task: to drag Iran into the 20th century in the same way his neighbor Mustafa Kemal Atatürk was modernizing Turkey. Literacy, transport infrastructure, the health system, industry, and agriculture had all been neglected and were pathetically underdeveloped. Like Atatürk, Reza Shah aimed to improve the status of women and to that end he made wearing the chador (black cloak) illegal. Like Atatürk, too, he insisted on the wearing of Western dress and moved to crush the power of the religious establishment. However, Reza had little of the subtlety of Atatürk and his edicts made him many enemies. Some women embraced his new dress regulations, but others found them impossible to accept.

"On his return to Iran on 1 February 1979, Khomeini told the exultant masses of his vision for a new Iran, free of foreign influence and true to Islam: 'From now on it is I who will name the government'. Ayatollah Khomeini soon set about proving the adage that 'after the revolution comes the revolution'. His intention was to set up a clergy-dominated Islamic Republic, and he achieved this with brutal efficiency" (*Lonely Planet*, 2013).

Prior to World War I, the Tiflis Datum of 1914 was established by the Russians and later was transformed to the Pulkovo 1931 Datum referenced to the Bessel 1841 ellipsoid. No parameters are currently available on this ancient system in Iran (*Army Map Service Geodetic Memorandum 1302*).

The British Survey of India surveyed the Trans-Iranian Arc from Iran to India. According to John W. Hager, "Reading the diary of that operation is fascinating. They ran out of spare tires for their vehicles and took to cutting out the good portions of the tires (or tyres) and putting them back together with nuts and bolts. I would guess that was Nahrwan Datum. Nahrwan was also used by the Iranian Oil Exploration and Producing Company (I.O.E.P.C.), the successor (1954) to the Anglo-Persian Oil Company. It mainly covers the area from Iraq in the area Ahvāz-Abādan east to Shīrāz-Būshehr. The connection to Nahrwan was tenuous at best. Nahrwan, code NAH, located at Nahrwan S.E. Base, $\Phi_o = 33°\ 19'\ 10.87''\ N \pm 0.20''$, $\Lambda_o = 44°\ 43'\ 25.54''\ E \pm 0.69''$, $\alpha_o = 10°\ 55'\ 51.8''$ to North End Base, Clarke 1880 (*where a = 6,378,249.145 m, and 1/f = 293.465 – Ed.*).

"The next was Būshehr Datum, code BUS, at camp 279, $\Phi_o = 28°\ 29'\ 26''\ N$, $\Lambda_o = 51°\ 05'\ 49''\ E$, Clarke 1880. The area of coverage was east of Shīrāz – Būshehr to Kermān – Mināb. This was a temporary datum until Final Datum 58 was adopted.

"And then there was Final Datum 58, code FIN, at 49/87 Manyur (*Ahwaz – Ed.*), $\Phi_o = 31°\ 23'\ 59.190''\ N$, $\Lambda_o = 48°\ 32'\ 31.380''\ E$, Clarke 1880. Adopted in 1958 covering the combined area of Nahrwan and Būshehr datums. The station designation (49/87) looks suspiciously like a Survey of India designation. In actuality, Final Datum 58 was derived from Nahrwan. There were several 1st and 2nd order nets that ran the full east–west extent and went north to intersect with the trans-Iranian arc (Nahrwan and European 1950). Tellurometer and Geodimeter (*Electronic Distance Measuring equipment brands – Ed.*) were coming into their own and results from those instruments may have been included in the Final Datum 58 adjustment. Tellurometer traverses were being used more and more by the I.O.E.P.C. (Iranian Oil Exploration and Producing Companies). In 1955, 1956, and 1957, the country was photographed using SHORAN controlled photography. The SHORAN control network, Trans-Iranian Arc, selected arcs in the I.O.E.P.C. area, and new 2nd order triangulation, base lines and astros were all combined and adjusted to produce a network on European 1950 Datum."

The U.S. Army Map Service let a contract in 1963 for first order surveying in Iran. The work included establishment of 3,750 km of precise electronic traverse, determination of 1st order astronomic positions and Laplace azimuths, 4,400 km of third order traversing, and the location of 250 horizontal picture points for photogrammetric mapping (*Surveying in Iran by A.V. Cocking, Surveying and Mapping, December 1969, pp. 655–660*).

A curious variety of datums and coordinate systems have been used in Iran over the years, some by official government interests, most by the oil exploration and development industry. European Datum 1950 (*some offer the ED77 designation, albeit more by wishful thinking than by fact – Ed.*) in which the origin is at Helmertturm, Potsdam, Germany where $\Phi_o = 52°\ 22'\ 53.9540''\ N$, $\Lambda_o = 13°\ 04'\ 01.1527''$ East of Greenwich. The defining azimuth to station Golmberg is: $\alpha_o = 154°\ 47'\ 32.19''$, and the ellipsoid is the International 1924 (Hayford 1909) where $a = 6,378,388\ m$, and $1/f = 297$.

The Iran–Iraq Zone Lambert Conformal Conic is on the Final Datum 58, where the Central Meridian, $\lambda_o = 45°\ E$, latitude of origin, $\varphi_o = 32°\ 30'\ N$, scale factor at origin, $m_o = 0.9987864078$, False Easting = 1,500 km, False Northing = 1,166,200 m.

The "RassadIran" Datum is referenced to the International 1924 ellipsoid, and is derived from a chain originating at Kangan, near Tāherī, and is likely just a local snippet of ED50. For a tiny area of coastal Iran, someone cooked up a Hotine Rectified Skew Orthomorphic projection where $\varphi_o = 27°\ 31'\ 07.7837''\ N$, $\lambda_o = 52°\ 36'\ 12.7410''\ E$, False Easting = 658,377.437 m, False Northing = 3,044,969.194 m, $\gamma_o = +0°\ 34'\ 17.9803''$, and $m_o = 0.999895934$. This origin point represents the location of "TOTAL 1," an on-site ITRF96 point, such that the three-parameter shift from RassadIran to the ITRF96 is: $\Delta X = -133.688$ m, $\Delta Y = -157.575$ m, and $\Delta Z = -158.643$ m. The ITRF96 coordinates of TOTAL 1 are: $\varphi = 27° 31' 03.8822''$ N, $\lambda = 52° 36' 13.1239''$ E, and $H_o = h_o = 84.81$ m. The three-parameter transformation for all of Iran from ED50 to WGS84 is: $\Delta X = -192.359$ m ± 0.3162 m, $\Delta Y = +263.787$ m ± 0.3162 m, and $\Delta Z = -24.450$ m ± 0.3162 m *(Strategy for Computation of Transformation Parameters from Local Datums to Regional Datum, F. Tavakoli, National Cartographic Center of IRAN, PCGIAP, July)*.

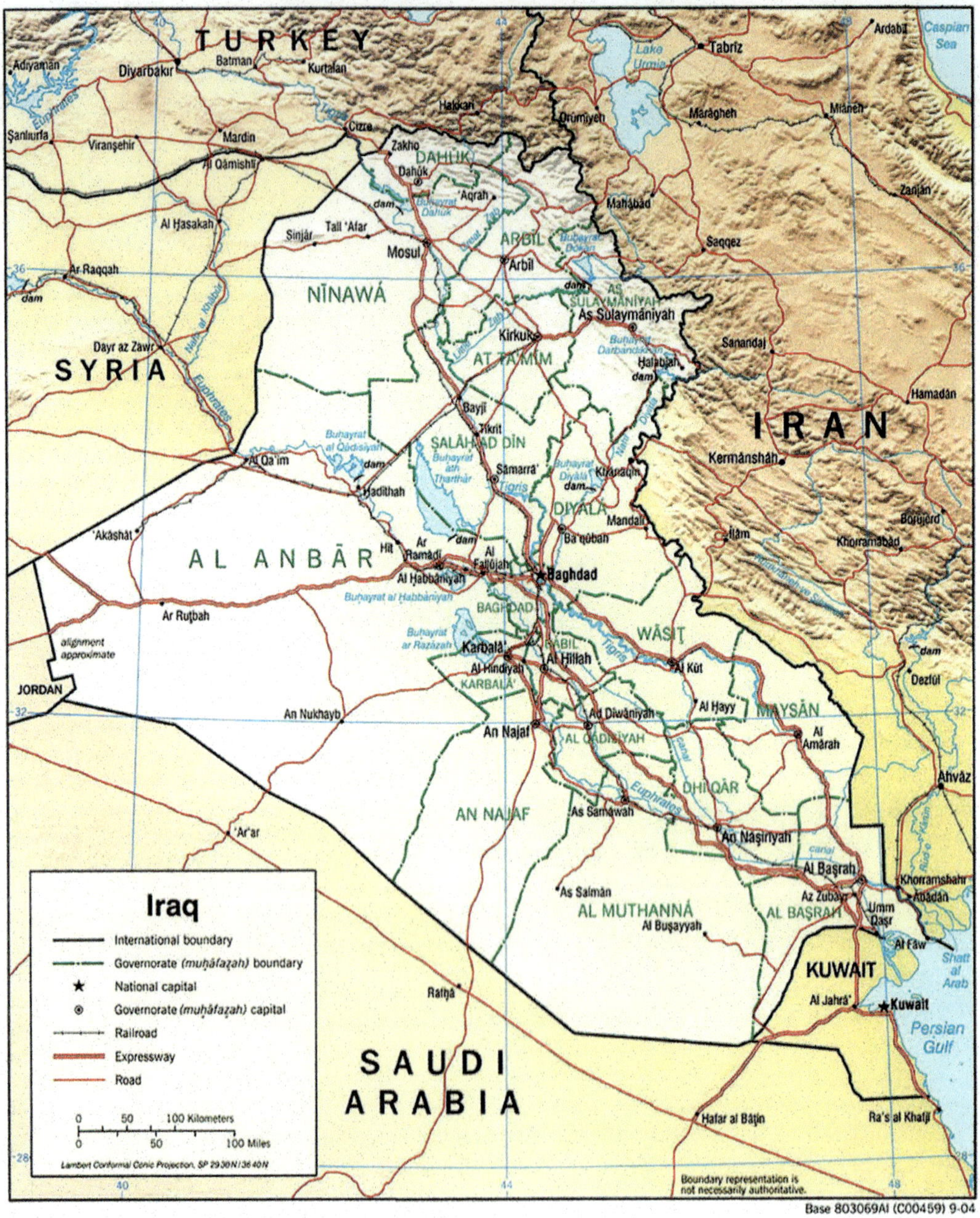

Base 803069AI (C00459) 9-04

Republic of Iraq

Iraq is slightly more than three times the size of New York State. Bordered by Iran (1,599 km), Jordan (179 km), Kuwait (254 km), Saudi Arabia (811 km), Syria (599 km), and Turkey (367 km). The terrain is mostly broad plains; reedy marshes along Iranian border in south with large, flooded areas; mountains along borders with Iran and Turkey. The highest point is *Cheekha Dar* (Kurdish for "Black Tent") (3,611 m), the lowest point is the Persian Gulf (0 m), and the mean elevation is (312 m).

"By 6000 B.C., Mesopotamia had been settled, chiefly by migrants from the Turkish and Iranian highlands. Sumer is the ancient name for southern Mesopotamia. Historians are divided on when the Sumerians arrived in the area, but they agree that the population of Sumer was a mixture of linguistic and ethnic groups that included the earlier inhabitants of the region. Sumerian culture mixed foreign and local elements. The Sumerians were highly innovative people who responded creatively to the challenges of the changeable Tigris and Euphrates rivers. Many of the great Sumerian legacies, such as writing, irrigation, the wheel, astronomy, and literature, can be seen as adaptive responses to the great rivers. The precariousness of existence in southern Mesopotamia also led to a highly developed sense of religion. Cult centers such as Eridu, dating back to 5000 B.C., served as important centers of pilgrimage and devotion even before the rise of Sumer. Many of the most important Mesopotamian cities emerged in areas surrounding the pre-Sumerian cult centers, thus reinforcing the close relationship between religion and government" (*Country Studies, Library of Congress 2014*).

"In A.D. 637 the Arab armies of Islam swept north from the Arabian Peninsula and occupied Iraq. Their most important centers became Al-Kufa, Baghdad, and Mosul. In 749, the first Abbasid caliph was proclaimed at Al-Kufa, and the Abbasids would go on to make Iraq their own. The founding of Baghdad by Al-Mansur saw the city become, by some accounts, the greatest city in the world. In 1258 Hulagu – a grandson of the feared Mongol ruler Genghis Khan – laid waste to Baghdad and killed the last Abbasid caliph. Political power in the Muslim world shifted elsewhere. By 1638, Iraq had come under Ottoman rule. After a period of relative autonomy, the Ottomans centralized their rule in the 19th century, where after Iraqi resentment against foreign occupation crystalized even as the Ottomans undertook a massive program of modernization. The Ottomans held on until 1920, when the arrival of the British saw Iraq come under the power of yet another occupying force, which was at first welcomed then resented by Iraqis. Iraq became independent in 1932 and the period that followed was distinguished by a succession of coups, countercoups and by the discovery of massive reserves of oil. On 14 July 1958 the monarchy was overthrown in a military coup and Iraq became a republic.

"In 1979 Saddam Hussein replaced Al-Bakr as president, the revolution in Iran took place and relations between the two countries sank to an all-time low. Saddam, increasingly concerned about the threat of a Shiite revolution in his own country, declared that Iraq wanted a return to exclusive control over the Shatt al-Arab River. Full-scale war broke out on 22 September 1980, with Iraqi forces entering Iran along a 500 km front. The eight years of war that followed were characterized by human-wave infantry advances and the deliberate targeting of urban residential areas by enemy artillery, all for little territorial gain. A million lives were lost and the economic cost to Iraq alone is estimated at more than US$100 billion" (*Lonely Planet*, 2014). "The acting Iraqi National Intelligence Service Director General Mustafa al-Kadhimi became prime minister in May 2020 after the previous prime minister resigned in late 2019 because of widespread protests demanding more employment opportunities and an end to corruption. His mandate as prime minister was to guide Iraq toward an early national legislative election, which was held in October 2021" (*World Factbook*, 2022).

Thanks to John W. Hager, "The British divided up the trig list production in the Middle East early in WW II. I don't know the exact boundaries, but I believe that Turkey, Iraq, Iran, the Arabian

Peninsula and areas east went to the Survey of India (S.I). The rest went to Middle East Land Forces (M.E.L.F.).

"The principal datum for Iraq was Nahrwan (code NAH) observed at Nahrwan S. E. Base in 1929 or 1930. $\Phi_o=33°\ 19'\ 10.87''\ N\pm0.20''$, $\Lambda_o=44°\ 43'\ 25.54''\ E\pm0.69''$, $\alpha_o=10°\ 55'\ 51.8\pm1.16''$ to North End Base, Clarke 1880 (ellipsoid where $a=6,378,249.2$ m and $1/f=293.4660208$. – *Ed*.)

"Another datum of importance was Fao (code FAO) at Fao Roof where $\Phi_o=29°\ 58'\ 23''$ N and $\Lambda_o=48°\ 28'\ 55''$ E, Everest (ellipsoid where $a=6,377,301.243$ m, $1/f=300.8017$ – *Ed*.). Also given is Fao Flagstaff which is probably derived from Fao Roof. Here $\Phi_o=29°\ 58'\ 23.82''$ N, $\Lambda_o=48°\ 28'\ 56.88''$ E. Reference is *Triangulation in Arabia, Iraq & Persia*; International (ellipsoid where $a=6,378,388$ m, $1/f=297$ – *Ed*.) NH-39-M Fao, Survey of India 1946. The description is "Mark engraved on floor of roof of chief clerk's quarters midway between chimney and pillar and 18 feet 4 inches from edge of roof to north-west. Astronomical observations for time, latitude and azimuth were also taken here."

"Another is Iraqi Date Society at $\Phi_o=30°\ 31'\ 51.1''$ N, $\Lambda_o=47°\ 50'\ 03.7''$ E. This was taken from British Admiralty (BA) chart 3846 (21 Oct 1926). Probably used on DA 3842–3845. Without access to these charts, I can only surmise that this is the approaches to Basra up the Shatt al Arab.

"The last I have for Iraq is M79 (code MSN) where $\Phi_o=35°\ 33'\ 15.19$ N, $\Lambda_o=44°\ 18'\ 21.77''$ E. (Probably established by W. E. Browne in 1927–31). Looking at an atlas, my guess is that it had to do with the oil fields around Kirkuk. *(See further discussion below – Ed.)*

"Final Datum 58 (1958) (code FIN) was derived from Nahrwan by the Iranian Oil Exploration and Producing Companies (I.O.E.P.C.) for southwest Iran east to Būshehr. When the surveys tied into the Būshehr datum area, the whole of the I.O.E.P.C. area east to the Strait of Hormuz was adjusted to Final Datum 58 and European Datum (1950). I felt the connection to Nahrwan was poor, but it did provide a unified system for the I.O.E.P.C. I don't think that Final Datum 58 was ever applied to Iraq" (*John W. Hager, personal communication, 5 March 2010*).

The origin point for Final Datum 1958 is Maniyur where $\Phi_o=31°\ 23'\ 59.19$ N, $\Lambda_o=48°\ 32'\ 31.38''$ E, and the network was included in the Nahrwan 1967 Adjustment. With respect to the *unverified* transformation parameters **from** Nahrwan Datum of 1967 (Iraq Datum) **to** WGS84 Datum are: $\Delta X=+65$ m, $\Delta Y=-334$ m, $\Delta Z=+267$ m (*from unofficial source, 2005*). A variation on Nahrwan is Karbala Datum which is the realization in the city of Bagdad.

"The triangulation conducted by W.E. Browne of the Iraq Petroleum Company in the years 1927–31 consists chiefly of a main arc running from Mosul to Jabal Hamrin South and three independent local systems in the immediate vicinity. Browne's triangulation, for the most part, has very good configuration, which is composed mainly of braced quadrilaterals and polygons. He has utilized a good number of short bases (about 700 meters long) and rather frequent astronomical azimuths. Some of his bases, however, were measured with an unstandardized steel tape. Browne's main arc from Mosul to Jabal Hamrin South was originally surveyed in seven sections, and finally consolidated into one continuous arc. However, this consolidation was not rigorously done. The seven sections were put together by average geographic differences at stations common to two sections. The origin of this arc (Mosul to Jabal Hamarin South) is station M79, whose geographics are above. The task of rigorously adjusting Browne's survey of Iraq datum and eventually to new Egyptian datum is seriously hindered by the lack of information at this headquarters *(Army Map Service – Ed.)* dealing with Browne's survey" (*Tentative Procedure for the Adjustment of W.E. Browne's Triangulation in Iraq to New Egyptian Datum, Edmund A. Early, Army Map Service Geodetic Memorandum, 7 July 1948*).

"The Clarke 1880 New Egyptian Datum geographics of the origin of the Nahrwan system, (South End Base) are: $\varphi_o=33°\ 19'\ 17.4838''$ N, $\lambda_o=44°\ 43'\ 21.1053''$ E. In the adjustment of Iraq to Syria at the border, a number of difficulties were encountered. The border overlap does not consist of a continuous group of common points, but of 2 separate groups of positions, which on the Syrian side at least are not connected by a direct arc of triangulation. In the northern group of common points, use was made, on the Syrian side of first order positions only, found in *fascicule V* of the latest French lists *(Coordonnees Rectangulaires Lambert et Altitudes de Points Géodésiques du Levant)*. Iraq values of secondary quality were taken from British trig lists for Iraq *(Triangulation in Iraq, Syria, and Turkey, No. 37)*. The following 9 points were used in the northern connection least squares solution: Tell Hanta, Tell Safra, Tell Hamidi, Tell Brak, Tell Aarbib, Kapes Darh, Melzbate, Tell Ahmar, and Tell Hamdoun. In the southern connection between Iraq and Syria, 4 positions were found to be common to both Syrian and British lists. These positions are Chaafu, Kamiyse, Abou Kemal Minaret, and Tell Hedkouk" *(Conversion of Clarke 1880 Iraq and Caucasus Zone Nahrwan datum Lambert grid coordinates in Iraq and Iran to Clarke 1880 New Egyptian datum UTM grid coordinates in Zones 37, 38, and 39, Patricia K. Hayden, Army Map Service Geodetic Memorandum, 6 August 1948)*.

"The primary triangulation of Iraq was executed between the years 1930–35. It should be considered as having a secondary order of precision. The primary triangulation is divided into two series: the North Series and the South Series *(as detailed in the above paragraph – Ed.)*. Both series are based on the Nahrwan Base near Bagdad. *(By examination of the area just east of Bagdad with Google Earth™, one can see the NW to SE irrigation canal segment that the baseline paralleled – Ed.)*. The North Series closes on a measured base at Kirkuk and the South Series on that at Batha. Discrepancies between the measured and triangulated lengths of the Kirkuk and Batha bases are said to have been adjusted these discrepancies are given as follows: Kirkuk – 1/20,000, Batha – 1/15,000. The initial values of the triangulation for latitude and longitude were taken at the south end of the Nahrwan Base. Astronomic observations were then made at Kirkuk and Batha Bases for latitude, longitude, and azimuth *(LaPlace stations – Ed.)*. Average triangle closure of the North Series is 1.48″ with a maximum of 5.31″ and the closure of the South Series is 1.61″ with a maximum of 5.39″.

"The Iraq Zone Grid B is one of the two overlapping grid systems decided upon at the Cairo Conference in 1940 between the Middle East Command and the Survey of India. The specifications of the Iraq Grid B are as follows: Spheroid: Clarke 1880, Latitude of Origin: 32° 30′ N, Longitude of Origin: 45° E of Gr., Scale Factor at Parallel of Origin: 0.9987864078, False Northing: 1,166,200 meters, False Easting: 1,500,000 meters. Limits of Zone: North – 300,000 Northing line of Caucasus Zone, East – Meridian of 60° E, South – Parallel of 28° N., West to 57° E., South on this meridian to 550,000 Northing line of this Zone, West on this line to 39° E., West – Meridian of 39° E.

"The Caucasus Zone Grid C is the second overlapping grid in Iraq decided upon in 1940. The specifications of the Caucasus Zone Grid C are as follows: Spheroid: Clarke 1880, Latitude of Origin: 39° 30′ N, Longitude of Origin: 45° E of Gr., Scale Factor at Parallel of Origin: 0.998461538, False Northing: 675,000 meters, False Easting: 2,155,500 meters. Limits of Zone: North – 300,000 Northing line of Caspian Zone, East – Meridian of 60° E, South – 300,000 Northing line of this Zone, West – Meridian of 39° E" *(Iraq. Conversion of Nahrwan Datum to ED UTM, Issac C. Lail, Army Map Service Geodetic Memorandum, 9 November 1957)*.

"The Appendix by Major Thomas gives the history and details of organization of the work done by his party, which consisted of 75 persons of all ranks and 18 vehicles. The main difficulties encountered were mechanical breakdowns; bogging of vehicles; difficulties in climbing mountains due to ice, snow and precipices; variable weather conditions consisting of icy winds of cyclonic intensity, clouds and haze; and the difficulty of maintaining scattered parties with water, petrol, kerosene, rations and clothing. In a few instances parties were benighted and had to spend an uncomfortable night on high peaks without food, warmth, or shelter, and with temperatures well below freezing point. In one case, where a night was thus spent at 13,000 feet, two of the party suffered next day from snow blindness. Two lives were nearly lost during the ascent of a mountain station in which the surface consisted of loose shale and slate, making a foothold extremely precarious and dangerous" *(Notice: Trans-Persian Trigonometrical Connection Between Iraq and India, J.C., Empire Survey Review, No. 68, pp. 277–279)*.

Recent work has established GPS reference stations in Iraq and is referred to as the Iraqi Geospatial Reference System *(Iraq on the Map: Installing Reference Stations for Accurate Engineering, Anas Malkawi, GPS World, March 1, 2011)*.

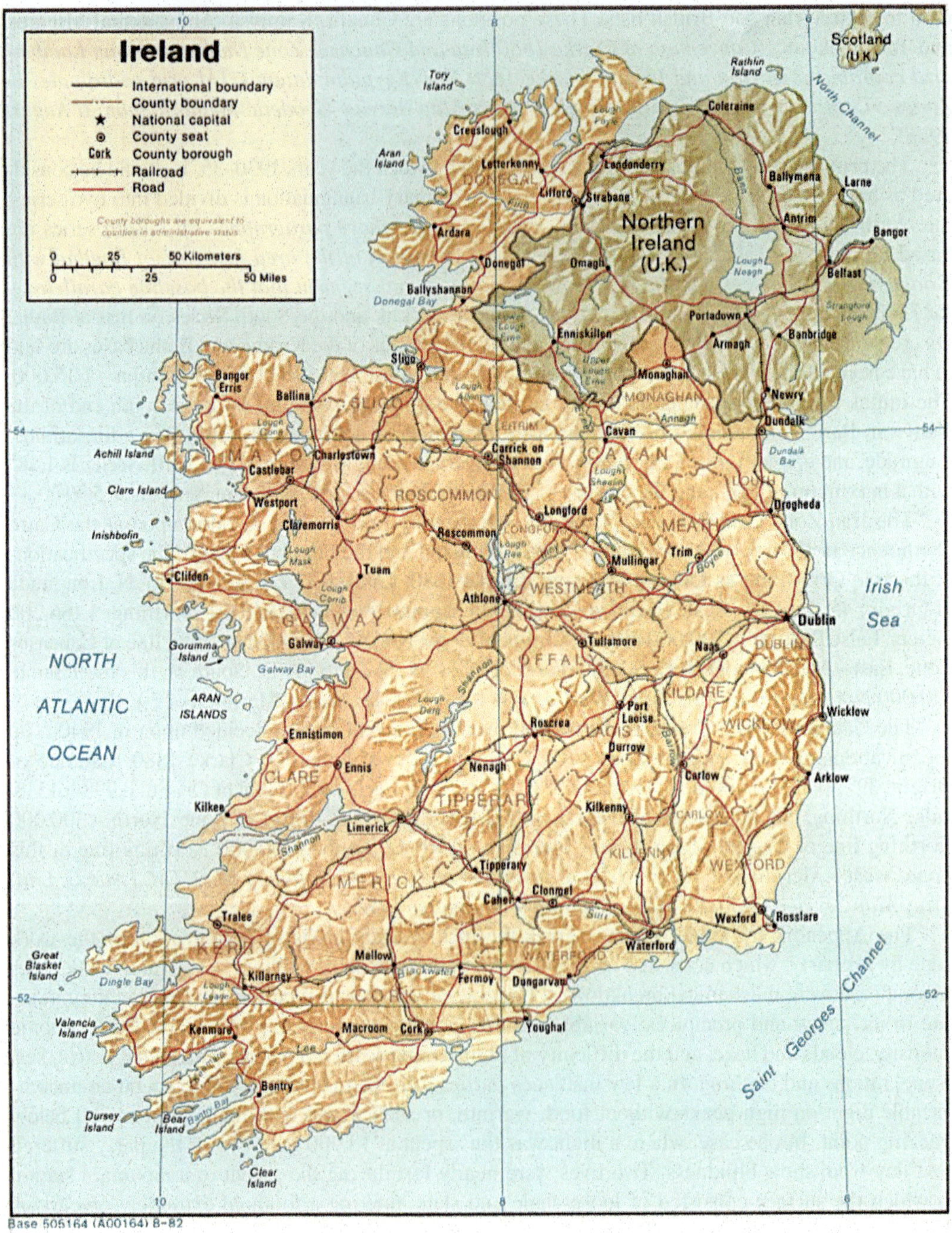

Base 505164 (A00164) B-82

Ireland

Ireland is in Western Europe and is occupying five-sixths of the island of Ireland in the North Atlantic Ocean, west of Great Britain. It is slightly larger than West Virginia and is bordered by the United Kingdom (490 km). The terrain is mostly flat to rolling interior plain surrounded by rugged hills and low mountains with sea cliffs on west coast. The highest point is *Carrauntoohil* (1,041 m), the lowest point is the Atlantic Ocean (0 m), and the mean elevation is (118 m).

"Celtic tribes arrived on the island between 600 and 150 B.C. Invasions by Norsemen that began in the late 8th century were finally ended when King Brian Boru defeated the Danes in 1014. Norman invasions began in the 12th century and set off more than seven centuries of Anglo-Irish struggle marked by fierce rebellions and harsh repressions. The Irish famine of the mid-19th century was responsible for a drop in the island's population by more than one quarter through starvation, disease, and emigration. For more than a century afterward, the population of the island continued to fall only to begin growing again in the 1960s. Over the last 50 years, Ireland's high birthrate has made it demographically one of the youngest populations in the EU.

"Ireland was neutral in World War II and continues its policy of military neutrality. Ireland joined the European Community in 1973 and the euro-zone currency union in 1999. The economic boom years of the Celtic Tiger (1995–2007) saw rapid economic growth, which came to an abrupt end in 2008 with the meltdown of the Irish banking system. As a small, open economy, Ireland has excelled at courting foreign direct investment, especially from US multi-nationals, which helped the economy recover from the financial crisis and insolated it from the economic shocks of the COVID-19 pandemic" (*World Factbook*, 2022).

Boundary maps in Ireland were made to accompany the "*terriers*" (property records) of the surveys in 1636–1640 by order of Lord Strafford, Viceroy of Ireland. In 1654–1659, the "*Down Survey*" comprised maps of the "townlands" averaging 300 acres each and baronies totaling over two-thirds of the surface of Ireland, about 20,000,000 acres. The Ordnance Survey was established in 1791 to produce national mapping, this included all of Ireland. The need for an accurate map of Ireland was brought to the fore in the 1800s by problems with a local tax, known as the "*County Cess*." In 1824, the committee chaired by Thomas Spring Rice recommended to the (British) House of Commons that a survey of Ireland was required to provide a definitive indication of acreages and ratable values for the purpose of establishing local taxes in Ireland. That same year, Colonel (later Major General) Thomas Colby of the Ordnance Survey received orders to proceed with the work of triangulation and Six Inch (6″ = 1 mile) topographical surveys for all of Ireland.

"Reaction to the Engineers by the local people was mixed but generally they were regarded with suspicion. A particular nuisance was the removal by local people of the poles, set up as targets on mountains, before the surveyors had a chance to observe them and, in one case, the observers were attacked. However, by contrast, in Glenomara, County Clare, the people climbed the mountain with them in a great crowd, with flutes, pipes and fiddles, treating the building of the trigonometrical station as a festive occasion."

The initial baseline for the Irish survey was selected by Colby at the Plain of Magelligan near Lough Foyle in Londonderry. Colby decided to use the principle of compensation for the manufacture of the baseline measuring apparatus. The principle was to use two metal bars, one of brass, and one of iron, which were placed 1⅛ inches apart but joined rigidly to each other at their centers. The bars were allowed to expand or contract freely, with pivoted steel tongues fixed to both bars at their ends. These were marked with silver pins, and although the length of the bars changed with temperature, the distance between the two pins remained constant. Six sets of bars were made, each just over 10 feet long. The baseline of nearly 8 miles was measured over 1827–1828. The Lough

Foyle Base was remeasured in 1960 using electronic distance measuring equipment, with a difference found to Colby's original measurement of only 1 inch!

The One Inch Map ($1'' = 1$ mile) of Ireland was cast on the ellipsoidal Bonne projection in 1850 which was the projection in vogue throughout Europe at the time. Also used for the ½ inch series and some 9-inch sheets, the Latitude of Origin (φ_o) = 53° 30′ North, and the Central Meridian (λ_o) = 8° 00′ West of Greenwich. The Airy 1830 ellipsoid was used where $a = 6,377,563.396$ m and $1/f = 299.3$. The radius of the mean parallel used for the Bonne was 13,361,612.2 feet.

The topographic surveying demanded greater accuracy than the methods used for the One Inch Map. Colby issued an instruction (specifications) in what became known as the *"Colonel's Blue Book."* Note that in the United States, the current (1999) specifications for note keeping for acceptance of data by the Federal Geodetic Control Committee are based on the *"Blue Book."* However, Colby's annual reports were also termed the same. In 1838, with the Irish survey on a firm footing, Colby returned to England and turned his attention to the survey of Great Britain. Sir Thomas A. Larcom, KCB was the Officer in Charge at Mountjoy, Dublin from 1828 to 1846. The final cost was £820,000 (more than twice the original estimate of time and money), but the Survey of Ireland served as a model for the remainder of Great Britain.

In 1858, Captain Alexander Ross Clarke (the same fellow that computed his ellipsoids of 1858, 1866, and 1880) selected the observations to be used in the adjustment. Clarke's interlocking network of "well-conditioned" triangles is now known as the "Principal Triangulation of Ireland (1824–1832)." Clarke rigorously adjusted the observations by the method of least squares in 21 independently computed but connected blocks with the aid of an average of eight computers (persons). Note that in 1881, Colonel A. R. Clarke, R.E. received a forced retirement from the Ordnance Survey of Great Britain rather than accept a post to the island of Mauritius. Clarke was 52 years old at the time. For the remaining 30 years of his life, he published no further scientific work.

In the meantime, some form of framework was required on which to control the new mapping at the scale of 6 inches to 1 mile. Six Inch Map control was therefore based on a network of secondary and tertiary blocks of triangulation, begun in 1832 and completed in 1841, just ahead of the chain survey teams who were surveying the detail. One Inch Map control began in 1852, and 25-Inch Map control began in 1888. Although these lower-order control points included five of Clarke's points, they were based only on provisional coordinates. In Northern Ireland, the coordinates of the origins of Counties Down and Armagh were in general agreement with the values deduced from the Principal Triangulation. In the other four counties they were about 80 feet too far south and 25 feet too far west. Unlike the aphylactic sheet system used in the United States with the Polyconic Projection, each Irish County had its own central meridian as a true "Grid" about a single point of origin. There were 26 County Cassini-Soldner Grids in what is now the Republic of Ireland, and an additional 6 were in Northern Ireland. Of these coordinate origins, 16 were based on church towers and spires, 2 were observatories, 2 were "round towers" (ancient defensive stone structures from various wars), and a number were on mountains and at monuments. Printed sheets had neither grid nor graticule. This system of local (county) Cassini-Soldner coordinates was also used in England and was later adopted for the "Meridional Circuits" of New Zealand. An observation on the practicality of this seemingly haphazard method of mapping stated: *"When the map of Ireland is picked up and shaken, it is only the mathematician who hears the rattle."* This system of provisional coordinates resulted in the County Coordinates being in active use for over 100 years. The Irish Grid, based on a single Cassini-Soldner projection for the entire island was used from 1936 to 1956.

The Ordnance Survey of Northern Ireland (U.K.) was set up in 1921, and the Irish Free State occupying five-sixths of the number of counties was established in 1922. The boundary between the two was settled in 1925. Eire declared itself completely independent in 1937 and was renamed the Republic of Ireland in 1949. The funds for first- and second-order re-triangulations of Northern Ireland were authorized in 1947 and the survey was completed in 1956. The first-order re-triangulation of Ireland was carried out between 1962 and 1964. Ordnance Survey of Northern Ireland (Belfast) did not adopt the 1965 adjustment values for subsequent mapping, whereas Ordnance

Survey of Ireland (Dublin) did. A fascinating result of the adjustments was the "Airy Modified" ellipsoid where the semi-major axis was changed to 6,377,340.189 m! This was done to best accommodate previous published maps and coordinates, and, thus, the ellipsoid was reduced by 35 parts per million. In the days when Grid-to-Graticule Tables were used, this was very important. The Gauss-Krüger Transverse Mercator projection Irish National Grid was adopted as a replacement for the old Cassini-Soldner Irish Grid. The defining parameters include φ_o and λ_o being the same as the old Bonne (and Cassini-Soldner) Grids, False Easting=200km, False Northing=250km, and the Scale Factor at Origin (m_o)=1.000035. (Notice the "35" in the scale factor.)

A subsequent adjustment was performed in 1975 which was designed to provide a common coordinate system for the whole of Ireland for mapping purposes. The Northern Ireland primary stations were held completely fixed. The Ireland Datum of 1975 origin is at Slieve Donard (new) where Φ_o=54° 10′ 48.262″ North, and Λ_o=05° 55′ 11.898″ West of Greenwich. Various other datums used for scientific purposes include: the Ordnance Survey of Great Britain Scientific Network of 1970 (OGSB SN 1970), OGSB (SN) 80, ED50, ED87, and WGS84. The latest major campaign, IRENET 95, provides European Terrestrial Reference Frame (ETRF 89) coordinates for a network throughout Ireland.

Compatibility between the existing Irish National Grid and ETRF 89 has been developed in three levels. The first level will provide a simple Easting and Northing coordinate shift between the old Grid and the new Grid and have an accuracy of better than ±5 m relative to the new geodetic framework. The second level will provide a seven-parameter Helmert (geocentric) transformation estimated to be better than ±2 m relative to the new geodetic framework. The third level will be a polynomial algorithm with an accuracy of better than ±0.2 m relative to the new geodetic framework. A new National Grid is contemplated based on the GRS80 ellipsoid and may be the UTM (Zones 29 & 30).

The Malin Head Vertical Datum of 1970 is the current system for all 1:50,000 mapping in Ireland. Earlier maps used the low water mark of the spring tide on the 8th of April 1837 at Poolbeg Lighthouse, Dublin Bay. The Malin Head (County Donegal) Datum is approximately 2.71 m above the Poolbeg Lighthouse Datum. All large-scale mapping in Northern Ireland uses mean sea level at Belfast which is 0.037 m below the Malin Head Datum.

The Ordnance Survey of Ireland now has a Co-ordnate (*sic*) Converter that will interactively convert the following co-ordinate reference systems: WGS84 (World Geodetic System 1984)/ETRF89 (European Terrestrial Reference Frame 1989), Irish Grid, ITM (Irish Transverse Mercator), UTM (Universal Transverse Mercator). OSGM15 and OSTN15 have updated transformations for UK and Ireland. The Ordnance Survey of Great Britain (OSGB), Ireland (OSI) and Land & Property Services (LPS – Formerly OSNI) have collaborated again to improve the OSGM02 geoid model covering the United Kingdom and Ireland. A new Geoid model OSGM15 was launched on the 26th of August 2016. The polynomial transformation for Ireland and Northern Ireland has not changed however there are solutions available for download. Grid Inquest II took over from Grid Inquest I on the 26th of August 2016.

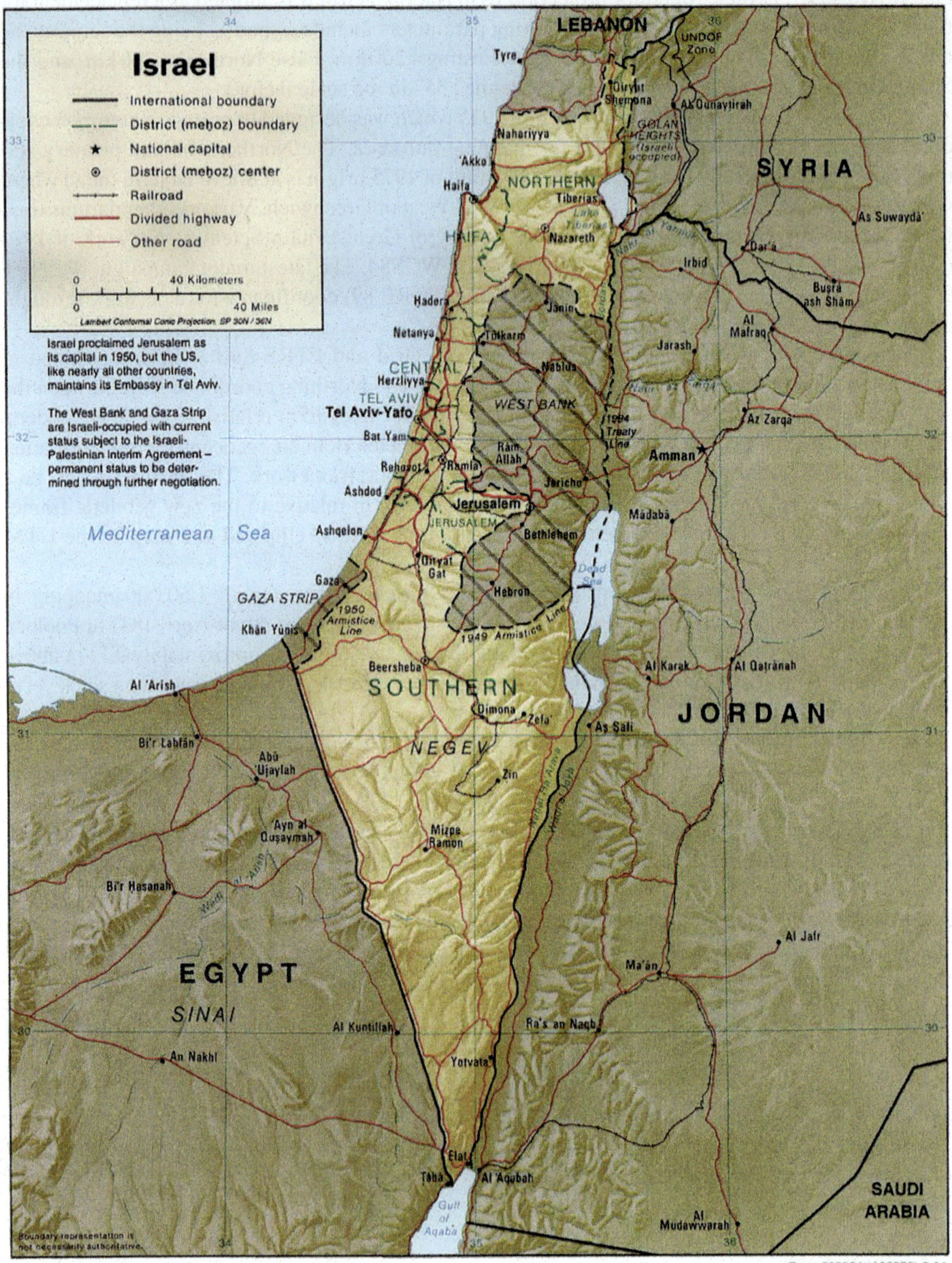

Base 802634 (A00853) 9-01

The State of Israel

Israel is hilly in the North, and its highest peak is *Mitspe Shlagim* 2,224 m; note this is the highest named point, the actual highest point is an unnamed dome slightly to the west of *Mitspe Shlagim* at 2,236 m; both points are on the northeastern border of Israel, along the southern end of the Anti-Lebanon mountain range. It is bounded on the North by Lebanon (81 km), on the West by the Mediterranean Sea, on the East by Syria (83 km) and Jordan (327 km), by Egypt to the Southwest (208 km), and by the Negev Desert to the South with the Gulf of Aqaba at its extreme Southern port. The State of Israel was established by decree of the United Nations on May 15, 1948.

Archaeological evidence has uncovered the existence of man in what is now modern Israel during the Paleolithic Period (Old Stone Age). During the subsequent Neolithic Period (New Stone Age), humans cultivated crops and had built towns such as Jericho by 7,000 B.C.

"Israel has emerged as a regional economic and military powerhouse, leveraging its booming high-tech sector, massive defense industry, and concerns about Iran to foster partnerships around the world, even with some of its former foes. The State of Israel was declared in 1948, after Britain withdrew from its mandate of Palestine. The UN proposed partitioning the area into Arab and Jewish states, and Arab armies that rejected the UN plan were defeated. Israel was admitted as a member of the UN in 1949 and saw rapid population growth, primarily due to migration from Europe and the Middle East, over the following years. Israel fought wars against its Arab neighbors in 1967 and 1973, followed by peace treaties with Egypt in 1979 and Jordan in 1994. Israel took control of the West Bank and Gaza Strip in the 1967 war, and subsequently administered those territories through military authorities. Israel and Palestinian officials signed a number of interim agreements in the 1990s that created an interim period of Palestinian self-rule in the West Bank and Gaza. Israel withdrew from Gaza in 2005. While the most recent formal efforts to negotiate final status issues occurred in 2013–2014, the US continues its efforts to advance peace. Immigration to Israel continues, with more than 20,000 new immigrants, mostly Jewish, in 2020.

"The Israeli economy has undergone a dramatic transformation in the last 25 years, led by cutting-edge, high-tech sectors. Offshore gas discoveries in the Mediterranean, most notably in the Tamar and Leviathan gas fields, place Israel at the center of a potential regional natural gas market. However, longer-term structural issues such as low labor force participation among minority populations, low workforce productivity, high costs for housing and consumer staples, and a lack of competition, remain a concern for many Israelis and an important consideration for Israeli politicians. Former Prime Minister Benjamin Netanyahu dominated Israel's political landscape from 2009 to June 2021, becoming Israel's longest serving prime minister before he was unseated by Naftali Bennett, after Israel's fourth election in two years. Bennett formed the most ideologically diverse coalition in Israel's history, including the participation of an Arab-Israeli party. Under the terms of the coalition agreement, Bennett would remain as prime minister until August 2023, then Alternate Prime Minister and Foreign Minister Yair Lapid would succeed him. Israel signed normalization agreements – brokered by the US – with Bahrain, the United Arab Emirates, and Morocco in late 2020 and reached an agreement with Sudan in early 2021" (*World Factbook*, 2022).

The first serious mapping of Israel on modern lines was undertaken in 1798 by Napoleon as an extension of his survey of Egypt. A Topographic Section was formed which consisted of four officers, an astronomer, and four "intelligent soldiers." Bases were measured at Alexandria and Cairo by the "*Service Topographique de l'Armee d'Egypte*," and topographic maps were compiled with a 10 km Grid with an origin at the great pyramids of Giza. The coastline depicted on these early French topographic maps was actually based on British Admiralty Charts. Survey work on the ground was completed late in 1801, and by the end of 1803, compilation in Paris had reached a stage

DOI: 10.1201/9781003307785-96

where the maps could be engraved on copper plates. The sheets were printed in 1808, but Napoleon ordered that they should remain under seal as state secrets. The maps were not finally published until 1817.

In 1865, Captain C. W. Wilson, RE (later Major General Sir Charles Wilson), surveyed the City of Jerusalem at a scale of 1:2,500. The success of Wilson's survey led directly to the establishment of an association called the *Palestine Exploration Fund* (PEF). November of 1871 was the beginning of the PEF surveying and mapping activities, and by 1874 Lieutenant H. H. Kitchener, RE (later Field Marshal Lord Kitchener of Khartoum), arrived to assist Lieutenant C. R. Conder in the first successful mapping expedition of the entire region since Napoleon's attempt. Conder's report speaks of oak forests and bears, wolves, wild pig, cheetah, deer, antelope, and a great variety of game birds. The fauna were the same as reported in the Bible except for the lion, last recorded by the Crusaders in the 12th century. The surveying and mapping were completed by 1878 and comprised 26 sheets, all based on a single Cassini-Soldner projection with a Central Meridian $(\lambda_o) = 34°\ 56'$ East of Greenwich. The ellipsoid of reference was the Clarke 1866. There was no Grid associated with this sheet series where each map measured 15' from north to south and 22' from east to west. Sir Arden-Close remarked, "It is interesting to note that the field work was done by prismatic compass and that no plane-table was used, though, in general, the country lends itself remarkably well to plane-tabling. The reason given by Conder was that the members of the party moved everywhere on horseback, and that a plane table is an inconvenient thing to carry on a horse." Conder and Kitchener supervised Stanford's engraving of the final maps in 1881.

In 1883–1884, Kitchener accompanied an expedition and carried out a triangulation from a base at Aqaba to the Dead Sea, connecting it with the original Palestine control framework near Beersheba. In December 1913, Captain S. F. Newcomb, RE was given charge of a large surveying and archeological expedition which started in Gaza and was to cover the entire Sinai Peninsula. One member of that expedition included T. E. Lawrence, later known as Lawrence of Arabia! Close remarked, "It was doubtless the experience which Lawrence gained on this expedition that justified his being commissioned into the Directorate of Military Survey in September 1914, where he was immediately involved in the compilation of the map of Sinai." Lawrence's cartographic drafting work of Israel between Gaza and Aqaba was performed in Cairo where he also interviewed prisoners, processed intelligence data from agents behind enemy lines, and he produced a handbook on the Turkish Army.

Although a gridded version (referenced to the Bessel 1841 ellipsoid) of Conder's one inch to the mile map series was used by the Egyptian Expeditionary Force during World War I, there was also a need for larger-scale maps of Palestine for artillery and tactical purposes. The British Forces had no town maps of some of the key towns of Palestine. Towns situated beyond the front line, such as Gaza, Beersheba, Ramleh, and others, were photographed by the Aerial Squadrons and maps of those towns were made. The first map, that of Gaza, was produced on January 25, 1917, being probably the first town map ever made by the use of aerial photographs. Other maps, which were produced during 1918, like those of Nablus and El-Kerek were maps which demonstrated a new solution to the problem of the use of aerial photographs for the purpose of mapping towns situated in hilly areas. Town maps for the Palestine Front were an immediate necessity, not an academic exercise, and the war served as an immediate catalyst. The new *Survey of Palestine* department (now the *Survey of Israel*) was established by the Occupied Enemy Territory Administration after the war, and therefore inherited some good topographic maps. They were then able to concentrate on improving the triangulation network and connecting it with the French triangulation in Syria, as well as carrying out cadastral surveys for land settlement.

Under the post-World War I Mandate, Lebanon and Syria (the *Levant*) were protectorates of France, and Israel and Jordan collectively called Palestine and Trans-Jordan were protectorates of Great Britain. A result of that is the northern reaches of Israel are also covered by the Levant Zone. This is a particularly difficult Grid system because it was developed by General Mehmed Sevki Pacha, Director of the Turkish Map Service based on the French Army Truncated Cubic Lambert (partially) Conformal Conic. Earlier Turkish Ottoman topographic maps of the Levant, northern Israel, and Palestine were on the Turkish Bonne Grid. The joins and transformations among the Levant Zone, the Turkish Bonne, and the Palestine Grid (mentioned below) are the source of decades of computational heartburn for cartographers; not to mention my Grad students' homework problems …

Mr. John W. Hagar offered that the Palestine Datum of 1928 has its origin at station Number 2 where $\Phi_o = 31°\ 18'\ 06.27''$ North, $\Lambda_o = 34°\ 31'\ 42.02''$ East of Greenwich, the ellipsoid of reference is the Clarke 1880 where $a = 6,378,300.789$ m, $1/f = 293.466004983713280$, and elevation $= 98.9$ m. The Cassini-Soldner Civil Grid of 1933 (origin adopted is the principal point 82′M (Jerusalem) having the geographic coordinates $\varphi_o = 31°\ 44'\ 02.749''$ N, $\lambda_o = 35°\ 12'\ 39.29''$ East of Greenwich $+ 04.200''$ E $= 35°\ 12'\ 43.490''$. The addition of $04.200''$ to the longitude is in accordance with the decision in 1928 to adopt the French value for the longitude at the points of junction 73′M and 98′M in the north, and to correct all Palestine longitudes accordingly. Palestine longitudes were originally based on those of Egypt at the Transit of Venus station, and a correction of $3.45''$ was indicated to the Egyptian longitudes. Imara Base (1′M or 5′DM) is the original false origin of the Grid coordinates (*i.e.*, FN = FE = 100 km) for the Cassini-Soldner Civil Grid. Final implemented Cassini-Soldner False Easting is then 170,251.555 m, False Northing = 126,867.909 m.

A Military version of this system based on the Gauss-Krüger Transverse Mercator is identical to the Civil Cassini-Soldner Grid except for the False Easting at False Origin where 1,000,000 meters is added, and for coordinates used for the southern Sinai while it was still occupied by the Israelis, an additional 1,000,000 meters was added south of the South False Origin! Of course, that is now defunct.

Mr. John W. Hagar further observed that: "I don't normally pay much attention to the limits of a Grid but here it might be of interest. North is the 150,000-meter Northing Grid line of the Levant Zone. This was redefined, probably as a line of latitude, when the Levant Zone was eliminated. East is the meridian of 39° E. South is a loxodrome from 19° N, 39° E to 26° 45′ N, 35° E, thence west on parallel of 26° 45′ N to 70,000-meter Easting Grid line. West is the 70,000-meter Easting Grid line. You will see that the Grid is not symmetrical east and west and thus can be extended into the Sinai which is precisely what Israel did when they occupied the Sinai."

I have noticed that ellipsoidal loxodromes will warm a professor's heart, but they generally drive cartographers (and Grad students) nuts! About 30 years ago, I was teaching my once every other year grad course in coordinate systems. On the occasion of a 3-day weekend, I assigned a computational problem for intersecting ellipsoidal loxodromes with Grid lines of a constant value as are found for offshore oil leases in the Gulf of Mexico as well as the Palestine Grid limits. The problem is particularly involved because it is iterative between isometric space and in this Louisiana instance, Gauss-Krüger space for UTM. On the first day of class after the holiday, I had the students pass their homework solutions up to me. They then nervously asked me to show the solution on the blackboard. I complied, filling about three blackboards with the computational solution. One of the students in class then asserted, "Yes sir, that is correct." Stunned, I thanked the lady grad student for the validation! (She later got an "A" in the course). The point to all of this is that there are many "British Grids" still extant throughout the world that have boundaries defined by ellipsoidal loxodromes. Furthermore, shipping lanes called Safety Fairways and Traffic Separation Schemes (TSS) are defined by end points connected by ellipsoidal loxodromes. This is one of the realities of contemporary computational cartography the GIS Mapping Scientist may encounter in the real world.

My first job in mapping was as a junior map draftsman for Offshore Navigation, Inc. (now out of business) in New Orleans during the early 1960s. In later decades, ONI retained me for consulting work for various coordinate system puzzles they encountered from time-to-time. In the early 1980s, ONI received a curious message from Israel, and I was asked to solve the problem. Four points were given: Zikhron Yaakou, Tel Aviv, Ashdod, and Khan Yunis. Coordinates were furnished in Latitude and Longitude and in X, Y coordinates. No other explanation was offered to me (or to ONI). Objective: figure out how one set of coordinates related to the other and "solve the problem, Cliff." Knowing what the Datums and Grids of the Palestine and particularly Israel used, I started trying the permutation of systems. The trial-and-error solution yielded the fact that someone in Israel offered the coordinates of the four points in geodetic coordinates referenced to the European Datum of 1950 (International 1924 ellipsoid), and the coordinates of the same four points in Cassini-Soldner Grid coordinates in meters referenced to the Palestine Datum of 1928 on the Clarke 1880 ellipsoid. Therefore, Palestine 1928 to European Datum 1950 is: $\Delta X = -76$ m, $\Delta Y = +64$ m, $\Delta Z = +442$ m. For instance, Ashdod on Palestine 1928, $\phi = 31°\ 50'\ 07.039''$ N, $\lambda = 34°\ 38'\ 17.396''$ E, and on ED 50, $\phi = 31°\ 50'\ 01.8994''$ N, $\lambda = 34°\ 38'\ 13.6922''$ E. (No elevations were offered.)

In October of 1989, Dr. Ron K. Adler, Director General of the *Survey of Israel* offered that the "new" Usrlurim Datum origin at station Urim was: $\Phi_o = 31°\ 20'\ 42.687''$ North, $\lambda_o = 34°\ 28'\ 02.835''$ East of Greenwich. The Israel New Datum is referenced to the GRS 1980 ellipsoid where $a = 6,378,137$ m, and $1/f = 298.2572215381489$. The Seven-Parameter Datum shift from the Israel New Datum to WGS84 Datum are: $\Delta X = -23.500$ m, $\Delta Y = -18.190$ m, $\Delta Z = -17.530$ m, Scale $= +5.43 \times 10^6$, $R_x = -0.30$ arc seconds, $R_y = -1.84$ arc seconds, and $R_z = +1.64$ arc seconds. The new Grid system is defined as being a Gauss-Krüger Transverse Mercator where $\phi_o = 31°\ 44'\ 03.817''$ N, $\lambda_o = 35°\ 12'\ 16.261''$ East of Greenwich, False Northing $= 626,907.39$ m, False Easting $= 219,529.584$ m. For example, on the New Israel Grid: $X = 186,691.878$ m, $Y = 666,264.4$ m, $Z = 82.545$ m (height above ellipsoid), the corresponding WGS 84 Datum coordinates are: $\phi = 32°\ 05'\ 21.16923''$ N, $\lambda = 34°\ 51'\ 26.51726''$ E., $h = 82.526$ m. Thanks for the help on this column go to Mr. Russell Fox of the UK Ordnance Survey, Mr. John W. Hagar (retired from DMA/NIMA), Mr. Mal Jones of Perth, Australia, and others.

Israel has implemented a nationwide Kinematic GPS Continuously Operating Reference Station network. The entire country is now on a new coordinate system named Israel 2005 (IL05). A new Grid has also been implemented for cadastral purposes based on the Transverse Mercator; the selected scale factor at origin is 1.0000067. There is a 50 km shift in the Y axis and a 500 km shift in the X axis, implemented so as to keep the grid lines unchanged on large-scale maps (*Establishment of National Grid Based on Permanent GPS Stations in Israel,* Steinberg, G., and Even-Tzur, G, Surveying and Land Information Science, Vol. 65, No. 1, 2005, pp. 47–52). See also: *Kinematic Datum Based on the ITRF as a Precise, Accurate, and Lasting TRF for Israel,* Ronen, H., Even-Tzur, G., DOI: 10.1061/(ASCE)SU.1943-5428.0000228. © 2017 American Society of Civil Engineers.

A fascinating account of all surveying and mapping activities in a significant portion of the 20th century is in: *A Survey of Palestine under the British Mandate 1920–1948,* 2010, Gavish, Dov, ISBN 978-0-415-59498-1, 337 pages.

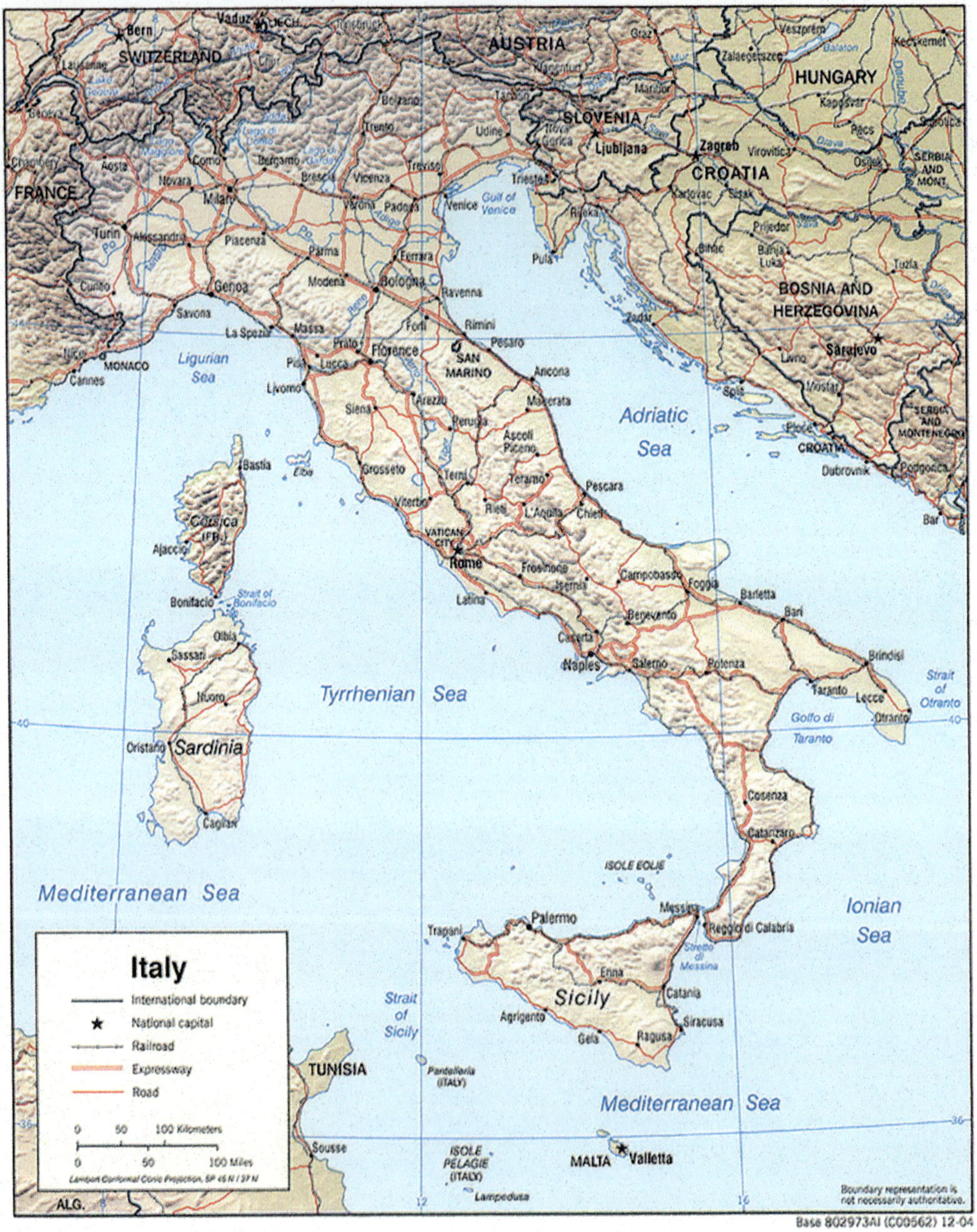

Base 802973AI (C00562) 12-04

Italian Republic

Italy has a coastline of 2,700 km that includes the Ligurian Sea, the Tyrrhenian Sea, the Mediterranean Sea, the Ionian Sea, the Adriatic Sea, and the Gulf of Venice. It borders the following countries: Austria (430 km), France (488 km), Holy See (Vatican City) (3.2 km), San Marino (39 km), Slovenia (232 km), and Switzerland (740 km).

According to *Lonely Planet*, "The Etruscans were the first people to rule the peninsula, arriving somewhere between the 12th and 8th century B.C. They were eventually subsumed within the mighty Roman Empire, leaving little cultural evidence, other than the odd tomb." Thanks to the *CIA Factbook*, "Italy became a nation-state in 1861 when the city-states of the peninsula, along with Sardinia and Sicily, were united under King Victor Emmanuel II. An era of parliamentary government came to a close in the early 1920s when Benito Mussolini established a Fascist dictatorship. His disastrous alliance with Nazi Germany led to Italy's defeat in World War II. A democratic republic replaced the monarchy in 1946 and economic revival followed. Italy was a charter member of NATO and the European Economic Community. It has been at the forefront of European economic and political unification, joining the Economic and Monetary Union in 1999."

The Italian map *Grande Carta d' Italia* was created in 1862 by the decision of the parliament of the United Kingdom of Italy. The topographic survey for the new map was cast on the Bonne projection centered on the intersection of the 40°N parallel and the meridian of the Capodimonte Observatory where $\Lambda_o = 14°\ 15'\ 27.91''$ East of Greenwich. The *Istituto Topografico Militare* was created in 1862 and renamed *Istituto Geografico Militare (IGM)* in 1882. In 1875, it was decided to extend the topographic survey from the former Kingdom of Naples to the entire territory of the United Kingdom of Italy. The primary datum origin of the triangulation system around Rome was observed by Prof. Respighi in 1874 at the vertical axis of the round turret of Monte Mario. The geodetic coordinates of the origin are: $\varphi_o = 41°\ 55'\ 25.42''$ N and $\lambda_o = 12°\ 27'\ 14.00''$ East of Greenwich, (actually $\lambda_o = 00°\ 00'\ 00''$ since it was considered the national Prime Meridian), and with an azimuth to Monte Gennaro where $\alpha_o = 62°\ 38'\ 20.03''$. These geodetic coordinates were derived from the astronomic observatory of Capodimonte near Naples. In 1875, the projection of the *Carta d' Italia* was changed to the Polyhedric projection, *Projezione Naturale*. The coordinates of Castello Monte Mario remained unchanged until the adjustment of the European Datum of 1950, even though the station was astronomically observed in 1904–1905 and in 1940. In 1886, the Italian Cadastre was formed, and the projection used to this day for Italian cadastral plans is the Cassini-Soldner. (The mapping equations for the Polyhedric projection and for the Cassini-Soldner projection are in the *Manual of Photogrammetry*, 5th & 6th editions.)

The origin of the *Genova* Datum of 1874 for Northern and Central Italy was observed where $\Phi_o = 44°\ 25'\ 08.48''$ N and $\Lambda_o = 8°\ 55'\ 21.08''$ East of Greenwich and with an azimuth to *Monte del Telegrafo* where $\alpha_o = 117°\ 31'\ 08.86''$. The origin for *Castanea delle Furie* Datum for South Italy and Sicily was observed where $\Phi_o = 38°\ 15'\ 53.380''$ N and $\Lambda_o = 15°\ 31'\ 18.435''$ East of Greenwich and with an azimuth to *Milazzo* where $\alpha_o = 271°\ 09'\ 16.26''$. The origin for *Guardia Vecchia* Datum for Sardinia was observed where $\Phi_o = 41°\ 13'\ 21.15''$ N and $\Lambda_o = 9°\ 23'\ 59.21''$ East of Greenwich ($3°\ 03'\ 13.29''$ West of Rome, *Monte Mario*), and with an azimuth to *La Curi* where $\alpha_o = 156°\ 51'\ 01.34''$. All of these 19th-century Italian datums were referenced to the Bessel 1841 ellipsoid where $a = 6,377,397.15$ m, and $1/f = 299.1528$.

The baselines were measured with the Bessel base apparatus consisting of bimetallic bars of iron and tin as follows: *Foggia* (1859–1860) 3,939.4206 m ± 1/1,319,000; *Catania*, Sicily (1865) 3,692.1800 m ± 1/587,000; *Crati* (1871) 2,919.5530 m ± 1/751,000; *Lecce* (1872) 3,044.2301 m ± 1/836,000; *Udine*

(1874) 3,248.5785 m±1/1,504,000; *Somma* (1878) 9,999.5380 m±1/2,288,000; *Ozieri*, Sardinia (1879) 3,402.2287 m±1/1,890,000; *Piombino* (1895) 4,621.5696 m±1/945,000.

Considering the shape and topography of Italy, it was decided to cover the entire kingdom with a closed first-order net. By adapting to the terrain, the sizes, and shapes of the classically observed triangles vary considerably. For instance, the triangles in Liguria, along the French boundary, in Puglia, and in Sicily all have an average side length of 22 km, while the average side length of the triangles in central Italy measure 55 km. The smallest first-order side, *Monte Trazzonara-Trasconi* in the province of Puglia has a length of 12.5 km, and the largest side of *Monte Capanna-Punta Maggiore di Monte Nieddu* connecting the islands of Elba and Sicily has a length of 232 km. The locations of the baselines enumerated above were selected in such a manner that the connecting chains consisted of 20–25 triangles and did not exceed a length of 400 km.

After 1890, the first-order net *Rete Geodetica Italiana Fondamentale* was revised, partially resurveyed, and destroyed marks were restored. The revision was sufficiently accurate north of the parallel of Rome to warrant an update, and the Genova Datum of 1908 was based on the new coordinates of the origin (pillar on the terrace of the observatory of the Naval Hydrographic Institute) as: $\varphi_o = 44° 25' 08.235''$ N and $\lambda_o = 8° 55' 15.709''$ East of Greenwich and with a new azimuth to *Monte del Telegrafo* where $\alpha_o = 117° 31' 08.91''$. The Bessel 1841 remained as the ellipsoid of reference. The geodetic coordinates referring to the Genova Datum of 1908 were published in 1919.

To this first-order net were attached: the first-order net in *Venezia Giulia* (Littoral) extended from the sides *Udine – Monte Kanin* and *Udine – Aquileia* and was observed in 1930–1031. All the first-order stations except for Učka (*Monte Maggiore*) are collocated with the *K. und k. III* Austro–Hungarian Military triangulation. The first order net in Venezia Tridentina (South Tyrol) was observed for the purpose of the boundary survey between Italy and Austria and was executed in 1921–1923. In 1944–1945, the net was rigorously adjusted to the Genova Datum of 1908. During World War II, the eastern coast of the Adriatic Sea was occupied by the Italian Army, and the Dalmatian chain was established in 1942. This chain of 37 stations represents a first-order link between the Udine and Foggia baselines. The Italian first order net was tied in 1876 with the triangulation of Tunis, in 1900 with Malta, in 1928 with the triangulation of France with a new connection projected in 1955 and observed in 1957, in 1941 with the triangulation of Greece, in 1951 with Corsica, and in 1953 with the triangulations of Switzerland, Austria, and Yugoslavia.

After World War I, the Austro–Hungarian Empire disintegrated, and Italy succeeded in pushing its northeastern frontiers to the watershed separating the Adriatic Basin from the Danubian Basin. In order to assure a more favorable boundary from a military aspect, Italy annexed parts of the former Austrian provinces of *Tyrol, Carinthia, Carniola, Dalmatia*, and the *entire Littoral province* including the cities of *Trieste* and *Fiume (Rijeka)*. The classical geodetic network of Italy has merged with the cadastre. The cadastre is comprised of two basic types, the Tyrol and the Trentino, and a major part of the cadastre corresponding to the remainder of Italy, The cadastral plans have traditionally been cast on the Cassini-Soldner projection as mentioned earlier, but they have been essentially locally referenced to church spires or other prominent features. The result has been a system of local Cassini-Soldner coordinate systems that total 849 origins! Recent work in the cadastral offices has been to compute transformations of the local systems into the national triangulation network established by the IGM.

Prof. Boaga recast Gauss' original equations for the Transverse Mercator into a truncated series specifically for the Italian Peninsula, and he computed tables in 1945. The grid system is based on the datum recomputed for Italy during World War II, named Roma 1940 Datum. The datum origin is at the original castle at *Monte Mario*, where $\Phi_o = 41° 55' 25.51''$ N±0.027'' and $\Lambda_o = 00° 00' 00'' = 12° 27' 08.40''$ East of Greenwich, and with an azimuth to *Monte Soratte* where $\alpha_o = 06° 35' 00.88'' \pm 0.12''$. The ellipsoid of reference is the Hayford 1909 (International 1924) where $a = 6,377,388$ m and $1/f = 297$. The Italian Geodetic Commission determined the new coordinates of Monte Mario in 1940 and adopted the International ellipsoid in 1942. The Gauss-Boaga Transverse

Mercator (mapping equations are published in the *Manual of Photogrammetry*, 5th & 6th editions) is comprised of two zones: West Zone (I) from 6° to 12° 27′ 08.40″ East of Greenwich (meridian of Monte Mario), and East Zone (II) from 11° 57′ 08.40″ (meridian 30′ West of Monte Mario) to 18° 30′ East of Greenwich. An overlapping zone of 30′ is limited by the meridians of 11° 57′ 08.40″ and 12° 27′ 08.40″ East of Greenwich to coincide with the column of the 1:100,000 scale sheets of the *Carta d' Italia* confined between the −0° 30′ and 0° 00′ (*Monte Mario*) meridians. The scale factor at origin (m_o) = 0.9996, the False Easting for Zone I (West) = 1,500,000 m and for Zone II (East) = 2,520,000 m.

After World War II, the U.S. Army Map Service established the European Datum of 1950 (ED50) and recomputed all classical triangulation of Europe, North Africa, Asia Minor, and the Middle East. Subsequent *military* mapping of the affected countries has been on ED50, referenced to the International ellipsoid. Civilian surveying and mapping have usually been retained on local native datums until recent decades since the advent of the GPS satellites. According to *DMA Technical Report 8350.2*, the transformation parameters from ED50 to WGS84 for Sardinia are: $\Delta a = -251$ m, $\Delta f \times 10^4 = -0.14192702$, $\Delta X = -97$ m ± 25 m, $\Delta Y = -103$ m ± 25 m, $\Delta Z = -120$ m ± 25 m, and this is based on a solution of two points published in 1991. Furthermore, from ED50 to WGS84 for Sicily: $\Delta X = -97$ m ± 20 m, $\Delta Y = -88$ m ± 20 m, $\Delta Z = -135$ m ± 20 m, and this is based on a solution of three points published in 1991. According to the European Petroleum Studies Group (EPSG), peninsular Italy transformation parameters from Rome40 to WGS84 are: $\Delta X = -225$ m, $\Delta Y = -65$ m, $\Delta Z = +9$ m. Note that no accuracy estimates are given, nor are the number of points used in a solution given. Therefore, the user should exercise caution in implementing these values.

On the other hand, *official* transformation parameters for Italy are now published by the International Association of Geodesy (IAG), the *Bundesamt für Kartographie und Geodäsie* (German Federal Office for Cartography and Geodesy), and Eurographics. Note that the European sign convention for rotations is *opposite* from the United States (and Australian) standard. Therefore, the U.S. standard sign convention for rotations is listed in the following parameters. In Sardinia, for Rome 1940 to ETRS89: $\Delta X = -168.6$ m, $\Delta Y = -34.0$ m, $\Delta Z = +38.7$ m, $R_x = +0.374″$, $R_y = +0.679″$, $R_z = +1.379″$, $\Delta s = -9.48$ ppm. In Sicily, for Rome 1940 to ETRS89: $\Delta X = -50.2$ m, $\Delta Y = -50.4$ m, $\Delta Z = +84.8$ m, $R_x = +0.690″$, $R_y = +2.012″$, $R_z = -0.459″$, $\Delta s = -28.08$ ppm. In Peninsular Italy, for Rome 1940 to ETRS89: $\Delta X = -104.1$ m, $\Delta Y = -49.1$ m, $\Delta Z = -9.9$ m, $R_x = -0.971″$, $R_y = +2.917″$, $R_z = -0.714″$, $\Delta s = -11.68$ ppm. These seven-parameter transformations cannot be truncated to just three-parameter translations only, without complete recalculation of the least squares solutions for only three parameters. Do not truncate the above-published rotation and scale change parameters. The transformations with the above seven parameters for the different regions of Italy are expected to yield positions of about 3–4 m accuracy. An example test point published for Peninsular Italy is for Rome 1940: $\varphi = 40°\ 26′\ 12.48″$ N, $\lambda = 17°\ 49′\ 32.52″$ E transforms to ETRS89: $\varphi = 40°\ 26′\ 14.80″$ N, $\lambda = 17°\ 49′\ 32.58″$ E. Andrew Glusic of the U.S. Army Map Service originally published most of the historical details contained herein in an internal document in 1959. See also: Geodetic Datums of the Italian Cadastral Systems, Timar, Gabor, Baiocchi, Valerio, and Lelo, Keti, *Geographia Technica*, No. 1, 2011, pp. 82–90.

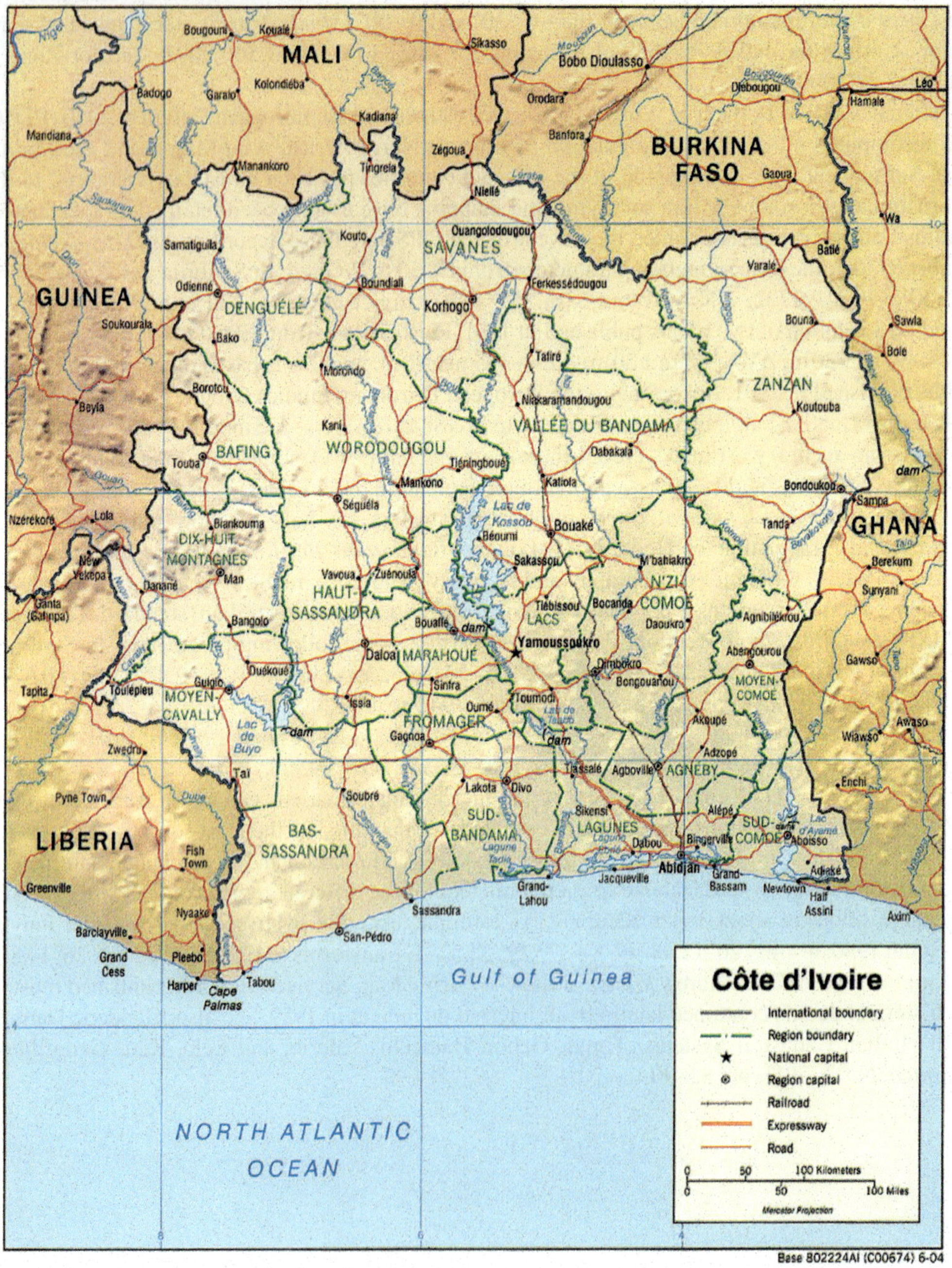

Base 802224AI (C00674) 6-04

Republic of Côte D'Ivoire

Côte d'Ivoire lies on the West African coast on the Gulf of Guinea. Its outline is roughly that of a square 560 km on a side, with an area of 322,460 km² – nearly the same size as New Mexico. It is bounded on the east by Ghana (668 km), on the north by Burkina Faso (584 km), and Mali (532 km), and on the west by Guinea (610 km). The entire southern border is the Gulf of Guinea coastline (515 km). The *Côte d'Ivoire* is mostly flat to undulating plains with mountains in the northwest. The lowest point is the Gulf of Guinea (0 m), and the highest point is Mont Nimba (1,752 m). The republic achieved independence from France on August 7, 1960, and the national capital since 1983 is Yamoussoukro, although Abidjan remains the administrative center.

According to the Library of Congress Country Study, the first recorded history of *Côte d'Ivoire* is found in the chronicles of North African traders who from early Roman times conducted a caravan trade across the Sahara in salt, slaves, gold, *etc*. The first Europeans to explore the West African coast were the Portuguese. The earliest recorded French voyage to West Africa took place in 1483. The first West African French settlement, Saint Louis, was founded in the mid-17th century in Senegal, while at about the same time the Dutch ceded to the French a settlement at *Ile de G*orée of Dakar. A French mission was established in 1687 at Assini, and it became the first European outpost in that area. Assini's survival was precarious, however, and only in the mid-19th century did the French establish themselves firmly in *Côte d'Ivoire*. The colonies of French Sudan and *Côte d'Ivoire* were included in the federation of French West Africa when it was constituted on June 16, 1895. A French decree of October 17, 1899, transferred a number of administrative units from French Sudan to other members of French West Africa, including Bquna, Kong, and Odienné to *Côte d'Ivoire*. The remainder of French Sudan was organized into the civil territory of Upper Senegal and Middle Niger along with two and later three (after December 20, 1900) military territories having headquarters at Timbuctou, Bobo Dioulasso, and Zinder. On March 23, 1902, the Governor General of French West Africa issued an arête modifying the boundary between *Côte d'Ivoire* and the Second Military District headquartered at Bobo Dioulasso.

There was no topographic survey of the French Equatorial Africa or *"Afrique Equatorial Français"* (*AEF*) until after World War II. In December 1945 the *Cabinet du Directeur, Institut Géographique National* in Paris issued an instruction concerning the systems of projection to be utilized in French West Africa. The instruction detailed that a Gauss (Gauss-Schreiber Transverse Mercator) system of projection was to be used for the group of regular map compilations and related works that included geodesy, topography, photogrammetry, and cartography for a range of scales that included 1:200,000. All of the map series were to be cast on the International Map of the World sheet system based on the graticule. During the war, the Ivory Coast Belt Transverse Mercator (*"Fuseau Côte d'Ivoire"*) was used such that the ellipsoid of reference was the Hayford 1909 (International 1909). The unit of measurement was the meter, the Central Meridian, $(\lambda_o) = 6° 30'$ West of Greenwich, and the Latitude of Origin, (φ_o), by definition was the equator. The Scale Factor at Origin $(m_o) = 0.999$, the False Northing = 100 km, and the False Easting = 100 km. D. R. Cockburn and W. L. Barry of Army Map Service translated the *IGN* Instruction dated September 20, 1950, "1. The General Directorate has decided to abandon the projection systems now in use in the French Overseas Territories and Departments and to utilize henceforth, in all these territories, a new projection system called the Universal Transverse Mercator (*Mercator Transverse Universelle*), having a unique definition." See: Republic of Bénin.

The practice of the *Institut Géographique National* (IGN) at the time for West Africa was to provide mapping control in the form of astronomical points ("Astro Stations") at approximately 60 km intervals. Curiously, the local datum known as Abidjan Datum, however, used the Clarke 1880 ellipsoid of reference where $a = 6{,}378{,}249.145$ m, and $1/f = 293.465$. *"The reader is cautioned to not confuse the Abidjan Datum of 1948 with the Blue Nile Datum of 1958 which is commonly misnamed for its origin point at Adindan, Egypt!"* Thanks to Mr. Roger Lott, retired Chief Surveyor of British Petroleum, the European Petroleum Studies Group (EPSG) database lists the Abidjan Datum of 1948 (surveyed in 1948 by Vantroys according to *Annals Hydrographique*, 1961) origin as: $\Phi_o = 05°\ 18'\ 51.01''$ N, $\Lambda_o = -04°\ 02'\ 06.04''$ West of Greenwich. The EPSG database lists the transformation from Abidjan to WGS84 as: $\Delta X = -124.76$ m, $\Delta Y = +53$ m, $\Delta Z = +466.70$ m. The Locodjoro 1965 Datum is listed by EPSG as an *alias* of the same datum.

A datum listed in *Annals Hydrographique* for the western *Côte d'Ivoire* port is the *Tabou Datum of 1950* where $\Phi_o = 04°\ 24'\ 40''$ N, $\Lambda_o = 07°\ 21'\ 29''$ West of Greenwich. The local survey grid, presumably based on the Hatt Azimuthal Equidistant projection, has its origin at the East End of the *Tabou baseline* where $X = 50{,}000$ m and $Y = 50{,}000$ m and the Y axis is defined as being parallel to the meridian. The datum origin point has local grid coordinates of $X = 50{,}228.36$ m, and $Y = 50{,}460$.

REFERENCES: I

Annales Hydrographiques, 2ᵉ Série, Tome Quatorziéme, année 1892, Paris, pp. 351–355.
Annals Hydrographique, 1961.
Army Map Service Geodetic Memorandum 1302.
CIA Factbook, 2005.
Cocking, A.V., Surveying in Iran, *Surveying and Mapping*, December 1969, pp. 655–660.
Conversion of Clarke 1880 Iraq and Caucasus Zone Nahrwan datum Lambert grid coordinates in Patricia K. Hayden, Iraq and Iran to Clarke 1880 New Egyptian datum UTM grid coordinates in Zones 37, 38, and 39, Army Map Service Geodetic Memorandum, 6 August 1948.
Coordonnees Rectangulaires Lambert et Altitudes de Points Géodésiques du Levant, fascicule V.
Country Studies, Library of Congress, 2014.
D. R. Cockburn and W. L. Barry of Army Map Service translated the Institut Géographique National (IGN) Instruction dated 20 September 1950.
DMA Technical Report 8350.2.
Dr. Ron K. Adler, Personal Communication, October 1989.
Edmund A. Early, Tentative Procedure for the Adjustment of W.E. Browne's Triangulation in Iraq to New Egyptian Datum, Army Map Service Geodetic Memorandum, 7 July 1948.
Gavish, Dov, *A Survey of Palestine under the British Mandate 1920–1948*, 2010, ISBN 978-0-415-59498-1, 337 pages.
Geodætisk Instituts Publikationer VII, Island Kortlægning, 1900.
Geodetic Report, Survey of India, CPT. G. Bomford, R.E., 1928.
Geodetic Report, Survey of India, CPT. G. Bomford, R.E., 1930.
Geodetic Work in India – War and Post-War, B.L. Gulatee, *Empire Survey Review*, No. 77, Vol. X, 1950.
GPS-mælingar í grunnstöðvaneti 1993.
Gunnar Þorbergsson, Personal Communication, November 2004.
International Association of Geodesy (IAG), the Bundesamt für Kartographie und Geodäsie (German Federal Office for Cartography and Geodesy), and Eurographics.
Issac C. Lail, Iraq Conversion of Nahrwan Datum to ED UTM, Army Map Service Geodetic Memorandum, 9 November 1957.
John W. Hagar, Personal Communication, July 2000.
John W. Hager, Personal Communication, 05 March 2010.
Library of Congress Country Study, 2009.
Lonely Planet, 2005.
Lonely Planet, 2013.
Lonely Planet, 2014.
Mal Jones of Perth, Australia, Personal Communication, 2000.
Malkawi, Anas, Iraq on the Map: Installing Reference Stations for Accurate Engineering, *GPS World*, March 1, 2011.
Manual of Photogrammetry, 5th & 6th editions.
Notes on East Indies Maps, Theater Area T, Army Map Service, pp. 149–157, March 1945.

Notice: Trans-Persian Trigonometrical Connection between Iraq and India, J.C., *Empire Survey Review*, No. 68, pp. 277–279.

Quilty, L.J., A Report on the State of Geodesy in Java, Jakarta, 1995.

Ronen, H., Even-Tzur, G., Kinematic Datum Based on the ITRF as a Precise, Accurate, and Lasting TRF for Israel, DOI: 10.1061/(ASCE)SU.1943-5428.0000228. © 2017 American Society of Civil Engineers.

Russell Fox of the UK Ordnance Survey, Personal Communication, 2000.

Schepers, J.H.G., and Schulte, Capt. F.C.A., *Geodetic Survey in the Netherlands East Indies*, Topographische Dienst, 1931, 17 pages.

Steinberg, G., and Even-Tzur, G, Establishment of National Grid Based on Permanent GPS Stations in Israel, *Surveying and Land Information Science*, Vol. 65, No. 1, 2005, pp. 47–52.

Subarya, C., Matindas, R.W., *Datum Geodesi Nasional 1995 (DGN-95) Yang Geosentrik*, Bakosurtanal, 27 pages, 1995.

Survey of India through the Ages, by Lt. General S. M. Chadha, delivered at the Royal Geographical Society on 8 November 1990.

Tavakoli, F., Strategy for Computation of Transformation Parameters from Local Datums to Regional Datum, National Cartographic Center of IRAN, PCGIAP.

Timar, Gabor, Baiocchi, Valerio, and Lelo, Keti, Geodetic Datums of the Italian Cadastral Systems, *Geographia Technica*, No. 1, 2011, pp. 82–90.

Triangulation in Arabia, Iraq & Persia, NH-39-M Fao, Survey of India 1946.

Triangulation in Iraq, Syria, and Turkey, No. 37.

World Factbook and NGA GeoNames Server, 2013.

World Factbook, 2009.

World Factbook, 2022.

J

Jamaica 395
Japan 401
Hashemite Kingdom of Jordan 405

Caribbean Sea
Jamaica
Parish boundary
National capital
Parish capital
Railroad
Road
HANOVER
SAINT JAMES
TRELAWNY
SAINT ANN
SAINT MARY
WESTMORELAND
SAINT ELIZABETH
MANCHESTER
CLARENDON
SAINT CATHERINE
SAINT ANDREW
PORTLAND
SAINT THOMAS
Kingston
Montego Bay
Falmouth
Negril
Savanna-la-Mar
Black River
Mandeville
May Pen
Spanish Town
Port Antonio
Ocho Rios
Port Maria
Morant Bay
Jamaica Channel
Caribbean Sea
Base 802855AI (C00236) 11-02

Jamaica

The island of Jamaica is slightly smaller than Connecticut, and it has a 1,022 km coastline. Jamaica claims a 12-nautical-mile territorial sea, and a continental shelf of 200 nautical miles or to the edge of the continental margin. The terrain is mostly mountainous with a narrow discontinuous coastal plain. The lowest point is the Caribbean Sea, and the highest point is Blue Mountain Peak (2,256 m).

Inhabited by Arawak Indians when Columbus discovered the island in 1494, a Spanish colony was established in 1509. The island was captured by British naval officers William Penn and Robert Venables in 1655, the Spanish being finally expelled in 1658. At that time the total population of the island was estimated at 3000, chiefly the slaves of the eight hidalgos who were the "lords proprietors," opposed to further European immigration. By then, the native Indian population had been wiped out by the conquistadors, just as the Arawak had previously done to the other peaceful race – the Lucayan of the Bahamas. Of the indigenous population nothing remains but the name of the island, "*Xayamaca*," the "Land of Springs," and numerous examples of native craft that were left in caves. Jamaica and the Cayman Islands were ceded to Britain in 1670 under the Treaty of Madrid and Jamaica remained a British Colony until it became independent in 1962 within the Commonwealth. "Deteriorating economic conditions during the 1970s led to recurrent violence as rival gangs affiliated with the major political parties evolved into powerful organized crime networks involved in international drug smuggling and money laundering. Violent crime, drug trafficking, and poverty pose significant challenges to the government today. Nonetheless, many rural and resort areas remain relatively safe and contribute substantially to the economy" (*World Factbook*, 2022).

"The island does not lend itself to camping except in the Blue Mountains, so one of the bugbears of the profession is the long walks to and from the field each day. This fact is best illustrated by the experiences of surveyors working in the '*Cockpits*,' who have found their days to be each divided into three parts – three hours for the walk to work over trails which have such picturesque names as the '*Devil's Staircase*,' leading through the '*District of Look Behind*,' six hours behind the instrument and four hours for the return journey to his car or to his lodging" (*Survey Review*, 1955).

The earliest topographic maps of Jamaica were four sheets of Kingston and of Lower St. Andrew at the scale of 2½ inches to the mile made by the Military Authorities in 1909–1912 and known locally as the Pomeroy Map. All other maps were based on compass surveys until the late 1930s.

The Department of Lands and Surveys was formed in 1938, and "the layout of the trigonometrical survey was complete by the end of the year. All lines had been cleared; at the 38 primary stations, 4-foot concrete piers had been built and somewhat smaller piers at 20 secondary stations. At one-third of the primary stations, the observations were completed with the two geodetic Tavistock theodolites in use; they showed triangular closures averaging about 1″. *(Tavistock was the name of the town where the manufacturer was located. The instruments were actually made by Cooke, Troughton and Simms – Ed.).* All angular observations were made between sunset and midnight to projectors; sixteen pointings on eight zeros exhibited an average range of 4″. It was hoped to complete the measurement of a 3½-mile base in Westmorland by January 1939. Invar wires were to be used. An azimuth was observed in Trelawny, with a probable error of about ±1″. The fixing of latitude, longitude and check azimuths had not been undertaken at the date of reporting." "The azimuth observations were made at Maxwell Hall hill, Kempshot (*sic*) on the 28th and 30th of March 1938. Using *Ursa Minor* at or near elongation the azimuth of the side Maxwell Hall-Etingdon was determined. In all, thirty-two pointings were made, and the probable error of the result is 1″. The azimuth of the side was calculated (and later corrected) to be $\alpha_o = 81° \ 16' \ 36.25''$." Although observing Polaris must be done at a specific time of night at elongation, it's a particularly good choice, especially when exact standard time is difficult or impossible to obtain. The resultant observed "true" azimuth is quite reliable and accurate.

"*Datum Latitude and Longitude:* — In accordance with the original instructions from the Geographical Section *(of)* the General Staff, the Flagstaff in Fort Charles, Port Royal, the latitude and longitude of which had previously been determined to be $\Phi_o=17°\ 55'\ 55.8''$ N, $\Lambda_o=76°\ 50'\ 37.26''$ West of Greenwich, was taken as the triangulation origin. It has been connected to the triangulation by intersections from: Wareika, Montpelier, Red Hills, Rodney's Look-out, Plumb Point."

"*Datum Level:* — A line of level (*sic*) was run from the Plumb Point station to the Admiralty Datum in the swimming bath at Port Royal. The mark is reported to be ... 3.50 ft. above mean sea level. The distance from Plumb Point to the Port royal bath is 4½ miles and the disagreement between the outward and return leveling was .086 of a foot, a very good result obtained by Sergeant D.K. Black, R.E. (*I. J. Harris, Capt., R.E., 20th May 1939*)." Note that ordinary differential leveling would allow a total error of ±0.106 feet even by 21st-century standards! The Clarke 1880 ellipsoid was referenced (and modified for Jamaica) such that $a=6,378,249.136$ m and $1/f=293.46631$. The Lambert Conical Orthomorphic was chosen as the projection for Jamaica, and quite interestingly was designed to be a tangent conic, where $m_o=1.0$. Using feet as the unit of measure until 1980, the defining parameters of the "Jamaica Foot Grid" are: central meridian, $\lambda_o=77°$ W, latitude of origin, $\varphi_o=18°$ N, False Easting=550,000 feet, and False Northing at the latitude of origin=400,000 feet, all on the Fort Charles Flagstaff Datum of 1938. The first complete coverage of the island with aerial photography was flown in 1941–1942. The British Directorate of Overseas Surveys (DOS) compiled 12 topographic maps at 1:50,000 scale of the island in 1947 and this was the first DOS project undertaken by the new agency. Brigadier Martine Hotine (the father of the Rectified Skew Orthomorphic Oblique Mercator projection) was the *only* Director of the DOS for its entire existence. Of course, the units were feet for this edition, even though the scale was metric.

In April of 1969, Keith A. Lee wrote: "In the original scheme only one azimuth had been observed by the Royal Engineers; that line Maxwell Hall to Etingdon being situated near the western end of the island. It was considered prudent to determine the azimuth of another line and preferably one in the eastern part of the island. For reasons of accessibility and convenience the line Coopers Hill to Nutfield was chosen. On the night 24/25 October 1968, the azimuth of this line was determined by observations on Polaris. The instrument used was Wild T3 No. 53112 and the observations were made on the 16 circle positions recommended in 'The Retriangulation of Great Britain.' Not having the equipment for a proper determination of astronomical co-ordinates, the listed geodetic position of the stations was used in computing the azimuth. It was noted that this computed azimuth was 10.01″ larger than the listed geodetic azimuth of the Royal Engineers'. The observation was thought to be of a fairly high standard – the *p.e.* being ±0.29″ – and although a systematic swing was present in the individual determinations it was certainly not of the order of 10″." Further investigations by Lee proved that the original reference azimuth of the old Fort Charles Flagstaff Datum of 1938 was in error as a combination of mathematical blunder and probably deflection of the vertical. Note that the reference azimuth quoted with the definition of the old datum was the corrected value that was carried in the readjustment discussed in the following paragraphs. Lee did recommend that the original origin point for the datum be retained, and it was.

From Watson almost 40 years after Captain Harris: "In 1969 it was decided to re-adjust the primary triangulation of Jamaica and to incorporate into the new adjustment all additional and relevant observations made since the previous adjustment. In particular, many electronic distance measurements and some additional astronomical azimuth information were now to be included. As the same time, it was decided to change spheroids and carry out all computations on the Clarke 1866 spheroid instead of the current Clarke 1880 spheroid! These decisions were agreed between the Jamaica Survey Department and the Directorate of Overseas Surveys." Although quite unusual when readjusting a datum to go to an *older* ellipsoid of reference, in this case the reason was that the Inter-American Geodetic Survey had extended the North American Datum of 1927 into the

Caribbean. In the case of Hong Kong, the ellipsoid was changed to an older one to allow for a "better fit" for a readjustment.

"The bulk of the triangulation was observed and adjusted by a team of Royal Engineers during the period 1937–45. Finally, during 1951–62 a team of surveyors from the Directorate of Overseas Surveys, assisted by members from the local department, measured by Tellurometer a good selection for the primary lines." *(Tellurometer was a brand name of electronic distance meter once manufactured in South Africa. – Ed.)*

"The local department would undertake the daunting task of producing abstracts and/or Photostat copies of all observed angles and lengths held by them. These abstracts together with the observed angles and lengths held by DOS would constitute the data for the new adjustment. Historically, this task involved using angles observed between 1937 and 1970 by the local department on different types of survey using different types of instruments. In addition, from 1960 onwards, length measurements by Tellurometer and other electronic distance measuring equipment were carried out by both DOS and the local department. DOS observed Tellurometer traverses all-round the island between primary stations and produced a network of Tellurometer traverses in the Cockpit part of the country. The local department extended and strengthened existing minor work all over the island by triangulation and Tellurometer traverse.

"During the next few years, the local department prepared and sent batches of data and annotated diagrams on map sheets. In 1971–72, after consultation with the local department and DOS, a detachment of Royal Engineers, under the code name 'Calypso Hop,' observed additional angles and lengths in some of the weaker points of the framework. Finally in 1974, a start was made on the secondary and minor adjustment by producing fresh diagrams showing all the information supplied by the local department and held by DOS. Missing data, as revealed by the diagram, and obvious errors which arose during the plotting of the data, were referred back to Jamaica, which department continued to send further information as their diligent searches of their records brought them to light. Close and friendly co-operation with the local department helped considerably with this huge task of collating and collecting the data.

"Naturally the moment had to come when a halt had to be called to any further additions to the data to be adjusted. Any weaknesses in the framework would be revealed in the results of the adjustment, and these weaknesses could be strengthened in the future, though, here, it must be added that serious consideration should be given to trying to hold adjusted co-ordinates for say 20 to 30 years before attempting a complete re-adjustment. Too much localized chopping and changing leads to confusion and a lack of continuity. This may appear to be a rather pragmatic approach, but surveyors and engineers are pragmatic people and surveys have still to be started and computed from the best results available at the time.

"It must be remembered that no co-ordinates are ever final, and, if certain small areas of the adjustment are weaker than the mainly strong whole, additional field work can be carried out in the future to strengthen those weaker areas and if absolutely necessary, small controlled re-adjustments carried out. Following on from the adjustments, some 291 offset points were fixed by azimuth and Tellurometer length and, finally, 36 additional points were computed mainly by the method of intersection. Geographical co-ordinates and Lambert Conical Orthomorphic projection coordinates were computed for 1,392 points" *(W. Watson, April 1977).*

The Jamaica Datum of 1969 retained the original origin point at the Fort Charles Flagstaff, where $\Phi_o = 17°\ 55'\ 55.8''$ N, $\Lambda_o = 76°\ 50'\ 37.26''$ W, the reference azimuth $\alpha_o = 81°\ 16'\ 36.25''$ were used as corrected by Lee, and the ellipsoid of reference was now the Clarke 1866, where $a = 6,378,206.4$ m and $b = 6,356,583.6$ m. The new unit of measure was the meter, the re-defined "Jamaica Metre Grid" retained the original parameters except for the change in ellipsoid, the False Easting was now 250 km and the False Northing at the latitude of origin was now 150 km. Deflection of the vertical at the flagstaff was equated to zero, as was the geoid-ellipsoid separation. However, the new grid did not appear on official topographic maps until 1981.

Control densification work continued by the Jamaica Survey Department, and DOS with its library was incorporated into the Ordnance Survey of Great Britain. A combined adjustment of the original 1969 DOS data along with the control densification and trigonometric levels was performed in 1984 with excellent results. Minor problems with the original 1969 adjustment were corrected, and the final adjustment was performed in blocks with a PDP 11/34 minicomputer. (That's the same model of computer many of us old photogrammetrists used to have that powered our analytical stereoplotters.)

On November 16, 1994, the Government of Jamaica signed a formal treaty with the Government of the Republic of Cuba on the delimitation of the maritime boundary between the two states. Based on the mutually agreed principle of the equidistance method, a list of 106 points was filed with the Secretary General of the United Nations. Although the points were plotted on navigation charts, it was explicitly noted in the treaty that the charts were intended for illustrative purposes only, and the lines connecting the tabulated points were defined as geodesics on the North American Datum of 1927 and the Clarke 1988 ellipsoid. (A geodesic in this case is defined as the shortest distance on the surface of the ellipsoid of revolution.)

Jamaica filed a formal declaration with the Secretary General of the United Nations under the United Nations Convention on the Law of the Sea regarding its claim to territorial waters based on Jamaica's archipelagic basepoints as of October 14, 1996. The list of coordinates is comprised of 28 points described in latitude and longitude to the closest integer arc second "referenced to the North American Datum of 1927 (NAD27) and based on Clarke's (1866) spheroid with a semimajor axis of 6,378,206.4 metres and a flattening of 1/294.978." Since the text of the declaration includes reference to those basepoints plotted on navigation charts; the projection is a normal Mercator and therefore the "straight lines" connecting the basepoints are ellipsoidal loxodromes or rhumb lines. Readers may recall that it was Snellius, the father of geodesy that coined the word "loxodrome."

NIMA lists the transformation parameters for Jamaica from the NAD27 to WGS84 as a mean of 15 stations observed throughout the northern portion of the Caribbean as: $\Delta a=-69.4$, $\Delta f \times 10^4 = -0.37264639$, $\Delta X=-3$ m± 3 m, $\Delta Y=+142$ m± 9 m, and $\Delta Z=+183$ m± 12 m. In addition, in 1996 the U.S. National Geodetic Survey (NGS) observed several existing triangulation stations at the Norman Manley International Airport in Kingston, Jamaica. Mr. Dave Doyle of the NGS graciously provided the local coordinates of a number of points at the airport. In particular for "Airport 5," $\varphi=17°\ 56'\ 06.776''$ N, $\lambda=76°\ 47'\ 44.836''$ W, Northing$=142,844,681$ m, Easting$=271,626.526$ m, and the observed NAD83 geodetic coordinates are: $\varphi=17°\ 56'\ 16.27529''$ N, $\lambda=76°\ 47'\ 41.29986''$ W. Computing the three-parameter geocentric transformation from JAD69 to NAD83 for this single point yields $\Delta X=+58$ m, $\Delta Y=+209$ m, and $\Delta Z=+392$ m. The large differences with NIMA mean values, however, have a simple explanation. NIMA values for Jamaica are for the transformation from NAD27 to WGS84; they are not from JAD69 to WGS84. Although both classical datums are referenced to the Clarke 1866 ellipsoid, note that the origin coordinates for JAD69 are exactly the same as for the Jamaica Datum of 1938: it's at the Fort Charles Flagstaff! The reader will recall that the origin for NAD27 is at Meades Ranch in Kansas, and that is quite a long distance from Jamaica.

Prof. Glendon G. Newsome of the University of Technology in Kingston and Prof. Bruce R. Harvey of the University of New South Wales wrote a paper on transformations in Jamaica, and they used four fiducial points on the island. The three parameters from JAD69 to WGS84 they solved are: $\Delta X=+65.33$ m± 0.96 m, $\Delta Y=+212.46$ m± 1.49 m, and $\Delta Z=+387.63$ m± 0.69 m. Newsome and Harvey acknowledge the critical need for a good geoid for Jamaica, and they hope for the day when a new national datum may be established. Thanks also go to Prof. Hugart Brown retired from the Metro State College of Denver for his kind help over the years with Jamaican survey history. Thanks to Mr. Russell Fox of the Ordnance Survey of the United Kingdom and his soon to be closed International Library for all of his help.

In 2008, the Ministry of Agriculture established a GPS RTK Virtual Reference System for the island consisting of a five-station network (http://www.spatialvision.com/ministry-of-agriculture-jamaica-signs-major-contract-with-spatial-innovision-to-deliver-the-national-gps-infrastructure/).

In 2012, the Jamaica VRS and Cadastral Surveying was discussed by Prof. G. G. Newsome, Prof. G. Peake, and Mr. R. Douglas in the September issue of *Coordinates* magazine (https://mycoordinates.org/the-jamaica-vrs-and-cadastral-surveying-2/).

In 2015, *The Spatial Active Global Geomatrix* was presented by Mr. Siburn Clark, RICS and Mr. Douglas Nelson at the 4th Annual Caribbean Valuation & Construction Conference in which they discussed the 13 CORS sites in Jamaica of the VRS system (https://www.researchgate.net/publication/290814255_Spatial_Global_Active_Geomatrix).

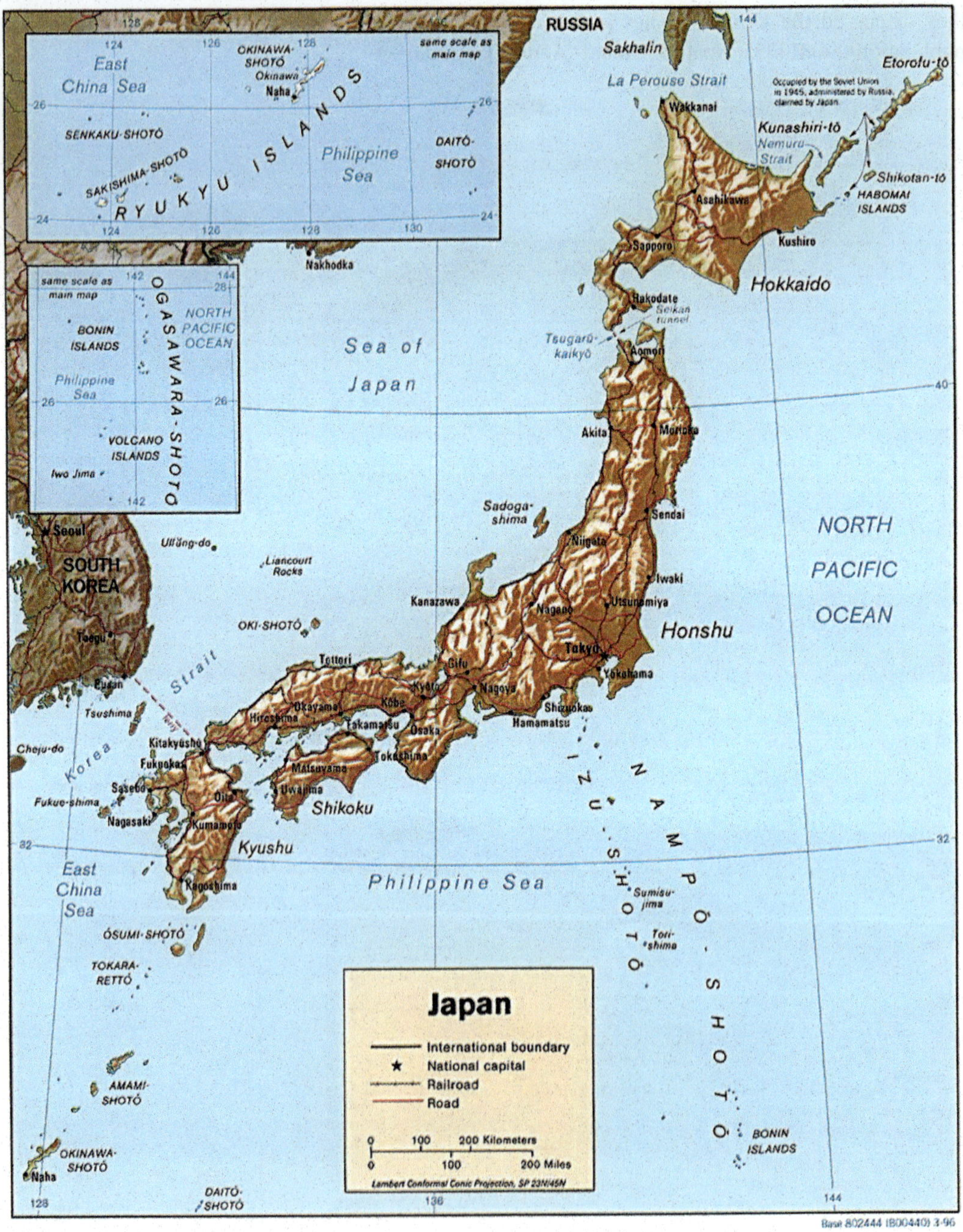

Base 802444 (B00440) 3-96

Japan

Comprised of four main islands: *Honshū, Shikoku, Kyūshū,* and *Hokkaidō,* Japan also includes the Bonin Islands (*Ogasawara-Guntō*), *Daitō-Shotō, Minami-Jima, Okino-tori-Shima, Ryukyu Islands* (*Nansei-Shotō*), and Volcano Islands (*Kazan-Rettō*). Japan is mostly mountainous, the lowest point is *Hachirō-Gata* (lake) at −4 m, the highest is *Fuji-Yama* at 3,776 m, and both places are on *Honshū*.

Japan was inhabited by humans as early as 30,000 B.C. Its written history began in the 5th century A.D., after it adopted handwriting from the Chinese culture. Buddhism was introduced *circa* 552, and Japan closely imitated Chinese institutions during the 6th–9th centuries. First visited by the Portuguese in 1542–1543, European influences had little effect on religion, Shintoists and Buddhists represent 84% of the current population and Christianity is practiced by less than 1%. For several hundred years, the Portuguese had far greater influence than the later English and Dutch trading companies. Commodore Mathew Perry, USN secured the first commercial treaty in 1853. The "Land of the Rising Sun" is a constitutional monarchy with a parliamentary government. There are 47 prefectures, Japan's independence dates back to 660 B.C. (traditional founding by Emperor Jimmu), and its constitution is dated May 3, 1947.

"During the late 19th and early 20th centuries, Japan became a regional power that was able to defeat the forces of both China and Russia. It occupied Korea, Formosa (Taiwan), and southern Sakhalin Island. In 1931–32 Japan occupied Manchuria, and in 1937 it launched a full-scale invasion of China. Japan attacked US forces in 1941 – triggering America's entry into World War II – and soon occupied much of East and Southeast Asia. After its defeat in World War II, Japan recovered to become an economic power and an ally of the US. While the emperor retains his throne as a symbol of national unity, elected politicians hold actual decision-making power. Following three decades of unprecedented growth, Japan's economy experienced a major slowdown starting in the 1990s, but the country remains an economic power. In March 2011, Japan's strongest-ever earthquake, and an accompanying tsunami, devastated the northeast part of *Honshū* Island, killed thousands, and damaged several nuclear power plants. Prime Minister Shinzo Abe was reelected to office in December 2012 and has since embarked on ambitious economic and security reforms to improve Japan's economy and bolster the country's international standing. In November 2019, ABE became Japan's longest-serving post-war prime minister" (*World Factbook*, 2022).

In 1869, the Survey Division of the Ministry of Civil Services was established. In 1888, the Imperial Land Survey of the Army General Staff absorbed the Survey Division and carried out the fundamental surveying and mapping of the entire Japanese Empire. The Tokyo Datum of 1892 was established at the *Azabu* origin, where $\Phi_o = 35° 39' 17.5148''$ North, $\Lambda_o = 139° 44' 30.097''$ East of Greenwich. The defining azimuth was determined from the old Tokyo Observatory at *Azabu* to station *Kanou-Yama* (Kanou Mountain) as: $\alpha_o = 156° 25' 28.442''$. The Tokyo Datum of 1892 was referenced to the Bessel 1841 ellipsoid where the semi-major axis $(a) = 6,377,397.155$ m and the reciprocal of flattening $(1/f) = 299.1528128$. The pervasive Tokyo Datum of 1918 reflected a new determination of Longitude at *Azabu* Observatory such that the origin was re-defined as: $\Phi_o = 35° 39' 17.5148''$ North, $\Lambda_o = 139° 44' 40.5020''$ East of Greenwich. The orientation was not changed, and the deflection of the vertical at the observatory was defined to be nil. The ellipsoid height of mean sea level in Tokyo Bay was defined to be −24.414 m. For military use, the Japanese Imperial Land Survey (JILS) used a conformal double transverse Mercator Grid known as the Gauss-Schreiber Transverse Mercator from about 1892 to 1921. (That's the same formulae that the U.S. Coast & Geodetic Survey chose for the TM zones on the North American Datum of 1927.) The multiple belt system defined the East Belt origin as: $\varphi_o = 36° 03' 34.9523''$ N, $\lambda_o = 139° 44' 40.5020''$ E, where the X-axis began at the meridian crossing the transit circle at Tokyo Observatory. The East Belt covered the area south of the *Tsugaru Kaikyō* (Strait) and east of 135°, comprising all of *Honshū* east of *Kōbe* to 142°. The West belt origin was: $\varphi_o = 36°\ 03'\ 34.9523''$ N, $\lambda_o = 132°\ 04'\ 42.9183''$ E, where the meridian crossing the *Kammuri-Yama* First-Order station in *Hiroshima-Ken* was the X-axis origin, and the limits of the West belt were from 126° to 135°. The North Belt origin was: $\varphi_o = 45°\ 00'\ 00''$ N, $\lambda_o = 142°\ 15'\ 17.2085''$ E, where the meridian crossing the *Yubari-Dake* First-Order station in *Hokkaidō* was the X-axis origin and was used for computation in *Hokkaidō*. This North Belt was also used in *Karafutō*, the southern half of Sakhalin Island invaded during the Russo-Japanese War of 1904–1905 and occupied by the Japanese until after World War II. *(For an interesting story on that see Sakhalin Island – Ed.)* The southern belt covered the area south of 31°, where the Formosa Belt origin was $\varphi_o = 23°\ 40'\ 00''$ N, $\lambda_o = 120°\ 58'\ 25.9750''$ E, where the meridian crossing the *Koshizan* First-Order station (now called *Hu-tzu-shan* in Taiwan) was the X-axis origin. These grids did not use False Origins, and in all cases the Scale Factor at Origin (m_o) was equal to unity. The purpose of the JILS Grids was to compute the basic primary triangulations of these areas. Similar systems were implemented by the JILS in Manchuria, China, and Korea.

According to Mr. John W. Hager, "After 1921, the JILS Grids changed to the following: Latitude of Origin for Korea and zones 1 through 5 is 36° N, for Formosa 24° N. The Longitudes of Origin – Korea 128°, zone 1 132°, zone 2 136°, zone 3 140°, zone 4 144°, zone 5 148°, and Formosa 121°. All zones use a unity Scale Factor, and False coordinates were used, but the only one I have is for zone 2 where FN=6,000 km, and FE=3,600 km. Notes on the Japanese Artillery (JA) versus the JILS: where JILS X (northing)=0 meters=JA Y=250,000 meters, JILS Y (easting)=0 meters=JA X=200,000 meters. Further stated that Latitude of Origin=28° N, Longitude of Origin=129° 19' 23.6028" E, (Yuwan-Dake). This is marked as the New Tokyo (1918) value and with a further remark to see also 129° 19' 13.1978" E, which is the old Tokyo (1892). Convenient in that the Grid values are the same and only the geographic vary. The *Korean National Geography Institute* and *National Construction Research Institute* use the same gimmick by increasing the longitude of origins by 10.405"."

After World War II, the duties of the JILS were transferred to the Geographical Survey Institute (GSI) in the Ministry of Home Affairs. In 1947, a new Surveying Law defined 13 zones for cadastral mapping ("Public Survey"), on the Gauss-Schreiber Transverse Mercator with specific prefectures for each zone. For Zone 1, $\varphi_o = 33°\ 00'$ N, $\lambda_o = 129°\ 30'$ E, required for *Nagasaki and Saga*. For Zone 2, $\varphi_o = 33°\ 00'$ N, $\lambda_o = 131°\ 00'$ E, required for *Fukuoka, Ōita, Kumamoto, Miyazaki, and Kagoshima*. For Zone 3, $\varphi_o = 36°\ 00'$ N, $\lambda_o = 132°\ 10'$ E, required for *Yamaguchi, Shimane, and Hiroshima*. For Zone 4, $\varphi_o = 33°\ 00'$ N, $\lambda_o = 133°\ 30'$ E, required for *Kagawa, Ehime, Tokushima, and Kōchi*. For Zone 5, $\varphi_o = 36°\ 00'$ N, $\lambda_o = 134°\ 20'$ E, required for *Hyōgo, Tottore, and Okayama*. For Zone 6, $\varphi_o = 36°\ 00'$ N, $\lambda_o = 136°\ 00'$ E, required for *Fukui, Shiga, Mie, Nara, Wakayama, Kyōto, and Ōsaka*. For Zone 7, $\varphi_o = 36°\ 00'$ N, $\lambda_o = 137°\ 10'$ E, required for *Ishikawa, Toyama, Gifu, and Aichi*. For Zone 8, $\varphi_o = 36°\ 00'$ N, $\lambda_o = 138°\ 30'$ E, required for *Niigata, Nagano, Gumma, Yamanashi, and Shizuoka*. For Zone 9, $\varphi_o = 36°\ 00'$ N, $\lambda_o = 139°\ 50'$ E, required for *Fukushima, Tochigi, Ibaraki, Chiba, Saitama, Kanagawa, and Tōkyō*. For Zone 10, $\varphi_o = 40°\ 00'$ N, $\lambda_o = 140°\ 50'$ E, required for *Aomori, Akita, Yamagata, Iwate, and Miyagi*. For Zone 11, $\varphi_o = 45°\ 00'$ N, $\lambda_o = 140°\ 15'\ 17.2085''$ E, where the central meridian is 2° west of the Longitude of Yubari-dake First-Order Trig. station in Hokkaido. Zone 11 is required for the following sub-prefectures of *Hokkaidō*: *Iburi, Usu, Abuta, Hiyama, Shiribeshi, Otaru, Hakodate, and Ō Shima*. For Zone 12, $\varphi_o = 45°\ 00'$ N, $\lambda_o = 142°\ 15'\ 17.2085''$ E, where the central meridian is the Longitude of *Yubari-dake* First-Order Trig. station in *Hokkaidō*. Zone 12 is required for the following sub-prefectures of *Hokkaidō*: *Sapporo, Ishikari, Abashiri (Mombetsu), Asahigawa, Kamikawa, Soya, Urakawa, Muroran, Yūbari* (excluding *Usu, and Abuta*), *Sorachi, and Rumoi*. For Zone 13, $\varphi_o = 45°\ 00'$ N, $\lambda_o = 144°\ 15'\ 17.2085''$ E, where the central meridian is the Longitude of *Yubari-dake* First-Order Trig. station in *Hokkaidō*. Zone 13 is required for the following sub-prefectures of *Hokkaidō*: *Nemuro, Kushiro, Abashiri (excluding Mombetsu), Kasai, and Kushiro*. As usual, there was no false origin, and the Scale Factor at Origin was equal to unity for all of these grids for the "Public Survey."

Thanks to Professor Kazuo Kobayashi of the Survey College of Kinki, there have been some new additions to the "Grids of the Public Survey." Mr. John W. Hager later pored over some maps to identify the likely areas of coverage. For Zone 14, $\varphi_o=26°\ 00'$ N, $\lambda_o=142°\ 00'$ E, and appears to be for the Bonin Islands (*Ogasawara-Guntō*), and Volcano Islands (*Kazan-Rettō*). For Zone 15, $\varphi_o=26°\ 00'$ N, $\lambda_o=127°\ 00'$ E, and appears to be for *Okinawa Guntō* and *Amami Guntō*. For Zone 16, $\varphi_o=26°\ 00'$ N, $\lambda_o=124°\ 00'$ E, and appears to be for *Sakishima Guntō*. For Zone 17, $\varphi_o=26°\ 00'$ N, $\lambda_o=131°\ 00'$ E, and appears to be for the Daito Islands (*Kita Daito Jima and Okino Daito Jima*). For Zone 18, $\varphi_o=20°\ 00'$ N, $\lambda_o=136°\ 00'$ E, and appears to be for *Parece Vela*. For Zone 19, $\varphi_o=26°\ 00'$ N, $\lambda_o=154°\ 00'$ E, and appears to be for Marcus (*Minami Tori Shima*).

In 1985, Mr. Toshiyuki Shiina, General Manager of the Geodetic Surveying Department of the Aero Asahi Corporation, wrote to me of the special Grid he devised for the Seikan Tunnel Project. (The project was the longest tunnel under the seabed in the world at the time.) Since the tunnel spanned two Public Coordinate Grids between the Main Island of Honshū and the northern island of Hokkadō, he chose to define his own special purpose Grid for the project based on the Gauss-Krüger Transverse Mercator projection. Mr. Shiina also chose a Scale Factor at Origin, $m_o=0.9999$ and the projection origin at: $\varphi_o=41°\ 10'$ N, $\lambda_o=140°\ 10'$ E.

The GSI reports that it is moving the country to the Japanese Geodetic Datum 2000 (JGD2000). In *TR8350.2* the transformation (for Japan) from Tokyo 1918 to WGS84 is: $\Delta a=739.845$, $\Delta f \times 10^4=0.10037483$, $\Delta X=-148$ m± 8 m, $\Delta Y=-507$ m± 5 m, $\Delta Z=-685$ m± 8 m.

Japan now has a GPS network established with a density of about 20 km used primarily for an earthquake warning system. The new published transformation parameters from Tokyo 97 Datum to ITRF94 are: $\Delta X=+147.414$ m, $\Delta Y=-507.337$ m, $\Delta X=-680.507$ m. Thanks to Prof. Kazuo Kobayashi, "the lowest point in Japan is the middle of the Seikan Tunnel at -256.5674 m. The history of Japanese Datum leveling above Mean Sea Level at Tokyo Bay at "1-chome Nagatacho, Chiyokaku, Tokyo in 1894=24.5000 m, in 1923=24.4140 m and in 2011=24.3900 m."

The Geospatial Information Authority of Japan (GSI) has published extensive information on their data *in English*: https://www.gsi.go.jp/ENGLISH/page_e30030.html; there is a detailed preface: https://www.gsi.go.jp/common/000001194.pdf; concept of the New Japanese Geodetic System: https://www.gsi.go.jp/common/000001195.pdf; the new Height System of Japan: https://www.gsi.go.jp/common/000001197.pdf; and Development of a new Geoid: https://www.gsi.go.jp/common/000001198.pdf.

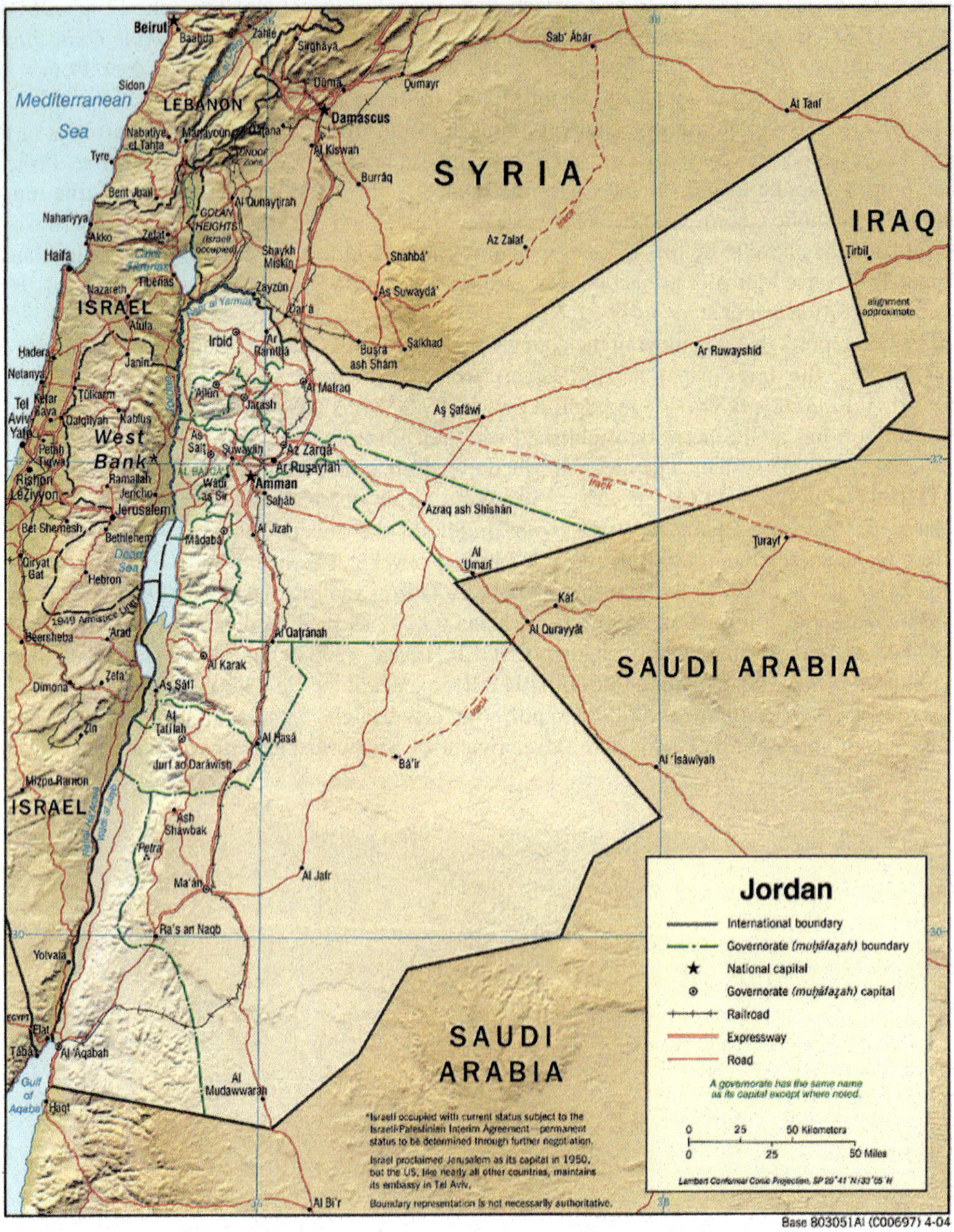

Base 803051AI (C00697) 4-04

Hashemite Kingdom of Jordan

The kingdom is slightly smaller than Indiana, and borders Iraq (181 km), Israel (335 km), Saudi Arabia (744 km), and Syria (375 km). Jordan is mostly desert plateau in the east, and highland in the west. The Great Rift Valley separates the East and West banks of the Jordan River, the lowest point is the Dead Sea (−408 m), and the highest point is *Jabal Ram* (1,734 m).

According to *Webster's New Geographical Dictionary*, "Created (as Transjordan) 1921 out of former Turkish territory and proclaimed an independent state 1923 under Emir Abdullah ibn al-Husayn, but a mandate under British protection; mandate revoked 1946 and by treaty of March 1946 became an independent kingdom."

"The country has had four kings. Jordan's long-time ruler, King Hussein (1953–99), successfully navigated competing pressures from the major powers (US, USSR, and UK), various Arab states, Israel, and Palestinian militants, which led to a brief civil war in 1970 referred to as 'Black September' and ended in King Hussein's ouster of the militants from Jordan. Jordan's borders also have changed. In 1948, Jordan took control of the West Bank and East Jerusalem, eventually annexing those territories in 1950 and granting its new Palestinian residents Jordanian citizenship. In 1967, Jordan lost the West Bank and East Jerusalem to Israel in the Six-Day War but retained administrative claims until 1988 when King Hussein permanently relinquished Jordanian claims to the West Bank. King Hussein signed a peace treaty with Israel in 1994 after Israel and the Palestine Liberation Organization signed the Oslo Accords in 1993.

"Jordanian kings continue to claim custodianship of the holy sites in Jerusalem by virtue of their Hashemite heritage as descendants of Prophet Mohammad and agreements with Israel and Jerusalem-based religious and Palestinian leaders. Israel has authorized the Jordanian Islamic Trust, or *Waqf*, to administer affairs in the *Al Haram ash Sharif*/Temple Mount holy compound, and the Jordan-Israel peace treaty reaffirmed Jordan's 'special role' in administering the Muslim holy shrines in Jerusalem.

"King Hussein died in 1999 and was succeeded by his eldest son, Abdallah II, who remains the current king. In 2009, King Abdallah II designated his son Hussein as the Crown Prince. During his reign, Abdallah II has contended with a series of challenges, including the Arab Spring influx of refugees from neighboring states, and a perennially weak economy" (*World Factbook*, 2022).

The *Office of the Geographer, U.S. Department of State* provides background on the Jordan – Saudi Arabia Boundary: "The new boundary passes through a desert or near desert area of limited economic potential; the frontier is virtually rainless (under 2 cm everywhere) and almost devoid of population. Scattered wells and access to the Gulf of Aqaba (for Jordan) are the primary points of strategic value.

"The frontier region is a tilted plateau with the highest elevation in the west and the lower in the east. The interior district is composed of sand areas interspersed with eroded lava flows, which are dissected by dry streambeds. A few peaks attain elevation of 1,000 m, approximately 300 m above the average elevation of the plateau. The general alignment of the drainage pattern is toward the south and east reflecting the slope of the plateau. Soil, except in the wadi bottoms, is very thin or non-existent.

"Close to the shores of the Gulf of Aqaba, the border traverses the escarpment marking the edge of the Dead Sea – Jordan – Aqaba fault valley. The escarpment, in places, looms 1,400 m above the level of the adjacent sea. However, many dry watercourses dissect the escarpment into a series of broken blocks. A narrow coastal plain fringe the shores of the Gulf with relatively easy access to the interior furnished by the numerous wadi beds."

DOI: 10.1201/9781003307785-102

"After the end of World War I, Great Britain received a League of Nations mandate for Palestine encompassing Jordan. The British soon divided the mandate for administrative purposes along the Jordan River – Wadi Araba line. However, the precise southern limits of Palestine and Trans-Jordan were indefinite. At the time, Britain claimed access to the Gulf of Aqaba, while the Arabians considered Ma'an, about 80 km to the north, to be within their domain based on its inclusion in the Ottoman *Vilayet* of the *Hejaz*. The adjacent parts of Arabia were occupied by the independent sultanates of the *Nejd* and *Hedjaz*, which later became the core of the Kingdom of Saudi Arabia (1932). In the middle of the 1920s, Britain began a series of negotiations with the Arabian sultanates to settle the southern limits of both the Trans-Jordan and the Syrian mandates. The eastern and central sectors were agreed upon in 1925 with the disputed Kaf region being assigned to the *Nejd*. The western sector of Jordan with the *Hejaz*, however, could not be agreed upon. Britain ultimately delimited the boundary unilaterally to include within Trans-Jordan a narrow outlet on the Gulf of Aqaba. This action was rejected by the *Hejaz* and later by the combined kingdom of Saudi Arabia. King Ibn Saud, however, did agree to maintain the *status quo* pending a solution to the dispute.

"In 1946, Trans-Jordan became independent as the Hashemite Kingdom of the Jordan, with the boundary problem still unsolved. Efforts were made in 1961–63 to settle the question, again to no avail. Finally in 1965, a mutually acceptable line was delimited. Jordan gained an enlarged coastline (19 kilometers) on the Gulf of Aqaba and 6,000 km^2 of territory in the interior. In turn, 7,000 km^2 of Jordanian-administered territory were ceded to Saudi Arabia. The new boundary came into effect on November 7, 1965. The TAP line (Trans-Arabian Pipeline) crosses the boundary at approximately 31° 13′ East. The pipeline is currently the most important economic feature crossing the boundary."

Early mapping (pre-World War I) of the area comprising the Kingdom of Jordan were small-scale compilations by the Ottoman Turks and were cast on the ellipsoidal Bonne projection. Large-scale mapping was confined to local planetable compilations for cadastral tax applications. This constituted the first *Tabu* (land registry office) in 1857 in the Greater Syrian town of *Bilad Ash Sham*. Modern mapping was accomplished by the British Army Middle East Land Forces (MELF) for the Trans-Jordan–Nejd Boundary 1:100,000 scale series. In 1937 the Trans-Jordanian Department of Lands and Surveys took over from the Survey of Palestine the responsibility for mapping the area east of the Jordan River and carried on the work under British direction. Work in the northern part of the area was done in cooperation with the Free French Forces in Syria and Lebanon. Before World War I, southwestern Palestine was mapped by the British War Office cooperatively with the Survey of Egypt. Although the British Mandate of Transjordan ended in 1946, the Department of Lands and Surveys, Transjordan and later Jordan continued to work, with varying degrees of closeness, with mapping services of the British Army. During and immediately following World War II, topographic mapping in Transjordan was conducted by survey units attached to MELF. Among the MELF publications covering parts of Transjordan were three series at scales of 1:25,000, three at 1:50,000, and one at 1:500,000. In 1914–1915, the British *Directorate of Military Survey* published the Africa 1:125,000 Sinai Peninsula series and in 1943 the 1:50,000 Transjordan Lava Belt series.

Thanks to Mr. John W. Hager, the military mapping of Jordan was referenced to the Palestine Datum of 1928 which has its origin at station number 2, where $\Phi_o = 31° 18′ 06.27″$ North, $\Lambda_o = 34° 31′ 42.02″$ East of Greenwich, the ellipsoid of reference is the Clarke 1880, where $a = 6,378,300.78$ m, $1/f = 293.466308$, and elevation $= 98.9$ m. The Gauss-Krüger (military) Grid (origin adopted is the principal point 82′M (Jerusalem) having the geographic coordinates $\phi_o = 31° 44′ 02.749″$ N, $\lambda_o = 35° 12′ 39.29″$ East of Greenwich $+ 04.200″$ E $= 35° 12′ 43.490″$. The addition of 04.200″ to the longitude is in accordance with the decision in 1928 to adopt the French value for the longitude at the points of junction 73′M and 98′M in the north, and to correct all Palestine (and Transjordan) longitudes accordingly. Palestine longitudes were originally based on those of Egypt at the Transit of Venus station, and a correction of 3.45″ was indicated to the Egyptian longitudes. Imara Base (1′M or 5′DM) is the original false origin of the Grid coordinates (i.e., FN = 126,876.909 m, FE = 170,251.555 m), and 1,000,000 m is added south of the false origin to avoid negative numbers.

After World War II, the U.S. Army Map Service (AMS) decided to eliminate all of the individual datums and grid systems of Europe, the Mid-East, and North Africa and to combine all into a single datum called the European Datum of 1950. The origin was at the *Helmetturm* (Helmert's Tower) in Pottsdam, Germany, where $\Phi_o = 52°\ 22'\ 51.446''$ North, $\Lambda_o = 13°\ 03'\ 58.928''$ East of Greenwich, and was referenced to the International 1924 ellipsoid where $a = 6,378,388$ m, and $1/f = 297$. AMS did convert all the Palestine and Jordan to the European Datum of 1950. The method used was a conformal transformation on the complex plane using UTM coordinates.

Thanks to a memo by Isaac C. Lail (AMS, 1957): "The Transjordan Triangulation Extension was initiated in 1937 by the Department of Lands and Surveys, Amman and is based on the major triangulation of Palestine. The connection for the northern arm is at the stations 52M, 53M, and 54M and to the south, the stations are 44M, 89M, and 92M. Using the sides between these points as a base the triangulation was extended eastwards into Transjordan. The stations were occupied, and their lengths of sides were accepted from the major triangulation of Palestine as the starting base. The mean triangle closure from the beginning of the extension to points 169M, 172M, and 176M was 7.9″. The triangles to the east of these points had larger closures of up to 13″. Therefore, this extension was considered to be of 3rd order accuracy. The Palestine–Transjordan Chain starts on the Palestine major stations 11M, 12M, and 13M and connects with the Egyptian stations E10, F10, G10, and H10 and then goes along the border of Egypt to the Gulf of Aqaba. From the gulf it goes north through Transjordan to the area of Jericho and closes on the Stations 92M and 44M of the Palestine major net. The chain had 121 triangles, which had a mean closure of 2.4″ with a maximum of 9.4″ and only 11 of the 121 triangles had a closure of over 5″. Thus, the chain was considered to have 2nd order accuracy. The bases, differences between measured and calculated lengths have errors as follows: 201M–202M = 1/35,000 and 157M–182M = 1/320,000. Three astronomical azimuths were taken as stations 196M, 205M, and 137M. The results were consistent but varied with computed values from the coordinates."

A fascinating account of all surveying and mapping activities in a significant portion of the 20th century is in *A Survey of Palestine under the British Mandate 1920–1948* (Dov Gavish, 2010, ISBN 978-0-415-59498-1, 337 pages).

"The British Army contracted a civilian surveyor, Kolomoytzeff in 1942 to observe a major triangulation connecting the Syrian Cadastral Primary Triangulation to the Existing work in Transjordan but it was not fully completed, and it is said that the computations were not carried out in a rigorous manner. The net started from two Syrian primary stations, *Tell El Aarar and Chababiye*. Another station, *Tell Koulib*, was used but not occupied. Three of the Transjordan Extension stations and four of the Palestine Transjordan Chain stations were reoccupied. Then to the east and south to 31° N, Kolomoytzeff observed 21 new stations."

Based on a solution I derived about 40 years ago, the local datum shift from Palestine 1928 Datum to European Datum 1950 is: $\Delta X = -76$ m, $\Delta Y = +64$ m, and $\Delta Z = +442$ m.

The Royal Jordan Geographic Center (RJGC) was established in Amman in 1975. The process of computerization of all Jordanian land records is complete and the new grid system, the Jordan Transverse Mercator (JTM), is based on 6° belts with a central meridian of 37° East and a Scale Factor at Origin $(m_o) = 0.9998$. "The National Geodetic Network is a high accuracy Doppler-based geodetic network," and the ellipsoid of reference is reported to (strangely) remain as the International 1924! No transformation parameters are presently offered by the government.

However, thanks to Prof. Steven H. Savage of Arizona State University, the JTM has a False Easting of 500 km, a False Northing of −3,000 km, and the three-parameter transformation to WGS84 is: $\Delta X = -86$ m, $\Delta Y = -98$ m, and $\Delta Z = -11$ m.

REFERENCES: J

Background Notes on the Jordan – Saudi Arabia Boundary, Office of the Geographer, U.S. Department of State.

Dov Gavish, *A Survey of Palestine under the British Mandate 1920–1948*, 2010, ISBN 978-0-415-59498-1, 337 pages.

file:///Users/cliff/Downloads/ClarkeNelson2015SpatialActiveGeomatrixRICSIPTICaribbeanValuationConstructionConference.pdf

G. G. Newsome, G. Peake, and R. Douglas, Coordinates, September 2015. https://mycoordinates.org/the-jamaica-vrs-and-cadastral-surveying-2/

http://www.spatialvision.com/ministry-of-agriculture-jamaica-signs-major-contract-with-spatial-innovision-to-deliver-the-national-gps-infrastructure/

https://www.gsi.go.jp/common/000001194.pdf

https://www.gsi.go.jp/common/000001195.pdf

https://www.gsi.go.jp/common/000001197.pdf

https://www.gsi.go.jp/common/000001198.pdf

https://www.gsi.go.jp/ENGLISH/page_e30030.html

I. J. Harris, Capt., R.E., 20th May 1939.

Isaac C. Lail, Army Map Service, Geodetic Memo, 1957.

John W. Hager, Personal Communication, 2002.

John W. Hager, Personal Communication, November 2006.

Prof. Glendon G. Newsome, Prof. Bruce R. Harvey, GPS Coordinate Transformation Parameters for Jamaica, Date unknown.

Prof. Kazuo Kobayashi, Survey College of Kinki, Personal Communication, January 2002.

Russell Fox Personal Communication, Ordnance Survey Library, April 2003.

Survey Review, 1955.

Toshiyuki Shiina, Aero Asahi Corp., Personal Communication, 1985.

W. Watson, Typed manuscript, Directorate of Overseas Surveys, April 1977.

Webster's New Geographical Dictionary.

World Factbook, 2022.

K

Republic of Kazakhstan 411
Republic of Kenya 415
Kiribati 421
The Republic of Korea 427
State of Kuwait 431
Kyrgyz Republic 435

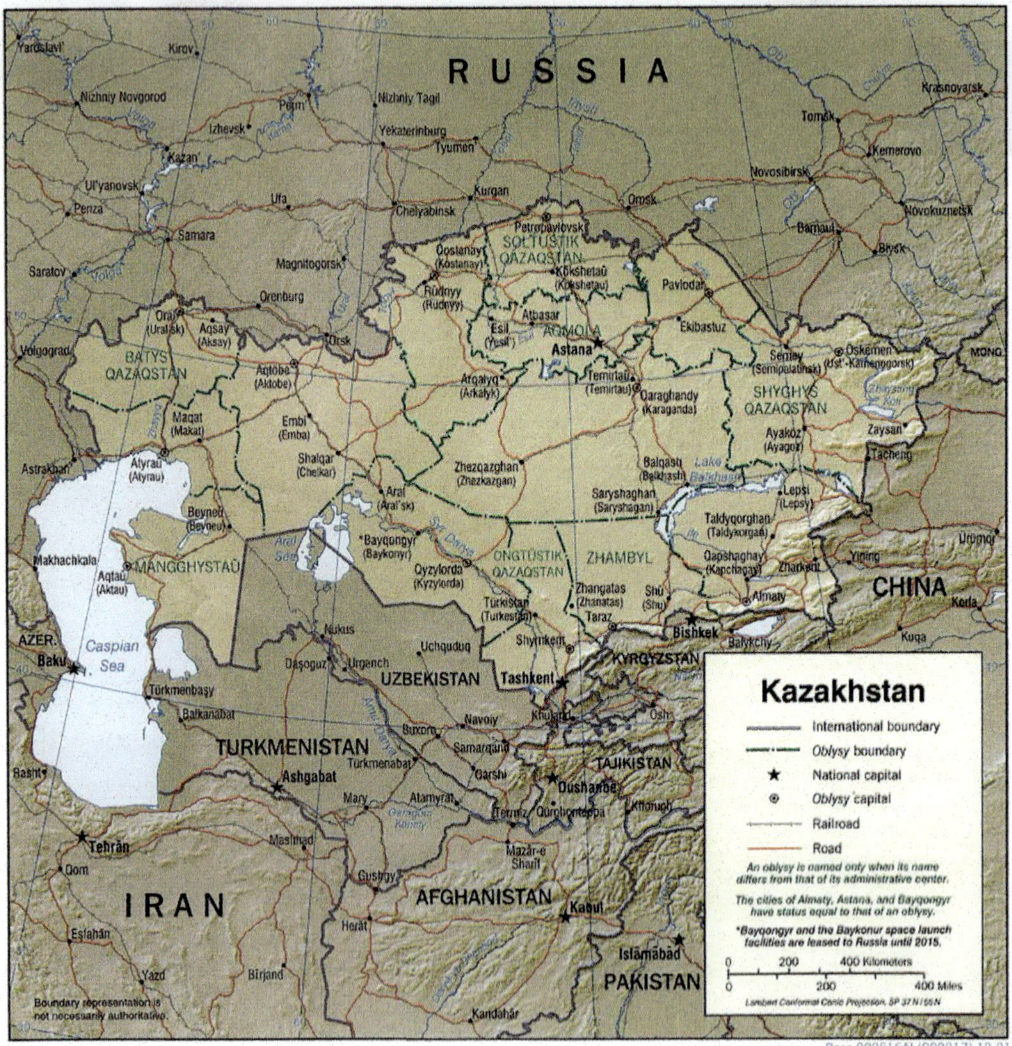

Base 802516AI (C00217) 12-01

Republic of Kazakhstan

With an area of about 2,717,300 km², Kazakhstan is more than twice the combined size of the other four Central Asian states and is slightly less than four times the size of Texas. The country borders China (1,533 km), Kyrgyzstan (1,224 km), Russia (6,846 km), Turkmenistan (379 km), and Uzbekistan (2,203 km). The lowest point is *Vpadina Kaundy* (−132 m), and the highest point is Khan Tangiri Shyngy *(Pik Khan-Tengri)* (6,995 m). Because of a catastrophic idea of the former Soviet Union, the two main rivers that flowed into the Aral Sea have been diverted for irrigation; it is now drying up and leaving behind a harmful layer of chemical pesticides and natural salts; these substances are then picked up by the wind and blown into noxious dust storms (*World Factbook*, 2010).

"Humans have inhabited present-day Kazakhstan since the earliest Stone Age, generally pursuing the nomadic pastoralism for which the region's climate and terrain are best suited. The earliest well-documented state in the region was the Turkic Kaganate, which came into existence in the sixth century A.D. The Qarluqs, a confederation of Turkic tribes, established a state in what is now eastern Kazakhstan in 766. In the eighth and ninth centuries, portions of southern Kazakhstan were conquered by Arabs, who also introduced Islam. The Oghuz Turks controlled western Kazakhstan from the ninth through the eleventh centuries; the Kimak and Kipchak peoples, also of Turkic origin, controlled the east at roughly the same time. The large central desert of Kazakhstan is still called *Dashti-Kipchak*, or the *Kipchak Steppe*. In the late ninth century, the Qarluq state was destroyed by invaders who established the large Qarakhanid state, which occupied a region known as Transoxania, the area north and east of the Oxus River (the present-day Syrdariya), extending into what is now China. Beginning in the early eleventh century, the Qarakhanids fought constantly among themselves and with the Seljuk Turks to the south. In the course of these conflicts, parts of present-day Kazakhstan shifted back and forth between the combatants. The Qarakhanids, who accepted Islam and the authority of the Arab Abbasid caliphs of Baghdad during their dominant period, were conquered in the 1130s by the Karakitai, a Turkic confederation from northern China. In the mid-twelfth century, an independent state of Khorazm along the Oxus River broke away from the weakening Karakitai, but the bulk of the Karakitai state lasted until the invasion of Chinggis (Genghis) Khan in 1219–21. After the Mongol capture of the Karakitai state, Kazakhstan fell under the control of a succession of rulers of the Mongolian Golden Horde, the western branch of the Mongol Empire. The horde, or *zhuz*, is the precursor of the present-day clan, which is still an important element of Kazak society. By the early fifteenth century, the ruling structure had split into several large groups known as khanates, including the Nogai Horde and the Uzbek Khanate. The present-day Kazaks became a recognizable group in the mid-fifteenth century, when clan leaders broke away from Abul Khayr, leader of the Uzbeks, to seek their own territory in the lands of Semirech'ye, between the Chu and Talas rivers in present-day southeastern Kazakhstan. The first Kazak leader was Khan Kasym (r. 1511–23), who united the Kazak tribes into one people. In the sixteenth century, when the Nogai Horde and Siberian khanates broke up, clans from each jurisdiction joined the Kazaks. The Kazaks subsequently separated into three new hordes: the Great Horde, which controlled Semirech'ye and southern Kazakhstan; the Middle Horde, which occupied north-central Kazakhstan; and the Lesser Horde, which occupied western Kazakhstan. Russian traders and soldiers began to appear on the northwestern edge of Kazak territory in the seventeenth century, when Cossacks established the forts that later became the cities of Oral (Ural'sk) and Atyrau (*Gur'yev*)" (*Library of Congress Country Study*, 2010).

"Non-Muslim ethnic minorities departed Kazakhstan in large numbers from the mid-1990s through the mid-2000s and a national program has repatriated about a million ethnic Kazakhs (from Uzbekistan, Tajikistan, Mongolia, and the Xinjiang region of China) back to Kazakhstan.

As a result of this shift, the ethnic Kazakh share of the population now exceeds two-thirds. Kazakhstan's economy is the largest in the Central Asian states, mainly due to the country's vast natural resources" (*World Factbook*, 2022).

In the early 19th century, the Russian penetration began into the area between the Caspian Sea and Iran, Afghanistan, and China. In 1839, the Russian Army moved into the area of Turkestan which was for centuries an object of struggle between Turko–Iranian and Chinese influences and completed its occupation in 1857. In 1867, the *Military Topographic Department of Turkestan* was formed and was headed by a major or brigadier general. The Astronomic and Physical Observatory of Tashkent was founded officially in 1878. The city of Tashkent, now the capital of Uzbekistan became the capital of Turkestan, and by 1888 the region of Transcaspia, now in southwest Kazakhstan, was incorporated as a new province. Geodetic triangulation performed between 1896 and 1929 was based on base lines measured with Jäderin apparatus (invar steel and brass wires) at Kazalinsk and Arys in Kazakhstan and oriented on the Tashkent Datum. In order to make a topographic survey at 1:4,000 scale of the City of Tashkent, a traverse net was established in 1885 by Pomerantsev. The revised coordinates of the Tashkent Observatory SW pillar used for the city survey are: $\Phi_o = 41°\ 19'\ 30.42''$ North, $\Lambda_o = 38°\ 58'\ 00.99''$ East of Pulkovo (or $69°\ 17'\ 39.54''$ East of Greenwich). The defining azimuth at the point of origin to the North Stone Pillar (probably the mire) of the Observatory is: $\alpha_o = 00°\ 52'\ 08.25''$, and the ellipsoid of reference is the Bessel 1841 where $a = 6,377,397.155\,m$, and $1/f = 299.1528128$. The Indian Datum of 1916 origin at Kalianpur Hill Station is: $\Phi_o = 24°\ 07'\ 11.26''$ North, $\Lambda_o = 77°\ 39'\ 17.57''$ East of Greenwich. The defining azimuth at the point of origin to station Surantal is: $\alpha_o = 190°\ 27'\ 05.10''$. The ellipsoid of reference is the Everest 1830, where $a = 6,377,276.345\,m$, and $1/f = 300.8017$.

The Kazalinsk Datum of 1891 origin at the finial Cross on the Town Church is: $\Phi_o = 45°\ 45'\ 46.450''$ North, $\Lambda_o = 62°\ 06'\ 01.66''$ East of Greenwich. The defining azimuth at the point of origin to station Sulutan is: $\alpha_o = 20°\ 34'\ 07.34''$. The ellipsoid of reference again is Bessel 1841. The three-parameter shift from Kazalinsk 1891 to Tashkent 1895 is: $\Delta X = +530\,m$, $\Delta Y = -160\,m$, $\Delta Z = -104\,m$. (The fit of four points agrees to about 20 m in each geocentric component.)

The main chain from the Aral Sea to India together with attached triangulations represents a geodetic work not planned in advance but executed differently by time, purpose, and accuracy. The enormous difficulties were faced by triangulators during the field work in the regions with dense overgrowth along the *Syr Darya River* and in *Fergana Valley*, but particularly in the glaciers of the *Pamir Mountains* meant that the survey procedures many times deviated from the basic standards of that time. Consequently also, accuracy dropped far below the claimed Second Order. There are no published observations or sides and angles of the triangle net, but only geographic coordinates which, for the triangulation established in 1896–1924, are fragmentarily included in the Catalog of Kazakhstan. During the revolution (1917–1921), many records of the chains which should be attached to the adjusted Main Chain were lost. The remaining data is buried in the "secret" crates held at the U.S. Army Documents Depository in Mineral Wells, Texas. Supposedly, this is done to avoid offending the deceased officials of the long-defunct U.S.S.R.!

In the *Zapiski Catalog, Vol. 41, part. III, Section 5, page 13*, the 1880 Triangulation in Semiryechenska Province details 37 trigonometric points of the local system oriented on the astronomic point *Gorohudzir* that was determined in 1879 by the chronometric expedition of Schwartz with: $\Phi_o = 44°\ 07'\ 41.87''$ North, $\Lambda_o = 49°\ 26'\ 41.01''$ East from Pulkovo ($\Lambda_o = 79°\ 46'\ 19.56''$ East of Greenwich). Near the astro point a base line was measured and its astronomic azimuth was determined by observation of α Polaris. However, these points were not included in the Catalog of COL. Gedeonov in the *Zapiski Vol. 53, part II, pp. 229–288*. *(Note that these references to "Zapiski Catalog" are the Annals of the Czarist Russian Topographic Brigades starting with volume 1 in 1837 by General Schubert. – Ed.)*

In the *Anuario of the (Argentine – Ed.) Instituto Geográfico Militar, 3er Volumen, 1914*, a history of the Russian Geodetic and Topographic activities of the 19th century is offered in *Spanish*. An annex of that Argentine *Anuario* provides a map of all Czarist Russian Geodetic work until 1912.

A major First-Order chain commenced in Kazakhstan at the Aral Sea near the town of Aral, it followed the *Syr Darya River* through Kazalinsk and the now present *Baykonur Cosmodrome* which was the old *Tyuratam* Ballistic Missile Silo Farm – *(a discomforting sight 50+ years ago when I saw it imaged by the Corona Program – Ed.)*, and finally down to Tashkent, Uzbekistan. An independent semi-circular arc of First-Order triangulation had been surveyed in the vicinity of Akmolinsk *(now Aqmola – Ed.)*, a mineral-rich region of the Kazakh steppes, and dozens of astronomical observations had been completed throughout all of Kazakhstan and along the border with China.

"The main 1901 chain from *Orenburg* to Tashkent through Kazakhstan was based on a Second-Order triangulation in support of the railroad and the irrigation project of the *Syr Darya Valley*. The section from the Aral Sea to Kazalinsk is 163 km long and contains 68 triangles, from Kazalinsk to *Arys* is 675 km long and contains 138 triangles, and from *Arys* to *Osh* at the border between Uzbekistan and Kyrgyzstan is 547 km long and contains 72 triangles. From the Second-Order stations of the main chain, Third-Order stations such as churches, mosques, smokestacks, water towers, monuments and mountain peaks were included. This totaled 358 Second-Order stations and 105 Third-Order points. All the stations were provided with permanent markers, but they were not all uniform. At the beginning of the triangulation survey in 1901, ceramic cylinders were used; later stones and stone pillars with lead bolts and chiseled crosses were utilized. Some triangulators used subsurface markers consisting of bricks with chiseled crosses, and some covered surface markers with heaps of stones or with soil. End points of base lines were marked with both surface and subsurface marks, where bolts of lead or brass with engraved crosses were set into stone pillars or bricks. The base lines were measured with the standard Jäderin brass and invar apparatus. Sections of the main chain were computed and adjusted independently where different starting points were used. The northwestern section of the *Syr Darya Valley* observed in 1901 to *Ob* was computed from the old 1888 triangulation surveyed by Zalesky. The Kazalinsk base line measured from station *Sulutem* to *Karmyz* was measured in 1907 and was 8,863.3094 m ±0.00152 m, or 1:4,900.000. The *Arys* base line exit side measured from *May Tyube* to *Kur Say* was measured in 1913 and was 8,521.440 m ±0.049 m or 1:173,907" (*Triangulation in Turkestan and its Connection with India, Army Map Service Technical Report Number 21, Andrew M. Glusic, 100 pp., 1957*).

In 1998, geodetic surveys commenced in the Caspian Sea area of Kazakhstan, and the Russian System 42 Datum was used where the origin is at Pulkovo Observatory: $\Phi_o = 59°\ 46'\ 18.55''$ North, $\Lambda_o = 30°\ 19'\ 42.09''$ East of Greenwich, the defining azimuth at the point of origin to Signal A is: $\alpha_o = 317°\ 02'\ 50.62''$, and the reference ellipsoid is the Krassovsky 1940 where $a = 6,378,245$ m, and $1/f = 298.3$. The local grid system used in that area of Kazakhstan is the Gauss-Krüger Transverse Mercator where the Central Meridian $(\lambda_o) = 56°\ 46'$ E, False Northing Latitude of Origin $(\varphi_{FN}) = 00°\ 08'\ 00''$ N, False Easting = 300 km, and the Scale Factor at Origin $(m_o) =$ unity. The three-parameter transformation **from** System 42 **to** WGS 84 in the Caspian Sea area of Kazakhstan is: $\Delta X = +14.471$ m, $\Delta Y = -132.753$ m, $\Delta Z = -83.454$ m, and is based on an occupation at five stations: *Bolat, Yesim Sevirne, Dlinnaya Dolina, Daralsai,* and *Aul*. The seven-parameter position vector transformation **from** System 42 **to** WGS 84 is: $\Delta X = +43.822$ m, $\Delta Y = -108.842$ m, $\Delta Z = -119.585$ m, $R_x = +1.455''$, $R_y = -0.761''$, $R_z = +0.737''$, and $\delta s = +0.549$ ppm. A test point provided for this transformation is **from** System 42: $\varphi = 46°\ 30'\ 00''$ N, $\lambda = 50°\ 00'\ 00''$ E **to** WGS 84: $\varphi = 46°\ 30'\ 00.321''$ N, $\lambda = 49°\ 59'\ 55.513''$ E. This solution is based on a mean of 13 stations along the entire Kazakhstan coastline of the Caspian Sea and is considered good to ±2 m. Thanks go to Mr. Phil Smart, M.R.I.C.S. of RPS Energy.

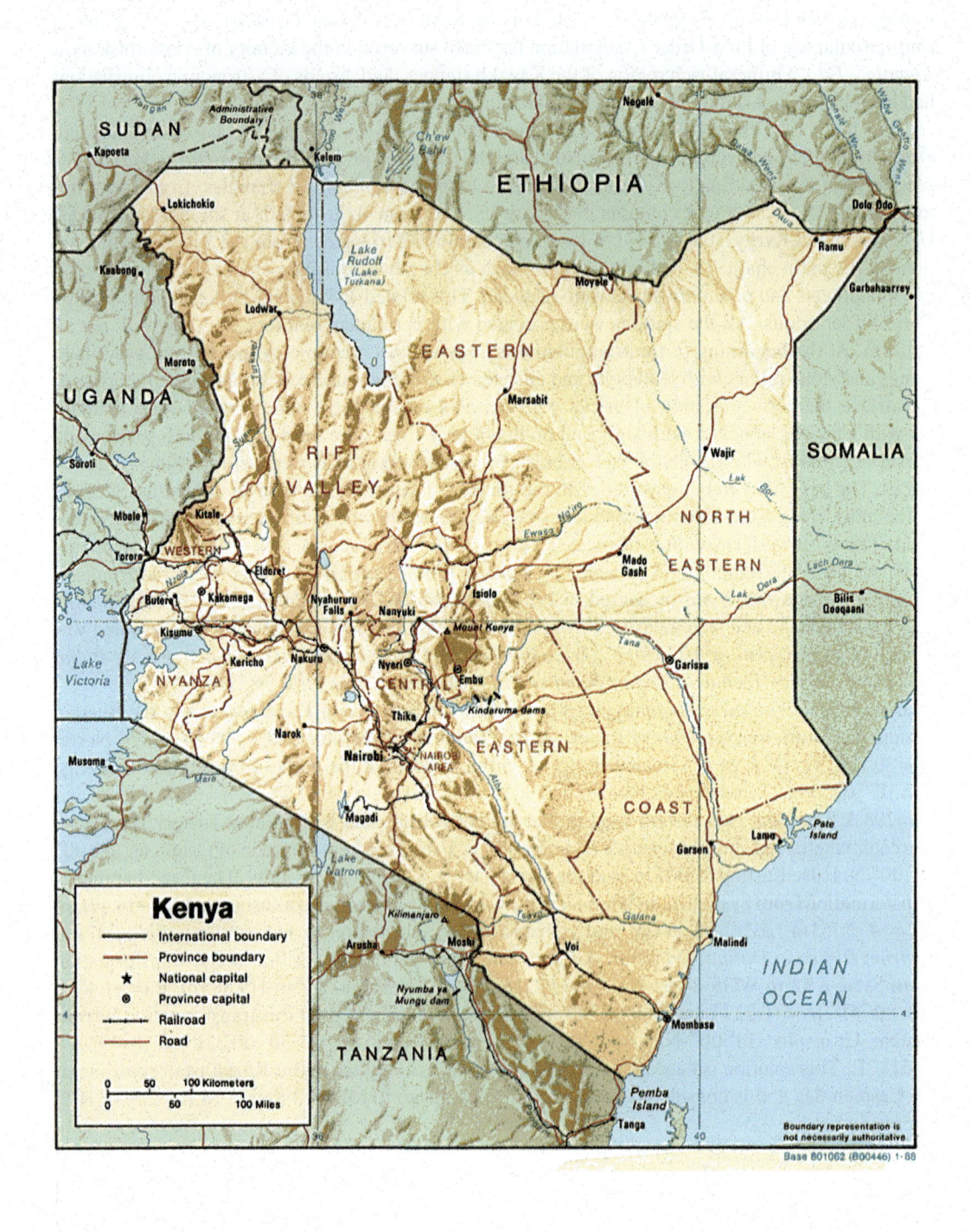

Base 801062 (B00446) 1-88

Republic of Kenya

Kenya is bordered on the north by Ethiopia (861 km), on the east by Somalia (682 km), on the southeast by the Indian Ocean (536 km), on the south by Tanzania (769 km), on the west by Uganda (933 km), and on the northwest by Sudan (232 km). The lowest point in Kenya is the Indian Ocean, and the highest point is Mount Kenya at 5,199 m. Comprised of the Nairobi Area and seven provinces – Central, Coast, Eastern, Northeastern, Nyanza, Rift Valley, and Western; Kenya is slightly larger than twice the size of Nevada.

The coast of Kenya was long dominated by Arabs and was seized in the 16th century by the Portuguese. The Europeans were expelled by the Omanis; the coast then came under the rule of the Sultan of Zanzibar, and later leased in 1867 to the British East Africa Association. The British extended their holdings into the interior and fixed an initial southern boundary with the German East Africa Company in 1886. The former Colony of British East Africa gained independence on 12 December 1963.

Thanks to Mr. Morgan W. Davis, "The history of surveying in East Africa begins with the domination of those lands by European powers in the late 1800s. The British entered into a number of agreements defining spheres of influence in 1886, 1890, 1891, and 1894. The present-day boundaries between Kenya and its neighbors are a result of legal descriptions hashed out in negotiations and subsequent triangulation and boundary surveys. Negotiations between the United Kingdom and Germany in 1886 and 1890 established spheres of influence north and south respectively, of a line beginning at the Indian Ocean near Vanga and extending to the eastern shore of Lake Victoria:

> The line of demarcation starts from the mouth of the River Wanga or Umbe (Umba), runs direct to Lake Jipe, passes thence along the eastern side and round the northern side of the lake and crosses the Lumi River …
>
> After which it passes midway between the territories of Taveita and Chaga, skirts the northern base of the Kilimanjaro range, and thence is drawn direct to the point on the eastern side of Lake Victoria Nyaza (Lake Victoria) which is intersected by the 1st degree of south latitude.

The line between the Indian Ocean and Lake Jipe was surveyed by a plane table. Most of the mapping in East Africa between 1890 and 1910 was a result of the boundary commissions. Basic topographic mapping of varying quality was accomplished along the narrow zones of the surveyed boundaries. There was little opportunity to extend mapping to the interiors of the colonies. An important outcome of this early phase was the consolidation of the War Office as the authority on boundary surveys and maps in Africa. Both the Foreign Office and the Colonial Office relied heavily on the expertise of the War Office on technical matters related to surveying and mapping, as well as for help in wording legal descriptions in negotiations. The Topographic Section of the General Staff of the War Office played a crucial role, as well in the policies and activities of survey departments in the colonies.

The colonial Survey Committee was created in an attempt to organize the mapping effort in the British East African colonies. The first meeting was held on August 14, 1905. They recommended that there be two survey departments, standardized topographic map scales at 1:62,500, 1:125,000, 1:250,000, and 1:1,000,000. In a 1907 meeting, they adopted the Clarke 1858 ellipsoid for Africa. They decided on the spelling of place names on maps. The Committee continued to be an important governing body up to the World War II years. Major E. H. Ellis was appointed Inspecting Officer to the departments in the Uganda and East Africa Protectorates (Kenya) in order to help expedite work. He submitted a comprehensive report in February 1907 in which he noted that a topographic section had not been constituted. He insisted that a full section of two officers and six to eight surveyors be formed. He recommended map sheets covering 45′ longitude and 30′ latitude or 30′ × 30′, at 1:125,000 scale for developed areas, and 1½° longitude by 1° latitude in undeveloped areas, utilizing the rectangular polyconic projection. *(This was the same specification utilized during the same era by the British Survey of India. – Ed.)*

In late 1908 one officer, three NCOs, and a civilian were assembled to begin fieldwork on 1:125,000 sheets for Kijabe and Nairobi, and a special 1:62,500 sheet for Nairobi. Mapping continued until the outbreak of World War I. The Africa Series GSGS 1764 in 33 sheets at 1:250,000 scale covered both Uganda and the East Africa Protectorates. The maps were published in monochrome, principally between 1905 and 1907. These were provisional sheets with a paucity of detail. The maps were later also published in color. Each sheet covered 1½° longitude and 1° latitude with a graticule spacing of 30′. They were reprinted in 1939–1941 during the East African Campaign. It was not until 1953 and thereafter that Series GSGS 1764 was replaced at the same scale by Series GSGS 4801 and subsequently Series Y503.

After World War I, the War Office was no longer available to do work in the African colonies. German East Africa had been assigned to Great Britain as a mandate from the League of Nations in 1919 and was renamed *Tanganyika*. The Arc of the 30th meridian was proposed as the foundation of triangulation in the East African colonies. Observations on a portion of the arc in western Uganda had been taken prior to 1914, and the triangulation net in Uganda was tied to it. Surveying on the arc had been done in northern Rhodesia, and it was felt that it was important to close the gap in the arc in Tanganyika.

Martin Hotine *(years later promoted to Brigadier – Ed.)* surveyed the arc of the 30th meridian in Tanganyika between 4½° and 9° South during the years 1931–1933. Depletion of funds in late 1933 left a gap in the arc between 1½° and 4½° South. From July 1936 to August 1937, a survey was conducted wholly within *Tanganyika* to fill the gap, consisting of observation angles and some azimuths. Uganda had withdrawn from the project due to fears that if their portion of the arc was connected to South Africa, they would be forced to recompute their already completed surveys on a new projection and grid system.

This leads to a major theme of discussion during the years between the two great wars – that of a common datum and projection for all of British Africa. Debate raged over this topic until the exigencies of war during the Second World War permitted the British military to force a solution. In a memorandum circulated in 1926, it was assumed that a common datum could be chosen, utilizing a meridional orthomorphic projection from Khartoum to Cape Town. The Clarke 1880 ellipsoid was suggested. During the second Conference of Empire Survey Officers (1931), it was assumed that all colonial governments would adopt the Transverse Mercator projection because it was already accepted by Egypt, South Africa, and two of the West African territories. The width of the zones could not be agreed upon. Kenya saw little prospect in adopting the proposal because its cadastral work was computed on Clarke 1858 and the Cassini projection. Extension and re-computation of its triangulation was more urgent than conversion of its completed surveys to a new datum. In January 1934, GSGS proposed a coordinated projection and grid zone embracing South Africa, South Rhodesia, Sudan, Egypt, and the Central and East African territories. They recommended the Clarke 1880 ellipsoid and the Transverse Mercator projection on a 6° grid. The same parameters were recommended in a meeting of a sub-committee of the Colonial Survey Committee on October 3, 1935. Brigadier M. N. MacLeod insisted on the adoption of the meter as the map unit. Each time a new recommendation would be put forward for a common set of map parameters, one or more colonial governing bodies would shoot it down for various reasons.

Lord Hailey wrote *An African Survey* (1938) after his tour of Africa in 1935. His views were taken up by the Colonial Survey Committee in 1939, at which time they once again recommended a 6° grid and the adoption of the meter. Whittingdale replied that a 2°-system was more appropriate for topographic mapping and military surveys. Huntley showed the military advantages of the 2° grid (artillery), and that it was inconvenient for cadastral surveyors to apply the corrections that a 6° grid would necessitate. *(The same reason for practicality continues to this day for NOT using the UTM grid for civil GIS and surveying applications. – Ed.)* South Africa totally opposed the change from 2° to 6° zones. There was general agreement on the adoption of the meter on map grids.

A policy for military mapping was defined in July 1940, which utilized the Clarke 1880 ellipsoid and the Transverse Mercator projection with 5° zones. The central meridians were placed at 32°

30′E, 37° 30′E, and 42° 30′E. A scale factor reduction of 0.05% was introduced to provide correct scale on two parallel meridians approximately 1° 49′ on either side of the central meridian. The scale error at the central meridian was about 1:2,000, and it was about 1:2,200 at the edges of the grid zones. The grid was originally in yards but was later changed to meters. This became known as the East African War System, and it was eventually applied to an area bounded by 19°N, 15°E, 12°S and the Indian Ocean. The *Directorate of Colonial Surveys* was born on 1 March 1946, with Brigadier Martine Hotine as its first (and only) Director. An allowance of £2 million was approved for this centralized organization of geodetic and topographic surveys. For the first time in the eastern African colonies, two problems, which had plagued the survey effort from the earliest days, were addressed: lack of funds and the lack of a centralized organizing body. In 1947, fieldwork for basic topographic mapping was commenced. The first 1:50,000 scale sheets of Series Y731 were produced for the *Kenya–Ethiopia Boundary Commission*. At least 470 sheets were produced, virtually all of which were contoured, and 64 sheets along the Ethiopian border."

The 1:100,000 scale sheets of Series Y633 were produced between 1958 and 1968, mainly by the *Survey of Kenya* and the *Directorate of Military Survey*. A general map series at 1:250,000 (Y503) has been derived from the 1:50,000 and 1:100,000 scale sheets. The Survey of Kenya produced 42 of the 50 sheets needed to cover the country. Kenya is covered by seven sheets of the 1:1,000.000 International Map of the World. The most commonly used geodetic parameters for maps produced by the Kenyan authorities are Arc Datum 1960 referenced to the Clarke 1880 (modified) ellipsoid, Transverse Mercator projection with coordinates on the UTM grid.

In the 1970s, first-order EDM traverses were run between stations adjusted on the Arc 1960 Datum and Clifford triangulation. In 1972–1973 the *Survey of Kenya*, in conjunction with the U.S. Defense Mapping Agency and the Directorate of Military Survey of the U.K. made the first experimental Doppler satellite survey in Kenya. Recently the *Kenya Institute of Surveying and Mapping (KISM)* took GPS observations on existing control points. A. S. Lwangasi of the University of Nairobi reported the "results of a datum transformation carried out on 25 control points from Arc 1960 Datum to WGS 84 Datum: $\Delta X = -179.1$ m ± 0.7 m, $\Delta Y = -44.7$ m ± 0.7 m, $\Delta Z = -302.6$ m ± 2.2 m."

In a letter dated July 20, 1989, Mr. J. R. R. Aganyo wrote for the Director of Surveys of the Survey of Kenya that the old Cassini-Soldner used in Kenya has the following parameters: grid name: Cassini-Soldner; years used: since introduction of cadastral surveys; central meridians: 33°, 35°, 37°, 39° East; unit: English Foot where 1 foot = 0.30480 International meters, exactly; ellipsoid: Clarke 1858 where $a = 20,926,348$ English feet and $1/f = 294.26$.

In January 2000, Mr. Russell Fox of the U.K. Ordnance Survey sent a memo to me that was written by the famous H. F. Rainsford on 28 September 1961 with the (then) *Directorate of Overseas Surveys*: "Since there appears to be some confusion of thought about the 'origin' of the trigonometric data lists produced by this Directorate, the purpose of this paper is to clarify the position so far as possible. Up to the present date, all trig Lists have included in the preamble the words – 'New 1950 Arc Datum.' This denotes that the results in the list are based on the Arc of the 30th meridian, which was computed by the D.O.S., from South Rhodesia to Uganda, in the 1950 (*circa*). The values of the stations accepted as a starting point in South Rhodesia had been computed continuously from South Africa. These Arc results have been held fixed since 1950, and it is hoped that they will remain so far as long as possible in the future, since they are used not only by the D.O.S., but also by the Congo and Portuguese Africa, and they provide a uniform system from the Cape to the Equator.

The South African datum is an arbitrary one, as at no station were the Astronomic and Geodetic latitude, longitude and Azimuth made coincident. On the Arc itself the (A–G) values vary (sometimes quite abruptly) between:

latitude +20″ and −30″
longitude +12″ and −10″
azimuth +15″ and −08″

The only astronomic elements that have been held fixed on the Arc adjustment are – in South Africa one latitude, longitude, and azimuth (but each at a different station) and an astronomic azimuth at Kicharere in Uganda, just south of the Equator.

The Year 1950 was used in the title as a convenient epoch mainly to distinguish from previous systems such as the '1935 Arc Datum.' (Original also emphasized) Tanganyika was the first East African territory in which geodetic trig. control was computed based on the Arc and used for control of topographic surveys. It was known that some of this trig. was not up to primary standards, but it was the only work available, and it was hoped that re-computation based on the Arc would produce results of sufficient accuracy for the purpose required.

Since Laplace Azimuths had not been available for the Arc computation nor in *Tanganyika*, the Tanganyika trig. was computed without holding fixed any azimuths, which were, in any case, of doubtful value. When the trig. computation reached *Malindi* in Kenya from the Arc it was found that the (A–G) azimuth was approximately 20″.

It was then decided that a new approach was necessary. Put in new primary circuits based on the Arc, and observe frequent astronomic stations and tellurometer lengths, much closer together than the old, measured bases. The trig. circuits *(were)* to be adjusted to the fixed (or nearly fixed) scale and azimuth checks. This policy has been carried out and results have already been circulated for:

The Lake Circuit
Uganda Primary
Kenya Primary

All these results have been headed, as before, 'New 1950 Arc Datum', because the fundamental datum, which is the Arc, has not been changed. Whenever the new coordinates differ from the previous; this is due to a re-computation (including new observations) of part of the trig. system.

To avoid any further misunderstanding in the future it is proposed to change the heading of trig. lists now to 'New *1960* Arc Datum'. Most of the Tanganyika main trig. has still to be recomputed and a letter will be sent to each territory indicating the particular trig. chains which have already been recomputed and circulated under the 1950 heading.

The Figure of the Earth used is the Modified Clarke 1880, for which $a = 6378249.145$ and $1/f = 293.465$ in International Metres. The geodetic tables used are *Latitude Functions Clarke 1880 Spheroid*, Army Map Service, but most D.O.S. computations are now done on the Electronic Computer, which computes its own geodetic factors *'ab initio'*. Co-ordinates are also produced on the U.T.M. projection."

Thanks go to Mr. Washington Abuto wherein his letter of November 24, 1997, for the Director of Surveys of the *Survey of Kenya* enclosed a paper detailing much of Kenya's history of Grids and Datums. That paper, authored by Mr. Mahinda, served as the basis of much of the specific geodetic history quoted in Mr. Davis' graduate-level term paper of 1999.

AFREF Newsletter No. 5 (2008) reports a Molodensky-Badekas seven-parameter transformation solution for Kenya based on 20 common points between WGS84 and Arc 1960 Datum. Although the parameters are listed, no coordinates are listed for the Arc60 Datum origin point necessary for a Molodensky-Badekas model, so the model is likely Bursa-Wolfe, instead. Furthermore, no guidance is provided regarding which direction the parameters are intended to be used, nor are any test points provided. https://www.rcmrd.org/newletters

AFREF Newsletter No. 11, (2010) reports on the KENREF (Kenya Reference Frame) proposal that will include CORS at 21 primary stations and eventually 71 stations at secondary locations (https://www.rcmrd.org/newletters).

Renewing Kenya Vertical Datum by Davin N. Siriba (2020), https://issuu.com/iskkenya/docs/isk_journal_april_2019__9_/s/10844529.

Assessment of EGM2008 using GPS/levelling and free-air gravity anomalies over Nairobi County and its environs by Patroba Achola Odera, *South African Journal of Geomatics, Vol. 5, No. 1, February 2016*, http://sajg.org.za/index.php/sajg/article/viewFile/356/218.

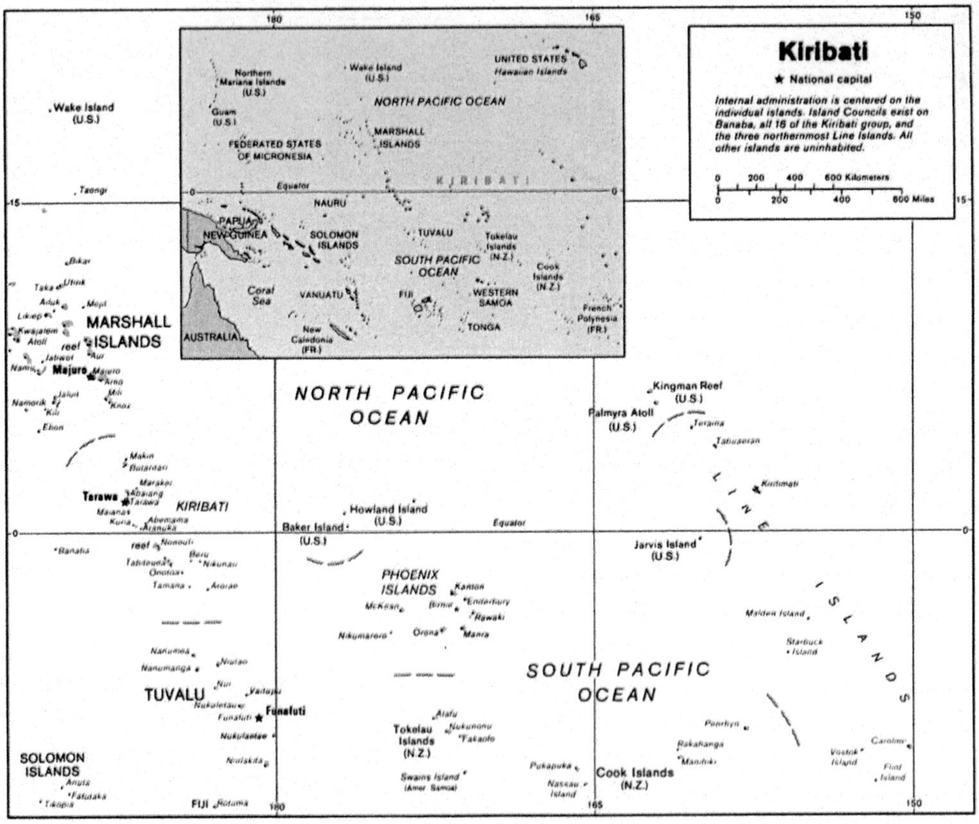
Kiribati
★ National capital
Internal administration is centered on the individual islands. Island Councils exist on Banaba, all 16 of the Kiribati group, and the three northernmost Line Islands. All other islands are uninhabited.
0 200 400 600 Kilometers
0 200 400 600 Miles
Northern Mariana Islands (U.S.)
Guam (U.S.)
Wake Island (U.S.)
UNITED STATES
Hawaiian Islands
NORTH PACIFIC OCEAN
MARSHALL ISLANDS
FEDERATED STATES OF MICRONESIA
Equator
KIRIBATI
NAURU
PAPUA NEW GUINEA
SOLOMON ISLANDS
TUVALU
Tokelau Islands (N.Z.)
SOUTH PACIFIC OCEAN
Cook Islands (N.Z.)
Coral Sea
VANUATU
FIJI
WESTERN SAMOA
TONGA
French Polynesia (FR.)
New Caledonia (FR.)
AUSTRALIA
Wake Island (U.S.)
Taongi
Bikar
Taka
Utirik
Aduk
Mejit
Likiep
Kwajalein Atoll
MARSHALL ISLANDS
reef
Jabwot
Aur
Majuro
Majuro
Arno
Jaluit
Mili
Namorik
Kili
Knox
Ebon
NORTH PACIFIC OCEAN
Kingman Reef (U.S.)
Palmyra Atoll (U.S.)
Teraina
Tabuaeran
Kiritimati
LINE ISLANDS
Makin
Butaritari
Marakei
Tarawa
Abaiang
Tarawa
KIRIBATI
Maiana
Kuria
Abemama
Aranuka
Howland Island (U.S.)
Baker Island (U.S.)
Equator
Jarvis Island (U.S.)
Banaba
reef
Nonouti
Beru
Tabiteuea
Nikunau
Onotoa
Tamana
Arorae
PHOENIX ISLANDS
Kanton
McKean
Birnie
Enderbury
Rawaki
Nikumaroro
Orona
Manra
Malden Island
Starbuck Island
Nanumea
Nanumanga
Niutao
Nui
TUVALU
Vaitupu
Nukufetau
Funafuti
Funafuti
Nukulaelae
Niulakita
SOUTH PACIFIC OCEAN
Atafu
Tokelau Islands (N.Z.)
Nukunonu
Fakaofo
Swains Island (Amer. Samoa)
Penrhyn
Rakahanga
Manihiki
Pukapuka
Nassau Island
Cook Islands (N.Z.)
Caroline
Vostok Island
Flint Island
SOLOMON ISLANDS
Anuta
Fatutaka
Tikopia
FIJI
Rotuma
180
165
150
15
0

Kiribati

Kiribati consists of a three-island group: the Gilbert Islands, the Line Islands, and the Phoenix Islands. These groups of islands straddle the equator and the International Date Line, about halfway between Hawaii and Australia. In 1995, The Republic of Kiribati proclaimed that all of its territories lie in the same time zone as the Gilbert Islands group (GMT+12). The total land area equals 717 km^2, about four times the size of Washington, DC. The country is composed of mostly low-lying coral atolls surrounded by extensive reefs. The lowest point is the Pacific Ocean, the highest point is on *Banaba Island* (81 m). Twenty of the 33 islands are inhabited, and the capital is Tarawa.

The islands were originally settled by Austronesians thousands of years ago. Around the 14th century *A.D.*, the islands were invaded by Fijians and Tongans. The first recorded European encounter with Kiribati was by the Spanish explorer Quiros in 1606. By 1820, all of the islands had been charted. At that time, the Russian hydrographer A. I. Krusenstern gave the group the name Gilbert Islands. Until 1870, many British and American whaling vessels sought sperm whales in Gilbertese waters. Starting in 1850, trading vessels passed through seeking at first coconut oil and later copra. In the 1860s, slave ships known as "*blackbirders*" carried off islanders to work in plantations in Peru, Fiji, Tahiti, Hawaii, and Australia. The male population was decimated, and European disease (measles) took a further large toll on the "I'Kiribati" people. The Ellice group (now Tuvalu) and the Gilbert Islands became a British Protectorate in 1892. In 1975, the Ellice Islands seceded from the colony and became the independent nation of Tuvalu. On July 12, 1979, Kiribati obtained its own independence from the United Kingdom and became a republic within the Commonwealth.

According to Russell Fox of the Ordnance Survey, "We revised our Tarawa 1:50,000 map in 1997 and produced new 1:25,000 photomapping of the Line and Phoenix Islands in 1995/96 using Australian Army aerial photography, but we did not do any geodetic survey work in that decade. I believe that a GPS survey, aerial photography, and new mapping (on WGS84) project was planned for Tarawa for 1998/99 under Australian aid. The following notes summarize what we know about Kiribati. Background: The Republic of Kiribati (pronounced 'kiribass') comprises the Gilbert, Phoenix and Line Island groups in the central Pacific Ocean. Total land area 717 sq km, total sea area 5.2 million sq km, greatest extent 4000 km W-E by 2000 km N-S. The modern republic has its genesis in a British protectorate proclaimed in 1892, which became the Gilbert and Ellice Islands Colony in 1916. Independence was granted to the Gilbert, Phoenix and Line Islands in 1979 as the Republic of Kiribati. The Ellice Islands had previously seceded, in 1975, as the Republic of Tuvalu. The International Date Line passes through Kiribati, but the Government of Kiribati has (legitimately) legislated that the Line and Phoenix Islands will observe the same date as the Gilbert Islands. Survey history: The Royal Navy carried out hydrographic surveys, based on astro fixes, from the early nineteenth century onward. Little survey and mapping work was done by the British colonial authorities before WW2. Japanese military occupation of the Gilbert Islands in 1941 lead to extensive aerial photography and mapping by the US armed forces, who drove out the enemy in fierce fighting. Post-WW2: Aerial photography: Gilbert Islands RNZAF 1962/63; USN1964; Fiji Lands, Mines and Survey Department 1968/69. The 1968/69 cover was used subsequently by the Directorate of Overseas Surveys (DOS) for mapping. Line Islands RAF 1950s. Phoenix Islands RNZAF 1962/63. All Kiribati, 1984/85, by the Royal Australian Survey Corps, as part of Operation Anon, one of the Australian Army's Pacific Doppler campaigns. Survey: DOS control surveys 1967–73, by Tellurometer traversing, covered all the islands in the Gilbert group, with each atoll on its own astro datum and local Transverse Mercator grid. International Spheroid (DOS's default spheroid for the Pacific) was used. Christmas Island was surveyed by the RNZN, British Military Survey, and contract staff between 1941 and the 1960s, and by the Kiribati Survey Department in 1979/80. DOS then brought those surveys together, computing on Christmas Island 1967 Astro Datum, International Spheroid, Kiritimati Local [TM] Grid. The first modern survey work in the Phoenix and Line Islands was the precise Doppler campaign of 1984/85. Mapping: DOS produced 1:25,000 (a few sheets were done at 1:12,500 & 1:10,000), photomap series of all the islands between 1972 and 1996. The larger atolls, Tarawa and Kiritimati (Christmas Island) were also mapped at 1:50 000 on single sheets. Generally, two editions were produced, one showing the local TM grid, and the other (for military use) UTM grid. Early editions of the Christmas Island sheet showed UTM Grid only. Southern Tarawa was mapped at 1:2500 and 1:1250."

All of these classical Datums are referenced to the international 1924 ellipsoid where $a = 6,378,388$ m, and $1/f = 297$. Starting with specifics for the Gilbert Islands Group, for the *Abaiang* Datum of 1962, the Datum origin is at the first-order station Flagstaff of Government Station, also called HMS Cook Astro "H" where $\Phi_o = 01°\ 49'\ 25.029''$ N and $\Lambda_o = 173°\ 01'\ 25.830''$ East of Greenwich. According to a British Admiralty Report of Survey file, "The Government flagstaff on a coral rock plinth in the centre of the Government Station – *Taburao*." The local grid is based on the Transverse Mercator projection with the central meridian, $\lambda_o = 172°\ 55'$ E, the False Easting = 20 km, and the False Northing = zero. The Scale Factor at Origin is unity ($m_o = 1.0$). From *Abaiang* Datum of 1962 to WGS 84 Datum, the converted Doppler solution is $\Delta X = +254.8$ m, $\Delta Y = -322.4$ m, and $\Delta Z = -270.0$ m. The Operation ANON 1984–85 solution is $\Delta X = +254.3$ m, $\Delta Y = -323.4$ m, and $\Delta Z = -275.6$ m.

For the *Abemama* Datum of 1944, the Datum origin is at Signal Station, southwestern tip of Steve Island (Station "Flag"), where $\Phi_o = 00°\ 27'\ 36''$ N and $\Lambda_o = 173°\ 49'\ 11''$ East of Greenwich. For the *Abemama* Datum of 1959, the Datum origin is at Cook Astro Point where $\Phi_o = 00°\ 24'\ 19.02''$ N, $\Lambda_o = 173°\ 55'\ 36.57''$ East of Greenwich, and $H_o = 2.14$ m. The local grid is based on the Transverse Mercator projection with the central meridian, $\lambda_o = 173°\ 51'$ E, the False Easting = 20 km, and the False Northing = 100 km. The Scale Factor at Origin is unity ($m_o = 1.0$). From Abemama Datum of 1959 to WGS 84 Datum, the converted Doppler solution is $\Delta X = +289.4$ m, $\Delta Y = +656.2$ m, and $\Delta Z = +303.4$ m. Note that the UTM coordinates of local traverses depend on the position of Observation Spot that was connected by traverse to C where $\varphi_o = 00°\ 24'\ 29.02''$ N and $\lambda_o = 173°\ 55'\ 36.57''$ E.

For the *Arorae* Datum of 1965, the Datum origin is at *Arorae Astro Observation Spot* where $\Phi_o = 02°\ 38'\ 36.7''$ S and $\Lambda_o = 176°\ 49'\ 33.3''$ East of Greenwich. The local grid is based on the Transverse Mercator projection with the central meridian, $\lambda_o = 176°\ 49'$ E, the False Easting = 10 km, and the False Northing = 500 km. The Scale Factor at Origin is unity ($m_o = 1.0$). From *Arorae* Datum of 1965 to WGS 84 Datum, the converted Doppler solution is $\Delta X = +221.4$ m, $\Delta Y = -34.4$ m, and $\Delta Z = -21.6$ m.

For the *Beru* Datum of 1970, the Datum origin is at third-order station BRZ 10 where $\Phi_o = 01°\ 19'\ 29.9632''$ S, $\Lambda_o = 175°\ 59'\ 16.9134''$ East of Greenwich, and $H_o = 1.73$ m. The local grid is based on the Transverse Mercator projection with the central meridian, $\lambda_o = 175°\ 59'$ E, the False Easting = 10 km, and the False Northing = 300 km. The Scale Factor at Origin is unity ($m_o = 1.0$). From Beru Datum of 1970 to WGS 84 Datum, the converted Doppler solution is $\Delta X = +179.9$ m, $\Delta Y = -595.3$ m, and $\Delta Z = +6.96$ m. The Operation ANON 1984–1985 solution is $\Delta X = +181.3$ m, $\Delta Y = -585.6$ m, and $\Delta Z = -7.2$ m.

For the *Butaritari* Datum of 1965, the Datum origin is at third-order station BTZ 26 where $\Phi_o = 03°\ 15'\ 40.629''$ N, $\Lambda_o = 172°\ 41'\ 45.8381''$ East of Greenwich, and $H_o = 1.87$ m. The local grid is based on the Transverse Mercator projection with the central meridian, $\lambda_o = 172°\ 50'$ E, the False Easting = 20 km, and the False Northing = zero. The Scale Factor at Origin is unity ($m_o = 1.0$). From Butaritari Datum of 1965 to WGS 84 Datum, the converted Doppler solution is $\Delta X = +253.8$ m, $\Delta Y = +6.1$ m, and $\Delta Z = +528.2$ m. The Operation ANON 1984–1985 solution is: $\Delta X = +254.2$ m, $\Delta Y = +3.2$ m, $\Delta Z = +544.2$ m.

For the *Kuria* Datum of 1962, the Datum origin is at HMS Cook Astro, Kuria 1962 where $\Phi_o = 00°\ 13'\ 00.4''$ N and $\Lambda_o = 173°\ 23'\ 06.8''$ East of Greenwich. The local Kuria and *Aranaka* Grid is based on the Transverse Mercator projection with the central meridian, $\lambda_o = 173°\ 30'$ E, the False Easting = 30 km, and the False Northing = 100 km. The Scale Factor at Origin is unity ($m_o = 1.0$). From Kuria Datum of 1962 to WGS 84 Datum, the converted Doppler solution is based on a fourth parameter where a Z-axis rotation is performed first (algebraically added to the longitude), where $R_z = 102.84765''$ and then the standard three-parameter shift is performed where $\Delta X = +219.1$ m, $\Delta Y = -24.9$ m, and $\Delta Z = +137.0$ m. The Operation ANON 1984–1985 solution is $R_z = 102.804''$, $\Delta X = +218.6$ m, $\Delta Y = -24.8$ m, and $\Delta Z = +140.1$ m.

For the *Little Makin* Datum of 1972, the Datum origin is at station *Bikati Astro* where $\Phi_o = 03°\ 16'\ 19.90''$ N and $\Lambda_o = 172°\ 40'\ 36.21''$ East of Greenwich. The local grid is based on the Transverse Mercator projection with the central meridian, $\lambda_o = 172°\ 50'$ E, the False Easting = 20 km, and the False Northing = zero. The Scale Factor at Origin is unity ($m_o = 1.0$). From Little Makin Datum of 1972 to WGS 84 Datum, the converted Doppler solution is $\Delta X = +243.4$ m, $\Delta Y = +221.1$ m, and $\Delta Z = -104.1$ m. The Operation ANON 1984–85 solution is $\Delta X = +239.5$ m, $\Delta Y = +189.9$ m, and $\Delta Z = -121.6$ m.

For the *Maiana* Datum of 1965, the Datum origin is at *Maiana Astro 1965*. The local grid is based on the Transverse Mercator projection with the central meridian, $\lambda_o = 173°\ 02'$ E, the False Easting = 20 km, and the False Northing = zero. The Scale Factor at Origin is unity ($m_o = 1.0$). From Maiana Datum of 1965 to WGS 84 Datum, the converted Doppler solution is $\Delta X = +215.5$ m, $\Delta Y = -27.9$ m, and $\Delta Z = -159.7$ m.

For the *Marakei* Datum of 1969, the Datum origin is at SW Point A where $\Phi_o = 01°\ 58'\ 58''$ S and $\Lambda_o = 173°\ 15'\ 22''$ East of Greenwich. The local grid is based on the Transverse Mercator projection with the central meridian, $\lambda_o = 173°\ 16'$ E, the False Easting = 10 km, and the False Northing = zero. The Scale Factor at Origin is unity ($m_o = 1.0$). From Marakei Datum of 1969 to WGS 84 Datum, the converted Doppler solution is $\Delta X = +188.1$ m, $\Delta Y = -237.6$ m, and $\Delta Z = -185.6$ m. The Operation ANON 1984–85 solution is $\Delta X = +188.7$ m, $\Delta Y = -229.9$ m, and $\Delta Z = -189.3$ m.

For the *Nikunau* Datum of 1965, the Datum origin is at third-order station NKZ 7 where $\Phi_o = 01°\ 23'\ 28.9196''$ S and $\Lambda_o = 172°\ 28'\ 46.4327''$ East of Greenwich. The local grid is based on the Transverse Mercator projection with the central meridian, $\lambda_o = 176°\ 27'$ E, the False Easting = 10 km, and the False Northing = 300 km. The Scale Factor at Origin is unity ($m_o = 1.0$). From Nikunau Datum of 1965 to WGS 84 Datum, the converted Doppler solution is $\Delta X = +229.8$ m, $\Delta Y = -38.4$ m, and $\Delta Z = -311.7$ m. The Operation ANON 1984–85 solution is $\Delta X = +230.3$ m, $\Delta Y = -17.7$ m, and $\Delta Z = -315.3$ m.

For the *Nikunau* Datum of 1959, the Datum origin is at Government Flagstaff where $\Phi_o = 01°\ 20'\ 44.37''$ S and $\Lambda_o = 176°\ 26'\ 31.36''$ East of Greenwich. The local grid is based on the Transverse Mercator projection with the central meridian, $\lambda_o = 176°\ 27'$ E, the False Easting = 10 km, and the False Northing = 300 km. The Scale Factor at Origin is unity ($m_o = 1.0$). From Nikunau Datum of 1965 to WGS 84 Datum, the converted Doppler solution is $\Delta X = +229.8$ m, $\Delta Y = -38.4$ m, and $\Delta Z = -311.7$ m. The Operation ANON 1984–85 solution is $\Delta X = +230.3$ m, $\Delta Y = -17.7$ m, and $\Delta Z = -315.3$ m.

For the *Nonouti* Datum of 1965, the Datum origin is at third-order station NNZ 14 where $\Phi_o = 00°\ 39'\ 57.7067''$ S, $\Lambda_o = 174°\ 26'\ 52.1428''$ East of Greenwich, and $H_o = 2.62$ m. The local grid is based on the Transverse Mercator projection with the central meridian, $\lambda_o = 174°\ 20'$ E, the False Easting = 20 km, and the False Northing = 200 km. The Scale Factor at Origin is unity ($m_o = 1.0$). From Nonouti Datum of 1965 to WGS 84 Datum, the converted Doppler solution is $\Delta X = +221.9$ m, $\Delta Y = -99.5$ m, and $\Delta Z = -926.8$ m. The Operation ANON 1984–1985 solution is $\Delta X = +221.2$ m, $\Delta Y = -96.3$ m, and $\Delta Z = -935.1$ m. Note that the UTM coordinates of local traverses in 1963 of the Nonouti Survey depend on position A being $\Phi_o = 00°\ 40'\ 16.4''$ S and $\Lambda_o = 174°\ 27'\ 28''$ E.

For the *Onotoa* Datum of 1970, the Datum origin is at third-order station ONZ 7. The local grid is based on the Transverse Mercator projection with the central meridian, $\lambda_o = 175°\ 33'$ E, the False Easting = 10 km, and the False Northing = 350 km. The Scale Factor at Origin is unity ($m_o = 1.0$). From Onotoa Datum of 1970 to WGS 84 Datum, the converted Doppler solution is $\Delta X = +244.4$ m, $\Delta Y = +197.9$ m, and $\Delta Z = -243.1$ m. The Operation ANON 1984–1985 solution is $\Delta X = +245.2$ m, $\Delta Y = +203.7$ m, and $\Delta Z = -248.4$ m.

For the *Tabiteuea* Datum of 1959, the Datum origin is at third-order station TBZ 1 Astro where $\Phi_o = 01° 28' 05.6''$ S and $\Lambda_o = 175° 03' 15.0''$ East of Greenwich. The local grid is based on the Transverse Mercator projection with the central meridian, $\lambda_o = 174° 53'$ E, the False Easting = 30 km, and the False Northing = 300 km. The Scale Factor at Origin is unity ($m_o = 1.0$). From Tabiteuea Datum of 1959 to WGS 84 Datum, the converted Doppler solution is $\Delta X = +250.8$ m, $\Delta Y = +235.6$ m, and $\Delta Z = -377.2$ m.

For the *Tamana* Datum of 1962, the Datum origin is at Astro Observation Spot where $\Phi_o = 02° 30' 09.0''$ S and $\Lambda_o = 175° 58' 45.8''$ East of Greenwich. The local grid is based on the Transverse Mercator projection with the central meridian, $\lambda_o = 175° 59'$ E, the False Easting = 10 km, and the False Northing = 400 km. The Scale Factor at Origin is unity ($m_o = 1.0$). From Tamana Datum of 1962 to WGS 84 Datum, the converted Doppler solution is $\Delta X = +202.3$ m, $\Delta Y = -181.6$ m, and $\Delta Z = +128.2$ m. The Operation ANON 1984–1985 solution is $\Delta X = +206.8$ m, $\Delta Y = -159.8$ m, and $\Delta Z = +95.7$ m.

For the *Tarawa* Datum of 1966, the Datum origin is at first-order station Tarawa SECOR AMS 1966 where $\Phi_o = 01° 21' 42.13''$ N and $\Lambda_o = 172° 55' 47.27''$ East of Greenwich. The local grid is based on the Transverse Mercator projection with the central meridian, $\lambda_o = 173° 02'$ E, the False Easting = 20 km, and the False Northing = zero. The Scale Factor at Origin is unity ($m_o = 1.0$). From Tarawa Datum of 1966 to WGS 84 Datum, the converted Doppler solution is $\Delta X = +176.0$ m, $\Delta Y = -421.8$ m, and $\Delta Z = +282.3$ m. The Operation ANON 1984–1985 solution is $\Delta X = +184.0$ m, $\Delta Y = -356.3$ m, and $\Delta Z = +287.5$ m. For the Betio Anchorage Survey of 1959, note that the UTM coordinates of local traverses depend on the position C being $\Phi_o = 01° 19' 42.98''$ N and $\Lambda_o = 172° 58' 31.747''$ E.

For the Line Island group, according to John W. Hager, the origin is at Station Beacon on *Kiritimati* (Christmas Island) where $\Phi_o = 01° 59' 08''$ S and $\Lambda_o = 157° 29' 00''$ West of Greenwich. Established by N. J. Till, hydrographic surveyor, M.N.Z.I.S. in April–September 1941. English Harbor Observation Spot on *Tabuaerau* (Fanning Island) is the Datum origin where $\Phi_o = 03° 51' 23''$ N and $\Lambda_o = 159° 21' 50''$ West of Greenwich.

For the Phoenix Island Group, Birnie Island Astro is where $\Phi_o = 03° 35' 07.875''$ S and $\Lambda_o = 171° 31' 03.194''$ West of Greenwich. Kanton 1939 Datum at the American Eclipse Expedition Pier, USS Bushnel where $\Phi_o = 02° 49' 07.2''$ S and $\Lambda_o = 171° 42' 53.5''$ West of Greenwich. The position is described as a monolith on the western side of the island. Kanton 1963 is where $\Phi_o = 02° 47' 20.0''$ S±0.3″, $\Lambda_o = 171° 39' 49.0''$±0.3″ West of Greenwich, and the reference azimuth is $\alpha_o = 75° 15' 19.15''$±0.15″ to CAN AZ1 from south (USC&GS second order). Canton Astro 1966 at Canton SECOR Astro is where $\Phi_o = 02° 46' 28.99''$ S±0.04″, $\Lambda_o = 171° 42' 53.5''$±0.05″ West, and the azimuth is $\alpha_o = 385° 51' 02.65''$±0.11″ (*sic*) from SECOR RM-1 to TBM #3 (SECOR AZ MK) from South, referenced to the International 1909 ellipsoid. *Enderbury Island Astro* is where $\Phi_o = 03° 08' 30.140''$ S, $\Lambda_o = 171° 05' 34.95''$ West of Greenwich, and the reference azimuth is $\alpha_o = 179° 04' 00.69''$ to Bas. The Gardner Island Astro Datum is where $\Phi_o = 04° 40' 18.85''$ S, $\Lambda_o = 174° 32' 27.71''$ West of Greenwich, and the reference azimuth is $\alpha_o = 252° 25' 24.41''$ to station Line. The *Hull Island Astro* Datum is where $\Phi_o = 04° 29' 15.263''$ S, $\Lambda_o = 172° 10' 15.188''$ West of Greenwich, and the reference azimuth is $\alpha_o = 001° 50' 22.20''$ to station Base. The *McKean Island Astro* is where $\Phi_o = 03° 35' 51.375''$ S, $\Lambda_o = 174° 07' 37.522''$ West of Greenwich, and the reference azimuth is $\alpha_o = 005° 04' 59.26''$ to station North. The *Phoenix Island Astro* is where $\Phi_o = 03° 43' 13.375''$ S, $\Lambda_o = 170° 42' 56.004''$ West of Greenwich, and the reference azimuth is $\alpha_o = 309° 23' 37.76''$ to station South. *Sydney Island Astro* is where $\Phi_o = 04° 26' 57.975''$ S, $\Lambda_o = 171° 15' 43.885''$ West of Greenwich, and the reference azimuth is $\alpha_o = 009° 45' 57.97''$ to station Nee. For both the Line Islands Group and the Phoenix Islands Group, the ellipsoid of reference is the Clarke 1866 unless otherwise noted.

Thanks again go to Russell Fox of the UK Ordnance Survey; John W. Hager, retired from AMS/DMA/NIMA; and Richard W. Stevenson, head of the Reference and Bibliography Section, and Gary Fitzpatrick, senior reference librarian, both of the Library of Congress; and David Llewellyn, senior draftsman, Lands and Survey Division, Bairiki, Tarawa, Republic of Kiribati.

A number of geodetic surveys have been performed on various islands by *Geoscience Australia*:

- ftp://ftp.ga.gov.au/geodesy-outgoing/gnss/pub/SPSLCMP/PreviousLevellingSurvey Reports/KIRIBATI%20LEVEL%20SVY%202007.pdf
- ftp://ftp.ga.gov.au/geodesy-outgoing/gnss/pub/SPSLCMP/14-7573-4%20Record%20 2014-18%20Kiribati%202013%20WCAG.pdf
- https://researchdata.edu.au/kiribati-levelling-survey-report-2009/1230349
- ftp://ftp.ga.gov.au/geodesy-outgoing/gnss/pub/SPSLCMP/14-7573-3%20Record%20 2014-17%20Kiribati%202012%20WCAG.pdf

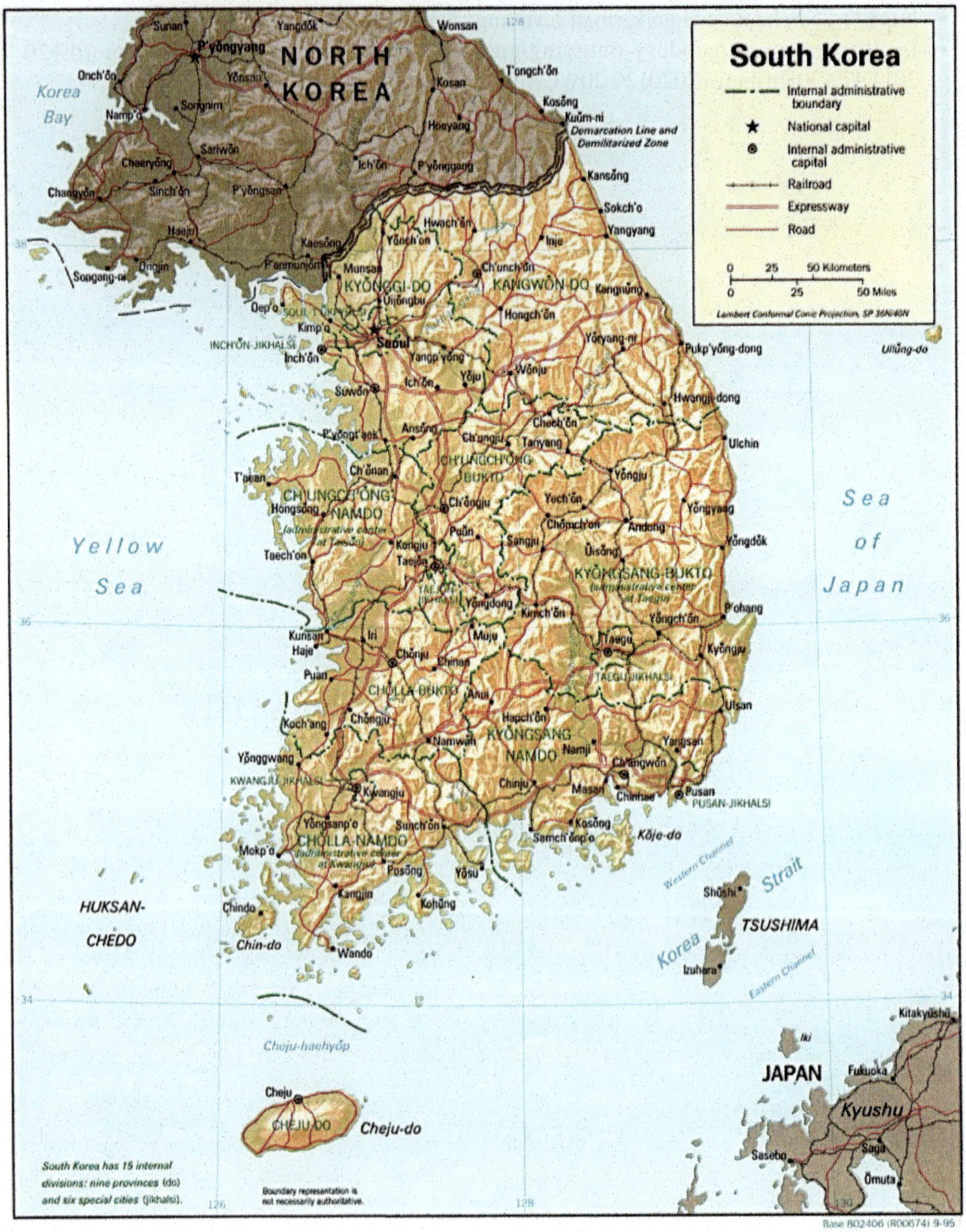

Base 802406 (R00674) 9-95

The Republic of Korea

Located in eastern Asia, southern half of the Korean Peninsula bordering the Sea of Japan and the Yellow Sea. It is slightly smaller than Pennsylvania, slightly larger than Indiana. The terrain is mostly hills and mountains with wide coastal plains in west and south. The highest point is *Hallasan* (1,950 m), the lowest point is the Sea of Japan (0 m).

The ancient kingdom of *Chosǒn* was established in the northern part of the peninsula during the third millennium B.C. Buddhism was introduced in the 4th century A.D., which the Koreans then carried to Japan. The Yi dynasty ruled the kingdom from the capital Seoul from 1392 to 1910. Korea was forced to grant a treaty opening its ports to Japan in 1876. In resisting Chinese control, Korea was the focus of the Chinese–Japanese War 1894–1895, and further rivalry over Korea resulted in the Russo–Japanese War 1904–1905. Korea became a Japanese protectorate in 1905 and was annexed as a province in 1910. In 1945 after World War II, Korea was brought under the temporary protection of the USSR and the United States. The two zones of occupation were divided at the 38th parallel, and that line continues to divide the peninsula's politics.

The first geodetic datum used in Korea was the Tokyo Datum of 1892 where $\Phi_o = 35° 39' 17.515''$ North, $\Lambda_o = 139° 44' 30.097''$ East of Greenwich. The defining azimuth was determined from the old Tokyo Observatory at *Azabu* to station *Kanoyama* (Kano Mountain) as $\alpha_o = 156° 25' 28.44''$. The Tokyo Datum of 1892 is referenced to the Bessel 1841 ellipsoid where the semi-major axis $(a) = 6,377,397.155$ m and the reciprocal of flattening $(1/f) = 299.1528128$. Geodetic surveys started in Korea in 1910, and 13 baseline measurements were completed by 1918. 400 First Order, 2,401 Second Order, and 31,646 Third Order triangulation stations were established in all of the Korean Peninsula during this time. The number of stations in South Korea totaled 16,089. There were five tide stations used for the establishment of the vertical datum to local mean sea level, and 6,629 km of leveling were run. The original triangulation was the Datum for the Old Cadastral Grids based on the Gauss – Schreiber Transverse Mercator formulae. The three belts were East, Central, and West with Central Meridians $(\lambda_o) = 129°, 127°$, and $125°$ East of Greenwich, respectively. The False Northing Latitude of Origin was 38° North, the False Northing = 500 km, the False Easting = 200 km. For all three belts the Scale Factor at Origin $(m_o) = 1.0$. Note that these narrow belts used the simpler Gauss-Schreiber formulae for the Transverse Mercator the same as the U.S. Coast and Geodetic Survey later did in the 1930s for the NAD 1927 State Plane Coordinate Systems. The Tokyo Datum of 1918 was later adopted for use in Korea, but the Cadastral Grids were *not* changed to accommodate that new system. The only difference between the two datums is that the 1918 re-determination of Longitude is: $\Lambda_o = 139° 44' 40.502''$ East of Greenwich, which is an increase of 10.405''. The latitude was the same for both years. In Korea, the geographic coordinates of the triangulation stations are on the Tokyo Datum of 1918, but the Grid coordinates are on the Tokyo Datum of 1898. This is a little-known fact of Korean mapping that has led some cartographers to question their own sanity!

The number of triangulation stations lost or destroyed during World War II and the Korean War was about 12,000 or approximately 80% of the marks in South Korea. The rearrangement and reconstruction of the triangulation stations was begun in 1957. Rearrangement in this context means that work is done to build the stone marker to coincide with the recovered footing of the original triangulation station. Reconstruction work refers to rebuilding the footing and marker at the original site or nearby and to perform the triangulation observations anew. All of this work was completed by 1986. In 1975, electro-optical distance meters (EDM) were introduced to the observational techniques. The re-survey has been of 1,291 existing First and Second Order stations as well as 14,798 Third- and Fourth-Order stations. This has resulted in a new geodetic network being established where the accuracy specification is ±3 cm horizontal, ±5 cm vertical.

The Korean Datum of 1985 origin is at station Suwon on the grounds of the National Geographic Institute (NGI) where $\Phi_o = 37°\ 16'\ 31.9034''$ North, $\Lambda_o = 127°\ 03'\ 05.1451''$ East of Greenwich. The defining azimuth to station *Donghak-san* is: $\alpha_o = 170°\ 58'\ 18.190''$. The ellipsoid of reference is the WGS 84 where $a = 6,378,137$ m, $1/f = 298.257223563$. The 40 Laplace stations established for this new datum were planned for a density of one station per 5,000 km². By 1996, 37 Laplace station observations were completed. The civilian Cadastral Grid system will continue on the new datum with no change in parameters other than the ellipsoid.

The leveling net in Korea was initially established between 1910 and 1915, but the Korean War virtually destroyed all existing marks in the south. Since 1960, NGI started the re-survey of its network and by 1986 completed the First-Order net. This is composed of 16 loops and 38 routes with a total length of 3,400 km and 2,030 benchmarks spaced at 2–4 km intervals. The primary benchmark of the Republic of Korea is at Inha University in the city of Inchon. Second-Order benchmarks total 4,035 points along 7,600 km of leveling routes.

NGI has utilized satellite surveying techniques since 1979. Two Magnavox 1502 receivers were employed at Pusan, Kyeonju, and Cheju-do through cooperation with Japan until 1982. Twenty islands have been occupied for Transit satellite observations as of 1991. Since 1991, NGI has been using GPS receivers to strengthen the classical network. Plans to establish 20 permanent GPS observation stations in a Continuously Operating Reference Station (CORS) network started with the first station called SUWN using a TurboRogue™ SNR-8000 receiver on March 15, 1995.

Gravity surveys have been conducted by NGI since 1975. The original design plan called for 25 First-Order stations and 2,000 Second-Order stations. By 1996, 1,709 Second-Order relative gravity stations had been observed at existing benchmarks. The initial point for the Korean gravimetric datum is also at station Suwon in connection with the international gravity datum at the Geographical Survey Institute of Japan. Auxiliary First-Order stations are located at Seoul National University ($\varphi = 37°\ 27.1'$ N, $\lambda = 126°\ 57.0'$ E), Kyeongbuk National University ($\varphi = 35°\ 53.2'$ N, $\lambda = 128°\ 36.9'$ E), Pusan National University ($\varphi = 35°\ 13.0'$ N, $\lambda = 129°\ 05.0'$ E), and the Korean Standards Institute ($\varphi = 36°\ 23.1'$ N, $\lambda = 127°\ 22.4'$ E).

On August 20, 1888, Russia and Korea signed an agreement providing for freedom of navigation on the Tumen River for coasting vessels of both nationalities. The treaty also spoke of the river as "their common frontier." However, since Japan annexed Korea in 1910 the status of that common boundary is not clear. By 1914, Japan submitted a plan for the delimitation of the boundary by utilizing the Rule of the Thalweg, but World War I and the Russian Revolution prevented any action.

The Korea "Demarcation Line" mentioned in the first paragraph represents the partitioning of Korea affected by the July 27, 1953 Panmunjom Agreement ending the Korean hostilities. Approximately 238 km long (148.5 miles), the line follows a sinuous path over rugged terrain.

The China–Korea boundary received attention in November of 1961 when magazines published in both countries carried features on the *Ch'ang-pai* mountain range and specifically "*The Pond of Heaven*" which is a volcanic lake. Both countries claimed the same lake. The first attempt by both countries to define their common border in this region dates back to 1713, but disputes and confusion obviously continue.

On January 30, 1974, the governments of Japan and Korea signed two maritime agreements that established a continental shelf boundary in the northern part of the maritime region adjacent to the countries. The boundary is defined by a series of ellipsoidal loxodromes (rhumb lines) between points referenced to the Tokyo Datum of 1918. On September 20, 1978, the Republic of Korea promulgated a system of "straight baselines" by Presidential Decree No. 9162. These straight lines are also defined as loxodromes on the Tokyo Datum of 1918, and Korea defers its boundary claim of territorial waters in narrow water bodies such as the Korean Strait and Cheju Hachyop.

DMA/NIMA lists the three-parameter datum shift from Tokyo Datum of 1918 to WGS 84 Datum in the Republic of Korea as $\Delta X = -147$ m ± 2 m, $\Delta Y = +506$ m ± 2 m, $\Delta Z = +687$ m ± 2 m, and is based on a 29-station solution. However, my solution for nine well-distributed First Order points (*N'pyong, Haynam, Namhai, Bangejin, Sokcho, Hansan, Sangha, Uljin,* and *Kangwha*) is:

$\Delta X = -323$ m ± 2 m, $\Delta Y = +309$ m ± 2 m, $\Delta Z = +653$ m ± 2 m. The actual rms fit of my solution to test points is Latitude = ±2.20 m, Longitude = ±1.26 m, and Height = ±7.87 m. On the other hand, using a seven-parameter Molodensky model (with the Datum origin) with these co-located positions yields: $\Delta X = -325.89$ m, $\Delta Y = +324.13$ m, $\Delta Z = +664.51$ m, scale = -7.60×10^6, $R_z = +01.75''$, $R_y = -9.20''$, $R_x = -7.39''$. The actual rms fit of my seven-parameter solution to test points is: Latitude = ±1.69 m, Longitude = ±0.84 m, and Height = ±0.65 m. For example, test point "N'pyong 21" on Tokyo 1918 Datum: X = 498,278.75, Y = 269,683.64, $\varphi = 37°\ 58'\ 54.538$ N, $\lambda = 127°\ 47'\ 35.747$ E, H = 784.65 m. On WGS 84 Datum: $\varphi = 37°\ 59'\ 04.483$ N, $\lambda = 127°\ 47'\ 38.404$ E, h = 812.007 m. Thanks for the geodetic history of Korea and data go to Mr. Heungmuk Cho of the Geodesy Division, National Geography Institute in Suwon-shi, Republic of Korea.

"Korean Geodetic Datum 2002 (KGD2002), has been adopted since 1st January 2003, replacing the Tokyo datum which has been used in the country since early 20th century. The new datum is based on the International Terrestrial Reference System (ITRS) and uses the Geodetic Reference System 1980 (GRS80) ellipsoid. The KGD2002 uses International Terrestrial Reference Frame 2000 (ITRF2000) at epoch 1st January 2002, which has a geocentric origin. The datum origin of KGD 2002 is located at NGII and the coordinates are: $\varphi_o = 37°\ 16'\ 33.3659''$ N, $\lambda_o = 127°\ 03'\ 14.8913''$ E, $h_o = 91.253$ m. Subsequently, the 1st order geodetic control stations consisting of 14 GPS CORS stations were readjusted for determining the KGD2002 coordinates. All of the works had been completed by the end of 2002.

"The *National Geographical Information Institute (NGII)* of Korea together with a number of surveying contractors has held GPS observation campaigns over the geodetic network since 1996. During these campaigns, about 11,000 points were observed until the end of 2007." (*Implementation of the New Korean Geocentric Datum and GPS CORS Management*, Young-Jin Lee, Hung-Kyu Lee, Chan-Oh Kwon and Jun-Ho Song, Integrating Generations, FIG Working Week 2008, Stockholm, Sweden, June 14–19, 2008.)

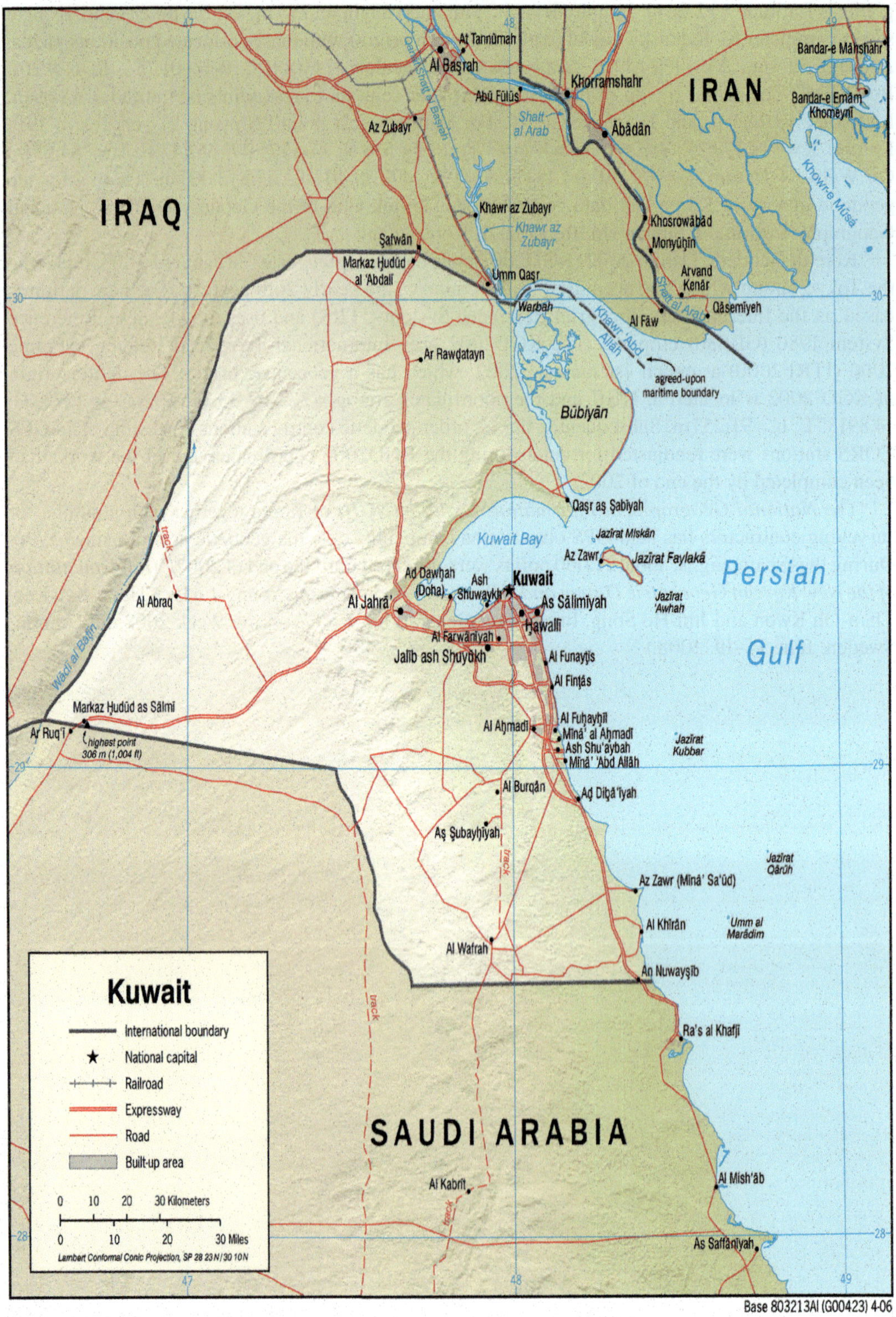

Base 803213AI (G00423) 4-06

State of Kuwait

Bordered by Iraq (240 km) and Saudi Arabia (222 km), Kuwait's coastline is 499 km long, the lowest point is the Persian Gulf (0 m), and the highest point is unnamed (306 m). Kuwait is slightly smaller than New Jersey.

"Evidence of the first proper settlement in the region dates from 4500 *B.C.*, and shards of pottery, stone walls, tools, a small, drilled pearl and remains of what is perhaps the world's earliest seafaring boat indicate links with the Ubaid people who populated ancient Mesopotamia. The people of Dilmun also saw the potential of living in the mouth of two of the world's great river systems and built a large town on Failaka Island, the remains of which form some of the best structural evidence of Bronze Age life in the world. A historian called Arrian, in the time of Alexander the Great, first put the region on the map by referring to an island discovered by one of Alexander's generals en route to India. Alexander himself is said to have called this, the modern-day island of Failaka, Ikaros, and it soon lived up to its Greek name as a Hellenistic settlement that thrived between the 3rd and 1st centuries *B.C.* With temples dedicated to Artemis and Apollo, an inscribed stele with instructions to the inhabitants of this high-flying little colonial outpost, stashes of silver Greek coins, busts, and decorative friezes, Ikaros became an important trading post on the route from Mesopotamia to India. While there is still a column or two standing proud among the weeds, and the odd returning Kuwaiti trying to resettle amid the barbed wire, there's little left to commemorate the vigorous trading in pearls and incense by the Greeks. There's even less to show for the Christian community that settled among the ruins thereafter" (*Lonely Planet*, 2010).

"Kuwait has been ruled by the Al Sabah dynasty since the 18th century. The threat of Ottoman invasion in 1899 prompted Amir Mubarak Al Sabah to seek protection from Britain, ceding foreign and defense responsibility to Britain until 1961, when the country attained its independence. Kuwait was attacked and overrun by Iraq in August 1990. Following several weeks of aerial bombardment, a US-led UN coalition began a ground assault in February 1991 that liberated Kuwait in four days. In 1992, the Amir reconstituted the parliament that he had dissolved in 1986. Amid the 2010–11 uprisings and protests across the Arab world, stateless Arabs, known as Bidoon, staged small protests in early 2011 demanding citizenship, jobs, and other benefits available to Kuwaiti nationals. Other demographic groups, notably Islamists and Kuwaitis from tribal backgrounds, soon joined the growing protest movements, which culminated in late 2011 with the resignation of the prime minister amidst allegations of corruption. Demonstrations renewed in late 2012 in response to an amiri decree amending the electoral law that lessened the voting power of the tribal blocs" (*World Factbook*, 2022).

The *Directorate of Military Survey* of the British military was responsible for all mapping of Kuwait until its independence in 1961. The 1:100,000 scale map series of Kuwait and the Neutral Zone (Kuwait–Saudi Arabia) was published from 1958 to 1961. The coordinate system used was the Iraq Zone Grid where the Lambert Conformal Conic projection used a Central Meridian (λ_o) = 45° E, Latitude of Origin (φ_o) = 32° 30′ N, Scale Factor at Origin (m_o) = 0.998786408, False Easting = 1,500 km, False Northing = 1,166,200 m. The ellipsoid of reference is the Clarke 1880 (IGN) where $a = 6,378,249.2$ m, and $1/f = 293.4663077$. I was informed by Mr. Gener Bajao that the first major geodetic Datum of the Persian Gulf area established by W.E. Browne of the Iraq Petroleum Company in 1927–1931 at the South End Base at station Nahrwan (East of Baghdad) such that: $\Phi_o = 33°$ 19′ 10.87″ North, $\Lambda_o = 44°$ 43′ 25.54″ East of Greenwich. Subsequent adjustments have resulted in the name Nahrwan Final Datum 1958.

The Kuwait Aminoil Grid of 1951 is based on the Azimuthal Equidistant Projection (Postel Projection) and the ellipsoid of reference is the Clarke 1866 where $a = 6,378,206.4$ m and $b = 6,356,583.6$ m. The Central Meridian $(\lambda_o) = 48° 20' 53.2''$ East, Latitude of Origin $(\varphi_o) = 28° 33' 48.5''$ North, False Easting = 105,600 U.S. Survey Feet, False Northing = 264,000 U.S. Survey Feet. This curious system is likely the result of a geophysical exploration of the area by an American Oil Company that did not have a command of international geodetic surveying practices – not exactly an uncommon occurrence in the 1950s. Control is likely a series of temporary radio-location towers used for offshore positional control. The Kuwait Oil Co. Datum does not seem to have a known origin; it may be an ersatz system.

Sometime after World War II, the U.S. Army Map Service and the British Directorate of Military Survey (DMS) recomputed all of the classical geodetic datums of the Middle East on the European Datum of 1950 by using a two-dimensional transformation on the complex conformal plane of the Universal Transverse Mercator (UTM) Grid. Of course, Kuwait was included in this new datum coverage and ED50 is referenced to the International 1924 ellipsoid where $a = 6,378,388$ m and $1/f = 297$. Large-scale topographic maps of Kuwait at 1:50,000 scale are on ED50 with a UTM Grid, published by NGA. A new edition published after the war is referenced to the WGS84 Datum.

The KUDAMS mapping project of Kuwait was initiated in 1983 using the Kuwait Transverse Mercator Grid where the Central Meridian $(\lambda_o) = 48°$ E, Scale Factor at Origin $(m_o) = 1.0$, and the False Easting = 500 km. The datum used for the KUDAMS project is the Ain el Abd Datum of 1970 located in the "Arq" Oil Field of Saudi Arabia near the Neutral Zone where $\Phi_o = 28° 14' 06.171''$ N, $\Lambda_o = 48° 16' 20.906''$ East of Greenwich, and the ellipsoid of reference is the International 1924. The three-parameter transform to WGS84 Datum is: $\Delta X = +294.7$ m, $\Delta Y = +200.1$ m, $\Delta Z = -525.5$ m, thanks to the EPSG Database.

The United Nations prepared a report on the Iraq–Kuwait Boundary Survey and orthophoto mapping project which consisted of a number of points determined on the WGS 84 Datum for the physical monuments established by personnel from New Zealand and Sweden. As of October 1994, the point lists were still secret *(IKBDC/Doc.20, 15 February 1993)*. From what can be determined from the UN work, they apparently did not attempt to recover the points established by the 30th Engineer Battalion (Topographic), 18th Airborne Corps in 1990 (*Desert Storm Surveying*, Funk & Lafler, POB, Vol. 17, No. 1, October–November 1991).

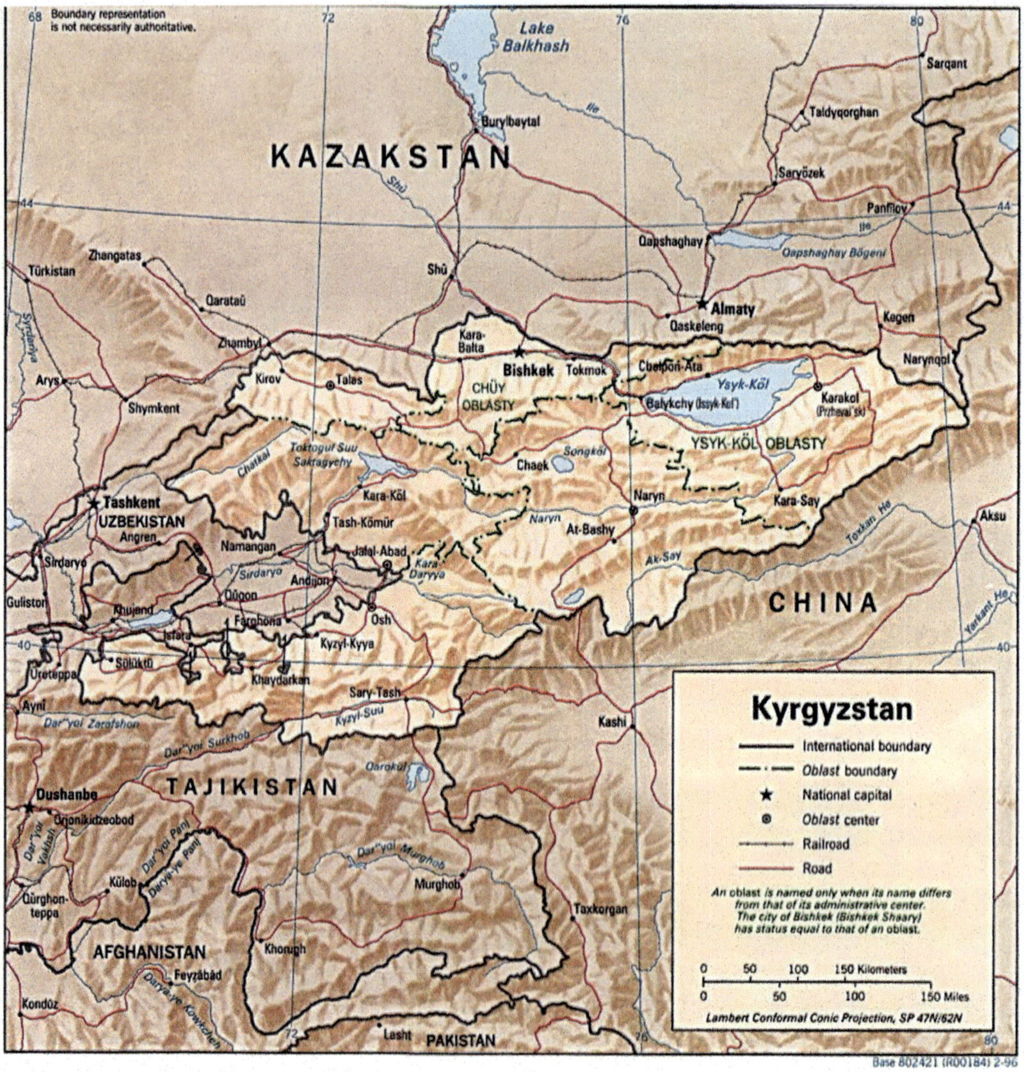

Base 802421 (R00184) 2-96

Kyrgyz Republic

Located in central Asia, west of China, south of Kazakhstan, Kyrgyzstan is bordered by China (1,063 km), Kazakhstan (1,212 km), Tajikistan (984 km), and Uzbekistan (1,314 km). The terrain has peaks of the Tien Shan mountain range and associated valleys, and basins encompass the entire country. The highest point is *Jengish Chokusu (Pik Pobedy)* (7,439 m), the lowest point is *Kara-Daryya (Karadar'ya)* (132 m), and the mean elevation is (2,988 m).

"The modern nation of Kyrgyzstan is based on a civilization of nomadic tribes who moved across the eastern and northern sections of present-day Central Asia. In this process, they were dominated by, and intermixed with, a number of other tribes and peoples that have influenced the ultimate character of the Kyrgyz people. Stone implements found in the Tian Shan mountains indicate the presence of human society in what is now Kyrgyzstan as many as 200,000 to 300,000 years ago. The first written records of a Kyrgyz civilization appear in Chinese chronicles beginning about 2000 B.C. The Kyrgyz, a nomadic people, originally inhabited an area of present-day northwestern Mongolia. In the fourth and third centuries *B.C.*, Kyrgyz bands were among the raiders who persistently invaded Chinese territory and stimulated the building of the original Great Wall of China in the third century *B.C.* The Kyrgyz achieved a reputation as great fighters and traders. In the centuries that followed, some Kyrgyz tribes freed themselves from domination by the Huns by moving northward into the Yenisey and Baikal regions of present-day south-central Siberia. The first Kyrgyz state, the Kyrgyz Khanate, existed from the sixth until the thirteenth century *A.D.*, expanding by the tenth century southwestward to the eastern and northern regions of present-day Kyrgyzstan and westward to the headwaters of the Ertis *(Irtysh)* River in present-day eastern Kazakstan. In this period, the khanate established intensive commercial contacts in China, Tibet, Central Asia, and Persia. In the meantime, beginning about 1000 *B.C.*, large tribes collectively known as the Scythians also lived in the area of present-day Kyrgyzstan. Excellent warriors, the Scythian tribes farther west had resisted an invasion by the troops of Alexander the Great in 328–27 *B.C.* The Kyrgyz tribes who entered the region around the sixth century played a major role in the development of feudalism.

"The Kyrgyz reached their greatest expansion by conquering the Uygur Khanate and forcing it out of Mongolia in *A.D.* 840, then moving as far south as the Tian Shan range – a position the Kyrgyz maintained for about 200 years. By the twelfth century, however, Kyrgyz domination had shrunk to the region of the Sayan Mountains, northwest of present-day Mongolia, and the Altay Range on the present-day border of China and Mongolia. In the same period, other Kyrgyz tribes were moving across a wide area of Central Asia and mingling with other ethnic groups. The Mongols' invasion of Central Asia in the fourteenth century devastated the territory of Kyrgyzstan, costing its people their independence and their written language. The son of Chinggis (Genghis) Khan, Dzhuchi, conquered the Kyrgyz tribes of the Yenisey region, who by this time had become disunited. For the next 200 years, the Kyrgyz remained under the Golden Horde and the Oriot and Jumgar khanates that succeeded that regime. Freedom was regained in 1510, but Kyrgyz tribes were overrun in the seventeenth century by the Kalmyks, in the mid-eighteenth century by the Manchus, and in the early nineteenth century by the Uzbeks. The Kyrgyz began efforts to gain protection from more powerful neighboring states in 1758, when some tribes sent emissaries to China. A similar mission went to the Russian Empire in 1785. Between 1710 and 1876, the Kyrgyz were ruled by the Uzbek Ququn *(Kokand)* Khanate, one of the three major principalities of Central Asia during that period. Kyrgyz tribes fought and lost four wars against the Uzbeks of Ququn between 1845 and 1873. The defeats strengthened the Kyrgyz willingness to seek Russian protection. Even during this period, however, the Kyrgyz occupied important positions in the social and administrative structures of the khanate, and they maintained special military units that continued their earlier tradition of military organization, some Kyrgyz advanced to the position of khan. In 1876 Russian troops defeated the Quqon Khanate and occupied northern Kyrgyzstan. Within five years, all Kyrgyzstan had become part of the Russian Empire, and the Kyrgyz slowly began to integrate themselves into the economic and political life of Russia. In the last decades of the nineteenth century, increasing numbers of Russian and Ukrainian settlers moved into the northern part of present-day Kyrgyzstan. Russian specialists began large-scale housing, mining, and road construction projects and the construction of schools. In the first years of the twentieth century, the presence of the Russians made possible the publication of the first books in the Kyrgyz language; the first Kyrgyz reader was published in Russia in 1911. Nevertheless, Russian policy did not aim at educating the population; most Kyrgyz remained illiterate, and in most regions traditional life continued largely as it had before 1870.

"By 1915, however, even many Central Asians outside the intelligentsia had recognized the negative effects of the Russian Empire's repressive policies. The Kyrgyz nomads suffered especially from confiscation of their land for Russian and Ukrainian settlements. Russian taxation, forced labor, and price policies all targeted the indigenous population and raised discontent and regional tension. The Kyrgyz in Semirech'ye Province suffered especially from land appropriation. The bloody rebellion of the summer of 1916 began in Uzbekistan, then spread into Kyrgyzstan and elsewhere. Kazaks, Turkmen, Uzbeks, and Kyrgyz participated. An estimated 2,000 Slavic settlers and even more local people were killed, and the harsh Russian reprisals drove one-third of the Kyrgyz population into China. Following a brief period of independence after the 1917 Bolshevik Revolution toppled the empire, the territory of present-day Kyrgyzstan was designated the Kara-Kyrghyz Autonomous Region and a constituent part of the Union of Soviet Socialist Republics (Soviet Union) in 1924. In 1926 the official name changed to the Kyrgyz Autonomous Republic before the region achieved the status of a full republic of the Soviet Union in 1936. In the late 1980s, the Kyrgyz were jolted into a state of national consciousness by the reforms of Soviet leader Mikhail S. Gorbachev and by ethnic conflict much closer to home. As democratic activism stirred in Kyrgyzstan's cities, events in Moscow pushed the republic toward unavoidable independence" *(Library of Congress Country Studies, 2014)*.

"Preliminary results from the legislative election in November 2021 suggest that pro-government parties will hold a majority in the *Jogorku Kengesh* (Kyrgyzstan's legislature). Continuing concerns for Kyrgyzstan include the trajectory of democratization, endemic corruption, a history of tense, and at times violent, interethnic relations, border security vulnerabilities, and potential terrorist threats" *(World Factbook, 2022)*.

"After the decision of the International Geodetic Union to connect the Indian and Russian triangulations (London, 1909), the survey to accomplish this task started with measurement of the base line near the town of *Osh*. The same year of 1909 the second order chain *Osh-Alay Valley* was already in work. LTC Cheykin in 1910 extended the chain to Pamirsky Post and in 1911 to the boundary of Afghanistan. In 1912 the baseline of *Kyzyl Rabat* was measured and the connection with the triangulation of India established. This section includes 87 stations and represents the most uniform part of the "main chain." The coordinates of the stations of section *Osh-Beik Pass* are for "Datum NW (II) base point of the *Osh Base Line* where $\Phi_o = 40° 37' 16.670'' N \pm 0.10''$, $\Lambda_o = 72° 56' 11.175''$ East of Greenwich $\pm 01.380''$, $\alpha_o = 151° 54' 01.860'' \pm 1.26''$ to SE (I) base point and referenced to the Bessel 1841 ellipsoid where $a = 6,377,397.155$ m, $1/f = 299.1528$.

"The coordinates of *Osh* Datum resulted from the adjusted chain, but because the chain was later connected with Tashkent Observatory (Uzbekistan), should be considered as preliminary. The relation with the coordinates of the "main chain" is established by stations 3284 (I) Osh, NW (II) base point-datum, 86 Beik *(Sarblock)*, and *87 Ak Turuk-Tau (Kutek)* for which final coordinates are published in *Geodezist 1934*, vol. 11/12, Pgs. 45 and 54. Final adjustment on Tashkent Datum for 3284 (I) Osh, NW (II) base point-datum: $\Phi_o = 40° 36' 55.740'' N$, $\Lambda_o = 72° 56' 22.880'' E$. For 86 Beik *(Sarblock)* on Osh Datum: $\Phi_o = 37° 19' 09.93'' N$, $\Lambda_o = 75° 04' 24.56'' E$, and on Tashkent Datum: $\Phi_o = 37° 18' 48.28'' N$, $\Lambda_o = 75° 04' 35.36'' E$. For 87 Ak Turuk-Tau *(Kutek)* on Osh Datum:

$\Phi_o = 37°\ 17'\ 43.94''$ N, $\Lambda_o = 74°\ 59'\ 55.53''$ E, and on Tashkent Datum: $\Phi_o = 37°\ 17'\ 22.29''$ N, $\Lambda_o = 75°\ 00'\ 06.35''$ E. By means of these three stations a transformation of coordinates on Osh Datum to Tashkent Datum is possible. The coordinates of Osh Datum never appeared in any Russian publication because they were computed just for the International Geodetic Union.

"In 1928 the Academy of Sciences of the USSR and the Emergency Society for German Sciences came to the agreement to send an exploring expedition in the regions of the Northwest Pamirs. In 1932 *Dr. Richard Finsterwalder*, who was responsible for the survey and mapping, published his *Scientific Records of Alai-Pamirs Expedition*. The task of the expedition required an adequate topographical map on which would be based the studies of the expedition after the explorations in the field were completed. For this purpose, Dr. Finsterwalder had chosen as the most suitable the terrestrial photogrammetric survey which for the first time found its application at a large scale in an exploring expedition" (*Triangulation in Turkestan and its Connection with India, Andrew M. Glusic, Army Map Service Technical Report No. 21, February 1957, 100 pages*).

"The State Geodetic Networks (GGS) of the Central Asian countries were parts of the former USSR's net. Its regional part was based on the first Central Asian geodetic net realized in 1885–1946 based on the ellipsoid Bessel 1841. The next GGS in the region was developed in 1946–1988 within the State program developed by Prof. F. Krassovsky following the "Fundamentals of Construction of the State Geodetic Network of USSR" issued in 1954 and 1961 on the ellipsoid Krassovsky 1940 *(where $a = 6,378,245\ m$ and $1/f = 298.3 - Ed.$).* Its class 1 network has chains of approximately equilateral triangles with sides of 20–25 km located roughly in the direction of the Earth's meridians and parallels at intervals of 200–250 km with 800–1000 km perimeter. The area that bordered by the class 1 triangulation chain is covered with solid nets of class 2 triangles with sides of about 10–20 km. The network of geodetic points can be made more dense by construction of class 3 and class 4 triangles. The main cartographic projection used in the Central Asian region was the Gauss-Krüger projection with 6° and 3° zones. The main Datum, Geodetic and Cartesian coordinates were Pulkovo 1942 (SK-42) *(the origin point is at Pulkovo Observatory where $\Phi_o = 59°\ 46'\ 18.55''$ North, $\Lambda_o = 30°\ 19'\ 42.09''$ East of Greenwich. The defining azimuth at the point of origin to Signal A is: $\alpha_o = 317°\ 02'\ 50.62'' - Ed.$),* based on the ellipsoid Krassovsky 1940. There are other coordinate systems used in the Soviet Union as Pulkovo 63, SK-90 *etc.*, but Pulkovo 42 has been used as a base for all other following reference systems.

"The Kyrgyz Government has announced the National Reference System '*Kyrg06*' in 2010 (Decree No. 235 from October 7, 2010). The National Reference System is ordered to use for geodetic and topographic activities, engineering survey, construction, and exploitation of buildings and structures, land surveying and maintaining the state cadaster. Kyrg-06 is based on the International Terrestrial Reference Frame ITRF-2005, UTM projection with five 3° zones" (*GNSS Application Trends in Central Asia, Akylbek Chymyrov, Nyugat-magyarországi Egyetem, Geoinformatikai Kar, Székesfehérvár, 2014*).

REFERENCES: K

AFREF Newsletter No. 11, (2010). https://issuu.com/iskkenya/docs/isk_journal_april_2019__9_/s/10844529
AFREF Newsletter No. 5, 2008. https://www.rcmrd.org/newletters
Akylbek Chymyrov, GNSS Application Trends in Central Asia, Nyugat-magyarországi Egyetem, Geoinformatikai Kar, Székesfehérvár, 2014.
Andrew M. Glusic, Army Map Service Technical Report No. 21, February 1957, 100 pages.
Andrew M. Glusic, Triangulation in Turkestan and its Connection with India, Army Map Service Technical Report Number 21, pp., 1957.
Argentine Anuario, Instituto Geográfico Militar, 3er Volumen, 1914.
COL. Gedeonov, Zapiski Vol. 53, part II, pp. 229–288.
David Llewellyn, Lands and Survey Division, Republic of Kiribati, Personal Communication July 2002.
EPSG Database, 2010.
ftp://ftp.ga.gov.au/geodesy-outgoing/gnss/pub/SPSLCMP/14-7573-3%20Record%202014-17%20Kiribati%202012%20WCAG.pdf.

ftp://ftp.ga.gov.au/geodesy-outgoing/gnss/pub/SPSLCMP/14-7573-4%20Record%202014-18%20Kiribati%202013%20WCAG.pdf.
ftp://ftp.ga.gov.au/geodesy-outgoing/gnss/pub/SPSLCMP/PreviousLevellingSurveyReports/KIRIBATI%20LEVEL%20SVY%202007.pdf.
Funk & Lafler, *Desert Storm Surveying*, POB, Vol. 17, No. 1, October–November 1991.
Gen. Schubert, Zapiski Catalog, volume 1 in 1837.
Heungmuk Cho, Geodesy Division, National Geography Institute in Suwon-shi, Republic of Korea, Personal Communication October 1999.
https://researchdata.edu.au/kiribati-levelling-survey-report-2009/1230349.
IKBDC/Doc.20, 15 February 1993.
John W. Hager, Personal Communication July 2002.
Kazakhstan Library of Congress Country Study, 2010.
Library of Congress Country Studies, 2014.
Lonely Planet, 2010.
Lord William Malcolm Hailey, An African Survey, 1938.
Morgan W. Davis, Personal Communication May 2003.
Patroba Achola Odera, *South African Journal of Geomatics*, Vol. 5, No. 1, February 2016. http://sajg.org.za/index.php/sajg/article/viewFile/356/218
Phil Smart, M.R.I.C.S., RPS Energy, Personal Communication, March 2010.
Richard W. Stevenson, and Gary Fitzpatrick, Library of Congress, Personal Communication July 2002.
Russell Fox, U.K. Ordnance Survey, Personal Communication July 2002.
Russell Fox, U.K. Ordnance Survey, Personal Communication, January 2000.
Washington Abuto, Personal Communication, 24 November 1997.
World Factbook, 2010.
World Factbook, 2022.
Young-Jin Lee, Hung-Kyu Lee, Chan-Oh Kwon and Jun-Ho Song, Implementation of the New Korean Geocentric Datum and GPS CORS Management, Integrating Generations, FIG Working Week 2008, Stockholm, Sweden 14–19 June 2008.
Zapiski Catalog, Vol. 41, part. III, Section 5, page 13.

L

Lao Peoples Democratic Republic 441
The Republic of Latvia 445
The Lebanese Republic 451
Kingdom of Lesotho 455
Republic of Liberia 459
Great Socialist People's Libyan Arab Jamahiriya 465
Principality of Liechtenstein 469
Republic of Lithuania 473
Grand Duchy of Luxembourg 477

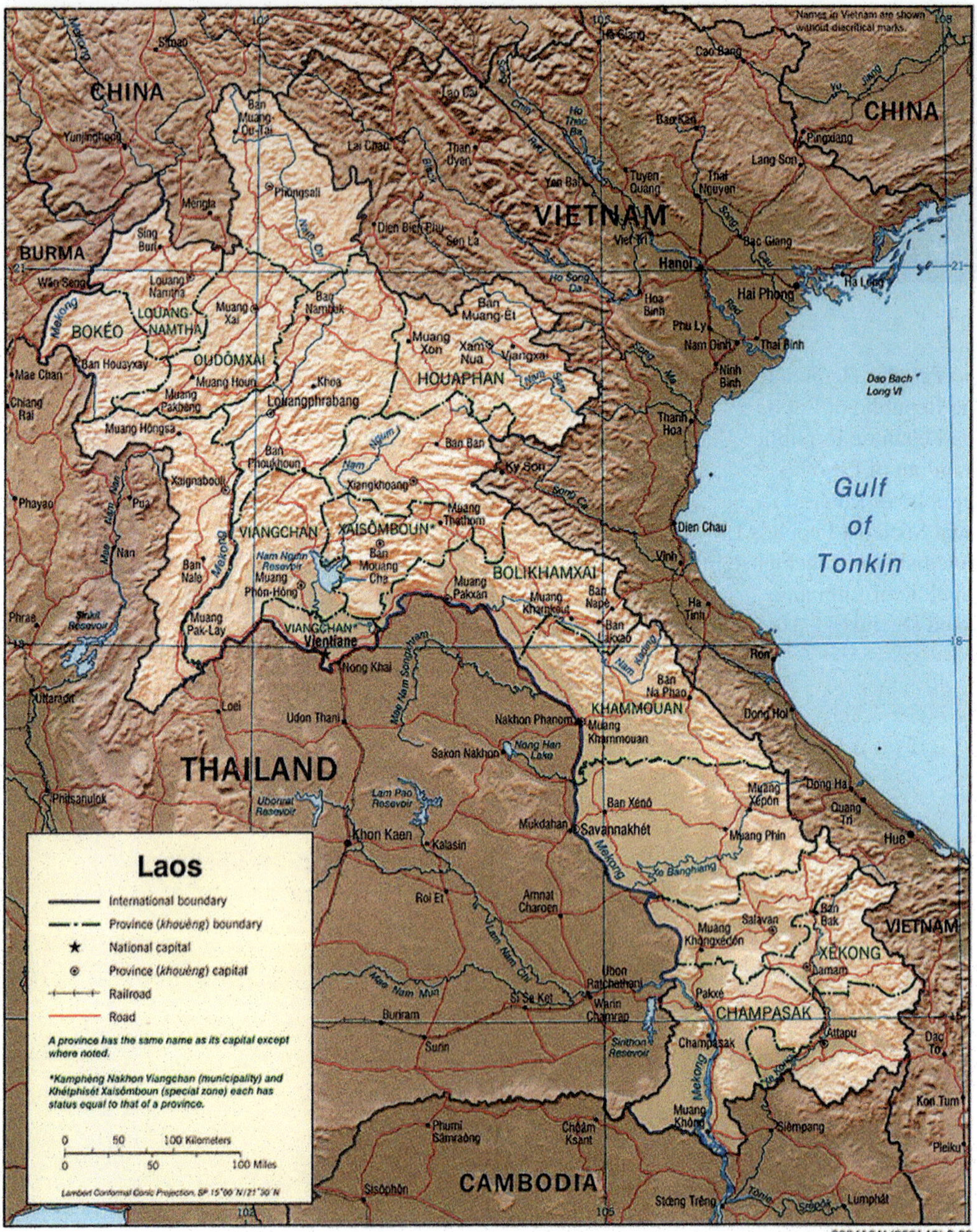

802412AI (C00140) 8-03

Lao Peoples Democratic Republic

The Lao PDR is landlocked and bordered by Burma (235 km), Cambodia (541 km), China (423 km), Thailand (1,754 km), and Vietnam (2,130 km). Slightly larger than Utah, its terrain is comprised of mostly rugged mountains with some plains and plateaus; the lowest point is the Mekong River (70 m), and the highest point is Phou Bia (2,817 m).

"Historical research shows that the rudimentary structures of a multiethnic state existed before the founding of the Kingdom of Lan Xang in the 13th century A.D. These pre-13th century structures consisted of small confederative communities in river valleys and among the mountain peoples, who found security away from the well-traveled rivers and overland tracks where the institutions and customs of the Laotian people were gradually forged in contact with other peoples of the region. During these centuries, the stirring of migrations as well as religious conflict and syncretism went on more or less continuously. Laos's short-lived vassalage to foreign empires such as the Cham, Khmer, and Sukhothai did nothing to discourage this process of cultural identification and, in fact, favored its shaping. In the thirteenth century … the rulers of Louangphrabang (Luang Prabang) constituted a large indigenous kingdom with a hierarchical administration. Even then, migratory, and religious crosscurrents never really ceased. The durability of the kingdom itself is attested to by the fact that it lasted within its original borders for almost four centuries. Today, the Lao People's Democratic Republic (Lao PDR, or Laos) covers only a small portion of the territory of that former kingdom" (Library of Congress Country Study).

"In the 2010s, the country benefited from direct foreign investment, particularly in the natural resource and industry sectors. Construction of a number of large hydropower dams and expanding mining activities have also boosted the economy. Laos has retained its official commitment to communism and maintains close ties with its two communist neighbors, Vietnam and China, both of which continue to exert substantial political and economic influence on the country. China, for example, provided 70% of the funding for a $5.9 billion, 400-km railway line between the Chinese border and the capital Vientiane, which opened for operations in December 2021. Laos financed the remaining 30% with loans from China. At the same time, Laos has expanded its economic reliance on the West and other Asian countries, such as Japan, Malaysia, Singapore, Taiwan, and Thailand" (*World Factbook* 2022).

The first mapping of Indochina of any importance was performed by the *Bureau Topographique*, which was set up as a section of the General Staff after the French assumed control of the area in 1886. This organization started triangulation work in Indochina and compiled topographic maps at 1:100,000, 1:200,000, 1:500,000, and 1:1,000,000 scales as well as boundary and river maps. The 1:500,000 series, which was completed in 1899, was the most important of these early maps. The major mapping organizations were the *Service Géographique de l'Indochine – SGI*; the *Institut Géographique National – IGN*, in Paris; and the *Service Cartographique des Forces Terrestres d'Extrême Orient – F.T.E.O.* (Cartographic Service of the Far East Land Forces) (*Foreign Maps, TM 5–248*). France commenced the first substantial geodetic network in Laos in 1902. It took the form of a first-order triangulation network that comprised 47 stations. Although some of these stations were as far apart as 100 km, the average distance between them was 30 km. Unfortunately, many of the stations were not marked with durable materials. By 1955, ~65% of the points had been destroyed. During the late 19th and early 20th centuries, the British Survey of India (SOI) extended their classical triangulation into neighboring Burma and the French triangulation in the Lao PDR was subsequently connected. A series of regional adjustments then "established" the Indian Datum in Laos.

As a sidebar, I need to explain the presence of the "Indian Datum" in the Lao PDR. A good personal friend of mine of 40+ years is Dr. Muneendra Kumar, retired Chief Geodesist of the United States National Geospatial Intelligence Agency – NGA. "Muni" tells me that there is no such thing as the "Indian Datum." There is only a series of piecemeal regional "adjustments" of field observations and there has never (ever!) been a unified classical datum adjustment. Dr. Kumar should know – he retired from the Survey of India before he immigrated to the United States, and he once personally walked 5,000 km in 1 year while performing a classical triangulation in Nepal as a junior SOI officer! Therefore, take the moniker "Indian Datum" with various dates with a grain of salt – these flavors only represent local adjustments "cobbled together" to assure a smooth-appearing consistency of various published map series. The legendary "origin" of the Indian Datum as defined in 1900 and labeled as Indian 1916 origin at Kalianpur Hill Station: $\Phi_o = 24° 07' 11.26''$ N, $\Lambda_o = 77° 39' 17.57''$ East of Greenwich, the initial azimuth to Surantal from south is: $\alpha_o = 190° 27' 05.10''$. The ellipsoid of reference is the Everest 1830 where $a = 6,377,276.345$ m, and $1/f = 300.8017$. John W. Hager tells me that "In 1954, the triangulation of Thailand was adjusted to Indian 1916 based on ten stations on the Burma border. In 1960, the triangulation of Cambodia and Vietnam was adjusted holding fixed two Cambodian stations connected to the Thailand adjustment of stations from the Cambodian-Vietnam adjustment. North Vietnam was also adjusted to this system but with lower standards." Note that the Lao PDR was the connecting link.

Thanks to Andrew Dyson, Jones, Rohde, Lloyd, and Sougnatti in *GPS World, March 1999*, "between 1963 and 1975, AMS produced three series of 1:50,000 scale maps based on the remaining French control network. Most of the associated computations were performed on the Indian Datum of 1960 using the Everest Spheroid (*sic*). Some later maps were produced based on the Indian Datum of 1975. In 1982, in cooperation with the Soviet Union, the Lao National Geographic Department (NGD) initiated a new geodetic survey to provide control for small-scale mapping … this survey defined a local geodetic datum, referred to as the *Vientiane Datum of 1982*."

The defining parameters are as follows: $\Phi_o = 18° 01' 31.6301''$ N, $\Lambda_o = 102° 30' 56.6999''$ East of Greenwich, ellipsoid height (h_o) = 223.56 m, and the ellipsoid of reference is the Krassovsky 1940, where $a = 6,378,245$ m and $1/f = 298.3$. The ellipsoidal height was defined as being equal to the mean sea level height for the origin station. The geoid-ellipsoid separation at Vientiane (Nongteng) was therefore zero.

The Lao Datum 1993 was created following completion of a national GPS survey in cooperation with the Socialist Republic of Vietnam. It is defined as follows: Origin Station is Pakxan (35203), where $\Phi_o = 18° 23' 57.0056''$ N, $\Lambda_o = 103° 38' 41.802''$ East of Greenwich, ellipsoid height (h_o) = 177.600 m, and the ellipsoid of reference again, is the Krassovsky 1940.

The cobbled series of local adjustments of SOI triangulations for various purposes (as explained to me by Dr. Kumar) have resulted in a number of transformations published by the Laos government involving "Indian Datum of 19XX," such as **from** "Lao National Datum 1997" **to** "Indian Datum 1954 (Vientiane Area): $\Delta X = -168.711$ m ±0.034 m, $\Delta Y = -951.115$ m ±0.034 m, $\Delta Z = -336.164$ m ±0.034 m. **From** Lao National Datum 1997 **to** Indian Datum 1960: $\Delta X = -153$ m, $\Delta Y = -1012$ m, $\Delta Z = -357$ m. **From** Lao National Datum 1997 **to** Lao Datum 1993: $\Delta X = +0.652$ m ±0.15 m, $\Delta Y = -1.619$ m ±0.15 m, $\Delta Z = -0.213$ m ±0.15 m. Finally, **from** Lao National Datum 1997 **to** Vientiane Datum 1982: $\Delta X = +2.227$ m ±0.79 m, $\Delta Y = -6.524$ m ±1.46 m, $\Delta Z = -2.178$ m ±0.79 m. According to an official government statement, "The parameter generation assumed that Vientiane Datum 1982 spheroidal (*sic*) heights were the same as mean sea level heights in all parts of Laos." As this assumption is incorrect, **extreme caution** (*sic*) should be exercised when interpreting transformed height information."

Thanks go to Dr. Muneendra Kumar, Mr. John W. Hager, and Mr. Malcolm A. B. Jones for their patience explaining these intricacies to me over the decades. Publications of the Australian Government's Land Titling Project by Mr. Andrew Jones made this a whole lot easier for me to describe in detail.

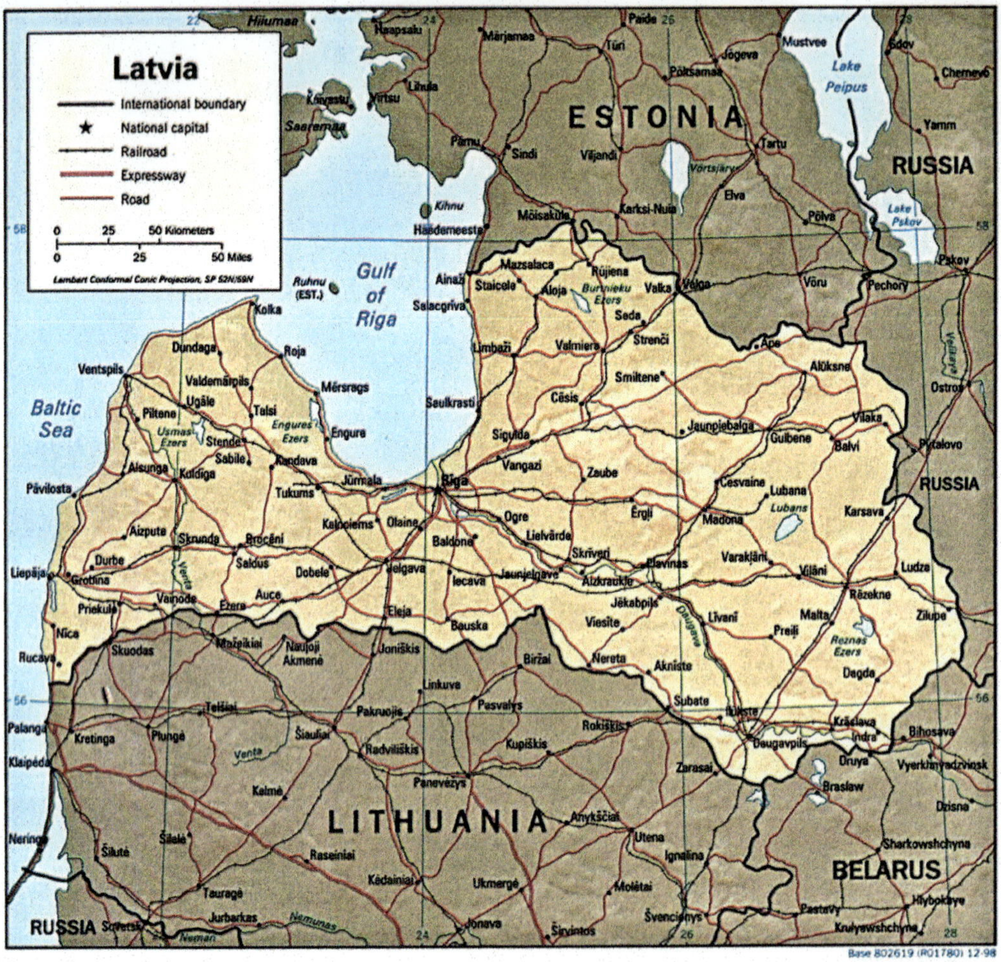

Base 802619 (R01780) 12-98

The Republic of Latvia

Latvia shares borders with Estonia to the north (339 km), Russia to the east (217 km), Belarus to the southeast (141 km), and the Baltic Sea and Gulf of Riga to the west comprise a coastline of 531 km. Slightly larger than West Virginia, the country is mostly low coastal plain with the highest point being Gaizinkalns at 312 m.

Between 2500 and 1500 *B.C.*, the Finno-Ugric and proto-Baltic tribes settled on Baltic shores. The closest ethnic relatives of the Latvians are the ancient Prussians, the Galinds, the Jatvings, and the Lithuanians. The first settlers in the territory of Latvia were the Livonians or "Libiesi." The Livonians were once concentrated in the northern part of Latvia, but today only about 100 individuals speak their ancient language which nevertheless has contributed to a prominent Latvian dialect. By the 12th century, the natives were split into a number of tribal groups, all practicing nature religions. The Knights of the Sword (Livonian Order) were crusaders that forcibly converted Latvia to Christianity in the 13th century. For centuries, Latvia has been under Swedish, Polish, German, and Russian rule. In 1918, Latvia proclaimed independence from Czarist Russia. By 1940, Latvia was occupied by the Soviet Union and was soon overrun by Nazi Germany. Soviet forces reoccupied in 1944–1945, and Latvia remained under Soviet rule until 1991 when it was admitted into the United Nations. In May 1994, the Latvian National Independence Movement finished first in Latvia's first post-Soviet local elections; the ex-communists fared the worst.

Survey activities in Latvia began with Tenner's first-order network of 1820–1832 in Semgallen and Courland and were published by Czarist Russia in 1843 and 1847. Tenner later supplemented his primary net with lower-order stations. The Tenner chains were originally computed on the Walbeck 1819 ellipsoid, where the semi-major axis (a) = 6,376,895 m and $1/f = 302.7821565$; they were later recomputed on the Bessel 1841 ellipsoid. Between 1878 and 1884, Schulgin further increased the density of lower-order stations in the area originally surveyed by Tenner. However, the majority of these latter station monuments did not survive into the 20th century, and they were ignored by the Russians. The Tenner net in the east did not extend further north than the Sestukalns-Geisenkalns side, and the Struve primary net extended north from this side through Yuryev and over the Gulf of Finland. The Russian Western Frontier surveys were executed mainly by Yemel'yanov and Nikifirov between 1904 and 1912. These chains formed a major part of the modern (early) 20th-century network of Eastern Latvia with some of the first-order stations being old Tenner or Struve stations. This Russian survey covered much of western and central Livonia in 1904–1905, and eastern Livonia and Lettgallen were covered with the 1906 and 1912 nets. Lower-order nets to fourth order supplemented the primary chains. These Czarist Russian surveys were computed on the Pulkovo 1904 Datum or the Yuryev II Datum, both of which are referenced to the Bessel 1841 ellipsoid, where $a = 6,377,397.155$ m and $1/f = 299.1528128$.

After Latvia achieved Republic status, the new Latvian Survey Office began to unify these various nets in stages, before commencing with their own survey activities in 1924. The results of this program of unification are published in *Latvijas Valsts Trigonometrieskais*. The Pulkovo 1904 Datum points were recomputed into Yuryev II Datum based on 14 available first-order Scharmhorst points and this was then put into the Senks Soldner Grid. The 1905 Russian net had already been computed on the Yuryev II Datum and was accepted into the Old Gaisenkalns Soldner Grid, where $\varphi_o = 56° 52' 15.184'' N$, $\lambda_o = 25° 57' 34.720''$ East of Greenwich. No False Origin was used. Similarly, the 1906 and 1912 nets were already on Yuryev II Datum, and those form the Old Vitoleieki Soldner Grid, where $\varphi_o = 56° 40' 08.64'' N$, $\lambda_o = 27° 15' 11.79''$ East of Greenwich, and no False Origin was used. Tenner's survey had been calculated on the Walbeck 1819 ellipsoid and was now recomputed on the Bessel 1841 ellipsoid and controlled by the Yemel'yanov-Nikifirov survey of the Puci-Sarmes side. This had the effect of introducing a slight swing to the orientation of the Tenner triangulation. The Latvian Kulkigan and Puci systems cover this net. For most of Eastern Latvia, the earlier Russian surveys could be used to form the basis of the Latvian triangulation as the station centers could be found. In Courland and Semgallen the original triangulation dated back to 1820 and few of the old stations could be recovered. This necessitated a completely new survey, which was started in 1924 from the Liepaja-Paplaka base and extended in stages to the Puci-Sarmes side and thence to Riga and Jelgava. The origin of this system is the Yuryev II Datum value of Riga St. Petri Church "tower ball" which is the same station as the Tenner first-order point. *(The datum origin is monumented with a plaque in the floor of the vestibule of the church that is directly under the tower ball or finial – Ed.)* The Yuryev II Datum azimuth to Jelgava Church was adopted for the orientation of the net. Although Riga Church and Courland value is the same as Scharnhorst value and the azimuth identical, the coordinates of Jelgava Church vary slightly by 0.002″ in each axis. This was due to the scale of Liepaja and Jelgava bases, which were adopted for the Courland System in preference to the less reliable Scharnhorst scale. This became known as the "Provisional Courland System (datum)." The Provisional Courland System was divided into two Cassini-Soldner Grids: The "Riga System" with its origin at Riga St. Petri Church, where $\varphi_o = 56° 56′ 53.919″$ N, $\lambda_o = 24° 06′ 31.898″$ East of Greenwich, and the "Vardupe System" with its origin at the Provisional Courland station Vardupe, where $\varphi_o = 56° 51′ 32.961″$ N, $\lambda_o = 21° 52′ 03.462″$ East of Greenwich. No False Origin was used for either Grid. The Provisional Courland System was immediately adjusted and computed before the triangulation of central and east Latvia was completed. This Provisional Courland System was first adjusted within itself and then adjusted to the Latvian part of the Baltic Ring. The lower-order control as far east as 24° 20′ East of Greenwich was adjusted and computed in terms of this system.

The General Latvian Triangulation Net of first-order stations covers practically all of Latvia, including Courland. The lower-order control east of 24° 20′ East of Greenwich was computed in terms of the General Latvian Triangulation Net. Actually, there is a small overlap around 24° 20′ East of Greenwich for which the coordinates of all stations, of all orders, were computed in both the Provisional Courland System and the General Latvian Triangulation Net. The 1924 net was adjusted in stages to fit eight bases that were the following: Puci-Sarmen Jelgava, Jekabpils-Dabarkalns, Garmaniski-Viski, Kangari-Jamilova, Kirbbisi-Akija, Duorno Sielo-Dziedzinka (Polish Base), Arula-Urkaste (Estonian Base), and Liepaja-Paplaka. The chain Puci-Sarmen to Jekabpils-Dabarkalns forms the backbone of the modern net from which the adjustment started. The origin of the General Latvian Triangulation Net is Riga Petrikirche (top of the Riga Church spire), where $\Phi_o = 56° 56′ 53.919″$ N, $\Lambda_o = 24° 06′ 31.898″$ East of Greenwich, and the reference azimuth to Mitau German Church: $\alpha_o = 215° 24′ 04.38″$. The value for Riga Petrikirche approximates the Dorpat II System (datum). The Latvian control was computed in terms of four Cassini-Soldner Grids. The Grid names and the coordinates of the respective origins are as follows: Vardupe Cassini-Soldner Grid, where $\varphi_o = 56° 51′ 32.961″$ N, $\lambda_o = 21° 52′ 03.462″$ E, the Riga Cassini-Soldner Grid, where $\varphi_o = 56° 56′ 53.919″$ N, $\lambda_o = 24° 06′ 31.898″$ E, the Gaisinkalns Cassini-Soldner Grid, where $\varphi_o = 56° 52′ 15.031″$ N, $\lambda_o = 25° 57′ 34.920″$ E, and the Vitolnieki Cassini-Soldner Grid, where $\varphi_o = 56° 40′ 08.447″$ N, $\lambda_o = 27° 15′ 12.252″$ E. These grid systems cover zones of about 1½° to 2° wide and overlap slightly. The Vardupe Grid is computed from the geographics of the Provisional Courland System, while the Gaisinkalns and Vitolnieki Grid values correspond to the General Latvian Triangulation Net. The Riga Cassini-Soldner Grid coordinates are computed from both the Provisional Courland System geographics and the General Latvian Triangulation Net. However, care is taken in the Latvian "Trig" Lists to show from which geodetic system the Riga Cassini-Soldner coordinates are computed.

During these inter-war years, the Russians were also actively re-computing their survey information in the Baltic States. Prior to 1932 the Russian horizontal control of the Baltic States was always referenced to Dorpat Observatory at Tarbu in Estonia. In 1932, the Russians set up Pulkovo Observatory 1932 as their horizontal datum and origin reference to the Bessel 1841 ellipsoid, and later revised this to Pulkovo 1942, now properly termed "System 42" (datum) referenced to the Krassovsky 1940 ellipsoid.

The Republic of Latvia

Thanks to E. A. Early of the U.S. Army Map Service, "In 1942 the German Army undertook the conversion of the Latvian Soldner coordinates to DHG Pulkovo" (Deutches Heeres Gitter – German Army Grid). "The first phase of the conversion embodied the change of projection from Soldner to Gauss-Krüger. The Latvian Geodetic Engineer Mensin set up formulas and tables to convert the four Latvian Soldner systems to the German Gauss-Krüger system. However, upon checking these formulas as the boundaries of the Soldner systems inadmissible gaps were discovered. Mensin's formulas were then abandoned, and new ones were derived following the method given in Jordan-Eggert's *Handbuch der Vermessungskunde*. Since there were no reliable geodetic connections to the Pulkovo system available at that time, the conversion of the Latvian system to the Pulkovo 1932 system could only be approximated. The value of the datum point of the general Latvian triangulation net approximates the Dorpat II system value. The necessary formulas converting Dorpat II system values to the Pulkovo 1932 system were available in the official Russian work of Brigade Engineer O. A. Sergjew, *Making and Editing of Military Maps*, Moscow 1939. In the absence of better data, these formulas were taken as a basis for the conversion to Pulkovo 1932 datum. The German Preliminary DHG Pulkovo 1932 coordinates resulting from this conversion were published in the form of *Ausgabe Koordinatenkartei* by the Kriegs-Karten und Vermessungsamt Riga, in 1943.

In 1943, extensive surveys were executed along the Latvian–Russian border for the final connection of the Latvian triangulation with the Pulkovo system. At the conclusion of these surveys, the Latvian system (already in terms of the preliminary DHG Pulkovo 1932 system) was converted to the Pulkovo 1932 system by a rigid-field adjustment. The Russian Pulkovo 1932 system coordinates used in this adjustment were taken from Russian Catalogs. As a consequence of this adjustment new conversion constants were computed to convert from the four Latvian Soldner systems to the Final DHG Pulkovo 1932 system. As mentioned previously, the triangulation of Latvia is not completely uniform since the triangulation in Courland is based on the Provisional Courland System adjustment. Only the first-order stations in Courland are available in terms of the General Latvian Triangulation Net. The lower-order trig in Courland was converted from the Provisional Courland System to the General Latvian Triangulation Net by a graphical adjustment (triangle by triangle) based on the comparison of first-order values. After the lower-order trig in Courland was converted to terms of the General Latvian Triangulation Net, the Final DHG Pulkovo 1932 coordinates for all Latvia were computed. These coordinates were published in 1943 as a second edition *Ausgabe Endwerte Koordinatenkartei* by the Kriegs-Karten und Vermessungsamt, Riga. In the fall of 1944, the publication of the *Koordinaten-Verzeichnis* (trig books) was begun. The coverage of these books is scanty. Also, it is noted that there are differences of up to a meter, as some stations, between the *Koordinaten-Verzeichnis* values and those from the second edition *Koordnatenkartei*. It appears that this difference is accounted for by the fact that some of the Latvian traverse points (as included in the second edition *Koordinatenkartei*) were resurveyed by the German Army and consequently were listed in the *Koordinaten-Verzeichnis* books by the German Survey values." To convert from DHG Pulkovo 1932 Datum Grid coordinates to European Datum 1950 coordinates on the UTM Grid, zone 34 use the following: (UTM Northing)=0.9996056758 * (DHG Northing)+0.0000176163 * (DHG Easting)+828.01, and (UTM Easting)=0.9996056758 * (DHG Easting)+0.0000176163 * (DHG Northing)+365.98. The NIMA published values for that general region of Europe from European Datum 1950 to WGS 84 are: $\Delta X = -87$ m ± 3 m, $\Delta Y = -95$ m ± 3 m, and $\Delta Z = -120$ m ± 3 m. The NIMA published values for System 42 Datum (in Latvia) to WGS 84 Datum are: $\Delta X = +24$ m ± 2 m, $\Delta Y = -124$ m ± 2 m, and $\Delta Z = -82$ m ± 2 m.

"(The) origin of Latvian Coordinate System LKS-92 definition was based on two GNSS campaigns in 1992 and 2003. There are two continuously operating reference networks in Latvia: LatPos and EUPOS®-Riga. GNSS stations of these networks have fixed coordinate values in LKS-92. At the Institute of Geodesy and Geoinformatics of the University of Latvia both LatPos and EUPOS®-Riga station daily coordinate values are calculated. The coordinate differences between epochs 1989.0 and 2018.5 were obtained for LatPos and EUPOS®-Riga stations, expressed in ITRF14.

ITRF reflects the motion of Eurasian plate in global frame of the Earth and ETRF89 system reflects the intraplate motion. Mean yearly coordinate components in ETRF89 were analysed. Comparison of LatPos and EUPOS®-Riga station coordinate components in ETRF89, LKS-92 and ETRF2000 coordinate systems was performed. Future of Latvian coordinate system LKS-92 is discussed." *Baltic Journal of Modern Computing*, Vol. 7 (2019), No. 4, 513–524, https://doi.org/10.22364/bjmc.2019.7.4.05.

According to the decision of IAG Reference Frame Sub-commission for Europe (EUREF), the EVRF2007 solution as the vertical reference has to be deployed in EU countries. The new height system *LAS-2000,5* had been enacted as the European Vertical Reference System's *EVRF2007* realization in Latvia and the new geoid model LV'14 had been introduced by Latvian authority Latvian Geospatial Information Agency. However, the appreciation of the quality of quasi-geoid model LV'14 is rather contradictious among the users in Latvia. The independent estimate and comparison of the two Latvian geoid models developed till now has been performed by the Institute of Geodesy and Geoinformatics. Previous geoid model LV98 which was developed for *Baltic-1977* height system almost 20 years ago is outdated now. Preparatory actions were described in order to fulfill the task of comparison of the geoids in two different height systems. The equations and transformation parameters are presented in this article for the normal height conversion from *Baltic-1977* height system to the Latvian realization named *LAS-2000,5*.

Changing the national height system and geoid model in Latvia, Janis Balodis, Katerina Morozova, Gunars Silabriedis, Maris Kalinka, Kriss Balodis, Ingus Mitrofanovs, Irina Baltmane, Izolde Jumare, Institute of Geodesy and Geoinformatics, University of Latvia, Riga, Latvia, 2020, (unpublished manuscript).

Because of the groundbreaking work the University of Latvia's Institute of Geodesy and Geoinformatics published on their research to develop a one-centimeter geoid model and the invention of a relatively inexpensive Digital Zenith Camera (DZC) by Dr. Ansis Zariņš, I traveled to Riga during the summer of 2019. LSU's Center for GeoInformatics had acquired an absolute gravity meter along with a couple of relative gravity meters and various related instruments and vehicles with the same objective for the State of Louisiana. Since their DZC only works at night with clear skies for star shots, days were open for the staff to take me around Riga and environs. A major point of interest for me was the origin of the Yuryev II Datum and the General Latvian Triangulation Net Datum Origin is Riga Petrikirche (top of the Riga Church spire). The Latvian Geodesists then took me inside of the church, and a plaque in the floor of the vestibule, directly under the plumb line of the spire displayed the actual datum origin. By the way, their DZC worked flawlessly, and LSU purchased one for €100 K, including training. Our DZC is currently kept busy here in Louisiana observing the deflection of the vertical at all of our GPS CORS sites.

Lebanon

- —— International boundary
- —·— Governorate (*mohafazat*) boundary
- ★ National capital
- ⊙ Governorate (*mohafazat*) capital
- ══ Expressway
- ┼┼┼ Railroad
- —— Road

0   15 Kilometers
0   15 Miles

Lambert Conformal Conic Projection, SP 33¹0 N/34³0 N

Boundary representation is
not necessarily authoritative.

The Lebanese Republic

Lebanon is almost ¾ the size of Connecticut and is comprised of a narrow coastal plain; the Bekaa Valley separates Lebanon and the Anti-Lebanon Mountains. Bordered by Israel (79 km) to the south and by Syria (375 km) to the east and north; the western side of the republic is the Mediterranean Sea (225 km). Lebanon has a territorial sea claim of 12 nautical miles, and the highest point in the country is *Qurnat as Sawda'* at 3,088 m.

There is evidence of human habitation in Lebanon for several thousand years from mid-3rd millennium *B.C.* that has been under the control variously of Sumerians, Akkadians, Amorites, Egyptians, Assyrians, and Babylonians. Once part of the Persian Empire, Alexander the Great conquered the region in the 4th century *B.C.*, and it later flourished under the Roman Empire. Lebanon was overrun by Muslim Arabs in 635–636 *A.D.* and remained under the Turks during the Crusades until the British and French invaded during World War I because Turkey was an ally of Germany.

"Following World War I, France acquired a mandate over the northern portion of the former Ottoman Empire province of Syria. The French demarcated the region of Lebanon in 1920 and it gained independence in 1943. Since then, Lebanon has experienced periods of political turmoil interspersed with prosperity built on its historical position as a regional center for finance and trade, although that status has significantly diminished since the beginning of Lebanon's economic crisis in 2019, which includes simultaneous currency, debt, and banking crises. The country's 1975–90 civil war, which resulted in an estimated 120,000 fatalities, was followed by years of social and political instability. Sectarianism is a key element of Lebanese political life. Neighboring Syria has historically influenced Lebanon's foreign policy and internal policies, and its military occupied Lebanon from 1976 until 2005. Hizballah – a major Lebanese political party, militia, and international terrorist organization – and Israel continued attacks and counterattacks against each other after Syria's withdrawal and fought a brief war in 2006. Lebanon's borders with Syria and Israel remain unresolved" (*World Factbook*, 2022).

In 1799, Napoleon Bonaparte commenced his military campaign for the conquest of Egypt and "Upper Egypt" (the Palestine and Greater Syria). *La Carte d'Egypte et de Syrie* was published by the Dépôt de la Guerre beginning in 1808. The ersatz Datum was based on astronomical observations in Cairo and Jerusalem and was referenced to the Plessis ellipsoid, where the semi-major axis, $a = 6,375,738.7$ m and the reciprocal of flattening $1/f = 334.29$. Much of the coast was actually based on published British Admiralty charts of the time. (See: The State of Israel.) The projection was the ellipsoidal Bonne, the "standard" for France and most of Europe at the time. "Le centre de la projection correspond à l'axe de la grande pyramide du Nord, à Memphis." (The center of the projection corresponds to the axis of the great pyramid of the North at Memphis.)

The French *Expédition du Liban* (1860–1861) was made after the massacre of Christians in Syria (and Lebanon) occurred during May and June 1860. After enforcing the peace, one topographic brigade remained to perform some exploratory mapping. Some minor triangulation was performed from Tyre to Tripoli along the coast. This resulted in one reconnaissance sheet at 1:100,000 scale, and one 1:200,000 sheet being published in 1862.

The early maps of Turkey including the Levant area were on the Bonne projection also, but the projection origin was the finial of the dome of the Aya Sofia Mosque. The Ottoman Turkish ellipsoidal Bonne of Syria was used from 1909 to 1923 (and the territory of what is now Lebanon), projection Latitude of Origin $(\varphi_o) = 28° 58' 50.8188"$ N, and the Central Meridian $(\lambda_o) = 39° 36'$ East of Greenwich. The geodetic network was calculated on the Clarke 1880 (IGN) ellipsoid, where $a = 6,378,249.2$ m and $1/f = 293.4660208$, and according to the SGA, the Datum Origin was at the South End of the Base of *Makri Keuī*, near Constantinople (Istanbul). The 1:200,000 general map of Asia Minor was published in 1911 under the direction of General Mehmed Sevki Pacha, Director of the Turkish Map Service.

The French established the Bureau Topographique du Levant in 1918, and after 1920, the chain of triangulation was extended eastward along the northern border of Syria with Turkey to Iraq. The French geodetic triangulation parties were quite impressed with the Bekaa Valley and the vast bounty of orange and banana harvests. Planimetric compilation was aided by aerial photography flown by a French military aviation squadron of the 39th Regiment. The topographic brigade was commanded by Lieutenant Colonel G. Perrier, and he organized the observations for the establishment of an astronomical origin for a Datum in the Bekaa Valley of Lebanon that would serve Syria as well. The baseline was measured, and the South End of the Base at Bekaa was the fundamental origin for the astronomical observations. The Latitude of the pillar was observed by Captain Volontat in 1920 with a prismatic astrolabe, where $\Phi_o = 33°\ 45'\ 34.1548''$ N. An azimuth was obtained at the same pillar with a microscopic theodolite by Captain Volontat, by observing Polaris at elongation. The direction was defined to a pillar constructed at the Ksara Observatory, where $\alpha_o = 28°\ 58'\ 50.8188''$. Longitude was also observed by Volontat at the same observatory, where $\Lambda_o = 35°\ 53'\ 25.26''$ East of Greenwich. (The longitude was then geodetically transferred to the South End of the Bekaa Base.)

In November 1997, Colonel George Massaad, then the Director of Geographic Affairs of the Lebanese Army, sent a photograph to me of the fundamental point at Bekaa South Base. The point is monumented by a stone pyramid that is over 2 m high, it is ~2 m², has an (apparently) bronze tablet describing the significance of the monument, it is straddled by a great iron skeleton target obviously over 4 m high, and the entire structure is enclosed by a formal iron fence! The monument recalls the aviation accident that took the lives of Captains Govin and Renaud of the Geodetic Section of the *Service Géographique de l'Armée* on July 15, 1924, at *Muslimié*, near Aleppo. Shortly after World War II, the U.S. Army Map Service computed the coordinates of the origin of the Bekaa geodetic system on the New Egyptian Geodetic Datum as: $\varphi = 35°\ 45'\ 34.2205''$ N, $\lambda = 35°\ 54'\ 36.4962''$ E. The geodetic coordinates of station Ksara are: $\varphi = 33°\ 49'\ 25.58''$ N, $\lambda = 35°\ 53'\ 25.26''$ E. The Bekaa Valley Datum of 1920 is referenced to the Clarke 1880 (IGN) ellipsoid as previously defined. A check baseline was measured at Bab in Syria, and another astronomical position was observed (Laplace Station), where $\Phi_o = 36°\ 13'\ 48.77''$ N, $\Lambda_o = 37°\ 30'\ 30.195''$ East of Greenwich, and the reference azimuth from Bab to Cheikh Akil signal is $\alpha_o = 179°\ 58'\ 33.152''$. The triangulation was computed on the Clarke 1880 ellipsoid, Levant Zone Grid, Lambert Conical Orthomorphic projection.

The Levant Lambert Zone (1920) is based on the French Army Truncated Cubic formulae where the developed meridional arc is expressed in series form and is truncated at terms higher than the cubic. Furthermore, another idiosyncrasy of the French Army formulae is that the Lambert (fully) Conformal Conic utilizes one of the principal radii of the ellipsoid called the Radius of Curvature in the Plane of the Meridian (ρ). The French Army instead substitutes the Length of the Ellipsoid Normal Terminated by the Semi-Minor Axis (υ). Although not strictly conformal, this is the system that was commonly used by the French in all their colonies (before World War II) that utilized the Lambert Conic projection. The Levant Lambert Zone, also known as the Syrian North Lambert Zone Latitude of Origin (φ_o) = 34° 39′ N, the Central Meridian (λ_o) = 37° 21′ East of Greenwich. The Scale Factor at Origin (m_o) = 0.9996256 (secant conic), and the False Easting and False Northing = 300 km. The scale of the triangulation was governed by the two bases (Bekaa and Bab) which had an internal precision of one part in two million. In the case of the initial azimuth of the Bekaa Base, a large number of observations were made in order to determine the mean azimuth. The maximum range of the observations was 48″ which does not represent good geodetic accuracy. Then azimuth, Latitude, and Longitude were measured at the Bab Base at Aleppo in Syria, and the differences from the geodetic values mathematically carried through the chain from Bekaa are as follows (Astronomic−Geodetic): $\Delta\varphi = -6.318''$, $\Delta\lambda = +10.789''$, $\Delta\alpha = +21.125''$. Thus the SGA decided not to apply a Laplace correction to the azimuths, assuming the 21″ was due to an error at the origin and not over the network. That 21″ error was later verified by the U.S. Army Map Service (AMS), in the 1950s. AMS computed an azimuth between two stations in the area utilizing the geodetic coordinates of the station in terms of the European Datum Mediterranean Loop and the Bekaa Valley Datum values.

The Tripoli Lambert Grid of 1920 origin is based on the North End of the Tripoli Base, where Latitude of Origin (φ_o) = 34° 27′ 04.7″ N, the Central Meridian (λ_o) = 35° 49′ 01.6″ East of Greenwich. The Scale Factor at Origin (m_o) = 1.0 (tangent conic), and the False Easting and False Northing = zero. This quite obscure grid was probably used only for a hydrographic survey in the vicinity of Tripoli, and the South End of the Tripoli Base cartesian coordinates were published by the French as X = +1,257.02 m and Y = −1,197.29 m. Considering the tiny geographic extent of the survey, the Hatt Azimuthal Equidistant or the Roussilhe Oblique Stereographic equations would yield the same transformation results to cartesian coordinates.

In 1922, the *Travaux du Cadastre et d'Amelioration Agricole des Etats, de Syrie, des Alaouites et du Liban sous Mandat Français* established the *SCHÉMA DE LA PROJECTION STÉRÉOGRAPHIQUE* which was based on the Roussilhe Oblique Stéréographique projection. The Latitude of Origin (φ_o) = 34° 12′ N, the Central Meridian (λ_o) = 39° 09′ East of Greenwich, the Scale Factor at Origin (m_o) = 0.9995341 (secant plane), and the False Easting and False Northing = zero. This grid has caused some consternation in the literature because attempts to substitute the fully conformal formulae of Paul D. Thomas' "Conformal Projections" fail to yield correct transformation results. In fact, the Roussilhe (Russell) formulae were developed by the Hydrographer of the French Navy in the late 19th century, and this is a common Grid used on many hydrographic surveys by the French well into the 20th century.

The latest available transformation parameters from the Bekaa Valley Datum of 1920 to the WGS 84 Datum are: $\Delta X = -182.966$ m, $\Delta Y = -14.745$ m, $\Delta Z = -272.936$ m. The mean planimetric error for these parameters is 5 m. Example test point: Bekaa Datum Origin: $\varphi = 33°$ 45′ 34.1548″ N, $\lambda = 35°$ 54′ 37.1188″ E, H = 870.513 m. WGS84 Datum coordinates of the same point: $\varphi = 33°$ 45′ 33.8602″ N, $\lambda = 35°$ 54′ 40.6802″ E, h = 868.64 m. According to NIMA TR 8350.2, transformation parameters from the European Datum 1950 to WGS84 Datum are: $\Delta X = -103$ m, $\Delta Y = -106$ m, $\Delta Z = -141$ m for Lebanon.

According to the World Bank,

> The leading geospatial agency in Lebanon is the Directorate of Geographic Affairs of the Lebanese Army (GAD), which produces topographical mapping and geospatial data for both military and civil purposes. The Directorate's products can be purchased subject to case by case approval. In an attempt to open broader access through a National Spatial Data Infrastructure (NSDI) approach, the Ministry of Administrative Reforms (OMSAR) implemented a GIS portal few years ago and created a NSDI regulatory framework and coordination mechanism for access to geospatial data, but the implementation failed due to lack of funding or sustainable arrangements for coordination, access, sharing and dissemination of data. However, there was progress in standardization and for example the Lebanon-Syria coordinate reference system is used for all mapping in Lebanon and thus, the most important key standard for NSDI is being applied. (The World Bank Land Administration System Modernization Project (P159692))

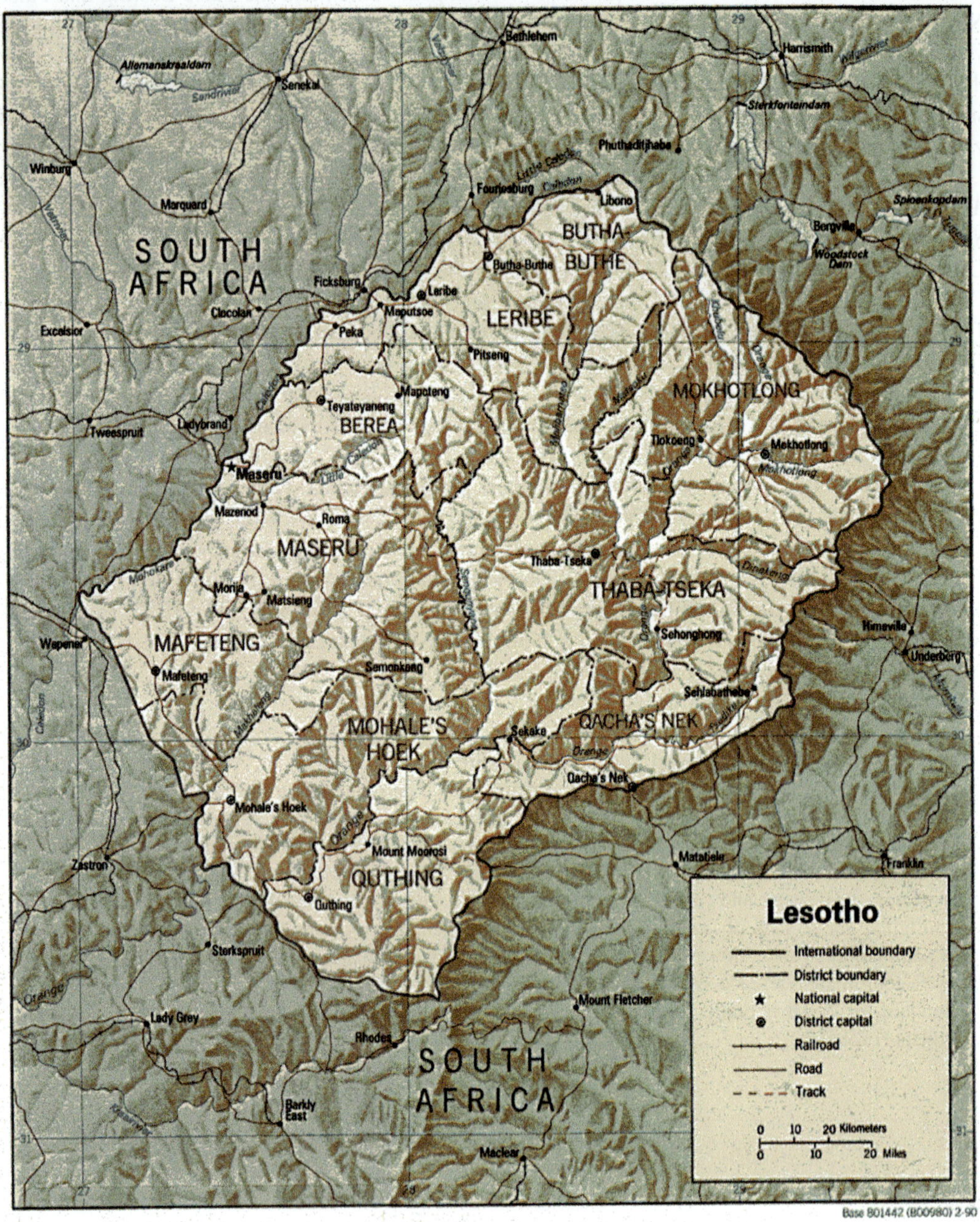

Base 801442 (B00980) 2-90

Kingdom of Lesotho

The Kingdom of Lesotho is completely surrounded by the Republic of South Africa. Its boundaries run with those of KwaZulu-Natal to the east, Eastern Cape to the south, and the Free State to the north and west. It lies between latitudes 28° and 31° South and longitudes 27° and 30° east. It covers an area of ~30,300 km² (*slightly smaller than Maryland – Ed.*), of which about one-quarter in the west is lowland country, varying in height above sea level from 1,500 to 1,600 m, the remaining three-quarters being highlands, rising to a height of 3,482 m at Thabana-Ntlenyana in the Maluti Range, which forms the eastern boundary with KwaZulu-Natal. The mountain ranges run from north to south and those in the central area, the Maluti, are spurs of the main Drakensberg, which they join in the north, forming a high plateau varying in height from 2,700 to 3,400 m. It is in this area where two of the largest rivers in Southern Africa, the Orange (Senqu) and the Tugela, and tributaries of the Caledon, have their source. This phenomenon has caused Lesotho to be called the "sponge" of Southern Africa.

"People described as Sotho have lived in Southern Africa since at least the 10th century *A.D.*, moving throughout the high veldt of the region. By the 16th century, the Sotho people had arrived in the area known now as Lesotho, marrying, and intermingling with the Khoisan people, and forming small chiefdoms. Extensive trade links were established between the groups, as well as with outside people. Grain and hides, for example, were traded for iron from the Transvaal area.

By the early 19th century, white traders were on the scene, exchanging their ever-reliable beads for cattle. In came the Voortrekkers (Boer pioneers), and suddenly the people of the area, now called Basotholand, had to recognize that constant expansion for 300 years was placing extreme pressure on the environment. At the same time, consolidation and expansion of the Zulu state was causing a chain-reaction of violence throughout southern Africa. Survival by the loosely organized southern Sotho society is attributed to the strong leadership of Moshoeshoe the Great.

The Basotho emerged as a people around 1820 when Moshoeshoe the Great gathered the tribes scattered by Zulu raids and established a stronghold at Butha-Buthe, and later on the mountain of Thaba-Bosiu, about 30 km from what is now Maseru. By 1840, his people numbered about 40,000. Worried by the Boers, Moshoeshoe the Great enlisted British support, but the British were equally worried about Moshoeshoe, and launched an unsuccessful attack on him. When the English left defeated, the Boers pressed their claims to the land, leading to the 1858 Free State-Basotho War (won by Moshoeshoe) and another in 1865 (in which Moshoeshoe lost much of the western lowlands). In 1868, under increasing pressure from the Boers, Moshoeshoe placed the region under the protection of the British government, but as part of the deal, lost even more land to the Boers.

The British signed over control to the Cape Colony in 1871 – a year after the death of Moshoeshoe the Great – and the new government wasted no time reducing the power of the chiefs. After another war in 1880, the land was again shuttled back to British control. This turned out to be a lucky break for the people of Lesotho. Had they remained part of the Cape Colony, they would have become part of the newly-formed Union of South Africa and, under apartheid, would have become a homeland" (*Lonely Planet*, 2008).

"The original Lesotho consisted of the high plains of the Mohokare (Caledon) valley and adjacent areas. Modern Lesotho has lost much of the western part of this land but has gained the high mountain ranges in the east, known as the Maloti. The present boundaries of Lesotho follow in part a series of rivers, the Tele, the Senqu, the Makhaleng and the Mohokare. Between the Makhaleng and Mohokare, the south-western boundary follows a beaconed boundary fence, while between the sources of the Mohokare and Tele, the long eastern and southern boundaries follow a high mountain watershed. This section of the boundary is for much of its distance the Continental Divide between the Atlantic and Indian Oceans, and it is seldom far from dramatic escarpment cliffs which make access to Lesotho on this side extremely difficult.

The soils in the mountain area are of basaltic origin, and those in the lowlands are derived mainly from the underlying cave sandstone. In the lowlands, the soil has been cropped continuously for upwards of 100 years. Because of the absence of fuel, practically all cattle manure is burnt, so that little or no organic matter is returned to the land. Thus, with increasing population, both human and livestock, excessive demands have been made on the soil which has lost its structure and has become seriously eroded. The soils in the mountains have been brought into cultivation comparatively recently and are rich, but shallow. With uncontrolled grazing, the areas above the arable land, in many places, became denuded of the grass cover, and the rush of surface water caused serious gully erosion on the arable land situated below. Several measures have been, and are being, taken to control this erosion and restore and preserve the grass cover" (*Kingdom of Lesotho*).

"In the early days of the white settlement of the Cape of South Africa, surveys of land were largely dependent on the isolated efforts of surveyors who established coordinates, each limited to the area covered by a particular survey; so that, with the serious discrepancies arising from the absence of a common measure and a common orientation in contiguous surveys, their correlation for the purpose of compiling even approximately correct maps was a difficult if not impossible task. Eventually, however, after about 1883, part of the Cape Colony and Natal was provided with a skeleton of geodetic survey undertaken by Colonel Sir William Morris – a survey which owed its conception and completion to the energetic and persistent efforts of Sir David Gill. This fundamental triangulation made it possible for the first efforts to be made in these two provinces towards a unification of all survey work, by providing a precise and rigid system of secondary triangulation. The positions of the stations of this triangulation were defined in terms of the geographical coordinates of latitude and longitude, and from these were computed the corresponding rectangular spheroidal coordinates. Their use, however, in precise computations involves an amount of work which private (practicing) surveyors could not be asked to undertake, and therefore, when used in connection with farm (cadastral) survey work, they were regarded as plane coordinates, and sued as such. Consequently, the spheroidal systems had to be limited within narrow boundaries so that the neglect of the curvature would not impair their value for controlling property surveys. The use of these spheroidal coordinates as though they are plane coordinates is what is commonly called the Cassini-Soldner projection." The readers will recall that the Cassini-Soldner projection is one of the aphylactics; it is neither equal-area (authalic) nor conformal (orthomorphic) and is a nightmare to use for computing survey adjustments. (*See explanation at Hong Kong – Ed.*)

"The geodetic framework of the Cape Colony was completed in 1893, and in the years 1903–1906 was extended through the Orange Free State and Transvaal. The final results of this triangulation were published in the form of geographical and Cassini-Soldner coordinates; but the inherent disadvantages of these Cassini coordinates for purposes of cadastral survey decided the late Dr. van der Sterr to abandon all the existing coordinate systems upon his appointment as Director of Trigonometrical Survey in 1919. He then established a series of parallel systems of conform coordinates in which each system is limited to a width of two degrees of longitude and the several origins of the coordinates are defined by the point of intersection of the Equator with the odd numbered meridians" (*Trigonometrical Survey*, D. R. Hendrikz, Union of South Africa 1955).

In December of 1989, Mr. S. Mosisili, Chief Surveyor for the Lesotho Department of Land, Surveys and Physical Planning sent a letter to me regarding the mapping grid used in Lesotho. Repeated later in a 1995 *Hoofdirektoraat: Opmetings en Grondingligting* (Chief Directorate: Surveys and Land Information) letter to a protégé of mine, Mr. Robert M. Frost, P.L.S., "The coordinate system used in ... Lesotho is the Gauss Conform Transverse Mercator projection which uses 2° longitude belts based on all the odd meridians (... 27°, 29°, ... etc.). No arbitrary scale factors or false origins are applied to the coordinates: X is measured positive southwards from the equator and Y positive westwards from the nearest odd meridian. The unit of measurement since the 1970s is the international meter. The ellipsoid used in Lesotho is the Modified Clarke 1880 where

$a = 6,378,249.145$ m, and $b = 6,356,514.967$ m. This ellipsoid, together with the coordinates of the initial point (*i.e.,* origin) of the geodetic survey, define the Cape or South African Datum. The initial point of the Cape Datum is Buffelsfontein near the coastal city of Port Elizabeth. The coordinates of Buffelsfontein are: $\Phi_o = 33° 59' 32.000''$ S, $\Lambda_o = 25° 30' 44.622''$ E, $h = 280.1$ m."

The *seminal* paper by J. Rems and Dr. Charles L. Merry in *Survey Review,* **30**, 236 (April 1990) entitled, *Datum Transformation Parameters in South Africa,* goes into exquisite detail of the individual characteristics of each of the geodetic networks of the nations of South Africa. With regard to "(b) Lesotho: It is somewhat meaningless in a country of this small size to determine a full seven-parameter transformation. Even the scale factor is hopelessly correlated with the translations. The only purpose of using more than a simple three-parameter transformation would be to model systematic distortions in the geodetic network. There is little improvement in σ_o in going from a three- to a seven-parameter transformation, so this argument falls away in the case of Lesotho. Again, it appears that the three-parameter model would be adequate for all transformations." Parameters derived by Rems & Merry for Lesotho from Cape Datum to WGS 84 Datum are: $\Delta X = -136.0$ m ± 0.4 m, $\Delta Y = -105.5$ m ± 0.4 m, $\Delta Z = -291.1$ m ± 0.4 m. Details on recent developments in research should be directed to Professor Merry at the University of Cape Town.

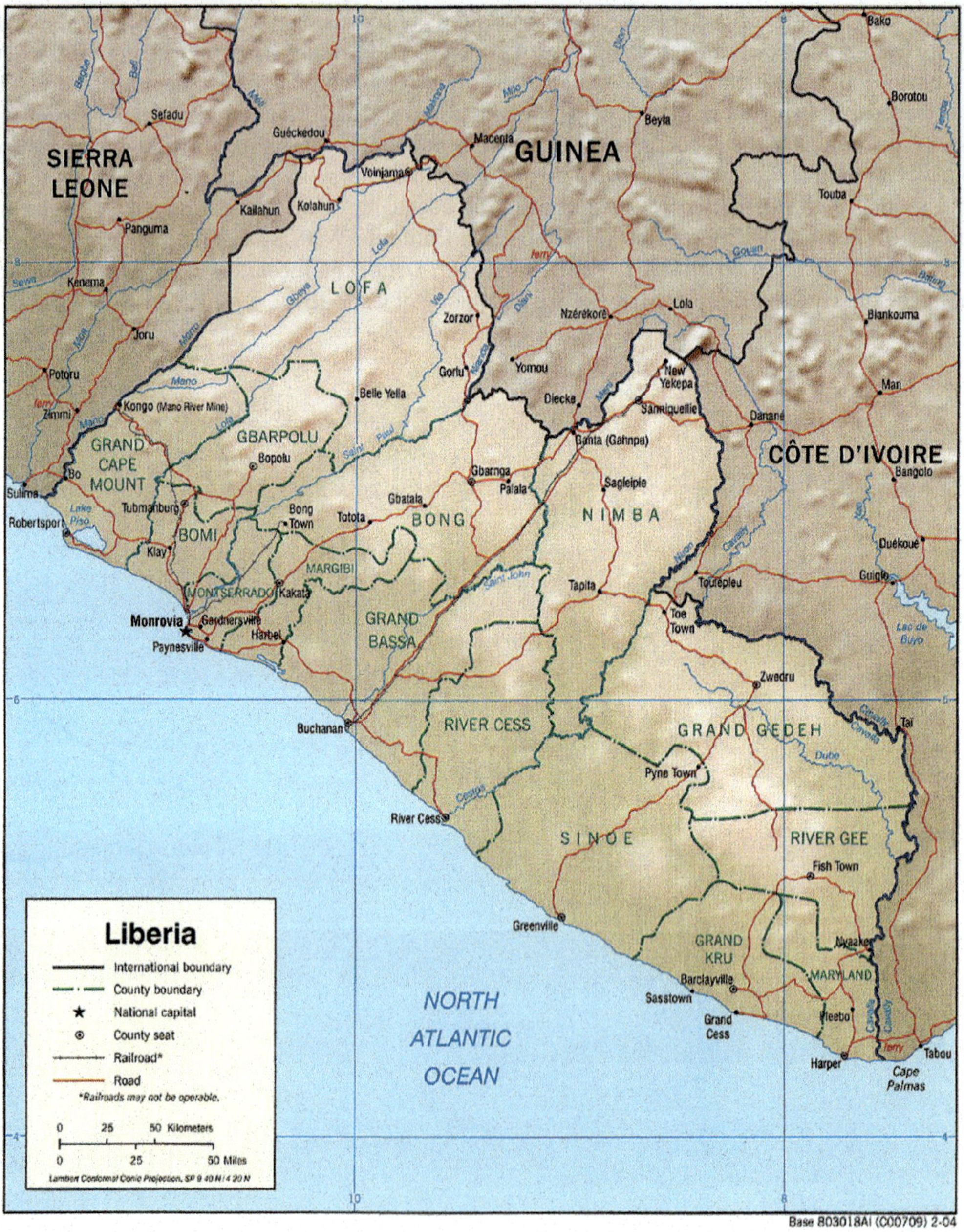

Base 803018AI (C00709) 2-04

Republic of Liberia

Established as a republic in 1847, Liberia has been subject to political turmoil starting in 1980 and has only recently begun the process of rebuilding the social and economic structure. Slightly larger than Tennessee, Liberia is bordered by Guinea (563 km), *Côte d'Ivoire* (716 km), and Sierra Leone (306 km). With a coastline if 579 km, Liberia is mostly flat to rolling coastal plains rising to rolling plateau and low mountains in the northeast. The lowest point is the Atlantic Ocean (0 m), and the highest point is Mount Wuteve (1,380 m) (*World Factbook*, 2010).

"It is generally believed that before 1822 there were 16 different tribes living in what was called the 'Pepper Coast', 'Grain Coast' or 'Malaguetta Coast'. One of them was exclusively living in what is nowadays Liberia: The Bassa, the other 15 tribes were dispersed in the region. According to the traditions of many African tribes their ancestors were Pygmies, or persons of small size, and memories of them still live on in numerous stories and legends. Although no trace of their existence was ever discovered in West Africa, they are well-known to the peoples of this sub region. In Liberia they are called 'Jinna.' No recorded history can prove their existence, but they still play an important role in the oral history and the religious life of some of Liberia's tribes. When the Golas, who are supposed to be the oldest of the Liberian tribes, travelled from the interior of Central Africa to this West African region they reportedly met these small-sized peoples, who were bushmen and who *'dwelt in caves and the hollows of large trees, and lived on fruits and roots of wild trees'*, according to Liberian historian Abayomi Karnga.

"A second group of peoples is reported to have arrived in the region about 6,000 B.C. Though their origin is not very clear they most likely came from the Western Sudan. These newly arrived people defeated the Golas and other tribes such as the Kissi and established an empire under the leadership of King Kumba, after whom they were called. The Kumbas comprised distinct groups which developed into different tribes after the death of their leader; the Kpelle, the Loma, the Gbande, the Mende, and the Mano, all belonging to the same linguistic group. They were chiefly agriculturalists but also developed arts such as pottery, weaving, and basket making. Their blacksmiths were able to make spears, arrowheads, hoes, knives, rings, and iron rods. These iron rods were used as a medium of exchange.

"The third group of peoples who arrived and settled in the region which is now known as Liberia migrated to this part of West Africa quite recently. They were the Kru, Bassa, Dei, Mamba, and Grebo tribes. They came from what is now the Republic of Ivory Coast. Population pressure – due to the mass emigration of tribes from the Western Sudan where the mediaeval empires had declined after their conquest by the Moroccan army – had resulted in tribal wars. The Kru arrived in the early sixteenth century. They came by sea, as did – later – a part of the Grebo. Those Grebos who took the sea-route were later called 'seaside Grebos' in order to distinguish them from their kinsmen who decided to travel by land, the safer way. Those who braved the dangerous waves still feel superior to these so-called 'bush-Grebos'. All the peoples of this group belong to the same linguistic group.

"The last group of tribes to arrive from 'over land' was the Mandingo-group, comprising the Vai and the Mandingo tribes. The Vai also migrated to the West African central region in the 16th century and had probably the same motivation as the tribes of the third group. They crossed the western part of the actual republic of Liberia, clashed with the Gola whom they subsequently defeated, and – later – moved to the coast where they settled. The Vai form the first tribe of this region which was Muslim, unlike the tribes previously mentioned which were all animists. It was one of the few tribes of Black Africa who developed its *own script*. About the 17th century the Mandingos began to arrive in Liberia. They were Muslims too. They too originated from the Western Sudan. They left this region after the Empire of Mali – of which they formed a part – was considerably reduced by the Emperor of Gao, Askia Mohammed, in the 16th century.

"Documents reveal that the first white men who landed on this part of the West African coast were probably Hanno the Carthaginian and his sailors; in the year 520 *B.C.* Hanno the Carthaginian may have reached the coast near Cape Mount, where he encountered the Golas. Trade started, but the contacts between the two races were limited to only a few trade visits. It was not until the 14th century that further and more frequent contacts were established. About 1364 the Normans settled (temporarily) at a few places on the coast of Liberia and started trading with the coastal tribes from whom they bought ivory, pepper, gold, and camwood. The Portuguese also frequented the Liberian coast as from this period and soon even controlled the trade. In fact, they had a monopoly for over a century, before they were replaced by other European maritime powers (France, England, Holland). From the 15th century onwards the Portuguese, Spanish, Dutch, English and French sailors and traders became common and accepted visitors of the West African coast. Their influence on the history of the area was considerable and their impact on the lives of the peoples of the coast quite distinct. The Portuguese *e.g.*, named regions, mountains, and rivers, and some of these names are still in use. In the Liberian area the Portuguese gave names such as (from west to east) Gallinhas River, St. Paul's River, Mesurado River, St. John River, Cestos River, Sanguin River and Cavalla River, and named the promontories Cabo do Monte (at present Cape Mount), Cabo Mesurado (Cape Mesurado) and Cabo das Palmas (Cape Palmas).

"The coast between Sierra Leone and Ivory Coast was called 'Malagueta Coast' or 'Pepper Coast' after one of its main products, the Malagueta pepper. The English and the Dutch preferred the name 'Grains of Paradise', referring to the same product. Later they abbreviated this name, and it became the 'Grain Coast'. The names of the coastal regions to the east of Cape Palmas were also based on their main commercial products: the Ivory Coast, the Gold Coast and (later) the Slave Coast. The spices, gold, and ivory, sought by the Europeans were exchanged for textiles, alcoholic beverages, general merchandise and, later, when this trade degenerated into the barbarous slave trade, horses and weapons. The trade in slaves soon ousted the more common trade and by the end of the sixteenth century all European powers (of that period) were engaged in this historical and inhumane commerce, an activity that resulted in the diaspora of the Black race and the loss for Africa of an estimated 20 million people. The damage for Africa was even greater than the mere loss of a large number of its population because the captured sons and daughters of its soil were its most able and most productive workers. When the slave trade was over Africa was left with hostilities between tribes, with a disrupted social and family life and with famines. The resulting weakness and dissensions greatly facilitated the imposition of foreign rule. Historical documents leave no doubt that the people of the coastal area had reached an admirable standard of living. They made many products which were of a higher quality than those produced in pre-industrial Western European countries – as cited by Sir Harry Johnston in his famous two volume work on Liberia (1906).

"The traditions of the Norman traders who visited Liberia in the fourteenth century and the authentic records of the Portuguese commerce with that country before 1460 and 1560, reveal a condition of civilization and well-being amongst the untutored natives which is somewhat in contrast to what one finds in the same coast at the present day; still more in contrast with the condition of the Liberian coast lines in the early part of the nineteenth century, suggesting that the rapacity of the Europeans, combined with the slave trade, did much to brutalize and impoverish the coastal tribes of Liberia during the two hundred years between 1670 and 1870. They seem to have been well furnished with cattle, (in Northern, perhaps not in Southern Liberia), with sheep, goats, and fowl, to have carried on a good deal of agriculture, and not to have been such complete savages as were the natives of the still little-known parts of Portuguese Guinea, or the people of the Ivory Coast, who were wild cannibals. (...) Having cast a glance at the principal commercial products of these countries when they were first discovered by Europeans, it may be interesting to note the trade goods which Europe was able to offer to the Blacks from the fourteenth to the seventeenth century.

To begin with a negative statement, there were no cotton goods, no calicoes in the holds of these vessels such as there would be nowadays. Strange to say, it was the natives of the Gambia and other rivers of Northern Guinea, and Cape Mount in Liberia that impressed the Europeans with the excellence of their cotton fabrics, and actually sent some cotton goods to Portugal. (…) It is possible that no cotton goods were exported from Europe to West Africa till the end of the seventeenth and the beginning of the eighteenth centuries. Since that time the cotton goods of Lancashire, of Germany, and of Barcelona have almost killed the local industries of weaving and dyeing. (…) As early as the time of Ca'da Mosto (middle of the fifteenth century) cannon were taken on the ship and gunpowder was fired to astonish and frighten the Negroes; but there seems to have been no sale of gunpowder till the close of the fifteenth century.

"The Europeans were not only responsible for the naming of places, the exchange of products, the introduction of fire-arms and the carrying off of millions of Africans, but also for the introduction and spreading of hitherto unknown diseases, such as dysentery, syphilis and certain parasites. In the region which is now called Liberia the (slave) trade thus contributed to the impoverishment of and the hostilities between tribes. Up till the present day the inter-tribal relations are affected by the events of this period. The Golas, Krus, Kpelles and Kissis were notorious slave traders conniving with unscrupulous Europeans who looted the coastal areas. Besides this, the northern tribes of the Mano and the Gio were feared because of their cannibalism, a practice which was also not uncommon among the Greboes and the Krus. It was in this environment of slave trade, suspicion, fear and open discord and hostilities that the first colonists arrived, aboard an American ship, in 1820. They came from the U.S.A. where their African ancestors had been sold to white masters. With the arrival of these black and colored people an experiment in 'black colonialism' started" (*van der Kraaij, The Open Door Policy of Liberia – An Economic History of Modern Liberia, Bremen, 1983, 2 volumes, pp. 1–5*).

"In 1926, the then Firestone Tire & Rubber Company leased 1,600 square miles of jungle in Liberia, West Africa, with the goal of producing its own rubber. Selected then for its location, soil and climatic conditions, Liberia is now home to the world's largest single natural rubber operation. Today, Firestone Liberia operates on a much smaller area of land with close to 8 million rubber trees planted on 200 square miles at its Harbel location (named after Firestone founder Harvey Firestone and his wife Idabelle)" (*Firestone Rubber Co., 2010*).

Thanks to John W. Hager in December 2010, there was a Firestone Datum, but that may just be another name for Jidetaabo (*NGA Geonames Server*), where Jidetaabo Astronomic Station is at: $\Phi_o = 04° 34' 40.33'' N \pm 3.7''$ *probable error*, $\Lambda_o = -07° 38' 55.74''$ West of Greenwich $\pm 2.16''$ *probable error*, and a reference azimuth to an unlisted point is $\alpha_o = 359° 42' 41.5''$, with the reference ellipsoid likely the Clarke 1866, where $a = 6,378,206.4$ m, and $b = 6,356,583.6$ m. *Google Earth*™ shows this location to be in the vicinity of a road intersection, and the *NGA Geonames Server* shows this location to be in the middle of a number of Firestone properties. With the reference azimuth being mere minutes of arc from a cardinal direction, there are a number of properties in the vicinity of Jidetaabo that resemble the Rectangular Survey of the Public Lands in the United States – not likely a coincidence.

Furthermore, thanks to Dave Doyle of the U.S. National Geodetic Survey (*Personal communication, December 2010*), a 1934 survey of the Firestone Plantations Company property at Cape Palmas Liberia by the U.S. Coast & Geodetic Survey lists the point of beginning at a concrete monument (corner No. 1), where $\varphi = 4° 34' 49.221'' N$, $\lambda = -07° 39' 21.895'' W$. Quoting from the U.S.C. & G.S. report, "The Astronomic datum determined at Gedetarbo (*sic*) by meridian telescope is the one upon which the initial position N. Cor. (Corner No. 1) is based. The position North Cor. (Corner No. 1) given in this description has been determined by second order triangulation, which depends upon the position of Astronomic station and azimuth from East Base to West Base. It may be defined in terms of the position of East Base as follows: $\Phi_o = 04° 34' 40.187'' N$, $\Lambda_o = -07° 38' 55.739'' W$, and azimuth to West Base is $\alpha_o = 86° 35' 41.5''$ *(C.F. Maynard, 3/12/31)*."

(*Hager, op. cit.*) continued: "Liberia 1964 at Robertsfield (*yes, one word - Ed.*) Astro, $\Phi_o = 6° 13' 53.02'' \pm 0.07''$ N, $\Lambda_o = 10° 21' 35.44'' \pm 0.08''$ W, $\alpha_o = 195° 10' 10.57'' \pm 0.14''$ to Robertsfield Astro Azimuth Mark from south, Clarke 1880, elevation $(H_o) = 8.2331$ meters.

Liberian Government (no code assigned). Possibly this is Jidetaabo 1928, correct time frame.

C. F. Maynard identified Du Plantation as being on Liberian Government datum, but I have no position. Further on C. F. Maynard, he and J. A. Bond published *The Triangulation of the Philippine Islands*, Vol. I, Department of Commerce and Communications, Bureau of Coast and Geodetic Surveys, Manila, 1927. He was evidently in the Philippines in 1927 and then in Liberia in 1928 or 1929. That is not the U. S. Coast and Geodetic Survey but the Philippine.

"Zigida at Zigida: $\Phi_o = 8° 02' 01.33''$ N, $\Lambda_o = 9° 34' 09.07''$ W, Clarke 1866. The Clarke 1866 is logical in that those surveys were done by Americans for an American company. Clarke 1880 for Liberia 1964 is in line with the recommendations of the I.G.G.U. for use in Africa.

"The U.S. Coast and Geodetic Survey created the Hotine RSO for a map series at (maybe) 1:250,000 or even smaller. I can't remember if it was a photo mosaic or a line map, but the photo coverage was incomplete. The grid tables are dated 1952 so it is possible that the Gedetarbo 1928 datum was used. But the ellipsoids don't agree, not that that would be important at that map scale. I have no idea why they chose the International ellipsoid. In the 1964 era, a SHORAN trilateration net was established based on the Liberia 1964 datum. Photography was flown with the photo nadir points determined by SHORAN, and 1:250,000 maps of the entire country and 1:50,000 maps of selected areas compiled."

For the Liberian Rectified Skew Orthomorphic projection, the Latitude of Origin $(\varphi_o) = 6° 35'$ N, Longitude of Origin $(\lambda_o) = 9° 25'$ W, Scale Factor at Origin $(m_o) = 0.99992$, False Northing $= 0$, False Easting $= 1,500,000$ m, Azimuth of Initial Line at Origin $(\gamma_o) = 126° 21' 47.451''$ (from South), $N = 0.6x + 0.8y$, $E = -0.8x + 0.6y + 1,500,000$; $x = 0.6N - 0.8E + 1,200,000$, $y = 0.8N + 0.6E - 900,000$ (*Personal communication, John W. Hager, 15 August 2005*).

A partial trig list furnished by Hager included a couple of undated points that appear to be in the near vicinity of the Liberian Government Datum of 1929 origin where at W. Base, $\varphi = 6° 27' 55.113''$ N, $\lambda = 10° 27' 04.868''$ W, and at E. Base, $\varphi = 6° 27' 57.845''$ N, $\lambda = 10° 26' 21.903''$ W.

(*Doyle, op. cit.*) Liberia Project No. 13-017-448, 1953 Annual Report by George E. Morris, Jr., Cdr., U.S.C.&G.S., "The project is divided into two principal tasks. The first is to provide the field inspection on the aerial photographic and airborne magnetometer surveys of Liberia under a Technical Cooperation Project. The second one is to organize and operate the Liberian Cartographic Service for surveying and mapping in Liberia. The second task includes the training of Liberian citizens in field and office procedures."

According to **TR 8350.2, from** Liberia 1964 Datum **to** WGS 84 Datum, (Clarke 1880), $\Delta X = -90$ m, ± 15 m, $\Delta Y = +40$ m, ± 15 m, $\Delta X = +88$ m, ± 15 m, and is based on collocation at four points as of 1987.

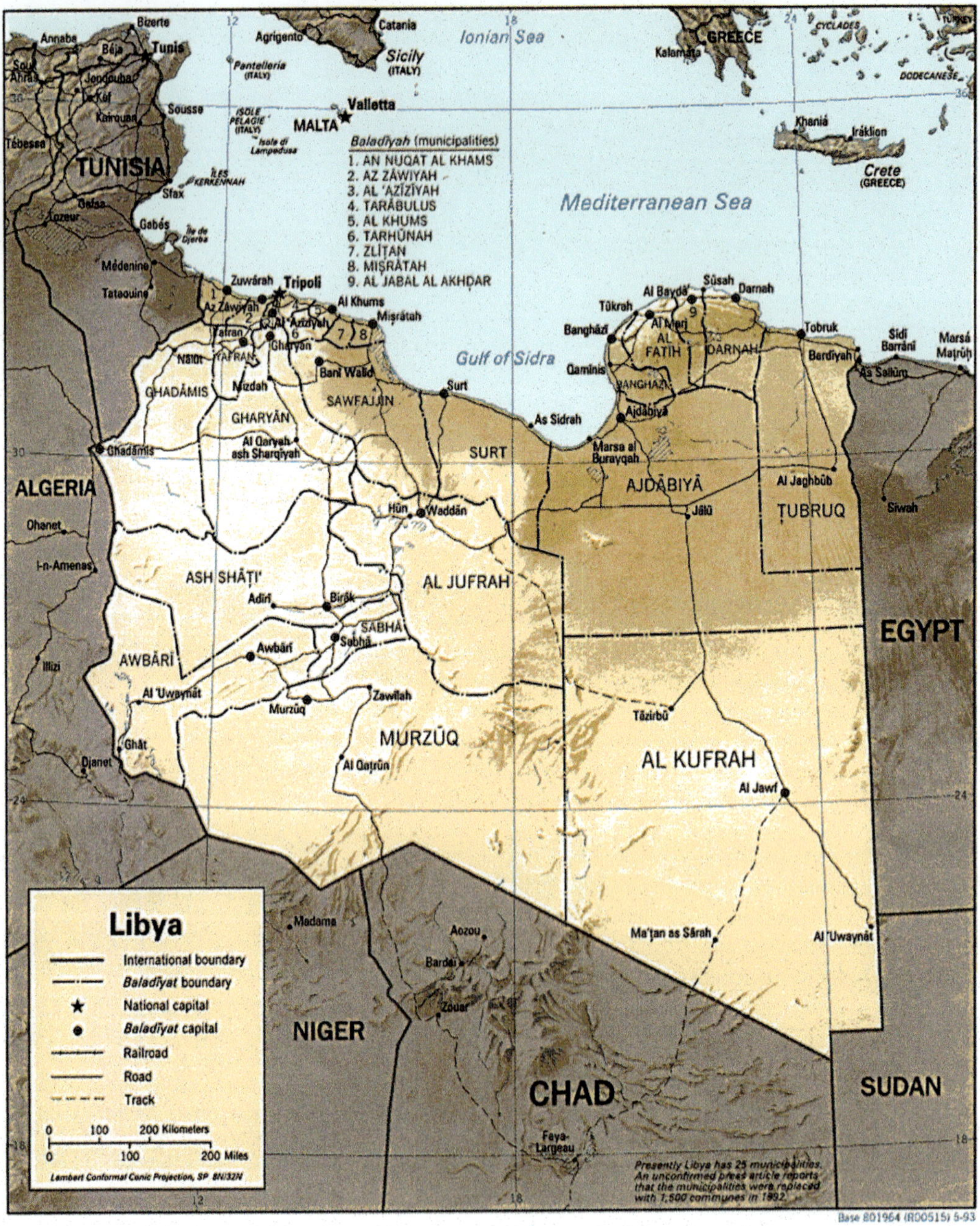

Base 801964 (R00515) 5-93

Great Socialist People's Libyan Arab Jamahiriya

The climate of Libya is Mediterranean along the coast and is extreme desert in the interior. The lowest point is Sabkhat Ghuzayyil (–47 m), and the highest is Bikku Bitti (2,267 m). Libya borders Algeria, 982 km, Chad, 1,055 km, Egypt, 1,115 km, Niger, 354 km, Sudan, 383 km, and Tunisia, 459 km. Slightly larger than Alaska, the country's northern border is the Mediterranean Sea with a 1,770 km coastline.

"Until Libya achieved independence in 1951, its history was essentially that of tribes, regions, and cities, and of the empires of which it was a part. Derived from the name by which a single Berber tribe was known to the ancient Egyptians, the name *Libya* was subsequently applied by the Greeks to most of North Africa and the term *Libyan* to all of its Berber inhabitants. Although ancient in origin, these names were not used to designate the specific territory of modern Libya and its people until the twentieth century, nor indeed was the whole area formed into a coherent political unit until then. Hence, despite the long and distinct histories of its regions, modern Libya must be viewed as a new country still developing national consciousness and institutions" (*Library of Congress Country Studies*).

Its territorial sea claim is 12 nautical miles, but the unique claim is that approx. one-third of its coastline comprises the Gulf of Sirte and they claim a closing line defined by 32° 30′ North! That is contrary to the *International Law of the Seas* definition of a bay – from which a territorial sea boundary is then measured from a line connecting the headlands of such a bay. A true bay must have headlands that are separated by no more than one-third of the depth of the bay. The Gulf of Sirte is truly a gulf and is not a bay. The only way a nation can claim a bay is through the exception of the Historical Bay Rule, but the international community must recognize that for at least 100 years! Some such Historical Bays do exist throughout the world as a result of that accepted convention. Since Libya has claimed such a closing line for the Gulf of Sirte, the United States has sent its Navy into the claimed territorial waters to upset the legal prescription of 100 years. More than once Libya has sent fighter planes to defend its claim and armed conflicts have occurred with the U.S. Navy. Spectacular when it happens, the purpose is to maintain the concept of *Freedom of the Seas* as promulgated by the United States and the rest of the international community of nations. Once upset by open and notorious occupation, the prescription of 100 years then restarts from zero.

Libya became an independent nation in 1951 and because no government organization existed that could accomplish mapping on a national scale, local native mapping was nonexistent until the Survey Department of Libya (SDL) was established in 1968. The Italian *Instituto Geografico Militare* (IGN) (Military Geographic Institute) performed the official topographic mapping prior to World War II. It established a second- and third-order geodetic network of control that covers all of the Libyan coastal area except for part of the Gulf of Sirte, where it used a fourth-order network. IGM produced topographic maps at scales ranging from 1:25,000 to 1:100,000, but the areas covered are limited to the coast and a few important oases of the interior. The Italian *Comando Superiore Forze Armate Africa Settentrionale Ufficio Topocartografico* (High Command, Armed Forces, North Africa, Topo cartographic Office) compiled a 1:1,000,000 scale map that was the first "large scale" map of the entire country. The *Carta Topografica della Cirenaica* was produced at the scale of 1:50,000 by IGM from 1920 to 1936. These monochrome sheets cover the coastal area between Bengasi and Derna. They were based on 1920–1923 triangulation and planetable surveys. Two of the sheets were revised in 1936 and were standard topographic maps intended for civilian use. A marginal note on each sheet states the graphical shift necessary to adjust the sheets to the survey of 1933–1934. The *Carta della Tripolitania* was produced at the scale of 1:50,000 by IGM in 1933–1940, and covers scattered areas of northwest Libya. The *Carta della Libia* was produced at the scale of 1:100,000 by the IGM from 1915 to 1938. All of these maps produced by the IGM have no grid. However, the U.S. Army Map Service reported two grid systems created for Libya by the British Geographical Section, General Staff (GSGS) during the 1940s that are referenced on the Clarke 1880 ellipsoid, where $a = 6,378,249.145$ m and $1/f = 293.4660208$. The Libya Zone is based on the Lambert Conical Orthomorphic projection where the Central Meridian $(\lambda_o) = 18°$ East of Greenwich, the Latitude of Origin $(\varphi_o) = 31°$ North, the Scale Factor at Origin, $(m_o) = 0.99938949$, the False Easting $= 1,000,000$ m, and the False Northing $= 550,000$ m. The South Libya Zone is also based on the Lambert Conical Orthomorphic projection where the Central Meridian $(\lambda_o) = 18°$ East of Greenwich, the Latitude of Origin $(\varphi_o) = 23°$ North, the Scale Factor at Origin, $(m_o) = 0.99907$, the False Easting $= 600,000$ m, and the False Northing $= 800,000$ m. An example South Libya Zone computational test point is: $\varphi = 21° \ 13' \ 43.397''$ N, $\lambda = 12° \ 19' \ 13.071''$ E, $X = 210,785.725$ m, $Y = 415,440.430$ m. Note that both of these zones are truly secant Lambert Conformal Conic projections with two standard parallels, but the British Convention of defining parameters is used. There are many commercially available software packages that incorrectly classify this sort of thing as a *single parallel* Lambert zone. Do not be misled by projection nomenclature blunders misappropriated by ignorant computer programmers – instead, use the *Manual of Photogrammetry*, 5th or 6th edition as the definitive reference for the mathematics and the proper terminology.

Since North Africa has the highest density of geodetic enigmas, puzzles, rumors, *etc.*, in January of 2006 I turned to Mr. John W. Hager for some historical insights of Libyan geodetic history: "The Italians did the pre-WWII surveys and it was done on the Bessel (*1841 – Ed.*) ellipsoid, not (on) the International. Misurata Marina Datum (1930) was based on an Astronomic station at $\Phi_o = 32° \ 22' \ 20.35''$ N, $\Lambda_o = 15° \ 12' \ 45.86''$ E, initial azimuth $(\alpha_o) = 180° \ 00' \ 38.31''$ to an unknown station and unknown whether measured from north or south. Bessel ellipsoid. I do not find anything for Sirte 1930 and would suggest that this may be Brega (1930) at Magnaghi's Astronomical Pier $\Phi_o = 30° \ 25' \ 00.21''$ N, $\Lambda_o = 19° \ 34' \ 18.60''$ E. The Italians published their data in a series of pamphlets. The U.S. Army Map Service probably adjusted these on European 1950 but there is some question as to how good the adjustment was. There was a grand scheme to adjust a loop around the Mediterranean. The problem was that there is one missing triangulation figure straddling the Libya – Egypt border. In the west, the triangulation observed by the French from Spain through Morocco, Algeria, and Tunisia is good. The Tunisian triangulation is connected to Italy through Pantelleria and Sicily. It was adjusted by IGN (French *Institut Géographique National – Ed.*) under contract from AMS in, I would guess, the 1950s. Thus, Libya is cantilevered from Tunisia and the adjustment is weak. In 1956 or 1957 AMS contracted for SHORAN-controlled photography used to produce 1:250,000 scale maps of Libya on European Datum 1950. The error in the positioning of the photography could have been up to about 62 m. I never had any indication that the actual error was anywhere near that."

During a 1970s Annual Meeting of the American Congress on Surveying and Mapping, a geodesist presented a paper on a new grid system that had been implemented in Libya. Termed the "Libyan Grid System," the projection was defined as a Gauss-Krüger Transverse Mercator with 2° "bands" where the first band termed zone #4 has a Central Meridian $(\lambda_o) = 7°$ East of Greenwich and ends with zone #14 that has a Central Meridian $(\lambda_o) = 27°$ East of Greenwich. The Scale Factor at Origin, $(m_o) = 0.9999$, and the False Easting of each band $= 200,000$ m. Of course, the Latitude of Origin is defined to be the Equator, and no False Northing was specified. An example Libyan Grid System computational test point for zone 6 is: $\varphi = 23° \ 00'$ N, $\lambda = 10° \ 25'$ E, $X = 140,197.96$ m, $Y = 2,544,435.31$ m, but I don't remember if this was included in the paper, or it was something that I personally cooked up.

The Association of Petroleum Surveying & Geomatics (APSG) is a very collegial group of petroleum-oriented geodetic surveyors that publishes a (free) database of parameters and transformations through the EPSG website. Intended to service the needs of "Oil Patch Surveyors" worldwide, it is a very nice collection of tidbits and useful parameters. A 2006 presentation of Libyan geodetic data and transformations included a couple of transforms that are pertinent to the current topic. Entitled "Esso (International 1924) to WGS72, $\Delta X = -69$ m, $\Delta Y = -91$ m, $\Delta Z = -147$ m, but no accuracy estimates are offered, nor the number of points used to derive this "mystery" transformation. Another datum relation listed is the "European Libyan Datum of 1979" (ELD79) to WGS84 which is reportedly referenced to the International 1929 ellipsoid is, where $\Delta X = -69$ m, $\Delta Y = -96$ m, $\Delta Z = -152$ m. Some parameter sets show a precision to the millimeter, but such is not worth publishing without a concomitant indicator of expected accuracy". The source referenced is EPSG.

Attempts over the years to request information from the Libyan government regarding geodetic transformation relations have never been answered, so how close these parameters conform to actual conditions is unknown. Thanks again to the APSG and to Mr. John W. Hager.

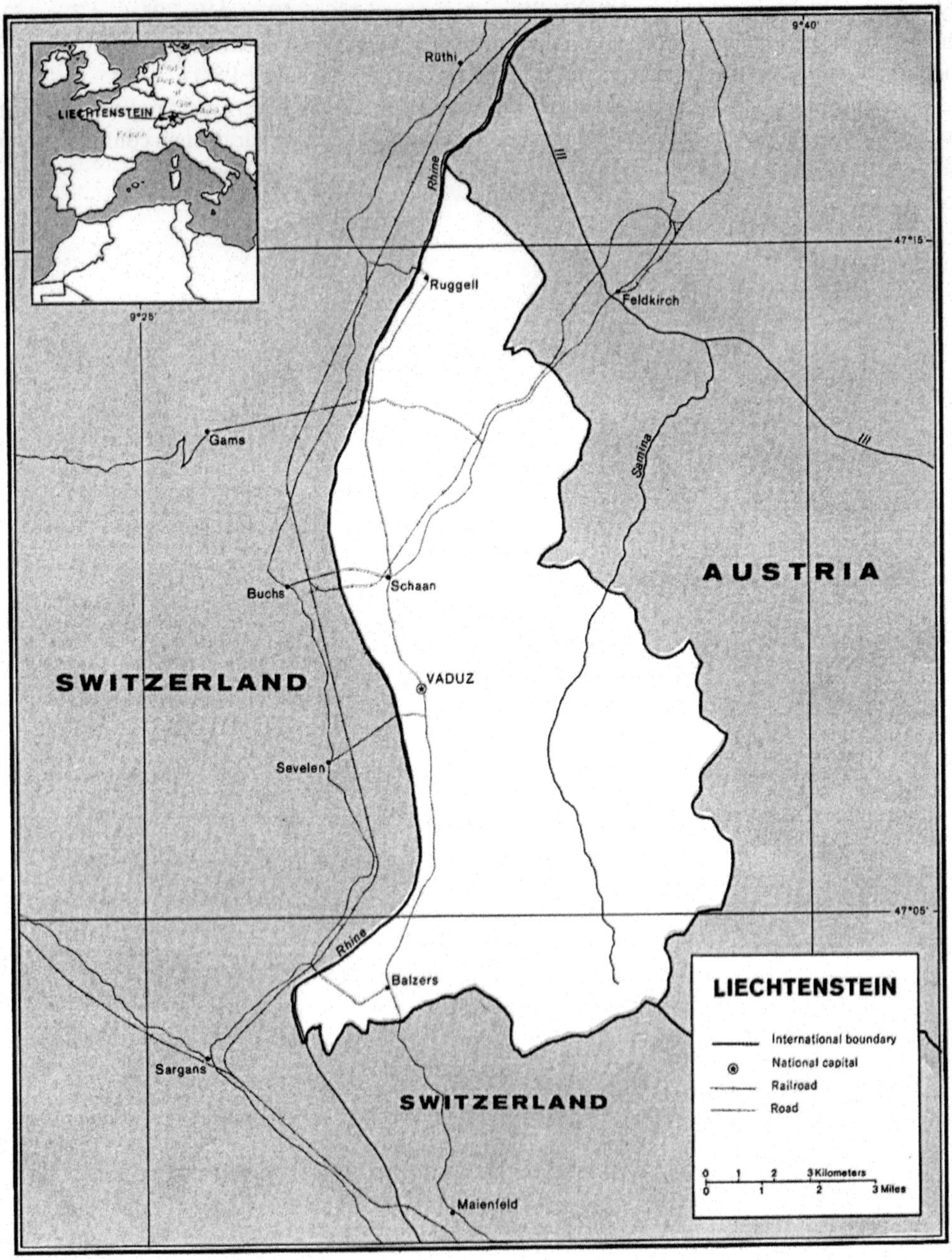

LIECHTENSTEIN
Rüthi
Ruggell
Feldkirch
Gams
Buchs
Schaan
VADUZ
Sevelen
Balzers
Sargans
Maienfeld
Rhine
Samina
Ill
AUSTRIA
SWITZERLAND
SWITZERLAND
9°25'
9°40'
47°15'
47°05'
LIECHTENSTEIN
International boundary
National capital
Railroad
Road
0 1 2 3 Kilometers
0 1 2 3 Miles

Principality of Liechtenstein

About 0.9 times the size of Washington, DC, the Principality of Liechtenstein is bordered by Austria (34.9 km) and by Switzerland (41.1 km). The lowest point is Ruggeller Riet (430 m) and the highest point is Vorder-Grauspitz (2,599 m). The terrain is mostly mountainous (Alps) with Rhine Valley in the western third.

"The Liechtenstein family, of Austrian origin, acquired the fiefs of Vaduz and Schellenberg in 1699 and 1713 respectively, and gained the status of an independent principality of the Holy Roman Empire in 1719 under the name Liechtenstein. The French, under Napoleon, occupied the country for a few years. Napoleon was the founder of the Rhine Confederation in 1806 and accepted Liechtenstein as a member. Liechtenstein considers itself therefore to be a sovereign state since 1806. In 1815 within the new German Confederation, Liechtenstein could prove its independence once more. In 1868, after the German Confederation dissolved, Liechtenstein disbanded its army of 80 men and declared its permanent neutrality, which was declared during both world wars.

"In 1919, Liechtenstein and Switzerland concluded an agreement whereby Switzerland assumed representation of Liechtenstein's diplomatic and consular interests in countries where Switzerland maintains representation and Liechtenstein does not. According to an agreement concluded with Austria in 1979, Liechtenstein citizens may seek consular assistance from Austrian representatives abroad in countries in which neither Liechtenstein nor Switzerland maintain representation. After World War II, Liechtenstein became increasingly important as a financial center, resulting in more prosperity" (*U.S. Department of State Background Notes,* 2011).

Complete topographic map coverage consists of four sheets at 1:10,000 scale. The *Fürstliche Regierung Liechtenstein* is responsible for all official mapping of the principality. The datum and grid is the Swiss Grid. "On the 1:10,000 maps, the north and west edges have the full grid values. The south and west edges have the grid values less the false coordinates, *i.e.*, FE = 600 km, FN = 200 km" (*John W. Hager, 11 July 2001*). Dr. Christoph Brandenberger informed me that these false coordinates were originally from the Swiss military (Bonne Grid) and are now adopted for civilian use (*Personal communication, January 2001*).

The Old Berne Observatory Datum of 1898 published an Astronomical Latitude $(\Phi_o) = 46°\ 57'\ 08.66''$ N, based on observations executed by E. Plantamour in 1875 and an Astronomical Longitude $(\Lambda_o) = 7°\ 26'\ 22.5''$ East of Greenwich. The defining azimuth to station Rötifluh was $(\alpha_o) = 11°\ 12'\ 05.24''$. The ellipsoid height and deflection of the vertical are not defined and therefore are forced to zero at the origin. The reference ellipsoid is the Bessel 1841, where $a = 6,377,397.155$ m and $1/f = 299.15281285$ (*op. cit., Brandenberger 2001*). For the Grid of the Swiss National Maps and of Liechtenstein, the value $\varphi_o = 46°\ 57'\ 07.90''$ N was chosen based on more recent measurements (1937), and the Central Meridian $(\lambda_o) = 7°\ 26'\ 22.5''$ East of Greenwich. The radius of the Gaussian Sphere evaluated at the Grid origin for the Bessel 1841 ellipsoid is $R = 6,378,815.9036$ m. The Grid Scale Factor at Origin $(m_o) = 1.00072913843$, and the False Origin is the same as for the old Swiss Bonne Grid of 600 and 200 km. Conformal doubles became the "carte du jour" projections of Europe during the early 20th century, and the cylinder, the cone, and the plane were all used as developable surfaces. The "oblique" for this Swiss system is really a misnomer; it's merely transverse at an oblique latitude. The Rosenmund projection is truly unique in the world for a national grid. The combination of the Bern Observatory (horizontal) Datum of 1898 with the Pierre du Niton (vertical) Datum of 1902 or LN02, and the Rosenmund projection and Grid of 1903 have collectively been known since as the "CH1903 System" (*Convention Helvetica 1903 System*), of Switzerland.

In 1988, a new network of 104 GPS station observations began, and the resultant adjustment has become the new national (terrestrial) reference system of Switzerland and is called the CHTRS95. The local reference *frame* realized for the old CH1903 Datum is called LV95. The "CH1903+" was held fixed at a new fundamental point, Zimmerwald Z_o where $\varphi_o = 46° 52' 42.27031''$ N, $\lambda_o = 7° 27' 58.41774''$ East, and $X_o = 191,775.0616$ m, $Y_o = 602030.7698$ m, all still referenced to the Bessel 1841. The new point was chosen because the original location no longer exists, and the original coordinates of triangulation point Gurten was kept to maintain orientation. The deflection of the vertical is now defined at Zimmerwald Z_o: $\zeta_o = +2.64''$, $\eta_o = +2.73''$, and $H_o = 897.8408$ m. Transforming CH1903+ Datum to CHTRS95 Datum (WGS84 ellipsoid) then is accomplished by $\Delta X = +674.374$ m, $\Delta Y = +015.056$ m, $\Delta Z = +405.346$ m (*op. cit., Brandenberger 2001*).

The European Datum of 1950 was computed for Switzerland by the U.S. Army Map Service in the 1950s, and to transform from EU50 to WGS84, $\Delta X = -87$ m, $\Delta Y = -96$ m, $\Delta Z = -120$ m. To transform from EU79 to WGS84, $\Delta X = -86$ m, $\Delta Y = -98$ m, $\Delta Z = -119$ m. These parameters are according to NGA's *TR 8350.2*, January 3, 2000.

According to the Swiss Federal Office of Topography (**SWISSTOPO**), the seven-parameter Datum shift from CH1903+ to WGS84 is: $\Delta X = +660.077$ m ± 4.055 m, $\Delta Y = +013.551$ m ± 4.816 m, $\Delta Z = +369.344$ m ± 3.914 m, $\alpha = 2.484$ cc ± 0.417 cc, $\beta = 1.783$ cc ± 0.455 cc, $\gamma = 2.939$ cc ± 0.411 cc, and $M = 1.000000566 \pm 0.00000566$ (*op. cit., Brandenberger 2001*).

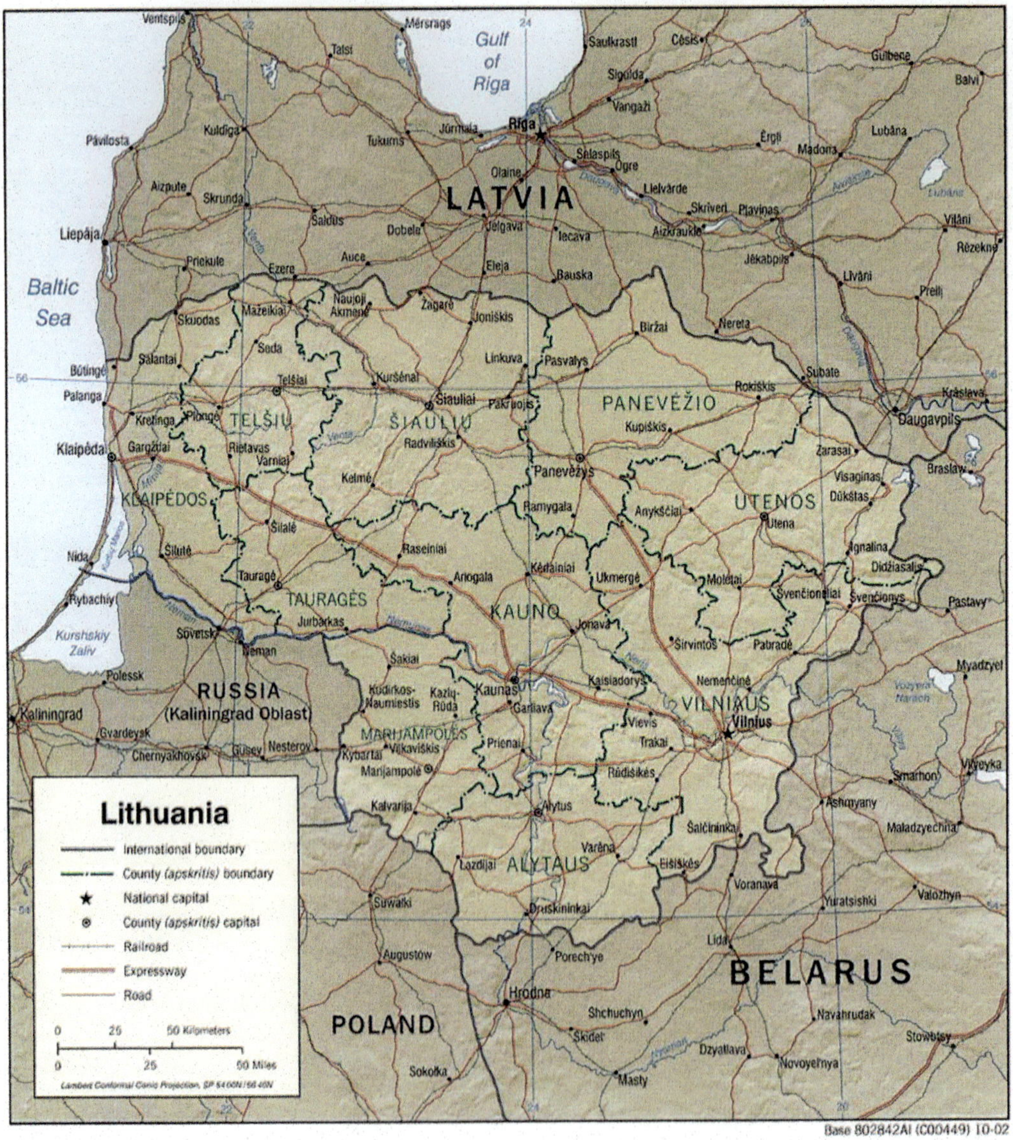

Base 802842AI (C00449) 10-02

Republic of Lithuania

Slightly larger than West Virginia, Lithuania is bordered by Belarus (502 km), Latvia (453 km), Poland (91 km), and Russia (Kaliningrad Oblast) (227 km). The country is mostly lowland with many scattered small lakes; the lowest point is the Baltic Sea (0 m), and the highest point is Juozapine Kalnas (292 m).

"Lithuanian lands were united under *Mindaugas* in 1236; over the next century, through alliances and conquest, Lithuania extended its territory to include most of present-day Belarus and Ukraine. By the end of the 14th century Lithuania was the largest state in Europe. An alliance with Poland in 1386 led the two countries into a union through the person of a common ruler. In 1569, Lithuania and Poland formally united into a single dual state, the Polish-Lithuanian Commonwealth. This entity survived until 1795, when its remnants were partitioned by surrounding countries. Lithuania regained its independence following World War I but was annexed by the USSR in 1940 – an action never recognized by the US and many other countries. On 11 March 1990, Lithuania became the first of the Soviet republics to declare its independence, but Moscow did not recognize this proclamation until September of 1991 (following the abortive coup in Moscow). The last Russian troops withdrew in 1993. Lithuania subsequently restructured its economy for integration into Western European institutions; it joined both NATO and the EU in the spring of 2004" (*CIA World Factbook*, 2008).

The earliest geodetic survey of Lithuanian lands was executed by Zhylinskiy between 1889 and 1890 and was based on the Warsaw 1875 Datum referenced to the "Adjustment Spheroid," where $a = 6,380,880$ m and $1/f = 263.597$. "The (*Lithuanian – Ed.*) original coordinates derive from lower-order surveys of the years 1879–1880, 1886–1901, 1902–1903, 1905–1909, 1911–1912, 1917, and 1930. In the vicinity of the meridian of Dorpat, Struve's Arc (variously called the Russo-Scandinavian Arc or Gradmessung, or Tenner's Arcs), for which results were published in *L'arc de Meridian entre le Danube et la Mer d'Iesicle* (sic), Struve, W., St. Petersburg, 1860, provided the sides later utilized in lieu of base and principal azimuth lines. These were used either by the Imperial government, but provinces (*e.g.,* Courland, Livonia) or by guberniyas (*e.g.,* Wilno). To this end Struve's arc sometimes was locally extended. The configuration of first-order lines is shown by a diagram supplied with the *Catalog of Points Determined by the Triangulation of the West Border States 1886–1901* in which are given also the principal points of the local systems associated with Struve's arc, and disagreements due to varying plans of adjustment" (*Army Map Service Geodetic Memo No. 880*, 30 November 1951). In 1938, *Bulletin Géodésique*, No. 57 published the Lithuanian National Report by Colonel Ing. A. Krikščiūnas which contained the astronomical coordinates of the Kaunas Observatory as determined by the method of *Horrebow-Talcott* being: $\Phi = 54° 53' 43.99''$ N ±0.05″, $\Lambda = 1^h 35^m 29.504^s$ ±0.0014 East of Greenwich. (*Note that the dual-bubble method was simultaneously published by Horrebow, a Captain in the British Royal Engineers and by Talcott, a Captain in the U.S. Army Corps of Engineers – Ed.*) The historical grid systems used locally in Lithuania on the Potsdam Datum were the Lithuanian Gauss-Krüger Transverse Mercator Belts, where the Central Meridian for the West Belt is $\lambda_o = 21°$ E with a False Easting of 1,500 km, for the Central Belt is $\lambda_o = 24°$ E with a False Easting of 2,500 km, and for the East Belt is $\lambda_o = 27°$ E with a False Easting of 3,500 km. All three belts have a scale factor at origin $m_o = 1.0$, and all are referenced to the Bessel 1841 ellipsoid, where $a = 6,377,397.155$ m and $1/f = 299.1528$.

"The Soviet occupation of Lithuania before WWII referenced topographic maps to the Pulkovo 32 System where: $\Phi_o = 59° 46' 18.71''$ N, $\Lambda_o = 30° 19' 39.55''$ East of Greenwich and retained the Bessel 1841 ellipsoid of reference. Soviet maps published after WWII were converted to the single coordinate system, System 42. The origin of the system is the center of Pulkovo Observatory with corrected coordinates $\Phi_o = 59° 46' 18.55''$ N, $\Lambda_o = 30° 19' 42.09''$ East of Greenwich and azimuth from point Bugry to Sablino $\alpha_o = 121°\ 06'\ 38.79''$. The Krassovsky ellipsoid with dimensions $a = 6,378,245$ m and $1/f = 298.3$ replaced the Bessel 1841 ellipsoid. The projection and military grid remained the same. The mean sea level of the Baltic Sea at the Kronstadt tidal gauge (*St. Petersburg – Ed.*) was accepted as zero altitude. Coordinates are not reduced to sea level but to the reference ellipsoid, which is tangent to the geoid only at Pulkovo" (*Foreign Maps, TM 5-248*, 1963). The military grid mentioned is the Russia Belts with identical parameters to the UTM except that $m_o = 1.0$ and referenced to the Krassovsky ellipsoid.

As reported in the first paragraph, Lithuania was the first country to declare its independence from the Soviets in 1990. By 1994, the Lithuanian Coordinate System (*Lietvos Koordinačių Sistema*) or LKS 94 was established, and the entire Soviet *System 42* was dumped. Introduced on September 30, 1994, the LKS 94 coincides with EUREF 89 Coordinate System ETRS89 and was transferred with points: Akmeniškiai (0311), Meškonys (0312), Šašeliai (0408), and DainavÚlÚ (0409). The reference ellipsoid is the GRS 1980, where $a = 6,378,137$ m and $1/f = 298.257\ 222\ 101$ ... "Lithuanian GPS Network coordinates are computed in the Coordinate System LKS94. Coordinates of triangulation and traverse points were recomputed in the new Coordinate system. This System will be used in various activities: geodetic, cartographic, land managing, cadastre and creation of Information Systems" (*National Report of Lithuania, Kumentaitis, Z. and Petroškevičius, P., Munich 1995*).

The current national mapping is based upon the Gauss-Krüger implementation of the Transverse Mercator projection and one zone is in use, with a central meridian of 24° east. The False easting is 500,000, False Northing 0, Latitude of Origin 0 degrees north. The Central Meridian scale factor is 0.9998 not the 1.0 value often associated with eastern European Gauss-Krüger projections. The national mapping agency is Valstybiné Geodezijos ir Kartografijos Tarnyba, Ukmerges 41, LT-2600 Vilnius.

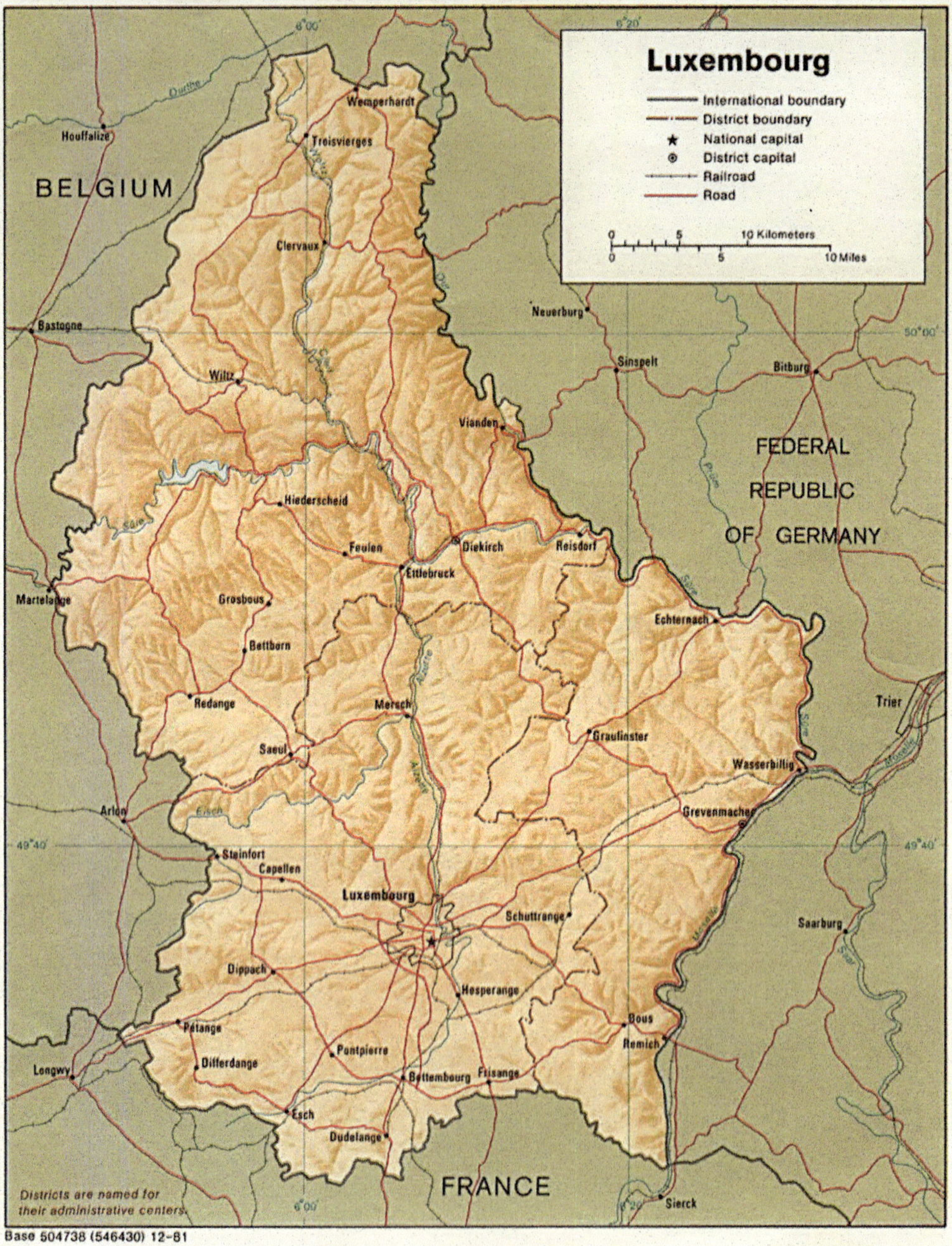

Base 504738 (546430) 12-81

Grand Duchy of Luxembourg

Bordered on the south by France (73 km), on the north and west by Belgium (148 km), and on the east by Germany (138 km), the Grand Duchy is comprised mostly of rolling uplands with shallow valleys and forms part of the plateau of Ardennes. There are uplands rising to low mountains in the north with a steep slope down to the Moselle basin flood plain in the southeast and is watered by the *Sûre* and *Alzette* rivers. The lowest point is the Moselle River (133 m) and the highest point is *Buurgplaatz* (559 m). Luxembourg is slightly smaller than Rhode Island and is completely landlocked.

Included in the Roman Empire from 50 B.C., it was later a part of the Frankish kingdoms of Austrasia and of Charlemagne. According to the CIA's *World Factbook*, "Founded in 963 A.D., Luxembourg became a grand duchy in 1815 and an independent state under the Netherlands. It lost more than half of its territory to Belgium in 1839 but gained a larger measure of autonomy. Full independence was attained in 1867. Overrun by Germany in both World Wars, it ended its neutrality in 1948 when it entered into the Benelux Customs Union and when it joined NATO the following year. In 1957, Luxembourg became one of the six founding countries of the European Economic Community (later the European Union), and in 1999 it joined the euro currency area."

J. Hansen produced maps at a scale of 1:50,000 between 1883 and 1906. During World War I, the British produced a series of topographic maps with the Nord de Guerre Zone shown by full lines. The French Nord de Guerre Zone (1914–1948) was based on the French Army Truncated Cubic Conic, where the Latitude of Origin was $\varphi_o = 49°\ 30'\ 00''$, the Central Meridian was $\lambda_o = 7°\ 44'\ 13.95''$ East of Greenwich, the Scale Factor at Origin $(m_o) = 0.999509082$, the False Easting was 500 km, and the False Northing was 300 km. The ellipsoid of reference was the Plessis Reconstituted, where $a = 6,376,523.994$ m and $1/f = 308.624807$. The datum used at the time was the New Triangulation of France (NTF 1887). The 1:50,000 scale series produced by J. Hansen and by the French Institut Géographique National (IGN) 1:50,000 and 1:100,000 series were expressed in degrees with longitudes referred to Paris ($\lambda_o = 2°\ 20'\ 13.95''$ East of Greenwich). The old IGN 1:20,000, 1:25,000, and 1:50,000 series *also* have coordinates in grads referred to Paris.

The national datum of Luxembourg (LUREF) was established in 1930 with its fundamental point (LaPlace astronomical point) at Habay-la-Neuve in Belgium, and the ellipsoid of reference is the Hayford International 1924, where $a = 6,378,388$ m and $1/f = 297$. The Luxembourg Transverse Mercator grid was adopted in 1940 and uses the Gauss-Krüger equations, where the Latitude of Origin, $\varphi_o = 49°\ 50'$ N, the Central Meridian, $\lambda_o = 6°\ 10'$ East of Greenwich, the Scale Factor at Origin is unity, the False Easting = 80 km, and the False Northing = 100 km. The Administration du Cadastre et de la Topographie (ACT) offers a worked example: $\varphi = 49°\ 34'\ 17.60287''$ N, $\lambda = 5°\ 55'\ 50.69323''$ E; X (Northing) = 70,910.00 m, Y (Easting) = 62,935.00 m.

The National Luxembourg Height Datum, *Nivellement General du Luxembourg* (*NG-L*) has its initial reference point at Wemperhardt (near the northern border with Belgium) where the elevation is 528.030 m. This is based on the reference Tide Gauge at Pegel, Amsterdam. "The NG-L is based on geometric levelling (*sic*) only, it counts 3800 points which corresponds to a density of 1.8 points/km^2."

I wrote a letter of inquiry to the ACT in October 1997 regarding the Luxembourg TM as well as the transformation parameters from LUREF to WGS84/EUREF89. Mr. Patrice Schonckert, Director of ACT, replied the same month and he verified that the projection parameters were unchanged, but that the datum transformation parameters were released "only to recognized users." The transformation parameters from LUREF to WGS84/EUREF89 are now public.

The Bursa-Wolfe Transformation Model: $\Delta X = -192.986$ m, $\Delta Y = +13.673$ m, $\Delta Z = -39.309$ m, $R_x = 0.4099''$, $R_y = 2.9332''$, $R_z = -2.6881''$, and Scale Factor = 0.43 ppm. Although the International Association of Geodesy (IAG), the Bundesamt für Kartographie und Geodäsie (German Federal

Office for Cartography and Geodesy), and Eurographics publish the parameters also, they use the European sign convention for rotations which is *opposite* from the United States (and Australian) standard. To my surprise, the Luxembourg government (ACT) publishes their parameters above with the United States standard!

Curiously, ACT also uses the American standard rotation convention for the Molodensky-Badekas model: $\Delta X = -265.983$ m, $\Delta Y = +76.918$ m, $\Delta Z = +20.182$ m, $R_x = 0.4099''$, $R_y = 2.9332''$, $R_z = -2.6881''$, and Scale Factor $= 0.43$ ppm, and the Rotation Origin is: $X_o = 4,098.647.674$ m, $Y_o = 442,843.139$ m, and $Z_o = 4,851,251.093$ m. Since the Molodensky-Badekas model is usually employed (by me) when the datum origin or fundamental point is in another country or at a great distance away from the region of interest, I transformed the rotation origin geocentric coordinates to geodetic coordinates with the expectation that I would obtain the coordinates of Habay-la-Neuve in Belgium. No such luck; the rotation origin offered by ACT is merely the projection origin of the Luxembourg TM grid.

REFERENCES: L

Abayomi Karnga, Liberian historian, 2011.
Andrew Dyson, Jones, Rohde, Lloyd, and Sougnatti, GPS World, March 1999.
Andrew Jones, Australian Government Land Titling Project, 2007.
Army Map Service Geodetic Memo No. 880, 30 November 1951.
Ausgabe Endwerte Koordinatenkartei, Kriegs-Karten und Vermessungsamt, Riga, 1943. Koordinaten-Verzeichnis, 1944.
Ausgabe Koordinatenkartei, Kriegs-Karten und Vermessungsamt, Riga, 1943.
Baltic J., Coordinate analysis of Latvian CORS stations, *Modern Computing*, vol. 7, no. 4, 513–524 (2019). doi: 10.22364/bjmc.2019.7.4.05.
Catalog of Points Determined by the Triangulation of the West Border States 1886–1901.
Changing the national height system and geoid model in Latvia, Janis Balodis, Katerina Morozova, Gunars Silabriedis, Maris Kalinka, Kriss Balodis, Ingus Mitrofanovs, Irina Baltmane, Izolde Jumare, Institute of Geodesy and Geoinformatics, University of Latvia, Riga, Latvia, 2020 (unpublished manuscript).
CIA World Factbook, 2005.
CIA World Factbook, 2008.
COL George Massaad, Director of Geographic Affairs, Lebanese Army, Personal Communication, November 1997.
COL Ing. A. Krikščiūnas, Bulletin Géodésique, No. 57, Lithuanian National Report, 1938.
Dave Doyle, U.S. National Geodetic Survey, Personal Communication, December 2010.
Dr. Christoph Brandenberger, Personal Communication, January 2001.
Dr. Muneendra Kumar, retired Chief Geodesist of the United States National Geospatial Intelligence Agency – NGA, Personal Communication, 2007.
E. A. Early, U. S. Army Map Service, c.a. 1950.
EPSG, 2006.
Firestone Rubber Co., 2010.
Foreign Maps, 5-248.
Google Earth, 2011.
J. Rems and Dr. Charles L. Merry, Datum Transformation Parameters in South Africa,
John W. Hager, Personal Communication, 11 July 2001.
John W. Hager, Personal Communication, 15 August 2005.
John W. Hager, Personal Communication, 2006.
John W. Hager, Personal Communication, 2007.
John W. Hager, Personal Communication, December, 2010.
Kingdom of Lesotho.
Kumentaitis, Z. and Petroškevičius, P., National Report of Lithuania, Munich 1995.
Library of Congress Country Studies, Libya 2006.
Library of Congress Country Study, Lao, 2007.
Liechtenstein U.S. Dept. of State Background Notes, 2011.
Lonely Planet, 2008.
Malcolm A. B. Jones, Personal Communication, 2007.

Manual of Photogrammetry, 5th or 6th editions.

NGA Geonames Server, 2011.

O. A. Sergjew, Making and Editing of Military Maps, Moscow 1939.

Patrice Schonckert, Director, Administration du Cadastre et de la Topographie, Personal Communication, October 1997.

S. Mosisili, Chief Surveyor for the Lesotho Department of Land, Surveys and Physical Planning, Personal Communication, December 1989.

S. Mosisili, Chief Surveyor, Lesotho Department of Land, Surveys and Physical Planning, Sir Harry Johnston, Liberia, 2 vols., 1906.

Struve, W., L'arc de Meridian entre le Danube et la Mer d'Iesicle (sic), St. Petersburg 1860.

Survey Review, 30, 236, April 1990.

The World Bank Land Administration System Modernization Project (P159692).

TR 8350, 2011.

Travaux du Cadastre et d'Amelioration Agricole des Etats, de Syrie, des Alaouites et du Liban sous Mandat Français, 1922.

Trigonometrical Survey, D. R. Hendrikz, Union of South Africa 1955.

van der Kraaij, The Open Door Policy of Liberia – An Economic History of Modern Liberia, Bremen, 1983, 2 volumes, pp. 1–5.

World Factbook, 2010.

World Factbook, 2022.

M

Macau Special Administrative Region	483
Republic of Macedonia	487
The Republic of Madagascar	493
Republic of Malawi	497
Malaysia	501
Republic of Maldives	505
Republic of Mali	509
Republic of Malta	513
Republic of the Marshall Islands	517
Department of Martinique	521
Islamic Republic of Mauritania	523
The Republic of Mauritius	527
United Mexican States	531
Federated States of Micronesia	535
Republic of Moldova	539
Principality of Monaco	543
Mongolia	547
Montenegro	551
Montserrat	553
The Kingdom of Morocco	555
The Republic of Moçambique	559

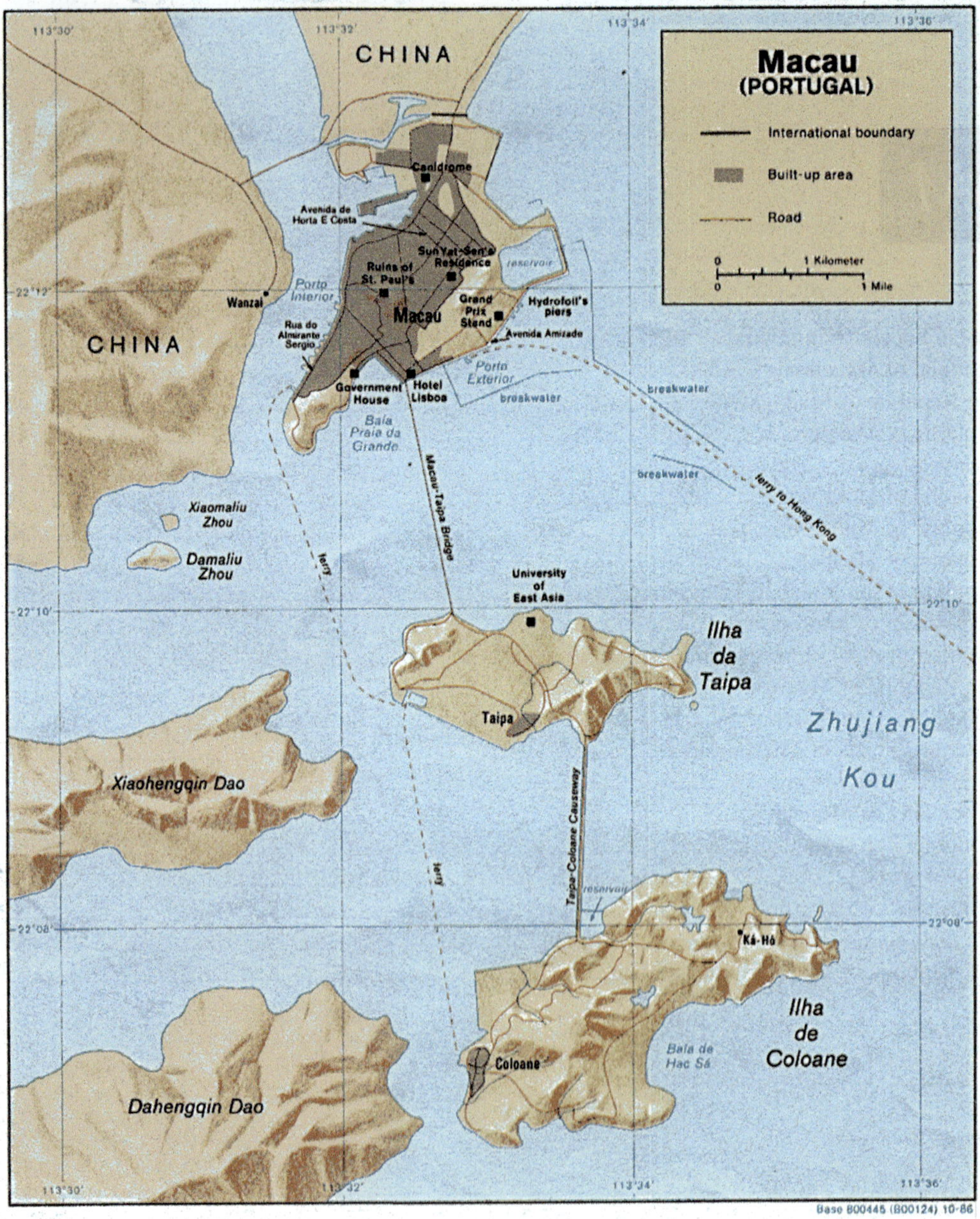

Base 800445 (B00124) 10-86

Macau Special Administrative Region

Macau S.A.R. is situated in the South China Sea, consisting of the Macau Peninsula and the two small islands of Taipa and Colôane about 64 km west of Hong Kong. It is less than one-sixth the size of Washington, DC. The terrain is generally flat, the highest point is Alto Coloane (172 m), and the lowest point is the South China Sea (0 m).

Visited by Portuguese traders in 1513, it was settled by the Portuguese in 1587. Declared a Portuguese territory in 1849, its claim was recognized by China in an 1887 treaty. In 1951, the status was changed from a colony to an overseas territory of Portugal. On April 13, 1987, Portugal agreed to return Macau to China on December 20, 1999. *Ilha da Taipa* is connected to the mainland by a bridge, and to *Ilha de Colôane* by a causeway. Ferries run between Macau and the islands as well as to Hong Kong.

In 1907, General Castelo Branco established a triangulation network in the Macau Peninsula. The earliest baseline was set in *Avenida Conselheiro Borja*. The coordinates of that baseline was (0,0) for "Base A," and (−673.73, 84.84 m) for "Base B"; the total distance was 679.056 m. The original astronomical Datum was observed in 1912 at the Hospital *Conde de S. Januario* on the "colina" (mountain), *Colina de Viscounde de S. Januario*, and was observed again by the Portuguese military in 1965. The coordinates appear to have been lost, but the Datum was transferred to *Guia Farol* (lighthouse) ($\varphi_o = 22°\ 11'\ 51.557''$ North, $\lambda_o = 113°\ 32'\ 48.063''$ East of Greenwich).

The Macau Datum origin was later again transferred to *Monte da Barra*, where $\varphi_o = 22°\ 11'\ 03.139''$ North, $\lambda_o = 113°\ 31'\ 43.625''$ East. This was a plane system and did not use an ellipsoid. The grid was based on an Azimuthal Equidistant projection.

In 1919–1920, *Capitão de Artilharia José Soares Zilhão* (Capitan of Artillery) and *Capitão-Tenente de Marinha Justino Henrique Herz* (Naval Full Lieutenant) established a triangulation network in Macau, Taipa, and Colôane. The azimuth of the initial Branco line was found to be in error by 001° 42′ 06″ eastwards. This network initially served as control for the large-scale hydrographic survey of the Port of Macau. The network was expanded and later included more stations. The above information was provided by "Direcção das Obras Dos Portos de Macau." The Zilhão/Herz hydrographic survey obviously included work on heighting. Only one tide gauge was existing on the peninsula at the time so additional gauges were established on Taipa and Colôane. The vertical datum was defined as the "hydrographic zero plane at 0.7 m below the level of the maximum tide low point of spring tides." This level was marked on all three gauges, and heights with respect to the zero plane were transferred by leveling to the trig stations *Passagem* (1920) and *Monte de Artilharia* (1920) on Taipa and Colôane, respectively.

According to "Repartição Téchnica das Obras de Macau-1943," the baseline was updated such that the coordinates of "Base A" were (20,000 m North, 20,000 m East) and the coordinates of "Base B" were (19,915.18 m, 20,673.54 m). The distance was 678.8545 m. The work was supervised by Engineer Chau of the Sanitation Brigade of the Macau Public Works Department.

A topographic survey of Macau was carried out by Lt. Numes of the Macau Military Command of the Portuguese Army in 1955–1956. The task was to produce 1:2,000 scale mapping of Taipa and Colôane. According to Miss S. Fletcher of the U.K. Geodesy Division of Her Majesty's Defence Geographic and Imagery Intelligence Agency, "An assessment of the earlier survey work (1907, 1920, 1940s) was carried out to establish the best control on which to base the new survey network required. As already mentioned, it was found that there was conflict between measurement and orientation of the baseline used. The coordinates differenced could not be resolved by shifts or rotations of the coordinate axes. It was decided that as Chau's results did not actually include a check on baseline orientation, and that the change to orientation made in 1920 still maintained the original baseline length, coordinates for control points based on Macau would be taken from the 1920 results. Height control was taken from the trig stations *D.Maria* (Macau), *Monte da Barra* (Macau), *Passagem* (Taipa), and *Monte de Artilharia* (Colôane), since all had previously been geodetically leveled with respect to the vertical datum (the Hydrographic Zero Plane of 1920). The network contained two control stations on Macau (*Monte da Barra* and *Guia Terraco*), 35 stations on Taipa and 43 on Colôane. The average length of side on the islands was about 500 m. Trig stations were marked with concrete pillars. The two islands are joined together by an observed braced quadrilateral figure, as is the connection between Taipa and the Macau peninsula. The latter was observed using a Universal Wild T4 theodolite; most of the other observations were made using Wild T3 theodolites. Leveling was carried out to height a number of trig points using an NK2 precise level; the remaining points were measured by Trig Heighting (vertical and horizontal angles). Plane tabling was also undertaken. Two baselines were measured: Base A (NW) – Base B (SE) on Taipa and Base C (Sul) – Base D (Norte) on Colôane. These were used to provide control for scale within the observed network and also to check the relative positions of the two control points on the Macau Peninsula as determined by the 1920 survey. This corresponded to an average of 0.023 m which was within the accuracy required for the purpose of this survey. The Taipa baseline is recorded as being measured by nine sets of observations (forward and reverse) with a 100-meter invar tape. The accuracy is quoted to ±0.0021 m (1.48 sigma).

In 1965–1966, a Hydrographic Survey was carried out by Captain Fernandes. The first part of this hydrographic survey was to carry out triangulation observations in Macau for the planning of major dredging works in the port. Although the 1920 and 1940s work provided coordinates for various trig stations on the Macau Peninsula, they were found not to have a high enough order of accuracy for what was required and discrepancies between the two data sets were in the order of tens of centimeters. The results of the 1955–1956 topographic survey were assessed as providing a suitable base for this re-observation work as it included a well-measured baseline, adequate observation procedures, and adjustment of results by least squares. The source coordinates for *Monte da Barra* and *Guia* (Terraco) were taken as fixed (originally dating back to 1920 results) and stated as: *Monte da Barra* (X=19207.44 m, Y=16878.29 m), *Guia* (Terraco) (X=21026.58 m, Y=18367.08 m). The same coordinate system as for the 1955–1956 survey was used. Observations were carried out using a Wild T2 theodolite with tripod. Only one trig station (Observatorio) made use of forced centering. Third order observation procedures were employed (four sets) with mean misclosures of sets in the order of 1″ (of arc). Confusion over the definition of the vertical datum had arisen over the years and so another task of this survey was to try to rectify this. The heights of trig stations in this survey are quoted in terms of a newly determined Mean Sea Level. Two sets of heights are quoted: "pont. Mira" – height to the top of the pillar/reference mark, and "terreno" – height of ground level."

In May 1981 the No. 512 Specialist Team, Royal Engineers of the U.K. Military Survey observed and computed the coordinates of two Doppler TRANSIT stations in the (then) Province. (This was the same era as when they re-surveyed Hong Kong). They were based on existing trig points at *Monte da Barra* and *Colôane Alto*. These stations were observed before the Macau Survey Department carried out the network re-adjustment such that the 512 STRE position of *Monte da Barra* was held to: $\varphi_o = 22° 11′ 03.14″$ N, $\lambda_o = 113° 31′ 43.63″$ E, $H_o = 74.45$ m. This marked the beginning of Macau's use of an ellipsoid for its network and plane coordinate system. The ellipsoid of reference was once thought to be the Clarke 1858 and cast on the Cassini-Soldner Grid (same as Hong Kong). Not so, the ellipsoid of reference is the International 1924 where the semi-major axis, $a = 6,378,188$ m and the reciprocal of flattening, $1/f = 297$.

In 1986, a team of Portuguese geodetic surveyors carried out astronomic observations for the baseline Monte da Barra–Colôane Alto, but results were not received in Macau until late 1991. The results showed that the latest local survey network (1981) still contained a swing of approximately

1′ 19.73″ of arc. That fact stands out like a sore thumb in a later paragraph here where the seven-parameter transformation values are listed.

The subsequent WGS 72 coordinates of *Monte da Barra* are: $\varphi_o = 22°\ 10'\ 58.696''$ N, $\lambda_o = 113°\ 31'\ 53.827''$ E, $h_o = 70.29$ m. The line between the two TRANSIT points checked with the new adjustment results to within 0.036 m, or 4.7746 parts per million. The WGS 72 ellipsoid semi-major axis, $a = 6,378,135$ m, and the reciprocal of flattening, $1/f = 298.26$.

Geographic Engineer and Director Adelino M. L. Frias dos Santos of the *Direcção dos Serviços de Cartografia e Cadastro* wrote to me in 1997 regarding the systems of Macau. Thanks also to Geographic Engineers Law Sio Peng and Lou Kuan Hou, they cleared up some of the long-standing mysteries of Macau. As mentioned previously, the original astronomical Datum at the *Hospital Conde de S. Januario* on the "colina" (mountain), *Colina de Viscounde de S. Januario*, was observed again by the Portuguese military in 1965 such that at station *Observatório*, $\Phi_o = 22°\ 11'\ 45.46''$ N, $\Lambda_o = 113°\ 32'\ 39.00''$ E. The old references to *Guia* Lighthouse have been confused by others because there are two OTHER stations: *Guia II* (*Chapa da Guia*) X = 21,052.91 m E, Y = 18,366.28 m N, and *Guia Terraço* (*Guia Marco*) X = 21,026.88 m E, Y = 18,367.24 m N. Thanks to Mr. John W. Hager, the projection used for *Observatório Metereológico* Datum of 1965 is the Azimuthal Equidistant with origin at Point "Base W," where $\varphi_o = 22°\ 12'\ 44.63''$ N, $\lambda_o = 113°\ 32'\ 11.29''$ E, and False Easting = False Northing = 20 km.

According to *Instituto Hidrográfico Portugal*, the transformation vector used for the conversion from WGS 72 Datum to *Observatório Metereológico* Datum 1965 is: $\Delta X = +187.77$ m, $\Delta Y = +297.54$ m, $\Delta Z = +160.35$ m. According to Engineers Peng and Hou, the 512 STRE returned in 1991 and established a strong geodetic link with the Hong Kong and Macau networks based on the WGS 84 Datum and ellipsoid. "Six primary geodetic stations were occupied in Macau (including *Monte da Barra, Hotel Oriental, Taipa Grande Taipa Pequeña, Colôane Alto*, and *Monte Ká Hó*). Three years later, we observed the coordinates of fourteen stations (including the six primary stations) with the Global Positioning System and transformed the WGS 84 coordinate system into the local coordinate system. The Molodensky-Badekas transformation system (seven-parameter transformation) was adopted to transform the WGS 84 coordinate system into the local coordinate system." Those parameters are Rotation Origin at: $X_o = -2,362,038.81$ m, $Y_o = 5,417,496.26$ m, $Z_o = 2,390,633.71$ m, and $\Delta X = +203.010$ m, $\Delta Y = +302.491$ m, $\Delta Z = +155.296$ m, $R_x = +32.876850''$, $R_y = -76.963371''$, $R_z = -32.622853''$, Scale = -8.204889 ppm. Thanks to Mr. Paul Gosling of the Geodetic Branch of the Defence Geographic Centre and to Mr. Giles André of the Defence Geographic and Imagery Intelligence Agency for their substantial contributions.

Republic of Macedonia

Slightly larger than Vermont, Macedonia is bordered by Albania (151 km) Bulgaria (148 km), Greece (246 km), and Kosovo (159 km). Macedonia is a mountainous territory covered with deep basins and valleys; three large lakes, each divided by a frontier line; country is bisected by the Vardar River which is the lowest point (50 m), and the highest point is Golem Korab (*Maja e Korabit*) (2,764 m) (*World Factbook*, 2012 and *NGA GeoNames Server*, 2012).

Macedonia is the birthplace of Alexander III the Great (356–323 *B.C.*). "The history of the ancient Macedonian kingdom begins with Caranus, who was the first known king (808–778 *B.C.*). The Macedonian dynasty Argeadae originated from Argos Orestikon, a city in located in southwestern Macedonia region of Orestis. By 65 *B.C.* Rome conquered the Seleucid Macedonian kingdom in Asia under its last king Antiochus VII. Finally, the defeat of Cleopatra VII in 30 *B.C.*, brought an end to the last of the Macedonian descendants in Egypt, and with it, the last remains of the Macedonian Empire that was once the mightiest in the world disappeared from the face of the earth. In the 9th century (*AD – Ed.*), while the Byzantine Empire was ruled by the Macedonians Emperors of the Macedonian Dynasty, the Macedonian brothers Cyril and Methodius from the largest Macedonian city of Salonica, created the first Slavonic alphabet, founded the Slavic literacy, and promoted Christianity among the Slavic peoples. Macedonia remained a Byzantine territory until the Ottoman Turks conquered it in 1389. In the 19th century, Greece, Serbia, and Bulgaria freed themselves from the Turkish rule and actively began conspiring against the Macedonians displaying territorial aspirations on their land. In 1912, Greece, Serbia, and Bulgaria joined forces and defeated the Turkish army in Macedonia. 100,000 Macedonians also participated and helped in the Turkish evacuation, but the victors did not reward them. The Treaty of London (May 1913), which concluded the First Balkan War, left Bulgaria dissatisfied with the partition of Macedonia among the allies which resulted after the war. Bulgaria's attempt to enforce a new partition in a Second Balkan War failed, and the Treaty of Bucharest (August 1913) confirmed a pattern of boundaries that (with small variations) has remained in force ever since" (*historyofmacedonia.org*, 2012).

"In the wake of the First World War, Vardar Macedonia (the present-day area of the Republic of Macedonia) was incorporated into the newly formed Kingdom of Serbs, Croats, and Slovenes. Throughout much of the Second World War, Bulgaria and Italy occupied Macedonia. Many citizens joined partisan movements during this time and succeeded in liberating the region in late 1944. Following the war, Macedonia became one of the constituent republics of the new Socialist Federal Republic of Yugoslavia under Marshall Tito. During this period, Macedonian culture, and language flourished. As communism fell throughout Eastern Europe in the late 20th century, Macedonia followed its other federation partners and declared its independence from Yugoslavia in late 1991" (*Background Notes, U.S. Department of State*, 2011).

"On the arrival of the Topographical Section, R.E., under the command of Mr (local Captain) Meldrum, a base was measured about January 1916, on the Hortiach plateau. This base was 1355.1 m long. No records of the base are available, but CMS Crusher, R.E., who took part in the measurement, states that it was measured twice with a 300-foot steel tape. He cannot remember what the difference between the two measurements was, but it was small and the mean of the two values was accepted. The tape had not been compared with any standard since leaving Southampton and had experienced rough usage on the Gallipoli peninsula. It was stretched along the ground, which had not been prepared in any way, but the alignment was mostly over shortish grass. The tension was applied by hand and, where possible, the handle of the tape was placed on the ground and an arrow stuck in to mark the position. No corrections of any sort were applied, nor were any astronomical observations made for latitude or azimuth.

"The base was extended by a number of small triangles to the side Beaz Tash – Kotos and from thence to Salonika and to the Langaza Plain, where it effected a junction with the French triangulation on the side Two Peaks – Deve Kran. Observations were made with a 5″ micrometer theodolite, measures being taken on one zero only. Very few of the stations were pooled, and the triangular error for 44 triangles (3 of which were not closed) averaged 36.6″, the greatest being 124″.

"Co-ordinates of the points were computed on the rectangular system, the French value of the bearing of the side Two Peaks – Deve Kran (36° 24′ 31″) being used in the computations. The White Tower, Salonika, was used as the initial station and the French co-ordinates for it (X = 144,810 m, Y = 123,359 m) were accepted. [The French values were obtained from a system of triangulation executed by that nation emanating from a base measured in the Vardar Plain, with an astronomically determined azimuth. Arbitrary values for the X and Y co-ordinates were given to one end of the base. Subsequently the azimuth was found to be considerably in error, and to save recomputation this error was considered as if due to convergency and the co-ordinates of the point which, taken as an origin, would give this convergency were calculated. This point (X = 173,000 m, Y = 129,000 m) is therefore the true origin of the system].

"When the triangulation reached the Langaza Plain a second base was measured. The ground was rough for a good part of its length. One measure was made but it was not utilized in the computations, presumably on account of the roughness of the measurement, although the base was 3562.4 m long (nearly three times the length of the original base) and was connected to the triangulation by an almost ideal system of extensions. The difference between the measured length and that derived through the triangulation from the Hortiach Plateau base was 4.7 m (a proportion of 1/760 approx.). On completion of this work the triangulation was carried eastwards from the Hortiach base with a view to making a connection with a naval system in the neighbourhood of Stavros, to which it joined on the side Four Tree Hill – Stavros Point. The naval value for this side was 7449.6 m and the army value 7444.7 m (value for this side in new triangulation 7448 m).

"No azimuth check was obtained as the naval work was based on geographical coordinates and no attempt seems to have been made to convert them into the rectangular system used by the army. In this extension most of the stations had been poled before observations were begun. The average triangular error for 23 triangles was 31″ (10 additional triangles were not closed).

This being completed, a reconnaissance was commenced towards Kavalla and Drama. Stations were not poled. The naval value for the side of commencement (Four Tree Hill – Stavros Point) was used in computing the triangles, but in computing the co-ordinates of the points the army values of the co-ordinates of the two stations of the initial side were used.

"As the naval value of this side differed by about 1 in 1,500 from the army value a considerable error was introduced into this work, which was increased owing to the poor shape of the initial triangle. The average triangular error of this work for 28 triangles was 40.4″ (8 triangles in addition were not closed).

"More or less simultaneously with this work, another party commenced a system of triangulation in the Struma Valley. A base 6,584.1 m long was measured but the tape had lost a handle and the alignment lay practically entirely through standing corn 4 to 5 ft high. No use (except for very temporary purposes) appears to have been made of the measure, though the computations are not clear. This triangulation carried westwards and then south, made a junction with the triangulation in the Langaza Plain on the side Deve Kran – Jagladzik, the original value of which appears to have been used for the computations of the new work. The series was also carried down the Struma Valley, effecting a junction with the Drama work on the side Bare Hill – Karakol. The value of the length of this side is given below:

Deduced from the Struma series: 4,677.2 m
Deduced from the Drama series: 4,679.2 m

The difference in the common bearing was 12′ 02″, while the co-ordinates of the common points as deduced from the two series are:

Small Peak: Struma series: X = 196,331 Y = 151,221 Drama series: X = 196,473 Y = 151,220
Tree 2: Struma series: X = 193,624 Y = 157,909 Drama series: X = 193,789 Y = 157,917

"Such was the state of the triangulation when Mr. Meldrum left the section in August 1916. 2nd Lieutenant (acting Captain) McNamara, R.E., took over from him and under his direction the triangulation was extended into the Doiran area, which had previously been in the French sector. Subsequently it was also extended from the neighbourhood of Salonika to Katerini.

In January 1917, the Topographical Section was absorbed into the Survey Company, which was placed under the command of Major Wood, R.E. Difficulties at once began to crop up about the value of points. When new triangulation was commenced from any part of the old, and attempts made to tie up with the other work based on a different side, no agreement could be obtained. Differences were large, in the case of the Struma Valley amounting to as much as 200 m in the co-ordinates. As there had been a good deal of overlapping and no attempt had been made to adjust differences, most points had two, three or even four values. It was apparent that no agreement was to be expected from the old triangulation, nor was it of sufficient accuracy to extend from, if there was an advance, or to attempt any counter battery work such as was being done in France. It depended on a short base extended by badly shaped triangles, the triangular errors were often large (up to 2 minutes) and often only two angles of the triangle had been observed. It was therefore decided to start a new net as a framework on which to recompute the old triangulation.

"THE NEW TRIANGULATION

In June 1917, a base 6½ kilometres (approx.) long was measured on the Lanagza Plain with a Jäderin apparatus with Invar wires borrowed from the French. Two wires were used for the measurement, the third being kept as a standard. Comparisons were made before and after work each day and necessary corrections were applied to the measures to reduce them to that of the standard wire. The base was measured twice, and the difference, after corrections for temperature and level, between the two measures was 5 cm. Observations for latitude and azimuth were also made. The triangulation net was designed to cover the entire British area, and the observations were made with a 5" micrometer theodolite – four measures on each of four zeros were made of each angle. The stations were marked with large stones buried in the ground with a triangle cut on their upper surfaces. Cairns with poles were erected over the marks. Observations were as a rule taken to the pole, but on long rays, luminous signals were employed. The average triangular error of the network was 7.4" for 64 triangles.

"The network was adjusted after the following method: the station error (*i.e.* the amount by which the sum of the angles at the station differed from 360°) was distributed proportionally to the size of the angles at the station. The triangular error was divided equally amongst the three angles, any excess being given to the largest or two larger angles. The triangles were then computed and the mean of the two values of a common side assumed to be correct. The angles of the triangles were then readjusted, so that on re-computation the values of the side common to both triangles was as near the original mean as it was possible to obtain, when using angles to one place of decimal of a second. The mean value of the common side thus obtained was accepted as the final value. The average correction thus applied to the angles was 7.2". The geographical co-ordinates of the points were then computed. The initial azimuth was the value obtained astronomically at the south end of the base. The initial latitude and longitude were those of the White Tower as given on the Admiralty chart. [This value of latitude, when carried through the triangulation, only differed from the value derived from observations for latitude made at the south end of the base by 1.8". It was not considered advisable to change the old value of latitude for such a small amount.

"The initial height is that at the top of the pillar close to the seashore in the grounds of the office of the French Topographical Service at Salonika taken at 1 metre (this is the approximate height above Mean Sea Level). The geographical co-ordinates have been converted to rectangular co-ordinates. The latitude and longitude of the point of origin of the French system (X = 173,000 metres, Y = 129,000 metres) was obtained by taking the naval values for the geographical coordinates of the White Tower (latitude 40° 37' 31", longitude 22° 57' 41") as equivalent to the French rectangular co-ordinates (X = 144,810, Y = 123,359) of that place and computing the geographical co-ordinates of the French point of origin. The rectangular co-ordinates of the point of origin and of the White Tower are taken from page 1 of the list of co-ordinates published by the French. All the old triangulation was then re-computed on the rectangular system using, wherever possible, the values of the sides of the new triangulation and the old computations were then destroyed. All new subsidiary triangulation has emanated from sides of the main network.

"The Serbians, extending from the stations Krechevo; 2 Peaks 2; White Tower; of the British triangulation carried a series to the neighbourhood of Monastir and effected a junction with the French and Serbian triangulation in that neighbourhood, which depended on a base measured near Monastir by the French. The agreement between the two systems was very good.

In 1918 when the British took over the area west of the Vardar River the subsidiary triangulation was extended to that region and affected a junction with some triangulation executed by the French there. Several of the stations were identical. A comparison of the values shows a consistent difference of 4 metres in "X" and 3 metres in 'Y' between the two systems. Heights agreed within two metres. During the hostilities in September 1918, and in the succeeding month, the triangulation was extended northwards to the Strumica Valley. Several of the German-Bulgar stations were utilized as stations of the British triangulation and a direct junction between the two systems was thus made.

"A synopsis of the German triangulation was obtained from Sofia and, though only values for rectangular co-ordinates were obtained, there is reason to suppose, from investigations made into their maps *etc*; that some of the work was computed on a spherical basis. It is understood that this work was based on a side of the Austrian geodetic triangulation in Albania. From information obtained from the German maps it is apparent that the geographical co-ordinates of the point of origin of their rectangular system is latitude 40° 45′ 00″, longitude 40° 30′ 00″ East of Ferro (= 22° 50′ 14″ East of Greenwich). Using, for the stations common to the two systems, the geographical co-ordinates obtained from the British triangulation and the German rectangular co-ordinates, the geographical co-ordinates of the German point of origin were computed. This was found to be latitude 40° 45′ 03.12″, longitude 22° 50′ 07.00″ (East of Greenwich). See below:

- Signal Allemand (*Drei Königseck No.613*) 40° 45′ 03.19″ 22° 50′ 06.97″
- Kara Bail (*Kara Bair Signal No.606*) 40° 45′ 03.10″ 22° 50′ 07.00″
- West Calkali (*Calkali Höhe No.573*) 40° 45′ 03.07″ 22° 50′ 07.09″
- Signal Boche (*Smeirkaja Signal No.1556*) 40° 45′ 03.10″ 22° 50′ 06.92″
- Means 40° 45′ 03.12″ 22° 50′ 07.0″

"The difference between these values and those accepted by the Germans gives a discrepancy between the two systems, *i.e.*, 3″ in latitude and 7″ in longitude. From the respective rectangular co-ordinates, the mean differences for the false bearings is 18′ 06″, as opposed to 17′ 59″, the convergency due to the distance between the origins, or a difference in azimuth between the two systems of only 7″. Similarly, the mean difference in lengths of sides is 1.85 metres for a mean side of 12,190 metres or 0.015%. The difference in height averages 1 m. The rectangular coordinates on the British system of the German point of origin are X = 134,212, Y = 137,353."

(*The British Triangulation in Macedonia 1916–1918,* by Lt Col H. Wood, R.E., 15 November 1918) https://salonikacampaignsociety.org.uk/publications-and-dvds/scs-digital-collection/.)

(*Personal communication, Mr. Keith Edmonds, Membership Secretary, Salonika Campaign Society, 8 April 2022.*)

"In the epoch between the two World Wars (1918–1941), the Military Geographic Institute (MGI) of the Kingdom of Yugoslavia (*Vojni Geografski Institit Kraljevine Jugoslavije*), the agency responsible for geodetic survey and mapping, made a great effort towards unification of the triangulation and production of a uniform map covering the entire national territory a considerable part of which never was surveyed. WWII prevented the completion of this project. After WWII, under different circumstances, the Geographic Institute of the Yugoslav People's Army (*Geografski Institut Jugoslovenske Narodne Armije-GIJNA*) has been quite successfully proceeding with the work initiated by its predecessor. From 1917–1924, the Clarke 1880 ellipsoid ($a = 6,378,249.145$, $1/f = 293.465$ – Ed.) was used by the Military Geographic Institute. In order to obtain geographic coordinates uniform with the geographic coordinates of the Austro–Hungarian Military Triangulation which covers the western part of Yugoslavia and because the Direction General of Cadaster and State Domains (*Generalna Direkcija Katastra I Državnih Dobara*) already had adopted the Bessel ellipsoid, the geographic coordinates of Serbian triangulation referring to Clarke 1880 ellipsoid were transformed to (the) Bessel (1841) ellipsoid ($a = 6,377,397.155$, $1/f = 299.1528128$ – Ed.). The degree survey which includes the arc along the 22° meridian observed in 1927–1930 was computed on the International (Hayford 1909) ellipsoid ($a = 6,378,388$, $1/f = 297$ – Ed.) for the purposes of the International Geodetic Association.

"The Military Geographic Institute of the Yugoslavian People's Army adopted for its new topographical survey and map compilation the Gauss-Krüger (Transverse Mercator) projection, but the maps compiled prior to WWII, still reproduced and predominantly used, are constructed in the polyhedric projection. (*Identical to the Local Space Rectangular projection discussed in the Manual of Photogrammetry, 5th & 6th editions, Chapter 3 – Ed.*)

"The Gauss-Krüger (Transverse Mercator) projection of 3° zones with starting meridian Greenwich, central meridian 15°, 18°, 21° (*only 21° for Macedonia – Ed.*) East of Greenwich and scale factor 0.9999, used since 1924 in the Yugoslav Cadastral survey, was in 1949 introduced in the construction of sheets of the Yugoslav military maps. (*False Easting = 500 km. – Ed.*) The sheets are cut in the graticule system of the International map. The l33, L34, K33 and K34 sheets at 1:1,000,000 scale with dimensions $4°_\varphi \times 6°_\lambda$ cover the area of Yugoslavia.

"The polyhedric projection. In an attempt to assure continuity with the Austro-Hungarian mapping the Military Geographic Institute of the Kingdom of Yugoslavia in the construction of the plane table sheets and maps used the polyhedric projection with starting meridian of Paris (20° Ease of Ferro or 2° 20′ 13, 98″ East of Greenwich). The sheets were cut along the meridians and parallels.

"In 1922, baselines were observed at Strumica (6,623.806 m±1/2,158,000) and at Prilep (5,982.555 m ±1/5,489,000) with Jäderin base apparatus. Concerning the accuracy of measurement of base lines, it should be mentioned that the ±ΔS/S fraction given with the lengths of base line represent so-called internal accuracy of the base line derived from the difference between the forward and backward measurements. This internal error does not include systematical errors caused by the apparatus and its constants (calibration of invar wires by the *Bureau International des poids et measures* in Sevres, France is carried out by a mean error up to ± 22 microns which would produce an accumulation of systematical error up to 0.92 mm/km). After a careful analysis of the Strumica base line the Federal Geodetic Administration published the following mean total relative error:

Strumica base line $E_r = \pm 9.4$ mm/6,623.806 m = ±1/7,000,000 or ±1.4 mm/km. The density of triangulation in Macedonia is 1 trig point to 1.6 km².

"The vertical control of Macedonia refers to the leveling datum Trieste, Molo Sartorio of which the elevation of starting benchmark within one year (1875) of observations of tidal gauge was defined to be +3.352 meters above the mean sea level of the Adriatic Sea" (*Mapping of the Countries in Danubian and Adriatic Basins, Andrew M. Glusic, Army Map Service Technical Report No. 25, June 1959*).

The Hermannskogel 1871 Datum origin is at: $\Phi_o = 48° \ 16′ \ 15.29″$ N, $\Lambda_o = 33° \ 57′ \ 41.06″$ East of Ferro, where Ferro = 17° 39′ 46.02″ East of Greenwich and azimuth to Hundsheimer is $\alpha_o = 107° \ 31′ \ 41.7″$ and is referenced to the Bessel 1841 ellipsoid. *Technical Report 8350.8* lists three-parameter transformation values for Yugoslavia (prior to 1990) for Slovenia, Croatia, Bosnia and Herzegovina, and Serbia but not for Macedonia.

According to a 2021 paper by Prof. Bashkim Idrizi of the University of Prishtina, **from** MGI 1901 **to** ETRS89, $\Delta X = +517.4399$ m, $\Delta Y = +228.7318$ m, $\Delta Z = +579.7954$ m, Rx = −4.045″, Ry = −4.304″, Ry = +15.612″, δs = −8.312 ppm. Also **from** MGI 1901 **to** WGS 84, $\Delta X = +521.748$ m, $\Delta Y = +229.489$ m, $\Delta Z = +590.921$ m, Rx = −4.029″, Ry = −4.488″, Ry = +15.521″, δs = −9.78 ppm**.**

https://www.proc-int-cartogr-assoc.net/4/45/2021/ica-proc-4-45-2021.pdf.

Madagascar
Faritany boundary
National capital
Faritany capital
Railroad
Road
0 50 100 Kilometers
0 50 100 Miles
Vila de Mocimboa da Praia
Grand Comore
Moroni
Foumbouni
Anjouan
Moheli
COMOROS
Dzaoudzi
Mayotte
(Administered by France, claimed by Comoros)
Îles Glorieuses
(FRANCE)
Cap d'Ambre
Antsiranana
Nosy Mitsio
Sosumau
Nosy Be
Hellville
Ambilobe
Vohimarina
ANTSIRANANA
Sambava
Pemba
MOZAMBIQUE
Nacala
Lumbo
Moçambique
Antsohihy
Andapa
Antalaha
Befandriana
Maroantsetra
Mozambique Channel
Mahajanga
Marovoay
Île Chesterfield
MAHAJANGA
Maevetanana
Andilamena
Nosy Boraha
Île Juan de Nova
(FRANCE)
Fenerive
Ambakireny
Ambatondrazaka
Maintirano
Toamasina
TOAMASINA
INDIAN
OCEAN
Îles Barren
Ankavandra
Tsiroanomandidy
Antananarivo
Brickaville
ATANANARIVO
Moramanga
Miandrivazo
Ambatolampy
Antsirabe
Mahanoro
Morondava
Ambositra
Nosy Varika
Ambohimahasoa
Mananjary
Fianarantsoa
FIANARANTSOA
Morombe
Ambalavao
Manakara
TOLIARY
Ihosy
Île Europa
(FRANCE)
Farafangana
Toliary
Ampanihy
Faradofay
Ambovombe
40
45
50
15
20
25

The Republic of Madagascar

Madagascar is the fourth largest island in the world, its length is almost 1,600 km (995 miles) and its maximum width is 576 km (360 miles). Tsaratanana Massif in the northern part of Madagascar is the highest point at 2,880 m (9,449 ft). The island is obliquely oriented to the cardinal directions on a northeast to southwest angle, and this fact has had a whopping effect on the development of mathematical cartography for the island. The centrally located capitol of the Republic of Madagascar is Antananarivo, and other major cities include Toliara in the south, Mahajanga in the northwest, Antsiranana in the north, and Toamasina in the east. The port of Hell-Ville is on the island of Nosy Bé off the northwest coast.

The Portuguese discovered the island of Madagascar in 1500, which was in the same era as their explorations of the entire eastern coast of Africa. The French established some colonial stations temporarily in the 17th century and re-established their presence in the 18th century. The local native Hovas group signed a treaty with France in 1868, and France declared that Madagascar was a French protectorate in 1882. The French reorganized the administrative system of Madagascar in 1924 and independence was achieved in 1960.

Original survey observations were performed by Owen of the British Royal Navy in 1824–1825, and French Navy observations followed in 1830 and 1850. Geodetic triangulation for a hydrographic survey of the Bay of Diego Suarez was commenced by Colonial Engineer M. Grégoire in 1887. The baseline was measured for a length of 1,044.295 meters, and a repeat measurement differed by a millimeter. This is typical quality for the French in the late 19th century. The Antsifrana Datum of 1887 was established at the Antsirana pillar, where $\Phi_o = 12° 16' 25.5''$ South, $\Lambda_o = 46° 57' 36.2''$ East of Paris. The defining azimuth was determined from Antsirana to signal station Oronjia as: $\alpha_o = 79° 12' 19''$. Antsirana 1887 Datum is referenced to the Clarke 1880 ellipsoid, where $a = 6,378,249.145$ m and $1/f = 293.465$. The Hellville Datum of 1888 was established at the meridian pillar, where $\Phi_o = 13° 24' 20.7''$ South, $\Lambda_o = 45° 57' 05.0''$ East of Paris. The defining azimuth was determined from the observation tower in Hellville to the fire beacon at station Tany-Kely as: $\alpha_o = 197° 25' 18''$. Hellville 1888 Datum is also referenced to the Clarke 1880 ellipsoid. The Mojanga Datum of 1888 was established at the pillar in Mojanga, where $\Phi_o = 15° 43' 24.2''$ South, $\Lambda_o = 43° 58' 54.4''$ East of Paris. Although the Clarke 1880 ellipsoid was referenced, no record exists of the defining azimuth for this tiny Datum. Note that all three of these local Datums in northwest Madagascar also defined Hatt Azimuthal Equidistant Grids for the respective hydrographic surveys. The usual rectangular coordinate systems (*Systémés Usuel*) were implemented in quadrant systems instead of in false origin systems. Simultaneously, the French Army started using the Bonne pseudo-conic equal area projection for topographic surveys and plane table mapping of Madagascar. Note that the Bonne was the projection "de rigueur" in Europe at the time, and it was the basis of the *Depôt de la Guerre* map series of France. The quadrant system was used for the topographic maps of Madagascar as well as for computation of the triangulation.

The "Navy at the Antsirana pillar where: $\Phi_o = 12° 16' 20.3''$ South (determined by Driencourt in 1904), and $\Lambda_o = 46° 57' 36.2''$ East of Paris (no change from Favé in 1887). A later report (date unknown) gives a local Hatt Grid in the vicinity having an origin where: the Central Meridian, $(\lambda_o) = 46° 55' 34.669''$ East of Paris, the Latitude of Origin, $\varphi_o = 12° 16' 26.148''$ South, a False Easting of 80 km, and a False Northing of 30 km. Considering the coordinate precision reported, the International 1924 ellipsoid was presumably used, where $a = 6,378,388$ m and $1/f = 297$.

The *Nosy Vé* Datum of 1907 was established for the hydrographic survey of the waters off the port of *Tuléar* (Toliara) in southwest Madagascar. The origin is at the meridian pillar, where $\Phi_o = 23°$

39′ 00.20″ South, $\Lambda_o = 41°$ 16′ 44.2″ East of Paris. Hellville 1888 Datum and the Hatt Azimuthal Equidistant Grid also used the quadrant system.

In 1926, the Hellville 1888 Datum was re-defined in terms of longitude, where $\Lambda_o = 48°$ 16′ 28.95″ (±1.5″) East of Greenwich. Presumably, this change was identified as the Hellville Datum of 1926. Although the original meridian pillar was gone, the new origin was observed to have the coordinates of X = −28.1 m, y = −2.6 m (note: X, y sic.).

In 1928 Laborde, a colonel in the French artillery wrote that the Bonne projection chosen for the *Service Géographique de Madagascar* was as bad a choice as it had been for France. "Beginning in 1924, the defects of the Bonne Projection became totally inconvenient for use on the first part in geodetic work for the calculation of geodetic coordinates on the ellipsoid, and on the other part the rectangular topographic surveys accumulated a substantial error. In 1925, regions quite removed from the Bonne Axis of projection, notably the delta region of Mangoky in the southwest part of Madagascar, attained errors of 22′ for angles and $\pm^1/_{580}$ for lengths. It was decided to find another solution urgently so that the quality would not be compromised for the survey campaign of 1925. Pressed for time, a local provisional projection was adopted in which the distortions were practically negligible within the region of application. Furthermore, difficulty was encountered in recruiting specialized personnel, and in 1926, the triangulation operations were constrained for lack of officers. Since training had to be furnished to the personnel because of the isolation of the survey parties due to the topography; the field calculations had to be simple to employ and therefore rectangular coordinates were used. Since the Bonne Projection was useless, provisional local coordinates systems were improvised. Between 1925–26 four local projections were put into service based on the Gauss Conform Projection. (*Ed. note: The actual math model used for all four was the Gauss-Schreiber Transverse Mercator.*) The Laborde Projection, adopted in 1926, was put into full use for the reduction of the calculations of the 1926 geodetic survey campaign and for the topographic survey campaign of 1927."

The Laborde Projection is a triple conformal projection, designed to have two lines of zero scale error oblique to the meridian. The projection of the ellipsoid onto the plane is carried out in three steps; the first is the projection of the ellipsoid on the sphere. The second step is a Gauss-Schreiber Transverse Mercator projection of the sphere onto a cylinder that is symmetrically secant about the central meridian of the projection. The third step in forming the Laborde Projection is a conformal distortion of the plane through a rotation in the secant case. Beginning with the first step (the conformal projection of the ellipsoid on the sphere), Laborde followed Gauss. Gauss showed that the ellipsoid could be mapped conformally on the sphere in such a way that the distortions in the neighborhood of some chosen point are of the third order or higher. Also note that the indirect or double projection through the sphere is not exactly the same as the direct one in that they differ concerning the higher order terms of the expanded transformation formulae. The conformal projection of the ellipsoid on the sphere was used by the Prussian Land Survey in 1876–1923 as the foundation for a double projection, and it is currently used in computations for cadastral retracement. This triple projection of Laborde's using the Gaussian Sphere to achieve an Oblique Mercator is fundamentally different from the method used some 20+ years later by Brigadier Martin Hotine of the UK. Most commercial software houses that I have noticed are abysmally ignorant on this point.

In reference to the four provisional coordinate systems mentioned earlier by Laborde, he later commented on the relation of these local systems to the Laborde Projection. His discussion referred to the 1925–1926 local triangulation situated in the region of *Bemolanga-Morafenobe*. The work was carried out over 4,200 km² and was comprised of 127 geodetic triangulation points calculated on a provisional Gauss-Schreiber Transverse Mercator coordinate system. Two points of an old reconnaissance triangulation were chosen as a base for a local grid. This grid was "une projection conforme de Gauss limiteé" (a Gauss-conformal projection of limited extent). Laborde further commented that the transformation from the provisional system to the Laborde Oblique Mercator was one translation, one rotation, and a change of scale without any deformation, "sans acune déformation." This is an implementation of one of the most valuable properties of conformal projection: the ability to maintain the undistorted shape of surveys and mappings through simple three-parameter (translation in X, translation in Y, and rotation) transformations! In my coverage on Belgium, I discussed how the U.S. Army Map Service implemented the European Datum 1950 by such three-parameter transformations (on the complex plane) with the UTM as a unifying tool. Laborde did this back in the 1920s, and the Russians do it in an attempt to obfuscate the relation between their "System 42 Datum" (Pulkovo origin) and the WGS 84 Datum. The Russians still do it in the 21st century by using their screwy "System 1963" series of local coordinates ("maskirovka") which are merely translated and rotated Gauss-Krüger Grids. However, Laborde's implementation for Madagascar was successful in *simplifying* the task!

Laborde's command of the *Service Géographique de Madagascar* also resulted in the establishment of the unifying coordinate system for the colony. After long observation, the coordinates of the observatory in the capital of Antananarivo were used as the origin of the Madagascar Datum of 1925, where $\Phi_o = 18°\ 55'\ 02.10''$ South, $\Lambda_o = 47°\ 33'\ 06.45''$ East of Greenwich. Of course, the ellipsoid of reference is the International 1924. The defining parameters were published by the retired General Jean Laborde in the seminal work, "Traitè des Projections des Cartes Gèographiques a L'Usage des Cartographes et des Geodèsiens," by L. Driencourt & J. Laborde, Paris, 1932. Remember Driencourt? He performed the 1904 final astronomical observations for the longitude of Antsirana Datum of 1906. Driencourt later became the Hydrographer of the French Navy (long before the four-volume book was published). Those parameters are Latitude of Origin $(\Phi_o) = 21^G$ South (18° 54'), Central Meridian $(\lambda_o) = 21^G$ East of Paris (46° 26' 13.95'' East of Greenwich), Scale Factor at Origin $(m_o) = 0.9995$, False Easting = 400 km, False Northing = 800 km, Azimuth of Axis of Strength $(\alpha) = 21^G$, and the False Easting at the False Origin = 1,000 km. For example, for geodetic station Morondava, $\varphi = 20°\ 17'\ 34.2443''$ South, $\lambda = 44°\ 16'\ 52.6884''$ East of Greenwich, Easting (Y) = 174,896.590 m, Northing (X) = 644,433.000 m, (H = 31.421 m).

Back in September of 1992, the late John P. Snyder asked for a copy of my 1982 notes on the Laborde Projection of Madagascar. After John diddled a while with trying to see how close the Hotine Rectified Skew Orthomorphic (RSO) Oblique Mercator would fit, he concluded that he could match my transformations to only "about 0.08 meters for my part of Madagascar ... worse of course away from the island." This approximation is of interest for U.S. Geological Survey medium-scale cartographic applications, but it is useless for geodetic work and for large-scale engineering mapping. I have seen examples of much (apparent) hand wringing done by software organizations (U.S. and elsewhere), that have attempted to foist this sort of "work-around" on government agencies in Madagascar. The grid is commonly used at great distances from the mainland because of the existence and official publication of a *False Easting at False Origin!* The RSO is *not* a valid substitute for the Laborde of Madagascar.

The U.S. National Imagery and Mapping Agency (NIMA) publishes a parameter set for transforming from the Madagascar Datum of 1925 ("Tananarive 1925") to WGS 84 Datum. However, they offer no information on accuracy of, or the origin of the parameters. I tried a nine-point solution and I obtained slightly different results as: $\Delta X = -191.745$ m, $\Delta Y = -226.365$ m, and $\Delta Z = -115.609$ m. The resultant rms for Latitude = 0.215 m, for Longitude = 0.323 m, and for ellipsoid height (h) = 9.958 m. The WGS 84 Datum observed coordinates for Morondava are: $\varphi = -20°\ 17'\ 39.079425''$ South, $\lambda = 44°\ 16'\ 51.545430''$ East of Greenwich, and h = 31.421 m.

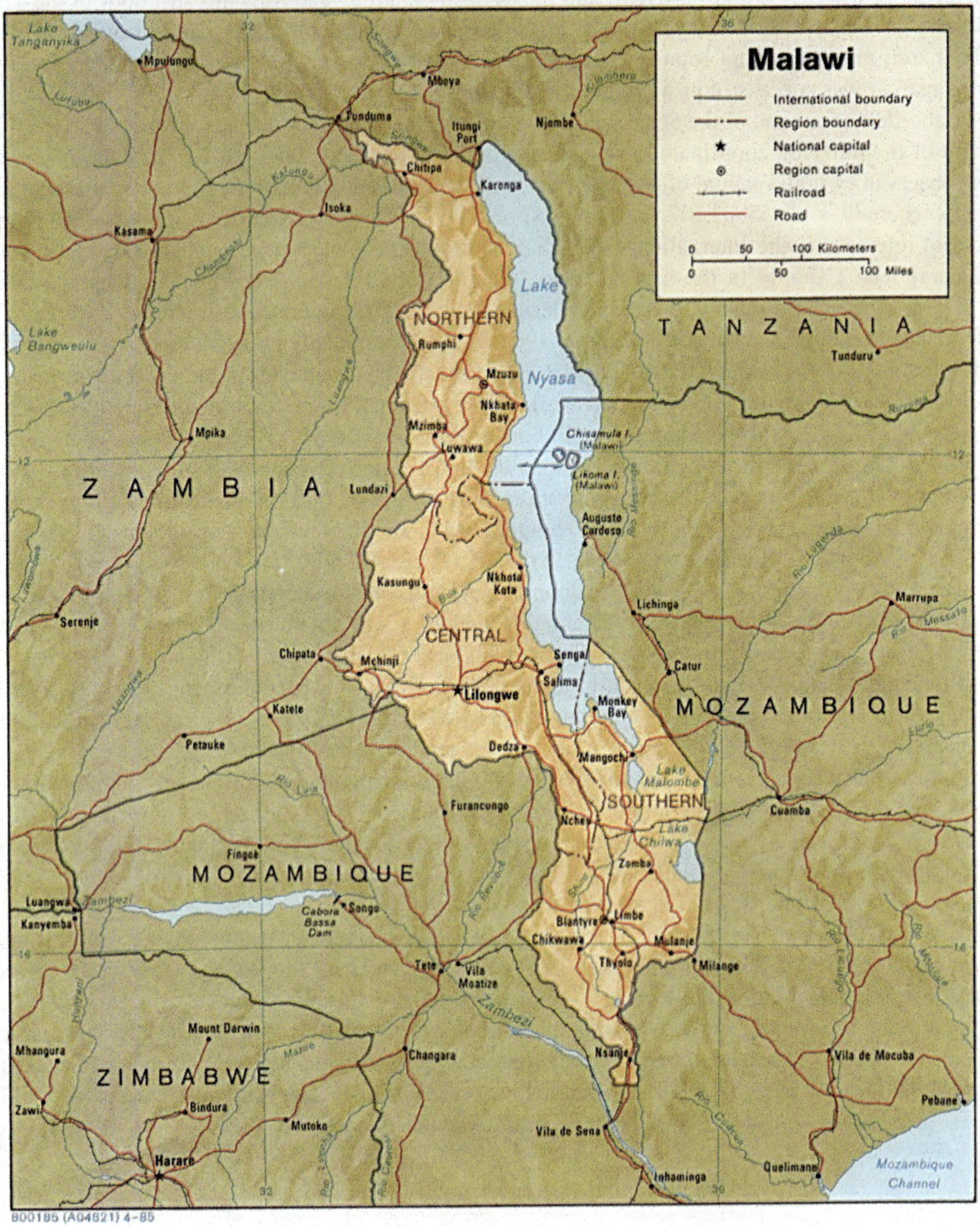

800185 (A04821) 4-85

Republic of Malawi

Bordered by Moçambique (1,569 km), Tanzania (475 km), and Zambia (837 km), the terrain consists of narrow elongated plateau with rolling plains, rounded hills, and some mountains. The lowest point is at the junction of the Shire River and the international boundary with Moçambique (37 m), and the highest point is Sapitwa (*Mount Mlanje*) (3,002 m) (*World Factbook*, 2011).

"Hominid remains and stone implements have been identified in Malawi dating back more than 1 million years, and early humans inhabited the vicinity of Lake Malawi 50,000 to 60,000 years ago. Human remains at a site dated about 8000 *B.C.* show physical characteristics similar to peoples living today in the Horn of Africa. At another site, dated 1500 *B.C.*, the remains possess features resembling Negro and Bushman people. Malawi derives its name from the Maravi, a Bantu people who came from the southern Congo about 600 years ago. On reaching the area north of Lake Malawi, the Maravi divided. One branch, the ancestors of the present-day Chewas, moved south to the west bank of the lake. The other, the ancestors of the Nyanjas, moved down the east bank to the southern part of the country. By *A.D.* 1500, the two divisions of the tribe had established a kingdom stretching from north of the present-day city of Nkhotakota to the Zambezi River in the south, and from Lake Malawi in the east, to the Luangwa River in Zambia in the west. Migrations and tribal conflicts precluded the formation of a cohesive Malawian society until the turn of the 20th century. In more recent years, ethnic and tribal distinctions have diminished. Regional distinctions and rivalries, however, persist. Despite some clear differences, no significant friction currently exists between tribal groups, and the concept of a Malawian nationality has begun to take hold. Predominately a rural people, Malawians are generally conservative and traditionally nonviolent. Although the Portuguese reached the area in the 16th century, the first significant Western contact was the arrival of David Livingstone along the shore of Lake Malawi in 1859. Subsequently, Scottish Presbyterian churches established missions in Malawi. One of their objectives was to end the slave trade to the Persian Gulf that continued to the end of the 19th century. In 1878, a number of traders, mostly from Glasgow, formed the African Lakes Company to supply goods and services to the missionaries. Other missionaries, traders, hunters, and planters soon followed. In 1883, a consul of the British Government was accredited to the 'Kings and Chiefs of Central Africa,' and in 1891, the British established the Nyasaland Protectorate (Nyasa is the Yao word for 'lake'). Although the British remained in control during the first half of the 1900s, this period was marked by a number of unsuccessful Malawian attempts to obtain independence. A growing European and U.S.-educated African elite became increasingly vocal and politically active – first through associations, and after 1944, through the Nyasaland African Congress (NAC). A new constitution took effect in May 1963, providing for virtually complete internal self-government. The Federation of Rhodesia and Nyasaland was dissolved on December 31, 1963, and Malawi became a fully independent member of the Commonwealth (formerly the British Commonwealth) on July 6, 1964" (*Department of State Background Notes*, 2011).

"The South African network formed a gridiron system of triangles arranged in eight circuits. From 1906 to 1907, Captain Gordon surveyed a gap of 2 degrees in the Arc, between Simms's chain in Southern Rhodesia and the newly completed triangulation in the Transvaal. The Transvaal triangulation was joined at points Pont and Dogola, and Simms's chain was connected at Standaus and Wedza. In the meantime, thanks to Gill's further effort, funds were obtained to continue Simms's chain northwards. Between 1903 and 1906, Dr Rubin carried out this work (Gill, 1933). The large difficulties, experienced by Rubin's party, were similar to that of Simms's: problems with haze and light, strong winds, lack of roads, transport and labor. In spite of these problems, Rubin managed to extend the chain of triangulation for almost 800 km; from Manyangau, in Southern Rhodesia, to

Mpange, in Northern Rhodesia (now Zambia). When Sir David Gill retired, in 1907, the Geodetic Survey of South Africa was completed, and the Arc of the 30th Meridian extended from the Cape to the southern shores of Lake Tanganyika. Between 1908 and 1909, a section of 2 degrees in amplitude was measured in Uganda: from the Semliki baseline, at latitude 1° North, across the Equator, to latitude 1° South. The work on the southern portion of the Arc was resumed only in 1931, when Major Hotine, picked up Rubin's points in Northern Rhodesia and, in 1933, brought the triangulation, through Tanganyika (now Tanzania), up to Urundi (now Burundi). Later, the Tanganyika's Survey Division continued the survey of the 400 km of the Arc (Rowe, 1938), between Kigoma and Uganda, across Urundi and Ruanda (now Rwanda). Since 1937, the Arc extended from the Cape to the Equator" (*Arc of the 30th Meridian, Tomasz Zakiewicz, Cairo, Egypt, April 16–21, 2005*). The Arc of the 30th Meridian is referenced to the Cape Datum of 1950 where the astronomic coordinates of the initial point of the Cape Datum near Port Elizabeth are for Buffelsfontein, where $\Phi_o = 33° 59' 32.000''$ S and $\Lambda_o = 25° 30' 44.622''$ E. The ellipsoid of reference is the Clarke 1880, where $a = 6,378,249.145$ m and $1/f = 293.4663077$.

The Moçambique – Malawi boundary is 1,569 km in length and is demarcated. It traverses Lake Nyasa for about 205 miles including lines around Likoma Island and Chisumulu Island, which are part of Malawi. Southward from Lake Nyasa to the Malosa River, the boundary extends along straight-line segments for 312 km passing through both Lake Chiuta and Lake Chilwa. It follows consecutively the thalwegs of the Malosa, Ruo, and Shire rivers downstream for 240 km. The boundary then continues northwestward to the Zambia tripoint utilizing features along the Shire-Zambezi and the Lake Nyasa-Zambezi drainage divides for most of the remainder of the distance.

The largest classical horizontal geodetic datum of Moçambique is the Tete Datum of 1960. Its origin is at the station at the NW end of the Tete Baseline (MGM 799), where $\Phi_o = 16° 09' 03.058''$ S, $\Lambda_o = 33° 33' 51.300''$ E. The reference azimuth to station Caroeira (MGM 40), $\alpha_o = 355° 50' 21.07''$ from south, and $H_o = 132.63$ m. The southern third of Malawi is covered by the Tete Datum chain of quadrilaterals. In 1995, a comprehensive adjustment of the entire geodetic network of Moçambique was initiated by Norway Mapping in collaboration with the government. The project was concluded in January of 1998, and the result was a 32-point constrained adjustment of 759 two-dimensional triangulation points throughout the country and the designation of a new datum called MOZNET/ITRF94, compatible with the WGS84 Datum. A seven-parameter Bursa-Wolf transformation was developed, but the national model yields residual errors as high as 30 m. Four "regional" models were developed, but these accuracies vary between 1 and 10 m, depending on the "region." The MOZNET 98 adjusted coordinates of Base Tete NW are $\varphi = 16° 09' 07.0480''$ S, $\lambda = 33° 33' 49.7778''$ E.

The parameters published by DMA/NIMA for the Malawi Arc 1950 Datum to WGS84 are: $\Delta X = -161$ m ± 9 m, $\Delta Y = -73$ m ± 24 m, $\Delta Z = -317$ m ± 8 m.

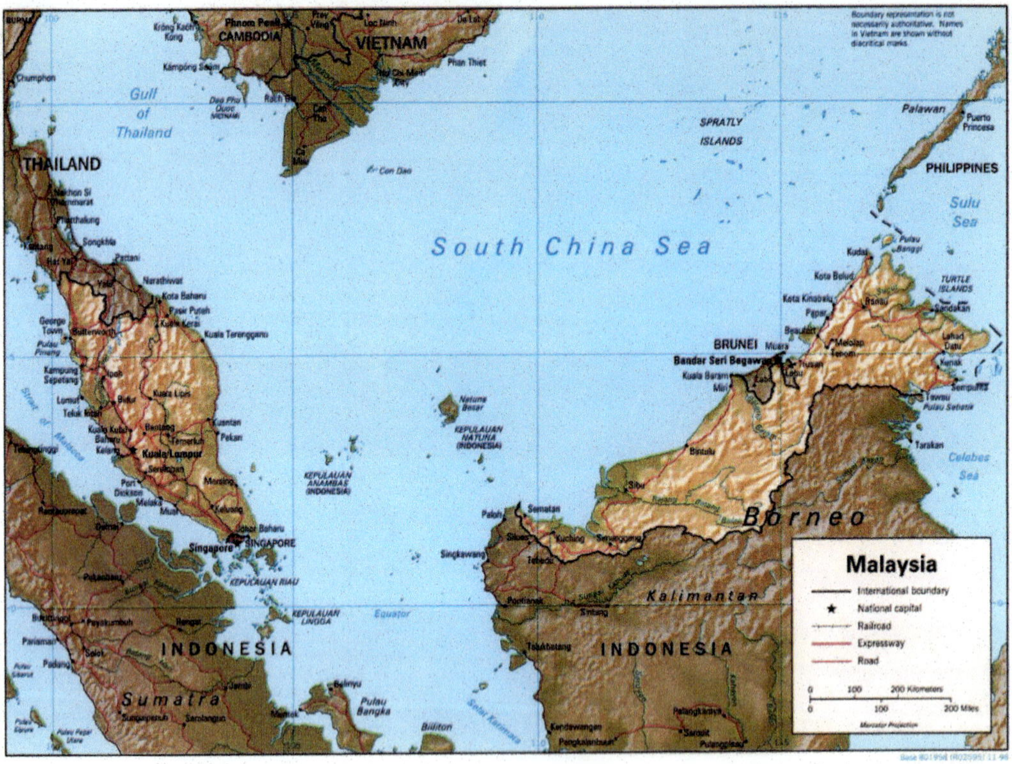

Base 801956 (R02595) 11-96

Malaysia

Bordered by Thailand (506 km), Brunei (381 km), and Indonesia (1,782 km), Malaysia is slightly larger than New Mexico and has a coastline totaling 4,675 km, Peninsular Malaysia (2,068 km) and East Malaysia (2,607 km). The lowest point is the Indian Ocean (0 m), and the highest point is Gunung Kinabalu (4,100 m) (*The 2008 World Factbook*).

The south Malay peninsula, Sumatra, central Java, and east Borneo, has been settled from *circa* 1st century *B.C.* by Hindu Pallavas from southeast India. In the early 15th century, the Malay states came under Chinese influence, but later Muslim traders gained ascendency in several states such as Malaka. After the early 16th century Malaya was dominated by the Portuguese who were succeeded by the British (*Merriam-Webster's Geographical Dictionary* 3rd Edition, 1997). Set up as the Union of Malaya in 1946, later formed the Federation of Malaya which became independent in 1957. Malaysia was formed in 1963 when the former British colonies of Singapore and the East Malaysian states of Sabah and Sarawak on the northern coast of Borneo joined the Federation. Singapore seceded from the Federation in 1965.

"The Survey Department, Federated Malay States and Straits Settlements (*FMS&SS*) accomplished most of the basic surveying and mapping of Malaya between 1923 and 1945. Topographic sheets for most of the country were based on planetable surveys at 1:31,680 and 1:63,360 (*2 inches to a mile and 1 inch to a mile scales – Ed.*); for areas on the east coast, however, the sheets were based on sketch and reconnaissance surveys and poor planetable work. Numerous city plans, small-scale maps, and planimetric state maps also were produced. Most of the work done by this agency has been superseded or revised by World War II mapping" (*TM 5-248, Foreign Maps,* 1963).

"The first attempt at triangulation survey was made in Penang in 1832 by Lieutenant Woore of the British Royal Navy. About 1880, the governor of the Straits Settlements, Sir Frederick Weld conceived the idea of introducing the Torrens System of registration of land titles into the Straits Settlements. Accordingly in 1882, he commissioned Sir W. E. Maxwell, who was at the time the Commissioner of Lands in the Straits Settlements to go to Australia to report on the suitability of the Torrens Act. Passed by the South Australian Legislature in 1858 with a view to its introduction in the Straits Settlements. Maxwell returned an enthusiastic supporter of the system and brought in legislation to affect its introduction. Bitter opposition to his proposals was aroused and lengthy arguments ensued. Eventually it was the Malay States that took the initiative to introduce a modified Torrens System. Maxwell realized that the Torrens System could not be operated successfully unless it was based and maintained on a survey sufficiently accurate to permit the re-identification of property boundaries. Arising from this need, the various States began to organize and form their own Survey Departments. The history of the Department of Survey and Mapping is therefore the history of these various State Survey Departments and their gradual amalgamation into a federal organization as it is today. In 1885, the Johore Survey Department began under Dato Yahya Bin Awal-ed-din as Chief Surveyor, trained as a surveyor by Major McCallum in Singapore in 1868. Also in 1885, H.G. Deanne, a contract surveyor from Ceylon was appointed by the Public Works Department, Perak, to carry out the trigonometrical survey of Perak. He measured the 4.6-mile Larut baseline and carried out astronomical determinations for latitude and azimuth near Taiping. This trigonometrical survey in Perak together with the Penang and Province Wellesley triangulations and Malacca Triangulation (1886–1888), laid the foundation of the existing control framework. … by the end of 1901, the major triangulation of Perak and Selangor had been completed and work had been in progress in Negeri Sembilan since 1899. This period also witnessed the commencement of trigonometrical surveys in various parts of the country." The two earliest datums extant in Malaysia were the Perak System and the Bukit Asa System, both referenced to the Everest 1830 ellipsoid. However, the quality of

DOI: 10.1201/9781003307785-125

the early works were so inconsistent that it was decided to re-observe the principal triangles of the general triangulation with the object of bringing the work up to modern standards. This triangulation scheme in Peninsular Malaysia was known as the Primary or Repsold Triangulation which was completed in 1916. The datum origin is at Kertau, where astronomical latitude and azimuth were determined. The longitude of the datum is derived from the value $\Lambda_o = 100°\ 20'\ 44.4''$ E. for Fort Cornwallis Flagstaff, Penang, obtained by Commander M. Field in 1893 by the exchange of telegraphic signals at Singapore, Old Transit Circle. The longitude accepted for Singapore was that fixed in 1881 by Lt. Commander Green with respect to the Transit circle of Madras Observatory, the corresponding longitude of the latter being taken as $\Lambda_o = 80°\ 14'\ 51.51''$ E (*Geographical Positions in Malaya and Siam, A. G. Bazley, Empire Survey Review No. 30, pp. 450–457*). At that time, there were two grids used for Peninsular Malaysia, the Malay Cassini-Soldner Grid, and the Johore Cassini-Soldner Grid, both referenced to the Indian Datum of 1916, the ellipsoid of revolution was the Everest 1830, where $a = 6,974,310$ Indian Yards and $1/f = 300.80$. The Malay Grid projection origin was $\varphi_o = 4°\ 00'$ N, $\lambda_o = 102°$ E, Scale Factor at Origin (m_o) = 1.0, False Easting = 500,000 yards, and False Northing = 300,000 yards. The Johore Grid projection origin was $\varphi_o = 2°\ 02'\ 33.30''$ N, $\lambda_o = 103°\ 33'\ 45.93''$ E, Scale Factor at Origin (m_o) = 1.0, False Easting = 450,000 yards, and False Northing = 300,000 yards (*Personal communication, DMATC Form 5000–1, 6 December 1974*). In 1948, it was replaced by a new system known as the Malayan Revised Triangulation (MRT). This was followed by a lengthy process of additional measurements and re-computation until 1968. As a result, this system is then referred to as MRT68. On the other hand, the geodetic network used in Borneo is called the Borneo Triangulation (BT68).

The Malayan Revised Triangulation has been used for geodetic, mapping, cadastral, and several other activities since 1948 in Peninsular Malaysia. This network consists of 77 geodetic, 240 primary, 837 secondary, and 51 tertiary stations. This network is based on the conventional observations with many of the triangulation points dated as far back as 1885. The MRT 1948 has been adopted as a result of the re-computations of the earlier network together with the Primary (Repsold) Triangulation carried out between 1913 and 1916. The reference ellipsoid used for the MRT 1948 is the Modified Everest, where $a = 6,377,304.063$ m and $1/f = 300.8017$, and the MRT 1948 datum origin is *also* at Kertau, Pahang, where $\Phi_o = 03°\ 27'\ 50.71''$ North, $\Lambda_o = 102°\ 37'\ 24.55''$ East of Greenwich. The map projection for Peninsular Malaysia is the Rectified Skew Orthomorphic (RSO) developed by the late Brigadier Martin Hotine specifically for Malaysia and Borneo. The Hotine RSO defining parameters for MRT 1948 consist of: Conversion Factor (1 chain = 20.11678249 m *from Chaney & Benoit, 1896*), projection origin $\varphi_o = 4°\ 00'$ N, $\lambda_o = 102°\ 15'$ E, Scale Factor at Origin (m_o) = 0.99984, basic or initial line of projection passes through the Skew Origin at an azimuth of $(\gamma_o) = -\sin^{-1}(-0.6)$ or $323°\ 01'\ 32.8458''$, False Easting = 804,671 m, False Northing = zero. The Bursa-Wolf transformation from WGS84 to MRT48 (Kertau) listed by (Aziz, W., Sahrum, M., and Teng, C., in *Cadastral Reform in Malaysia: A vision to the 2000s*) is as follows: $\Delta X = +379.776$ m, $\Delta Y = -775.384$ m, $\Delta Z = +86.609$ m, scale = not given, $R_X = +2.59674''$, $R_Y = +2.10213''$, $R_Z = -12.11377''$ (*Note that the rotations as published are the left-handed convention favored by most Europeans and are opposite the convention used by the U.S. and Australian militaries – Ed.*) According to *TR 8350.2*, the transformation parameters **from** "Kertau 1948 Datum" **to** WGS 84 Datum are: $\Delta X = -11$ m ± 10 m, $\Delta Y = +851$ m ± 8 m, $\Delta Z = +5$ m ± 6 m, and the solution is based on six satellite stations. (*Notice the whopping magnitude of the ΔY component ... likely the result of errors in time transfers to determine longitudes. – Ed.*)

The Geodetic Datum of Malaysia 2000 (GDM 2000) has re-defined the Cassini-Soldner (cadastral) Grid origins of the various states of Peninsular Malaysia as follows: Johor State is @ GPS Station GP58, *Institut Haiwan Kluang*, where $\varphi_o = 02°\ 07'\ 18.04708''$ N, $\lambda_o = 103°\ 25'\ 40.57045''$ E, Cassini Northing = +8,758.320 m, Cassini Easting = −14,810.562 m, Offset Northing = +8,758.320 m, Offset Easting = −14,810.562 m. North Sembilan & Melaka State is @ GPS Station GP10, K. Perindustrian Senawang, Seremban, where $\varphi_o = 02°\ 40'\ 56.45149''$ N, $\lambda_o = 101°\ 58'\ 29.65815''$ E, Cassini Northing = −4,240.573 m, Cassini Easting = +3,673.785 m, Offset Northing = −3,292.026 m, Offset Easting = +3,915.790 m. Pahang State is @ GPS Station GP31, Sek. Ren. Keb. Kuala Mai, Jerantut, where $\varphi_o=03°\ 46'\ 09.79712''$ N, $\lambda_o=102°\ 22'\ 05.87634''$ E, Cassini Northing$=+6,485.858$ m, Cassini Easting$=-7,368.228$ m, Offset Northing$=+6,485.858$ m, Offset Easting$=-7,368.228$ m. Selangor State is @ GPS Station 251D, Felda Soeharto, K. Kubu Baharu, where $\varphi_o=03°\ 41'\ 04.73658''$ N, $\lambda_o=101°\ 23'\ 20.78849''$ E, Cassini Northing$=+56,464.049$ m, Cassini Easting$=-34,836.161$ m, Offset Northing$=+503.095$ m, Offset Easting$=-13,076.704$ m. Terengganu State is @ GPS Station P253, Kg. Matang, Hulu Terenganu, where: $\varphi_o=04°\ 58'\ 34.62672''$ N, $\lambda_o=103°\ 04'\ 12.99225''$ E, Cassini Northing$=+3,371.895$ m, Cassini Easting$=+19,594.245$ m, Offset Northing$=+3,371.895$ m, Offset Easting$=+19,594.245$ m. P. Pinang & S. Pereai State @ GPS Station P314, TLDM Georgetown, where $\varphi_o=05°\ 25'\ 17.46315''$ N, $\lambda_o=100°\ 20'\ 39.75707''$ E, Cassini Northing$=+62.283$ m, Cassini Easting$=-23.414$ m, Offset Northing$=+62.283$ m, Offset Easting$=-23.414$ m. Kedah & Perlis State is @ GPS Station TG35, Gunung Perak, Kuala Muda, where $\varphi_o=05°\ 57'\ 52.82177''$ N, $\lambda_o=100°\ 38'\ 10.93600''$ E, Cassini Northing$=0$ m, Cassini Easting$=0$ m, Offset Northing$=0$ m, Offset Easting$=0$ m. Perak State is @ GPS Station TG26, Gunung Larut Hiijau, Taiping, where $\varphi_o=04°\ 51'\ 32.62688''$ N, $\lambda_o=100°\ 48'\ 55.47811''$ E, Cassini Northing$=+133,454.779$ m, Cassini Easting$=-1.769$ m, Offset Northing$=+0.994$ m, Offset Easting$=-1.769$ m. Finally, for Kelantan State is @ GPS Station P243, B. Polis Melor, Kota Bharu, where $\varphi_o=05°\ 58'\ 21.15717''$ N, $\lambda_o=102°\ 17'\ 42.87001''$ E, Cassini Northing$=+8,739.894$ m, Cassini Easting$=+13,227.851$ m, Offset Northing$=+8,739.894$ m, Offset Easting$=+13,227.851$ m. The new GDM2000 RSO for Peninsula Malaysia retains the exact same parameters as for MRT 48 *except* for the basic or initial line of projection that passes through the Skew Origin at an azimuth of $(\gamma_o)=-\sin^{-1}(-0.6)$ or $323°\ 01'\ 32.86728''$, and the fact that the new ellipsoid of revolution is the GRS 80, where $a=6,378,137$ m and $1/f=298.2572221$.

The Borneo Triangulation network in Sabah and Sarawak consists of the Borneo West Coast Triangulation of Brunei and Sabah (1930–1942), and was established with the origin at Bukit Timbalai, Labuan Island in 1948, where $\Phi_o=05°\ 17'\ 03.55''$ North, $\Lambda_o=115°\ 10'\ 56.41''$ East of Greenwich. The reference ellipsoid used for the BT 1948 is the (*other*) Modified Everest, where $a=6,377,298.556$ m and $1/f=300.8017$. The BT68 results from the readjustment of the primary control of East Malaysia (Sabah, Sarawak plus Brunei) made by the Directorate of Overseas Surveys, United Kingdom (DOS), and the old Borneo West Coast Triangulation of Brunei and Sabah (1930–1942), the Borneo East Coast Triangulation of Sarawak and extension of the West Coast Triangulation in Sabah (1955–1960), and some new points surveyed between 1961 and 1968. The Hotine RSO defining parameters for BT 1948 and for BT68 consist of: Conversion Factor (1 chain$=20.11676512$ m *from Sears, Jolly, & Johnson,* 1927), projection origin $\varphi_o=4°\ 00'$ N, $\lambda_o=115°\ 00'$ E, Scale Factor at Origin $(m_o)=0.99984$, basic or initial line of projection passes through the Skew Origin at an azimuth of $(\gamma_o)=\sin^{-1}(0.8)$ or $53°\ 19'\ 56.9537''$, False Easting$=$False Northing$=$zero. The new GDM2000 RSO for Borneo retains the exact same parameters as for BT 1948 and BT 68 *except* for the basic or initial line of projection that passes through the Skew Origin at an azimuth of $(\gamma_o)=-\sin^{-1}(-0.8)$ or $53°\ 18'\ 56.91582''$, and the fact that the new ellipsoid of revolution is the GRS 80. According to *TR 8350.2*, the transformation parameters **from** "Timbalai 1948 Datum" **to** WGS 84 Datum (Sarawak and Sabah) are: $\Delta X=-679$ m ±10 m, $\Delta Y=+669$ m ±10 m, $\Delta Z=-48$ m ±12 m, and the solution is based on eight satellite stations.

Except where otherwise explicitly cited, the historical details and current transformation parameters have been obtained from *A Technical Manual on the Geocentric Datum of Malaysia (GDM2000),* Department of Survey and Mapping Malaysia, Kuala Lumpur, August 2003. Other Malaysian datums known to exist at one time or another include: Blumut, Johore, BS76, East Malaysia, Bukit Panau, Fort Cornwallis Flagstaff, Panang, Gun Hill, Negri Sembilan, Gunung Hijau Larut, Perak, Gunung Gajah Trom, Trengganu, Jesselton, North Borneo, Kuching Survey Origin, Sarawak 1930, Mount Robertson, Perak Kedah, Semporna, Sinyum, Pahang, Timbalai 1936, Timbalai 1938, and Timbalai 1968. Probably most of these "datums" were temporarily used for local cadasters as a result of triangulation chains observed within individual Malaysian States before comprehensive datum adjustments could be performed. The remainder were likely used for local hydrographic harbor surveys.

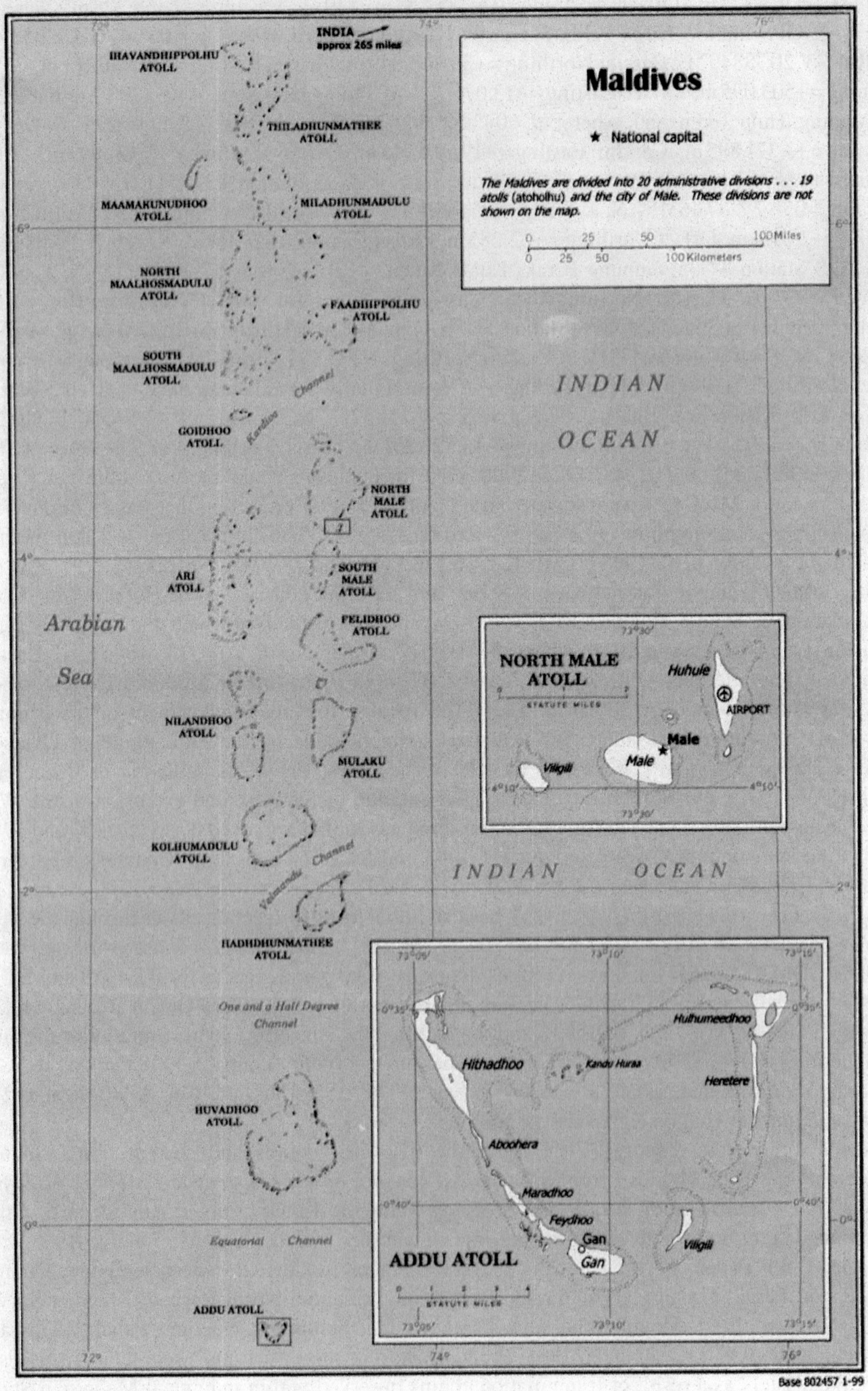

Base 802457 1-99

Republic of Maldives

The Maldives are about 1.7 times the size of Washington, DC, and is comprised of 1,190 islands grouped into 26 atolls (200 inhabited islands, plus 80 islands with tourist resorts). It is an archipelago with a strategic location astride and along major sea-lanes in the Indian Ocean. The coastline is 644 km in length, and 80% of the land area is <1 m above local mean sea level. The highest point is an unnamed location on Wilingili island in the Addu Atoll at 2.4 m (7.9 ft). There are 19 administrative divisions consisting of the following atolls: Alifu, Baa, Dhaalu, Faafu, Gaafu Alifu, Gaafu Dhaalu, Gnaviyani, Haa Alifu, Haa Dhaalu, Kaafu, Laamu, Lhaviyani, Maale (Male), the City of Male, Meemu, Noonu, Raa, Seenu, Shaviyani, Thaa, and Vaavu.

"Upon Conversion (*to Islam – Ed.*), in 1153 A.D., the Maldives was declared a Sultanate and the ruling monarch adopted the name of al *Sultan Muhammad Ibn Abdullah Siri Bavanaadittiya Mahaa Radhun*, popularly known as *Dharumavantha Rasgefaanu*" (*Republic of Maldives Official Website*). The Maldives was under European control from the 16th century and came under British protection in the 19th century. It became independent in 1965, and proclaimed itself a republic in 1968.

During World War II, the Maldives was still part of the British Indian Ocean Territory, and the British Grid that covered the entire area was the "Maldive-Chagos Belt." The defining Gauss-Krüger Transverse Mercator parameters include Latitude of Origin, (φ_o)=Equator, by definition, Central Meridian, (λ_o)=73° 30′ East of Greenwich, Scale Factor at Origin (m_o)=unity, False Easting=417,000 Indian yards, False Northing=1,450,000 Indian yards, the ellipsoid of reference was the Everest 1830, where a=6,974,310.60 Indian yards, $1/f$=300.8017, and 1 Indian yard=0.914399205 m). The limits of the Maldive-Chagos Belt were: North – Parallel of 8° N, East – Meridian of 76° East, South – Parallel of 10° South, and West – Meridian of 72° East. The standard British Military Grid color was not specified, but the color found published for this Belt was black. An example computation test point is: φ=6° 31′ 22.717″ S, λ=72° 00′ 36.426″ E, X=236,791.02 yds, Y=66,963.14 yds. (*Note that my own computation yielded X=236,791.016 yds., Y=66,963.138 yds. – Ed.*)

The name of the local Maldives astronomic datum is unknown, but in July 2004, Ruth Adams of the UK Hydrographic Office stated that "The difference between the original astro graticule and WGS84 Datum is over 3 miles" (*Hydrographic Journal*). The difference is primarily in the East-West component; that being a function of time-keeping chronometer error.

A recent issue of *Coordinates* magazine contained an article by V. Raghu Venkataraman, *et al.*, about the National Remote Sensing Agency (NRSA), Hyderabad, India, that performed a "Large scale topographical mapping for the Republic of Maldives." General photogrammetric details were discussed but no details of the local Grid parameters or transformation parameters from the local datum(s) to WGS84 were offered.

Thanks to John W. Hager, "I was able to garner some of the inscrutable details of the Maldives: For Gan 1970 located at AR1, Φ_o=0° 41′ 48.123″ S, Λ_o=73° 09′ 49.038″ E, International ellipsoid. For Gan located at Cag 11T, Φ_o=0° 42′ 02.204″ S, Λ_o=73° 09′ 44.854″ E, International ellipsoid. … I made a further note on Gan 1970 that I was not sure that it was the datum point, but another point connected to the datum point. The station designation, 'Cag,' has the earmark of being British, something to do with Chagos. I have two names but the same coordinates. First is Maradu (1923) at station "D", Φ_o=0° 39′ 43.62 S, Λ_o=73° 07′ 38.114″ E, Everest 1830 ellipsoid. This was from "Addu Atoll Survey," Aug. - Sept. 1942. Obviously, the position, presumably an astro, is much earlier. For the second, Manira at station Maralu (027), with the same coordinates but International ellipsoid.

I have a note questioning whether Manira is a typo for Mahira and also that the position for Mahira (8) is $\Phi_o = 0°\ 34'\ 27.56''$ S, $\Lambda_o = 73°\ 12'\ 44.77''$ E."

According to the NGA Technical Report *TR8350.2*, the three-parameter transformation from the local native Maldives Datum on the island of Gan ("Gan 1970" referenced on the International 1909 ellipsoid) to WGS84 Datum is: $\Delta X = -133$ m, $\Delta Y = -321$ m, $\Delta Z = +50$ m; the relation is based on only one observed point, and the uncertainty of each translation component is stated to be ±25 m. Thanks again to Mr. John W. Hager for his help on this enigma!

Mali
International boundary
Region boundary
National capital
Region capital
Railroad
Road
Track
0 100 200 Kilometers
0 100 200 Miles
Lambert Conformal Conic Projection, SP 8N/32N
ALGERIA
MAURITANIA
TOMBOUCTOU
KIDAL
GAO
MOPTI
KAYES
KOULIKORO
SÉGOU
SIKASSO
SENEGAL
GUINEA
BURKINA
NIGER
BENIN
NIGERIA
TOGO
GHANA
COTE D'IVOIRE
SIERRA LEONE
LIBERIA
NORTH ATLANTIC OCEAN
Bight of Benin
Chenachane
Aguêraktem (well)
Taoudenni
Poste Maurice Cortier
Bordj le Prieur
Tessalit
Ti-n-Zaouâtene
Kidal
Tidjikdja
Kifa
'Ayoun el 'Atroûs
Néma
Tombouctou
Goundam
Lac Faguibine
Bourem
Gao
Ménaka
Lac Niangay
Nampala
Nioro du Sahel
Kayes
Nayé
Mourdiah
Niono
Mopti
Bandiagara
Djenné
Ségou
Koulikoro
Bamako
Kita
San
Koutiala
Bougouni
Sikasso
Kédougou
Labé
Siguiri
Kankan
Kabala
Makeni
Kenema
Macenta
Nzérékoré
Tubmanburg
Monrovia
Kakata
Buchanan
Man
Daloa
Yamoussoukro
Bouaké
Korhogo
Ferkessédougou
Adzopé
Bobo-Dioulasso
Gaoua
Nouna
Ouahigouya
Ouagadougou
Koudougou
Pô
Dori
Ayorou
Abala
Niamey
Dosso
Fada-N'Gourma
Kamba
Kandi
Dapaong
Natitingou
Bolgatanga
Tamale
Parakou
Wawa
Kainji Dam
Sokodé
Blitta
Ilorin
Techiman
Sunyani
Kumasi
Lake Volta
Akosombo Dam
Kpalimé
Ho
Lomé
Porto-Novo
Cotonou
Ibadan
Abeokuta
Lagos
Niger
Bani
Aoul
Edjérir

Republic of Mali

Slightly less than twice the size of Texas, Mali is bordered by Algeria (1,376 km), Burkina Faso (1,000 km), Guinea (858 km), Côte d'Ivoire (532 km), Mauritania (2,237 km), Niger (821 km), and Senegal (419 km). The terrain is mostly flat to rolling northern plains covered by sand, savanna in the south, and rugged hills in the northeast. The lowest point is the Senegal River (23 m), and the highest point is *Hombori Tondo* (1,155 m) (*The World Factbook*, 2010).

"Mali is the cultural heir to the succession of ancient African empires – Ghana, Malinké, and Songhai – that occupied the West African savannah. These empires controlled Saharan trade and were in touch with Mediterranean and Middle Eastern centers of civilization. The Ghana Empire, dominated by the Soninke or Saracolé people and centered in the area along the Malian-Mauritanian frontier, was a powerful trading state from about *A.D.* 700 to 1075. The Malinke Kingdom of Mali had its origins on the upper Niger River in the 11th century. Expanding rapidly in the 13th century under the leadership of Soundiata Keita, it reached its height about 1325, when it conquered Timbuktu and Gao. Thereafter, the kingdom began to decline, and by the 15th century, it controlled only a small fraction of its former domain. The Songhai Empire expanded its power from its center in Gao during the period 1465–1530. At its peak under Askia Mohammad I, it encompassed the Hausa states as far as Kano (in present-day Nigeria) and much of the territory that had belonged to the Mali Empire in the west. It was destroyed by a Moroccan invasion in 1591. Timbuktu was a center of commerce and of the Islamic faith throughout this period, and priceless manuscripts from this epoch are still preserved in Timbuktu. The United States and other donors are making efforts to help preserve these priceless manuscripts as part of Mali's cultural heritage. French military penetration of the *Soudan* (the French name for the area) began around 1880. Ten years later, the French made a concerted effort to occupy the interior. The timing and resident military governors determined methods of their advances. A French civilian governor of *Soudan* was appointed in 1893, but resistance to French control did not end until 1898, when the *Malinké* warrior Samory Touré was defeated after 7 years of war. The French attempted to rule indirectly, but in many areas, they disregarded traditional authorities and governed through appointed chiefs. As the colony of French *Soudan*, Mali was administered with other French colonial territories as the Federation of French West Africa. In 1956, with the passing of France's Fundamental Law (*Loi Cadre*), the Territorial Assembly obtained extensive powers over internal affairs and was permitted to form a cabinet with executive authority over matters within the Assembly's competence. After the 1958 French constitutional referendum, the *Republique Soudanaise* became a member of the French Community and enjoyed complete internal autonomy. In January 1959, *Soudan* joined Senegal to form the Mali Federation, which became fully independent within the French Community on June 20, 1960. The federation collapsed on August 20, 1960, when Senegal seceded. On September 22, *Soudan* proclaimed itself the Republic of Mali and withdrew from the French Community" (*Background Note: Mali, U.S. Department of State*, 2010).

"In 2012, rising ethnic tensions and an influx of fighters - some linked to *Al-Qa'ida* - from Libya led to a rebellion and military coup. Following the coup, rebels expelled the military from the country's three northern regions, allowing terrorist organizations to develop strongholds in the area. With French military intervention, the Malian Government managed to retake most of the north. However, the government's grasp in the region remains weak with local militias, terrorists, and insurgent groups continuously trying to expand control. In 2015, the Malian Government and northern rebels signed an internationally mediated peace accord. Despite a June 2017 target for implementation of the agreement, the signatories have made little progress. Extremist groups were left out of the peace process, and terrorist attacks remain common. In May 2021, Colonel Assimi

Goita arrested the interim president in a second coup in an effort to slow election preparations, claiming that the transition government needed to prioritize improving security before elections occur. In June 2021, Mali's military transitional leaders appointed Choguel Kokalla Maiga as prime minister" (*World Factbook*, 2022).

There are no known geodetic triangulation arcs in the entire country of Mali. Mapping has been accomplished of local cities such as Bamako and Timbuktu (*Tombouctou*) by the local *L'institut Géographique National du Mali*, but the majority of topographic mapping is based on local astronomic control (Astro points) used to orient photomosaics and to produce form line topography such as the *Fond Topographique ou planimétrique*. Currently published maps of Mali are referenced to the Clark 1880 ellipsoid, where $a = 6,378,249.145$ m and $1/f = 293.465$, and the grid shown is the UTM. There is no known geodetic datum other than some recent references to WGS 84. The Mali IGN does picture one Wild Heerbrugg GPS receiver in its inventory. Other grid systems that may have covered portions of Mali are the South Sahara Lambert Zone and the *Fuseau Côte d'Ivoire* TM Belt. The parameters of the South Sahara Lambert Zone are: Central Meridian $(\lambda_o) = 5°\ 00'$ W, Latitude at Origin $(\varphi_o) = 20°\ 00'$ N, Scale Factor at Origin $(m_o) = 0.999071$, False Easting $= 1,600$ km, False Northing $= 600$ km. The *Fuseau Côte d'Ivoire* TM Belt parameters are: Central Meridian $(\lambda_o) = 6°\ 30'$ W, Latitude at Origin $(\varphi_o) = 0°\ 00'$ N, Scale Factor at Origin $(m_o) = 0.999$, False Easting $=$ False Northing $= 1,000$ km. The Mali IGN does not seem to be an active participant in the AFREF geodetic program for the African continent.

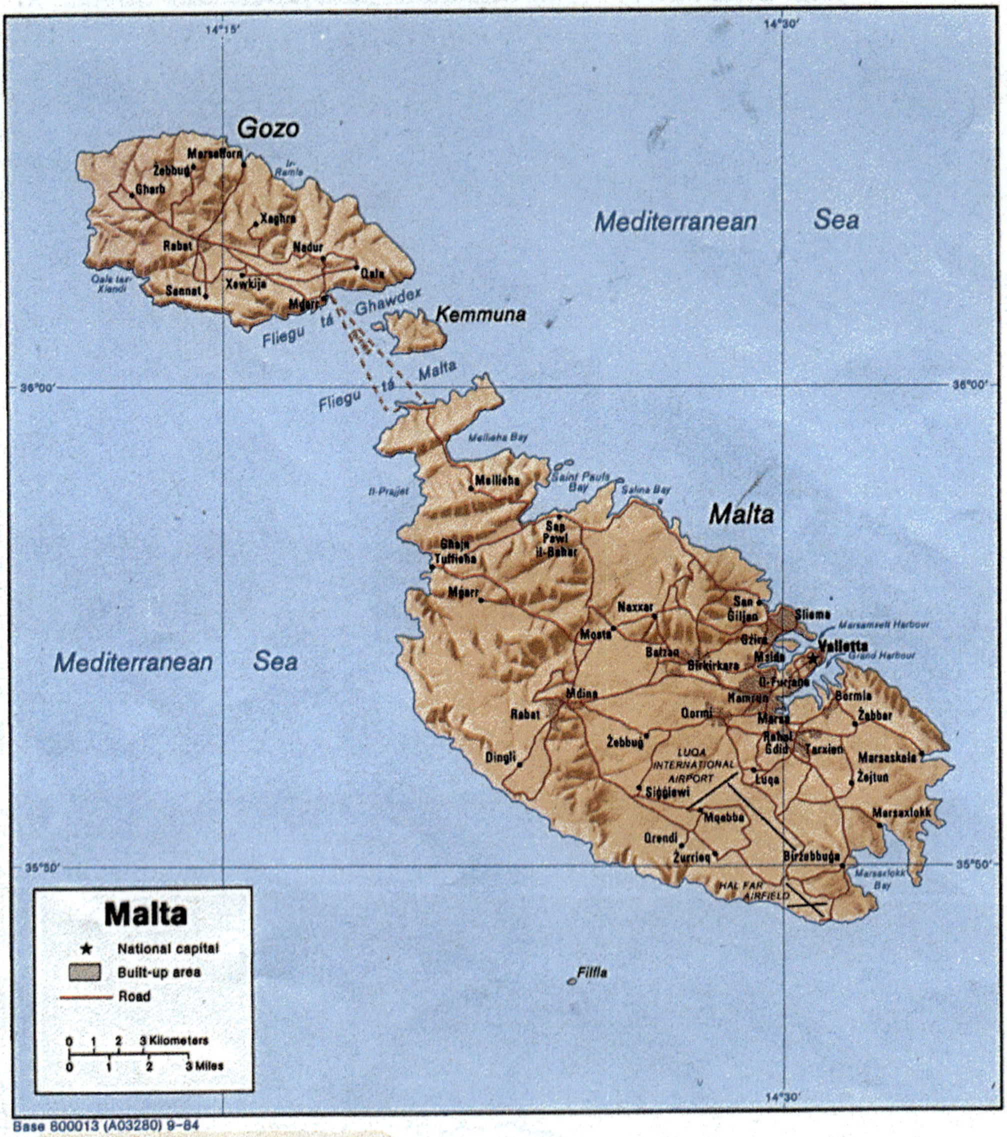

Base 800013 (A03280) 9-84

Republic of Malta

Malta is slightly less than twice the size of Washington, DC, the lowest point is the Mediterranean Sea (0 m), and the highest point is *Ta'Dmejrek* (253 m) which is near Dingli. The country comprises an archipelago, with only the three largest islands (Malta, Ghawdex or Gozo, and Kemmuna or Comino) being inhabited (*World Factbook*, 2010).

"Malta was an important cultic center for earth-mother worship in the 4th millennium *B.C.* Archeological work shows a developed religious center there, including the world's oldest freestanding architecture, predating that of Sumer and Egypt. (*For example, Ġgantija near Xagħram on Gozo and Ħaġar Qim and Mnajdra on the southwest coast of the main island – Lonely Planet, 2010*). Malta's written history began well before the Christian era. The Phoenicians, and later the Carthaginians, established ports and trading settlements on the island. During the second Punic War (218 *B.C.*), Malta became part of the Roman Empire. During Roman rule, in *A.D.* 60, Saint Paul was shipwrecked on Malta. In 533 *A.D.* Malta became part of the Byzantine Empire and in 870 came under Arab control. Arab occupation and rule left a strong imprint on Maltese life, customs, and language. The Arabs were driven out in 1090 by a band of Norman adventurers under Count Roger of Normandy, who had established a kingdom in southern Italy and Sicily. Malta thus became an appendage of Sicily for 440 years. During this period, Malta was sold and resold to various feudal lords and barons and was dominated successively by the rulers of Swabia (now part of Germany), Aquitaine (now part of France), Aragon (now part of Spain), Castile (now part of Spain), and Spain. In 1522, Suleiman II drove the Knights of St. John out of Rhodes, where they had established themselves after being driven out of Jerusalem. They dispersed to their commanderies in Europe, and in 1530 Charles V granted them sovereignty over the Maltese islands. For the next 275 years, these famous "Knights of Malta" made the island their domain. They built towns, palaces, churches, gardens, and fortifications and embellished the island with numerous works of art. Over the years, the power of the Knights declined, and their rule of Malta ended with their peaceful surrender to Napoleon in 1798. The people of Malta rose against French rule, which lasted two years, and with the help of the British evicted them in 1800. In 1814, Malta voluntarily became part of the British Empire. Malta obtained independence on September 21, 1964, became a Republic on December 13, 1974. The last British forces left in March 1979. Malta joined the European Union (EU) on May 1, 2004" (*U.S. Department of State Background Note, May 2010*).

Prof. Peter Dare, Chairman of the Department of Geodesy & Geomatics Engineering at the University of New Brunswick performed a GPS survey of Malta in 1993 with a group of students from the University of East London. The historical details of early surveys in Malta have been extracted from Prof. Dare's GPS Survey Report. "A survey was carried out in 1896 as a basis for 1:2,500 scale mapping. In 1900 a connection from Sicily to the islands of Malta and Gozo was observed by the Italian government. The Royal Engineers carried out a retriangulation in 1928 due to the large number of discrepancies in the previous surveys. Although good station descriptions were made of these points, unfortunately most of these points can no longer be located. Coastal defense surveys were carried out by the British Navy and Army during the second world war. It is believed that coordinates for these points are accurate to about 0.2 m" (Directorate of Military Survey, 1956).

"A complete retriangulation of Malta took place during 1955 and 1956 again by the military personnel (Logan, I., *A Critical Report: The Provision of Ground Control for Aerial Triangulation and Photogrammetric Mapping of Malta – 1968*, RICS, London, 1973). This was to provide control for new 1:2,500 mapping and to provide a triangulation network for the future. Primary and secondary points are thought to be accurate to about 0.05 m, while tertiary points (fixed by intersection and resections) are accurate to about 0.5 m. During 1956 additional triangulation points were added and these are also thought to be accurate to about 0.5 m (Directorate of Military Survey, 1956).

In late June 1968 the Directorate of Overseas Surveys (DOS) started fieldwork on the island of Malta to provide control for aerial photography so that new contoured maps at a scale of 1:2,500 could be produced. This was completed by January 1, 1969. A total of 85 control points were created for the aerotriangulation control. In addition, two 1955 primary points that had been destroyed were replaced by two new ones and three other 1955 points have since been classified as third order. This DOS work is documented in Logan (1973). The new system is now known as the Malta Precise Network 1993 (Malta PN′93)." The Holland office of Racal Survey previously determined the transformation from European Datum 1950 (ED50) to WGS 84 as: $\Delta X = -83.7$ m, $\Delta Y = -81.9$ m, and $\Delta Z = -131.6$ m. Prof. Dare found these shift parameters to agree well with his 1993 GPS survey, although he determined seven-parameter transformations also with Leica SKI™ software.

Thanks to Mr. John W. Hager, the Malta Datum of 1928 origin is at Point de Vue, where $\Phi_o = 35°$ 52′ 55.95″ N, $\Lambda_o = 14°$ 24′ 15.32″ W, and the defining azimuth (from North) to Gargur is: $\alpha_o = 48°$ 20′ 37.4″, and the height, $H_o = 204.0$ m. Note that this old datum from the War Office Triangulation of 1928 is referenced to the International ellipsoid, where $a = 6,378,388$ m and $1/f = 297$.

Hager points out a curiosity in that Eastern Telegraph Company Longitude Pillar is *also* at Point de Vue, where $\Phi_o = 35°$ 55′ 159.28″ N, $\Lambda_o = 14°$ 29′ 22.35″ W, and the defining azimuth (from North) to Gargur is the same: $\alpha_o = 48°$ 20′ 37.4″, but the ellipsoid of reference is the Clarke 1880! That curiosity begins to make sense when we look at what was done in World War II.

The Geographical Section, General Staff of the British Army (GSGS) in World War II established a Maltese Lambert Zone. The tangent zone parameters are: Central Meridian, $\lambda_o = 14°$ 27′ 48.92″ West of Greenwich, Latitude of Origin, $\varphi_o = 35°$ 55′ 30.46″ N, Scale Factor at Origin, $m_o = 1.0$ (*by definition for a tangent zone*), False Easting = 143,156.33 UK Yards, and False Northing = 129,524.33 UK Yards, where 1 m = 1.093614249 UK Yards, and the Clarke 1880 is the ellipsoid of reference, where $a = 6,378,249.145$ m and $1/f = 293.465$.

On the other hand, the European Datum of 1950 was computed for Malta after World War II, and it is referenced to the International ellipsoid, the same as the original Malta Datum of 1928. According to *TR 8350.2*, the transformation in Malta from ED50 to WGS84 is: $\Delta X = -107$ m ±25 m, $\Delta Y = -88$ m ±25 m, $\Delta Z = -149$ m ±25 m, and this is based on one station collocated as of 1991. However, it is likely *not as reliable* as the parameters confirmed by Prof. Peter Dare in 1993.

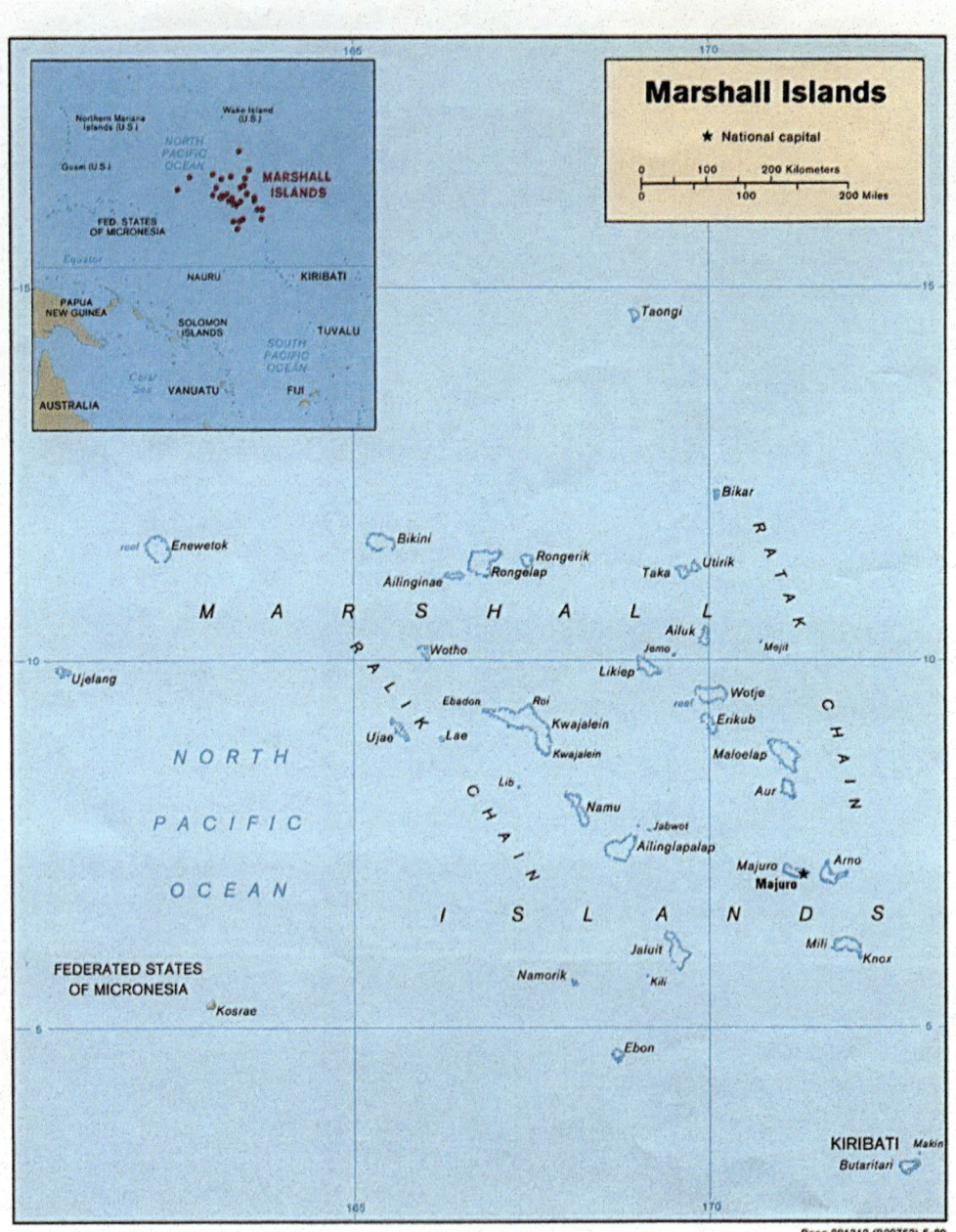

Marshall Islands

★ National capital

Republic of the Marshall Islands

The Marshall Islands terrain is low coral limestone and sand islands. The lowest point is the Pacific Ocean (0 m), the highest point is an unnamed location on Likiep (10 m). About the same size as Washington, DC, the coastline is 370.4 km long (*World Factbook*, 2014).

"Little is clearly understood about the prehistory of the Marshall Islands. Researchers agree on little more than that successive waves of migratory people from Southeast Asia spread across the Western Pacific about 3,000 years ago and that some of them landed on and remained on these islands. The Spanish explorer *de Saavedra* landed there in 1529. They were named for English explorer John Marshall, who visited them in 1799. The Marshall Islands were claimed by Spain in 1874. Germany established a protectorate in 1885 and set up trading stations on the islands of Jaluit and Ebon to carry out the flourishing copra (dried coconut meat) trade. Marshallese *iroij* (high chiefs) continued to rule under indirect colonial German administration. At the beginning of World War I, Japan assumed control of the Marshall Islands. Their headquarters remained at the German center of administration, Jaluit. U.S. Marines and Army troops took control from the Japanese in early 1944, following intense fighting on Kwajalein and Eniwetok atolls. In 1947, the United States, as the occupying power, entered into an agreement with the UN Security Council to administer Micronesia, including the Marshall Islands, as the Trust Territory of the Pacific Islands. On May 1, 1979, in recognition of the evolving political status of the Marshall Islands, the United States recognized the constitution of the Marshall Islands and the establishment of the Government of the Republic of the Marshall Islands. The constitution incorporates both American and British constitutional concepts" (*Background Notes, U.S. Department of State,* 2014).

"U.S. nuclear testing took place between 1946 and 1958 on the islands of Bikini and Eniwetok. The people of Bikini were removed to another island, and a total of 23 U.S. atomic and hydrogen bomb tests were conducted. Despite cleanup attempts, the islands remain uninhabited today because of nuclear contamination. The U.S. paid the islands $183.7 million in damages in 1983, and in 1999, the U.S. approved a one-time $3.8-million payment to the relocated people of Bikini atoll. Kwajalein atoll is the site of an American military base and has been used for missile defense testing since the 1960s" (*InfoPlease,* 2014).

Dr. Helmut Schmid was one of the original German V-2 Rocket Scientists from Peenemünde that were brought to the United States at the end of World War II. While most of the Germans were taken to White Sands Missile Proving Grounds, Dr. Schmid was moved to Aberdeen Proving Grounds in Maryland. His protégé was Dr. Duane C. Brown, the Father of American Computational Analytical Photogrammetry. While at Aberdeen, Dr. Schmid designed the WILD Heerbrugg BC-4 ballistic camera for tracking the flights of guided missiles at night. (*I personally own one that I converted for terrestrial photogrammetry. Every Spring semester I have it taken out for my students to see how the "old timers" used to do things. Personally, I haven't been able to pick the thing up for the last 30 years. – Ed.*) Some of the BC-4 camera systems were transferred to the U.S. Coast & Geodetic Survey for the BC-4 Photogrammetric Geodesy program of the 1960s and early 1970s and were directed by Dr. Schmid. (*I once was introduced to Dr. Schmid. He was not impressed with a U.S. Army Captain; I got a grunt for an acknowledgement. – Ed.*) The remainder of the BC-4 camera systems were transferred to the Kwajalein atoll for missile defense testing, their original purpose.

Thanks to Mr. Ed Carlson of NOAA, "There are at least 2 existing Marshall Island horizontal datums; the islands of Kwajalein, Eniwetok, and Roi-Namur defined by the U.S. Air Force in 1959 referenced to the International ellipsoid ($a = 6,378,388$ m, $1/f = 297$ – *Ed.*), and densified by the U.S.

Coast & Geodetic Survey (USC&GS), now NGS, during 1960–63, and the island of Majuro defined by the U.S. Geological Survey (USGS) referenced to the Clarke 1866 ellipsoid (a = 6,378,206.4 m, b = 6,356,583.8 m – *Ed.*). Differences in positions (latitude and longitude) between these datums and a geocentric reference system defined by GPS can vary in excess of over 320 meters across the country."

Thanks to Mr. John W. Hager, "Ailinginae (1952) at 6 Astro, $\Phi_o = 11°\ 06'\ 48.5''$ N, $\Lambda_o = 166°\ 28'\ 43.4''$ E, $\alpha_o = 279°\ 18'\ 43.0''$ to 7 from south, International, $H_o = 2.1$ m. Ailinglapalap (1951), at 2 Astro, $\Phi_o = 7°\ 17'\ 22.3''$ N, $\Lambda_o = 168°\ 44'\ 39.1''$ E, International, $H_o = 7$ ft. Ailuk (1951), at 1 Astro, $\Phi_o = 10°\ 12'\ 50.3$ N, $\Lambda_o = 169°\ 58'\ 46.8$ " E, $\alpha_o = 230°\ 03'\ 19.9''$ to 2 from south, International. Aon Island, code APB, $\Phi_o = 11°\ 13'\ 04.79''$ E, $\Lambda_o = 169°\ 45'\ 53.04''$ E. Arno (1951), at 2 Astro, $\Phi_o = 7°\ 03'\ 01.3$ N, $\Lambda_o = 171°\ 33'\ 28.1''$ E, $\alpha_o = 178°\ 56'\ 24.4''$ to 2a from south, International, $H_o = 2.3$ m. Arumenii Island, code ARB, $\Phi_o = 12°\ 14'\ 53.5''$ N, $\Lambda_o = 170°\ 07'\ 44.4''$ E. Aur (1951), at 1a, Astro Point, $\Phi_o = 8°\ 18'\ 54.9''$ N, $\Lambda_o = 171°\ 09'\ 10.8''$ E, $\alpha_o = 327°\ 24'\ 21.8''$ to 26a from south, International, $H_o = 1.5$ m. Bikar (1952), at Astro Point, $\Phi_o = 12°\ 14'\ 25.11''$ N, $\Lambda_o = 170°\ 07'\ 53.93''$ E, $\alpha_o = 321°\ 56'\ 28.7''$ to eight from south, International, $H_o = 5.8$ ft. Bikini (1946), code BIK, at BIK, $\Phi_o = 11°\ 37'\ 54.5''$ N, $\Lambda_o = 165°\ 31'\ 17.5''$ E, $\alpha_o = 118°\ 45'\ 55.5''$ to MON from south, Clarke 1866. Scaled from Japanese Chart 458 (1:50,000, 1926), which was copied as H.O. Chart 6032 (1944) by U.S.S. Sumner and Bowditch. Stellar α_o by Sumner. Used by Holmes & Narver, Inc. May also be defined as AIR (Beacon "H"), $\Phi_o = 11°\ 30'\ 24.906''$ N, $\Lambda_o = 165°\ 24'\ 55.168''$ E, $\alpha_o = 268°\ 00'\ 01.4''$ to ENYU from south. Also called USN 1946 and BIK (assumed). Bokku Island, code BOK, $\Phi_o = 9°\ 30'\ 49''$ N, $\Lambda_o = 170°\ 00'\ 04''$ E. Bonto, code BON, $\Phi_o = 8°\ 32'\ 48''$ N, $\Lambda_o = 171°\ 06'\ 53''$ E. Dalap 1968 Astro, at Dalap 1968, $\Phi_o = 7°\ 05'\ 14.0''$ N, $\Lambda_o = 171°\ 22'\ 34.5''$ E, Clarke 1866. Department of the Interior, Geological Survey coordinate list. Dodo Island Observation Spot, code DOA, $\Phi_o = 7°\ 07'\ 38''$ N, $\Lambda_o = 171°\ 41'\ 06''$ E. Ebon (1951), at Astro Point, $\Phi_o = 4°\ 35'\ 08.9''$ N, $\Lambda_o = 168°\ 41'\ 14.7''$ E, $\alpha_o = 223°\ 10'\ 46.8''$ to 1 from south, International. Eniirikku 1919, code ENG, $\Phi_o = 11°\ 29'\ 49''$ N, $\Lambda_o = 165°\ 20'\ 22''$ E. Trig List, Washington, DC, Feb. 1970. Eni Jun Island, code ENJ, $\Phi_o = 8°\ 36'\ 05''$ N, $\Lambda_o = 171°\ 02'\ 44''$ E. Eniwetok Astro 1944, code ENI, at Eniwetok Astro Pier, $\Phi_o = 11°\ 33'\ 23.480''$ N, $\Lambda_o = 162°\ 21'\ 10.250''$ E, $\alpha_o = 58°\ 44'\ 57.70$ to North Base from south, Clarke 1866, U.S.S. Bowditch. Also Privilege at $\Phi_o = 11°\ 21'\ 51.439$ N, $\Lambda_o = 162°\ 21'\ 14.735''$ E, or… 51.383'' N and… 14.736'' E, or Eniwetok of Holmes & Narver… 51.466'' N and… 14.736'' E. Eniwetok-Wake 1960, code ENW, at Wake (8), $\Phi_o = 19°\ 16'\ 19.606''$ N, $\Lambda_o = 166°\ 39'\ 21.798''$ E, Hough. Readjustment of USAF 1959 Datum (*q.v.*). Essentially the mean values based on the absolute positions on Wake and Eniwetok. Enuebing (1919), code ETS, at Eniibing, $\Phi_o = 7°\ 17'\ 25.50''$ N, $\Lambda_o = 168°\ 45'\ 11.90$ E. From a tabulation of geographic positions by J.H.O. Enyvertok (Eniwetok) Island, code ENH, at Observation spot, $\Phi_o = 11°\ 17'\ 43''$ N, $\Lambda_o = 167°\ 28'\ 27''$ E. Erappu Channel, code ERA, $\Phi_o = 10°\ 19'\ 34.9''$ N, $\Lambda_o = 169°\ 54'\ 23.5''$ E. Jaluit Astro, code ASK, as 61a, $\Phi_o = 5°\ 55'\ 19.5''$ N, $\Lambda_o = 169°\ 38'\ 37.2''$ E, $\alpha_o = 18°\ 30'\ 07.0''$ to 59 from south, International, $H_o = 2.2$ m. Jemo (1951), code ASD, at A (Astro), $\Phi_o = 10°\ 04'\ 47.5''$ N, $\Lambda_o = 169°\ 31'\ 18.6''$ E, International, $H_o = 9.0$ ft. Kili, code KKM, at Kili Island W. Coast Observation Spot (1923), $\Phi_o = 5°\ 38'\ 47''$, $\Lambda_o = 169°\ 07'\ 00''$ E. Kwajalein Astro 1952, code ASJ, at Station 42, Ennylabegan I., $\Phi_o = 8°\ 47'\ 19.2'' \pm 0.1''$ N, $\Lambda_o = 167°\ 37'\ 26.8'' \pm 0.2''$ E, $\alpha_o = 304°\ 33'\ 39.0'' \pm 0.5''$ to 43 from north, International, height = 2.39 ft. Observed by 71st Engineer Survey Liaison Detachment with 60° astrolabe. Kwajalein Astro U.S. Navy 1944, code KWA, at Astro, $\Phi_o = 8°\ 44'\ 06.33''$ N, $\Lambda_o = 167°\ 44'\ 28.63''$ E. Prismatic astrolabe used with chronograph checked by radio time signals. Approximately 300 stars computed. Observed by LT. E. V. Mohl, U.S.N.R., U.S.H.O., March 1944. Lae (1952), at 2. Astro Point, $\Phi_o = 8°\ 55'\ 31.6''$ N, $\Lambda_o = 166°\ 15'\ 58.6''$ E, $\alpha_o = 64°\ 07'\ 56.1''$ to 1 from south, International, $H_o = 5.7$ ft. Leuen Anchorage Observation Spot, code LET, $\Phi_o = 7°\ 45'\ 32''$ N, $\Lambda_o = 168°\ 13'\ 23''$ E. Likiep, code LIK (?), at South Pass Reef. $\Phi_o = 9°\ 50'\ 22''$ N, $\Lambda_o = 169°\ 13'\ 23''$ E. Likiep, code LIK (?), at 1 Astro Point, $\Phi_o = 9°\ 50'\ 24.1''$ N, $\Lambda_o = 169°\ 18'\ 51.3''$ E, $\alpha_o = 126°\ 00'\ 14.9''$ to 2 from south, International, $H_o = 5.0$ ft. Lotj Island, code LAD, at Observation Spot, $\Phi_o = 8°\ 55'\ 18''$ N, $\Lambda_o = 166°\ 12'\ 58''$ E. Majuro Astro, 1944, code MAJ, at Majuro Astro, 1944, $\Phi_o = 7°\ 04'\ 25.73''$ N, $\Lambda_o = 171°\ 19'\ 18.08''$ E. U.S.S. Bowditch, prismatic astrolabe used with chronograph checked by radio time signals. Approximately 300 stars computed. Majuro 1951, code ATR, at Astronomic Station No. 2, $\Phi_o=7°\ 05'\ 02.2''$ N, $\Lambda_o=171°\ 22'\ 25.2''$ E, $\alpha_o=83°\ 53'\ 22.9''$ to α mark (1) from south, International, $H_o=9.3$ ft. Maleolap (1952), at 1. Astropoint, $\Phi_o=8°\ 54'\ 15.5''$ N, $\Lambda_o=170°\ 50'\ 41.1''$ E, $\alpha_o=40°\ 04'\ 27.4''$ to 2 from south, International, $H_o=7.9$ ft. Mejit (1919), code MEC, at Astro Station B1, $\Phi_o=10°\ 16'\ 54''$ N, $\Lambda_o=170°\ 52'\ 36''$ E. Mellu Island, code MEB, $\Phi_o=11°\ 21'\ 38''$ N, $\Lambda_o=166°\ 59'\ 13''$ E. Mili (1951), at 17a Astro Point, $\Phi_o=6°\ 01'\ 45.9''$ N, $\Lambda_o=171°\ 56'\ 50.6''$ E, $\alpha_o=134°\ 23'\ 10.7''$ to 13 from south, International, $H_o=2.2$ m. Namorik (1951), code ATQ, at 1a Astro Point, $\Phi_o=5°\ 36'\ 34.0''$ N, $\Lambda_o=168°\ 06'\ 08.3$ E, $\alpha_o=348°\ 14'\ 35.1''$ to 2 from south, International, $H_o=0.6$ m. Namu (1951), at 1. Astro, $\Phi_o=7°\ 45'\ 47.6''$ N, $\Lambda_o=168°\ 13'\ 14.2''$ E, $\alpha_o=303°\ 01'\ 41.7''$ to 2 from south, International, $H_o=4.5$ ft. Pokaakku, code POK, $\Phi_o=14°\ 34'\ 03''$ N, $\Lambda_o=168°\ 57'\ 24''$ E. Rikuraru, code RIK, $\Phi_o=10°\ 01'\ 28.34''$ N, $\Lambda_o=169°\ 00'\ 47.77''$ E. Rongelap, at Observation Spot, $\Phi_o=11°\ 08'\ 55''$ N, $\Lambda_o=166°\ 53'\ 35''$ E. Rongelap (1952), code ROP, at 25. Astro, $\Phi_o=11°\ 09'\ 02.5''$ N, $\Lambda_o=166°\ 52'\ 03.2''$ E, $\alpha_o=75°\ 03'\ 50.1''$ to 24 from south, International, $H_o=2.5$ m. Rongelap 1959, $\Phi_o=11°\ 26'\ 43.62$ N, $\Lambda_o=167°\ 03'\ 27.56''$ E. Rongerik (1952), at 1a. Astro station, $\Phi_o=11°\ 22'\ 27.8''$ N, $\Lambda_o=167°\ 30'\ 59.0''$ E, $\alpha_o=174°\ 00'\ 56.2''$ to 2a from south, International, $H_o=1.9$ m. Rube Point, code RUA, at Observation Spot, $\Phi_o=4°\ 35'\ 13''$ N, $\Lambda_o=168°\ 13'\ 23''$ E. Takowa Island, code TAA, at Observation Spot, $\Phi_o=6°\ 13'\ 36''$ N, $\Lambda_o=171°\ 48'\ 14''$ E. UK requested change from Takowaka. Taongi Astro 1952, code TAO, at 2. Astro, $\Phi_o=14°\ 37'\ 43.5''$ N, $\Lambda_o=169°\ 01'\ 04.0''$ E, $\alpha_o=20°\ 00'\ 23.5''$ to 1 from south, International, $H_o=5.6$ ft. Taroa Island Trig Station, $\Phi_o=8°\ 42'\ 44''$ N, $\Lambda_o=171°\ 13'\ 48''$ E. Ujelang (1952), at 3. Astro, $\Phi_o=9°\ 46'\ 07.1''$ N, $\Lambda_o=160°\ 59'\ 10.2''$ E, International, elevation$=5.2$ ft. Ujelang, at Observation Spot, $\Phi_o=9°\ 46'\ 29''$ N, $\Lambda_o=160°\ 57'\ 43''$ E. USAF 1959, at WAKE (8), $\Phi_o=19°\ 16'\ 19.637''$ N, $\Lambda_o=166°\ 39'\ 21.745''$ E, $\alpha_o=273°\ 29'\ 34.503''$ Bikini (4) to Rolap (2), Hough. Preliminary determination of position of Initial Point. Utirik (1951), at 2. Astro, $\Phi_o=11°\ 13'\ 08.5''$ N, $\Lambda_o=169°\ 50'\ 42.5''$ E, $\alpha_o=73°\ 04'\ 55.0''$ to 1. from south, International, $H_o=5.9$ ft. Wotho (1952), at 3. Astro, $\Phi_o=10°\ 10'\ 24.3''$ N, $\Lambda_o=166°\ 00'\ 03.3''$ E, $\alpha_o=306°\ 04'\ 31.2''$ to 1 from south, International, $H_o=9.7$ ft. Wotje Atoll 1952, code WOJ, at 2 Astro Point, $\Phi_o=9°\ 27'\ 58.6''$ N, $\Lambda_o=170°\ 14'\ 09.7''$ E, $\alpha_o=320°\ 12'\ 34.7''$ to 42 azimuth mark from south, International, $H_o=6.4$ ft. Wotje, at Observation Spot, $\Phi_o=9°\ 27'\ 31''$ N, $\Lambda_o=170°\ 14'\ 32''$ E. I never was aware of any classified datums in the area. There could have been, but it would have been on a need-to-know basis. There could well be duplicates in this list JWH."

Majuro Atoll has a local grid system based on the Azimuthal Equidistant projection with origin at station Dalap, where latitude of origin, $\phi_o=7°\ 05'\ 14.0''$ N, central meridian, $\lambda_o=171°\ 22'\ 34.5''$ E, scale factor at origin, $m_o=1.0$ *(by definition)*, False Easting$=85$ km, and False Northing$=40$ km. Reference ellipsoid is the Clarke 1866 *(Herbert W. Stoughton, Ph.D., P.E., P.L.S., C.P., FASPRS, 1 April 2008)*.

According to *TR 8350.2*, **from** Wake-Eniwetok 1960 Datum **to** WGS84, $\Delta X=+102$ m, ± 3 m, $\Delta Y=+52$ m, ± 3 m, $\Delta Z=-38$ m, ± 3 m; and **from** Wake Island Astro 1952 **to** WGS84, $\Delta X=+276$ m, ± 25 m, $\Delta Y=-57$ m, ± 25 m, $\Delta Z=+149$ m, ± 25 m.

Dominica, Guadeloupe, and Martinique

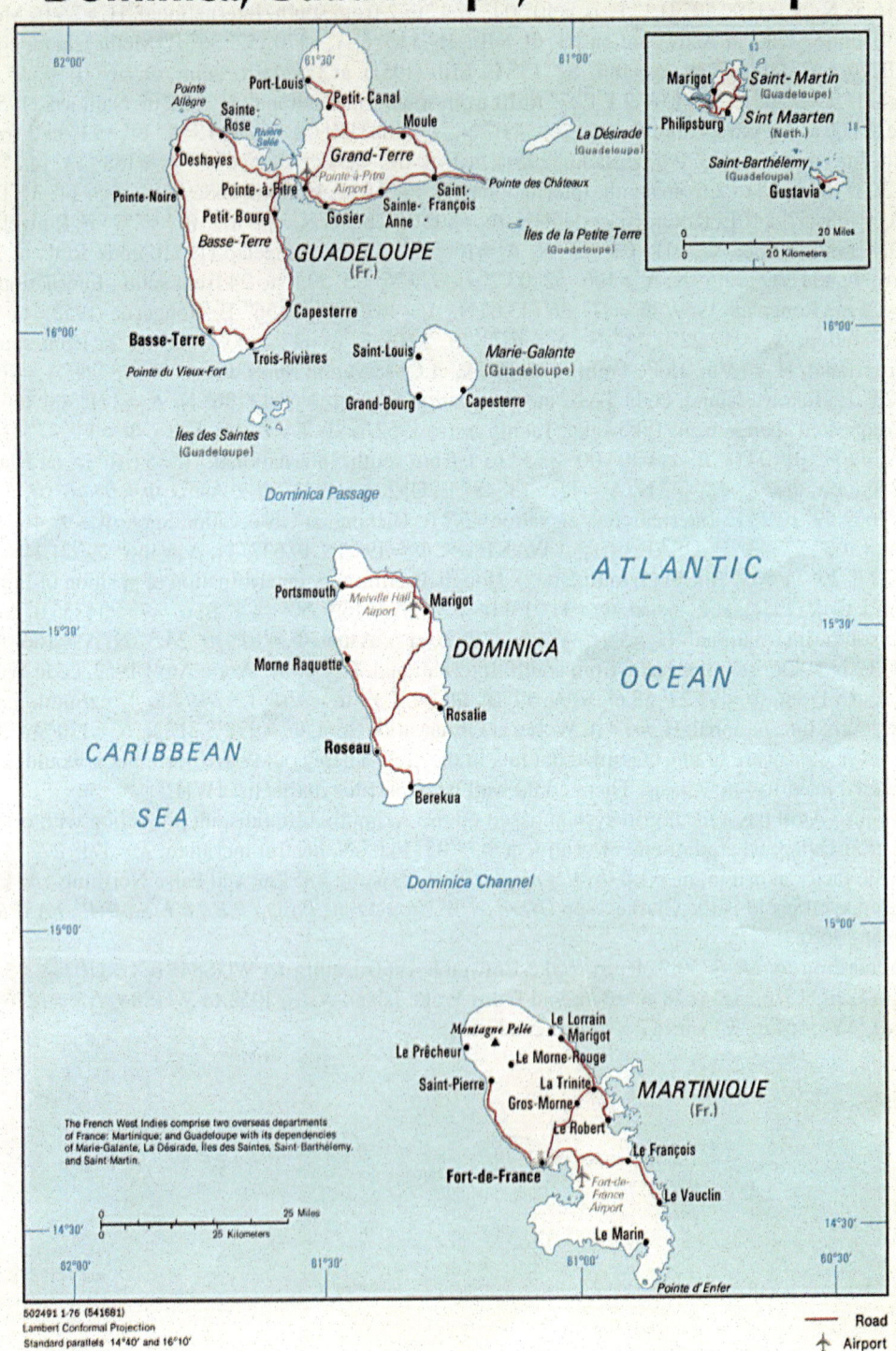

Department of Martinique

Martinique is mountainous with an indented coastline; dormant volcano; Caribbean island between the Caribbean Sea and North Atlantic Ocean, north of Trinidad and Tobago.

Subtropical tempered by trade winds; moderately high humidity; rainy season (June to October); vulnerable to devastating cyclones (hurricanes) every 8 years on average.

At the 1502 sighting of Martinique by Columbus, it was inhabited by the Carïb Indians who called the island *Mandinina,* or "Island of Flowers." Pierre Gelain d'Esnambuc landed on the northwestern side of the island and established a settlement and fort in 1635 that would later become the island's first capital, Saint Pierre. The following year, French King Louis XIII signed a decree authorizing the use of slaves in the French West Indies. Within 5 years, settlers had colonized the land to the south to Fort-de-France, where they constructed a fort on the rise above the harbor. Clearing forests for sugar plantations caused conflicts and eventually warfare with the native Carïbs until the last surviving Indians were forcibly removed in 1660. Occupied by the British several times over the centuries, Martinique was made a full Department of France by the 1946 Constitution. Martinique is the 1763 birthplace of Empress Josephine.

In 1902, a blast from Mont Pelée (a still-active volcano) laid waste to Saint Pierre with a burst of superheated gas and burning ash. The sole survivor of the town's 30,000 inhabitants was a prisoner in the local jail. Although part of the town has since been rebuilt, the capital was permanently moved to Fort-de-France. In 1876, the astronomic determination of the Fort-de-France Flagstaff was: $\Phi_o = 14°\ 35'\ 58.1''$ N, $\Lambda_o = 61°\ 04'\ 29.5''$ West of Greenwich, and the ellipsoid of reference was the Clarke 1880, where $a = 6,378,249.145$ m and $1/f = 293.465$. A later determination at the same point observed in 1939 yielded: $\Phi_o = 14°\ 35'\ 59.76''$ N, $\Lambda_o = 61°\ 04'\ 14.90''$ West of Greenwich, and the ellipsoid of reference was the International 1924, where $a = 6,378,288$ m and $1/f = 297$. By 1976, the coordinates of the Fort-de-France Flagstaff were published by the French Navy (*Annales Hydrographiques, 5ème Série – Vol. 4, Fasc. 1–1976 No. 743*), where $\Phi_o = 14°\ 35'\ 54.8''$ N, $\Lambda_o = 61°\ 04'\ 12.78''$ West of Greenwich, again referenced to the International 1924 ellipsoid. However, the local datum is named "Fort Desaix 1925–1926" after a cadastral survey performed by the Topographic Engineer, M. Jarre where Fort Desaix was his chosen coordinate system origin. The Fort Desaix monument is ~1.7 km north of Fort Saint Louis and Fort-de-France. Subsequent geodetic surveys appear to have ignored the Fort Desaix monument in favor of Fort-de-France monumentation. The geodetic survey performed during the Hydrographic Mission of Martinique of 1938–1939 and the geodetic survey performed during the *Service Hydrographique de Océanographique de la Marine* in 1984 (SHOM, 1984) collocated only three published points that were observed also in 1953 by the Institute Geographic National (IGN): Crèvecœur Borne (20), Vauclin Borne (70), and Caravelle Station (210). While the 1938–1939 survey employed a curious local Cartesian system that listed the Desaix origin point as having the coordinates of (x = −28,258.41 m, y = +34,202.88 m), Crèvecœur Borne (x = +22,498.93 m, y = −18,414.83 m), Vauclin Borne (x = +19,282.08 m, y = −06,056.30 m), and Caravelle (x = +19,615.82 m, y = +17,230.61 m); the SHOM 84 survey utilized the UTM Grid (Zone 20), where Crèvecœur Borne (X = 730795.38 m, Y = 1,598,424.38 m, Z = 201 m), Vauclin Borne (X = 733,894.08 m, Y = 1,611,512.18 m, Z = 6 m), and Caravelle (X = 727,587.87 m, Y = 1,634.081.73 m, Z = 155 m).

IGN has published a three-parameter transformation from the Fort Desaix Datum of 1925–1926 (International 1924 ellipsoid) to WGS 84 Datum as: $\Delta X = +186$ m, $\Delta Y = +482$ m, $\Delta Z = +151$ m. The seven-parameter transformation is listed as: $\Delta X = +126.93$ m, $\Delta Y = +547.94$ m, $\Delta Z = +130.41$ m, $\delta s = +13.8227$ ppm, $R_x = +2.7867''$, $R_y = −5.1612''$, $R_z = −0.8584''$. Note that the sense of the rotations shown herein conform to the American-Australian standard rather than the left-handed sense favored by many Europeans. The IGN guarantees the seven-parameters to be good to better than six meters, as the level is sufficient for the most part of cartographic applications. (*Ce niveau est suffisant pour la plupart des applications cartographiques.*)

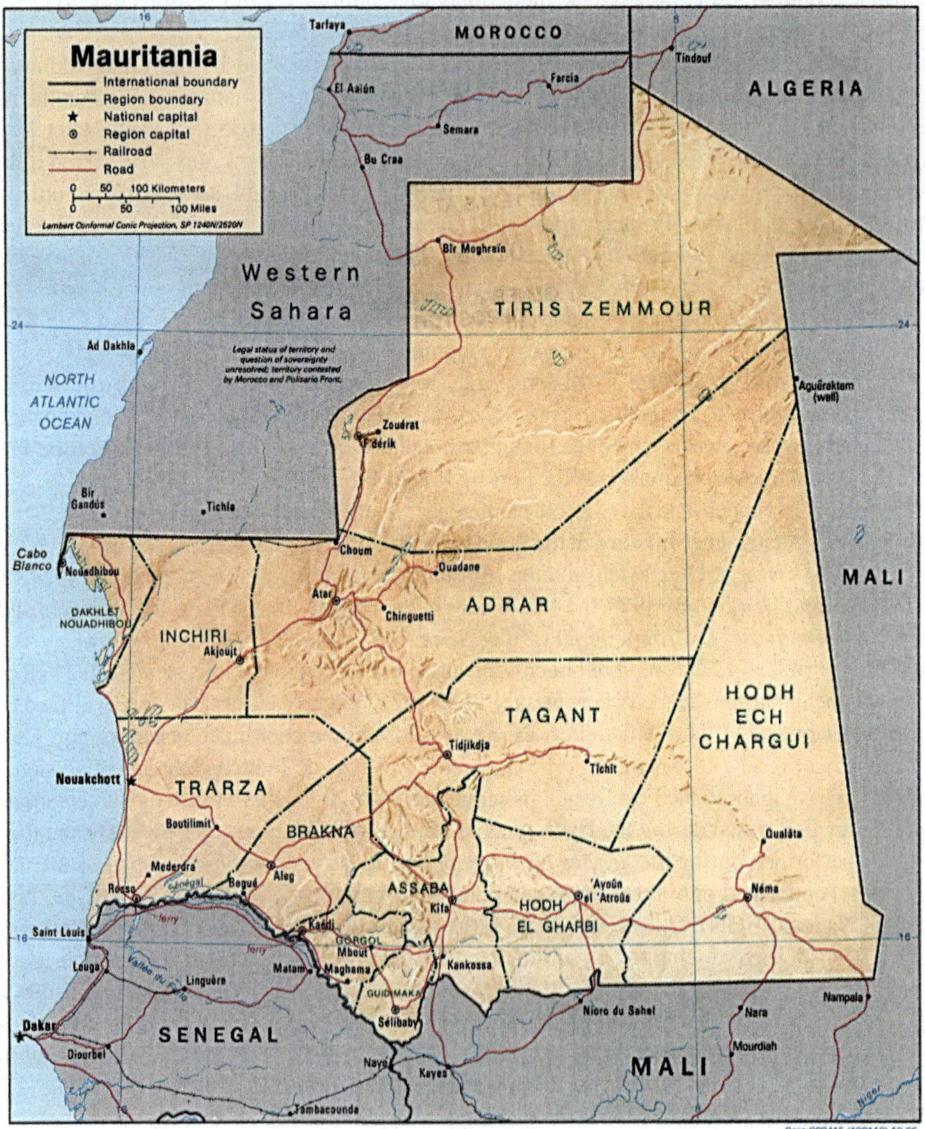

Base 802415 (A00110) 10-95

Islamic Republic of Mauritania

Bordered by Algeria (463 km), Mali (2,237 km), Senegal (813 km), Western Sahara (1,561 km), and the Atlantic Ocean (754 km), Mauritania is slightly larger than three times the size of New Mexico. The lowest point is *Sebkhet Te-n-Dghamcha* (–5 m), and the highest point is *Kediet Ijill* (915 m) (*World Factbook*, 2009).

"Mauritania's mineral wealth has been exploited since Neolithic times. Archeological evidence of a copper mining and refining site near Akjoujt in west-central Mauritania dates from 500 *B.C.* to 1,000 *B.C.* Modern exploitation of copper at Akjoujt and the more important ore deposits between Fdérik and Zouîtât began after independence. Berbers moved south to Mauritania beginning in the third century *A.D.*, followed by Arabs in the eighth century, subjugating and assimilating Mauritania's original inhabitants. From the eighth through the fifteenth century, black kingdoms of the western Sudan, such as Ghana, Mali, and Songhai, brought their political culture from the south. The divisive tendencies of the various groups within Mauritanian society have always worked against the development of Mauritanian unity. Both the Sanhadja Confederation, at its height from the eighth to the tenth century, and the Almoravid Empire, from the eleventh to the twelfth century, were weakened by internecine warfare, and both succumbed to further invasions from the Ghana Empire and the Almohad Empire, respectively. The one external influence that tended to unify the country was Islam. The Islamization of Mauritania was a gradual process that spanned more than 500 years. Beginning slowly through contacts with Berber and Arab merchants engaged in the important caravan trades and rapidly advancing through the Almoravid conquests, Islamization did not take firm hold until the arrival of Yemeni Arabs in the twelfth and thirteenth centuries and was not complete until several centuries later. Gradual Islamization was accompanied by a process of Arabization as well, during which the Berber masters of Mauritania lost power and became vassals of their Arab conquerors. From the fifteenth to the nineteenth century, European contact with Mauritania was dominated by the trade for gum arabic. (*Note that this was an essential ingredient used by Photogrammetric Technicians for decades when constructing aerial mosaics from contact paper prints – Ed*). Rivalries among European powers enabled the Arab-Berber population, the Maures (Moors), to maintain their independence and later to exact annual payments from France, whose sovereignty over the Senegal River and the Mauritanian coast was recognized by the Congress of Vienna in 1815. Although penetration beyond the coast and the Senegal River began in earnest under Louis Faidherbe, governor of Senegal in the mid1800s, European conquest or 'pacification' of the entire country did not begin until 1900. Because extensive European contact began so late in the country's history, the traditional social structure carried over into modern times with little change" (*Library of Congress Country Studies*, 1988).

The first hydrographic survey of the coasts of Mauritania and Senegal was performed by the corvette *La Bayadère* and the escort vessel *Le Lévrier*, commanded by the French Naval Captain Roussin and the Hydrographic Surveyor Givry in 1817. In 1910, Commandant L. V. Lebail established two signal points at the Cape Blanc Lighthouse. The first published coordinates of the Mauritanian coastline were observed by Naval Ensign Yayer in 1935 with an S.O.M. astrolabe for five points: Chauve (cement monument) $\Phi = 16° 46' 33'' N$, $\Lambda = 16° 21' 10'' W$, Nouakchott (post at Officers' magazine) $\Phi = 18° 05' 53'' N$, $\Lambda = 15° 57' 23'' W$, Angel (cement monument) $\Phi = 18° 38' 29'' N$, $\Lambda = 16° 07' 54'' W$, Mahara (wood target in sand dune) $\Phi = 19° 06' 33'' N$, $\Lambda = 16° 17' 06'' W$, and Timiris (house in the village of Memrhar) $\Phi = 19° 21' 21'' N$, $\Lambda = 16° 31' 08'' W$, with all longitudes referenced to Greenwich (*Mission Hydrographique du Sénégal et de la Mauritanie, par M. P. Bonin, Annales Hydrographiques, 3e série, Tome Quinziéme, année 1937*). The following year, an extensive survey of Cape Blanc was performed and the original signal points of Labail's

1910 work were recovered and included in the survey. The Goëland Datum of 1910 originated at *Signal Goëland* (sea gull) of L. V. Lebail's 1910 observation, where $\Phi_o = 20° 54' 46.72''$ N, $\Lambda_o = 17° 03' 07.09''$ W, and the ellipsoid of reference was the Germain 1865, where $a = 6,377,397.2$ m and $1/f = 299.15$. *Note that this ancient datum was later revised by the French and re-computed on a different ellipsoid in 1961.* Apparently because of World War II, the results of this later survey were not published until after the war (*Annales Hydrographiques, 3e série, Tome Seiziéme, années 1938-1919,* **1946**). From January 20 to February 24, 1955, a hydrographic survey of the Noakchott region was performed based on new triangulation that was extended from Yayer's original work of 1935 (*Mission Hydrographique de la Côte Ouest d'Afrique, par M. Pierre Mannevy, pp. 43–51*).

In 1961, the French Navy returned to Mauritania and performed second order geodetic surveys of the coast. Interestingly, they recovered some of the old monuments from earlier surveys and re-determined coordinates, both astronomically and geodetically compensated for the local deflection of the vertical as well as re-computed these positions on a different ellipsoid. The Goëland Datum of 1910, now re-computed on the Clarke 1880 ellipsoid, where $a = 6,378,249.145$ m, and $1/f = 293.465$ changed the astronomic coordinates to: $\Phi_o = 20° 54' 46.7238''$ N, $\Lambda_o = 17° 03' 08.1820''$ W, and the geodetic coordinates were now: $\varphi_o = 20° 54' 43.3490''$ N, $\lambda_o = 17° 03' 06.7295''$ W.

A now-infamous monument and Datum used by oil companies for exploration and production for the coastal area of Mauritania north of Cape Timiris is "Jouik 1961." Apparently unknown to the "oil patch" is that although the UTM coordinates of Jouik were published in 1961, the point was originally observed by Weber of the *Institut Géographique National (IGN) in 1956*, where $\Phi_o \equiv \varphi_o = 20° 30' 29.2626''$ N, $\Lambda_o \equiv \lambda_o = 16° 13' 57.2743''$ W, the ellipsoid of reference being of course, the Clarke 1880 (*Mission Hydrographique de L'Atlantique Sud (6 Février 1961-18 Février 1963) par M. Antoine Demerliac, Annales Hydrographiques, 4e série, Tome Quatorziéme, années 1967–1968, pp. 3–19*). According to version 6.18 of the EPSG database, the transformation parameters **from** Jouik 1961 **to** WGS84 are: $\Delta X = -80.01$ m, $\Delta Y = +253.26$ m, $\Delta Z = +291.19$ m, with the transformation being reported accurate to ±1 m and was based on a five-point solution in 2002 by "Woodside" (*the centimeter precision is dubious – Ed.*).

Other points of great interest in coastal Mauritania published in 1961 include the coordinate reference system originated at Cape Saint Anne and observed by Yayer in 1935, where $\Phi_o = 20° 41' 10.2266''$ N, $\Lambda_o = 16° 40' 49.7288''$ W, and the geodetic coordinates were now: $\varphi_o = 20° 41' 06.6298''$ N, $\lambda_o = 16° 40' 50.2925''$ W. Also, in 1935 Yayer observed Nouamrhar (*Maison de Timiris*), where $\Phi_o = 19° 21' 21.0036''$ N, $\Lambda_o = 16° 31' 07.9843''$ W, and the geodetic coordinates were now: $\varphi_o = 19° 21' 20.8220''$ N, $\lambda_o = 16° 31' 06.2062''$ W.

Also, according to version 6.18 of the EPSG database, the transformation parameters **from** Nouakchott 1965 **to** WGS84 are: $\Delta X = +124.5$ m, $\Delta Y = -63.5$ m, $\Delta Z = -281$ m, with the transformation being reported accurate to ±5 m and was based on a seven-point solution in 1992 within Nouakchott City. The whopping differences in magnitude and sign between the transformation parameters for Cape Saint Anne and for Nouakchott are to be expected for local datums established by independent astronomical observations. Obviously, there is not a unified triangulation network that covers even the coastal regions of Mauritania much less the remainder of the republic.

On the other hand, according to version 6.18 of the EPSG database, the Mauritania 1999 Datum is based on a unified solution of 35 GPS stations of which 8 were originally observed by *IGN* in 1962. No further data seems to exist at present in the public domain.

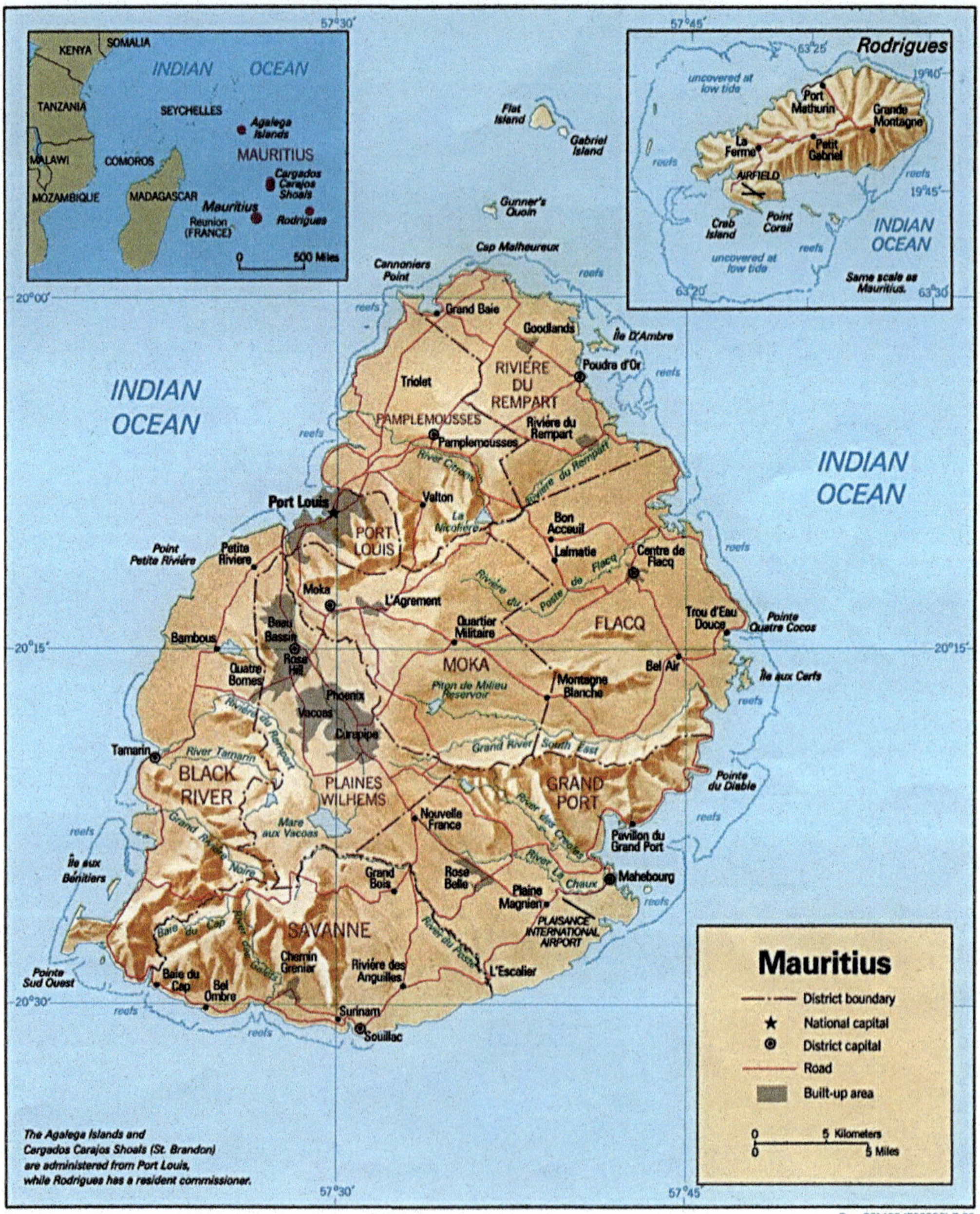

Base 801498 (E00303) 7-90

The Republic of Mauritius

The Republic of Mauritius is an island nation in the Indian Ocean that includes Rodrigues Island, the Agalega Islands, and the Cargados Carajos Shoals (Saint Brandon). Mauritius was discovered by the Portuguese in the early 16th century, it was occupied by the Dutch from 1598 to 1710, and was held by the French from 1715 until 1810 when it was captured by the British. Mauritius was formally ceded to the United Kingdom in 1814, it became independent in 1968, and became a republic in 1992. This nation has a land area of 1,850 km², which is almost 11 times the size of Washington, DC.

The first triangulation was carried out by the astronomer, *Monsieur L'Abbe La Caille* (literally Mister Abbot of the Quail), when he was sent to the island by the French East India Company in 1753. He observed the position of his observatory (Mabile's house) that was situated 4,730 ft east and 2,610 ft north of the Port Louis Time Ball, as: $\Phi_o = 20° 09' 42''$ South, $\Lambda_o = 55° 08' 15''$ East of Paris. He determined the summit of *Le Pouce* as one of ~90 stations. Note that "Le Pouce" literally means "the thumb" in English, which is an apropos description of its shape! I guess that the Grid system used by LaCaille was based on the Bonne projection, which was in vogue with the French at that time. Unfortunately, the records of LaCaille's survey did not survive the ravages of time.

In 1874, Lord Lindsay led an expedition to Mauritius in order to observe the transit of Venus (across the local meridian). The purpose was to observe a network of precisely determined longitudes in the Indian Ocean. The points on Mauritius included Belmont (primary), Pamplemousses, and Solitude. Others in the network were the island of Rodrigues and the French islands of *Réunion* and *St. Paul*. Interestingly, the datum origin in Egypt (different British network and decade) is still named "Venus."

In 1876, the Surveyor General of the Colony of Mauritius, M. Connal, published coordinates of the second triangulation and used the Datum origin as *Le Pouce*, where $\Phi_o = 20° 11' 49.156''$ South, $\Lambda_o = 57° 31' 39.60''$ East of Greenwich. The ellipsoid of reference was the Bessel 1841 where the semi-major axis $(a) = 6,377,397.155$ m and the reciprocal of flattening $(1/f) = 299.1528128$. This point was also the projection origin of the first Cassini-Soldner Grid System of Mauritius and was used with a quadrant system (north-south, east-west), rather than with the more "modern" False Eastings and False Northings.

In 1902, Captain Harrison of the British Colonial Survey Section produced a series of "1 inch" (to the mile) maps also using *Le Pouce* as the origin but based on the Clarke 1858 ellipsoid. Harrison's mapping, presumably also using the Cassini-Soldner Grid, was based on two networks of triangulation that were never connected to form a consistent system. The basis of all subsequent control and mapping prior to the 1990s was based on yet another survey.

In 1934, Captain V. E. H. Sanceau, R.E., performed the re-triangulation of Mauritius. He originally intended to use Connal's work as the basis of his survey in that "more complete records existed than of the work of the Colonial Survey Section." Eventually, only the latitude and longitude of *Le Pouce* and the azimuth Le Pouce – Corps de Garde (LaCaille's Rock) were retained. The only additional points originally used in Connal's survey that were occupied by Sanceau were: Signal Mountain, *Lagrave* (the grave), *Montagne Blanche* (white mountain), *Nouvelle Decouverte* (new discovery), *Montagne Cocotte* (chuck-chuck or hen mountain), *Montagne Bambou* (bamboo mountain), and the West Peak of the *Fayence* Mountains. The new triangulation of Mauritius established the *Le Pouce* Datum of 1934 with an origin of: $\Phi_o = 20° 11' 42.25''$ South, $\Lambda_o = 57° 31' 18.58''$ East of Greenwich. The ellipsoid of reference was the Clarke 1880 where $a = 6,378,249.145$ m and $1/f = 293.465$. Sanceau's triangulation also established a new Grid based on the Lambert Conical Orthomorphic (Conformal Conic) where the scale factor at origin was 1.0 (tangent conic), and the

False Easting (FE) and False Northing (FN) were 3,500,000 ft each. The unit of measure was the yard where 1 m = 1.09362311 yards. Captain Sanceau also produced "The 1″ Topographical Map."

In 1943, the United States Army, Office of the Chief of Engineers issued tables on the Mauritius Zone with the FN and FE being 1,166,666.67 yards. It was noted that the "new" FN and FE of the Ordnance Survey maps were equal to 1,000,000 m on the Blue Grid.

According to D. Ramasawmy, Chief Surveyor, "During 1962–1964 the Directorate of Overseas Surveys (DOS) UK undertook a Tellurometer traverse using most of Captain Sanceau's trigonometric points and at the same time establishing about 60 new additional points. During the period 1967–1969 with the assistance of DOS, the control points were increased by a further 80 points. During the same period, a primary level network was established for the island.

Formerly, all color maps of Mauritius were produced by DOS and later on, the French *Institut Géographique National* (IGN) also produced some maps. The scale of these maps was 1:100,000 (1 sheet covering the entire island) and 1:25,000 (13 sheets covering the whole island). It was only during the early seventies that the first 1:2,500 maps of the urban areas were produced by DOS from 1968 aerial photography. IGN did also some 1:2,500 mapping of the North from 1973 photography and part of the southern part from 1974 photography in 1981. From 1981 onwards most of the 1:2,500 mapping is being done by the Survey Department, Mauritius."

From July 11, 1994 to September 16, 1994, Professor Peter Dare of the University of East London led a group of students for a GPS survey of Mauritius. Seventeen old points were occupied, and the transformation developed by Dare was published as a three-parameter shift from the Mauritius 1994 Datum (WGS 84 ellipsoid) to Le Pouce 1934. Those parameters are: $\Delta X = +770.126$ m, $\Delta Y = -158.383$ m, $\Delta Z = +498.232$ m, and the r.m.s. of each component is ± 0.250 m. Prof. Dare remarked that the shift components were in "acceptable agreement" with the WGS 72 parameters published by the British Hydrographic Office in 1982. Note that Dare used the "Leica SKI™" software for his students' analysis, and a **left-handed** coordinate system is supported by that software.

In November 1997, the Chief Surveyor requested assistance from the University of New Orleans with respect to the relation between the Mauritius Datum of 1994 and the *Le Pouce* Datum of 1934. He also requested software for a new Grid System. Mr. Ramasawmy suggested a "UTM projection," with false coordinates of 1,000,000 m in each component. The "UTM projection" was not used because of my "well-known" disdain for military Grids being used in standard (or especially in "modified") form for civilian applications. Based on my direction, Dr. David Fabre, then a UNO Graduate Student, prepared the transformation software for the Republic of Mauritius Oblique Stereographic Grid. Dr. Fabre also "volunteered" to perform a geodetic analysis to determine a seven-parameter Bursa-Wolfe datum shift. The seven (**right-handed**) parameters from Mauritius 1994 to Le Pouce 1934 are: $\Delta X = -91.824$ m, $\Delta Y = -292.222$ m, $\Delta Z = -115.604$ m, $\delta s = 23.15822 \times 10^6$, $R_x = +1.01''$, $R_y = -19.74''$, $R_z = -22.14''$. The residuals with respect to: Latitude = ± 0.12 m, Longitude = ± 0.15 m, height = ± 0.50 m.

As a computational "test point" for the reader, coordinates of Fort George, Martello Tower are provided. For Lindsay's 1874 position: $\varphi = 20°~08'~45.92''$ South, $\lambda = 57°~29'~26.39''$ East. For Connal's 1876 Cassini-Soldner position: N.18,484 ft, W.12,686 ft. For LT. Coghlan's 1877 position: $\varphi = 20°~08'~45.92''$ South, $\lambda = 57°~29'~26.39''$ East. For Sanseau's 1934 position: $\varphi = 20°~08'~36.2866''$ South, $\lambda = 57°~29'~02.1478''$ East, h = +11.6 m. For Dare's 1996 position: $\varphi = 20°~08'~47.3129''$ South, $\lambda = 57°~29'~27.4235''$ East, h = -7.4 m.

The grid developed by Fabre is based on the Oblique Stereographic Double projection, as similarly used by the Canadians in the Maritime Provinces. The origin of the Mauritius Stereo Grid is at *Le Pouce* with a scale factor at origin equal to unity, False Easting = 2,000,000 m, and False Northing = 500,000 m. The rationale for the disparate numbers for the false origin is that I am a proponent of the U.S. Coast & Geodetic Survey (NGS) preference for a different number of digits in each rectangular component to avoid blunders of reporting. Note that a reason for choosing a stereographic projection is that it is ideally suited for regions that are circular in shape.

Everything has changed with reference to the coordinate systems of Mauritius since 1999: "The Ministry of Housing and Lands (MHL) identified the need to develop a new geodetic datum to support and maintain the Digital Cadastral Database (DCDB) in line with Government initiatives to modernize land administration in Mauritius.

In 2008, Landgate (Australia) undertook a rigorous assessment of all available geodetic survey data comprising the 1934 triangulation survey, and GPS surveys conducted in 1994 and 2001. This analysis revealed that these surveys, either in combination or singularly, were not of sufficient accuracy for the basis of a new geodetic datum. In addition, new GPS observations were required to accurately position the Island of Mauritius in a global sense. Landgate recommended a new GPS survey, with stringent survey practices, to deliver the required accuracy. This survey was conducted by MHL in July 2008 and Landgate processed the data with assistance from Geoscience Australia (GA) and determined the new datum through rigorous network adjustment.

The new Geodetic Datum for Mauritius is named Geocentric Datum of Mauritius 2008 (GDM2008) and derivative mapping coordinates based on Universal Transverse Mercator (UTM) is named Map Grid of Mauritius 2008 (MGM2008). The datum is based on the World Geodetic System 1984 (WGS84) reference ellipsoid. The reference coordinates in International Terrestrial Reference Frame (ITRF) 2005@2020 were derived for two datum points namely *Grand Bassin* and The Mount and the GPS surveys were adjusted on these two datum points by least square (*sic – Ed.*) adjustments.

Based on the above adjusted primary control points, the Geodetic Project Team of the Cadastre Unit designed, observed, and processed a secondary GPS network consisting of 61 control points spread over the whole island. This network included 17 primary control points which were held fixed during adjustment and the resulting average error ellipse of the adjusted points for both the semi-major and semi-minor axes are 6 mm. The whole exercise was carried out between 2010 and 2012.

"To cater for plane surveys, a plane coordinate system has been devised that minimizes the effect of projection scale factor. It is based on GDM2008 datum and Lambert Conformal Conic Two Parallel map projection and is referred to as the Local Grid Mauritius 2012 (LGM2012)" (MHL-2013).

Mexico
International boundary
State (estado) boundary
National capital
State (estado) capital
Railroad
Road
0 100 200 300 Kilometers
0 100 200 300 Miles
Lambert Conformal Conic Projection, SP 1900N/2920N
UNITED STATES
NORTH PACIFIC OCEAN
Gulf of Mexico
Gulf of California
Bahía de Campeche
Golfo de Tehuantepec
BAJA CALIFORNIA
BAJA CALIFORNIA SUR
SONORA
CHIHUAHUA
COAHUILA DE ZARAGOZA
NUEVO LEÓN
TAMAULIPAS
SINALOA
DURANGO
ZACATECAS
SAN LUIS POTOSÍ
NAYARIT
JALISCO
COLIMA
MICHOACÁN DE OCAMPO
GUERRERO
OAXACA
VERACRUZ-LLAVE
TABASCO
CHIAPAS
CAMPECHE
YUCATÁN
QUINTANA ROO
GUATEMALA
BELIZE
HONDURAS
EL SALVADOR
NIC.
Los Angeles
San Diego
Tijuana
Mexicali
Ensenada
Yuma
Phoenix
Tucson
Albuquerque
Amarillo
Little Rock
Atlanta
Birmingham
Montgomery
Fort Worth
Dallas
Abilene
Odessa
Shreveport
Jackson
Mobile
New Orleans
Austin
Houston
Galveston
San Antonio
Laredo
Corpus Christi
Brownsville
Matamoros
Ciudad Juárez
El Paso
Heroica Nogales
San Quintín
Isla Cedros
Rosarito
Hermosillo
Guaymas
Santa Rosalía
Ciudad Obregón
Chihuahua
Delicias
Hidalgo del Parral
Nuevo Laredo
Monclova
Reynosa
Monterrey
Saltillo
Torreón
Gómez Palacio
Los Mochis
Escollos Alijos
Loreto
La Paz
Culiacán
Durango
Mazatlán
Ciudad Victoria
Ciudad Mante
Tampico
Ciudad Madero
Zacatecas
San Luis Potosí
Aguascalientes
Islas Tres Marías
Tepic
Puerto Vallarta
Guadalajara
León
Guanajuato
Irapuato
Celaya
Querétaro
Pachuca
Poza Rica
Zamora
Morelia
México
Toluca
Cuernavaca
Puebla
Jalapa
Veracruz
Córdoba
Orizaba
Islas Revillagigedo
Manzanillo
Colima
Lázaro Cárdenas
Chilpancingo
Acapulco
Oaxaca
Salina Cruz
Coatzacoalcos
Minatitlán
Villahermosa
Ciudad del Carmen
Campeche
Progreso
Mérida
Cancún
Isla Cozumel
Chetumal
Belize City
Belmopan
Tuxtla Gutiérrez
San Cristóbal de Las Casas
Tapachula
Guatemala
Puerto Barrios
San Pedro Sula
Tegucigalpa
San Salvador
Puerto San José
1. AGUASCALIENTES
2. DISTRITO FEDERAL
3. GUANAJUATO
4. HIDALGO
5. MÉXICO
6. MORELOS
7. QUERÉTARO DE ARTEAGA
8. TLAXCALA
Boundary representation is not necessarily authoritative.
Base 802540 (R01257) 3-97

United Mexican States

Slightly less than three times the size of Texas, Mexico is bordered by Belize (250 km), Guatemala (962 km), and the United States (3,141 km). The lowest point is *Laguna Salada* (–10 m), and the highest point is *Vocal Pico de Orizaba* (5,700 m). "According to tradition, Francisco Gonzalez Bocanegra, an accomplished poet, was uninterested in submitting lyrics to a national anthem contest; his fiancée locked him in a room and refused to release him until the lyrics were completed" for the National Anthem of Mexico (*World Factbook*, 2012).

"Archaeological evidence testifies to the presence of early hunters and gatherers in Mexico around 10000 to 8000 *B.C.* During the next few thousand years, humans domesticated indigenous plants, such as corn, squash, and beans. With a constant food supply assured, people became permanent settlers. Leisure time became available and was used for developing technical and cultural skills. Villages appeared as the number of people and food supplies increased. By 1500 *B.C*, the early inhabitants were producing handmade clay figurines and sophisticated clayware. Between 200 *B.C.* and *A.D.* 900, Mesoamerica was the scene of highly developed civilizations. Archaeologists have designated this Classic Period as the Golden Age of Mexico. This era was a time when the arts and sciences reached their apex, when a writing system developed, and when a sophisticated mathematical system permitted the accurate recording of time. Religion was polytheistic, revering the forces of nature in the gods of rain, water, the sun, and the moon. The most important deity was *Quetzalcóatl*, the feathered serpent and the essence of life, from whom all knowledge derived. Metals came into use only by the end of the period, but despite this handicap, impressive architectural structures in the pyramids at *Teotihuacán* near Mexico City, the Pyramid of the Niches at El Tajín in the state of Veracruz, and the Temple of the Sun at Palenque in present-day Chiapas were built and survive to this day. The last nomadic arrivals in the Valley of Mexico were the Mexica, more commonly known as the Aztec. Although recent linguistic and archaeological work suggests the Aztec may have come from northwest Mexico, their origins are obscure.

"According to legend, the Aztec came from *Aztlán*, a mythical place to the north of the Valley of Mexico around *A.D.* 1100. Lured by stories of the riches of the Aztec, a Spanish adventurer, Hernán (sometimes referred to as Fernando or Hernando) Cortés, assembled a fleet of 11 ships, ammunition, and over 700 men and in 1519 set sail from Cuba to Mexico. The party landed near present-day Veracruz in eastern Mexico and started its march inland. Superior firepower, resentment against the Aztec by conquered tribes in eastern Mexico, and considerable luck all aided the Spanish in their conquest of the Aztec. The Aztec and their allies had never seen horses or guns, the Spanish had interpreters who could speak Spanish, Maya, and *Náhuatl* (the Aztec language), and perhaps what was most important, Cortés unwittingly had the advantage of the legend of *Quetzalcóatl*, in which the Aztec are said to have believed that a white god would arrive in ships from the east in 1519 and destroy the native civilizations. Unwilling to confront the mysterious arrival whom he considered a god, the Aztec emperor, *Moctezuma* II (anglicized as Montezuma), initially welcomed the Spanish party to the capital in November 1519. Montezuma soon was arrested, and the Spanish took control of *Tenochtitlán*. The Aztec chieftains staged a revolt, however, and the Spanish were forced to retreat to the east. The Spanish recruited new troops while a smallpox epidemic raged through *Tenochtitlán*, killing much of the population, possibly including Montezuma. By the summer of 1521, the Spanish were ready to assault the city. The battle raged for 3 weeks, with the superior firepower of the Spanish eventually proving decisive. The last emperor, *Cuauhtémoc*, was captured and killed. In the nineteenth century, the legend of *Cuauhtémoc* would be revived, and the last Aztec emperor would be considered a symbol of honor and courage, the first Mexican national hero.

"The 11-year (1810–1821) period of civil war that marked the Mexican wars of independence was largely a byproduct of the crisis and breakdown of Spanish royal political authority throughout the American colonies. A successful independence movement in the United States had demonstrated the feasibility of a republican alternative to the European crown" (*Country Studies, Library of Congress*, 2012).

"In 1898, the Mexican Geodetic Commission under the direction of Angel Anguiano agreed to collaborate with Canada and the United States to measure the arc of the 98°W meridian comprising some 1,100 km within Mexico. The survey was to encompass some 80,000 km. and cover parts of the states of Guerrero, Oaxaca, Puebla, Tlaxcala, San Luis Potos, Veracruz, and Tamaulipas; following the Sierra Madre Oriental and crossing the state of Oaxaca. The arc consisted of 76 vertices and reached heights of 2,500 m above sea level; the lengths of the sides varied from 10 to 130 km. The baselines measured were in Oaxaca, Tecamachalco, Apam, Rio Verde and La Cruz. The works were initiated in 1901 and concluded in 1915. The U.S. Coast & Geodetic Survey reported to the International Geodetic Association in Hamburg that the participation of Mexico was of the same quality as that of the Americans" (translated from INEGI, 1997 – Ed.). This initial geodetic survey in Mexico was included in the adjustment of the North American Datum of 1927 (NAD27) where the origin is at Meades Ranch, Kansas (and part of the 98.W meridian arc), where $\Phi_o = 39°\,13'\,26.686''$ N, $\Lambda_o = 98°\,32'\,30.506''$ W, the reference azimuth to station Waldo is $\alpha_o = 75°\,28'\,09.64''$, and the ellipsoid of reference is the Clarke 1866 where $a = 6,378,206.4$ m and $b = 6,356,583.8$ m. "In the early 1930s most of the Vera Cruz area was covered by 1:100,000 – scale sheets. Aerial surveys had scarcely begun. In 1938 or 1939 the *Comisión Geográfico Militar* was created (in 1954 to become the *Departamento Cartográfico Militar*). When the United States Air Force decided to obtain aerial mapping photographs of Mexico in 1942, Mexico agreed, stipulating that it was to receive, among other products, contact prints of all photographs taken. The United States made World Aeronautical Charts (WACs 1:1,000,000 scale) from its prints; Mexico considered its copies of the photographs to be state secrets, and the prints were not available for civilian use" (*Topographic Mapping of the Americas, Australia, and New Zealand*, Mary Lynette Larsgaard, 1984).

In July 1953, the Inter-American Geodetic Survey (IAGS) signed an agreement with the Mexican government for a cooperative geodetic survey and mapping program. Geodetic surveys continued to be adjusted to the NAD27 up to 1993. "In close cooperation with the National Geodetic Survey (NGS) and from 1991 onwards, Mexico's *Instituto Nacional de Estadística Geografía e Informática* (INEGI) has transitioned from the classical, mainly two-dimensional methods to the more accurate three-dimensional GPS-based satellite methodologies."

(*GPS High Accuracy Geodetic Networks in Mexico*, Tomás Soler, et al., *Journal of Surveying Engineering*, May 1996). The seven-parameter transformation from NAD27 to ITRF92, Epoch 1988.0 is as follows: $\Delta X = -60.2$ m ± 3.8 m, $\Delta Y = +141.5$ m ± 2.2 m, $\Delta Z = +158.2$ m ± 4.4 m, $\delta\varepsilon = +1.04''$ $\pm 0.1''$, $\delta\psi = +0.22'' \pm 0.1''$, $\delta\omega = -1.97'' \pm 0.1''$, $\delta s = -0.13 \times 10^{-6} \pm 0.28$ (*op. cit.*, Soler, 1996). Of course, the ellipsoid of reference is the GRS80, where $a = 6,378,137$ m and $1/f = 298.257222101$. Note that Dr. Soler is Chief Technical Officer, Spatial Reference System Division of the U.S. National Geodetic Survey. Now on the ITRF2000, epoch 2004.0; OPUS is now available for all of Mexico.

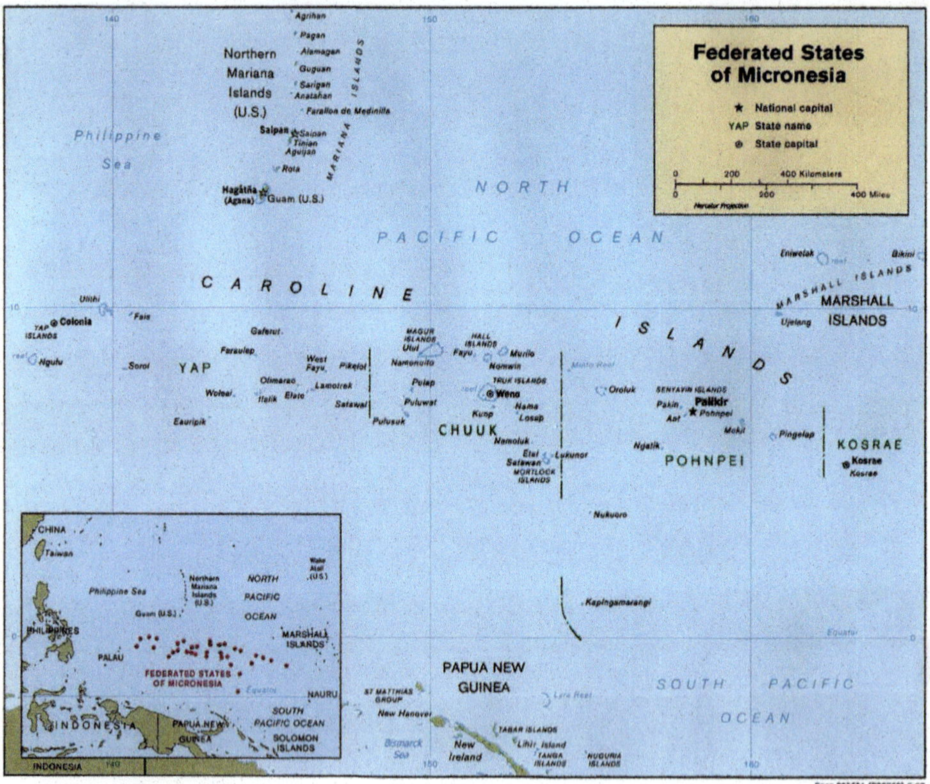

Base 802631 (800768) 5-99

Federated States of Micronesia

Four times the size of Washington, DC (land area only), the terrain is comprised of islands that vary geologically from high mountainous islands to low, coral atolls. Volcanic outcroppings appear on Pohnpei, Kosrae, and Chuuk. The lowest point is the Pacific Ocean (0 m), and the highest point is *Dolohmwar (Totolom)* (791 m) (*World Factbook*, 2014).

"The ancestors of the Micronesians settled there over 4,000 years ago. A decentralized chieftain-based system eventually evolved into a more centralized economic and religious empire centered on Yap. Nan Madol, consisting of a series of small artificial islands linked by a network of canals, is often called the Venice of the Pacific. It is located near the island of Pohnpei and used to be the ceremonial and political seat of the Saudeleur Dynasty that united Pohnpei's estimated 25,000 people from about 500 until 1500 A.D., when the centralized system collapsed" (*Princeton University*, 2014).

"In 1525 Portuguese navigators in search of the Spice Islands (Indonesia) came upon Yap and Ulithi. Spanish expeditions later made the first European contact. At that time, Spain withdrew from its Pacific insular areas and sold its interests to Germany, except for Guam which became a US insular area. German administration encouraged the development of trade and production of copra. In 1914 German administration ended when the Japanese navy took military possession of the Marshall, Caroline, and Northern Mariana Islands. Japan began its formal administration under a League of Nations mandate in 1920. During this period, extensive settlement resulted in a Japanese population of over 100,000 throughout Micronesia. The indigenous population was then about 40,000. Sugar cane, mining, fishing, and tropical agriculture became the major industries. World War II brought an abrupt end to the relative prosperity experienced during Japanese civil administration. By the War's conclusion most infrastructure had been laid waste by bombing, and the islands and people had been exploited by the Japanese Military to the point of impoverishment. The United Nations created the Trust Territory of the Pacific Islands (TTPI) in 1947. Ponape (then including Kusaie), Truk, Yap, Palau, the Marshall Islands, and the Northern Mariana Islands, together constituted the TTPI. The United States accepted the role of Trustee of this, the only United Nations Trusteeship to be designated as a 'Security Trusteeship,' whose ultimate disposition was to be determined by the UN Security Council. As Trustee the US was to 'promote the economic advancement and self-sufficiency of the inhabitants.' The President of the US appointed a High Commissioner of the TTPI, and he, in turn, appointed an administrator for each of the 'Districts' mentioned above. The TTPI remained under the civil administration of the US Navy Department until 1951, when authority passed to the Department of the Interior. In 1979, upon implementation of the FSM Constitution, the US recognized the establishment of the FSM national and state governments. Self- sufficiency, however, remained a dim prospect, in part because private-sector growth had never been encouraged by the TT Administration. On July 12, 1978, following a Constitutional Convention, the people of four of the former Districts of the Trust Territory, Truk (now Chuuk), Yap, Ponape (now Pohnpei) and Kusaie (now Kosrae) voted in a referendum to form a Federation under the Constitution of the Federated States of Micronesia (FSM). United Nations observers certified this referendum as a legitimate act of self- determination. Thereby, the people reasserted their inherent sovereignty which had remained dormant, but intact, throughout the years of stewardship by the League of Nations and the United Nations. Upon implementation of the FSM Constitution on May 10, 1979, the former Districts became States of the Federation, and in due course adopted their own State constitutions. Nationwide democratic elections were held to elect officials of the National and four State governments" (*Government of the Federated States of Micronesia*, 2014).

Yap Astro 1951 (West Base) is where $\Phi_o = 9° 32' 48.15''$ N ±0.2″, $\Lambda_o = 138° 10' 07.48''$ E ±0.1″, $\alpha_o = 175° 36' 30.91''$ to Yap Occultation, and $\alpha_o = 45° 06' 05.5''$ to Azimuth Mark (East Base) measured from North. Originally referenced to the Modified Clarke 1866 ellipsoid, where $a = 6,378,450.047$ m and $1/f = 294.9786982$. In 1965, Army Map Service established Yap SECOR (*SEquential Collation of Ranges - Ed*) Astro station at: $\Phi_o = 9° 32' 48.15''$ N ±0.4″, $\Lambda_o = 138° 10' 07.48''$ E ±0.4″, $\alpha_o = 128° 11' 40.2''$ to SECOR AZ MK measured from south. AMS later published a set of four 1:25,000 scale maps on the International 1924 ellipsoid (*John W. Hager, Personal communication 15 October 2000*). All the Caroline Islands of Micronesia have local Grid systems based on the Oblique Azimuthal Equidistant projection referenced to the Modified Clarke 1866 ellipsoid with origins at their respective Astro stations. For Yap @ Yap SECOR, the geodetic coordinates at the origin are as listed above, and the Grid coordinates at the origin are $X_o = 39,987.92$ m, $Y_o = 60,022.98$ m. For Palau @ Arakabesan Island, $\Phi_o = 7° 21' 04.3996''$ N, $\Lambda_o = 134° 27' 01.6015''$ E, and the Grid coordinates at the origin are: $X_o = 50,000.00$ m, $Y_o = 150,000.00$ m. For Pohnpei @ Distad (USE), $\Phi_o = 6° 57' 54.2725''$ N, $\Lambda_o = 158° 12' 33.4772''$ E, and the Grid coordinates at the origin are: $X_o = 80,122.82$ m, $Y_o = 80,747.24$ m. For Truk Atoll @ Truk SECOR RM1, $\Phi_o = 7° 27' 22.3600''$ N, $\Lambda_o = 151° 50' 17.8530''$ E, and the Grid coordinates at the origin are: $X_o = 60,000.00$ m, $Y_o = 70,000.00$ m (*Map Projections – A Working Manual*, John P. Snyder, USGS Professional Paper 1395, September 10, 1987, page 200).

According to *TR8350.2*, the only transformation parameters offered for the Caroline Islands, Federated States of Micronesia are from Kosrae Island at Kusaie Astro 1951 to WGS 84, where $\Delta X = +647$ m ±25 m, $\Delta Y = +1777$ m ±25 m, $\Delta Z = -1124$ m ±25 m; furthermore, note that the local datum is considered to be referenced to the International 1924 ellipsoid, where $a = 6,378.388$ m and $1/f = 297$.

The coordinate systems of Micronesia have apparently been originally established with a modified version of the Clarke 1866 ellipsoid by the U.S. Army Corps of Engineers Army Map Service soon after World War II. The U.S. Geological Survey apparently considered the coordinate systems of Micronesia to be on the standard version of the Clarke 1866 ellipsoid. Maps published by Army Map Service and presumably its successor agencies now consider the coordinate systems of Micronesia to be on the International 1924 ellipsoid. Since each island appears to have its own Astro origin for a coordinate system, and each island is relatively small, the difference in which reference ellipsoid is used for coordinate transformations is practically nil.

Thanks to Mr. John W. Hagar (*Personal communication, May 2014*), the astronomical observations of many of the islands of Micronesia have resulted in the following individual datums. For Ulithi Island: Asor Astro 1966 Datum, where $\Phi_o = 10° 02' 07.83''$ N, $\Lambda_o = 139° 45' 55.70''$ E, International 1924 ellipsoid. For Nomoi Island: Astronomic Station No. 1 (1951) Datum, where $\Phi_o = 5° 20' 05.5''$ N, $\Lambda_o = 153° 43' 55.2''$ E, International 1924 ellipsoid. On Yap Island also, East Fayu Astro 1920 Datum, where $\Phi_o = 8° 34' 48''$ N, $\Lambda_o = 151° 21' 40''$ E, observed by the United Kingdom. Fais Astro (1951) Datum observed at Fais Astro station, where $\Phi_o = 9° 45' 03.8''$ N, $\Lambda_o = 140° 31' 02.8''$ E, $\alpha_o = 328° 07' 54.6''$ to azimuth mark from south, $H_o = 18.7$ m, International 1924 ellipsoid. For Gaferut Island, Faraulep 1920 Datum at station F11, where $\Phi_o = 8° 35' 45''$ N, $\Lambda_o = 144° 33' 16''$ E, Bessel 1841 ellipsoid and likely observed by the Japanese. For Helen Island 1927 Datum at station South Point, where $\Phi_o = 2° 58' 35.7''$ N, $\Lambda_o = 131° 48' 51.4''$ E. For Truk Island, Iben Astro 1947 Datum, where $\Phi_o = 7° 29' 13.05''$ N, $\Lambda_o = 151° 49' 44.42''$ E, observed at Japanese Station #26. This was observed by USS Maury (AGS 16) 1947 and referenced to the Clarke 1866 ellipsoid. For Ifalik 1921 Datum, where $\Phi_o = 7° 14' 57''$ N, $\Lambda_o = 144° 27' 01''$ E, Clarke 1866 ellipsoid. For Kaarappu (1918) Datum, where $\Phi_o = 6° 47' 45.10''$ N, $\Lambda_o = 158° 01' 09.54''$ E, observed by the United Kingdom. For Kapingamarangi (1951) Datum observed at Astro Station 3, where $\Phi_o = 1° 01' 54.1''$ N, $\Lambda_o = 154° 47' 53.6''$ E, $\alpha_o = 218° 42' 40.3''$, and referenced to the International 1924 ellipsoid. For Kusaie 1951 Datum, where $\Phi_o = 5° 20' 25.3''$ N, $\Lambda_o = 163° 02' 48.0''$ E, $\alpha_o = 123° 07' 40.4''$ to 2 (AUX) from south, $H_o = 358.7$ ft. For Kusaie 1962 Datum observed at Allen Sodano Light, where $\Phi_o = 5° 21' 48.80''$

N ±0.10″, $\Lambda_o = 162°\ 58'\ 03.48″$ E ±0.04″, $\alpha_o = 90°\ 40'\ 08.87″$ ±0.19″ to Kusaie 1951 from south, $H_o = 456$ ft, and both are referenced to the International 1924 ellipsoid. For Lamotrek 1951 Datum observed at an Astro station, $\Phi_o = 7°\ 27'\ 11.8″$ N, $\Lambda_o = 146°\ 22'\ 43.3″$ E, $\alpha_o = 196°\ 52'\ 03.7″$ 1 to 2 from south, the Astro position was reduced to station 1, where $\Phi_o = 7°\ 27'\ 13.9″$ N, $\Lambda_o = 146°\ 22'\ 42.3″$ E, referenced to the International 1924 ellipsoid. For Le Le Island, $\Phi_o = 5°\ 20'\ 07″$ N, $\Lambda_o = 163°\ 01'\ 23″$ E, observed by the United Kingdom. For Losap 1924 Datum observed at an Astro Station, $\Phi_o = 6°\ 53'\ 38″$ N, $\Lambda_o = 152°\ 44'\ 01″$ E. From H.O. chart 81288, Magur Island Datum origin is, where $\Phi_o = 8°\ 59'\ 25″$ N, $\Lambda_o = 150°\ 07'\ 03″$ E. For Manto 1918 Datum with origin at Point F-2, where $\Phi_o = 7°\ 02'\ 57.86″$ N, $\Lambda_o = 157°\ 49'\ 54.38″$ E, observed by the United Kingdom. For Mokil Island, Manton Datum with origin at B Manton Astro, $\Phi_o = 6°\ 41'\ 06″$ N, $\Lambda_o = 159°\ 46'\ 31″$ E. For Murilo Island 1951 Datum with origin at Murilo Astro, 19, where $\Phi_o = 8°\ 36'\ 28.3″$ N, $\Lambda_o = 152°\ 14'\ 13.6″$ E, $\alpha_o = 277°\ 44'\ 18.3″$ to 2 Azimuth Mark from south, the reference ellipsoid is the International 1924. For Namoluk 1924 Datum at origin station Astro F, Namorukku, where $\Phi_o = 5°\ 55'\ 24″$ N, $\Lambda_o = 153°\ 06'\ 57″$ E. For Ngatik 1951 Datum origin at Astronomic Station No. 8: $\Phi_o = 5°\ 47'\ 18.4″$ N, $\Lambda_o = 157°\ 20'\ 31.4″$ E, $\alpha_o = 52°\ 56'\ 55.4″$ to Azimuth Mark from south, referenced to the International 1924 ellipsoid. For the Ngulu Island 1951 Datum origin at Ngulu Astro (1): $\Phi_o = 8°\ 17'\ 42.7″$ N, $\Lambda_o = 137°\ 29'\ 32.5″$ E, $\alpha_o = 258°\ 30'\ 04.7″$ to 2 from south, referenced to the International 1924 ellipsoid. For Nomoi Island 1916 Datum origin at Satawan Island Astro: $\Phi_o = 5°\ 20'\ 03.40″$ N, $\Lambda_o = 153°\ 44'\ 07.40″$, referenced to the Bessel 1841 ellipsoid and likely a Japanese observation. For Nukuoro 1951 Datum at origin Astronomic Station No. 4: $\Phi_o = 3°\ 49'\ 43.3″$ N, $\Lambda_o = 154°\ 57'\ 35.9″$ E, $\alpha_o = 229°\ 33'\ 35.7$ to Azimuth Mark from south and referenced to the International 1924 ellipsoid. For Onari Datum: $\Phi_o = 8°\ 45'\ 28″$ N, $\Lambda_o = 150°\ 19'\ 37″$ E, from H.O. chart 81288. For Oroluk 1951 Datum, $\Phi_o = 7°\ 37'\ 29.5″$ N, $\Lambda_o = 155°\ 09'\ 12.4″$ E, $\alpha_o = 223°\ 12'\ 29.4″$ to AP3 from south, referenced to the International 1924 ellipsoid. For Pingelap 1951 Datum origin No. 1 Astro: $\Phi_o = 6°\ 12'\ 30.9″$ N, $\Lambda_o = 160°\ 42'\ 13.4″$ E, $\alpha_o = 321°\ 51'\ 16.0″$, referenced to the International 1924 ellipsoid. For Pisaras Island Datum: $\Phi_o = 8°\ 35'\ 05″$ N, $\Lambda_o = 150°\ 24'\ 11″$ E. For Ponape Astro 1962 Datum, $\Phi_o = 6°\ 58'\ 35.49″$ N ±0.09″ (P.E.), $\Lambda_o = 158°\ 11'\ 16.14″$ E ±0.10″ (P.E.), $\alpha_o = 276°\ 20'\ 47.65″$ ±0.18″(P.E.) from Ponape RM1 to DISTAD from south, referenced to the International 1924 ellipsoid. Note that the Japanese observation (on the Bessel 1841 ellipsoid) is: $\Phi_o = 7°\ 00'\ 23.20″$ N, $\Lambda_o = 158°\ 13'\ 49.80″$ E. For Pulap 1920 Datum origin at Astronomic Station B1: $\Phi_o = 7°\ 38'\ 20.99″$ N, $\Lambda_o = 149°\ 24'\ 58.80″$ E. For Pulusuk 1921 Datum at Astro station: $\Phi_o = 6°\ 42'\ 15″$ N, $\Lambda_o = 149°\ 18'\ 47″$ E. For Puluwat 1951 Datum origin at Astro Sta. 1, Kihoko: $\Phi_o = 7°\ 21'\ 25.3″$ N, $\Lambda_o = 149°\ 11'\ 37.2″$ E, $\alpha_o = 001°\ 15'\ 41.8″$ to 3. α mark (Kinan) from south, referenced to the International 1924 ellipsoid. For Sorol 1920 Datum origin at Sorol Island Astro: $\Phi_o = 8°\ 08'\ 00″$ N ±5.6, $\Lambda_o = 140°\ 24'\ 30″$ E±2.5″ referenced to both the Clarke 1866 and the International 1924 ellipsoids. The Sumner Astro 1944 Monument at: $\Phi_o = 10°\ 05'\ 25.06″$ N, $\Lambda_o = 139°\ 42'\ 15.78″$ E was observed by U.S.S. Sumner, Oct.–Nov.1944. Compared to Ulithi 1951 Datum origin at Ulithi Astro (32): $\Phi_o = 10°\ 04'\ 52.8″$ N ±0.92, $\Lambda_o = 139°\ 44'\ 22.5″$ ±5.03″ E, $\alpha_o = 121°\ 28'\ 11.0″$ to 31 from north and referenced to the Clarke 1866 ellipsoid, but incorrect time signals were applied. The correct longitude should be 139° 44′ 22.3″. The United Kingdom in 1994 gave values of $\Phi_o = 10°\ 04'\ 51.0″$, $\Lambda_o = 139°\ 44'\ 28.2″$. (No explanation.) For the Truk Astro 1951 Datum origin at Astro Station No. 2: $\Phi_o = 7°\ 21'\ 37.7″$ N, $\Lambda_o = 151°\ 53'\ 34.3″$ E, $\alpha_o = 51°\ 58'\ 17.1″$ to 76 from south, referenced to the International 1924 ellipsoid. For Ulul Astro Station 1916 Datum origin at Obs. Spot: $\Phi_o = 8°\ 34'\ 59.0″$ N, $\Lambda_o = 149°\ 39'\ 27.9″$ E in the Namonuito area, note that USHO Chart 5416 (1945) Minto Reef 1:50,000 Series W756 Sheet 5046 II, Ed. 1-AMS Minto Reef is based on this. No further explanation was offered. For West Fayu 1921 Datum origin point at Astro Station: $\Phi_o = 8°\ 05'\ 17″$ N, $\Lambda_o = 146°\ 44'\ 29″$ E, and was observed by the United Kingdom. For Woleai Island 1951 Datum origin at Occultation Station 1955 Astro: $\Phi_o = 7°\ 22'\ 17.8″$ N, $\Lambda_o = 143°\ 54'\ 04.2″$ E, $\alpha_o = 296°\ 10'\ 42.0″$ to 17 (Az. Mark) from south, $H_o = 4.2$ ft, referenced to the International 1924 ellipsoid. Finally, for Yorupikku $\Phi_o = 6°\ 41'\ 12″$ N, $\Lambda_o = 143°\ 04'\ 47″$ E, observed by the United Kingdom.

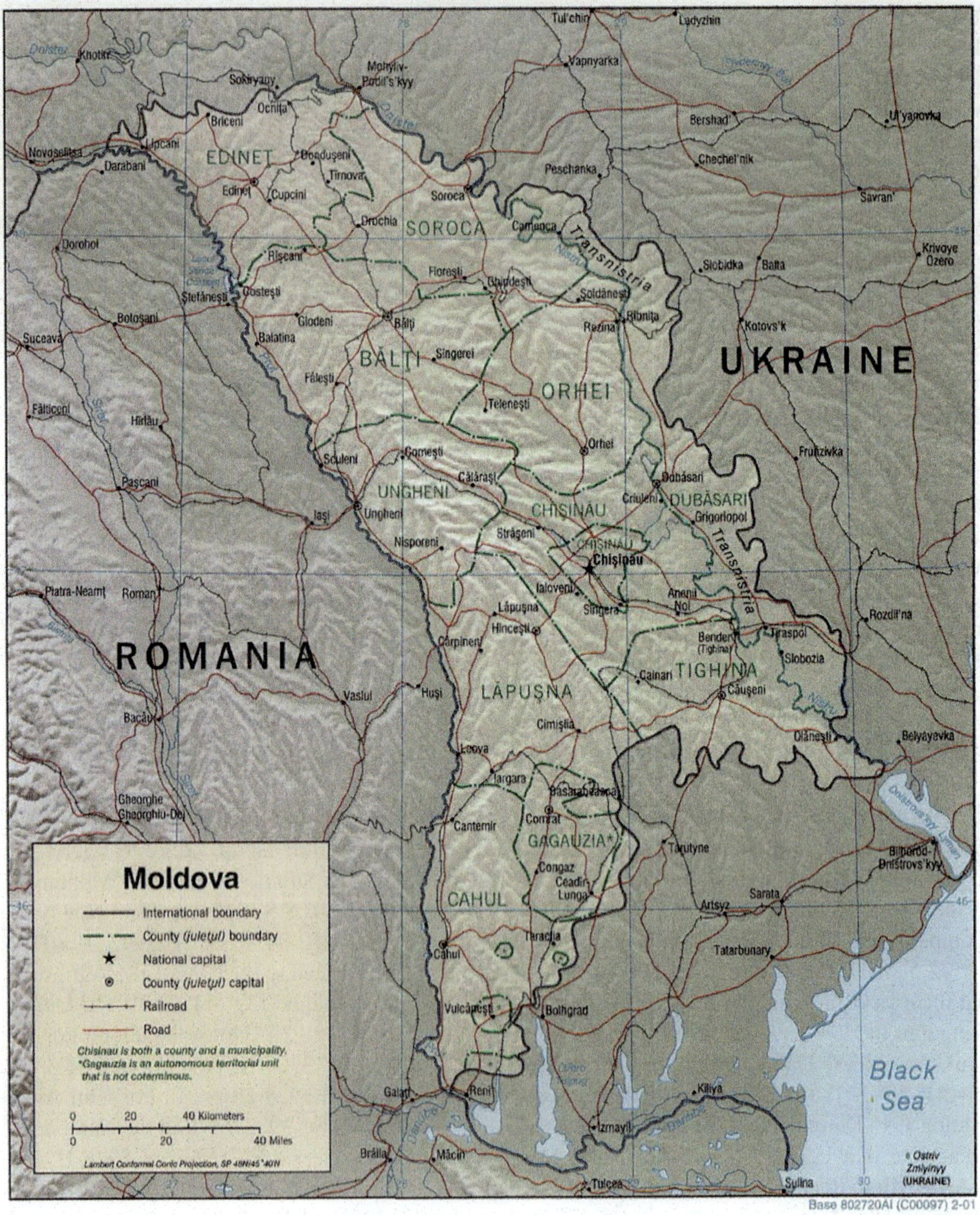

Base 802720AI (C00097) 2-01

Republic of Moldova

Slightly larger than Maryland, Moldova is bordered by Romania (450 km) and Ukraine (940 km). Moldova's terrain is comprised of rolling steppe with a gradual slope to the Black Sea; the lowest point is the Dniester (2 m) and the highest point is Dealul Bălănești (430 m) (*NGA GeoNames Server*).

"The history of the Republic of Moldova is the history of two different regions that have been joined into one country, but not into one nation: Bessarabia and Transnistria. Bessarabia, the land between the Prut and Nistru rivers, is predominantly ethnic Romanian in population and constitutes the eastern half of a region historically known as Moldova or Moldavia (the Soviet-era Russian name). Transnistria is the Romanian-language name for the land on the east bank of the Nistru River; the majority of the population there is Slavic–ethnic Ukrainians and Russians – although Romanians are the single largest ethnic group there. To a great extent, Moldova's history has been shaped by the foreigners who came to stay and by those who merely passed through, including Greek colonists, invading Turks and Tatars, officials of the Russian Empire, German and Bulgarian colonists, communist apparatchiks from the Soviet Union, soldiers from Nazi Germany, Romanian co-nationalists, and twentieth-century Russian and Ukrainian immigrants. Each group has left its own legacy, sometimes cultural and sometimes political, and often unwelcome. Moldova's communist overlords, the most recent 'foreigners,' created the public life that exists in Moldova today. Independence has brought about changes in this public life, but often only on the surface. What further changes Moldova makes will depend partly on how much time it has before the next group of 'foreigners' comes to call. Moldova's Latin origins can be traced to the period of Roman occupation of nearby Dacia (in present-day Romania, Bulgaria, and Serbia), *ca. A.D.* 105–270, when a culture was formed from the intermingling of Roman colonists and the local population. After the Roman Empire and its influence waned and its troops left the region in *A.D.* 271, a number of groups passed through the area, often violently: Huns, Ostrogoths, and Antes (who were Slavs). The Bulgarian Empire, the Magyars, the Pechenegs, and the Golden Horde (Mongols) also held sway temporarily. In the thirteenth century, Hungary expanded into the area and established a line of fortifications in Moldova near the Siretul River (in present-day Romania) and beyond. The region came under Hungarian suzerainty until an independent Moldovan principality was established by Prince Bogdan in 1349. Originally called Bogdania, the principality stretched from the Carpathian Mountains to the Nistru River and was later renamed Moldova, after the Moldova River in present-day Romania. During the second half of the fifteenth century, all of southeastern Europe came under increasing pressure from the Ottoman Empire, and despite significant military victories by Stephen the Great (*Stefan cel Mare*, 1457–1504), Moldova succumbed to Ottoman power in 1512 and was a tributary state of the empire for the next 300 years. In addition to paying tribute to the Ottoman Empire and later acceding to the selection of local rulers by Ottoman authorities, Moldova suffered repeated invasions by Turks, Crimean Tatars, and Russians. In 1792 the Treaty of Iasi forced the Ottoman Empire to cede all of its holdings in what is now Transnistria to the Russian Empire. An expanded Bessarabia was annexed by, and incorporated into, the Russian Empire following the Russo-Turkish War of 1806–12 according to the terms of the Treaty of Bucharest of 1812. Moldovan territory west of the Prut River was united with Walachia. And in the same year, Alexandru Ioan Cuza was elected prince of Walachia and the part of Moldova that lay west of the Prut River, laying the foundations of modern Romania. These two regions were united in 1861" (*Library of Congress Country Studies*, 1995).

"Moldova was incorporated into the Soviet Union at the close of World War II. Although the country has been independent from the USSR since 1991, Russian forces have remained on

Moldovan territory east of the Dniester River supporting a Transnistrian separatist region with a Slavic majority population of mostly Ukrainians and Russians" (*World Factbook*, 2013).

"The triangulation consisted of a net of the first order, supplemented by auxiliary nets of the second- and third-orders. In respect of the Moldavian district, the Dobrudzha, and Muntenia as far as the meridian of Zimnicea, was calculated on Bessel's ellipsoid (where $a = 6,377,397.155$ m, $1/f = 299.1528128$ – Ed.); whereas for computing the triangulation westwards of this meridian, Clarke's ellipsoid of 1880 (where $a = 6,378,249.145$ m, $1/f = 293.465$ – Ed.) was employed" (*Memorandum on the General State of Geodetic Work in Romania – Brief Historical Review, by General Radu Bodnârescu and Colonel Virgil Joan, translated by the U.S. Army Map Service, RHO/AMS Memo 318, 7 October 1960*). The geocentric datum shift parameters from the local datum to the WGS84 datum were computed using control points coordinates scaled from both the old and from the modern maps, in vicinities near the zone centers" (*Rectification of the Romanian 1:75,000 Map Series, Prior to World War I*, Timár, G. and Mugnier, C., Acta Geodaetica et Geophysica Hungarica, vol. 45(1), pp. 89–96 (2010) DOI: 10.1556/AGeod.45.2010.1.13).

"One side of the Russian triangulation for Bessarabia was taken as the basis for calculating the Moldavian triangulation. This had been taken as far as one side of the Austrian triangulation in Bukovina. Simultaneously, *i.e.*, in 1874, an Austrian and a Romanian officer surveyed the difference in longitude between Iași and Czernowitz. The triangulation work was continued southwards, without attempting to fit in with a side of the Austrian or Russian triangulation, and without measurement of a geodetic baseline" (*ibid. Bodnârescu and Joan*).

"Bucharest (1920) (code BUC), Militari (east end of base) $\Phi_o = 44°\ 26'\ 07.2823''$ N, $\Lambda_o = 26°\ 01'\ 00.2060''$ E, $\alpha_o = 96°\ 43'\ 22.8''$ to Ciorogârla (west end of base) from south. International ellipsoid. Positioned determined from West Pillar of the Military Observatory (*Observatorul Militar Astronomic din Dealul Piscului*). Also called New Romanian. Reference: *Vornetvure Zum Planheft Sudosteuropa Nordlicher Teil Rumenien*.

"Constanta (code COS), Minaret of the Main mosque in Constanta (*Kyustendyhi*) (*Kyustendja*), $\Phi_o = 44°\ 10'\ 31''$ N, $\Lambda_o = 28°\ 39'\ 30.55''$ E, $\alpha_o = 305°\ 15'\ 01.7''$ East Pyramid to West Pyramid of the *Kyustendyi* Base, Walbeck. That came out of the Zapiski vol. 43.

"Observatorul Militar Astronomic din Dealul Pisculua, Bucuresti (code OAR), $\Phi_o = 44°\ 24'\ 34.20''$ N$\pm 0.06''$ (1895), $\Lambda_o = 26°\ 06'\ 44.98''$ E$\pm 0.075''$ (1900.7), $\alpha_o = 127°\ 01'\ 53.005''$ (1895) to Cotroceni. The latitude error value of $\pm 0.06''$ came from "Determinari Astronomice," *Institutul Geografic Militar*, Romania. The same latitude value but $\pm 0.04''$ was published by *Dragomir, General-locotenent ing, Vasile, Rotaru, Colonel Dr. ing. Marian, Mărturii Geodezice*, Bucuresti, 1986, pg. 151. The astro position was transferred geodetically to Bucharest (BUC).

"Old Romanian (code ORU), Bucharest Meridian, $\Phi_o = 44°\ 26'\ 06.63''$ N $\pm 0.04''$ (1899), $\Lambda_o = 23°\ 46'\ 32.00''$ E of Paris, $\alpha_o = 26°\ 16'\ 45.95''$ E, Clarke 1880. Used on old maps, pre-WW II, at 1:20,000 and 1:100,000. Taken from a Planheft, possibly the one listed above for BUC.

"Saint Anna (code STD), at South Base Point of Arad Base, $\Phi_o = 46°\ 18'\ 47.63''$ N, $\Lambda_o = 39°\ 06'\ 54.19''$ E. of Ferro, $\alpha_o = 21°\ 27'\ 08.17''$ E. of Greenwich. Zach ellipsoid ($a = 6,376,480$, $1/f = 310$). Derived geodetically from Vienna. Probable reference is *Positions Rechnungen fur die neue Special Karte der Monarchie*, 1887 Vol. I, Part 1 and 2" (*John W. Hager, Personal communication, March 2013*).

"The Austrian military geographic survey invested a tremendous work into the geodetic and cartographic works in parts of Old Romania, mainly in two waves. First in the second half of the 18th century in the frame of the First Military Survey, then during the Crimean War in Oltenia, Muntenia and Northern Dobrogea, developing a full first-order triangulation net from the Transylvanian border to the Danube and the Black Sea. The network was on the Walbeck ellipsoid and its Molodensky-type displacement parameters are: $\Delta X = +1317$ m; $\Delta Y = +73$ m; $\Delta Z = +357$ m, the error of these figures is under 20 meters. The low error shows that the network was very precise at the time of the survey. The Austrian work in the Danube Principalities preceded the main part of the Second Military Survey in Transylvania (or connected to it), perhaps giving a good training opportunity also for that measurement" (*Habsburg geodetic and cartographic activities in the Old Romania,* Gábor Timár, Dept. of Geophysics and Space Sciences, Eötvös University, H-1117 Budapest, Pázmány Péter sétány 1/a, Hungary).

System 42 of the Soviet era represents a significant volume of geodetic control for Moldova in which the origin point is at Pulkovo Observatory, where $\Phi_o = 59° 46' 18.55''$ North, $\Lambda_o = 30° 19' 42.09''$ East of Greenwich. The defining azimuth at the point of origin to Signal A is: $\alpha_o = 317° 02' 50.62''$. System 42 is referenced to the Krassovsky 1940 ellipsoid, where $a = 6,378,245.0$ m and $1/f = 298.3$. The seven-parameter transformation from System 42 to the European Terrestrial Reference Frame 1989 locally referred to as MOLDREF89 is: $\Delta X = -617.880$ m, $\Delta Y = -253.456$ m, $\Delta Z = -315.690$ m, $R_1 = 5.79748''$, $R_2 = -2.44443''$, $R_3 = -5.1534''$, $\delta s = -13.51806$ (*Development of new geodetic infrastructure in Republic of Moldova,* V. Chiriac, *et al.,* EUREF 2011, Chisinau, 25–28 May 2011). MOLDPOS is the national real-time kinematic GPS network, and the country has implemented a first-order gravity network enforced by three absolute gravity stations and a national quasigeoid (*Regional quasigeoid solution for the Moldova area from GPS levelling data,* A. Marchenko, I. Monin).

The Transversal Mercator for Moldova (TMM) is used for large-scale mapping and has a central meridian $\lambda_o = 28° 24'$ E, $m_o = 0.99994$. No information is offered for a False Easting, but it may be 500 km since UTM is their adopted standard for small scales.

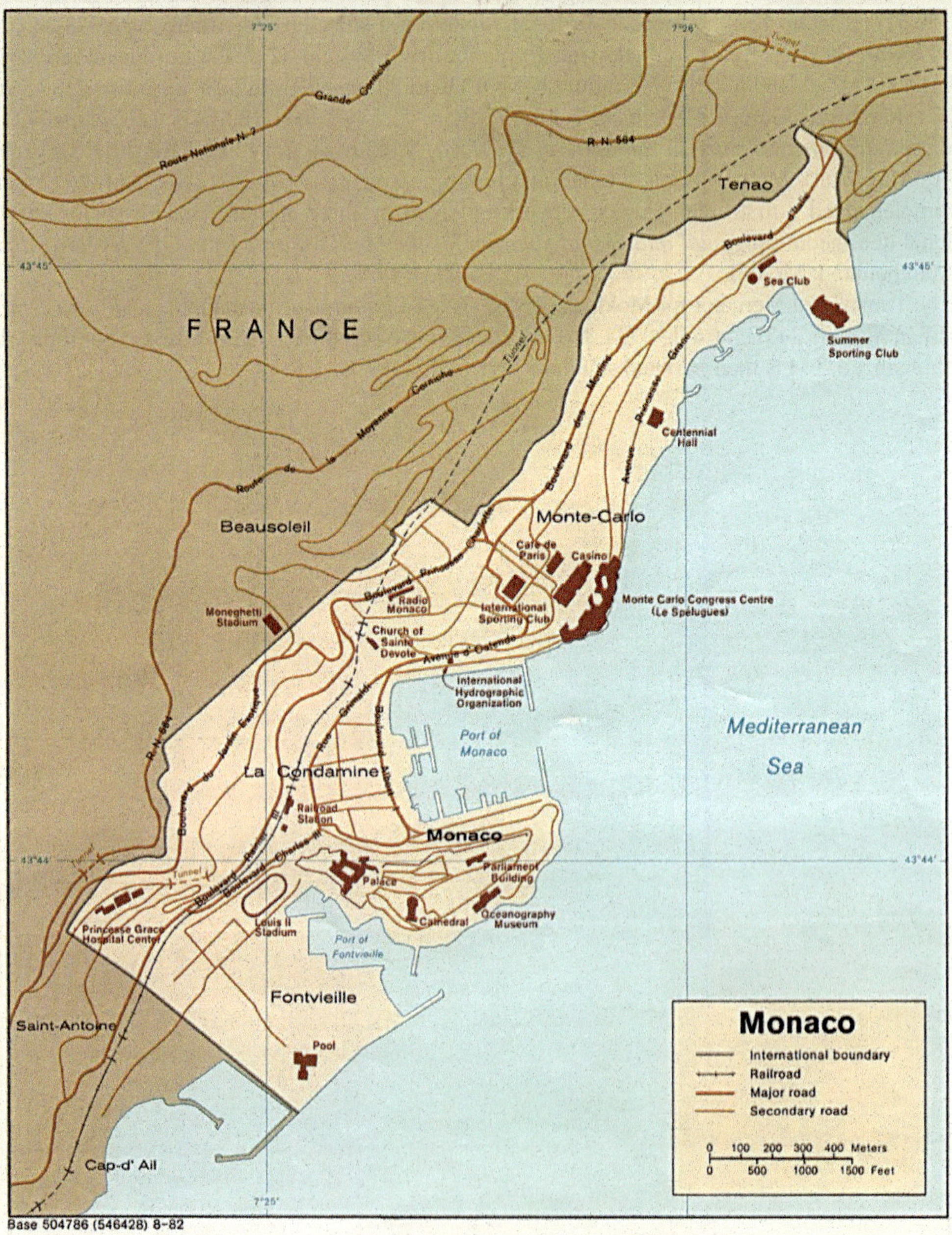

Base 504786 (546428) 8-82

Principality of Monaco

Monaco is about three times the size of the National Mall in Washington, DC, and it has a 6-km border with France; the lowest point is the Mediterranean Sea (0 m), and the highest point is Mont Agel (140 m).

"Since ancient times, Monaco has always been at the crossroads of history. Initially the Ligurians, the ancient people who first settled Monaco, were concerned with the strategic location of the Rock of Monaco. Evidence of the Ligurian occupation of Monaco was found in a cave in the Saint Martin's Gardens. Originally a mountain-dwelling people, they were known for their hard work and their frugality, two traits by which Monegasque citizens are known for today. Founded by the Phocaeans of Massalia during the 6th century, the colony of Monoikos became an important port of the Mediterranean coast. Julius Caesar stopped in Monoecus after the Gallic Wars on his way to campaign in Greece. After the collapse of the Western Roman Empire in 476, Monaco was ravaged by Saracens and barbarian tribes. After the Saracens were expelled in 975, the depopulated area was reclaimed by the Ligurians. In 1215, construction began on a fortress atop the Rock of Monaco by a detachment of Genoese Ghibellines. With the intention of turning the Rock of Monaco into a military stronghold, the Ghibellines created a settlement around the base of the Rock to support the garrison. To draw in residents from Genoa and other surrounding cities, the Ghibellines offered land grants and tax exemptions to newcomers. Civil strife in Genoa between the Guelph and Ghibelline families resulted in many taking refuge in Monaco, among them the Guelph family. Son of Otto Canella, Consul of Genoa in 1133, Grimaldo began the House of Grimaldi, the future ruling family of Monaco. In 1297, François Grimaldi ('Malizia', translated from Italian as 'The Cunning') disguised as a Franciscan monk alongside his cousin Rainier I and his men captured the fortress atop the Rock of Monaco. At his death in 1309, François Grimaldi was succeeded by his cousin, Rainier I. His son, Charles Grimaldi, who would come to be known as Charles I, is considered by historians to be the real founder of the Principality. In 1612, Hercule's son Honoré II, was first given the title of 'Prince of Monaco', which became the official title of the ruler of Monaco and would be passed on to his successors. After Napoleon abdicated the throne in 1814, Monaco was returned to its previous state under the new rule of Honoré IV.

"However, the Principality was re-established as a protectorate of the Kingdom of Sardinia by the Congress of Vienna in 1815. Monaco remained a protectorate until 1860 when, by the Treaty of Turin at the time of Italy's unification, Monaco was ceded to France. With unrest in Menton and Roquebrune, the Prince gave up his claim to the two towns (which made up 95% of the Principality at the time) in return for four million francs. Both the transfer of these two cities and Monaco's sovereignty were recognized by the Franco-Monegasque Treaty of 1861. In spite of the four-million-franc indemnity, Monaco's reduced size and loss of the income it would have gained from Roquebrune and Menton prevented the Principality from escaping its difficult financial predicament. Albert I, previously devoted to scientific research in the fields of oceanography and paleontology, assumed the role of Prince of Monaco in 1889. With an outstanding reputation, a seat in the Academy of Sciences, and various discoveries which are too numerous to discuss, he established the Oceanographic Museum, which is one of the top centers for oceanography to this day. Jacques-Yves Cousteau was the Director of the Oceanographic Museum from 1957 to 1988" (*Embassy of Monaco, 2015*).

The coordinate systems of Monaco are one and the same as France. The Meridian Survey of Picard was (1669–1671), followed by the Meridian Survey of Cassini (1683–1718), the Triangulation of Cassini (1733–1770), the Meridian Survey of France (1739–1740); the Triangulation of the Geographic Engineers (1792–1884) applied the Meridian of Delambre and Méchain, and the New

Meridian of France (1870–1896). The New Triangulation of France (NTF) had a principal network that consisted of 800 points of first order spaced at 30 km intervals, 5,000 points of second order spaced at 10 km intervals, and 60,000 points of third and fourth orders spaced at 3 km intervals. The origin of the datum had its fundamental point located at the Cross of the Panthéon in Paris, where $\Phi_o = 48° 50' 46.52"$ N, $\Lambda_o = 02° 20' 48.62"$ (East of Greenwich) and referenced to the Clarke 1880 ellipsoid, where $a = 6,378, 249.145$ m and $1/f = 293.465$. Monaco is on the French Lambert Zone III (Sud); the Latitude of Origin was $\varphi_o = 44° 06' 00"$, the Central Meridian was $\lambda_o = 2° 20' 13.95"$ East of Greenwich (zero degrees from the Paris Meridian), the Scale Factor at Origin (m_o) = 0.999877501, and the False Easting was 600 km, and the False Northing was 200 km. After World War II, the U.S. Army Map Service developed the European Datum of 1950 (ED50) such that its fundamental point was located at the Helmert Tower in Potsdam, where $\Phi_o = 52° 22' 53.954"$ N, $\Lambda_o = 13° 04' 01.153"$ (East of Greenwich) and referenced to the International 1924 ellipsoid, where $a = 6,378, 388$ m and $1/f = 297$. For the ED50, the UTM Grid is used. The current system for France and Monaco is the *Réseau Géodésique Française* 1993 (RGF93) associated with the European Terrestrial Reference System of 1989 (ETRF89) and referenced to the GRS80 ellipsoid, where $a = 6,378, 137$ m and $1/f = 298.257222101$. For all of France (& Monaco) the "Lambert-93" parameters are Latitude of Origin $\varphi_o = 46° 30'$ N, the Central Meridian is $\lambda_o = 3°$ East of Greenwich, the Northern Standard Parallel $\varphi_N = 49°$ N, the Southern Standard Parallel $\varphi_S = 44°$ N, and the False Easting is 700 km and the False Northing is 6,600 km.

Standard transformations published by the *Institut Géographique National* (IGN) for France and for Monaco are: **from** NTF **to** WGS84, $\Delta X = -168$ m, $\Delta Y = -60$ m, $\Delta Z = +320$ m; **from** NTF **to** ED50, $\Delta X = -84$ m, $\Delta Y = +37$ m, $\Delta Z = +437$ m; **from** ED50 **to** WGS84, $\Delta X = -848$ m, $\Delta Y = -97$ m, $\Delta Z = -117$ m. A bilinear grid interpolation system is also offered to the public free of charge by the IGN **from** NTF **to** RGF93.

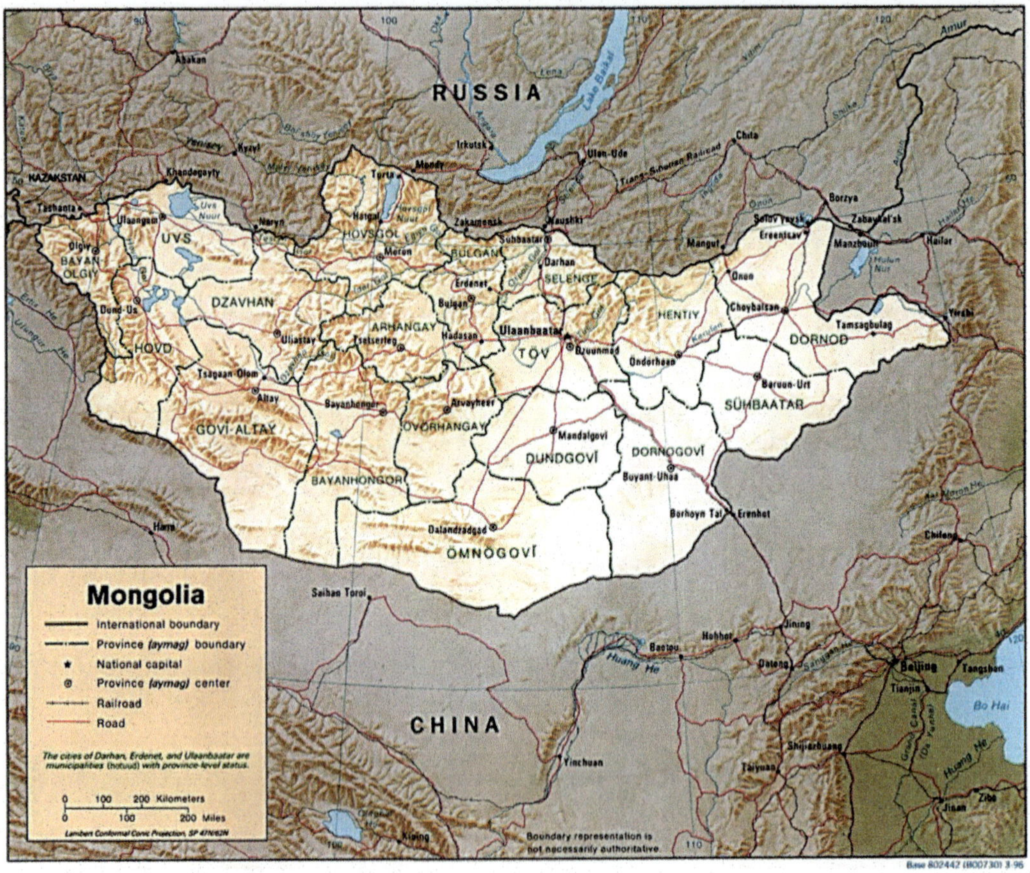

Base 802442 (B00730) 3-96

Mongolia

The current capital of Mongolia is Ulaanbaatar, and the area of the country is slightly smaller than Alaska; its borders are with China (4,677 km) and with Russia (3,485 km). Mongolia has a continental desert climate with large daily and seasonal temperature ranges; its terrain is comprised of a vast semi-desert and desert plains, grassy steppe, mountains in the west and southwest, and the Gobi Desert is in the south central. The lowest point is *Hoh Nuur* at 518 m, and the highest point is *Nayramadlin Orgil* (*Huyten Orgil*) at 4,374 m.

Although the region was inhabited since early times by nomadic peoples, the Mongol tribe made its entrance into history during the 13th century under the leadership of Genghis Khan. The Mongol Empire, with its original capital at Karakorum and later at Beijing, stretched from the Danube River in Eastern Europe to China. In the 14th century, the former empire was broken up and absorbed into China under the Yüan dynasty, originally established by Kublai Khan, grandson of Genghis in 1279. By 1368, the Ming dynasty supplanted the Yüan, shattering the Mongol unity. The former Outer Mongolia eventually gained its independence from China on July 11, 1921, as the Mongolian People's Republic. The date of the country's present constitution is February 12, 1992, when it was renamed Mongolia.

In 1918, the Chinese General Staff compiled a monochrome series at 1:100,000 and 1:300,000 covering the Mongolia–China border areas. The sheets are based on original Chinese surveys. Relief is shown by contours and roads are classified by vehicular limitations. In the 1940s, the Survey Department, Ministry of National Defense, compiled 1:500,000 and 1:1,000,000 series covering all of Mongolia. The Kwantung Army Headquarters produced sheets for military use in 1942–1943 for two areas in eastern Mongolia from sheets originally produced by the Japanese in1935 and 1942, Russian maps dated 1906 and 1933, and a rough survey made by the Japanese in 1912. Relief is shown by contours of form-line accuracy and hill shading. No grid system was used.

Virtually all of the maps produced by the Japanese for Mongolia cover the eastern part. Most were produced by the Japanese General Staff. A 1:100,000 monochrome series published in 1913–1914 covers part of Mongolia east of 106°; in the late 1930s a 1:200,000 series was compiled from Russian maps to cover northeastern Mongolia. During the period of 1923–1943, a 1:500,000 series was compiled for eastern Mongolia. From this series and Russian maps, a 1:200,000 series was produced by the Kwantung Army Headquarters mentioned previously. No grid system was used.

Mapping of Mongolia by the Russians was originally conducted during the 1930s. The *Upravleniye Topografov* (Military Topographic Administration) was formed in 1932 and compiled a 1:200,000 series of small, scattered areas and 1:500,000 and 1:100,000 series for more extensive areas in eastern Mongolia. Relief is shown by form-lines and contours. Geodetic surveys of Mongolia were conducted from 1939 to 1946, and the primary triangulation of the country is comprised of eight north-south arc chains and three east-west arc chains. I count 27 baselines and 54 LaPlace stations on a diagram published by the government in 1999. Thanks to a letter that year from B. Munkhzul, Geodetic Engineer for the State Administration of Geodesy and Cartography, the basic classical geodetic network of Mongolia is comprised of second-order accuracy, with third- and fourth-order points used to densify the network. Including the benchmarks based on the Kronstadt Datum (*Kronshtadsky futshtok*) (sic), there are 27,500 geodetic monuments in Mongolia. The Russian "System 42" Datum is referenced to the Krassovsky 1940 ellipsoid, where $a = 6,378,245$ m and $1/f = 298.3$. The origin is at Pulkovo Observatory: $\Phi_o = 59° \ 46' \ 18.55''$ North, $\Lambda_o = 30° \ 19' \ 42.09''$ East of Greenwich, and the defining azimuth at the point of origin to Signal A is: $\alpha_o = 317° \ 02' \ 50.62''$. The Grid system used in Mongolia for mapping from classical triangulation is the standard Russian Belts such that the False Eastings are equal to 500 km at the central meridians, and the scale factor at the central meridians is equal to unity. The Gauss-Krüger Transverse Mercator uses 6° belts with identical zones as the UTM.

In 1954–1955, the U.S. Army Map Service (AMS) compiled sheets of a 1:250,000 polychrome series with the Universal Transverse Mercator Grid. It covers scattered areas of Mongolia along the Russian and China borders. In 1942–1944, AMS copied a few sheets of a Russian 1:1,000,000 series, and in 1949–1958 compiled a 1:1,000,000 polychrome series for the remaining three-fourths of the country.

Mongolia appears to be the most geodetically advanced country in central Asia. Their national mapping staff was educated in Moscow until 1981 when geodetic and photogrammetric education was offered at the Mongolian Technical University. With the assistance of Swedesurvey, Mongolia has established a new national datum called "MONREF 97." This new datum is based on the International Terrestrial Reference Frame (ITRF 2000) epoch 1997.8. Essentially, this is cartographically identical with the World Geodetic System (WGS 84). The GPS observations were carried out and financed by MONMAP Engineering Services Co., Ltd., Ulaanbaatar in co-operation with the Ministry of Defense, Mongolia. The processing of the GPS observations, development of transformation formulae, and recommendations for a new grid system were performed by Swedesurvey and financed by the Swedish International Development Agency. The new MONREF 97 system will replace the old Russian "System 42," but the Baltic height system of elevations will not be replaced. MONREF 97 is comprised of 38 points at 34 different locations and is similar in concept to the High Accuracy and High Precision Reference Networks of each state in the United States. MONREF 97 is based on two national GPS campaigns carried out in the autumn of 1997. Trimble 4000 SSi receivers were used for the observations and "Bernese 4.2" software was used for the adjustment. Since that software package produces results contrary to U.S. military *and* civilian convention and usage, the standard U.S. rotations will be given herein.

It is fascinating to see that when Mongolia decided to change things, they even changed their grid system in a most surprising way. The Russian "System 42" Datum (locally termed "MSK42") used the identical projection parameters as the Universal Transverse Mercator (UTM) Grid, but with a *different* scale factor at origin. Mongolia has chosen to eschew that old system and has adopted the UTM for their new national Grid! However, it is interesting to note that the national mapping agency of Mongolia recognizes that although UTM may be convenient for national use, individual cities and smaller regions are encouraged to use systems with more sensible scale factors at origin and projections better suited for their shapes. Mongolia is covered by UTM zones 46 through 50.

The published datum shift parameters are offered in a variety of different models that are intriguing. The most familiar is the standard military three-parameter transformation where for MSK42 to MONREF 97 (WGS84) $\Delta a = -108$, $\Delta f = +0.000000480812$, $\Delta X = +13$, $\Delta Y = -139$, $\Delta Z = -74$. Other transformation models include the seven-parameter Bursa-Wolfe where for MSK42 to WGS84: $\Delta X = -78.042$ m, $\Delta Y = -204.519$ m, $\Delta Z = -77.450$ m, $R_x = -1.774''$, $R_y = +3.320''$, $R_x = -1.043''$, $\delta = -4.95105766$ ppm. Unfortunately, no test points were provided for these transformation parameters, but the three-parameter model will give a clue. Another datum shift method published by the Mongolian government is the two-dimensional Helmert transformation that works with the Russian Gauss-Krüger Transverse Mercator and the UTM. The parameters are: X_0 (translation in X), a (X coefficient), Y_0 (translation in Y), b (Y coefficient), δ (scalar), and α (rotation). There is a separate set of parameters published for each UTM zone, and this is the identical technique used by AMS for the computation of the European Datum 1950.

A fourth technique for performing datum shifts from MSK42 to MONREF 97 is a series of Gauss-Krüger projection parameters to transform directly from MSK42 Latitude and Longitude to MONREF 97 UTM coordinates. A fifth and final technique published by the Mongolian government is a table of differences in Latitude and differences in Longitude (all in meters) that serves as a system for implementing bi-linear interpolation akin to the NADCON technique published by the U.S. National Geodetic Survey. Since there is a paucity of gravity observations in Mongolia, the new datum is not a true three-dimensional system. There is great hope to someday have a reliable geoid model for the entire country that will enable GPS Leveling techniques to be implemented.

Zone transformation parameters for UTM zones were recalculated in 2005 and in 2008. A complete airborne gravity survey of Mongolia was carried out in two fall campaigns 2004–2005 by the Danish National Space Center. Absolute gravity was observed in 2006–2007. A Mongolian geoid height model was produced with 16 cm accuracy for the whole country and 2–5 cm accuracy for the city of Ulaanbaatar. Transformation parameters and the geoid height model are accessible from ALAGaC web page for the public (Munkhtsetseg, D., *Geodetic Network and Geoid Model of Mongolia*, isprs.org/proceedings/XXXVIII/7-C4/121_GSEM2009).

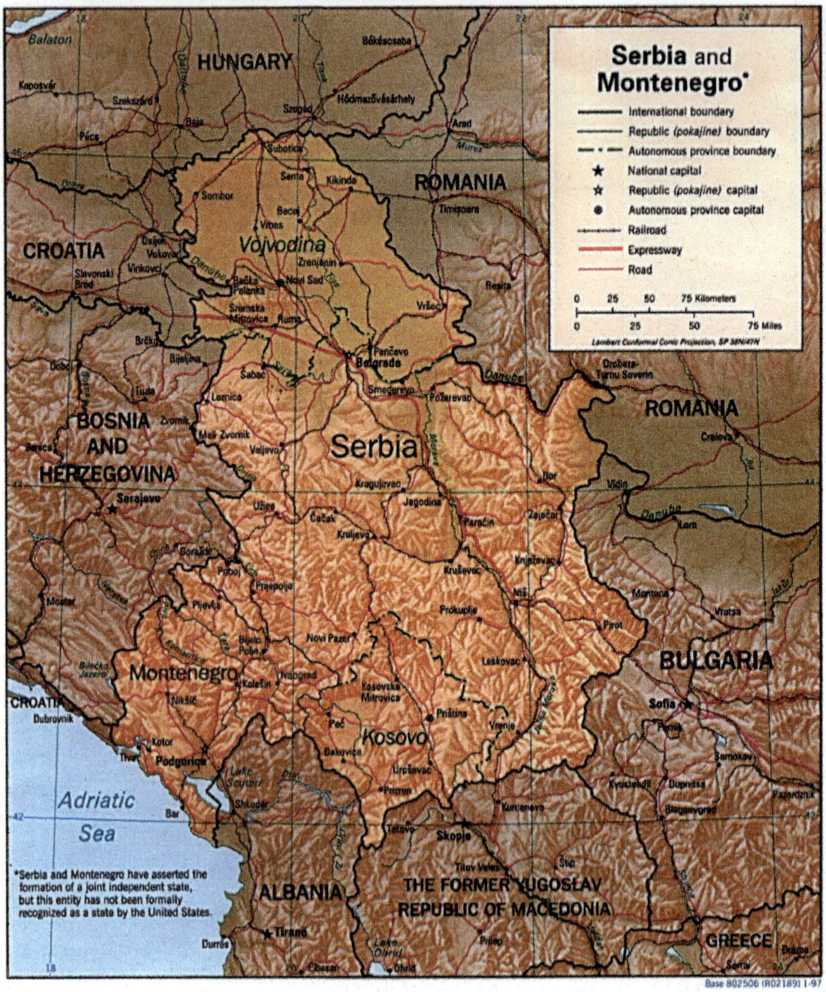

Base 802506 (R02189) 1-97

Montenegro

"Slightly smaller than Connecticut, Montenegro is bordered by: Albania (172 km), Bosnia and Herzegovina (225 km), Croatia (25 km), Kosovo (79 km), and Serbia (124 km). The terrain is highly indented coastline with narrow coastal plain backed by rugged high limestone mountains and plateaus; the lowest point is the Adriatic Sea (0 m), and the highest point is *Bobotov Kuk* (2,522 m)" (*World Factbook*, 2013).

"The use of the name *Crna Gora* (Montenegro) began in the 13th century in reference to a highland region in the Serbian province of *Zeta*. The later medieval state of *Zeta* maintained its existence until 1496 when Montenegro finally fell under Ottoman rule. Over subsequent centuries, Montenegro, while a part of the Ottoman Empire, was able to maintain a level of autonomy. From the 16th to 19th centuries, Montenegro was a theocracy ruled by a series of bishop princes; in 1852, it was transformed into a secular principality. Montenegro was recognized as an independent sovereign principality at the Congress of Berlin in 1878. After World War I, during which Montenegro fought on the side of the Allies, Montenegro was absorbed by the Kingdom of Serbs, Croats, and Slovenes, which became the Kingdom of Yugoslavia in 1929; at the conclusion of World War II, it became a constituent republic of the Socialist Federal Republic of Yugoslavia. When the latter dissolved in 1992, Montenegro federated with Serbia, first as the Federal Republic of Yugoslavia and, after 2003, in a looser State Union of Serbia and Montenegro. In May 2006, Montenegro invoked its right under the Constitutional Charter of Serbia and Montenegro to hold a referendum on independence from the state union. The vote for severing ties with Serbia barely exceeded 55% – the threshold set by the EU – allowing Montenegro to formally restore its independence on 3 June 2006.

"In 1879–1880 the Triangulation of Montenegro was realized by Russian military geodesists" (A. Gaina, *Astronomy, Geodesy and Map-Drawing in Moldova since the Middle Ages till the WWI*, Serb. Astron. J., No. 162 (2000), 121–125).

The Ferro prime meridian was widely used in Europe as the standard, proposed in 1634 as the "Westernmost point of the Old World." Commonly considered as being 20° west of the Paris meridian, and according to the French *Bureau International de l'Heure*, the civilian Greenwich-Ferro difference is $\Delta\lambda = 17°\ 39'\ 45.975''$. The military used the value of Theodor Albrech where $\Delta\lambda = 17°\ 39'\ 46.020''$, which was the official shift used for Yugoslavian cartography (*The Ferro Prime Meridian*, Gábor Timár, Geodézia És Kartográfia, pp. 3–7, 2007).

The Active (GPS) Geodetic Reference Network of Montenegro is called MONTEPOS (*Republic Geodetic Authority of Serbia – AGROS*, N. Tesla, N. Matic, V. Milenkovic, U.N./Turkey/European Space Agency Workshop, September 14–17, 2010, Istanbul). Cadastral mapping in Montenegro to-date still employs the Yugoslavia Reduced Gauss-Krüger Transverse Mercator. The scale factor at origin ($m_o = 0.9999$), the central meridian of the belt (C.M. = λ_o = 18° East of Greenwich), and the False Easting at C.M. = 500 km.

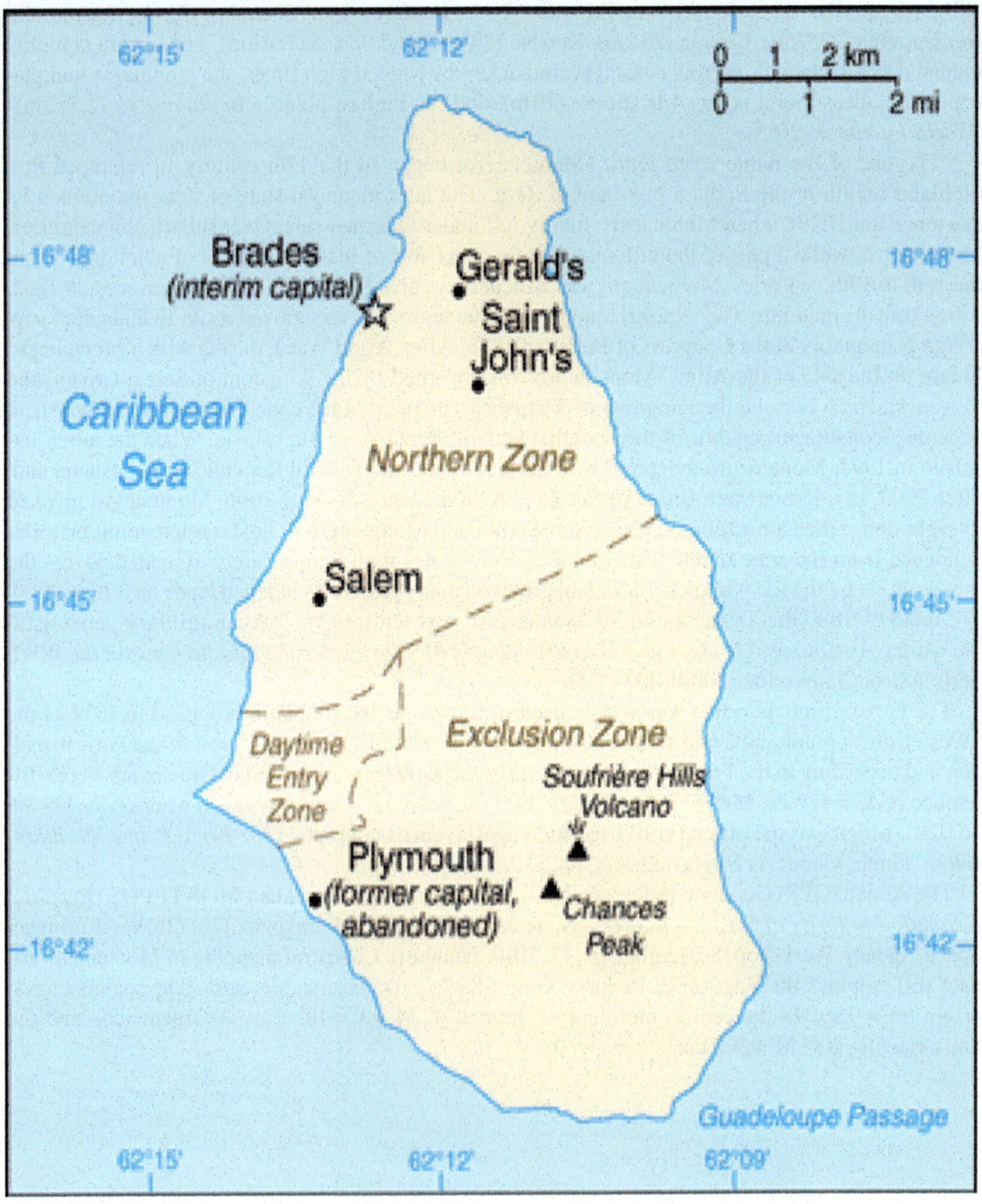
62°15'
62°12'
0 1 2 km
0 1 2 mi
16°48'
16°48'
Brades
(interim capital)
Gerald's
Saint John's
Caribbean Sea
Northern Zone
Salem
16°45'
16°45'
Daytime Entry Zone
Exclusion Zone
Soufrière Hills Volcano
Plymouth
(former capital, abandoned)
Chances Peak
16°42'
16°42'
Guadeloupe Passage
62°15'
62°12'
62°09'

Montserrat

Montserrat is about 0.6 times the size of Washington, DC, the lowest point is the Caribbean Sea (0 m), and the highest point is Chances Peak in the *Sourfiere Hills* (914 m).

On November 11, 1493, Christopher Columbus discovered the island and named it after *Santa Maria de Montserrate*, after the mountain abbey outside of Barcelona, Spain. "English and Irish colonists from St. Kitts first settled on Montserrat in 1632; the first African slaves arrived three decades later. The British and French fought for possession of the island for most of the 18th century, but it finally was confirmed as a British possession in 1783. The island's sugar plantation economy was converted to small farm landholdings in the mid-19th century. Much of this island was devastated and two-thirds of the population fled abroad because of the eruption of the *Soufriere Hills* Volcano that began on 18 July 1995" (*World Factbook*, 2011).

"In early 2007, Montserrat government authorities warned residents and visitors of volcanic activity and an increase in pyroclastic flows in Tyres Ghaut, Gages Valley and behind Gages Mountain. After placing sections of the lower Belham Valley off limits in January 2007 due to the danger posed by growth of the volcano's dome, the government of Montserrat lifted those restrictions in September. Access to all areas on the southern flanks of the Belham Valley east of the Belham Bridge and areas south remains prohibited, as is south of Jack Boy Hill to Bramble Airport and beyond. The Government of Montserrat has issued several recent proclamations and warnings urging residents and visitors to be vigilant and to be prepared to move at short notice. It last erupted in January 2009, requiring the large-scale evacuation of residents. Since that evacuation, residents have returned to their homes. On October 4, 2009, a series of eruptions began, but no one has been required to evacuate as of yet" (*U.S. Department of State*, 2011).

Originally mapped by the British Directorate of Overseas Surveys (DOS) from aerial photography taken from 1952 to 1959. There is one sheet at 1:25,000 scale of the island. The DOS performed a geodetic control survey and established the local datum origin at point M-36, the Montserrat Island Astro in 1958. The ellipsoid of reference is the Clarke 1880, where $a = 6,378,1249.145$ m and $1/f = 293.465$. DOS established a local British West Indies (BWI) Transverse Mercator Grid for the island where the Central Meridian (λ_o) = 62° W, Latitude of Origin (φ_o) = Equator, the Scale Factor at Origin (m_o) = 0.9995, the False Easting = 400 km, and the False Northing = nil (zero at the Equator). The shift **from** Montserrat Island Astro 1958 Datum **to** WGS 84 Datum is: $\Delta X = +174$ m ±25 m, $\Delta Y = +359$ m ±25 m, and $\Delta Z = +365$ m ±25 m.

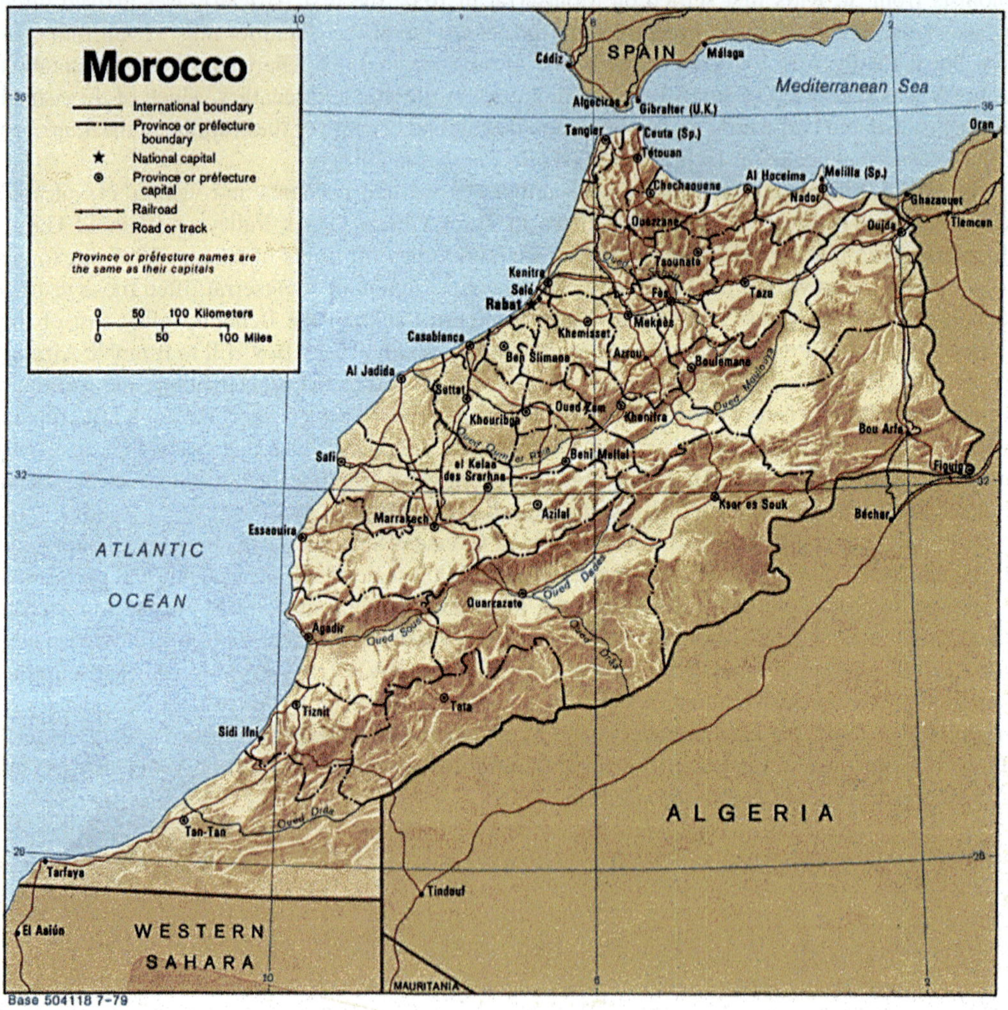

Base 504118 7-79

The Kingdom of Morocco

Slightly larger than twice the size of California, Morocco is bordered by Algeria (1,941 km); Mauritania (1,564 km); Spain (Ceuta) (8 km); and Spain (Melilla) (10.5 km). The terrain is mountainous northern coast (Rif Mountains) and interior (Atlas Mountains) bordered by large plateaus with intermontane valleys, and fertile coastal plains; the south is mostly low, flat desert with large areas of rocky or sandy surfaces. The highest point is *Jebel Toubkal* (4,165 m); the lowest point is *Sebkha Tah* (−59 m), and the mean elevation is (909 m).

The Roman province of Mauritania was invaded by Muslims in the 7th century A.D., leading to the founding of the present country of Morocco as an independent kingdom in the 9th century. Later centuries saw occupation by the Portuguese, English, Spanish, and French. In 1912, two protectorates were established: French Morocco, on the Atlantic coast which included most of the present country, and Spanish Morocco, on the Mediterranean coast. In recent years, the kingdom has annexed much of the former Spanish Sahara. The geodetic history of the entire country is a combination of Spanish and French influences. The coordinate systems of Morocco are among the more difficult to comprehend and unusual in the entire world.

The first-order Frontier Triangulation of Spanish Morocco was attached to Spain by measurements of the *Instituto Geográfico y Catastral de Madrid* in 1923–1929. It was composed entirely of one meridional arc and was originally referenced to the Madrid Datum on the Struve ellipsoid. This was locally computed on the Beni Meyimel Gauss-Krüger Transverse Mercator Grid of 1923 that was adjusted in the late 1950s to the European Datum of 1950. Station *Beni Meyimel* on ED50 is: $\varphi_o = 35° 45' 34.223''$ North, $\lambda_o = 5° 41' 09.5168''$ West of Greenwich. The local Spanish Grid origin for *Puerto de Tánger* (Port of Tangier) was at *Faro* (Lighthouse) *de Malabata*, where $\varphi_o = 35° 48' 47.7''$ North, $\lambda_o = 5° 44' 57.0''$ West of Greenwich. Neither of these Grids had false origins. The same triangulation arc was later completely recomputed with respect to the namesake of the capital of the former Spanish Morocco, the *Tétouan* Datum of 1929, where $\Phi_o = 35° 35' 41.493''$ North, $\Lambda_o = 5° 19' 19.545''$ West of Greenwich. The azimuth to *Tétouan* Southwest Base is: $\alpha_o = 231° 44' 50.67''$.

There were two types of classical triangulation in French Morocco. The original triangulation, the "Reconnaissance Triangulation," was started about 1910 and covered a major portion of the country. It was accomplished by Topographic Engineers attached to French Foreign Legion troops engaged in the conquest of Morocco. The work was done on a yearly basis to satisfy military needs, and the officers in charge did their own computations. Errors were cumulative because values for the last two or three primary stations were accepted by the extending party the following season. The work was accomplished in approximately 40 different portions and was not as well coordinated as the previous 50 years' worth of French efforts in adjacent Algeria. The Reconnaissance Triangulation is referenced to the *Agadir* Datum of 1921, where $\Phi_o = 30° 25' 25.0''$ North, $\Lambda_o = 9° 37' 48.06''$ West of Greenwich. The origin is the minaret of a mosque in the 16th-century Port of *Agadir*. The solar azimuth to *Signaux de Founti* is: $\alpha_o = 109° 19' 36''$. A subsequent Polaris observation (two pointings) was not adopted.

The French North African ellipsoidal Bonne projection was used for the Reconnaissance Triangulation, and this Grid continues to influence mapping systems in North Africa to this day. Proper computation of these Grid coordinates must be performed from the series form of the equations for millimeter accuracy. Texts that suggest the universal use of spherical formulae with the equivalent authalic radius and latitudes are wrong! The Latitude of Origin (φ_o) = 35° 06' North, and the Central Meridian (λ_o) of zero degrees from Paris corresponds to 2° 20' 13.95'' West of Greenwich. The False Origin is 100 km for both Eastings and Northings. This Grid has perplexed novice cartographers (and many "old pros") for decades. Even more confusing, some implementations of this

Grid display no False Origin. The active use of this Grid was terminated in 1942, but it continues to confound, as I shall explain later.

The second type of French triangulation commenced in 1922 and is known as the *Triangulation Réguliere* (regular triangulation). This work was carried out entirely by Topographic Brigades of the *Service Géographique de l'Armee* (Army Geographic Service), sent from Paris to Morocco. All computations were performed in Paris. The *Réguliere* Datum of 1922 is based on the origin at the astronomic station *Merchich* 1921 (south of Casablanca), where $\Phi_o = 33°\ 26'\ 59.672''$ North, $\Lambda_o = 7°\ 33'\ 27.295''$ West of Greenwich. The azimuth to station *Mohammed El Kebir* is: $\alpha_o = 46°\ 52'\ 11.291''$ (from south), and the ellipsoid of reference is the Clarke 1880, where $a = 6,378,249.145$ m and $1/f = 293.465$. During the same era, of seven baselines that were measured for the classical triangulation, *Ber Rechid* and *Agourai* were locally adopted as temporary datum origins. The *Ber Rechid* 1920 Datum is defined, where $\Phi_o = 33°\ 17'\ 41.6''$ North, $\Lambda_o = 9°\ 56'\ 02.05''$ West of Paris. The azimuth from North Base to South Base is: $\alpha_o = 169°\ 34'\ 15.3''$. The geodetic definition of the *Agourai* Datum seems to be lost, but it may exist somewhere in Rabat or Paris. The remaining baselines of the *Réguliere* Datum of 1922 included *Marrakech, Guercif, Taroudant, Bon Denib*, and *Sidi Ben Zekri*, where "sidi" is an Arabic term of obeisance, similar to "my lord." In the 1950s, the U.S. Army Map Service (AMS) found the *Réguliere* classical triangulation to be intrinsically correct.

The Lambert (partially conformal) Conic projection is used for the *Réguliere* Datum of 1922, and this Grid also confounds cartographers to this day. There are two original zones: for Zone *Nord*, the Latitude of Origin $(\varphi_o) = 33°\ 18'$ North (or 37^G where 100 grads $= 90°$), the Central Meridian $(\lambda_o) = 5°\ 24'\ (6^G)$ West of Greenwich, and the Scale Factor at Origin $(m_o) = 0.999625769$. For Zone *Sud*, the Latitude of Origin $(\varphi_o) = 29°\ 42'$ North, the Central Meridian $(\lambda_o) = 5°\ 24'$ West of Greenwich, and the Scale Factor at Origin $(m_o) = 0.999615596$. The False Origin is 500 km for Eastings and 300 km for Northings *for both zones*. The formulae used with this Grid have a long and lurid history dating back to the 19th century, courtesy of the French way of doing things.

During the 19th century, France was the center of cartographic research in the mathematics of map projections. Because projection table computations were performed by hand, all formulae were commonly truncated past the cubic term to ignore infinite series terms considered *at the time*, too small to warrant the extra effort. For instance, the Lambert Conformal Conic projection was used only to the cubic term in the formulae for the tables of the developed meridional distances. This resulted in French Army projection tables that have become part of the arcane lore of computational cartography. This "secret" has been explicitly detailed in the French literature for a century, but apparently ignored by readers of English-only papers. Standard Lambert formulae will not work for Morocco, and the use of fully conformal formulae will yield computational errors that exceed 15 m! Note that mathematical elegance is not what matters in a country's coordinate transformations; what matters is computational conformity to local *legal* standards. In Morocco, one had better use the French Army Truncated Cubic Lambert Conic projection. Watch out for that "shrink-wrap" software in perfect conformity with United States Standards.

The *Maroc Nord* and *Maroc Sud* zones have an eastern limit to their Grids that are defined by the 448,000 m Easting (X) coordinate of the North African ellipsoidal Bonne projection. The intersection of this Grid limit with the graticule was a graphical simplicity but a computational nightmare until the paper of Karl Rinner was used at AMS. In the 1930s, Professor Rinner published a paper (in German) detailing his development of the formulae for the ellipsoidal Bonne projection in series form. AMS developed a reversion of his series, and computational algorithms were developed for the North African Bonne in Morocco among other places.

In the past 50 years or so, a couple of other places on the Atlantic coast have been rumored to be datums in Morocco. Those places are *Modagor* and *Sidi Ifni*. Neither of these are datums; rather, they are local Grids based on the *Réguliere* Datum of 1922. In particular, the *Sud d'Ifni* 1955 Hatt Azimuthal Equidistant Grid is defined at the origin point (Camp monument at *Maison Guizol*), where $(\varphi_o) = 29°\ 02'\ 06''$ North, the Central Meridian $(\lambda_o) = 10°\ 30'\ 06''$ West of Greenwich, and the Scale Factor at Origin $(m_o) = 1.0$. The False Origin is 50 km each for Eastings and Northings.

The orientation of the Grid is defined to "Fourmi" as: $\alpha_g = 40° 19' 47''$. The Mogador 1957 Lambert Grid origin is: $(\varphi_o) = 31° 30'$ North, the Central Meridian $(\lambda_o) = 10° 00'$ West of Greenwich, and the Scale Factor at Origin $(m_o) = 0.999932968$. The False Origin is 200 km each for Eastings and Northings. In 1921, the position of *Mogador Ma*t on the Reconnaissance Triangulation was: $\varphi = 31° 30' 36.7''$ North, $\lambda = 9° 46' 31.2''$ West of Greenwich. The same point on the *Réguliere* Datum of 1922 is: $\varphi = 31° 30' 13.178''$ North, $\lambda = 9° 46' 23.643''$ West of Greenwich. The *Mosque d'Agadir* has also been recomputed on the *Réguliere* Datum of 1922 as: $\varphi = 30° 25' 54.785''$ North, $\lambda = 9° 37' 36.811''$ West of Greenwich.

The Kingdom of Morocco added two more zones to their national Grid system to cover the annexed former Spanish Sahara. Maintaining a single national Central Meridian $(\lambda_o) = 6^G$ West of Greenwich; for Zone III the Latitude of Origin $(\varphi_o) = 29^G$ North, the Scale Factor at Origin $(m_o) = 0.999616304$, and the False Origin is 1,200 km Easting, 400 km Northing. For Zone IV the Latitude of Origin $(\varphi_o) = 25^G$ North, the Scale Factor at Origin $(m_o) = 0.999616437$, and the False Origin is 1,500 km Easting, 400 km Northing. The National Imagery and Mapping Agency (NIMA) lists the shift from "MERCHICH Datum" (*sic*) to WGS 84 Datum as: $\Delta X = +31$ m ± 5 m, $\Delta Y = +146$ m ± 3 m, $\Delta Z = +47$ m ± 3 m. This relation is based on nine stations somewhere in Morocco.

In 2004 a team of Spanish and Moroccan geodesists performed a GPS campaign to map the Strait of Gibraltar and produce a map at the scale of 1:25,000. Referenced to the GRS80 ellipsoid and projected on the Lambert Conformal Conic secant projection, the point of origin is at $\varphi_o = 35° 57'$ N, $\lambda_o = 5° 37' 30''$ West of Greenwich, and a scale factor at origin of $m_o = 0.999995266$ (*Geodetic works carried out in the Strait of Gibraltar,* Almazan Garate, J.L., *et al.*, Human and Social Sciences at the Common Conference, November 18–22, 2013, pp. 301–306). In 2011, The *Cartographie et Geodesie Nationale du Maroc* reported that the fundamental network of Morocco consists of ~9,000 points, a permanent GPS network of 15 stations, a new national leveling network covering 13,806 km, and a unification of the system of coordinates (*Cartographie et Geodesie Nationale,* Hmamouchi, Y., FIG Working Week 2011, Marrakech, Morocco, 18–22 May 2011).

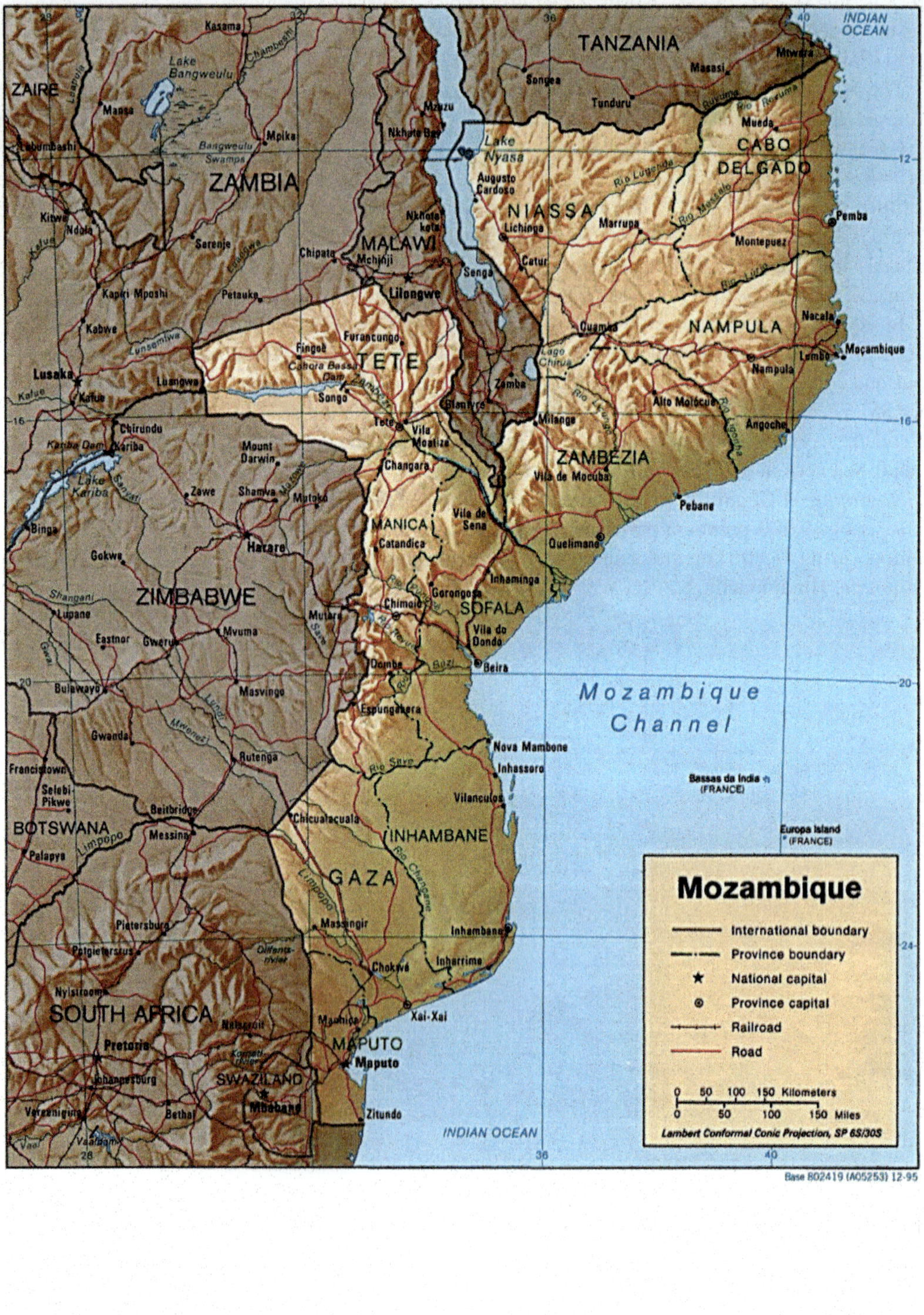

Base 802419 (A05253) 12-95

The Republic of Moçambique

Moçambique is slightly more than five times the size of Georgia; slightly less than twice the size of California. It is bordered by Malawi (1498 km); South Africa (496 km); Eswatini (108 km); Tanzania (840 km); Zambia (439 km); and Zimbabwe (1,402 km). The terrain is mostly coastal lowlands, uplands in center, high plateaus in northwest, and mountains in west. The highest point is *Monte Binga* (2,436 m), the lowest point is the Indian Ocean (0 m), and the mean elevation is (345 m).

The first Portuguese explorer to arrive in Moçambique was Pero de Covilha who was dispatched in 1487 to find a route to India, which he reached via Egypt and Aden. On his return trip in 1489, he visited several places on the East Coast of Africa, including the ancient port of Sofala. The flourishing Arab port city on the Island of Moçambique was visited by Vasco de Gama in 1498, following the rounding of the Cape of Good Hope the previous year by Bartolomeo Dias. Soon a Portuguese trading port, the region remained under the control of Portugal through a complicated series of arrangements until its independence in June of 1945. That port city remained the capital until 1907 when the capital was moved to the southern port of *Marques* (*Maputo*). Moçambique has ten provinces, some of which are the namesakes of local datums: *Cabo Delgado, Gaza, Inhambane, Manica, Maputo, Nampula, Niassa, Sofala, Tete,* and *Zambezia*.

Before the peace accord of October 1992, Moçambique had been devastated by civil war and was one of the poorest countries on the globe. Prospects have subsequently improved, and with its solid economic performance in 1996–1997, Moçambique has begun to exploit its sizable agricultural, hydropower, and transportation resources. The restoration of electrical transmission lines to South Africa and the completion of a new transmission line to Zimbabwe (permitting the giant *Cahora Bassa* hydropower plant to export large amounts of electricity) will greatly improve foreign exchange receipts.

Land surveying and boundary surveys were authorized by the Portuguese crown in 1857. Topographic mapping for Moçambique was designed by the Portuguese "Junta das Investigações do Ultramar" (Board of Overseas Research) in Lisbon, Portugal. The Junta coordinated the activities of the geographic mission that established horizontal and vertical control for photogrammetric mapping accomplished by *Serviços Geográficos e Cadastrais* (SGC) in Maputo. Control surveys for systematic mapping was initiated in 1931 by the SGC with the assistance of the Junta, and all 61 sheets at 1:250,000 scale were completed by 1955.

The Dutch ceded South Africa to the British in 1814. The last two decades of the 19th century saw extensive African development by the European powers. The geodetic survey of the 30th Meridian West of Greenwich became a symbol of the progress of documenting the British Colonial Empire and its neighbors in Africa. This was the same rationale as had already been established by Everest 100 years prior in India. The classical triangulation of the 30th Meridian arc has its roots both in South Africa that extended northward from the Cape and also southward from British East Africa (Kenya and Uganda).

The Cape Datum has its initial origin point in Port Elizabeth, South Africa, at station Buffelsfontein, where $\Phi_o = 33° 59' 32.000''$ South, $\Lambda_o = 25° 30' 44.622''$ East of Greenwich. The ellipsoid of reference is the Clarke 1880, where $a = 6,378,249.145$ m, $1/f = 293.465$. The azimuth to station Zuurberg is: $\alpha_o = 004° 15' 26.311''$, and the geoidal height (H_o) at the origin is defined by implication to be zero. The deflection of the vertical at Buffelsfontein was initially assumed to be zero, and although non-zero values were subsequently published in the middle 1960s, this has been ignored. The Cape Datum theoretically covers a substantial portion of Southeast Africa but has been subjected to several regional adjustments that effectively separate into a variety of individual local systems. The countries surrounding Moçambique that are on the Cape Datum include South Africa and Swaziland.

The Arc Datum has the same origin as the Cape Datum, but its initial fieldwork started in Uganda and Kenya. Individual country adjustments of this meridional chain (or arc) of quadrilaterals have resulted in the countries surrounding Moçambique that are on the Arc Datum to include Tanzania which is on the Arc Datum of 1960. This particular adjustment is carried further north into the Sudan. Malawi, Zambia, and Zimbabwe are on the Arc Datum of 1950. Note that Zimbabwe was formerly Southern Rhodesia, Zambia was formerly Northern Rhodesia, Malawi was formerly Nyasaland, and Tanzania was formerly Tanganyika, which had been formerly German East Africa. After World War I, the British assumed colonial administration of Tanganyika. The consequence of this was that Moçambique was now surrounded by British colonies, and its borders were subsequently entirely surveyed by the British Royal Engineers.

The Moçambique–Tanzania boundary initially delimited German and Portuguese spheres of influence in East Africa. In accordance with the terms of a German-Portuguese declaration signed at Lisbon in 1886, a boundary was established between the Indian Ocean and Lake Nyasa. Subsequent treaties delimited the boundary in additional detail, and a joint expedition in 1907 changed the boundary slightly in the west and demarcated the land segment by pillars. Following World War I, the former German territory of Tanganyika was made a British mandate, and during 1936 and 1937 an exchange of notes between Portugal and the United Kingdom determined the sovereignty of the island in the Rio Rovuma (Ruvuma River) which forms more than 90% of the total boundary. With a total length of 470 miles, streams comprise 445 miles of the distance of which the Rio Rovuma and its tributaries account for all but about one mile. The Rule of the Thalweg (thread of the stream) is specified for this riparian boundary. The parameters published by DMA/NIMA for the Tanzania Arc 1960 Datum to WGS84 are: $\Delta X = -175$ m ± 6 m, $\Delta Y = -23$ m ± 9 m, $\Delta Z = -303$ m ± 10 m.

The Moçambique – Malawi boundary is approximately 975 miles in length and is demarcated. It traverses Lake Nyasa for about 205 miles including lines around Likoma Island and Chisumulu Island, which are part of Malawi. Southward from Lake Nyasa to the Malosa River, the boundary extends along straight-line segments for 195 miles passing through both Lake Chiuta and Lake Chilwa. It follows consecutively the thalwegs of the Malosa, Ruo, and Shire rivers downstream for 150 miles. The boundary then continues northwestward to the Zambia tripoint utilizing features along the Shire-Zambezi and the Lake Nyasa-Zambezi drainage divides for most of the remainder of the distance. The parameters published by DMA/NIMA for the Malawi Arc 1950 Datum to WGS84 are: $\Delta X = -161$ m ± 9 m, $\Delta Y = -73$ m ± 24 m, $\Delta Z = -317$ m ± 8 m.

The Moçambique–Zambia boundary commences in the Northwest at the tripoint (with Malawi) where the Lake Nyasa–Zambezi River drainage divide meets at the 14th parallel in accordance with the agreement of June 11, 1891, between the United Kingdom and the Kingdom of Portugal. The tripoint was determined to be located at $\phi = 14°\ 00'\ 00''$ S, $\lambda = 33°\ 14'\ 32''$ E by a joint boundary commission in 1904. The boundary proceeds to the southwest along the Zambesi River drainage divide until it meets the River Aroangwa or Luangwa. Thence, the boundary follows along that river until it meets the Zambesi River to tripoint (with Zimbabwe) Beacon Number 1, where $\phi = 15°\ 37'\ 27''$ S, $\lambda = 30°\ 25'\ 20.3''$ E. Note that Zambia is where the Livingstone Memorial is located. ("Dr. Livingstone, I presume …?") The parameters published by DMA/NIMA for the Zambia Arc 1950 Datum to WGS84 are: $\Delta X = -147$ m ± 21 m, $\Delta Y = -74$ m ± 21 m, $\Delta Z = -283$ m ± 27 m.

The Moçambique – Zimbabwe boundary is about 765 miles in length. It is demarcated throughout this distance by pillars or rivers. The tripoint (with Zambia) Beacon Number 1 addressed in The Northern and Southern Rhodesia Order in Council 1963 indicated that their common boundary joined the Moçambique tripoint at the *medium filum acquae* (median line) of the Zambesi River. In the central part, the boundary traverses an area of numerous escarpments and peaks including the Inyanga Mountains and the Chimanimani Mountains. The parameters published by DMA/NIMA for the Zimbabwe Arc 1950 Datum to WGS84 are: $\Delta X = -142$ m ± 5 m, $\Delta Y = -96$ m ± 8 m, $\Delta Z = -293$ m ± 11 m.

The Moçambique – South Africa boundary, which is about 305 miles in length, consists of two discontinuous parts. The longer part extends southward from the confluence of the Limpopo River and the Luvuvhurivier River for 255 miles along straight-line segments to the northern tripoint with

Swaziland and the *M'Pundweni* Beacon. The remainder of the boundary from the southern tripoint with Swaziland to the Indian Ocean follows the *Maputo* River (*Great Usutu*) downstream for about 17 miles and then continues by straight-line segments for another 33 miles. The parameters published by DMA/NIMA for the South Africa Cape Datum to WGS84 datum shift are not as accurate as those are according to Rens & Merry. Professor Charles Merry of the University of Cape Town found a substantial improvement with a four-Parameter transformation, such that $\Delta X=-190.0\,m \pm 7.7\,m$, $\Delta Y=-137.9\,m \pm 3.7\,m$, $\Delta Z=-257.4\,m \pm 4.7\,m$, scale$=+11.0\times 10^{-6}$, and the aggregate positional rms$=\pm 2.9\,m$. Note that the values I quoted are the most reliable for points *within* Moçambique.

The Moçambique – Swaziland boundary has two tripoints with South Africa. Northward from the Maputo River it extends along the summit of the Lebombo Mountains for approximately 66 miles to *M'Pudweni* Beacon. The boundary consists of various straight-line segments demarcated by trigonometrical and/or boundary beacons. The original treaty in this area dates back to 1869, and portions remained in dispute until 1927. The *M'Ponduine* Geodetic Station was used as the origin of a local boundary grid that, based on an ellipsoidal polyhedric projection, has the origin as: $\phi_o=25°\,56'\,47.19''$ S, $\lambda_o=31°\,58'\,40.46''$ E. No false origin was used, and the *M'Uguene* Geodetic Station given as a computational check is: $\phi=26°\,07'\,25.98''$ S, $\lambda=32°\,16'\,31.58''$ E and X$=-29{,}757.40\,m$, Y$=+19{,}691.48\,m$. There is no doubt that this boundary Grid is on the Cape Datum, Clarke 1880 ellipsoid. Professor Merry's solution for Swaziland is a three-parameter shift to WGS84 Datum, where $\Delta X=-136.8\,m \pm 0.5\,m$, $\Delta Y=-106.6\,m \pm 0.5\,m$, $\Delta Z=-293.0\,m \pm 0.5\,m$, and the aggregate positional rms$=\pm 0.7\,m$.

There are two actively used classical geodetic Datums in Moçambique. Both are referenced to the Clarke 1866 ellipsoid, a favorite of the Portuguese in Africa – they used it in Angola, too! The most famous is the *Observatorio Campo Rodrigues* Datum with origin near *Lourenço Marques* (*Maputo*), where $\Phi_o=25°\,58'\,06.99''$ South, $\Lambda_o=32°\,35'\,37.75''$ East of Greenwich. One of Professor Merry's solutions for a three-parameter shift to WGS84 Datum for points only within Moçambique is: $\Delta X=-134.2\,m$, $\Delta Y=-111.7\,m$, $\Delta Z=-333.9\,m$, and the aggregate positional rms$=\pm 2.1\,m$. The second classical geodetic Datum used is the Tete Datum with origin at: $\Phi_o=16°\,10'\,54.50''$ South, $\Lambda_o=33°\,36'\,08.09''$ East of Greenwich. Merry's solution for that shift to WGS 84 Datum is: $\Delta X=-86.2\,m$, $\Delta Y=-104.7\,m$, $\Delta Z=-224.7\,m$, and the aggregate positional rms$=\pm 0.16\,m$. The government of Moçambique adopted the use of the UTM Grid in 1954, and, with the exception of the M'Ponduine Polyhedric Grid, no other Grids are used. There are reports of the existence of another Datum by the name of *Madzanzua,* but no other information was available.

"The 1st order Moçambique geodetic network still presents zones with low precision, such as the north coast. The conjunction of GPS and terrestrial observation s of well distributed points within the territory would by a good solution to improve the equality of this network. The prediction of re-observation of more than 30 well distributed points within the whole territory using GPS, in order to obtain a valid group of vectors is included in the Portugal and Moçambique cooperation protocol. This will make the conversion to ITRF94 datum more consistent and will allow the integration of all networks in the same system" (*Adjustment of the Classical Terrestrial Geodetic Network of Moçambique Tied to ITRF,* Santos, P., et al., Workshop – AFREF I, 5th FIG Regional Conference, Accra, Ghana, March 8–11, 2006).

"The adoption of MozNet is a decisive step for the modernization of the geo-referencing activities in Moçambique. Moçambique has now a network of permanent GPS stations that permit the definition and realization of a geocentric datum. This network still needs to be densified. The difference between MozNet/ITRF94 and MozNet/ITRF2008 is around 2 cm between the extreme north and south of the country which is negligible for practical purposes. Since the differences between MozNet/ITRF94 and MozNet/ITRF2008 are small, it was decided to adapt MozNet/ITRF94 as the new official reference system realized by the permanent network (MozPerm) and the already observed points" (*Validation and Implementation of MozNet – the geocentric reference frame of Moçambique,* Fernandes, R.M.S., et al., MozNet – Luxembourg, 13–17 October 2014).

Dr. Fernandes informs me that 2 years ago an aerial gravimetric survey was completed of the entire country in support of a national geoid (*Personal communication, 28 March 2017*).

REFERENCES: M

Mali–Niger Boundary, International Boundary Study No. 150 – January 13, 1975, Bureau of Intelligence and Research, Department of State, USA.

Algeria–Niger Boundary, International Boundary Study No. 99 – May 1, 1970, Bureau of Intelligence and Research, Department of State, USA.

C. U. Ezeigbo, Definition of nigerian geodetic datum from recent doppler observations, *Survey Review*, vol. 30, 237, July 1990.

Captain J. Calder Wood, The survey framework of Nigeria, *Empire Survey Review*, vol. III, no. 21, 386–395, July 1936.

Captain F. S. Spence, Himalayan survey, *Geographical Journal*, vol. 153, no. 2, 223–230, July 1987.

Chad–Niger Boundary, International Boundary Study No. 73 – August 1, 1966, Bureau of Intelligence and Research, Department of State, USA.

Charles L. Merry and J. Rens, *Survey Review*, vol. 30, 236, April 1990.

Collaborative Mapping in Nicaragua, 1957.

David Johanson, U.S. Naval Oceanographic Office, Personal Communication, 2006.

Dr. Roy Ladner, Niue ties to the NGD91, University of New Orleans, 1999.

EUREF Permanent GNSS Network.

Final Report of the 12th Parallel Survey in the Republic of Niger, IGN, 1969.

Foreign Maps, TM 5-248, 1963.

Geographic Notes, Department of State, Office of the Geographer September 8, 1986.

German South West Africa – British Bechuanaland Boundary Survey of 1898–1903, EPSG Database, 2006.

http://euref.eu/symposia/2019Tallinn/05-16-Netherlands.pdf.

http://gnss1.tudelft.nl/pub/vdmarel/reader/CTB3310_RefSystems_1-2a_print.pdf.

http://www.epncb.eu/_networkdata/siteinfo4onestation.php?station=DELF00NLD.

http://www.kartverket.no/en/data/Open-and-Free-geospatial-data-from-Norway/, June, 2017. Informe Detallado de la Comisión Técnica de Demarcación de la Frontera entre Guatemala y Honduras, 2009.

Institut Géographique National (IGN), SGC 1312, December 1945.

Instituto Nicaragüense de Estudios Territoriales (INETER), 2009.

International Pndaries Research Unit, Durham University, 22 July 2010.

L. P. Lee, First-order Geodetic Triangulation of New Zealand 1909–1949 and 1973–1974.

Lambert Conical Orthomorphic Projection Tables, Niger Zone, RESTRICTED, Office of the Chief of Engineers, Washington, DC, 1943.

Lonely Planet, July 2006.

M. L. Larsgaard, *Topographic Mapping of the Americas*, McGraw Hill: Australia, and New Zealand, 1934.

Mal Jones, Personal Communication, April 2005.

Memoria de la Dirección General de Cartografía, Guatemala, September 1957.

National Report of the Netherlands 2019- EUREF.

Nepal Country Study, Library of Congress, 1991.

Nicaragua Library of Congress Country Studies, 2009.

Niger – Country Studies, U.S. Department of State, 2011.

Niger – *Lonely Planet*, 2011.

Niger–Nigeria Boundary, International Boundary Study No. 93 – December 15, 1969, Bureau of Intelligence and Research, Department of State, USA.

Nigeria Library of Congress Country Study – 2008.

Niraj Manandhar, Study of Geodetic datum of Nepal, China and Pakisthan (*sic – Ed.*) and its transformation to World Geodetic System, *Nepalese Journal on Geoinformatics*, vol. 10, 2008.

Professor Michael Barnes, MSc MRICS MCInstCES MRIN, Lecturer, Geomatics, Geospatial & Mapping Sciences, School of Geographical & Earth Sciences, University of Glasgow, Glasgow, Personal Communication, 2022.

Reference Systems for Surveying and Mapping.

S. Izundu Agajelu, On conformal representation of geodetic positions in Nigeria, *Survey Review*, vol. 29, 223, January 1987.

The Geodetic Survey of Nepal 1981–1984, Directorate of Military Survey, UK, November 1985.

Heuvelink Hendrik, *Triangulation du Royaume des Pays-Bas*, Nabu Press: Charleston, SC, vol. 1, 1903, vol. 2, 1921.

World Factbook, 2009.

World Factbook, 2013.

World Factbook, 2022.

N

Republic of Namibia 565
Kingdom of Nepal 569
The Kingdom of the Netherlands 573
New Zealand 577
Republic of Nicaragua 583
Republic of Niger 587
Federal Republic of Nigeria 591
Niue 595
The Kingdom of Norway 599

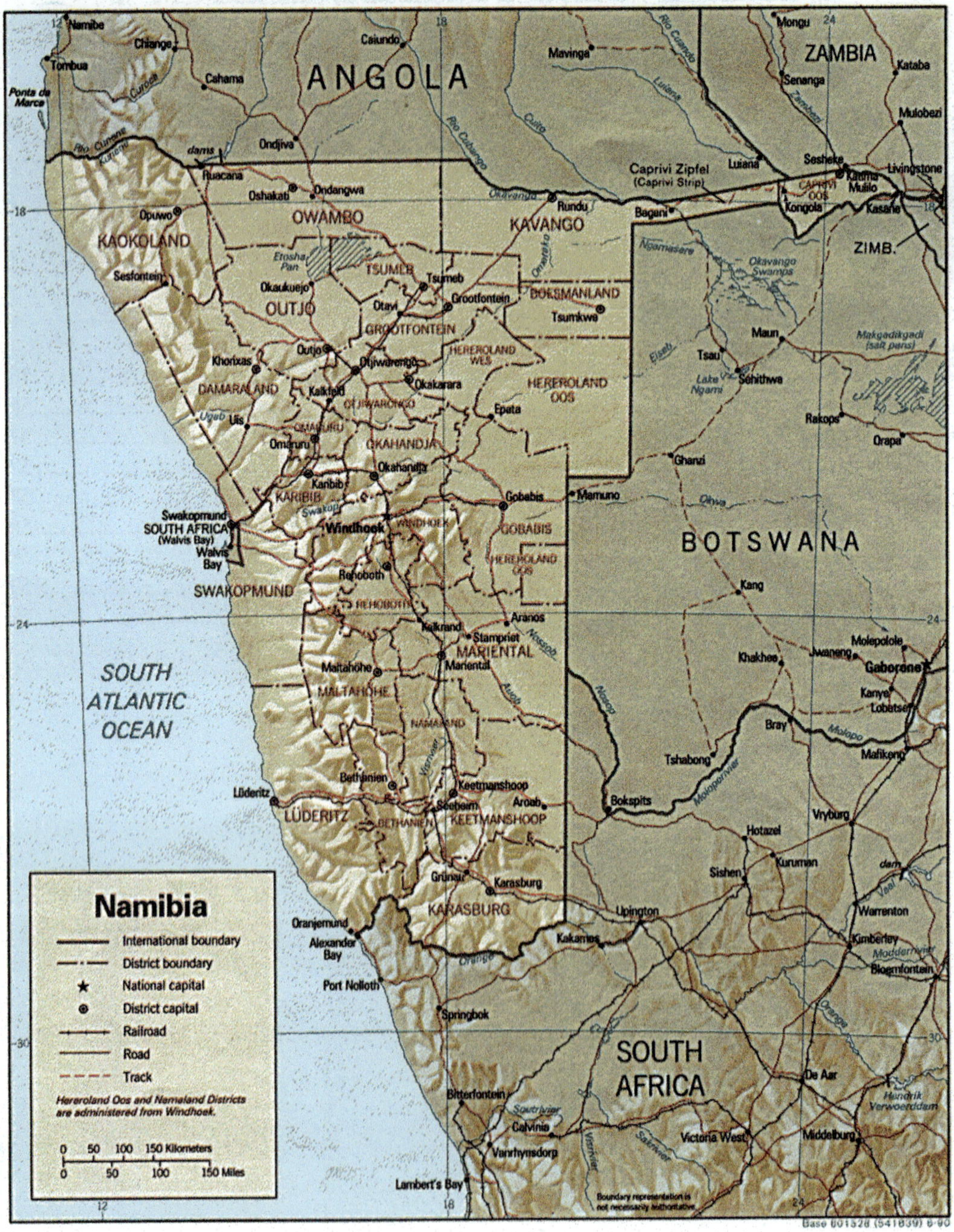

Base 801528 (541839) 6-90

Republic of Namibia

Namibia is bordered on the West by the South Atlantic Ocean (1,572 km), and by the following countries in clockwise order from the north: Angola (1,376 km), Zambia (233 km), the Caprivi Strip (Zipfel) extends to Zimbabwe (Tripoint), Botswana (1,360 km), and South Africa (855 km). The country is slightly more than half the size of Alaska; the terrain is mostly high plateau with the Namib Desert along the coast and the Kalahari Desert in the east. The lowest point is the South Atlantic Ocean (0 m), and the highest point is *Konigstein* (2,606 m).

According to *Lonely Planet*, "Southern Africa's earliest inhabitants were the San, a nomadic people organized in extended family groups who could adapt to even the severest terrain. San communities later came under pressure from Khoi-Khoi groups. The Khoi-Khoi were tribal people who raised livestock rather than hunted, and who were among the first pottery makers in the archaeological record books. They came from the south, gradually displacing the San, and remained in control of Namibia until around 1500 A.D. Descendants of the Khoi-Khoi and San people still live in the country, but few have retained their original lifestyles. Between 2300 and 2400 years ago, the first Bantus appeared on the plateaus of south-central Namibia. Their arrival marked the first tribal structures in southern African societies. Other tribes either retreated to the desert or the swamps of the Okavango Delta or were enslaved into Bantu society. Because Namibia has one of the world's most barren and inhospitable coastlines, it was largely ignored by European explorers. The first European visitors were Portuguese mariners seeking a way to the Indies in the late 15th century, but they confined their activities in Namibia to erecting stone crosses at certain points along the coast as navigational guides. It wasn't until the last-minute scramble for colonies towards the end of the 19th century that Namibia was annexed by Germany, except for the enclave of Walvis Bay, which was taken in 1878 by the British for the Cape Colony. In 1904, the Herero people, who were Bantu-speaking cattle herders, launched a rebellion, but it was brutally put down. Meanwhile, in the south, diamonds had been discovered east of Lüderitz by a South African laborer. In the blink of an eye, the German authorities branded the entire area between Lüderitz and the Orange River a *sperrgebiet*, or 'forbidden area'. German rule came to an end during WWI when German forces surrendered to a South African expeditionary army fighting for the Allies." Namibia gained her independence on 21 March 1990 from the Union of South Africa.

The Angola-Zambia-Namibian "Triune" point was reported on 21 October 1964 as Beacon No. 9 (corresponding to Beacon No. 32 of the Kwando series), located at: $\varphi = 17°\ 28'\ 29.28''$ S, $\lambda = 23°\ 25'\ 47.604''$ E.

The first geodetic surveys and small-scale topographic mapping were accomplished between 1885 and 1915 by three German agencies: the *Königliche Preußische Landesaufnahme* (Royal Prussian Survey Office); its civilian successor, *Reichsamt für Landesaufnahme* (Empire Land Survey Office); and the *Generalstab des Heeres* (General Staff of the Army). These agencies produced maps at scales of 1:100,000, 1:200,000, and 1:1,000,000; they varied greatly in the amount of detail and were inaccurate. However, since the original classical trigonometric surveys were observed by the Germans, the ellipsoid of reference was, and still is, the Bessel 1841 where $a = 6,377,483.865$ m, and $1/f = 299.1528128$. Note that the strange value for a is because of the conversion between the International (SI) meter and the "Namibian Legal Meter." (Thanks to David Johanson of the U.S. Naval Oceanographic Office for that tidbit.) The classical geodetic system of Namibia (onshore and offshore) is the Schwarzeck Datum of 1903 where the origin is: $\Phi_o = 22°\ 45'\ 35.820''$ S, $\Lambda_o = 18°\ 40'\ 34.549''$ E, fixed during the German South West Africa – British Bechuanaland Boundary Survey of 1898–1903. (Thanks to the EPSG Database for that gem of information.) According to Professor

Charles L. Merry and Mr. J. Rens, "the initial point *Schwarzeck* is near Gobabis. ... The geoidal height and deflections of the vertical at this initial point are assumed to be zero."

The national Grid system of Namibia, the "Southwest Africa Transverse Mercator Belts" includes Central Meridians (λ_o) = 13° E, 15° E, 17° E, 19° E, 21° E, 23° E, and 25° E. The Scale Factor at Origin is equal to unity (1.0), the False Northing Latitude of Origin is 22° S, and there is no reported False Easting or False Northing, a common practice of the late 19th-century Europeans. My guess is the grid system has been renamed since the republic has achieved its independence.

Again, according to Professor Charles L. Merry and Mr. J. Rens in *Survey Review*, 30, 236 (April 1990), "In 1982 and 1983 the African Doppler Survey (ADOS) was carried out as a project supported by the International Association of Geodesy, in which precise absolute positions were determined in most African countries. ... three (*sites were located – Ed.*) in Namibia. In 1980 the South African Surveys and Mapping Directorate commenced a systematic satellite Doppler survey of South Africa and Namibia, using the translocation technique. ... Over the past few years, a number of estimates of geoidal heights for southern Africa have been made. We will now use the most recent one. However, the geoid is needed in the determination of these same transformation parameters. This apparent impasse can be resolved by iteration. In the case of the Cape Datum one of the ADOS points is close to the initial point where the geoidal height is known to be zero, for the Namibian Datum the initial point is an ADOS point. In each case preliminary translation components can be determined at the initial point and applied to the (*Conventional Terrestrial System – Ed.*) CTS-based geoid; this transformed geoid can then be used in the process of determining a final set of transformation parameters for the appropriate datum."

"Turning now to Namibia, the results for this datum are summarized ... Although the rotations are not as large as those in Zimbabwe and are barely significant, they do serve to model distortions in the geodetic network and hence provide an improved fit between this network and the CTS. Again, it must be emphasized that these rotations have no physical interpretation. As in all the countries investigated, except for South Africa, the scale factor plays no major role. Although the seven-parameter set does improve the fit, it is by no means as remarkable an improvement as that experienced in Zimbabwe and it is debatable whether the extra effort is worth it. Consequently, we recommend that (*the*) three-parameter transformation shown: $\Delta X = +616.6 \, m \pm 1.3 \, m$, $\Delta Y = +103.0 \, m \pm 1.3 \, m$, $\Delta Z = -256.6 \, m \pm 1.3 \, m$. ... As in Zimbabwe, the Namibian networks suffer from significant distortions but in this case a seven-parameter transformation provides little improvement over a three-parameter transformation."

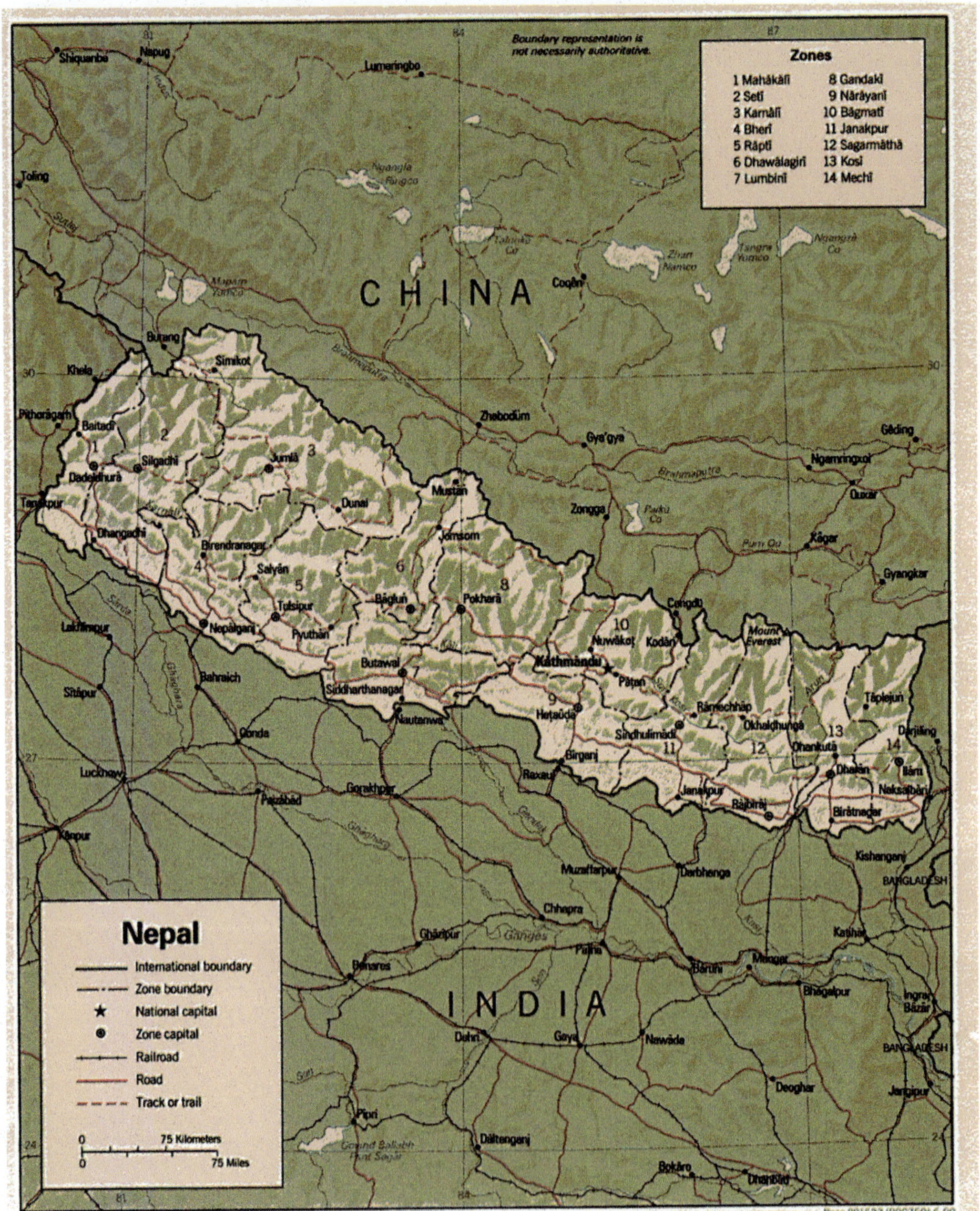

Base 801533 (B00750) 6-90

Kingdom of Nepal

Slightly larger than Arkansas, the terrain is comprised of Tarai or flat river plain of the Ganges in the south, central hill region, and rugged Himalayas in the north. The lowest point is *Kanchan Kalan* (70 m); the highest point is *Mount Everest* (8,850 m – the highest point in Asia). Nepal is bordered by China (1,236 km) and India (1,690 km) (*World Factbook*, 2013).

"Nepal has been a kingdom for at least 1,500 years. During most of that period, the *Kathmandu Valley* has been Nepal's political, economic, and cultural center. The valley's fertile soil supported thriving village farming communities, and its location along trans-Himalayan trade routes allowed merchants and rulers alike to profit. Since the fourth century, the people of the Kathmandu Valley have developed a unique variant of South Asian civilization based on Buddhism and Hinduism but influenced as well by the cultures of local Newar citizens and neighboring Tibetans. One of the major themes in the history of Nepal has been the transmission of influences from both the north and the south into an original culture. During its entire history, Nepal has been able to continue this process while remaining independent. The long-term trend in Nepal has been the gradual development of multiple centers of power and civilization and their progressive incorporation into a varied but eventually united nation. The Licchavi (4th to 8th centuries) and Malla (12th to 18th) kings may have claimed that they were overlords of the area that is present-day Nepal, but rarely did their effective influence extend far beyond the Kathmandu Valley. By the sixteenth century, there were dozens of kingdoms in the smaller valleys and hills throughout the Himalayan region. It was the destiny of Gorkha, one of these small kingdoms, to conquer its neighbors and finally unite the entire nation in the late eighteenth century. The energy generated from this union drove the armies of Nepal to conquer territories far to the west and to the east, as well as to challenge the Chinese in Tibet and the British in India. Wars with these huge empires checked Nepalese ambitions, however, and fixed the boundaries of the mountain kingdom. Nepal in the late twentieth century was still surrounded by giants and still in the process of integrating its many localized economies and cultures into a nation state based on the ancient center of the Kathmandu Valley. Nepal took a fateful turn in the mid-nineteenth century when its prime ministers, theoretically administrators in service to the king, usurped complete control of the government and reduced the kings to puppets. By the 1850s, a dynasty of prime ministers called Rana had imposed upon the country a dictatorship that would last about 100 years. The Ranas distrusted both their own people and foreigners – in short, anyone who could challenge their own power and change their position. As the rest of the world underwent modernization, Nepal remained a medieval nation, based on the exploitation of peasants and some trade revenues and dominated by a tradition-bound aristocracy that had little interest in modern science or technology. After the revolt against the Ranas in 1950, Nepal struggled to overcome its long legacy of underdevelopment and to incorporate its varied population into a single nation. One of the early casualties of this process was party-based democracy. Although political parties were crucial in the revolution that overthrew Rana rule, their constant wrangling conflicted with the monarchy's views of its own dignity and with the interests of the army. Instead of condoning or encouraging a multiparty democracy, King Mahendra Bir Bikram Shah Dev launched a coup in late 1960 against Bishweshwar Prasad (B.P.) Koirala's popularly elected government and set up a system of indirect elections that created a consultative democracy. The system served as a sounding board for public opinion and as a tool for economic development without exercising effective political power. Nepal remained until 1990 one of the few nations in the world where the king, wielding absolute authority and embodying sacred tradition, attempted to lead his country towards the twenty-first century" (*Country Study, Library of Congress*, 1991).

The first survey of Nepal was conducted by the Survey of India (SOI) in 1924. The first mapping was issued by the British Military Survey and later by SOI in 1927 which was compiled from plane table and alidade surveys with a 500 ft contour interval. The Colombo Plan of 1950 resulted in SOI surveys and mapping of the country for 10 years. During that time, Muneendra Kumar was a Survey Officer with the SOI and led triangulation parties into the mountains of Nepal. He later retired from the SOI, immigrated to the United States, and received his Ph.D. in Geodesy under Prof. Ivan I. Mueller at Ohio State University. After serving as Chief of Vertical Control Surveys at the U.S. National Geodetic Survey when we first started our personal friendship in 1977, he transferred to the National Imagery and Mapping Agency and finally retired as Chief Geodesist from the National Geospatial-Intelligence Agency. While working his triangulation party in Nepal, Dr. Kumar once walked 5,000 km in one year!

The first SOI surveys in the Kingdom of Nepal were referenced to the Indian Datum of 1916 where the origin is a station Kalianpur Hill where $\Phi_o = 24°\ 07'\ 11.26''$ N, $\Lambda_o = 77°\ 39'\ 17.57''$ East of Greenwich, and the ellipsoid of reference is the Everest 1830, where $a = 6,974,310.600$ Indian Yards, and $1/f = 300.8017$, $\xi = -0.29''$, $\eta = +2.87''$, $N = 0$ m. A Czechoslovakian team observed 7 Laplace (Astro) stations with 14 azimuth stations in 1977 through the assistance of the United Nations Development Program.

"Between 1982 and 1985 soldiers from the Military Survey branch of the Royal Engineers* undertook a major project in Nepal. The aim was to establish a first order geodetic network – a series of precisely coordinated trigonometrical points across the whole country – on which future lower order surveys and mapping could be based. The project, which was called Exercise High Trig, was carried out in conjunction with the Nepal Survey Department. A total of 14 Doppler points were fixed, ... a further agreement was to provide a comprehensive network of gravity observations.

"In the first year, work started in the east of the country with the project headquarters situated in the British Gurkha cantonment ad *Dharan*. A total of 16 survey stations were occupied in the first year and the distances and angles between them were measured.

"In the second year, work continued westward through the central region. The headquarters was now located in Maya House, Kathmandu, a fairly palatial accommodation hired through the British Embassy. Eventually this was supplemented by a tented advance camp at *Pokhara* in the *Annapurna* region. We capitalized on the groundwork of the previous year, and blessed by excellent weather, progress was rapid, and more work was achieved than expected. In all 36 stations were occupied, four of these being re-occupations of the previous year's stations in order to tie together the two pieces of work. The project was completed during the 1984/85 season, finishing in the 'far west' of the country" (Himalayan Survey, Captain, F. S. and Spence, R. E., *The Geographical Journal*, vol. 153, no. 2, , 223–230, July 1987).

The Nepal Datum of 1981 origin is at Station 12/157 *Nagarkot* where $\Phi_o = 27°\ 41'\ 31.04''$ N, $\Lambda_o = 85°\ 31'\ 20.23''$ East of Greenwich, and the ellipsoid of reference is the Everest 1830, where $a = 6,377,276.345$ m, and $1/f = 300.8017$, $\xi = -37.03''$, $\eta = -21.57''$, $N = 0$ m., in which the deflections of the vertical are derived from an astronomical position observed by the Czechoslovak Geodetic Institute (*The Geodetic Survey of Nepal 1981–84, Directorate of Military Survey*, UK, November 1985).

The University of Colorado and the Massachusetts Institute of Technology established a precise GPS network in Nepal during 1991. The classical origin point at Nagarkot on the WGS84 Datum is: $\Phi_o = 27°\ 41'\ 33.778''$, $\Lambda_o = 85°\ 31'\ 16.384''$.

"The topographical maps of Nepal are prepared and published in two parts: one of the Eastern Nepal and other of the Western Nepal. In both cases the ground controls for the topographical map preparation were established by GPS technique. Since the maps thus prepared have to be based on Everest Ellipsoid 1830, Nepal Datum the GPS established ground control points has to be transformed from WGS-84 to Everest Ellipsoid 1830.

"In the process of the determination of the transformation parameter (the common points between the reference frame *i.e.*, WGS-84 and Everest 1830) first order points based on Nepal datum, Everest 1830 only 11 points were used as the common points.

"The values of the transformation (**WGS-84 to Local**) parameters are as follows: $\Delta X = -293.17$ m, $\Delta Y = -726.18$ m, $\Delta Z = -245.36$ m. ... 1-sigma accuracy of the determination of transformation parameter is 0.26 m" (Study of Geodetic Datum of Nepal, China and Pakisthan (sic – Ed.) and its transformation to World Geodetic System, Niraj Manandhar, *Nepalese Journal on Geoinformatics*, 10, 2008).

* Professor Michael Barnes, MSc MRICS MCInstCES MRIN, Lecturer, Geomatics, Geospatial & Mapping Sciences, School of Geographical & Earth Sciences, University of Glasgow, Glasgow G12 8QQ was a Geodetic Surveyor attached to the 19 Topographic Squadron of 42 Survey Engineer Regiment for the Survey of Nepal in 1981–1984.

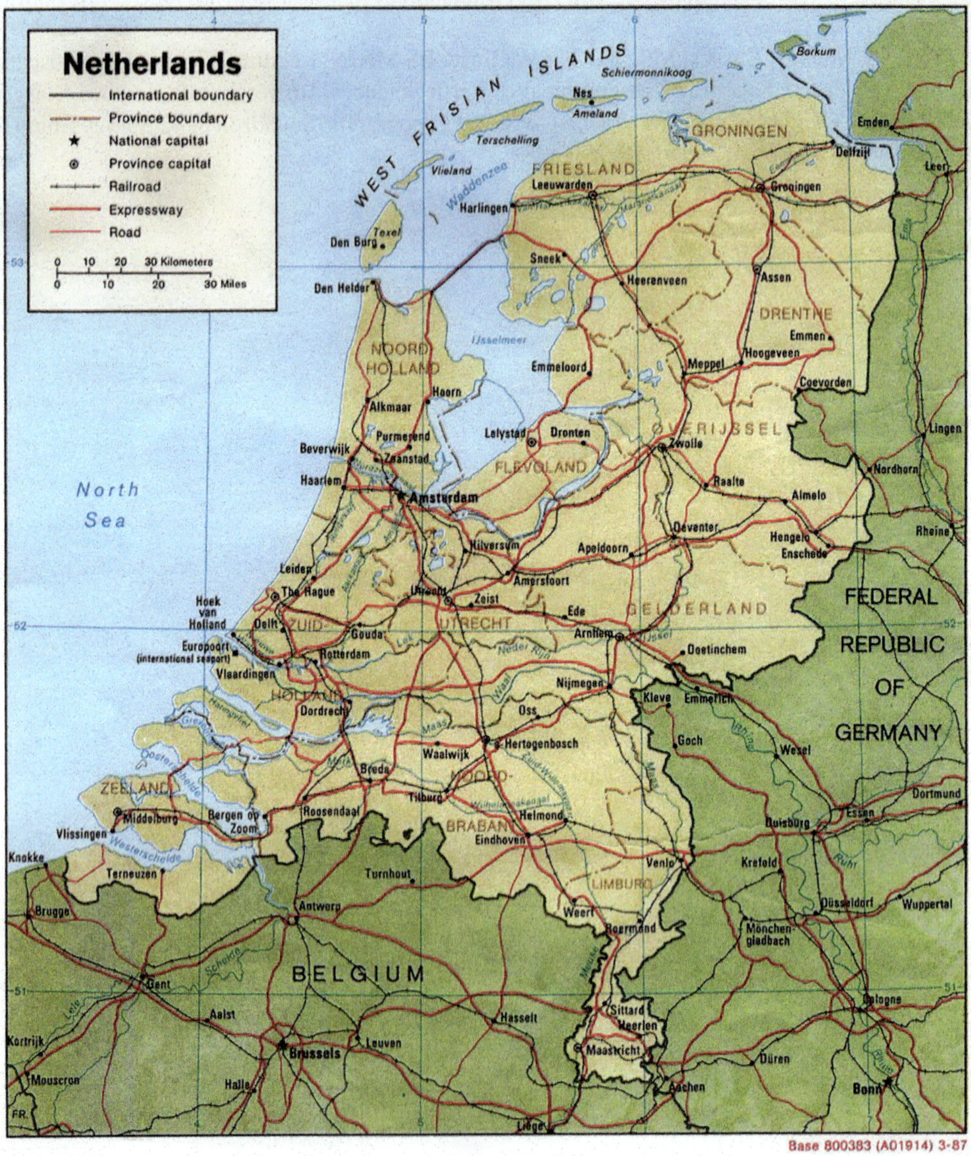

Base 800383 (A01914) 3-87

The Kingdom of the Netherlands

With an area slightly less than twice *the* size of New Jersey, the kingdom borders the North Atlantic Ocean to its north and west (451 km), Belgium to its south (450 km), and Germany to its east (577 km). The terrain is mostly coastal lowland and reclaimed land (lowest point is *Zuidplaspolder* at −7 m); there are some hills in the southeast and the highest point is *Vaalserverg* (322 m).

The kingdom was established in 1815 by French Emperor Napoleon, and initially controlled Belgium and Luxembourg. Inhabited since Paleolithic time, the region has been subjected to influences from early Celts, Germanic peoples, and Romans. The Netherlands is located at the mouths of three major European rivers: the *Rhine*, the *Maas* or *Meuse*, and the *Schelde*.

Willebrord Snell van Roijen (*Snellius*) was born in 1580 in Leiden, Netherlands, and studied law at the University of Leiden, although he taught mathematics there while he studied law. Snell's father was a professor of mathematics, and in 1613, he succeeded his father as professor of mathematics at the University of Leiden. The following year, he innovated the first classical triangulation based on his discovery of the Law of Sines. Using a brass Quadrant with a radius of 60 cm and by measuring an initial baseline of 328 m (total of five bases), he observed a quadrangle starting at Leiden and proceeded to complete a chain of quadrilaterals between *Bergen op Zoom* and *Alkmaar*. Later *Snellius* measured noontime shadows of towers in order to determine the length of a degree of the meridian. In 1617, he published *Eratosthenes Batavus De Terrae Ambitus Vera Quanitate*, which detailed his proposed techniques that established the science of Geodesy. He also discovered optical refraction (Snell's Law) and explored the *loxodrome*, a word he coined.

Before 1790, there was no systematic large-scale mapping in the Netherlands, either national or regional. The first mapping efforts from 1791 to 1794 and 1807 to 1811 were made to support cadastral mapping for taxation. In 1811, Napoleon I decreed that the entire country be surveyed and registered for the establishment of a cadastre. The Dutch Cadastre was established in 1832 and remained in the Ministry of Finance until 1973. Between 1809 and 1822, complete topographic mapping coverage of the Netherlands at a scale of 1:115,200 was achieved based on the first primary triangulation, published in 1861, on older cartographic materials, and on plane table foils. The Topografisch Bureau was founded in 1815 under the Ministry of Defense and was renamed *Topografisch Dienst Nederlands* (TDN) in 1931. The old triangulation and mapping was cast on the ellipsoidal Bonne projection where the latitude of origin ($\varphi_o = 51°\ 30'$ N), the central meridian was based on the (then) Prime Meridian at Amsterdam, and the scale factor at origin was unity. The ellipsoid of reference was the Bessel 1841 where the semi-major axis (a) = 6,377,397.155 m, and the reciprocal of flattening $1/f = 299.1528128$. (I think the old Amsterdam Datum of 1802 was first referenced to the Krayenhoff 1827 ellipsoid.) The kilometric grid squares on the old series are numbered over each 1:50,000 scale sheet 0–40 West to East, and 50–75 South to North, and a reference coordinate is given by sheet number and square number. The graticule was referenced to Amsterdam. TDN has used aerial photogrammetry since 1932.

A new triangulation was surveyed from 1886 to 1913 and published in *Triangulation du Royaume des Pays-Bas*, the first volume in 1903 and the second volume in 1921. The origin of the *Rijksdrihoeksmeting van het Kadaster* of 1918 Datum (RD 1918) at the *Lieve Vrouwe* (Holy Virgin) Church tower in *Amersfoort* is: $\Phi_o = 52°\ 09'\ 22.178''$ N, $\Lambda_o = 5°\ 23'\ 15.5''$ East of Greenwich (0° 30' 15.522″ East of Amsterdam), and $h_o = 0$ m. The ellipsoid of reference for RD 1918 is the Bessel 1841, and the Schreiber Stereographic double projection was chosen for the kingdom. The Stereographic

Grid origin is at the RD 1918 origin; the False Easting is 155 km, the False Northing is 463 km, and the scale factor at origin is unity. The result of this particular choice for the false origin is that no East to West reference can be more than 280 km, and no North to South reference can be <300 km. A comparison was made by the Dutch between values of 12 trigonometric stations along the border, as determined by the new Dutch triangulation on the Bessel ellipsoid and the new Belgian, values on the Hayford ellipsoid. The Dutch triangulation lies 2.95 and 10.0 centesimal seconds South and East respectively of the Belgian triangulation. According to Jacob A. Wolkeau of the U.S. Army Map Service, these differences were mainly due to residual differences in the respective Datum origins and the reduction of the triangulations on different ellipsoids of reference. Comparisons made by the Belgians showed varying differences of 4.4″ to 5.2″ in latitude and 6.1″ to 10.2″ in longitude. The figures used in the 1950s British map series revision of the Dutch maps to the adjacent Belgian maps were 4.6″ in latitude and 10.2″ in longitude determined from a number of primary stations along the border. The transformation of the RD1918 Datum to WGS84 Datum is defined by a seven-parameter Bursa-Wolfe shift using the *standard NGA right-hand rotation sign convention*: $\Delta X=+565.036$ m, $\Delta Y=+49.914$ m, $\Delta Z=+465.839$ m, $R_x=+0.4094″$, $R_y=-0.3597″$, $R_x=+1.86854″$, $\delta s=-4.0772$ ppm. A test point offered by Maarten Hooijberg in *Practical Geodesy*, 1997 is: $\varphi_{RD1918}=51° 59′ 13.3938″$, $\lambda_{RD1918}=4° 23′ 16.9953″$, $h_{RD1918}=30.696$ m → $\varphi_{WGS84}=51° 59′ 09.9145″$, $\lambda_{WGS84}=4° 23′ 15.9533″$, $h_{WGS84}=74.312$ m. Incidentally, Hooijberg also states that the WGS84 coordinates of *Lieve Vrouwe* tower are: $\varphi_{WGS84}=52° 09′ 18.62″$, $\lambda_{WGS84}=5° 23′ 13.9327″$, $h_{WGS84}=43.348$ m. European software commonly utilizes a left-handed rotation sign convention that appears to be mysteriously favored by NATO countries in lieu of the standard that is recognized and used by the United States, Australia, and most (I guess) of the Western Hemisphere. The best explanation I have heard for this curious disparity is a certain senior geodesist in the U.S. National Geodetic Survey once mused, "They probably didn't understand the math!"

The RD 1918 was recomputed on the European Datum of 1950 (EU50) and referenced to the Hayford 1909 (International 1909) ellipsoid where $a=6,377,388$ and $1/f=297$. To transform from EU50 to WGS84, NIMA lists the three-parameter values as $\Delta X=-87$ m ± 3 m, $\Delta Y=-96$ m ± 3 m, $\Delta Z=-120$ m ± 3 m, and the mean solution was based on 52 stations. To transform from EU79 to WGS84, NIMA lists the three-parameter values as $\Delta X=-86$ m ± 3 m, $\Delta Y=-98$ m ± 3 m, $\Delta Z=-119$ m ± 3 m, and the mean solution was based on 22 stations.

The first known elevation benchmark in Europe was the Amsterdam City Watermark (*Amsterdams Peil* or *AP*). Although now lost, the original was set in 1684! In 1707, the city watermark (*AP*) was already indicated on a water gauge near *Bilderham* – 25 km from the original mark. Extensions of precise levels were not performed until Napoleonic times. General C. R. T. Krayenhoff supervised the extension of the vertical datum from Amsterdam to the rivers *Rhine, Meuse,* and *Ijssel,* and along the coast of the *Zuiderzee*. Ramsden's leveling instrument was used which utilized a spirit level vial attached to a telescope. Krayenhoff's initial point was the water gauge at *Amstel Lock*. The *AP* was decreed to be the general datum plane of the Netherlands in 1818 and by 1860, the *AP* was published as the datum reference for some 550 benchmarks of the Prussian Railroad system. In 1876, a level loop was run among the original five stones and the deviation from the mean height for these 200-year-old benchmarks amounted to a maximum of 4 mm! Because of subsequent confusion with various levelings carried into Germany, the Dutch introduced new leveling results in January of 1891 as *Normaal Amsterdams Peil,* or *NAP*. At that same time, the Germans changed the name for their usage to *Normal-Null*. By 1928, the German and Dutch levels were compared anew with the result being $NN = NAP - 0.021$ mm. The *NAP* is now extended from Lapland to Gibraltar and from Scotland to Sicily; based on the original Dutch work in 1684.

- Reference Systems for Surveying and Mapping
    - http://gnss1.tudelft.nl/pub/vdmarel/reader/CTB3310_RefSystems_1-2a_print.pdf

- National Report of the Netherlands 2019- EUREF
  - http://euref.eu/symposia/2019Tallinn/05-16-Netherlands.pdf

- EUREF Permanent GNSS Network
  - http://www.epncb.eu/_networkdata/siteinfo4onestation.php?station=DELF00NLD

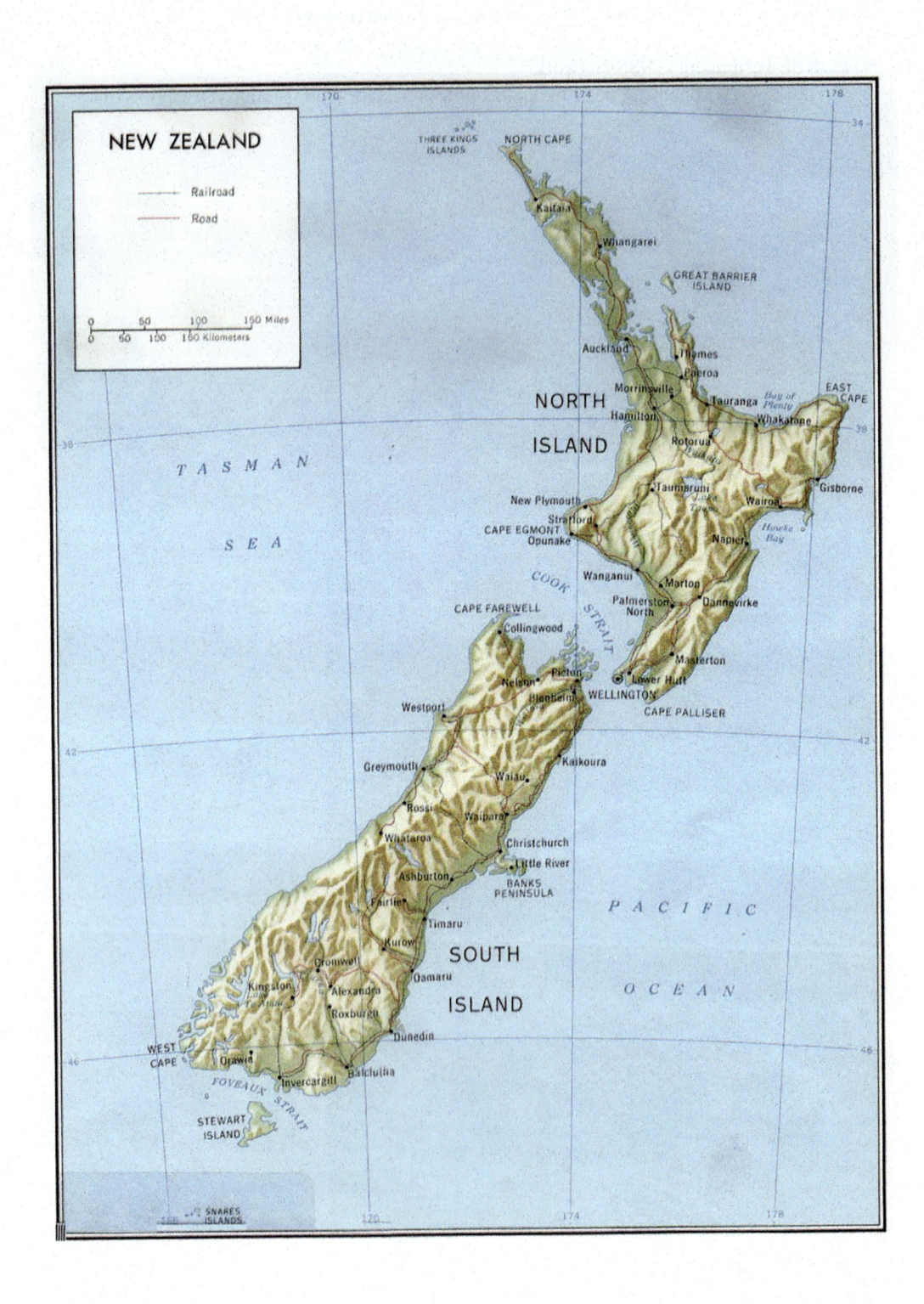

NEW ZEALAND

New Zealand

New Zealand is almost twice the size of North Carolina; about the size of Colorado, it is predominately mountainous with large coastal plains. Includes *Antipodes Islands*, *Auckland Islands*, *Bounty Islands*, *Campbell Island*, *Chatham Islands*, and *Kermadec Islands*. The highest point is *Aoraki/Mount Cook* (3,724 m); the lowest point is the Pacific Ocean (0 m), and the mean elevation is (388 m).

New Zealand is the southernmost extent of colonization by Polynesians. It is believed that the colonization took place in at least two major waves, the first of which is thought to have occurred around 950 A.D. and the second around 1200–1400 A.D. The first group consisted of hunters and depended for survival on the flightless "Moa" bird that is now extinct. The second Polynesian group was more agrarian, and archeological evidence indicates that the two cultural groups overlapped. Abel Janszoon Tasman, a captain for the Dutch East Indies Company, was the first European to discover the Maori of New Zealand. He sighted a "large land, uplifted high" near the modern-day town of *Hokitika* on the West Coast of South Island on December 13, 1642. He sailed northwards to a bay which he subsequently named Murderer's Bay (now renamed Golden Bay), after four of his men were attacked and killed by Maori warriors. The entire region was once thought by Tasman to be part of Tierra del Fuego, so he named it *Staten Landt*. Soon after Tasman's voyage, it was discovered that it was not Staten Landt, and it was renamed *Nieuw Zeeland*.

In the late 1800s modern surveying had begun. Captain James Cook, son of a Scottish migrant farmhand, was apprenticed to a Quaker ship owner, and he learned his trade in the difficult waters of the North Sea. He studied mathematics at night during off-seasons, and later in Nova Scotia he mastered surveying with the plane table and alidade. He was chosen to command a scientific voyage to the Pacific Ocean. He was commissioned Lieutenant prior to his first survey, and he was promoted to the rank of Captain before his third and last survey voyage. On the first voyage (1768–1771), his charter was to first go to Tahiti and observe the transit of Venus. These observations were to be later combined with other simultaneous observations (by others) in different locations in order to establish the magnitude of one astronomical unit (AU), which is defined as the mean distance of the Earth from the Sun. (*See Mauritius*) After making these observations, Cook was to sail south and discover or not the supposed Antipodes or great Southern Continent. The East Coast of North Island was sighted October 7, 1769. Cook circumnavigated North and South Island while taking almost continual observations. It took 6 months to produce an "astonishing" chart of New Zealand that mapped 2,400 miles of coastline. On this first voyage, Cook did not sail with John Harrison's chronometer but instead used Dr. Maskelyne's Lunar Almanac, published in 1767, which he used to fix his longitudes. However, Cook did use Harrison's chronometer on his two subsequent voyages (1772–1775, 1776–1780); he verified his longitudes to within 30 minutes. Cook's chart was used for the next 100 years as the settlement of New Zealand began. Thanks to Mrs. Peggy Haeger for the above research done in the late 1990s for a graduate-level course on coordinate systems that I used to teach at the University of New Orleans.

According to L. P. Lee in *First-order Geodetic Triangulation of New Zealand 1909–1949 and 1973–1974*, "In New Zealand, as indeed in many a similar young country," … "the settlement survey proceeded well in advance of the triangulation which should have controlled it. Even the triangulations were done in reverse order, the smaller networks being observed first, the larger and more accurate networks later. Thus, the procedure has been one of successive approximations, and each stage has led to a revision or recalculation of the work done earlier."

The first use of triangulation to control local surveys was by Felton Mathew, the first Surveyor-General in 1840–1841; the limited area covered was near Auckland. In 1849 another small triangulation was begun near Christchurch. Both of these were superseded by more accurate surveys. In 1852, six Provincial Governments were formed to administer New Zealand, and later this increased to nine provinces; each of which had its own survey department. (A special department of the General Government conducted Surveys of native Maori lands.) Henry Jackson based his land surveys on triangulation in Wellington Province beginning in 1865. The principal sections were in *Wellington*, *Wairarapa*, and *Rangitikei*, covering the three districts where settlements were located, and each section was erected upon its own baseline, with several check bases included for verification.

The specifications for surveying the New Zealand public lands with steel tapes originated with experiments on the Thames goldfields in 1869. The specifications were finally written to require steel tapes only (no chains) by James McKerrow in 1886, which preceded both the Swedish Jäderin steel wire apparatus in 1887 and the U.S. Coast & Geodetic Survey steel tapes in 1891!

"In the Province of *Otago*, a control net of triangulation was not used, but a uniform system to govern the orientation of surveys was introduced by J.T. Thomson in 1856. The province was divided into four large districts called 'meridional circuits.' Within each circuit an initial station was selected, and true meridian was determined by astronomical observation. Bearings, but not distances, were carried outwards by traverse from the initial station to the boundaries of the circuit, following the chief valleys suitable for settlement. Points on these traverses were called 'geodesic stations,' and were usually from 15 to 25 km apart, providing a series of reference points by which any survey within the meridional circuit could be oriented in terms of the meridian of the initial station. Any further control was merely local, being based upon a small triangulation net extending only over the region where it was immediately required, so that a large number of independent triangulations came to be distributed throughout the settled are of the Province. Each such triangulation was regarded as the control for an area around it called a 'survey district,' and each meridional circuit was eventually divided into many such districts, often irregular in shape: although the later additions tended to be bounded by lines parallel to and perpendicular to the meridian of the initial station. Local surveys could be coordinated with reference to the geodesic stations within a survey district."

The original datum for New Zealand was the Mt. Cook Datum of 1883, located in the city of Wellington. Thompson eventually adopted the meridional circuit system for all of New Zealand in 1877 as modeled by his original system earlier used in Otago. A total of 28 Meridional Circuits were established: 9 in the North Island and 19 in the South Island. Those Meridional Circuits with their original initial origins are: *Mt. Eden/Mt. Eden, Bay of Plenty/Maketu, Poverty Bay/ Patutahi, Taranaki/Huirangi, Tuhirangi/Thuirangi, Hawkes Bay/Hawkes Bay, Wanganui/Mt. Stewart, Wairarapa/Opaki, Wellington/Mt. Cook, Collingwood/Parapara, Nelson/Botanical Hill, Karamea/Karamea, Marlborough/Goulter Hill, Buller/ Buller Initial, Grey/Grey Initial, Amuri/ Isolated Hill, Hokitika/Hokitika Initial, Okarito/Abut Head, Mt. Pleasant/Mt. Pleasant, Gawler/ Gawler Downs, Jacksons Bay/Mt. Eleanor, Timaru/Mt. Horrible, Lindis Peak/Lindis Peak Initial, Mt. Nicholas/Mt. Nicholas, Mt. York/Mt. York, Observation Point/Observation Point, North Taieri/North Taieri,* and *Bluff/Observation Spot*. Computations on these meridional circuits were performed on the plane with the point of origin being the initial point. With the geodetic coordinates known for each initial point, the survey computations were equivalent to using the Polyhedric projection, which is the same as the Local Space Rectangular commonly, used in computational photogrammetry.

In 1901 a new secondary triangulation was started in order to bring all the different nets of triangles into harmony in the Wellington and Taranaki districts. The North Island geodetic triangulation of 1921–1938 started actual field observations in 1923 and continued until being suspended during the Great Depression of the early 1930s. It resumed in 1936. The South Island geodetic triangulation of 1938–1942 started with the observations across *Cook Strait* in a quadrilateral with one line measuring 120 km from *Papatahi* to *Attempt Hill*. Final fieldwork was observed from 1947 to 1949 including baseline-measuring equipment obtained on loan from *Tanganyika* (now *Tanzania*), which was used for three South Island bases. In 1948 12 LaPlace stations were observed with time signals transmitted from Dominion Observatory, especially for this purpose. When the computations were completed, the "New Zealand Geodetic Datum 1949" (NZGD49) was established where the initial station of origin was: *Papatahi Trig Station* $\Phi_o = 41° 19' 08.9000''$ S, $\Lambda_o = 175° 02' 51.0000''$ E of Greenwich, azimuth to Kapiti No. 2 $\alpha_o = 347° 55' 02.500''$, and the ellipsoid of reference is the International 1924 where $a = 6,378,388$ m and $1/f = 297$. *Papatahi* is a centrally situated station of the main net and one of the corner stations of the subsidiary net containing *Kelburn*. The values of deflection of the vertical for the north-south component at 65 latitude stations and deflection of the vertical for the east-west component at 39 azimuth stations from the first adjustment were known, and the latitude and azimuth at Papatahi were chosen so as to make the means of these differences equal to zero. The longitude adopted for Papatahi was that derived from Kelburn. The stations were coordinated on the National Grids, each island being on an independent Transverse Mercator projection, which had been selected by H.E. Walshe just before World War II. The New Zealand North Island Belt Latitude of Origin $\varphi_o = 39°$ S, Central Meridian, $\lambda_o = 175° 30'$ E, Scale Factor at Origin, m_o = unity, False Northing = 400,000 yards, False Easting = 300,000 yards where 1 foot = 0.304799735. The New Zealand South Island Belt Latitude of Origin, $\varphi_o = 44°$ S, Central Meridian, $\lambda_o = 171° 30'$ E, Scale Factor at Origin, m_o = unity, False Northing = 500,000 yards, False Easting = 500,000 yards. No further first-order work was contemplated for about 25 years, and in 1972–1974, some re-observation and extension was done with theodolites and Geodimeter Model 8 electronic distance meters.

When the metrication of surveys was begun in 1973, a one-projection coordinate system was adopted for topographic maps (the New Zealand Map Grid), but for cadastral surveying it was decided to retain the meridional circuit systems, but the Polyhedric coordinates were replaced by Transverse Mercator coordinates referred to the old origins.

In August 1998, "Land Information New Zealand" (LINZ) approved the adoption and implementation of a new geocentric datum, New Zealand Geodetic Datum 2000 (NZGD2000). The new coordinates of points changed by approximately 200 m relative to the old datum, NZGD49. A one-projection coordinate system was adopted for 1:50,000 scale and 1:250,000 scale topographic maps (the New Zealand Transverse Mercator 2000) that replaces the NZMG. The NZTM2000 Latitude of Origin, $\varphi_o = 0°$, Central Meridian, $\lambda_o = 171°$ E, Scale Factor at Origin, $m_o = 0.9996$, False Northing = 10,000,000 m, and False Easting = 1,600,000 m. For cadastral surveys in terms of NZGD2000 the 28 new meridional circuits replace the existing circuits, which were in terms of NZGD49. The new circuits are referred to as "<name> Circuit 2000," to distinguish them from the old circuits. The origins of latitude and longitude of the NZGD2000 circuit projections are almost the same as their NZGD49 equivalents being rounded down to the nearest arc second. The central meridian scale factors at origin of the NZGD2000 circuit projections are the same as those of their NZGD49 equivalents. The false origin coordinates of NZGD2000 circuit projections are 100 km *greater* than their NZGD49 equivalents, being 800 km N and 400 km E. This is to reduce the risk of confusion between the NZGD2000 and NZGD49 projections. The NZGD2000 circuit projections are based on the GRS80 ellipsoid of revolution where $a = 6,378,137$ m and $1/f = 298.257222101$. The SI standard for the meter has been adopted. The NZGD2000 circuit projections have a scale factor at origin of unity except for North Taieri 2000 (0.99996) and *Mt. Eden* 2000 (0.9999).

The Circuit Parameters are as follows: *Mount Eden* 2000- $\varphi_o = 36° 52' 47''$ S, $\lambda_o = 174° 45' 51''$ E, $m_o = 0.9999$; *Bay of Plenty* 2000- $\varphi_o = 37° 45' 40''$ S, $\lambda_o = 176° 27' 58''$ E, $m_o = 1.0$; *Poverty Bay* 2000- $\varphi_o = 38° 37' 28''$ S, $\lambda_o = 177° 53' 08''$ E, $m_o = 1.0$; *Hawkes Bay* 2000- $\varphi_o = 39° 39' 03''$ S, $\lambda_o = 176° 40' 25''$ E, $m_o = 1.0$; *Taranaki* 2000- $\varphi_o = 39° 08' 08''$ S, $\lambda_o = 174° 13' 40''$ E, $m_o = 1.0$; *Tuhirangi* 2000- $\varphi_o = 39° 30' 44''$ S, $\lambda_o = 175° 38' 24''$ E, $m_o = 1.0$; *Wanganui* 2000- $\varphi_o = 40° 14' 31''$ S, $\lambda_o = 175° 29' 17''$ E, $m_o = 1.0$; *Wairarapa* 2000- $\varphi_o = 40° 55' 31''$ S, $\lambda_o = 175° 38' 50''$ E, $m_o = 1.0$; *Wellington* 2000- $\varphi_o = 41° 18' 04''$ S, $\lambda_o = 174° 46' 35''$ E, $m_o = 1.0$; *Collingwood* 2000- $\varphi_o = 40° 42' 53''$ S, $\lambda_o = 172° 40' 19''$ E, $m_o = 1.0$; *Nelson* 2000- $\varphi_o = 41° 16' 28''$ S, $\lambda_o = 173° 17' 57''$ E, $m_o = 1.0$; *Karamea* 2000- $\varphi_o = 41° 17' 23''$ S, $\lambda_o = 172° 06' 32''$ E, $m_o = 1.0$; *Buller* 2000- $\varphi_o = 41° 48' 38''$ S, $\lambda_o = 171° 34' 52''$ E, $m_o = 1.0$; *Grey* 2000- $\varphi_o = 42° 20' 01''$ S, $\lambda_o = 171° 32' 59''$ E, $m_o = 1.0$; *Amuri* 2000- $\varphi_o = 42° 41' 20''$ S, $\lambda_o = 173° 00' 36''$ E, $m_o = 1.0$; *Marlborough* 2000- $\varphi_o = 41° 32' 40''$ S, $\lambda_o = 173° 48' 07''$ E, $m_o = 1.0$;

Hokitika 2000- $\varphi_o=42°$ 53′ 10″ S, $\lambda_o=170°$ 58′ 47″ E, $m_o=1.0$; *Okarito* 2000- $\varphi_o=43°$ 06′ 36″ S, $\lambda_o=170°$ 15′ 39″ E, $m_o=1.0$; *Jacksons Bay* 2000- $\varphi_o=43°$ 58′ 40″ S, $\lambda_o=168°$ 36′ 22″ E, $m_o=1.0$; *Mount Pleasant* 2000- $\varphi_o=43°$ 35′ 26″ S, $\lambda_o=172°$ 43′ 37″ E, $m_o=1.0$; *Gawler* 2000- $\varphi_o=43°$ 44′ 55″ S, $\lambda_o=171°$ 21′ 38″ E, $m_o=1.0$; Timaru 2000- $\varphi_o=44°$ 24′ 07″ S, $\lambda_o=171°$ 03′ 26″ E, $m_o=1.0$; *Lindis Peak* 2000- $\varphi_o=44°$ 44′ 06″ S, $\lambda_o=169°$ 28′ 03″ E, $m_o=1.0$; *Mount Nicholas* 2000- $\varphi_o=45°$ 07′ 58″ S, $\lambda_o=168°$ 23′ 55″ E, $m_o=1.0$; *Mount York* 2000- $\varphi_o=45°$ 33′ 49″ S, $\lambda_o=167°$ 44′ 19″ E, $m_o=1.0$; *Observation Point* 2000- $\varphi_o=45°$ 48′ 58″ S, $\lambda_o=170°$ 37′ 42″ E, $m_o=1.0$; *North Taieri* 2000- $\varphi_o=45°$ 51′ 41″ S, $\lambda_o=170°$ 16′ 57″ E, $m_o=0.99996$; and *Bluff* 2000- $\varphi_o=46°$ 36′ 00″ S, $\lambda_o=168°$ 20′ 34″ E, $m_o=1.0$. Thanks to Mr. Graeme Blick of LINZ for a copy of L.P. Lee's monograph on the history of geodetic triangulation in New Zealand and to Mr. Mal Jones of Perth Australia for his continuing generous help.

Nicaragua

- International boundary
- Department boundary
- ★ National capital
- ⊙ Department capital
- Railroad
- Road

Lambert Conformal Conic Projection, SP 9N/17N

Republic of Nicaragua

Slightly smaller than the state of New York, Nicaragua is bordered by Costa Rica (313 km) and Honduras (940 km). The terrain is extensive Atlantic coastal plains rising to central interior mountains, narrow Pacific coastal plain interrupted by volcanoes. The highest point is *Mogoton* (2,085 m), the lowest point is the Pacific Ocean (0 m), and the mean elevation is (298 m).

Two basic culture groups existed in pre-colonial Nicaragua. In the central highlands and Pacific coast regions, the native peoples were linguistically and culturally similar to the Aztec and the Maya. Most people of central and western Nicaragua spoke dialects of *Pipil*, a language closely related to *Nahuatl*, the language of the Aztec. Most of Nicaragua's Caribbean lowlands area was inhabited by tribes that migrated north from what is now Colombia. The various dialects and languages in this area are related to *Chibcha*, spoken by groups in northern Colombia. When the Spanish arrived in western Nicaragua in the early 1500s, they found three principal tribes, each with a different culture and language: the *Niquirano*, the *Chorotegano*, and the *Chontal*. The *Chontal* were culturally less advanced than the *Niquirano* and *Chorotegano*, who lived in well-established nation-states. Occupying the territory between *Lago de Nicaragua* and the Pacific Coast, the *Niquirano* were governed by chief *Nicara*o, or Nicaragua, a rich ruler who lived in *Nicaraocali*, now the city of *Rivas*. The *Chorotegano* lived in the central region of Nicaragua. These two groups had intimate contact with the Spanish conquerors, paving the way for the racial mix of native and European stock now known as mestizos. The *Chontal* (the term means foreigner) occupied the central mountain region (*Library of Congress Country Studies,* 2009).

"Former Sandinista President Daniel Ortega was elected president in 2006, 2011, 2016, and most recently in 2021. Municipal, regional, and national-level elections since 2008 have been marred by widespread irregularities. Democratic institutions have weakened under the Ortega administration as the president has garnered full control over all branches of government, especially after cracking down on a nationwide anti-government protest movement in 2018. In the lead-up to the 2021 presidential election, most of the prominent opposition candidates (were either arrested or forced into exile leaving only five lesser-known candidates of mostly small parties allied to Ortega's Sandinistas to run against him" (*World Factbook*, 2022).

The government has 15 geographic departments: *Boaco, Carazo, Chinandega, Chontales, Estelí, Granada, Jinotega, Leon, Madriz, Managua, Masaya, Matagalpa, Nueva Segovia, Rio San Juan, Rivas* and the two autonomous regions of *Atlántico Norte* and *Atlántico Sur*. The national holiday is Independence Day, September 15, 1821 (*The World Factbook*, 2009).

The first map of the region was prepared by Christopher Columbus in 1502 and covered the Caribbean coast. Nicaragua was noted for its paucity of surveying and mapping for many years into the 20th century. "Conventional mapping was impossible in the undeveloped eastern two-thirds of the country with its constant cloud cover, vast areas, lack of roads, flat terrain combined with heavy jungle, and extensive coastal swamps, all making plane table techniques unworkable" (*Collaborative Mapping in Nicaragua,* 1957). "The Military Intelligence Division, General Staff, U.S. Army, published during 1929–1934, a hachured 1:250,000 scale map series covering the entire country" (*Foreign Maps, TM 5-248, 1963*). "In 1946 the *Oficina de Geodesía* (Office of Geodesy) was organized as a part of the *Ministerio de Guerra, Marina y Aviación* (Ministry of War (Army), Navy and Air Force) and an agreement was made between Nicaragua and the United States to establish basic geodetic control in Nicaragua" (*Topographic Mapping of the Americas, Australia, and New Zealand,* M. L. Larsgaard, 1934). An old publication of the U.S. Army Map Service Inter-American Geodetic Survey (IAGS) lists the first director of the *Oficina de Geodesía* in 1946 as

General Anastasio Somoza DeBayle, who later rose to become the infamous Dictator of Nicaragua for decades! Initial geodetic surveys commenced in 1949.

The oldest geodetic datum of Central America is the *Ocotepeque Datum of 1935* which was established at *Base Norte* in Honduras where $\varphi_o = 14°\ 26'\ 20.168''$ North, $\lambda_o = 89°\ 11'\ 33.964''$ West of Greenwich, and $H_o = 806.99$ m above mean sea level. The defining geodetic azimuth to Base Sur is: $\alpha_o = 358°\ 54'\ 21.790''$ (*Memoria de la Dirección General de Cartografía, Guatemala, September 1957*), and the ellipsoid of reference is the Clarke 1866 where $a = 6,378,206.4$ m and $1/f = 294.9786982$. The corresponding astronomic observations at that mountainous location are: $\Phi_o = 14°\ 26'\ 13.73''$ North ($\pm 0.07''$), $\Lambda_o = 89°\ 11'\ 39.67''$ West ($\pm 0.045''$), and the defining astronomic azimuth to Base Sur is: $\alpha_o = 358°\ 54'\ 20.37''$ ($\pm 0.28''$) (*Informe Detallado de la Comisión Técnica de Demarcación de la Frontera entre Guatemala y Honduras*). The difference between these two sets of coordinates is due to the local gravimetric deflection of the vertical.

The IAGS developed a series of map projections for each of the Central American countries during the late 1940s through the 1950s. Each of these coordinate systems was based on the Lambert Conformal Conic projection with two standard parallels, similar in treatment as the Coast & Geodetic Survey did for those applicable states in the United States. The two Lambert Conformal Conic zones for the Republic of Nicaragua are: Nicaragua North (*Norte*) and South (*Sud*) Zones – Both zones use the same Central Meridian $(\lambda_o) = 85°\ 30'\ 00''$ West of Greenwich and False Easting of 500 km. Zone *Norte* has a Latitude of Origin $(\varphi_o) = 13°\ 52'$ North, the False Northing = 359,891.816 m, and the scale factor at origin $(m_o) = 0.99990314$. Zone *Sud* has a Latitude of Origin $(\varphi_o) = 11°\ 44'$ North, the False Northing = 288,876.327 m, and the scale factor at origin $(m_o) = 0.9992228$. Sometime in the 1960s, the IAGS extended the North American Datum of 1927 (Clarke 1866 ellipsoid) into Central America. Because of that work, the published three-parameter datum shift from NAD27 to WGS84 for the region that includes Costa Rica through Nicaragua is: $\Delta X = 0$ m ± 8 m, $\Delta Y = +125$ m ± 3 m, $\Delta Z = +194$ m ± 5 m, and the relation is based on a solution of 19 stations. Boundary disputes with neighboring countries were finally settled in 1960 (and re-monumented with Costa Rica in 2004) except for a continuing squabble with Colombia regarding *Isla de San Andres* and *Isla de Providencia* in the Caribbean Sea (*Geographic Notes, Department of State, Office of the Geographer September 8, 1986*). The national mapping agency (now civilian since 1981) of Nicaragua is the *Instituto Nicaragüense de Estudios Territoriales (INETER)*.

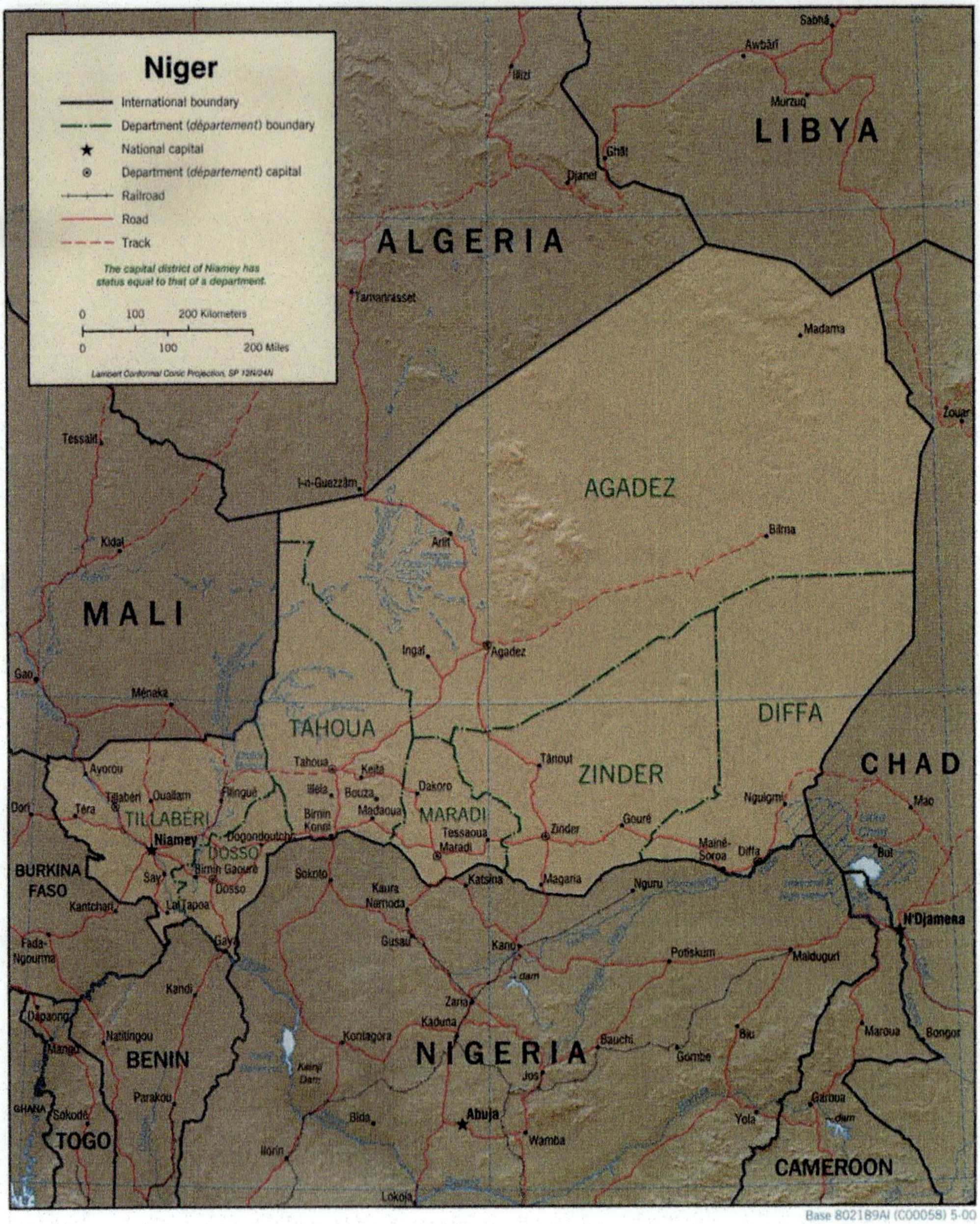

Base 802189AI (C00058) 5-00

Republic of Niger

Bordered by Algeria (956 km), Benin (266 km), Burkina Faso (628 km), Chad (1,175 km), Libya (354 km), Mali (821 km), and Nigeria (1,497 km), Niger has predominately desert plains and sand dunes, flat to rolling plains in the south and hills in the north. The lowest point is the Niger River (200 m), and the highest point is *Idoûkâl-n-Taghès* (2,022 m).

"Considerable evidence indicates that about 600,000 years ago, humans inhabited what has since become the desolate Sahara of northern Niger. Long before the arrival of French influence and control in the area, Niger was an important economic crossroads and the empires of *Songhai, Mali, Gao, Kanem,* and *Bornu,* as well as a number of Hausa states claimed control over portions of the area. During recent centuries, the nomadic Tuareg formed large confederations, pushed southward, and, siding with various *Hausa states,* clashed with the *Fulani* Empire of *Sokoto,* which had gained control of much of the *Hausa* territory in the late 18th century. In the 19th century, contact with the West began when the first European explorers – notably Mungo Park (British) and Heinrich Barth (German) – explored the area searching for the mouth of the Niger River. Although French efforts at pacification began before 1900, dissident ethnic groups, especially the desert Tuareg, were not subdued until 1922, when Niger became a French colony" (*Niger – Country Studies, U.S. Department of State,* 2011).

"Before the Sahara started swallowing Niger around 2500 *B.C.*, it supported verdant grasslands, abundant wildlife and populations thriving on hunting and herding. Long after the desert pushed those populations south, Niger became a fixture on the trans-Saharan trade route. Between the 10th and 18th centuries, West African empires, such as the *Kanem-Borno, Mali,* and *Songhaï* flourished here, trafficking gold, salt, and slaves. The French strolled in late in the 1800s, meeting stronger-than-expected resistance. Decidedly un-amused, they dispatched the punitive *Voulet-Chanoîne* expedition, destroying much of southern Niger in 1898–99. Although *Tuareg* revolts continued, culminating in Agadez's siege in 1916–17; the French had control.

French rule wasn't kind. They cultivated traditional chiefs' power, whose abuses were encouraged as a means of control, and the enforced shift from subsistence farming to high-density cash crops compounded the Sahara's ongoing migration. In 1958 France offered its West African colonies self-government in a French union or immediate independence. Countless votes conveniently disappeared, enabling France to claim that Niger wished to remain within its sphere of influence. Maintaining close French ties, Niger's first president, Hamani Diori, ran a repressive one-party state. After surviving several coups, he was overthrown by Lieutenant Colonel Seyni Kountché after food stocks were discovered in ministerial homes during the Sahel drought of 1968–74. Kountché established a military ruling council. Kountché hit the jackpot in 1968 when uranium was discovered near Arlit. Mining incomes soon ballooned, leading to ambitious projects, including the 'uranium highway' to *Agadez* and *Arlit.* By the 1990s, Nigerians were aware of political changes sweeping West Africa and mass demonstrations erupted, eventually forcing the government into multiparty elections in 1993. However, a military junta overthrew the elected president, Mahamane Ousmane, in 1996. In 1999, during widespread strikes and economic stagnation, president Mainassara (1996 coup leader) was assassinated, and democracy re-established. Peaceful elections in 1999 and 2004 witnessed victory for Mamadou Tandja" (*Niger – Lonely Planet,* 2011).

"Burkina Faso and Mali have jointly submitted a boundary dispute to the International Court of Justice (ICJ) for binding adjudication. The two states had signed an agreement to send their dispute to the ICJ on 24 February 2009 (entering into force on 20 November 2009) but the application was filed into the Court's Registry on 20 July 2010. Burkina Faso and Niger have asked the ICJ to determine the course of their boundary between two identified endpoints, the survey pillar at

Tong Tong (14° 25′ 04″ N, 00° 12′ 47″ E) in the north to the *Boutou* curve in the south (12° 36′ 18″ N, 01° 52′ 07″ E). This constitutes the vast majority of the central section of their boundary and the two parties made clear that the northern and southern extremities of the boundary (from *Tong Tong* to the tripoint with Mali, and *Boutou* to the tripoint with Bénin) have already been demarcated by a Joint Technical Commission" (*International Boundaries Research Unit, Durham University, 22 July 2010*).

During World War II, the Niger Zone was established as a "British Grid" where the Latitude of Origin $(\varphi_o) = 13°$ N, Central Meridian $(\lambda_o) = 0°$ (Equator), Scale Factor at Origin $(m_o) = 0.99932$, False Easting = 1,800 km, False Northing = 500 Km. The limits of the Grid were: North, 16° N, East, 1° 30′ East; South: 10° N; and West, 14° West. The ellipsoid of reference was the Clarke 1880; the datum: ersatz (*Lambert Conical Orthomorphic Projection Tables, Niger Zone, RESTRICTED, Office of the Chief of Engineers, Washington, DC, 1943*). (Later declassified – Ed.)

In December of 1945, the *Institut Géographique National* (IGN) issued SGC 1312 in which *l'A.E.F.* and *l'A.O.F.* (French East and West Africa) would employ a series of Gauss-Krüger Belts that included Niger. By September 20, 1950, that was rescinded in favor of the UTM Grid proposed by the U.S. Army Map Service.

The Chad–Niger Boundary is 1,175 km in length. In the south the Nigeria tripoint is situated in Lake Chad at about 13° 42′ 53″ N, 13° 38′ 20″ E, and in the north the Libya tripoint is situated northwest of the *Tibesti* at 23° N and 15° E. Northward the boundary traverses Lake Chad, crosses typical sandy and gravelly surfaces, and continues through a region of rocky ridges and steep-sided hills. There are no known pillars demarcating the boundary (*Chad–Niger Boundary, International Boundary Study No. 73 – August 1, 1966*, Bureau of Intelligence and Research, Department of State, USA).

The Niger–Nigeria Boundary is 1,497 km in length. From the tripoint with Dahomey on the median line of the Niger River, it extends northward and then eastward to the Republic of Chad tripoint at 13° 42′ 29″ N, and approximately 13° 38′ E. In the extreme eastern sector, the boundary follows the thalweg of the eastward flowing *Komadugu Yobe* for more than 272 km and then continues for about 25.6 km in Lake Chad to the Chad tripoint. The boundary is demarcated by pillars and the *Komadugu Yobe* (*Niger–Nigeria Boundary, International Boundary Study No. 93 – December 15, 1969*, Bureau of Intelligence and Research, Department of State, USA).

Located in the Sahara, the Algeria–Niger boundary is about 956 km in length. Northeastward from the tripoint with Mali, it consists of three straight-line sectors of 174.4, 227.2, and 548.8 km, respectively. The boundary is un-demarcated and traverses sparsely populated areas (*Algeria–Niger Boundary, International Boundary Study No. 99 – May 1, 1970*, Bureau of Intelligence and Research, Department of State, USA).

The Mali–Niger boundary extends for approximately 816 km between the Upper Volta and Algeria tripoints. The line is not demarcated by pillars. Although it follows several valleys, more than two-thirds of the boundary consists of straight-line segments (*Mali–Niger Boundary, International Boundary Study No. 150 – January 13, 1975*, Bureau of Intelligence and Research, Department of State, USA).

The only datum known for Niger has its origin at Point 58 east of the town of *Dosso* and near the border with Nigeria where $\Phi_o = 12°\ 52′\ 44.045″$ N, $\Lambda_o = 3°\ 58′\ 37.040″$ E of Greenwich. Thanks to John W. Hager, "Azimuth is 97° 30′ 04.237″ to C. F. L. 1 from north. Elevation = 266.71. Astro observed by *IGN* in 1968. This was used as a temporary datum pending the adjustment of the 12th Parallel to Adindan. It was for the section of the 12th Parallel in Niger and Upper Volta. Reference is *Final Report of the 12th Parallel Survey in the Republic of Niger.*" Surveyed in 1969 by the French *IGN*, the ellipsoid of reference is the Clarke 1880 where $a = 6{,}378{,}249.145$ m, $1/f = 293.465$. This was used as the basis for computation of the 12th Parallel traverse conducted 1966–1970 from Senegal to

Chad and connecting to the Adindan triangulation in Sudan. Remarkably, references to this datum origin point appear to have interchanged the values for Latitude and Longitude, incorrectly placing the point somewhere in Cameroon! According to *TR 8350.2*, the datum shift **from** Point 58 Datum **to** WGS84 Datum is: $\Delta X = -106$ m ± 25 m, $\Delta Y = -129$ m ± 25 m, and $\Delta Z = +169$ m ± 25 m, and is based on two points collocated in 1991.

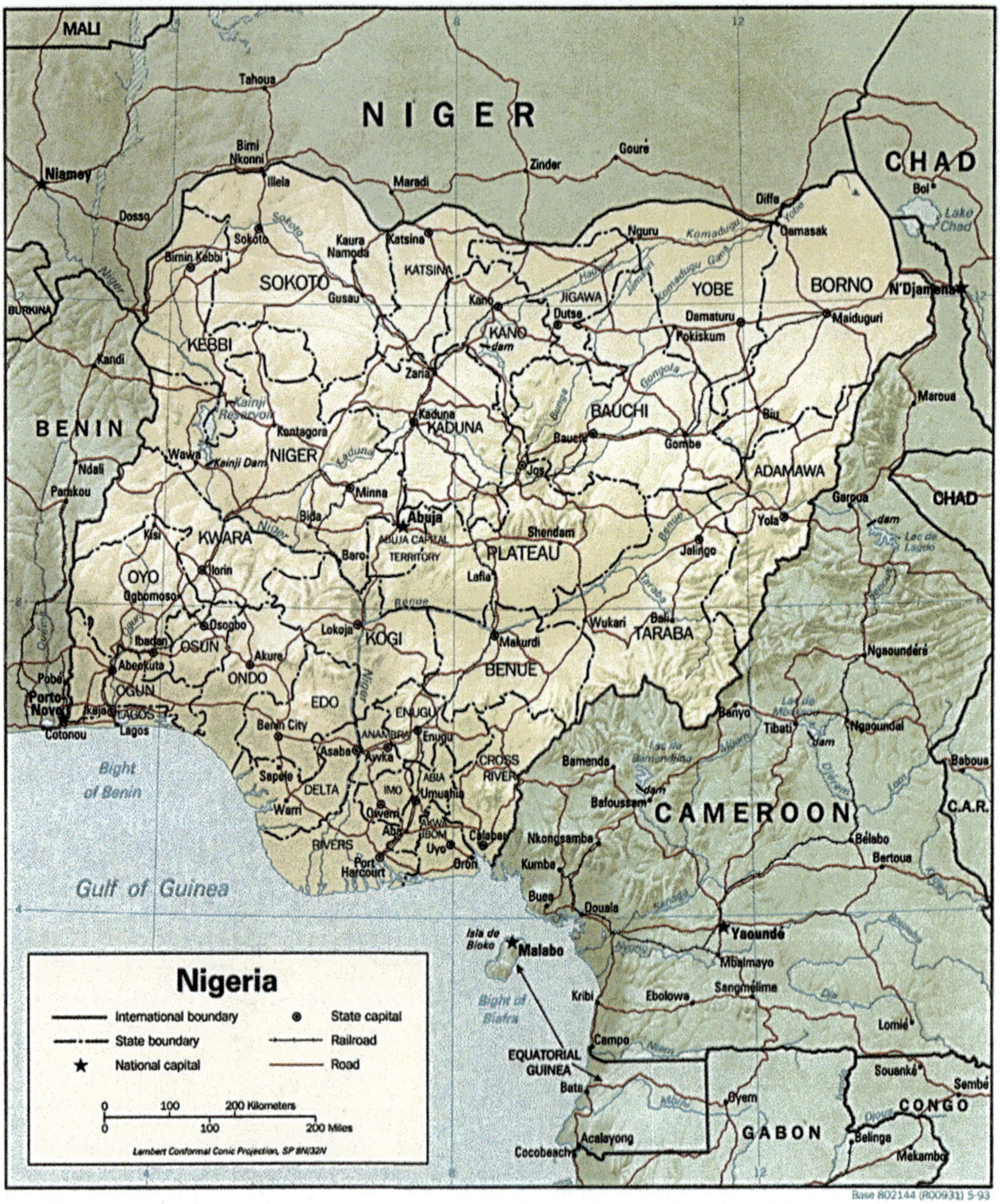

Base 802144 (R00931) 5-93

Federal Republic of Nigeria

Bordered by Benin (773 km), Cameroon (1,690 km), Chad (87 km), Niger (1,497 km), and the Gulf of Guinea (853 km), the lowest point is the Atlantic Ocean (0 m), and the highest point is *Chappal Waddi* (2,419 m). The border with Bénin, formerly Dahomey, is a curious combination of physical monuments and riparian calls consisting of both *Rule of the Thalweg* (thread of the stream) and of *medium filium aquae* (geographic center of the stream) principles. Slightly more than twice the size of California and Africa's most populous country, Nigeria is composed of more than 250 ethnic groups; the following are the most populous and politically influential: *Hausa* and *Fulani* 29%, *Yoruba* 21%, *Igb*o (*Ibo*) 18%, *Ijaw* 10%, *Kanuri* 4%, *Ibibio* 3.5%, *Tiv* 2.5%.

The earliest settlements in what is now Nigeria are indicated by stone tools dating back 12,000 years. "Microlithic and ceramic industries were developed by pastoralists in the savanna from at least the fourth millennium *B.C.* and were continued by grain farmers in the stable agricultural communities that subsequently evolved there. To the south, hunting and gathering gradually gave way to subsistence farming on the fringe of the forest in the first millennium *B.C.* The cultivation of staple foods, such as yams, later was introduced into forest clearings. The stone ax heads, imported in great quantities from the north and used in opening the forest for agricultural development, were venerated by the *Yoruba* descendants of Neolithic pioneers as 'thunderbolts' hurled to earth by the gods. The primitive iron-smelting furnaces at *Taruga* dating from the fourth century *B.C.* provide the oldest evidence of metalworking in West Africa, while excavations for the *Kainji* Dam revealed the presence of ironworking there by the second century *B.C.* The transition from Neolithic times to the Iron Age apparently was achieved without intermediate bronze production. Some scholars speculate that knowledge of the smelting process may have been transmitted from the Mediterranean by Berbers who ventured south. Others suggest that the technology moved westward across the Sudan from the Nile Valley, although the arrival of the Iron Age in the Niger River valley and the forest region appears to have predated the introduction of metallurgy in the upper savanna by more than 800 years. The usefulness of iron tools was demonstrated in the south for bush cutting and in the north for well digging and the construction of irrigation works, contributing in both regions to the expansion of agriculture. The earliest culture in Nigeria to be identified by its distinctive artifacts is that of the *Nok* people. These skilled artisans and ironworkers were associated with *Taruga* and flourished between the fourth century *B.C.* and the second century *A.D.* in a large area above the confluence of the Niger and Benue rivers on the *Jos Plateau*. The *Nok* achieved a level of material development not repeated in the region for nearly 1,000 years. Their terra-cotta sculpture, abstractly stylized and geometric in conception, is admired both for its artistic expression and for the high technical standards of its production" (*Library of Congress Country Study* – 2008).

"British influence and control over what would become Nigeria grew through the 19th century. A series of constitutions after World War II granted Nigeria greater autonomy; independence came in 1960. Following nearly 16 years of military rule, a new constitution was adopted in 1999, and a peaceful transition to civilian government was completed. The government continues to face the daunting task of reforming a petroleum-based economy, whose revenues have been squandered through corruption and mismanagement, and institutionalizing democracy. In addition, Nigeria continues to experience longstanding ethnic and religious tensions. Although both the 2003 and 2007 presidential elections were marred by significant irregularities and violence, Nigeria is currently experiencing its longest period of civilian rule since independence. The general elections of April 2007 marked the first civilian-to-civilian transfer of power in the country's history. residential and legislative elections were held in early 2019 and deemed broadly free and fair despite voting irregularities, intimidation, and violence" (*World Factbook*, 2022).

TABLE N.1
The Values of Latitude and Longitude Projected to the North Terminal of the Minna Base Are as Follows

	Latitude	Longitude
Observed at *Minna* L 40	9° 38′ 11″.700	6° 30′ 53″.850
From *Kano* K 2	07″.278	31′ 00″.278
Naraguta N 1	12″.768	31′ 00″.370
Lafia Beri Beri N 26	03″.940	30′ 56″.713
Zaroa N 144	08″.665	*31′ 02″.580*
Mean	9° 38′ 08″.87	6° 30′ 58″.76
Adopted values	9° 38′ 09″.000	6° 30′ 59″.000

The first geodetic surveys of Nigeria were performed by the British Royal Engineers in 1910–1912 (*TM 5–248, Foreign Maps 1963*). The majority of geodetic work in Nigeria was performed beginning in 1928, and most of the older work done was discarded with the exception of the four bases at *Eruwa*, *Kano*, *Naraguta*, and *Udi*. The azimuths at *Eruwa*, *Kano*, and *Naraguta* were re-observed in 1928 and *Kano* again and *Udi* were re-determined in subsequent years. Difficulties with local fauna "can be illustrated by the fact that during the survey of one of the reserves a locally engaged laborer was charged by a herd of elephants and, being unable to blaze his trail whilst escaping, was bushed for two days." The following portions of triangulation executed prior to 1931 with the 6-inch Troughton & Simms micrometer theodolite were retained, with base nets being excluded from the percentages quoted: *Naraguta – Kano* 1929–1931, 11 stations or 36% of the chain; *Minna – Udi* 1930, 17 stations or 30% of the chain; *Ilorin – Eruwa* 1927–1928, 26 stations or 100% of the chain. The ellipsoid of reference chosen for Nigeria was the Clarke 1880, where $a=20,926,202$ feet, $1/f=293.465$, and the meter$=1.09362311$ yard. The mean datum was obtained at the North Terminal of the *Minna* Base by projecting the values of latitude and longitude obtained at *Kano*, *Naraguta*, *Lafia*, *Beri Beri*, and *Zaria* through the triangulation and obtaining the resulting mean value at *Minna*. Of the above-mentioned points two lie to the east of *Minna* and the other two to the north–northeast. A better mean value at *Minna* would have been obtained if it had been possible to carry up coordinates from the south and west also, but at the time the datum was adopted in 1928 no triangulation to the south and west of Minna was available (Table N.1).

The points at *Lafia Beri Beri* and *Zaria* are *not* now part of the primary framework; both are classed as secondary points. The geoid height of *Minna*, L 40 is $H_o=279.6\,m$.

"The spheroidal (*sic*) tables for the Clarke 1880 figure in use in Nigeria are based on those computed by Morris for the Geodetic Survey of South Africa, and the unit of length of the tables is the British foot. This in identifying the unit of length of the triangulation with the unit of length in the tables as altered scale of the Clarke Spheroid is achieved, the fractional amount, represented by the logarithmic correction 0.00000031, being 1/138,161. Hence for the triangulation of Nigeria all computations of position incorporate this linear change of 1,138,161. It is, however, in conformity with the usage in the greater part of British and French Africa" (Capt. J. Calder Wood, *Empire Survey Review*, vol. III, no. 21, July 1936).

The geodetic and cadastral surveys of colonial Nigeria appear to be of exceptional quality, particularly considering the enormous financial strains of funding such work during the worldwide Great Depression of the 1930s. With British and other European surveyors providing the technical and supervisory expertise, the local native population was pushed to the background with little education and training provided. Ethnocentrism among the Europeans prevailed for several decades until local Nigerians were able to gain advanced university educations in the UK and other European institutions of higher learning. With local Nigerian universities producing highly educated surveyor professionals, the level of high-quality geodetic surveys gained ground in the latter half of the 20th century. Some resentment of prior survey qualities, both geodetic and cadastral, seemed to be reflected in some of the geodetic literature, but such appears to now be on the wane.

A curiosity of the colonial period is the Nigerian Colony Coordinate System which is based on the "Notes on Projection (NOP)" Modified Transverse Mercator. An excellent paper on the topic is "S. Izundu Agajelu, On conformal representation of geodetic positions in Nigeria" (*Survey Review*, vol. 29, 223, January 1987), with the following excerpts quoted: "The 'classical' method of map projection in Nigeria, is based on the *Notes on Projection* (*NOP*) published by the Federal Surveys of Nigeria in the 1920s, in which reference is made to examples worked in 1927 on the basis of logarithmic and interpolating tables." "Further the *NOP* tables were developed in feet, whereas Nigeria went metric early in the 1970s." "The system is discussed in this paper primarily because Nigeria has recently gone over to UTM for representing its control positions. Some of the major differences between the UTM and the Modified Transverse Mercator System (MTM) used in *NOP* are as follows; (1) The UTM has the scale factor at the central meridian as 0.9996 while the MTM system has the value of 0.99975. (2) The central meridians in the UTM system in Nigeria are 3°E. 9°E. and 15°E, corresponding to zones 31, 32, and 33 respectively, while those of MTM are 4° 30′, 8° 30′, and 12° 30′, for the West, Mid and East belts respectively. (3) In UTM the Eastings of the central meridians are fixed as a false value of 500,000m while in the Nigerian system, the three central meridians have the values 230,728.266m, 670,553.984m, 1,110,369.702m, respectively. (4) In the UTM the maximum angular distance of a point in a belt, from the central meridian of the belt is 3° while in the MTM, it is 2°." In Agajelu's Conclusions he states: "(1) Contrary to the description at page 10 of *NOP*, it has been proved conclusively in this paper, that the Nigerian system is based not on the double projection method, but on direct projection from the ellipsoid to the map plane. (2) The direct projection formulae referred to in 1 above are based on power series expansion truncated at the third term. This truncation limits the accuracy of representation of certain geodetic positions to the order 1 part in 10^8. This is certainly not desirable since we want a representation good to 1 part in 10^9. (3) For the Nigerian latitude zone, meridional distances can be computed either by the e – series truncated at terms containing e^6 or by the n – series truncated at terms containing n^4. (4) Assuming the description in page 10 of *NOP*, the radius of the conformal sphere which is 4,000th part less than the actual radius of the earth (Clarke 1880 ellipsoid) is 6,354,798.792 m. The corresponding mean radius for latitude 9° used in computing the scale corrections in *NOP* is 6,365,757.688 m."

The solution to the problems enumerated above is simple to state but financially difficult to implement. The United States National Geodetic Survey recomputed all of its horizontal geodetic observations on the new GRS1980 ellipsoid for the North American Datum of 1983. The same solution is available to the Republic of Nigeria if they are interested in doing it the proper way.

The relation of the *Minna* 1928 Datum to the WGS72 Datum was investigated in (C.U. Ezeigbo, Definition of Nigerian geodetic datum from Recent Doppler Observations, *Survey Review*, vol. 30, 237, July 1990) where in his conclusion he states the three-parameter shift is: **From** *Minna* 1928 Datum **to** WGS 72 Datum where $\Delta X=-92.9$ m, $\Delta Y=-116.0$ m, $\Delta Z=+116.4$ m, and this is based on an 11-station solution good to ±2.5 m. According to *TR 8350.2*, the three-parameter shift: **From** *Minna* 1928 Datum **to** WGS 84 Datum is where $\Delta X=-92$ m ±3 m, $\Delta Y=-93$ m ±6 m, $\Delta Z=+122$ m ±5 m, and this is based on a 6-station solution.

The African Reference Framework (AFREF) project is an African initiative with international support designed to unify the coordinate reference systems in Africa using Global Navigation Satellite Systems and in particular, the GPS as the primary positioning tool. The project is a work in progress as of 2009.

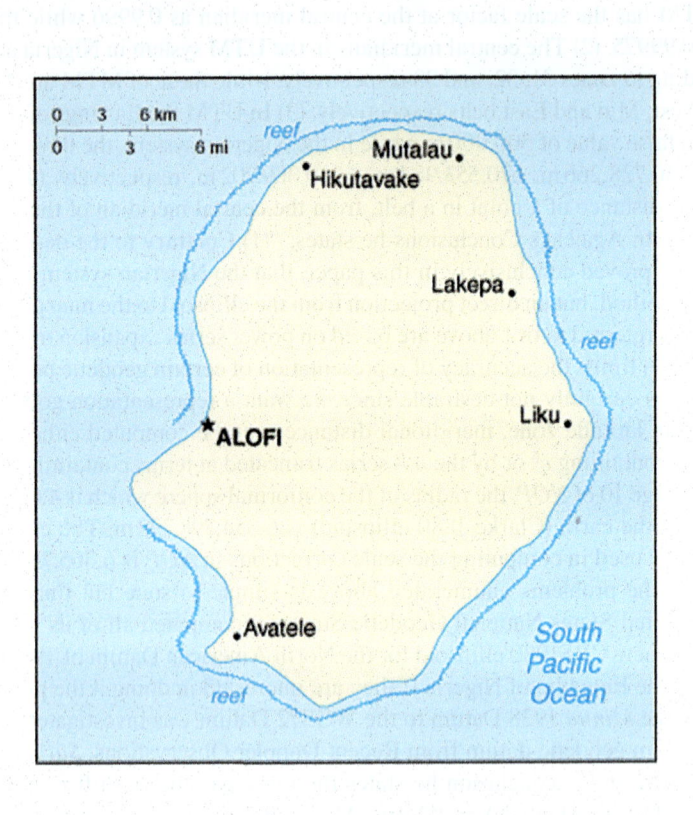

Niue

Niue Island is situated in the South Pacific Ocean approximately 480 km east of *Tonga*, 550 km southeast of *Samoa* and 900 km west of *Rarotonga*. The name NIUE is composed of "NIU," a coconut tree, and "E," behold. The island is isolated and does not form part of any recognized island group. Niue has an area just over 250 km². It is approximately 21 km long and 18 km wide, and the main road that roughly follows the coastline is about 64 km in length. The island is an elevated coral outcrop with a coral reef fringing a precipitous and broken coastline. The general formation takes the shape of two terraces, the lower coastal terrace being about 28 m above sea level; the upper terrace that forms the bulk of the island is about 69 m above sea level. Apart from the rise from the lower to the upper terrace, there are no hills. The island has no running streams or surface water. There are no good harbors, and the best anchorage is at *Alofi*, being an open roadstead. Niue is on the edge of the hurricane belt. Hurricanes and winds of high velocity are sometimes experienced between December and March. The climate is mild and equable; the mean annual temperature over the last 30 years has been 24.7°C, and the annual rainfall for the same period 2,047 mm. Occasionally droughts do occur, but the rainfall is generally well distributed over the entire year.

It is believed that Niue has been inhabited for over a thousand years. Some authorities believe that there were two principal migrations to the island, one from Samoa and one from Tonga and a smaller migration from *Pukapuka*. There were probably also contacts with *Aitutaki* and *Rarotonga*. In 1774, the English Capitan Cook landed on Niue and received a less than friendly reception. Cook named it *Savage Island*. In 1830 the Reverend John Williams attempted to land Christian Island teachers, but the party was repulsed. The next 15 years continued to have little success with attempts to bring Christianity to the island until 1846. Nukai Peniamina who was a Niuean teacher of the London Missionary Society landed and succeeded in converting the inhabitants, and the first European missionary landed in 1851. After requests from the Niueans, a British protectorate was declared over the island in 1900, and the island was formally annexed by New Zealand in 1901. In 1974 Niue gained independence in free association with New Zealand.

According to correspondence in 1998 from the Department of Justice, Lands & Survey Niue, "The island is so flat that we do not have a triangulation system. What we have instead is a series of control traverses along the main roads around and across the island. They are primarily for cadastral survey. The very first survey of Niue was established in 1903 (or there about) by Harzard, a surveyor from New Zealand. He established the initial or origin station Tomb Point and also observed the Latitude and Longitude, however there is no record or information of the Grid System used." (Note: Tomb Point Datum of 1903 origin coordinates are $\Phi_o = -19°\ 01'\ 42''$ South, $\Lambda_o = -169°\ 55'\ 15''$ West of Greenwich.) "What we have in our records is just a single plan showing all the marks including bearings and distances with no Grid lines. In the late 1930s to the early 1940s another visiting surveyor (A. A. Bailey) established the traverses along all the main roads as aforementioned. This surveyor managed to find only a few of the old marks placed by Harzard back in 1903. This survey is based on a plane Grid system with the coordinates in terms of Tomb Point and stellar observations made every 20 to 25 stations. The false origin of Tomb Point was 0.0 chain North and 0.00 chain East. In 1980 with the introduction of metrication, the false origin of 15 000 m N and 5 000 m E were adopted in order to keep all coordinates on the island positive. All cadastral surveys are now based and in terms of the above false origin. In the late 70's to the 80's general roads were re-traversed by modern equipment – EDM, Total Station theodolite, *etc.* … these traverses were also in terms of Tomb Point. The mapping Grid is now officially known as Niue Map Grid. In 1990, Niue took another step further by completing a GPS survey of the whole Island. Please note that the GPS survey stations selected include some of the existing marks established in the 60's, and from Plan 461 known as the Niue North Control. This latter control was established in the mid 80's incorporating some of Bailey's marks as part of the traverse regime. We are now in the process of re-adjusting some of the road traverses / controls in terms of the NGD 91. But we found that some of the controlled closure from GPS-to-GPS system is quite large considering the distances between marks. We need to investigate and analyze all available data to pinpoint the problem. There is the possibility that the GPS survey for some of the stations are unreliable or questionable."

The astronomic origin coordinates observed by A. A. Bailey for the Tomb Point Datum of 1945 are $\Phi_o = 19°\ 03'\ 10''$ South, $\Lambda_o = 169°\ 55'\ 22''$ West of Greenwich, $H_o = 18.86\,m$ above sea level, and the defining astronomic azimuth to Namoui is $\alpha_o = 356°\ 10'\ 42''$. The ellipsoid of reference is the International 1924 where the semi-major axis, $a = 6,378,388\,m$ and the reciprocal of flattening, $1/f = 298$. The plane coordinate system used for this Datum is presumably based on the Transverse Mercator projection. Considering the size of Niue, the choice of projection is rather moot for this old Datum since the coordinates were only used for cadastral purposes and never intended for geodetic applications.

The traverse of 1965 was performed by Lee, and a few points were added to the list of cadastral control. The bulk of new North Control was observed in the middle 1980s by J. G. Kammanankada and was adjusted with an H-P 41 CV pocket computer! The GPS Survey performed in 1991 was by the Department of Survey and Land Information of New Zealand, and 11 existing points were co-located in that survey. The new definitions are as follows: The Niue Geodetic Datum 1991 (NGD91) is based on origin point "SW Pacific Mark" where the geodetic coordinates are: $\varphi_o = -19°\ 04'\ 54.704''$ South, $\lambda_o = 169°\ 55'\ 29.383''$ West of Greenwich, the ellipsoid height is $h_o = 88.00\,m$, and the Geodetic Reference System 1980 is the ellipsoid of reference where $a = 6,378,137\,m$, and $1/f = 298.257222101$. The new Niue Map Grid is based on the Transverse Mercator projection with the origin at Tomb Point with NGD91 coordinates of the Grid Latitude of Origin $\varphi_o = -19°\ 03'\ 13.96''$ South, the Grid Central Meridian $\lambda_o = 169°\ 55'\ 15.15''$ West of Greenwich, the scale factor at origin is equal to unity, the False Easting $= 5\,km$, and the False Northing $= 15\,km$. No parameters were offered for performing transformations between the old cadastral coordinate system and the new Niue Map Grid. Although it is technically correct to consider the NGD91 as the very first geodetic datum, the local surveyors were left in a quandary as how to relate all the cadastral surveys of the 20th century to the new coordinate system.

Dr. Roy Ladner was in need of just one more Engineering Course at the University of New Orleans to complete his requirements for a Ph.D. back in the fall of 1999. I suggested that he re-adjust all the surveys of Niue with ties to the NGD91 as a special topic course in engineering, and he agreed. The software used to accomplish the adjustment was "ADJUST v.4.41" available from the National Geodetic Survey. All original observations were tabulated by Dr. Ladner consisting of 808 stations for the 1940s traverse and an additional 180 stations for the 1980s traverse. The result was two constrained adjustments that provided the Niue Lands & Survey Department with a tool to allow correlation of the NGD91 with the historical records of the cadastral surveys of the 20th century. Dr. Ladner's report consisted of a final written report as well as all input data, ADJUST software, utility software written specifically for Niue, *etc.*, on a CD disk that was transmitted to the Lands & Survey Department. Although Niue is moving towards Tonga at a rate of about 20 cm per year, a three-parameter Datum Shift from the Tomb Point Datum 1945 to NGD91 is $\Delta X = +138\,m$, $\Delta Y = -178\,m$, and $\Delta Z = +92\,m$ (data as of 1991). Hurry and transform before the parameters change!

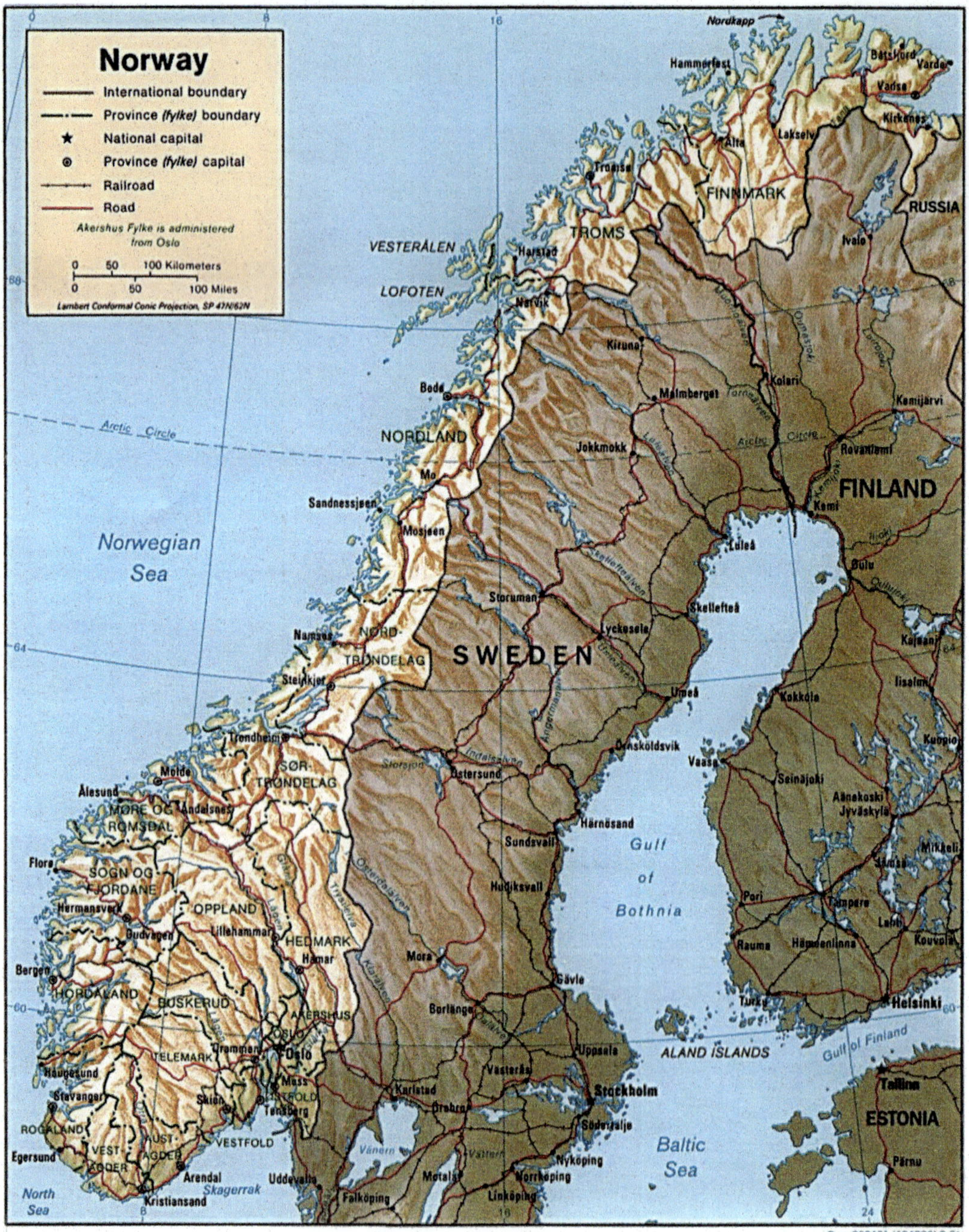

Base 802431 (A04936) 2-96

The Kingdom of Norway

The Kingdom of Norway occupies the western part of the Scandinavian Peninsula. It is slightly larger than twice the size of Georgia and is slightly larger than New Mexico. It is bounded by Finland (709 km), Sweden (1,666 km), and Russia (191 km). The terrain is glaciated; mostly high plateaus and rugged mountains broken by fertile valleys; small, scattered plains; coastline deeply indented by fjords and Arctic tundra in the north. Because of the numerous fjords and small coastal islands, the Kingdom has one of the longest coastlines in the world. Norway claims the islands of Svalbard and Jan Mayen in the Norwegian Sea. Its highest point is *Galdhopiggen* (2,469 m), its lowest point is the Norwegian Sea (0 m), and its mean elevation is (460 m).

Norway was settled in the Middle Stone Age (*circa* 7000 *B.C.*), and by the 9th century *A.D.*, the Norse expeditions began which colonized the islands of Scotland, Ireland, Iceland, and Greenland. Trondheim was the Norwegian capital until 1380. Kristiania, founded in 1050, became the capital in the 14th century and was renamed Oslo in 1924.

The earliest modern map of Norway was the map of Scandinavia drawn by Claudius Clavus in Italy in about 1425. Several other maps were compiled of the entire peninsula, but the first national cartographer of Norway was Melchior Ramus who mapped the southern coast from 1689 to 1693. German foresters were employed in the 18th century to map the land resources of the Kingdom after the Scandinavian wars. The excellent quality of the work and the need for military maps of Norway after the many years of war with Sweden prompted the establishment of the *Norges Graundsers Oppmäling* (Norwegian Border Survey) on December 14, 1773. Attached to the military, the initial attempts of the NGO at mapping by plane table and alidade without basic control were inevitably deemed unreliable. In January of 1779, General Von Huth directed that subsequent mapping be based on astronomically determined points and classical triangulation surveys. Initial longitude determinations were based on fire signals, gunpowder explosions, and pendulum clocks. This was found too inaccurate, and in the winter of 1779–1780, a baseline was measured on *Lake Storsren* using wooden survey bars. By 1784, a triangulation arc was surveyed between *Kongsvinger* and *Verda*l. Additional triangulation work continued, and the survey was adjusted in 1810. The geographical position of *Bergen* was compared to another determination from a triangulation arc from *Lindesnes*. The difference in longitude was 9″ and that error was considered satisfactory at the time. From 1791 to 1803 a series of hydrographic charts were published from the surveys of LT. F. C. Grove of the Royal Danish Navy. Printed in Copenhagen from copper plates, the "Grove Charts" were used for navigation for about 100 years.

A re-organization of surveying and mapping within the government in 1805 combined military and economic objectives in the same department. The *Norske Topographiske Oppmäling* (Norwegian Topographic Survey, or NTO) passed among several ministries including Defense, Interior, Finance, Customs and Trade, and finally back to Interior. In 1823, all operating funds were suspended. The first 50 years of survey and mapping work had not been reproduced in large quantities because maps were generally considered a military secret. The first map printed was a County map of *Smaalenenes* that was published in 1826 at a scale of 1:200,000. In fact, two of the NTO surveyors obtained permission to have it engraved and printed in Paris at their own expense. In 1833, the organization's name was changed to *Norges Geografiske Opmäling* (Geographical Survey of Norway, or NGO). By 1854, NGO obtained the expertise and equipment to publish their own maps. The first series of published maps were cast on the Cassini-Soldner projection and were referenced to the Svandberg 1805 ellipsoid where $a = 6,376,797.0$ m and $1/f = 304.2506$. Presumably, this was based on the Oslo Observatory Datum (of 1810?) where $\Phi_o = 59° 54′ 44.00″$ North, $(\Lambda_o) = 10° 43′ 22.5″$ East of Greenwich. However, the defining azimuth to *Husbergoen* was not observed until 1869. The sheets had no printed grid and were cast on the graticule such that they measured 20′ of latitude and 1° of longitude. The sheets started at 58° North and were evenly spaced longitudinally from Oslo Observatory. This lasted until 1844 when the ellipsoid was changed to the Bessel 1841, and then the series was continued through 1890. This strange series of map sheets was actually a common design for the time. (The Grove Charts were based on the Cassini-Soldner projection also.) In 1891, the Cassini-Soldner projection was replaced by the polyhedric projection. As I have pointed out, the polyhedric is mathematically equivalent to the system commonly used for computational photogrammetry, which is the local space rectangular. The U.S. equivalent was merely based on a different aphylactic projection, the polyconic.

Prof. Hamsteen became the new director of the survey in 1832, and a new triangulation commenced in 1834 from *Kristiania* (Oslo) to *Trondheim* covering most of southern Norway. The Russian-Scandinavian geodetic survey extended from Swedish Lapland to *Fuglenes* at *Hammerfest* in Norway from 1845 to 1850. The *Fuglenes* Datum of 1850 origin is where $\Phi_o = 70° 40′ 11.23″$ North, $(\Lambda_o) = 23° 40′ 12.8″$ East of Greenwich. The ellipsoid of reference is the Bessel 1841, but some of the original manuscript computations must surely be on the Walbeck 1819, which was the standard for the Russians at that time. The defining azimuth was to station *Jedki*. By 1874, the NGO was transferred from the General Staff to the Army Department. The Triangulation Division started a Second-Order triangulation in 1880 and adopted a network adjustment philosophy in 1903. Computations in geodetic coordinates were introduced in 1891. The Gauss-Krüger Transverse Mercator system for Norway was initiated in 1916 and is based on the Oslo Observatory Datum of 1844. The ellipsoid of reference is the Bessel 1841 where the semi-major axis $(a) = 6,377,397.155$ m and the reciprocal of flattening $(1/f) = 299.1528128$.

Terrestrial photogrammetric experiments started in 1907, and phototheodolites were purchased in 1910. In 1920, a Zeiss Stereoautograph was purchased for the compilation of terrestrial photography, although the plane table and alidade was still used to supplement detail for holidays in the photo coverage. All topographic mapping was performed with aerial photographs after 1933. By 1936, a Zeiss C-1 Universal Stereoplanigraph was in NGO production.

The military Grids used for Norway through World War II were the Northern European Lambert Conformal Conic Zones I–III. These were complicated systems that had Grid boundaries defined by other Grid systems, by the graticule, and by numerous ellipsoidal loxodromes. Although easy to compute nowadays, these Grids were part of the nightmare for U.S. Army Topographic Engineers in the 1940s.

After World War II, the Oslo Observatory Datum of 1844 was transformed to the European Datum of 1950 (ED50) through the Northern Block adjustment of 1949–1951. The U.S. Coast & Geodetic Survey performed the computations, and the modern Norwegian triangulation south of approximately 62° latitude was connected to the Central European Net via the principal triangulation arcs of Sweden and Denmark. Flare triangulation over the *Skagerrak* was used between Denmark and Norway. The Norwegian triangulation north of approximately 62° latitude was subsequently included in the European Datum as a supplementary adjustment of the Northern Block from September 1951 through June 1952. The only exception to the standard Universal Transverse Mercator (UTM) Grid zones in the entire world is for the southern half of Norway, where the zone exceeds ±3° from the Central Meridian! This exception accommodates the southern half of the Kingdom in one "V 32" zone. During this same period, Norway adjusted their new datum NGO 1948, and they retained the Bessel 1841 ellipsoid with the point of origin at the Oslo Observatory.

Svalbard is a group of nine main islands north of Norway that officially became a territory of the Kingdom on August 14, 1925. The most famous island of the group is *Spitzbergen*, and the main economic activity is coal mining. The coordinate systems of *Svalbard* include the Thumb Point Datum of 1948 origin where $\Phi_o = 79° 03′ 58.97″$ North, $(\Lambda_o) = 2° 48′ 15.36″$ East of Greenwich and is referenced to the Bessel 1841 ellipsoid. The Hedgehog Datum origin is where $\Phi_o = 76° 57′ 53.210″$ North, $(\Lambda_o) = 17° 19′ 52.500″$ East of Greenwich and is referenced to the Clarke 1880 ellipsoid. The New *Spitzbergen* Datum of 1948 consisted mainly of re-computing Hedgehog Datum coordinates on the Bessel 1841 ellipsoid. Other observations that were recomputed in that same epoch included the Isaksson triangulation of 1909–1910, the Polish 1934 Expedition triangulation, the Cadastral system of the *Svalbard* Commissioner, several points of the Italian Hydrographic Expedition, and the *Dahl* Grid system. Note that this strategic island group was the subject of a treaty 5 years prior to becoming a Norwegian territory, and the signatories totaled 41 sovereign nations. Considering the number of countries participating, the number of local datums for so small an area is not surprising.

Jan Mayen is an island northeast of Norway that was annexed on May 8, 1929. The island is slightly larger in area than Washington, DC, there are no permanent inhabitants, and the only occupation is for manning a weather station and a Loran C base station. In 1949, the *Norsk Polarinstitutt* established the only classical coordinate system for the island as the *Nordlaguna* Datum of 1949 origin where $\Phi_o = 71°\ 00'\ 46.79''$ North, $(\Lambda_o) = 08°\ 27'\ 51.3''$ West of Greenwich. The local *Nordlaguna* Grid is based on the Datum origin with a Gauss-Krüger Transverse Mercator where the scale factor at origin, $m_o = 1.0$, the central meridian (C. M.), $\lambda_o = 08°\ 30'$ West of Greenwich, the False Easting (Y axis) at C. M. $= 50$ km, and the False Northing (X-axis) $= 7,800$ km. Note that the transposition of the axes labels is common in Europe.

In January of 1986, all mapping activities in the Kingdom were consolidated under the new banner of *Staatens Kartverk* (Norwegian Mapping Authority) in *Hønefoss*. By 1990, the concern for establishing a coordinate system that was consistent for European offshore hydrocarbon exploitation resulted in the "North Sea Formulae." The published parameters to shift from ED50 to ED87 are the standard 15-coefficient multiple regression equations published by the U.S. National Imagery and Mapping Agency (NGA). The published parameters for transforming from the "new" European Datum of 1987 to the World Geodetic System of 1984 Datum have a specific subset of relations referred to as the "WGS84*SEA." The parameters are: $\Delta X = -82.981$ m, $\Delta Y = -99.719$ m, $\Delta Z = -110.709$ m, $R_x = -0.5076$, $R_y = +0.1503$, $R_z = +0.3898$, scale $= -0.3143$ where the rotations are in micro radians and the scale factor is in parts per million (10^{-6}). For a published computational example; from ED87 coordinates: $\varphi = 52°$, $\lambda = 02°$; WGS84*SEA coordinates are: $\varphi = 51°\ 59'\ 57.0927''$, and $\lambda = 01°\ 59'\ 55.1400$. The number of each country's terrestrial points used in the solution was based on a proportion of comparative North Sea coastline lengths. This transformation system was a consensus of agreement among Norway (26), United Kingdom (43), Denmark (12), Germany (9), and the Netherlands (13).

If the preceding paragraph sounds complicated, fasten your seatbelts! A transformation algorithm was developed by the Norwegians to go from either the NGO1948 Datum or the ED50 to the WGS84 Datum. This series of transformations were developed as "block" transformations based on specific areas. All of the shifts are on the conformal plane of the UTM Grid, which was the rationale used by the USC&GS (on contract to AMS) to compute and adjust the Northern Block of the ED50. With 12 different blocks defined for the Scandinavian Peninsula, the implementation is non-trivial. The Norwegian government is commercially marketing a datum shift solution similar to the Canadians and the British. A software package is available (for a price) that will implement such coordinate transformations. The package is called "WSKTRANS." Another government commercial offering in the positioning field is a differential GPS correction service called "SATREF." The Norwegians now tout their Stamnett system, and a veritable alphabet soup of acronyms now appear in annual national reports such as EUREF-NOR94, EUREF-NOR95, and EUREF-NOR96 campaigns. When the geodesists start comparing positions at the sub-centimeter level, the accuracy for aerial photogrammetric applications is definitely more than adequate!

Whether you are a professional developer or a "regular" user, this provides you with information on how to download and use the Norwegian Mapping Authority's open and free maps and geospatial data.

The following data sets from the Norwegian Mapping Authority are available as free downloads:

- Topographical land data from 1:50 000 to 1:5,000,000 (N50, N250, N500, N1000, N2000, and N5000 vector and raster images)
- Administrative/property boundaries
- Road networks including addresses

- Place name data
- Historical maps
- National elevation models

To download the data sets you must register (except for historical maps). Data sets must be used according to the Norwegian Mapping Authority's terms of use.

You can download data sets from our main download page here or from our historical maps site (texts in Norwegian only). We advise you to read this entire article, especially if you are unsure about further procedures and the knowledge required to make full use of our data sets.

If you have any questions or comments, please find contact info by using the "veiviseren" (Wizard).

If you are interested in purchasing other geospatial data, please contact the Norwegian Hydrographic Service for marine geospatial data or one of our distributors for topographical land or property data.

For Professionals

The map data sets are released in a form that is as close to the original as possible. This provides professional developers with as much free rein as possible to develop their own products and services based on the Norwegian Mapping Authority's data.

Topographic vector data, boundaries, road data, and place names are provided in SOSI and GeoJSON formats. The elevation models are provided as DEMs. N50-N5000 topographical land data can be found in ESRI File Geodatabase format, and in addition the N50 data is available in raster format as MrSID.

At the moment, professional developers and GIS workers will benefit the most from many of our open data sets. It is our goal that the map data can be used to develop new solutions, which can be useful to many users. (http://www.kartverket.no/en/data/Open-and-Free-geospatial-data-from-Norway/, June 2017.)

REFERENCES: N

Algeria–Niger Boundary, International Boundary Study No. 99 – May 1, 1970, Bureau of Intelligence and Research, Department of State, USA.
C. U. Ezeigbo, Definition of Nigerian geodetic datum from recent Doppler observations, *Survey Review*, vol. 30, 237, July 1990.
Captain J. Calder Wood, *The survey framework of Nigeria, Empire Survey Review*, vol. III, no. 21, July 1936.
Captain, F. S., Spence, R. E., Himalayan survey, *Geographical Journal*, vol. 153, no. 2, 223–230, July 1987.
Chad–Niger Boundary, International Boundary Study No. 73 – August 1, 1966, Bureau of Intelligence and Research, Department of State, USA.
Charles L. Merry. J. Rens, *Datum Transformation Parameters in South Africa, Survey Review*, vol. 30, 236, April 1990.
Collaborative Mapping in Nicaragua, 1957.
David Johanson, U.S. Naval Oceanographic Office, Personal Communication, 2006.
Dr. Roy Ladner, Niue ties to the NGD91, University of New Orleans, 1999.
EUREF Permanent GNSS Network.
Final Report of the 12th Parallel Survey in the Republic of Niger, IGN, 1969.
Foreign Maps, TM 5-248, 1963.
Geographic Notes, Department of State, Office of the Geographer September 8, 1986.
German South West Africa – British Bechuanaland Boundary Survey of 1898–1903, EPSG Database, 2006.
Heuvelink Hendrik, *Triangulation du Royaume des Pays-Bas*, Nabu Press: Charleston, SC, vol. 1, 1903, vol. 2, 1921.
http://euref.eu/symposia/2019Tallinn/05-16-Netherlands.pdf.
http://gnss1.tudelft.nl/pub/vdmarel/reader/CTB3310_RefSystems_1-2a_print.pdf.

References: N

http://www.epncb.eu/_networkdata/siteinfo4onestation.php?station=DELF00NLD.
http://www.kartverket.no/en/data/Open-and-Free-geospatial-data-from-Norway/, June, 2017.
Informe Detallado de la Comisión Técnica de Demarcación de la Frontera entre Guatemala y Honduras, 2009.
Institut Géographique National (IGN), SGC 1312, December 1945.
Instituto Nicaragüense de Estudios Territoriales (INETER), 2009.
International Boundaries Research Unit, Durham University, 22 July 2010.
L. P. Lee, First-order Geodetic Triangulation of New Zealand 1909–1949 and 1973–1974.
Lambert Conical Orthomorphic Projection Tables, Niger Zone, RESTRICTED, Office of the Chief of Engineers, Washington, DC, 1943.
Lonely Planet, July 2006.
M. L. Larsgaard, *Topographic Mapping of the Americas*, McGraw-Hill: Australia, and New Zealand, 1934.
Mal Jones, Personal Communication, April 2005.
Mali–Niger Boundary, International Boundary Study No. 150 – January 13, 1975, Bureau of Intelligence and Research, Department of State, USA.
Memoria de la Dirección General de Cartografía, Guatemala, September 1957.
National Report of the Netherlands 2019- EUREF.
Nepal Country Study, *Library of Congress*, 1991.
Nicaragua Library of Congress Country Studies, 2009.
Niger – Country Studies, U.S. Department of State, 2011.
Niger – *Lonely Planet*, 2011.
Niger–Nigeria Boundary, International Boundary Study No. 93 – December 15, 1969, Bureau of Intelligence and Research, Department of State, USA.
Nigeria Library of Congress Country Study – 2008.
Niraj Manandhar, Study of Geodetic datum of Nepal, China and Pakisthan (*sic – Ed.*) and its transformation to World Geodetic System, *Nepalese Journal on Geoinformatics*, 10, 2008.
Professor Michael Barnes, MSc MRICS MCInstCES MRIN, Lecturer, Geomatics, Geospatial & Mapping Sciences, School of Geographical & Earth Sciences, University of Glasgow, Glasgow, Personal Communication, 2022.
Reference Systems for Surveying and Mapping.
S. Izundu Agajelu, On conformal representation of geodetic positions in Nigeria, *Survey Review*, vol. 29, 223, January 1987.
The Geodetic Survey of Nepal 1981–1984, Directorate of Military Survey, UK, November 1985.
World Factbook, 2009.
World Factbook, 2013.
World Factbook, 2022.

O

Sultanate of Oman 607

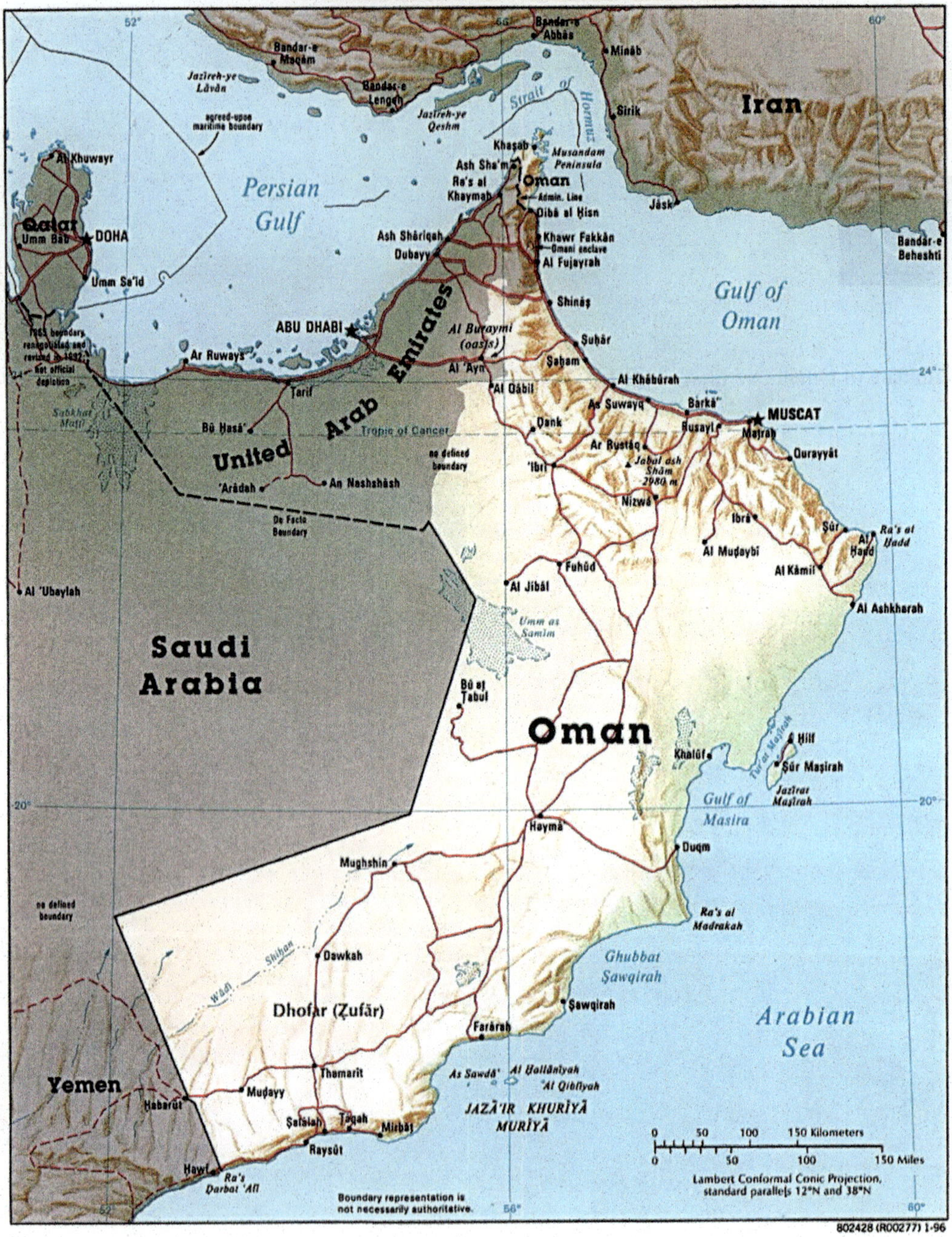

802428 (R00277) 1-96

Sultanate of Oman

The Sultanate is bordered on the west by Saudi Arabia (676 km), United Arab Emirates (410 km), Yemen (288 km), and on the north, east, and south by the Gulf of Oman and the Arabian Sea (2,092 km). The terrain is comprised of a vast central desert plain with rugged mountains in the north and south. The lowest point is the Arabian Sea (0 m), and the highest point is *Jabal Shams* (2,980 m).

"The Persian Gulf lies between two of the major breadbaskets of the ancient world, the Tigris-Euphrates area (Mesopotamia, meaning 'between the rivers') in present-day Iraq and the Nile Valley in Egypt. Mesopotamia, a part of the area known as the Fertile Crescent, was important not only for food production but also for connecting East to West. Mesopotamia became the linchpin of ancient international trade. The fertile soil between the Tigris and the Euphrates produced a large surplus of food; however, it did not support forests to produce the timber necessary to build permanent structures. The region also lacked the mineral resources to make metals. Accordingly, the early inhabitants of Mesopotamia were forced to go abroad and trade their food for other raw materials. They found copper at Magan, an ancient city that lay somewhere in the contemporary state of Oman and, via Magan, traded with people in the Indus Valley for lumber and other finished goods. The people of Magan were both middlemen and suppliers because the city was a source of copper as well as a transit point for Indian trade. Over time, other cities developed that were exclusively entrepôts, or commercial way stations. One of the best known of these cities was Dilmun. Dilmun probably lay on what is now the island state of Bahrain. Excavations on the island reveal rich burial mounds from the Dilmun period (*ca.* 4000 to 2000 B.C.). The area that constitutes the present-day Persian Gulf states was on the immediate periphery of the rise of Islam. In A.D. 610, Muhammad – a merchant of the Hashemite branch of the ruling Quraysh tribe in the Arabian town of Mecca – began to preach the first of a series of revelations that Muslims believe was granted him by God … Shia Muslims hold the fundamental beliefs of other Muslims. In addition to these tenets, however, Shia believe in the *imamate*, which is the distinctive institution of Shia Islam. The term *Kharijites* also became a designation for Muslims who refused to compromise with those who differed from them. Their actions caused the Sunni community to consider them assassins. In the eighth century, some *Kharijites* began to moderate their position. Leaders arose who suppressed the fanatical political element in *Kharijite* belief and discouraged their followers from taking up arms against Islam's official leader. *Kharijite* leaders emphasized instead the special benefits that Kharijites might receive from living in a small community that held high standards for personal conduct and spiritual values. One of these religious leaders, or imams, was *Abd Allah ibn Ibad*, whose followers founded communities in parts of Africa and southern Arabia. Some of *Abd Allah's* followers, known as *Ibadis*, became the leaders of Oman" (*Library of Congress Country Studies*).

"In 1970, Qaboos bin Said Al-Said overthrew his father, and has since ruled as sultan. Sultan Qaboos has no children and has not designated a successor publicly; the Basic Law of 1996 outlines Oman's succession procedure. Sultan Qaboos' extensive modernization program opened the country to the outside world, and the sultan has prioritized strategic ties with the UK and US. Oman's moderate, independent foreign policy has sought to maintain good relations with its neighbors and to avoid external entanglements.

"Inspired by the popular uprisings that swept the Middle East and North Africa beginning in January 2011, some Omanis staged demonstrations, calling for more jobs and economic benefits and an end to corruption. In response to those protester demands, Qaboos in 2011 pledged to implement economic and political reforms, such as granting Oman's bicameral legislative body more power and authorizing direct elections for its lower house, which took place in November 2011.

Additionally, the Sultan increased unemployment benefits, and, in August 2012, issued a royal directive mandating the speedy implementation of a national job creation plan for thousands of public and private sector Omani jobs. As part of the government's efforts to decentralize authority and allow greater citizen participation in local governance, Oman successfully conducted its first municipal council elections in December 2012. Announced by the sultan in 2011, the municipal councils have the power to advise the Royal Court on the needs of local districts across Oman's 11 governorates. Sultan Qaboos, Oman's longest reigning monarch, died on 11 January 2020. His cousin, Haytham bin Tariq bin Taimur Al-Said, former Minister of Heritage and Culture, was sworn in as Oman's new sultan the same day" (*World Factbook*, 2022).

The only known classical geodetic datum in Oman is the *Fahud* Geodetic Datum of 1954, which was established by the Iraq Petroleum Corporation (IPC) for the Petroleum Development Oman (PDO) operation. Beginning in 1964, Hunting Surveys, Ltd., established an extensive control network based on the *Fahud* Datum of 1954, and this enterprise took over 8 years to develop. Although the work was performed to a 3rd-Order geodetic accuracy specification, the operational demands of a production oil field precluded the rigorous adjustment of the network. As a result, the classical network retained a number of relatively high distortions of up to 10 meters in magnitude for horizontal coordinate positions and up to 5 meters in elevation! A major Doppler satellite campaign was observed between September 1983 and August 1984. During this period, 106 existing stations in the *Fahud* Datum of 1954 network were occupied and observed. Subsequent to the initial period of survey work in Oman, all survey stations to the North of latitude 22° were given the prefix "NO," and those to the south the prefix "SO." This numbering system was later extended into the province of *Dhofar*, where the prefix "D" is used instead. The *Fahud* Datum of 1954 uses the Clarke 1880 as its ellipsoid of reference, where $a = 6,378,300.782$ m and $1/f = 293.4663077$. The origin point of the *Fahud* Datum is near station "NO68-024," where the Astronomical Latitude, $\Phi_o = 22°\ 17'\ 31.182''$ North, and the Astronomical Longitude, $\Lambda_o = 56°\ 29'\ 18.820''$ East of Greenwich. The original astronomical origin point at F.N. Fort (station "F1" with defining azimuth to station "F2" of the baseline) has been destroyed. The baseline was located on a gravel plain on the southern side of an airstrip constructed for the survey. A total of 32 stations were triangulated in *Fahud*, including F15 and F5, and 15 stations were triangulated in *Natih*, including N1. Original coordinates of the triangulation were published on the WWII Trucial Coast/Qatar Grid on the Cassini-Soldner projection. The Central Meridian, $\lambda_o = 50°\ 45'\ 41''$ E, the False Northing Latitude of Origin, $(\varphi_{FN}) = 25°\ 22'\ 56.5''$ N, and *both* the False Eastings and False Northings are 100 km. Of course, the Scale Factor at Origin by definition is equal to unity. Also used in the region was the Mecca-Muscat Zone on the Lambert Conical Orthomorphic projection where the Central Meridian, $\lambda_o = 45°$ East of Greenwich, the Latitude of Origin, $\varphi_o = 23°$ N, a Scale Factor at Origin, $m_o = 0.99907$, and a False Easting, F.E. = 1,000 km at the Central Meridian. The Oman Transverse Mercator projection has a Central Meridian, $\lambda_o = 54°$ East of Greenwich, a Scale Factor at Origin, $m_o = 0.9997$, and a False Easting, F.E. = 500 km at the Central Meridian. Notice the curious similarity in the digits used for the Scale Factor at Origin values for the Mecca-Muscat Lambert zone and the Oman Transverse Mercator belt – I'll bet that's not a coincidence.

"The astronomical observations were approximately located on a 25 km grid in order to provide control for relative gravity surveys. Six to ten pairs of circum-meridian observations and 8–13 pairs of time stars were generally observed for the astrofixes. By 1957, triangulation networks had been established in the *Buraimi* area, the *Duqm/Huqf* area, and in the vicinity of *Fahud* covering over 2 degrees of latitude between *Dhank* and *Nihayda*. After the completion of the initial survey, the IPC Photo survey section in Tripoli, Lebanon used February 1953 aerial photographs to produce a *Fahud* Topographical Map at a scale of 1:25,000 on the UTM grid. A subsequent geological map was cast on the Trucial Coast grid in July 1956.

"During 1964, a survey was conducted by the British Army Directorate of Military Survey (DMS), which involved a cantilevered Telurometer traverse from Dubai in the United Arab Emirates through *Buraimi* to *Fahud* (Series 13B). The traverse, to extend available control, was carried out by 13 Field Survey Squadron, during April–May 1964. In 1964, Hunting Surveys Ltd. Was contracted by PDO to install new control by Telurometer traversing for mapping and oil exploration purposes. The specification called for station intervals of about 20 km, with azimuth stations: and at each station, for 3 sets of simultaneous reciprocal vertical angles. Distances were to be observed in both forward and reverse directions with agreement better than 1/50,000. Polaris azimuths were to be observed at 100 km intervals. (This was amended to 60 km in later years). Closure between azimuths was specified to be less than 2″ per station" (*M.H.B. Jensen, December 1984*).

The original *Fahud* Height Datum was based on reciprocal trigonometric heighting from a single MSL determination at *Mina al Fahal* (formerly known as *Saih al Maleh*) in Muscat. Station NO68-024 had an orthometric height of 323.50 m above mean sea level at *Mina al Fahal*. According to the *Ordnance Survey of Great Britain:* "PDO Height Datum of 1993—Misclosure between Muscat and *Salalah* less than 5 meters with differences from of up to 5 meters from old *Fahud* Datum. The PHD93 adjustment was initially known as the Spine (*sic – Ed*). supersedes *Fahud* Vertical Datum from 1993."

The transformation **from** WGS84 Datum **to** PDO Survey Datum 1993 listed by Wilfred of Medco Energy in 2005 is: $\Delta X = +180.624$ m, $\Delta Y = +225.516$ m, $\Delta Z = -173.919$ m, $\delta S = -16.71006$ ppm, $R_x = -0.80970''$, $R_y = -1.89755''$, $R_z = +8.33604''$. No information was offered regarding transformation accuracy nor the sense of the coordinate system rotations, but fortunately a test point was provided, where PSD93 (Clarke 1880) $\varphi = 20°\ 20'\ 56.967''$ North, $\lambda = 56°\ 31'\ 15.467''$ East and WGS84 $\varphi = 20°\ 20'\ 58.495''$ North, $\lambda = 56°\ 31'\ 25.456''$ East.

The Honorable Gary A. Grappo used to be the U.S. Ambassador to the Sultanate of Oman. Ambassador Grappo holds a master's degree in geodesy from Purdue University.

REFERENCES: O

M.H.B. Jensen, Personal Communication, December 1984.
Oman Library of Congress Country Studies, 2007.
World Factbook, 2022.

P

Islamic Republic of Pakistan 613
Republic of Panamá 617
Independent State of Papua New Guinea 621
Republic of Paraguay 625
Republic of Perú 629
The Republic of the Philippines 633
The Republic of Poland 637
Kingdom of Portugal 641

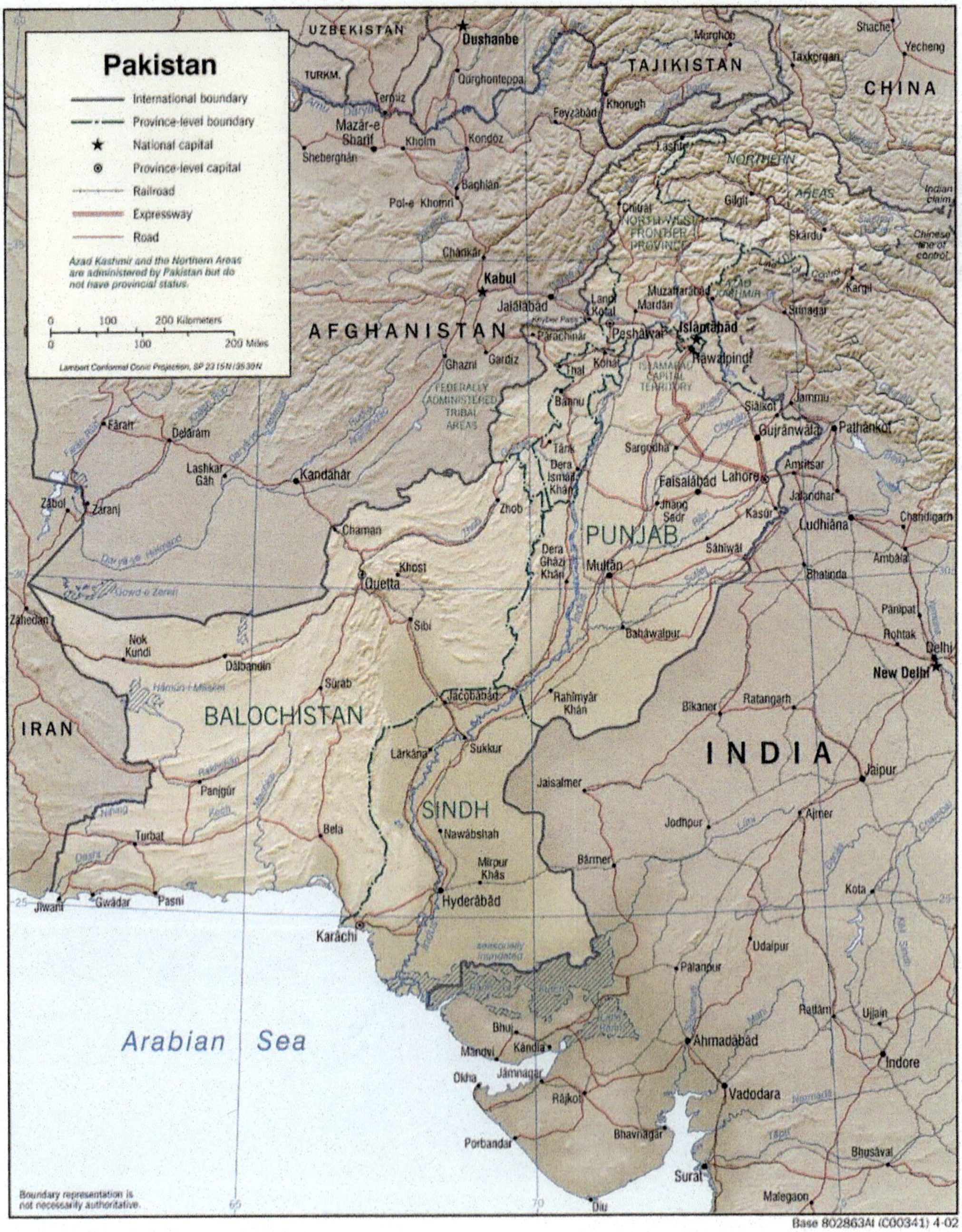

Base 802863AI (C00341) 4-02

Islamic Republic of Pakistan

Pakistan occupies a position of great geostrategic importance, bordered by Iran (959 km) on the west, Afghanistan (2,670 km) on the northwest, China (438 km) on the northeast, India (3,190 km) on the east, and the Arabian Sea (1,046 km) on the south. The total land area is estimated at 803,940 km^2. The terrain is divided into three major geographic areas: the northern highlands, the *Indus River plain* in the center and east, and the *Balochistan Plateau* in the south and west. The highest point is *K2* (*Mt. Godwin-Austen*) (8,611 m), the lowest point is the Arabian Sea (0 m), and the mean elevation is (900 m) (*World Factbook*, 2022).

Pakistan has a history that can be dated back to the *Indus Valley* civilization (ca. 2500–1600 B.C.), the principal sites of which lay in present-day *Sindh* and *Punjab* provinces. Pakistan was later the entryway for the migrating pastoral tribes known as Indo-Aryans, or simply Aryans, who brought with them and developed the rudiments of the religio-philosophical system of what later evolved into Hinduism. They also brought an early version of Sanskrit, the base of *Urdu, Punjabi*, and *Sindhi* languages that are spoken in much of Pakistan today. Hindu rulers were eventually displaced by Muslim invaders, who, in the 10th, 11th, and 12th centuries, entered northwestern India through the same passes in the mountains used earlier by the Indo-Aryans. The culmination of Muslim rule in the *Mughal* Empire (1526–1858, with effective rule between 1560 and 1707) encompassed much of the area that is today Pakistan. *Sikhism*, another religious movement that arose partially on the soil of present-day Pakistan, was briefly dominant in Punjab and in the northwest in the early 19th century. All of these regimes subsequently fell to the expanding power of the British, whose empire lasted from the 18th century to the mid-20th century, until they too left the scene, yielding power to the successor states of India and Pakistan. The Muslim-majority areas in northwestern and eastern India were separated and became Pakistan, divided into the West Wing and East Wing, respectively. The placement of two widely separated regions within a single state did not last, and in 1971 the East Wing broke away and achieved independence as Bangladesh. The pride that Pakistan displayed after independence in its long and multicultural history has disappeared in many of its officially sponsored textbooks and other material used for teaching history (although the Indus Valley sites remain high on the list of the directors of tourism). As noted anthropologist Akbar S. Ahmed has written in *History Today*, "In Pakistan the Hindu past simply does not exist. History only begins in the seventh century after the advent of Islam and the Muslim invasion of *Sindh*" (*Library of Congress Country Study*, 2009).

The boundary with Iran, some 800 km in length, was first delimited by a British commission in 1893, separating Iran from what was then British Indian Balochistan. In 1957 Pakistan signed a frontier agreement with Iran, and since then the border between the two countries has not been a subject of serious dispute.

"Pakistan's boundary with Afghanistan is about 2,250 kilometers long. In the north, it runs along the ridges of the *Hindu Kush* (meaning Hindu Killer) mountains and the Pamirs, where a narrow strip of Afghan territory called the *Wakhan Corridor* extends between Pakistan and Tajikistan. The *Hindu Kush* was traditionally regarded as the last northwestern outpost where Hindus could venture in safety. The boundary line with Afghanistan was drawn in 1893 by Sir Mortimer Durand, then foreign secretary in British India, and was acceded to by the Amir of Afghanistan that same year. This boundary, called the Durand Line, was not in doubt when Pakistan became independent in 1947, although its legitimacy was in later years disputed periodically by the Afghan government as well as by *Pakhtun* tribes straddling the Pakistan-Afghanistan border. On the one hand, Afghanistan claimed that the Durand Line had been imposed by a stronger power upon a weaker one, and it favored the establishment of still another state to be called *Pashtunistan* or *Pakhtunistan*.

On the other hand, Pakistan, as the legatee of the British in the region, insisted on the legality and permanence of the boundary. The Durand Line remained in effect in 1994.

"In the northeastern tip of the country, Pakistan controls about 84,159 square kilometers of the former princely state of *Jammu and Kashmir*. This area, consisting of *Azad Kashmir* (11,639 square kilometers) and most of the Northern Areas (72,520 square kilometers), which includes *Gilgit* and *Baltistan*, is the most visually stunning of Pakistan. The Northern Areas has five of the world's seventeen highest mountains. It also has such extensive glaciers that it has sometimes been called the 'third pole.' The boundary line has been a matter of pivotal dispute between Pakistan and India since 1947, and the *Siachen Glacier* in northern *Kashmir* has been an important arena for fighting between the two sides since 1984, although far more soldiers have died of exposure to the cold than from any skirmishes in the conflict.

"From the eastern end of the Afghanistan-Pakistan border, a boundary of about 520 kilometers runs generally southeast between China and Pakistan, ending near the *Karakoram Pass*. This line was determined from 1961 to 1965 in a series of agreements between China and Pakistan. By mutual agreement, a new boundary treaty is to be negotiated between China and Pakistan when the dispute over *Kashmir* is finally resolved between India and Pakistan.

"The Pakistan-India cease-fire line runs from the *Karakoram Pass* west-southwest to a point about 130 kilometers northeast of *Lahore*. This line, about 770 kilometers long, was arranged with United Nations (UN) assistance at the end of the Indo-Pakistani War of 1947–48. The cease-fire line came into effect on January 1, 1949, after eighteen months of fighting and was last adjusted and agreed upon by the two countries in the Simla Agreement of July 1972. Since then, it has been generally known as the Line of Control" (*Note Lahore is the name of a grid system.*)

"The Pakistan-India boundary continues irregularly southward for about 1,280 kilometers, following the line of the 1947 Radcliffe Award, named for Sir Cyril Radcliffe, the head of the British boundary commission on the partition of Punjab and Bengal in 1947. Although this boundary with India is not formally disputed, passions still run high on both sides of the border. Many Indians had expected the original boundary line to run farther to the west, thereby ceding Lahore to India; Pakistanis had expected the line to run much farther east, possibly granting them control of Delhi, the imperial capital of the Mughal Empire.

"The southern borders are far less contentious than those in the north. The *Thar Desert* in the province of *Sindh* is separated in the south from the salt flats of the *Rann of Kutch* by a boundary that was first delineated in 1923–24. After partition, Pakistan contested the southern boundary of *Sindh*, and a succession of border incidents resulted. They were less dangerous and less widespread, however, than the conflict that erupted in *Kashmir* in the Indo-Pakistani War of August 1965. These southern hostilities were ended by British mediation, and both sides accepted the award of the Indo-Pakistan Western Boundary Case Tribunal designated by the UN secretary general. The tribunal made its award on February 19, 1968, delimiting a line of 403 kilometers that was later demarcated by joint survey teams. Of its original claim of some 9,100 square kilometers, Pakistan was awarded only about 780 square kilometers. Beyond the western terminus of the tribunal's award, the final stretch of Pakistan's border with India is about 80 kilometers long, running west and southwest to an inlet of the Arabian Sea" (*Library of Congress Country Study*, 2009).

"In response to Indian nuclear weapons testing, Pakistan conducted its own tests in mid-1998. India-Pakistan relations improved in the mid-2000s but have been rocky since the November 2008 *Mumbai* attacks and have been further strained by attacks in India by militants believed to be based in Pakistan. Imran Khan took office as prime minister in 2018 after the Pakistan *Tehreek-e-Insaaf* (PTI) party won a plurality of seats in the July 2018 general elections. Pakistan has been engaged in a decades-long armed conflict with militant groups that target government institutions and civilians, including the *Tehreek-e-Taliban* Pakistan (TTP) and other militant networks" (*World Factbook*, 2022).

Islamic Republic of Pakistan

Geodetic control and mapping of India, Pakistan, and Bangladesh was introduced by the British in the 18th and 19th centuries during the Great Trigonometrical Survey of India (*Mapping an Empire, The Geographical Construction of British India 1765–1843*, M.H. Edney 1997). The arduous process of mapping the subcontinent of India was finally put on a firm footing by Sir George Everest, and Pakistan still relies on the Everest 1830 ellipsoid (modified) for its reference surface as well as the Indian Datum of 1916 with origin at Kalianpur Hill in India, where $\Phi_o=24°\ 07'\ 11.26''$ N, $\Lambda_o=77°\ 39'\ 17.57''$ East of Greenwich, and the ellipsoid of reference is the Everest 1830, where $a=6,974,310.600$ Indian Yards and $1/f=300.8017$. The original topographic mapping of Pakistan by the British Survey of India (SOI) was based on the Lambert Conical Orthomorphic projection, India Zone I, where the Central Meridian, $\lambda_o=68°$ E, the Latitude of Origin $\varphi_o=32°\ 30'$ N, the Scale Factor at the Latitude of Origin $m_o=823/824=0.998786408$, and False Easting $=3,000,000$ Indian Yards, and the False Northing $=1,000,000$ Indian Yards. The scale factor at origin, m_o, should be of particular interest. The British way of doing things (back then) has been overlooked by most authors in recent years, and one penchant of the Old Masters was to express m_o as a fraction with one component being a prime number, in this case 823. Who were the Old Masters? Names such as DeGraff-Hunter, McCaw, Bomford, and Hotine – some of the most illustrious names in Geodesy for the 20th century! The limits of Zone I are: *North Limit*: 2,160,000 yd Northing line of India Zone 0; *East Limit*: Meridian of 78° E, southwards to 31° N, thence along this parallel to 81° E, thence along this meridian to 29° N, thence along this parallel to 79° E, thence along this meridian to 28° N; *South Limit*: Parallel of 28° N; to the *West Limit*: Meridian of 60° E to the point of beginning (*Lambert Conical Orthomorphic Projection Tables, India Zone I*, issued by Office of the Chief of Engineers, Washington, DC, Second Edition 1944). Also used in Pakistan, particularly in the offshore areas is India Zone IIA, where the Central Meridian, $\lambda_o=74°$ E, the Latitude of Origin $\varphi_o=26°$ N, the Scale Factor at the Latitude of Origin $m_o=823/824=0.998786408$, and False Easting $=3,000,000$ Indian Yards, and the False Northing $=1,000,000$ Indian Yards. The original limits of Zone IIA were: *North Limit*: Parallel of 28° N, eastwards to 79° E, thence along this meridian to 29° N, thence along this parallel to 82° E.; *East Limit*: Meridian of 82° E, southwards to 28° N, thence along this parallel to 83° E, thence along this meridian to 26° N, thence along this parallel to 82° E, thence along this meridian to 22° N; *South Limit*: Parallel of 22° N, westwards to 72° N, thence along this meridian to 20° E, thence along this parallel to 60° N; to the *West Limit*: Meridian of 60° E, northwards to 24° N, thence along an (*ellipsoidal – Ed.*) loxodrome to 25° N, 57° E, and finally, thence along the meridian of 57° E, to 28° N, being the point of beginning (*Lambert Conical Orthomorphic Projection Tables, India Zone IIA,* issued by Office of the Chief of Engineers, Washington, DC, 1943).

The Indian Datum of 1960, as it is called in Pakistan, is a local adjustment that has changed the original parameters of the ellipsoid of reference for Everest 1830 (modified for Pakistan), where $a=6,377,309.613$ m and $1/f=300.8017$ as a result of adoption of a new yard to meter conversion factor (*TR 8350.2, Appendix A.1*). The Lambert Conical Orthomorphic Zones I and IIA parameters have been changed to meters as the basic unit for Pakistan. *TR 8350.2* does not list any transformation parameters from Indian 1960 to WGS84. However, offshore oil exploration surveys in Pakistan have yielded a couple of three-parameter transformations that are plausibly similar **from** Indian 1960 Datum **to** WGS 84 Datum: for "Middle Indus Region," $\Delta X=+270$ m, $\Delta Y=+664$ m, $\Delta Z=+225$ m, as determined at Station *Routi*, and for "Bela North Block," $\Delta X=+272$ m, $\Delta Y=+670$ m, $\Delta Z=+231$ m, as determined at Station *Chitrawala*. There are no accuracy estimates available for either of these transformations, so they should be used with caution. Thanks to Dr. S. M. Shuaib of Karachi, for his letter of 18 March 1985 and the reference materials received regarding Pakistan's Offshore Lease Block limits.

Caribbean Sea
Golfo de los Mosquitos
COSTA RICA
Bocas del Toro
Almirante
BOCAS DEL TORO
Palmar Sur
Golfito
San Andrés
Bajo Boquete
La Concepción
David
CHIRIQUI
Pedregal
Puerto Armuelles
Golfo de Chiriquí
Soloy
Tolé
VERAGUAS
Santa Fe
Santiago
COCLE
El Copé
Penonomé
El Valle
Aguadulce
Chitré
HERRERA
Las Tablas
LOS SANTOS
Los Asientos
El Tigre
Isla de Coiba
Salud
COLON
Colón
Portobelo
El Porvenir
La Chorrera
PANAMA
Panama
Balboa
Chepo
Cañita
SAN BLAS
Bay of Panama
ARCHIPIÉLAGO DE LAS PERLAS
Gulf of Panama
La Palma
Yaviza
DARIÉN
Turbo
COLOMBIA
NORTH PACIFIC OCEAN
Panama
International boundary
Province boundary
National capital
Province capital
Railroad
Road
San Blas is a territory (comarca).
0 25 50 75 Kilometers
0 25 50 75 Miles
Lambert Conformal Conic Projection, SP 9N17N

Republic of Panamá

Panamá is bordered on the west by Costa Rica (330 km), and on the East by Colombia (225 km); its coastline is 2,490 km by the Caribbean Sea to the north and by the Pacific Ocean to the south. Panamá is slightly smaller than South Carolina, its lowest point is the Pacific Ocean (0 m), and the highest point is *Volcán de Chiriquí* (3,475 m). The soils of *Chiriquí* Province are volcanic in origin, and the range-fed cattle on the local grasses there develop a distinctive flavor to their flesh. My late Uncle Gus (Johnny Gustin) owned the only gourmet restaurant (*Restaurante Sky Chef*), in Panamá City in the 1950s; one item on his menu was a *Chiriquí* Porterhouse steak. Although tough as shoe leather, they were sensationally popular for their unique taste!

Explored by Christopher Columbus in 1502, Panamá was occupied by various American Indian groups including the peaceful *San Blas* fishermen and the headhunter warriors of *Darién*. The Spanish explorer Vasco Nuñez de Balboa established the first successful colony at *Darié*n in 1510, and Balboa was the first European to discover the Pacific Ocean in 1513. (The Panamanian unit of currency is the Balboa.) The City of Panamá was originally founded in 1519 but was later rebuilt a few miles to the west after the British pirate Henry Morgan sacked and burned it in 1671. I used to live on a cove in the Bay of Panamá where I could see the ruins of Old Panamá City from the front yard of my parents' home. The Isthmus of Panamá has been a major center of commerce and trade since its discovery because it provides the shortest land bridge between the Caribbean Sea and the Pacific Ocean. When the French Geodesists traveled to Quito to measure the size and shape of the Earth in the 17th century, they packed their scientific instruments onto mules to traverse the Isthmus of Panamá. The French originally attempted to dig a canal across the isthmus in the 19th century, but failed to conquer the female *Anopheles* mosquito, the carrier of Malaria. After the United States purchased the rights from the French, the Americans conquered both the *Anopheles* mosquito (through the efforts of Dr. Gorgas of the U.S. Army), and the "Cucaracha rock" (an awful gooey glop when wet and nearly rock-hard when dry), of the Culebra Cut. The Panamá Canal was completed in 1914 after the country achieved independence from Colombia, and the Canal Zone concession was granted to the United States. The U.S. Canal Zone was later ceded back to Panamá by a 1977 treaty and then invaded in 1989 to capture General Manuel Noriega (1934–2017). During that invasion, U.S. Navy SEALs engaged in a battle at the then new *Paitia* Airfield – an area that had been used for anti-aircraft emplacements during World War II, later abandoned and then used by my boyhood cronies and I when we stalked and hunted each other in the tall grass with *Daisey Red Ryder© BB* rifles!

"An ambitious expansion project to more than double the Canal's capacity – by allowing for more Canal transits and larger ships – was carried out between 2007 and 2016" (*World Factbook*, 2022).

The oldest known datum for the republic is the Panamá-Colón Datum of 1911 where the origin is at Balboa Hill: $\Phi_o = 09°\ 04'\ 57.637''$ N, $\Lambda_o = -79°\ 43'\ 50.313''$ West of Greenwich and the azimuth (likely from south), to station *Salud* is: $\alpha_o = 185°\ 02'\ 39.54''$. The ellipsoid of reference is the Clarke 1866, where $a = 6,378,206.4$ m and $b = 6,356,583.8$ m (*Definitions of Terms used in Geodetic and Other Surveys*, Hugh C. Mitchell, U.S. Coast & Geodetic Survey, Special Publication No. 242, 1948). The U.S. Army Corps of Engineers established this datum during surveys that were performed between 1904 and 1935. The map projection associated with that old datum is the Panamá Polyconic (American Polyconic math model) where the Latitude at Origin, $\phi_o = 08°\ 15'$ N, the Central Meridian, $\lambda_o = -81°$ W, the Scale Factor at Origin, $m_o = 1.0$, the False Easting = 1,000,000 yards and the False Northing = 1,092,972.10 yards. This polyconic is probably associated with the pre-World War II artillery grid system known as the World Polyconic Grid, and after that war the Inter-American Geodetic Survey (IAGS) of the U.S. Army Map Service started entering into cooperative agreements with all of the countries of Latin America. Through these agreements, the IAGS carried the North American Datum of 1927 through the West Indies and Central America into Panamá. The origin of the NAD27 is at triangulation station Meades Ranch, Kansas, where $\Phi_o = 39°\ 13'\ 26.686''$ N, $\Lambda_o = 98°\ 32'\ 30.506''$ West of Greenwich and the azimuth (from south), to station Waldo is: $\alpha_o = 75°\ 28'\ 09.64''$. Of course, the ellipsoid of reference is the Clarke 1866. As was the custom of the IAGS for nations in Central America, the Lambert Conformal Conic was employed as the local grid system or each country in one or two zones. For the Republic of Panamá, a single zone was developed since there is little north-south extent in the predominately east-west country. The Panamá Lambert Conformal Conic grid system is defined where the Latitude at Origin, $\phi_o = 08°\ 25'$ N, the Central Meridian, $\lambda_o = -80°$ W, the Scale Factor at Origin, $m_o = 0.99989909$, the False Easting = 500 km and the False Northing = 294,865.303 m. Reports of the existence of an Ancon and a Balboa datum are likely erroneous, considering that the origin of the Panamá-Colón Datum of 1911 is at Balboa Hill, and the access road up the hill is from the town of Ancon, Canal Zone where Ancon Hill is pretty small in comparison.

In 1946 aerial photography began to be used for Panamanian mapping; a 1:20,000 scale map of the Canal Zone was made by the Army Map Service (*PAIGH, Centro* 1952, p. 419; *UN 1979*, p. 193). The *Sección de Cartografía* was established in the same year; it would eventually be renamed the *Instituto Geográfico Nacional* "Tommy Guardia" (*UN 1979*, p. 21; *Twenty* 1966, p. 108). As late as 1949, there were features that did not appear on then current maps, such as a range of mountains 120 miles long and over 5,200 ft high only 60 miles from the Canal Zone (Conly 1956, p. 338; *Cartography in the Americas,* 1951. p. 53). In 1954, the *Sección* became the *Oficina de Cartografía*, and in 1955, it was renamed the *Dirección de Cartografía* (*Twenty* 1966, p. 108). Between 1958 and 1960, the U.S. Army Map Service began a 1:25,000 scale map series, which was <25% complete in 1963 (Birch, 1964, p. 11). By 1964, the AMS series at 1:50,000 scale covered the south coast, the peninsula west of Panamá City, and islands in the Gulf of Panamá. In 1962, the *Dirección* began work on a 1:10,000 scale map series.

One of the major reasons for such sparse and patchy cartographic coverage in such a relatively small country was the almost continuous cloud cover, which for 20 years precluded obtaining satisfactory aerial photography. As the same time the vegetation, fauna, and physiography mitigated against the use of planetable mapping. Up to 1965, about 47 1:50,000 scale sheets had been issued. In 1967, side looking airborne radar (SLAR) cut through the clouds and provided imagery that could be used in the compilation of 1:250,000 scale sheets covering 55% of the country that included the *Darién* Province. By late 1968, all 12 of the 1:250,000 scale sheets, 109 of the 199 1:50,000 scale sheets, and 47 of the 700 1:25,000 scale sheets had been completed.

In the late 1950s, I used to go to Sidney Townsend's house in Balboa for her Friday night "Bee-Bop" dances while I was a student at Balboa High School. Sidney's father was a Panama Canal Company Land Surveyor, and he usually stayed upstairs while the high school kids danced downstairs.

One evening, I walked up the stairs and listened to a surveying story related by Mr. Townsend. As I recall, he said that he went on a surveying expedition in the 1930s into *Darién* Province. He took along a large contingent of helpers that was comprised mainly of local Panamanian laborers. One of the men was a fellow with a distinctively shaped scar on one cheek of his face. He remembered him because the fellow disappeared one night, and Mr. Townsend assumed that the fellow got tired of the hard work and just walked back to the city. Years passed by, and Mr. Townsend said that one afternoon after World War II, he was walking along *Avenida Central*, the main shopping street in downtown Panamá City, and he noticed one glass display case that had a number of shrunken heads. One head startled him as he recognized the scar on one cheek of the face! Mr. Townsend told me he purchased the head from the shopkeeper, and he took it to the late fellow's Parish Priest for a proper Catholic burial. I suspect that Mr. Townsend might have been associated with the monumentation and survey of the 14 boundary markers for the Panamá-Colombia Boundary Treaty signed on June 17, 1938 (*International Boundary Study No. 62 – January 30, 1966*, The Geographer, Bureau of Intelligence and Research, U.S. Department of State).

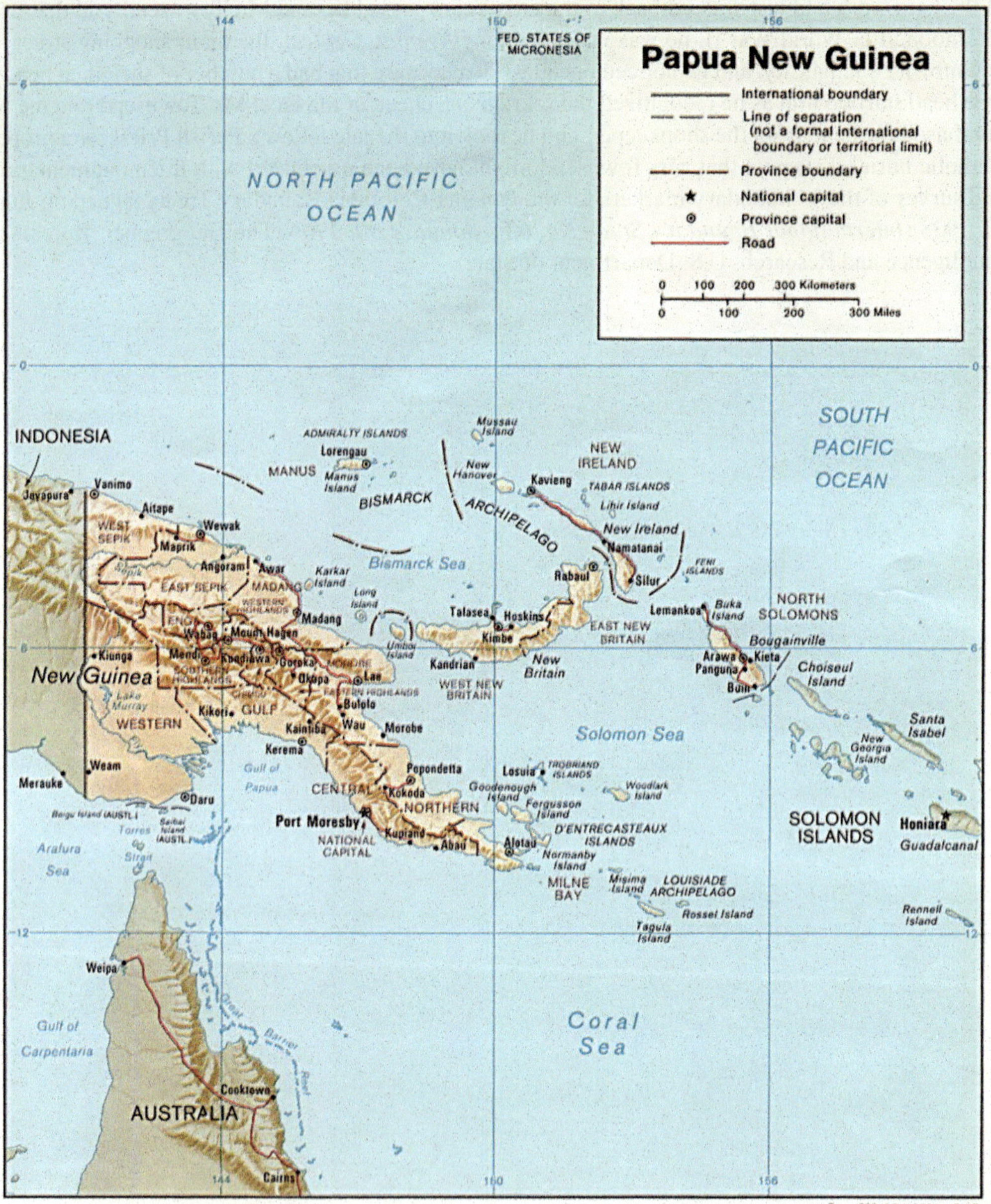

Base 800495 (A05601) 3-87

Independent State of Papua New Guinea

Papua New Guinea is slightly larger than California, bordering Indonesia (824 km) with a coastline of (5,152 km), the terrain is mostly mountains with coastal lowlands and rolling foothills. The highest point is *Mount Wilhelm* (4,509 m), the lowest point is the Pacific Ocean (0 m), and the mean elevation is (667 m).

During the 16th century, Portuguese and Spanish navigators visited the island. Annexed by Queensland in 1883, the region became a British Protectorate in 1884 and was annexed by Great Britain in 1888 as British New Guinea. Administration was passed to Australia in 1905 and the name was changed to the Territory of Papua. In 1949, it was united with the Territory of New Guinea to form Papua New Guinea.

The Independent State of Papua New Guinea became independent in 1975. The country is comprised of the Eastern part of New Guinea, the island of Bougainville, and the Bismarck Archipelago; a total area of 462,840 km² which is slightly larger than California. According to the *CIA Factbook,* the "natural hazards include active volcanism; situated along the Pacific 'Ring of Fire;' the country is subject to frequent and sometimes severe earthquakes; mud slides; tsunamis." On July 18, 1998, a tsunami took the lives of 2,200 north shore residents of Papua New Guinea.

"A secessionist movement in Bougainville, an island well-endowed in copper and gold resources, reignited in 1988 with debates about land use, profits, and an influx of outsiders at the *Panguna Copper Mine*. Following elections in 1992, the PNG government took a hardline stance against Bougainville rebels and the resulting civil war led to about 20,000 deaths. In 1997, the PNG government hired mercenaries to support its troops in Bougainville, sparking an army mutiny and forcing the prime minister to resign. PNG and Bougainville signed a truce in 1997 and a peace agreement in 2001, which granted Bougainville autonomy. An internationally monitored nonbinding referendum asking Bougainvilleans to choose independence or greater self-rule occurred in November 2019, with 98% of voters opting for independence" (*World Factbook*, 2022).

The first Australian Engineer Officer for mapping was posted to *Rabaul* on New Britain in 1914. Topographic mapping of the area began during World War II and consisted mainly of inch to the mile compilations with classical triangulation control. The Australian military mapping installations consisted of drafting and computation sections quartered in tents. Map printing services in Queensland were transferred to the U.S. Army 69th Engineer Topographic Battalion's lithographic detachment in Port Moresby. Supplemented by reconnaissance aerial photo mosaics, additional mapping control continued through the 1950s with assistance from the Royal Australian Survey Corps and the U.S. Army (*Australia's Military Map-Makers,* 2000).

The oldest "Astro station" serving as a local datum is *Paga Hill 1939* near Port Moresby, where $\Phi_o = 9°\ 29'\ 00.31''$ S, $\Lambda_o = 147°\ 08'\ 21.66''$ E of Greenwich, and the ellipsoid of reference is the Bessel 1841, where $a = 6,378,397.155$ m., and $1/f = 299.1528$. The grid system commonly associated with the Paga Hill Datum of 1939 is the 1943 Southern New Guinea Lambert Zone where the Latitude of Origin, $\varphi_o = 8°$ S, Central Meridian, $\lambda_o = 150°$ E, Scale Factor at Origin, $m_o = 0.9997$, False Northing = 1,000 km, False Easting = 3,000 km. The original limits of the Zone were for the North: Parallel of 7° S, east to 153° 30′ E, thence north along this meridian to 5° S, thence east along this parallel to 165° E. East: Meridian of 165° E. South: Parallel of 12° S, west to 145° E, thence west along this parallel to 141° E, thence south along this meridian to 11° S, thence west along this parallel to 137° E. West: Meridian of 137° E. Recent source data for Paga Hill Datum of 1939 now state the ellipsoid of reference as: International 1924, where $a=6,378,388$ m and $1/f=297$. When this supposed change occurred is unknown.

Thanks to John W. Hager for the following: other astro positions in Papua New Guinea include: Brown Island, East New Britain Province $\Phi_o=5°\ 01'\ 40''$ S, $\Lambda_o=151°\ 58'\ 54''$ E; Cay, *Panaeati & Deboyne Is.*, Milne Bay Province $\Phi_o=14°\ 41'$ S, $\Lambda_o=152°\ 22'$ E; *Dedele Point*, Central Province $\Phi_o=10°\ 14'$ S, $\Lambda_o=148°\ 45'$ E; *Dobodura* Astro Fix, Northern Province, $\Phi_o=8°\ 45'\ 50.13''$ S, $\Lambda_o=148°\ 22'\ 38.8''$ E; *Dumpu, Madang Province*, $\Phi_o=5°\ 50'\ 34.4''$ S, $\Lambda_o=145°\ 44'\ 29.55''$ E; *Guadagasal* Astro Fix, Gulf Province, $\Phi_o=7°\ 15'\ 33.6''$ S, $\Lambda_o=146°\ 58'\ 42.0$ E; *Guasopo B, Woodlark Is.*, Milne Bay Province, $\Phi_o=9°\ 13'\ 39''$ S, $\Lambda_o=152°\ 57'\ 03''$ E; *Hetau Island Naval Astro, Buka Is.*, North Solomons Province, $\Phi_o=5°\ 09'\ 57''$ S, $\Lambda_o=154°\ 31'\ 12''$ E.

Hong Astro (1947), *Manus Is., Manus Province*, West Base, $\Phi_o=1°\ 58'\ 03.930''$ S, $\Lambda_o=147°\ 22'\ 03.320''$ E, azimuth $\alpha_o=111°\ 55'\ 58.00''$ to Az. Mk. from north, Clarke 1866 ellipsoid, elevation $=6.0$ ft; *Jammer Bay, Milne Bay Province*, $\Phi_o=9°\ 58'\ 28''$ S, $\Lambda_o=152°\ 11'\ 15''$ E; *Kavieng, New Ireland Province*, $\Phi_o=2°\ 36'$ S, $\Lambda_o=150°\ 50'$ E; *Keila Island* Astro, East New Britain Province, $\Phi_o=4°\ 48'\ 28''$ S, $\Lambda_o=152°\ 11'\ 15''$ E; *Kieta*, North Solomons Province, Ashton, $\Phi_o=6°\ 12'\ 42.68''$ N, $\Lambda_o=155°\ 37'\ 43.69''$ E; *Koiaris*, North Solomons Province, *Koiaris Astro* 1947, $\Phi_o=6°\ 18'\ 06.11''$ S, $\Lambda_o=155°\ 11'\ 47.32''$ E, azimuth $\alpha_o=322°\ 19'\ 42.4''$ to Az. Mark #1 from south, International ellipsoid, established by 657th Eng. Astro. Det., March 1947; *Losuia, Milne Bay Province, Losuia*, $\Phi_o=8°\ 32'\ 33.825''$ S, $\Lambda_o=151°\ 03'\ 59..466''$ E; *Matupi*, East New Britain Province, *Matupi Astronomic Station* 1957, $\Phi_o=4°\ 14'\ 12.210'''$ S, $\Lambda_o=152°\ 11'\ 26.54''$ E, International ellipsoid, Elevation $=2.4$ m; *Popondetta*, Astro fix, $\Phi_o=8°\ 46'\ 07.76''$ S, $\Lambda_o=148°\ 12'\ 51.55''$ E; St. Matthais, New Ireland Province, South Base, $\Phi_o=1°\ 40'\ 30''$ S, $\Lambda_o=149°\ 54'\ 54''$ E; *Salankaua, Morobe Province*, $\Phi_o=6°\ 33'\ 28.4''$ S, $\Lambda_o=147°\ 51'\ 07.2''$ E; *Torokina*, North Solomons Province, Naval astro station, $\Phi_o=6°\ 12'\ 18''$ S, $\Lambda_o=155°\ 02'\ 02.5''$ E; *Wabutina, Milne Bay Province, Wabutin* (spelling may be Wabutima), $\Phi_o=8°\ 30'\ 54.628''$ S, $\Lambda_o=151°\ 03'\ 24.947''$ E; *Wau, Morobe Province*, $\Phi_o=7°\ 20'\ 28.12''$ S, $\Lambda_o=146°\ 42'\ 55.6''$ E; *Wewak*, $\Phi_o=3°\ 32'\ 52''$ S, $\Lambda_o=143°\ 37'\ 37''$ E.

The various local astro datums listed above represent the fixes used for navigational charts. In regard to how these various datums are related to the WGS 84 Datum, the Australian Maritime Safety Authority comments: "For some charts, particularly in Papua New Guinea, the correction to be applied to GPS cannot be calculated and these charts display a specific warning to this effect. Use of GPS alone on these charts is hazardous."

For the most part, cartographic products of Papua New Guinea have been on the Australian Geodetic Datum of 1966 with its origin at **Johnston** Cairn, where $\Phi_o=25°\ 56'\ 54.5515''$ S, $\Lambda_o=133°\ 12'\ 30.0771''$ E, $h_o=571.2$ m, and the ellipsoid of reference is the Australian National Spheroid: $a=6,378,160$ m and $1/f=298.25$. A new system is the Papua New Guinea Geodetic Datum 1994 (PNG94), which is a geocentric datum defined by a widespread network of geodetic stations around PNG. There are three permanent GPS base stations operating in PNG. The Papua New Guinea Map Grid 1994 (PNGMG) is the UTM grid on the GRS80 ellipsoid. According to the Department of Surveying and Land Studies of the Papua New Guinea University of Technology, "A very approximate relationship between AGD66 and PNG94 coordinates is as follows: PNG94 Latitudes are approximately 5″ north of AGD66 latitudes, PNG94 Longitudes are approximately 4″ east of AGD66 longitudes, PNGMG Eastings are approximately 120 m greater than AMG66 Eastings, and PNGMG Northings are approximately 160 m greater than AMG66 Northings."

There is a caveat to this approximate relation between AGD66 and PNG94. Again, according to the Department of Surveying and Land Studies, "Tectonic motion is unaccounted for in the realization of the datum. Relative motion between different tectonic regions in PNG is often in *excess of 8 cm per year*. There are inconsistencies of up to 12 m between tabulated PNG94 coordinates and those derived from high precision GPS survey network adjustments...."

Thanks to Mr. John W. Hager for his patience with my requests and his generous help.

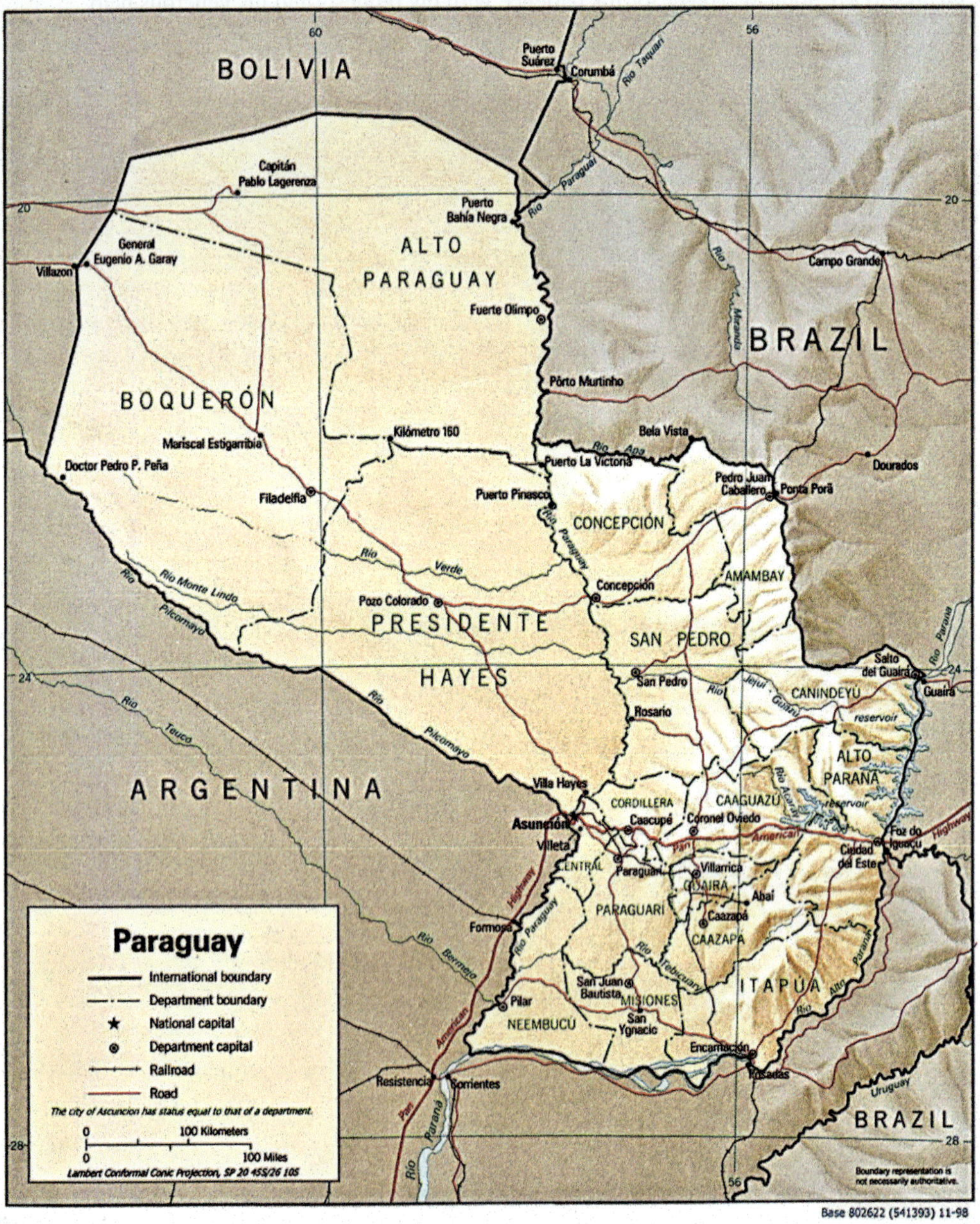

Base 802622 (541393) 11-98

Republic of Paraguay

Slightly smaller than California, Paraguay is landlocked and is bordered by Argentina (1,880 km), Bolivia (750 km), and Brazil (1,365 km). The lowest point in the republic is the junction of *Rio Paraguay* and *Rio Parana* (46 m) and the highest point is *Cerro Pero or Cerro Tres Kandu* (842 m). Paraguay is comprised of grassy plains and wooded hills to the east of Rio Paraguay; the *Gran Chaco* region west of Rio Paraguay is mostly low, marshy plain near the river and dry forest and thorny scrub elsewhere (*CIA Factbook*).

Discovered by Spanish conquistadors in 1524, Paraguay had been populated by semi-nomadic *Chaco Indian* tribes for millennia. "Although few relics or physical landmarks remain from these tribes, the fact that 90 percent of Paraguayans still understand the indigenous Guaraní language is testament to Paraguay's Indian lineage. The Spanish conquistadors arrived in 1524 and founded Asunción in 1537. Paraguay's colonial experience differed from that of neighboring countries such as Bolivia and Argentina, because it did not have what the Spanish were searching for – gold or other large mineral deposits" (*Library of Congress Country Studies*). First influenced by the Jesuits, indigenous society was converted to Catholicism until the Jesuits were expelled in 1767. A succession of dictators eventually resulted in only 79 persons owning half of the country's land by 1900! After wars in the 18th and 19th centuries including the disastrous War of the Triple Alliance (1865–1970) – between Paraguay and Argentina, Brazil, and Uruguay – Paraguay lost two-thirds of all adult males and much of its territory. More war in the 20th century combined with a continued succession of dictators has left the country lacking in substantive economic growth. The current democratically elected administration seems to have garnered some international support for progress, but Paraguay's legacy of political instability may persist.

"Following the *Chaco War of 1932–35* with Bolivia, Paraguay gained a large part of the *Chaco* lowland region. (*The war started over the publication of a **postage stamp** showing a disputed area of the northern plains called the Chaco that purported to belong to Paraguay. Bolivia objected and the war ensued. – Ed.*) The 35-year military dictatorship of Alfredo Stroessner ended in 1989, and Paraguay has held relatively free and regular presidential elections since the country's return to democracy" (*World Factbook*, 2022).

Although some small-scale maps were privately compiled of Paraguay in the 18th and 19th centuries, the majority of mapping was produced by the Office of Municipal Engineers in Asunción (*Oficina de Ingenieros Municipales de Asunción*) until the establishment of the Military Geographic Institute (*Instituto Geográfico Militar – IGM*) in 1941. Until 1962, IGM was dedicated to low-order surveys with the exception of a period of 4 years when with the help of an Argentine Technical Mission, IGM reproduced four topographic map sheets at a scale of 1:50,000 for the metro area of Asunción. These sheets were cast on the Paraguay Gauss-Krüger Transverse Mercator projection where the national Grid has four Belts: (1) $\lambda_o=63°$ W, False Easting F.E.=4,500 km, (2) $\lambda_o=60°$ W, False Easting F.E.=5,500 km, (3) $\lambda_o=57°$ W, False Easting F.E.=6,500 km, (4) $\lambda_o=54°$ W, False Easting F.E.=7,500 km. All four Belts have a scale factor at origin (m_o)=unity and the False Northing=10,002,288.299 m. The ellipsoid of reference is the International 1924, where $a=6,378,388$ m and $1/f=297$. Note that this is an extension of the Argentine TM Grid.

In January 1962, with the help of the Inter-American Geodetic Survey (IAGS), IGM concentrated its forces for the establishment of basic vertical and horizontal control and for the training of its personnel in the various phases of geodesy and cartography. In October 1965, IGM acquired a photographic laboratory. *Chua* Datum is defined by the 1966 astronomic observation reduced to the *Bureau International de l'Heure* (*BIH*) pole, specifically, $\Phi_o=19°$ 45′ 41.16″ North, $\Lambda_o=48°$ 06′ 07.56″ West of Greenwich. Although used in Paraguay, astro station *Chua* is actually located about

720 km East in Brazil just west of the *Uberaba airport* and north of the *Grande River*. Curiously, the South American Datum of 1969 (SAD69) origin is defined by the same astronomic observation reduced to the Conventional International Origin (CIO) pole, specifically, $\Phi_o = 19°\ 45'\ 41.34''$ North, $\Lambda_o = -48°\ 06'\ 07.80''$ West. Furthermore, the SAD69 geodetic value for *Chua* is: $\varphi_o = 19°\ 45'\ 41.6527''$ North, $\lambda_o = 48°\ 06'\ 04.0639''$ West. The ellipsoid of reference remains the same, thanks to John W. Hager, 1992.

The classical triangulation arcs of Paraguay consist of *Brazil – Alegre* (1:1,350,000 error), *Pastoreo – Alegre* (1:448,000 error), *Asunción – 5H07* (1:1,130,000 error), 5H07 – *Paraguari* (1:875,000 error), *Paraguari – Caaguazú* (1:493,000 error), *Caaguazú – Cerrito* (1:311,000 error), *Cerrito – Meza* (1:297,000 error), *Caaguazú – Pastoreo* (1:543,000 error), *Cerrito – Pilar* (1:425,000 error), *5H01 – Alberdi* (1:420,000 error), *Alberdi – Pilar* (1: 540,000 error), *Pilar – Diaz* (1: 675,000 error), *Asunción – San Pedro* (1:306,000 error), and *San Pedro – Caaguazú* (1:297,000 error) (*DMATC Geodetic Memorandum No. 1683, May 1973*).

In 1977, James W. Walker of the Defense Mapping Agency Topographic Center (DMATC – same place as the original Army Map Service where I was stationed about 10 years earlier), published two sets of transformation parameters for Paraguay. The Seven-Parameter Transformation **from** NWL9D **to** Paraguay's Chua Datum is: $\Delta X = +160.35\,\text{m}$, $\Delta Y = -229.67\,\text{m}$, $\Delta Z = +28.88\,\text{m}$, $\delta s = -0.52029 \times 10^5$, $R_x = +1.62926''$, $R_y = +1.58638''$, $R_z = -1.47279''$, and the gross total positional error was estimated at ±3.64 m. The corresponding transformation **from** NWL9D **to** SAD69 in Paraguay is: $\Delta X = +80.68\,\text{m}$, $\Delta Y = +8.83\,\text{m}$, $\Delta Z = +36.14\,\text{m}$, $\delta s = -0.80233 \times 10^5$, $R_x = +1.27144''$, $R_y = +0.56022''$, $R_z = -2.28029''$, and the gross total positional error was estimated at ±3.56 m.

James A. Slater published a paper describing "A New GPS Geodetic Control Network for Paraguay" (*S&LIS, 1993*), in which a mapping project was sponsored by the World Bank for the establishment of approximately 170 GPS stations. No transformation parameters were offered in the paper, though. The old *TR8350.2* parameters **from** *Chua Astro* (Paraguay) **to** WGS84 are: $\Delta X = -134\,\text{m}, \pm 6\,\text{m}$, $\Delta Y = +229\,\text{m}, \pm 9\,\text{m}$, and $\Delta Z = -29\,\text{m}, \pm 5\,\text{m}$, and that is based on six stations.

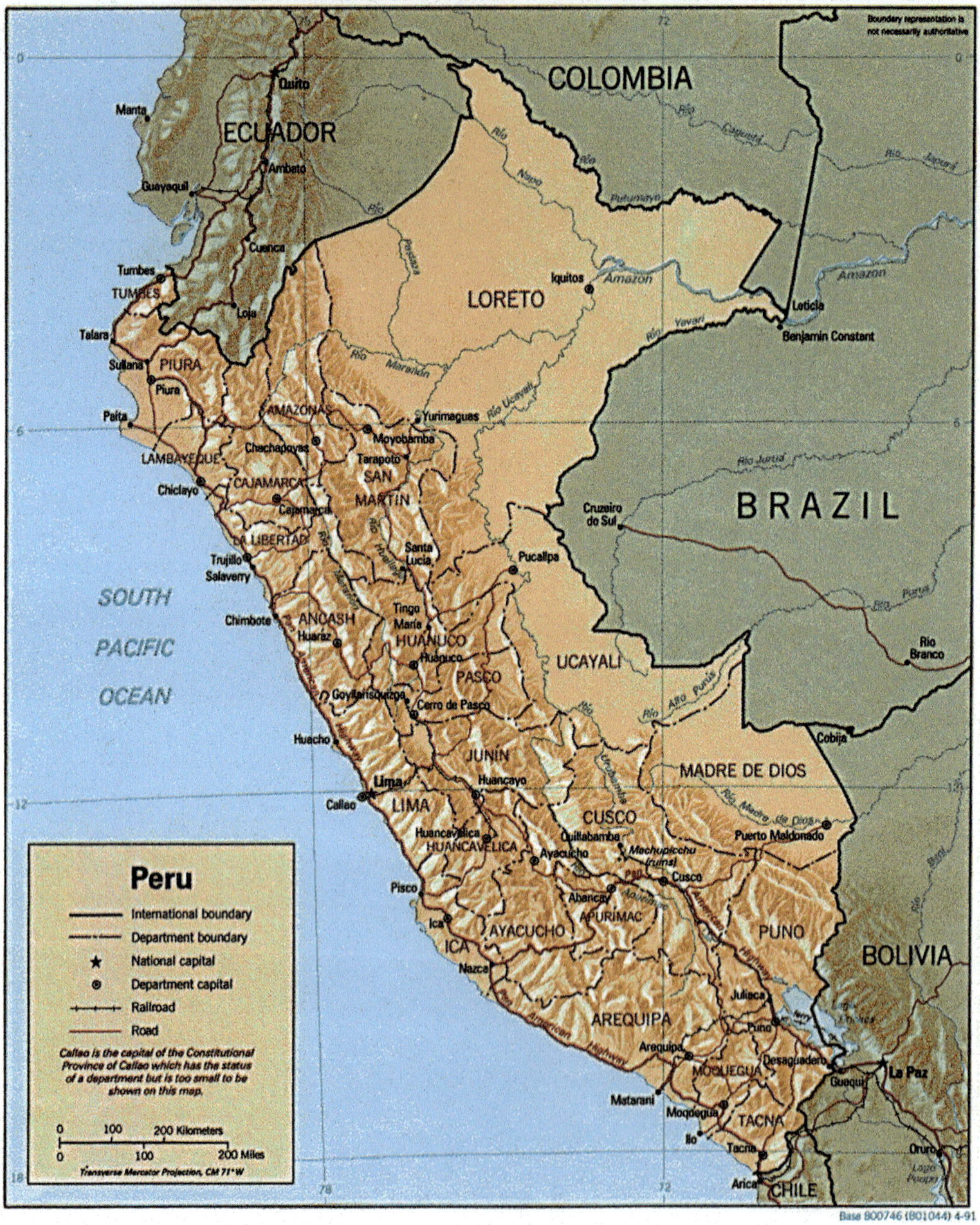

Base 800746 (801044) 4-91

Republic of Perú

Perú borders Bolivia (900 km), Brazil (1,560 km), Chile (160 km), Colombia (1,496 km), Ecuador (1,420 km), and the Pacific Ocean (2,414 km). Perú is slightly smaller than Alaska, its lowest point is the Pacific Ocean (0 m), and the highest point is *Nevado Huascaran* (6,768 m); the terrain is comprised of the western coastal plain (costa), high and rugged Andes in the center (sierra), and the eastern lowland jungle of the Amazon Basin (selva). The climate varies from tropical in the east to dry desert in the west; it is temperate to frigid in the Andes Mountain range.

The seat of the Inca Empire was established in *Cuzco, Perú* about 1230 A.D. and by the mid-15th century, ruled parts of Colombia, Ecuador, Perú, Bolivia, Chile, and Argentina. The Spaniard Francisco Pizarro conquered the Incas in 1533 with the help of Diego de Almagro. Lima was founded in 1535, and initially became the seat of the Viceroyalty of Perú. The Argentinean leader, José Francisco de San Martín, declared the Perúvian independence from Spain in 1821 and remaining Spanish forces were defeated in 1824 (*Webster's Geographical Dictionary*).

"After a dozen years of military rule, Peru returned to democratic leadership in 1980, but experienced economic problems and the growth of a violent insurgency. President Alberto Fujimori's election in 1990 ushered in a decade that saw a dramatic turnaround in the economy and significant progress in curtailing guerrilla activity. Nevertheless, the president's increasing reliance on authoritarian measures and an economic slump in the late 1990s generated mounting dissatisfaction with his regime, which led to his resignation in 2000. A caretaker government oversaw a new election in the spring of 2001, which installed Alejandro Toledo Manrique as the new head of government – Peru's first democratically elected president of indigenous ethnicity. The presidential election of 2006 saw the return of Alan Garcia Perez who, after a disappointing presidential term from 1985 to 1990, oversaw a robust economic rebound. Former army officer Ollanta Humala Tasso was elected president in June 2011, and carried on the sound, market-oriented economic policies of the three preceding administrations. Poverty and unemployment levels have fallen dramatically in the last decade, and today Peru boasts one of the best performing economies in Latin America. Pedro Pablo Kuczynski Godard won a very narrow presidential runoff election in June 2016. Facing impeachment after evidence surfaced of his involvement in a vote-buying scandal, President Kuczynski offered his resignation on 21 March 2018. Two days later, First Vice President Martin Alberto Vizcarra Cornejo was sworn in as president. On 30 September 2019, President Vizcarra invoked his constitutional authority to dissolve Peru's Congress after months of battling with the body over anticorruption reforms. New congressional elections took place on 26 January 2020 resulting in the return of an opposition-led legislature. President Vizcarra was impeached by Congress on 9 November 2020 for a second time and removed from office after being accused of corruption and mishandling of the COVID-19 pandemic. Because of vacancies in the vice-presidential positions, constitutional succession led to the President of the Peruvian Congress, Manuel Merino, becoming the next president of Peru. His ascension to office was not well received by the population, and large protests forced his resignation on 15 November 2020. On 17 November, Francisco Sagasti assumed the position of President of Peru after being appointed President of the Congress the previous day. Jose Pedro Castillo Terrones won the second round of presidential elections on 6 June 2021 and was inaugurated on 28 July" (*World Factbook*, 2022).

La Cartografía Nacional was established in 1859 when the President of the Republic, Don Ramón Castilla placed don Mariano Felipe Paz in charge of preparing a general map of Perú. From 1901 to 1909, a geodetic chain of triangulation was surveyed from *Viviate* to *Piura* by consulting French Army Officers with the intention of accomplishing a meridional arc of high precision in the equatorial region. In 1904, Lieutenant Colonel Pablo Berthon of the *Misión Militar Francesca* organized a topographic section of the Military Academy in *Chorrillos*. On March 6, 1906, President José Pardo approved the topographic service. Numerous topographic sheets were compiled by planetable and alidade of *Lima, Ancón,* and *Tumbes,* the latter probably based entirely on the previous chain observed by the French Army Officers. In 1912, the name was changed to the *Servicio Geográfico del Ejercito* (Army Geographic Service), and the Major General and Chief of Staff of the Army ordered fieldwork extended into the areas of *Puno, Moquegua, Canta,* and *Chiclayo*. In 1916, the *Servicio Geográfico* was reorganized under one Director with a topographic section and a cartographic section. The work then concentrated on topographic fieldwork (by planetable and alidade) of nine sheets of Arequipa at a scale of 1:200,000 and was based on a geodetic network referenced to the Clarke 1880, where $a = 6,378, 249.145$ m and $1/f = 293.465$. These were also the first Perúvian maps to be printed in color.

On May 10, 1921, the National Map of Perú was commenced by planetable and alidade methods, and the ellipsoid of reference for the rest of the 20th century was changed to the International (Hayford, 1909), where $a = 6,378,388$ m and $1/f = 297$. The National Map of Perú was to be compiled at a scale of 1:200,000, based on the Greenwich Prime Meridian, and the projection was to be the Polyhedric. The reader may recall that the ellipsoidal Polyhedric projection can be easily approximated by the "Local Space Rectangular" projection when constrained to a plane. (See the *Manual of Photogrammetry*, 5th & 6th editions.) The plan was to produce 261 map sheets that were cast on a standard graticule of 48′ of latitude and 36′ of longitude.

On February 2, 1944, the name of the organization was changed to *Instituto Geográfico Militar (IGM) del Perú*. In 1957, IGM was authorized to change compilation methods to employ aerial photography and stereophotogrammetry to continue with the National Map. On April 9, 1958, IGM initiated photogrammetric operations with negatives flown by the Inter-American Geodetic Survey (IAGS) of the U.S. Army Map Service. The new specification for mapping was to be based on the Provisional South American Datum of 1956 with origin in the town of *La Canoa, Anzoátegui Province*, Venezuela, where $\Phi_o = 08° 34′ 17.170″$ North, $\Lambda_o = 63° 51′ 34.880″$ West of Greenwich, and the defining azimuth to station Pozo Hondo $(\alpha_o) = 40° 22′ 45.96″$. The standard sheets were to be at 1:100,000 scale, cast on the UTM grid, and the graticule was to be 30′ of latitude and 30′ of longitude. The vertical datum was defined as mean sea level.

With the cooperation of the IAGS, a concerted campaign of geodetic triangulation was commenced that included the following chains with lengths and positional errors at the terminal points: *Tarqui – Piura* (225 km), 1:352,000; *Piura – Chimbote* (450 km), 1:273,000; *Chimbote – Salinas* (250 km), 1:309,000; *Salinas – Conchan* (125 km), 1:321,000; *Conchan – Marcona* (360 km), 1:324,000; *Huancayo – Pucusana – Pucara* (150 km), 1:333,000; *Cerro de Pasco – San Francisco – Chonta* (150 km), 1:789,000; *Marcona – Majes* (310 km), 1:310,000; *Majes – Sama* (225 km), 1:336,000; *Sama – Chaca* (120 km), 1:300,000; *Puquio –Huahupasa – Huancaramis* (140 km), 1:237,000; *Cerro Apopata – Camane Avabacas* (140 km), 1:269,000; *Ayabacas – Anta* (300 km), 1:283,000; *Anta – Ayacucho* (240 km), 1:202,000; *Ayacucho – Huancayo* (175 km), 1:438,000; and Pichus – Cero Jampito – Cerro de Pasco (125 km), 1:291,000 (*DMATC GM No. 1684*). In 1980, the agency name was changed to *Instituto Geográfico Nacional*.

The old transformation from NWL9D (WGS72) to PSAD56 for Perú expressed with seven parameters are as follows: $\Delta X = +282.57$ m, $\Delta Y = -185.85$ m, $\Delta Z = +401.38$ m, $\delta s = -2.69414 \times 10^5$, $R_x = -0.31989″$, $R_y = -0.39589″$, $R_z = +2.29014″$. The new three-parameter transformation from PSAD56 to WGS84 as published in 1991 in *NIMA/NGA TR 8350.2* is: $\Delta X = -279$ m ± 6 m, $\Delta Y = +175$ m ± 8 m, $\Delta Z = -379$ m ± 12 m, and was based on six collocated points. Curiously, IGN published different parameters from WGS84 to PSAD56 in 2005 as follows: $\Delta X = 303.55$ m, $\Delta Y = -265.41$ m, $\Delta Z = +358.42$ m. Note that the datum shift direction is *different* than that published by NIMA/NGA, and that the Peruvian-published parameters do not have an associated accuracy statement. Transformation parameters have been offered by the U.S. Military for Perú that utilize the South American Datum of 1969, but there seems to be zero recognition of that relationship (or use) by Perú.

Republic of Perú

In addition to the standard UTM grid used for mapping associated with military applications, there are four Transverse Mercator Grid systems used in Perú. The first three are on the PSAD56. The Perú Transverse Mercator East Zone is defined by a Central Meridian $(\lambda_o) = 70°$ 30′ West of Greenwich, Latitude of Origin (φ_o) = Equator, False Northing Latitude of Origin $(FN\varphi_o) = 9°$ 30′ South, Scale Factor at Origin $(m_o) = 0.99933$, False Easting = 1,324,000 m, and False Northing = 1,040,084.558 m. The Perú Transverse Mercator Central Zone is defined by a Central Meridian $(\lambda_o) = 76°$ 00′ West of Greenwich, Latitude of Origin (φ_o) = Equator, False Northing Latitude of Origin $(FN\varphi_o) = 9°$ 30′ South, Scale Factor at Origin $(m_o) = 0.99933$, False Easting = 720,000 m, and False Northing = 1,039,979.159 m. The Perú Transverse Mercator West Zone is defined by a Central Meridian $(\lambda_o) = 80°$ 30′ West of Greenwich, Latitude of Origin (φ_o) = Equator, False Northing Latitude of Origin $(FN\varphi_o) = 6°$ 00′ South, Scale Factor at Origin $(m_o) = 0.99933$, False Easting = 222,000 m, and False Northing = 1,426.834.743 m. The Perú International Petroleum Company (IPC) Transverse Mercator (October 27, 1950) is defined by a Central Meridian $(\lambda_o) = 74°$ 38′ 03″ West of Greenwich, Latitude of Origin (φ_o) = Equator, False Northing Latitude of Origin $(FN\varphi_o) = 9°$ 08′ 08″ South, Scale Factor at Origin (m_o) = unity, False Easting = 870,000 m, and False Northing = 1,080,000 m. The IPC Grid is obviously on an *ersatz datum* (a technical term for "baloney"), because the ellipsoid of reference is the Clarke 1866, and is obviously something cooked up with the *wrong set of projection tables*! (That's not unheard of, especially when a geodesist has had to "winter over" on a remote island and forgot to pack the *proper* book of tables.) *Ayabacas Base* and *Sama River Base* are the names of two datums that apparently were old "astro" stations observed in a past century, probably the 20th.

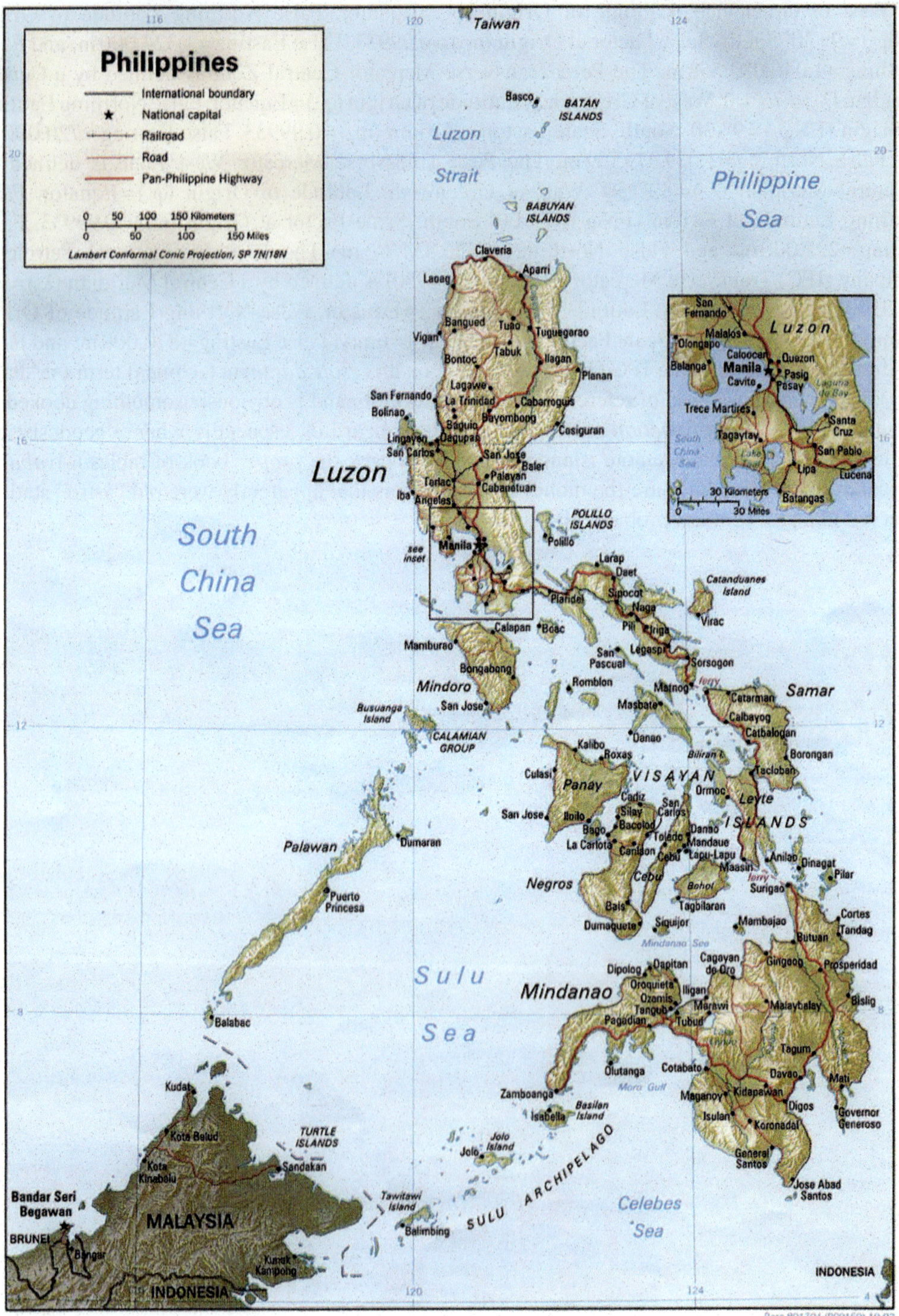

Base 801704 (R00150) 10-93

The Republic of the Philippines

The Philippines is slightly less than twice the size of Georgia, slightly larger than Arizona. Its territorial maritime claim is an irregular polygon extending up to 100 nm from coastline as defined by the 1898 treaty; since the late 1970s, it has also claimed a polygonal-shaped area in South China Sea as wide as 285 nm. The terrain is mostly mountains with narrow to extensive coastal lowlands, the highest point is *Mount Apo* (2,954 m), the lowest point is the Philippine Sea (0 m), and the mean elevation is (442 m).

Chinese traders visited the Philippines in the 10th century A.D., and Muslims settled in the southern part of the islands during the 15th century. The Portuguese navigator Ferdinand Magellan discovered the islands in 1521, and the first successful European settlement was made by the Spanish under the explorer Miguel López de Legazpi in 1565. The battle of Manila Bay was fought during the Spanish-American War, and the *Treaty of Paris* with Spain ceded the Philippines to United States control in 1898. The treaty limits of the Philippines consisted of an enclosing box with appropriate zigzags to accommodate the island of Borneo.

The U.S. Coast and Geodetic Survey (USC&GS) executed the earliest topographic work of significance from 1901 to 1942. Numerous local datums were established at "Astro stations" that include: *Bancalan I., Cagayan Sulu I., Davao, Mindanao I., Iligan, Mindanao I., Iloilo, Panay I., Legaspi, Luzon I., Misamis Oriental, Mindanao I., Ormoc, Leyte I., Tacloban, Leyte I., Vigan, Luzon I.,* and *Zamboanga, Mindanao I.* The results of the basic trigonometric surveys were reported on topographic and hydrographic field sheets that have never been published. A 1:200,000 series of city plans and topographic sheets were the only series printed before World War II.

The Luzon Datum of 1911 is defined by its origin near *San Andres Point* on *Marinduque Island* in the Southern *Tagalog* Region. That point is at station *Balanacan* (a port name), where $\Phi_o = 13°$ 33′ 41.000″ North, $\Lambda_o = 121°$ 52′ 03.000″ East of Greenwich, and the geoid/spheroid separation $H_o - h_o = 0.34$ m. The defining geodetic azimuth to station Baltasar is: $\alpha_o = 009°$ 12′ 37.000″, the ellipsoid of reference is the Clarke 1866, where $a = 6,378,206.4$ m and $1/f = 294.9786982$.

The first grid system used in the Philippines was devised by the USC&GS in 1919, which was also the first Grid used in the United States. The "Grid System for Progressive Maps in the United States" was the defining design for the World Polyconic Grid (WPG), the predecessor of the Universal Transverse Mercator (UTM) Grid. During the 1930s, the Work Projects Administration for the City of New York extended some of the specifications for the "Grid System for Military Maps." The first use of false easting was introduced to the American Grid System at the Central Meridian (λ_o), where False Easting = 1,000,000 U.S. Survey yards. May of 1943 marked the first use of the WPG by the U.S. Army Map Service (AMS), and the Philippines were based on a special local meridian (122°) of that Grid as computed and tabulated by the USC&GS. Although the WPG overprint color was purple for most of the world, the color used for the Philippines was black. The 1:200,000 maps had a 10,000-yard grid interval and the WPG was used in the Philippines until 1952.

Other Grids found in the Philippines include the Netherlands East Indies (NEI) Equatorial Zone British Metric Blue Grid based on the Lambert Conical Orthomorphic projection. The NEI Equatorial Zone Black or Brown Grid in U.S. yards is also found there (numerically the same Grid with different units), but the occurrences for both are only on Cagayan Sulu Island and in the *Sibutu Island* group. These two grids commonly found in Malaysia and Indonesia flabbergasted me when I first encountered them some decades ago. Besides *Bukit Rimpah Datum* being referenced to the Bessel 1841 ellipsoid, the Scale Factor at the Latitude of Origin (m_o) = 0.997, False Easting = 900,000.000 m (984,250.000 yards), False Northing = 3,900,000.00 m (4,265,083.333 yards). The Central Meridian (λ_o) = 110°, and the Latitude of Origin (φ_o) = 00° = the equator! I wasted my time trying to get these parameters to work with Lambert Conformal Conic equations until I realized that the Lambert Conformal Conic with a Latitude of Origin at the equator decomposes to a Normal Mercator! The NEI transformations worked fine after that...

Straight baselines are a method for a sovereign nation to define its territorial waters by defining points along its coast from which lines are drawn (on a Mercator projection). This then establishes a line (ellipsoidal loxodrome or rhumb line) from which that nation's claim to territorial limits is measured. The straight baseline is a "new" development in international law. It had its inception in 1951 with the decision in the *Anglo–Norwegian Fisheries* case in which the International Court of Justice upheld Norway's method of delimiting an exclusive fisheries zone. This was by drawing straight baselines along the Norwegian coast above the Arctic Circle, independent of the low water mark. This established a new system of baselines from which the territorial sea could be measured, provided certain geographic situations were satisfied. This system with certain modifications was approved by the 1958 Geneva conference on the Law of the Sea. (*See Shore and Sea Boundaries*, Aaron Shalowitz, USC&GS, U.S. Govt. Printing Office, Vols. 1–3, various dates of publications.)

On June 17, 1961, the Government of the Philippines approved Republic Act No. 3046, "An Act to Define the Baselines of the Territorial Sea of the Philippines." The Philippine Government adopted the so-called "archipelago principle" in drawing a series of 80 straight baselines about the external group of islands. This baseline system in effect closed the important Surigao Strait, Sibuto Passage, Balabac Strait, and Mindoro Strait as well as the more internal passages through the Philippine Islands. The largest body of water enclosed was the Sulu Sea, and other significant seas enclosed include the Moro, Mindanao, and Sibuyan. Furthermore, the Indonesian island of Pulau Miangas and the Indonesian Straight Baselines were enclosed within the Philippine territorial sea! The United States (and Indonesia) did not recognize this declaration of territorial waters, and this represents a thorny issue in international boundary claims. Furthermore, to become recognized under international law; there must be an absence of formal protests from other nations, and the territorial waters declaration must be ratified by at least 22 other nations. (*According to Shalowitz, there may be a 100-year prescriptive period.*)

From 1947 to 1962, a national civil Grid was used on the Luzon Datum of 1911, and it was known as the Philippine Transverse Mercator Grid with four zones. All four Gauss-Schreiber (*a double projection – Ed.*) zones had a False Easting at the Central Meridian of 500 km, all four had a Scale Factor at Origin $(m_o) = 0.99995$, and all four had a False Northing Latitude of Origin of: $(N_{FN}) = 04°\ 00'\ 00''$ North. Zones II, III, and IV had Central Meridians $(\lambda_o) = 121°, 123°,$ and $125°$, respectively. For some reason unfathomable to me, Zone I had a Central Meridian $(\lambda_o) = 118°\ 20'$ East of Greenwich. Of course, the military 1:50,000 mapping was based on the Gauss-Krüger Transverse Mercator projection with the UTM grid.

In 1962, a new national civil Grid was introduced for the Luzon Datum of 1911, and it was changed to the Gauss-Krüger Transverse Mercator projection. The previous Grid Scale Factor at Origin and False Easting was retained, the False Northing Latitude of Origin was changed to the equator, and the Central Meridians $(\lambda_o) = 117°, 119°, 121°, 123°,$ and $125°$ for Zones I to V. A civilian edition of the 1:50,000 topographic series was produced in association with U.S. agencies starting in 1961 and completed in the 1970s. All 967 sheets have been published in color.

In 1987, the Philippine Bureau of Coast and Geodetic Survey was made a part of the National Mapping and Resource Information Authority (NAMRIA). A total of 467 GPS stations were observed which included 330 first-order stations, 101 second-order stations, and 36 third-order stations. This series of new observations was adjusted and published as the Philippine Reference System of 1992 (PRS92). According to NAMRIA, this included the development of a Philippine geoid model, the establishment of an EDM calibration baseline, and the determination of the seven Bursa-Wolf transformation parameters between the Luzon Datum of 1911 and WGS 84. Those parameters from WGS84 to PRS92 are published as: $\Delta X = +127.623\,m$, $\Delta Y = +67.245\,m$, $\Delta Z = +47.043\,m$, $\delta s = +1.06002 \times 10^6$, $R_x = +3.07''$, $R_y = -4.90''$, and $R_z = -1.58''$. Since this work was done with the assistance of the Australian government, the sign of the rotation parameters is assumed to be the standard right-handed system favored in Australian and U.S. practice.

"NAMRIA is spearheading the establishment of the Philippine Active Geodetic Network (AGN) as part of the implementation of the PRS92 Project. As envisioned, the AGN will be composed of stations strategically located all over the country which continuously provide geographic data. To date, all six Active Geodetic Stations (AGS) have already been set up: four are roof-based (NAMRIA Main Building in *Taguig, Urdaneta City Hall in Pangasinan*, Registry of Deeds Building in *Tagaytay*, and Basa Air Base in *Floridablanca, Pampanga*) while two are ground-based (*Nueva Ecija University of Science and Technology* in *Cabanatuan* City and 415th Police Provincial Mobile Group in Candelaria, Quezon). All six are ready for interconnection to the network. For the Data and Control Center (DCC), the installation of the Center's ICT requirements has been completed. Support systems needed in the operation and maintenance of the DCC, such as fire suppression and access control systems, have already been configured and tested. The Horizontal Control Network established 2,367 second order and third order GCPs and established 5,286 fourth-order GCPs. For the Vertical Control Network, 17,410 km of level lines were surveyed. About 65 first order GCPs were recovered and observed from 2009 until June 2010. Processing and adjustments of GCPs using the Active Geodetic Station (AGS) – PageNET in Taguig City as reference control were finished in July 2010. These zero-order points are now available for use as reference points in geodynamic studies. Around 362 first order GCPs have been contracted out in six clusters for recovery and re-observation. This activity will update the coordinates of the first order GCPs established under the Natural Resources Management Development Project (NRMDP) in 1989. Some 300 GCPs were recovered and observed by contractors and GPS data are being evaluated. Nine clusters of benchmarks in loops were recovered and observed using the Global Positioning System to determine the relationships of the Mean Sea Levels (MSL) of the major islands in the Philippines to the National Vertical Datum. As of May 2011, all of the nine clusters of benchmarks in loops were observed, processed and MSL relationships observed. Some 80 first-order gravity stations will be established nationwide for scientific applications and for the formulation of a Philippine geoid model. As of May 2011, established are 80 first order and 68 second-order gravity stations. Around 5,000 kilometers of level line will be surveyed to connect the GCPs to the existing benchmarks to compute the Geoid undulations. The activity is essential in the computation of the Geoid model of the Philippines. The project is ongoing. The NAMRIA PRS92 Data Integration Task Group continued to undertake the quality control of cadastral datasets for Ilocos province-Region I and the National Capital Region (NCR). All plotted lots are currently being converted from local to grid coordinates. In compliance with the agreements reached during the PRS92 Coordinating Conference held on 18 April 2011, the Task Group conducted a retraining program and lecture on the derivation of local transformation parameters of Regions III, VII, VIII, IX, and XIII. This is also in pursuance of the commitment of NAMRIA to Provide relevant technical assistance and support for the full adoption of PRS92 and enhancement on the capability of the regional offices. On the other hand, the surveying arm of the Task Group is presently establishing image control points (ICPs) at the northern portion of *Bohol* province. Said ICPs will be used for the transformation of satellite imageries into PRS92. A total of 48 ICPs for four map sheets covering the Province of *Bohol* is set to be established. Similarly, a total of 70 photo control points (PCPs) were established in the Province of *Aklan*. Additional 12 PCPs were established in *Casiguran, Aurora*. At present, the processing of Global Positioning Systems (GPS) raw data and of PCP Description Sheets was finished and they are ready for submission. A total of 161 PCPs covering *Casiguran, Eastern Laguna de Bay, Northwestern Panay*, and *Pampanga* is set to be established for use in the large scale topographic base mapping under the PRS92 project. There are three developed systems currently installed in 18 regions of the country. These are the regional Geodetic Network Information System (GNIS), the Land Survey Data Management System (LSDMS), and the Metadata Entry for environment and natural resources datasets. As of May 2011, Regions 1,2,5,7,8,11, and CAR were visited for the update of regional GNIS and LSDMS, and installation of the four-parameter derivation program and online synchronization module of both systems" (*NAMRIA, 22 March, 2017*).

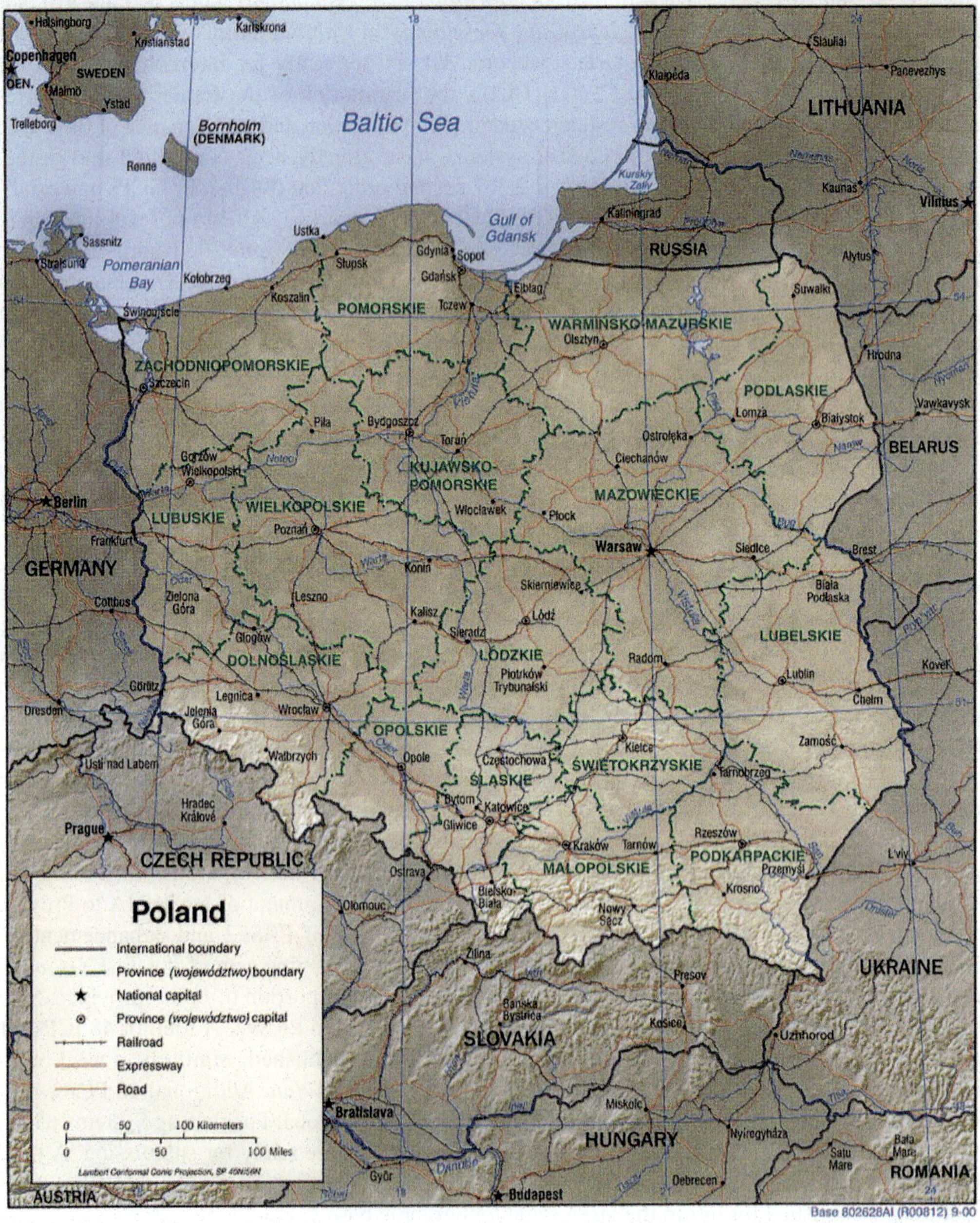

Base 802628AI (R00812) 9-00

The Republic of Poland

The north and central regions of Poland are essentially flat and are characterized by morainic topography. This lack of natural barriers on the North European Plain has been a major reason for so many invasions of Poland throughout history. The southern boundary is mountainous with the highest peak being *Rysy* at (2,499 m); the lowest point is *Raczki Elblaskie* at (–2 m).

In the 10th century, Poland was a Slavic duchy. The crown became elective after 1572, and various wars caused much loss of territory. Napoleon partly reestablished (1807–1815) the kingdom, which was later to become closely aligned with Russia. Poland was declared a republic in 1918, but after nearly another century of being overrun and controlled by others, the new constitution was dated October 16, 1997.

"Free elections in 1989 and 1990 won Solidarity control of the parliament and the presidency, bringing the communist era to a close. A 'shock therapy' program during the early 1990s enabled the country to transform its economy into one of the most robust in Central Europe. Poland joined NATO in 1999 and the EU in 2004" (*World Factbook*, 2022).

Early mapping of Poland was instituted by the Prussians for the western half of the present country, and ~17% of the southeast was mapped by the Austro–Hungarian Empire. The remainder of Poland was surveyed and mapped by czarist Russia. The dates of this early mapping activity go back to 1816. The early *Prussian Landesaufnahme* characteristically used the Cassini-Soldner projection in its spherical form that was based on equivalent (Gaussian) spheres referenced to the Bohnenberger ellipsoid and the Zach ellipsoids, and later the Bessel 1841 ellipsoid. (See also the Republic of Hungary and the Czech Republic.) The Prussians and the Austrians introduced the concept of the cadaster, or system of surveys and land registration for ownership and taxation. The Austro–Hungarian surveyors had similar preferences for ellipsoids, but the Russians were a different story.

The tsarist Russians performed surveys and topographic mapping of Poland in the 19th and early 20th centuries, but these works were for military purposes only. They did nothing with respect to individual land ownership registration, and they preferred the sazhen for their unit of measurement. In the years between the two world wars, this source material was responsible for some very strange-looking contour maps of Poland when the unit of measurement was changed from sazhens to meters where 1 sazhen = 2.134 m. (The only time I have seen similar strange values for contours is when I have graded some of my sophomores' campus topo maps!) The Russians preferred the Walbeck 1819 ellipsoid, where $a = 6,376,896$ m and the reciprocal of flattening $(1/f) = 302.78$. Some of these old maps also referred longitudes to Ferro in the Canary Islands; a practice dropped after World War II.

New geodetic triangulation started after the founding of the republic, and the origin of the Polish National Datum of 1925 (PND 1925) is at station *Borowa Gora* (*gora* is Polish for mountain), where (Astronomic Latitude of Origin) $\Phi_o = 52°\ 28'\ 32.85''$ North, and (Astronomic Longitude of Origin) $\Lambda_o = 21°\ 02'\ 12.12''$ East of Greenwich. The ellipsoid of reference is the Bessel 1841, where $a = 6,377,397.155$ and $1/f = 299.1528128$. Instruments used by the Polish military included theodolites manufactured by Bamberg, Fennel, Wild-Heerbrugg, and Aerogeopribor. Over 120 triangles were observed from 1927 to 1935, with the average angular error of figure not exceeding 0.56 arc seconds. (Post-World War II observations with T-3 theodolites yielded errors not exceeding 0.46 arc seconds.) Baselines were measured at *Grodno, Kobryn, Warsaw, Lomza, Luniniec, Mir,* and *Braslaw*. Laplace stations (astro shots) were observed at *Varsovic* (initial point *Borowa Gora*), *Borkowo, Kopciowka,* and *Skopowka*.

The 1:25,000, 1:100,000, and 1:300,000 maps produced by the *Wojskowy Instytut Geograficzny* (WIG) or Military Geographical Institute, were products of cartometric datum shifts (*physical cut and paste – once called "paneling"*) to PND 1925 and are cast on the Polish Stereographic Grid. The old Polish Stereographic Grid is based on the mathematical model by Rousilhe, who was the Hydrographer of the French Navy. All ellipsoidal oblique stereographic projections developed and used worldwide before World War I are based on Rousilhe's work that was originally published in *Annals Hydrographique*. The development of the projection for nearby Romania (STEREO 70) was done by the Bulgarian geodesist, Hristow in the late 1930s. The PND 1925 WIG Military Stereographic Latitude of Origin $(\phi_o) = 52°\ 00'$ N, Central Meridian $(\lambda_o) = 22°\ 00$ E, the Scale Factor at Origin $(m_o) = 1.0$, the False Easting $= 600$ km, and the False Northing $= 500$ km. The PND 1925 Grids developed for the Cadastre around the same time were cast on the Gauss-Krüger Transverse Mercator where the Scale Factor at Origin $(m_o) = 1.0$, the False Easting $= 90$ km, the False Northing $= minus\ 5,700$ km *at the equator*, and the Central Meridians $(\lambda_o) = 15°$, $17°$, $19°$, and $21°$ East of Greenwich.

During World War II, the *Generalstab des Herres, Reichsamt fur Landesaufnahme* (German Army) produced topographic maps of Poland cast on the *Deutches Herres Gitter* (DHG) Grid, which is identical to the UTM except for the Scale Factor at Origin $(m_o) = 1.0$. Of course, the Datum used was the PND 1925, as was the equivalent treatment of Poland by the USSR with the Russia Belts TM that had the same defining parameters as the DHG except for the Datum and ellipsoid. The Russian coverage during the war had *double* Grids in Poland that exhibited unresolved horizontal Datum discrepancies ranging from 160 to 250 m. After the war, the USSR converted their *military* topographic coverage of the Warsaw Pact countries to the System 42 Datum that has its origin at Pulkovo Observatory and is referenced to the Krassovsky ellipsoid. In the Republic of Poland, their preferred terminology of that Datum is "Polish National 1942" or "PN 42."

Large-scale topographic maps of Poland published in the latter part of the 20th century are on the "UKLAD 65 System," the parameters of which have been a closely held secret. In the past few years, little information has dribbled out of Poland on these "Strefa" or Zones. In February 2000, Mr. Wojtek Hanik sent the de-classified parameters to me! There are five *Strefa* comprising the UKLAD 65 System, four are based on the "Quasi-Stereographic" Grid (Rousilhe Oblique Stereographic), and the fifth is a Gauss-Krüger Transverse Mercator Grid. *Strefa 1* covers the following provinces: *Biala Podlaska, Eastern Bielsko, Chelm, Kielce, Krakow, Krosno, Lodz, Lublin, Nowy Sacz, Piotrkow, Premysl, Radom, Rzeszow, Sieradz, Tarnobrzeg, Tarnow,* and *Zamosc*. The UKLAD 65 *Strefa 1* Quasi-Stereographic Grid Latitude of Origin $(\phi_o) = 50°\ 37'\ 30''$ N, Central Meridian $(\lambda_o) = 21°\ 05'\ 00''$ E, the Scale Factor at Origin $(m_o) = 0.9998$, the False Easting $= 4,637$ km, and the False Northing $= 5,467$ km. *Strefa 2* covers the following provinces: *Bialystok, Ciechanow, Lomza, Olsztyn, Ostroleka, Plock, Siedlce, Skierniewice, Suwalki,* and *Warszawa*. The UKLAD 65 *Strefa 2* Quasi-Stereographic Grid Latitude of Origin $(\phi_o) = 53°\ 00'\ 07''$ N, Central Meridian $(\lambda_o) = 21°\ 30'\ 10''$ E, the Scale Factor at Origin $(m_o) = 0.9998$, the False Easting $= 4,603$ km, and the False Northing $= 5,806$ km. *Strefa 3* covers the following provinces: *Bydgoszcz, Elbag, Gdansk, Koszalin, Slupsk, Szczecin, Torun,* and *Wloclawek*. The UKLAD 65 *Strefa 3* Quasi-Stereographic Grid Latitude of Origin $(\phi_o) = 53°\ 35'\ 00''$ N, Central Meridian $(\lambda_o) = 17°\ 00'\ 30''$ E, the Scale Factor at Origin $(m_o) = 0.9998$, the False Easting $= 3,703$ km, and the False Northing $= 5,627$ km. *Strefa 4* covers the following provinces: *Gorzow, Jelenia Gora, Kalisz, Konin, Legnica, Leszno, Opole, Pila, Poznan, Walbrzych, Wroclaw,* and *Zielona Gora*. The UKLAD 65 *Strefa 4* Quasi-Stereographic Grid Latitude of Origin $(\phi_o) = 51°\ 40'\ 15''$ N, Central Meridian $(\lambda_o) = 16°\ 40'\ 20''$ E, the Scale Factor at Origin $(m_o) = 0.9998$, the False Easting $= 3,703$ km, and the False Northing $= 5,627$ km. *Strefa 5* covers the following provinces: *Western Bielsko, Czestochowa,* and *Katowice*. The UKLAD 65 *Strefa 5* Gauss-Krüger Transverse Mercator Grid Central Meridian $(\lambda_o) = 18°\ 57'\ 30''$ E, the Scale Factor at Origin $(m_o) = 0.999983$, the False Easting $= 237$ km, and the False Northing $= minus\ 4,700$ km.

For small-scale mapping, the GUGiK 80 Quasi-Stereographic (Rousilhe) projection is used where the Latitude of Origin $(\phi_o) = 52°\ 12'$ N (approx.), Central Meridian $(\lambda_o) = 19°\ 10'$ E. The scale

The Republic of Poland

factor at a point is designed to be equal to unity at a distance of 215 km from the projection origin. These mysterious parameters of which some are still held, and some are now public reflect the history of the nation. Those countries that have a long and recent history of war, occupation, or blood spilled at borders will be particularly sensitive about releasing Grid and/or Datum relation parameters. The release of some of this previously secret data may be an indication of the republic's confidence in the future.

"The ETRS89 was introduced in Poland technically by the GNSS technique in the last years of the 20th century and by law in 2000. On 2 June 2008, the Head Office of Geodesy and Cartography in Poland (GUGiK) commenced operating the multifunctional precise satellite positioning system named ASG-EUPOS. The ASG-EUPOS network defines the European Terrestrial Reference System ETRS89 in Poland. A close connection between the ASGEUPOS stations and 18 Polish EUREF Permanent Network (EPN) stations controls the realization of the ETRS89 on Polish territory. In 2010–2011 GUGiK integrated the ASG-EUPOS with the existing geodetic networks (horizontal and vertical) using GNSS and spirit levelling. Those actions resulted in developing and then legal introduction in 2012 new technical standards: to the National Spatial Reference System (PSOP) and to establish and maintain the geodetic (horizontal and vertical), gravity and magnetic control in the country. Thus, the geodetic, gravimetric, and magnetic system in Poland has been associated with the European one (previous and current). This allowed for the next step of networks integration in Poland, namely, in 2013 started integration of national geodetic control with gravimetric control. Modern geodetic, gravimetric, and magnetic networks in Poland are to be fully consistent with the European system. In 2011, following the initiative by the Section of Geodetic Networks and the Section of Earths' Dynamics of the Committee on Geodesy of the Polish Academy of Sciences, a new research network 'Polish Research Network for Global Geodetic Observing System' (acronym GGOS-PL) has been established" (*POLISH NATIONAL REPORT ON GEODESY 2011–2014*, Jarosław Bosy, and Jan Kryński, Vol. 64, Warsaw, 2015).

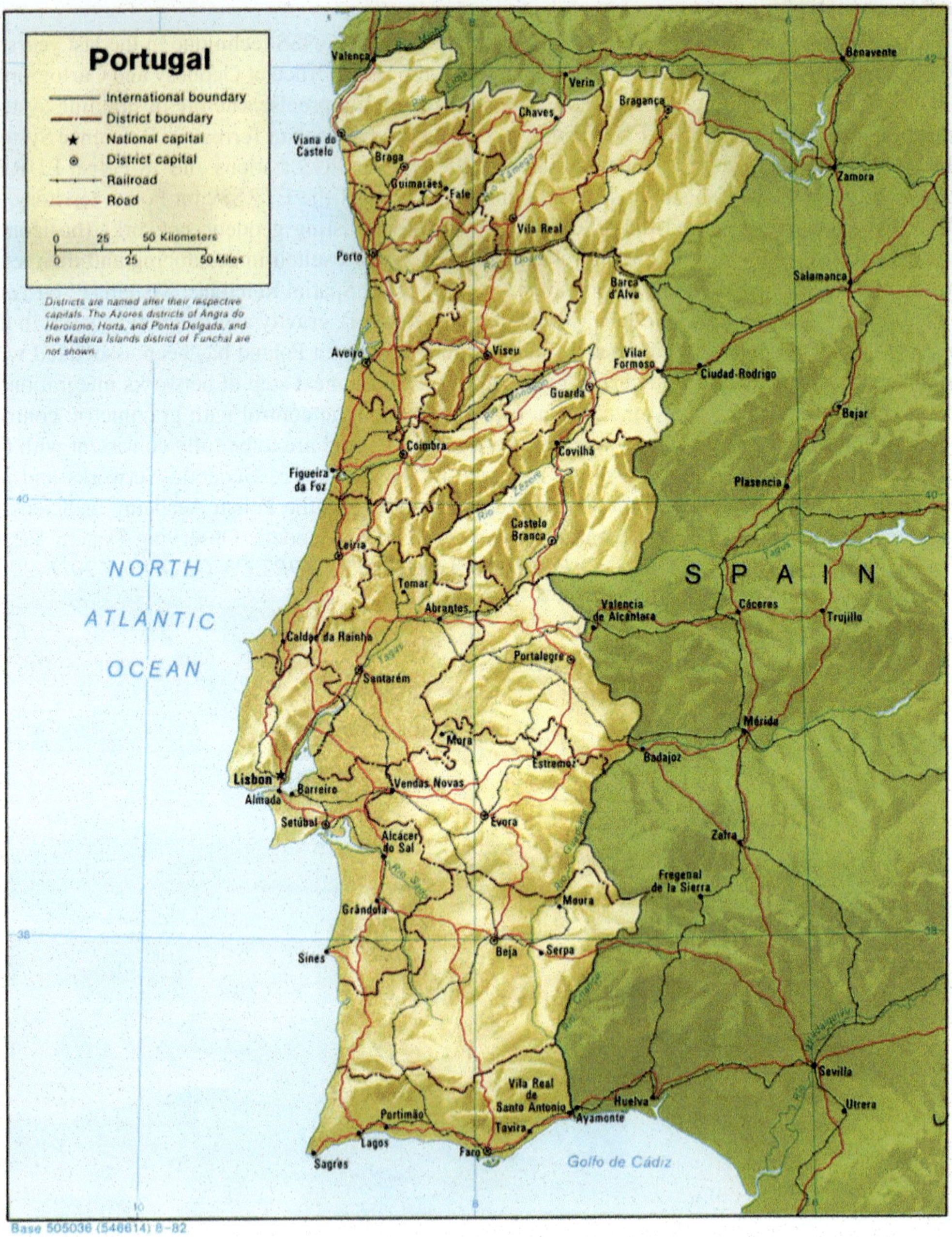

Base 505036 (546614) 8-82

Kingdom of Portugal

Portugal is slightly smaller than Indiana; it is mountainous north of the Tagus River and has rolling plains in the south. With its only border being Spain to the east and north, the North Atlantic Ocean is on Portugal's west and south. Portugal holds sovereignty over the Azores and Madeira Islands; both archipelagos are strategic locations along the western sea with approaches to the Strait of Gibraltar. The highest point is *Ponta do Pico* on *Ilha do Pico* in the Azores at 2,351 m.

Inhabited by late Stone Age people, many megalithic tombs date from 2500 B.C. Portugal is one of the oldest nation states in Europe, and its foundation in 1139 pre-dates Spain by nearly 350 years. The Romans arrived in 216 B.C. and named the entire peninsula Hispania. However, the region between the Douro and Tagus Rivers was named Lusitania by the Celt-Iberian inhabitants. Later overrun by Germanic tribes and then by the Moors, eventually the kingdom of Portucale, comprising León and Castile, was declared independent by King Afonso Henriques.

According to the Portuguese *Instituto Geográfico do Exército* (Army Map Service), in 1420, aware of the navigational importance of cartography, Prince Henry the Navigator commissioned the master cartographer Jácome of Majorca to teach apprentice cartographers in Portugal the art of preparing navigational charts. One of the oldest existing maps, thought to have been prepared by the cartographer Pedro Reinel, dates from 1500. Portuguese cartographers of the period were considered the most skilled in the world, their maps providing the most accurate representation of the Earth. Conscious of the importance of cartography in tackling the country's economic problems, Queen Mary of Portugal created the Royal Military Archive in 1802, in order to house the different national cartography departments. This body was the precursor of the Military Geographic Institute. At the end of the 18th century, having closely followed the progress of this science, it was the Portuguese military cartographers that established and developed the geodesic (*sic*) network in Portugal. Modern Portuguese cartography dates back to 1778 when work began on the first fundamental geodetic triangulation network, which lasted until 1848. The survey work for the 37 sheets which make up the *Carta Geral do Reino* (General Map of the Kingdom), also known as The Chorographic Map of Portugal, at a scale of 1:100,000, lasted from 1853 to 1892. Produced al- most entirely by Army officers, the map series was awarded the "Lettre de Distinction" during the Paris International Congress of Geographic Sciences in 1875. Based on the *Castello de São Jorge* Datum in Lisbon, the ellipsoid of reference was the Bessel 1841, where $a = 6,377,397.155$ m and the reciprocal of flattening $(1/f) = 299.1528128$. The projection *de jour* in Europe at that time was the ellipsoidal Bonne, and for this series the Latitude of Origin $(\varphi_o) = 38° 42' 43.631'' $ N, and the Central Meridian $(\lambda_o) = 9° 07' 54.806''$ West of Greenwich. The scale factor at origin was equal to unity, and there was no False Easting or False Northing.

In 1881, the Army Staff Department published the *Carta Itinerária da 1a Divisão Militar* (Itinerary Map of the 1st Army Division) at a scale of 1:250,000. In 1891 the Army Staff Department began work on the publication of the *Carta dos Arredores de Lisboa* (Map of the Outskirts of Lisbon) at a scale of 1:20,000. This was the first military topographic map, which later became known as the Map of Portugal. The ellipsoidal Bonne projection was used for this series also, and the Latitude of Origin $(\varphi_o) = 39° 40'$ N and the Central Meridian $(\lambda_o) = 8° 07' 54.806''$ West of Greenwich. The scale factor at origin was equal to unity, False Easting = 200 km and False Northing = 400 km. The two baselines were observed in 1886–1889 and were at Batel-Montijo and at Melriça.

In 1911, the Military Cartographic Section of the Army General Staff was created and continued the work on the Map of the Outskirts of Lisbon at 1:20,000 scale, and the Map of Portugal at 1:250,000 scale. Later, work was suspended on these two series during the restructuring process of the services which started in 1926 and ended in 1932 with the creation of the *Serviços Cartográficos do Exército* (Army Cartographic Service).

The Regulatory Decree dated November 24, 1932 established the *Serviços Cartograficos do Exercito* (SCE) (Army Cartographic Service) under the aegis of the Army Staff (EME), and the heirs of Portuguese cartographic traditions. The initial priority of the SCE was to publish the Military Map of Portugal, the Itinerary Map of Portugal, and other documentation necessary for the defense of Portugal. The first sheet, produced at a scale of 1:25,000, covered the Abrantes region. Thus, in 1937, the SCE adopted purely classical methods in their survey activities. From then on, photogrammetric processes were initiated, and by 1940 were exclusively used in medium-scale mapping.

In 1995 the survey for the Military Map of Continental Portugal was completed at a scale of 1:25,000 with a total of 639 sheets. The Gauss-Krüger Transverse Mercator Grid was used for this series where the Latitude of Origin (φ_o) = the equator (by definition) and the Central Meridian (λ_o) = 8° 07′ 54.862″ West of Greenwich. The scale factor at origin was equal to unity, False Easting = 300 km, and False Northing = 200 km. This is based on the Castello de Sâo Jorge Datum, where Φ_o = 38° 42′ 43.631″ N and the Central Meridian Λ_o = 9° 07′ 54.8446″ West of Greenwich. The ellipsoid of reference is the Hayford 1909 or the International (Madrid) 1924, where a = 6,378,388 m and $1/f$ = 297. This series was then followed by the survey of the islands of Azores and Madeira with 51 sheets at the same scale. The Gauss-Krüger Transverse Mercator Grids for the islands of the Azores Archipelago are quite similar in that the parameters are φ_o = 0°, the scale factor at origin is equal to unity, False Easting = zero, and False Northing Latitude of Origin (φ_{FN}) = 38° 45′ N. For *Ilha de Sâo Jorge*, the Central Meridian (λ_o) = 28° West of Greenwich; for *Ilha do Faial*, λ_o = 28° 42′ W; for Ilha do Pico, λ_o = 28° 20′ W; and for Ilha Graciosa, λ_o = 28° W. For the Madeira Archipelago, the *Ilha de Porto Santo* Gauss-Krüger Transverse Mercator φ_o = 0°, the scale factor at origin is equal to unity, False Easting = zero, False Northing Latitude of Origin (φ_{FN}) = 33° 03′ 23.9412″ N, and λ_o = 16° 20′ 01.2304″ West of Greenwich. For the *Ilha de Madeira e Desertas*, φ_o = 0°, the scale factor at origin is equal to unity, False Easting = zero, False Northing Latitude of Origin (φ_{FN}) = 32° 45′ N, and λ_o = 16° 55′ West of Greenwich. Thanks go entirely to Mr. Jorge Teixeira Pinto, Director of Geodetic Services, *Instituto Portugués de Cartografía e Cadastro*.

"Portugal has adopted a new map projection, based on the ETRS89 datum, as a national coordinate system for topographic mapping purposes. This new map projection is designated as PT-TM06 (*EPSG:3763*). The Portuguese Geographic Institute (*IGP*) has observed with GPS more than 1000 points of the national geodetic network and provides their coordinates. Older reference systems are still in use, in particular Datum 73 (*epsg:4207*) and Datum Lisboa (*epsg:4274*), both based on the Hayford ellipsoid. Usual projections of these datums are identified with epsg codes *epsg:27493* (D73) and *epsg:20790* (DLisboa). Coordinates of more than 8000 points (1st, 2nd and 3rd order points of the geodetic network are provided by IGP) as well as Bursa-Wolf parameters to convert from these to ETRS89. Coordinates in these projections can be transformed to ETRS89 using PROJ.4 command cs2cs. The RMSE of coordinates transformed using these parameters are of approximately 0.40 m (for D73) and 1.50 m (for DLisboa).

"Using around 900 of the points provided by IGP, datum shift grids were calculated, using standard kriging interpolation, for the differences: Datum 73 to ETRS89 (*pt73_e89.gsb*) and Datum Lisboa to ETRS89 (*ptLX_e89.gsb*). These files are in format NTv2 and can be placed in the directory pointed to by environment variable PROJ_LIB. Tests made with a set of around 140 points not involved in the generation of the grids provided RMS errors of 5 cm in the case of Datum 73 and 8 cm for Datum Lisboa. The grid files can be also easily configured in ArcGIS (as a Custom Geographic Transformation) and other commercial and open-source GIS programs" (Gonçalves, J., 2009. Paper presented at the 6th National Conference of Geodesy and Cartography, https://citeseerx.ist.psu.edu/viewdoc/download?doi=10.1.1.720.4243&rep=rep1&type=pdf).

REFERENCES: P

Aaron Shalowitz, Shore and Sea Boundaries, USC&GS, U.S. Govt. Printing Office, Vols. 1–3, various dates of publication.
Australia's Military Map-Makers, 2000.
CIA Factbook, 2007.
DMATC Geodetic Memorandum No. 1683, May 1973.
Dr. S. M. Shuaib of Karachi, Personal Communication, 18 March 1985.
Gonçalves, J., Conversões de Sistemas de Coordenadas Nacionais para ETRS89 Utilizando Grelhas, *6th National Conference of Geodesy and Cartography*, 2009.
https://www.fc.up.pt/pessoas/jagoncal/coordenadas/index_en.htm.
Hugh C. Mitchell, Definitions of Terms used in Geodetic and Other Surveys, U.S. Coast & Geodetic Survey, Special Publication No. 242, 1948.
International Boundary Study No. 62 – January 30, 1966, The Geographer, Bureau of Intelligence and Research, U.S. Department of State.
James A. Slater, A New GPS Geodetic Control Network for Paraguay, S&LIS, 1993.
Jarosław Bosy, and Jan Kryński, *Polish National Report on Geodesy 2011–2014*, Vol. 64, Warsaw, 2015.
J. Chris McGlone, *Manual of Photogrammetry*, 5th & 6th editions. ASPRS.
John W. Hager, Personal Communication, 1992.
John W. Hager, Personal Communication, February 2005.
Jorge Teixeira Pinto, Director of Geodetic Services, Instituto Portugués de Cartografía e Cadastro, Personal Communication, March 2002.
Lambert Conical Orthomorphic Projection Tables, India Zone I, issued by Office of the Chief of Engineers, Washington, DC, Second Edition, 1944.
Lambert Conical Orthomorphic Projection Tables, India Zone IIA, issued by Office of the Chief of Engineers, Washington, DC, 1943.
M. H. Edney, Mapping an Empire, The Geographical Construction of British India 1765–1843, 1997.
NAMRIA, 22 March 2017.
NIMA/NGA TR 8350.2, 1991.
PAIGH, Centro 1952, p. 419; UN 1979, p. 193.
Pakistan Library of Congress Country Study, 2009.
Paraguay Library of Congress Country Studies, 2007.
TR 8350.2, Appendix A.1.
UN 1979, p. 21; Twenty 1966, p. 108.
Webster's Geographical Dictionary.
World Factbook, 2022.

Q

State of Qatar 647

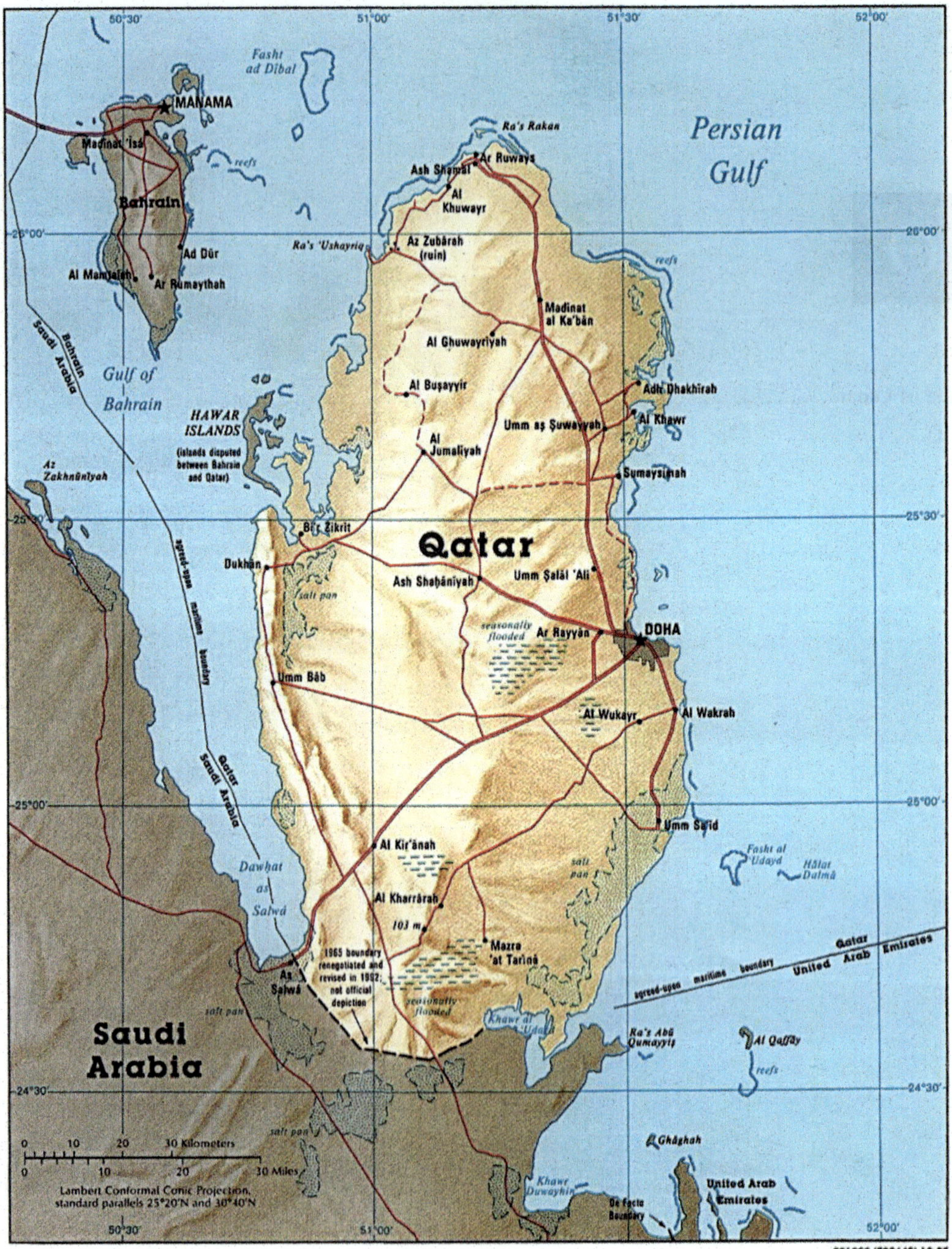

801898 (E00442) 12-95

State of Qatar

Having a coastline on the Persian Gulf of 563 km, Qatar is slightly smaller than Connecticut and borders Saudi Arabia (60 km). International boundaries have been signed with Bahrain, Iran, Saudi Arabia, and Abu Dhabi of the United Arab Emirates. The lowest point is the Persian Gulf (0 m), and the highest point is *Tuwayyir al Hamir* (103 m). The climate is arid and mild, with pleasant winters and very hot, humid summers. The terrain is mostly flat and barren desert covered with loose sand and gravel (*CIA World Factbook*).

"Human habitation of the Qatar Peninsula dates as far back as 50,000 years, when small groups of Stone Age inhabitants built coastal encampments, settlements, and sites for working flint, according to recent archaeological evidence. Other finds have included pottery from the Al Ubaid culture of Mesopotamia and northern Arabia (*circa* 5000 B.C.), rock carvings, burial mounds, and a large town that dates from about 500 B.C. at *Wusail*, some twenty kilometers north of *Doha*" (*Library of Congress Country Studies*).

"Ruled by the al-Thani family since the mid-1800s, Qatar transformed itself from a poor British protectorate noted mainly for pearling into an independent state with significant oil and natural gas revenues. During the late 1980s and early 1990s, the Qatari economy was crippled by a continuous siphoning off of petroleum revenues by the Amir, who had ruled the country since 1972. His son, the current Amir Hamad bin Khalifa al-Thani, overthrew him in a bloodless coup in 1995. Oil and natural gas revenues enable Qatar to have one of the highest per capita incomes in the world.

"Recently, Qatar's relationships with its neighbors have been tense, although since the fall of 2019 there have been signs of improved prospects for a thaw. Following the outbreak of regional unrest in 2011, Doha prided itself on its support for many popular revolutions, particularly in Libya and Syria. This stance was to the detriment of Qatar's relations with Bahrain, Egypt, Saudi Arabia, and the United Arab Emirates (UAE), which temporarily recalled their respective ambassadors from Doha in March 2014. Tamim later oversaw a warming of Qatar's relations with Bahrain, Egypt, Saudi Arabia, and the UAE in November 2014 following Kuwaiti mediation and signing of the Riyadh Agreement. This reconciliation, however, was short-lived. In June 2017, Bahrain, Egypt, Saudi Arabia, and the UAE (the 'Quartet') cut diplomatic and economic ties with Qatar in response to alleged violations of the agreement, among other complaints" (*World Factbook*, 2022).

It seems that the oldest Astronomical Station recorded for Qatar is at *Om, Al Da'asa* the origin of the 1941 datum and coincident with the Cassini-Soldner Grid referenced to the Helmert 1906 ellipsoid, where $a = 6,378,200$ m and $1/f = 298.3$. The Latitude of Origin for the *Al Da'asa 1941 Datum* is $(\varphi_o) = \Phi_o = 25°\ 22'\ 56.50''$ N, the Central Meridian $(\lambda_o) = \Lambda_o = 50°\ 45'\ 41.00''$ East of Greenwich, the Scale Factor at Origin by definition is unity, the False Northing = the False Easting = 100 km; sometimes the False Easting and False Northing are found to have been doubled. This was apparently the first plane coordinate system used for the *Dukhan Oil Field*. Another Astro Station local datum is reported to be at *Al Buşayyir*, but there does not seem to be any coordinates available for the 1946 datum. (The ellipsoid of reference is likely the Clarke 1880.)

The Qatar National Datum of 1970 has its origin at Station *G2, Balad Ibrahim*, where $\Phi_o = 25°\ 16'\ 10.6570''$ N, and $\Lambda_o = 51°\ 36'\ 20.6227''$ East of Greenwich, and the ellipsoid of reference is the International 1924 where: $a = 6,378,388$ m and $1/f = 297$. The projection associated with this datum is the Gauss-Krüger Transverse Mercator for the Qatar National Local Grid where the Latitude of Origin $(\varphi_o) = 24°\ 27'$ N, the Central Meridian $(\lambda_o) = 51°\ 13'$ E, the Scale Factor at Origin $(m_o) = 0.99999$, and the False Northing = the False Easting = 300 km.

I am informed by Mr. Gener Bajao that for offshore Qatar, the UTM Grid is used as referenced to the first major geodetic Datum of the Persian Gulf area established by W.E. Browne of the Iraq Petroleum Company in 1927–1931 at the South End Base at station *Nahrwan* (East of *Baghdad*) such that $\Phi_o = 33°\ 19'\ 10.87''$ North, $\Lambda_o = 44°\ 43'\ 25.54''$ East of Greenwich, and the Clarke 1880 is the ellipsoid of reference, where $a = 6,378,300.782$ m and $1/f = 293.4663077$.

"Hunting Surveys, Ltd. performed a survey of Qatar that was completed in 1972. First Order Tellurometer (EDM) traverses were performed over the entire country that comprised 92 traverse stations forming 10 closed loops and 2 junction figures. The 937-line kilometers were measured with an average distance between stations of 8.76 km. Traverse observations at each station consisted of 10 or 12 rounds of horizontal angles, half of which were measured in the evening and half in the following morning. Eight sets of Tellurometer measurement, 4 in each direction were taken at different times on different days. Three sets of simultaneous reciprocal vertical angles were measured in the middle of the day. The average standard deviation for the observed horizontal angles was 0.41″ and the worst value obtained for a full set of observations was 0.86″. The average standard deviation for the measured distances was 1:876,000 and the worst value obtained was 1:198,000. Standard deviations were not calculated in the case of vertical angles where spreads of up to 10 seconds of arc were accepted. LaPlace azimuth control was provided by 16-star Black's azimuth programs at 12 traverse stations spaced at approximately 60–70 km intervals around the network. The average standard deviation for the azimuths obtained was 0.41″ and the worst value obtained was 0.63″."

"The Qatar Local system coordinates were converted to *Ain el Abd Datum* by means of a datum and spheroid change by Cartesian shifts. The comparison of the values obtained from the two survey systems is given for Station 'G16,' the common station. ... The Qatar value: $\varphi = 24°\ 40'\ 21.6345''$ N, $\lambda = 50°\ 51'\ 37.0981''$ E, and the Saudi value: $\varphi = 24°\ 40'\ 22.4297$ N, $\lambda = 50°\ 51'\ 35.7352$ E'" (*National Geodetic Survey Final Report, September 1972, Hunting Surveys Limited*).

Thanks to a complete report on the Qatar National Datum of 1995 that I received from Mr. Alistair Strachan back in July of 1996, the three-parameter transformation from WGS84 Datum to QND 95 is: $\Delta X = -127.781$ m, $\Delta Y = -283.375$ m, and $\Delta Z = +21.241$ m, and the Qatar National Local Grid on the QND95 is where the Latitude of Origin $(\varphi_o) = 24°\ 27'$ N, the Central Meridian $(\lambda_o) = 51°\ 13'$ E, the Scale Factor at Origin $(m_o) = 0.99999$, the False Northing $= 300$ km, and the False Easting $= 200$ km. The ellipsoid of reference is again the International 1924. A coordinate transformation package called "QTRANS" is available from the government that includes the local geoid relations. Although a seven-parameter transformation is available from the Qatar Geographic Information Systems Centre, the three parameters given above are estimated to be good to about one meter. Extensive geodetic documentation is available freely from the government and nothing geodetic appears to be considered a military secret – a most enlightened philosophy that should be practiced by many larger nations throughout the world.

REFERENCES: Q

CIA World Factbook, December 2007.
National Geodetic Survey Final Report, September 1972, Hunting Surveys Limited.
Qatar Library of Congress Country Studies, 2007.
W.E. Browne, Iraq Petroleum Company, 1927–1931.
World Factbook, 2022.

R

Român ia 651
Russia Federation 657
Republic of Rwanda 663

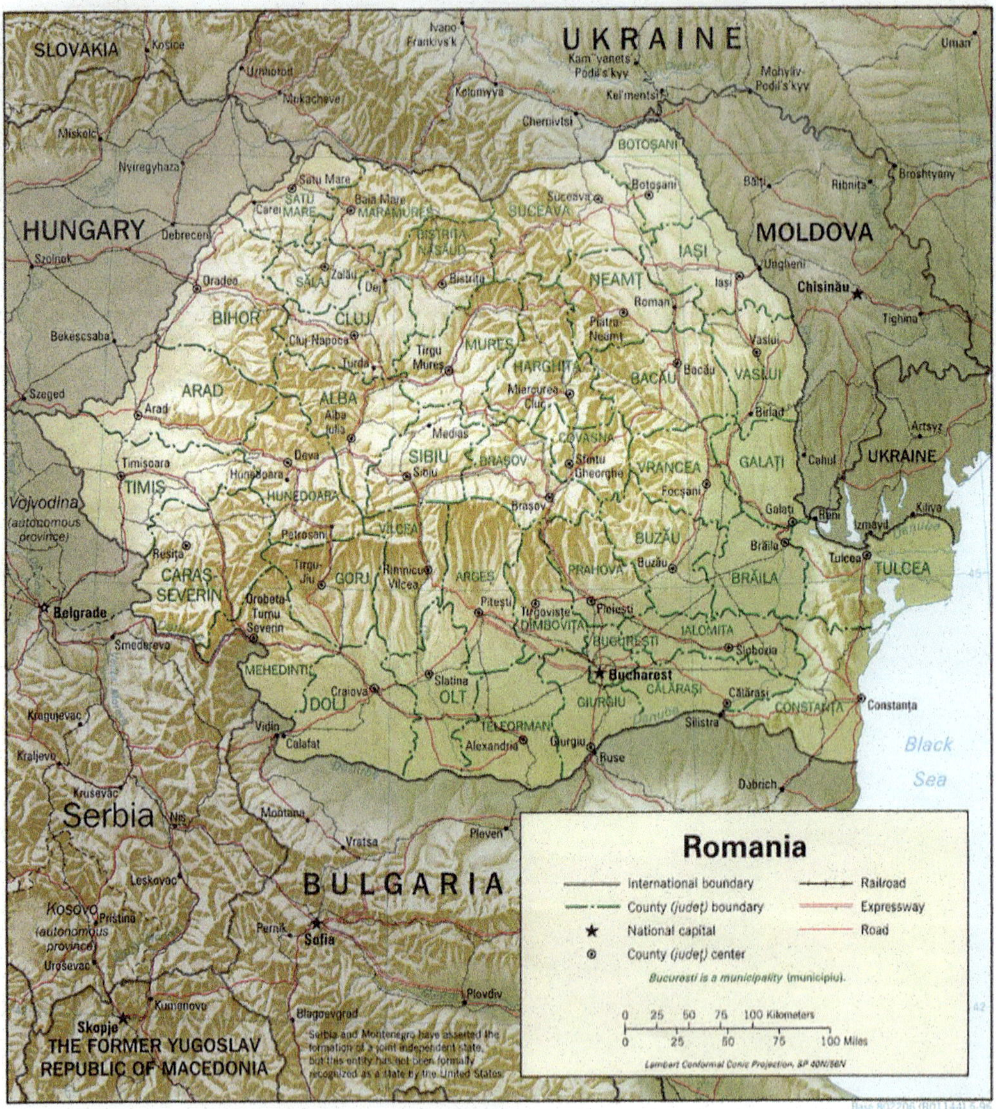

Base 802206 (R01144) 5-96

România

România is situated in Central Europe, in the northern part of the Balkan Peninsula and its territory is marked by the Carpathian Mountains, the southern border is the Danube River, and the eastern border is the Black Sea. Bordering countries are Ukraine (601 km) to the north, Moldova (683 km) to the northeast, Bulgaria (605 km) to the south, Serbia (531 km) to the southwest, and Hungary (424 km) to the northwest. The highest point is *Moldoveanu* (2,543 m), and the lowest point is sea level.

Traces of human presence date back to the Lower Paleolithic Period (~2 million years B.C.), and relatively stable populations in România were found beginning with the Neolithic Period (6–5,000 years B.C.). Strabo, a famous geographer, and historian in the age of Emperor Augustus wrote that "the *Dacians* have the same language as the *Getae*"; the basic difference is that the former inhabited the mountains and *Transylvania*, while the latter settled in the *Danube River valley*. The Romanians are the only descendants of the Eastern Roman stock, and their language is one of the romance languages. România has been referred to as a "Latin Island in a sea of Slavs."

"The principalities of *Wallachia* and *Moldavia* – for centuries under the suzerainty of the Turkish Ottoman Empire – secured their autonomy in 1856; they were *de facto* linked in 1859 and formally united in 1862 under the new name of România. The country gained recognition of its independence in 1878. It joined the Allied Powers in World War I and acquired new territories – most notably *Transylvania* – following the conflict. In 1940, Romania allied with the Axis powers and participated in the 1941 German invasion of the USSR. Three years later, overrun by the Soviets, România signed an armistice. The post-war Soviet occupation led to the formation of a communist 'people's republic' in 1947 and the abdication of the king. The decades-long rule of dictator Nicolae Ceausescu, who took power in 1965, and his Securitate police state became increasingly oppressive and draconian through the 1980s. Ceausescu was overthrown and executed in late 1989. Former communists dominated the government until 1996 when they were swept from power. România joined NATO in 2004 and the EU in 2007" (*World Factbook*, 2022).

The Austro–Hungarian Empire was surveyed at a time when topographic surveys were based partially on geometry and partially upon art. The topographer, more an artist than a surveyor, proudly made the map by himself. Therefore, the proverb: "There are plenty of surveyors, but few topographers." Maps are no longer produced by ingenious topographer-artists, but by a team of specialists like those that comprise the ASPRS. The Second Topographical Survey (in România) of the Austro–Hungarian Empire was the *Franziszeische Aufnahme* of 1806–1869 that utilized the Cassini-Soldner Grid of *Vizakna, Sibiu (Hermannstadt)* Observatory for Transylvania, România. The original coordinates of the Cassini Grid origin used from 1817 to 1904 were: $\varphi_o = 45°\ 50'\ 25.430''$ North, $\lambda_o = 41°\ 46'\ 32.713''$ East of Ferro, geodetically determined from Vienna. As per the European geodetic convention of the time, no false origin was employed, and coordinates were computed with quadrant signs. This Grid was cast on the von Zach 1812 ellipsoid, where $a = 6,376,385$ m and $1/f = 310$.

While still a part of the Austro–Hungarian Empire, Constantin Barozzi (1833–1921) helped in the Viennese measurement of the *Slobozia geodetic base* and in the triangulation chain that connected *Dobrogea* to *Transylvania*. By 1870, Barozzi was appointed as the chief (and founder) of the Scientific War Depôt, the national military mapping department of România. The third topographic survey of the Austro–Hungarian Empire (*Neue Aufnahme*) was conducted from 1869 to 1896 and was based on the St. Anna Datum of 1840 located in *Arad*, România. The origin was at *Arad South Base*: $\Phi_o = 46°\ 18'\ 47.63''$ North, $\Lambda_o = 39°\ 06'\ 54.19''$ East of Ferro (geodetically determined from Vienna). This Datum was originally reference to the Zach 1812 ellipsoid, but the Austrians introduced the Bessel 1841 ellipsoid as a new standard for the empire in 1869, where $a = 6,377,397.155$ m and $1/f = 299.1528$. The defining azimuth to *Kurtics* was determined astronomically, but the angular value was not published and is now lost. The *Arad 1840 Base* length was 8,767.578 m. The metric system was legally established in 1872, and the 1:25,000 mapping scale was introduced along with the polyeder (polyhedric) projection to eliminate inconsistencies in map sheet lines. Remember that in past columns I have pointed out that the polyhedric projection is mathematically equivalent to the local space rectangular (LSR) coordinate system that is commonly used in computational photogrammetry. Of course, back then they did not transform first to the Earth-centered Geocentric Coordinate System and then perform a 3×3 rotation secant or tangent to the surface of the ellipsoid the way we do now. (It's trivial with Fortran or C, but mind-boggling with tables of logarithms.)

A chain from *St. Anna base line* to *Sibiu (Hermannstadt)* Observatory was computed in 1846. The computations again were based on the Zach 1812 ellipsoid and *St. Anna* was used as a starting point. From these computations for *Sibiu Observatory*, the following values were obtained: $\varphi = 45°$ 50′ 28.95″ North, $\lambda = 41°$ 46′ 39.00″ East of Ferro, and azimuth to *Presbe*: $\alpha = 359°$ 16′ 33.78″. If these values are compared with the values derived in 1870 from Vienna University using the Bessel 1841 ellipsoid: $\varphi = 45°$ 50′ 25.97″ North, $\lambda = 41°$ 46′ 31.66″ East of Ferro, and azimuth to *Presbe*: $\alpha = 359°$ 16′ 43.24″. The result of this "bust" was that the cartographic adjustments were a nightmare for nearly a century when trying to fit other Datums to Românian sheets.

Thanks to Mr. John W. Hager, the *Constanta Datum* established by the Russians in the 19th century originated at the minaret of the main mosque in *Constanta*, where $\Phi_o = 44°$ 10′ 31″ North, $\Lambda_o = 29°$ 39′ 30.55″ East of Greenwich. The reference azimuth from East Pyramid to West Pyramid on the *Kyustendyi (Constanta) Base* $\alpha_o = 305°$ 15′ 01.7″, and the ellipsoid used was the Walbeck, where $a = 6,376,896$ m and $1/f = 302.78$.

The Austrians were at work in *Bukovina* and *Transylvania* through this time and established the *Kronstadt base* in 1876 that measured 4,130.141 m, ±1/2,000,000. A famous name in East European coordinate systems, *Kronstadt* is merely the Austrian name for the central Românian Alpine city of *Braşov*. Two ellipsoidal Bonne Grids were established in 1870, both cast on the Bessel 1841 ellipsoid. The West Românian Bonne $\varphi_o = 45°$ 00′ North, $\lambda_o = 26°$ 06′ 41.18″ East of Greenwich, the East Românian Bonne $\varphi_o = 46°$ 30′ North, $\lambda_o = 27°$ 20′ 13.35″ East of Greenwich. Both of these Grids had the scale factor at origin equal to unity and had no false origin.

During World War I, the necessity for a conformal Grid system became evident for artillery fire control, and Professor Cholesky developed a Lambert Conformal Conic Cadrilaj (Grid) for România. Cholesky imposed two conditions on the design:

i. The projection distance between parallels 45^G and 50^G shall equal the distance between parallels 50^G and 55^G, and both these distances shall be 500 km.
ii. The scale factor at the central parallel shall equal the reciprocal of the scale factors at the projection limits.

Thanks to Mr. Giles André of the Defence Geographic Centre in Middlesex, Placinţeanu in 1940 and Negoescu in 1942 attempted to re-create and extend the original Lambert-Cholesky Tables. The closest computational cartographic fit yielded the following parameters: Clarke 1880 ellipsoid (modified for this Grid) $a = 6,378,249.2$ m, $e^2 = 0.00680348764$, $m_o = 0.99844674$, $\varphi_o = 45°$ 02′ 29.216″ North, $\lambda_o = 24°$ 18′ 44.99″ East of Greenwich, False Easting = 500 km, False Northing = 504,599.11 m. These parameters are not a reflection of a Lambert-Cholesky projection, but for a projection in sympathy with Cholesky's over România.

After World War I, the new origin point for the *Bucharest Datum of 1920* was based on the *Bucharest Military Observatory* and the astronomical coordinates were transferred to point *Militari*, situated at the end of the Bucharest baseline, where $\Phi_o = 44°$ 26′ 07.2832″ North, $\Lambda_o = 26°$ 01′ 00.207984″ East of Greenwich. The reference azimuth to *Ciorogârla* (west end of base) $\alpha_o = 96° 43' 22.8''$ (measured from south), and the new ellipsoid used was the Hayford 1909 (later termed the International 1924), where $a = 6,378,388$ m and $1/f = 297$.

With 10 years of triangulation progress, the New Românian Datum of 1930 was established and referenced to the International 1924 ellipsoid. The origin was at *Observatorul Militar Astronomic din Dealul Piscului, Bucuresti,* called "Dealul Piscului" in the West, where $\Phi_o = 44° 24' 34.20''$ North ±0.06" (1895), $\Lambda_o = 26° 06' 44.98''$ East of Greenwich ±0.075" (1900.7). The reference azimuth to *Cotroceni* originally observed in 1895 was $\alpha_o = 127° 01' 53.005''$. A Rousilhe Stereographic Grid was developed with its projection center near the geographic center of the country. The defining parameters are $\varphi_o = 51^G = 45° 54'$ North, $\lambda_o = 28^G 2138.51 = 25° 23' 32.8772''$ East of Greenwich, FE = FN = 500 km, $m_o = 0.9996666666$. Since the origin is near *Kronstadt* (now *Braşov*), it came to be known as the "Stereographic Projection with Kronstadt as Central Point."

After World War II, România was introduced to the Soviet "System 42" Datum with its origin at Pulkovo Observatory, where $\Phi_o = 59° 46' 18.55''$ North, $\Lambda_o = 30° 19' 42.09''$ East of Greenwich. The reference azimuth to Signal A is: $\alpha_o = 317° 02' 50.62''$, and the new ellipsoid used is the Krassovsky 1940 where $a = 6,378,245$ m, and $1/f = 298.3$. The standard Grid used in the former U.S.S.R. and most former satellite countries is the Russia Belts Gauss-Krüger Transverse Mercator where the Grid parameters are identical to the UTM except that the Russia Belts have a scale factor at origin on the Central Meridian equal to unity. However, România did not go along with that for its internal use even though the Soviets used their Russia Belts for military topographic mapping of the country. The "Stereo 70" was developed for România based on the Hristow Oblique Stereographic projection. The projection center was selected as $\varphi_o = 45°$ North, $\lambda_o = 25°$ East of Greenwich, and associated parameters were FE = FN = 500 km, and $m_o = 0.999750$. Example computation point is: $\varphi = 44° 30' 30'' N$, $\lambda = 26° 03' 03'' E$ such that on the Stereo 70 Grid: Northing (X) = 334,794.541 m, Easting (Y) = 583,553.824 m. As a computational exercise using the identical geodetic coordinate but computing the Grid on the Kronstadt Stereo with its different ellipsoid, Northing (Y) = 345,588.461 m, Easting (X) = 552,344.592 m. Note that the Kronstadt Stereo unit of angular measurement is Grads ($100^G = 90°$) so: $\varphi = 49^G 45^C 37^{CC}.037 N$, $\lambda = 28^G 94^C 53^{CC}.704 E$ for the identical point. NGA lists the transformation from System 42 in România to WGS 84 Datum as: $\Delta X = +28$ m ±3 m, $\Delta Y = -121$ m ±5 m, $\Delta Z = 177$ m ±3 m, and this is based on four co-located points computed in 1997. The U.S. National Geodetic Survey assisted the government of România with the establishment of the România High Accuracy Reference Network, and for station *Dealul Piscülci* on System 42: $\varphi = 44° 24' 22.383'' N$, $\lambda = 26° 06' 44.126'' E$, $H = 89.275$ m, and on EUREF89: $\varphi = 44° 24' 22.71021'' N$, $\lambda = 26° 06' 38.74635'' E$, $h = 124.520$ m.

The International Boundary Treaties of România indicate that the border of the country is largely monumented or along streams and rivers. Interestingly, the boundary with Hungary distinguishes between navigable streams that are defined by the thread of the main channel or thalweg, and non-navigable streams that are defined by the geographical center or *medium filium aquae*. Thanks for help go to Mr. Russell Fox of the Ordnance Survey, Mr. Giles André of the Defence Geographic Centre, Mr. Dave Doyle of the National Geodetic Survey, and Mr. John W. Hager.

The successful development of countries in the 21st century is going to be largely dependent on the free availability of spatial data. Although there is little difficulty in a national government charging fair market value for data, the total suppression of spatial data distribution for military purposes of security is quite out of date and futile. Military operations in the recent past have clearly demonstrated that existing native topographic maps and existing native geodetic control have absolutely nothing to do with successful coordination of fire support. Secrecy of positional data and topographic maps only hurts the local economy and discourages foreign investment. It is hoped that the formerly "closed societies" will move to a more enlightened philosophy regarding GIS technology.

The transformation package offered by the Romanian government is TransDatRO: *http://www.ancpi.ro/index.php/en/download-2*. Transformations among the Transylvanian grid coordinates are possible courtesy of Dr. Gábor TIMÁR at: *http://sas2.elte.hu/tg/viteaz_en.htm*.

"Transylvania had an unsettled history in the 20th century. As a result of the repeated changing of the region between Hungary and Romania and of the Soviet dominance between 1945–1990, several map grid systems have been introduced. Besides the old Hungarian 'Marosvásárhely' Stereographic system and the Romanian Stereo-70 system, the Soviet-type Gauss-Krüger projection has been also introduced. These grids can be handled together with the world's quasi-standard, the UTM system, using this small MS Excel table.

"*Viteaz* is this MS Excel table that allows you to convert UTM, Soviet-type Gauss-Krüger, Romanian Stereo-70 and old Hungarian 'Marosvásárhely' Stereographic coordinates. You can enter your input in any of the listed projections and will be given the result in all the three other systems.

"Notice: I've written it for my pleasure – you can use it free, without any restriction. Distribute it freely but if you put it to your website, please place a link to this site, too.

Please, drop me a mail: timar@ludens.elte.hu with your new version if you modify the table."

This page is maintained by Gábor TIMÁR, Space Research Group, Dept. of Geophysics, Eötvös University of Budapest. Last modified: 12 June 2013.

Romania has 74 GPS RTK Continuously Operating Reference Stations at approximately 70 km spacing. The National (*relative – Ed.*) Gravity Network of the first and second orders (about 270 points) was observed by the Ministry of Defense – Topography and Cartography Directorate. The country has four points determined by Absolute Gravity.

Astrogeodetic leveling was part of a research project culminating in November 2015. Results showed accuracies of (*xi, eta*) of ±0.500″ and ±0.745″, respectfully with a Topcon M505AX with a CCD chip. * In comparison, in 2018 LSU purchased a Digital Zenith Camera that has an accuracy of ±0.100″ in *both* components – after one hour's observation – automatically from the University of Latvia, Institute for Geodesy & GeoInformatics.

- *https://iag.dgfi.tum.de/fileadmin/IAG-docs/NationalReports2015/IUGG_2011-2014_Final_Report_ROMANIA.pdf*

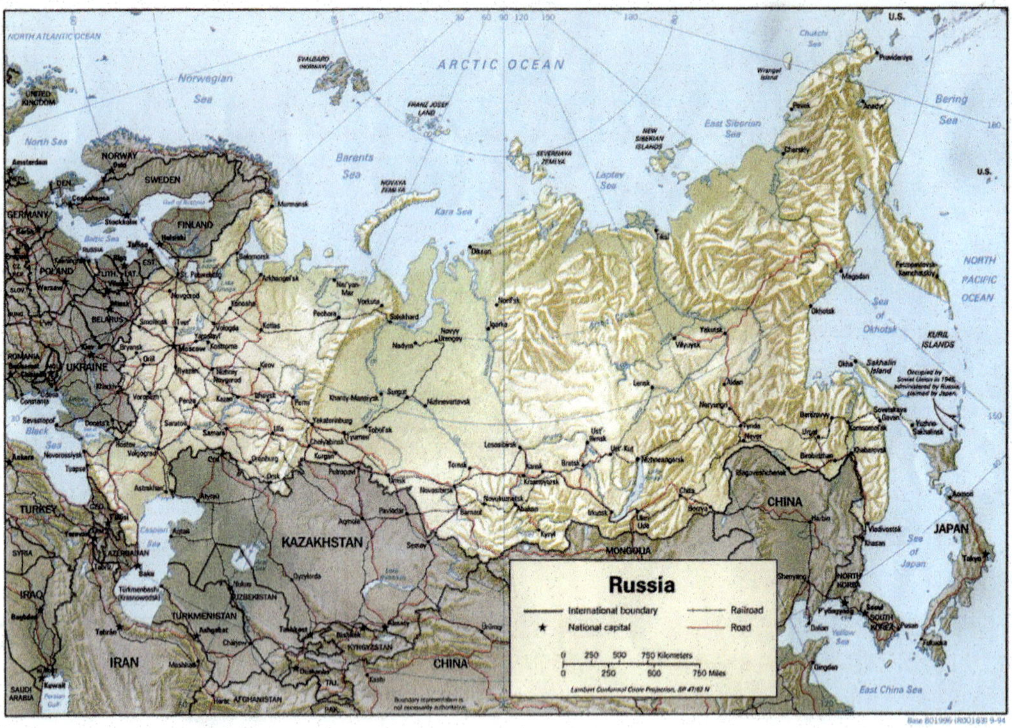

Base 801996 (R00163) 9-94

Russia Federation

Russia is bordered by Azerbaijan (338 km), Belarus (1,312 km), China (4,179 km), Kazakhstan (7,644 km), Korea (18 km), Latvia (332 km), Lithuania (261 km), Mongolia (3,452 km), Norway (191 km), Poland (210 km), and Ukraine (1,944 km). The terrain has a broad plain with low hills west of Urals; vast coniferous forest and tundra in Siberia; uplands and mountains along southern border regions. The lowest point is the Caspian Sea (–28 m); the highest point is *Gora El'brus (Гора Эльбрус)* (5,633 m), which is the highest point in Europe. Russia is the largest country in the world in terms of area but is unfavorably located in relation to major sea lanes of the world; despite its size, much of the country lacks proper soils and climates (either too cold or too dry) for agriculture; Lake Baikal, the deepest lake in the world, is estimated to hold one-fifth of the world's fresh water (*World Factbook*, 2015).

"Russia has its cultural origins in *Kyivan Rus*, the kingdom located in what is today Ukraine and Belarus. From here the Slavs expanded into modern European Russia. The birth of the Russian state is usually identified with the founding of Novgorod in A.D. 862, although until 1480 Russia was overrun by the Mongols. It was not until the Romanov dynasty (1613–1917) that Russia became the vast nation it is today – territorial expansion from the 17th to 19th centuries saw the country increase in size exponentially to include Siberia, the Arctic, the Russian Far East, Central Asia and the Caucasus. Peter the Great dragged the country kicking and screaming out of the Dark Ages, setting up a navy and building a new capital, St Petersburg, in 1703. Catherine the Great continued Peter's progressive policies to create a world power by the mid-18th century.

"The 19th century saw feverish capitalist development undermined by successively autocratic and backwards tsars. The most prominent example was Nicholas II, whose refusal to countenance serious change precipitated the 1917 revolution. What began as a liberal revolution was hijacked later the same year in a coup led by the Bolsheviks under Lenin, which resulted in the setting up of the world's first communist state.

"The Communist Party held power from 1917 until 1991, during which time Russia became a superpower, having created the Union of Soviet Socialist Republics (USSR) and absorbing some 14 neighboring states between 1922 and 1945. The terror of Stalin, the reforms of Khrushchev and the stagnation during the Brezhnev era finally led to Mikhail Gorbachev's period of reform known as *perestroika* in 1985. Within six years the USSR had collapsed alongside communism and reformer Boris Yeltsin led Russia into a new world of cutthroat capitalism" (*Lonely Planet*, 2015).

"Russia shifted toward a centralized authoritarian state under President Vladimir Putin (2000–2008, 2012-present) in which the regime seeks to legitimize its rule through managed elections, populist appeals, a foreign policy focused on enhancing the country's geopolitical influence, and commodity-based economic growth. Russia faces a largely subdued rebel movement in Chechnya and some other surrounding regions, although violence still occurs throughout the North Caucasus" (*World Factbook*, 2022).

"A U.S. telecommunications engineer arrested and charged with espionage in Russia will be allowed to leave under an agreement with the Russian security service, the U.S. State Department said Tuesday. State Department deputy spokesman James Foley said Richard Bliss' employer, the telecommunications company Qualcomm, and Russian security officials reached an agreement that would allow Bliss to depart for the United States possibly as early as Tuesday. 'We welcome his release,' Foley said. Bliss, 29, a field technician, was arrested Nov. 25 in the southern Russian city of Rostov-on-Don while installing a wireless telecommunications system for the Russian company *Rostovelektrosvyaz*. Interfax, quoting unidentified federal security officials, confirmed that Bliss would be allowed to go home for the holidays but could be called back to Russia if necessary.

There were no other details. One of Bliss' local attorneys said the technician was due to leave Russia late Tuesday night. 'He's flying out at 2300 hours,' Valery Petrayev said by telephone from Rostov, where Bliss has been held since his arrest. Petrayev said Bliss had been allowed home until Jan. 10 but might not be recalled then. He said the release was a sign that the Federal Security Service, or FSB, would not bring formal spying charges. 'They let him go for Christmas, New Year, until Jan. 10 for the moment, maybe longer. It depends on whether the FSB comes up with any evidence that he was spying,' he said. 'Who sends a spy home for Christmas? It's funny!' Petrayev added" (*Moscow Times, 24 December 1997*). (*As I recall, President Bill Clinton personally telephoned President Boris Yeltsin and explained that the young Bliss was ignorant of the fact that recording GPS coordinates of potential cell phone towers on hill tops was considered an act of espionage when Russian Triangulation Stations were ALSO on the same hill tops – Ed.*)

"There are three distinct periods of mapping activity in the U.S.S.R.; these coincide with the country's history and territorial changes. In the first period, between 1845 and 1920, the Russian Empire extended deeply into Europe and included the territories of Estonia, Latvia, and Lithuania, and large parts of Poland, Finland, and other neighboring countries. For the first time in the history of mapping into the Russian Empire, topographic maps were based on an accurate geodetic survey. The principal mapping agency for this period was the *Korpus Voyennykh Topografov* – KVT (Corps of Military Topographers) under the Russian Imperial General Staff.

"The second period started in 1920, after the Bolshevik Revolution, when the Russian Empire became the Soviet Union and lost large parts of its European territory to the independent states: Estonia, Latvia, Lithuania, Poland, and others. This period ended in 1945. During this time the Soviets established two new mapping organizations: the *Voyenno-Topograficheskoye Upravleniye* – VTU (Military Topographic Administration) under the *General'nyy Shtab Krasnoy Armii* (General Staff of the Red Army) and the *Glavnoye Upravleniye Geodezii i Kartografi* – GUGK (Main Administration of Geodesy and Cartography), first under the Council of Ministers and later under the Ministry of Interior. These two organizations surveyed the territory east of the new border to the Astrakhan-Moscow-Leningrad line and published maps at the scales of 1:25,000, 1:50,000, and 1:100,000. Only small portions of the Far East, around *Lake Khanka*, were covered by topographic maps at these scales.

"The third period started in 1945 after the end of World War II. The western boundary of the U.S.S.R. again moved westward, and Estonia, Latvia, and Lithuania were again engulfed by the Soviets, together with a large portion of former Polish territory. GUGK and VTU remained the two most important mapping organizations, and their main task was to consolidate the entire triangulation net of the country and establish a new system of coordinates based on the newly computed Krassovsky ellipsoid. All the existing maps were converted to the new system of coordinates and a photogrammetric compilation of unmapped territory was started. By 1956 the entire Soviet Union had been covered by topographic maps at the scale of 1:100,000" (*TM 5–248, Foreign Maps, Army Map Service, October 1963*).

"a. *Verst Maps*. These maps have heterogeneous coordinate systems based on the local astronomic positions or on Pulkovo Observatory, with vertical datum based on the nearest sea or lake level, established by barometric leveling. Horizontal control was computed on the Bessel 1841 spheroid ($a = 6,377,397.155\,m$, $1/f = 299.1528$ – *Ed.*) and the polyhedral projection (*same as the Local Space Rectangular as defined in the Manual of Photogrammetry, 5th & 6th editions – Ed.*). Longitudes were referenced to Pulkovo or Ferro (*Ferro was sometimes considered $17°\ 39'\ 59.411''$ West of Greenwich – Ed.*) and in some instances, Greenwich. Verst maps have no military grid.

"b. *Soviet Maps Produced before World War II*. These maps retained the Bessel spheroid ($a = 6,377,397$ m (sic); $\alpha = 1:229.15$ (sic)) but adopted the Gauss-Krüger projection and Greenwich as the prime meridian. There were two main systems of coordinates: Pulkovo 32 System ($\Phi_o = 59°\ 46'\ 18.71''$; $\Lambda_o = +30°\ 19'\ 39.55''$) (*Note: in TM 5–248, Astronomic Latitude of Origin (Φ_o) is abbreviated "B_o" which is German for Breite or Latitude, and Astronomic Longitude of Origin (Λ_o), is abbreviated "L_o" which is German for Lange or Longitude – Ed.*) and Svobodnyy 1935

($\Phi_o=51°$ 25′ 36.55″; $\Lambda_o=+128°$ 11′ 34.77″). The Pulkovo 1932 System was used in European U.S.S.R., and the Svobodnyy System was used east of the meridian. Discrepancy between the two systems reached 270 m. in northing and 790 m in easting. In Central Asia, the *Tashkent* 1875 System of coordinates was used with origin at the *Tashkent Observatory* ($\Phi_o=41°$ 19′ 31.35″; $\Lambda_o=+69°$ 17′ 40.80″), and in the Caucasus the *Baku* 1927 System was used, the origin at the Tower of Khan Court ($\Phi_o=40°$ 21′ 57.90″; $\Lambda_o=+49°$ 50′ 27.57″). All topographic maps larger than 1,500,000 show geographic values of sheet corners and graticules and carry the Gauss-Krüger military grid. Gauss-Krüger rectangular coordinates are computed for each 6° zone, the first zone starting at Greenwich; zones are numbered eastward. The origin of each zone is the intersection of the central meridian and the equator. A false easting of 500,000 meters is added to the central meridian to avoid negative values. Rectangular coordinates should be read up and right. (*Note that other than using a scale factor at origin equal to unity and the eastward numbering of the zone numbers, the Russia Belts are the same as the Universal Transverse Mercator UTM Grid System that uses a scale factor at origin equal to 0.9996 and zone numbers increasing westward starting at Greenwich – Ed.*)

"c. *Soviet Maps Published after World War II.* These maps were converted to the single coordinate system Pulkovo 1942 – (*now properly referred to as* "System 42" – *Ed.*) The origin of the system is the center of Pulkovo Observatory with corrected coordinates $\Phi_o=59°$ 46′ 18.55″; $\Lambda_o=+30°$ 19′ 42.09″ and azimuth from point *Bugry* to *Sablino* 121° 06′ 38.79″. The Krassovsky ellipsoid with dimensions a=6,378,245 m (*sic*) and $\alpha=1$:298.3 (*sic*) replaced the Bessel spheroid. The projection and military grid remained the same. The mean sea level of the Baltic Sea at the *Kronstadt* tidal gage was accepted as zero altitude. Coordinates are not reduced to sea level but to the reference ellipsoid, which is tangent to the geoid only at Pulkovo" (*ibid, TM-5-246, 1963*). The current NATO transformation parameters **from** System 42 (*European Russia – Ed.*) to WGS 84 are: $\Delta X=+28.0$ m, $\Delta Y=-130.0$ m, and $\Delta Z=-95.0$ m.

"The Russian Far East has a different story, however since the Russian professional literature on surveying and mapping does not include a comprehensive study of the Far East. Local datums established astronomically for chains surveyed in different regions included: *Khabarovsk* 1907 Datum where: $\Phi_o=48°$ 30′ 22.61″, $\Lambda_o=+135°$ 08′ 04.08″, $\alpha_o=48°$ 44′ 49.10″ to II Base Point *Matveyevka*. The length of the old 1907 Khabarovsk base line measured along the string, served as initial values in the primary computing of the 1st order chain *Khabarovsk – Iman – Vladivostok* and to it attached 2nd and 3rd order nets established in 1907–1909. The *Svobodnyy* 1935 Datum had its origin at station *Chernigovskiy* (coordinates mentioned above) and base line measurements were performed with the four-wire *Jäderin apparatus* as follows: Iman 1907, 9,837.648 m; *Knevichi* 1909, 5353.511 m; *Khabarovsk* 1909, 6906.296 later abandoned; *Khabarovsk* 1912, 6201.515 m; *St. Olga* 1912, 5443.905 m; N. *Tambovskoye* 1914, 4644.222 m; *Ferma* 1910, 6058.629 m and Ferma 1914, 6058.722. The *Iman* 1914 Datum origin at 359 Northern Terminal of the *Iman base line* defined on the Bessel 1841 ellipsoid is where: $\Phi_o=45°$ 53′ 07.54″, $\Lambda_o=+133°$ 41′ 40.23″, $\alpha_o=168°$ 40′ 14.12″ to 351 *Iman*, South Base Point. The *Harbin* 1905 Datum origin is at *New Harbin* 1904 where: $\Phi_o=45°$ 44′ 49.00″, $\Lambda_o=+126°$ 37′ 39.70″, $\alpha_o=65°$ 02′ 34.00″ to *Huanshanchzuiza*. The *Vladivostok* 1914 Datum origin at station 61 *Vladivostok* 1903 is where: $\Phi_o=43°$ 07′ 13.10″, $\Lambda_o=+131°$ 53′ 33.92″. The *Ferma* 1910 Datum origin at 750 *Ferma* I is where: $\Phi_o=52°$ 40′ 54.10″, $\Lambda_o=+140°$ 03′ 17.33″, $\alpha_o=265°$ 35′ 26.36″ to 747 *Verpli* (II Southeastern Terminal). The Langr 1909 Datum origin referenced to the Clarke 1880 ellipsoid where: a=6,378,249.145 m, 1/f = 293.465 is at 371(898) *Langr East Base* Point where: $\Phi_o=53°$ 18′ 59.12″, $\Lambda_o=+141°$ 27′ 38.68″, $\alpha_o=134°$ 35′ 49.7″ to West Base Point Terminal. Thanks to the customary use of *"maskirovka,"* the custom of obfuscating cartographic/geodetic data by the KGB/FSB, the *Langr 1909 Datum* origin was used for the "secret" coordinate system of the nearby offshore Sakhalin Island. The *De Kastri 1910 Datum* was referenced to the Clarke 1880 ellipsoid and the origin point at 235 *De Kastri Bay* is where: $\Phi_o=51°$ 27′ 50.7″, $\Lambda_o=+140°$ 48′ 41.4″, $\alpha_o=257°$ 24′ 56.47″ to 638 Southern Terminal.

"By the decree of the Council of Ministers of USSR No. 760 of 7 April 1946 an order was given to unify the entire USSR triangulation into one uniform system oriented at Pulkovo defined in 1942. The decree prescribed also transformation of coordinates and the recasting of maps referring to local datums into a newly established uniform system. (Connection between Pulkovo and *Svobodnyy* was established in 1935). In 1954 the transformation of coordinates of local triangulations into Pulkovo System 42 was completed" (*Analysis of Surveys in the Russian Far East, Army Map Service Technical Report No. 31, Andrew M. Glusic, September 1960, 114 pages*).

Rwanda and Burundi

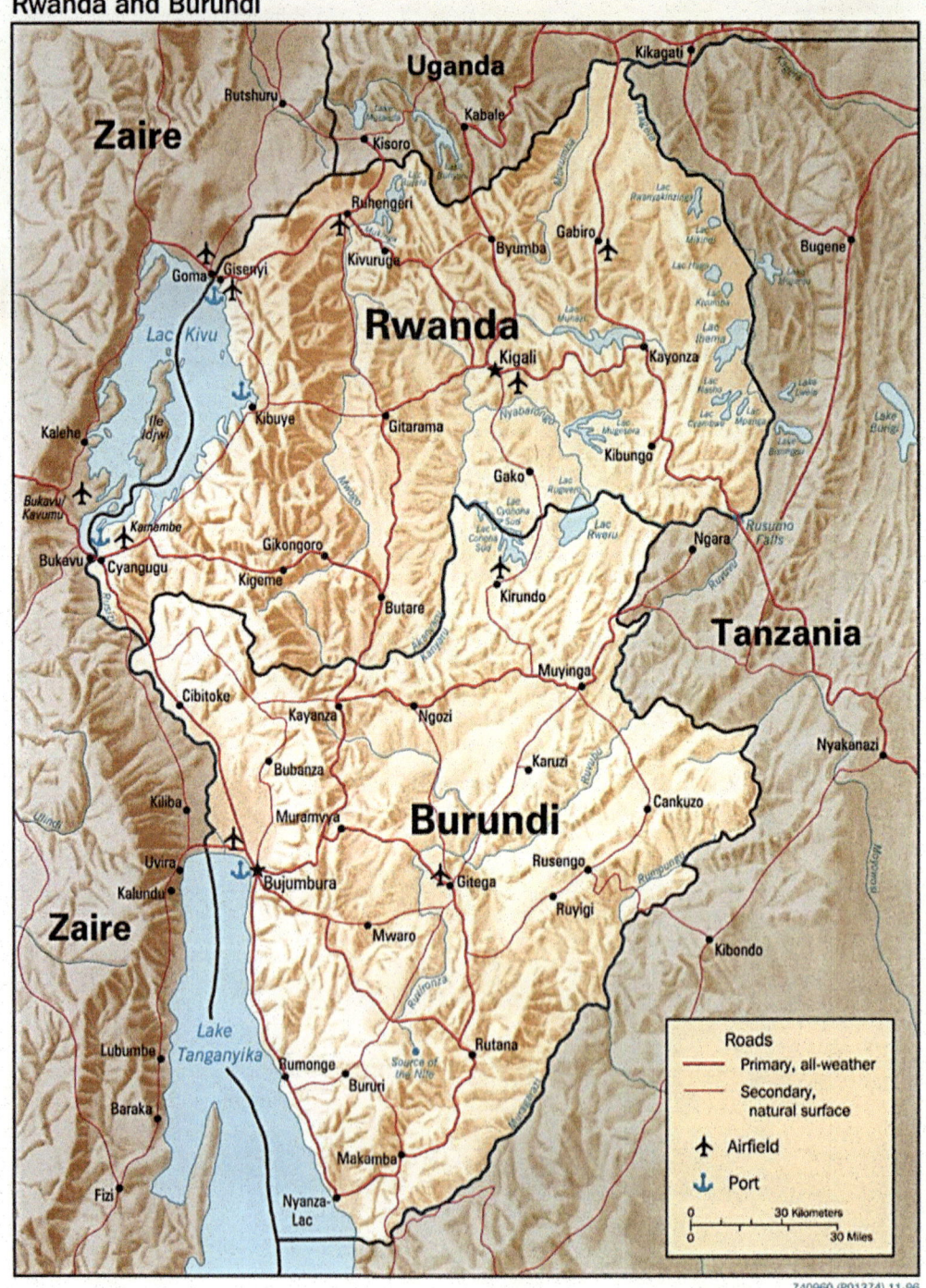

Republic of Rwanda

Slightly smaller than Maryland, Rwanda is bordered by Burundi (290 km), Democratic Republic of the Congo (217 km), Tanzania (217 km), and Uganda (169 km). The lowest point is the *Rusizi River* (950 m), and the highest point is *Volcan Karisimbi* (4,519 m). Rwanda is comprised of "mostly grassy uplands and hills; relief is mountainous with altitude declining from west to east" (*World Factbook*, 2022).

"The original Rwandans, the Twa Pygmies, were gradually displaced by bigger groups of migrating *Hutu* tribes' people from 1000 A.D. Later came the *Tutsi* from the north, arriving from the 16th century onwards. The authority of the Rwandan *mwami* (king) was far greater than that of his opposite number in Burundi, and the system of feudalism that developed here was unsurpassed in Africa outside Ethiopia. *Tutsi* overlordship was reinforced by ceremonial and religious observance. The Germans took the country in 1890 and held it until 1916, when their garrisons surrendered to Belgian forces during WWI. During Belgian rule, the power and privileges of the *Tutsi* increased, as the new masters found it convenient to rule indirectly through the *mwami* and his princes. However, in 1956, *Mwami Rudah*igwa called for independence from Belgium and the Belgians began to switch allegiance to the *Hutu* majority. The *Tutsi* favored fast-track independence, while the *Hutus* wanted the introduction of democracy first" (*Lonely Planet*, 2013).

"In 1959, three years before independence from Belgium, the majority ethnic group, the *Hutus*, overthrew the ruling *Tutsi* king. Over the next several years, thousands of *Tutsis* were killed, and some 150,000 driven into exile in neighboring countries. The children of these exiles later formed a rebel group, the Rwandan Patriotic Front (RPF), and began a civil war in 1990. The war, along with several political and economic upheavals, exacerbated ethnic tensions, culminating in April 1994 in a state-orchestrated genocide, in which Rwandans killed up to a million of their fellow citizens, including approximately three-quarters of the *Tutsi* population. The genocide ended later that same year when the predominantly *Tutsi* RPF, operating out of Uganda and northern Rwanda, defeated the national army and *Hutu* militias, and established an RPF-led government of national unity. Approximately 2 million *Hutu* refugees – many fearing *Tutsi* retribution – fled to neighboring Burundi, Tanzania, Uganda, and Zaire. Since then, most of the refugees have returned to Rwanda, but several thousand remained in the neighboring Democratic Republic of the Congo (DRC, the former Zaire) and formed an extremist insurgency bent on retaking Rwanda, much as the RPF tried in 1990. Rwanda held its first local elections in 1999 and its first post-genocide presidential and legislative elections in 2003. Rwanda in 2009 staged a joint military operation with the Congolese Army in DRC to rout out the Hutu extremist insurgency there and Kigali and Kinshasa restored diplomatic relations. Rwanda also joined the Commonwealth in late 2009. President Paul Kagame won reelection in 2010, and again in 2017 after changing the constitution to allow him to run for a third term" (*World Factbook*, 2022).

"Extending from west to east between the *Rusizi* (*Ruzizi*) and *Kagera rivers*, the Burundi–Rwanda boundary has a length of about 180 miles. Most of the boundary follows streams, of which the *Ruwa* (*Luhwa*), *Kanyaru* (*Akanyaru*), and *Kagera* are the principal rivers. It also passes through *Lac Cohoha* (*Lac Cyohoha–Sud*) and *Lac Rweru* (*Lac Rugwero*) and utilizes a number of straight-line segments between streams. There are no known boundary pillars demarcating the boundary" (*International Boundary Study, Burundi–Rwanda Boundary, No. 72* – June 1, 1966, U.S. Department of State).

"The Congo (Leopoldville)–Rwanda boundary has a length of approximately 135 miles. Southward from the Uganda tripoint on *Volcan Sabinyo* (*Sabinio*), it crosses an area of volcanic mountains for about 37 miles, passes through *Lac Kivu* for 71 miles, and follows the thalweg of the *Ruzizi* (*Rusizi*) for 27 miles. The Burundi tripoint is located at the junction of the *Ruzizi* and *Luhwa* (*Ruwa*) rivers. A tripoint with Uganda was established initially in 1892 following a compromise agreement between the United Kingdom and Germany relative to their respective spheres of influence west of Lake Victoria. Article I, Paragraph 1 of the agreement provided that the boundary between British and German territories should follow the parallel of one-degree south latitude across *Lake Victoria* and thence should continue westward, deflected only to include *Mount Mufumbiro* in the British sphere, to its intersection with the 30th meridian of longitude east of Greenwich. At this point the 30th meridian was the self-defined limit of the Congo Free State" (*International Boundary Study, Democratic Republic of the Congo (Zaire)–Rwanda Boundary, No. 52* – June 15, 1965, U.S. Department of State).

"The Rwanda-Tanzania boundary is demarcated throughout by the *Kagera* or by pillars. In the south the Burundi tripoint is located at the confluence of the *Mwibu* and *Kagera*, and in the north the Uganda tripoint is located at the confluence of the *Kagitumba* (*Kakitumba*) and *Kagera*. Except for straight-line segments between pillars in the central part, it follows the thalweg or median of the *Kagera*. The boundary is about 135 miles in length (*International Boundary Study, Rwanda–Tanzania Boundary, No. 69* – May 2, 1966, U.S. Department of State).

"The Rwanda-Uganda boundary has a length of approximately 105 miles, of which about 40 percent consists of streams. Eastward from the Congo (Leopoldville) tripoint on *Sabinio* (*Volcan Sabinyo*) peak, it passes through an area of volcanic mountains for about 29 miles, follows the thalweg of the *Vigaga* (*Kirurumu*) for 8 miles, crosses a second area of high peaks (including 3 miles of swamps) for 37 miles, and follows down-stream successively the thalwegs of the *Lubirizi, Muvogero* (*Tshinzinga*), and *Kakitumba* (*Kagitumba*) river for 31 miles to the Tanzania tripoint. The tripoint is located at the junction of the thalweg of the *Kakitumba* and the midstream of the *Kagera*. The boundary is demarcated throughout either by pillars or streams" (*International Boundary Study, Rwanda – Uganda Boundary, No. 54* – July 1, 1965, U.S. Department of State).

Thanks to John W. Hager, the classical geodetic origin of the Cape Datum/Arc 1950 Datum is at Buffelsfontein (in Port Elizabeth, Union of South Africa), where $\Phi_o=33°\ 59'\ 32.000''$ S, $\Lambda_o=25°\ 30'\ 44.622''$ E, azimuth to *Zuurberg* measured from South: $\alpha_o=183°\ 58'\ 15.00''$, $\xi=-3.46''$, $\eta=-0.59''$, $h_o=280.1\,m$, and the ellipsoid of reference is the Clarke 1880, where $a=6,378,249.145\,m$, and $1/f=293.465$. Rwanda is covered by the Congo Oriental Transverse Mercator Belt (*Fuseau*) 30° where $\lambda_o=30°$ E, $m_o=1.0$, False Easting$=220\,km$, False Northing$=565\,km$. The published relation between Arc 1950 Datum and WGS84 Datum near Rwanda (in Burundi) by the NGA is as follows: $\Delta X=-153\,m\ \pm20\,m$, $\Delta Y=-5\,m\ \pm20\,m$, $\Delta Z=-292\,m\ \pm20\,m$, and this solution was based on three points in 1991.

REFERENCES: R

Andrew M. Glusic, Analysis of Surveys in the Russian Far East, Army Map Service Technical Report No. 31, September 1960, 114 pages.
Dave Doyle of the National Geodetic Survey, Personal Communication, April 2001.
Gábor Timár at: http://sas2.elte.hu/tg/viteaz_en.htm.
Giles André of the U.K. Defence Geographic Centre, Personal Communication, April 2001.
Giles André, Defence Geographic Centre, Middlesex, Personal Communication, April 2001.
https://iag.dgfi.tum.de/fileadmin/IAG-docs/NationalReports2015/IUGG_2011-2014_Final_Report_ROMANIA.pdf.
International Boundary Study, Burundi–Rwanda Boundary, No. 72 – June 1, 1966, U.S. Department of State.
International Boundary Study, Democratic Republic of the Congo (Zaire)–Rwanda Boundary, No.52 – June 15, 1965, U.S. Department of State.
International Boundary Study, Rwanda–Tanzania Boundary, No 69 – May 2, 1966, U.S. Department of State.
International Boundary Study, Rwanda–Uganda Boundary, No 54 – July 1, 1965, U.S. Department of State.
J. Chris McGlone, *Manual of Photogrammetry*, 5th & 6th editions. ASPRS.
John W. Hager, Personal Communication, April 2001.
John W. Hager, Personal Communication, March 2013.
Lonely Planet, 2013.
Lonely Planet, 2015.
Moscow Times, 24 December 1997.
Russell Fox of the Ordnance Survey, Personal Communication, April 2001.
TM 5–248, Foreign Maps, Army Map Service, October 1963.
World Factbook, 2015.
World Factbook, 2022.

S

Sakhalin Island	669
The Islands of Samoa	675
Democratic Republic of São Tomé and Príncipe	679
Kingdom of Saudi Arabia	683
Republic of Senegal	689
Republic of Serbia	693
Republic of Seychelles	697
Republic of Sierra Leone	701
Republic of Singapore	705
Slovak Republic	709
Republic of Slovenia	713
Solomon Islands	717
Federal Republic of Somalia	721
Republic of South Africa	725
The Kingdom of Spain	731
Democratic Socialist Republic of Sri Lanka	735
Federation of Saint Kitts and Nevis	739
Saint Lucia	741
Saint Vincent and the Grenadines	743
Republic of the Sudan	749
Republic of Suriname	755
Kingdom of Swaziland	759
The Kingdom of Sweden	763
The Swiss Confederation	769
The Syrian Arab Republic	773

Japan-USSR: Northern Territories

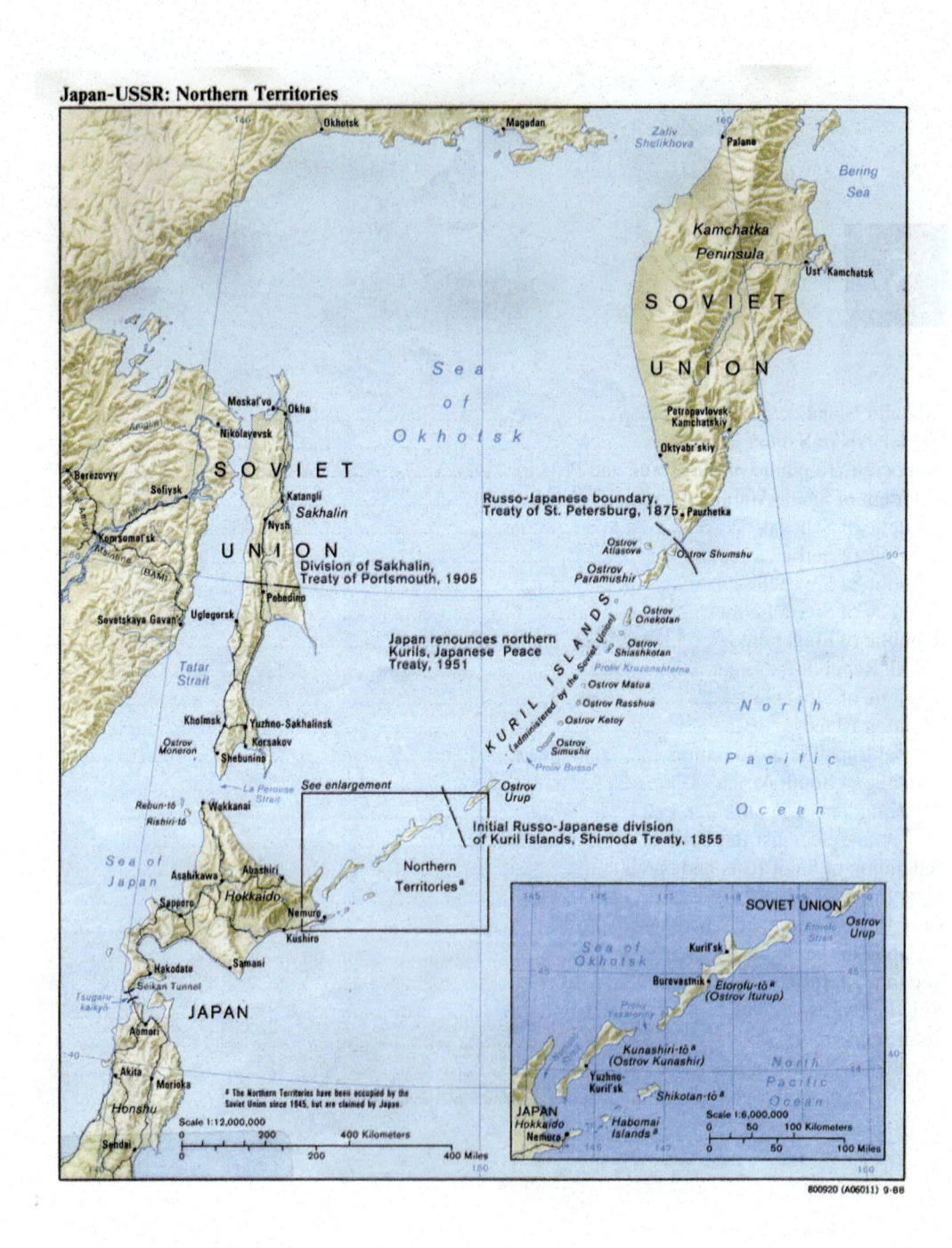

Okhotsk
Magadan
Zaliv Shelikhova
Palane
Bering Sea
Kamchatka Peninsula
Ust'-Kamchatsk

SOVIET UNION

Petropavlovsk-Kamchatskiy
Oktyabr'skiy

Sea of Okhotsk

Moskal'vo
Okha
Nikolayevsk

Berezovyy
Sofiysk
Katangli
Sakhalin
Nysh

Russo-Japanese boundary, Treaty of St. Petersburg, 1875
Pauzhetka

SOVIET UNION

Ostrov Atlasova
Ostrov Shumshu
Ostrov Paramushir

Komsomol'sk
Amur
Magistral (BAM)
Pobedino

Division of Sakhalin, Treaty of Portsmouth, 1905

Sovetskaya Gavan'
Uglegorsk

Japan renounces northern Kurils, Japanese Peace Treaty, 1951

Ostrov Onekotan
Ostrov Shiashkotan
Proliv Kruzenshterna
Ostrov Matua
Ostrov Rasshua
Ostrov Ketoy
Ostrov Simushir
Proliv Bussol'

KURIL ISLANDS
(administered by the Soviet Union)

Tatar Strait

Kholmsk
Yuzhno-Sakhalinsk
Korsakov
Ostrov Moneron
Shebunino

North Pacific Ocean

La Perouse Strait
See enlargement
Ostrov Urup

Rebun-tō
Rishiri-tō
Wakkanai

Initial Russo-Japanese division of Kuril Islands, Shimoda Treaty, 1855

Sea of Japan

Asahikawa
Abashiri
Northern Territories*
Sapporo
Hokkaido
Nemuro
Kushiro

Hakodate
Samani
Tsugaru-kaikyō
Seikan Tunnel

Inset (Enlargement)

SOVIET UNION
Sea of Okhotsk
Kuril'sk
Ostrov Urup
Etorofu Strait
Burevestnik
Etorofu-tō* (Ostrov Iturup)
Proliv Yekateriny
Kunashiri-tō (Ostrov Kunashir)
Yuzhno-Kuril'sk
Shikotan-tō*
North Pacific Ocean
JAPAN
Hokkaido
Nemuro
Habomai Islands*

Scale 1:6,000,000
0 50 100 Kilometers
0 50 100 Miles

JAPAN

Aomori
Akita
Morioka
Honshu
Sendai

* The Northern Territories have been occupied by the Soviet Union since 1945, but are claimed by Japan.

Scale 1:12,000,000
0 200 400 Kilometers
0 200 400 Miles

Sakhalin Island

Sakhalin is separated from the mainland by the narrow and shallow *Mamiya Strait* or *Strait of Tartary*, which often freezes in winter in its narrower part, and from *Hokkaido* (Japan) by the *Soya Strait* or *Strait of La Pérouse*. Sakhalin is the largest island of Russia, being 948 km long, and 25–170 km wide, with an area of 78,000 km². Nearly two-thirds of Sakhalin is mountainous. The Western Sakhalin Mountains peak in *Mount Ichara*, (1,481 m) while the highest peak of the Eastern Sakhalin Mountains is *Mount Lopatin*, (1,609 m) is also the island's highest mountain. *Susuanaisky* and *Tonino-Anivsky ranges* traverse the island in the south, while the swampy Northern-Sakhalin plain occupies most of its north.

"Sakhalin was inhabited in the Neolithic Stone Age. Among the indigenous people of Sakhalin are the *Ainu* and the *Nivkh*, as well as others. The Chinese in the Ming dynasty knew the island as *Kuyi*, the Ming sent 400 troops to Sakhalin in 1616, after a newfound interest because of northern Japanese contacts with the area, but later withdrew as it was considered there was no threat to China from the island. A Japanese settlement in the southern end of Sakhalin of Ootomari was established in 1679 in a colonialization attempt. The 1686 Nerchinsk Treaty between Russia and China, which defined the *Stanovoy Mountains* as the border, made no explicit mention of the island. However, as the Chinese governments did not have a military presence on the island, people from both Japan and Russia attempted to colonize the island, albeit from different ends. Sakhalin became known to Europeans from the travels of Ivan Moskvitin and Martin Gerritz de Vries in the 17th century, and still better from those of Jean-François de La Pérouse (1787) and Ivan Krusenstern (1805). On the basis of it being an extension of Hokkaido, geographically and culturally, Japan unilaterally proclaimed sovereignty over the whole island in 1845, as well as the *Kuril Islands*, as there were competing claims from Russia. In 1855, Russia and Japan signed the Treaty of Shimoda, which declared that both nationals could inhabit the island: Russians in the north and Japanese in the south, without a clear boundary between. Following the Opium War, Russia forced the Qing to sign the unequal Treaty of Aigun and Convention of Peking, under which China lost claim to all territories north of *Heilongjiang* (*Amur*) and east of *Ussuri*, including Sakhalin, to Russia. A *katorga* (penal colony) was established by Russia on Sakhalin in 1857, but the southern part of the island was held by the Japanese until the 1875 Treaty of Saint Petersburg, when they ceded it to Russia in exchange for the *Kuril Islands*."

In the first days of 1904, the famous American novelist, Jack London heard of Japan mobilizing for war against Russia. Obtaining employment from Hearst newspapers, London sailed for *Yokohama* with a contingent of other reporters on January 7th. After many columns had been posted and printed along with the first photographs of the conflict ever published in the United States by the Hurst newspapers, London was exiled from his attachment with the Japanese Army. His Courts Martial death sentence (fist fight on the battle front with a man that was stealing the feed of London's horse) was pardoned by the Imperial Japanese Army through the personal intercession of President Teddy Roosevelt. President Roosevelt later served as a mediator at the subsequent peace conference, which was held at Portsmouth, New Hampshire (August 9 to September 5, 1905). In the resulting Treaty of Portsmouth, Japan gained control of the *Liaotung Peninsula* (and *Port Arthur*) and the South Manchurian railroad (which led to Port Arthur), plus half of Sakhalin Island below 50° N reverted to Japan. "South Sakhalin was administrated by Japan as Karafuto-chō, with the capital *Toyohara*, today's *Yuzhno-Sakhalinsk*, and had quite a large number of migrants from Japan and Korea. *Karafuto* plays a major part in the history of the development of the coordinate system of Sakhalin Island.

"In August 1945, (*40 years later – Ed.*), according to the Yalta Conference (*Franklin D. Roosevelt, Winston Churchill, and Joseph Stalin – Ed,*), agreement, the Soviet Union took over the control of Sakhalin. The Soviet attack on South Sakhalin started on 11 August 1945, as a part of Operation August Storm, four days before the Surrender of Japan, after the bombing of *Hiroshima*. (Japan had quite a presence here and developed much infrastructure.) It was not until the 113th Rifle Brigade and the 365th Independent Naval Infantry Rifle Battalion from *Sovietskaya Gavan* landed on *Tōrō*, a seashore village of western Sakhalin on 16 August that the Soviets broke the Japanese defense line. The Soviets completed the conquest of Sakhalin on 25 August 1945 by occupying the capital, *Toyohara* "(Cambridge Encyclopedia, Vol. 65)."

"In order to cover the Pacific Coast, particularly the *Gulf of Tatary* and *Tatar Strait* with nautical charts, the Chief Hydrographic Administration was formed under Major General Zhdanko, the Expedition to Pacific Ocean (*Ekspeditsiya Vostochnoga Okeana*), which in 1909–1914 in this area carried out astronomic and magnetic observations, triangulation and topographical survey of the Coast at 1:8,400, 1:16,800, 1:21,000, 1:42,000 and 1:84,000 scale. In the period of six years 42 astro stations and 165 trig points were determined and 4400 km² (*of*) coastal area (*was*) covered by topographical survey. The triangulation on continental coast extends from *Sakhalin Bay* ($\varphi = 53°30'$ N) to *Cape Syurkum* ($\varphi = 53°04'$ N) and on the western coast of Sakhalin Island from *Baikal Bay* ($\varphi = 53°30'$ N) to *Cape Tyk* ($\varphi = 51°44'$ N). This triangulation consisting of 54 stations and 111 intersected points, in respect to its shape and manner in which the observations were carried out, is rather an exploratory than regular type of net. The acute angles and the sides with the lengths from 1–50 and more kilometers forming a net in which merely 35 triangles are closed and (*a*) large majority of trig points determined by intersection and even by resection would entirely justify such a statement. The stations of the low western coast of Sakhalin largely are determined by resection (Pothenot, Hansen) by observing the trig points located on the dominating continental coast. The astronomic stations were marked with wooden posts having (*an*) inscription consisting of (*the*) capital letter A and of (*a*) number showing the year of establishment. The trig stations were marked with a bottle set 2 feet below the ground level as (*a*) subsurface marker (*see Finland – for my comments concerning 'standard' Russian survey markers – Ed.*); the bottle was covered by 4 bricks tied with cement together. Above markers of stations, wooden quadripode and tripode signals (pyramids) were erected. The intersected points, which were not clearly visible permanent objects, were marked by wooden poles with target boards without permanent surface and subsurface marks. In the determination of astronomic positions, 10″ Hildebrand universal instruments and 6–11 portable chronometers were used. Only station 179 was determined by a sextant. The latitudes were determined by (*the*) Pevtsov method, the longitudes by (*the*) Tsinger method and the azimuths by observation of Polaris or (*the*) Sun. In the observation of triangulation 10″ Hildebrand and 10″ little Kern universal instruments were used. The observations were carried out by (*the*) directional method of Struve at 3–8 positions of the horizontal circle. In the computing (*of*) the angular condition of the closed triangles (*it*) was satisfied by $-\frac{1}{3}$ ($\varepsilon \pm \nu$) and (*the*) geographic coordinates (*were*) computed by Gaussian formulae on (*the*) Clarke 1880 ellipsoid (*where: $a = 6,378,249.145$, and $1/f = 293.465$*).

"The northern part of the coastal triangulation covering the continental coast north of *De Kastri Bay* and (*the*) northwestern coast of Sakhalin consisting of 125 trig points was carried out in 1909–1913. In 1909 the base line on the *Langr Island* (*Ostrov Baydukov – Ed.*) was measured by 4-meter wooden rods along the string (*plumb bob string lines – Ed.*) and with the length $d = 4,965.1859 \text{ m} \pm 1/_{70,000}$ served for the derivation of scale of the northern part of the net. In the same year for the purpose of the orientation of (*the*) northern part of the net, the Eastern Terminal of the base line was astronomically determined as follows: 371 (898) Langr, East Base Point $\Phi_o = 53°18'59.12'' \pm 0.9''$, $\Lambda_o = 141°27'38.68'' \pm 2.3''$ East of Greenwich, with the astronomic azimuth determined in (*the*) Western Terminal to (*the*) Eastern Terminal $\alpha_{902\text{WBP}-898\text{EBP}} = 134°35'48.7'' \pm 4.7''$. The position of (*the*) Eastern Terminal and the azimuth of the base line in the Western Terminal defined on (*the*) Clarke 1880 ellipsoid constitute the *Langr Datum* (*of 1909*).

"Independently of the northern part of the net, in 1910 in the *De Kastri Bay* (*Zaliv Chikhachëva –
Ed.*), a small triangulation consisting of 9 trig points was established. In order to provide the scale
for this triangulation, a base line with wooden rods along the string was measured. The length and
accuracy of the base line are as follows: d = 693.2565 m ± $1/_{18,152}$. As the initial point in the computing of geographic coordinates, the astronomic station 235 *De Kastri Bay*, Eastern Observatory,
determined in 1868 by Staritskiy and Onatsevich, was used with the following position: $\Phi_o = 51°\ 27'$
$50.7''$, $\Lambda_o = 140°\ 48'\ 41.4''$ East of Greenwich. For the orientation of the net served the azimuth from
the astronomic station 235 to 608 Southern Terminal of the base line determined in 1910 by observation of (*the*) Sun having the following value: $\alpha = 257°\ 24'\ 56.47''\ \pm 2.5''$. (*I do not believe a Solar
Azimuth accuracy of ±2.5" was likely observed in 1910; more likely ±25.0" – Ed.*) The position of
astronomic station 235 De Kastri Bay and the azimuth to 608 Southern Terminal defined on (*the*)
Clarke 1880 ellipsoid constitute (*the*) *De Kastri Datum* (*of 1910*).

"In 1912, stations 599 *Arbat Mountain* and 613 *De Kastri Mountain* were observed and included
into (*the*) 1910 triangulation. In 1913, the northern triangulation referring to *Langr datum* was
extended from the side 682 *Lazarev –* 684 *Pogobi* to *De Kastri Bay* and stations 613 *De Kastri
Mountain*, 600 *De Kastri, Kloster Kamp*, 599 *Arbat Mountain* and 596 *Cape Soborniy* of the local
1910–1912 net were observed and included into (*the*) *Langr system*. The side 600–613 common
to *Langr and De Kastri* system differs for $^{\Delta S}/_S = 1/_{2,500}$. In 1914, (*the*) De Kastri triangulation was
extended from the side 5 *De Kastri Mountain*, South Signal – 8 Kloster Kamp, Lighthouse (these
two stations are not included in (*the*) Far East Catalogue) to the side 520 *Topazovaya Mountain –
503 Lysaya* (*Syurkum*) *Mountain*. In the computing of the net consisting of 31 stations the starting
values used were those of 1910, hence the 1914 net together with *De Kastri* 1910–1912 triangulation
constitutes a homogeneous system of 40 trig points. Using the coordinates of the mentioned four
stations (594, 599, 600, and 613), which were in 1913 included into (*the*) *Langr system*, the *De Kastri
system* can be incorporated into (*the*) *Langr system*" (*Andrew W. Glusic, AMS Technical Report
No. 31*, Analysis of Surveys in the Russian Far East, *1960*).

During the 2000s, I was contacted by Marathon Upstream Sakhalin Services, Ltd. regarding
a little problem that they had with Sakhalin Island coordinates. It seems that the company was
founded jointly with the Russian government in order to design, construct, and build a pipeline
from the offshore petroleum platforms east of Sakhalin to the island and then south to the port of
Korsakov which is below the winter "ice line" of the northern Pacific Ocean. Marathon had initially
informed the Russians that a large-scale topographic survey was necessary in order to plan the
routing of the pipeline and that Marathon needed their permission to perform such a survey and
mapping project. The Russians responded that such a mapping project was unnecessary, 1:10,000
scale topographic maps already existed, and that they would provide them to Marathon. Well, the
Russians kept their word. They shipped complete topographic coverage of the southern half of the
island (south of 50° N), at 1:10,000 scale map sheets as printed on 11" × 17" paper. There was just
one problem … there was a kilometer grid overprinted on each sheet, but there were little rectangular holes in each paper map sheet that had been physically cut in all four margins with a razor blade
where the coordinates had presumably been printed! The values of whatever coordinate system
existed were literally cut out, and no information was offered regarding the kilometer grid system
of Sakhalin Island and its relation to any datum of latitude and longitude! Thanks to the KGB, but
likely its successor (the FSB *Федеральная служба безопасности*), that was a secret. About that
time, I received a telephone call from Marathon Oil in Houston, Texas.

I dropped by the office shortly thereafter and was greeted by Roger Holeywell, the Chief
Cartographer of Marathon Oil Company. I was briefed on their conundrum and shown the maps in
question. I noticed that in addition to the screwy-looking 11" × 17" paper maps with little rectangular holes cut on all four margins, the contour maps had spot heights shown at hills and mountain
tops with little dots and elevations shown. To me, some of those dots obviously were triangulation stations from past surveys; not all were photogrammetric spot heights. Being vaguely familiar
with the history of the ping-pong possession of the island between the Imperial Russians and the

Imperial Japanese, I thought that perhaps those spot height dots might offer an introduction to a geodetic Rosetta Stone of the island's coordinate system. Marathon had already contracted with a company to digitize the Sakhalin topographic maps, along with the dots and their elevation values. I was authorized to contact the scanning company and they later delivered digital files of the spot heights in an arbitrary Cartesian coordinate system with the elevations listed as each spot height's identification.

Being familiar with the Russian penchant for obfuscation for the sake of obfuscation, (*maskirova*), I expected that the "secret" FSB coordinate system was a conformal one and was probably based on a Gauss-Krüger Transverse Mercator projection. Question was, what were the coordinates of the origin? Holeywell already realized that the standard 1935 Russia Belts Transverse Mercator Grids did not match the major "rack" of the angle of the local grid system to the true meridian, and that was why I was called in for a consult. What was the "shift" of the coordinate system, what was the datum, and where was the origin located? Was there any relation to any "unclassified" coordinate system that Marathon Surveyors could use with their dual-frequency GPS receivers? (The Russians are notorious about *not* allowing foreign Geodetic Surveyors to collocate at *ANY* Russian triangulation stations!) Local coordinate systems are OK, but absolute ties are absolutely *verboten*!

The Marathon cartographic scanning contractors provided the local X-Y coordinates of the "spot heights" with the elevations listed as the IDs in an ASCII text list. My research partner at the time, the late Dr. Michael E. Pittman (retired from the University of New Orleans), took the text list and matched the IDs of some of the spot heights with a *Karafuto* Trig List that had corresponding identical spot heights. *Karafuto* Trig List control – what was that? Well, after World War II, Japan lost the war in the Pacific. The meeting in Yalta referenced in a paragraph above (Roosevelt, Churchill, and Stalin), resulted in the Japanese Geodetic Control being part of the "spoils of war" acquired by U.S. Forces, and the southern half of Sakhalin Island was returned to Soviet possession – as well as the creation of the "Iron Curtain." Also, *Geodetic Surveying in Japan during 1927–1929* by Rikuti Sokuryobu was previously published in the IUGG 1930 Proceedings in Japan and it detailed the progress of Japanese Geodetic Triangulation in *Karafuto* actually since 1905 – remember the Portsmouth Treaty? The *Karafuto* Trig List geodetic coordinates as transformed to the Japanese Transverse Mercator Grid (North Belt) were compared with the Marathon contractor's ASCII list of spot heights and 20 points were found to match! Those 20 points were indeed triangulation stations and were not just photogrammetric spot heights. The four-parameter transformation equations derived were: $X = 1.0002x - 0.0007y + 3861.32$, $Y = 0.0007x + 1.0002y - 2242.79$. The angle of the X-Y local grid system on the paper maps with respect to true north indicated that the central meridian was at some significant longitudinal distance from the central axis of Sakhalin Island. On close examination, the convergence angle between grid north and true north was suspiciously close to the value of the origin of the *Langr Datum of 1909* – over on the mainland. When the actual origin longitude was used for a central meridian, the system matched all *Karafuto* control to the digitized spot heights from the Russian $11'' \times 17''$ paper maps! Mr. Holeywell was happy.

SAMOA ISLANDS

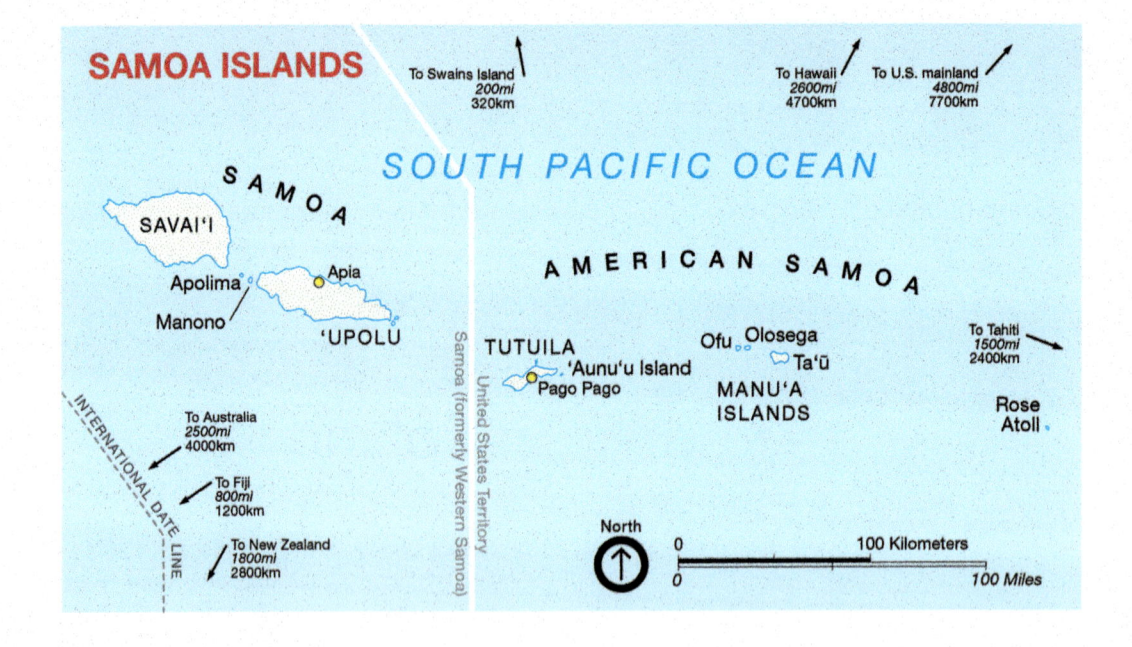

To Swains Island
200mi
320km

To Hawaii
2600mi
4700km

To U.S. mainland
4800mi
7700km

SOUTH PACIFIC OCEAN

SAMOA

SAVAI'I
Apolima
Manono
Apia
'UPOLU

AMERICAN SAMOA

TUTUILA
'Aunu'u Island
Pago Pago

Ofu Olosega
Ta'ū
MANU'A ISLANDS

Rose Atoll

To Tahiti
1500mi
2400km

INTERNATIONAL DATE LINE

Samoa (formerly Western Samoa)
United States Territory

To Australia
2500mi
4000km

To Fiji
800mi
1200km

To New Zealand
1800mi
2800km

North

0   100 Kilometers
0   100 Miles

The Islands of Samoa

American Samoa is slightly larger than Washington, DC. The lowest point is the Pacific Ocean, and the highest point is *Lata Mountain* (964 m). Samoa is slightly smaller than Rhode Island, the lowest point is the Pacific Ocean, and the highest point is *Mount Silisili* (1,857 m).

Thanks to *American Samoa Tourism 2009* and to *Samoan Sensation 2011*, the islands of independent Samoa are comprised of nine islands: *Savai'i* and *Upolu* with the most inhabitants, and with only two others, *Manono* and *Apolima* being inhabited. The other five are: *Fanuatapu, Namu'a, Nuutele, Nuulua, and Nuusafee*. American Samoa, a territory of the United States is comprised of five volcanic islands: *Tutuila, Ta'u, Ofu, Olosega, Aunu'u*, and of two atolls: *Rose Atoll* and *Swain's Island*. Polynesians first migrated to the Samoan Islands more than 3,000 years ago. In 1722, the Dutch explorer Jacob Roggeveen became the first European to arrive in Samoa, but for most of the 18th-century European influence was limited to occasional trading with ships that passed through the islands. In 1872, the U.S. Navy established a harbor in *Pago Pago* on the island of *Tutuila*. In 1899 Western Samoa was passed into German hands while the United States took control of what is now American Samoa. In 1914, New Zealand took control of Western Samoa, and it was the first Polynesian state to gain independence in 1962.

"During the German's Administration of Samoa early in the 20th century, German surveyors started the first ever surveying system in 1914 and used as the basis for all surveys in Samoa. This triangulation system was established and called the Observatory Origin. Between 1921 and 1927 the circuit traverse was carried out for the entire area of *Apia* based on the *Observatory Origin* established in 1914. In 1941 the German Triangulation was recalculated and shifted to the north pier of some meters from the first origin established in 1914 and called it the *Lemuta Origin*. This shift from the first origin in 1914 to the one established in 1941 formed the now so called *Lemuta Datum* as the datum used for all cadastral survey within the country up till now. All control traverses based on this datum called *Lemuta* were completed between 1953 and 1954" (Polutea, *et al., Improving Samoa's Geographic Information Services through the Upgrade of its National Geodetic Survey Network*, Government of Samoa, 2004).

"The Lemuta datum was established in 1953 by traverse survey, and has an arbitrary origin located near the Observatory on *Mulenu'u Peninsular*. The Observatory datum, which also has its origin somewhere near the Observatory, is an even older datum that is sometimes used. A preliminary attempt at this was done in 1995 by New Zealand resident surveyors without the benefit of high precision GPS to eliminate errors in the primary control. It is understood from anecdotal information that there may be differences in the Lemuta datum [sic] over different parts of the Islands" (*Infrastructure Asset Management Project – Phase II*, Government of Samoa, 2004). Thanks to John W. Hager, "*Apia Observatory Transit House* is the same as *Lemuta*. Lemuta at Apia Observatory Transit Pillar, $\Phi_o = $ S 13° 48′ 26″ ±0.4″ probable error (1911), $\Lambda_o = $ W 171° 46′ 30″, ±2″ probable error (1921), referenced to the International ellipsoid where $a = 6,378,388$ meters and $1/f = 297$. Latitude was observed by the German Government, surveyor Lammert. Longitude observed by New Zealand government. The pillar is in the garden of the observatory. Based on the above values, *Matautu Point F.S.* is $\varphi = $ S 13° 49′ 01.6″, $\lambda = $ W 171° 45′ 09.6″. Reference is a Hydrographic Report, *Apia Harbor*. I observed a grid on 1:20,000 topo sheets for Western Samoa which I called the Lemuta Datum Grid. TM, International, $\varphi_o = $ S 13° 48′ 26″, $\lambda_o = $ W 171° 46′ 30″, FN = FE = 0, scale factor = 1.0."

Samoa used the Western Samoa Integrated Grid, WSIG, which was adopted in 1988. The grid was based on the WGS72 Datum with the Western Samoa Integrated Grid, (WSIG): Projection: Transverse

Mercator, ellipsoid: WGS72 where $a=6,378,135$ meters and $1/f=298.26$, Central Meridian: 188° E (–172°) Scale Factor=unity, False Easting=700,000, and False Northing=7,000,000.

"The Samoan Geodetic Network is comprised of 34 survey marks of which 19 new marks have been established. The network comprises 26 primary stations and 8 secondary stations. The horizontal datum for Samoa is defined as the Samoan Geodetic Reference System 2005 (SGRS2005). The following high precision fundamental geodetic stations referred to the GRS80 ellipsoid where $a=6,378,137$ meters and $1/f=298.257222101$ are: 102 – *Faleolo* CGPS $\varphi=$S 13° 49' 55.95916", $\lambda=$W 171° 59' 58.32189", h=47.600m, and 104 – *Fagalii* CGPS $\varphi=$S 13° 50' 57.14900", $\lambda=$W 171° 44' 18.34120", h=76.875 m. This datum is defined within the International Terrestrial Reference Frame 2000 (ITRF2000) at the epoch of 2016.00. The Samoan Map Grid is defined as UTM Zone 2. The shift **from** WGS72 **to** SGRS2005 is: $\Delta X=0.460$m, $\Delta Y=-13.615$m, $\Delta X=5.786$m. The SGRS2005 coordinates of Observatory Doppler are: $\varphi=$S 13° 48' 52.903267", $\lambda=$W 171° 46' 50.92481", H=1.41 m" (*The Samoan Geodetic Network,* Government of Samoa, December 2005).

Again thanks to John W. Hager, "America Samoa 1962 (code AMA) at *Betty 13 ecc* (USGS), $\varphi_o=14°$ 20' 08.34" S±0.13" ($\Phi_o=$same), $\lambda_o=170°$ 42' 52.25"±0.08" ($\Lambda_o=...$ 52.15"), $\alpha_o=184°$ 15' 07.47"±0.21" to *Betty 13* eccentric Azimuth mark measured from south, Clarke 1866 (*where $a=6,378,206.4$ meters, $b=6,356,583.8$ meters – Ed.*), $H_o=5.43$ meters. Reference is *United States Department of the Interior, Geological Survey, Horizontal Control Data*. This was also the position of Doppler 10655.

"*Rose Atoll (1939)* (no code) at Observation Spot, $\Phi_o=14°$ 32' 51.91" S, $\Lambda_o=168°$ 08' 33.7" W, $\alpha_o=37°$ 25' 11.29" to *Lan* from south. Astro observation by astrolabe, using chronograph and radio time signals, U.S.S. *Bushnell*. Taken from NOS chart 83484. *Swains* (1939) (code ASP), at Astro Pier (U.S.S. *Bushnell*), $\Phi_o=11°$ 03' 35.97" S, $\Lambda_o=171°$ 04' 24.99" W, $\alpha_o=232°$ 54' 21.76" *Gagie* to *Teleapa* from South. Astro observation by astrolabe, using chronograph and radio time signals, U.S.S. *Bushnell. Swains Island* (code SWS) at Observation Spot, $\Phi_o=11°$ 03' 18" S, $\Lambda_o=171°$ 05' 53" W. The existence of this datum was from a request from the UK in 1994 for a datum code. *Tutuila 1962* (code TUT), at *Tutuila Astro Sta.* USGS (unadjusted) or AMS 10–62 Astro Station, $\Phi_o=14°$ 19' 50.530" S, $\Lambda_o=170°$ 42' 45.767" W, $\alpha_o=89°$ 02' 33.6" to *Betty 13 ecc.* From the same source file as American Samoa 1962 there are values of *Tutuila 1962* based on American Samoa 1962 datum of $\varphi_o=14°$ 20' 08.334" S, $\lambda_o=170°$ 42' 51.882" W, $\alpha_o=89°$ 02' 35"." According to *TR 8350.2*, **from** American Samoa Datum of 1962 **to** WGS84 Datum: $\Delta X=-115$ m±25 m, $\Delta Y=+118$ m±25 m, $\Delta X=+426$ m±25 m.

The current new datum for American Samoa is NAD83 (2002–HARN) where the U.S. National Geodetic Survey has computed NADCON tables for Ofu, Olosega, and Ta'u Islands listed as the **eshpgn Grid Pair File Prefix, and for *Tutuila,* and *Aunu'u Islands* listed as the **wshpgn Grid Pair File Prefix, thanks to Dave Doyle of the National Geodetic Survey. The NADCON files state: "… American Samoa never went through the intermediate step of island datum to NAD83 (86). Those islands were adjusted directly from their old island datums to HPGN/HARN."

Sao Tome and Principe

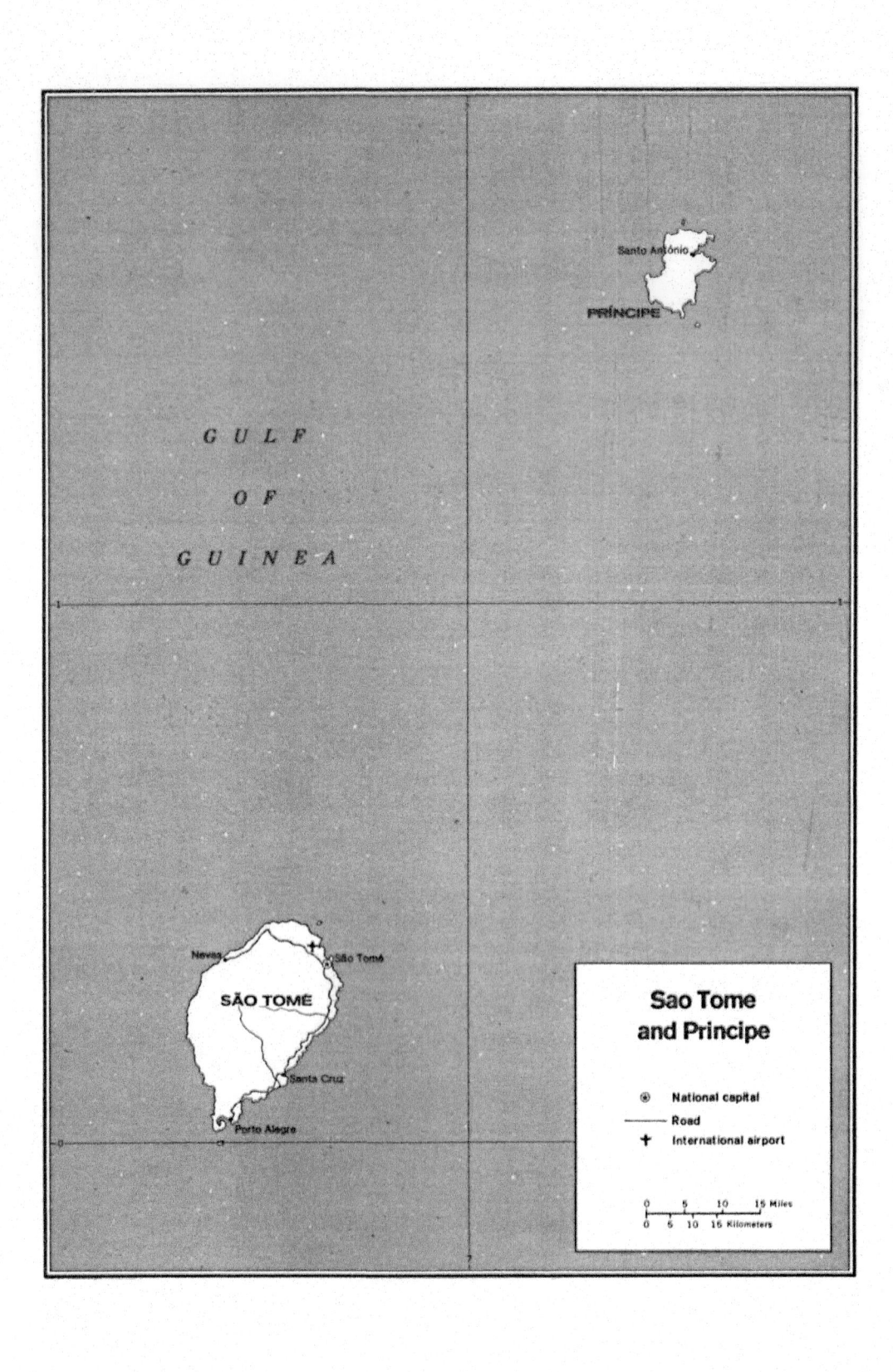

- ⊛ National capital
- —— Road
- ✢ International airport

Democratic Republic of São Tomé and Príncipe

São Tomé and Príncipe is the smallest country in Africa, being more than five times the size of Washington, DC. It has zero land boundaries; its coastline is 209 km long and is measured from claimed archipelagic baselines with a territorial sea of 12 nautical miles. The terrain is volcanic, mountainous and the lowest point is the Atlantic Ocean (0 m); the highest point is *Pico de São Tomé* (2,024 m) (*World Factbook*, 2013).

"Before being 'discovered' and colonized by the Portuguese in the late 15th century, the islands of São Tomé and Príncipe were comprised of rainforests dense with vegetation and birdlife, but, most likely, no people (though there is a legend that present-day Angolares were really the first inhabitants of the land). The islands' volcanic soil proved good for cultivation, and, under Portuguese rule, by the mid-16th century the islands were the foremost exporter of sugar, though the labor-intensive process required increasing amounts of slaves from Africa. When the price of sugar fell and slave labor proved difficult to control, the islands increasingly looked towards the slave trade to bolster the economy, becoming an important weigh station for slave ships heading from Africa to Brazil. In the 19th century two new cash crops, coffee, and cocoa, overtook the old sugar plantations. By the early 20th century São Tomé was one of the world's largest producers of cocoa. In 1876 slavery was outlawed but was simply replaced with a similar system of forced labor for low wages. Contract workers came in from Mozambique, Cape Verde and other parts of the Portuguese empire. During these times there were frequent uprisings and revolts, often brutally put down by the Portuguese. In 1953 the *Massacre of Batepá*, in which many Africans were killed by Portuguese troops, sparked a full-fledged independence movement. Portugal held on, however, until the fall of the fascist government in 1974, after which it got out of its colonies in a hurry. São Tomé & Príncipe achieved independence on 12 July 1975. The Portuguese exodus left the country with virtually no skilled labor, an illiteracy rate of 90%, only one doctor and many abandoned cocoa plantations. An economic crisis was inevitable. Manual Pinto da Costa, who was the first president and, until then, a moderate, was forced to concede to many of the demands of the more radical members of his government. The majority of the plantations were nationalized four months after independence, legislation was passed prohibiting any one person from owning more than 100 hectares of land, and a people's militia was set up to operate within workplaces and villages. The country remained closely aligned with Angola, Cuba and communist Eastern Europe until the demise of the Soviet Union, when Santoméans began to demand multiparty democracy. The first multiparty elections were held in early 1991 and led to the inauguration of the previously exiled Miguel Trovoada as the new president in April of that year. Elections in 2001 brought Fradique de Menezes to power. De Menezes pledged to use revenues from increased tourism and exploitation of the country's newly discovered offshore oilfields to improve the standard of living and modernize the islands' infrastructure. Grand changes seemed imminent. But complications with extracting the oil in addition to possible overestimations of the oil deposits have delayed economic progress, and there is a palpable growing restlessness in the deeply indebted and impoverished nation. A brief and bloodless coup attempt was peacefully resolved in 2003 while the president was out of the country. De Menezes was re-elected in 2006 in internationally observed, peaceful elections. São Tomé presently scrapes by on US$25 million a year of foreign aid and US$5 million in cocoa exports" (*Lonely Planet*, 2013).

"Carlos Vila Nova was elected president in early September 2021 and was inaugurated 2 October 2021. New oil discoveries in the Gulf of Guinea may attract increased attention to the small island nation" (*World Factbook*, 2022).

In 1915, Admiral Gago Coutinho was Chief of the Geodetic Mission of São Tomé and Príncipe. Since 1898 he had lived for 20 years in the African bush, sleeping in camping tents, working for boundary demarcations and geodetic triangulations in Timor, Moçambique, Angola, and São Tomé. His triangulation reconnaissance related by him was that "it was necessary to use techniques similar to navigation like to set a course with the compass, ... recognize land and make observations with the sextant at the top of the trees, to sound, ..., to find the most highest and suitable point." In 1909 the Colonial Office of Portugal ordered 4 theodolites from Filotécnica Salmoiraghi under specifications by Gago Coutinho. To provide the highest accuracy the horizontal circle was to be covered and the scales were to be engraved in platinum to avoid the oxidation of the African climate. During the first field campaign, Gago Coutinho noticed that the instrument was far from ideal, its principal defects being the ocular thumbscrew and the circles which at the time were very difficult if not impossible to engrave accurately. Some improvements made them operational and were being used in the São Tomé and Príncipe Geodetic Mission (1915–1917), also directed by Gago Coutinho, and the first Cape Verde geographical mission (1918–1921). To scale the geodetic network, a 10 km baseline was measured every 200 km. Coutinho was the first Portuguese Surveyor to use invar wires, and the most used lengths were 8, 24, and 48 m. He found that after 10 years of use, the invar wires reached their final lengthening of a third of a millimeter. (*This is a curious characteristic of the quench-annealed invar alloy of steel in that it can, (but not always), "jump" to a different (longer) length under field conditions. – Ed.*) Using appropriate procedures, the possible accuracy was 1/1,000,000 (*Geodetic Field Operations for Cartography – an Overview Over the First Portuguese Geodetic Mission in the Colonial Territories 1907–1910*, Paula Santos and Ana Roque, FIG 29 May 2012).

It is thought that the *São Tomé* Datum origin is at triangulation station *Fortaleza*, where $\Phi_o = 00°$ $20'$ $49.02''$ N, $\Lambda_o = 06°$ $44'$ $41.85''$ E and is referenced to the International 1924 ellipsoid, where $a = 6,377,388$ m and $1/f = 297$. (*Now the National Museum – Ed.*) The *Príncipe* Datum origin is at triangulation station *Morro do Papagaio*, where $\Phi_o = 01°$ $36'$ $46.87''$ N, $\Lambda_o = 07°$ $23'$ $39.56''$ E and is also referenced to the International 1924 ellipsoid (*Limits in the Seas No. 98- São Tomé and Príncipe*, U.S. Department of State, November 1, 1983).

However, according to Mr. John W. Hager, "*São Tomé* (code PST) at *Pico São Tomé* is at latitude, $\varphi_o = 00°$ $16'$ $10.72''$ E, longitude, $\lambda_o = 06°$ $32'$ $48.953''$ E, azimuth, $\alpha_o = 223°$ $33'$ $26.903''$ to *Mukinkí*, Clarke 1866 (*where $a = 6,378,206.4$ m, $1/f = 294.9786982$ – Ed.*), altitude, $H_o = 2,023.74$ meters. Astro latitude is $\Phi_o = 00°$ $16'$ $19.0''$. International ellipsoid values are latitude, $\varphi_o = 00°$ $16'$ $10.74''$ E, longitude, $\lambda_o = 06°$ $32'$ $48.97''$ E. References: *Ilha De. S. Tomé Resultados Finais 1915–1918*, Gago Coutinho, (Carlos Viegas), Capitão de Mar e Guerra, Lisboa, 1920. *International Hydrographic Bureau Special Publication No. 24c, Geographic Positions*, p. 52, Monaco, January 1951. The IHB cites Clarke 1866 to 1951. Maps of 1957 (and later) are on International. IHB also gives azimuths to (2) *Mukinkí*, (1) *Fortaleza*, (19) *Ponta-Furada*, and (7) *Rôlas*.

"Now for Principe. Morro do Papagaio (code MDP) at *Morro do Papagaio* is at latitude, $\varphi_o = 01°$ $36'$ $46.87''$ N, longitude, $\lambda_o = 07°$ $23'$ $39.65''$ E, azimuth, $\alpha_o = 183°$ $11'$ $59.62'' \pm 0.26''$ *Abade* to *Ponta Capitão* from south, International ellipsoid, altitude, $H_o = +680.45$. Reference: *Triangulaçã e levantamento da carta da Ilha do Príncipe; Longitudes em s. Tomé; Megnetismo*, Lisboa, 1934.

"Google Earth is a bust – mostly clouds. If I remember correctly, the 1:25,000 topographic maps (dated 1957) of São Tomé show a large amount of narrow-gauge railroads. They are in the cocoa plantations and went up and down hills. In some places they were connected by zig-zag lines which indicated that they were too sinuous to map. I couldn't locate any examples on the imagery. Re the 1:25,000 maps I mentioned earlier, map grid - Projection: Mercator, Ellipsoid: International Origin: São Tomé, latitude 00° 16' 10.74" n, longitude 6° 32' 48.97" E False Northing=False Easting=0. *n.b.* north and east are negative. I can't make sense of my notes. I also have the latitude of origin as the mid-latitude of the area, that is, 00° 12' 00.00" N. This gives an origin value of N=29,816.61

and E=0. My suspicion is that the map sheets had the grid and geographic values of the control printed on them. If only... I also have a note sheet for Príncipe but no information. Suspicion, again, that I expected a similar grid there but did not have the time to get a map sheet from the library and analyze it. *JWH*."

Current aeronautical information states that the international airports are on the WGS84 Datum.

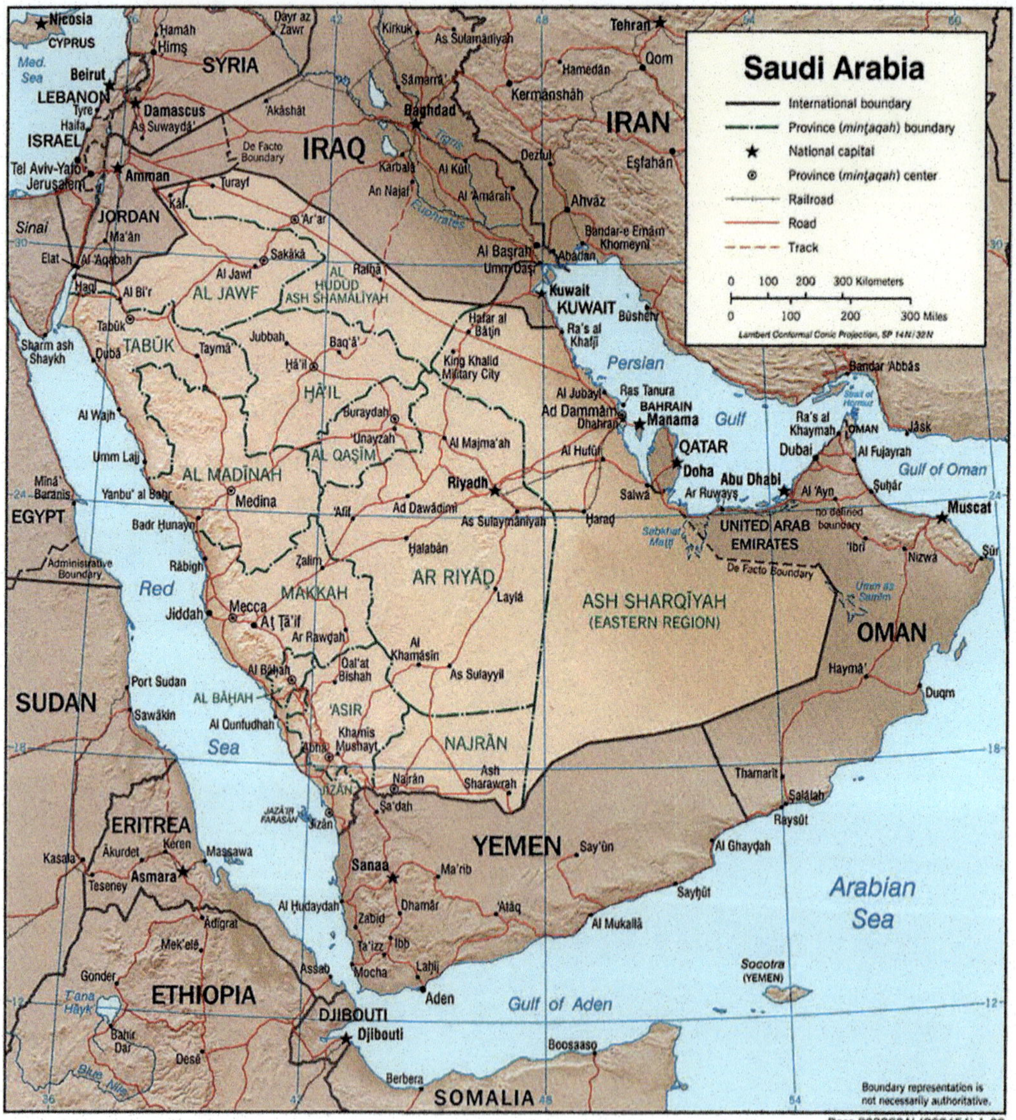

Base 802922AI (C00454) 1-03

Kingdom of Saudi Arabia

The kingdom is bordered by Iraq (814 km), Jordan (728 km), Kuwait (222 km), Oman (676 km), Qatar (60 km), UAE (457 km), and Yemen (1,458 km). With a coastline of 2,640 km, Saudi Arabia's lowest point is the Persian Gulf (0 m); although they prefer to call it the "Arabian Gulf," and the highest point is *Jabal Sawdā´* (3,133 m). With regard to international boundary treaties with Saudi Arabia, land agreements appear to exist with Jordan, UAE, Iraq, and Oman. For Limits in the Seas, continental shelf boundaries have been signed with Iran and Bahrain. In both cases, the principle of equidistance has been followed, more or less (*Office of the Geographer, U.S. Department of State*).

"The first concrete evidence of human presence in the Arabian Peninsula dates back 15,000 to 20,000 years. Bands of hunter-gatherers roamed the land, living off wild animals and plants. As the European ice cap melted during the last Ice Age some 15,000 years ago, the climate in the peninsula became dry. Vast plains once covered with lush grasslands gave way to scrubland and deserts, and wild animals vanished. River systems also disappeared, leaving in their wake the dry riverbeds (*wadis*) that are found in the peninsula today. This climate change forced humans to move into the lush mountain valleys and oases. No longer able to survive as hunter-gatherers, they had to develop another means of survival. As a result, agriculture developed – first in Mesopotamia, then the Nile River Valley, and eventually spreading across the Middle East. The development of agriculture brought other advances. Pottery allowed farmers to store food. Animals, including goats, cattle, sheep, horses and camels, were domesticated, and people abandoned hunting altogether. These advances made intensive farming possible. In turn, settlements became more permanent, leading to the foundations of what we call civilization – language, writing, political systems, art and architecture. Located between the two great centers of civilization, the Nile River Valley and Mesopotamia, the Arabian Peninsula was the crossroads of the ancient world. Trade was crucial to the area's development; caravan routes became trade arteries that made life possible in the sparsely populated peninsula. The people of the peninsula developed a complex network of trade routes to transport agricultural goods highly sought after in Mesopotamia, the Nile Valley and the Mediterranean Basin. These items included almonds from Taif, dates from the many oases, and aromatics such as frankincense and myrrh from the Tihama plain. Spices were also important trade items. They were shipped across the Arabian Sea from India and then transported by caravan. The huge caravans traveled from what is now Oman and Yemen, along the great trade routes running through Saudi Arabia's *Asīr Province* and then through *Makkah* and *Al Madīnah*, eventually arriving at the urban centers of the north and west. The people of the Arabian Peninsula remained largely untouched by the political turmoil in Mesopotamia, the Nile Valley and the eastern Mediterranean. Their goods and services were in great demand regardless of which power was dominant at the moment – Babylon, Egypt, Persia, Greece or Rome. In addition, the peninsula's great expanse of desert formed a natural barrier that protected it from invasion by powerful neighbors" (*Embassy of Saudi Arabia, 2008*).

During the early part of the 20th century, the Survey of India, while under British rule, extended some of their triangulation arcs into the Arabian Peninsula – as I recall from my own readings in recent years. In 1910, the British War Office published topographic maps of Arabia in four sheets at 32 miles to the inch (approximately 1:2,000,000 scale). The British Directorate of Military Survey produced maps at a scale of 1:253,440 (1 inch=4 miles), along the Red Sea from surveys in 1915–1917 (*Foreign Maps, TM 5–248, 1963*). In 1933, Saudi Arabia granted the first concession for oil exploration to an American company – ESSO, and the Arabian American Company was founded: ARAMCO.

The earliest geodetic surveying of any importance were the ARAMCO surveys that started in the 1930s. The original datum is *Umm Er Rus* at point 506 *Umm Er Rus*, $\Phi_o = 26°\ 19'\ 04''$ North, $\Lambda_o = 50°\ 07'\ 50''$ East, and the defining astronomic azimuth to Station 511 *Midra Jinubi* from south is: $\alpha_o = 82°\ 34'\ 32.64''$. The ellipsoid of reference is the Clarke 1866, where $a = 6,378,206.4$ m and $1/f = 294.9786982$, and elevation $= 149.7$ (units unknown). The Grid system associated with this datum is the ARAMCO Gauss-Krüger Transverse Mercator Zones 1–10 where each zone is a 2° belt with central meridians (λ_o) from 37°E to 55°E, scale factor at origin (m_o) $= 0.9999$, False Easting $= 150$ km, False Northing $= 100$ km, and False Northing Latitude at Origin $= 12°$N for each zone. According to John W. Hager, "As best I can remember, the ARAMCO surveys ran from south of the former Kuwait-Saudi Neutral Zone to about the Qatar-Saudi boundary. There were several arcs, but they were never adjusted as a whole."

Thanks again to John W. Hager, "In 1954 there was a SHORAN survey in northeast Saudi Arabia by the name of *Carmine* performed by Worldwide Aerial Surveys. (Worldwide was a combination of Aero Services Corporation and Fairchild Aerial Surveys created to handle large projects.) The purpose was to provide mapping photography for 1:250,000 scale mapping. From my experience with similar work in Iran, I would estimate that the trilateration network yielded positions of the ground control stations to better than 10 meters and positions of the photo nadir points at better than 50 m. The boundaries between Saudi Arabia and Kuwait, and Saudi Arabia and Iraq were in dispute and there were two Neutral Zones as a result. Finally, all the parties got together and resolved the problem. One result was point: *Ain el Abd* (HIRAN 2) $\varphi = 28°\ 14'\ 06.171''$ N, $\lambda = 48°\ 16'\ 20.906''$ E."

Another datum according to John W. Hager is the *Selwa Datum* at ARAMCO Station 915, $\Phi_o = 24°\ 44'\ 06.270''$N, $\Lambda_o = 50°\ 48'\ 08.381''$E, and the defining astronomic azimuth to an unknown station is: $\alpha_o = 351°\ 05'\ 24.456''$. The ellipsoid of reference is also the Clarke 1866, elevation $= 5.7295$ (units unknown) at *Khor Ed Duan*. Mr. Hager continued, "About 1957, Aero Services Corporation out of Philadelphia (later Aero Service Division of Western Geophysical in Houston), did a trilateration net using the MRA 1 Tellurometer. It ran along the coast from about latitude 27°N south to *Fort Tarut* and *Ra's Tann rah*. The following season the trilateration net was continued south with the purpose of establishing the boundary between Saudi Arabia and Bahrain. This portion ran into trouble with numerous ambiguities in the readings. The fact that a number of the lines were over water probably did not help matters. Army Map Service, with advice from the Tellurometer Corp. was able to straighten things out and get a satisfactory solution."

Again, thanks to Mr. Hager, "*Ain el-Abd* (1964) at Point 'A', *Ain el-Abd*, $\Phi_o = 28°\ 14'\ 06.968''$N $\pm 0.076''$ (P.E.), $\Lambda_o = 48°\ 16'\ 27.868''$E $\pm 0.044''$(P.E.), and the defining astronomic azimuth from north to 127(AMI) is: $\alpha_o = 307°\ 07'\ 34.85''\ \pm 0.08''$(P.E.), **Clarke 1880** ellipsoid, height $= 52$ meters. The point is described as ARAMCO "Q" or AMINOIL 43, a 2-inch iron pipe embedded in concrete set in the ground, located beside a natural spring named *Ain el-Abd*." Considering the probable errors of the observations and the choice of ellipsoid, my guess is that this point was originally an Astro Station by the British Royal Navy.

Established in 1925 and used during World War II for portions of the Arabian Peninsula, the Mecca-Muscat Zone was established by the British General Staff, Geographic Section and was known as one of the "British Grids" where the projection was the fully conformal *secant* Lambert Conical Orthomorphic. The Latitude of Origin (φ_o) $= 45°$N, the Central Meridian (λ_o) $= 23°$E, the Scale Factor at Origin (m_o) $= 0.99907$, the False Easting $= 1,000$ km, and the False Northing $= 600$ km. This ersatz datum used the Clarke 1880 ellipsoid, where $a = 6,378,300.78$ meters, $1/f = 293.466308$. I am informed by NGA that the current version of *GeoTrans* will now accommodate the British Definition parameters for defining a Lambert zone. From every public source available, it appears that all unclassified military mapping is on the Clarke 1880 ellipsoid, and the UTM Grid is utilized.

On 12 October 1966, the first comprehensive geodetic survey of Saudi Arabia was commenced by a consortium of British, French, Dutch, Japanese, and Saudi Arabian companies which completed a 15,000 km survey in 45 months. For this survey, the *Rub'al Khali – the Empty Quarter* – was excluded. The network incorporated 702 traverse stations and about 2,500 benchmarks. All were permanently monumented, paneled, and recorded on strip photography flown at 1:30,000 scale. All the routes were traversed to First-Order specifications and double-run leveled. All the routes were leveled to Second-Order standards except for the line connecting *Jeddah, Al Riyādh, Az Zahrān (Dhahran)*, and *Ain el Abd,* along which First-Order leveling and relative gravity observations with Worden meters were taken. The traversing and leveling lines followed the same routes as closely as the terrain permitted, and common stations or height connections were established at frequent intervals rarely exceeding 100 km. In addition, a flare triangulation net was planned between the 47 primary stations using five of the traverse sections as baselines. Six tide gauges were installed; four beside the Red Sea and two in the Persian Gulf, and simultaneous tidal observations were taken for a year initially. Two standardization bases were established with Geodimeters to calibrate the Tellurometers used for traversing. In all, 189 Laplace stations were observed for azimuth control and geoid computations.

"Field operating conditions in Saudi Arabia still preserve some unique features. The size of the country and the limited supplies obtainable in provincial towns made the logistic support of mobile field parties a vital task. The scheduled services of Saudi Arabian Airlines were used to fly in personnel, fresh food, and spares. For most of the length of the Red Sea coast there is a narrow coastal plain with patches of *sebkah* salt flats. Behind this the mountains of *Al Hejāz* and *Asīr* rise to a ridge elevation of between 1,000 and 3,000 m. East of the ridge the igneous rocks and lava flows of the Western Shield gradually become submerged beneath the sand plain with just occasional hill ranges outcropping. Further east a series of westward facing sedimentary escarpments, the largest of these being the *Tuwaiq* escarpment of Jurassic limestone, break up the monotony of the plains. These escarpments in turn dip beneath the sands of the *Rub'al Khali, the Dahna*, and the *Great Nefud*, which provide the conventional image of the desert with huge longitudinal sand dunes often vivid orange in color. Beside the Persian Gulf, low sandy hills descend gently to the *sebkahs* along the coast. Across this obstacle course a network of paved roads is being extended at the rate of 1,000 km a year, but only a few thousand kilometers had been completed at the time of the survey. The climate is harsh and immoderate but not often disagreeable. The summer months are predictably hot and fortunately – because of the heat haze – quite unsuitable for geodetic observations. The winter can be surprisingly cold with light frosts in the early mornings, and snow on occasions. In spring and autumn, storms bring rain and flash floods. In November 1967 the rainfall all over the country was exceptionally heavy and brought all the field parties to a temporary halt. It is always windy, a hot sand blast in summer, penetratingly cold in winter. Sandstorms blew tents away and even brought a tower down. The diurnal temperature range is enormous, and one removes layers of clothing progressively during the morning only to replace garment after garment during the afternoon" (J. Leatherdale, *The Geodetic Survey of Saudi Arabia,* Chartered Surveyor, December 1970).

In a later paragraph of Leatherdale's paper on page 273, he provides insight to the common question regarding the proper ellipsoid to associate with the *Ain El Abd Datum of 1970.* "The initial computations of the traverse net were made on European Datum as defined by the geographical coordinates of the HIRAN 2 station at *Ain El Abd* near the Kuwait Neutral Zone border. Deviation and separation were not defined. Trigonometric heights were calculated and adjusted to the leveling net at all benchmark connections. Distances were reduced first to sea level and later to spheroid. Initial values for the geodetic positions of the traverse stations were calculated without adjustment for azimuth or closure, by the Gauss mid-latitude formula extended to fifth order terms." On page 274 Leatherman continues, "At an early stage we expected that the separation of the reference spheroid of European Datum from the geoid in Saudi Arabia would be inconveniently large. It was agreed that the adjustment should be carried out in terms of European Datum and that the adjusted coordinates would be transformed to a new 'best fitting datum' for the International spheroid in Saudi Arabia by the method given by Weightman in *Bulletin Géodésique, No. 85, 1967.* The geoid-spheroid separation varied from −6 m to −68 m and the range of the deviation components was 32″ in meridian and 45″ in prime vertical. Three alternative definitions of best fitting were considered: to minimize the separations, or the deviations, or both simultaneously. It was considered more meaningful and more useful to minimize the separations. Being smoothed quantities, they provide a better representation of the overall pattern, and by minimizing the separations, the scale errors are also minimized. The adopted best fitting datum has been designated the *Ain El Abd (1970) Datum*. Any readjustment of European Datum positions in this area would affect the relationship between the datums but not the final coordinates." The ellipsoid of reference for the *Ain El Abd Datum of 1970* is the International 1924, where $a = 6,378,388$ m and $1/f = 297$.

In 1995, Mr. J. Anthony Cavell, PLS, CFed, now Deputy Director of the LSU Center for GeoInformatics (C4G), developed an oblique Mercator Grid (Hotine Rectified Skew Orthomorphic projection), for Saudi Arabia. The defining parameters are: Central Scale Factor $(k_o) = 0.99919$, Center Latitude $(\varphi_o) = 22°\ 30'$N, Central Line through: $\varphi_1 = 17°$N, $\lambda_1 = 51°$E, $\varphi_2 = 31°$N, $\lambda_2 = 36°$E, False Easting $(x_o) = 4,000$ km, False Northing $(y_o) = 500$ km, and the ellipsoid of reference is the WGS 84, where $a = 6,378,137$ m, $1/f = 298.257223563$. A test point is: $\varphi = 16°$N, $\lambda = 43°$E; $X = 1,455.596.409$ m, $Y = 2,162,102.853$ m.

According to *TR 8350.2*, the transformation parameters **from** the *Ain el Abd 1970 Datum* **to** the WGS 84 Datum are: $\Delta X = -143$ m ± 10 m, $\Delta Y = -236$ m ± 10 m, $\Delta Z = +7$ m ± 10 m. Nine stations were used to derive the parameters which were published in 1991.

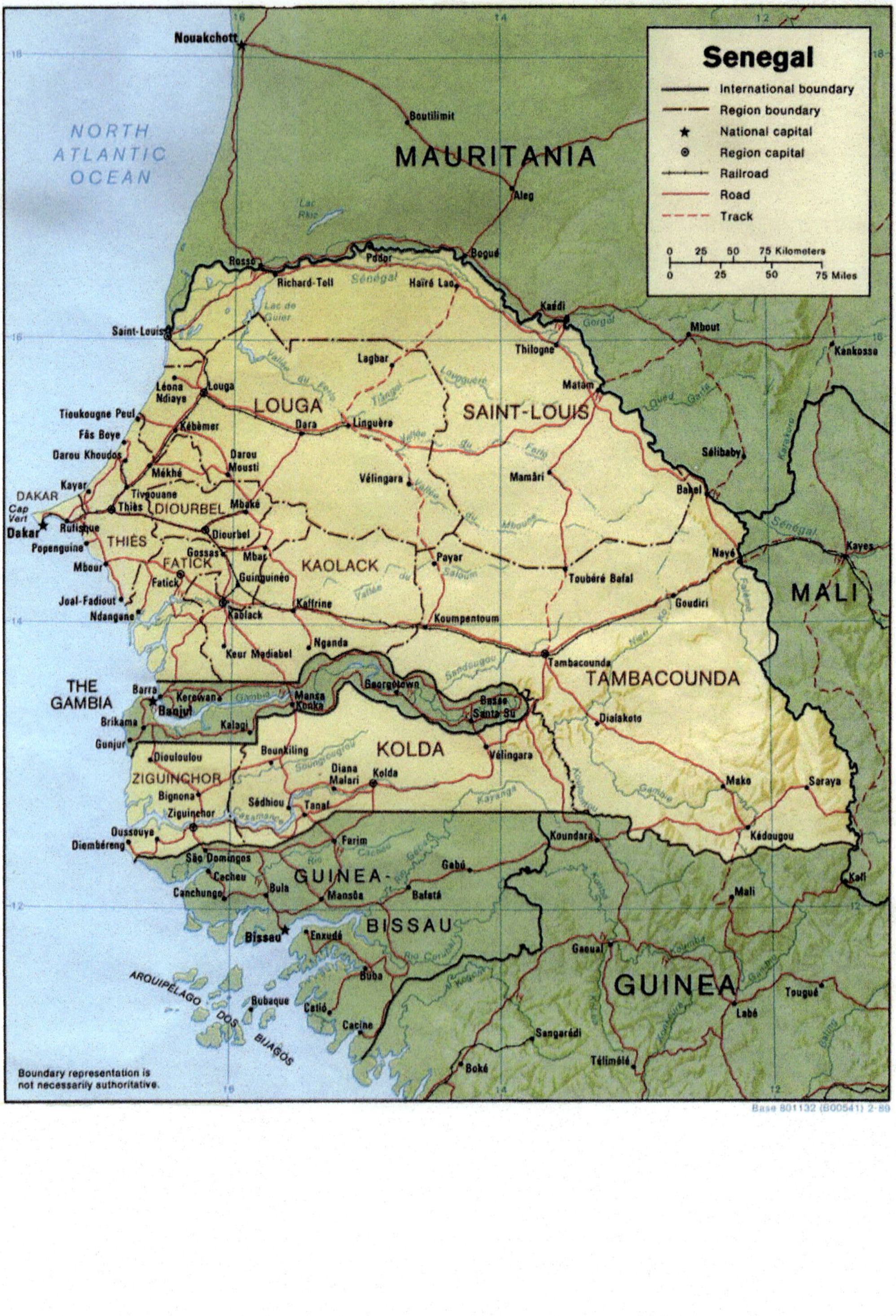

Base 801132 (B00541) 2-80

Republic of Senegal

Slightly smaller than South Dakota, Senegal borders the Gambia (740km), Guinea (330km), Guinea–Bisseau (338km), Mali (419km), and Mauritania (813km). The lowest point is the Atlantic Ocean (0m), and the highest point is an unnamed feature near *Nepen Diakha* (581 m) (*World Factbook*, 2010).

"Senegal was part of the Ghana Empire in the 8th century and the Djolof kingdom in the area between the Senegal River and modern-day Dakar. The Toucouleur people, among the early inhabitants of Senegal, converted to Islam in the 11th century, although their religious beliefs retained strong elements of animism. The Portuguese had some stations on the banks of the Senegal River in the 15th century, and the first French settlement was made at *Saint-Louis* in 1659. *Gorée Island* became a major center for the Atlantic slave trade through the 1700s, and millions of Africans were shipped from there to the New World. The British took parts of Senegal at various times, but the French gained possession in 1840 and made it part of French West Africa in 1895. Dakar was built as the administrative centre, and as early as 1848 Senegal had a (French) deputy in the French parliament. In 1946, together with other parts of French West Africa, Senegal became an overseas territory of France. On June 20, 1960, it formed an independent republic federated with Mali, but the federation collapsed within four months. Although Senegal is neither a large nor a strategically located country, it has nonetheless played a prominent role in African politics since its independence. As a black nation that is more than 90% Muslim, Senegal has been a diplomatic and cultural bridge between the Islamic and black African worlds. Senegal has also maintained closer economic, political, and cultural ties to France than probably any other former French African colony" (*InfoPlease,* 2010).

"Since the 1980s, the Movement of Democratic Forces in the *Casamance* – a separatist movement based in southern Senegal - has led a low-level insurgency. Several attempts at reaching a comprehensive peace agreement have failed. Since 2012, despite sporadic incidents of violence, an unofficial cease-fire has remained largely in effect. Senegal is one of the most stable democracies in Africa and has a long history of participating in international peacekeeping and regional mediation. The Socialist Party of Senegal ruled for 40 years until Abdoulaye Wade was elected president in 2000 and re-elected in 2007. Wade amended Senegal's constitution over a dozen times to increase executive power and weaken the opposition. In 2012, Wade's decision to run for a third presidential term sparked public backlash that led to his defeat to current President Macky Sall. A 2016 constitutional referendum limited future presidents to two consecutive five-year terms. The change, however, does not apply to Sall's first term. In February 2019, Sall won his bid for re-election; his second term will end in 2024. One month after the 2019 election, the National Assembly voted to abolish the office of the prime minister. Opposition and civil society organizations criticized the decision as a further concentration of power in the executive branch at the expense of the legislative and judicial branches" (*World Factbook* 2022).

The earliest geodetic reference in Senegal is to the longitude of a light on the east jetty of Dakar in 1885 by telegraphic determination from San Fernando Observatory (Cadiz, Spain) by Bouquet de la Grye which also gave a position for a chimney at the Dakar dry dock. The earliest geodetic survey in Senegal seems to have been in the City of Dakar by Rounguere and Laurent, 1903–1904 where the origin (at the old jetty) was: $\Phi_o = 14°\ 40'\ 27''$ N, $\Lambda_o = 17°\ 25'\ 22''$ West of Greenwich. The next survey apparently was done by the French Navy in the 1909–1910 Hydrographic Mission to West Africa (*Mission Hydrographique de L'Afrique Occidentale*) by *Capitaine de Frégate* M. Lebail. A survey of the *Casamance River* estuary near the southern border with Guinea–Bissau was controlled by a short triangulation base line of 323.003 m measured from 23 February to 27 May 1909. The origin of the Hatt Azimuthal Equidistant Grid was $\Phi_o = 12°\ 35'\ 05.10''$ N, $\Lambda_o = 16°\ 42'\ 47.45''$ West of Greenwich, no False Origin was used (Jacob M. Wolkau, *AMS memo 27 Oct. 1947*). In Dakar, an astronomic observation pillar was constructed near the Armory, and from astrolabe observations over four months, the latitude was determined to be: $\Phi_o = 14°\ 40'\ 40.6''$ N. An updated position for the 1903–1904 latitude of the old jetty was determined to be: $\Phi_o = 14°\ 40'\ 26.8''$ N. A baseline of 641.60 meters was measured, and a reference azimuth from the East Terminal of the Baseline to *Signal Gorée* was: $\alpha_o = 137°\ 14'\ 43.8''$. Another Hatt Azimuthal Equidistant Grid was used for the hydrographic survey of *Cape Vert*, and the origin point was the West Terminal of the Baseline, and no False Origin was used.

The French 1922 report to the IUGG gave general details of their work up to then. An additional Astro position was determined at *Saint-Louis*: $\Phi_o = 16°\ 01'\ 31''$ N, $\Lambda_o = 16°\ 30'\ 23''$ West of Greenwich.

The 1930–1931 Hydrographic Mission of *Saloum* (*Mission Hydrographique du Saloum*) by *Lieutenant de Vaisseau* Tromeur was performed because of the steadily increasing shipping tonnage with the river port of *Kaolack*. A Hatt Azimuthal Equidistant Grid was used for the hydrographic survey of the *Saloum River*, and the origin point chosen was one earlier established by Lebail in 1910, where $\Phi_o = 13°\ 51'\ 11.22''$ N, $\Lambda_o = 16°\ 45'\ 29.97''$ West of Greenwich was the South Terminal of the *Saloum 1,102.8675-m Baseline*, and the reference azimuth was: $\alpha_o = 001°\ 25'\ 06.2''$. A secondary coordinate system used an origin at the South Terminal of the 1,896.1760-m *Baseline at Diamniayo*, where $\Phi_o = 14°\ 03'\ 19.36''$ N, $\Lambda_o = 16°\ 35'\ 13.00''$ West of Greenwich, and the azimuth from the South Terminal to the North Terminal was: $\alpha_o = 047°\ 25'\ 41''$. All angles observed were with a WILD Heerbrugg Universal Theodolite (T-4), and to this date (1931), no mention of any ellipsoid of revolution was given for Senegal. However, in later memoranda of the French *Institut Geographique National (IGN)*, the use of the Clarke 1880 ellipsoid in all of French Africa was implicit.

The definitive geodetic survey of Dakar was that of *Lieutenant de Vaisseau* Bonin as reported in *Annals Hydrographiques de 1937*. The origin was defined as the South Terminal of the Base of the Route d'Yof (*Terme Sud de la Base de la Route d'Yof*), where $\Phi_o = 14°\ 43'\ 53.2''$ N, $\Lambda_o = 17°\ 29'\ 18.4''$ West of Greenwich, and the ellipsoid of reference was the Clarke 1880 using the *specific French definition* parameters, where $a = 6,378,249.2$ m and $1/f = 293.4660208$. In (*Wolkau, ibid*), "However, these geographics are not consistent with the geographics of *Gouvernement* as cited. The geographics of *Terme Sud de la base de la route d'Yof* that are consistent with these of *Gouvernement* are: $\Phi_o = 14°\ 43'\ 53.1''$ N, $\Lambda_o = 17°\ 29'\ 19.1''$ West of Greenwich. It appears that this inconsistency was noted by the French and corrected in their city plan of *Port Da Dakar*, but their published rectangular coordinates were not corrected. These positions are regarded as local UTM coordinates and made to refer to the 15th meridian which is central to UTM Belt 28." In 1938, Gougenheim densified the triangulation of Dakar, using the same Hatt Grid as Bonin, but with a False Easting = False Northing of 50 km.

In 1944, the U.S. Lake Survey, New York Office, Corps of Engineers published tables for the Senegal Belt on the Transverse Mercator projection. The ellipsoid of reference is the Clarke 1880 with the *standard U.S. Army* parameters being: $a = 6,378,249.145$ m and $1/f = 293.465$. The latitude of origin is $\varphi_o = 13°$N, Central Meridian, $\lambda_o = 16°$ West of Greenwich, Scale Factor at origin, $m_o = 0.99975$, False Easting = 400 km, False Northing = 500 km, Limits of Belt North: Parallel of 16°N; East: Meridian of 14° West; South: Zero-meter Northing line of Senegal Belt; West: Meridian of 19° West. A test point is provided, where $\varphi = 15°\ 47'\ 39.616''$ N, $\lambda = 14°\ 02'\ 29.729''$ W, $X = 609,817.26$ m, $Y = 810,047.94$ m.

From October 1946 to August 1948, *Capitaine de Corvette* M. Paul Bonnin led the *Mission de Triangulation de L'A.O.F.* Utilizing WILD Heerbrugg T-2 and T-3 theodolites with Bilby towers, the triangulation party started about 30 km southeast of *Dakar* and occupied a couple of existing triangulation stations named *Tiao* and *Niangol*. According to *Instruction N°. 1312 S.G.C. de l'Institut Géographique National*, dated December 12, 1945, a new Grid was used to calculate the triangulation; *Fuseau Senegal*, the parameter of which were: latitude of origin, $\varphi_o = $ Equator, Central

Meridian, $\lambda_o = 13°\ 30'$ West of Greenwich, Scale Factor at origin, $m_o = 0.999$, False Easting = False Northing = 1,000 km, and surprisingly the ellipsoid of reference was the International 1909, where $a = 6,378,388$ m and $1/f = 297$. Coordinates of these two points are: Tiao: $\varphi = 14°\ 39'\ 32.270''$ N, $\lambda = 16°\ 59'\ 14.945''$ W, X = 624,488.42 m, Y = 2,622,547.50 m, and Niangol: $\varphi = 14°\ 37'\ 28.384''$ N, $\lambda = 17°\ 09'\ 28.763''$ W, X = 606,046.48, Y = 2,619,027.27 m (*Annales Hydrographiques*, 3ᵉ série, Tome vingt-et-unième, Année 1949, pp. 49–68).

From January 5, 1949, to May 12, 1950, Mannevy densified Bonin's triangulation in the region south of *Dakar* and used *Fuseau Senegal* as the Grid system for his computations and published coordinates.

On September 20, 1950, *SGC 1312* was rescinded by the *IGN* in Paris in favor of the UTM Grid for Senegal, using the Clarke 1880 (*AMS version*) as the ellipsoid of reference.

A detailed triangulation and hydrographic survey of the mouth of the *Casamance River* was performed from February 17 to June 6, 1951, and from February 20 to March 20, 1952, led by M. Alain Le Fur. Astro station *Djogué* (in the vicinity of the *Djogué Lighthouse or Phare*), was observed with an astrolabe, and the final coordinates were: $\Phi_o = 12°\ 34'\ 14.5''$ N, $\Lambda_o = 16°\ 44'\ 28.5''$ West. The triangulation was computed and published on *Fuseau Senegal*. From October 1, 1953, to November 1, 1954, M. Albert Sauzay, Principal Hydrographic Engineer performed a survey of the *Saloum River* in a continuation of Bonin's work and Sauzay continued with computations and publication of his Trig List on the *Fuseau Senegal* Grid.

Re-observed in 1958, Astro station *Djogué* was updated to $\Phi_o = 12°\ 34'\ 14.3''$ N, $\Lambda_o = 16°\ 44'\ 25.5''$ West, and was re-published on the UTM Grid, Zone 28 by M. Jean Bourgoin, Principal Hydrographic Engineer (*Mission Hydrographique de la Côte Ouest d'Afrique*). The local hydrographic survey of the mouth of the *Saloum River* was a continuation on the local *Djogué Datum*, Clarke 1880 ellipsoid. From January to March of 1960, M. André Comolet-Tirman extended a triangulation traverse from the Saloum River to the border with Gambia. Published in UTM coordinates, Zone 28, the *Djogué Datum origin* was updated *again* to: $\Phi_o = 12°\ 34'\ 14.7''$ N, $\Lambda_o = 16°\ 44'\ 25.5''$ West, X = 310,918.9 m, Y = 1,390,166.1 m.

In January 1962, M. Antoine Demerliac measured a baseline in the city of *Thiès*, east of *Dakar*. The average of 4 invar measurements was 3,710.5040 m. This compared favorably with a Tellurometer measurement of 3,710.5093 m (*Annales Hydrographiques*, 4ᵉ Série, Tome quatorzième, Années 1967–1968).

Thanks to John W. Hager, "the International Hydrographic Bureau published another list of coordinates in 1961, the France section included a number of points for Senegal. In 1977, the Canadian Hydrographic Service conducted a LORAN-C and Satnav-based survey off Senegal and The Gambia. The 'Datum' for the shore-based LORAN-C stations was given as Clarke 1880!" In a note from Mr. Russell Fox of the Ordnance Survey, Southampton to Mr. Malcolm A. B. Jones of Perth, Australia of December 7, 1999, "I believe that *Yoff* 200 and Point 58 datums were *IGN* adjustments of the Senegal-Mali section of the 12th Parallel Survey. *Yoff* 200 was used in the *Senegambia* adjustment, linking the Senegal and Gambia trig networks. Adindan Datum itself was the US DMA adjustment of the entire 12th Parallel survey. I believe the main Senegal Datum is *Dakar-Hann IGN 1952*, using Clarke 1880 (SGA-IGN) spheroid." In a related note from Mr. Jean-Pierre Pirat of the French *IGN* to Mr. Malcolm A. B. Jones later that month, "The geodetic Systems that one finds in Senegal were established during the measurement of the geodetic advance of the 12th parallel in 1968 (Senegal-Sudan crosses). They are composed of the Adindan system as you say it in your message, but also *Yoff* datum and Point 58 datum. *Yoff* (or *Yoff-200 datum*): the datum is the astronomical point of *Yoff* (n°200 in the repertory of the 12th Parallel), spheroid Clarke 1880. Adindan datum: the datum is the astronomical point of Adindan located as the border Sudan-Egypt, spheroid Clarke 1880. (*Actually, that is the Blue Nile Datum with origin at Adindan. – Ed.*) It is the system retained for the international geodetic connections. The parameters of passage (datum shifts) of these local systems toward World System WGS84 are respectively: *Yoff-200* towards WGS84: Tx = −31, Ty = +173, Tz = +90; Adindan (Senegal) towards WGS84 $T_x = -128$, $T_y = -18$, Tz = +224, Point 58 towards WGS84: no comment."

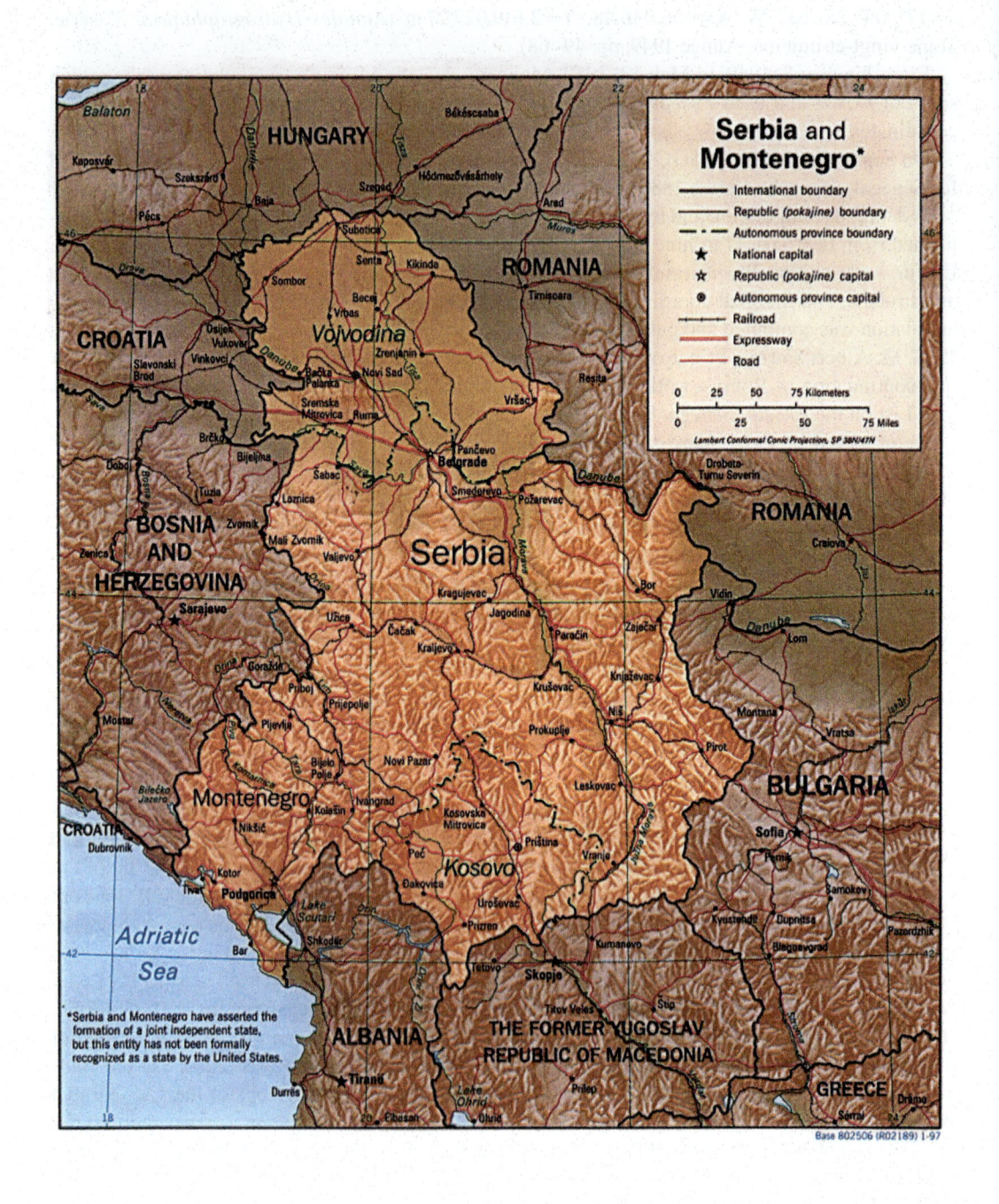

*Serbia and Montenegro have asserted the formation of a joint independent state, but this entity has not been formally recognized as a state by the United States.

Republic of Serbia

Slightly smaller than South Carolina, Serbia is bordered by: Bosnia and Herzegovina (302 km), Bulgaria (318 km), Croatia (241 km), Hungary (151 km), Kosovo (352 km), Macedonia (62 km), Montenegro (124 km), and Romania (476 km). With rich fertile plains to the north; limestone ranges and basins to the east; ancient mountains and hills to the southeast; the lowest points are the *Danube River* and *Trgoviški Timok* (35 m); the highest point is *Midžor* (2,169 m) (*World Factbook & NGA GeoNames Search*, 2013).

"Serbia's history has been punctuated by foreign invasions, from the time the Celts supplanted the Illyrians in the 4th century *B.C.*, through to the arrival of the Romans 100 years later, the Slavs in the 6th century *A.D.*, the Turks in the 14th century, the Austro-Hungarians in the late 19th and early 20th centuries, and the Germans briefly in WWII. A pivotal nation-shaping event occurred in *A.D.* 395 when the Roman Emperor Theodosius I divided his empire giving Serbia to the Byzantines, thereby locking the country into Eastern Europe. This was further cemented in 879 when Saints Cyril and Methodius converted the Serbs to the Orthodox religion. Serbian independence briefly flowered from 1217 with a golden age during Stefan Dušan's reign (1346–55). After his death Serbia declined and at the pivotal Battle of Kosovo in 1389 the Turks defeated Serbia, ushering in 500 years of Islamic rule. Early revolts were crushed but one in 1815 led to *de facto* Serbian independence that became complete in 1878" (*Lonely Planet*, 2013).

"The Kingdom of Serbs, Croats, and Slovenes was formed in 1918; its name was changed to Yugoslavia in 1929. Various paramilitary bands resisted Nazi Germany's occupation and division of Yugoslavia from 1941 to 1945 but fought each other and ethnic opponents as much as the invaders. The military and political movement headed by Josip 'Tito' Broz (Partisans) took full control of Yugoslavia when German and Croatian separatist forces were defeated in 1945. Although Communist, Tito's new government and his successors (he died in 1980) managed to steer their own path between the Warsaw Pact nations and the West for the next four and a half decades. In 1989, Slobodan Milosevic became president of the Republic of Serbia, and his ultranationalist calls for Serbian domination led to the violent breakup of Yugoslavia along ethnic lines. In 1991, Croatia, Slovenia, and Macedonia declared independence, followed by Bosnia in 1992. The remaining republics of Serbia and Montenegro declared a new Federal Republic of Yugoslavia (FRY) in April 1992 and under Milosevic's leadership, Serbia led various military campaigns to unite ethnic Serbs in neighboring republics into a 'Greater Serbia.' These actions were ultimately unsuccessful and led to the signing of the Dayton Peace Accords in 1995. Milosevic retained control over Serbia and eventually became president of the FRY in 1997. In 1998, an ethnic Albanian insurgency in the formerly autonomous Serbian province of Kosovo provoked a Serbian counterinsurgency campaign that resulted in massacres and massive expulsions of ethnic Albanians living in Kosovo. The Milosevic government's rejection of a proposed international settlement led to NATO's bombing of Serbia in the spring of 1999, to the withdrawal of Serbian military and police forces from Kosovo in June 1999, and to the stationing of a NATO-led force in Kosovo to provide a safe and secure environment for the region's ethnic communities. FRY elections in late 2000 led to the ouster of Milosevic and the installation of democratic government" (*World Factbook*, 2013).

"President Aleksandar Vucic has promoted an ambitious goal of Serbia joining the EU by 2025. Under his leadership as prime minister, in 2014 Serbia opened formal negotiations for accession" (*World Factbook* 2022).

"With the establishment of Austrian Cadaster in 1817, Cadastral triangulation with its 10 plain rectangular systems (Cassini-Soldner projection) was started. The Military topographical survey in order to utilize reduced cadastral planimetry, adopted all 10 cadastral systems. For the northern part of Serbia (*north of Belgrade – Ed.*), The *Budapest, Gellérthegy Observatory* was used for Hungary where from 1817 to 1874, $\Phi_o=47°\ 29'\ 15.97''$ North, $\Lambda_o=36°\ 42'\ 51.57''$ East of Ferro; since 1879, $\Phi_o=47°\ 29'\ 14.93''$ North, $\Lambda_o=36°\ 42'\ 51.69''$ East of Ferro; in the *Hermannskogel system of K.u.K. III MT*, $\Phi_o=47°\ 29'\ 14.07147''$ North, $\Lambda_o=36°\ 42'\ 56.2316''$ East of Ferro. The triangulation of each system was carried out independently. The scale and the orientation were based either on a directly measured base line and directly determined azimuth or on a side and an azimuth of II Military triangulation. Each of the systems was computed independently with the adjustment carried our empirically. The coordinates were computed in Cassini-Soldner projection, and represent an average value obtained from individual values computed in different ways after the removal of the angular and side differences empirically. Thus, the unstable relationships among the positions of origins and among the orientations of systems caused the discrepancies along the boundaries of provinces (systems) which made the constructions of a uniform map impossible. The stations were marked by wooden posts. Stone markers were placed 30–50 years later where the remnants of wooden posts and signals were found. At the time of placing the stone markers, 12% of the stations were not found.

"Topographical surveys were carried out by plane table method. The instructions for the survey changed gradually (1807, 1817, 1833, and 1860) and the manner of drafting also was subject to many changes; thus, the plane table sheets were not completed uniformly. Prior to 1860 elevations, except for trig points, were not measured. The instructions of 1860 provided the measurement of elevation for detail points as well as the introduction of contour lines. The scale remained 1:28,800 as in the first topographic survey because some of the Second topographical survey sheets were used as a base for the Third topographical survey it is important to know the areas where the topographical survey was based on reduced cadastral planimetry. The cadastral survey was not yet completed as the time this topographical survey was under way and the cadastral maps were not used in Serbia and *Banat* (*Vojvodina*) surveyed topographically in 1864–1866" (*Mapping of the Countries in Danubian and Adriatic Basins,* Andrew M. Glusic, Army Map Service Technical Report No. 25, June 1959). Later editions were re-ambulated and compiled at 1:75,000 scale.

"For a long period of time in the Austro-Hungarian Army, the pace (1 pace=0.75 meters) was used as the unit of length measurement. The range of the guns and rifles was determined in paces. This was the main reason that the 1:75,000 scale was adopted in 1872" (*op. cit. Glusic,* 1959).

"When Russia occupied the *Danube Principalities* (the Old Romania) in 1853, the Turco-Russian tension led to the Crimean War. During this war, Austria forced its former ally, Russia, to give up this occupation, and Austrian armies took their place between 1854 and the Paris conference resolving the peace at the end of Crimean War. The MGI the time and made a whole geodetic survey in *Wallachia* (*Oltenia and Muntenia*) and *Northern Dobrogea*. This survey was the first systematic geodetic triangulation in Romania (Timár, G. (2008): *Habsburg geodetic and cartographic activities in the Old Romania,* Studii şi Cercetări, Seria Geologie-Geografie [Complexul Muzeal Bistriţa-Năsăud] 13: 93–102). The Central European Arc Measurement, later the European Arc Measurement (*Mittel-Europäische Gradmessung* and *Europäische Gradmessung*), respectively; brokered by the Prussians in the 1860s, had this reason behind the curtain. These projects aimed to determine the real and accurate size and shape of the Earth. However, not all European countries were prepared to make the necessary measurements and provide them to the community. This could be a possibility for the Austrians to offer their help to the Turkish Empire to make the necessary measurement. Whatever was the background, the Turkish authorities entered the Austrian survey teams to the European parts of the empire, and the survey was made between 1871 and 1875 in the whole European territory of the Turkish Empire.

"The tension between Russia and Turkey broke out in a new war in 1878. The Congress of Berlin, resolving the peace after the hostilities, let the Serbs and the Bulgarians to form their own countries. As Serbia sought for any possibility to unite all southern Slavs into a country (in the later Yugoslavia), it was a 'natural' enemy of the actual ruler of many claimed territories, the Habsburg

Empire. The tension was increased further by the occupation of *Bosnia-Herzegovina* by the Austro-Hungarian Monarchy. No further Austrian surveys were made prior to the WWI here" (*The Austro-Hungarian Triangulations in the Balkan Peninsula (1853–1875)*, Béla Kovács and Gábor Timár).

The Ferro prime meridian was widely used in Europe as the standard, proposed in 1634 as the "Westernmost point of the Old World." Commonly considered as being 20° west of the Paris meridian, and according to the French *Bureau International de l'Heure*, the civilian Greenwich-Ferro difference is $\Delta\lambda = 17°\ 39'\ 45.975''$. The military used the value of Theodor Albrech where $\Delta\lambda = 17°\ 39'\ 46.020''$, which was the official shift used for Yugoslavian cartography (*The Ferro Prime Meridian*, Gábor Timár, Geodézia És Kartográfia, pp. 3–7, 2007).

The Active Geodetic Reference Network of Serbia (AGROS) was founded in 2005. With average distance between stations being 70 km, the entire Serbian territory has 32 Global Navigation Satellite Systems (GNSS) sites. International cooperation covers connections to GNSSnet.hu (Hungary), MAKPOS (Macedonia), BULIPOS (Bulgaria), CROPOS (Croatia), MONTEPOS (Montenegro), and ROMPOS (Romania) (*Republic Geodetic Authority of Serbia – AGROS,* N. Tesla, N. Matic, V. Milenkovic, U.N./Turkey/European Space Agency Workshop, 14–17 September 2010, Istanbul).

Cadastral mapping in Serbia to-date still employs the Gauss-Krüger Transverse Mercator, the Cassini-Soldner, and the Oblique Stereographic projections in various scales ranging from 1:500 to 1:5,000 (*Land Administration in Post Conflict Serbia*, I. Aliksic, Z. Gospavic, Z. Popovic, 2003).

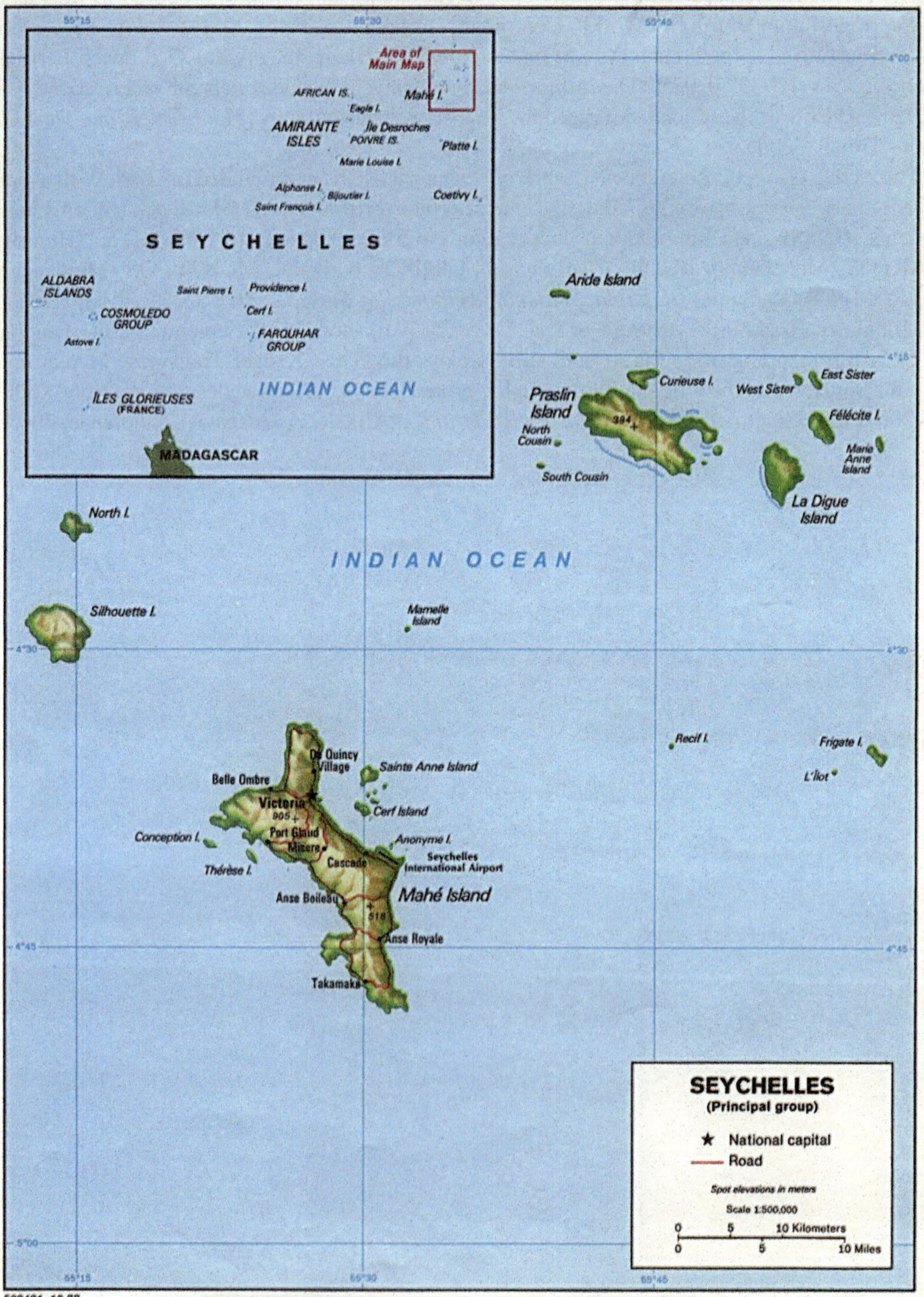
Area of Main Map
AFRICAN IS.
Eagle I.
Mahé I.
AMIRANTE ISLES
Île Desroches
POIVRE IS.
Platte I.
Marie Louise I.
Alphonse I.
Bijoutier I.
Saint François I.
Coetivy I.
SEYCHELLES
ALDABRA ISLANDS
Saint Pierre I.
Providence I.
COSMOLEDO GROUP
Cerf I.
Astove I.
FARQUHAR GROUP
INDIAN OCEAN
ÎLES GLORIEUSES (FRANCE)
MADAGASCAR
Aride Island
Curieuse I.
East Sister
West Sister
Praslin Island
394
Félicité I.
North Cousin
Marie Anne Island
South Cousin
La Digue Island
North I.
INDIAN OCEAN
Silhouette I.
Mamelle Island
Recif I.
Frigate I.
L'Ilot
De Quincy Village
Sainte Anne Island
Belle Ombre
Victoria
905
Cerf Island
Port Glaud
Conception I.
Anonyme I.
Misere
Cascade
Seychelles International Airport
Thérèse I.
Anse Boileau
Mahé Island
518
Anse Royale
Takamaka
SEYCHELLES
(Principal group)
National capital
Road
Spot elevations in meters
Scale 1:500,000
0 5 10 Kilometers
0 5 10 Miles

503481 10-77

Republic of Seychelles

The Seychelles are comprised of an archipelago in the Indian Ocean, northeast of Madagascar, and have a total land area of 455 km² or 2.5 times the size of Washington, DC, with a coastline of 491 km. The lowest point is the Indian Ocean (0m), and the highest point is *Morne Seychellois* (905m). The Mahé Group is granitic with a narrow coastal strip that is rocky and hilly. The other islands are coral, flat elevated reefs. In total, there are 41 granitic islands and about 75 coralline islands.

The Seychelles islands first appeared on Portuguese charts in 1505 but remained uninhabited for more than a century and a half after they became known to Western explorers. The islands may have had visitors from Arabia much earlier, but there is no known record. In 1742 the French Governor of Mauritius, Mahé de Labourdonnais, sent an expedition to the islands. A second expedition in 1756 reasserted formal possession by France and bestowed upon the islands their present name in honor of the French Finance Minister under King Louis XV. The French being the first to colonize the Seychelles, maintained control for a 50-year period. Official land surveys for property subdivision are known from the 1780s and 1790s. A triangulation may have been observed, as in Mauritius, but no permanently marked stations survived into the 20th century and no old trig records were kept. The new colony barely survived its first decade and did not really flourish until 1794, when Queau de Quincy became Commandant. After being captured and freed several times during the French Revolution and the wars of Napoleon, the islands passed officially to the British under the Treaty of Paris in 1814. From the date of its founding until 1903, the Seychelles was regarded as a dependency of Mauritius. However, in 1888, a separate Administrator nominated Executive and Legislative Councils which were established for the archipelago. Nine years later the Administrator was given the full powers of a Governor, and on August 31, 1903, the Seychelles became a separate British Crown Colony. The coralline islands were added to the colony in 1903 and the whole territory became an independent republic within the Commonwealth in 1976.

The colony's administration continued the property surveying that had begun in French times, but surveys were done in isolation in the absence of a triangulation network. The early settlers received free grants of land, known as "habitations," and in 1787 a survey plan of *Mahé* was made which showed 29 such habitations of irregular shape and size. The survey resulted from formal orders issued by the Commandant, who was also charged with the task of drawing up deeds and rectifying or disallowing all bad titles. There is virtually no record available in Seychelles that throws much light upon these early surveys and the methods by which they were made. An overall plan of *Mahé* was made in 1829 and portrays all the properties for which information was then available, but the representation is in graphical form only. The earliest land measurements employed in Seychelles are the French foot or pied, the toise which contains 6 pieds, the perche of 20 pieds and l'Arpent common which has 40,000 pieds². In addition to these measurements the meter and English foot have also been used.

The British Royal Navy was very active in Seychelles waters during the 19th century, suppressing the slave trade between Africa and Arabia (slave trading and slavery having been abolished within the British Empire in 1807 and 1833, respectively). Therefore, extensive Admiralty hydrographic surveys were carried out throughout the islands, based on coastal astro fixes and small local trig schemes for harbor surveys. The HMS Stork surveyed Mahé and the adjacent islands in 1890. The positions were not reliable and a survey by HMS Enterprise in 1931 of Curieuse Bay showed that latitudes should be decreased by 47″ and longitudes increased by 34″. Mahé itself required a correction of about 10″ in latitude in 1943.

The first modern, permanently marked, triangulation of the Seychelles Group was carried out by a British military unit, the East Africa Survey Group, in 1943. The object was to provide a mapping and artillery grid for coastal batteries. The HMS Challenger observed an astro station on South East Island in 1943, and that has served as the exclusive local datum for the islands on the Seychelles bank, that is, those islands north of 5°S and between 55°E and 56°E. The South East Island Datum of 1943 originates on an astrofix taken by the crew of HMS Challenger, where $\Phi_o = 4°\ 40'\ 39.460''$ S and $\Lambda_o = 55°\ 32'\ 00.166''$ E, and the ellipsoid of reference is the Clarke 1880, where $a = 6378249.145$ m and $1/f = 293.465$. The Seychelles Belt Grid is based on the Gauss-Krüger Transverse Mercator projection where the central meridian, $\lambda_o = 53°$ E, the latitude of origin $\varphi_o = 10°$ S, the scale factor at the latitude of origin $m_o = 0.9995$, False Easting = 500 km, and False Northing = 1,100 km. The 1944 Belt limits are for the north being the Equator, for the east the meridian of 57° E, for the south the parallel of 19° 30' S, and for the west the meridian of 52° E northwards to 11° S, thence along this parallel to 50° E, and thence along this meridian to the Equator.

A number of hydrographic survey datums have been established over the decades in the Seychelles that include the Victoria Lighthouse Datum, where $\Phi_o = 4°\ 36'\ 57.835''$ S, $\Lambda_o = 55°\ 28'\ 12.347''$ E, but the UK Military Survey notes that the "Co-ordinates were computed from rectangular co-ordinates and depended on Challenger's Observation Spot at Port Victoria being $\varphi = 4°\ 37'\ 12.41''$ S, $\lambda = 55°\ 27'\ 19.24''$ E, based on Clarke 1880. From the co-ords of the observation spot, the position of Victoria Lighthouse should be $\varphi = 4°\ 36'\ 57.894''$ S, $\lambda = 55°\ 28'\ 12.523''$ E, hence positions should be moved 0.059" S and 0.176" E." The Mahé Island Datum of 1964 origin is at station "SITE," where $\Phi_o = 4°\ 40'\ 14.644''$ S, $\Lambda_o = 55°\ 28'\ 44.488''$ E. the Mahé Island Datum of 1971 has its origin at the same place and coordinates of the 1964 datum, and "a third datum called Mahé 1971 is also used in the published literature. The datum does not seem to have been used outside Mahé and its origin is unclear" (*UK Mil Svy*). The Bird Island Datum is based on an astrolabe observation, where $\Phi_o = 3°\ 43'\ 08.403''$ S, $\Lambda_o = 55°\ 11'\ 58.079''$ E, and the Aldabra Atoll Datum of 1878 origin is at station "North Base," where $\Phi_o = 9°\ 23'\ 23''$ S, $\Lambda_o = 46°\ 14'\ 25''$ E and has a local grid based on the Azimuthal Equidistant projection (*also termed a "Cassini" and a "Gnomonic" by UK Mil Svy*), where the central meridian, $\lambda_o = 46°\ 14'\ 24.8''$ E, the latitude of origin $\varphi_o = 9°\ 23'\ 22.7''$ S, the scale factor at the latitude of origin $m_o = 1.0$, and False Easting = False Northing = 10 km. The grid was surveyed in 1964 by HMS Owen, and the point of grid origin is at Grand Post Observation Spot (North Base, N).

A number of transformation solutions have been developed from the local South East Island Datum of 1943 to WGS84 Datum. According to *TR 8350.2*, the three-parameter shift to WGS84 is, where $\Delta X = +41$ m, $\Delta Y = -220$ m, $\Delta Z = -134$ m, ±25 m, and is based on one point. According to the UK Hydrographic Office in January 1999, the seven-parameter shift from SEI43 to WGS84 is where: $\Delta X = -76.269$ m ±0.3 m, $\Delta Y = -16.683$ m ±0.3 m, $\Delta Z = +68.562$ m ±0.3 m, k = -13.868 ppm, $R_x = -6.275''$, $R_y = 10.536''$, $R_z = -4.286''$. Presumably, the sense of the rotation elements is counterclockwise as employed by most European nations, the *opposite* of the standard American-Australian convention. A separate UK Hydrographic Office report containing transformation parameters from SEI43 to WGS84 is reported to be based on a solution of 10 collocated stations such that for the three-parameter shift: $\Delta X = -43.635$ m, $\Delta Y = -179.785$ m, $\Delta Z = -267.721$ m and for the seven-parameter shift: $\Delta X = +30.768$ m, $\Delta Y = -129.010$ m, $\Delta Z = -91.673$ m ±0.3 m, k = -10.901 ppm, $R_x = -1.98470''$, $R_y = +7.51328''$, $R_z = +0.64532''$.

"Directorate of Overseas Surveys (DOS) work in the Seychelles began in 1959–61 with the aerial photography of the Seychelles Group and Aldabra Atoll, and the computation of the United Nations Survey Team's 1958–59 trig survey of Mahé and nearby islands. The geodetic system adopted by DOS was Local Datum, Clarke 1880 (Modified) Spheroid, and UTM Grid. The UN main and minor trig was computed in two blocks and co-ordinates were issued in 1962. The first map series – Mahé 1:10 000 in eleven contoured sheets – was published in 1963, followed by Aldabra 1:25 000, Mahé 1:50 000, large scale photomaps and dyeline series of the small islands in the Seychelles Group and 1:1250 & 1:2500 series of Mahé.

"Surveys in the Admirantes and Farquhar Groups began in 1974, with new astrofixes and connections to the observation sites of old Admiralty surveys and to new Doppler stations placed by 512STRE (512 Specialist Team, Royal Engineers). The DOS team also observed photo control, surveyed potential airfield sites and collected place names. DOS computed the Admirantes Group on WGS72 Datum and UTM Grid and started the Admirantes and Farquhar Groups photomap series in 1976.

"DOS conducted a Doppler campaign in the Aldabra and Farquhar Groups during 1981/82, with the purpose of bringing those islands onto WGS72 Datum, thereby aiding the definition of the Seychelles' Exclusive Economic Zone. Azimuths were also observed on each island. UTM gridded, WGS72 editions of the Aldabra and Farquhar photomaps were published in 1992/93.

"EDM traversing for control densification was carried out on *Mahé, Praslin, La Digue* and nearby small islands during 1976–78. In 1977 inter-island Tellurometer connections were observed by DOS between the outer and main islands of the Seychelles Group, allowing the extension of South East Island Datum to all islands in the group. Clarke 1880 (Modified) Spheroid and UTM Grid were retained. The connections were computed, and co-ordinates issued, in 1978. The three-parameter shift from SE Island Datum to WGS84 Datum is ΔX –42.9 m, ΔY –178.5 m, ΔZ –277.8 m, based on four common points by DOS" (*Russell Fox, retired Chief Librarian, Ordnance Survey*). In my opinion, the transformation parameters listed in *TR 8350.2* appear to be grossly incorrect for "MAHE 1971" to WGS84 because of the very close agreement in different surveys and adjustments carried out over several years by the various United Kingdom survey units.

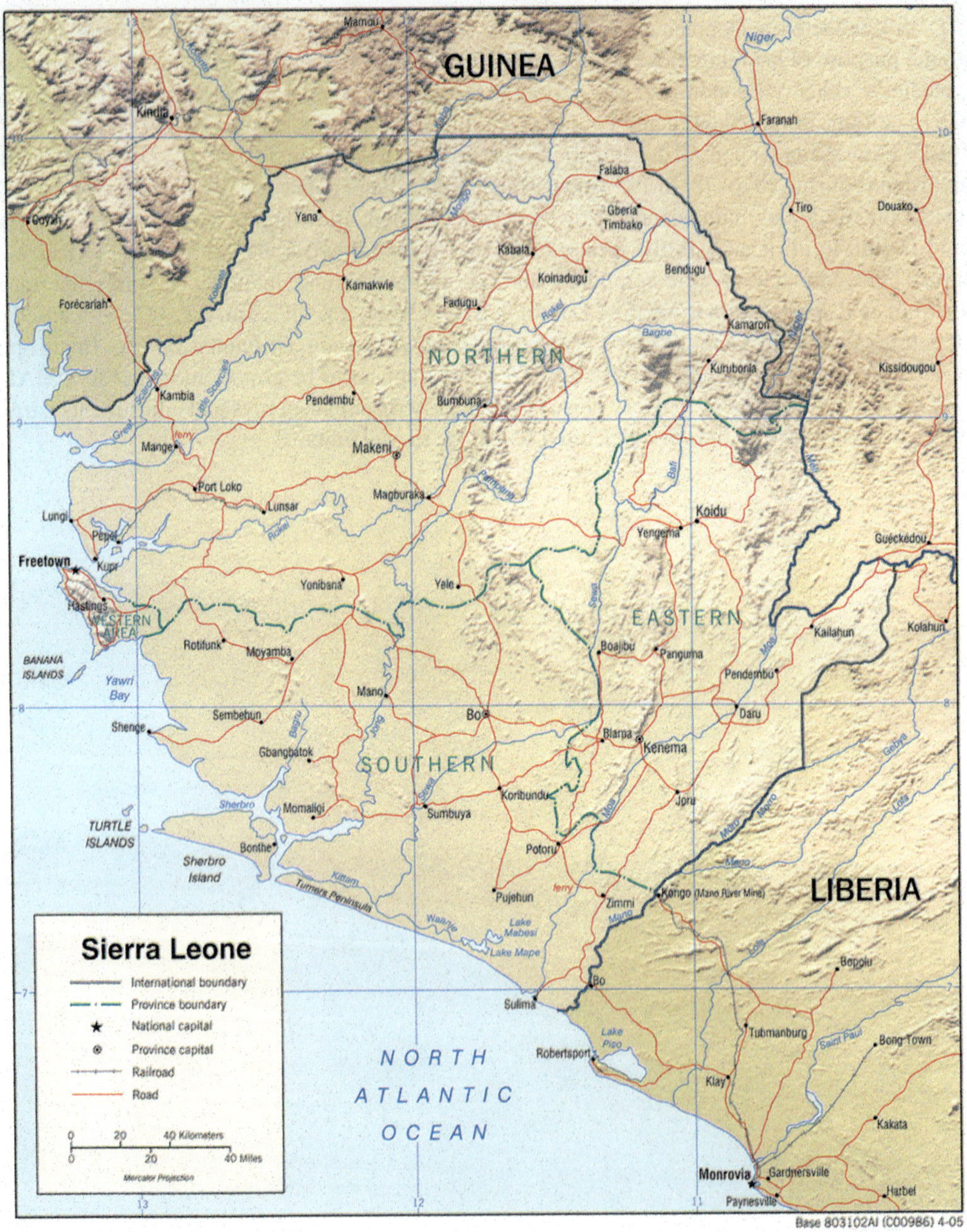

Base 803102AI (C00986) 4-05

Republic of Sierra Leone

Slightly smaller than South Carolina, Sierra Leone is bordered by Guinea (652 km) and Liberia (306 km). With a coastline of 402 km, the terrain is comprised of a coastal belt of mangrove swamps, wooded hill country, upland plateau, and mountains in the east. The lowest point is the Atlantic Ocean (0 m), and the highest point is *Loma Mansa* (1,948 m) (*World Factbook*, 2011).

"European contacts with Sierra Leone were among the first in West Africa. In 1652, the first slaves in North America were brought from Sierra Leone to the Sea Islands off the coast of the southern United States. During the 1700s there was a thriving trade bringing slaves from Sierra Leone to the plantations of South Carolina and Georgia where their rice-farming skills made them particularly valuable. In 1787 the British helped 400 freed slaves from the United States, Nova Scotia, and Great Britain return to Sierra Leone to settle in what they called the 'Province of Freedom.' Disease and hostility from the indigenous people nearly eliminated the first group of returnees. This settlement was joined by other groups of freed slaves and soon became known as Freetown. In 1792, Freetown became one of Britain's first colonies in West Africa. Thousands of slaves were returned to or liberated in Freetown. Most chose to remain in Sierra Leone. These returned Africans – or Krio as they came to be called – were from all areas of Africa. Cut off from their homes and traditions by the experience of slavery, they assimilated some aspects of British styles of life and built a flourishing trade on the West African coast.

In the early 19th century, Freetown served as the residence of the British governor who also ruled the Gold Coast (now Ghana) and the Gambia settlements. Sierra Leone served as the educational center of British West Africa as well. Fourah Bay College, established in 1827, rapidly became a magnet for English-speaking Africans on the West Coast. For more than a century, it was the only European-style university in western Sub-Saharan Africa. The colonial history of Sierra Leone was not placid. The indigenous people mounted several unsuccessful revolts against British rule and Krio domination. Most of the 20th century history of the colony was peaceful, however, and independence was achieved without violence. The 1951 constitution provided a framework for decolonization. Local ministerial responsibility was introduced in 1953, when Sir Milton Margai was appointed Chief Minister. He became Prime Minister after successful completion of constitutional talks in London in 1960. Independence came in April 1961, and Sierra Leone opted for a parliamentary system within the British Commonwealth. Sir Milton's Sierra Leone Peoples Party (SLPP) led the country to independence and the first general election under universal adult franchise in May 1962." Significant political unrest has plagued the republic since then (*Background Notes*, U.S. Department of State, 2011).

"An Anglo–French convention of June 28, 1882, delimited a boundary between the territories of Sierra Leone and Guinea from the Atlantic Ocean inland along the drainage divide of the *Great Scarcies* and *Melikhoure* to an indefinite point in the interior. On August 10, 1889, France and the United Kingdom signed an arrangement extending the Guinea–Sierra Leone boundary northward to the 10th parallel and then eastward to the 13th meridian west of Paris (10° 39′ 46.05″ West of Greenwich). In order to determine the boundary between British and French spheres of influence west and south of the upper Niger river, an Anglo–French agreement of June 26, 1891, stated that the 13th meridian west of Paris was to be followed where possible from the 10th parallel to *Timbekundu* (the source of the *Timbe* or *Niger*). The boundary commissions were to keep both banks of the Niger in the French sphere, but the line could be deflected by agreement east or west of the meridian. A Franco–Liberian convention of December 8, 1892, delimited the western sector of the Guinea–Liberia boundary as the parallel of *Timbekundu* (9° 05′ N.) to the 13th meridian west of Paris.

"An agreement of January 21, 1895, between the United Kingdom and France established the present Guinea–Sierra Leone boundary from the Atlantic Ocean to *Timbekundu*. The agreement stated that the sector from the 10th parallel to Timbekundu would follow the drainage divide or *"watershed separating the basin of the Niger on the one hand from the basins of the Little Scarcies and other rivers, falling westward to the Atlantic Ocean on the other hand."* In accordance with the Anglo–French agreement of January 21, 1895, British and French commissions between December 1895 and May 1896 surveyed and demarcated the boundary as contained in a *proces-verbal* of April 9–30, 1896. The demarcation was accepted by the French and British Governments by notes exchanged at Paris on June 14 and 16, 1898. In accordance with *proces-verbaux* of March 12, 1903, the Guinea–Sierra Leone boundary was demarcated by pillars between *Tembikundu* and the 13th meridian west of Paris. The demarcation was approved by an exchange of notes between France and the United Kingdom on March 22 and April 5, 1904, respectively. In the meantime, on June 25, 1903, an Anglo–Liberian *proces-verbal* delimited the Liberia–Sierra Leone boundary as the 13th meridian west of Paris southward from the *Wulafu (Ou Lafou)* to the *Mano River*. A Franco–Liberian agreement of September 18, 1907, stated that it was physically impossible to apply the theoretical lines of their 1892 convention and that natural topographical lines should be used where possible. The agreement redrew the Guinea–Liberia boundary and transferred a large strip of Liberian territory to French Guinea. In implementation of the 1907 agreement, a commission delimited the boundary as approved by a second agreement on January 13, 1911. Located between the *Makona* and *Mauwa rivers* to the east of the 13th parallel west of Paris, the *Liberian Kailahan area* was transferred to Sierra Leone by an Anglo–Liberian convention of January 21, 1911. The transfer of the area moved the tripoint with French Guinea eastward to the left bank of the *Makona* and the center of the *Dundugbia*. By an exchange of notes on July 6, 1911, the United Kingdom and France agreed to extend the Guinea–Sierra Leone boundary to the tripoint with Liberia by utilizing the *Wulafu, Meli,* and *Makona rivers*. An Anglo–French agreement of September 4, 1913, reconfirmed the 1903 demarcation east of Timbekundu and delimited in detail the present Guinea–Sierra Leone boundary from the *Wulafu* to the Liberian tripoint on the left bank of the *Makona* at the center of the *Dundugbia*" (*Guinea – Sierra Leone Boundary, International Boundary Study No. 136* – July 2, 1973, U.S. Department of State).

"An Anglo–Liberian convention of November 11, 1885, established the *Mano River* as the boundary between Sierra Leone and Liberia from the Atlantic Ocean to a somewhat indefinite point in the interior. To the north a *proces verbal* of June 25, 1903, demarcated a line from the right bank of the *Wulafu (Ou Lafou),* at the point where the 13th meridian west of Paris intersects the river, and then followed this meridian to the *Mano*. A Franco–Liberian agreement of September 18, 1907, indicated it was physically impossible to apply the theoretical lines of their boundary convention of December 8, 1892, and that as far as possible natural topographical lines should be used to prevent future disputes. Article I of the 1907 agreement stated that the Franco (Guinean)–Liberian boundary would begin at 'The left bank of the Makona River, from the point where that river enters Sierra Leone to a point to be determined, approximately five kilometers south of Bofosso.' Thus, the sector formerly a part of the Liberia–Sierra Leone boundary between the *Oudalfou* and *Makona* became a part of the Guinea–Sierra Leone boundary. An Anglo–Liberian convention of January 21, 1911, readjusted the northern and southern parts of the 1903 meridian sector of the Liberia–Sierra Leone boundary in accordance with natural features and tribal divisions. In the north the *Kailahun area* east of the meridian between the *Makona* and *Mauwa* was transferred by Liberia to Sierra Leone in exchange for an area west of the meridian between the *Morro* and *Mano*. Article 1 of the 1911 convention affords the transfer of territory as follows:

    a. The line marking the western boundary of the Republic shall start from the meeting point on the *Moa River* of the *Tengea* and *Kunyo* sections of the *Kissi* country and shall be continued in a southerly direction to a point on the *Maia River,* so that it corresponds, between these two points, with the western boundary of the Tengea section and the eastern boundaries of the *Kunyo and Tungi* sections of that country.

b. From this point the boundary shall follow the course of the Maia, Makwoi, and Mauwa Rivers to the point where the Mauwa River intersects the provisional line laid down by the Anglo–Liberian Boundary Commission of 1902–3.

c. From this point the boundary shall follow the provisional line mentioned above until it reaches the point where that line meets the Morro River.

d. From this point the boundary shall follow the Morro River to the junction of that river with the Mano River.

e. From this point the boundary shall follow the provisional line to the seacoast.

In accordance with the 1911 convention, a demarcation of the boundary between the *Makona and Mauwa* by a joint commission in 1913–14 was approved by an agreement of June 19–26, 1917. An exchange of notes between Liberia and the United Kingdom on January 16–17, 1930 approved a later demarcation of the meridian boundary between the *Mauwa and Morro*" (*Liberia – Sierra Leone Boundary, International Boundary Study No. 129* – October 13, 1972, U.S. Department of State).

Among the coordinate systems utilized in the country are the Sierra Leone Colony Datum of 1924 with origin point at *Kortright*, where $\Phi_o = 08°\ 28'\ 44.4''N$, $\Lambda_o = 13°\ 13'\ 03.81''W$, and the reference ellipsoid was the War Office (or McCaw), where $a = 20,926,201$ feet, $1/f = 296$, and the Grid system was the Colony Coordinates or New Sierra Leone Peninsula Transverse Mercator where the Central Meridian, $\lambda_o = 12°\ 00'W$, the Latitude of Origin, $\varphi_o = 6°\ 40'N$, Scale Factor at Origin, $m_o = $ Unity, False Easting $= 500,000$ feet, and False Northing $=$ Nil (*Private communication to Adam I. Alimi, (UNO Graduate Student) from Russell D. Fox, Overseas Surveys Directorate, Ordnance Survey, 13 April 1988*). According to John W. Hager, this Grid has also been used with a False Easting $= 166,666.66$ Yards.

The Sierra Leone Datum of 1960 origin point is at SLX2 Astro station, where $\Phi_o = 08°\ 27'\ 17.567''N$, $\Lambda_o = 12°\ 49'\ 40.186''W$, reference azimuth, $\alpha_o = 142°\ 41'\ 34.5''$, and the reference ellipsoid was the Clarke 1880, where $a = 6,378,249.145$ meters, $1/f = 293.465$, and the Grid system was the UTM where the Central Meridian, $\lambda_o = 12°\ W$.

"Precise Ephemeris Doppler fixes were observed at four existing trigonometrical stations in Sierra Leone during 1984 as part of the African Doppler Survey (ADOS), a project initiated by the International Association of Geodesy and the African Cartographic Association. The following lists the results and, for comparison, the corresponding Sierra Leone 1960 Datum positions: ADOS stn # ASL001, Local stn # 62-X-2, *Mile 6 Village*, (WGS72 Datum): $\varphi = 08°\ 20'\ 31.268''N$, $\lambda = 12°\ 59'\ 08.727''W$, $h = 63.2\,m$, (Sierra Leone Datum 1960): $\varphi = 08°\ 20'\ 30.993''N$, $\lambda = 12°\ 59'\ 07.608''W$, $H = 29.0\,m$; ADOS stn # ASL002, Local stn # DOS 4, *Kabala Town*, (WGS72 Datum): $\varphi = 09°\ 34'\ 45.254''N$, $\lambda = 11°\ 33'\ 41.228''W$, $h = 564.4\,m$, (Sierra Leone Datum 1960): $\varphi = 09°\ 34'\ 45.354''N$, $\lambda = 11°\ 33'\ 40.223''W$, $H = 530.6\,m$; ADOS stn # ASL003, Local stn # 71-X-1, *Kailahun Rest House*, m (WGS72 Datum): $\varphi = 08°\ 16'\ 22.103''N$, $\lambda = 10°\ 34'\ 19.965''W$, $h = 358.5$, (Sierra Leone Datum 1960): $\varphi = 08°\ 16'\ 21.831''N$, $\lambda = 10°\ 34'\ 19.053''W$, $H = 325.2\,m$; ADOS stn # ASL004, Local stn # 118-X-2, *Kwaje*, (WGS72 Datum): $\varphi = 06°\ 59'\ 31.112''N$, $\lambda = 11°\ 28'\ 13.849''W$, $h = 100.0\,m$, (Sierra Leone Datum 1960): $\varphi = 06°\ 59'\ 30.467''N$, $\lambda = 11°\ 28'\ 12.887''W$, $H = 68.0\,m$" (*op. cit., Adam I. Alimi, Russell D. Fox, 1988*).

"Survey Beacon SLS 28/57/107 at $\varphi = 07°\ 58'\ 01''N$, $\lambda = 11°\ 47'\ 38''W$. This position was scaled from a large-scale map and is listed as the origin of the Grid where $FN = 80,000$ feet, and $FE = 40,000$ feet. This position appears to be in the neighborhood of *Bo*" (*Personal communication - John W. Hager, 7 December 2011*).

Technical Report 8350.2 lists the three-parameter transformation **from** Sierra Leone Datum of 1960 **to** WGS84 as: $\Delta X = -88\,m$, $\Delta Y = +4\,m$, $\Delta Z = +101\,m$.

MALAYSIA
Johor Bahru
Sungai Johor
Pulau Ubin
Pulau Tekong
Singapore Changi International Airport
Singapore Strait
Main Strait
INDONESIA
Pulau Batam
Teluk Jodoh
Teluk Tering
agreed-upon maritime boundary
Singapore
Railroad
Mass rapid transit system (MRT)
Expressway
Major road
Minor road
Ferry
Built-up area
Scale 1:150,000

Republic of Singapore

A Malay city of importance in the 13th century, it was destroyed by the Japanese in the 14th century and remained a ruin until re-founded by Sir Thomas Stamford Raffles in 1819 for the British East India Company. Singapore became the capital of the Straits Settlements in 1833 (*Webster's Geographical Dictionary*). It joined the Malaysian Federation in 1963 but separated 2 years later and became independent on August 9th. The islands of Singapore have a total area of 693 km^2, and the republic is slightly larger than 3½ times the size of Washington, DC. The terrain of Singapore is lowland and has a gently undulating central plateau containing water catchments and a nature preserve. The lowest point is the *Singapore Strait* (0 m), and the highest point is *Bukit Timah* (166 m) (*CIA Factbook*).

"The Singapore Survey Department commenced its operation in 1826. It was then under the administration of the Assistant Engineer of the British Army Garrison who was the Surveyor and Registrar of Titles. Officers of the local Garrison conducted Land Survey work until 1847. For the next 20 years, the supervisory control was passed to the Surveyor-General of Bengal (Survey of India). In 1879, the department was incorporated with the Public Works Department and was under the direction of MacCallum the Colonial Engineer, assuming office as Surveyor-General. In 1881, he instituted the practice of marking boundaries with granite stones. Some of those original monuments can still be found in rural areas. From 1942 to 1945, the department functioned as part of the Property Department of the Municipal Administration. In 1946, the department was separated from the survey departments of the Federation of Malaya. From 1 June 1959, the administration of the department came under the charge of the Chief Surveyor, Singapore. Except for a brief period from 16 September 1963 to 9 August 1965 when Singapore was part of Malaysia, the department was under the portfolio of the Minister for Law. The department was re-organized in June 1965 in preparation for fulfillment of its role as an autonomous agency in April 1996. With the re-organization, four technical sections, namely Control Survey Section, Inspection Survey Section, Survey Services Section, Mapping and Records Section together with Administration Section were formed to carry out its functions" (*Singapore Ministry of Law*).

Geodetic triangulation in Singapore by British Royal Engineers was initiated in 1904 as part of the "Primary Triangulation of Malaya," published in 1917. According to A. G. Bazley in "*Geographical Positions in Malaya and Siam*," Empire Survey Review, No. 30, "Malaya: "The geographical positions of points were dependent upon latitude and azimuth determinations at *Bukit Asa* (*Asa Hill near the Lower Pierce Reservoir – Ed.*), and on the longitude of Fort Cornwallis Flagstaff in *Penang*, the latter being supposed to be 100° 20′ 44.4″ E. This value was obtained by Commander (later Admiral) Mostyn Field in H.M.S. *Egeria* (in) 1893, by the exchange of telegraphic signals with Mr. Angus Sutherland at Singapore, Old Transit Circle. The longitude, 103° 51′ 15.75″ E., accepted for Singapore in order to arrive at this determination of Fort Cornwallis Flagstaff, was based upon that of an Observation Spot, 103° 51′ 15.00″ E., fixed in 1881 by Lieutenant Commander Green, United States Navy, by meridian distance from the transit circle of *Madras Observatory*, the corresponding longitude of the latter being taken as 80° 14′ 51.51″ E.

"In 1894–6 another value of the longitude of *Madras*, 80° 14′ 47.06″ E., was determined via *Potsdam, Teheran, Bushire, etc.*, which value was later increased by 1.635 seconds of arc as a result of Albrecht's adjustment of European longitudes, the accepted longitude of *Madras* thereby becoming 80° 14′ 48.69″ E.; this is 2.82 seconds of arc, if the Green meridian distance across the *Bay of Bengal* and the latest astronomical longitude are accepted. After applying this correction, the geodetic longitude of the mark at Singapore Government Offices becomes 103° 51′ 07.96″ E., whereas the astronomical longitude depending upon the same position of *Madras* is 103° 51′ 05.28″ E., the respective meridian distances between Government Offices and Fort Cornwallis Flagstaff being: Geodetic ... 3° 30′ 26.38″; Astronomical ... 3° 30′ 23.70″.

"The possibility of a junction with Siam on the north and with the Dutch East Indies on the south, combined with the fact that it had in many cases been found difficult to account for the differences between modern standard traverses and the triangulation used to control them, resulted in the old 1917 triangulation being superseded by another which does attain to the degree of accuracy expected of a modern survey."

The Geodetic Survey of Malaya. – The new triangulation, details of which were published in 1931, extends from Kedah, including Penang, to Singapore, with a secondary chain through *Kelantan*. The latitudes and azimuths depend upon new determinations at *Kertau*, but the longitudes depend upon the value of *Kertau* from the F.M.S. (1917) (*possible meaning is "Fundamental Malayan Survey of 1917"–Ed.*) data – which, as already stated, was based on Fort Cornwallis Flagstaff–the value being 100° 20′ 44.4″ E. If the same (1917) figures for *Kertau* are accepted, however, the new triangulation places the flagstaff in longitude 100° 20′ 44.09″.

"The principal station at Singapore in the new survey is *Mt Faber*, situated 50.62″ south and 2′ 00.88″ west of Government Offices, while at Fort Cornwallis the trigonometrical station is 0.403″ north and 0.631″ east of the flagstaff; allowing for these connections we find that the new geodetic meridian distance between Singapore Government Offices and Fort Cornwallis Flagstaff is 3° 30′ 26.02″.

"In addition to the different datums for latitude and longitude used in the 1931 triangulation, between the old and the new geodetic work there are differences in azimuth varying from 1 to 3 minutes, together with small differences of varying amounts in scale, on the average about 1 foot in 40,000 feet. The latitudes and longitudes of the 1931 survey appear to be respectively 2.5 and 9 seconds of arc too great and are to be investigated further."

The *Kertau Datum* was revised for West Malaysia and Singapore the same year that the Timbalai Datum was published for East Malaysia by the Survey of India. The origin point for the *Revised Kertau Datum of 1948* is: $\Phi_o = 03° 27′ 50.71″$ N, $\Lambda_o = 102° 37′ 24.55″$ East of Greenwich; the deflection of the vertical components are $\xi_o = 3.47″$, $\eta_o = -10.90″$. The ellipsoid of reference is the Everest 1830 (revised), where $a = 6,377,304.063$ m and $1/f = 300.8017$. The plane coordinate system associated with this datum is the Singapore Cassini-Soldner Grid such that the origin is the Flagstaff at the Empress Place Building, where the Latitude of Origin, $(\phi_o) = 1° 17′ 15.08″$ N, the Central Meridian, $(\lambda_o) = 103° 51′ 10.78″$ E, the scale factor at origin, $(m_o) = 1.0$, and both the False Easting and False Northing = 30 km. The datum shift parameters listed in *NGA TR8350.2* **from** Kertau *1948 Datum* **to** WGS84 Datum are: $\Delta X = -11$ m ± 10 m, $\Delta Y = +851$ m ± 8 m, and $\Delta Z = +5$ m ± 6 m, and six points were used in the 1987 solution. NATO lists the datum shift parameters as a seven-parameter Bursa-Wolf transform, *herein modified to the Standard American Rotation Convention* as: $\Delta X = -366.94$ m, $\Delta Y = +719.29$ m ± 8 m, and $\Delta Z = -88.93$ m, $\delta s = +9.093$, $R_x = +2.498″$, $R_y = +2.142″$, $R_z = -12.057″$; however, NATO does **not** state the accuracy of the transformation parameters, nor do they indicate the number of collocated points used for their published solution!

There's a new geocentric datum in town for Singapore, originally termed the SVY-95 Datum and now renamed as "SVY-21." The "origin" is listed as being at BASE7: $\phi_o = 1° 22′ 02.915414″$ N, $\lambda_o = 103° 49′ 31.975227″$ E, $h_o = 26.824$ m, $H_o = 17.113$ m; however, no explanation is offered regarding how a geocentric datum can have a datum origin! The plane coordinate system is now based on a Transverse Mercator projection, where $\phi_o = 1° 22′ 00″$ N, $\lambda_o = 103° 50′ 00″$ E, Scale Factor at Origin (m_o) presumably is unity, False Easting = 128,001.642 m, and False Northing = 138,744.572 m. Presumably, the mapping equations are the Gauss-Krüger, but it really doesn't make a difference considering the size of Singapore. The parameters have been chosen so that coordinates in this new datum will be close to the Cassini-Soldner coordinates but with an additional 100,000 m to both the Northings and Eastings. The resulting "1" prefix is intended to easily distinguish between the Cassini-Soldner coordinates and the Integrated Survey Network (ISN) coordinates.

The new SVY-21 Datum geodetic coordinates of the Flagstaff at the Empress Place Building are: $\phi = 1° 17' 15.5294''$ N, $\lambda = 103° 51' 10.8046''$ E. The Flagstaff is one of three fiducial points held for the readjustment of the new datum. Bin and Chai published a paper titled, *"Improving Cadastral Survey Controls using GPS Surveying in Singapore"* in Survey Review, 33, 261 (July 1996). In that paper, the authors state that they used the Leica "SKI" software package to derive auxiliary parameters to convert GPS vectors into the Revised Kertau Datum. Since that Swiss package uses the European Rotation convention, those *four* parameters converted to the American Standard Rotation convention **from** *Revised Kertau Datum of 1948* **to** SVY-21 Datum are as follows: $\delta s = +0.9475$ ppm ± 0.23 ppm, $R_x = -6.4227'' \pm 0.34''$, $R_y = +3.8310'' \pm 0.21$, $R_z = +7.2881'' \pm 0.16''$. "The average difference for 58 stations in both latitude and longitude is 0.0006″ or 0.018 m." Note that the equations for all the transformations described herein are in the *Manual of Photogrammetry*, 5th and 6th editions.

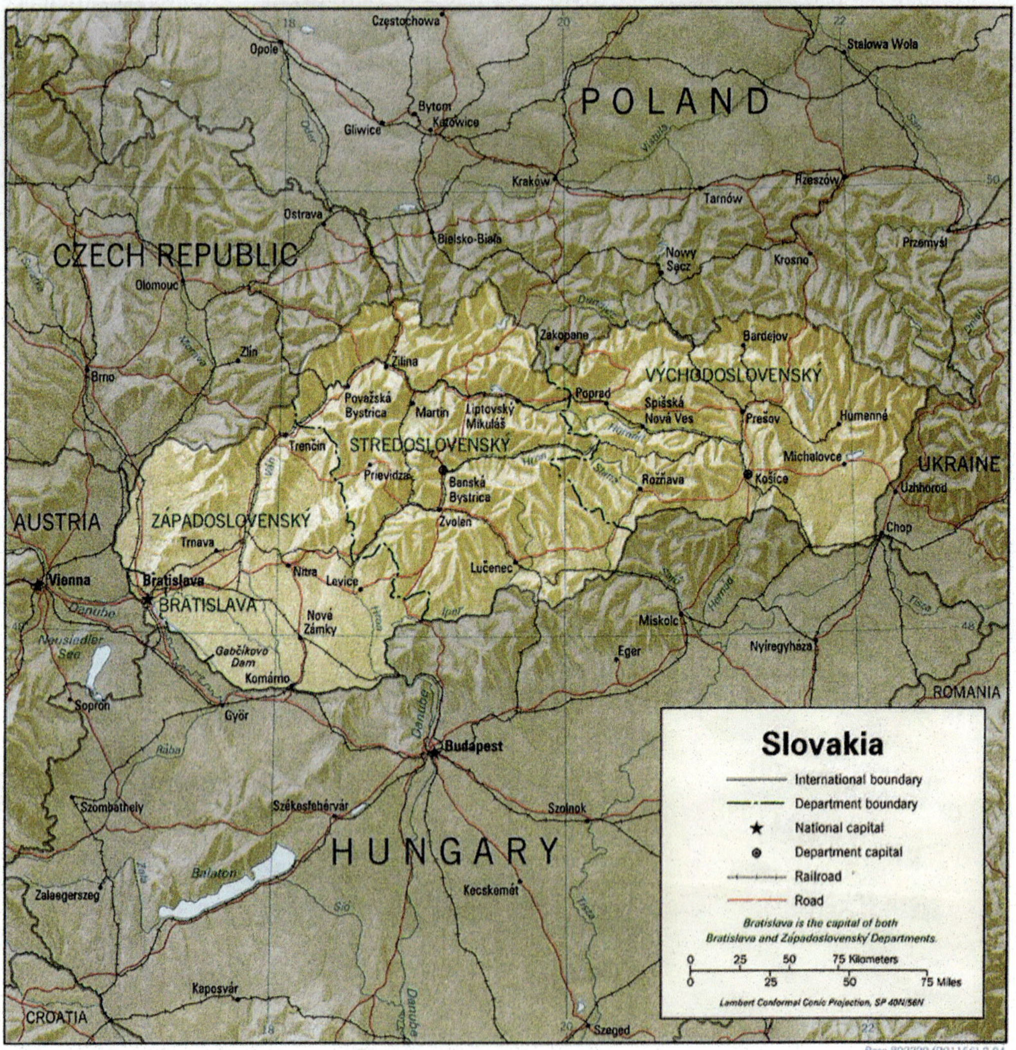

Base 802220 (R01166) 2-94

Slovak Republic

About twice the size of New Hampshire, Slovakia is bordered by: Austria (91 km, Czech Republic (197 km), Hungary (676 km), Poland (420 km), and Ukraine (90 km). The terrain consists of rugged mountains in the central and northern part and lowlands in the south; the lowest point is the *Bodrok River* (94 m), the highest point is *Gerlachovsky Štít* (2,655 m) (*World Factbook*, 2011).

"Slovakia's roots can be traced to the 9th century state of Great Moravia. Subsequently, the Slovaks became part of the Hungarian Kingdom, where they remained for the next 1,000 years. Following the formation of the dual Austro-Hungarian monarchy in 1867, language and education policies favoring the use of Hungarian (Magyarization) resulted in a strengthening of Slovak nationalism and a cultivation of cultural ties with the closely related Czechs, who were themselves ruled by the Austrians. After the dissolution of the Austro-Hungarian Empire at the close of World War I, the Slovaks joined the Czechs to form Czechoslovakia. Following the chaos of World War II, Czechoslovakia became a Communist nation within Soviet-dominated Eastern Europe. Soviet influence collapsed in 1989 and Czechoslovakia once more became free. The Slovaks and the Czechs agreed to separate peacefully on 1 January 1993. Slovakia joined both NATO and the EU in the spring of 2004 and the euro zone on 1 January 2009" (*World Factbook* 2022).

The territory was covered by the old cadastral triangulations with origins at *Gusterberg, St. Stephan Tower (Vienna), Gellerthegy, Pschow*, and coordinates referring to *Vienna University Datum, St. Anna Datum*, and *Hermannskogel Datum*. A new first-order net was started in 1936 and was completed in 1956. The basic cadastral trigonometric net was connected with the first-order nets of Austria, Germany, Poland, and Romania (through the *Carpatho-Ukraine* and Slovakia). Between 1918 and 1932, the *Military Geographic Institute* (MGI) applied the Lambert conformal conic projection for triangulation computations and mapping. This was based on the Hermannskogel Datum of 1871 referenced to the Bessel 1841 ellipsoid of revolution, where the semi-major axis $(a) = 6,377,397.155$ m and the reciprocal of flattening $(1/f) = 299.1528128$. The Hermannskogel 1871 Datum has its origin with $\Phi_o = 48°\ 16'\ 15.29''$ N, $\Lambda_o = 33°\ 57'\ 41.06''$ *Est de l'Ile de Fer* (East of Ferro Island in the Canaries), where Ferro = $17°\ 39'\ 46.02''$ East of Greenwich and azimuth to station Hundsheimer is $\alpha_o = 107°\ 31'\ 41.7''$. The secant Lambert Grid had the standard parallels of $\varphi_N = 50°\ 15'$ N and $\varphi_S = 48°\ 30'$ N, a Central Meridian $(\lambda_o) = 35°\ 45'$ East of Ferro, a False Easting = 1,000 km, and a False Northing = 500 km. This point corresponds to the center of the southern sheet line of the 1:75,000 sheet titled "4260 Vsetin." "Only 3% of the state territory was mapped. The author of this projection was Captain Benes, and this projection is practically Lambert" (*Prof. Bohuslav Veverka, personal communication, February 2003*).

Ing. Josef Krøvák (commonly spelled *Krovák* or *Krovak*) prepared the Conformal Oblique Conic Projection of Czechoslovakia in 1922 for the preparation of cadastral (tax) maps and topographic maps of medium scales for the civil geodetic service of Czechoslovakia. The "starting meridian" was termed Ferro where the MGI usage *in Slovakia* used the relation: Ferro = $17°\ 40'$ East of Greenwich (*op. cit., Veverka, 2003*). The Krovak Projection is a double projection in that the oblique conic is projected from the Gaussian Sphere where the radius = 6,380,703.6105 m. The Gaussian Sphere was "invented" by Carl Friederich Gauss and is also commonly known as the "conformal sphere." It is simply the geometric mean of the ellipsoidal normal (at a point) terminated by the semi-minor axis and the radius of the ellipsoid (at the same point) in the plane of the meridian. To be succinct, it's $[\sqrt{\nu\rho}]$ evaluated in this case at $\varphi = 49°\ 30'$ N. For the Slovak Republic, they used the Bessel 1841 ellipsoid at that latitude. Whenever one sees the term "double projection," the generating sphere is usually the Gaussian Sphere. The oblique cone has a pole centered at $\varphi_o = 59°\ 42'\ 42.6969''$ N, $\lambda_o = 42°\ 30'$ East of Ferro (southwest of *Helsinki, Finland*). The spherical cartographic coordinates are transformed into the rectangular plane coordinates of the uniform cadastral system. For this purpose, the reduced (0.9999) Gaussian Sphere is projected on the surface of an oblique cone touching the sphere around the central cartographic parallel, having a cartographic latitude of 78° 30' N, with the vertex in the extended axis connecting the center of the sphere at the rotation angle of 30° 17' 17.3031". This is still in use as of 2000 and is known as *Systém - Jednotné Trigonometrické Sítì Katastrální* or S-JTSK (System of the Unified Slovak Trigonometrical Cadastral Net). The Czechs state (*Prof. Ing. Bohuslav Veverka, Prague, November 1997*) that the "scale, location and orientation of the S-JTSK on the surface of the Bessel's ellipsoid was derived from the results of the historical Austro/ Hungarian military surveys in the years 1862–98."

"The average density of control points of first- through fifth order is 1 per 2.7 km². The State Astronomic and Geodetic Network contains 66 points in S-JTSK and in S-42/83. The State Astronomic and Geodetic Network in S-42/83 has been adjusted together with the astronomic and geodetic networks of the East European Countries" (*Geodesy, Cartography and the Cadastre of the Real Estates in the Slovak Republic, 2000*).

After World War II, the agencies responsible for geodetic, topographic, and cartographic activities in the Second Czechoslovakian Republic were in a stage of re-organization up to the end of 1953. During the years 1953–54, those agencies were subsequently organized according to the pattern established in the USSR. The *Ustrední Správa Geodesie a Kartografie* – USGK, (Central Administration of Geodesy and Cartography) was established. The *Zakladni Trigonometricka Sit* – ZTS (Basic Trigonometric Net) (*op. cit., Veverka, 2003*) included the first-order net of the Protectorate and the first-order net established in 1949–1953 in Slovakia (*op. cit., Veverka, 2003*). The adjustment of the net was carried out by the method of Pranis-Praniévitch on the Krassovsky 1940 ellipsoid, where $a=6,378,245$ meters and $1/f=298.3$. The Datum is defined as "System 42" where the origin is at Pulkovo Observatory: $\Phi_o=59°\ 46'\ 18.55''$ North, $\Lambda_o=30°\ 19'\ 42.09''$ East of Greenwich. The defining azimuth at the point of origin to Signal A is: $\alpha_o=317°\ 02'\ 50.62''$. The "Russia Belts" Grid System is used with the System 42 Datum; identical to UTM except that the scale factor at origin is unity.

A provisional military version used since 1952 for 1:25,000 scale mapping (Gauss-Krüger – since 1953) (*op. cit., Veverka, 2003*) is a modification of the Russia Belts system in that the False Northing at origin was $\varphi=49°\ 30'$ North, the False Northing = 200 km, the False Easting = 500 km, and the scale factor at origin $(m_o)=0.99992001$. Everything else remained the same as the standard Gauss-Krüger Transverse Mercator Grid. Boundary treaties with adjacent countries refer to ancient datums and grids that include the old double stereographic projections of the 19th and early 20th centuries. The S-JTSK Krovak Projection is alive and well in the Slovak Republic for the 21st century. "Differences between S-42 and S-52 are max. 2 meters" (*op. cit., Veverka, 2003*).

TR 8350.2 provides two three-parameter transformations for all of the former Czechoslovakia. **From** S-42 (referenced to the Krassovsky 1940 ellipsoid) **to** WGS84: $\Delta X=+26\,m\ \pm3\,m$, $\Delta Y=-121\,m\ \pm3\,m$, $\Delta Z=-78\,m\ \pm2\,m$, and **from** S-JTSK (referenced to the Bessel 1841 ellipsoid) **to** WGS84: $\Delta X=+589\,m\ \pm4\,m$, $\Delta Y=+76\,m\ \pm2\,m$, $\Delta Z=+480\,m\ \pm3\,m$.

The Slovak Republic has published (*BKG 2001–2003*) a full seven-parameter Molodensky transformation **from** S-JTSK **to** ETRS89: $\Delta X=+559.0\,m$, $\Delta Y=+68.7\,m$, $\Delta Z=+451.5\,m$, $R_x=+7.920''$, $R_y=+4.073''$, $R_z=4.251''$, $\delta s=+5.71''$, $X_o=3,980,912.082\,m$, $Y_o=1,392,955.999\,m$, $Z_o=4,767,344.572\,m$. The Molodensky offset origin point appears to be southeast of *Banská*, in a geometrically central part of the republic. A newer version using the Bursa-Wolfe seven-parameter transformation **from** S-JTSK **to** ETRS89: $\Delta X=+485.0\,m$, $\Delta Y=+169.5\,m$, $\Delta Z=+483.8\,m$, $R_x=+7.786''$, $R_y=+4.398''$, $R_z=4.103''$, $\delta s=0$. This was computed in 2006 with about 700 identical points and when checked with three test points, the transformation accuracy was about 1 m.

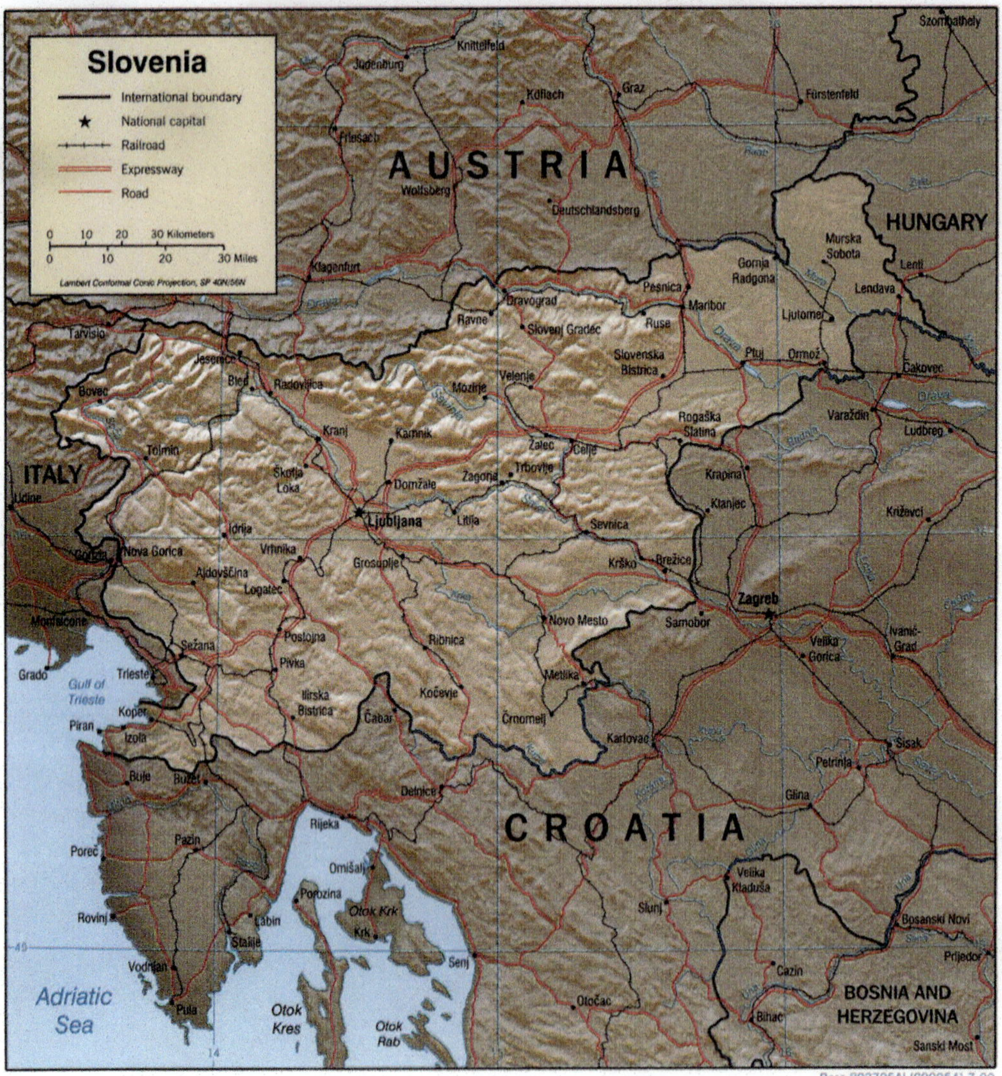

Base 802705AI (C00054) 7-00

Republic of Slovenia

Slovenia is bordered by Austria (330 km), Croatia (455 km), Hungary (102 km), and Italy (199 km). Slightly smaller than New Jersey, Slovenia has a 46.6 km coastline on the Adriatic Sea (0 m), and the highest point is *Triglav* (2,864 m), near the Italian/Austrian tripoint (*World Factbook, NGA Geonames Server and Google Earth*, 2011).

"Slovenia was originally settled by Illyrian and Celtic peoples. It became part of the Roman Empire in the first century B.C. The Slovenes were a south Slavic group that settled in the region in the 6th century A.D. During the 7th century, the Slavs established the state of *Samu*, which owed its allegiance to the Avars, who dominated the Hungarian plain until Charlemagne defeated them in the late 8th century" (*InfoPlease,* 2011).

"According to the 16th century French political philosopher, Jean Bodin, Slovenes practiced the unique custom of the Installation of the *Dukes of Carinthia* for almost 1,000 years, until the late 14th century. According to some scholars, Bodin's account of how Slovene farmers contractually consented to be governed by the duke influenced Thomas Jefferson's drafting of the Declaration of Independence. From as early as the 9th century, Slovenia had fallen under foreign rulers, including partial control by Bavarian dukes and the Republic of Venice. With the exception of Napoleon's 4-year tutelage of parts of Slovenia and Croatia – the 'Illyrian Provinces' – Slovenia was part of the Habsburg Empire from the 14th century until 1918. Nevertheless, Slovenia resisted Germanizing influences and retained its unique Slavic language and culture. In 1918, Slovenia joined with other southern Slav states in forming the Kingdom of Serbs, Croats, and Slovenes as part of the peace plan at the end of World War I. Renamed in 1929 under a Serbian monarch, the Kingdom of Yugoslavia fell to the Axis powers during World War II. Following communist partisan resistance to German, Hungarian, and Italian occupation and elimination of rival resistance groups, socialist Yugoslavia was born under the helm of Josip Broz Tito. On June 25, 1991, the Republic of Slovenia declared its independence" (*U.S. Dept. of State Background Notes,* 2011).

"Historical ties to Western Europe, a growing economy, and a stable democracy have assisted in Slovenia's postcommunist transition. Slovenia acceded to both NATO and the EU in the spring of 2004; it joined the euro zone and the Schengen zone in 2007" (*World Factbook* 2022).

Thanks to Michael Rittri, "the MGI (Ferro) datum (Prime Meridian) covered both Austria and former Yugoslavia, but when the countries switched to the Greenwich meridian, Austria assumed that Ferro was 17° 40′ 00″ west of Greenwich, while Yugoslavia assumed it was 17° 39′ 46.02″ west (the Albrecht value). So, when using MGI with Greenwich, one must distinguish between MGI (Greenwich, Austria) and MGI (Greenwich, Yugoslavia), since longitudes differ by 13.98″, about 300 meters. The MGI datum was finalized in 1901, but the switch to Greenwich occurred later (in the 1920s?). A new adjustment was made in Yugoslavia in 1948. This datum is known as D48, at least in Slovenia, and it was (probably) designed to be a minor improvement, in the sense that some distortion was removed, but that coordinates are not changed more than a couple of meters (compared to MGI 1901). This D48 system was definitely adopted in Slovenia. Slovenia has later made a densification of the European ETRS89 datum; this is known as D96 in Slovenia, or 'Slovenia Geodetic Datum 1996' in the EPSG database (EPSG: 6765).

"I think the datum transformations currently published by EPSG from 'MGI' to WGS84 will work fine, if you respect their Area of Use. Just remember that if the Area of Use is in Austria, then 'MGI' is really the Austrian MGI. And if the Area of Use is in former Yugoslavia, then 'MGI' is really the Yugoslavian MGI (or possibly the 1948 version of the Yugoslavian MGI)."

The most common grid found on the D48 datum referenced to the Bessel 1841 ellipsoid of revolution, where $a = 6,377,397.155$ meters and $1/f = 299.1528128$ is the Yugoslavia Reduced Gauss-Krüger

Transverse Mercator and is still used in Slovenia. The scale factor at origin (m_o=0.9999), the central meridian of the belt C.M.=λ_o=15° East of Greenwich and the False Easting at C.M.=500 km. The D48 datum, a readjustment of the Hermannskogel 1871 datum is still referenced to the Bessel 1841 ellipsoid. Recent adjustments and observations with GPS receivers have resulted in a new datum realization for Slovenia to the European Terrestrial Reference System of 1989 (ETRS89). The seven-parameter Helmert transformation published for Slovenia from SI_D48 datum to ETRS89 datum is where: ΔX=+426.9 m, ΔY=+142.6 m, ΔZ=+460.1 m, Scale=+17.1 × 10^{-6}, Z-rotation (ω)=−12.42 seconds, Y-rotation (ψ)=+4.49 seconds, and X-rotation (ξ)=+4.91 seconds. A test point provided is from SI_D48: φ=45° 36′ 54.72″N, λ=14° 01′ 59.52″E, X=5,052,752.959 m, Y=424,607.776 m, ETRS89: φ=45° 36′ 53.733″N, λ=14° 01′ 42.771″E, X=5,051,723.46 m, Y=424,258.16 m.

A complete series of datum transformations have been published in the open literature for Slovenia that are grouped according to specific areas within the country. Although the language used is Slovenian, the parameters are easily discerned. One series of parameters are grouped into three separate regions of the country, another series are grouped into seven separate (smaller) regions. Obviously, the smaller the region the greater, the transformation accuracy.

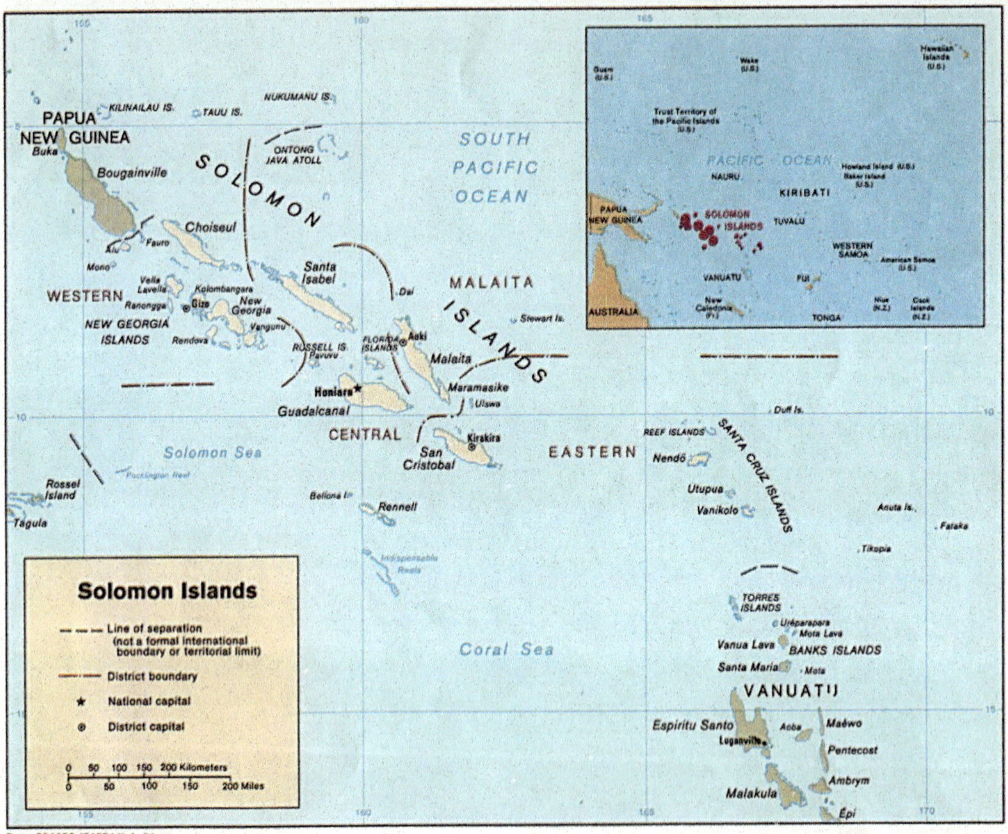

Base 504633 (545714) 4-81

Solomon Islands

Slightly smaller than Maryland, the Solomon Islands have a coastline of 5,313 km, and its terrain is comprised of mostly rugged mountains with some low coral atolls. The lowest point is the Pacific Ocean (0 m), and the highest point is *Mount Popomanaseu* (2,310 m) (*World Factbook*, 2014).

"The Solomon Islands were initially settled by at least 2000 *B.C.* – well before the archaeological record begins – probably by people of the Austronesian language group. Pottery of the Lapita culture was in use in Santa Cruz and the Reef Islands about 1500 *B.C.* Material dating to about 1000 *B.C.* has also been excavated at *Vatuluma Cave* (*Guadalcanal*), on *Santa Ana Island*, and on the outlying islands of *Anuta* and *Tikopia*.

"The first European to reach the islands was the Spanish explorer Álvaro de Mendaña de Neira in 1568. Subsequently, unjustified rumors led to the belief that he had not only found gold there but had also discovered where the biblical king Solomon obtained the gold for his temple in Jerusalem. The islands thus acquired the name *Islas de Solomón*. Later Spanish expeditions to the southwest Pacific in 1595 and 1606 were unable to confirm the discoveries reported by Mendaña. Geographers came to doubt the existence of the group, and it was not until the late 18th century, after further sightings by French and English navigators, that the Solomons were accurately charted. After the settlement of Sydney by the English in 1788, naval and commercial shipping began increasingly to pass through the Solomons' waters.

"Roman Catholic missionaries failed to establish a settlement in the 1840s but did so in 1898. Anglican missionaries, who had been taking islanders to New Zealand for training since the 1850s, began to settle in the Solomons in the 1870s. Other missions arrived later.

"By the late 19th century the islands were being exploited for labor to work the plantations of Fiji and other islands and of Queensland, Australia. About 30,000 laborers were recruited between 1870 and 1910. To protect their own interests, Germany and Britain divided the Solomons between them in 1886; but in 1899 Germany transferred the northern islands, except for Buka and Bougainville, to Britain (which had already claimed the southern islands) in return for recognition of German claims in Western Samoa (now Samoa) and parts of Africa. The British Solomon Islands Protectorate was declared in 1893, partly in response to abuses associated with labor recruitment and partly to regulate contacts between islanders and European settlers, but mainly to forestall a threat of annexation by France. Colonial rule began in 1896. Although generally humane, administrators were more concerned with promoting the interests of European traders and planters than those of the islanders, and islanders were punished harshly for offenses against colonial law and order. The murder of government tax collectors by members of the Kwaio ethnic group on *Malaita* in 1927 was answered with a savage punitive expedition, backed by an Australian warship, that burned and looted villages and killed many of the Kwaio. Together with some of his associates, Basiana, the leader of the tax collectors' killers, was hanged, and his young sons were forced to witness the execution.

"With the outbreak of World War II in the Pacific, the Japanese began occupying the protectorate early in 1942, but their advance farther southward was stopped by U.S. forces, which invaded on August 7. Fighting in the Solomons over the next 15 months was some of the most bitter in the Pacific; the long Battle of *Guadalcanal* was one of the crucial conflicts of the Pacific war. Throughout the campaign the U.S. forces and their allies were strongly supported by the islanders. After the war, because of the proximity of an airfield and the availability of flat land and of the military's buildings, *Honiara* on *Guadalcanal* became the new capital, replacing Tulagi.

"Another result of the war was to stimulate political consciousness among the islanders and so inspire a nationalist movement known as Maasina Rule, which lasted from 1944 to 1952. Subsequently, in response to the worldwide movement for decolonization, the Solomons set out on the path of constitutional development. The country was formally renamed Solomon Islands in 1975, and independence was attained on July 7, 1978" (*Encyclopaedia Brittanica, 2014*).

Thanks to Mr. John W. Hager, "the Astro Stations observed in the Solomon Islands include *Cruz*, at *"Cruz Astro 1947,"* $\Phi_o=9°$ 25' 27.61" S, $\Lambda_o=159°$ 59' 10.14" E, $\alpha_o=99°$ 46' 39.3" to Az. Mark from south, International ellipsoid, elevation=2.20 feet. Reported under water, 23 Oct. 1961. Established by the 657th Eng. Astro. Det. March 1947. CZ-X-6, $\Phi_o=11°$ 34' 13.3920" S, $\Lambda_o=166°$ 52' 55.8300" E, International ellipsoid, $H_o=10.5$ feet. Area is *Santa Cruz Islands, Islands of Ndeni (Nendo), Utupua, Vanikoro (Vanikolo)*. I don't know whether or not it includes the *Reef Islands*. CZ-X-8, $\Phi_o=9°$ 47' 43.8000" S, $\Lambda_o=167°$ 06' 24.3000" E, International ellipsoid. *Duff Islands*. GUX 1 (1960), at *GU.X.1 Jean*, $\Phi_o=9°$ 27' 05.272" E, $\Lambda_o=159°$ 58' 31.752" E, $\alpha_o=300°$ 16' 31.045" to GU.X. 2 from north, International ellipsoid, $H_o=576.25$ feet. Also listed at DOS Astro Gux 1. Source is *"Co-ordinate List British Solomon Islands Protectorate,"* *Honiara, Guadalcanal*, 23 Oct. 1961. *Swallow Islands*, at 1966 SECOR Astro, $\Phi_o=10°$ 18' 21.42" S, $\Lambda_o=166°$ 17' 56.79" E. International ellipsoid".

The official datum of the Solomon Islands is *GUX 1 ASTRO FIX*, and the only Grid system ever used is the UTM Zone 57, all referenced to the International ellipsoid, where $a=6,378,388$ m and $1/f=298$. This appears to be the "working system" for cadastral applications. Current mapping is referenced to the WGS84 Datum. According to *TR 8350.2*, the transformation parameters **from** GUX 1 ASTRO Datum **to** WGS84 Datum are: $\Delta X=+252$ m± 25 m, $\Delta Y=-209$ m± 25 m, $\Delta Z=-751$ m± 25 m. The magnitude of the parameters that would apply to the *Swallow Islands* likely would be significantly smaller, perhaps by an order of magnitude or more, because of the superior timing of the SECOR satellites over spring-wound chronometers used for the classical Astro Stations.

In September of 2013, the Solomon Islands reported on the condition of their geodetic control to the FIG Small Island Developing States Symposium in Fiji.

See also: *Crustal Motion Studies in the Southwest Pacific: Geodetic Measurements of Plate Convergence in Tonga, Vanuatu and the Solomon Islands*, David A. Philips, http://www.soest.hawaii.edu/gg/academics/theses/Phillips_D_2003_Dissertation.pdf.

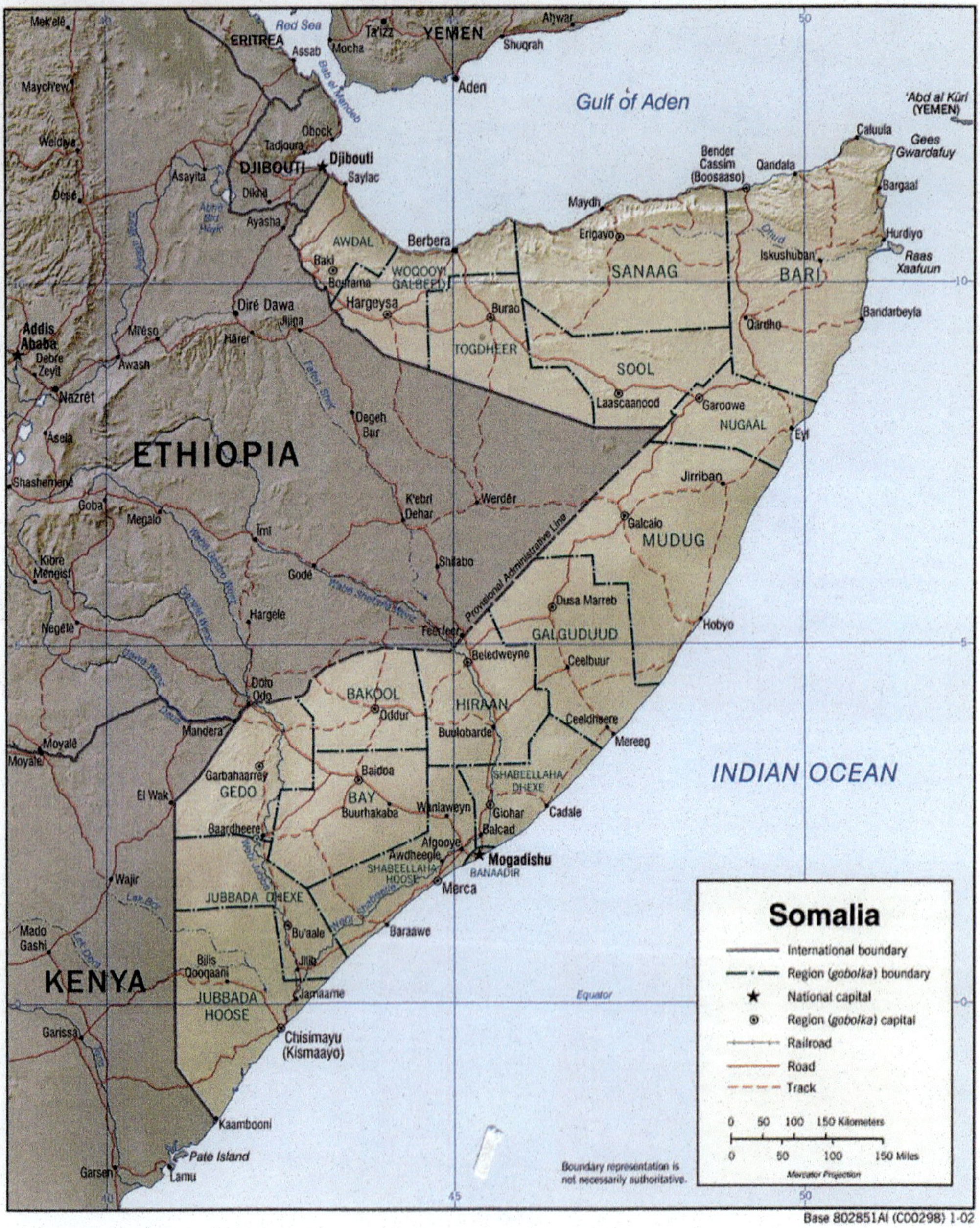

Base 802851AI (C00298) 1-02

Federal Republic of Somalia

Slightly smaller than Texas, Somalia is bordered by Djibouti (58 km), Ethiopia (1,600 km), and Kenya (682 km). The terrain consists of mostly flat to undulating plateau, rising to hills in the north; the lowest point is the Indian Ocean (0 m), and the highest point is *Shimbiris* (2,416 m).

"Located in the horn of Africa, adjacent to the Arabian Peninsula, Somalia is steeped in thousands of years of history. The ancient Egyptians spoke of it as 'God's Land' (*the Land of Punt*). Chinese merchants frequented the Somali coast in the tenth and fourteenth centuries and, according to tradition, returned home with giraffes, leopards, and tortoises to add color and variety to the imperial menagerie. Greek merchant ships and medieval Arab dhows plied the Somali coast; for them it formed the eastern fringe of Bilad as Sudan, '*the Land of the Blacks*.' More specifically, medieval Arabs referred to the Somalis, along with related peoples, as the Berberi. By the eighteenth century, the Somalis essentially had developed their present way of life, which is based on pastoral nomadism and the Islamic faith. During the colonial period (approximately 1891 to 1960), the Somalis were separated into five mini-Somalilands: British Somaliland (north central); French Somaliland (east and southeast); Italian Somaliland (south); Ethiopian Somaliland (the *Ogaden*); and, what came to be called the Northern Frontier District (NFD) of Kenya. In 1960 Italian Somaliland and British Somaliland were merged into a single independent state, the Somali Republic. In its first nine years the Somali state, although plagued by territorial disputes with Ethiopia and Kenya, and by difficulties in integrating the dual legacy of Italian and British administrations, remained a model of democratic governance in Africa; governments were regularly voted into and out of office" (*Library of Congress Country Studies,* 2013).

"In 1969, a coup headed by Mohamed Siad Barre ushered in an authoritarian socialist rule characterized by the persecution, jailing, and torture of political opponents and dissidents. After the regime's collapse early in 1991, Somalia descended into turmoil, factional fighting, and anarchy. In May 1991, northern clans declared an independent Republic of Somaliland that now includes the administrative regions of *Awdal, Woqooyi Galbeed, Togdheer, Sanaag,* and *Sool*. Although not recognized by any government, this entity has maintained a stable existence and continues efforts to establish a constitutional democracy, including holding municipal, parliamentary, and presidential elections. The regions of *Bari, Nugaal,* and *northern Mudug* comprise a neighboring semi-autonomous state of *Puntland*, which has been self-governing since 1998 but does not aim at independence; it has also made strides toward reconstructing a legitimate, representative government but has suffered some civil strife. *Puntland* disputes its border with Somaliland as it also claims portions of eastern *Sool* and *Sanaag*" (*World Factbook* 2013).

"By 2012, Somali powerbrokers agreed on a provisional constitution with a loose federal structure and established the central government in *Mogadishu*. Since then, four interim regional administrations have been established and there have been two presidential elections. However, significant and fundamental governance and security problems remain" (*World Factbook* 2022).

There are a number of datum origins in Somalia, corresponding to areas in which the various colonial powers have had dominion over the past couple of centuries. Thanks to John W. Hager, in the north, corresponding to the "old British Somaliland is *Berbera Pier* (*the port city of Berbera on the Gulf of Aden – Ed.*) where: $\Phi_o = 10°\ 26'\ 24.0''$ N, $\Lambda_o = 45°\ 00'\ 39.0''$ E, and the ellipsoid of reference is the Clarke 1880 where: $a = 6,378,249.145$ m, and $1/f = 293.465$."

Proceeding south, the next datum origin found is *Dolo*, Northwest of *Mogadishu*, in the vicinity of *Dolo Bay*, where $\Phi_o = 04°\ 10'\ 36.60''$ N, $\Lambda_o = 42°\ 50'\ 00.15''$ E; *unknown ellipsoid, probably Clarke 1880*. "The Ethiopia – Somalia boundary consists of three distinct sectors. The thalweg of the *Dewa* (*River – Ed.*) constitutes a 22-mile sector between the Kenya tripoint and the confluence of the *Dewa* with the *Ganālē-Doryā River* at *Dolo*, from which junction the rivers form the *Giuba*. Between *Dolo* and 8°N, 48°E, the second sector is delimited by a provisional administrative line for 509 miles. The final sector, between 8°N, 48°E and the tripoint with the French Territory of the *Afars and Issas* at *Madaha Djalêlo*, is 463 miles long and is demarcated by boundary pillars" (*International Boundary Study No. 153, Ethiopia – Somalia*, U.S. Dept. of State, November 5, 1975). Dolo was likely established by the Anglo–Ethiopian Boundary Commission, 1932–1935, as reported by G.T. McCaw in *Empire Survey Review, Nos. 25–26, 1937*.

Continuing south, the next datum origin found is generally considered the major system of Somalia: *Afgooye Datum*, thanks to John W. Hager; established by the U.S.S.R. sometime in 1962–1968 at station "BM Ts 30," where $\Phi_o=02°\ 06'\ 12.14''$ N, $\Lambda_o=45°\ 09'\ 55.46''$ E, $h_o=128.210$ m, and the ellipsoid of reference is the Krassovsky 1940, where $a=6,378,245$ m and $1/f=298.3$. According to *TR8350.2*, the three-parameter transformation from *Afgooye Datum* to WGS84 Datum is: $\Delta X=-43$ m ±25 m, $\Delta Y=-163$ m ±25 m, $\Delta Z=+45$ m ±25 m. Out of curiosity, I transformed the *Afgooye Datum* coordinates of station "BM Ts 30" into WGS84 Datum: $\varphi=02°\ 06'\ 13.78''$ N, $\lambda=45°\ 09'\ 52.73''$ E, and I input the coordinates into Google Earth©. John W. Hager informs me that the *Afgooye Datum* origin is "Located on top of a square water tank. A Doppler station was established there." Those coordinates will indeed locate the water tank on Google Earth©, including its adjacent shadow! Note that the location is actually in *Lafoole*, southeast of the town of *Afgooye*. My son, André informs me that the tank is now gone.

Continuing southwest, the next datum origin is *Fortino di Trevis*, where $\Phi_o=01°\ 43'\ 19.10''$ N, $\Lambda_o=44°\ 46'\ 32.38''$ E which is in the Northeast part of the town of *Merca*; *unknown ellipsoid*.

Continuing southwest, the last datum origin is *Perduchi a Giumbo*, where $\Phi_o=00°\ 14'\ 50.70''$ S, $\Lambda_o=42°\ 37'\ 27.10''$ E which is in *Jidka Aaran*, between *Luglow* and Goob Weyn and Northeast of *Kismaayo*; *unknown ellipsoid*.

The most common legacy Grids found in Somalia are the East Africa Belts where the Central Meridians (λ_o) are J=42° 30′ E, K=47° 30′ E, L=52° 30′ E; the Scale Factor at Origin $(m_o)=0.9995$, False Easting=400 km, and False Northing=4,500 km, referenced to the Clarke 1880 ellipsoid. Of course, for the *Afgooye Datum*, the likely Grid used is the Russia Belt 8 Gauss-Krüger Transverse Mercator where $\lambda_o=45°$ E, $m_o=1.0$, and False Easting=500 km.

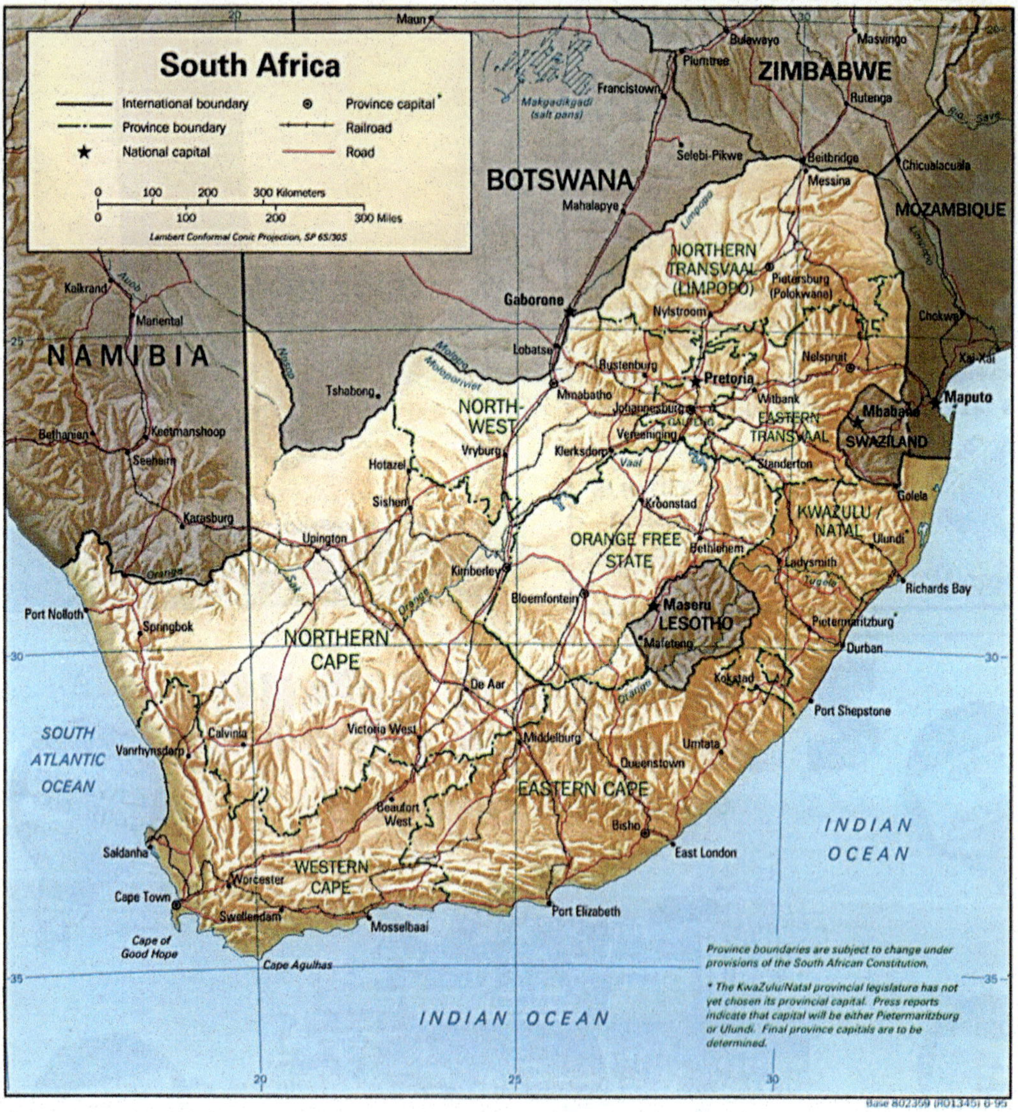

Base 802359 (R01345) 6-95

Republic of South Africa

Slightly less than twice the size of Texas, South Africa is bordered by Botswana (1,840 km), Lesotho (909 km), Moçambique (491 km), Namibia (967 km), Swaziland (430 km), and Zimbabwe (225 km). The terrain is comprised of a vast interior plateau rimmed by rugged hills and narrow coastal plain; the lowest point is the Atlantic Ocean (0 m), and the highest point is *Njesuthi* (3,3408 m) (*World Factbook*, 2012).

According to *Lonely Planet*, "South Africa's history extends back to around 40,000 B.C. when the San people first settled Southern Africa. By A.D. 500, Bantu-speaking peoples had arrived from West Africa's Niger Delta. Competing colonial European powers began settling here in small numbers from the 17th century, mostly in the Cape. Widespread colonial settlement of South Africa began in the 19th century. From 1836, groups of Boers dissatisfied with British rule in the Cape Colony trekked off into the interior in search of freedom. In a decade of migration known as the Great Trek, increasing numbers of *Voortrekkers* (Fore-trekkers – pioneers) abandoned their farms and crossed the *Senqu (Orange) River*. Reports from early missions told of vast, uninhabited – or at least poorly defended – grazing lands. Tensions between the Boers and the government had been building for some time, but the reason given by many trekkers for leaving was the 1833 act banning slavery. The Great Trek coincided with the *difaqane* (forced migration) and the Boers mistakenly believed that what they found – deserted pasture lands, disorganized bands of refugees and tales of brutality – was the normal state of affairs. This gave rise to the Afrikaner myths that the Voortrekkers moved into unoccupied territory or arrived at much the same time as black Africans. The Great Trek's first halt was at *Thaba 'Nchu*, near present-day *Bloemfontein*, where a republic was established. Following disagreements among their leadership, the various Voortrekker groups split, with most crossing the *Drakensberg* into *Natal* to try and establish a republic there. As this was Zulu territory, the Voortrekker leader Piet Retief paid a visit to King Dingaan and was promptly massacred by the suspicious Zulu. This massacre triggered others, as well as a revenge attack by the Boers. The culmination came at the Battle of Blood River (1838) in Natal. While the Boers sustained some injuries, more than 3000 Zulu were killed, reportedly causing the Ncome River to run red. After this victory (the result of vastly superior weapons), the Boers felt their expansion really did have that long-suspected stamp of divine approval. The 16 December victory was celebrated as the Day of the Vow until 1994, when it was renamed the Day of Reconciliation. Several short-lived Boer republics sprang up but soon the only serious contenders were the Orange Free State and the Transvaal. The republics' financial position was always precarious, and their economies depended entirely on cattle. Most trade was by barter. Just when it seemed that the republics, with their thinly spread population of fiercely independent Boers were beginning to settle into stable states, diamonds were discovered near Kimberley in 1869. Britain stepped in quickly and annexed the area. The Boers were disturbed by the foreigners, both black and white, who poured in following the discovery and were angry that their impoverished republics were missing out on the money the mines brought in. Long-standing Boer resentment became a full-blown rebellion in the Transvaal and the first Anglo–Boer War, known by Afrikaners as the War of Independence, broke out. It was over almost as soon as it began, with a crushing Boer victory at the *Battle of Majuba Hill* in 1881, and the republic regained its independence as the *ZAR* (*Zuid-Afrikaansche Republiek* – South African Republic). With the discovery of a huge reef of gold in the *Witwatersrand* (the area around *Johannesburg*) in 1886 and the ensuing explosive growth of Johannesburg (Jo'burg) itself, the ZAR was suddenly host to thousands of *uitlanders* (foreigners), black and white. This only intensified the Boers' grievances that had begun during the earlier diamond rush. In 1899 the British demanded that voting rights be given to the 60,000 foreign whites on the *Witwatersrand*. Paul Kruger

(*ZAR* president 1883–1900) refused and demanded that British troops be withdrawn from the republic's borders, leading to the second Anglo–Boer War. The conflict was more protracted than its predecessor, as the British were better prepared. By mid-1900, *Pretoria*, the last of the major Boer towns had surrendered. Yet resistance by Boer *bittereinders* (bitter enders) continued for two more years with guerrilla-style battles, which in turn were met by scorched-earth tactics by the British. In May 1902, the Treaty of *Vereeniging* brought a superficial peace. Under its terms, the Boer republics acknowledged British sovereignty."

"Nelson (Rolihlahla) Mandela, South Africa's most popular anti-apartheid leader, had witnessed the rise and decline of apartheid firsthand. In the mid-1980s, after more than twenty years in prison for opposing apartheid, he assumed a central role in helping to end it. Government and opposition leaders met for talks – tentative ones at first, and then with greater confidence and amid more publicity – and they agreed on a general approach to political reform. Four years of difficult and uneven progress, amid escalating violence and competing political pressures, finally paid off in 1994, when South Africa held its first multiracial democratic elections. And while both sides could claim some of the success in achieving this historic goal, both sides also faced even greater challenges in trying to establish a stable multiracial society in the decades ahead" (*U.S. Department of State, Country Studies, 2012*).

"Cyril Ramaphosa has made some progress in reigning in corruption, though many challenges persist. In May 2019 national elections, the country's sixth since the end of apartheid, the ANC won a majority of parliamentary seats, delivering Ramaphosa a five-year term" (*World Factbook* 2022).

"The pursuit of the figure of the earth has a long and interesting history in South Africa. The year 2001 marked the 250th anniversary since the prominent astronomer-geodesist Abbe de LaCaille who set foot on South African soil to catalogue the Southern stars by their celestial co-ordinates of right ascension and declination. Shortly after his arrival at the Cape, LaCaille set out to measure a meridian of arc in the southern hemisphere as no such measurement existed. Abbe de LaCaille measured a triangulation arc northwards from Cape Town, to determine the figure of the earth and obtained a result which indicated that the curvature of the earth was less at southern latitudes than at corresponding northern ones. This perplexity was later to be verified by Sir Thomas Maclear, Her Majesty's Astronomer at the Cape. Sir George Everest visited the Cape in 1820 and inspected the site of LaCaille's meridian arc. His experience in the Himalayas led him to believe that the presence of considerable mountain masses in the Cape could have caused some anomalous disturbance thus falsifying the astronomical latitude determinations made by LaCaille. Sir Thomas Maclear was tasked to verify LaCaille's meridian arc and commenced such in 1840, completing the task in 1848. The arc was extended southward beyond the possible gravitational effect of Table Mountain to Cape Town and northwards towards Namaqualand. Indeed, it was confirmed that the predictions made by Everest of a large disturbance of gravity at LaCaille's northern zenith sector station amounting to more than eight seconds of arc, accounted for the error in the observations made by LaCaille. The Cape coastal triangulation of Captain Bailey and Henry Fourcade was a chain of triangles tied to the southern end of Maclear's arc extending eastwards from Cape Town to the then Kei River frontier of the Cape Colony. Sir David Gill, Her Majesty's Astronomer at the Cape in 1879 began to study the general question of the Geodetic Survey of South Africa. Gill found Bailey's field records to be erroneous and inconsistent. However, the concept of the scheme of triangles was adopted by Gill. It was Gill's dream to commence an arc following the 30th meridian and stretching from Cape to Cairo, through the *Levant* and terminating at *Nordkapp*. A chain of triangles forming the backbone of the 30th meridian was to provide the geodetic control for countries traversed by the arc. Gill being appointed the honorary scientific adviser of the Geodetic Survey set sail and landed in Durban in 1883 under the command of Captain William Morris. The field party set out to measure the Pietermaritzburg base, which was then extended by triangulation to the geodetic chain. The geodetic chain later was extended northwards towards Newcastle. The chain was carried south-west from *Pietersmaritzburg* to Port Elizabeth, then northwards from *Port Elizabeth* to *Kimberly*. These geodetic operations were completed before the Anglo–Boer War of 1899–1902, after which the work was extended over the Orange Free State and *Transvaal* up to the former Rhodesia where the 30th meridian also commenced also under the instruction of Sir David Gill. Astronomical observations of latitude and longitude were made at frequent intervals in order to position the geodetic chains on the earth, and azimuth observations to orient the work to the earth's axis of rotation. Sir David Gill had taken great care in choosing a datum point free of 'considerable deviation of the plumb-line' for the geodetic survey. Differences between astronomic and geodetic measurements showed that his triangulation on the datum and the chosen Clarke 1880 ellipsoid was in good agreement with the figure of the earth in South Africa. From the 1920s onward the Trigonometrical Survey undertook extensive geodetic surveys. In the mid 1930s the Kaitob base and the Mtubatuba base were measured. Land Surveyor, H. S. K. Simpson played a key role in these surveys. These were the last of the taped baselines before the advent of the EDM (electromagnetic distance measurement). (*Note that the Tellurometer EDM was invented and manufactured in South Africa – Ed.*) Recent additions to the geodetic framework include a looped chain of triangulation passing through the Mtubatuba baseline which was attached to the northern Natal section of the 30th meridian. The northern *Transvaal* section of the 30th meridian arc was extended eastwards toward the *Moçambican* border; a loop of geodetic triangulation running parallel to the Botswana border was attached to the western side of the 30th meridian arc. In the northern Cape, Surveyors Leipoldt and Heatlie connected the northern end of Maclear's arc to the Port Elizabeth-Kimberly chain, while Surveyor Connan split this area in two with a north-south chain. Meanwhile, Mr D. P. M. Rosseau, took charge in the geodetic surveys in Namibia. The points of the geodetic survey were too far apart for a surveyor to connect to this system and therefore a densification of the Geodetic survey network was required. The *Natal* Trigonometrical Survey established a secondary triangulation spreading throughout Natal emanating from the 30th arc of meridian. Many farm beacons were thus established in terms of a spheroidal rectangular or Cassini-Soldner projection co-ordinates. This was the original co-ordinate system of the Geodetic survey which was later to be superseded by the Gauss Conform system thirty years later. The Cassini-Soldner co-ordinate calculations were found to be awkward by Oscar Schreiber, who later influenced Van Der Sterr in using the Gauss Conform projection. The work of Johannes Jacobus Bosman, Director of Secondary Triangulation in the Cape Colony is also worth mentioning. Bosman, under the direction of Gill established a chain connecting the northern end of Maclear's arc eastwards over the *Kalahari Desert* to connect with the geodetic chain near *Kimberly*. Colonel Winterbotham conducted a minor triangulation based on the geodetic survey of the *Orange Free State*. By 1919 a considerable amount of trigonometrical control became available for cadastral and mapping purposes. New concrete beacons were built over the old centre points, points were re-observed and co-ordinates were recomputed in the Gauss Conform system. This revision was initiated in 1919 by the newly appointed Director of Trigonometrical Survey, Willem Cornelis van Der Sterr. The present structure housing the Chief Directorate: Surveys and Mapping is named after him. The Primary triangulation scheme continued under the direction of Van Der Sterr and computations were carried out in *Mowbray* under the watchful eyes of geodesist Oscar Shreiber. Primary order triangulations of 40 km sides were reconnoitred to fill the open spaces encircled by the loops of geodetic chains. The interpolation followed into the primary points of the secondary order triangulation nets. Thereafter, the tertiary stations followed with these points being intersected by rays observed to and from the surrounding fixed secondary stations. Sub-tertiary intersections were conducted in urban areas, often to church spires, in order to provide control for street traverses which connect the underground reference marks placed at street intersections and upon which urban surveys are based. This highlights the description of the unified trigonometrical system upon which all mapping, cadastral and engineering surveys are based" (*Chief Directorate: Surveys & Mapping, Dept. of Land Affairs, Mobray, Cape Town, South Africa, 2001*).

Thanks to John W. Hager, the classical geodetic origin of the Cape Datum/Arc 1950 Datum is at *Buffelsfontein* (in *Port Elizabeth*), where $\Phi_o = 33°\ 59'\ 32.000''$ S, $\Lambda_o = 25°\ 30'\ 44.622''$ E, azimuth to Zuurberg measured from South: $\alpha_o = 183°\ 58'\ 15.00''$, $\xi = -3.46''$, $\eta = -0.59''$, $h_o = 280.1$ m, and the ellipsoid of reference is the Clarke 1880, where $a = 6,378,249.145$ m and $1/f = 293.465$. The Gauss

Conform Transverse Mercator Grid system has 2° wide central meridians at: 17°E, 19°E, ..., 31°E, 33°E.

"No arbitrary scale factors or false origins are applied to the co-ordinates; X is measured positive southwards from the equator and Y positive westwards from the nearest odd meridian. The unit of measurement since the 1970s is the International metre" (*Chief Directorate: Surveys and Land Information, 1995, Personal Communication*).

The new geodetic datum in South Africa is termed "*Hartbeeshoek 94 Datum*" and is referenced to the GRS80 ellipsoid, where $a = 6,378,137$ m and $1/f = 298.257222101$. The three-parameter transformation **from** Cape Datum **to** Hartbeeshoek 94 Datum is where: $\Delta X = +134.7$ m, $\Delta Y = +110.9$ m, $\Delta Z = +292.7$ m (*op. cit. Chief Directorate, 1995*). High-accuracy transformations are available from the Chief Directorate: Surveys and Land Information in the form of software that operates in somewhat similar fashion to the NADCON package of the U.S. National Geodetic Survey. The government of South Africa has also developed a high-accuracy geoid model for their country, and a GPS Real Time Network provides full-country coverage for high-precision surveys.

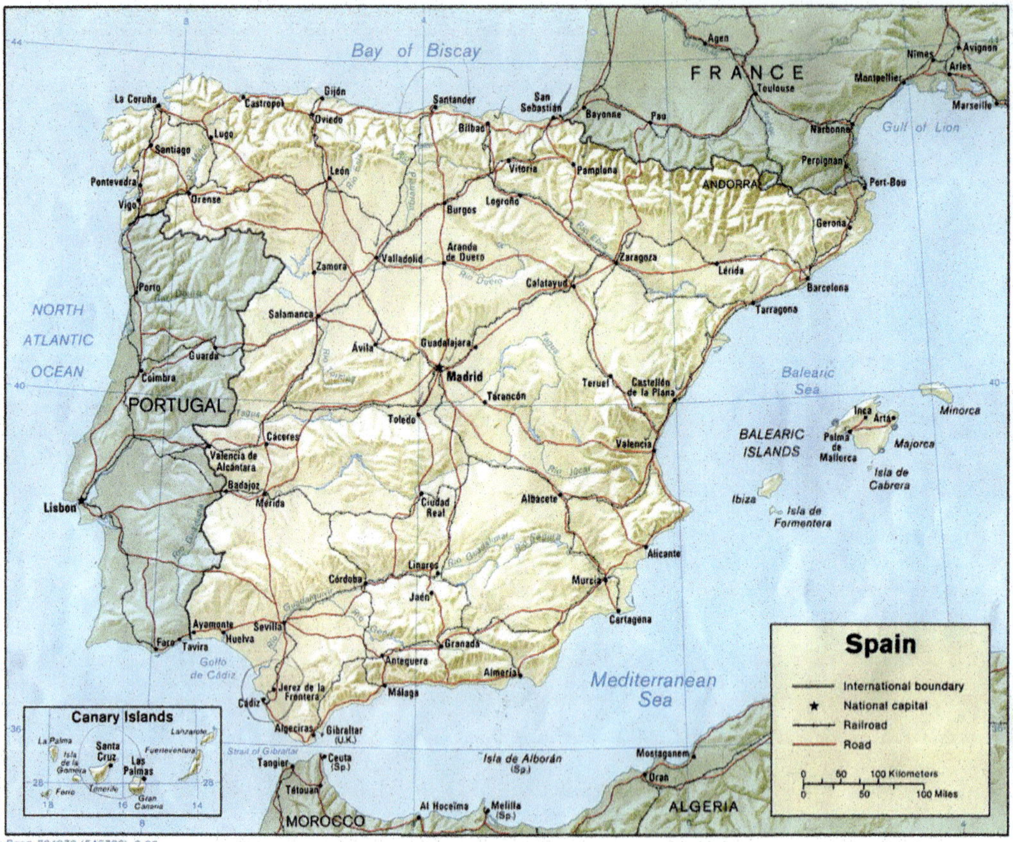

Base 504978 (546706) 3-82

The Kingdom of Spain

Spain is almost five times the size of Kentucky and is slightly more than twice the size of Oregon. Bordered by Andorra (63 km); France (646 km); Gibraltar (1.2 km); Portugal (1,224 km); Morocco (Ceuta) (8 km) and Morocco (Melilla) (10.5 km). The terrain is large, flat to dissected plateau surrounded by rugged hills and the *Pyrenees Mountains* in north. The highest point is *Pico de Teide* (*Tenerife*) on Canary Islands (3,718 m), the lowest point is the Atlantic Ocean (0 m), and the mean elevation is (660 m).

The Greeks and the Phoenicians settled Spain on the southern and eastern coasts. The Mediterranean coastal region was ruled by *Carthage*, which ceded it to Rome in 201 *B.C.* Invaded by Vandals in 409 *A.D.*, the Muslims of North Africa conquered Toledo from 711 to 719. The last of the Moors were expelled from *Granada* in 1492. "Spain remained neutral in World War I and II but suffered through a devastating civil war (1936–39). A peaceful transition to democracy following the death of dictator Francisco Franco in 1975, and rapid economic modernization (Spain joined the EU in 1986) gave Spain a dynamic and rapidly growing economy and made it a global champion of freedom and human rights. More recently, Spain has emerged from a severe economic recession that began in mid-2008, posting four straight years of GDP growth above the EU average. Unemployment has fallen, but remains high, especially among youth. Spain is the Eurozone's fourth largest economy. The country has faced increased domestic turmoil in recent years due to the independence movement in its restive *Catalonia* region" (*World Factbook* 2022).

Spain's cadastre began in 1854, but the Law of Cadastre established the way to perform the work in 1906. This was essentially a rural method for taxation until 1964 when urban areas of the Kingdom were brought into the cadastre.

The Spanish first-order triangulation of 1858–1885 consists of ten principal arcs: the meridional arcs of *Salamanca, Madrid, Pamplona,* and *Lerida*; the parallel arcs of *Palencia, Madrid,* and *Badajoz*; and the diagonal arcs of the North, East, and South coasts. The origin of the *Madrid Observatory Datum of 1853* is where: $\Phi_o = 40°\ 24'\ 29.70''$ North (after 20 years of observations), $\Lambda_o = 03°\ 41'\ 14.546''$ West of Greenwich. The defining azimuth was determined from the observatory to station *Hierro* as: $\alpha_o = 154°\ 31'\ 06.90''$ *from south* and is referenced to the Struve 1860 ellipsoid, where $a = 6,378,298.3$ meters and $1/f = 294.73$. However, the Spaniards did not use Greenwich as a Prime Meridian at the time; they used Madrid. Every country in Europe used their own Royal Observatory as their national prime meridian back then. The original triangulation comprised 235 triangles, which was observed with Repsold, Pistor, Brunner, and Kern theodolites. A polyhedric system of projection has been reported for the early topographic maps. In 1875, the Spanish Military adopted the Lambert Conformal Secant Conic Grid that has a Latitude of Origin $(\varphi_o) = 40°\ 00'$ North, a Scale Factor at Origin $(m_o) = 0.9988085293$ and a Central Meridian $(\lambda_o) = 03°\ 41'\ 14.546''$ West of Greenwich. The False Northing and False Easting were both equal to 600 km. That Grid is still in use. However, in 1880, the Military adopted a Bonne projection for the *Itinerario Militar* (military route map) with a Latitude of Origin $(\varphi_o) = 40°\ 45'$ North, a Scale Factor at Origin $(m_o) = 1.0$ and a Central Meridian $(\lambda_o) = 04°\ 26'\ 14.55''$ West of Greenwich. That Bonne base remained in use until 1951 when the UTM was adopted for all military applications in Spain.

During World War II, the Iberian Peninsula Lambert tangent Zone was used by the Allied Forces. This Grid had a Latitude of Origin $(\varphi_o) = 40°\ 00'$ North, and a Central Meridian $(\lambda_o) = 03°\ 41'\ 14.546''$ West of Greenwich. The False Northing = 600 km, and the False Easting = 530 km. A later triangulation densification in 1930–1933 was observed with Wild T3 theodolites. During the war, the Germans developed a series of computations for the Spanish Triangulation Network referenced to the Clarke 1880 ellipsoid. After the war, those German tables were useful in the re-computation of the classical triangulation of the Kingdom onto the European Datum of 1950 (ED50) which was referenced to the International ellipsoid. The subsequent transformation of the Spanish Net to ED50 was based on a two-step two-dimensional Helmert transformation. The initial step computed by the Axis (Germans) was still on a Lambert Conformal Conic, but referenced to the Clarke 1880, which was the standard for France. That series of co-located points that started computations in the *Pyrenees Mountains* was continued into a substantial portion of the *Iberian Peninsula*. After the war, the U.S. Army Map Service decided that the best way to merge the Madrid Observatory Datum of 1858 into the new ED50 was to use the existing German data as an intermediate step. In computing the new French datum values of the Spanish triangulation, use was made of the German adjustment of 1938. Values of the New Triangulation of France (NTF) were available for the following 11 Spanish triangulation stations: *Forceral, Canigou, Licuses, Rouge, Cabére, Maupas, Anie, Orhi, Baigura, La Rune,* and *Biarritz*. Computations to the Madrid Datum of 1858 were performed by V.U.K.A. 631 (April 1943) and by *Kriegskarten und Bermessungsamt*, Paris (August 1943).

Other Datums of Spain include San Fernando Observatory, where $\Phi_o = 36° 27' 54.6''$ North, $\Lambda_o = 06° 12' 17.8''$ West of Greenwich. Another version of the same point is where: $\Phi_o = 36° 27' 41.52''$ North, and $\Lambda_o = 06° 12' 19.5''$ West of Greenwich. This same point on the "*Salamanca Datum*" is where: $\Phi_o = 36° 27' 54.65''$ North, and $\Lambda_o = 06° 12' 18.00''$ West of Greenwich.

Pico de las Nieves Datum of 1934 in the Canary Islands is where the origin is: $\Phi_o = 27° 57' 41.273''$ North ±0.130'', and $\Lambda_o = -15° 34' 10.524''$ West of Greenwich ±0.380''. The defining azimuth was observed to station *Iselta* as: $\alpha_o = 212° 34' 48.30''$ *from south*, and the elevation is 1949.14 m. It is referenced to the International ellipsoid of 1909 (Madrid 1924), where $a = 6,378,388$ m and the reciprocal of flattening $(1/f) = 297$. Also found on the Canary Islands is the *Abona Datum* that is referenced to the International 1909 ellipsoid. The origin of that Datum is where: $\Phi_o = 28° 08' 22.93''$ North ±0.102'', and $\Lambda_o = 16° 25' 01.92''$ West of Greenwich ±0.105''. The defining azimuth was observed to station *Roja* as: $\alpha_o = 221° 48' 27.48''$, and the elevation is 16.82 m. The Royal Spanish "*Servicio Geografico del Ejercito*" (Army Map Service) Grid of 1945 for the Canaries is on the International 1909 ellipsoid. The Latitude of Origin $(\varphi_o) = 28° 30'$ North, and a Central Meridian $(\lambda_o) = 12° 00'$ West of Madrid (therefore, $\lambda_o = 15° 41' 14.546''$ West of Greenwich). The Scale Factor at Origin $(m_o) = 0.9999244799$, the False Northing (FN), and the False Easting (FE) = 400 km.

Interestingly, the British Grid system for the Canary Islands used during World War II by the Allies was also based on a secant Lambert Conformal Conic projection. The Latitude of Origin $(\varphi_o) = 27° 00'$ North, and a Central Meridian $(\lambda_o) = -17° 00'$ West of Greenwich. The Scale Factor at Origin $(m_o) = 0.9993817$, the False Northing (FN) = 100 km, and the False Easting (FE) = 200 km. This Grid was referenced to the Clarke 1880 ellipsoid presumably because the Canary Islands are so close to Africa. That ellipsoid is/was the (British) standard for that continent. (Some of the World War II-era British Grids were based on ersatz Datums "cooked up" for reasons of expediency during the war.)

The *Balearic Islands* in the Mediterranean Sea are said to be on the Madrid Datum of 1858, but I do not think that is true. Ten years later, station *Mola* was observed for the *Balearic Islands Datum of 1868*, where $\Phi_o = 38° 39' 53.17''$ North and $\Lambda_o = +5° 13' 19.94''$ East of Madrid (therefore, +1° 32' 05.394'' East of Greenwich). The longitude is also defined as −0° 48' 11.26'' West of Paris, which leaves a discrepancy of 2.71'' in Longitude between Greenwich and Paris. This is not a large error for an "Astro" shot done on an island in 1868, but it is quite large for the relation between two national observatories. The defining azimuth was observed to station *Furnas* as: $\alpha_o = 177° 39' 11.03''$ *from south,* and to station Camp Vey as: $\alpha_o = 160° 15' 40.48''$ *from south.* The elevation of *Mola* is 192.25 m, and of course the ellipsoid of reference is the Struve 1860.

NIMA lists the three-parameter datum shift from the European Datum 1950 (in Spain) to WGS84 as: $\Delta X = -84$ m ± 5 m, $\Delta Y = -107$ m ± 6 m, $\Delta Z = -120$ m ± 3 m, and is based on an 18-station solution.

The National Geodetic Network of GNSS Reference Stations (ERGNSS) Descriptive information: Currently 75 stations make up the network, of which 24 are integrated into the European network of *EUREF* and 3 (YEBE, MELI, and LPAL) in the global network of the IGS (*International*

GNSS Service). In addition, in order to make a more efficient management of expenditure and with a view to having an adequate density of stations throughout the national territory, the IGN shares the ownership of some stations of the ERGNSS with other institutions, such as Autonomous Communities and Ports of the State, through collaboration agreements. The stations are equipped with multi-frequency geodetic receivers, with reception capacity of other constellations, such as GLONASS or GALILEO and geodetic antennas, almost all "Choke ring" type and with calibration of the phase center variation (part of them with individual calibrations).

"The network of permanent stations of the IGN is the basic reference that gives access to the ETRS89 system directly. The objectives of this network can be summarized as: Obtaining very precise coordinates and speed field in all points of the network. Contribution to the definition of the new Global Reference Systems (ITRF) in the national territory. To be fundamental points of the European EUREF network of permanent stations (EPN) for the densification of global frameworks and definition of the ETRS89 system. Use of continuous data records for studies of Geodynamics, troposphere, ionosphere, meteorology, *etc*. Define a fundamental network as support for real-time applications of differential corrections (DGNSS) and RTK. To provide GNSS users, publicly, data for geodetic, cartographic, topographic and positioning work in general.

"In order to make a more efficient expenditure management and with a view to having an adequate density of stations throughout the national territory, the IGN shares with other institutions some of its stations of the ERGNSS network. The institutions that share these facilities are Autonomous Communities and Ports of the State through collaboration agreements. In the following list the institutions that share stations are mentioned. Services of the ERGNSS network: Post-process services: in the form of RINEX files of all stations, with different intervals (hours at 1, 5, 10 and 30 seconds or daily to 30 seconds), accessible by the user in different ways:

- Directly through an *FTP* server.
- Through this web page in the utility *"Geodetic Data"*.
- Through the *Geodetic Applications Program (PAG)*.

"Real-time positioning services: by connecting to the different 'streams' of data available in the Real-Time Services section. They allow instant precise positioning for the user by connecting through the Internet to the real-time servers of the IGN. For more detailed information, consult the real-time positioning services page.

"GNSS data analysis center: The IGN is a GNSS Data Analysis Center that performs continuous data processing for several projects from both the permanent stations of the ERGNSS network and from other networks, as well as being a EUREF Local Analysis Center (acronym IGE) from September 2001. The different continuous calculation processes for these projects are carried out in Bernese 5.2, they are automated. For any query about the geodesy services provided by the National Geographic Institute and the GNSS Permanent Station Network, the contact is *buzon-geodesia@fomento.es*" (*www.ign.es*, January 2018).

Sri Lanka

★ National capital
- - - Province boundary
⊙ Province capital
─┼─ Railroad
── Road

Base 802514 6-00

Democratic Socialist Republic of Sri Lanka

Sri Lanka is slightly larger than West Virginia, the lowest point is the Indian Ocean (0 m), and the highest point is *Pidurutalagala* (2,524 m) (*The World Factbook*, 2009). "Extensive faulting and erosion over time have produced a wide range of topographic features, making Sri Lanka one of the most scenic places in the world. Three zones are distinguishable by elevation: the Central Highlands, the plains, and the coastal belt" (*Library of Congress Country Studies*, 2009).

"The first Sinhalese arrived in Sri Lanka late in the 6th century *B.C.* probably from northern India. Buddhism was introduced in about the mid-third century *B.C.*, and a great civilization developed at the cities of *Anuradhapura* (kingdom from *circa* 200 *B.C.* to *circa A.D.* 1000), and *Polonnaruwa* (from about 1070 to 1200). In the 14th century, a south Indian dynasty established a Tamil kingdom in northern Sri Lanka. The coastal areas of the island were controlled by the Portuguese in the 16th century and by the Dutch in the 17th century. The island was ceded to the British in 1796, became a crown colony in 1802, and was united under British rule by 1815. As *Ceylon*, it became independent in 1948; its name was changed to Sri Lanka in 1972.

"Prevailing tensions between the Sinhalese majority and Tamil separatists erupted into war in July 1983. Fighting between the government and Liberation Tigers of Tamil Eelam (LTTE) continued for over a quarter century. Although Norway brokered peace negotiations that led to a ceasefire in 2002, the fighting slowly resumed and was again in full force by 2006. The government defeated the LTTE in May 2009.

"During the post-conflict years under President Mahinda Rajapaksa, the government-initiated infrastructure development projects, many of which were financed by loans from China. His regime faced significant allegations of human rights violations and a shrinking democratic space for civil society. In 2015, a new coalition government headed by President Maithripala Sirisena of the Sri Lanka Freedom Party and Prime Minister Ranil Wickremesinghe of the United National Party came to power with pledges to advance economic, governance, anti-corruption, reconciliation, justice, and accountability reforms. However, implementation of these reforms has been uneven. In October 2018, President Sirisena attempted to oust Prime Minister Wickremesinghe, swearing in former President Rajapaksa as the new prime minister and issuing an order to dissolve the parliament and hold elections. This sparked a seven-week constitutional crisis that ended when the Supreme Court ruled Sirisena's actions unconstitutional, Rajapaksa resigned, and Wickremesinghe was reinstated. In November 2019, Gotabaya Rajapaksa won the presidential election and appointed his brother, Mahinda, prime minister" (*World Factbook* 2022).

"The Triangulation of Sri Lanka commenced in 1857 with the measurement of a base (one side of a triangle) in *Negombo* on the West Coast (latitude 7° 10′) and at *Batticaloa* on the East Coast (latitude 7° 40′). Both bases are in low, flat country; and brick towers up to 70 feet high had to be built over the terminals to enable observations to be taken to surrounding points. The bases were measured in 1857 and 1859 respectively for the Topographical Survey. The *Negombo base* was measured with a 100-foot heavy iron chain along the ground, and this chain was compared frequently with two 100-foot 'standards' which had been laid out at either end of the line by means of a beam compass from a standard brass scale 1-yard long; the origin of this scale is unknown, but its graduations were evidently taken to be correct at 62° F. The *Batticaloa base* was measured with the same chain laid on planks and trestles, the 100-foot standards having been laid out with the chain itself before measurement; the ground here was very rough, and the base was only intended as a check line. Each base was measured once, and each measurement took about six weeks. Computed from

Negombo the triangulated value of the *Batticaloa base* was 2.25 feet shorter than its observed value. The triangulation was adjusted to both bases" (*Notes on the Base Lines of the Ceylon Triangulation*, J.E. Jackson, Empire Survey Review, Vol. 3, 129, pp. 129–130).

"Plane co-ordinates were computed using the Transverse Mercator Projection, with *Pidurutalagala* ($\varphi_o = 7°\ 00'\ 01.729''$ N, $\lambda_o = 80°\ 46'\ 18.160''$ E – *Ed.*), as origin. A re-computation of this Triangulation, with a few additional triangles observed, was carried out in 1888. Systematic large scale cadastral surveys had commenced in many parts of the country and during the first few years of this century (*19th – Ed.*), a good deal of triangulation of first order was carried out. Great difficulty was encountered in fitting the new work to the old.

"In 1929 it was decided to re-compute the Triangulation a second time for which purpose the two bases were re-measured more precisely and two astronomical azimuths observed which added greatly to the accuracy of the scheme. This re-computation was completed in 1933. The Triangulation is not of a geodetic or primary order, but the results are the best which is possible from the existing observations made during different periods" (*The Geodetic Horizontal Control Network*, Sri Lanka Ministry of Land, 2009). This re-computation by J.E. Jackson of the classical triangulation of Sri Lanka used the origin point in *Kandavelu* at: $\Phi_o = 7°\ 14'\ 06.838''$ N, $\Lambda_o = 79°\ 52'\ 36.670''$ E with a reference azimuth to station *Halgastota*: $\alpha_o = 176°\ 41'\ 33.18''$, the reference ellipsoid is the Everest 1830, where $a = 6,377,276.345$ m and $1/f = 300.8017$. If the reader compares the previous sentence to commonly published data, a discrepancy will be found. The actual name of this datum is "The *Jackson Datum of 1929*," and the origin point is *Kandavelu* which translated to English is *Hell Valley*, and is *not Kandawala* (*Dr. Muneendra Kumar, personal communication, 25 September 2003*).

The original Ceylon Belt Transverse Mercator Grid used the Indian Yard as the unit of measurement where 1 yard = 0.914399205 meters. The Latitude of Origin (φ_o) and the Central Meridian (λ_o) is at *Mount Pidurutalagala* (listed above), the Scale Factor at Origin (m_o) = unity, the False Northing = the False Easting = 176,000 yards. The limits of the Ceylon Belt are North: parallel of 10° N; East: Meridian of 84° E; South: parallel of 4° N; and West: Meridian of 78°E, northwards to 8° N, thence along this parallel to 79° 30' E, thence along this meridian to 10° N. A notation in the projection tables is as follows: "Within approximately 3° of from the central meridian the formulae given on page iii are wholly adequate, but as $\Delta\lambda$ increases beyond 3° there is an increasing departure from conformality due to the omission from the formulae of terms with coefficients of higher than the fifth power. This condition does not seriously disturb the inter-relation of the grid coordinates in a limited area but becomes apparent in the failure of the inverse solution to give geographic coordinated identical with those from which the grid coordinates were computed. The magnitude of this discrepancy varies with the latitude as well as with $\Delta\lambda$." (*British General Staff, Geodetic Section, Survey Computations, London, H.M. Stationery Office Second Edition, 1932, Reprinted 1941, page 98., War Department Corps of Engineers, U.S. Lake Survey – New York Office Military Grid Unit, 1944*).

A test point offered for the Ceylon Belt is as follows: $\varphi = 6°\ 49'\ 53.769''$ N, $\lambda = 82°\ 14'\ 41.813''$ E, X = 354,083.63 Indian Yards, Y = 155,849.30 Indian Yards, Convergence Angle (γ) = +0° 10' 31.02", Scale Factor at a Point (m) = 1.0003443. According to TR8350.2, *from Kandavelu (Jackson 1929) Datum to* WGS84 Datum, $\Delta X = -97$ m ±20 m, $\Delta Y = +787$ m ±20 m, $\Delta Z = +86$ m ±20 m.

A new local datum is named the *Sri Lanka Datum of 1999*. The new "Grid Co-ordinate System is: The Transverse Mercator projection on Everest ellipsoid (1830) with the following parameters is used to compute new grid co-ordinate system, Central meridian, E 80° 46' 18.16710", Latitude of Origin N 7° 00' 1.69750", Scale factor 0.9999238418, False Northing 500,000 m, False Easting 500,000 m, Pidurutalagala trigonometrical station is used as the latitude of origin and central meridian" (*Sri Lanka Ministry of Land, 2009*). Curiously, the GRS80/WGS84 ellipsoid is eschewed for the antiquated Everest 1830 without explanation.

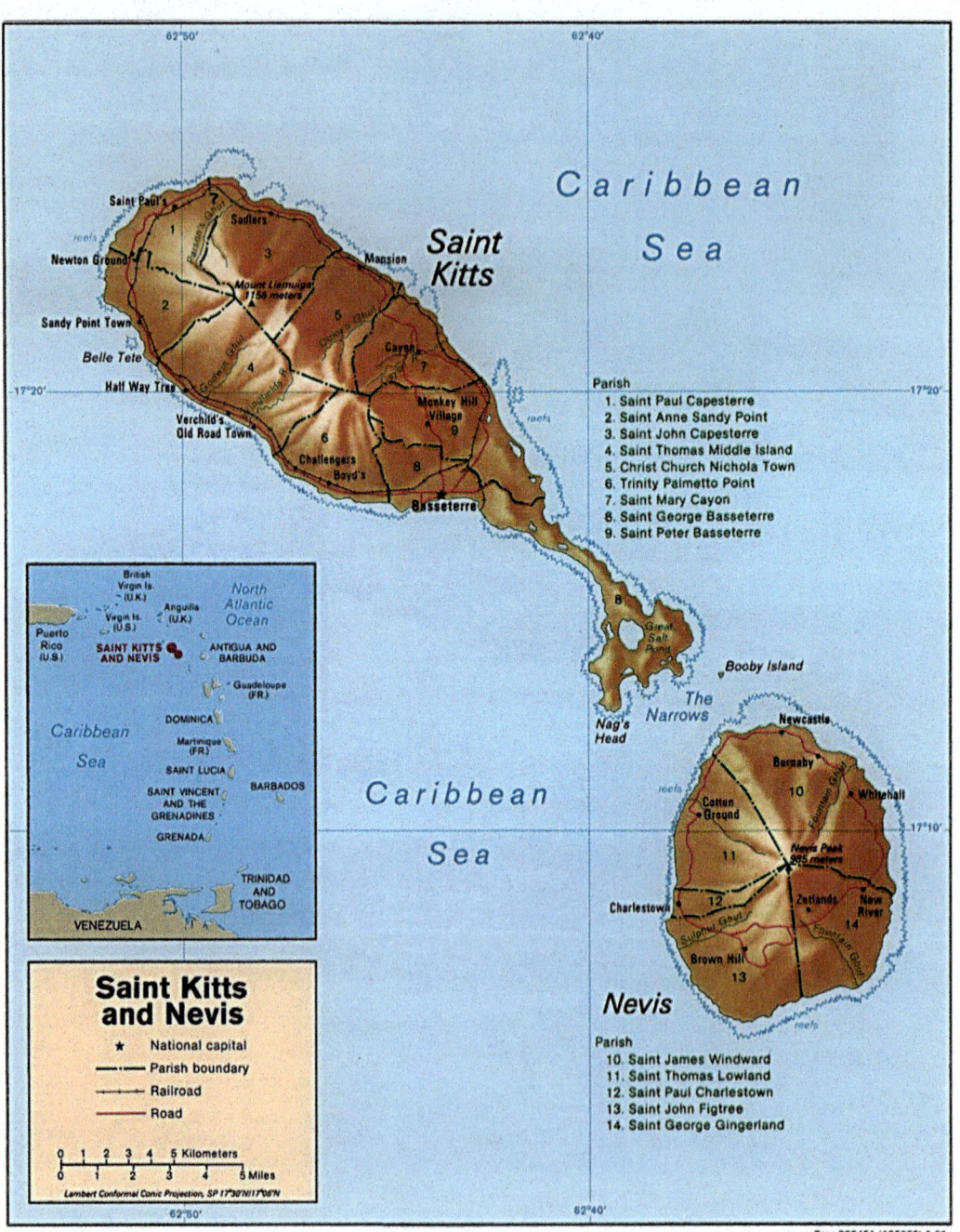

Saint Kitts and Nevis

Saint Kitts
Parish
1. Saint Paul Capesterre
2. Saint Anne Sandy Point
3. Saint John Capesterre
4. Saint Thomas Middle Island
5. Christ Church Nichola Town
6. Trinity Palmetto Point
7. Saint Mary Cayon
8. Saint George Basseterre
9. Saint Peter Basseterre

Nevis
Parish
10. Saint James Windward
11. Saint Thomas Lowland
12. Saint Paul Charlestown
13. Saint John Figtree
14. Saint George Gingerland

Legend
- ★ National capital
- Parish boundary
- Railroad
- Road

0 1 2 3 4 5 Kilometers
0 1 2 3 4 5 Miles
Lambert Conformal Conic Projection, SP 17°30'N/17°06'N

Base 802451 (A03232) 3-96

Federation of Saint Kitts and Nevis

With an area of about 1.5 times the size of Washington, DC, the lowest point is the Caribbean Sea (0 m), and the highest point is *Mt. Liamuiga* or *Mt. Misery* (1,156 m). With coastlines in the shape of a baseball bat and ball, the two volcanic islands are separated by a 3-km-wide channel called *The Narrows*; on the southern tip of long, baseball bat-shaped *Saint Kitts* lies the *Great Salt Pond; Nevis Peak* sits in the center of its almost circular namesake island and its ball shape complements that of its sister island (*World Factbook*, 2009).

"At the time of European discovery, Carïb Indians inhabited the islands of St. Kitts and Nevis. Christopher Columbus landed on the larger island in 1493 on his second voyage and named it after St. Christopher, his patron saint. Columbus also discovered *Nevis* on his second voyage, reportedly calling it Nevis because of its resemblance to a snowcapped mountain (in Spanish, '*Nuestra Señora de las Nieves*' or Our Lady of the Snows). European settlement did not officially begin until 1623–24, when first English, then French settlers arrived on *St. Christopher's Island*, whose name the English shortened to *St. Kitts Island*. As the first English colony in the Caribbean, *St. Kitts* served as a base for further colonization in the region.

The English and French held *St. Kitts* jointly from 1628 to 1713. During the 17th century, intermittent warfare between French and English settlers ravaged the island's economy. Meanwhile *Nevis*, settled by English settlers in 1628, grew prosperous under English rule. *St. Kitts* was ceded to Great Britain by the Treaty of Utrecht in 1713. The French seized both *St. Kitts* and *Nevis* in 1782. The Treaty of Paris in 1783 definitively awarded both islands to Britain. They were part of the colony of the *Leeward Islands* from 1871–1956, and of the West Indies Federation from 1958–62. In 1967, together with Anguilla, they became a self-governing state in association with Great Britain; Anguilla seceded late that year and remains a British dependency. The Federation of *St. Kitts* and *Nevis* attained full independence on September 19, 1983" (*Background Note, Bureau of Western Hemisphere Affairs, U.S. Dept. of State, 2009*).

"In 1998, a referendum on *Nevis* to separate from Saint Kitts fell short of the two-thirds majority vote needed" (*World Factbook* 2022).

Although local cadastral surveys of the British West Indies date back to the 19th century, the first known geodetic observations of *St. Kitts* and *Nevis* were in the middle of the 20th century. The origin of the local 1955 datum at *Fort Thomas* is Station K 12, where $\Phi_o = 17°\ 17'\ 17.37''$ N, $\Lambda_o = 62°\ 44'\ 08.295''$ W, the azimuth from North to *Station Upper Bayford* is: $\alpha_o = 13°\ 53'\ 02.7''$, and the reference ellipsoid is the Clarke 1880, where $a = 6,378,249.145$ m and $1/f = 293.465$. There is no published relation between the *Ft. Thomas Datum of 1955* and WGS 84 Datum, but the U.S. National Geodetic Survey (NGS) did perform a number of high-precision GPS observations on the island of *St. Kitts* in 1996. Although the NGS indeed occupied *one* of the local cadastral control points, they neglected to research the local coordinates of the point. The point occupied was KT 8, and the adjusted NAD 83 coordinates observed are: $\varphi = 17°\ 17'\ 58.85758''$ N, $\lambda = 62°\ 41'\ 43.83677''$ W, $h = 85.287$ m. Once the local BWI (*pronounced "bee-wee"*) coordinates are obtained, the transformation to WGS 84 will be a trivial computational exercise for local orienteering purposes. The BWI Transverse Mercator Grid for *St. Kitts* and *Nevis* is defined as: Central Meridian $(\lambda_o) = 62°$ W, Scale Factor at Origin $(m_o) = 0.9995$, False Easting = 400 km, False Northing = null.

The U.S. Army Map Service, Inter-American Geodetic Survey (IAGS) performed cooperative geodetic surveys of all Latin America and the Caribbean after World War II, and carried the North American Datum of 1927 throughout Central America and the Caribbean Islands. The approximate transformation from NAD 27 to WGS 84 for that area of the Caribbean is: $\Delta X = -3$ m ± 3 m, $\Delta Y = +142$ m ± 9 m, and $\Delta Z = +183$ m ± 12 m, and the solution is based on 15 stations in that region of the Caribbean. Thanks go to Mr. John W. Hager for the *Fort Thomas* geodetic reference.

Saint Lucia

- ★ National capital
- Quarter boundary
- Road

Saint Lucia

Now an independent state of the Windward Islands, *St. Lucia* has a total land area of 606 square kilometers, it has a coastline of 158 km, and the highest point is *Mount Gimie* (950 m). The island's terrain is mostly mountainous with some broad fertile valleys, and is 3½ times the size of Washington, DC.

The cannibal Carïbs replaced (perhaps ate?) all of the Arawak inhabitants of St. Lucia around 800–1,300 A.D. These tribes called the island of St. Lucia *"Ioüanalao"* and *"Hewanorra"* meaning "there is where the iguana is found" long before it was named by Christopher Columbus during his fourth voyage to the West Indies in 1502. Columbus did not land on the island, and the first attempts to settle on the island by the French and by the English were violently repulsed during most of the 17th century. Finally, the town of Soufrière was permanently founded at the end of that century, and in 1746, France officially recognized it as the first town on St. Lucia. The island was alternately held by the British and the French seven times over 150 years before it was finally ceded in the Treaty of Paris to the British in 1814.

"In the mid-20th century, Saint Lucia joined the West Indies Federation (1958–1962) and in 1967 became one of the six members of the West Indies Associated States, with internal self-government. In 1979, Saint Lucia gained full independence" (*World Factbook* 2022).

According to the Organization of American States (OAS), "An estimated 7,000 farmers in Saint Lucia actively cultivate bananas, the largest source of the country's employment and foreign exchange. Yet over 75 percent of these farmers own 10 acres of land or less. This is not unusual in Saint Lucia, where 92.7 percent of all farmers control on 24 percent of the land. At the opposite end of the scale, about 50 percent of all cultivable land is controlled by only 0.17 percent of the farmers, most of whom are absentee owners. Skewed land distribution has long been recognized as a major constraint to agrarian reform and the alleviation of rural poverty." This is a common theme in much of the world; I have been involved in land titlelization projects in South America for the same reasons, and photogrammetry with GPS control is the common thread to implementing the solution. The Institute of Surveyors St. Lucia was founded in September of 2003 and is the latest development of a long history of surveying and mapping of the island dating back to the 1771 survey of the *Bexton* and *Forrestierre* areas.

The British Directorate of Colonial Surveys (DCS), later the Directorate of Overseas Surveys and now the Ordnance Survey of the United Kingdom first published 1:25,000 scale topographic maps of the island in 1958.

The first astronomic position was at *Vieux Fort* (Old Fort), on *Vieux Fort Bay* located at the southern end of the island, and likely observed with an astrolabe. The current geodetic system is known as *St. Lucia Lighthouse 1955 Datum*, and the origin point at DCS 3 (near the southern town of *Laborie*) is: $\Phi_o=13°\ 42'\ 35''$ N, $\Lambda_o=60°\ 56'\ 37''$ W, the elevation $H_o=222.19$ m, the reference azimuth to *Cannelles* (DCS 9) $\alpha_o=30°\ 28'\ 52''$, and the ellipsoid of reference is the Clarke 1880 (modified), where $a=6,378,249.145$ m and $1/f=293.465$. The British West Indies (BWI) Grid for St. Lucia is on the Gauss-Krüger Transverse Mercator projection, where the central meridian $\lambda_o=-62°$ W, the scale factor at origin $m_o=0.9995$, the False Easting $=400$ km and there is no False Northing. The datum origin coordinates on the BWI Grid are $X=514218.711$ m and $Y=1515586.182$ m.

In 1998, Mr. John N. Wood of St. Lucia College in the Caribbean provided some local control on St. Lucia so that I could work up a three-parameter datum shift for his island. Coupling his classical survey data from DCS with GPS data observed by the U.S. National Geodetic Survey (NGS) and provided to me by Mr. David Doyle, I found three points in common between the two data sets. The shift from *St. Lucia Lighthouse 1955 Datum* to the WGS 84 Datum is then: $\Delta X=-134$ m, $\Delta Y=+119$ m, $\Delta Z=+314$ m. The fit of these three points was excellent and is probably good to better than a meter for the northern third of the island around the capitol city of *Castries*.

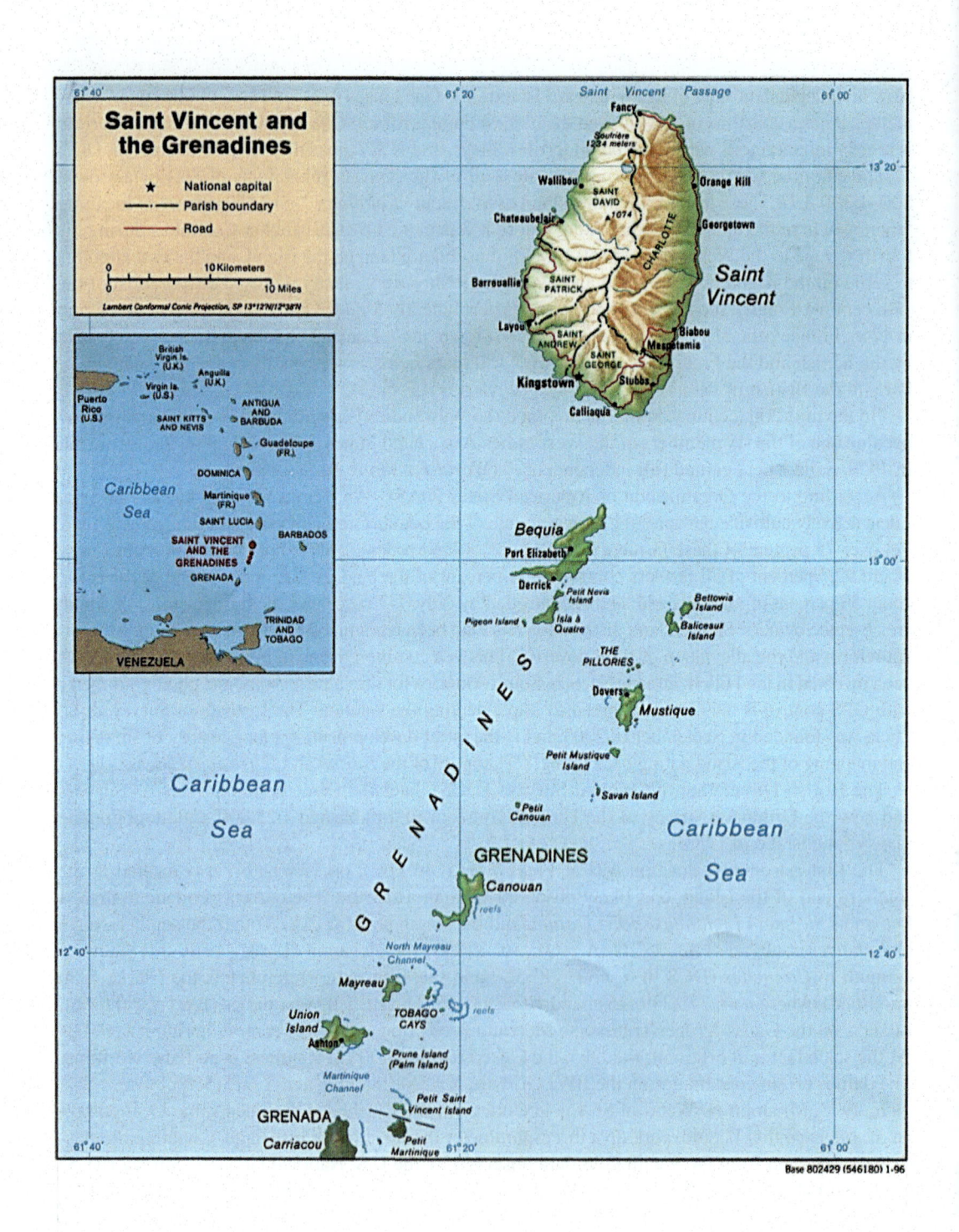

Saint Vincent and the Grenadines

- ★ National capital
- Parish boundary
- Road

Saint Vincent and the Grenadines

Slightly less than twice the size of Washington, DC, there are 32 Grenadine islands and cays of which the largest are *Bequia, Mustique, Canouan,* and *Union.* Of a total 389 km², the area of *Saint Vincent* is 344 km². Some of the smaller islands are privately owned – probably not by retired cartographers. Part of the *Windward Islands,* the name dates back to the 18th century when English ships bound for Jamaica followed the trade-wind passage and stopped at islands along the way. The islands constitute a north-south chain in the southern section of the *Lesser Antilles* and share a volcanic rock formation.

The cannibal warrior Carïbs arrived in Saint Vincent around the 14th century, and they strongly resisted colonial settlers until the 18th century. Christopher Columbus probably sighted the island on January 22, 1498 (St. Vincent's Day). In 1673 the first African slaves were shipwrecked in the *Grenadines*, but they managed to get to St. Vincent, intermarry with the Carïbs, and became known as the "Black Carïbs." In 1795 the Carïbs unsuccessfully rose in revolt against the British, and 5,000 or more of them were deported to *Roatan Island* off the coast of Honduras. The hurricane of 1898 and the volcanic eruption of 1902 were disastrous to the economy. In 1958 Saint Vincent joined the West Indies Federation, it received a new constitution in 1960, and it became a state in association with the United Kingdom in 1969. Independence for Saint Vincent and the Grenadines was achieved on October 27, 1979. *Encyclopedia Britannica* says, "In 1979 the *Soufrière volcano* (1,234 m) erupted once again, damaging agriculture and the tourist trade. Hurricane Allen virtually wiped out the all-important banana crop in 1980."

The earliest geodetic survey of Saint Vincent was of *Fort Charlotte (lighthouse) in 1946* by the Hydrographic Service of the British Admiralty on 4 December. The coordinates of *Fort Charlotte (V.1)* are $\Phi_o = 13°\,09'\,24''$ N and $\Lambda_o = 61°\,14'\,43''$ West of Greenwich, the reference azimuth from "V. 1" to "V. 3" is $\alpha_o = 107°\,30'\,13.42''$, the elevation of "V.2" is: $H_o = 370.36$ feet, and the baseline length (measured in 1945 by the Royal Engineers (R.E.) by catenary) from "V.30" to "V.32" is 2,347,504 m. (*Invar tapes or wires are commonly calibrated for a standard length by being supported only at the ends of the tape or wire with a specific tension, thus the sag forms a catenary curve – Ed.*) The reference ellipsoid is the Clarke 1880, where $a = 6,378,249.136$ m and $1/f = 293.46631$, the same parameters as for Jamaica. Courtesy of the *UK Military Survey*, "The height of "V.2" was established by the R.E. party by leveling from a Bench Mark on a step of the Aquatic Club, Kingstown; the height of the Bench Mark was established by the R.E. party from mean tide gauge reading taken over a period of four months." (*Note that a Metonic cycle is 18.67 years! – Ed.*)

The coordinate system used by the Lands and Surveys Department of Saint Vincent and the Grenadines is the British West Indies (BWI) Grid which is based on the Transverse Mercator projection where the Central Meridian $\lambda_o = 62°$ W, the Scale Factor at Origin $m_o = 1999/2000 = 0.9995$, and the False Easting is 400 km. Note that the unit of measurement for this BWI Grid is the meter where 1 meter = 3.2828456 feet. The strange conversion factor is likely due to an earlier colonial length standard that was used for property surveys, a common quirk of old British Colonies throughout the world.

In 1996, the United States National Geodetic Survey (NGS) performed a GPS survey of selected points on the island of Saint Vincent. The NGS occupied a number of existing survey control marks previously set by the Lands and Surveys Department (L&SD), and I was successful in obtaining the classical *St. Vincent Datum of 1946* coordinates of four of the six points collocated with GPS observations. Unfortunately, I do not have a record of who it was that sent the information to me on official L&SD stationery. I ran a solution for the three parameters of geocentric translation for those four points; two were listed as First Order and two were listed as Second Order. The resultant relation I derived from the St. Vincent Datum of 1946 to WGS 84 is $\Delta X = +196$ m, $\Delta Y = +332$ m, and $\Delta Z = +275$ m. I estimate the horizontal accuracy to be good to about 1 meter for the island of Saint Vincent. Because I had zero data on collocated points on any of the other islands, my guess is that the three-parameter shift values listed above are likely good to no more than a few meters for the remainder of the islands to the south because of the usually superb quality of work produced by the Royal Engineers. Thanks to Mr. Dave Doyle of NGS for the NAD83 coordinates of Saint Vincent.

New 2019 publication on Limits in the Seas from the U.S. Dept. of State: *https://www.state.gov/wp-content/uploads/2019/10/LIS-144.pdf*, 2007 Update on history details from Mr. Russell Fox, Librarian for the Directorate of Overseas Surveys.

DOS WORK IN ST VINCENT AND THE GRENADINES

Summarized from DOS Annual Reports and notes held by Russell Fox (OS International Library Manager 1994–2004)

- DOS is used here as shorthand for the Directorate of Colonial Surveys (DCS, 1946–57), Directorate of Overseas Surveys (DOS, 1957–1984), Overseas Surveys Directorate, Ordnance Survey (OSD, 1984–1991), and Ordnance Survey International (1991–2004).

Other abbreviations:

SVSD: St Vincent Survey Department
SVG: St Vincent Government
RE: Royal Engineers
RN: Royal Navy

SURVEYS AND COMPUTATIONS

1947/1948: RE 1945 primary trig of St V. computed by DCS; co-ords issued.

1951/1952: DCS Senior Surveyor W H Young, and two surveyors recruited from Trinidad by the SVG, identified existing trigs on air photos and established 93 new trigs to provide extra control for 25K mapping, road traversing and cadastral surveys. Extra height control was fixed for 25K contouring.

1952/1953: St Vincent. Control for 25K mapping completed. Closer control established for cadastral surveys. Many existing property beacons were tied in during this work. Subsequent investigation showed gross error in the existing property surveys. Brigadier Hotine, Director of Colonial Surveys, visited St V. in Dec 1952 and decided that a complete cadastral re-survey of the island was necessary. Mr. Young was instructed to commence this work while a suitably qualified surveyor could be recruited to continue the work permanently.

1953/1954: St Vincent. Cadastral surveys continued. Little cadastral re-survey work was possible owing to ad hoc surveys urgently required by SVG and carried out up to May 1954. The cadastral re-survey was postponed until trained staff were available. Five Vincentians were selected for the two-year survey training course in Trinidad. *[Mr. Young probably left St V. in mid-1954]*. All trig computations were completed at DCS, and lists prepared for issue.

1957: Mr. Young visited St V. in May and Sep to advise on local surveys and do 25K field completion.

1962/1963: The Grenadines. Tellurometer traversing & trig observed from St V. to Grenada by DOS/SVSD/RN. 115 stations built, 96 observed, 91 lines measured. Preliminary investigation of existing control in the Grenadines completed at DOS.

1964: The Grenadines computations were completed at DOS. Co-ords and heights computed for 121 stations. Two sets of co-ords were produced, one on St V. datum and the other on Grenada datum.

1965: A DOS surveyor spent one week in St V. assisting SVSD in premarking control for large scale mapping and measuring 11 Tellurometer lines to fix additional stations.

1971–1973: St Vincent. A DOS field party of four surveyors established a dense network of fourth order traverse stations for cadastral purposes. A perimeter Tellurometer traverse, with a cross-island link, was observed to check the scale of the existing trig. Additional control was observed for the Kingstown 2.5 K mapping. Photo control for 5 K mapping of the west coast was observed. In all, some 700 new stations were established. Field revision was carried out for the 25 K third edition mapping. During November 1971 the Soufriere volcano became active. DOS surveyors and geologists from Trinidad made periodic observations of the level of the crater lake and lava islands in it. Geodimeter observations were made between Chateaubelair and Soufriere to detect earth movements associated with recent eruptions. At DOS the perimeter Tellurometer traverse was computed, revealing inconsistencies in the scale of the existing network. DOS prepared to carry out a comprehensive re-adjustment of the main surveys observed since 1944/1945 *[but the re-adjustment was not done until the late 1970s]*.

1972–1974: Grenadines. Extra control was established for 2.5K mapping in the islands. Sixty-five new stations were fixed by EDM traverse and trig. On Mustique, existing Fairey Surveys control points were re-occupied by DOS; only one new station was required there. DOS computed final co-ords and provisional heights for Bequia, Mustique, Cannouan, Mayreau, and Union islands.

1976/1977: DOS computed provisional co-ords and heights for St V. from the 1972 fieldwork.

1978–1980: St Vincent. Two DOS surveyors were detached from the St Lucia field party for a month to recce and photo-identify control for 2.5K mapping N of Kingstown and on the E coast.

St Vincent and the Grenadines: DOS re-computed all surveys in one adjustment comprising 1,528 observation equations in 416 unknowns. Final co-ords were produced for 209 points. Two further adjustments were carried out to incorporate the minor control and a final co-ordinate list of 1,177 points was issued.

1980–1982: The DOS Caribbean Map Revision/Field Completion team and an SVSD surveyor carried out field completion and 2.5K map revision in St Vincent and the Grenadines, plus 5K and 10K map revision on some islands of the Grenadines. Plan and height control for 2.5K mapping in the Grenadines was observed and then computed at DOS.

1982/1983: 2.5K field completion was completed on St Vincent and Union islands by two surveyors from DOS and SVSD. The DOS party closed down in 1983, SVSD continuing the field completion work (Tables S.1 and S.2).

TABLE S.1
Aerial Photography

DOS Contract No.	Year	Scale	Coverage
2	1950	18 K	St Vincent
3	1951	14–18 K	W & central St Vincent
RAF	1965	25 K	The Grenadines
85	1966	6–12.5 K	W & SE St V. and Mustique
RAF	1966	?	Partial coverage of the Grenadines
113	1970	12.5 K	Central & E St Vincent
116	1971	20 K	W St Vincent
116	1971	12.5 K	The Grenadines
RAF	1972	20 K	Partial coverage of St Vincent
RN	1972	?	Helicopter phy of Ronde and Bequia
163	1977	12.5 K	W, S & E St Vincent
186	1981	25 K	St Vincent
186	1981	12.5 K	Bequia, Mustique, and Union islands and parts of St Vincent

I believe that the RAF flew some photography in 1946 but I have no details.

TABLE S.2
DOS Mapping

Scale	Series	Year	Coverage
200 K	?	?	St Vincent & the Grenadines
50 K	DOS417	1961–91, 8 eds	St Vincent. Ed 8 has the St Vincent Grenadines on the reverse
50 K	DOS417	1991	St V. Grenadines
25 K	DOS317	1959–83, 5 eds	St Vincent in two sheets
25 K	DOS344	1967–1970	The Grenadines in 5(?) sheets; this was a dual scale series with DOS244
10 K	DOS244	1967–1970	The Grenadines in? sheets; dual scale series with DOS344
10 K	DOS217	1983–1988	St V Grenadines in five sheets
5 K	DOS144	1983	Parts of the Grenadines in 5 dyeline sheets
2.5 K	DOS017	1980–1991	Approx. 60% of St Vincent covered in 134 dyeline sheets
2.5 K	DOS044	1983	Parts of the Grenadines in 29 dyeline sheets

I may have omitted some printed, large-scale editions from the 1970s. I have not attempted to summarize here any DOS geological and land use mapping of St Vincent and the Grenadines. RDF, 15.7.07.

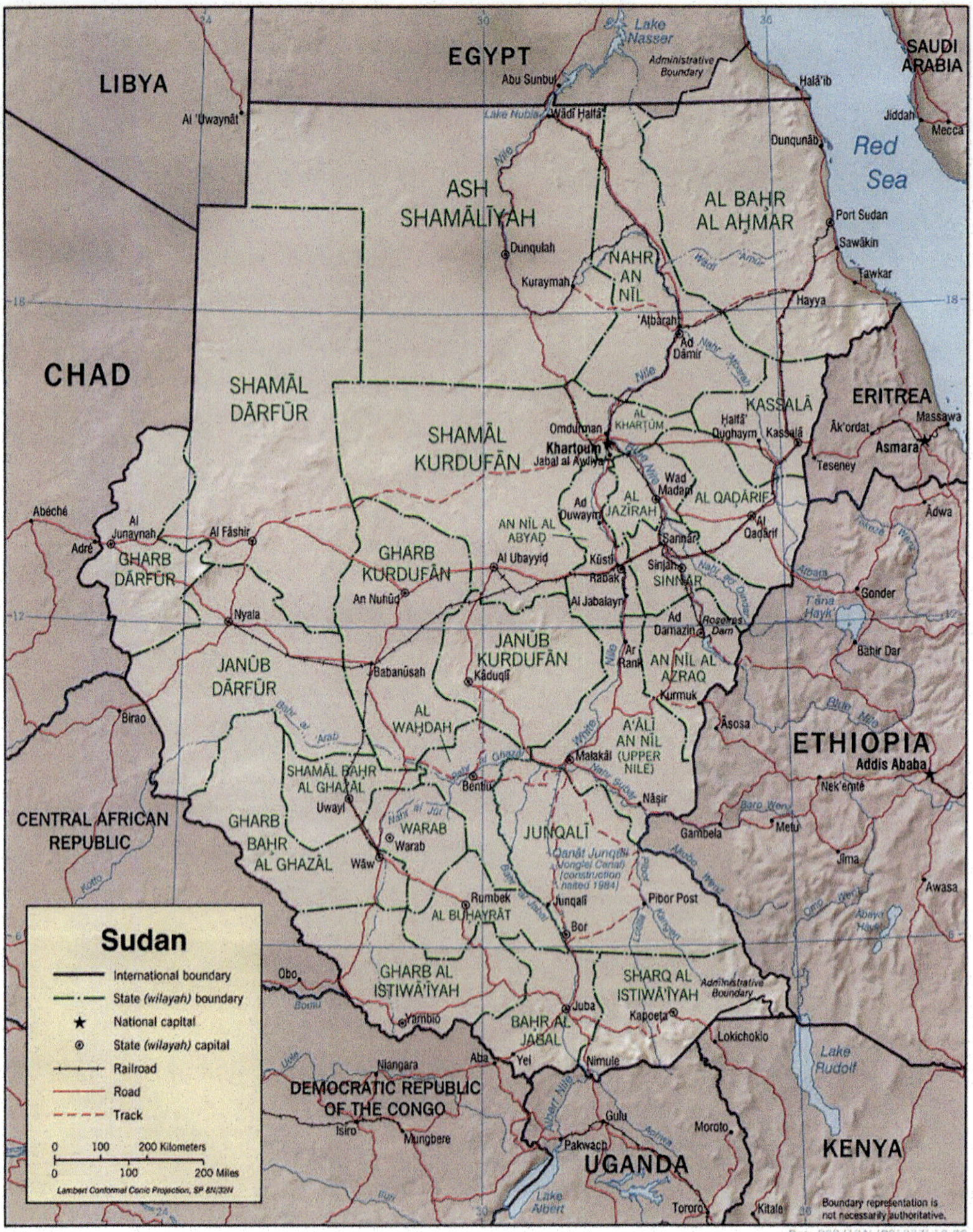

Base 802713AI (R01237) 10-00

Republic of the Sudan

Slightly less than one-fifth the size of the United States, Sudan is bordered by the Central African Republic (175 km)), Chad (1,360 km), Egypt (1,275 km), Eritrea (605 km), Ethiopia (769 km), Libya (383 km), and South Sudan (2,184 km) in which "Sudan-South Sudan boundary represents 1 January 1956 alignment; final alignment pending negotiations and demarcation; final sovereignty status of Abyei region pending negotiations between Sudan and South Sudan." Sudan is a generally flat, featureless plain with a desert that dominates the north (*World Factbook*, 2015).

"Archaeological excavation of sites on the Nile above Aswan has confirmed human habitation in the river valley during the Paleolithic period that spanned more than 60,000 years of Sudanese history. By the eighth millennium *B.C.*, people of a Neolithic culture had settled into a sedentary way of life there in fortified mud-brick villages, where they supplemented hunting and fishing on the Nile with grain gathering and cattle herding. Contact with Egypt probably occurred at a formative stage in the culture's development because of the steady movement of population along the Nile River. Skeletal remains suggest a blending of negroid and Mediterranean populations during the Neolithic period (eighth to third millennia *B.C.*) that has remained relatively stable until the present, despite gradual infiltration by other elements. Northern Sudan's earliest historical record comes from Egyptian sources, which described the land upstream from the first cataract, called *Cush*, as 'wretched.' For more than 2,000 years after the Old Kingdom (ca. 2700 – 2180 *B.C.*), Egyptian political and economic activities determined the course of the central Nile region's history. Even during intermediate periods when Egyptian political power in *Cush* waned, Egypt exerted a profound cultural and religious influence on the Cushite people.

"Over the centuries, trade developed. Egyptian caravans carried grain to *Cush* and returned to *Aswan* with ivory, incense, hides, and carnelian (a stone prized both as jewelry and for arrowheads) for shipment downriver. Egyptian traders particularly valued gold and slaves, who served as domestic servants, concubines, and soldiers in the pharaoh's army. Egyptian military expeditions penetrated *Cush* periodically during the Old Kingdom. Yet there was no attempt to establish a permanent presence in the area until the Middle Kingdom (*ca.* 2100 – 1720 *B.C.*), when Egypt constructed a network of forts along the Nile as far south as *Samnah*, in southern Egypt, to guard the flow of gold from mines in *Wawat*. Around 1720 *B.C.*, Asian nomads called Hyksos invaded Egypt, ended the Middle Kingdom, severed links with *Cush*, and destroyed the forts along the Nile River. To fill the vacuum left by the Egyptian withdrawal, a culturally distinct indigenous kingdom emerged at *Karmah*, near present-day *Dunqulah*. After Egyptian power revived during the New Kingdom (*ca.* 1570 – 1100 *B.C.*), the pharaoh Ahmose I incorporated Cush as an Egyptian province governed by a viceroy. Although Egypt's administrative control of Cush extended only down to the fourth cataract, Egyptian sources list tributary districts reaching to the Red Sea and upstream to the confluence of the *Blue Nile* and *White Nile rivers*. Egyptian authorities ensured the loyalty of local chiefs by drafting their children to serve as pages at the pharaoh's court. Egypt also expected tribute in gold and slaves from local chiefs. Once Egypt had established political control over *Cush*, officials and priests joined military personnel, merchants, and artisans and settled in the region. The Coptic language, spoken in Egypt, became widely used in everyday activities. The Cushite elite adopted Egyptian gods and built temples like that dedicated to the sun god Amon at *Napata*, near present-day *Kuraymah*. The temples remained centers of official religious worship until the coming of Christianity to the region in the sixth century. When Egyptian influence declined or succumbed to foreign domination, the Cushite elite regarded themselves as champions of genuine Egyptian cultural and religious values.

"By the eleventh century B.C., the authority of the New Kingdom dynasties had diminished, allowing divided rule in Egypt, and ending Egyptian control of *Cush*. There is no information about the region's activities over the next 300 years. In the eighth century B.C., however, *Cush* reemerged as an independent kingdom ruled from *Napata* by an aggressive line of monarchs who gradually extended their influence into Egypt. About 750 B.C., a Cushite king called Kashta conquered Upper Egypt and became ruler of *Thebes* until approximately 740 B.C. His successor, Painkhy, subdued the delta, reunited Egypt under the Twenty-fifth Dynasty, and founded a line of kings who ruled *Cush* and *Thebes* for about a hundred years. The dynasty's intervention in the area of modern Syria caused a confrontation between Egypt and *Assyria*. When the Assyrians in retaliation invaded Egypt, Taharqa (688 – 663 B.C.), the last Cushite pharaoh, withdrew and returned the dynasty to *Napata*, where it continued to rule *Cush* and extended its dominions to the south and east. Egypt's succeeding dynasty failed to reassert control over Cush. In 590 B.C., however, an Egyptian army sacked *Napata*, compelling the Cushite court to move to a more secure location at *Meroe* near the sixth cataract. For several centuries thereafter, the Meroitic kingdom developed independently of Egypt, which passed successively under Persian, Greek, and, finally, Roman domination. During the height of its power in the second and third centuries B.C., Meroe extended over a region from the third cataract in the north to Sawba, near present-day Khartoum, in the south. The pharaonic tradition persisted among a line of rulers at *Meroe*, who raised *stelae* to record the achievements of their reigns and erected pyramids to contain their tombs. These objects and the ruins of palaces, temples, and baths at *Meroe* attest to a centralized political system that employed artisans' skills and commanded the labor of a large work force. A well-managed irrigation system allowed the area to support a higher population density than was possible during later periods. By the first century B.C., the use of hieroglyphs gave way to a Meroitic script that adapted the Egyptian writing system to an indigenous, Nubian-related language spoken later by the region's people. Meroe's succession system was not necessarily hereditary; the matriarchal royal family member deemed most worthy often became king. The queen mother's role in the selection process was crucial to a smooth succession. The crown appears to have passed from brother to brother (or sister) and only when no siblings remained from father to son.

"Although *Napata* remained *Meroe's* religious center, northern Cush eventually fell into disorder as it came under pressure from the Blemmyes, predatory nomads from east of the Nile. However, the Nile continued to give the region access to the Mediterranean world. Additionally, Meroe maintained contact with Arab and Indian traders along the Red Sea coast and incorporated Hellenistic and Hindu cultural influences into its daily life. Inconclusive evidence suggests that metallurgical technology may have been transmitted westward across the savanna belt to West Africa from Meroe's iron smelteries. Relations between Meroe and Egypt were not always peaceful. In 23 B.C., in response to *Meroe's* incursions into Upper Egypt, a Roman army moved south and razed *Napata*. The Roman commander quickly abandoned the area, however, as too poor to warrant colonization. In the second century A.D., the Nobatae occupied the Nile's west bank in northern *Cush*. They are believed to have been one of several well-armed bands of horse- and camel-borne warriors who sold protection to the Meroitic population; eventually they intermarried and established themselves among the Meroitic people as a military aristocracy. Until nearly the fifth century, Rome subsidized the Nobatae and used *Meroe* as a buffer between Egypt and the Blemmyes. Meanwhile, the old Meroitic kingdom contracted because of the expansion of Axum, a powerful Abyssinian state in modern *Ethiopia* to the east. About A.D. 350, an Axumite army captured and destroyed Meroe city, ending the kingdom's independent existence.

"The emergence of Christianity reopened channels to Mediterranean civilization and renewed Nubia's cultural and ideological ties to Egypt. The church encouraged literacy in Nubia through its Egyptian-trained clergy and in its monastic and cathedral schools. The use of Greek in liturgy eventually gave way to the Nubian language, which was written using an indigenous alphabet that combined elements of the old Meroitic and Coptic scripts. Coptic, however, often appeared in ecclesiastical and secular circles. Additionally, early inscriptions have indicated a continuing knowledge of colloquial Greek in *Nubia* as late as the twelfth century. After the seventh century, Arabic gained importance in the Nubian kingdoms, especially as a medium for commerce. The Christian Nubian kingdoms, which survived for many centuries, achieved their peak of prosperity and military power in the ninth and tenth centuries. However, Muslim Arab invaders, who in 640 had conquered Egypt, posed a threat to the Christian Nubian kingdoms. Most historians believe that Arab pressure forced *Nobatia* and *Muqurra* to merge into the kingdom of *Dunqulah* sometime before 700. Although the Arabs soon abandoned attempts to reduce Nubia by force, Muslim domination of Egypt often made it difficult to communicate with the Coptic patriarch or to obtain Egyptian-trained clergy. As a result, the Nubian church became isolated from the rest of the Christian world.

"In January 1899, an Anglo-Egyptian agreement restored Egyptian rule in Sudan but as part of a condominium, or joint authority, exercised by Britain and Egypt. The agreement designated territory south of the twenty-second parallel as the Anglo-Egyptian Sudan. Although it emphasized Egypt's indebtedness to Britain for its participation in the re-conquest, the agreement failed to clarify the juridical relationship between the two condominium powers in Sudan or to provide a legal basis for continued British presence in the south. Britain assumed responsibility for governing the territory on behalf of the *khedive*" (*Library of Congress Country Study*, 2015).

"In the 21st century, Sudan faced conflict in *Darfur, Southern Kordofan,* and *Blue Nile* starting in 2003. Together, these conflicts displaced more than 3 million people; while some repatriation has taken place, about 2.28 million IDPs remained in Sudan as of December 2020. Sudan also faces refugee influxes from neighboring countries, primarily Central African Republic, Chad, Eritrea, Ethiopia, and South Sudan" (*World Factbook* 2022).

"Since 1899 cadastral mapping in the Sudan has been concentrated along the banks of the Nile from the Egyptian frontier to latitude 13° N, in the towns, and in the area of the *Gezira*, south of *Khartoum*, where cotton has been developed. These surveys were controlled by theodolite and steel tape either in the form of traverses or rectangulation. The early triangulation was used mainly to control topographical surveys, for it was essential to cover the whole country as rapidly as possible with a series of maps on scale 1:250,000. As a consequence, much of this early triangulation is of a relatively low order of accuracy, indifferently marked on the ground, and unfit for inclusion in a framework for medium and large-scale mapping. Brigadier Winterbotham's visit to the Sudan in 1929 led however to a revival of interest in the geodetic triangulation of the 30th Arc of Meridian. Lieut.-Col. S. L. Milligankk then Director of Surveys, made persistent efforts in face of the poverty of the country and the economic crisis to start the great work, and a length in 1935 a party under R.C. Wakefield went into the field. Progress was steady for five seasons and, by the time war intervened, two sections of the chain had been completed between the Egyptian frontier and latitude 13° 30′ N" (*Base Measurement in the Anglo–Egyptian Sudan*, D. F. Musey, M.A., Empire Survey Review, Vol. X, No. 72, 1949, pp. 66–74).

Geodetic work started in the Sudan in 1935 as part of the survey of the "Arc of the 30th Meridian" from Norway to South Africa. The geodetic work in the Sudan was an extension of previous work through Europe and ended finally in Egypt; the origin for the Egyptian work was considered in establishing an origin for the work in the Sudan. The fundamental origin of the Egyptian chain is "*Venus*" station near *Cairo*. In 1874 a number of expeditions were led by British scientists to various European colonies in Africa and the Indian Ocean in order to simultaneously observe the transit of Venus for the purpose of precisely determining differences in longitude. Helwân Observatory situated on *Az Zahra Hill* in the *Al Moqattam Hills, Qalyûbîya of Cairo* was utilized for the observations, and the station was termed "F$_1$," where $\Phi_o = 30° 01′ 42.8591″$ N, $\Lambda_o = 31° 16′ 33.6″$ East of Greenwich, the initial LaPlace azimuth being measured from Station O$_1$ (*Helwân*) to Station B$_1$ (*Saccara*), $\alpha_o = 72° 42′ 01.20″$ from South, and H$_o = 204.3$ m, based on mean sea-level at Alexandria. The Egyptian work was extended to *Adindan* which is just north of the Sudan border. The geographical positions of this station were computed according to the Hayford figure of 1909, where $a = 6,378,388$ m and $1/f = 297$.

"In *Mongalla* minor triangulation completed a chain from the western limit of the survey along the Sudan-Uganda boundary to the Nile and was connected at *Opari* to the 1927 survey. Various disconnected surveys, from Gwynn's of 1901 to Whitehouse's of 1926, have been incorporated, and the values of all trigonometrical stations have been re-computed. When smoke from grassfires caused the temporary cessation of the *Mongalla* survey, the inspector moved to *Upper Nile* and fixed astronomically some positions between the Nile and the *Pibor* at the request of the Egyptian Irrigation Department; latitude observations were made to a minimum of twenty pairs of stars, and longitude was determined by wireless time signals. Five new trigonometrical stations were fixed in the *Nuba Mountains*, and the position of *Dilling* was found by Astro-radio means to differ from that previously accepted by 7.01″ in latitude and 29.106″ in longitude. In a minor triangulation in the *Hadendowa District of Kassala* a base of 7,000 metres was measured near *Tendera Wells* and connected to *Jebels Asoteriba* and *Musma*r, which were fixed in 1902. A native surveyor was sent to make a plane-table survey in the *Qala' en Nahl* neighbourhood and measured a base of 3,500 meters along the Sennar-Gedaref railway, connecting up to *Jebel Beila* fixed by Gwynn in 1904" (*Sudan Government: Annual Report of the Survey Department for the Year 1930*, Empire Survey Review, pp. 139 & 140).

For the Sudan part of the arc, the origin of the geodetic work was in southern Egypt near *Abu Simbel*, south of *Lake Nasser*, at station *Adindan*, where $\Phi_o = 22°$ 10′ 07.1098″ N, $\Lambda_o = 31°$ 29′ 21.6079″ East of Greenwich, azimuth of line $Z_v - Y_v$: $\alpha_o = 58°$ 14′ 28.45″, deflection of the vertical: $\xi = +2.38″$, $\eta = -2.51″$, and the ellipsoid of reference was the Clarke 1880 (*modified*), where $a = 6,378,249.145$ m and $1/f = 293.465$. The *Blue Nile Datum of 1958* is the established classical datum of Sudan and much of North Africa. *Adindan* is the name of the origin, it is *not* the name of the datum; a most common mistake found in many "reference works."

See also: *The Arc of the Thirtieth Meridian between the Egyptian Frontier and Latitude 13° 45′ 1935–1950*, vols. 1 & 2, R.C. Wakefield and D.F. Munsey, Sudan Survey Department, 1950.

In the period between 1935 and 1945 all the geodetic work in Sudan was concentrated on the Arc of the 30th Meridian as it would be the basis for the whole program. During this period, the arc was carried south to latitude 13° 45′ N. Throughout this work, the English-made Tavistock theodolite was used in the observations with 16 sets of angles observed (*A Geodetic Datum for The Sudan*, M.O. Adam, Cornell University Thesis, June 1967, 95 pages).

The other part of the Arc of the 30th Meridian from latitude 13° 45′ N to the boundary with Uganda was observed together with a number of first- and second-order chains during the period from 1943 to 1952. Assistance on this work was given by the U.S. Army Map Service; one Party Chief was the late William Parkhurst that used a Wild Heerbrugg theodolite and moved from place to place via camel (*Personal communication, c.a.* 1963).

Based on a *very* recent analysis of 13 collocated points in Sudan, the three-parameter transformation **from** WGS84 datum **to** the *Blue Nile Datum of 1958* (*Adindan origin*) has updated the NGA values to $\Delta X = +162.6$ m ± 0.3025 m, $\Delta Y = +15.1$ m ± 0.3025 m, $\Delta Z = -204.5$ m ± 0.3025 m; RMS $= 1.0906$ m (*Common Point Coordinates Transformation Parameters Between Adindan (Sudan) New Ellipsoid and the World Geodetic System 1984 (GPS Datum) Coordinates Compared with Parameters of the American National Imagery and Mapping Agency (NIMA)*, Abdelrahim Elgizouli Mohamed Ahmed, International Journal of Advanced Research in Engineering and Applied Sciences, ISSN: 2278-6252, Vol. 2, No. 9, September 2014, pp. 26–39).

NORTH ATLANTIC OCEAN
GUYANA
French Guiana (FRANCE)
BRAZIL
NICKERIE
CORONIE
SARAMACCA
WANICA
COMMEWIJNE
MAROWIJNE
PARA
BROKOPONDO
SIPALIWINI
Paramaribo
Nieuw Nickerie
Totness
Friendship
Wageningen
Jenny
Boskamp
Groningen
Lelydorp
Nieuw Amsterdam
Moengo
Albina
Saint-Laurent du Maroni
Paranam
Onverwacht
Zanderij
Apoera
Bitagron
Brokopondo
Brownsweg
Afobaka Dam
Kamp 52
Bakhuis
Pokigron
Professor Doctor Ingenieur W. J. van Blommestein Meer
Grand Santi
Kajana
Cottica
Benzdorp
Apetina
Anapaike
Pelelu Tepu
Kwamalasamutu
Rosignol
New Amsterdam
Skeldon
Linden
Ituni
Takama
Suriname
International boundary
District (distrikt) boundary
National capital
District (distrikt) capital
Railroad
Road
The capital for the district of Sipaliwini has not yet been determined.
0 20 40 60 Kilometers
0 20 40 60 Miles
Lambert Conformal Conic Projection, SP 2N/6N
Boundary representation is not necessarily authoritative
801760 (B00447) 5-91

Republic of Suriname

The area of Suriname is slightly larger than Georgia; it borders Brazil (597 km), French Guiana (510 km), and Guyana (600 km). The terrain is mostly rolling hills with a narrow coastal plain with swamps. The lowest point is within the coastal plain at −2 m, the highest point is *Juliana Top* at 1,230 meters. Most of Suriname is tropical rain forest, and the majority of the 434,000 population lives within 30 km of the northern Atlantic Ocean coast.

Inhabited by Carïb and Arawak tribes prior to European settlement; the coast was sighted by Columbus in 1489. Spain officially claimed the area in 1593, but Portuguese and Spanish explorers gave the area little attention. Various attempts made to settle the area in the 17th century failed until the first permanent settlement was accomplished by the British Lord Willoughby of Parham, governor of Barbados in 1651. Suriname became a Dutch colony in 1667 according to the Treaty of Breda. The colony did not flourish, however, and there were numerous uprisings by the imported slave population as well as conflicts between the native tribes and the whites. Many of the slaves fled to the interior and established the five major Bush Negro tribes in existence today – the *Djuka, Saramaccaner, Matuwari, Paramaccaner,* and *Quinti*. Suriname became independent on 25 November 1975. The backbone of the Suriname economy is the export of alumina, produced since 1941 after investments made by ALCOA. The majority of aluminum (75%) used by the United States during World War II originated in Suriname.

"The Netherlands granted the colony independence in 1975. Five years later the civilian government was replaced by a military regime that soon declared Suriname a socialist republic. It continued to exert control through a succession of nominally civilian administrations until 1987, when international pressure finally forced a democratic election. In 1990, the military overthrew the civilian leadership, but a democratically elected government - a four-party coalition - returned to power in 1991. The coalition expanded to eight parties in 2005 and ruled until August 2010, when voters returned former military leader Desire Bouterse and his opposition coalition to power. President Bouterse was reelected unopposed in 2015" (*World Factbook* 2022).

In April and May of 1886, observations were made by Captain Haan of the Steamer Surinam for the mean difference in longitude between the stone stair landing of *Paramaribo* and the flagstaff of *Rickett Battery in Bridgetown, Barbados* by three time-transfers. In May of 1886, similar observations were made between Paramaribo and the upper signal of the *St. Marthe* and *St. Pierre Battery in Martinique*. With computations obtained from the U.S. Navy as well as additional Dutch measurements, the stone stair landing in *Paramaribo* was finally determined to have an astronomical longitude of $\Phi_o = 57°\ 29'\ 02.0''$ West of Greenwich. In *Annals Hydrographique*, the French remarked that in 1880, Dutch *Lieutenant de Vaisseau de Première Classe* Mulder, (Full Naval Lieutenant) observed a similar longitude that differed by only approx. 6 arc seconds; a remarkable feat, because it was all done with chronometers!

Topographic mapping began in 1947 when 1:40,000 scale aerial photography was flown by KLM Aerocarto. The initial control was based on the *Paramaribo Datum of 1947* where, according to J. B. Wekker, $\Phi_o = 5°\ 49'\ 25.40''$ North, $\Lambda_o = 55°\ 09'\ 09.20''$ West of Greenwich. This Datum was referenced to the Bessel 1841 ellipsoid, where $a = 6,377,397.155$ m and $1/f = 299.1528128$. For mapping, the Kaart van Suriname Rousilhe Stereographic Grid was used where the latitude of origin, $\varphi_o = 4°\ 07'$ N; the central meridian, $\lambda_o = 55°\ 41'$ W; the scale factor at origin was unity; False Easting = 300 km, and False Northing = 775 km. The map compilation was performed by *Centraal Bureau Luchkartering* (CBL). The area north of 4° N was mapped at 1:40,000 scale, and the entire country was mapped at 1:100,000 scale.

During the 1960s, a new primary triangulation network was based on HIRAN - SECOR – BC4 – PC-1000 – Doppler Transit observations in Suriname. (An aspect of these geodetic systems is that I am so old, I have had some association with all of them except for the PC-1000!) The new local system is known as the *Zanderij Datum of 1962*, where $\Phi_o = 5°\ 26'\ 53.45''$ North $\pm 0.10''$, $\Lambda_o = 55°\ 12'\ 19.04''$ East of Greenwich $\pm 0.10''$, and the reference azimuth from RM No. 1 to Az. Mk. measured from south $\alpha_o = 261°\ 59'\ 18.89''$. According to Mr. John W. Hager, "… the station name of HIRAN 14 AMS 1962 and latitude and longitude values (seconds only) of 53.25″ and 19.22″. This is a difference of 8.27 meters. The equipment in 1962 was quite bulky and I think that this later value represented the HIRAN antenna position and that the astro was located the 8.27 meters away. Another reason for having the astro point and the antenna some distance apart is that they would be making the HIRAN measurements simultaneous with or before they would complete the astro observations." The ellipsoid of reference for the Zanderij Datum of 1962 is the International 1924, where $a = 6,378,388$ m and $1/f = 297$. The projection adopted for this Datum is the Suriname Gauss-Krüger Transverse Mercator Grid, where the Central Meridian $\lambda_o = 55°\ 41'$ W, False Northing = zero, and False Easting = 500 km. Two scale factors at origin have been noticed with this Grid: $m_o = 0.99975$ and $m_o = 0.9999$, the latter observed on some 1:50,000 scale maps dated around 1978. The most common scale factor for the Suriname TM Grid is $m_o = 0.99975$.

In 1996, the U.S. National Geodetic Survey observed a number of positions with GPS receivers, one point being "008 Astro ECC 19," where $\varphi = 5°\ 26'\ 54.62257''$ N, $\lambda = 55°\ 12'\ 19.04''$ E. Although this is a different point than the old Datum origin, the similarity of the coordinates shows how close the NAD83 Datum is to the old Zanderij 1962 Datum. Proof of the pudding is the three-parameter shift values published by NGA in *TR8350.2*, January 3, 2000, where from Zanderij to WGS84: $\Delta X = -265$ m ± 5 m, $\Delta Y = +120$ m ± 5 m, $\Delta Z = -358$ m ± 8 m. The NIMA solution was based on five co-located points. Thanks for a lot of help on Suriname goes to Mr. John W. Hager and to Mr. Mark Nettles.

A fascinating account of a boundary recovery survey used in the Guyana-Suriname Maritime Boundary Delimitation by David H. Gray of Ottawa Canada is found at: https://pdfs.semanticscholar.org/6bc4/c38aa6bdbac5a7359a6714d90eb9e9193ae8.pdf.

Swaziland
International boundary
District boundary
National capital
District capital
Railroad
Road
0 10 20 Kilometers
0 10 20 Miles
SOUTH AFRICA
MOZAMBIQUE
HHOHHO
LUBOMBO
MANZINI
SHISELWENI
INDIAN OCEAN
Mbabane
Maputo
Manzini
Matsapha
Siteki
Piggs Peak
Hlatikulu
Nhlangano
Big Bend
Lavumisa
Witrivier
Nelspruit
Komatipoort
Ressano Garcia
Hectorspruit
Moamba
Barberton
Namaacha
Boane
Goba
Bela Vista
Amsterdam
Piet Retief
Paulpietersburg
Louwsburg
Pongola
Magudu
Jozini
Mkuze
Base 801485 (E00300) 4-90

Kingdom of Swaziland

Bordered by Moçambique (108 km) and South Africa (438 km), the lowest point is the *Great Usutu River* (21 m) and the highest point is *Emiembe* (1,862 m). Slightly smaller than New Jersey, the terrain is mostly mountains and hills; some moderately sloping plains (*World Factbook*, 2015).

"In eastern Swaziland archaeologists have discovered human remains dating back 110,000 years, but the ancestors of the modern Swazi people arrived relatively recently. During the great Bantu migrations into southern Africa, one group, the Nguni, moved down the east coast. A clan settled near what is now Maputo in Moçambique, and a dynasty was founded by the Dlamini family. In the mid-18th century increasing pressure from other Nguni clans forced King Ngwane III to lead his people south to lands by the Pongola River, in what is now southern Swaziland. Today, Swazis consider Ngwane III to have been the first king of Swaziland. The next king, Sobhuza I, withdrew under pressure from the Zulus to the Ezulwini Valley, which today remains the center of Swazi royalty and ritual. When King Sobhuza I died in 1839, Swaziland was twice its present size. Trouble with the Zulu continued, although the next king, Mswazi (or Mswati), managed to unify the whole kingdom. By the time he died in 1868, the Swazi nation was secure. Mswazi's subjects called themselves people of *Mswazi*, or *Swazis*, and the name stuck.

"The arrival of increasing numbers of Europeans from the mid-19th century brought new problems. Mswazi's successor, Mbandzeni, inherited a kingdom rife with European carpetbaggers – hunters, traders, missionaries, and farmers, many of whom leased large expanses of land. The Pretoria Convention of 1881 guaranteed Swaziland's 'independence' but also defined its borders, and Swaziland lost large chunks of territory. 'Independence' in fact meant that both the British and the Boers had responsibility for administering their various interests in Swaziland, and the result was chaos. The Boer administration collapsed with the 1899–1902 Anglo–Boer War, and afterwards the British took control of Swaziland as a protectorate. Swaziland became independent on 6 September 1968. The country's constitution was largely the work of the British. Currently, Swaziland's greatest challenge comes from the HIV/AIDS pandemic; the country has the world's highest HIV infection rate (almost 39% for adults between 15 and 49 years of age), and life expectancy has fallen as a result from 58 to 33 years" (*Lonely Planet*, 2015), (*The Kingdom of Swaziland*, D. Hugh Gillis, 1999, ISBN:0-313-30670-2).

"In 1897 a Portuguese-South African commission demarked with beacons or pillars the four principal points of the Moçambique-Swaziland boundary, which were designated by the 1888 commission as *Ipoye, Buchanan, Nunes da Silva,* and *Krogh*. The State Secretary of the South African republic agreed to the beaconing of the 1897 commission, except for the sector of the boundary between *Krogh Beacon* and *Mpundweni Beacon*. An Anglo-Portuguese commission re-demarcated the sector from *Krogh Beacon* to *Mpundweni Beacon* in 1925. The following gives the beacons from south to north with their locations which are based on a coordinate grid system from *M'Ponduine* Geodetic Station, where $\varphi = 25°\ 56'\ 47.19''$ S, $\lambda = 31°\ 58'\ 40.46''$ E, X=0.0 m, Y=0.0 m; *Muguene* Geodetic Station: X=−29,757.40 m, Y=+19.691.48 m; *Krogh*: X=−12,756.06 m, Y=+23,576.54 m; *Line*: X=−12,048.39 m, Y=+18,300.49 m; *Oribi*: X=−10,809.60 m, Y=+9,064.82 m; *Sikayana*: X=−10,976.64 m, Y=+6,718.89 m; *Xilungo*: X=−5,236.90 m, Y=+5,786.62 m; C: X=−3,268.72 m, Y=+5,038.19 m; D': X=−2,539.52 m, Y=+5,430.60 m; *Mpundweni Beacon (M'Ponduine Marco)*: X=+229.38 m, Y=+512.66 m, $\varphi = 26°\ 07'\ 25.98''$ S, $\lambda = 32°\ 16'\ 31.58''$ E" (*International Boundary Study, Mozambique-Swaziland Boundary, The Geographer, Bureau of Intelligence and Research, U.S. Department of State, No. 135,* June 4, 1973).

"At the suggestion of the Surveyor-General, *Pretoria*, it was eventually decided to compile a single sheet map at 1:500,000, utilizing not only the Irrigation Department's topographical map but also the cadastral survey of the Union *[of South Africa – Ed.]* currently under preparation at 1:250,000. The result was a single sheet map of Swaziland in four colors at 1:500,000 with a part of the imprint reading 'Drawn in the Trig Survey Office, *Pretoria*, for the Swaziland Administration.' The map is not dated but 500 copies were printed in March 1939. In the course of compilation, further small failings were found with the work of the 1905–11 survey, this time in connection with heights of peaks on the northwest boundary.

"Cartographic developments in South Africa in the 1930s and in the war years affected map provision in Swaziland. In 1935 the High Commissioner was informed of the intention of the Union Department of Irrigation to prepare a contour map of the Union including Swaziland, in 10 sheets. By 1951, ten preliminary plots covering a priority area in Swaziland had been issued. By 1954, twenty-two preliminary plots had been issued, the remaining eleven were nearing completion and contouring was receiving high priority. The heighting of Swaziland, together with a trigonometrical survey to cover the whole territory with a triangulation network tied to the Union system, took longer than estimated because of adverse observations conditions, but it was completed by December 1956" (*Geography and Imperialism 1820–1940*, Morag Bell, Manchester University Press, 1995, 338 pages).

The first Surveyor General of Swaziland was appointed in 1973. Prior to that, land registration surveys in the country came under the jurisdiction of the *Surveyor General of Transvaal*, South Africa, and all records were kept in *Pretoria*. The control framework of Swaziland was originally established by the then British Directorate of Overseas Surveys (DOS) and is generally strong (*Land Survey and Large-Scale Mapping in Sub-Saharan Africa*, UN Centre for Human Settlements (Habitat), Nairobi, 2001).

The classical system of Swaziland is part of the Cape Datum of 1950 where the origin point is *Buffelsfontein*, near *Port Elizabeth*, where $\Phi_o=33°$ $59'$ $32.000''$ S, $\Lambda_o=25°$ $30'$ $44.622''$ E, azimuth to *Zuurberg* measured from South: $\alpha_o=183°$ $58'$ $15.00''$, $\xi=-3.46''$, $\eta=-0.59''$, $h_o=280.1$ m, and the ellipsoid of reference is the Clarke 1880, where $a=6,378,249.145$ m and $1/f=293.465$. The Gauss Conform Transverse Mercator Grid system has 2° wide central meridians at: 31°E. "No arbitrary scale factors or false origins are applied to the co-ordinates; X is measured positive southwards from the equator and Y positive westwards from the nearest odd meridian. The unit of measurement since the 1970s is the International metre" (*Chief Directorate: Surveys and Land Information*, 1995, personal communication).

The new geodetic datum in South Africa is termed "*Hartbeeshoek 94 Datum*" and is referenced to the GRS80 ellipsoid, where $a=6,378,137$ m and $1/f=298.257222101$. The three-parameter transformation for Swaziland is based on four points **from** Cape Datum **to** Hartbeeshoek 94 Datum, where $\Delta X=+136.8$ m, $\Delta Y=+106.6$ m, $\Delta Z=+293.0$ m. "The seven- and four-parameter transformations provide little improvement over the three-parameter transformation and the scale factor cannot be separated from the translations" (*Datum Transformation Parameters in Southern Africa*, Rens, J, & Merry, C.L., Survey Review, 30, 236, April 1990).

Sweden
International boundary
National capital
Railroad
Road
0 50 100 Kilometers
0 50 100 Miles
Lambert Conformal Conic Projection, SP 47N/62N
Norwegian Sea
VESTERÅLEN
LOFOTEN
NORWAY
RUSSIA
FINLAND
ESTONIA
LATVIA
LITHUANIA
DENMARK
Gulf of Bothnia
Gulf of Finland
ALAND ISLANDS
Baltic Sea
Skagerrak
Kattegat
North Sea
Gotland
Öland
Hiiumaa
Saaremaa
Kiruna
Malmberget
Jokkmokk
Boden
Luleå
Piteå
Skellefteå
Storuman
Vilhelmina
Lycksele
Umeå
Örnsköldsvik
Östersund
Härnösand
Sundsvall
Ånge
Hudiksvall
Mora
Falun
Borlänge
Gävle
Uppsala
Västerås
Stockholm
Södertälje
Karlstad
Örebro
Nyköping
Norrköping
Linköping
Motala
Mariestad
Mellerud
Vänersborg
Uddevalla
Göteborg
Borås
Falköping
Jönköping
Nässjö
Värnamo
Västervik
Oskarshamn
Växjö
Kalmar
Visby
Varberg
Halmstad
Helsingborg
Hässleholm
Kristianstad
Karlskrona
Lund
Malmö
Ystad
Oslo
Trondheim
Bergen
Stavanger
Kristiansand
Narvik
Bodø
Mo
Mosjøen
Namsos
Steinkjer
Ålesund
Molde
Lillehammer
Hamar
Drammen
Porsgrunn
Arendal
Haugesund
Sandnessjøen
Harstad
Kolari
Kemijärvi
Rovaniemi
Kemi
Oulu
Kajaani
Kokkola
Vaasa
Seinäjoki
Jyväskylä
Tampere
Pori
Rauma
Turku
Helsinki
Lahti
Kotka
Tallinn
Pärnu
Riga
Ventspils
Liepāja
Jelgava
Šiauliai
Klaipėda
Frederikshavn
Ålborg
Holstebro
Viborg
Århus
Vejle
Esbjerg
Odense
Copenhagen
Roskilde
Korsør
Storsjön
Vänern
Vättern
Arctic Circle
Base 802435 (A04937) 2-96

The Kingdom of Sweden

Sweden is mostly flat with gently rolling lowlands; there are mountains in the west along the Norwegian border and the kingdom is slightly larger than California. The lowest point is the reclaimed bay of *Lake Hammarsjon*, near *Kristianstad* (–2.41 m); the highest point is *Kebnekaise* (2,111 m).

Inhabited during the Stone Age, several independent tribes lived in Sweden by the 9th century A.D. During that time, those adventurous tribes were among the Scandinavians that were known as the Vikings. Loosely united and converted to Christianity a couple centuries later, the Swedes conquered the Finns; they joined Norway and Denmark and finally broke away in 1537 under Gustav I Vasa. Sweden became a constitutional monarchy in 1809.

"An armed neutrality was preserved in both World Wars. Since then, Sweden has pursued a successful economic formula consisting of a capitalist system intermixed with substantial welfare elements. Sweden joined the EU in 1995, but the public rejected the introduction of the euro in a 2003 referendum. The share of Sweden's population born abroad increased from 11.3% in 2000 to 19.1% in 2018" (*World Factbook* 2022).

According to the *Lantmäteriet*, "The '*geometriska jordeböcker*' are the oldest large-scale maps in Sweden. One of the main tasks of the Land Survey following its establishment in 1628 was to carry out the mapping of villages and individual farms and their lands. It was primarily Crown farms that were the focus of interest. Cultivated fields and meadowland were mapped and information concerning yields and other information related to income and economic matters was collected. It is not clear whether the original purpose of the mapping was, in fact, to form the basis for taxation, but it can, definitely be seen as the predecessor to the Swedish land use maps (*Ekonomiska kartan*). The maps are unevenly distributed across Sweden. They are collected in large volumes sorted according to parish and district. The '*geometriska jordebö*cker' should not be confused with the Crown's standard '*jordeböcker*' which cover landed properties and contain fiscal information about them. The Crown's '*jordeböcker*' can be looked upon as being the first Swedish real property register and the '*geometriska jordeböcker*' as the first cadastral index maps. There are around sixty volumes for the period between 1630 and 1650. Most of the maps are at a scale of 1:5,000. We have only included in this series the volumes that have been scanned and are in digital format. To find older '*geometriska jordeböcker*' which are not yet scanned you should go to the series Cadastral Maps. In the Land Survey' map archives there are more than a hundred volumes of maps titled '*geometriska jordeböcker*' dating from the latter half of the 1600s and the early part of the 1700s. The maps are at varying scales, although most of them are large-scale maps. They mainly comprise farm maps that were produced for taxation purposes, maps to be used as the basis for the recruitment of and provision of material support of soldiers and maps needed for the organization of the return of land by the Church to the Crown."

During the early 18th century, the French scientist, Maupertuis joined with the Swedish astronomer, Celsius on the French expedition to *Lapland* for the determination of the length of a degree of the meridian arc. This was considered an expedition for insurance in case the sister trip to South America, was not conclusive in proving the shape of the earth. The area chosen for the chain of north-south triangles is now the southern land border between Sweden and Finland. Starting at the Lutheran church steeple in the city of *Torneå* (*Torne*) on the Gulf of Bothnia, the chain extended northwards along the *Torne River* to the (now) Finnish town of *Pello*. Maupertuis published his book on the Lapland expedition in 1737.

Soon after the commencement of the constitutional monarchy, the military survey of the kingdom was begun in 1811. Sweden gave up *Swedish Pomerania* in return for Norway, which entered into a personal union with Sweden (1814–1905). A civilian mapping authority for the compilation of an economic map was formed in 1859. The military and civilian mapping agencies were consolidated in 1894 and were known as the *Rikets Allmänna Kartverk* (RAK). After a series of consolidations and mergers, the current national mapping organization of the National Land Survey (*Statens Lantmäteriet*) was formed in 1985.

According to a personal communication from Dr. Lars Sjöberg of November 7, 1980, "The first systematic triangulation of Sweden started in 1805. All calculations were made on the ellipsoid, for Northern Sweden on Svanberg's ellipsoid and for Southern Sweden on Clarke 1880. For official maps (in general scale 1:100,000) Spens' projection, was used for Southern Sweden (up to Lat 61° 30') and a conform conic projection for Northern Sweden. In 1903 a new triangulation started in Southern Sweden. The calculations were made in plane coordinates (x, y) (Gauss-Hannover's projection and Bessel's ellipsoid). The scale in the net was determined from a Danish baseline, which was measured in 1838. A new measurement of the baseline was made in 1911 and that measurement differed significantly from the earlier one. In 1938, when 5 Swedish baselines and 6 azimuths had been measured, the scale and orientation of the nets (obtained from the above measurements) were compared. The measurements of the Swedish baselines agreed better with the 1911 measurements than with the observations of 1838. It was then decided to enlarge the net with a factor 1.00002 and turn it clockwise 0.00005 radians around a point in Southern Sweden. Up till then all calculations had been made in 6 different zones with the longitude of origin referring to '*Stockholms gamla observatorium*' (The Old observatory of Stockholm), which is 18° 03' 29.8" E of Greenwich. The longitude of origin for each zone was 6° 45' W, 4° 30' W, 2° 15' W, 0°, 2° 15' E, and 4° 30' E of *Stockholms Gamla Observatorium*. This system is still in use for large scale maps. *[Ed's. note: this letter from Dr. Sjöberg was dated 1980.]* In 1938 *Rikets Allmänna Kartverk* decided to reduce the number of projection zones to 3 namely 2° 15' W, 0°, and 4° 30' E for official maps with FE 1,500,000 m, 2,500,000 m, and 3,500,000 m, respectively. In 1945 RAK decided to use only one projection zone for official maps namely 2° 15' W with FE 1,500,000 m. Common for all zones are that latitude of origin is 0° and FN is 0. The scale factor along the central meridian m_o is 1.0000. For some official maps there is also a grid net in the UTM projection. This net is based upon the European Datum 1950 with $m_o = 0.9996$ and FE = 500,000 m."

I later wrote back to Dr. Sjöberg in July of the following year and inquired about the Spens projection. In Dr. Sjöberg's reply of August 7, 1981, "The Spens projection differs somewhat from the Lambert conic projection. Spens' projection satisfied the following conditions: 1. The scale factor (m_o) along the parallels $\varphi_1 = 65° 50' 20.4"$ and $\varphi_2 = 55° 21' 19.4"$ are equal. 2. The minimum scale factor between φ_1 and φ_2 equals m_o^{-1}. The first condition yields log n = 9.9407276−10 and $\varphi_o = 60° 44' 29.6"$ (These are Spens' results from 1817 used in the tables of Spens projection. The correct values are log n = 9.94072828−10 and $\varphi_o = 60° 44' 30.2"$.) From the second condition one obtains $m_o = 0.997903542$. The x-axis of the Spens projection is the meridian 5° W of the Old Observatory of Stockholm, directed southward. The origin is located at the parallel circle 72°. The Spens projection was described by P.G. Rosen (1876) in *Den vid Svenska Topografiska Kartverket använder projektionsmetoden*, 32 pp. As far as I know there is no word '*Gradblätterkarten*.' '*Karten*' means maps and '*Gradblätter*' 'degree maps.' However, I think you refer to the polyconic projection used for the old '*Generalstabskarten*' in the north of Sweden. This means that the conic projection is used repeatedly at each ½° parallel. Each map is made as a '*Gradblatt*' limited by parallel circles of every ½° and meridian of equidistance 1½°. Clarke's ellipsoidal parameters were used. The arc-triangulation in Lapland (*Tornedalen*) carried out 1730–1736 under the supervision of the Paris Academy was repeated in 1801–1803. From these latter measurements Svanberg computed the Earth dimensions (published 1805)."

According to a paper published (in German) by "*Professor im Generalstabe Karl D. P. Rosén*," Stockholm 1933, the Svandberg ellipsoid parameters used were $a = 6,376,797$ m and $1/f = 304.2506$. Similarly, the published parameters for the Clarke 1880 ellipsoid as used for the Northland projection were $a=6,378,249.2$ m and $1/f=293.465$. The specific formulae used in Sweden were discussed in 1951 by G. A. Rune in *Tabeller Till Gauss Hannoverska Projektion*, (Tables for Gauss's Hanoverian Projection) where he states (in English) in the Preface, "For facilitating the computation of the modern triangulation of Sweden, begun in 1903, the General Staff professor of that time Dr. Karl D. P. Rosén introduced the Gauss's Hanoverian projection, often called the Gauss-Krüger or, briefly, the Gauss's projection, a projection well-fitting Sweden with it marked extension in the meridian." Note that the defining parameters of the Bessel 1841 ellipsoid are: $a=6,377,397.155$ m and $1/f=299.1528128$. All of the Swedish classical datums have the same origin at the *Old Stockholm Gamla Observatory*, where $\Phi = 59°\ 20'\ 32.7''$ N and $\Lambda = 18°\ 03'\ 29.8''$ E. The triangulation of Sweden from 1903 to 1938 consisted of 170 triangles and was observed with Wanschaff and Hildebrand instruments achieved an average Ferrero's formula accuracy of $0.41''$. The later Swedish triangulation of 1939–1953 consisted of 222 triangles and was observed with Wild T-3 theodolites and achieved an average Ferrero's formula accuracy of $0.40''$.

That classical triangulation is defined as the RT 38 (*rikstrianguleringen* 1903–1950) datum. It has been replaced with RT 90 also called *Rikets Koordinatsystem* 1990, which is a local geodetic datum based on the Swedish third national triangulation (1967–1982) and is also referenced to the Bessel 1841 ellipsoid. The corresponding plane coordinate system is denoted **NT 90 2.5 gon V 0: −15** and is obtained by a Gauss-Krüger Transverse Mercator projection of the RT 90 latitudes and longitudes. The Central Meridian is $\lambda_o = 15°\ 48'\ 29.8''$ E, the scale factor at origin $m_o = 1.0$, and FE = 1,500 km. The central meridian was originally interpreted as "2.5 Gon West of the Old Observatory of Stockholm," but is now defined as relative to Greenwich (1 Gon = 0.9 degrees).

According to the *Lantmäteriet*, "The original map sheet system in Sweden is based on a grid in RT 90 2.5 gon V 0: −15 with the SW corner at (North.=6,100,000 m, East.=1,200,000 m), and NE corner at (North.=7700 000 m, East.=1900 000 m). This area is divided into 50 km squares, which are enumerated with 0–32 in South-North direction and lettered with A - N in West-East direction. Each 50 km square can be subdivided into four 25 km topographic map sheet squares, or subdivided into 100 5×5 km cadastral map sheets, which are enumerated from South to North by 0–9. and lettered from West to East by a - j. This original basic map sheet system has been modified in several ways for the modern series of maps, but the basic grid square notation is still frequently used, for instance in the numbering of geodetic control points. For larger scale mapping (>1:10000) there are six different zones of Transverse Mercator projections used in Sweden, in order to reduce the map projection errors. The other 5 zones apart from '2.5 gon V' differ only in the longitude of the central meridians, which are spaced by $2°\ 15'$. The boundaries of the projection zones are adjusted to follow administrative borders if possible. The coordinate system 'RT 90 5 gon V 61: -1' has the map projection parameters: Central meridian: $13°\ 33'\ 29''.8$ East Greenwich, False Easting: 100 000 m, False Northing: −6,100,000 m. Example of a point's coordinates in different coordinate systems: x (Northing)=6,200,000.000; y (Easting)=1,300,000.000 in 'RT 90 2.5 gon V 0:-15' x (Northing)=6,195,783.588; y (Easting)=1,440,736.999 in 'RT 90 5 gon V 0:-15' x (Northing)=95,783.588; y (Easting)=40,736.999 in 'RT 90 5 gon V 61:-1'."

"SWEREF 99 is a Swedish realization of ETRS 89. The processing of the GPS data was performed according to the EUREF guidelines and was based on observations made on permanent reference stations in Sweden (SWEPOS), Denmark, Finland (FinnRef) and Norway (SATREF) during the GPS-weeks 1014–1019 (June-July 1999).

SWEREF 99 coincides with WGS 84[G730] and WGS 84[G873] within some decimeters. Coordinates can be transformed from the Swedish coordinate datum RT 90, to SWEREF 99 through a seven-parameter transformation given below (estimated accuracy 7 cm, 1 sigma, 2D). The ellipsoid used with SWEREF 99 is GRS 80: $a=6378137$, $1/f=298.257222101$. SWEREF 99 replaces SWEREF 93 (the former realization of ETRS 89). If one ... prefers to define a transformation in the direction RT 90 to SWEREF 99, use the following parameters: $\Delta X = +414.1$ m, $R_x = +0.855$ arc seconds, $\Delta Y = +41.3$ m $R_y = -2.141$ arc seconds, $\Delta Z = +603.1$ m $R_z = +7.022$ arc seconds, and $\delta = 0.0$ ppm (scale = 1.0)."

For example: Latitude, longitude and height above the Bessel 1841 ellipsoid in RT 90: $\varphi=58°$ $00'$ $01.213296''$ $\lambda=17°$ $00'$ $11.683659''$ h=−5.397 m, latitude, longitude, and height above the GRS 80 ellipsoid in SWEREF 99: $\varphi=58°$ $00'$ $00.0''$, $\lambda=17°$ $00'$ $00.0''$, h=30,000 m. Thanks to Professor Lars Sjöberg now of the Geodesy Group at the Royal Institute of Technology.

Sweden operates the Nordic Geodetic Commission (NKG) Analysis Center and currently operates 90 GNSS station sites in cooperation with the EUREF permanent network. As of 2018, there were over 3,900 current subscriptions to the SWEPOS (Swedish National network of permanent GNSS stations operated by Lantmäteriet). Of interest is that SWEPOS offers not only dual-frequency correction services, but a single frequency DGNSS service has been offered since 2016, something not offered by the U.S. National Geodetic Survey. The new national height system RH2000 was implemented in 2005 and consists of about 50,000 passive benchmarks. So far 247 municipalities have implemented the replacement of RH2000 for their legacy height systems. Sweden had upgraded its FG-5 Absolute Gravity Meter to an FG-5X (Like LSU currently uses) and has observed at 14 sites as well as at 96 sites with A-10 Absolute Gravity Meters and at 200 sites with Relative Gravity Meters.

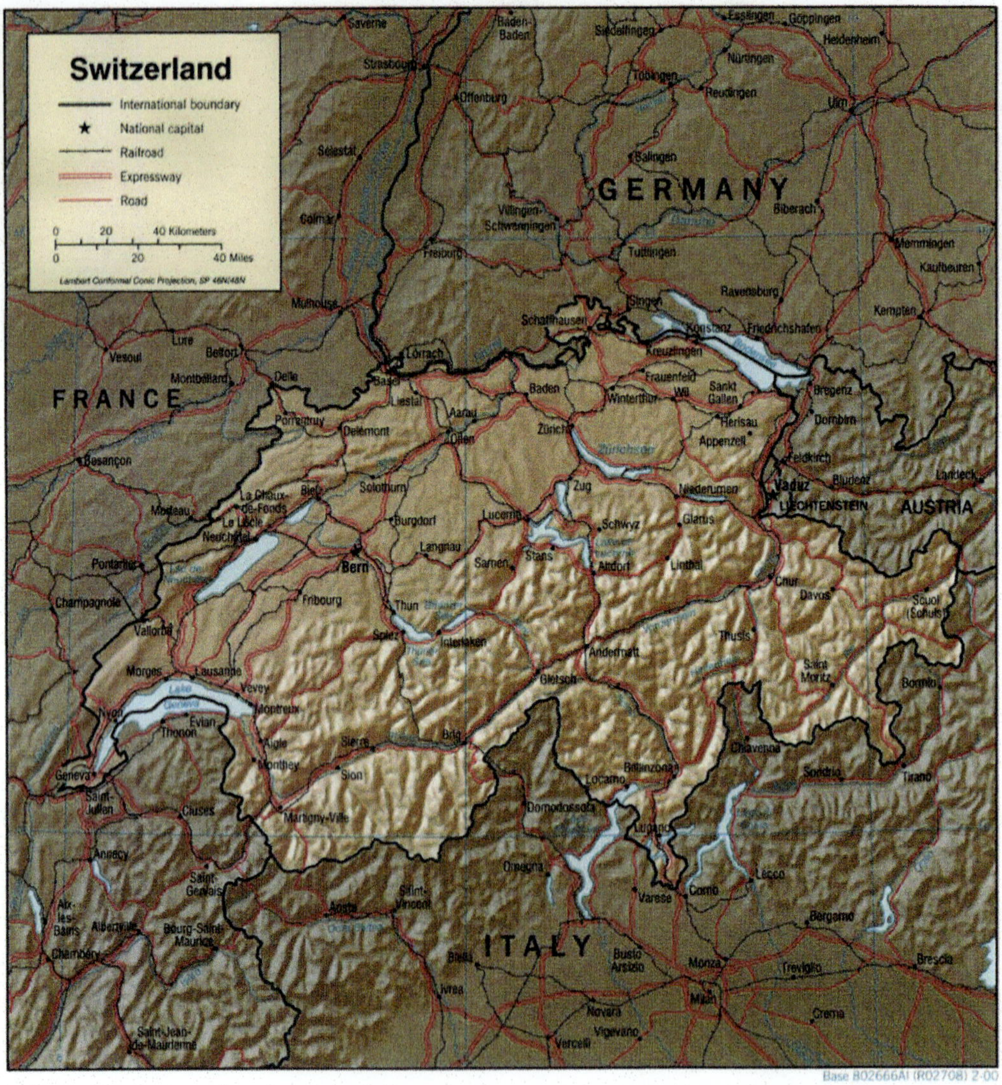

Base 802666AI (R02708) 2-00

The Swiss Confederation

Switzerland is slightly less than twice the size of New Jersey; it is bordered by Austria (158 km); France (525 km); Italy (698 km); Liechtenstein (41 km); and Germany (348 km). The terrain is mostly mountains (Alps in south, Jura in northwest) with a central plateau of rolling hills, plains, and large lakes. The highest point is *Monte Rosa* at 4,638 m, the lowest point is *Lake Maggiore* (195 m), and the mean elevation is (1,350 m).

Occupied by Helvetians who were conquered by Romans, the southwest was invaded by Burgundians and the northeast was invaded by Alamanni. In 1291 the Forest Cantons or provinces of Uri, Schwyz, and Unterwalden formed an anti-Hapsburg league that became the nucleus of the confederation. The perpetual neutrality of Switzerland was guaranteed by international agreement in 1815 (Congress of Vienna) and again in 1919 (Treaty of Versailles); its present constitution was adopted in 1874.

G. H. Dufour, later to become a general, founded the *Eidgenößisches Topographisches Bureau* (Topographical Bureau) in Geneva in 1838. Dufour decided to use the "*carte du jour*" projection of Europe for the time which was the ubiquitous Ellipsoidal Bonne originally used for topographic mapping by Cassini himself during the Napoleonic Campaigns. The Grid used for this Ellipsoidal Bonne has a Latitude of Origin $(\varphi_o) = 46°\ 57'\ 06.02''$ N, a Central Meridian $(\lambda_o) = 7°\ 26'\ 24.75''$ East of Greenwich ($5°\ 06'\ 10.80''$ East of Paris), a False Easting of 600 km, and a False Northing of 200 km. The *Berne Observatory Datum was circa 1840*, and the ellipsoid used was the Schmidt 1831, where $a = 6,376,804$ m and $1/f = 302.02$. Although my notes show a false origin for this old Grid, I suspect that the original use was with the traditional quadrant system. The false origin probably crept into use as the base was updated after the Rosenmund System was introduced in the 20th century. From 1845 to 1864, the publication of the first accurate map covering the whole of Switzerland was performed at the scale of 1:100,000, the slopes were shown by hatchures, and was known as the "Dufour Map." By this time, the classical triangulation of Switzerland comprised 40 triangles that had been observed with Kern instruments, and the average error of a figure was 0.86 arc seconds. The Swiss triangulation calculations were based on the Bessel 1841 ellipsoid after 1863, where $a = 6,377,397.155$ m and $1/f = 299.1528$. The office was transferred to *Berne* in 1868, and the publication of the original surveys at 1:25,000 (Swiss Central Plains) and at 1:50,000 (Alps) with contours was performed from 1870 to 1916. The Old Berne Observatory Datum of 1898 published an Astronomical Latitude $(\Phi_o) = 46°\ 57'\ 08.66''$ N, based on observations executed by E. Plantamour in 1875 and an Astronomical Longitude $(\Lambda_o) = 7°\ 26'\ 22.5''$ East of Greenwich. The defining azimuth to station *Rötifluh* was $(\alpha_o) = 11°\ 12'\ 05.24''$. The ellipsoid height and deflection of the vertical are not defined and therefore are forced to zero at the origin.

In 1900 the national mapping agency was renamed the *Eidgenößische Landestopographie*. The vertical Datum was defined as *Repère Pierre du Niton* 1902, a large rock in the harbor of *Geneva*, where $H_o = 373.600$ m ("*Gebrauchshöhe*") with a connection to the tide gauge in *Marseilles, France*. M. Rosenmund, an engineer with the bureau developed a new projection. The new system is an oblique conformal cylindrical double projection, similar in concept to what General Jean Laborde developed for Madagascar. For the Grid of the Swiss National Maps, the value $\varphi_o = 46°\ 57'\ 07.90''$ N was chosen based on more recent measurements (1937), and the Central Meridian $(\lambda_o) = 7°\ 26'\ 22.5''$ East of Greenwich. The radius of the Gaussian Sphere evaluated at the Grid origin for the Bessel 1841 ellipsoid is $R = 6,378,815.9036$ m. The Grid Scale Factor at Origin $(m_o) = 1.00072913843$, and the false origin is the same as previously listed for the old Swiss Bonne Grid of 600 and 200 km. Conformal doubles became the "*carte du jour*" projections of Europe during the early 20th century, and the cylinder, the cone and the plane were all used as developable surfaces. The "oblique" for this

Swiss system is really a misnomer; it's merely transverse at an oblique latitude. Laborde actually used the first oblique cylinder with a tilt at the origin, Rousilhe used the first oblique plane for the stereographic for most of his hydrographic surveys, and Krovak used the first oblique (tilted) cone for the Czech Republic. Rousilhe was the only system developed that became a widely used conformal double projection. The Rosenmund projection is truly unique in the world for a national grid. The combination of the Bern Observatory (horizontal) Datum of 1898 with the *Pierre du Niton* (vertical) Datum of 1902 or LN02, and the Rosenmund projection and Grid of 1903 have collectively been known since as the "*CH1903 System*" (Convention Helvetica 1903 System), of Switzerland.

The use of aerial photographs for map making commenced in 1930, although terrestrial photogrammetry was already in common use. By 1938, the first 1:50,000 series of the entire country was completed, and the office moved to a new building in Wabern, near Berne in 1941. The Swiss National Map series was completed at the 1:25,000 scale in 1979, and the first use of satellite receivers for national surveys was in 1979. In 1980 the office was renamed the *Bundesamt für Landestopographie*, and the first Swiss National map sheet was digitally updated in 1989.

In 1988, a new network of 104 GPS station observations began, and the resultant adjustment has become the new national (terrestrial) reference system of Switzerland and is called the *CHTRS95*. The local reference frame realized for the old CH1903 Datum is called LV95. The "CH1903+" was held fixed at a new fundamental point, Zimmerwald Z_o where $\varphi_o = 46° 52' 42.27031''$ N, $\lambda_o = 7° 27' 58.41774''$ East, and $X_o = 191,775.0616$ m, $Y_o = 602030.7698$ m, all still referenced to the Bessel 1841. The new point was chosen because the original location no longer exists, and the original coordinates of triangulation point *Gurten* were kept maintaining orientation. The deflection of the vertical is now defined at *Zimmerwald*: $\zeta_o = +2.64''$, $\eta_o = +2.73''$, and $H_o = 897.8408$ m. Transforming CH1903+ Datum to CHTRS95 Datum (WGS84 ellipsoid) then is accomplished by $\Delta X = +674.253$ m, $\Delta Y = +015.053$ m, $\Delta Z = +405.324$ m, according to the Swiss Federal Office of Topography.

The European Datum of 1950 was computed for Switzerland by the U.S. Army Map Service in the 1950s, and to transform from EU50 to WGS84, $\Delta X = -87$ m, $\Delta Y = -96$ m, $\Delta Z = -120$ m. To transform from EU79 to WGS84, $\Delta X = -86$ m, $\Delta Y = -98$ m, $\Delta Z = -119$ m. These parameters are according to NGA's TR 8350.2, January 3, 2000.

According to the Swiss Federal Office of Topography, the seven-parameter Datum shift from CH1903+ to WGS84 is: $\Delta X = +660.077$ m ± 4.055 m, $\Delta Y = +013.551$ m ± 4.816 m, $\Delta Z = +369.344$ m ± 3.914 m, $\alpha = 2.484$ cc ± 0.417 cc, $\beta = 1.783$ cc ± 0.455 cc, $\gamma = 2.939$ cc ± 0.411 cc, and $M = 1.000000566 \pm 0.000000052$. "Different applications of these transformation parameters, particularly in northern Switzerland, have shown that WGS-84 coordinates can be computed for all of Switzerland from the national coordinates with an accuracy better than 1 meter (1 sigma)."

Significant geodetic activities have been performed in Switzerland since 2001, and all is reported in National Reports to the International Union of Geodesy and Geophysics (IUGG). The accuracy of the known geoid in Switzerland is considered to be better than a few centimeters, largely due to an enormous number of observations for the deflection of the vertical (DOV). The European Reference Frame (EUREF) is supported, and the country is in full cooperation with all its neighbors.

- http://www.bernese.unibe.ch/publist/publist_2015.php
- https://iag.dgfi.tum.de/fileadmin/IAG-docs/NationalReports2007/Switzerland.pdf
- https://www.swisstopo.admin.ch/en/knowledge-facts/surveying-geodesy/reference-frames/transformations-position.html
- file:///Users/cjmce/Downloads/ch1903wgs84_e.pdf

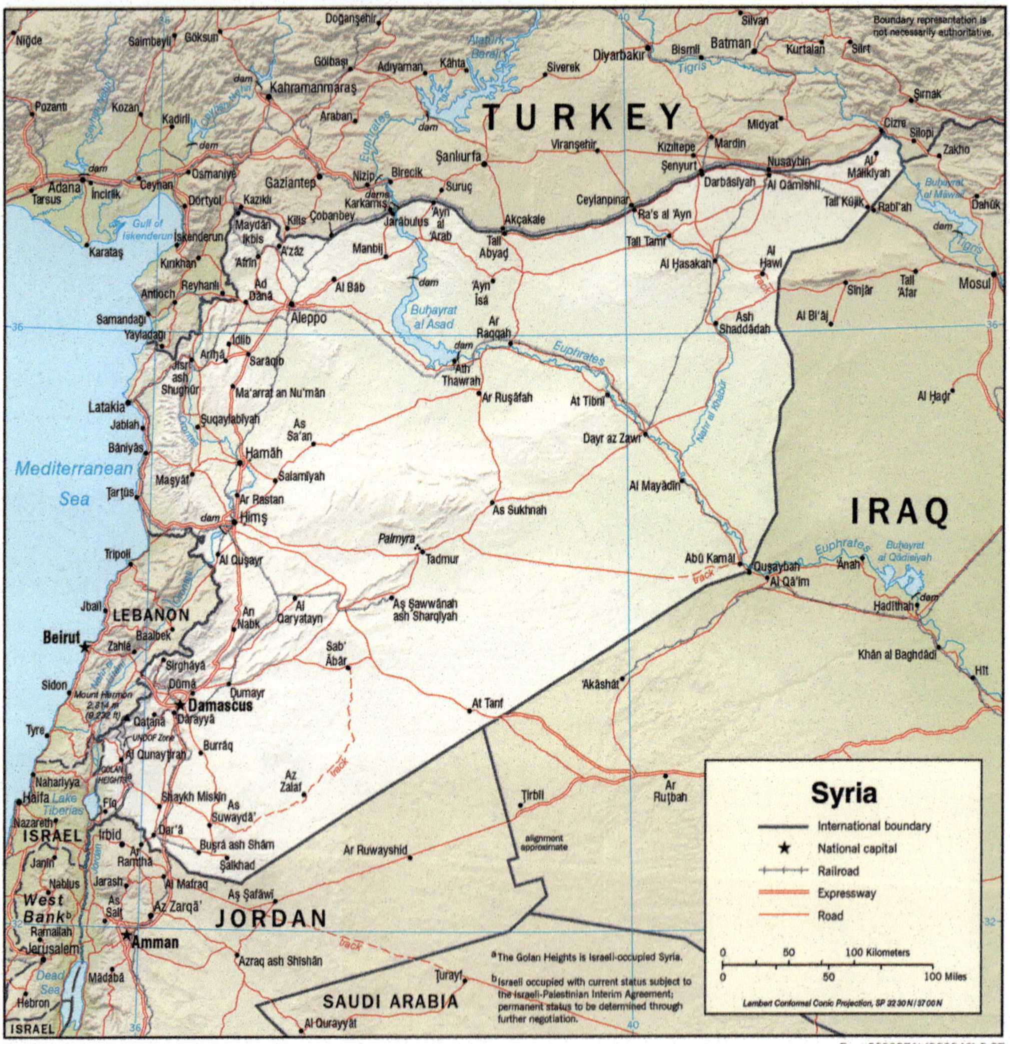

Base 803087AI (G00046) 5-07

The Syrian Arab Republic

Syria is bounded on the North by Turkey, on the East by Iraq, on the Southeast by Jordan on the Southwest by Israel and Jordan and the West by Lebanon and the Mediterranean Sea. The chief cities are Aleppo, Damascus, and Homs, all aligned more or less in a meridional arc on the western side of the country. The northeast region of the country has one city of geodetic significance: *Hassetché* (French) or *Al Hasakah*. Syria is slightly larger than North Dakota; its lowest point is near *Lake Tiberias* (−200 m), and its highest point is *Mount Hermon* (2,814 m).

Syria is an ancient country of Asia that once included modern Syria, Lebanon, Israel, and Jordan. There is evidence of human inhabitation for several thousand years from mid-3rd millennium B.C. that has been under the control variously of Sumerians, Akkadians, Amorites, Egyptians, Assyrians, and Babylonians. Once part of the Persian Empire, Alexander the Great conquered the region in the 4th century *B.C.*, and it later flourished under the Roman Empire. Syria was overrun by Muslim Arabs in 635–636 *A.D.* and was under the Ottoman Turks during the Crusades. During the first Crusade, the French established a presence in Antioch, and Godfrey of Bouillon decreed that a Barony be established there. After the Crusades, the Ottoman Turks resumed control for centuries until the British and French invaded during World War I because Turkey was an ally of Germany. Syria saw the end of Turkish rule in 1918. Part of the area (modern Lebanon and Syria) was mandated as a Protectorate by the French following World War I and the region was called the French *Levant* (Eastern Mediterranean). French troops finally left in 1946 when Syria became independent.

"The new country lacked political stability and experienced a series of military coups. Syria united with Egypt in February 1958 to form the United Arab Republic. In September 1961, the two entities separated, and the Syrian Arab Republic was reestablished. In the 1967 Arab-Israeli War, Syria lost the *Golan Heights* region to Israel. During the 1990s, Syria and Israel held occasional, albeit unsuccessful, peace talks over its return. In November 1970, Hafiz al-Asad, a member of the socialist Ba'ath Party and the minority Alawi sect, seized power in a bloodless coup and brought political stability to the country. Following the death of President Hafiz al-Asad, his son, Bashar al-Asad, was approved as president by popular referendum in July 2000. Syrian troops – stationed in Lebanon since 1976 in an ostensible peacekeeping role – were withdrawn in April 2005. During the July–August 2006 conflict between Israel and Hizballah, Syria placed its military forces on alert but did not intervene directly on behalf of its ally Hizballah. In May 2007, Bashar al-Asad's second term as president was approved by popular referendum.

"Influenced by major uprisings that began elsewhere in the region, and compounded by additional social and economic factors, antigovernment protests broke out first in the southern province of Dar'a in March 2011 with protesters calling for the repeal of the restrictive Emergency Law allowing arrests without charge, the legalization of political parties, and the removal of corrupt local officials. Demonstrations and violent unrest spread across Syria with the size and intensity of protests fluctuating. The government responded to unrest with a mix of concessions - including the repeal of the Emergency Law, new laws permitting new political parties, and liberalizing local and national elections - and with military force and detentions. The government's efforts to quell unrest and armed opposition activity led to extended clashes and eventually civil war between government forces, their allies, and oppositionists.

"International pressure on the Asad regime intensified after late 2011, as the Arab League, the EU, Turkey, and the US expanded economic sanctions against the regime and those entities that support it. In December 2012, the Syrian National Coalition, was recognized by more than 130 countries as the sole legitimate representative of the Syrian people. In September 2015, Russia launched a military intervention on behalf of the Asad regime, and domestic and foreign government-aligned forces recaptured swaths of territory from opposition forces, and eventually the country's second largest city, Aleppo, in December 2016, shifting the conflict in the regime's favor. The regime, with this foreign support, also recaptured opposition strongholds in the Damascus suburbs and the southern province of Dar'a in 2018. The government lacks territorial control over much of the northeastern part of the country, which is dominated by the predominantly Kurdish Syrian Democratic Forces (SDF), and a smaller area dominated by Turkey. The SDF expanded its territorial hold beyond its traditional homelands, subsuming much of the northeast since 2014 as it battled the Islamic State of Iraq and Syria. Since 2016, Turkey has been engaged in northern Syria and has conducted three large-scale military operations to capture territory along Syria's northern border in the provinces of Aleppo, *Ar Raqqah,* and *Al Hasakah*. Some opposition forces organized under the Turkish-backed Syrian National Army and Turkish forces have maintained control of northwestern Syria along the Turkish border with the Afrin area of Aleppo Province since 2018. In 2019, Turkey and its opposition allies occupied formerly SDF controlled territory between the cities of Tall Abyad to *Ra's Al 'Ayn* along Syria's northern border. The extremist organization Hay'at Tahrir al-Sham (formerly the Nusrah Front) in 2017 emerged as the predominate opposition force in Idlib Province, and still dominates an area also hosting additional Turkish forces. Negotiations between the government and opposition delegations at UN-sponsored Geneva conferences since 2014 and separately held discussions between Iran, Russia, and Turkey since early 2017 have failed to produce a resolution to the conflict. According to a September 2021 UN estimate, the death toll resulting from the past 10 years of civil war is more than 350,000, although the UN acknowledges that this is the minimum number of verifiable deaths and is an undercount. As of late 2021, approximately 6.7 million Syrians were internally displaced and approximately 14 million people were in need of humanitarian assistance across the country. An additional 5.7 million Syrians were registered refugees in Turkey, Jordan, Iraq, Egypt, and North Africa. The conflict in Syria remains one of the largest humanitarian crises worldwide" (*World Factbook* 2022).

In 1799, Napoleon Bonaparte commenced his military campaign for the conquest of Egypt and "Upper Egypt" (the Palestine and Greater Syria). *La Carte d'Egypte et de Syrie* was published by the *Dépôt de la Guerre* beginning in 1808. The five-sheet map series consisted of 47 *feuilles* (foils or manuscripts), one of which included *Acre, Nazareth, Jordan*, and the southern tip of present *Syria*. The ersatz Datum was based on astronomical observations in Cairo and Jerusalem and was referenced to the Plessis ellipsoid where the semi-major axis, $a = 6,375,738.7$ m and the reciprocal of flattening $1/f = 334.29$. Much of the coast was based on published British Admiralty charts of the time. (*See: The State of Israel.*) The projection was the ellipsoidal Bonne, the "standard" for France and most of Europe at the time. *"Le centre de la projection correspond à l'axe de la grande pyramide du Nord, à Memphis."* (The center of the projection corresponds to the axis of the great pyramid of the North at *Memphis*.)

In 1918, the *French Service Géographique de l'Armée (SGA)* was already compiling topographic features with planetable and alidade in Homs, Damascus, Sanameīn, and Haïfa. The geodetic network was calculated on the Clarke 1880 (IGN) ellipsoid, where $a = 6,378,249.2$ m and $1/f = 293.4660208$, and according to the *SGA*, the Datum Origin was at the South End of the Base of *Makri Keuī*, near Constantinople (Istanbul). The early maps of Turkey were on the Bonne projection, but the projection origin was the finial of the dome of the Aya Sofia Mosque. The Ottoman Turkish (ellipsoidal) Bonne of Syria (1909–1923) projection Latitude of Origin $(\varphi_o) = 28° 58' 50.8188''$ N, the Central Meridian $(\lambda_o) = 39° 36'$ East of Greenwich.

In Old Turkish, *"The Section and Arrangements Table was printed by the National Mapping Office* (of Turkey) *in the Year 1333 of the Moslem Calendar … the calculations were made because of the new lands that have been added. The shape and descriptions were based on Clarke 1880 measurements where the flattening was 1/293.46 and the half axis = 6,378,249.2 meters."* The angular unit of measurement was the grad, and the scale of the map was 1:200,000. False Northing and False Easting were equated to zero, per the standard practice of the time in Europe.

Old Turkish is written with the Arabic alphabet, while modern Turkish is written with the Roman alphabet.

Miranda H. Mugnier, personal communication (2022)

The French established the *Bureau Topographique du Levant* in 1918, and after 1920, the chain of triangulation was extended eastward along the northern border with Turkey to Iraq. Planimetric compilation was aided by aerial photography flown by a French military aviation squadron of the 39th Regiment. The Topographic Brigade was commanded by Lieutenant Colonel G. Perrier, and he organized the observations for the establishment of an astronomical origin for a Datum in the Bekaa Valley of Lebanon that would serve Syria as well. The baseline was measured, and the South End of the Base at Bekaa was the fundamental origin for the astronomical observations. The Latitude of the pillar was observed by Captain Volontat in 1920 with a prismatic astrolabe, where $\Phi_o = 33°\ 45'\ 34.1548''$ N. An azimuth was obtained at the same pillar with a microscopic theodolite by Captain Volontat, by observing Polaris at elongation. (*Good choice of method when your timekeeping is not radio-controlled. – Ed.*)

The direction was defined to a pillar constructed at the *Ksara Observatory*, where $\alpha_o = 28°\ 58'\ 50.8188''$. Longitude was also observed by the good Captain at the same observatory, where $\Lambda_o = 35°\ 53'\ 25.26''$ East of Greenwich. (The longitude was then geodetically transferred to the South End of the *Bekaa Base*.)

In November of 1997, Colonel George Massaad, then the *Director of Geographic Affairs of the Lebanese Army*, sent me a photograph of the fundamental point at *Bekaa South Base*. The point is monumented by a stone pyramid that is over 2 m high, is ~2 m^2, has an (apparently) bronze tablet describing the significance of the monument, is straddled by a great iron skeleton target obviously over 4 m high, and the entire structure is enclosed by a formal iron fence! The monument recalls the aviation accident that took the lives of Captains Govin and Renaud of the Geodetic Section of the *Service Géographique de l'Armée* on July 15, 1924, at *Muslimié*, near *Aleppo*. Shortly after World War II, the U.S. Army Map Service computed the coordinates of the origin of the Syrian geodetic system (*Bekaa*) on the New Egyptian Geodetic Datum as: $\varphi = 35°\ 45'\ 34.2205''$ N, $\lambda = 35°\ 54'\ 36.4962''$ E. The geodetic coordinates of station *Ksara* are: $\varphi = 33°\ 49'\ 25.58''$ N, $\lambda = 35°\ 53'\ 25.26''$ E. The *Bekaa Valley Datum of 1920* is referenced to the Clarke 1880 (IGN) ellipsoid as previously defined.

A check baseline was measured at *Bab*, and another astronomical position was observed (Laplace Station), where $\Phi_o = 36°\ 13'\ 48.77''$ N, $\Lambda_o = 37°\ 30'\ 30.195''$ East of Greenwich, and the reference azimuth from *Bab* to *Cheikh Akil* signal is where: $\alpha_o = 179°\ 58'\ 33.152''$.

The principal major classical triangulations of Syria are the *Aleppo Meridian Chain* and the *Coastal Series* (SGA) 1920–1929, the *Djerablous-Mennbidgi* Series (SGA) 1930, *Euphrates and Turkish Frontier Primary Series* (SGA) 1934–1937, *Hassetché* Series (SGA) 1932–1933, *Iraq Petroleum Company Frontier Series* (IPC/Syrian Petroleum Company), *Cadastral Central Series* (Bureau Cadastre) 1931, (IPC/SPC) 1934–1941, and the *Cadastral Abou Kemal Series*.

The *Aleppo* Chain was the backbone of the triangulation. The scale of the triangulation was governed by the two bases (*Bekaa* and *Bab*) which had an internal precision of one part in two million. The other primary work is connected by one or more connecting arcs. This series was based on the measurement of two bases (*Bekaa* and *Bab*) with invar wires with internal accuracy of about 1/25,000. But in the case of the initial azimuth of the *Bekaa Base*, a large number of observations were made in order to determine the mean azimuth. The maximum range of the observations was 48″ which does not represent good geodetic accuracy. Then a check azimuth, Latitude, and Longitude were measured at the Bab Base at Aleppo and the differences from the geodetic values carried through the chain from *Bekaa* are as follows (Astronomic−Geodetic): $\Delta\varphi = -6.318''$, $\Delta\lambda = +10.789''$, $\Delta\alpha = +21.125''$. Thus the *S.G.A.* decided not to apply a Laplace correction to the azimuths, assuming the 21″ was due to an error at the origin and not over the network. The 21″ error was verified by the U.S. Army Map Service by the computation of an azimuth between two stations in the area from utilizing the geographics of the station in terms of the *European Datum Mediterranean Loop* and *the Bekaa Valley Datum* geographics. The triangulation was computed on the Clarke 1880 ellipsoid, Levant Zone Grid, Lambert Conical Orthomorphic projection. The Levant Lambert Zone (1920) is based on the French Army Truncated Cubic formulae where the developed meridional arc is expressed in series form and is truncated at terms higher than the cubic. Furthermore, another idiosyncrasy of the French Army formulae is that the Lambert (fully) Conformal Conic utilizes one of the principal radii of the ellipsoid called the Radius of Curvature in the Plane of the Meridian (ρ). The French Army instead substitutes the Length of the Ellipsoid Normal Terminated by the Semi-Minor Axis (υ). Although not strictly conformal, this is the system that was commonly used by the French in all colonies (before World War II) that utilized the Lambert Conic projection. Failure to adhere strictly to the truncation (and above substitutions) will yield coordinates that although maybe strictly conformal, will differ from published (correct) coordinates in excess of ten meters! The Levant Lambert Zone, also known as the Syrian North Lambert Zone Latitude of Origin (φ_o) = 34° 39' N, the Central Meridian (λ_o) = 37° 21' East of Greenwich. The Scale Factor at Origin (m_o) = 0.9996256 (secant conic), and the False Easting and False Northing = 300 km.

The Syrian South Lambert Zone (1920), Latitude of Origin (φ_o) = 33° 18' N, the Central Meridian (λ_o) = 36° 00' East of Greenwich. The Scale Factor at Origin (m_o) = 0.999625769 (secant conic), the False Easting = 500 km, and the False Northing = 300 km.

In 1922, the *Travaux du Cadastre et d'Amélioration Agricole des États, de Syrie, des Alaouites et du Liban sous Mandat Français* established the Schéma de la Projection STEREOGRAPHIQUE which was base on the Rousilhe Oblique *Stéréographique* projection. The Latitude of Origin (φ_o) = 34° 12' N, the Central Meridian (λ_o) = 39° 09' East of Greenwich, the Scale Factor at Origin (m_o) = 0.9995341 (secant plane), and the False Easting and False Northing = zero. This Grid has caused some consternation in the literature because attempts to substitute the fully conformal formulae of Paul D. Thomas' "Conformal Projections" fail to yield correct transformation results. In fact, the Rousilhe formulae were developed by the Hydrographer of the French Navy in the late 19th century, and this is a common Grid used on many hydrographic surveys by the French well into the 20th century. Other variations of the Oblique Stereographic include the "Stereographic Double" projection used by Canada, and the Hristow "Stereo 70" used by Romania. Commercial software packages seem to universally ignore the different stereographic formulae (and others) that are legally defined in several countries around the world with the exception of my own academic stuff and that written by Mr. Roger Lott of British Petroleum, Inc. for the European Petroleum Studies Group (EPSG). For example, using a test point, φ = 37° 00' 00" N, λ = 42° 09' 00" E, X = 267,078.434 m, Y = 314,642.621 m, m = 1.0005837, γ = +1° 44' 49.8".

The only Syrian fundamental point is at *Hassetché*, where Φ_o = 36° 36' 23.100" N, Λ_o = 40° 46' 20.550" East of Greenwich, and the reference azimuth from to *Tell Amaa* is where: α_o = 172° 44' 59.4" from south, Clarke 1880 ellipsoid. All triangulation coordinates on this northeastern Syrian Datum have since been converted to the *Bekaa Valley Datum of 1920*. The *Hassetché* Series is the basic series for the triangulation in the eastern half of Syria. When the astronomical observations at the *Hassetché Datum* point (the South Terminal of the *Hassetché* Base) are compared to the geodetic value for this point based on the Bekaa Valley Datum, the differences (Astronomic – Geodetic) are as follows: $\Delta\varphi$ = –3.956", $\Delta\lambda$ = +10.508", $\Delta\alpha$ = +20.940". The *Hassetché* triangulation was computed on the Syrian Oblique *Stéréographique* Grid (1922).

Since 1989, the Syrians have used a new four-zone Grid based on the Gauss-Krüger Transverse Mercator projection. Being referenced to the European Datum of 1950 which uses the International 1924 ellipsoid, where a = 6,378,388 m and $1/f$ = 297; the Latitude of Origin by definition is the Equator, the False Easting of each zone is 200 km, the False Northing is zero at the Equator, the Scale Factor at Origin is 0.99974996, and the Central Meridians of the four zones are as follows: *Tretya* = 31° 30', *Sortya* = 34° 30', *Patya* = 37° 30', and *Sestya* = 40° 30'.

The latest available transformation parameters from the *Bekaa Valley Datum of 1920* in Syria to the WGS 84 Datum are: $\Delta X = -182.966$ m, $\Delta Y = -14.745$ m, $\Delta Z = -272.936$ m. The mean planimetric error for these parameters is 5 meters. Example test point: *Bekaa Datum Origin*: $\varphi = 33°\ 45'\ 34.1548''$ N, $\lambda = 35°\ 54'\ 37.1188''$ E, $H = 870.513$ m. WGS84 Datum coordinates of the same point: $\varphi = 33°\ 45'\ 33.8602''$ N, $\lambda = 35°\ 54'\ 40.6802''$ E, $h = 868.64$ m. Thanks go to Mr. John W. Hager for his assistance and to others.

REFERENCES: S

Jacob M. Wolkau, AMS memo 27 October 1947.
A. G. Bazley, Geographical Positions in Malaya and Siam, *Empire Survey Review*, No. 30.
Abdelrahim Elgizouli Mohamed Ahmed, Common Point Coordinates Transformation Parameters Between Adindan (Sudan) New Ellipsoid and the World Geodetic System 1984 (GPS Datum) Coordinates Compared with Parameters of the American National Imagery and Mapping Agency (NIMA) International Journal of Advanced Research in Engineering and Applied Sciences, ISSN: 2278–6252, vol. 2, no. 9, September 2014, pp. 26–39.
American Samoa Tourism, 2009.
André S. Mugnier, Personal Communication, 2013.
Andrew M. Glusic, Mapping of the Countries in Danubian and Adriatic Basins, Army Map Service Technical Report No. 25, June 1959.
Andrew W. Glusic, AMS Technical Report No. 31, Analysis of Surveys in the Russian Far East, 1960.
Annales Hydrographiques, 3ᵉ série, Tome vingt-et-unième, Année 1949, pp. 49–68.
Annales Hydrographiques, 4ᵉ Série, Tome quatorzième, Années 1967–1968.
Béla Kovács and Gábor Timár, The Austro-Hungarian Triangulations in the Balkan Peninsula (1853–1875).
Bin and Chai, Improving cadastral survey controls using GPS surveying in Singapore, *Survey Review*, vol. 33, 261, July 1996.
British General Staff, Geodetic Section, Survey Computations, London, H.M. Stationery Office Second Edition, 1932, Reprinted 1941, page 98.
buzon-geodesia@fomento.es, www.ign.es, January 2018.
Cambridge Encyclopedia, Vol. 65.
Capitaine de Corvette M. Paul Bonnin, Mission de Triangulation de L'A.O.F, October 1946 to August 1948.
Capitaine de Frégate M. Lebail, Mission Hydrographique de L'Afrique Occidentale, 1909–1910.
Chief Directorate: Surveys & Mapping, Department of Land Affairs, Mobray, Cape Town, South Africa, 2001.
Chief Directorate: Surveys and Land Information, 1995, Personal Communication.
CIA Factbook, 2005.
D. F. Musey, M.A., Base Measurement in the Anglo–Egyptian Sudan, *Empire Survey Review*, vol. X, no. 72, 1949, pp. 66–74.
D. Hugh Gillis, The Kingdom of Swaziland, 1999, ISBN:0-313-30670-2.
Dave Doyle, National Geodetic Survey, Personal Communication, July 2011.
Dave Doyle, NGS, Personal Communication, 2009.
David A. Philips, Crustal Motion Studies in the Southwest Pacific: Geodetic Measurements of Plate Convergence in Tonga, Vanuatu and the Solomon Islands.
Dr. Michael E. Pittman, Personal Communication, 2000.
Dr. Muneendra Kumar, Personal Communication, 25 September 2003.
Embassy of Saudi Arabia, 2008.
Encyclopaedia Brittanica, 2014.
EPSG database (EPSG: 6765).
Foreign Maps, TM 5-248, 1963.
G.T. McCaw, Empire Survey Review, Nos. 25–26, 1937.
Gábor Timár, The Ferro Prime Meridian, Geodézia És Kartográfia, pp. 3–7, 2007.
Gago Coutinho, (Carlos Viegas), Ilha De. S. Tomé Resultados Finais 1915–1918, Capitão de Mar e Guerra, Lisboa, 1920.
Geodesy, Cartography and the Cadastre of the Real Estates in the Slovak Republic, 2000.
Guinea – Sierra Leone Boundary, International Boundary Study No. 136 – July 2, 1973, U.S. Department of State.
http://www.soest.hawaii.edu/gg/academics/theses/Phillips_D_2003_Dissertation.pdf.

https://pdfs.semanticscholar.org/6bc4/c38aa6bdbac5a7359a6714d90eb9e9193ae8.pdf.
https://www.state.gov/wp-content/uploads/2019/10/LIS-144.pdf.
I. Aliksic, Z. Gospavic, Z. Popovic, Land Administration in Post Conflict Serbia, 2003.
InfoPlease 2011.
InfoPlease, 2010.
Infrastructure Asset Management Project – Phase II, Government of Samoa, 2004.
Instruction N°· 1312 S.G.C. de l'Institut Géographique National, 12 December 1945.
International Boundary Study No. 153, Ethiopia – Somalia, U.S. Dept. of State, November 5, 1975.
International Boundary Study, Mozambique-Swaziland Boundary, The Geographer, Bureau of Intelligence and Research, U.S. Department of State, No. 135, June 4, 1973.International Hydrographic Bureau Special Publication No. 24c, Geographic Positions, p. 52, Monaco, January 1951.
J. Anthony Cavell, PLS, CFed, Personal Communication, 1995.
J. B. Wekker, Personal Communication, February 2002.
J. Leatherdale, The Geodetic Survey of Saudi Arabia, Chartered Surveyor, December 1970.
J.E. Jackson, Notes on the Base Lines of the Ceylon Triangulation, Empire Survey Review, Vol. 3, 129, pp. 129–130.
John N. Wood, St. Lucia College in the Caribbean, 1998.
John W. Hager Personal Communication, 2009.
John W. Hager, Personal Communication 2012.
John W. Hager, Personal Communication, 2010.
John W. Hager, Personal Communication, 7 December 2011.
John W. Hager, Personal Communication, December 2013.
John W. Hager, Personal Communication, February 2002.
John W. Hager, Personal Communication, July 2008.
John W. Hager, Personal Communication, July 2011.
John W. Hager, Personal Communication, May 2014.
Kriegskarten und Bermessungsamt, Paris, August 1943.
Liberia – Sierra Leone Boundary, International Boundary Study No. 129 – October 13, 1972, U.S. Department of State.
Library of Congress Country Studies, 2009.
Lieutenant de Vaisseau Bonin, Annals Hydrographiques de 1937.
Lieutenant de Vaisseau de Première Classe Mulder, Annals Hydrographique, 1880.
Lieutenant de Vaisseau Tromeur, Mission Hydrographique du Saloum, 1930–1931.
Limits in the Seas No. 98- São Tomé and Príncipe, U.S. Dept. of State, November 1, 1983.
Lonely Planet, 2012.
Lonely Planet, 2013.
Lonely Planet, 2015.
M. Jean Bourgoin, Mission Hydrographique de la Côte Ouest d'Afrique, 1948.
M.O. Adam, A Geodetic Datum for the Sudan, Cornell University Thesis, June, 1967, 95 pages.
Manual of Photogrammetry, 5th and 6th editions.
Mark Nettles, Personal Communication, February 2002.
N. Tesla, N. Matic, V. Milenkovic, AGROS, U.N./Turkey/European Space Agency Workshop, 14–17 September 2010, Istanbul.
Note from Mr. Russell Fox of the Ordnance Survey, Southampton to Mr. Malcolm A. B. Jones of Perth, Australia of 7 December 1999.
Paula Santos and Ana Roque, Operations for Cartography – an Overview Over the First Portuguese Geodetic Mission in the Colonial Territories 1907–1910, FIG 29 May 2012.
Personal Communication, Russell Fox, retired Chief Librarian, Ordnance Survey, August 2007.
Polutea, et al., Improving Samoa's Geographic Information Services through the Upgrade of its National Geodetic Survey Network, Government of Samoa, 2004.
Primary Triangulation of Malaya, 1917.
Private Communication to Adam I. Alimi, (UNO Graduate Student) from Russell D. Fox, Overseas Surveys Directorate, Ordnance Survey, 13 April 1988.
Prof. Bohuslav Veverka, Personal Communication, February 2003.
R.C. Wakefield and D.F. Munsey, The Arc of the Thirtieth Meridian between the Egyptian Frontier and Latitude 13° 45′ 1935–1950, vols. 1 & 2, Sudan Survey Department, 1950.
Rikuti Sokuryobu, Geodetic Surveying in Japan during 1927–1929, IUGG 1930.
Roger Holeywell, Chief Cartographer, Marathon Oil Company, Personal Communication, 2000.

Russell D. Fox, Personal Communication, 7 July 2015.
Samoan Sensation, 2011.
Saudi Arabia Office of the Geographer, U.S. Department of State.
Sierra Leone Background Notes, U.S. Dept. of State, 2011.
Singapore Ministry of Law, 2006.
Slovenia U.S. Department of State Background Notes, 2011.
Somalia Library of Congress Country Studies, 2013.
South Africa U.S. Department of State, Country Studies, 2012.
Sri Lanka Ministry of Land, 2009.
St. Kitts & Nevis Background Note, Bureau of Western Hemisphere Affairs, U.S. Department of State, 2009.
Sudan Government: Annual Report of the Survey Department for the Year 1930, Empire Survey Review, pp. 139 & 140.
Sudan Library of Congress Country Study, 2015.
The Geodetic Horizontal Control Network, Sri Lanka Ministry of Land, 2009.
The Samoan Geodetic Network, Government of Samoa, December 2005.
Timár, G., Habsburg geodetic and cartographic activities in the Old Romania, Studii şi Cercetări, Seria Geologie-Geografie [Complexul Muzeal Bistriţa-Năsăud] 13: 93–102, 2008.
TR 8350.2, 1991.
TR 8350.2, 2014.
TR 8350.2.
Triangulaçã e levantamento da carta da Ilha do Príncipe; Longitudes em s. Tomé; Megnetismo, Lisboa, 1934.
United States Department of the Interior, Geological Survey, Horizontal Control Data.
V.U.K.A. 631, April 1943.
War Department Corps of Engineers, U.S. Lake Survey – New York Office Military Grid Unit, 1944.
Weightman, Bulletin Géodésique, No. 85, 1967.
William Parkhurst, Personal Communication, c.a. 1963.
World Factbook & NGA GeoNames Search, 2013.
World Factbook, 2009.
World Factbook, 2010.
World Factbook, 2011.
World Factbook, 2012.
World Factbook, 2013.
World Factbook, 2014.
World Factbook, 2015.
World Factbook, 2022.
World Factbook, NGA Geonames Server and Google Earth, 2011.
Yalta Conference (Franklin D. Roosevelt, Winston Churchill, and Joseph Stalin, August 1945.

T

Republic of China (Taiwan) 783
Republic of Tajikistan 787
United Republic of Tanzania 791
Kingdom of Thailand 795
Democratic Republic of Timor-Leste 799
Togolese Republic 803
Kingdom of Tonga 807
The Republic of Trinidad and Tobago 811
Tunisian Republic 815
Republic of Turkey 819
Turkmenistan 825
Tuvalu 827

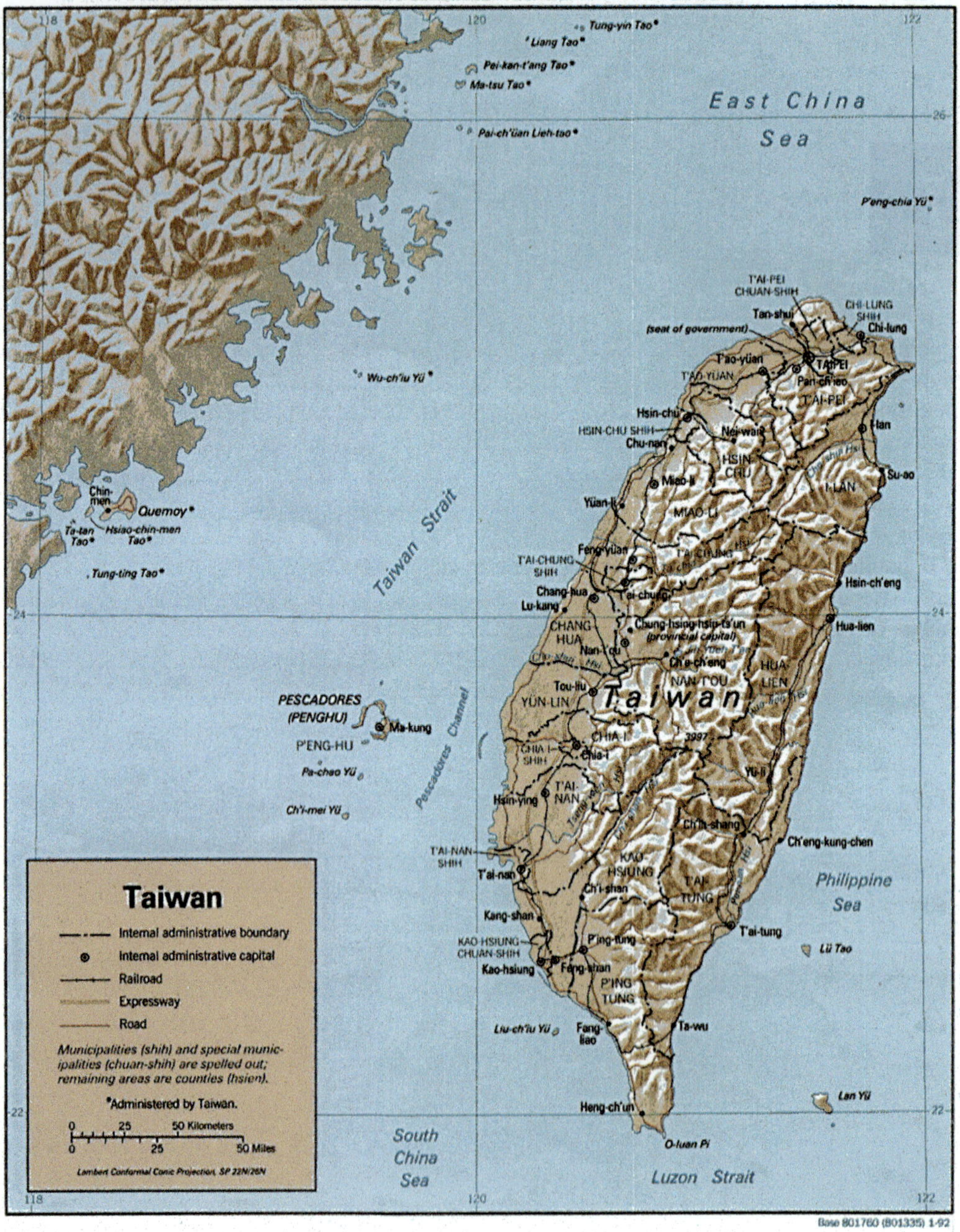

Base 801760 (801335) 1-92

Republic of China (Taiwan)

Taiwan is comprised of the main island, (formerly known as Formosa), the *Pescadores, Matsu,* and *Quemoy islands*. Slightly smaller than Maryland and Delaware combined, the terrain is mostly rugged mountains for the eastern two-thirds, flat to gently rolling plains in the west. The lowest point is the South China Sea (0 m); the highest point is *Yu Shan* (3,952 m) (*World Factbook*, 2010).

"There is evidence of human settlement in Taiwan dating as far back as 30,000–40,000 years ago; current prevalent thinking dates the arrival of the Austronesian peoples, ancestors of many of the tribal people who still inhabit Taiwan, between 4,000 and 5,000 years ago. For most of her long history, China seemed fairly indifferent to Taiwan. Early Chinese texts from as far back as *A.D.* 206 contain references to the island, but for the most part it was seen as a savage island, best left alone. Contact between China and Taiwan was erratic until the early 1400s, when boatloads of immigrants from China's Fujian province, disillusioned with the political instability in their homeland, began arriving on Taiwan's shores. When the new immigrants arrived, they encountered two groups of aboriginals: one, who made their homes on the fertile plains of central and southwestern Taiwan and the other, seminomadic, lived along the Central Mountain Range. Over the next century, immigration from Fujian increased, these settlers being joined by the Hakka, another ethnic group leaving the mainland in great numbers. By the early 1500s there were three categories of people on the island: Hakka, Fujianese and the aboriginal tribes. Today, Taiwan's population is mainly descended from these early Chinese immigrants, though centuries of intermarriage makes it likely a fair number of Taiwanese have some aboriginal blood as well" (*Lonely Planet*, 2010).

"During the 18th and 19th centuries, migration from *Fujian* and *Guangdong* provinces steadily increased, and Chinese supplanted indigenous peoples as the dominant population group. In 1895, a weakened Imperial China ceded Taiwan to Japan in the Treaty of Shimonoseki following the first Sino-Japanese war. During its 50 years (1895–1945) of rule, Japan expended considerable effort in developing Taiwan's economy. At the same time, Japanese rule led to the 'Japanization' of the island, including compulsory Japanese education and pressuring residents of Taiwan to adopt Japanese names. At the end of World War II in 1945, Taiwan reverted to Chinese rule. In October 1949 the People's Republic of China (P.R.C.) was founded on the mainland by the victorious communists. Chiang Kai-shek established a 'provisional' Republic of China (R.O.C.) capital in *Taipei* in December 1949" (*U.S. Department of State Background Notes,* 2010).

"Taiwan held legislative elections in 1992, the first in over forty years, and its first direct presidential election in 1996. In the 2000 presidential elections, Taiwan underwent its first peaceful transfer of power with the KMT loss to the DPP and afterwards experienced two additional democratic transfers of power in 2008 and 2016. Throughout this period, the island prospered, became one of East Asia's economic 'Tigers,' and after 2000 became a major investor in mainland China as cross-Strait ties matured. The dominant political issues continue to be economic reform and growth as well as management of sensitive relations between Taiwan and China" (*World Factbook*, 2022).

"Early records of survey work in Formosa date back to a Japanese Hydrographic Department survey of the west coast in 1897. Between 1900 and 1904 the Government General of Formosa under the supervision of the Japanese Government established approximately 2,000 stations. Japanese Imperial Land Survey initiated their work on a first-order net in Formosa in 1909. The datum for this and succeeding lower-order triangulation is *Koshizan 1906* and is defined at: $\Phi_o = 23° 58' 32.340''$ North, $\Lambda_o = 120° 58' 25.975''$ East of Greenwich, initial azimuth: $\alpha_o = 63° 46' 57.18''$ (south end of *Horisha base* to *Koshizan*), back azimuth: $\alpha = 243° 47' 21.611''$ (*Koshizan* to *south end of Horisha base*). Elevation = 555.34 meters. (*The reference ellipsoid was the Bessel 1841 where: $a = 6,377,397\ 155$ meters and the reciprocal of flattening $1/f = 299.1528128$ – Ed.*) Including the

2,000 stations established by the Government General of Formosa, the number of stations established and coordinated into Japanese Imperial Land Survey (*JILS – Ed.*) trig lists is 4,851. Japanese Hydrographic Department stations, which have been adjusted to Japanese Imperial Land Survey work, number 221. The grid system for Formosa is known as the Formosa Grid and is based on the Gauss-Schreiber projection: origin at 23° 40′ 00.000″ North, 120° 58′ 25.975″ East. (*The Gauss-Schreiber projection is a Transverse Mercator projected first to an equivalent sphere and then to the ellipsoid; the scale factor at origin is equal to unity, and there was no False Origin per standard Japanese practice of the time. – Ed.*) Control for this grid system is available for first, second and third order Government General of Formosa and Japanese Imperial Land survey stations. However, Station values of 4th order Government General of Formosa and Japanese Hydrographic Department surveys are in terms of geographical coordinates only. Level lines consisting of approximately 700 bench marks were established from 1904 through 1936 by the Japanese Imperial Land survey. The principal level line circles the island, with other lines joining this circuit. Datum point for the levels of Formosa is *Kirun tide gauge*" (*Army Map Service Geodetic Memorandum File 773.0303, 7 June 1950*). According to John W. Hager, "After 1921, the *JILS* Grid changed the Latitude of Origin for Formosa to $\varphi_o=24°$, and the Central Meridian for Formosa to $\lambda_o=121°$."

When the Republic of China (Taiwan) was established, the China Map Service was moved from the mainland to Taipei. In accordance with standard military practices of the U.S. Army Map Service (AMS), the Japanese Imperial Land Survey triangulation net was recomputed on the International 1924 ellipsoid, where $a=6,378,388$ m and $1/f=297$. The datum origin was retained but was translated to the Chinese *Huzi Shan (Tiger Cub Mountain)*. *The translation to English was provided by Professor Frank Tsai of Louisiana State University in 2010.* Various English spellings of the transliteration can be found in the literature, the one given here is the approved version on the NGA GeoNames server. When the International ellipsoid was adopted with the re-computation of the triangulation in 1950, the UTM Grid was also adopted for Taiwanese Military applications. The civilian cadastral applications continued to use the original Formosa Grid and origin as referenced on the Bessel 1841 ellipsoid for almost two more decades. During the Japanese occupation, a total of 3,873,468 cadastral parcels had been registered to the original Grid. Japanese geodetic control combined with cadastral control totaled 7,903 points. Note that the 1950 Taiwan re-computation on the International ellipsoid was accomplished with standard AMS techniques developed for use with mechanical calculators by performing a conformal 2D transformation on the complex plane; the identical way the European Datum of 1950 was computed. According to *TR 8350.2*, the three-parameter shift **from** "Hu-Tzu-Shan" (*sic*) **to** WGS84 is: $\Delta X=-637$ m ± 15 m, $\Delta Y=-549$ m ± 15 m and $\Delta Z=-203$ m ± 15 m, based on collocation at four points in 1991.

"A new conventional geodetic datum was computed based on geodetic observations carried out in 1978, and the new reference ellipsoid selected was the Geodetic Reference System 1967 ($a=6,378,160$ m, $1/f=298.25$ – *Ed.*). The origin point of this datum located at the *Hu-Tzu-Shan* (*Huzi Shan – Ed.*) astronomic station. The coordinates of this origin and the initial azimuth from this point to the reference point were both based on first-order astronomic observations. Four other basic assumptions for this datum were also made as follows:

1. The adopted center was the mass center of the earth.
2. Three axes were aligned parallel to the average terrestrial system.
3. Astronomic and geodetic coordinates for the origin point were kept the same.
4. The ellipsoid was assumed to be tangential to the geoid at the origin point.

Some defects can be clearly seen from the definition of this datum. The assumptions of (3) and (4) might not be valid as the separation between the ellipsoid and the geoid is significant. For the geodetic measurements made for this datum, the fundamental control networks extending from Hu-Tzu-Shan (*sic*) station to some 2,661 points of first-, second-, and third-order control stations were set up using triangulation and trilateration observations.

The coordinates, including latitude and longitude based on GRS67, grid coordinates based on the 2° – Zone Transverse Mercator (TM) projection, and height above the mean sea level for all geodetic control stations, were released by the Ministry of Interior in 1980. This set of coordinates was called the *Hu-Tzu-Shan (sic)* coordinate system, also named the TWD67 (Taiwan Geodetic Datum based on the GRS67)" (A Geocentric Reference System in Taiwan, Chang & Tseng, *Survey Review*, vol. 35, no. 273, July 1999, pp. 195–203). According to the National Land Surveying and Mapping Center the 2° – Zone Transverse Mercator (TM) projection has a Central Meridian of 121° East for *Taiwan Island, Ryukyu, Green Island,* and *Turtle Island*; 119° East for *Penghu, Kinmen,* and *Matsu*; 117° East for *Tungsha Islands*; and 115° East for *Nansha Islands*. The scale factor at origin is 0.9999, and the False Easting is 250 km for each belt. The specific projection type is defined to be the Gauß-Krüger Transverse Mercator. A shift published in an ESRI thread **from** TWD67 **to** WGS84 is as follows: $\Delta X = -764.558$ m, $\Delta Y = -359.515$ m, $\Delta Z = -180.510$ m, scale = 0.99998180, $R_X = +7.968''$, $R_Y = +3.5498''$, $R_Z = +0.4063''$. (No clue is offered regarding the sense of the rotations, nor is a test point given, nor is accuracy stated.)

The TWD67 was used for almost two decades. "Due to the occurrence of natural disasters, such as earthquakes and typhoons, and the many construction projects undertaken in the island during this period over 60% of the triangulation points, mostly below 500 meters high, were seriously damaged or even destroyed.

The final strategy was made by the working committee to establish a fundamental GPS network consisting of 8 permanent tracking stations, 105 first-order, and 621 second-order GPS control stations in the Taiwan area. The schedule was to set up 4 of the GPS control stations in 1993 and four others in 1994, to establish 105 first-order GPS control stations in 1995, and to implement 621 second-order GPS stations through 1996 to 1998, all by GPS" (*ibid., Chang & Tseng, 1999*). In 1995–1999, 1,967 third-order control points were established in Taiwan. A follow-up plan for Taiwan Province in 2000–2003 added an additional 2,500 third-order control points, and *Taiwan-Fukien* added 4,710 control points. The Taiwan Geodetic Datum of 1997 (TWD97) is referenced to the Geodetic Reference System 1980 ellipsoid, where $a = 6,378,137$ m and $1/f = 298.257222101$. TWD97 also uses a variety of Transverse Mercator belts in that the 2° system has Central Meridians at 121°E, 119°E, 117°E, and 115°E; the False Easting for each belt is 250 km, and the scale factor at origin, $m_o = 0.9999$. On the other hand, a 3° system has Central Meridians at 121°E and 118°E; the False Easting for each belt is 350 km, and the scale factor at origin, $m_o = 1.0$. The standard 6° UTM is also used with Central Meridians at 123°E and 117°E. The latitudinal extent of all of the Taiwan TWD97 Grids is from 21° 30′ N to 25° 30′ N.

Since the adoption of the TWD97, earthquakes have caused major terrestrial movements on the island; "the maximum dislocation of 9.8 meters is among the largest fault movements ever measured for modern earthquakes" (Establishment and maintenance of TWD97, Yang, Tseng, and Yu, *ASCE Journal of Surveying Engineering*, November 2001, pp. 119–132).

Attempts to develop a method for datum transformation between TWD67 and TWD97 have yielded some fascinating results. Using the traditional seven-parameter similarity transformation for Taiwan resulted in a transformation accuracy of ±0.5 m. On the other hand, using a 2D method on a conformal projection (like the U.S. Army Map Service did for Taiwan in 1950), yielded a transformation accuracy of ±0.16 m! (*Coordinate Transformation between Two Geodetic Datums of Taiwan by Least-Squares Collocation,* You & Hwang, *ASCE Journal of Surveying Engineering*, May 2006). Thanks to John W. Hager, other datums known to exist in Taiwan include for the *Pescadores (Penghu Qundao), Kobokutei (Japanese)* or *Hung-mu-ti (Chinese)*, where $\Phi_o = 23°~34'~19.069''$ N, $\Lambda_o = 119°~33'~45.695''$ E, Bessel 1841 ellipsoid until 1950, then International 1924 ellipsoid thereafter; *Mazutian* at South Point of Base, $\Phi_o = 26°~09'~20.95''$ N, $\Lambda_o = 119°~56'~57.45''$ E, $\alpha_o = 266°~45'~02.24''$ to Fu-wu (from north), International 1924 ellipsoid; *Quemoy (Jinmen Dao)* 1952, where $\Phi_o = 24°~25'~52.55'' \pm 0.56''$ N, $\Lambda_o = 118°~26'~20.69'' \pm 0.28''$ E, $\alpha_o = 61°~21'~38.89'' \pm 0.35''$ to East Point of Base (from north).

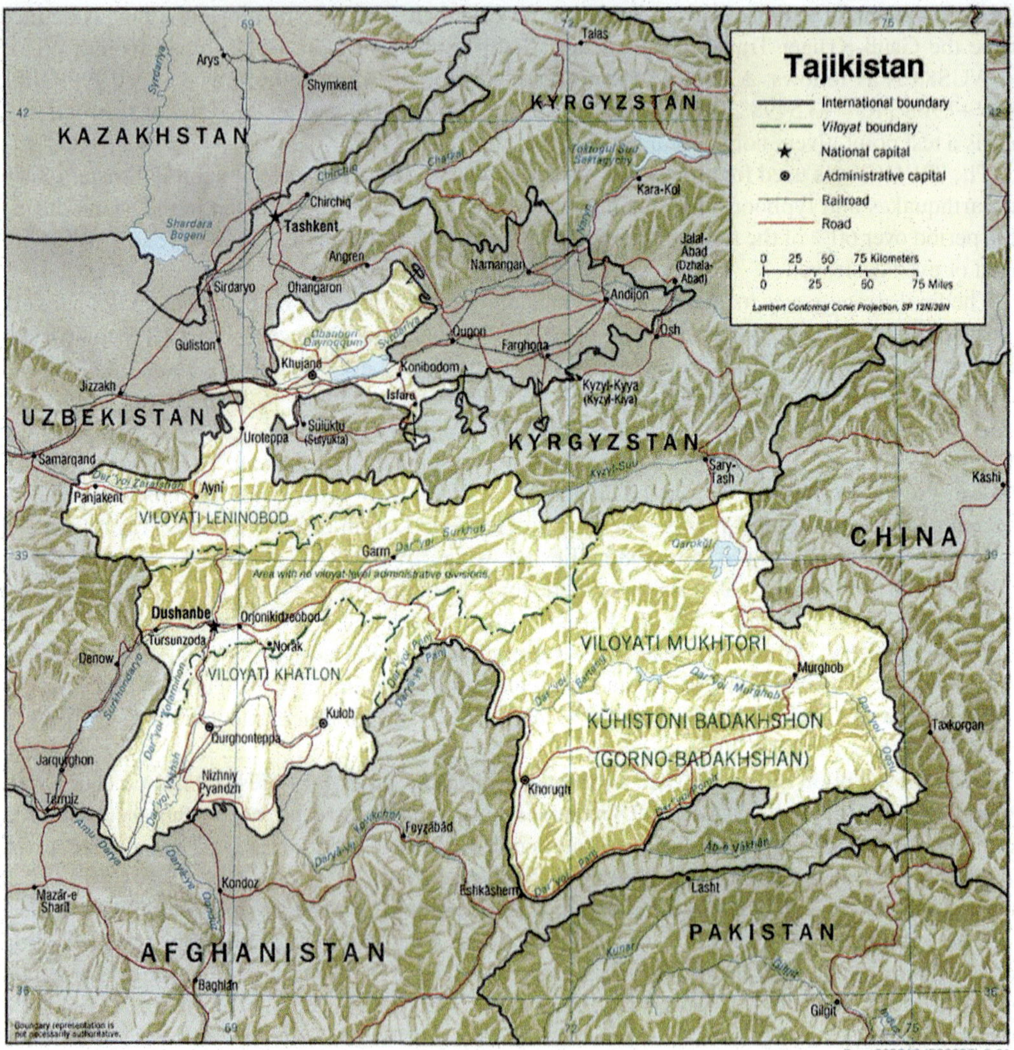

Base 802816 (R00627) 6-01

Republic of Tajikistan

Slightly smaller than Wisconsin, Tajikistan is bordered by Afghanistan (1,206 km), China (414 km), Kyrgyzstan (870 km), and Uzbekistan (1,161 km). The terrain consists of the Pamir and Alay Mountains that dominate the landscape; western Fergana Valley in the south; Kofarnihon and Vakhsh Valleys in the southwest. The lowest point is the *Syr Darya River* (*Sirdaryo*) at 300 m; the highest point is *Qullai Ismoili Somoni* at 7,495 m (*World Factbook*, 2014).

"The Tajiks, whose language is nearly identical with Persian, were part of the ancient Persian Empire that was ruled by Darius I and later conquered by Alexander the Great (333 *B.C.*). In the 7th and 8th centuries, Arabs conquered the region and brought Islam. The Tajiks were successively ruled by Uzbeks and then Afghans until claimed by Russia in the 1860s. In 1924, Tajikistan was consolidated into a newly formed Tajik Autonomous Soviet Socialist Republic, which was administratively part of the Uzbek SSR until the Tajik ASSR gained full-fledged republic status in 1929.

"Tajikistan declared its sovereignty in Aug. 1990. In 1991, the republic's Communist leadership supported the attempted coup against Soviet president Mikhail Gorbachev. Tajikistan joined with ten other former Soviet republics in the Commonwealth of Independent States on Dec. 21, 1991. A parliamentary republic was proclaimed, and presidential rule abolished in Nov. 1992. After independence, Tajikistan experienced sporadic conflict as the Communist-dominated government struggled to combat an insurgency by Islamic and democratic opposition forces. Despite continued international efforts to end the civil war, periodic fighting continued. About 60,000 people lost their lives in Tajikistan's civil war" (*Infoplease,* 2014).

"Though the country holds general elections for both the presidency (once every seven years) and legislature (once every five years), observers note an electoral system rife with irregularities and abuse, with results that are neither free nor fair. President Emomali Rahmon, who came to power in 1992 during the civil war and was first elected president in 1994, used an attack planned by a disaffected deputy defense minister in 2015 to ban the last major opposition political party in Tajikistan. In December 2015, Rahmon further strengthened his position by having himself declared 'Founder of Peace and National Unity, Leader of the Nation,' with limitless terms and lifelong immunity through constitutional amendments ratified in a referendum. The referendum also lowered the minimum age required to run for president from 35 to 30, which made Rahmon's first-born son Rustam Emomali, the mayor of the capital city of *Dushanbe*, eligible to run for president in 2020. In April 2020, Rahmon orchestrated Emomali's selection as Chairman of the Majlisi Milli (Tajikistan's senate), positioning Emomali as next in line of succession for the presidency. Rahmon opted to run in the presidential election in October 2020 and received 91% of the vote.

"The country remains the poorest in the former Soviet sphere. Tajikistan became a member of the WTO in March 2013. However, its economy continues to face major challenges, including dependence on remittances from Tajikistani migrant laborers working in Russia and Kazakhstan, pervasive corruption, and the opiate trade and other destabilizing violence emanating from neighboring Afghanistan. Tajikistan has endured several domestic security incidents since 2010, including armed conflict between government forces and local strongmen in the *Rasht Valley* and between government forces and criminal groups in *Gorno-Badakhshan Autonomous Oblast*. Tajikistan suffered its first ISIS-claimed attack in 2018, when assailants attacked a group of Western bicyclists with vehicles and knives, killing four" (*World Factbook*, 2022).

"In the early 19th century, the Russian penetration began into the area between the Caspian Sea and Iran, Afghanistan, and China. In 1839, the Russian Army moved into the area of Turkestan which was for centuries an object of struggle between Turko-Iranian and Chinese influences and completed its occupation in 1857. In 1867, the Military Topographic Department of Turkestan was formed and was headed by a major or brigadier general. The *Astronomic and Physical Observatory of Tashkent* was founded officially in 1878. The city of *Tashkent*, now the capital of Uzbekistan became the capital of *Turkestan*, and by 1888 the region of *Transcaspia*, now in southwest Kazakhstan was incorporated as a new province. The new government of Turkestan incorporated Khokanda, Tajikistan in 1875, and the *Fergana Valley*, Tajikistan in 1876. Triangulations from 1896 to 1929 were based on base lines measured with *Jäderin apparatus* (invar steel and brass wires) at *Kyzyl Ra*bat in southern Tajikistan and oriented primarily on *Osh Datum. Osh, II (NW) Base Point* determined in 1909 by Davidov is where: $\Phi_o = 40°\ 37'\ 16.67''\ N \pm 0.10''$, $\Lambda_o = 42°\ 36'\ 32.625''$ E of Pulkovo $\pm 0.092''$. Determined by a short passage of chronometer from Astro point *Osh, Old Church* with $\Lambda_o = 42°\ 28'\ 32.26''$ E of Pulkovo $\pm 0.45''$ determined in 1884 by telegraph, $\alpha_o = 152°\ 54'\ 01.86'' \pm 1.26''$ to I (SE) Base Point, and the ellipsoid of reference is the Bessel 1842', where $a = 6,377,397.155$ meters, and $1/f = 299.1528128$. The *Kyzyl Rabat base line* was at SW base point – NE base point measured in 1912 and was 8,395.683 m ± 0.00241 m, where $\Phi_o = 37°\ 26'\ 40.28''\ N \pm 0.42''$ N, Λ_o was not determined, and $\alpha_o = 75°\ 45'\ 22.72'' \pm 0.60''$ to NE Base Point (*Zapiski VTU, Vol. 69 Pt. 2 pl 235; Geodezist Vol. 11–12, p. 30*). Starting from the *Kyzyl Rabat base line*, there were eight triangles computed of the southeastern most section to the *Beik Pass*, where the main chain is tied to the triangulation of India.

"The main chain with the side *Beik (Sarblok)-Ak-turuk-tau (Kuhtek)* is on the *Beik Pas*s tied to the British Triangulation of India. The British chain of connection is composed of 7 geodetic quadrilaterals with diagonals, 2 central quadrilateral systems and 11 simple triangles – together 47 triangles which connect 33 stations. It is computed on the ellipsoid of Everest 1830 with the datum of Kalianpur. The *Indian Datum of 1916* origin at *Kalianpur Hill Station* is: $\Phi_o = 24°\ 07'\ 11.26''$ North, $\Lambda_o = 77°\ 39'\ 17.57''$ East of Greenwich. The defining azimuth at the point of origin to station *Surantal* is: $\alpha_o = 190°\ 27'\ 05.10''$. The ellipsoid of reference is the Everest 1830, where $a = 6,377,276.345$ meters, and $1/f = 300.8017$.

"In order to make the comparison, Russian coordinates of *Tashkent Datum* and Bessel ellipsoid for the tie points: *Beik (Sarblok)* $\varphi = 37°\ 18'\ 48.28''$ N, $\lambda = 44°\ 44'\ 56.81''$ E, $\alpha = 248°\ 12'\ 53.7''$ to *Ak-turuk-tau (Kuhtek)*, and *Ak-turuk-tau (Kuhtek)* $\varphi = 37°\ 17'\ 22.29''$ N, $\lambda = 44°\ 40'\ 27.80''$ E, $\alpha = 68°\ 10'\ 10.7''$ to Beik were converted by Helmert's formulae (Jordan-Eggert *Handbuch der Vermessungskunde* Vol. iii/1923 p. 737) to the ellipsoid of Everest (*Triangulation in Turkestan and its Connection with India*, Andrew M. Glusic, Army Map Service Technical Report No. 21, February 1957, 100 pages). (Note that the resultant disparities of Russian *Tashkent Datum* minus British *Kalianpur Datum for Beik (Sarblok)* were: $\Delta\varphi = -10.73''$, $\Delta\lambda = -5.22''$, $\Delta\alpha = -16.9''$, and for *Ak-tutuk-tau (Kuhtek)* were: $\Delta\varphi = -10.77''$, $\Delta\lambda = -5.29''$, $\Delta\alpha = -16.8''$ – *Ed.*).

Significant progress in geodetic surveying activities and cadastral mapping has been accomplished in the 21st century including high-resolution satellite imagery referenced to the ITRF2005 on the UTM Grid System (*Transfer of Technology for Cadastral Mapping in Tajikistan Using High Resolution Satellite Data*, Romuald Kaczynski, ISPRS Archives, Vol. XXXIX-B6, 2012 XXII ISPRS Congress, 25 August–1 September 2012, Melbourne, Australia).

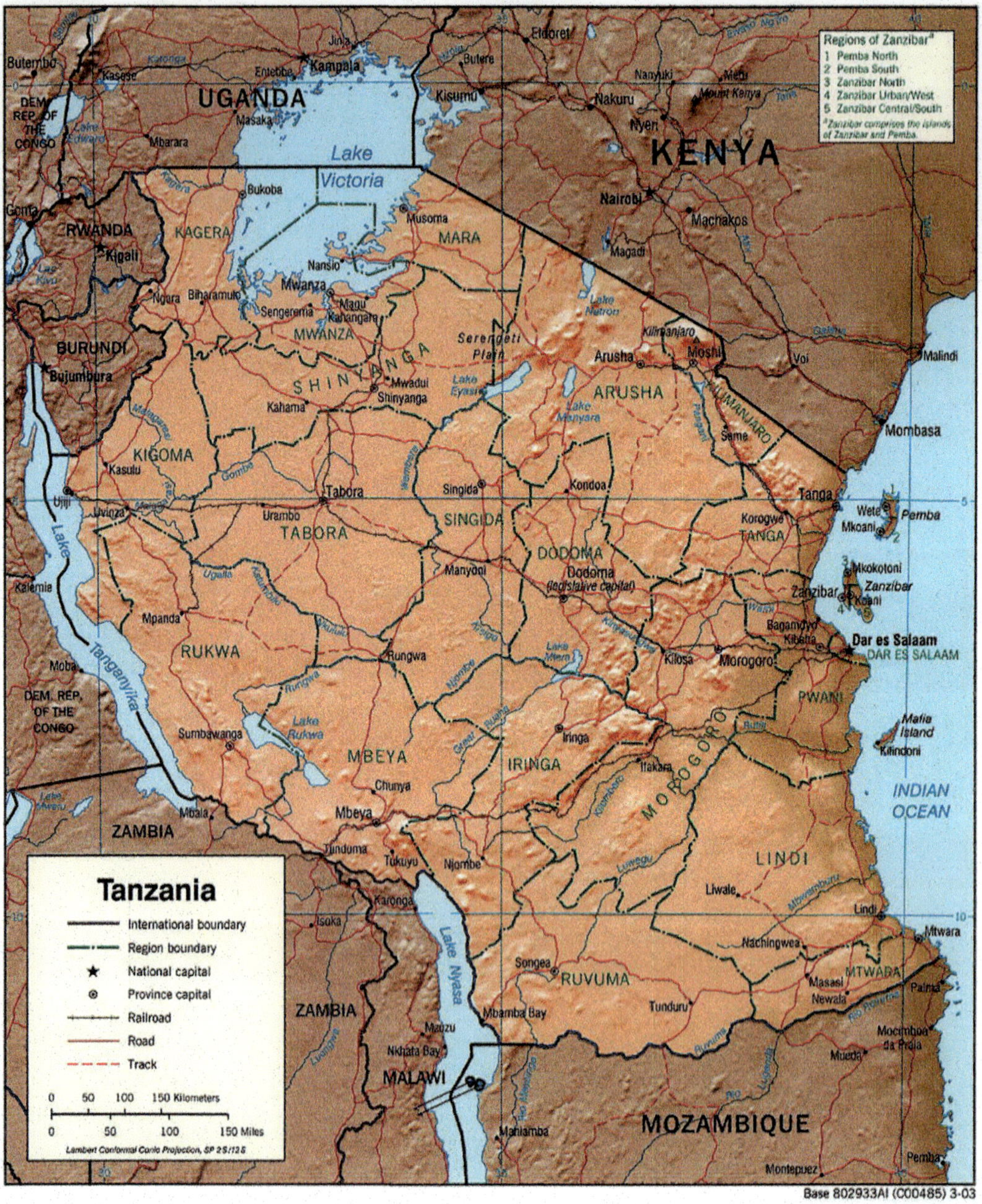

Base 802933AI (C00485) 3-03

United Republic of Tanzania

Tanzania is bordered by Burundi (451 km), Kenya (769 km), Malawi (475 km), Moçambique (756 km), Rwanda (217 km), Uganda (396 km), Congo (459 km), Zambia (338 km), and has a coastline of 1,424 km on the Indian Ocean. Slightly larger than twice the size of California, the terrain consists of plains along the coast with a central plateau and highlands in the north and south (*World Factbook*, 2022). The lowest point is the Indian Ocean (0 m), and the highest point is on *Mt. Kilimanjaro* recently determined by GPS observations at *Uhuru Peak* to be 5892 m, using the EGM96 geoid (*Professional Surveyor,* May 2000). The islands of *Zanzibar* and *Pemba* are significant agricultural sources of the clove spice. An acquaintance of mine from the University of New Orleans, Dr. Gil Richards, once told me that when the door of the airplane opens in *Zanzibar*, a flood of clove-laden air will fill the plane!

A Tanzanian gorge recently yielded a few bits of *Homo erectus*, but little is known about the country's really early history. Recorded history began around the 1st century *B.C.*, when various migrating tribes from West Africa first reached East Africa. While the country's coastal area had long witnessed maritime squabbles between Portuguese and Arabic traders, it wasn't until the middle of the 18th century that Arabic traders dared venture into the country's wild interior. European explorers began arriving in earnest in the mid-19th century, the most famous being Stanley and Livingstone. The famous phrase, "Dr Livingstone, I presume," stems from the duo's meeting at the town of *Ujiji* on *Lake Tanganyika*.

"During the late 19th century, Germany began colonizing *Tanganyika* – as the mainland was then known – by building railroads and commerce. The first railroad started construction in 1893 at Tanga on the coast and reached *Moshi* in 1911. A railroad from *Dar es Salaam* to *Morogoro* was begun in 1905 and was completed in 1907. However, the *tsetse fly* provided a significant obstacle to greater participation in establishing a large colony in German East Africa. After the Germans lost WWI, the League of Nations mandated the territory of *Tanganyika* to the British. The British had already conquered the offshore island of *Zanzibar*, which for centuries had been the domain of Arab traders" (*Lonely Planet*). *Tanganyika* became independent on December 9, 1961. The United Republic of *Tanganyika* and *Zanzibar* was constituted by a merger on April 27, 1964, and the name of the state was changed to the current United Republic of Tanzania on October 29, 1964.

"In 1964, a popular uprising overthrew the Sultan in Zanzibar and either killed or expelled many of the Arabs and Indians who had dominated the isles for more than 200 years. Later that year, Tanganyika and Zanzibar combined to form the United Republic of Tanzania, but Zanzibar retained considerable autonomy. Their two ruling parties combined to form the Chama Cha Mapinduzi (CCM) party in 1977. Nyerere handed over power to Ali Hassan Mwinyi in 1985 and remained CCM chair until 1990. Tanzania held its first multi-party elections in 1995, but CCM candidates have continued to dominate politics. Political opposition in Zanzibar has led to four contentious elections since 1995, in which the ruling party claimed victory despite international observers' claims of voting irregularities. In 2001, 35 people in Zanzibar died when soldiers fired on protestors following the 2000 election. John Magufuli won the 2015 presidential election and the CCM won a two-thirds majority in Parliament. He was reelected in 2020 and the CCM increased its majority in an election that was also critiqued by observers. Magufuli died in March 2021 while in office and was constitutionally succeeded by his vice president, Samia Suluhu Hassan" (*World Factbook*, 2022).

The international boundary between Tanganyika and the Belgian Mandated Territories of Ruanda and Urundi was demarcated by the Tanganyikan Survey Department in the years 1922–1924. The secondary triangulation in connection with this Boundary Commission was the first piece of triangulation work undertaken by the Department. Triangulation in the East and West *Usambara Mountains* was surveyed from 1894 to 1911 and was recomputed in 1931. The Anglo–German Boundary Commission surveyed Zanzibar from 1902 to 1906 and the First-Order survey was started based on the coordinates of *Vunta* (on the mainland), where $\varphi = -4°\ 28'\ 51.346''$ S, $\lambda = 37°\ 57'\ 37.954''$ E, and was referenced to the Clarke 1880 ellipsoid, where $a = 6,378,249.145$ m and $1/f = 293.465$. A Second-Order triangulation was performed in *East Usambara* in 1893, south of *Mombasa*, and west of *Tanga*.

Whitehouse performed some Third-Order triangulation along the southern shore of *Lake Victoria* in 1906–1907, which was based on the earlier Anglo–German Boundary Commission survey. Third-Order triangulation in 1898 was done by the *Anglo–German Nyasa Boundary Commission* and was based on a local Astro determination of latitude but used a value of longitude based on the Cape triangulation. The German–Portuguese Boundary Commission performed a survey in 1907 East of *Lake Nyasa* along the border with Moçambique (*Mitt. aus den Deutschen Schutzgebieten, Erganzungsheft 23, Berlin, 1910, pp. 49–56*). From 1922 to 1924, the Anglo–Belgian Boundary Commission for Ruanda–Urundi and Tanganyika performed a Second-Order triangulation that was part of the 30° Meridian Arc. This triangulation was carried across *Lake Tanganyika* and connected to a secondary chain originating from the Rhodesian Arc (now Zimbabwe). The *Morogoro Net* was surveyed from 1912 to 1914 and was a First-Order triangulation as was the *Pare-Moshi-Arusha Net* surveyed from 1913 to 1914. Both were based on original German field observations that were recomputed in 1931. (Originally published by Dr. von G. Pinkwart of Hanover in *Allgemeine Vermessungs-Nachrichten, Nos. 11 et sequitur,* 1926.) The connection with the Rhodesian and Uganda Arcs were performed from 1932 to 1933 and were First-Order triangulations organized by the British War Office under the supervision of Major Martin Hotine, Royal Engineers. Note that Hotine later would be promoted to Brigadier, and he would eventually direct the Directorate of Overseas Surveys (DOS) for its entire existence! Triangulation baselines measured with invar tapes included *Ardai* (1932) @ $37,478.1519$ ft ± 0.0324 ft; *Dodoma* (1933) @ $31,784.861$ ft ± 0.00865 ft; *Ulete* (1933) @ $20,684.480$ ft ± 0.04482 ft; *Seke* (1936) @ $37,948/116$ ft ± 0.03684 ft, etc. – the miniscule errors between forward and back chainings are mind-boggling!

"The Struve Arc is a chain of survey triangulations stretching from *Hammerfest* in Norway to the *Black Sea*, through 10 countries and over 2,820 km. These are points of a survey, carried out between 1816 and 1855 by the astronomer Friedrich Georg Wilhelm Struve, which represented the first accurate measuring of a long segment of a meridian. This helped to establish the exact size and shape of the planet and marked an important step in the development of earth sciences and topographic mapping. The original arc consisted of 258 main triangles with 265 main station points. Otto Stuve, son of F. G. W. Stuve, was keen to carry forward the ideas of his father. Sir David Gill took over as Her Majesty's Astronomer at the *Cape of Good Hope* in the late 1870s and had the dream of connecting with Struve's Arc. In 1879, Gil started the triangulation northward, and by 1892 all of its length in South Africa was complete together with two of the four baselines. 1897–1901 saw much of *Southern Rhodesia* (Zimbabwe) completed by Simms except for a small piece around 21°–22° south which was filled in later. 1903–1906 saw Rubin complete most of *Northern Rhodesia* (Zambia), and in 1908–1909 Jack and McCaw (*Hotine's Mentor – Ed.*) did a section in southern Uganda. Thus between 1879 and 1909 the great majority of the triangulation from South Africa to the Equator was completed. The gap was from 10°S northward to 1°S. This was partly completed by Hotine in 1931–1933 and the rest by the Tanganyika Survey Department in 1936–1938. Looking now from the Mediterranean Sea southwards Egypt was covered between 1907 and 1930 and much of the Sudan at varying periods between 1901 and 1951. This left one last section of around 600 miles in southern *Sudan* and *Northern Uganda* which was the most difficult of all because of the terrain in the *Sudd*. This was finally filled in with the assistance of the U.S. Army Map Service during 1952–1954" (*The Connection between the Struve Geodetic Arc and the Arc of the 30th Meridian,* Jr. R. Smith, FIG, September 2005). That historic connection was surveyed by William Parkhurst, my mentor who got me my first job as a Junior Map Draftsman at Offshore Navigation,

Inc. in 1963 in New Orleans. I recall later sitting in Bill's living room in Metairie, Louisiana and looking at the snapshots of him seated in a saddle on a camel during his surveys in the Sudan. (My late mother was a close friend of Bill's wife, Hazel – small world …).

A variety of astronomical observations have been performed in Tanzania, and the data has been graciously provided by John W. Hager: "for *Pemba Island* (1957) at Observation Spot+*Mpigangoma*, $\Phi_o = 5°\ 02'\ 41.656''$ S, $\Lambda_o = 39°\ 39'\ 02.465''$ E. Reference, East Africa - *West Coast of Pemba Island*, H.M.S. Owen, 1957 and H.M.S. Owen 1958, H.12/1299, 15 August 1958. For *Ras Shangani* Observation Spot β, $\Phi_o = 6°\ 09'\ 45.82''$ S, $\Lambda_o = 39°\ 11'\ 04.72''$ E., $\alpha_o = 358°\ 54'\ 39.2''$ to *Mangapwani Lighthouse* (G), Clarke 1880. H.M.S. 'Rambler' 1900. I have a pencil note on the ellipsoid "from 1951 or earlier? The British Admiralty usually used Clarke 1858, so I wonder (now) if I suspected a change in ellipsoid. Weti (code WET), at South End of *Weti Base*, $\Phi_o = 5°\ 03'\ 48.31''$ S, $\Lambda_o = 39°\ 43'\ 00.00''$ E, $\alpha_o = 71°\ 13'\ 50.35''$. I'll bet that the astro longitude wasn't very good or they were very lucky. *New Pillar Zanzibar* (code NPI) at Observation Spot New Pillar. $\Phi_o = 6°\ 09'\ 45.82''$ S, $\Lambda_o = 39°\ 11'\ 04.72'' \pm 07.5''$ E, $\alpha_o = 358°\ 54'\ 31.4''$ to Northern Lighthouse, Clarke 1858; possibly by Sir David Gill, 1905. The longitude was transferred to *Igurua* in Uganda. My original list had a *New Pillar datum* and a *Zanzibar datum*. I am now combining the two to have code NPI as *New Pillar Zanzibar*."

The grid systems known to exist in Tanzania include: the Tanganyika Territorial Transverse Mercator Belts (F-J) where the Central Meridians are from 27° 30′ to 42° 30′, the Latitude of Origin is the Equator, the Scale Factor at Origin is 0.9995, the False Easting is 133,333.333 m, and the False Northing is 3,333,333.333 m; the East Africa Transverse Mercator Belts G and H are where the Central Meridians are 32°30′ and 37°30′, same False Origins as the Territorial Belts, and the South African Belts have Central Meridians of 35° to 41° at 2° intervals with a Scale Factor at Origin of Unity and have False Eastings of 250 km and False Northings of 500 km. Current papers describing coordinate systems used in Tanzania refer exclusively to the standard UTM Grid with no reference to the older TMs used by the British.

TM 8350.2 offers transformation parameters **from** the Arc 1960 Datum as realized in Tanzania **to** WGS84 as: $\Delta X = -175$ m ± 6 m, $\Delta Y = -23$ m ± 9 m, and $\Delta Z = -303$ m ± 10 m. Thanks again to John W. Hager for the elusive Astro positions of Tanzanian Datum Origins and to the late Bill & Hazel Parkhurst for the memories.

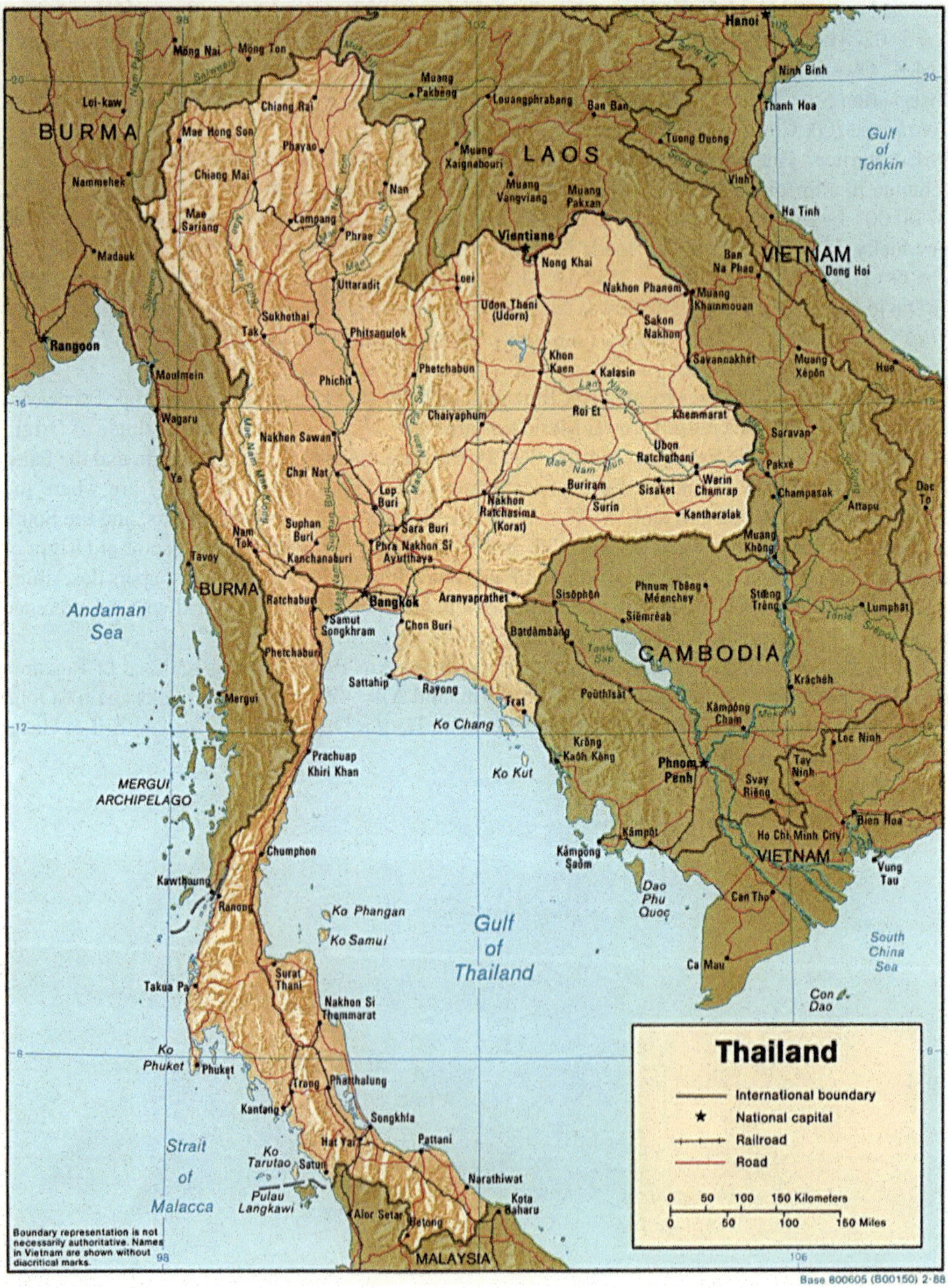

Base 800605 (B00150) 2-88

Kingdom of Thailand

Bordered by Burma (1,800 km), Cambodia (803 km), and Malaysia (506 km), Thailand is slightly more than twice the size of Wyoming. The lowest spot is the Gulf of Thailand (0 m), and the highest spot is *Doi Inthanon* (2,576 m) in the province of *Chiang Mai*. Thailand's coastline is 3,219 km long, and the territorial sea claimed is 12 nautical miles.

"Over the course of millennia, migrations from southern China peopled Southeast Asia, including the area of contemporary Thailand. Archaeological evidence indicates a thriving Paleolithic culture in the region and continuous human habitation for at least 20,000 years. The pace of economic and social development was uneven and conditioned by climate and geography. The dense forests of the *Chao Phraya Valley* in the central part of Thailand and the Malay Peninsula in the south produced such an abundance of food that for a long time there was no need to move beyond a hunting-and-gathering economy. In contrast, rice cultivation appeared early in the highlands of the far north and hastened the development of a more communal social and political organization. Excavations at *Ban Chiang*, a small village on the *Khorat Plateau* in northeastern Thailand, have revealed evidence of prehistoric inhabitants who may have forged bronze implements as early as 3000 B.C. and cultivated rice around the fourth millennium B.C. If so, the *Khorat Plateau* would be the oldest rice-producing area in Asia because the inhabitants of China at that time still consumed millet. Archaeologists have assembled evidence that the bronze implements found at the Thai sites were forged in the area and not transported from elsewhere. They supported this claim by pointing out that both copper and tin deposits (components of bronze) are found in close proximity to the *Ban Chiang* sites. If these claims are correct, Thai bronze forgers would have predated the 'Bronze Age,' which archaeologists had traditionally believed began in the Middle East around 2800 B.C. and in China about a thousand years later. Before the end of the first millennium B.C., tribal territories had begun to coalesce into protohistorical kingdoms whose names survive in Chinese dynastic annals of the period. *Funan*, a state of substantial proportions, emerged in the second century B.C. as the earliest and most significant power in Southeast Asia. Its Hindu ruling class controlled all of present-day Cambodia and extended its power to the center of modern Thailand. The Funan economy was based on maritime trade and a well-developed agricultural system; Funan maintained close commercial contact with India and served as a base for the Brahman merchant-missionaries who brought Hindu culture to Southeast Asia. On the narrow isthmus to the southwest of Funan, Malay city states controlled the portage routes that were traversed by traders and travelers journeying between India and Indochina. By the tenth century A.D. the strongest of them, *Tambralinga* (present-day *Nakhon Si Thammarat*), had gained control of all routes across the isthmus. Along with other city-states on the Malay Peninsula and Sumatra, it had become part of the Srivijaya Empire, a maritime confederation that between the seventh and thirteenth centuries dominated trade on the South China Sea and exacted tolls from all traffic through the *Strait of Malacca*. *Tambralinga* adopted Buddhism, but farther south many of the Malay city-states converted to Islam, and by the fifteenth century an enduring religious boundary had been established on the isthmus between Buddhist mainland Southeast Asia and Muslim Malaya. Although the Thai conquered the states of the isthmus in the thirteenth century and continued to control them in the modern period, the Malay of the peninsula were never culturally absorbed into the mainstream of Thai society. The differences in religion, language, and ethnic origin caused strains in social and political relations between the central government and the southern provinces into the late twentieth century" (*Library of Congress Country Study,* 2009).

A high spot in *Chiang Mai* was chosen by the U.S. Army Map Service (AMS) in the late 1960s as a site for a SECOR (Sequential Collation of Ranges) satellite receiving station in order to establish a precise geodetic position for a SHORAN (radiolocation) tower and beacon to provide positioning control to B-52 bombers for the bombing of North Vietnam. While I was the Company Commander of the enlisted detachment at AMS, I once had a soldier transported to me in chains and leg irons from the *Chiang Mai* SECOR site for a disciplinary action!

"The geodetic control network and basic topographic mapping in Thailand is the responsibility of the Royal Thai Survey Department (RTSD), which traces its origin to a small unit established in 1873 by King Rama V. (*Probably the origin of Rājburi Datum of 1878 – Ed.*) The first order control network consists (*as of 1985 – Ed.*) of 362 triangulation stations together with 200 first order traverse stations. The network includes 66 Laplace astronomic stations, 60 Doppler satellite stations and a large number of gravity stations. The geodetic coordinates are based on a 1975 adjustment. The RTSD has also surveyed a large number of second and third order stations as control for topographic mapping.

"The geodetic control was established on the pattern of the Survey of India geodetic network. It was made up of geodetic chains which formed a grid, with areas inside the grid which were not covered. Generally, the chains ran along the mountain ridges. This was unfortunate as it meant that geodetic control was generally not available on the flat plains, where the centers of population and agricultural activity were located, and where the major activity for the Land Titling project would occur. For mapping the RTSD uses the Universal Transverse Mercator projection which covers Thailand with two zones. The country is covered by basic topographic mapping at 1:50,000, with 830 maps on a $15' \times 15'$ format.

"Since 1903, the Department of Lands has been carrying out cadastral surveying and mapping for land titles. At that date, there was no overall geodetic network in existence, so a set of local systems was established, similar to the system in New Zealand. These systems are still used for cadastral mapping (*as of 1985 – Ed.*). There are 29 local systems, each with its local origin and each covering one of more of the 72 provinces. Initially, the origins chosen were prominent local points but in systems established later, an intersection of geographical graticule lines was selected for the origins. The systems are based nominally on Spherical Rectangular (Cassini) coordinates.

"The control for cadastral surveying was extended from each origin by traverse loops measured with theodolite and steel tape. Control stations consisting of buried numbered concrete blocks are placed about every 500 m. Azimuth control is determined by solar observations about every 10 km. Loop was added onto loop so that, even though the traversing was to a good standard, serious errors accumulated. The theoretical formulae are available to transform the local spherical rectangular coordinates to UTM on the national system. But the resulting positions are not reliable because of discrepancies in the positions and, particularly, the accumulation of traverse error. Position discrepancies of up to 80 m are common" (A Project for Upgrading the Cadastral System in Thailand, P.V. Angus-Leppan & I.P. Williamson, *Survey Review*, vol. 28, 215, January 1985, pp. 2–14 & April 1985, pp. 63–73). Records of the origin co-ordinates of those Cassini systems are no longer available (*Prof. Ian Williamson, University of Melbourne – personal communication, November 2010*).

Thanks to John W. Hager, "INDIAN 1916. At the commencement of operation of Survey of India in about 1802, *Madras* observatory, being the only institution equipped with precision instruments, was elected as the origin to which coordinates of all of the trigonometrical stations were to be referred: $\Lambda_o = 80°\ 18'\ 30''E$ (1805 – Lambton's value, used for the Atlas Sheets), and $\Lambda_o = 80°\ 17'\ 21''E$ (1815 – Warren's value, used for Standard Sheets and all other mapping) (*Gulatee, B. L., Deviation of the Vertical in India, Survey of India,* 1955), and (*Markham, Clements R., A Memoir on the Indian Surveys, London,* 1871). In 1840 Everest chose *Kalianpur Hill Station* as origin. The geodetic triangulation of India was first adjusted to form a self-consistent whole in about 1880 and that of Burma in 1916. These are also known as the "Published Values" (*Army Map Service Geodetic Memorandum No. 1600, Pakistan-India-North Burma Conversion of the Indian Datum "Published Values" to South Asia Datum* by Isaac C. Lail, Washington, 10 November 1966). Everest derived the weighted mean to be: $\Phi_o = 24°\ 07'\ 11.26''N$. This value was accepted as the geodetic latitude of the origin for the calculation of the primary triangulation of India in 1878. The longitude,

$\Lambda_o = 77°\ 41'\ 44.75''$, was determined by Everest in 1840 based on Warren's determination of the longitude of Madras Observatory in 1815. Operations undertaken in 1894-95-96 determined a correction of $-2'\ 33.94''$ in the longitude of *Madras*. To this is added a correction of $-6.76''$ to the value of *Kalianpu*r as determined in 1889. Except for this constant change of $-2'\ 27.18''$ in longitude, the 1900 definition of the datum point is identical with that of 1880 (*Account of the Operations of the Great Trigonometrical Survey of India; Volume XVII; ElectroTelegraphic Longitude Operations Executed During the Years 1894-95-96; The Indo–European Arcs from Karachi to Greenwich, Dehra Dun*, 1901). The origin was *Kalianpur Hill Station* as defined in 1900, where $\Phi_o = 24°\ 07'\ 11.26''$N, $\Lambda_o = 77°\ 39'\ 17.57''$E, $\alpha_o = 190°\ 27'\ 05.10''$ to *Surantal* from south, $H_o = 1,765$ ft, Everest 1830 ellipsoid.

"INDIAN 1937. Brigadier (then Major) Guy Bomford re-adjusted the geodetic triangulation adding in approximately 17 important new series, seven new base lines, recomputed values for the original 10 base lines, and including about 43 Laplace stations. The origin remains the same. As far as is known, no products have been produced on this system (Bomford, Major G., *Survey of India Professional Paper No. 28; The Readjustment of the Indian Triangulation*, Dehra Dun, 1939).

"INDIAN 1954 (Thailand). In 1954, the triangulation was adjusted by the U. S. Coast and Geodetic Survey (USC&GS) for the Army Map Service. Positions were computed in terms of the Indian 1916 adjustment based on ten stations along the Burma border. The defining parameters of the Indian 1916 Datum were retained.

"INDIAN 1960. The triangulation of Cambodia and Vietnam was adjusted to Indian Datum in 1960 holding fixed two Cambodian stations connected to the Thailand triangulation adjusted by the USC&GS in 1954 (INDIAN 1954). In turn, the primary triangulation of Laos was adjusted holding fixed four stations from the Cambodia Vietnam adjustment. North Vietnam was also adjusted to this system but with lower standards.

"INDIAN 1975. This is an adjustment of the primary network of South Burma and Thailand available to DMATC in 1975. Nine Doppler satellite positions were incorporated into this adjustment. Also included were 10 Geodimeter lengths, 4 Invar lengths, and 22 Laplace azimuths (*Geodetic Memorandum No. 1692, 1975 Adjustment of the Primary Triangulation of Thailand* by J. W. Walker, Washington, June 1976). Datum Origin: *Khao Sakaerang* where: $\Phi_o = 15°\ 22'\ 56.0487''$N, $\Lambda_o = 100°\ 00'\ 59.1906''$E, Geoid Height $= -20.46$ meters, Everest 1830 ellipsoid. This is Doppler Station 10084." Note that for the Everest 1830 ellipsoid, $a = 6,377,276.345$ m, $1/f = 300.8017$.

An earlier local datum origin west of *Bangkok* was "*Khao Luang (1878–1881)* at *Khao Luang Hill* where $\Phi_o = 13°\ 43'\ 30.34''$N, $\Lambda_o = 99°\ 32'\ 22.94''$E, $\alpha_o = 270°\ 15'\ 21.4''$ from *Khao Luang* to *Khao Khieo*, or $\alpha_o = 179°\ 44'\ 34.308'' \pm 0.168''$ from *Khao Ngem* to *Khao Ngu* (1910), $H_o = 1,382$ ft, Everest 1830 ellipsoid. An alternate spelling (transliteration) is *Khao Hluang*. (*This is also known as Rājburi Datum, the name being derived from the triangulation baseline – Ed.*) This was found in a Thai book (or booklet) 1918–1919 and from an Indian Publication published in 1947" (*ibid, Hager*). There are several pages of discussion on this old datum in *Survey Review*, by A.G. Bazley, October 1938, No.30, Vol. IV, pp. 450–457."

The current UTM coordinates are still referenced to Indian Datum 1975, using Everest ellipsoid, which was the national geodetic datum then and so adopted by Department of Lands when the department migrated from the Cassini-Soldner system. The shift parameters from Indian 1975 to WGS84 have been derived many times. The current ones are: $\Delta X = +204.4$ m, $\Delta Y = +837.7$ m, $\Delta Z = +294.7$ m (*Prof. Itthi Trisirisatayawong, Chulalongkorn University – personal communication, November 2010*). "It should be noted that there is no rotation and scale factor change. The latest national reference frame for the zero and first order geodetic network is tied to ITRF2005 using the GPS data observed in Nov. 2008 campaign" (*Prof. Chalermchon Satirapod, Chulalongkorn University – Personal communication, November 2010*).

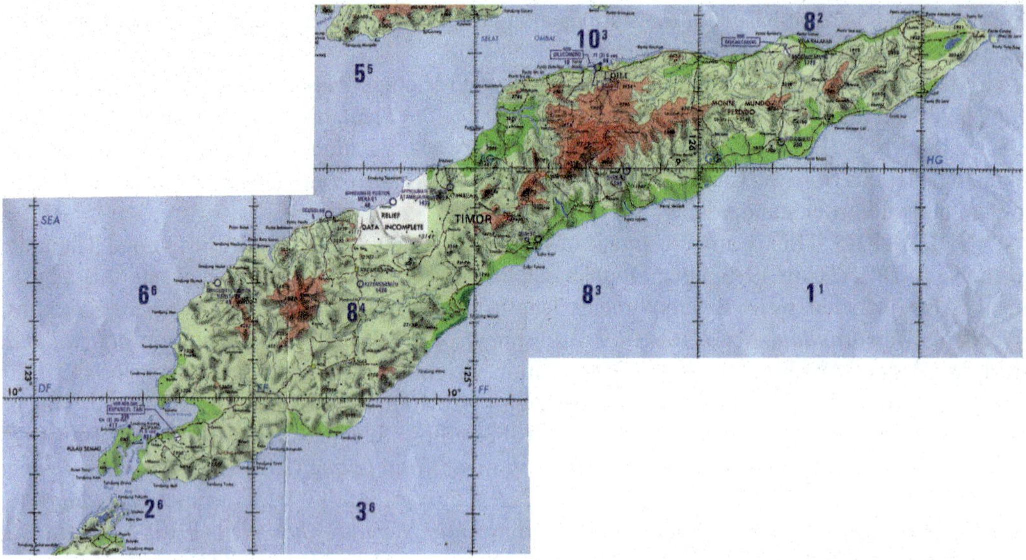

TIMOR
RELIEF
DATA INCOMPLETE
DILI
MONTE MUNDO PERDIDO
SEA
PULAU SEMAU

Democratic Republic of Timor-Leste

Bordered by Indonesia (253 km), the lowest point is the *Timor Sea, Savu Sea,* and *Banda Sea* (0 m); the highest point is *Foho Tatamailau* (2,963 m). The capital city is Dili.

"Little is known of Timor before A.D. 1500, although Chinese and Javanese traders visited the island from at least the 13th century, and possibly as early as the 7th century. These traders searched the coastal settlements for aromatic sandalwood, which was valued for its use in making furniture and incense, and beeswax, used for making candles. Portuguese traders arrived between 1509 and 1511, but it wasn't until 1556 that a handful of Dominican friars established the first Portuguese settlement at *Lifau* – in the present-day Oecussi enclave – and set about converting the Timorese to Catholicism.

"In 1642, Francisco Fernandes led a Portuguese military expedition to weaken the power of the Timor kings. Comprised primarily of Topasses, the 'black Portuguese' mestizos (people of mixed parentage) from neighboring Flores, his small army of musketeers settled in Timor, extending Portuguese influence into the interior. To counter the Portuguese, the Dutch established a base at *Kupang* in western Timor in 1653. The Portuguese appointed an administrator to Lifau in 1656, but the Topasses went on to become a law unto themselves, driving out the Portuguese governor in 1705. By 1749 the Topasses controlled central Timor and marched on *Kupang*, but the Dutch won the ensuing battle, expanding their control of western Timor in the process. On the Portuguese side, after more attacks from the Topasses in *Lifau*, the colonial base was moved east to *Dili* in 1769.

"The 1859 Treaty of Lisbon divided Timor, giving Portugal the eastern half, together with the north coast pocket of *Oecussi*; this was formalized in 1904. Portuguese Timor was a sleepy and neglected outpost ruled through a traditional system of *liurai* (local chiefs). Control outside Dili was limited and it wasn't until the 20th century that the Portuguese intervened in the interior.

"In 1941, Australia sent a small commando force into Portuguese Timor to counter the Japanese, deliberately breaching the colony's neutral status. Although the military initiative angered neutral Portugal and dragged Portuguese Timor into the Pacific War, it slowed the Japanese expansion. Australia's success was largely due to the support it received from the locals, for whom the cost was phenomenal. In 1942 the Portuguese handed control to the Japanese whose soldiers razed whole villages, seized food supplies and killed Timorese in areas where the Australians were operating. By the end of the war, between 40,000 and 60,000 Timorese had died.

"After WWII the colony reverted to Portuguese rule until, following the coup in Portugal on 25 April 1974, Lisbon set about discarding its colonial empire. Within a few weeks political parties had been formed in East Timor, and the Timorese Democratic Union (UDT) attempted to seize power in August 1975. A brief civil war saw its rival Fretilin (previously known as the Timorese Social Democrats) come out on top, declaring the independent existence of the Democratic Republic of East Timor on 28 November. But on 7 December the Indonesians launched their attack on Dili.

"Indonesia opposed the formation of an independent East Timor, and the leftist Fretilin raised the specter of Communism. The full-scale invasion of the former colony came one day after Henry Kissinger and Gerald Ford departed *Jakarta*, having tacitly given their assent. (Indeed, the Americans urged the Indonesians to conduct a swift campaign so that the world wouldn't see them using weapons provided by the USA.) Australia also sided with Indonesia, leaving the Timorese to face Indonesia alone.

"By 1976 there were 35,000 Indonesian troops in East Timor. Falintil, the military wing of Fretilin, fought a guerrilla war with marked success in the first few years, but weakened considerably thereafter. The cost of the brutal takeover to the East Timorese was huge; it's estimated that at least 100,000 died in the hostilities, and ensuing disease and famine" (*Lonely Planet*, 2015).

"Most of the country's infrastructure, including homes, irrigation systems, water supply systems, and schools, and nearly 100% of the country's electrical grid were destroyed (*including geodetic and cadastral records – Ed*). On 20 September 1999, Australian-led peacekeeping troops deployed to the country and brought the violence to an end. On 20 May 2002, Timor-Leste was internationally recognized as an independent state. In late 2012, the UN Security Council voted to end its peacekeeping mission in Timor-Leste and departed the country by the end of the year" (*World Factbook*, 2015).

"The Geographic Mission of Timor (MGT) was established in 1937 but had to stop their activities in 1941 because of the Japanese occupation during World War II. All the work done was lost except some pillar foundations at the most inaccessible points. MGT was reactivated in 1954 with the aim of providing a geodetic network to the territory with the accuracy required to create maps at the scale 1:50,000. For the implementation of the network, the International ellipsoid was adopted as reference. (*Note: $a = 6,378,388\,m$, $1/f = 297$ – Ed.*) Given the presumed existence of significant deviations of the vertical, it was considered that it would be unwise to tie to a single point. Thus, the coordinates of the vertices of the network were calculated using the astronomical values at three stations: *Lequi Levato* ($\Phi_o = 08°\ 30'\ 50''$ S, $\Lambda_o = 126°\ 21'\ 36''$ E, $\alpha_o = 09°\ 08'\ 24.73''$), *Burnuli* ($\Phi_o = 08°\ 53'\ 37''$ S, $\Lambda_o = 126°\ 20'\ 15''$ E, $\alpha_o = 175°\ 08'\ 38.42''$), and *Caitaba* ($\Phi_o = 09°\ 08'\ 19''$ S, $\Lambda_o = 125°\ 45'\ 15''$ E, $\alpha_o = 130°\ 52'\ 26.52''$), in *Relatórios da Missão Geográfica de Timor, IICT*. The datum was defined so the latitudes, longitudes and geodetic azimuths had minimum deviations from their corresponding astronomicals at these three points" (*Paula Santos, Instituto de Investigação Científica Tropical [IICT], personal communication 2015*).

"The leveling network and GPS points network had been established by BAKOSURTANAL during the Indonesian period. The concrete monuments of benchmarks and GPS points still exist in East Timor. However, documentary descriptions of benchmarks and GPS points, coordinates lists *etc.*, were lost during the violence following the independence vote in September 1999. The Indonesian Government's documents (statistic data, land registration data, resident registration data *etc.*,) were also lost during the violence following the independence vote in September 1999. Since land use data were destroyed during the violence after September 1999, many disputes have arisen in connection with land ownership within *Dili City*. In order to settle these disputes, there is an urgent need to develop large scale topographic maps, to conduct a land-use survey and to build a database system ensuring that the results of the land use survey are properly arranged on the newly developed large scale topographic maps" (*Study on Urgent Establishment of Topographic Mapping in East Timor – Creation of Topographic Information for Establishment of Cadastre in East Timor*, Toru Watanabe, Japan International Cooperation Agency, 2000).

Because of the urgent need for a reliable First-Order Geodetic network to support topographic mapping and cadastral mapping, the Portuguese government established a new independent system of control with new monuments in eight locations throughout East Timor. Collocation was not performed as the old classical coordinate records had been lost during the 1999 insurrection mentioned above. "The eight new stations (*estações*) representing the new **Rede Geodésica Fundamental Timor-Leste (RGFTL)** are: ABAC (*Bacau Airport*) $\varphi = 08°\ 29'\ 01.824''$ S, $\lambda = 126°\ 23'\ 46.465''$ E, $h = 577.373\,m$; CBBR (*Campo Barro*) $\varphi = 08°\ 44'\ 57.012''$ S, $\lambda = 125°\ 58'\ 16.818''$ E, $h = 1,173.132\,m$; CLMR (*Colméra Park*) $\varphi = 08°\ 33'\ 22.356''$ S, $\lambda = 125°\ 34'\ 10.106''$ E, $h = 53.664\,m$; MOLN (*Moleana*) $\varphi = 08°\ 55'\ 37.308''$ S, $\lambda = 125°\ 11'\ 02.319''$ E, $h = 183.054\,m$; OCSS (*Oecusse Airport*) $\varphi = 09°\ 11'\ 54.456''$ S, $\lambda = 124°\ 21'\ 08.758''$ E, $h = 47.459\,m$; RACA (*Town of Raca*) $\varphi = 08°\ 26'\ 33.360''$ S, $\lambda = 126°\ 59'\ 18.328''$ E, $h = 457.366\,m$; SAME (*Town of Same Heliport*) $\varphi = 09°\ 00'\ 06.984''$ S, $\lambda = 125°\ 39'\ 12.509''$ E, $h = 605.573\,m$; SUAI (*Fort Suai Loro*) $\varphi = 09°\ 21'\ 09.324''$ S, $\lambda = 125°\ 16'\ 33.707''$ E, $h = 46.803$ m" (*Relatório – Cálculo das Coordenadas*, Rui Fernandes, 31 Dezembro 2005). The new RGFTL is referenced to the International Terrestrial Reference Frame of 2000 (ITRF2000), epoch of 23 May 2003.

Thanks to Dr. Rui Fernandes for his help with getting me in touch with Ms. Santos of the IICT and also for his invaluable *on-site* assistance at our LSU Center for GeoInformatics for his expertise with JPL's GIPSY software.

Togo

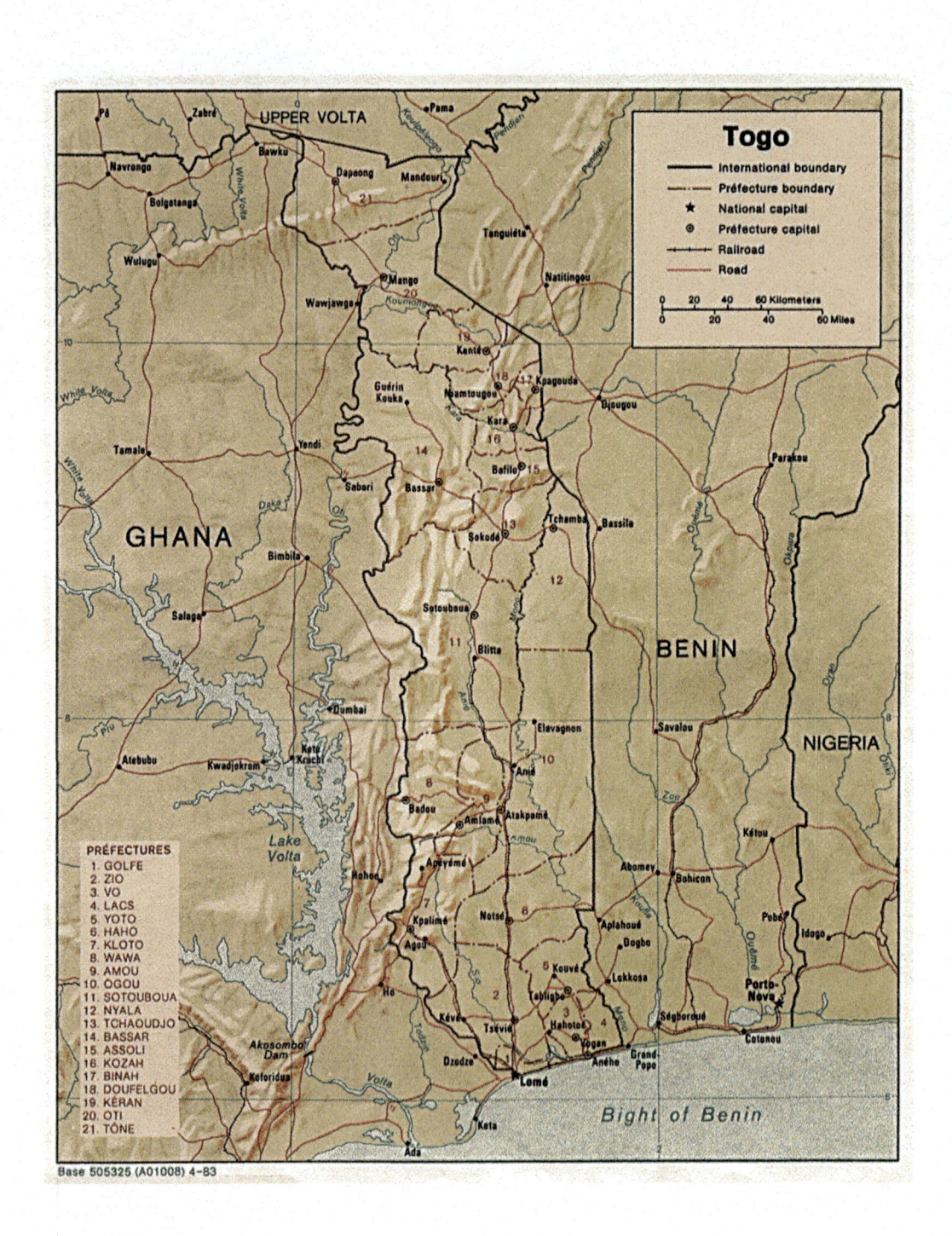

Legend:
- International boundary
- Préfecture boundary
- ★ National capital
- ⊙ Préfecture capital
- Railroad
- Road

Scale: 0 20 40 60 Kilometers / 0 20 40 60 Miles

PRÉFECTURES
1. GOLFE
2. ZIO
3. VO
4. LACS
5. YOTO
6. HAHO
7. KLOTO
8. WAWA
9. AMOU
10. OGOU
11. SOTOUBOUA
12. NYALA
13. TCHAOUDJO
14. BASSAR
15. ASSOLI
16. KOZAH
17. BINAH
18. DOUFELGOU
19. KÉRAN
20. OTI
21. TÔNE

Base 505325 (A01008) 4-83

Togolese Republic

Bordered by Benin (644 km), Burkina Faso (126 km), and Ghana (877 km), Togo is slightly smaller than West Virginia. The terrain is comprised of gently rolling savanna in the north; central hills; southern plateau; and a low coastal plain with extensive lagoons and marshes. The lowest point is the Atlantic Ocean (0 m), and the highest point is *Mont Agou* (986 m) (*World Factbook*, 2013).

"Togo's name comes from *togodo,* which means 'behind the lake' in Ewe – a reference to *Lake Togo*. The country was once on the fringes of several great empires and, when the Europeans arrived in the 16th century, this power vacuum allowed the slave-traders to use Togo as a conduit. Following the abolition of slavery, Germany signed a treaty in *Togoville* with local king Mlapa. Togoland, as the Germans called their colony, underwent considerable economic development, but the Togolese didn't appreciate the Germans' brutal 'pacification' campaigns. When the Germans surrendered at Kamina – the Allies' first victory in WWI – the Togolese welcomed the British forces. However, the League of Nations split Togoland between France and Britain – a controversial move that divided the populous of Ewe. Following a 1956 plebiscite, British Togoland was incorporated into the Gold Coast (now Ghana). French Togoland gained full independence in 1960" (*Lonely Planet,* 2013).

"A Franco–German convention of July 23, 1897, delimited a boundary between German Togoland and the French possessions of Dahomey and Sudan. The line passed northward from the Atlantic Ocean to the 11th parallel between Togoland and Dahomey.

The boundary then extended westward between Togoland and French Sudan as follows:

It shall then run westward along Lat. 11° N. to the *White Volta* so as in any case to leave *Pougno* to France and *Koun–Djari* [*Koundjouare*] to Germany. It shall then run along the thalweg of that river to Lat. 10° N., which it shall follow to its intersection with the meridian 03°52′ west of Paris (01°32′ west of Greenwich). The following year an Anglo–French convention of June 14, 1898, delimited the British Gold Coast–French Sudan boundary eastward from the *Black Volta* to the French Sudan–Togoland boundary. The extreme eastern sector of the boundary to the Togoland tripoint was as follows: … and shall then follow the thalweg of this river [*Nahau* or *Nouhao*] up or down stream, as the case may be, to a point situated 2 miles (3,219 metres) eastward of the road which leads from *Gambaga* to *Tenkrugu* (*Tingourkou*) [*Tenkodogo*], via *Bawku* (*Baukou*). Thence it shall rejoin by a straight line the 11th degree of north latitude at the intersection of this parallel with the road which is shown on Map No. 1 as leading from *Sansanne–Mango* to *Pama,* via *Jebigu* (*Djebiga*) [*Diabiga*]. After a joint survey in 1901–2, the Gold Coast–Togoland boundary was delimited in detail from the 9th parallel northward to the Gold Coast–Sudan–Togoland tripoint by an Anglo–German exchange of notes of June 25, 1904."

"An Anglo–French exchange of notes on March 18, 1904–April 25, 1904, redelimited the boundary between the Gold Coast and French Sudan, including the extreme eastern sector. Paragraph 41 of the delimitation schedule relative to the Togoland tripoint was amended in a second Anglo–French exchange of notes on May 24, 1906–July 19, 1906, which also included the demarcation of the boundary. The following is the revised paragraph:

"41. From this point the frontier runs in a straight line across an uninhabited country, covered by bush, in the direction of the point of intersection of the 11th parallel with the road from *Punio* (*Pounio*) to *Jebiga* (*Djebiga*) until it reaches the northern terminal point of the Anglo–German frontier, where it terminates."

Utilizing in part the boundary of the convention of July 23, 1897, a Franco–German declaration of September 28, 1912, delimited the French Sudan–Togoland boundary in detail and is the basis of the alignment of the present Togo–Upper Volta boundary.

Although the term continued in use in official treaties, the Sudan as part of French West Africa underwent a number of changes in size, administration, and names. In 1899 parts of French Sudan were transferred to French Guinea, Ivory Coast, and Dahomey. The remainder of French Sudan was organized into the civil territory of Upper Senegal and Middle Niger and initially into two and later three military territories with headquarters at *Tombouctou, Bobo Dioulasso,* and *Zinder.* During 1902 Upper Senegal and Middle Niger was renamed Senegambia and Niger which in turn was changed to the Colony of Upper Senegal and Niger in 1904.

"By a decree of March 1, 1919, the colony of Upper Volta was constituted by detaching various cercles from Upper Senegal and Niger. The remainder of Upper Senegal and Niger was renamed French Sudan on December 4, 1920. A decree of September 5, 1932 abolished the colony of Upper Volta and divided it among the colonies of Ivory Coast, Niger, and French Sudan. Following World War I, eastern Togoland became a League of Nations mandate under French administration and western Togoland became a British mandate. In 1946 the mandates were made United Nations trust territories and continued to be administered by France and the United Kingdom, respectively. During the mandate and trusteeship periods, French Togo had its own governmental structure, but British Togoland was administered by the United Kingdom as an integral part of the territory under the Government of the Gold Coast. In 1957 British Togoland was merged with the Gold Coast, including Ashanti and the Northern Territories, to form the new state of Ghana. On October 28, 1956, French Togo voted to become an autonomous republic within the French Union, and on April 27, 1960, it became an independent republic."

"The tripoint with Dahomey at Point No. 109 of the Franco–German delimitation of 1912 is by map measurement located at 11°00′ N. and approximately 0°55′E. The boundary then extends westward along the 11th parallel for about 16 miles to 0°40′E. It continues in a straight line southwestward for 12 miles to the *Sansargou River* at 10°56′N. and then follows the *Sansargo*u northward for 6 miles to 11°30′N. and 0°30′12″E. The remainder of the boundary consists of a straight line for 44 miles to the Ghana tripoint at pillar 148 of the Anglo–French demarcation of 1929 at approximately 11°08′13″N. and 0°08′09″W" (*International Boundary Study, No. 128–29 September 1972, Burkina Faso* (*Upper Volta*) *– Togo Boundary,* U.S. Department of State). See also *IBS Numbers 124 & 126.*

The local datum for Togo is considered to be *Lomé* which is referenced to the Clarke 1880, where $a = 6,378,249.145$ m and $1/f = 293.145$. However, transformation parameters from Lomé Datum to WGS84 Datum are not available in the literature. Using *GeoTrans,* I transformed the above tripoint coordinates to Clarke 1880 Geocentric coordinates. I also went to those same coordinates in Google Earth™ and could actually see nearby where that tripoint apparently is in Google Earth™! I then took those Google Earth coordinates of the apparent tripoint ($\varphi = 11°08′22.19″$ N, $\lambda = 00°08′10.42″$ W), and using *GeoTrans* I transformed those geodetic coordinates to WGS84 Geocentric coordinates. I then computed the shift parameters **from** *Lomé* **to** WGS84 to be: $\Delta X = -177$ m, $\Delta Y = +42$ m, $\Delta Z = +388$ m; the accuracy may be about ± 25 m.

The traditional Grids used in this part of French Africa include the Lambert Conformal Conic for Niger Zone (*Fuseau Niger*), where the central meridian (λ_o) = 0°, latitude of origin (φ_o) = 13° N, scale factor at origin (m_o) = 0.99932, False Easting = 1,800 km, and False Northing = 500 km; and the Guinea Zone (*Fuseau Guinea*) which has the same parameters as the Niger Zone except that the latitude of origin (φ_o) = 7° N. The Dahomey (Benin) Transverse Mercator Zone (*Fuseau Dahomey*) is where: the central meridian (λ_o) = 0° 30′ E, latitude of origin (φ_o) = 0°, scale factor at origin (m_o) = 0.9990, False Easting = 1,000 km, and False Northing = 1,000 km. However, all mapping since 1951 has been on the UTM Grid. The existing 1:50,000 scale topographic mapping coverage of the entire country is likely entirely controlled by French IGM Astros.

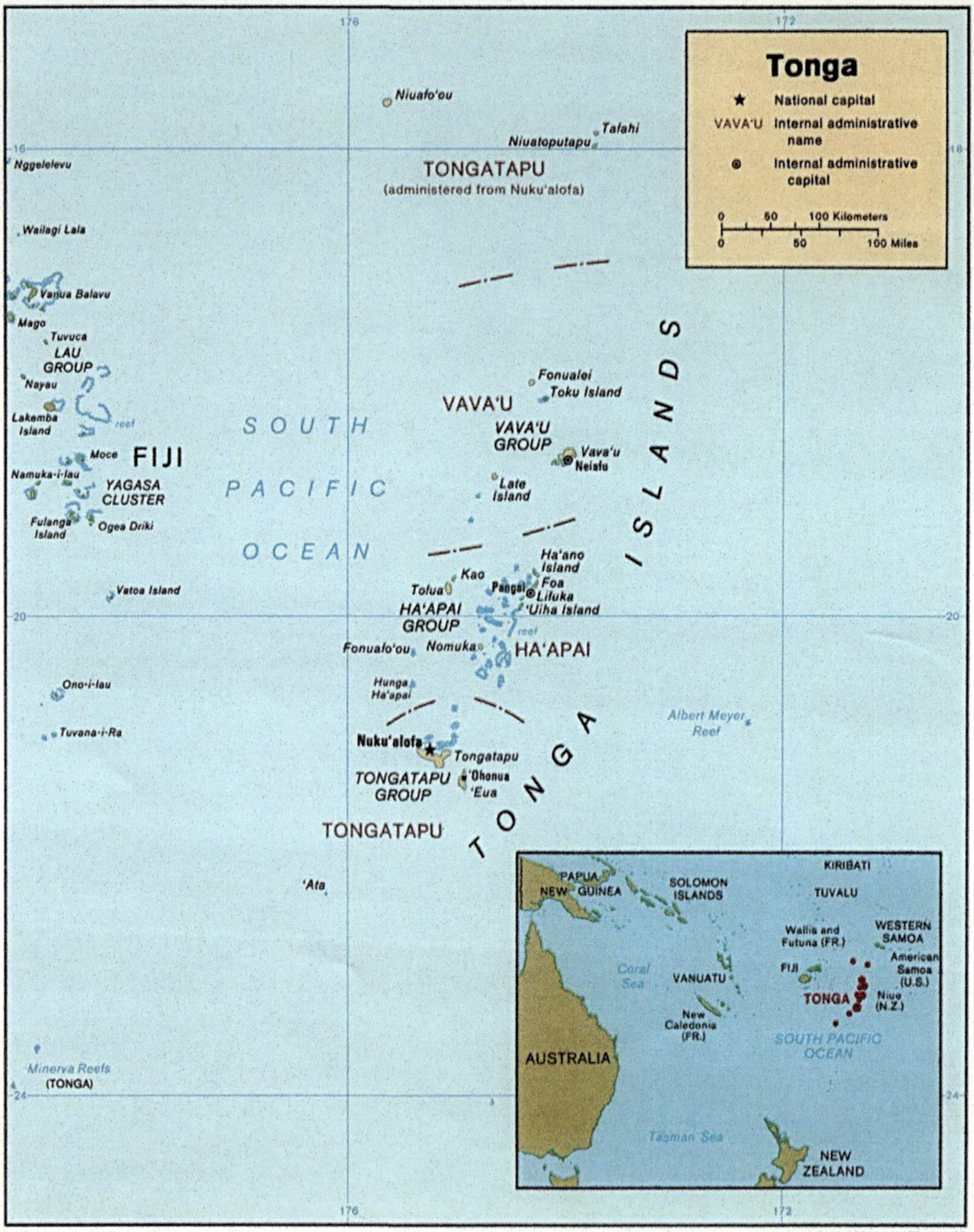

Base 800199 (A05044) 1-89

Kingdom of Tonga

Tonga is four times the size of Washington, DC, the terrain of the islands is mostly comprised of a limestone base formed from uplifted coral formations; others have limestone overlying a volcanic base. The lowest point is the Pacific Ocean (0 m); the highest point is an unnamed elevation on *Kao Island* (1,033 m). The Kingdom of Tonga is an archipelago of 169 islands, 36 of which are inhabited (*World Factbook*, 2014).

"Around 3000 years ago, the Lapita people from Southeast Asia migrated west via the Malay peninsula and the remote islands of the East Indies to settle in the scattered and pristine islands of the South Pacific. In Tonga, these original ancestors of today's Polynesian people founded settlements at *Toloa* – near the present-day location of *Fua'amotu* International Airport – and at *Heketa*, on the northeastern edge of *Tongatapu*. Three millennia later, reminders of these ancient times are dotted throughout the islands. The fascinating Ha'amonga a Maui trilithon still stands as an imposing legacy of early Tongan ingenuity. Eventually settling in the far-flung island groups of the Kingdom's archipelago, these early ancestors also developed a distinctive culture that still underpins traditional Tongan life in more contemporary times. Initial European contact with Tonga came in 1616, when the Dutch navigators Wilhelm Schouten and Jacob Le Maire discovered the Niuas, the small northern most islands of the Tongan archipelago. Contact with the local Niuas islanders was restricted to a minor altercation with a Tongan canoe. In 1643, the Dutch extended their exploration when Abel Tasman visited the Tongan Islands of *'Ata, 'Eua* and the largest island of *Tongatapu*. Unlike the first Dutch contact further north in 1616, Tasman's ships the Heemskerck and the Zeehaen stopped for water and replenishments, and Tasman also traded with the local communities. In 1773, the British explorer and navigator Captain James Cook visited Tonga's southern islands of *Tongatapu and 'Eua*. He returned in 1777 and spent two months exploring and charting the Tongan archipelago, with his legendary skill as a cartographer producing accurate charts still in regular use until recent times. During this voyage, a lavish feast for Cook and his men was presented by Chief Finau in the village of *Lifuka* in the *Ha'apai* island group. Cook was so impressed by Tongan hospitality he dubbed Tonga the 'Friendly Isles', not realizing the amiable and social nature of the locals actually concealed a plan to raid his boats and kill Cook and his crew! The conspiracy was only foiled at the eleventh hour after a dispute between Finau and other village nobles, and Cook sailed away oblivious of his intended fate. Ironically his positive and complementary name for the Kingdom of Tonga remains in common use. The northern island group *Vava'u* was discovered in 1781 by Spanish navigator, Don Francisco Antonio Mourelle, commander of the ship La Princesa. Mourelle named *Vava'u*'s well-protected harbor *Port of Refuge* and claimed the beautiful islands in the name of Spain. Over ensuring years early traders continued to visit Tonga and tensions grew between Europeans and Tongans. In 1806, this disquiet culminated in the sacking of the ship the Port-au-Prince in *Lifuka* in the *Ha'apai* island group. With the exception of a young cabin boy William Mariner, all the crew was killed. The lad was nurtured by Chief Finau in *Lifuka* for four years, learning the Tongan language and becoming immersed in the Kingdom's tradition and protocol. Mariner's book 'An Account of the Natives of the Tongan Islands' is now recognized as a significant insight into early Tongan life, customs and culture. Another navigator to visit Tongan waters was Captain William Bligh, and Fletcher Christian's infamous mutiny of the HMS Bounty actually occurred near the volcanic island of *Tofua* in the *Ha'apai* group. In 1845, the scattered and pristine islands of western Polynesia became united as the Kingdom of Tonga, and 30 years later officially became a constitutional monarchy and British Protectorate. The first King of this united Tonga was George Tupou I, and the modern Kingdom of Tonga is the only Pacific Island nation never to lose its indigenous governance or to be colonized. Located just west of the International

Date Line, Tonga is also the first country to experience the new day each morning. Tonga joined the Commonwealth of Nations in 1970, jettisoning the protectorate status in 1970, but still retaining its unique position as the only monarchy in Polynesia" (*2011 Tonga Visitors Bureau*).

"Old survey records indicated that surveying and mapping in Tonga began somewhere around in the early years of the 1900s onwards. Up until 1927, at least three to four expatriate surveyors from Australia and New Zealand, assisted by only few locally trained field surveyors were consistently employed by the Tonga Government to carry out the required survey works. These included coastal traverses, road and engineering surveys, land boundary surveys for hereditary estates, townships and new individually granted allotments (town and tax allotments – at standard legal sizes) were carried out in most parts of the main islands of Tonga – at a steady rate of progress due to shortage of trained personnel. This was the situation during and after the 1st World War in 1918. In 1927, it appeared that the size of the field survey forces, and their annual outputs were nowhere close to meeting the demands from a rapidly increasing population for more subdivided lands - tax and town allotments. However, during the period 1927–1957 a large amount of surveyed tax and town allotments were completed due to increasing number of Tongan overseas trained Surveyors (at technician level) that returned and were employed permanently by the Government of Tonga. Although the bulk of the survey works, completed for the period 1900–1957, were of good standard theodolite traverses, various and different meridians were used as origins and occasionally a true meridian was used. Bearings of contiguous surveys in an area were usually in the same terms, but no system of regular coordinates was set up as a survey and mapping basis for the completed works" (*A Report on the Geodetic Infrastructure of the Kingdom of Tonga*, Folau, V. & Malolo, T.L., Ministry of Lands, Environment, Climate Change and Natural Resources, Kingdom of Tonga, September 2013, Fiji).

Classical datum origins consist of "Astro Stations" in Tonga, all referenced to the International Ellipsoid, where $a = 6,378,388$ m and $1/f = 297$. *Vava'u* (Datum Origin?) is at *Trig B*, where $\Phi_o = 18°\ 37'\ 28.93''$ S, $\Lambda_o = 174°\ 03'\ 59.82''$ W, $\alpha_o = 64°\ 11'\ 47.0''$ to *Trig A* from North. *Nomuka Datum of 1959* Origin is where: $\Phi_o = 20°\ 15'\ 15.5''$ S, $\Lambda_o = 174°\ 48'??.?''$ W, $\alpha_o = 119°\ 28'\ 29.8''$ to *Toafa* from North. *Pangai Datum of 1961* Origin is at *Trig I*, where $\Phi_o = 19°\ 48'\ 20.7''$ S, $\Lambda_o = 174°\ 20'\ 56.3''$ W, $\alpha_o = 232°\ 02'\ 37.0''$ to *Trig II*. *Tongatabu Datum of 1942* (noted in the 1980s that this might not be used by the Kingdom of Tonga) Origin at South Station, where $\Phi_o = 21°\ 03'\ 23''$ S, $\Lambda_o = 175°\ 15'\ 36''$ W, $\alpha_o = 206°\ 06'\ 09.3''$ to *North Station*, observed by the U.S.S. Sumner (*John W. Hager, personal communication 5 May 2005*). The majority of the above datums are collectively known as the *Tonga Cadastral Survey Datum 1957–1961* (*TCSD57/61*).

"The above Survey Operations were executed by Leach and Browne, Registered Surveyors, Milford in Auckland, New Zealand. The Tonga Cadastral Survey Datum (TCSD 57/61) and as well as the Tonga Cadastral Survey Grid (TCSG 61)] was established at four island groups, each with its own circuit origin and azimuth determination. The four circuits are: *Tongatapu/'Eua, Nomuka, Ha'apai and Vava'u*. The purpose of the survey control network or circuit was to provide a homogeneous framework from which to survey and subdivide land parcels into town and tax allotments of 8¼ acres each. The area of 8¼ acres is based on a land parcel having the nominal dimensions of (100 fathoms × 100 fathoms). An instruction manual, titled "Instructions for the Cadastral Survey of the Kingdom of Tonga", was produced in 1958 by Leach and Browne, Reg. Surveyors, Milford in Auckland, NZ, which also established cadastral survey and office procedures for recording land registration information. A report title "General Notes on the Primary Control for the Cadastral Survey of Tonga" was produced and signed by D. L. Leach himself.

"The Tonga Cadastral Survey Grid 1961 (TCSG 61) Definition: Map Grid System – TCSG 61, Projection – Transverse Mercator based on the UTM Projection Zone One, Unit of Measurement: For UTM. Meter (1 m = 3.2808455 ft.) For TCSG 61, Link (100 links to 1 chain or 66 ft.), True Origin – the true origin of the UTM Zone One is the intersection of the Central Meridian (CM) of 177° West Longitude with the equator. The coordinates of this origin are 500,000 meters East and 10,000,000 meters North. False Origin – the origin of the TCSG 61 is a point on the 177° West Longitude, which has a UTM Northing coordinate value of 7,500,000 meters North. There is therefore a direct relation between the UTM Zone One projection coordinates in meters and the TCSG 61 projection coordinates in links. Scale Factor – the Scale Factor at the CM is ko=0.9996 (a reduction of 1=2,500).

"In remote and isolated islands or group of islands not covered during the Cadastral Survey Operations, Directorate of Overseas Surveyors established six more additional Survey Control Circuits and Datums (Horizontal and Vertical), where the Horizontal Datum in each case was either scaled from nautical chart or adopted from the Shoran Trilateration Survey. Where the positions (lat. & long.) of the horizontal datum were scaled from nautical charts, an astronomical azimuth was to be determined to another reference station within the circuit and, the height of the datum station (point of origin) was connected also to sea level so as to determine approximately the MSL height of the point of origin. For the Topographic Mapping Series, a total of 10 Survey Control Circuits (4 from the Tonga Cadastral Survey Operations 1957–1962 and 6 from DOS Mapping Control 1970–1971) were used to provide the horizontal and vertical control requirements for the production of the first Topographic Mapping Series of the Kingdom of Tonga, excluding *Telekitonga* and *Telekitokelau Islands* in the *Minerva Reefs*. The 10 independent Control Survey Circuits are as follows:

- *Niuafo'ou Circuit* - Datum scaled from Nautical Chart No. 968
- *Niuatoputapu Circuit* - Datum scaled from Nautical Chart No. 968
- *Fonualei/Toku Circuit* - Datum based on Shell Shoran Trilateration 1970
- *Vava'u Circuit* - Datum based on TCSD & TCSG
- *Late Circuit* - Datum based on Shell Shoran Trilateration 1970
- *Ha'apai Circuit* - Datum based on TCSD & TCSG
- *Nomuka Circuit* - Datum based on TCSD & TCSG
- *Hunga (Tonga & Hp)* - Datum based on Shell Shoran Trilateration 1970
- *TBU/'Eua Circuit* - Datum based on TCSD & TCSG
- *Ata Circuit* - Datum scaled from Nautical Chart No. 2421

"A survey operation was done by the *Shell International Petroleum Maatshappij N.V.,* The Hague, Netherlands in connection with geophysical seismic surveys required before the hydrocarbon exploration conducted in Tonga during the seventies. (*This was*) to establish a fourth order basic Geodetic Network over three major islands groups (*Tongatapu, Ha'apai and Vava'u*) in order to obtain coordinates on a common Datum for various base stations to be used for the positioning of pre-planned seismic surveys. Investigation of Tonga's survey record revealed that a large number of geodetic and cadastral controls existed. These controls were grouped in circuits, almost identical with islands groups, with each having its own astronomically determined reference station. Thus, a survey for a wider geodetic was made to tie up these circuits together. In spite of heavy seas and landing problems on some isolated islands, the two months field survey work was completed to acceptable standard and results at reasonable cost by trilateration using XR Shoran equipment. A total of 27 baselines were measured, 5 of these being TCSD & TCSMG lines. The Trilateration network, based on a single datum (assumed to be of the TCSD 57/61), consists of 11 stations of which 3 were newly established and 8 were existing stations. Two of the existing stations were adopted for the datum and for orientation" (*op. cit., Folau & Malolo*).

The current Datum of Tonga is the *Tonga Geodetic Datum of 2005* (*TGD2005*) where the ellipsoid of reference is the GRS80; $a=6,378,137\,m$, and $1/f=298.257222101$. The Tonga Map Grid is a modified version of the UTM Zone 1 Grid in which the False Origins have been changed to avoid confusion with the old cadastral grid system referenced to the International ellipsoid. The new False Origins are: False Northing=5,000,000 m, and False Easting=1,500,000 m. All other defining parameters of the UTM Zone 1 remain unchanged for the Tonga Map Grid. The Kingdom of Tonga has not published transformation parameters from the old cadastral Datums to TGD2005.

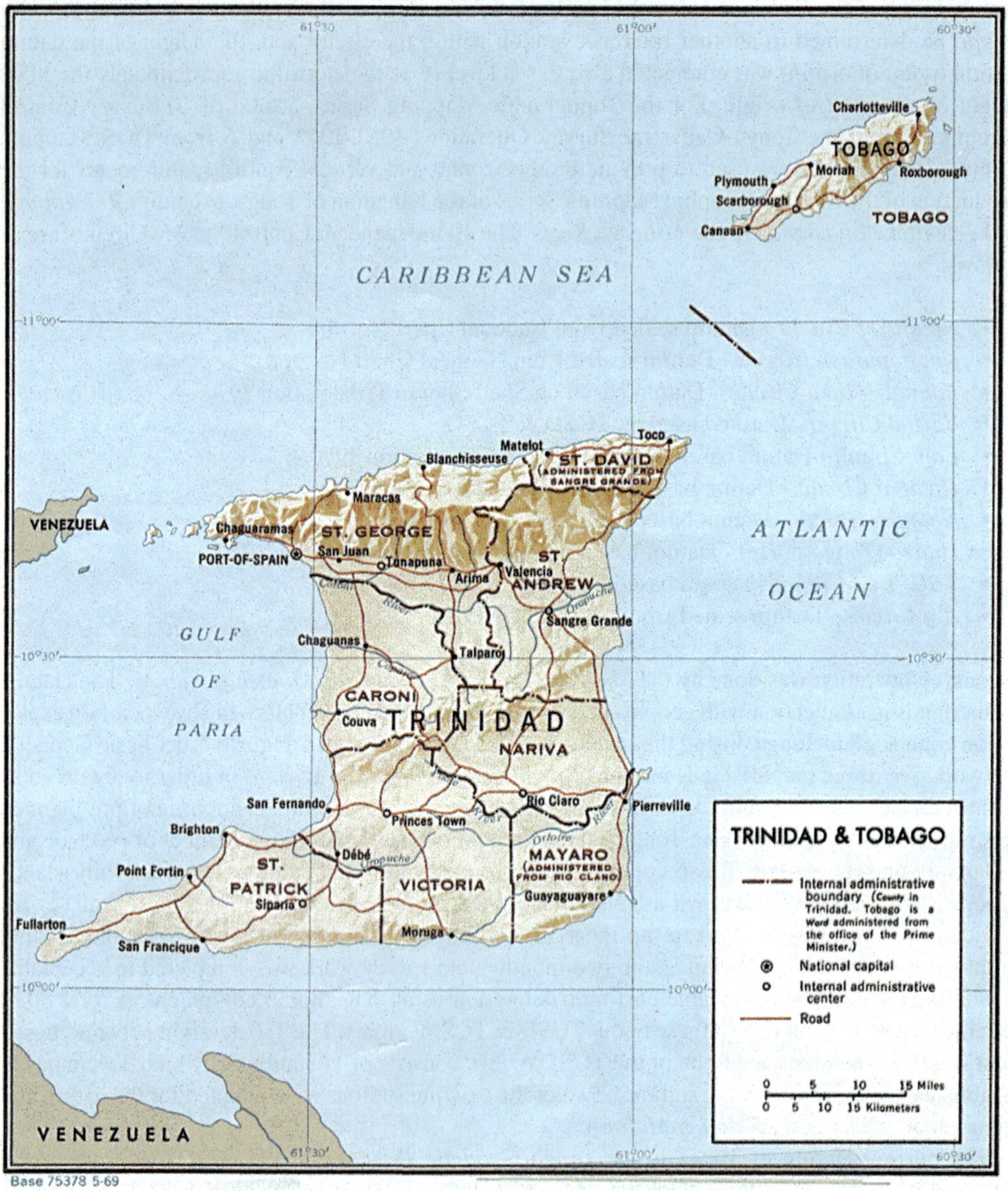

Base 75378 5-69

The Republic of Trinidad and Tobago

The highest point is *El Cerro del Aripo* at 940 m (3,084 ft). The island of Trinidad is mostly low plains with some hills and low mountains in the north; the island of Tobago is also low in the southern part and hilly to its north. My children and I found the shallow coral reefs to the south of Tobago to be spectacular!

Trinidad and Tobago were originally populated by the Igneri, a relatively peaceful Arawak subgroup, and by the cannibal Carïbs. Discovered by Columbus in 1498, the Spaniards established a colony on the island of Trinidad in 1577 but Sir Walter Raleigh destroyed it in 1595. Occupied by the British in 1797, it was formally established as a Colony in 1899 and was a member of the West Indies Federation from 1958 to 1962. Tobago, in contrast, was not colonized by Spain; it has had a most varied history and experienced a remarkable 31 changes of possession before being finally recaptured by the British in 1803. The first oil deposits were discovered in 1866, by 1908 crude oil production began, and in 1912 the first oil refinery was established. The first international boundary treaty between two sovereign nations that divided the mineral resources of the ocean bottom was signed between Venezuela and the United Kingdom in 1942. Both Trinidad and Tobago and Venezuela are parties to the Geneva Convention on the Continental Shelf. In 1954 marine drilling began off the West coast of the island in the *Gulf of Paria*. The Republic of Trinidad and Tobago became an independent state in 1962. In 1968 offshore petroleum discoveries off the East coast propelled Trinidad to the enviable status of the wealthiest nation in the Caribbean.

In 1787, the Spanish Governor of Trinidad signed the first Instructions for Surveyors. The last sentence reads: "All which shall be faithfully and punctually observed, as has been provided in these instructions, of which an attested copy, under my hand, is to be given to every surveyor, making him sign at the bottom of this original a receipt, **in order to convict him as necessary**." (*Signed*) Pedro de Ibarrarte.

The first triangulation of Trinidad was carried out from 1900 to 1903 by E. R. Smart, M.A., with third-order extensions carried out until discontinued in 1911. The *Orange Grove Base Line* was first laid out and then measured three times with a Gurley 300 ft. steel tape. The tape was certified at the U.S. National Bureau of Standards. The mean distance was 8,085.034 ft (~2,464 m). The triangulation was then carried westwards to *Battery, Port of Spain*, the Longitude of which had been accurately determined by telegraph by LCDR Green, U.S.N., in 1883 at *Orange Grove Meridian*, where $\Phi_o = 10° 37' 33.412"$ N, $\Lambda_o = 61° 22' 30.920"$ W. The triangulation then proceeded southwards to the *Harmony Hall Base Line*; a discrepancy being computed through the unadjusted triangulation of 8.4 inches was equal to nearly 1/11,500th part of the *Harmony Hall Base*. Latitude was determined at both bases by meridian altitudes of North and South stars, and the results showed reasonable agreement; Azimuth was also determined at both bases by observation of circumpolar stars at and near elongation. The adjustment of the triangulation of Trinidad was carried out by the usual approximate methods.

The triangulation of Tobago was carried out by Capt. G. M. Latham, R. E., and a party of the Royal Engineers in 1923. A base with a mean distance of 2,162.3741 ft (*sic*) (~659 m) was determined from three measurements with the Trinidad Base Measurement Apparatus. The main net of the Tobago triangulation was adjusted by the method of Least Squares.

"In 1925, on the advice of the Colonial Survey Committee, it was decided to adopt the Cassini Projection for Trinidad maps. A suitable sheet system based on a grid for the whole island has been evolved. The projection used for both Trinidad and Tobago is the Cassini, Clarke (1858) Spheroid. In Trinidad, the Standard Meridian is Long. 61° 20′ 00″ and the point of Origin is situated as Lat.

10° 26′ 30″ on that meridian, the Rectangular Coordinates of this point for convenience are called N. 325,000 links and E. 430,000 links. In Tobago the Standard Meridian is Long. 60° 41′ 09.632″ W, and the point of Origin (*Mount Dillon*, Trig Sta. L.) is situated as Lat. 11° 15′ 07.843″ N. on that meridian, the Rectangular Coordinates of this point for convenience are called N. 180,000 links and E. 187,500 links. *Brigadier H. St. J. Winterbotham, C.M.G., D.S.O., 15 October 1930.*" Note that 8,000 links = 1 mile, and 1 m = 4.971014137 links. Also, for the Clarke 1858 ellipsoid as used in Trinidad and Tobago: $a = 6,378, 293.6$ m and $1/f = 294.26$. The origin of the *Old Trinidad Datum of 1903* is defined at the *Harbour (sic) Master's Flagstaff*, where $\Phi_o = 10°$ 38′ 39.02″ North, $\Lambda_o = 61°$ 30′ 38.00″ West of Greenwich, and the corresponding Cassini-Soldner Grid coordinates are: N.436,366.91 links, E.333,604.30 links.

According to Mr. John W. Hagar, "In 1942 the Coast & Geodetic Survey put out Tables for Computation of Plane Coordinates on Lambert Grid: Trinidad. I would guess that these were part of the general contingency planning of WW II. Anyway, they are Lambert, Clarke 1866, yard, latitude of origin = 10° 25′ north (lower standard parallel = 10° 10′ north, upper standard parallel = 10° 40′ north), longitude of origin = 61° 30′ west, scale factor = 0.99999054, FE = 120,000 yards. I don't have a false northing but cannot conceive of a U.S. military grid without same. Never saw any maps or coordinates on this system."

Just across the *Gulf of Paria*, Ing. Dr. Adolfo C. Romero initiated the redefinition of classical triangulation for the northern expanses of the South American continent. The small Venezuelan town of *La Canoa*, just a few hundred miles from Trinidad was used as the origin of the *Provisional South American Datum of 1956* (PSAD 56), where $\Phi_o = 08°$ 34′ 17.170″ N, $\Lambda_o = 63°$ 51′ 34.880″ W. The ellipsoid of reference is the International (Hayford 1909 or Madrid 1924), where $a = 6,378,388$ m and $1/f = 297$. Earlier work in Venezuela had made connections to the island of Trinidad through the assistance of the Inter-American Geodetic Survey (IAGS), and classical triangulation on the *Loma Quintana Datum of 1911* was extant across the *Gulf of Paria*. The PSAD 56 coordinates were subsequently also carried over to Trinidad through the redefinition by Dr. Romero, the IAGS, and the U.S. Army Map Service.

The basic triangulation net of Trinidad and Tobago began to be re-observed and adjusted by approximate methods in 1963–1965. The U.S. Army Topographic Command accomplished a subsequent simultaneous least-squares adjustment of triangulation and traverse through the "USHER" computer program for horizontal control surveys using the Variation of Coordinates method. This is known as the *Naparima Datum of 1972*. The point of origin is station Naparima Hill, where $\Phi_o = 10°$ 16′ 44.8600″ N, $\Lambda_o = 61°$ 27′ 34.6200″ W, and the ellipsoid of reference is the International. When I visited Trinidad, I inquired where exactly, is the station on Naparima Hill? I was told that it was about 20 m past the cliff face in thin air. It seems that there was a rock quarry on the western face of the hill, and the station was blasted away shortly after the last observation period in the early 1960s. For comparison, the Trinidad Datum of 1903 coordinates for Naparima are: $\phi = 10°$ 16′ 37.737″ N, $\lambda = 61°$ 27′ 31.489″ W, H = 586 ft. Although the UTM Grid was adopted for use with the new Datum, local use of the Cassini Grids on the former Datum continued for various applications. These applications included the contractual description of offshore petroleum concessions ("Lease Blocks") and reporting regulations. As of 1999, the Cassini-Soldner Grid of Trinidad on the Old Trinidad Datum of 1903 was alive and well in the "oil patch."

In 1996, the U.S. National Geodetic Survey (NGS) made GPS observations of many countries in the Caribbean. Observations on the island of Trinidad included two points that happened to be on both the Old Trinidad Datum of 1903 and on the Naparima Datum of 1972. Thanks to Mr. David Doyle of NGS, I received the WGS84 coordinates of all stations observed. I performed a simple three-parameter analysis of those two points and derived the following relations: from Old Trinidad 1903 to WGS 84, $\Delta X = -33.250$ m, $\Delta Y = +232.675$ m, $\Delta Z = +484.542$ m, the fit was good to about 2 m in each component for the two points computed. From Naparima 1972 to WGS 84, $\Delta X = +0.332$ m, $\Delta Y = +369.359$ m, $\Delta Z = +172.897$ m, and the fit was good to better than a meter in each component for the two points computed.

In 1986, I was retained to study the relations of the various Grids and Datums of the area, and I traveled to the islands and mainland to research the original data. In 1987, I was invited to present the results of my analysis and development of the seven-parameter Bursa-Wolf Datum Shift relations of all coordinate systems in the Republic of Venezuela and the Republic of Trinidad and Tobago to the Minister of External Affairs (Secretary of State) in Port of Spain. A new International Boundary Treaty was subsequently signed and ratified by both republics. The defining Datum in the Treaty was the PSAD 56.

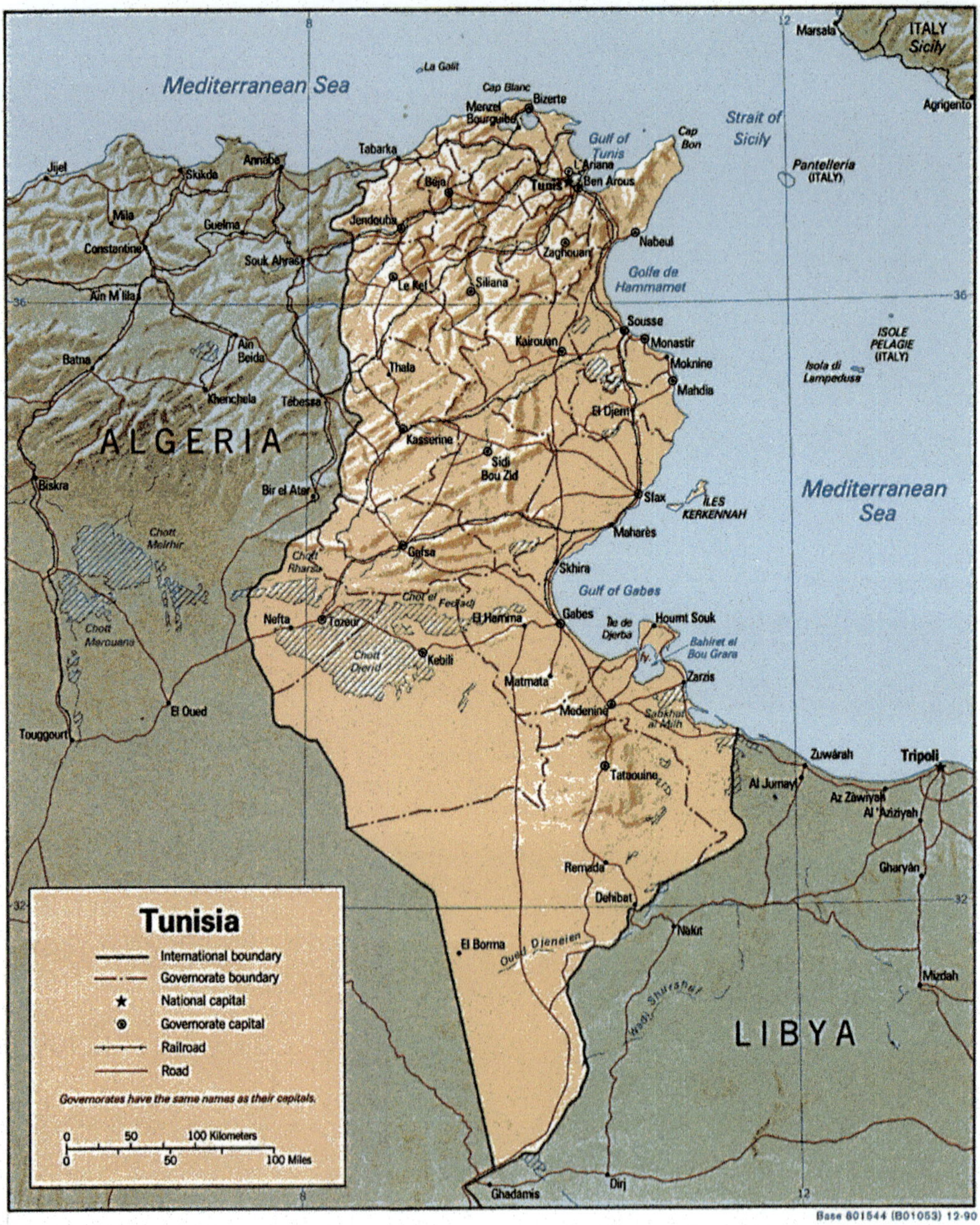

Base 801544 (B01053) 12-90

Tunisian Republic

The republic is located in Northern Africa and is bordered by the Mediterranean Sea (1,148 km), Algeria (965 km), and Libya (459 km). Tunisia is slightly larger than Georgia and has mountains in the north; it has a hot, dry central plain, and the semiarid south merges into the Sahara desert. The lowest point is *Shatt al Gharsah* (–17 m), and the highest point is *Jebel ech Chambi* (1,544 m) (*CIA World Factbook*, 2006).

Tunisia was a Roman province of Africa from the 2nd century B.C. to the 9th century A.D. when the Vandals overran it. Reconquered by the Byzantine Empire in the 6th century, it was then taken by Muslims in the 7th century. Both Libya and Tunisia came under Turkish suzerainty in the latter half of the 16th century. The former Barbary state engaged in piracy in the 19th century, was later the subject of dispute between France and Italy and was eventually invaded by France and became a French protectorate in 1881. "An agreement in 1886 between France and Turkey delimited a boundary between Tunisia and the Turkish vilayet of *Tripoli* in western Libya from the Mediterranean inland for a limited distance. A second agreement in 1892 delimited the boundary with greater accuracy than previously and inland as far as *Ghudämis*. On May 19, 1910, a Franco-Turkish convention delimited the present-day Libya-Tunisia boundary which was demarcated with pillars by a joint commission in 1910–1911" (*International Boundary Study No. 121 – April 7, 1972, Libya–Tunisia Boundary,* Department of State Bureau of Intelligence and Research, Issued by the Geographer).

The Tunisian government was reorganized in 1922, occupied by the Germans in 1942, captured by the U.S. and British forces in 1943, and was recognized by France as independent in 1956 (*Merriam-Webster's Geographical Dictionary,* 3rd Edition).

"The origins of the Topography and Cartography Office (*Office de la Topographie et de la Cartographie*), date as far back as the 15th of July 1886, when the Topographic Service was born, in application of the Real Estate Law dated 1st July 1995. This service became the Topographic Division on the 1st of January 1968, then the Topographic and Cartographic Department on the 5th of January 1970. The Topography and Cartography Office was established on the 25th of December 1974, by law no. 74–100, as a public enterprise with an industrial and commercial character, under the supervision of the Ministry of the Equipment, Housing and Land Management" (*Republic of Tunisia Topography and Cartography Office.*)

Prior to independence in 1956, official topographic mapping was carried out by the *Service Géographique de l'Armée* (Geographic Service of the Army) and later by the *Institut Géographique National – IGN* (National Geographic Institute), both of France. The maps ranged in scale from 1:10,000 to 1:1,000,000. Some of the mapping was delegated by *IGN* to the *Service Topographique* of Tunisia, and it performed local surveys for civil use, reconnaissance work, and it also produced cadastral maps. In 1915, the French Navy Hydrographic Surveyor, Commandant Léon Pélissier, established a local Grid system for the Triangulation of *La Porte du Lac de Bizerte* where the origin of the local grid was at *Dhebel Iskeul* and where $(\varphi_o) = 37° 07' 22.1''$ N, and the Central Meridian $(\lambda_o) = 7° 19' 00.3''$ East of Paris. (*Annals Hydrographique, Paris, le févirer 1926.*) Note that X and Y were set to null, as usual for that time. Chances are that the local grid was computed on the Hatt Azimuthal Equidistant projection but was computed on the Germain ellipsoid. (*Annals Hydrographique, 3ᵉ série Tome Quatriéme Année 1921.*)

The general triangulation of Tunisia was started in 1883 and was completed in 1908. There are two main triangulation chains in Tunisia: the sector of the North African parallel running through Tunisia – the west-east chain prolonging the Algiers parallel – to the end of *Cap Bon peninsula*; and the chain of the meridian of *Gabés*, stretching from *Tunis* to *Métameur*. The coordinates were computed on the *Colonne Voirol Datum of 1875* (of Algeria), commonly termed Voirol 75. The fundamental point is at the geodetic pillar of the *Colonne Voirol Observatory* (near *Algiers*), and the astronomical coordinates are: $\Phi_o=36°$ 45' 07.9" N ($40^G 8357.8^{cc}$), $\Lambda_o=3°$ 02' 49.45" East of Greenwich ($0^G 7887.3^{cc}$ East of Paris). The reference azimuth *from south* to *Melab el Kora* is: $\alpha_o=322°$ 16' 52.7", and the ellipsoid of reference is the Clarke 1880 (*IGN*), where $a=6,378,249.2$ m and $1/f=293.4660208$. An interesting ancient comment about observations at *Colonne Voirol* was that the baseline originally measured with the bi-metallic apparatus of Porro in 1854 was re-observed with superior accuracy subsequently obtained with the mono-metallic apparatus of INVAR steel in 1912. (*Section de Géodésie du Service Géographique de l'Armée, 1924,* Bulletin Géodésique Année 1925, No. 8.) There was no net adjustment made. Computation of coordinates was carried out progressively with the fieldwork. The results of these computations are contained in 34 folios entitled "Description Geometrique de la Tunisia." Between 1912 and 1926 the first-order net was adjusted with three bases included in the adjustment. Coordinates were computed on the *Carthage Datum of 1925*. The first-order complementary and some second-order control were adjusted on the *Carthage Datum in 1934*. The elements of the Carthage Datum of 1925 are at the origin point being the astronomic station *Carthage Cathedral* (*observed in 1878*), where $\Phi_o=36°$ 51' 06.5" N, $\Lambda_o=10°$ 19' 20.645" East of Greenwich (07° 59' 06.7" East of Paris). The reference azimuth *from south* (from *Carthage Cathedral*) to *Marsa* is: $\alpha_o=180°$ 01' 00.9", and the ellipsoid of reference is the Clarke 1880 (*IGN*) (*James W. Walker, U.S. Army Map Service, 22 November 1957, and Mémorial du Dépôt Général de la Guerre, t. XI, fascicule 3, p. 382*). The original French mapping was cast on the ellipsoidal Bonne projection – the ubiquitous projection *du jour* for the Europeans of the time. The North African (ellipsoidal) Bonne Grid Latitude of Origin: $(\varphi_o)=35°$ 06' N ($39^G 00$ N), the Central Meridian $(\lambda_o)=2°$ 20' 13.95" East of Greenwich, and some time before World War II, the False Easting and False Northing were changed from zero to 100 km for each. Interestingly, this old Bonne Grid still influences current mapping in that Grid limits of the Lambert Conic Grids are still defined by Bonne Grid values. The sheet boundaries of the new Lambert Grids are commonly computed by a reversion of the late Prof. Karl Rinner's Bonne power series formulae published in *Zeitschrift für Vermessungswesen* during the 1930s. That reversion allows cartographers to compute the intersection of a constant Bonne Grid value with a chosen arc of the parallel or of the meridian. Those intersections then were used to define the limits with the graticule of the Lambert Conic Grids computed in 1974 by Mr. John W. Hager now retired from NGA. The projection used for the topographic mapping on the *Carthage Datum of 1925* was the Lambert Conformal Conic in two different secant zones: North and South Tunisia separated by the parallel of $38^G 50^{cc}$ and having the following parameters for North Tunisia the Latitude of Origin $(\varphi_o)=36°$ North (40^G), the Central Meridian $(\lambda_o)=11^G$ East of Greenwich, and the Scale Factor at Origin $(m_o)=0.999625544$, and for South Tunisia the Latitude of Origin $(\varphi_o)=33°$ 18' North (37^G), the Central Meridian $(\lambda_o)=11^G$ East of Greenwich also, and the Scale Factor at Origin $(m_o)=0.999625769$. (Note that only the Central Meridians have different parameters from the Algerian Lambert zones.) The False Origin is 500 km for Eastings and 300 km for Northings *for both zones*, the same convention as used in the Kingdom of Morocco (*PE&RS*, June 1999) and in Algeria. In 1970, the French *IGM* developed on contract to the U.S. Army Topographic Command (TOPOCOM – my old duty station), some nomograms for converting from the Carthage 1925 Datum of Tunisian to the European Datum of 1950. "The nomograms involved not only a mathematical conversion due to the change in ellipsoid and geodetic system, but also an experimental correction of the imperfect early adjustments of local geodetic systems. The nomograms are based on hasty maps. The values of the corrections are to be taken as functions of the geographic coordinates, disregarding the alignment of the coast line and the triangulation points whose location is very much guesswork" (*COL du JEU, Chief Military Geographic Section, Ministry of State for National Defense of France, 16 June 1970*). A subsequent caveat discussed the imperative for extraordinary care in dealing with coordinate transformations in this region of North Africa and stressed that the maximum accuracy likely obtained under the best of circumstances is "several meters and this appears to be satisfactory for scales of 1:50,000 or less."

The first general precision leveling net of Tunisia was begun between 1887 and 1889. After fairly long interruptions, fieldwork resumed in 1903 and continued intermittently until 1909. It was completed from 1913 to 1914. In the process 2,118 km of roads and railways were leveled by the Army Geographical Service. Nevertheless, no absolute elevations were calculated at this time. In 1920 the calculations were made following the methods of the French general leveling survey, and the whole network was adjusted by the method of least squares, despite certain excessive closure errors. The altitude datum for this network was the Porte de France console at Tunis. Its altitude of 6.9 m was calculated from the daily readings of the tide gauge installed in the port of La Goulette in 1889. (*That satisfies the geodetic/hydrographic convention of a minimum of one Metonic cycle of 18.6 years. – Ed.*)

The numerous errors which had crept into the observations, which came to light when the altitudes of certain benchmarks near the Algerian-Tunisian frontier (calculated under both the Algerian and the Tunisian system) were compared, together with the appreciable closure errors, disturbed the officials of the Topographical Department and led them to check calculations already made. Eventually certain sections were resurveyed, and four fascicles of altitudes were published in December 1926. Faced with this confused situation, the Topographical Department decided in 1959 to renew the precision leveling net completely.

The new Tunisian precision leveling network surveyed by the Topographical Department consists of eleven first-order traverses and eleven peripheral zones, covering a distance of some 3,039,900 km and comprising 1,400 reference points and 2,324 benchmarks. Three hundred and thirty benchmarks belonging to the old network were incorporated into the new one. The fieldwork was begun on 15 February 1959 and completed on 28 February 1962."

Photogrammetry was initially employed in Tunisia in 1935 and was furnished by foreign governments. By 1964, the Department acquired a Zeiss C8 Stereoplanigraph equipped with an Ecomat. (*A beautiful instrument model that I used and taught with for about 20 years at the University of New Orleans. – Ed.*) "In addition, a WILD A8 Autograph and WILD B8 stereoplotter was put into service at *Tunis* by the end of 1965. In 1963 a laboratory equipped both for rectifying (with a Zeiss SEG V rectifier) and for printing aerial photographs (Zeiss KG30 contact printer, Zeiss reducer) was added to the existing cartographic reproduction laboratory (equipped with a Klimsch Super Autohorika)" (*Cartographic Activities in Tunisia – Selim Benghachame, Chief of the Topographical Service*).

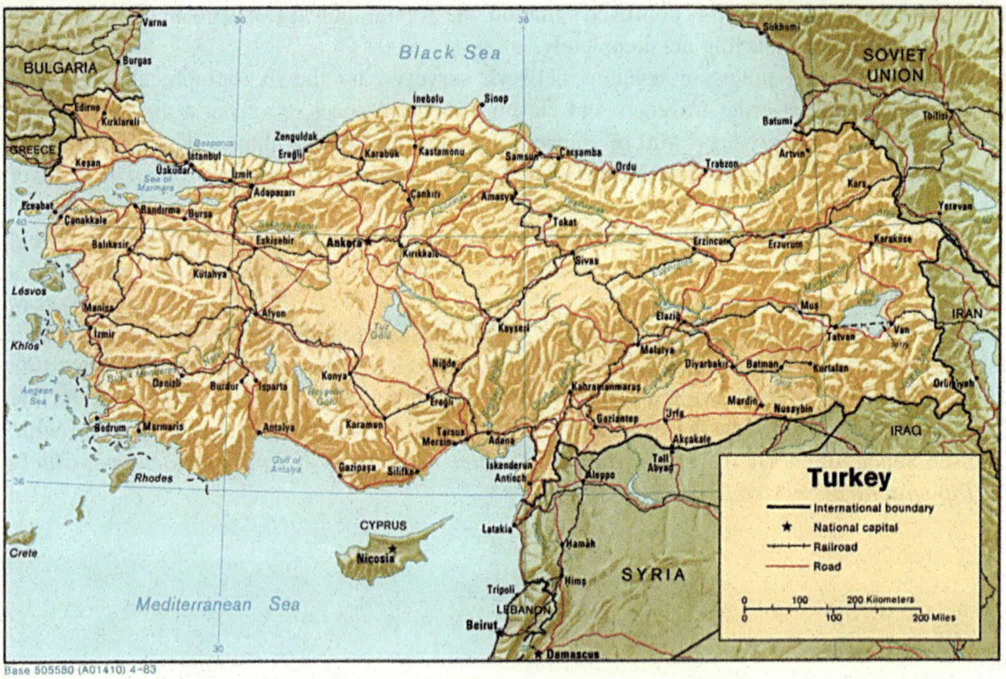

Base 505580 (A01410) 4-83

Republic of Turkey

The westernmost extremity of Turkey is Thrace, the European segment bounded by Bulgaria (240 km), and Greece (206 km), on the north by the Black Sea, on the Southwest by the Aegean Sea, and on the southeast by the Sea of Marmara. *Anatolia* is a peninsula surrounded on the north by the Black Sea, on the south by the Mediterranean Sea, Syria (822 km) and Iraq (352 km), on the west by the Aegean Sea, on the northwest by the Sea of Marmara, on the east by Iran (499 km), Azerbaijan (9 km), and Armenia (268 km), and on the northeast by Georgia (252 km).

The region was inhabited by an advanced Neolithic culture as early as 7,000 B.C., and metal instruments were in use by 2,500 B.C. The Hittite culture was replaced by several peoples until the 4th century B.C. when Alexander the Great of Macedonia conquered all of Anatolia. The Byzantine Empire ruled Anatolia until defeated by Seljuk Turkish forces in 1071 A.D. Troops of the Osmanli Dynasty moved into southern Europe and defeated Serbian forces at the battle of Kosovo in 1389. The Ottoman Empire flourished until multinational European forces drove Ottoman troops to cede substantial European territory in the Treaty of Karlowicz (1699). During World War I, the Turkish alliance with the Kaiser caused England, France, and Russia to declare war on the Ottoman Empire. After World War I, Atatürk, the founder of the present-day Republic repulsed a Greek attempt to expand Greece's postwar allotment of Ottoman territory. The Palestine was lost to the French and British Mandates, and the 1923 Treaty of Lausanne, negotiated between the Atatürk government and the Allies defined control of the Bosporus and the territorial extent of the new Republic of Turkey. During World War II, the government of Atatürk's successor, Ismet Inönü, maintained neutrality despite German pressure throughout the war. Turkey is now a member of NATO and of the UN and is becoming part of the European Union.

In 2004, the International Society of Photogrammetry held its convention in Turkey. Associated with that meeting, the General Command of Mapping in Ankara prepared a marvelous text available in a "pdf" file entitled *The Illustrated History of Turkish Cartography*. The oldest map in history was found at *Çatalhöyük* in central Anatolia and was dated as originating in 6,200 B.C. with Carbon$_{14}$ methods. The next major contribution to Turkish cartography was by Pirî Reis, born in the late 1460s. He was a pirate, naval admiral, captain of Egypt, cartographer, and master of oceanography. By today's standards, mapping of Turkey and the world by the Ottoman Turks was of a thematic nature through most of the 19th century. Large-scale mapping was accomplished by planetable and alidade, with control being established graphically through resection with the planetable. Much of Europe had been mapped in a similar fashion in the 18th century; with geodetic control being utilized mostly in the latter half of the 19th century.

The General Staff of the armed forces established the first Turkish surveying school in 1818. After the foundation of both *Mühendishane-i Berri-I Hümayun* (the Artillery School) and the Military Academy, the young officers were sent to France, England, and Prussia for education on western science and techniques. The first Cartographic Branch of the Armed Forces was formed in 1853. However, little geodetic surveying and mapping was performed for decades. Classical triangulation started in the *Vardar Bas*in and one year later, in May of 1896, a 7,235.520 m baseline was measured in *Eskişehir* with Brunner invar apparatus. One of the geodesists working on this project was Captain Mehmet Şevki, later to be promoted through the ranks to Major General and revered as the father of Turkish geodesy. Hydrographic surveys were beginning to be performed by the navy during these same years, and supporting triangulation was used for numerous harbor surveys and passages. In 1909, the survey office was upgraded to the Survey Commission. A series of topographic maps was started in 1911 and was based on the (1906–1923) Bonne projection, where the central meridian, $\lambda_o = 28°\ 58'\ 50.8188''$ East of Greenwich, the latitude of origin, $\varphi_o = 39°\ 36'\ 00''$ N, and the scale factor at origin, $m_o = 1.0$. Note that the longitude given above is but one of the various values published for the position of the center of the dome of the Aya Sofia mosque (originally built as the Cathedral of Constantinople), but the Turks actually used Aya Sofia as their Prime Meridian. (The relation of their Prime Meridian to Greenwich was immaterial to the usefulness of Turkish maps for Turkish applications.) The ellipsoid of reference was the Clarke 1880 *modified* by General Şevki, where $a = 6,378,249$ m and $b = 6,356,515$ m. Note that these parameters were published in 1333 of the Moslem calendar and were hand-written in *Old Turkish* (Arabic characters), the norm at the time, and the angular unit of measurement was the centesimal Grad ($360° = 400^G$).

Geodetic baselines measured before World War I included: *Bakirkoy (Istanbul)* 1909 (4396.703 m), *Edirne* 1911 (2919.240 m), *Adapazari* 1911 (3483.862 m), *Ezurum* 1912 (6127.396 m), and *Aleppo* 1915 (7477.857 m). The first-order triangulation performed during this period was mainly in support of the Balkan War of 1912. By then, Brigadier General Şevki was involved with triangulation and mapping that extended through much of European Turkey (*Edirne Region*) to the *Çanakkale Strait (Dardanelles)*. Remember that the Prime Meridian at the time for these topo sheets was the Aya Sofia mosque. Aerial photography was utilized by the Ottoman Empire during World War I (1914–1918), but it is thought to have been for reconnaissance purposes only. Note that during this same time period, the British were the first to use aerial photography for actual military mapping of Turkish fortifications in the Palestine during World War I in 1917.

The General Directorate of Mapping was established by law on May 2, 1925, in the newly formed Republic of Turkey. New classical triangulation necessitated the observation of baselines that included: *Ankara* 1925 (2050.426 m), *Izmir* 1928 (9536.311 m), *Dardanelles* 1929 (3250.321 m), *Eskişehir* 1931 (second observation), *Maltépé (Istanbul)* 1932 (4279.455 m), *Balikesir* 1933 (6106.972 m), and *Korkuteli (Bozova)* 1935 (6346.447 m). These post-World War I baselines were observed with the new French Carpentier invar apparatus. With the exception of the *Edirne baseline* that had an average error of 1/300,000, all the other baselines measured after the war had errors better than one part per million.

The first Turkish ventures into photogrammetry were in the 1930s, first with terrestrial applications and later with aerial. Foreign consultants carried out earlier experiments. In 1937, the first Zeiss C-5 Universal Stereoplanigraph was purchased, and German technicians were assigned to Ankara to assist in training as well as production. With the Werhmach's penchant for geodetic intelligence, Turkish military topographic sheets somehow found their way to Berlin. The new president of the republic, Ismet Inönü, insisted on maintaining neutrality during World War II. Nevertheless, there was great German interest in Turkey. The holiest place in all of Turkey is the Aya Sofia mosque. Being the national Prime Meridian, there was keen academic interest in the relation of the great dome with respect to Greenwich. The local datum for European Turkey was at the *Kandilli Observatory* where the geodetic coordinates are: $\varphi_o = 41° 03' 48.899''$ N and $\lambda_o = 29° 03' 55.2''$ East of Greenwich. This was also the origin for the new Gauss-Krüger Transverse Mercator projection for European Turkey (used from 1932 to 1946), and the relation of the great dome to Greenwich was necessary in order to recast the Bonne Grid to the TM Grid. The connecting survey between the Turkish first-order triangulation net and the mosque was assigned in 1934 to Major Niyazi, also an accomplished photogrammetrist. One of the first things the major had to do was find the center of the dome. "As it is known the upper sections of Aya Sofia are covered with lead plates. The lead layers near the pillar were removed, approximately 15 centimeters of earth underneath was dug, and a geodetic bronze rod was set with cement. The center of this little bronze rod is the head of Aya Sofia triangulation." Previous determinations of the position of Aya Sofia included: French Captain Gautier in 1816–1820 ($\Phi = 41° 00' 12''$ N, $\Lambda = 28° 59' 01''$), *Nautical Table London* in 1888 ($\Phi = 41° 00' 16''$ N, $\Lambda = 28° 58' 59''$), *Connaissance de Temps* in 1912 ($\Phi = 41° 00' 29.8''$ N, $\Lambda = 28° 58' 58''$), General Şevki ($\Phi = 41° 00' 26.56''$ N transferred from *Bakirkoy*), and Brigadier General Abdurraham ($\Phi = 41° 00' 31.034''$ N transferred from *Eskişehir*). Major Niyazi finally determined the coordinates of Aya Sofia as: ($\Phi = 41° 00' 30.0709''$ N, $\Lambda = 28° 58' 52.7238''$ West of Greenwich), but there was a problem. The major published his findings in *Old Turkish*! A certain couple of German Photogrammetric Technicians knew their photogrammetry and Zeiss C-5 Stereoplanigraphs very well, but they were apparently oblivious to geodetic publications published in *Old Turkish*. During World War II, there was a catastrophic military defeat of German and Axis forces in the European Turkish region of *Edirne*. The secret *Planheft* of the Wehrmach (German Army) mentioned that there was an unexplained shift in the longitudes (and Eastings) of native maps in the region of *Edirne*! I do not think this was just a "coincidence" that happened to affect indirect artillery fire control.

The following is another fascinating excerpt from Major Niyazi's description of the enormous field work associated with the Aya Sofia Dome Survey: "The geometric equipment of the post consisted on one large Hildebrandt theodolite, six heliotropes, five geodetic projectors, twelve storage batteries and one small Wild theodolite, while the participating staff consisted of assistants Lieutenants Hilki and Kemal, one permanent sergeant and twenty-one soldiers for general duty and heliotrope communications, besides myself. We also had tents and animals. Assigning three soldiers to each heliotrope we placed them at designated points, and to ensure the long stay of these troops on the mountaintops beside the first-order points, we gave them whatever supplies they needed. We also constructed caves as shelter in their long vigil. On some of these points we found caves already constructed in 1933, so we utilized them. In the matter of transportation, motorized transports first brought our equipment to the slope of the mountains, then horses, donkeys, ox and horse carts took it to the mountaintop. During the transportation of equipment, the officers of the post paid most attention to the transportation of the large theodolite and they saw to it that there was always an officer in charge of this instrument. After the instrument was taken out of the spring carriage one or more soldiers always carried it on their shoulders to the top of the mountain." The entire unit spent about a year in the field while observing the triangulation for the dome!

For Anatolia proper, the national system was the *Meşedağ Datum (near Ankara)*, where $\Phi_o = 39°$ 52′ 10.451″ N and $\Lambda_o = 32°$ 34′ 38.430″ East of Greenwich and $H_o = 3255$ m. Other local datums included the *Balikesir Datum of 1933* for the region south of the *Sea of Marmara (Mysia)*, where $\Phi_o \sim 39°$ 37′ N and $\Lambda_o \sim 27°$ 57′ E, and the *Söke Datum* for the region of southwest Turkey (*Caria*), where $\Phi_o \sim 37°$ 45′ N and $\Lambda_o \sim 27°$ 22′ 30″ E. I believe it is likely that all of these old datums were referenced to the Clarke 1880 ellipsoid as modified by General Şevki. The names of datums are often associated with the names of baseline terminals – not a coincidence.

According to a 1981 personal communication I had with Professor I. Kasim Yaşar of the Middle East Technical University in *Ankara*, the parameters of the (1932–1946) Gauss-Krüger Transverse Mercator were: "Latitude of Origin, $\varphi_o = $ Equator, Longitude of Origin, $\lambda_o = 27°$ E $-45°$ E in 3^g intervals with Zones for entire Turkey and was used with a scale factor at central meridian $m_o = 0.99992$ only for cadastral purposes and then abandoned. False Easting at origin = Zone number $\times 10^6 + 500,000$ meters, False Northing at origin = zero, Measurement unit = meter, and Ellipsoid name = International, Madrid 1924. Since 1946 this system was replaced with the identical system in all respects except that the zone width was changed to 6° and $m_o = 0.9996$ is still in use (*as of 1981 – Ed.*) and the system of projection is Gauss-Krüger for entire Turkey to a map scale 1/25,000. In addition to … this and only for cadastral maps of scale 1.5000, we employ again the Gauss-Krüger projection system in 3° interval width with the scale factor at the central meridian $m_o = 1.0000$."

The U.S. Army Map Service began computing the "Central European Adjustment" in the late 1940s after World War II. The adjustment was carried to Turkey through Greece and Bulgaria, in part using the Czarist Russian *Zapiski* journals of the 19th century in which the local origin was at the minaret of the main mosque in *Kyustendja* (now *Constanţa, Rômania*), where $\Phi_o = 44°$ 10′ 31″ North, $\Lambda_o = 28°$ 39′ 30.55″ East of Greenwich. Note that this longitude is a ***correction*** from that published for Rômania, thanks to Dr. Momchil Minchev of the Bulgarian Geoinformation Company. "The Central European Adjustment" was later renamed the *European Datum of 1950*. For the remainder of the 20th century, information on the geodetic foundation and the ED50 network in Turkey remained a military secret and its use was denied to all, *including to official civilian government surveyors!* The cadastral agencies were forced to establish their own networks, independent of the National Triangulation and National Leveling networks that were military secrets. In this respect, Turkey was following old European custom.

The times are changing for geodetic Turkey. *Official* transformation parameters for Turkey are now published by the International Association of Geodesy (IAG), the *Bundesamt für Kartographie und Geodäsie* (German Federal Office for Cartography and Geodesy), and Eurographics. Note that the European sign convention for rotations is *opposite* from the United States (and Australian) standard. Therefore, the U.S. standard sign convention for rotations is listed in the following parameters. In Turkey, for ED50 to ETRS89: $\Delta X = -84.1$ m, $\Delta Y = -101.8$ m, $\Delta Z = -129.7$ m, $R_x = +0.0''$, $R_y = +0.0''$, $R_z = +0.468''$, $\delta s = +1.05$ ppm. This seven-parameter transformation cannot be truncated to just a three-parameter translation only, without complete recalculation of the least squares solutions for only three parameters. Do not truncate the above-published rotation and scale change parameters! The transformation with the above seven parameters for Turkey is expected to yield positions of about two meters accuracy. An example test point published for Turkey on the European Datum 1950: $\varphi = 37°\ 08'\ 35.8''$ N, $\lambda = 28°\ 28'\ 25.32''$ E, which transforms to ETRS89: $\varphi = 37°\ 08'\ 32.07''$ N, $\lambda = 28°\ 28'\ 23.79''$ E. The mathematical equations for all the projections and datum transformations referred to herein are contained in the *Manual of Photogrammetry*, 5th and 6th editions.

The Turkish National Fundamental GPS Network (TNFGN – TUTGA) has been established in 2001 with some stations re-observed due to the 1999 earthquake. There are about 682 stations computed in ITRF 2005.0. The Turkish National Vertical Control Network (TNVCN-99) was established with the adjustment of 243 lines of 25,680 points with a total length of 29,316 km. The current geoid model is TG-09, and the General Command of Mapping is working on a plan for height modernization with a 1-cm Turkish geoid model (*2007–2011 Term Report of Turkish National Geodesy Commission*, General Command of Mapping, Ankara, 2011). Thanks to Cole Billeaud for research assistance.

Turkmenistan
International boundary
Welayat boundary
National capital
Welayat capital
Railroad
Road
0 200 Kilometers
0 200 Miles
Lambert Conformal Conic Projection, SP 12N/38N
KAZAKHSTAN
UZBEKISTAN
IRAN
AFGHANISTAN
Aral Sea
Caspian Sea
DASHHOWUZ WELAYATY
BALKAN WELAYATY
AHAL WELAYATY
LEBAP WELAYATY
MARY WELAYATY
Ashgabat
Turkmenbashi (Krasnowodsk)
Nebitdag
Gumdag
Gyzylarbat
Gyzylgaya
Bekdash
Yerbent
Dashhowuz
Urganch
Nukus
Qunghirot
Muynoq
Aqtau
Uchquduq
Nawoiy
Bukhoro
Samarqand
Qarshi
Charjew
Kerki
Gowurdak
Mary
Bayramaly
Tejen
Saragt
Gushgy
Esenguly
Bandar-e Torkeman
Sari
Gorgan
Shahrud
Bojnurd
Mashhad
Sabzevar
Neyshabur
Tayyebat
Gonabad
Deyhuk
Herat
Shindand
Delaram
Sheberghan
Meymaneh
Amu Darya
Garagum Kanaly
Murgap Deryasy
Sarygamysh Koli
Garabogazkol Aylagy
below sea level
Sumbar
Etrek Deryasy
Kashaf Rud
Base 802344 (R01131) 12-94

Turkmenistan

Bordered by Afghanistan (744 km), Iran (992 km), Kazakhstan (379 km), Uzbekistan (1,621 km), and the Caspian Sea (1,768 km); the terrain is flat to rolling sandy desert with dunes rising to mountains in the south and low mountains along the border with Iran. "Slightly larger than California, the lowest point is *Vpadina Akcha*naya (−81 m), *note: Sarygamysh Koli* is a lake in northern Turkmenistan with a water level that fluctuates above and below the elevation of *Vpadina Akchanaya* (the lake has dropped as low as −110 m) and the highest point is *Gora Ayribaba* (3,139 m)" (*World Factbook*, 2012).

"Present-day Turkmenistan covers territory that has been at the crossroads of civilizations for centuries. The area was ruled in antiquity by various Persian empires, and was conquered by Alexander the Great, Muslim crusaders, the Mongols, Turkic warriors, and eventually the Russians. In medieval times *Merv* (today known as *Mary*) was one of the great cities of the Islamic world and an important stop on the Silk Road" (*World Factbook*, 2012). "Merv suffered a number of attacks over the course of its history, but instead of being rebuilt on top of the older ruins, *Merv* slowly spread west. In total, five cities were constructed next to each other, largely because of the shifting rivers. The oldest section was the Erk Kala and in later centuries most people lived in the vast walled city called Sultan Kala. All of this was completely eradicated in 1221 under the onslaught of the Mongols. In 1218 Jenghiz Khan demanded a substantial tithe of grain from *Merv*, along with the pick of the city's most beautiful young women. The unwise Seljuq response was to slay the tax collectors. In retribution Tolui, the most brutal of Jenghiz Khan's sons, arrived three years later at the head of an army, accepted the peaceful surrender of the terrified citizens, and then proceeded to butcher every last one of the city's inhabitants, an estimated 300, 000 people" (*Lonely Planet*, 2012). "Annexed by Russia in the late 1800s, Turkmenistan later figured prominently in the anti-Bolshevik movement in Central Asia. In 1924, Turkmenistan became a Soviet republic; it achieved independence upon the dissolution of the USSR in 1991. Extensive hydrocarbon/natural gas reserves, which have yet to be fully exploited, have begun to transform the country. Turkmenistan is moving to expand its extraction and delivery projects. The Turkmen Government is actively working to diversify its gas export routes beyond the still important Russian pipeline network. In 2010, new gas export pipelines that carry Turkmen gas to China and to northern Iran began operating, effectively ending the Russian monopoly on Turkmen gas exports" (*World Factbook*, 2012).

By 1912, longitudes had been transmitted via telegraph from *Tashkent*, Uzbekistan to *Aşgabat, Merv,* and *Türkmenbaşy*. Also, a line of Astro Station observations had been performed from *Türkmenbaşy* up to the Aral Sea and then down southeast along the border with Uzbekistan to the tripoint with Afghanistan. These observations were presumably calculated on the Bessel 1841 ellipsoid, where $a = 6,377,397.15$ m and $1/f = 299.1528$. After the Russian revolution and the founding of the Soviet, geodetic work progressed towards standardization. The country is still on the old Soviet datum, "System 42" with origin at Pulkovo Observatory, where $\Phi_o = 59°\ 46'\ 18.55''$ North, $\Lambda_o = 30°\ 19'\ 42.09''$ East of Greenwich, the defining azimuth at the point of origin to Signal A is: $\alpha_o = 317°\ 02'\ 50.62''$ and the ellipsoid of reference is the Krassovsky 1940, where $a = 6,378, 245$ m and $1/f = 298.3$. The Russia Belts are a Grid System identical to UTM except that the scale factor at origin is $m_o = 1.0$. The closest transformation parameters available for Turkmenistan are the three-parameter transformation from System 42 to WGS 84 in the *Caspian Sea area* of Kazakhstan are: $\Delta X = +14.471$ m, $\Delta Y = -132.753$ m, $\Delta Z = -83.454$ m, and are based on an occupation at 5 stations: *Bolat, Yesim Sevirne, Dlinnaya Dolina, Daralsai,* and *Aul*.

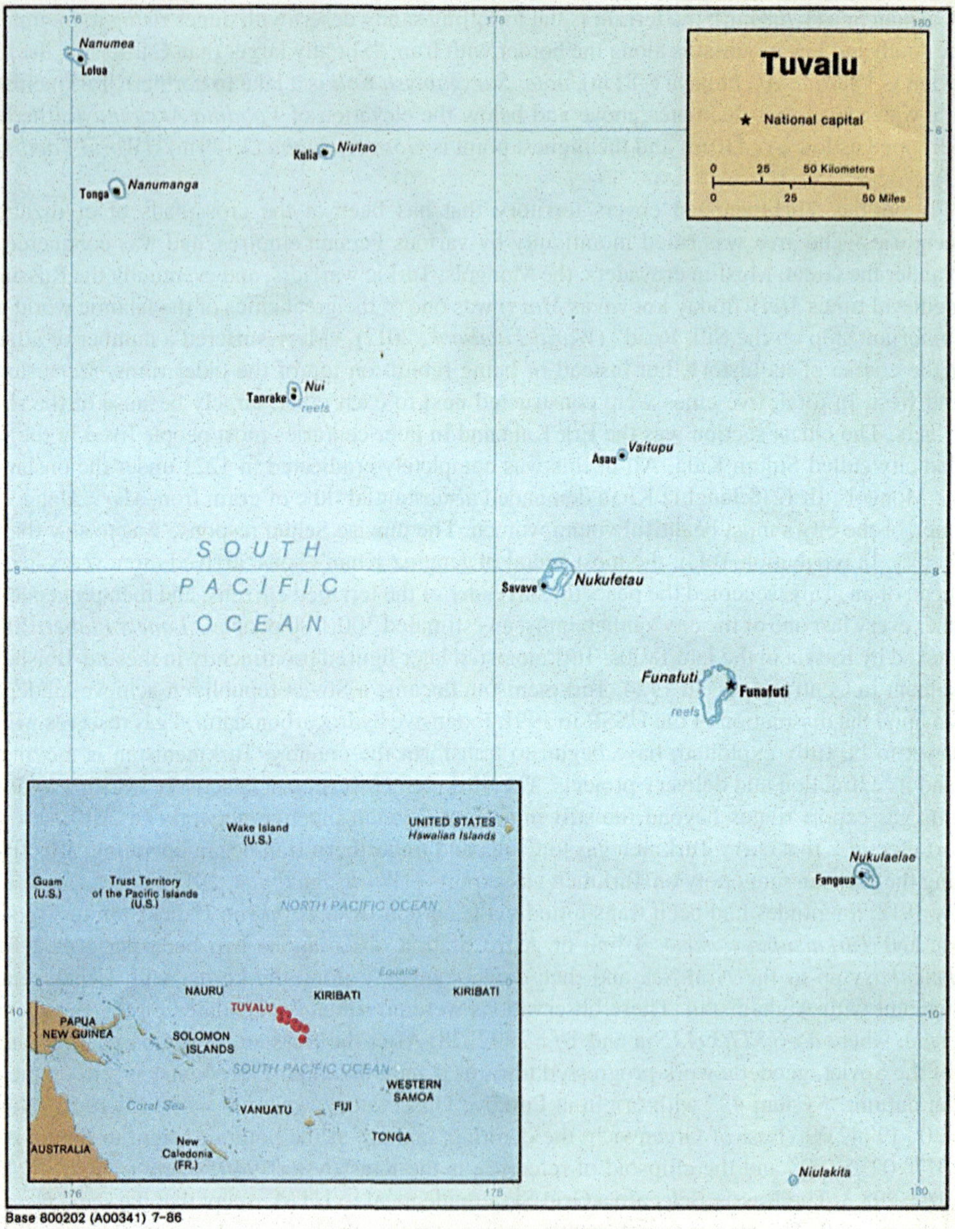

Base 800202 (A00341) 7-86

Tuvalu

Tuvalu comprises a chain, 580 km long, of nine coral atolls in the Pacific Ocean just west of the International Date Line. The total land area of the densely populated group is only 26 km² (10 mi²); however, the islands occupy 1.3 million km² (500,000 mi²) of ocean between *Kiribati* and the *Samoas*. Five of the islands are low-lying coral atolls; the highest point on these (and the highest point in *Tuvalu*) is just 4.6 m (15 ft) above sea level. The remaining four islands are pinnacles of land that rise up from the sea bed. On the islands, there are many reefs and salt-water ponds while the island of *Nanumea* hosts a freshwater pond, rare for an atoll. Made mostly of eroded coral, *Tuvalu* has poor soil, no streams or rivers, and few remaining outcrops of forest. Coconut palms grow in abundance across all the islands, but otherwise there is only enough soil to support subsistence agriculture for about three-quarters of the population. All other food is imported. Water needs are met by catchment and storage facilities because the porous, low-lying atolls are unable to hold ground water. The only land animals are the Polynesian rat, chickens, dogs, and pigs — all introduced species. Niulakita has no lagoon but has a swamp at its center. Because it has never had a permanent population, the southernmost island was not taken into account in the naming of the *Tuvalu* group. *Tuvalu* means "eight standing together." The climate is tropical, with an average temperature of 30°C (87°F) and little seasonal variation. The wet season is between October and March, and 350 cm (12 ft) of rain falls in a normal year. Cyclone (hurricane) activity is rare; there have been only four severe hits this century (but all since 1972).

Physiographically, *Tuvalu* belongs in Micronesia, but culturally, the islands belong to Polynesia. Language, traditions, and artifacts indicate that Polynesians from *Tonga* and *Samoa* in the southeast arrived in the island group early in the 14th century. In 1597, Don Alvaro de Mendaña y Neyra cruised through the coral atolls of Tuvalu. Further European contact came in the late 18th century, and the last of the islands were charted by 1826. They were named the *Ellice Islands* after the British member of parliament who owned the ship that first landed on *Funafuti Atoll* in 1819. In 1892, the islands became part of the *Gilbert* and *Ellice Islands* protectorate and, in 1916, became a Crown Colony. During World War II the United States used Tuvalu's northernmost atoll, *Nanumea*, as a base to repel the Japanese who were threatening the Gilbert Islands. Wrecks of air and sea craft remain on the island.

From the 1960s through 1977, Tuvaluans embarked on steady constitutional development. In 1974, the Polynesian Ellice Islanders voted to separate from the Micronesian Gilberts. They reverted to their pre-colonial name of Tuvalu and attained independence on October 1, 1978.

Hydrographic surveys were undertaken by the USS Sumner in 1943, the USS Hydro in 1944, and by the H.M.S. Cook from 1959 to 1963. Between 1962 and 1966 the HIRAN trilateration of the southwest Pacific was undertaken by the U.S. Air Force. A number of primary stations were established. These were originally expressed in terms of the *WGS 60 Datum, the Australian National Datum of 1966,* as well as being converted to the *Fiji Datum of 1956*. The British Directorate of Overseas Surveys (DOS) carried out survey work between 1968 and 1973, expressing their values in terms of local astronomic datums for individual islands. Some stations were linked to the HIRAN survey and hence were expressed in terms of *Fiji 56*. In 1974, the Royal Military Survey of the United Kingdom decided that where possible island areas should be positioned on *WGS 72*; hence, where possible the *Fiji 56* datums were converted to the *WGS 72 Datum* using cartesian shifts. In addition, shifts have been established between the local astro datums and Fiji 56 for some of the islands, but other values are held only in local datum terms for some island areas. In 1984 and 1985, the Australians carried out "Operation ANON." This provided 16 Doppler fixes to many points in the Tuvalu Group, yielding coordinates in terms of the WGS 72 Datum. The primary objective was to provide the government of Tuvalu with sufficient survey data to enable them to determine base points for the definition of their Exclusive Economic Zone.

The survey work for Operation ANON in 1984 and 1985 used 11 points from earlier surveys by the Australians. Because all field work was carried out on the WGS 72 Datum, the points were converted to the WGS 84 Datum using the standard NGA WGS72 to WGS84 transformation: $\Delta X = 0$ m, $\Delta Y = 0$ m, $\Delta Z = 4.5$ m, k= 0.219 ppm, and $R_z = 0.554''$.

On *Nanumea*, the northernmost of *Tuvalu's atolls*, the origin of the *Nanumea Sodano Astro Datum of 1966 at Laken Island* is $\Phi_o = 05° 39' 04.59''$ S, $\Lambda_o = 176° 04' 31.09''$ East of Greenwich, and the ellipsoid of reference is the International 1924, where $a = 6,378,388$ m and $1/f = 297$. The vertical datum is based on readings of the tide levels made on the ocean side (at the seaward end of the LST wreck) and in the lagoon (at NME 25), and these two staff gauges were connected by height traversing and also to the Post Office. From the *Nanumea Sodano Astro Datum to the WGS84 Datum*, $\Delta X = +225$ m, $\Delta Y = -114$ m, and $\Delta Z = -148$ m; the accuracy of this transformation is estimated to be ±2 m in each of Eastings and Northings. The *Nanumea* TM Local Grid is based on the Transverse Mercator projection where the Latitude of Origin is at the equator ($\varphi_o = 0°$), the Central Meridian $\lambda_o = 176° 06'$ E, the False Easting = 45 km, and the False Northing = 8,000 km. The Scale Factor at Origin is unity ($m_o = 1.0$).

On the island of *Nanumaga*, the origin of the *NMG 1 Astro Datum of 1974* is $\Phi_o = 06° 17' 15.04''$ S, $\Lambda_o = 176° 18' 52.86''$ East of Greenwich, and the ellipsoid of reference is the International 1924. The vertical datum is based on heights observed as part of the traversing. They have been related to mean sea level by 2 days of readings on a staff gauge set up in the boat channel. Two tide poles were erected on the reef opposite the Government flagstaff, and continuous observations were obtained for 2.5 days. From the NMG 1 Astro Datum of 1974 to the WGS84 Datum, $\Delta X = +204$ m, $\Delta Y = -31$ m, and $\Delta Z = +113$ m; the accuracy of this transformation is estimated to be between ±2 m and ±26 m in each of Eastings and Northings. The *Nanumaga* TM Local Grid is based on the Transverse Mercator projection where the Latitude of Origin is at the equator ($\varphi_o = 0°$), the Central Meridian $\lambda_o = 176° 19'$ E, the False Easting = 35 km, and the False Northing = 7,000 km. The Scale Factor at Origin is unity ($m_o = 1.0$).

On the *atoll of Nukufetau*, the location of the World War II airfield on *Motulalo Island*, the origin of the *NFT 1 Astro Datum of 1974* is $\Phi_o = 08° 01' 40.28''$ S, $\Lambda_o = 178° 18' 48.37''$ East of Greenwich, and the ellipsoid of reference is the International 1924. From the NFT 1 Astro Datum of 1974 to the WGS84 Datum, $\Delta X = +200$ m, $\Delta Y = -83$ m, and $\Delta Z = +96$ m; the accuracy of this transformation is estimated to be ±7 m in Eastings and ±21 m in Northings. The *Nukufetau* TM Local Grid is based on the Transverse Mercator projection, where the Latitude of Origin is at the equator ($\varphi_o = 0°$), the Central Meridian $\lambda_o = 178° 22'$ E, the False Easting = 40 km, and the False Northing = 6,000 km. The Scale Factor at Origin is unity ($m_o = 1.0$).

On the solitary coral island of *Niulakita*, I guess that the origin of the *Niulakita Astro Datum of 1965* is $\Phi_o = 10° 47' 21.6059''$ S, $\Lambda_o = 179° 27' 51.7081''$ East of Greenwich, and the ellipsoid of reference is the International 1924. From the Niulakita Astro Datum of 1965 to the WGS84 Datum, $\Delta X = +184$ m, $\Delta Y = -465$ m, and $\Delta Z = +119$ m; the accuracy of this transformation is estimated to be ±10 m in Eastings and ±19 m in Northings. The *Niulakita* TM Local Grid is based on the Transverse Mercator projection where the Latitude of Origin is at the equator ($\varphi_o = 0°$), the Central Meridian $\lambda_o = 179° 28'$ E, the False Easting = 15 km, and the False Northing = 3,000 km. The Scale Factor at Origin is unity ($m_o = 1.0$). A century ago, workers excavated guano here for commercial fertilizer. Later an Australian company used the island as a coconut plantation, and in 1944 the British government purchased the island and gave it to overpopulated Niutao, which relocated a few families there.

On the *atoll of Niutao*, the origin of the *NTO 1 Astro Datum of 1973* is $\Phi_o = 06°06'29.25''$ S, $\Lambda_o = 177°19'59.16''$ East of Greenwich, and the ellipsoid of reference is the International 1924. The vertical Datum at NTO 2 is based on a personal estimate of probable mean sea level! From the *NTO 1 Astro Datum of 1973* to the WGS84 Datum, $\Delta X = +219$ m, $\Delta Y = -198$ m, and $\Delta Z = -92$ m; the accuracy of this transformation is estimated to be ±15 m in Eastings and ±4 m in Northings. The *Niutao* TM Local Grid is based on the Transverse Mercator projection, where the Latitude of Origin is at the equator ($\varphi_o = 0°$), the Central Meridian $\lambda_o = 177° 20'$ E, the False Easting = 30 km, and the False Northing = 5,000 km. The Scale Factor at Origin is unity ($m_o = 1.0$).

The *atoll of Nukulaelae* is the easternmost of the Tuvalu islands and was the first island to accept Christianity. Because of rising sea level, *Nukulaelae* is threatened by saltwater seeping into the taro swamps. For Nukulaelae, the origin of the *Nukulaelae Astro Datum of 1965*, actually on Fanagua Island, is unknown, but the ellipsoid of reference is the International 1924. From the *Nukulaelae Astro Datum of 1965* to the WGS84 Datum, $\Delta X=+254$ m, $\Delta Y=-238$ m, and $\Delta Z=-234$ m; the accuracy of this transformation is estimated to be ±13 m in Eastings and ±14 m in Northings. The *Nukulaelae* TM Local Grid is based on the Transverse Mercator projection where the Latitude of Origin is at the equator ($\varphi_o=0°$), the Central Meridian $\lambda_o=179°\ 50'$ E, the False Easting=25 km, and the False Northing=2,000 km. The Scale Factor at Origin is unity ($m_o=1.0$).

On the *atoll of Vaitupu*, the educational center of Tuvalu, I guess that the origin of *Vaitupu Atoll Datum* is at point VTZ 1 Astro: $\Phi_o=07°\ 29'\ 24.710''$ S, $\Lambda_o=178°\ 41'\ 52.31''$ East of Greenwich, and the ellipsoid of reference is the International 1924. From the *Vaitupu Island Datum* to the WGS84 Datum, $\Delta X=+193$ m, $\Delta Y=+61$ m, and $\Delta Z=+201$ m; the accuracy of this transformation is estimated to be ±23 m in Eastings and ±26 m in Northings. The Vaitupu TM Local Grid is based on the Transverse Mercator projection, where the Latitude of Origin is at the equator ($\varphi_o=0°$), the Central Meridian $\lambda_o=178°\ 41'$ E, the False Easting=10 km, and the False Northing=1,000 km. The Scale Factor at Origin is unity ($m_o=1.0$).

On the capital *atoll of Funafuti*, the coordinates of the origin of the *UF5 Astro Datum of 1973* are unknown but the ellipsoid of reference is the International 1924. The vertical datum is based on an automatic tide gauge situated on the main jetty at *Fongafale*, and 48-hour readings were obtained. The gauge is run by the University of Hawaii. From the UF5 Astro Datum of 1973 to the WGS84 Datum, $\Delta X=+189$ m, $\Delta Y=+783$ m, and $\Delta Z=+256$ m. The Funafuti TM Local Grid is based on the Transverse Mercator projection where the Latitude of Origin is at the equator ($\varphi_o=0°$), the Central Meridian $\lambda_o=179°\ 08'$ E, the False Easting=50 km, and the False Northing=9,000 km. The Scale Factor at Origin is unity ($m_o=1.0$).

The only Micronesian community in Polynesian Tuvalu is on *Nui*. Nui is 255 km northwest of Funafuti Island. On the atoll of Nui, the coordinates of the origin of the *Nui Astro Datum of 1965* are $\Phi_o=07°\ 13'\ 40.09''$ S, $\Lambda_o=177°\ 09'\ 47.26''$ East of Greenwich, and the ellipsoid of reference is the International 1924. The vertical datum is based on ocean and lagoon tide levels read over four weekends. The ocean staff gauge was off the *Maneapa*, and the lagoon gauge was at the Government station. The two gauges were connected by height traverse to the local control and to the Post Office for height of the Meteorological Service mercury barometer. From the *Nui Astro Datum of 1965* to WGS84 Datum, $\Delta X=+259$ m, $\Delta Y=+217$ m, and $\Delta Z=-246$ m. The Nui TM Local Grid is based on the Transverse Mercator projection where the Latitude of Origin is at the equator ($\varphi_o=0°$), the Central Meridian $\lambda_o=177°\ 09'$ E, the False Easting=20 km, and the False Northing=4,000 km. The Scale Factor at Origin is unity ($m_o=1.0$). Thanks to Jane Resture for details of local culture. Thanks for all technical details of the complex coordinate systems of Tuvalu go entirely to Karen French, Geodetic Branch, Defence Geographic Centre of the British Military Survey.

The Pacific Islands Geospatial and Surveying Strategy 2017–2027 is a 10-year regional plan for developing geospatial and surveying capacity. The Tuvalu Geodetic Survey Project 2016-Phase I was supported by government funding and consisted of 4 weeks for field surveys on three islands: *Vaitupu, Nukufetau,* and *Funafuti*.

In 2017, the completion of Phase II of the Geodetic Survey for the four northern Islands (*Nanumea, Nanumaga, Niutao,* and *Nui*) included Geodetic survey, Cadastral survey, Topo survey, UAV/Drone survey, Tide Monitoring to establish mean sea level or MSL, lowest astronomical tide or LAT, and highest astronomical tide or HAT on these four islands. The year 2018 saw completion of the last Phase III of the Geodetic Project on *Nukulaelae, Niulakita,* and on *Funafuti* again.

- http://ggim.un.org/meetings/GGIM-committee/7th-Session/side_events/2%20-%20 Faatasi%20Malologa.pdf.
- https://tuvalutrustfund.tv/wp-content/uploads/2018/03/2018-TUVALU-NATIONAL-BUDGET.pdf.

REFERENCES: T

2007–2011 Term Report of Turkish National Geodesy Commission, General Command of Mapping, Ankara, 2011.

2011 Tonga Visitors Bureau.

A.G. Bazley, Geographical positions in Malaya and Siam, *Survey Review,* vol. IV, no. 30, pp. 450–457, October 1938.

Account of the Operations of the Great Trigonometrical Survey of India; Volume XVII; ElectroTelegraphic Longitude Operations Executed During the Years 1894-95-96; The Indo- European Arcs from Karachi to Greenwich, Dehra Dun, 1901.

Andrew M. Glusic, Triangulation in Turkestan and its Connection with India, Army Map Service Technical Report No. 21, February 1957, 100 pages.

Anglo–French exchange of notes on March 18, 1904–April 25, 1904.

Anglo–French exchange of notes on May 24, 1906–July 19, 1906.

Annals Hydrographique, 3e série Tome Quatriéme Année, 1921.

Annals Hydrographique, Paris, le févirer, 1926.

Army Map Service Geodetic Memorandum File 773.0303, 7 June 1950.

Bomford, Major G., Survey of India Professional Paper No. 28; The Readjustment of the Indian Triangulation, Dehra Dun, 1939.

Bundesamt für Kartographie und Geodäsie, 2005.

Capt. G. M. Latham, R. E., The triangulation of Tobago, 1923.

Cartographic Activities in Tunisia – Selim Benghachame, Chief of the Topographical Service, 1963.

Chang & Tseng, A Geocentric Reference System in Taiwan, *Survey Review*, vol. 35, no. 273, pp. 195–203, July 1999.

CIA World Factbook, 2006.

COL du JEU, Chief Military Geographic Section, Ministry of State for National Defense of France, 16 June 1970.

COL George Massaad, Director of Geographic Affairs of the Lebanese Army, Personal Communication, November 1997.

Cole Billeaud, Personal Communication, 2005.

Commandant Léon Pélissier, Triangulation of La Porte du Lac de Bizerte, 1915.

D. L. Leach, General Notes on the Primary Control for the Cadastral Survey of Tonga.

Dr. Gil Richards, University of New Orleans, 2007.

Dr. Lars Sjöberg, Personal Communication, 7 August 1981.

Dr. Lars Sjöberg, Personal Communication, 7 November 1980.

Dr. Momchil Minchev of the Bulgarian Geoinformation Company, Personal Communication, 2001.

Dr. Rui Fernandes, Relatório – Cálculo das Coordenadas, 31 Dezembro 2005.

Dr. von G. Pinkwart, Allgemeine Vermessungs-Nachrichten, Nos. 11 *et sequitur*, 1926, Hanover.

Folau, V. & Malolo, T.L., A Report on the Geodetic Infrastructure of the Kingdom of Tonga, Ministry of Lands, Environment, Climate Change and Natural Resources, Kingdom of Tonga, September 2013, Fiji.

Franco–German convention of July 23, 1897.

G. A. Rune, Tabeller Till Gauss Hannoverska Projektion, 1951.

Gulatee, B. L., Deviation of the Vertical in India, Survey of India, 1955.

http://ggim.un.org/meetings/GGIM-committee/7th-Session/side_events/2%20-%20Faatasi%20Malologa.pdf.

http://www.bernese.unibe.ch/publist/publist_2015.php.

https://iag.dgfi.tum.de/fileadmin/IAG-docs/NationalReports2007/Switzerland.pdf.

https://tuvalutrustfund.tv/wp-content/uploads/2018/03/2018-TUVALU-NATIONAL-BUDGET.pdf.

https://www.swisstopo.admin.ch/en/knowledge-facts/surveying-geodesy/reference-frames/transformations-position.html.

Infoplease, 2014.

Ing. Dr. Adolfo C. Romero, Provisional South American Datum of 1956 (PSAD 56), Personal Communication, December 1986.

International Boundary Study No. 121 – April 7, 1972, Libya-Tunisia Boundary, Department of State Bureau of Intelligence and Research, Issued by the Geographer.

International Boundary Study, No. 128–29 September 1972, Burkina Faso (Upper Volta) – Togo Boundary, U.S. Department of State.

Isaac C. Lail, Army Map Service Geodetic Memorandum No. 1600, PakistanIndiaNorth Burma Conversion of the Indian Datum "Published Values" to South Asia Datum, Washington, 10 November 1966.

J. W. Walker, Geodetic Memorandum No. 1692, 1975 Adjustment of the Primary Triangulation of Thailand, Washington, June 1976.
James W. Walker, U.S. Army Map Service, 22 November 1957.
John W. Hagar, Personal Communication, October 2000.
John W. Hager, Personal Communication, August 2001.
John W. Hager, Personal Communication, January 2007.
John W. Hager, Personal Communication, January 2008.
John W. Hager, Personal Communication, October 2010.
Jordan-Eggert, Handbuch der Vermessungskunde, vol. iii/1923, p. 737.
Karen French, Geodetic Branch, Defence Geographic Centre of the British Military Survey, Personal Communication, November 2001.
La Carte d'Egypte et de Syrie, Dépôt de la Guerre, 47 feuilles, 1808.
Land Survey and Large-Scale Mapping in Sub-Saharan Africa, UN Centre for Human Settlements (Habitat), Nairobi, 2001.
Leach and Browne, Reg. Surveyors, Instructions for the Cadastral Survey of the Kingdom of Tonga, 1958, Milford in Auckland, NZ.
Lonely Planet, 2008.
Lonely Planet, 2010.
Lonely Planet, 2012.
Lonely Planet, 2013.
Manual of Photogrammetry, 5th & 6th editions.
Markham, Clements R., A Memoir on the Indian Surveys, London, 1871.
Mémorial du Dépôt Général de la Guerre, t. XI, fascicule 3, p. 382.
Merriam-Webster's Geographical Dictionary, 3rd edition.
Miranda H. Mugnier, Personal Communication, 2022.
Mitt. aus den Deutschen Schutzgebieten, Erganzungsheft 23, Berlin, 1910, pp. 49–56.
Morag Bell, Geography and Imperialism 1820–1940, Manchester University Press, 1995, 338 pages.
P.G. Rosen, Den vid Svenska Topografiska Kartverket använder projektionsmetoden, 32 pp., 1876.
P.V. Angus-Leppan & I.P. Williamson, A Project for Upgrading the Cadastral System in Thailand, Survey Review, Vol. 28, 215, January 1985, pp. 2–14 & April 1985, pp. 63–73.
Paul D. Thomas, Conformal Projections.
Paula Santos, Instituto de Investigação Científica Tropical (IICT), Personal Communication 2015.
Pedro de Ibarrarte, Spanish Governor of Trinidad, 1787.
Planheft, Wehrmach, 1943.
Prof. Chalermchon Satirapod, Chulalongkorn University, Personal Communication, November 2010.
Prof. Frank Tsai, Louisiana State University, Personal Communication, 2010.
Prof. Ian Williamson, University of Melbourne, Personal Communication, November 2010.
Prof. Itthi Trisirisatayawong, Chulalongkorn University, Personal Communication, November 2010.
Prof. Karl Rinner, Zeitschrift für Vermessungswesen, 1930s.
Professor I. Kasim Yaşar, Middle East Technical University, *Ankara*, Personal Communication, 1981.
Professor im Generalstabe Karl D. P. Rosén, Stockholm 1933.
Rens, J, & Merry, C.L., Datum Transformation Parameters in Southern Afri*ca,* Survey Review, 30, 236, April 1990.
Republic of Tunisia Topography and Cartography Office, 2006.
Romuald Kaczynski, Transfer of Technology for Cadastral Mapping in Tajikistan Using High Resolution Satellite Data, ISPRS Archives, Vol. XXXIX-B6, 2012 XXII ISPRS Congress, 25 August – 01 September 2012, Melbourne, Australia.
Schéma de la Projection Stereographique, Travaux du Cadastre et d'Amélioration Agricole des États, de Syrie, des Alaouites et du Liban sous Mandat Français, 1922.
Section de Géodésie du Service Géographique de l'Armée, 1924, Bulletin Géodésique Année 1925, No. 8.
See also IBS Numbers 124 & 126.
Smith, Jr. R., The Connection between the Struve Geodetic Arc and the Arc of the 30th Meridian, FIG, September 2005.
South Africa Chief Directorate: Surveys and Land Information, Personal Communication, 1995.
Taiwan U.S. Departmnet of State Background Notes, 2010.
Thailand Library of Congress Country Study, 2009.
The Illustrated History of Turkish Cartography, 2004.

Toru Watanabe, Study on Urgent Establishment of Topographic Mapping in East Timor – Creation of Topographic Information for Establishment of Cadastre in East Timor, Japan International Cooperation Agency, 2000.
TR 8350.2, 2008.
TR 8350.2, 2010.
World Factbook, 2010.
World Factbook, 2012.
World Factbook, 2013.
World Factbook, 2014.
World Factbook, 2015.
World Factbook, 2022.
Yang, Tseng, and Yu, Establishment and maintenance of TWD97, *ASCE Journal of Surveying Engineering*, November 2001, pp. 119–132.
You & Hwang, Coordinate Transformation between Two Geodetic Datums of Taiwan by Least-Squares Collocation, *ASCE Journal of Surveying Engineering*, May 2006.
Zapiski VTU, vol. 69 Pt. 2 p. 235; Geodezist vol. 11–12, p. 30.

U

Republic of Uganda 835
Ukraine 839
United Arab Emirates 843
United Kingdom 847
United States of America 853
Oriental Republic of Uruguay 859
Republic of Uzbekistan 863

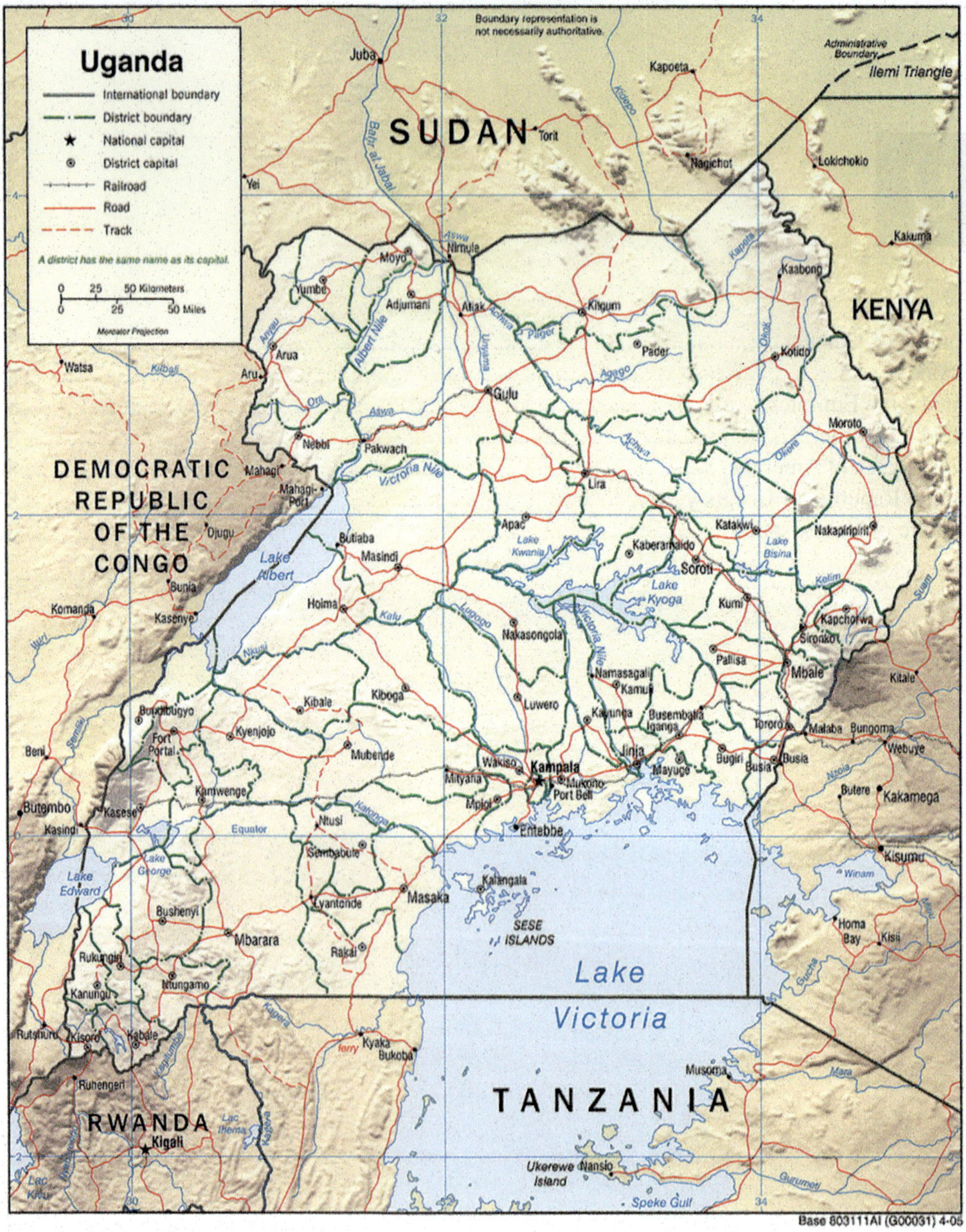

Base 803111AI (G00031) 4-05

Republic of Uganda

Uganda is bordered by the Democratic Republic of the Congo (Kinshasa) (765 km), Kenya (933 km), Rwanda (169 km), South Sudan (435 km), and Tanzania (396 km); slightly smaller than Oregon, the terrain is comprised mostly of plateau with a rim of mountains. The lowest point is Lake Albert (621 m), and the highest point is Margherita Peak on Mound Stanley (5,110 m).

"Uganda experienced two great waves of migration. The first brought the Bantu-speaking peoples from further west in Africa, and the second, the Nilotic people from Sudan and Ethiopia. These broad families are still geographically split today, the Bantu in the center and south of the country and the Nilotic peoples in the north. Until the 19th century, landlocked Uganda saw few outsiders compared with its neighbors. Despite fertile lands and surplus harvests, trading links with the great Indian Ocean ports were limited. During the reign of the Bugandan *kabaka* (king) Mwanga in the mid-19th century, contacts were finally made with Arab traders and early European explorers" (*Lonely Planet*, 2012). "The colonial boundaries created by Britain to delimit Uganda grouped together a wide range of ethnic groups with different political systems and cultures. These differences prevented the establishment of a working political community after independence was achieved in 1962. The dictatorial regime of Idi Amin (1971–1979) was responsible for the deaths of some 300,000 opponents; guerrilla war and human rights abuses under Milton Obote (1980–1985) claimed at least another 100,000 lives. The rule of Yoweri Museveni since 1986 has brought relative stability and economic growth to Uganda" (*World Factbook*, 2012).

Prior to becoming a protectorate of the British Empire, the concept of land ownership was unknown in Uganda. The British established the Mailo System in which land parcels were assigned to various levels of local royalty with the British Crown taking ownership of areas undesired by Ugandans. The sheer size of the number of parcels necessitated the training and employment of local Ugandans as surveyors, with mixed results with regard to the quality of the surveys and descriptions of calls. Consequently, a system of geodetic control was recognized as a requirement for a successful cadaster. The first datum established in Uganda Protectorate by the British was at origin point *Busowa M.T.S.* (*M.T.S. = Main Triangulation Survey – Ed.*) in 1907, where $\Phi_o = 00°\ 46'\ 01.492''$ N, $\Lambda_o = 32°\ 12'\ 59.324''$ East of Greenwich, and the azimuth to *Katwe Kangili M.T.S.* was $\alpha_o = 320°\ 01'\ 56.4''$ North by East (*clockwise – Ed.*). The ellipsoid of reference was the Clarke 1858, where $a = 6,378,235.6$ m and $1/f = 294.2606768$.

In 1929, a new datum was introduced with a new origin at *Igurua* (Arc of Meridian) (The "Arc of Meridian" refers to points included in the famous survey of the 30° East meridian which was from South Africa to Finland, Russia, and Norway – Ed.), where $\Phi_o = 00°\ 57'\ 18.173''$ S, $\Lambda_o = 30°\ 21'\ 52.756''$ East of Greenwich, and the azimuth from *Kicherere* (Arc of Meridian) to *Karamrani* (Arc of Meridian) was $\alpha_o = 340°\ 59'\ 52.11''$ North by East. The ellipsoid of reference was the Clarke 1858 again. "Rectangular spheroidal co-ordinates are plotted in 3 zones of Cassini-Soldner co-ordinates based on the following origins: – *Busowa M.T.S.* (1907 determination), latitude 0° 46' 01.492" North, longitude 32° 12' 59.324" East. *Busoga* Origin, latitude 0° 30' 00" North, longitude 33° 30' 00" East. Toro Origin, latitude 0° 30' 00" North, longitude 30° 30' 00" East. Apparently, the above particulars refer to work done before the years 1939–1940, when it was decided to replace the Cassini-Soldner co-ordinates by the Transverse Mercator system.

"The report does not include among the other geodetic data any particulars of the origin, *etc.*, of this system, and the only reference to it is a more or less casual remark to the effect that all the special Tables required for working on it were being computed by the Scientific Computing Service, Ltd., that most of the work had been done, although all the results had not yet been received, and that about £600 had been expended during the last year with this firm. These Tables have now been completed, and, through the courtesy of Dr. J.L. Comrie, the Managing Director of the firm, the present writer has been able to see a copy. From the introduction it appears that the Tables are based on the following fundamental data: – (1) Central Meridian, 32° 30′ East. (2) Origin of co-ordinates. Intersection of this meridian with the equator. (3) False co-ordinates of Origin: Northing=1,000,000 feet. Easting=1,350,000 feet. (4) Scale factor for reduction of scale error=0.9995. (5) Figure of the earth is Clarke's 1880 figure, for which the fundamental geodetic quantities are given in the R.G.S. Table V. Thus, the figure of the earth now adopted is different from that introduced in 1929 and it is presumed that the re-computation of the main triangulation already mentioned is based on the Clarke 1880 figure" (G.T. McCaw, *Empire Survey Review*, No. 52, pp. 257–262).

These Transverse Mercator parameters are generally attributed to Hume F. Rainsford, the author of *Survey Adjustments and Least Squares*, Constable, 1957 in which he discusses his association with the *Directorate of Overseas Surveys* (*DOS*) while in Uganda. Note that the Clarke 1880 ellipsoid is, where $a=6,378,249.145$ m, and $1/f =293.465$. "The circuits formed by several triangulation chains have been re-computed a number of times to rectify discrepancies in various circuits. The last re-computation was done in 1960, and the Uganda main triangulation was adjusted to fit the new co-ordinates of the 1960 re-computation of the 30th meridian arc. In Uganda, our 'datum' is now known as the '1960 Arc Datum.' The 1960 re-computation of the 30th meridian replaced the 1950 re-computation of the same arc which earlier, had been used to control the main triangulation network" (*The Uganda Triangulation Network: – Establishment and Current Status,* Y. Okia and J. Kitaka, Lands and Surveys, Entebbe, Uganda).

Further discussions of triangulation details of the region by Rainsford are covered in the chapter on Kenya. The Cape (Arc) Datum 1960 origin is at Buffelsfontein, where $\Phi_o=33°$ 59′ 32.000″ S and $\Lambda_o=25°$ 30′ 44.622″ E. The mean value to transform from Arc 1960 Datum to WGS84 in the general vicinity of Uganda (Tanzania & Kenya) is: $\Delta X=-160$ m, $\Delta Y=-8$ m, $\Delta Z=-300$ m.

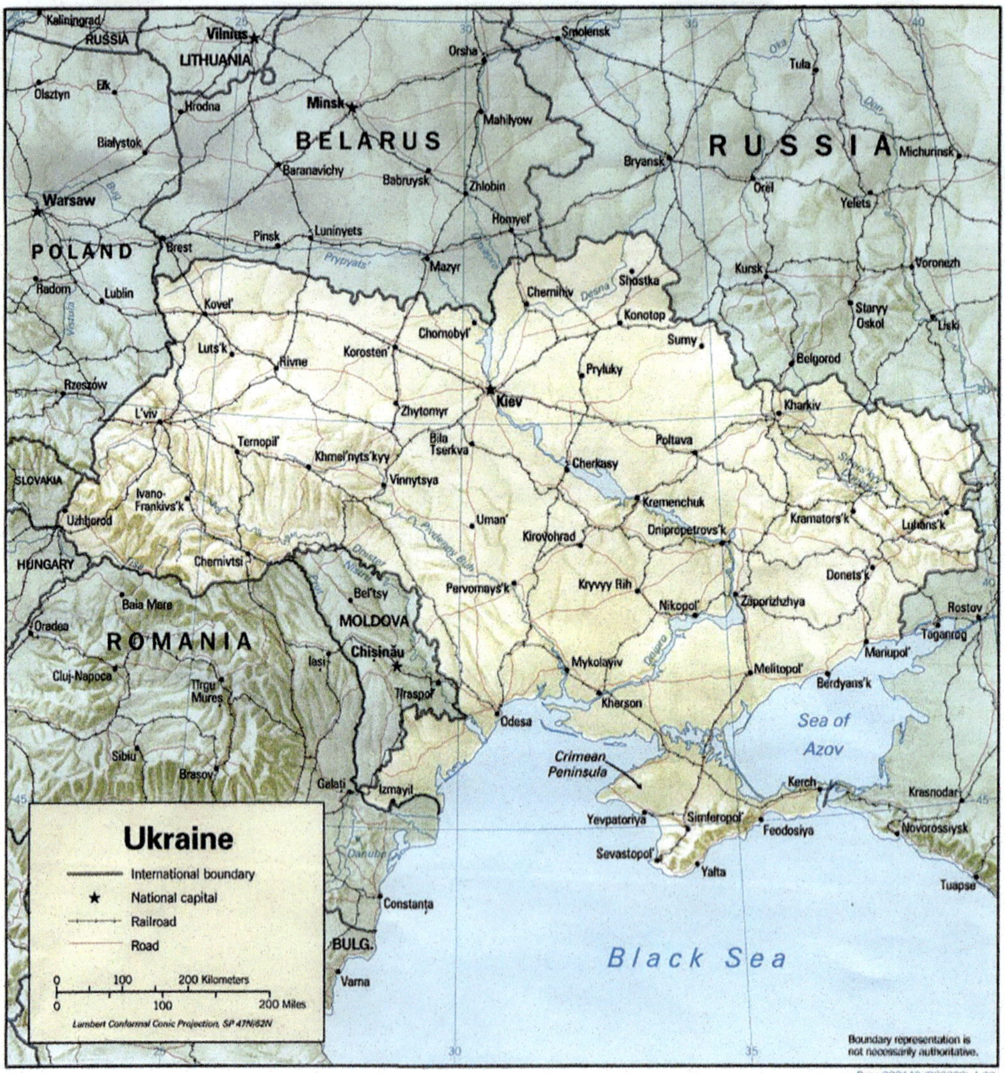

Base 802140 (R00339) 4-93

Ukraine

Ukraine is slightly smaller than Texas, and borders Belarus (891 km), Hungary (103 km), Moldova (939 km), Poland (526 km), România (169 km), Russia (1,576 km), and Slovakia (97 km). The coastline is 2,782 km along the Black Sea and the Sea of Azov. The climate is temperate continental, and most of Ukraine is steppes and plateaus, with the *Carpathian Mountains* in the west and the southeastern coast of the *Crimea* from *Sevastopol*, through *Yalta* and north to *Feodosiya*. The lowest point is the Black Sea (0 m), and the highest point is *Hora Hoverla* (2,061 m). The capitol is *Kiev*, and according to legend, the city was founded in 482 A.D. by a royal family of three brothers and one sister.

Paleolithic remains have been found in the region, but the oldest dwelling in *Kiev* is from the 25th century *B.C.*, about 4,500 years ago. "Ukraine was the center of the first Slavic state, *Kievan Rus*, which during the 10th and 11th centuries was the largest and most powerful state in Europe. Weakened by internecine quarrels and Mongol invasions, *Kievan Rus* was incorporated into the Grand Duchy of *Lithuania* and eventually into the Polish-Lithuanian Commonwealth. The cultural and religious legacy of *Kievan Rus* laid the foundation for Ukrainian nationalism through subsequent centuries. A new Ukrainian state, the *Cossack Hetmanate*, was established during the mid-17th century after an uprising against the Poles. Despite continuous Muscovite pressure, the Hetmanate managed to remain autonomous for well over 100 years. During the latter part of the 18th century, most Ukrainian ethnographic territory was absorbed by the Russian Empire. Following the collapse of czarist Russia in 1917, Ukraine was able to bring about a short-lived period of independence (1917–1920) but was reconquered and forced to endure a brutal Soviet rule that engineered two artificial famines (1921–22 and 1932–33) in which over 8 million died. In World War II, German and Soviet armies were responsible for some 7 to 8 million more deaths" (*World Factbook*, 2004). The republic achieved independence in 1991.

The czarist Russians performed surveys and topographic mapping of Ukraine in the 19th and early 20th centuries, but these works were for military purposes only. They did nothing with respect to individual land ownership registration, and they preferred the sazhen for their unit of measurement (*Paraphrased from Poland*). The existing classical triangulation net is a dense mesh to the west along the border with Poland, Hungary, România, and Moldova, primarily in the mountainous region and extending as far east as *Rivne, Ternopil'*, and *Chernivtsi*. A southern chain of figures reaches from the western city of *Izmayil*, through *Odessa* and *Kherson* to the *Crimea* where it includes *Feodosiya* and *Kerch*. There are seven other meridional arcs that are connected by three more-or-less continuous east-west chains. Although some first-order work is evident around *Kiev*, there is a very dense network about *Yalta* in the *Crimea*. There are a number of high-order local surveys evident in Ukraine, and I suspect that some of these locations may be coincident with now-empty underground silos; a once-favorite area for ICBM sites when the USSR had control of Ukraine.

The observations for the Horizontal State Geodetic Network (HSGN) of Ukraine began in 1923–1925, but it took over 30 years to complete both horizontal and vertical leveling work. Completed in 1970, the first-order network has been maintained while densification has continued for third- and fourth-order control. The HSGN consists of 19,538 points that include 547 first-order and 5,386 second-order points. The HSGN is on the "System 42" datum established (in 1942) by the USSR where the origin point is at *Pulkovo Observatory*, where $\Phi_o = 59° \ 46' \ 18.55''$ North, $\Lambda_o = 30° \ 19' \ 42.09''$ East of Greenwich. The defining azimuth at the point of origin to Signal A is: $\alpha_o = 317° \ 02' \ 50.62''$. System 42 is referenced to the Krassovsky 1940 ellipsoid where $a = 6,378,245.0$ m, and $1/f = 298.3$. The previously mentioned dense and continuous western network is entirely first-order in quality.

The remainder of the first-order network of the Ukraine is comprised of polygons, the lengths of each section being less than 200–250 km. There are 250 LaPlace (astronomic) stations in the HSGN and are located at each end of the first-order triangulation polygons and in the middle of each section. In continuous first-order chains, the LaPlace stations are spaced every ten triangles and the accuracy of the azimuths is ±1.2″. In second-order chains, LaPlace stations are located at baseline terminals. The accuracy of baseline distance measurements is not less than 4×10^{-6}. In general, a single Ukraine map sheet at a scale of 1:1,000,000 will contain about 35–70 LaPlace station points and about 20–30 baselines. The average density of HSGN points is 1 point in 30 km² but this varies in different regions. For instance, in the industrial region around Donbass the density goes up to 1 point in 5–10 km², while in the rural region around *Polissia* the density goes down to about 1 point in 40–50 km². The grid system associated with the Ukraine HSGN is the same as with all former countries of the Soviet Union – the Russia Belts which are identical to the UTM grid except that the scale factor at origin (m_o)=unity. For large-scale mapping, the width of the belts reduces to 3° rather than the standard 6° belt.

The Vertical State Geodetic Network (VSGN) consists of almost 11,000 km of first-order leveling, including 12,600 km of second-order leveling, 6,000 km of third-order leveling, and about 300,000 km of ordinary leveling. The average distance from any site in Ukraine to a first- or second-order level line does not exceed 40 km. The first-order VSGN is tied to the vertical networks of Poland, Slovakia, România, Hungary, Russia, and Belarus. The vertical datum is referenced to the *Kronstadt tide gauge* located at the *Baltic Sea*, near *St. Petersburg (Russia)*. Benchmark spacing in Ukraine is not in my files. The State Gravimetric Network is comprised of 80 first-order points and 20 second-order points with the fundamental point located in *Poltava*.

The NGA does not list datum transformation parameters from System 42 to WGS84 for Ukraine. My guess is that the parameters are pretty close to what they are for Moscow since the strategic importance of the country was so enormous to the USSR. Ukraine has now passed legislation that denotes WGS84 as the national datum of the republic.

Years ago, I sat in a hotel room in South America and watched "The Wall" being torn down. I was working on a U.S. A.I.D. project for land titlelization in which I designed the geodetic and photogrammetric aspects of the project for a canton in Ecuador. That process is a major project now in Ukraine, and GPS technology is an integral component of the social transformation. Those that read these columns of mine are aware that I often grouse on "La Ley" – "The Law" as it exists in much of Latin America in which a branch of the federal government is given the exclusive monopoly for geodetic surveying and topographic mapping of a country. That is a custom derived from the European way of doing things back in the 19th century. I don't care for the concept because it frustrates private commercial mapping in favor of some federal group, usually the military. Such an idea seems to be the current state of affairs in Ukraine and their federal government appears to have passed a similar 19th century era-type law. This may be a result of sociological/economic phenomena more than anything else, but it's disappointing to see such developments in new republics striving for excellence in a worldwide capitalistic environment. I wish them success in their endeavors to provide farmers with a title to the soil their forefathers have tilled for so many centuries; the geodetic and photogrammetric sciences will allow the technical aspects to flow smoothly.

I have to thank Dr. Momchil Minchev of Sofia, Bulgaria for his generous assistance in locating geodetic publications in English on Ukraine for me. The reports of Dr. Michael Cheremshynsky of the Ukraine Main Administration of Geodesy, Cartography and Cadastre of Ukraine in *Kiev* have made the technical details of the geodetic history possible for this article. Once again, Dr. Minchev has helped me unravel an enigma.

"2020 can be considered as the year of geospatial data in Ukraine with the Ukrainian geospatial community facing a historic moment of digitalization. We have introduced a 'single window' for natural resource management, which will help to save budget funds and develop territories, strengthen public control over the activities of state bodies and increase public confidence in the government.

"In April 2020, Ukraine's law on the National Spatial Data Infrastructure (NSDI) was finally adopted by the Ukrainian Parliament after more than 10 years and 4 attempts. Although Ukraine is not an EU Member State, the law is fully in line with INSPIRE and also reflects the main principles of EU open data policy" (*State Service of Ukraine for Geodesy, Cartography and Cadastre* [*StateGeoCadastre*] *2022*).

The Ukrainian government's geodetic website (*https://dgm.gki.com.ua/pererahunok-po-gelmertu-(po-kljuchu)-na-ploschini*) offers a Helmert-style datum conversion tool, and appears to have a completely open access portal to the nation's geodetic network, as typical for a free republic. The website pages are in Ukrainian and in English.

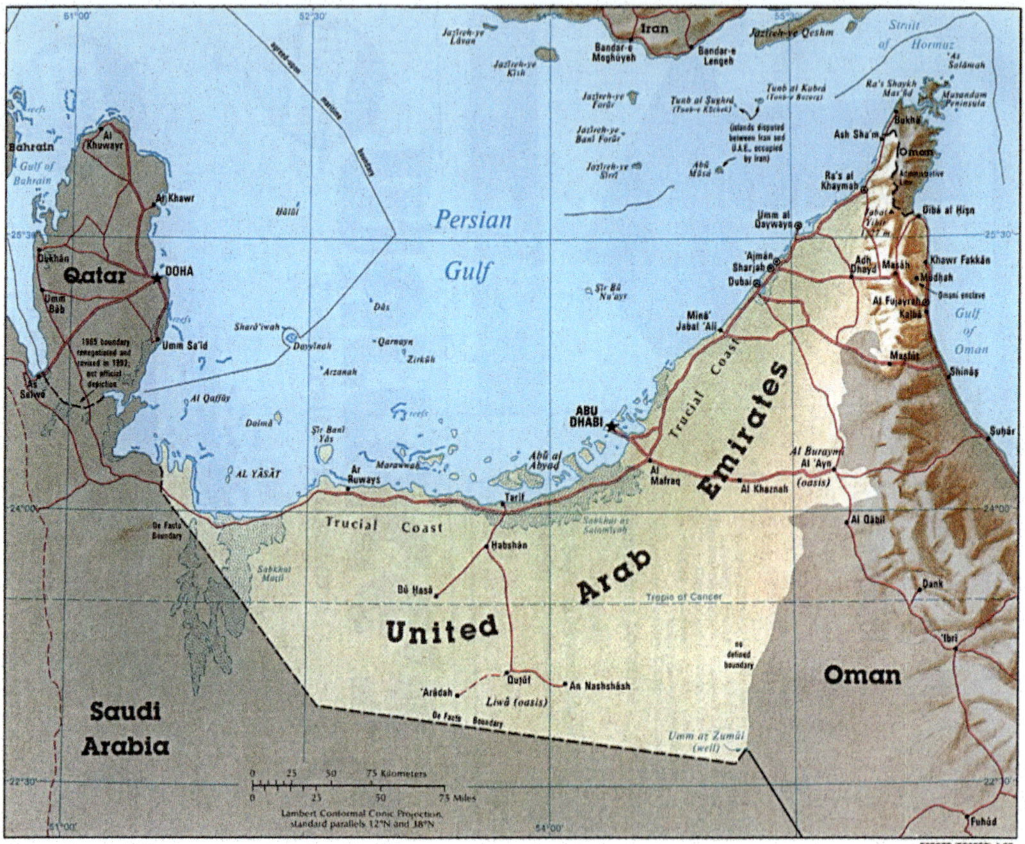

802372 (R00307) 4-95

United Arab Emirates

The United Arab Emirates covers an area slightly smaller than the state of Maine. Much of the interior of the UAE is desert and runs to the edge of the *Empty Quarter* of Saudi Arabia, the largest sand desert in the world. The northern and eastern sections are mountainous and green while the coastal areas are marked with salt flats.

Originally settled in the Bronze Age, the *Umm an-Nar's* culture established itself near modern *Abu Dhabi* in the 3rd century *B.C.*, and its influence extended to the interior of the Arabian Peninsula as well as along the coast to *Oman*. Some later settlements by the Greeks have been found, and in the Middle Ages most of the region was part of the Kingdom of *Hormuz*, which controlled trade in the Arabian Gulf. The Portuguese arrived in 1498 and stayed until 1633 until the British took control of the area. By 1820, the British had destroyed or captured all Qawasim pirate ships, and they imposed a General Treaty of Peace on nine Arab sheikhdoms in the area and installed a garrison. The area was known as the Trucial Coast until 1971. The seven emirates are *Abu Dhabi (Abu Zaby), Ajman, Dubai (Dubayy), Al Fujayrah, Ra's al Khaymah, Sharjah (Ash Shariqah),* and *Umm al Qaywayn*.

The first major geodetic Datum of the Arabian Gulf area was established by W.E. Browne of the Iraq Petroleum Company in 1927–1931 at the South End Base at station *Nahrwan* (East of Baghdad), such that $\Phi_o=33°\ 19'\ 10.87''$ North, $\Lambda_o=44°\ 43'\ 25.54''$ East of Greenwich, and the Clarke 1880 is the ellipsoid of reference, where $a=6,378,300.782$ m and $1/f=293.4663077$. The *Nahrwan Datum of 1929* is the most prevalent coordinate system of the entire Arabian Gulf area and is still found to this day.

The *Sir Bani Yas Island Datum of 1933* was established by the British Royal Navy in 1933, such that $\Phi_o=24°\ 16'\ 44.83''$ North, $\Lambda_o=52°\ 37'\ 17.63''$ East of Greenwich, and the Clarke 1880 is the ellipsoid of reference.

The *Ajman Datum of 1946* origin is such that $\Phi_o=25°\ 23'\ 50.19''$ North, $\Lambda_o=55°\ 26'\ 43.95''$ East of Greenwich and is referenced to the Helmert 1906 ellipsoid, where $a=6,378,200$ m and $1/f=298.3$. The first Grid was the World War II Trucial Coast/Qatar Grid on the Cassini-Soldner projection. The Central Meridian $\lambda_o=50°\ 45'\ 41''$ E, the False Northing Latitude of Origin $(\varphi_{FN})=25°\ 22'\ 56.5''$ N, and both the False Eastings and False Northings are 100 km. Of course, the scale factor at origin by definition is equal to unity.

In 1967, the Directorate of Military Surveys recomputed the *Mainland Trucial Coast* and *Qatar* triangulations on the International ellipsoid, European Datum 1950. Those coordinates were then transformed into *Nahrwan Datum 1929*. The Trucial Coast Transverse Mercator Grids became preferred to the old Cassini Grid. The Central Meridian $\lambda_o=55°\ 00'$ E, and the False Eastings have had values of 100 and 1,200 km. The False Northings are –2,000 km as measured from the Equator. The series covered the entire UAE. Topographic mapping was compiled in the 1960s by the British Directorate of Military Surveys, and a 1:50,000 scale map series (K763) comprising 155 sheets was completed in 1969.

Limited 1:50,000 mapping was produced in the early 1980s with the assistance of Syria, but coverage of Dubai was based on only 4 third-order points in the southeast corner of the Emirate. A military Survey Department was set up by the *Emirates* and new mapping was published between 1989 and 1991 as 138 orthophoto sheets on the *Nahrwan Datum of 1929* with the UTM Grid. A new GPS network was initiated for *Dubai* in 1991 with a new local Grid. The *Dubai* Local Transverse Mercator (DLTM) Grid is referenced to the WGS 84 ellipsoid, the Central Meridian $\lambda_o=+55°\ 20'$ E, and the False Easting = 500 km. The Northings are presumably measured from the Equator. Analysis of the old network indicated a potential positional error of the old classical control of up to 9 m horizontal. The first-order Geodetic GPS Network of Dubai is composed of 62 monumented points with distances between points ranging from 5 to 10 km. Of particular interest is that *Dubai* has completely abandoned the previous classical geodetic work extant in the *Emirate*. Zero effort was (apparently) made to relate the old to the new! I personally do not agree with this philosophy because I prefer to relate historical records to current and future work. However, I suspect that this unfortunate tack may be chosen from time-to-time for the sake of expediency.

Satellite positioning studies (by others) in the United Arab Emirates derived a set of Datum shift parameters from WGS72 Datum to Nahrwan Datum of 1929, where $\Delta X = +225.4$ m, $\Delta Y = +158.7$ m, $\Delta Z = +378.9$ m, based on observations of 8 stations. I personally would consider the tenths of a meter used in these parameters as very optimistic. Interestingly, NGA lists the transformation from *Nahrwan 1929* to WGS 84 as $\Delta X = -249$ m, $\Delta Y = -156$ m, $\Delta Z = -381$ m, ±25 m, based on two stations observed in 1987.

Significant developments have been implemented in the coordinate reference systems for two primary areas of the United Arab Emirates: in *Al Ain Region (Abu Dhabi)* and in *Dubai*. In *Al Ain Municipality*, Professor Kamal A. Abdalla of the University of Khartoum, Sudan spent a number of years as a consultant to the region, and in 2005 published a paper that stated, "The local geodetic network adopt(ed) Ras Ghantut datum and (is) based on the modified Clarke 1880 ellipsoid. The local control stations are non-homogeneous, unadjusted and have many limitations in terms of spatial data applications. While the global geodetic network is tied to the ITRF system, containing 33 well-distributed geodetic control stations. The transformation parameters between the global datum and the local datum were computed" (*Al Ain Local and Global Spatial Reference Systems, Map Middle East, 2005*).

In *Dubai*, a Leica SmartNet™ has been established in order to provide a virtual reference system (VRS) to the emirate (*Al Marzooqi, Y., Fashir, H., Babiker, T., Establishment & Testing of Dubai Virtual Reference System (DVRS) National GPS-RTK Network, GISdevelopment.net*). Furthermore, an anonymous paper (*An Absolute/Relative Gravity Base Net in the Emirate of Dubai*) details a new fundamental gravity network of nine relative and absolute gravity stations and that "These measurements were the first absolute gravity determinations in the whole South-West Asia in 1994."

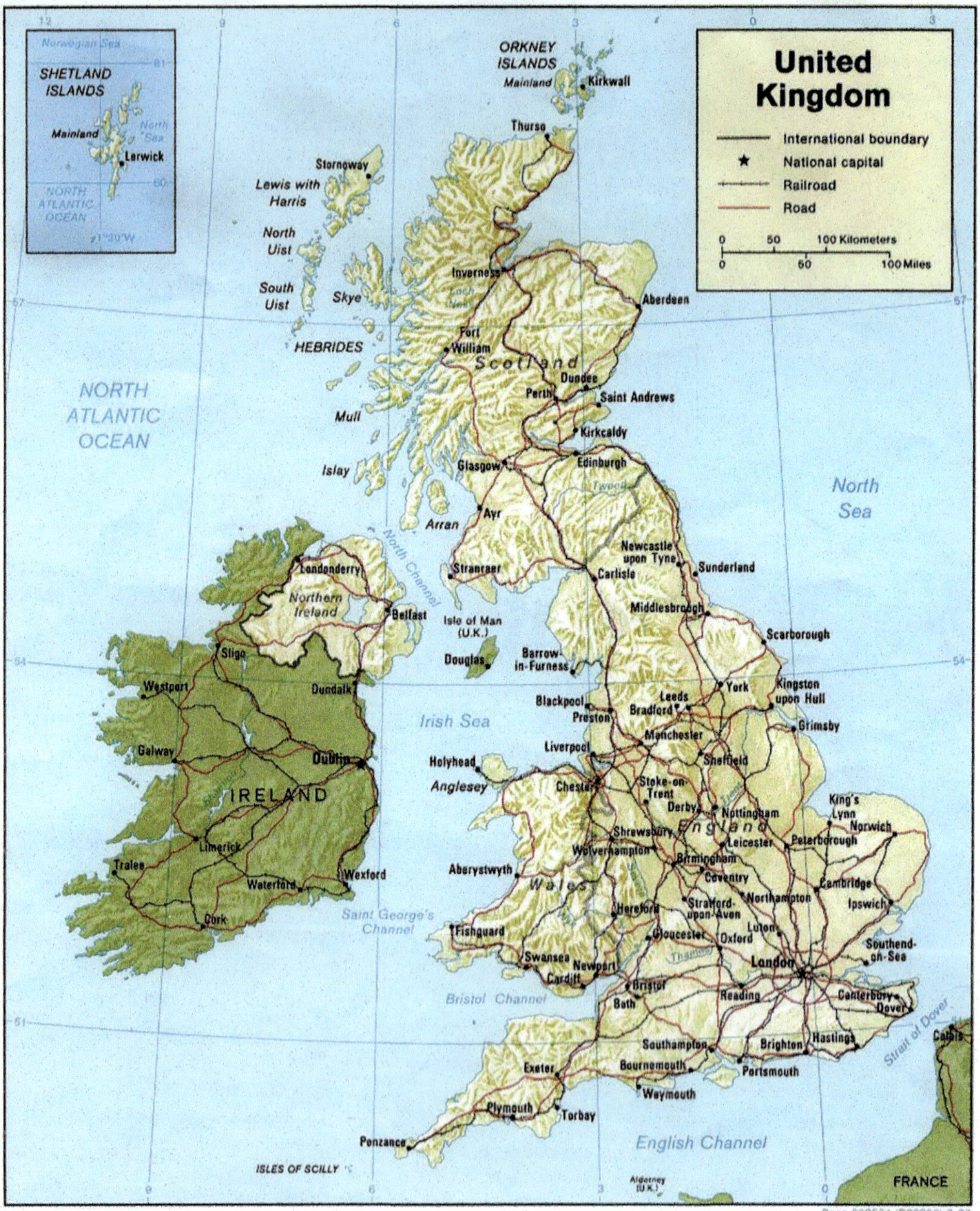

Base 800564 (B00326) 7-87

United Kingdom

The area of Great Britain is slightly smaller than Oregon; the kingdom has a 12,429 km coastline and the lowest point is *Fenland* (−4 m), the highest point is *Ben Nevis* (1,343 m).

Evidence from pre-Roman times includes Neolithic mound-tombs and henge monuments as well as Bronze Age Beaker culture tools, graves, and the famous Stonehenge monument. Brythonic-speaking Celtic peoples arrived during migrations of the first millennium B.C., according to *Webster's Geographical Dictionary*. England has existed as a unified entity since the 10th century. The union between England and Wales was begun in 1284 and formalized in 1536. England and Scotland agreed to join as Great Britain in 1707 and with Ireland in 1801. The present name of Great Britain and Northern Ireland was adopted in 1927. Great Britain is only 35 km from France and is linked by the "Chunnel." Because of the heavily indented coastline, no location in the country is more than 125 km from tidal waters!

Thanks to the web site of the Ordnance Survey, "England was squeezed between rebellion in Scotland and war with France when King George II commissioned a military survey of the Scottish Highlands in 1746. The job fell to William Roy, a far-sighted young engineer who understood the strategic importance of accurate maps. Walk into Ordnance Survey's Southampton headquarters and you'll see Roy's name engraved on the curved glass entrance doors, yet his vision of a national military survey wasn't implemented until after his death in 1790. By then Europe was in turmoil, and there were real fears that the French Revolution might sweep across the English Channel. Realizing the danger, the government ordered its defence ministry – the *Board of Ordnance* – to begin a survey of England's vulnerable southern coasts. In June 1791, the Board purchased a huge new Ramsden theodolite, and surveyors began mapping southern Britain from a baseline that Roy himself had measured several years earlier. The first one-inch (1″ = 1 mile) map of *Kent* was published in 1801, and a similar map of *Essex* followed – just as Nelson's victory at Trafalgar made invasion less likely! Within twenty years about a third of England and Wales had been mapped at the one-inch scale. If that seems slow in these days of aerial surveys and global positioning, spare a thought for Major Thomas Colby – later Ordnance Survey's longest serving Director General – who walked 586 miles in 22 days on a reconnaissance in 1819. In 1824, Parliament ordered Colby and most of his staff to Ireland, to produce a detailed six-inch to the mile valuation survey. Colby designed specialist measuring equipment, established systematic collection of place names, and reorganized the map-making process to produce clear, accurate plans. But Colby the perfectionist also traveled with his men, helped to build camps, and arranged mountaintop feasts with huge plum puddings at the end of each surveying season. Soon after the first Irish maps began to appear in the mid-1830s, the demands of the *Tithe Commutation Act* provoked calls for similar six-inch surveys in England and Wales. The government prevaricated but, by then, there was a new power in the land. This was the era of railway mania and if the one-inch map was unsuitable for calculating tithes, it was virtually useless for the new breed of railway engineers. To make matters worse, mapping of England and Scotland remained incomplete and, in 1840, the Treasury agreed that the remaining areas should be surveyed at the six-inch scale. Now, surveyors needed greater access than ever before; and so, in 1841, the Ordnance Survey Act gave them a legal right to 'enter into and upon any land' for survey purposes. A few months later Ordnance Survey's cramped Tower of London offices were at the centre of a national catastrophe when fire swept through the Grand Storehouse, threatening to engulf the Crown Jewels in the Martin Tower. Miraculously, the Jewels were saved, and most of Ordnance Survey's records and instruments were also carried to safety. But the blaze highlighted the Survey's desperate need for more office space and prompted a move to *Southampton*. The scene was now set for two decades of wrangling over scales. Throughout this period, Victorian reforming zeal was creating an acute need for accurate mapping. The issue was settled piecemeal until, by 1863, scales of six inches and twenty-five inches to the mile had been approved for mountain and moorland, and rural areas respectively. The one-inch map was retained, and detailed plans at as much as ten feet to the mile were introduced for built-up areas. By now, Major-General Sir Henry James – perhaps Ordnance Survey's most eccentric and egotistical Director General – was midway through his twenty-one-year term. James quickly saw how maps could be cheaply and quickly enlarged or reduced using the new science of photography, and he designed an elaborate glass studio at *Southampton* for processing photographic plates. James planted his name on everything he touched, and later claimed to have invented photo zincography, a photographic method of producing printing plates. In fact, the process had been developed by two of his staff. By 1895 the twenty-five-inch survey was complete. The twentieth century brought cyclists and motorists swarming onto the roads, and the new Director General, Colonel Charles Close, prepared to exploit this expanding leisure market. But by now, the tide of history was sweeping Ordnance Survey back to its roots. As Britain entered the First World War, surveyors, draughtsmen and printers from Ordnance Survey were posted overseas. Working in appalling conditions alongside the troops, surveyors plotted the lines of trenches, and for the first time, aerial photography was used to capture survey information. After the war, Sir Charles, as he now was, returned to his marketing strategy and appointed a professional artist to produce eye-catching covers for the one-inch maps. Ellis Martin's classic designs boosted sales to record levels, but the war had taken its toll; behind their bright new covers, the maps were increasingly out of date. In an uncanny echo of the mid-nineteenth century, a whole raft of new legislation brought demands for accurate, up-to-date mapping. Matters came to a head in 1935, and the Davidson Committee was established to review Ordnance Survey's future. That same year, a far-sighted new Director General, Major-General Malcolm MacLeod, launched the retriangulation of Great Britain. Surveyors began an Olympian task, building the now familiar concrete triangulation pillars on remote hilltops throughout Britain. Deep foundations were dug by hand, and staff dragged heavy loads of materials over isolated terrain by lorry, packhorse and sheer brute force. The Davidson Committee's final report set Ordnance Survey on course for the 21st century.

"The National Grid reference system was introduced, using the metre as its measurement. An experimental new 1:25,000 scale map was launched, leaving only the one-inch unscathed. It was almost forty years before this popular map (1:63,360) was superseded by the 1:50,000 scale series, first proposed by William Roy more than two centuries earlier. In 1939, war intervened once again. The Royal Artillery was now responsible for its own field surveys, but over a third of Ordnance Survey's civilian staff were called up, and its printing presses were kept busy with war production. It wasn't a soft option. Enemy bombing devastated *Southampton* in November 1940 and destroyed most of Ordnance Survey's city centre offices. Staff were dispersed to other buildings, and to temporary accommodation at *Chessington*. But the military appetite remained insatiable – the Normandy landings alone devoured 120 million maps! After the war, Ordnance Survey returned to Davidson's agenda; the retriangulation was completed, and metric maps began to appear along National Grid sheet lines. Aerial survey helped speed up the new continuous revision strategy, and up-to-date drawing and printing techniques were introduced. But the organization was still fragmented, scattered across southern England in a battered collection of worn-out buildings. All that changed in 1969, when Ordnance Survey move to its present, purpose-built headquarters on the outskirts of *Southampton*. Four years later, the first computerized large-scale maps appeared; the digital age had begun. Ordnance Survey digitized the last of some 230,000 maps in 1995, making Britain the first country in the world to complete a programmed of large-scale electronic mapping. Computers have transformed the map-making process, and electronic data is now routinely available to customers within 24 hours of being surveyed. The public still knows Ordnance Survey for its comprehensive range of printed leisure maps, yet electronic data now accounts for some 80% of Ordnance Survey's turnover. Independent estimates show that the national mapping agency's data now underpins up to £136 billion-worth of economic activity in Britain – everything from crime-fighting and conservation to marketing and mobile phones."

The original triangulation of Britain was carried out between 1783 and 1853 and is known as the "Principal Triangulation." Jesse Ramsden, a gifted but dilatory gentleman built the theodolite where the overall size of the horizontal circle measured 3 ft in diameter and was divided to a precision of a tenth of an arc second! (That's the same precision as the Wild Heerbrugg T-4 astronomical theodolite still manufactured as recently as the 1980s.) Major-General William Roy once wrote, "On one occasion he (Ramsden) attended at Buckingham Palace precisely as he supposed at the time named in the Royal mandate. The King remarked that he was punctual as to the day and hour, while late by a whole year!" The genius responsible for the final adjustment and computation of the *Principal Triangulation of 1783–1853* was Colonel Alexander Ross Clarke who also computed the Clarke ellipsoids of 1858, 1866, and 1880. The network selected by Clarke was an interlocking system of well-conditioned triangles. In 1967, the Ordnance Survey wrote, "This network was geometrically of great strength since it involved no fewer than 920 condition equations to find corrections to 1,554 observed directions subsequently used to fix 218 points." The One-Inch map series referred to a map scale of one-inch equals one mile. The One-Inch Scottish Bonne projection (1852–1936) was based on the Principal Triangulation and had a central meridian $\lambda_o = 4°$ W, a latitude of origin $\varphi_o = 57° 30'$ N, scale factor of unity, and with no false Origin. The One-Inch English and Welsh Cassini projection (1919–1936) was also based on the Principal Triangulation and had a central meridian $\lambda_o = 2° 41' 03.5620''$ W, a latitude of origin $\varphi_o = 53° 13' 17.2740''$ N, scale factor of unity, and with no false Origin.

The counties of the United Kingdom were based on local Cassini-Soldner projections that had the typical scale factor equal to unity, no false origin, and a single triangulation station as the projection origin. *System Ben Auler* ($\varphi_o = 56° 48' 50.3889''$ N, $\lambda_o = 4° 27' 49.7064''$ W) served the counties *Canna, Eigg, Muck, Rhum, Sanday,* and *Skye of the Inner Hebrides* and *Inverness of Scotland*. System Ben Cleuch ($\varphi_o = 56° 11' 08.8438''$ N, $\lambda_o = 3° 46' 05.2765''$ W) served the counties *Clackmannan* and *Perth of Scotland*. System Ben Clibrig (1839), ($\varphi_o = 58° 14' 07.8780''$ N, $\lambda_o = 4° 24' 35.3627''$ W) served the county of *Sutherland, Scotland*. System Black Down (1797, 1840), ($\varphi_o = 50° 41' 10.3186''$ N, $\lambda_o = 2° 32' 52.4856''$ W) served the counties *Dorset and Somerset of England*. System Bleasdale ($\varphi_o = 53° 54' 55.351''$ N, $\lambda_o = 2° 37' 20.752''$ W) served the county of *Lancashire, England*. System Brandon (1822), ($\varphi_o = 52° 24' 19.820''$ N, $\lambda_o = 0° 37' 21.040''$ E) served the counties *Durham* and *Northumberland of England*. System Broadfield ($\varphi_o = 55° 47' 59.8320''$ N, $\lambda_o = 4° 32' 20.5920''$ W) served the county of *Rentfrew, Scotland*. System Brown Carrick ($\varphi_o = 55° 24' 26.5714''$ N, $\lambda_o = 4° 42' 41.1291''$ W) served the counties *Ayr, Wigtown,* and *Kirkcudbright, Scotland*. System The Buck ($\varphi_o = 57° 17' 51.1940''$ N, $\lambda_o = 2° 58' 32.0297''$ W) served the counties *Aberdeen, Banff, East Lothian, Fife, Kincardine, Kinross, Midlothian,* and *West Lothian of Scotland*. System Cairn Glasher ($\varphi_o = 57° 20' 22.8895''$ N, $\lambda_o = 3° 50' 30.5116''$ W) served the county of *Nairn, Scotland*. System Cleisham (1840), ($\varphi_o = 57° 57' 50.6850''$ N, $\lambda_o = 6° 48' 41.4340''$ W) served the *Outer Hebrides*. System Craigowl ($\varphi_o = 56° 32' 52.4204''$ N, $\lambda_o = 3° 00' 48.5178''$ W) served the county of *Angus, Scotland*. System Cruach-na-Sleagh ($\varphi_o = 56° 07' 08.9328''$ N, $\lambda_o = 5° 43' 34.1145''$ W) served the counties *Argyle, Arran,* and *Bute of Scotland* and *Coll, Islay, Jura, Mull,* and *Tiree of the Inner Hebrides*. System Cryn-y-Brain (1852, 1853), ($\varphi_o = 53° 02' 16.8715''$ N, $\lambda_o = 3° 10' 22.2907''$ W) served the counties *Denbighshire* and *Flintshire of Wales*. System Danbury (Church) Spire (1844), ($\varphi_o = 51° 42' 57.9220''$ N, $\lambda_o = 0° 34' 32.9299''$ E) served the counties *Cambridgeshire* and *Isle of Ely, Huntingdonshire, Soke of Peterborough, Norfolk,* and *Suffolk of England*. System Derrington Great Law ($\varphi_o = 55° 26' 08.9646''$ N, $\lambda_o = 2° 26' 34.6838''$ W) served the county of *Berwick, Scotland*. System Ditchling (1793, 1845), ($\varphi_o = 50° 54' 04.0149''$ N, $\lambda_o = 0° 06' 21.7531''$ W) served the county of *Sussex, England*. System Dunnet Head (1838) ($\varphi_o = 58° 40' 10.1679''$ N, $\lambda_o = 3° 22' 13.4365''$ W) served the county of *Caithness, Scotland*. System Dunnose (1793, 1844), ($\varphi_o = 50° 37' 03.7288''$ N, $\lambda_o = 1° 11' 50.1015''$ W) served the counties *Berkshire, Buckingham, Derby, Hampshire, Isle of Wight, Leicester, Lincoln, Northampton, Nottingham, Oxford, Rutland, Warwick, Wiltshire, Worcester,* and *Stafford of England*. System Dunrig (1816, 1850), ($\varphi_o = 55° 47' 59.832''$ N, $\lambda_o = 4° 32' 20.592''$ W) served the county of *Peebleshire, Scotland*. System Finlay Seat

($\varphi_o=57°$ 34' 42.2883" N, $\lambda_o=3°$ 14' 26.5655" W) served the county of *Moray, Scotland*. System *Forest Hill* ($\varphi_o=54°$ 25' 30.700" N, $\lambda_o=2°$ 43' 41.593" W) served the county *Westmorland of England*. System *Foula* (1821), ($\varphi_o=60°$ 08' 26.2436" N, $\lambda_o=2°$ 05' 38.4808" W) served the counties *Orkney and Zetland, Scotland*. System *Hart Fell* (1816, 1847), ($\varphi_o=55°$ 24' 28.9742" N, $\lambda_o=3°$ 24' 00.1634" W) served the county of *Dumfries, Scotland*. System *Hensbarrow* (1790, 1843), ($\varphi_o=50°$ 22' 58.8532" N, $\lambda_o=4°$ 49' 05.1029" W) served the counties *Cornwall* and *Isles of Scilly, England*. System *Highgate* ($\varphi_o=51°$ 41' N, $\lambda_o=4°$ 50' W) served the county of *Pembroke, Wales*. System *High Pike* ($\varphi_o=54°$ 42' 19.3478" N, $\lambda_o=3°$ 03' 26.5277" W) served the county of *Cumberland, England*. System *Hollingbourne* ($\varphi_o=51°$ 16' 10.380" N, $\lambda_o=0°$ 39' 55.564" W) served the county of *Kent, England*. System *Lanark Church Spire* ($\varphi_o=55°$ 40' 20.8037" N, $\lambda_o=3°$ 46' 19.8904" W) served the counties *Dumbarton, Lanark, Stirling,* and *Roxburgh of Scotland*. System *Leith Hill Tower* (1792, 1822, 1824), ($\varphi_o=51°$ 10' 32.8895" N, $\lambda_o=0°$ 22' 10.9797" W) served the county of *Surrey, England*. System *Llangeinor* ($\varphi_o=51°$ 38' 26.6535" N, $\lambda_o=3°$ 34' 17.0360" W) served the counties *Anglesey, Brecknock, Caernarvon, Cardigan, Carmarthen, Glamorgan, Merioneth, Montgomery,* and *Radnor of Wales* and *Hereford, Monmouth,* and *Salop of England*. System *Nantwich Church Tower* ($\varphi_o=53°$ 04' 00.3405" N, $\lambda_o=2°$ 31' 09.1414" W) served the county of *Cheshire, England*. System *Rippon Tor* ($\varphi_o=50°$ 33' 57.0563" N, $\lambda_o=3°$ 46' 12.0872" W) served the county of *Devon, England*. System *St. Paul's* (1848), ($\varphi_o=51°$ 28' 15.9830" N, $\lambda_o=0°$ 22' 47.6011" W) served the counties of *Bedford, Essex, Hertsford, London,* and *Middlesex, England*. System *Sandhope Heights* ($\varphi_o=55°$ 30' 10.253" N, $\lambda_o=3°$ 02' 31.092" W) served the county of *Selkirk, Scotland*. System *Scourna-Lapach* (1846), ($\varphi_o=57°$ 22' 10.3858" N, $\lambda_o=5°$ 03' 31.5613" W) served the counties of *Ross* and *Cromarty, Scotland*. System *South Berule* (1846), ($\varphi_o=54°$ 08' 57.5699" N, $\lambda_o=4°$ 40' 05.2960" W) served the *Isle of Man, England*. System *York Minster* (1846), ($\varphi_o=53°$ 57' 43.1879" N, $\lambda_o=1°$ 04' 49.7302" W) served the counties *York* and *Yorkshire, England*.

The retriangulation of Great Britain was not based on a single origin point. However, it was not based on any new length measurements, either. The overall size of the network was constrained to agree with the old 18th-century Principal Triangulation using the old coordinates of 11 primary stations adjusted by Clarke. Therefore, the Ordnance Survey of Great Britain 1936 datum (OSGB36) does not have a single origin point. The ellipsoid of reference is the Airy 1830 where $a=6,377,563.396$ m and $b=6,356,256.910$ m. The overall size of the OSGB36 is based on the measurement of a single distance between two stations on Hounslow Heath in 1784 using 18-foot glass rods! The error incurred by using this scale constraint yielded an error in the length of the entire country of only 20 m! The English National Grid (1936–1950) was based on the Cassini-Soldner projection and had a central meridian $\lambda_o=1°$ 11' 50.1360" W, a latitude of origin $\varphi_o=50°$ 37' 03.7480" N, scale factor of unity, False Easting=500km, and False Northing=100km. Introduced during World War II, the English Belt Transverse Mercator actually came into use in 1950 and it continues to be used. Although the equations published appear to be a Gauss-Krüger expansion to the 5th derivative, they are cast as *latitude differences* from the False Northing latitude of true origin=49° N. The central meridian=2° W, and the scale factor at the central meridian, commonly mistaken for that of the UTM Grid is different in that by definition, $m_o=0.9996012717... \equiv \text{Log}_{10}$ $m_o=1.99982680$, exactly. I would love to find out where that number came from. The False Easting is 400 km, and the False Northing is −100 km.

The Ordnance Survey seven-parameter transformation from OSGB36 to WGS84 *modified to the standard American rotation convention* used by the National Geodetic Survey (NGS) and by the National Imagery and Mapping Agency is: $\Delta X=+446.448$ m, $\Delta Y=-125.157$ m, $\Delta Z=+542.060$ m, scale$=-20.4894\times10^{-6}$, $R_X=-0.1502"$, $R_Y=-0.2470"$, $R_Z=-0.8421"$. "OSGB36 is an inhomogeneous Terrestrial Reference Frame by modern standards. Do not use this transformation for applications requiring better than 5 metre (*sic*) accuracy in the transformation step, either vertically or horizontally. Do not use it for points outside Britain." The above transformation parameters were extracted from *A Guide to Coordinate Systems in Great Britain*, a "pdf" file downloaded from the

Ordnance Survey's website *www.gps.gov.uk*. Higher accuracy transformations are possible utilizing the free software available from the Ordnance Survey in the form of OSTN02, a bi-linear interpolation package similar in concept to the NADCON package of the NGS. The British Geoid Model, OSGM02 is available and is free from the Ordnance Survey. The stated vertical accuracy is 2 cm in mainland UK and 4 cm for other areas. I hope that other nations will follow the example of the United Kingdom, Australia, Mexico, and the United States in making critical national geodetic transformation models available at no charge to the public. I am indebted to Mr. Russell Fox retired from the Ordnance Survey and to Mr. John W. Hager, now retired from NIMA for their patience over the years in helping me to compile this mass of data on the UK.

The Ordnance Survey of the United Kingdom has published an up-to-date document that covers all of the current coordinate systems and recommended transformation methods and parameters as of 2020: *https://www.ordnancesurvey.co.uk/documents/resources/guide-coordinate-systems-great-britain.pdf.*

nationalatlas.gov™
Where We Are
GENERAL REFERENCE
CANADA
WASHINGTON
OREGON
IDAHO
MONTANA
WYOMING
NORTH DAKOTA
SOUTH DAKOTA
MINNESOTA
WISCONSIN
MICHIGAN
NEBRASKA
IOWA
NEVADA
UTAH
COLORADO
CALIFORNIA
ARIZONA
NEW MEXICO
KANSAS
MISSOURI
ILLINOIS
INDIANA
OHIO
PENNSYLVANIA
NEW YORK
MAINE
WEST VIRGINIA
VIRGINIA
KENTUCKY
TENNESSEE
NORTH CAROLINA
SOUTH CAROLINA
OKLAHOMA
ARKANSAS
TEXAS
LOUISIANA
MISSISSIPPI
ALABAMA
GEORGIA
FLORIDA
PACIFIC OCEAN
ATLANTIC OCEAN
GULF OF MEXICO
MEXICO
THE BAHAMAS
CUBA
HAWAII
ALASKA
Albers equal area projection
U.S. Department of the Interior
U.S. Geological Survey
The National Atlas of the United States of America®

United States of America

The United States is about half the size of Russia; about three-tenths the size of Africa; about half the size of South America (or slightly larger than Brazil); slightly larger than China; more than twice the size of the European Union. Bordered by Canada (8,891 km) (including 2,475 km with Alaska) and Mexico (3,111 km), the terrain is vast central plain, mountains in west, hills and low mountains in east; rugged mountains and broad river valleys in Alaska with rugged, volcanic topography in Hawaii. The highest point is *Denali* (6,190 m) (*Mount McKinley*) (highest point in North America), the lowest point is *Death Valley* (lowest point in North America) (−86 m), and the mean elevation is (760 m).

Thought to have originally been settled by Asian peoples during the Pleistocene (last ice age) that traveled over the land bridge now represented by the Aleutian Islands of Alaska, the ancient peoples now known as Native Americans or "Indians" thrived for thousands of years in North America. Northeastern coastal portions of North America were temporarily settled by Vikings in the first millennium *A.D.*, but their colonies did not survive. Again "discovered" by Christopher Columbus in 1492, the "New World" was successfully colonized by numerous Western European kingdoms, with North America primarily being colonized by Spanish, French, and English peoples – all at the eventual expense of the "Native Americans." The eastern seaboard coast was successfully colonized primarily by English peoples which consisted of the original 13 colonies or states. After the Declaration of Independence from England was signed on the 4th of July in 1776, a revolutionary war was fought and won by the colonists who named the new nation the United States of America.

The "Survey of the Coast" was formed by the U.S. Congress in 1806, and the first Superintendent of the Coast Survey was Ferdinand Hassler. He traveled back to his homeland, Switzerland, in order to purchase geodetic instruments crafted in Aarau. He then returned to commence the triangulation of the coast and harbors of the 13 original colonies as a foundation for the compilation of navigation charts. The new charts acted as an impetus for the maritime nations of Europe to make way to the new United States for commerce and trade. *Two* major surveys were commenced in the early years of the nation: one was geodetic in nature and foundation, and the other was (at least partially), the brainchild of President Thomas Jefferson, the Rectangular Survey of the Public Lands. The former as directed by Hassler was a coordinated classical geodetic triangulation of physically measured baselines that provided scale to chains of quadrilaterals for the geometric foundation of control for the planetable and alidade compilation of coastal charts with soundings. As a convenience to the casting of planetable sheets with a graticule *in the field*, Hassler invented the "American Polyconic" projection. (*The material used at the time for drafting media for the field compilation of valuable charts and maps was starched linen – I have a roll of drafting linen that continues to amaze my students.*) The latter type of survey served for the marking of the Range and Township system of property boundary lines with each 6 square mile Township being subdivided into (more or less) 36 aliquot equal-area sections: each measuring one mile by one mile and containing ~640 acres. The Rectangular Survey is oriented with the cardinal directions and consisted originally of traverse distances physically measured with Gunter's chains 66 ft in length and magnetic compasses controlled with solar observations. The original 13 colonies were already in the process of being surveyed according to the traditional European method of metes and bounds, but the remainder of the United States has since been surveyed with the rectangular system for property boundaries. The two basic surveys of the United States have only rarely been related or collocated, and the methodology of recovering original monuments and property boundaries is a separate, legally licensed profession of "Land Surveying" apart from the activities of the geodetic surveyor. Curiously, the original federal

Township maps of Louisiana still held at the State Land Office are pen-and-ink water-colored maps cast on paper, not linen.

Hassler's survey design was not restricted to only classical triangulations as the objective was to produce navigation charts in order to promote and encourage shipping trade and commerce. The hydrographic soundings of water depths needed vertical control referenced to local mean sea level, so tide gauges and geodetic leveling were also critical elements to the production of reliable charts that accurately charted the hazards to navigation. After his untimely death, Hassler's successors continued the original Congressional charter of the Survey of the Coast; later changing the name to the U.S. Coast & Geodetic Survey (USC&GS) and eventually to its present moniker, the National Geodetic Survey (NGS). While the coasts were being surveyed and charted, triangulations of the interior of the nation started being surveyed by the USC&GS also. After the Civil War in the 1860s, several catastrophic floods occurred in the lower Mississippi River valley. The U.S. Congress called out the Army, and the Civil Works Division of the Army Corps of Engineers was formed with the primary mission of flood control and maintenance of the channels of the navigable rivers and waterways of the United States. The Corps of Engineers undertook the General Survey of the Mississippi River in the 1870s and formed a continuous chain of quadrilaterals of second- and third-order triangulation from *St. Louis, Missouri* to the *Gulf of Mexico*. The Corps did not perform first-order surveys as that was the mission of the USC&GS. (*Note: This was at a time in our nation's history when there was **very little** duplication of effort by different arms of the federal government!*) An interesting side note to the Corps' triangulation activities was the original simultaneous development of an invention used in geodetic astronomy called the "Horobow-Talcott" level, a double-vial attachment used with astronomical theodolites. Horobow was a British Royal Engineer that published his new invention in a British technical paper almost simultaneously with Captain Talcott of the U.S. Army Corps of Engineers. Apparently, Horobow got the idea into print a bit earlier than Talcott, so he got "first billing." The Corps also conducted hydrographic surveys of the river depths of the navigable rivers and waterways which necessitated the establishment of tide gauges. One gauge in particular was at *Biloxi, Mississippi* which was used as the primary gauge for the establishment of the "Mean Low Gulf – (MLG)" hydrographic datum used as the reference to this day for dredging operations at the Mouth of the Mississippi River and upstream to *Baton Rouge, Louisiana*. MLG is nowadays merely applied as an "index" to whatever is the current vertical datum of the USA as determined by NGS which is –0.41 feet below the current "zero" of the *North American Vertical Datum of 1988 (NAVD88)*.

Hassler's first surveys were of the New England area of the country, and those triangulations were eventually adjusted into what was termed "The New England Datum." Subsequent adjustments of increasingly larger and larger areas of the country were named "The North American Datum." Finally, triangulation arcs spanned the breadth of the continent, and "The North American Datum of 1927" was published by the USC&GS. This spectacular geodetic feat was the first continental horizontal datum in the history of the world, and practically every nation of the world adopted the methods, techniques, standards, and specifications written and published by the USC&GS. Likewise, first-order geodetic levelings had accumulated thousands of kilometers of observations, and the first continental vertical datum was published as the "Sea Level Datum of 1929." Standards, specifications, and instruments were adopted by the rest of the world, including the famous "Fischer Level" instrument.

Topographic mapping duties were assumed by the U.S. Geological Survey (USGS) in the 19th century, and the planetable and alidade compilations were also cast on Hassler's Polyconic projection. This projection continued in the copperplate engravings of the topographic maps and charts published for decades and were eventually supplanted in the late 1960s. However, other projections were known in the federal mapping agencies for decades prior to that change but were not embraced, apparently out of typical federal agency inertia.

When the United States joined England in World War I, the first Americans to "hop the pond" and assist the British were a small group of mathematicians and geodesists of the USC&GS. Initially they assisted in the compilation of projection tables of the "Nord de Guerre" zone extended from France into Belgium. Apparently, notice of these tables were taken by some American Artillery Officers during the war because of the convenience of plane computations for the indirect control of cannon fire. After the war, the *North Carolina Geodetic Survey* wrote to the USC&GS and requested that a set of projection tables be compiled for their state. Soon thereafter, the idea took off and the State Plane Coordinate Systems were computed, established, and eventually legislated for every state of the United States. In the 1930s, much of the country was unemployed, and President Franklin Delano Roosevelt created the *Works Progress Administration* (*WPA*). A number of states had federally funded and founded geodetic surveys, including Louisiana. Triangulation traverse surveys were conducted with transit and steel tape by the WPA, and second-order closures were required to be better than one part in ten thousand (1:10,000). Consequently, one of the design criteria of the State Plane Coordinate Systems computed by the USC&GS was that the maximum systematic error of the conformal projections could not exceed 1:10,000. This rationale was enforced because the average civil engineer and land surveyor had very humble computational power at their disposal, and if they neglected to correct chained distances for the scale factor at a point, they could still produce excellent-quality (at the time) field survey closures using computations on the local State Plane Coordinate System Zone of their state.

As technology continued to progress including the invention of the electronic distance meter (EDM), local engineers, surveyors, and cartographers started discovering that the first-order triangulation stations had some significant errors in their published coordinates. The National Academy of Science published a paper documenting the apparent deficiencies of the NAD27 and recommended that Congress appropriate necessary monies to re-compute the old classical datum along with additional scale control through a proposed Transcontinental Traverse that would span the nation from east to west in several routes using first-order theodolites and EDM instruments. This was in the early 1970s, and the United States had entered the Space Age, including the invention and implementation of artificial geodetic satellites. The USC&GS had the BC-4 cameras (Army surplus) for intercontinental triangulation, the Army had the ANNA satellite and later the SECOR system for geodetic triangulation, but these systems took weeks or months of observations to acquire sufficient observations to compute a single latitude, longitude, and ellipsoid height! (*This was a significant improvement over the classical method that took years to decades.*) During the middle 1970s, the Navy TRANSIT satellite system and geoceiver were declassified for public use. The USC&GS acquired some of these electronic instruments including the famous Magnavox 1502 geoceiver that enabled observations referenced to the *World Geodetic System Datum of 1972*. The observations obtained at collocated Transcontinental Traverse stations showed that there were still significant residual systematic errors present! Nevertheless, the recomputed *North American Datum of 1983* included all of the original classical triangulation observations, the Transcontinental Traverse observations, and the occasional TRANSIT observations. The majority of the resultant errors of the NAD83 were now at least at the sub-meter level.

For a number of years into the middle 1980s, individual states entered into cooperative agreements with the NGS to observe a representative sampling of stations to establish the High Accuracy Reference Network (HARN) and similar names and acronyms with TRANSIT observations. The U.S. Air Force commenced the NAVSTAR system of artificial satellites that promised "real-time" positioning and dynamic locating called the Global Positioning System (GPS). It was found that a number of academics, some in geodesy but most in geophysics could obtain geodetic-quality positioning using a differential approach to positioning. The U.S. Coast Guard started installing Permanent GPS receivers with broadcast capabilities to commercial shipping activities, and the concept of Continuously Operating Reference System (CORS) sites was born. Ultra-precise positioning was now available to anyone with access to a dual-frequency receiver!

Problems had been developing with the vertical datum in the United States also, and observations were performed through several transcontinental levelings in what was termed "Basic Net A." The re-adjustment of the national vertical datum was accomplished and published in the 1990s as the North American Vertical Datum of 1988, but Congress had taken notice of the enormous cost of the new observations of "Basic Net A." Congress directed the NGS to come up with a solution to the cost of what had become ~$1,500 per mile of first-order differential leveling; a cost that would continue to skyrocket as the price of labor continued to rise with the cost of living in the United States.

In 1992, the NGS published the plan for the *National Height Modernization Program* that would employ the *first new methodology to measuring elevations in over 5,000 years*: **GPS Leveling**. By 1992, the Soviet Union was kaput. The possibility of WWIII breaking out any day with nuclear warfare had faded, and peace was breaking out everywhere (for a while, anyway). The mathematical model of the Earth's gravity field, the GEOID, was declassified by the U.S. Department of Defense as it was no longer needed to be Top Secret for the targeting and guidance of Inter-Continental Ballistic Missiles. Congress approved NGS's plan, and the installation of numerous GPS CORS sites commenced both by NGS and by private/public entities. The implementation of the GEOID is well-employed if not well understood in exquisite detail, and ellipsoid heights are now converted to equivalent orthometric elevations referenced to the NAVD88 now performed on a daily basis nationwide. The reliability of the transformation from "h" to "H" is not of as precise a level of accuracy as one would want; it is now in the decimeter range for much of the continental United States. However, the cure for the GEOID malady is in the works. GRAV-D is a current program of the NGS in that airborne gravity surveys are being flown currently for the entire United States with the plan to develop a three-dimensional datum, including a vastly improved and accurate GEOID for a new datum of the United States in 2022. It is hoped that the new datum will employ a centimeter-level (on the average) accurate GEOID for the entire continent.

Documentation on the mathematics and parameters of the State Plane Coordinate Systems of the *NAD27* can be found in "State Plane Coordinates by Automatic Data Processing" by Charles N. Claire, SP 62–4, 1968 at *https://geodesy.noaa.gov/library/*.

Documentation on the mathematics and parameters of the State Plane Coordinate Systems of the *NAD83* can be found in NOAA Manual NOS NGS 05, "State Plane Coordinates of 1983" by James E. Stem, 1989 at *https://geodesy.noaa.gov/library/*.

As of the time of this publication, the United States National Geodetic Survey is in the process of documenting the mathematics and parameters of the State Plane Coordinate Systems of the *North American Terrestrial Reference System of 2022*. Significant changes are in store.

For some areas of continuous crustal motion, it is expected that a quasi-geoid will be a necessary enhancement for the NAD22 as is being observed for Louisiana by LSU with terrestrial gravity observations (relative and absolute) and 100+ CORS sites throughout the state and the Gulf Coast operated and maintained by the *LSU Center for GeoInformatics*.

With respect to this history of geodetic operations in the United States, much is due to the copious publications of the USC&GS and the NGS, including in particular by the late Joseph Dracup, a wonderful man that was very kind to me.

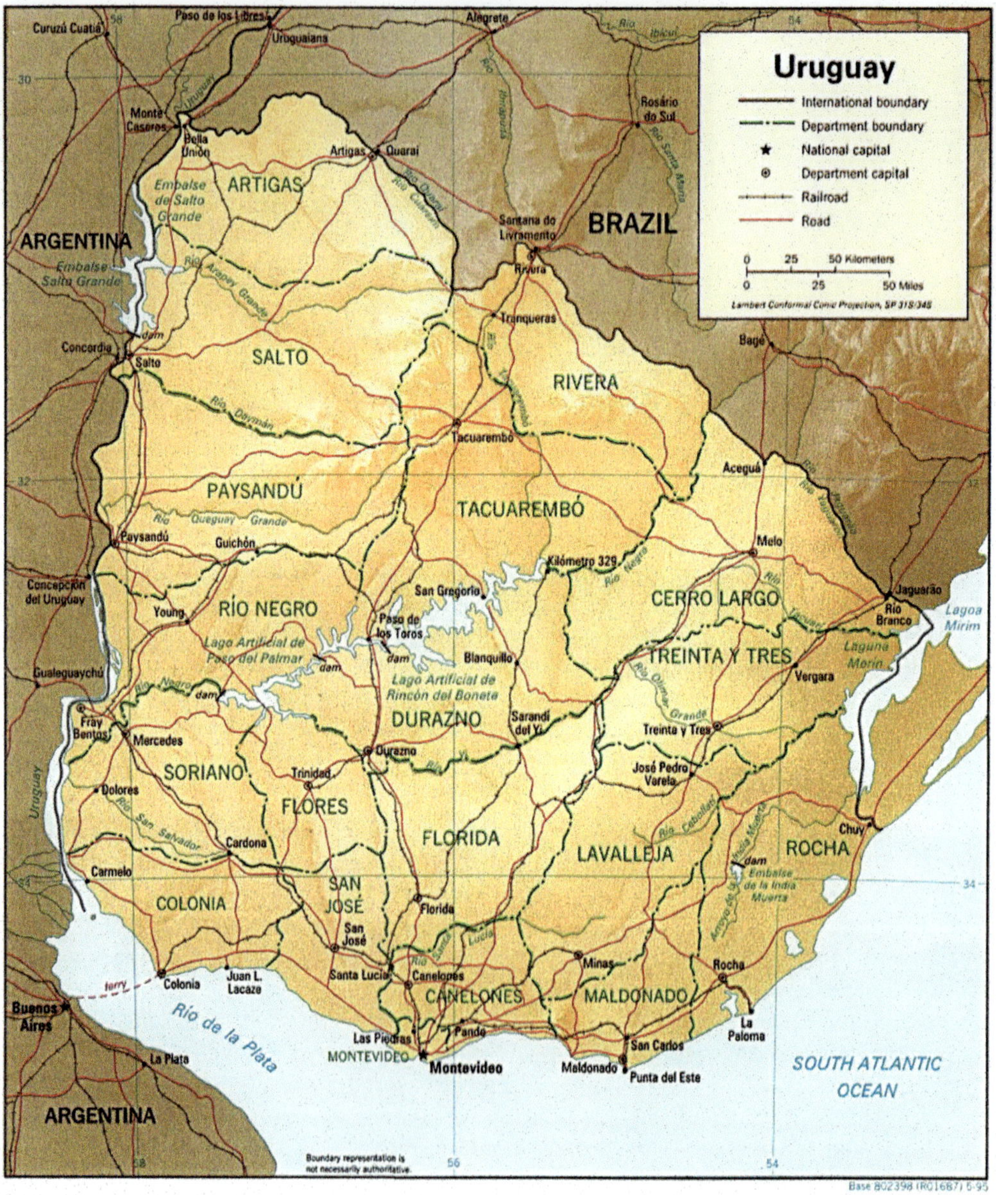

Base 802398 (R01687) 5-95

Oriental Republic of Uruguay

Uruguay is slightly smaller than the state of Washington and is the second-smallest country in South America after Suriname. Most (three-fourths) of the low-lying portion of the country is grassland and is ideal for raising cattle and sheep. The coastline is 660 km long, and Uruguay shares borders with Argentina and Brazil. The lowest point is the Atlantic Ocean (0 m) on the eastern border, and the highest point is *Cerro Catedral* (514 m).

The region was originally inhabited by *Charrúa* Indians and was visited by the Spanish explorer Juan Díaz de Solís in 1516. Colonia was founded by the Portuguese in 1680, and Montevideo was founded in 1726. Long associated with Brazil, the former *Banda Oriental, Cisplatine Province* gained independence on August 25, 1825. The economy of Uruguay is one of the two strongest of all of the South America countries.

The initial classical triangulations with Bamburg broken elbow astronomical theodolites and Huetz direction theodolites were performed in the central portion of the republic corresponding to the *Durazno* and *Tacuarembó* provinces in 1908. The *Servicio Geográfico Militar* (SGM) was founded in May of 1913 and was based in *Montevideo*. The following year, a triangulation network plan was designed, including four meridian arcs and five parallel chains. The classical triangulation net would take another 50+ years to complete. The Clarke 1880 ellipsoid was chosen for an initial figure of reference, where $a = 6,378,249.145$ m and $1/f = 293.465$. The grid system chosen in 1924 was based on the Polyhedric projection (same as Argentina), and the centesimal angle system of grads. The Bonne projection was considered as too cumbersome for survey computations. The initial reference system was called "Cerro de Montevideo," and it was considered a reconnaissance-quality system intended mainly for military purposes. This series of maps were produced at 1:100,000 scale. The initial instrument of choice for detailed cadastral observations was the Schraeder Tacheograph, a European instrument intended for cadastral surveys. The instrument intended to provide third- and fourth-order control was the Schasselons (*sic*) theodolite. Lt. Col. Silvestre Mato, Commander of the SGM estimated (optimistically) in 1924 that the initial mapping of the republic would take six brigades of Geographic Engineers 21 years to complete.

From 1912 to 1918, the *International Laguna (lagoon) Merím chain* was observed along the southern border with Brazil. At about the same time, the metropolitan area of *Montevideo* was observed between 1915 and 1917. Subsequently, the western border with Argentina along the *Uruguay River* was observed in a chain of figures from 1918 to 1937, and the eastern border north of the *Laguna Merím chain* with Brazil was observed from 1920 to 1940 to its junction with the Argentine border. During the 1930s, the coastal area was triangulated from the *Laguna Merím chain* to connect with the southern chain of the metropolitan area of *Montevideo*. In 1939, observations were made at what was to become the origin point for the national datum at *Yacaré*, in the northernmost *Department of Artigas*. In 1946, the SGM decided to change the reference figure to the International ellipsoid of 1930, where $a = 6,378,388$ m and $1/f = 297$. My guess is that Col. Floyd Hough of the U.S. Army Corps of Engineers Army Map Service had something to do with that. It was his urgings to the Pan American Institute of Geography and History in the late 1940s that prompted all of Latin America to eventually adopt the UTM Grid for military mapping along with the International ellipsoid. The founding of the Inter-American Geodetic Survey School in the Panama Canal Zone and the cooperative geodetic observation campaigns and military assistance programs implemented those changes. The *República Oriental del Uruguay* – U.S. Army Map Service (ROU-USAMS) adjustment of 1965 used the origin point at *Yacaré*, where $\Phi_o = 30° 35' 53.68''$ South, $\Lambda_o = 57° 25' 01.30''$ West of Greenwich, and the defining azimuth from *Yacaré* to *La Quisilla* is $\alpha_o = 87° 34' 26.28''$ (from south). The deflection of the vertical is defined to be zero at the origin point. This adjustment included 248 vertices, 8 LaPlace Astro stations, 325 triangles, and 15 geodetic bases at *Oeste (Uruguaiana), Guavitu, Tambores, Sudeste, Ubajay-Mudry, Molles, Cerro Chato, Artibas, Cerro Colorado, Florida, Agraciada, Tararidas, Carrasco, Maldonado,* and *Cerro*. This ROU-USAMS adjustment has commonly been called *Yacaré Datum* outside of Uruguay. The new national (civil) grid system associated with this datum is based on the Gauss-Krüger Transverse Mercator projection with a central meridian $\lambda_o = 55°\ 48'\ (62^G)$ West of Greenwich.

In 1969, Dr. Irene Fischer of the U.S. Army Map Service published the results of the *South American Datum of 1969 (SAD69)*. A total of 36 common stations along the Brazil-Uruguay border were included in the adjustment, and this datum is the exclusive reference system used by the Uruguay-Brazil Boundary Demarcation Commission. The *SAD69* origin is in Brazil at station *Chuá*, where $\Phi_o = 19°\ 45'\ 41.6527''$ South, $\Lambda_o = 48°\ 06'\ 04.0639''$ West of Greenwich, and the defining azimuth from *Chuá* to *Uberaba* is $\alpha_o = 271°\ 30'\ 04.05''$. The orthometric elevation $H_o = 763.28$ m, the deflection of the vertical at the origin is defined as $\xi_o = 0.31''$ and $\eta_o = -3.52''$, and ellipsoid height $h_o = 0$ m. The ellipsoid of reference is the International 1967, where $a = 6,378,160$ m and $1/f = 298.25$.

By 1990, the geodetic net of Uruguay was comprised of 420 first-order stations and some 3,000 stations of second, third, and fourth orders; all determined by various instruments and densification methods. Satellite surveying began in 1993 for topographic densification in Uruguay, and in 1995 geodetic-quality GPS receivers were used to participate in the South American Geocentric Reference System (SIRGAS) project. According to M.Sc.T.L.Eng. Barbato, F.D. of the University of the Republic, three LaPlace stations were initially occupied along with five other first-order stations in Uruguay. In 1997, three more stations in Uruguay were added to the SIRGAS campaign. The software used to initially process the GPS data was that developed by the University of Bern. Additional solutions were performed with the Bernese package and also with the Geomatics Canada package, "Geodetic adjustment using Helmert blocking of Space and Terrestrial data" (GHOST) and using geoidal heights from GeoidUru 1994 and GEMT2 models. The final República Oriental Uruguay 1998 (ROU 98) net adjustment was comprised of three SIRGAS-weighted position equations, 417 geodetic stations with geoid heights including 11 observed with GPS, 2337 horizontal directions, 66 distances, 11 astronomic azimuths, *etc.*, resulting in a variance of unit weight of 1.65. (That's very good!) According to Prof. Fabián Barbato, the general trend of the position change vectors from the ROU-USAMS 1965 Datum to the ROU 1995 Datum was 260° – a mean scalar of 2 m in Latitude and –37 m in Longitude. The Brazilian geodesists Piña and Costa (*Universidade Federal do Paraná*) published the transformation parameters from the ROU-USAMS 1965 Datum to the ROU 1995 Datum (Yacaré to WGS84) with both four-parameter and with seven-parameter models. Providing both sets of transformation parameters resolves any uncertainty regarding the sense of the rotations in the seven-parameter model by using the four-parameter set as blunder control. The four parameters are as follows: $\Delta X = -154$ m ±5 m, $\Delta Y = +162$ m ±7 m, $\Delta Z = +46$ m ±5 m, and the scale in ppm = -2.1365 ±1.5 ppm. The corresponding seven parameters are as follows: $\Delta X = -124$ m ±14 m, $\Delta Y = +184$ m ±11 m, $\Delta Z = +45$ m ±12 m, the scale in ppm = -2.1365 ±1.5 ppm, $R_x = -0.4384''$ ±0.001'', $R_y = +0.5446''$ ±0.001'', and $R_z = -0.9706''$ ±0.001''. Piña and Costa state that the average error ellipse accuracy of the classical stations on the ROU 98 Datum is better than 50 cm. In contrast, the TR8358.2 transformation parameters published by NIMA without accuracy estimators are $\Delta X = -155$ m, $\Delta Y = +171$ m, and $\Delta Z = +37$ m. Thanks go also for their kind assistance to Mr. Leonardo Ferrando of Montevideo and also to Mr. Rubén Rodríguez of Buenos Aires.

"Our country currently has 7 CORS, which make up the Active National Geodetic Network of the Oriental Republic of Uruguay (REGNA-ROU), developed and administered by the *Servicio Geográfico Militar*, within the framework of the framework of the fulfillment of one of its fundamental missions. Although the first advances in this regard were achieved between 1994 and 1995, with the installation of temporary stations, it is in the month of September 2006 that SGM, in agreement with the University of Memphis of the United States of America and within the framework of the Andes Project Centrals (CAP), installs its first CORS, near the city of Tacuarembó (located in rock mother), for monitoring the South American plate. This event was the milestone fundamental for the future implementation of the REGNA-ROU. At the beginning of 2007, with the incorporation of 2 new stations (UYRO in *Santa Teresa-Rochana* and UYMO in the *Fortaleza del Cerro de Montevideo*), the REGNA-ROU is definitively inaugurated. In the last 3 years another 4 stations (*UYDU in Sarandí del Yi – Durazno, UYPA in Paysandú, UYLP in La Paloma - Rocha and UYSO in Mercedes – Soriano*). In the near future, it is intended (*to*) achieve a distribution throughout the national territory where each season covering a 'buffer' 70 km (22 stations at least). Each of the stations is connected via GPRS/3G *routers*, to the central server (data management hardware and software GNSS), allowing monitoring and access to observations permanently and safe. Recorded data is managed by the Server and made available to the web in free and free form, such as geoservice through the Internet" (*http://www.igm.gub.ut/?mdocs-file-466*).

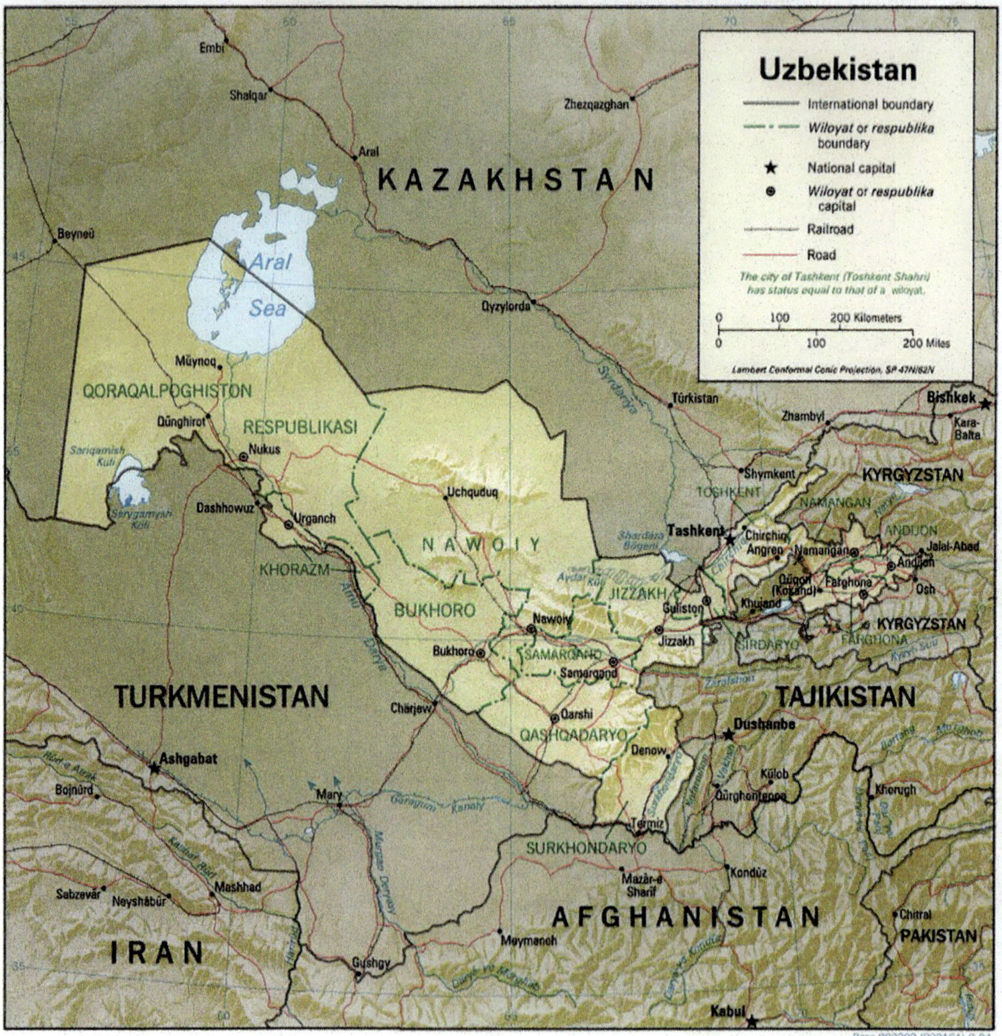

Base 802202 (R00161) 8-94

Republic of Uzbekistan

This ancient country of Uzbekistan (1200 *B.C.*) lies in the heart of Central Asia and is the center of the territory formerly known as Cimmeria, Bactria, and Turkestan. The oasis cities of *Samarkand, Tamerlane,* and *Tashkent* served as major cities on the *Great Silk Road* that connected China with Europe. Uzbekistan is one of the only two doubly landlocked countries in the entire world (the other is Liechtenstein).

The Fergana Valley was converted from its millennia-old agricultural tradition of varied crops to only cotton when the Aral Sea was diverted to irrigate the Soviet dream of a cotton-exporting capital. Uzbekistan gained its independence from the former Union of Soviet Socialist Republics on August 31, 1991, and now celebrates its independence holiday on September 1.

In 1839, the Russian Imperial Army moved into the area of *Turkestan* and completed its occupation in 1867. The city of *Tashkent* was selected as the capital of the newly formed Government General of Turkestan, which at the time comprised the two provinces of *Syr Darya* and of *Semiryechinsk*. Provinces added later were *Samarkand 1868, Bukhara'69, Amu Darya'73, Khokanda'75, Khiva'75, Fergana'76, Turkmenia'84*, and *Transcaspia 1888*. Colonization and railroad projects required topographic surveys, and the Military Topographic Department of Turkestan was formed in 1867. The *Physical Observatory of Tashkent* was officially founded by the Department in 1878.

The first classical triangulation was conducted from 1871 to 1895 with scale established at the baselines measured at *Miny Yuryukh, Tokacha, Samarkand, Chuamy, Nikolskiy,* and *Visokoye*. These were oriented primarily on Miny Yuryukh Datum and later on *Tashkent Datum*, retaining the astronomic azimuth from the NW to SE end point of the *Miny Yuryukh baseline*. That orientation was geodetically transferred to the *Tashkent Observatory Meridian Circle*. Triangulations of 1896–1929 were scaled to baselines measured by invar wires (*Jäderin base apparatus*) at *Kazalinsk, Arys, Osh,* and *Kyzyl Rabat*, and oriented on *Tashkent Datum, Kazalinsk Datum,* and *Osh Datum*.

Miny Yuryukh 1871 Datum was established at the NW Base Point (M), where $\Phi_o = 41° 17' 47.70''$ North, $\Lambda_o = 38° 57' 00.04''$ East of Pulkovo (where the Pulkovo Observatory is 79° 46' 19.56'' East of Greenwich). Scharnhorst executed the astronomic observations and referenced the datum to the Bessel 1841 ellipsoid of revolution where the semi-major axis $(a) = 6,377,397.155$ m and the reciprocal of flattening $(1/f) = 299.1528128$. The defining azimuth was determined at the point of origin to the SE Base point (N) as: $\alpha_o = 114° 26' 34.6''$. These Russian geodetic surveys were measured in units of sazhens. For instance, the *1874 Samarkand Baseline* was 1605 sazhens or 3,424 m. The early baselines were measured with wooden rods; however, no information is available on what kind of wood was used.

In the United States during the same time period, Magnolia wood was used for leveling rods because it was believed to have the lowest coefficient of expansion (from variations in temperature AND humidity) of all wood species. Note that the Americans (Coast & Geodetic Survey) still boiled their wood in paraffin just to make sure …

In order to make a topographic survey at 1:4,000 scale of the *City of Tashkent*, a traverse net was established in 1885 by Pomerantsev. The revised coordinates of the *Tashkent Observatory SW pillar* used for the city survey are: $\Phi_o = 41° 19' 30.42''$ North, $\Lambda_o = 38° 58' 00.99''$ East of Pulkovo (or 69° 17' 39.54'' East of Greenwich). The vertical datum of *Tashkent* was determined at the Meteorological Station of the Military Topographic Department located in the residence of Colonel Zhemchuznikov by means of barometric observations. The transfer of this vertical Datum point value to other horizontal datum origins with trigonometric leveling techniques produced geodetic problems of mind-boggling proportions!

The physical connection between the *Tashkent Datum of 1875* and the *Indian Datum of 1916* in the *Pamir region* was made by an exploratory triangulation party of the International Geodetic Union. The expedition was led by Professor Finsterwalder in the early part of the 20th century. The three-parameter shift from *Tashkent* to Indian is: $\Delta X = -223.632$ m, $\Delta Y = -281.310$ m, $\Delta Z = +304.059$ m. (The fit of two points agree to better than a meter.) The *Tashkent Datum of 1875* origin (at the Meridian Circle) is: $\Phi_o = 41° 19' 30.42''$ North, $\Lambda_o = 69° 17' 39.54''$ East of Greenwich. The defining azimuth at the point of origin to the North Stone Pillar (probably the mire) of the Observatory is: $\alpha_o = 00° 52' 08.25''$, and the ellipsoid of reference is the Bessel 1841. The *Indian Datum of 1916* origin at *Kalianpur Hill Station* is: $\Phi_o = 24° 07' 11.26''$ North, $\Lambda_o = 77° 39' 17.57''$ East of Greenwich. The defining azimuth at the point of origin to station *Surantal* is: $\alpha_o = 190° 27' 05.10''$. The ellipsoid of reference is the Everest 1830, where $a = 6,377,276.345$ m and $1/f = 300.8017$.

The *Kazalinsk Datum of 1891* origin at the finial Cross on the Town Church is: $\Phi_o = 45° 45' 46.450''$ North, $\Lambda_o = 62° 06' 01.66''$ East of Greenwich. The defining azimuth at the point of origin to station Sulutan is: $\alpha_o = 20° 34' 07.34''$. The ellipsoid of reference again is Bessel 1841. The three-parameter shift from *Kazalinsk 1891* to *Tashkent 1895* is: $\Delta X = +530$ m, $\Delta Y = -160$ m, $\Delta Z = -104$ m. (The fit of four points agrees to about 20 m in each geocentric component.)

The *Osh Datum of 1901* origin is defined at Point II, Northwest Base as: $\Phi_o = 40° 37' 16.670''$ North, $\Lambda_o = 72° 56' 11.175''$ East of Greenwich. The defining azimuth at the point of origin to Point II, Southeast Base is: $\alpha_o = 152° 54' 01.86''$. The ellipsoid of reference again is Bessel 1841. The three-parameter shift from *Osh 1901* to *Tashkent 1895* is: $\Delta X = -146.633$ m, $\Delta Y = +472.553$ m, $\Delta Z = +508.352$ m. (The fit of seven points agrees to 4.19 m in each geocentric component.) I decided to try a Bursa-Wolfe seven-parameter shift. The results yielded an average fit error in Latitude of 0.48 m, the error in Longitude was 1.09 m, and the error in Height was 0.09 m. The parameters that give this fit are: $\Delta X = +26.82$ m, $\Delta Y = -183.11$ m, $\Delta Z = -186.25$ m, $\delta s = 39.25 \times 10^6$, $R_z = -07.79''$, $R_y = +2.48''$, $R_x = +21.49''$.

Uzbekistan has only one Grid system, and that is based on the Russia Belts, a series of Gauss-Krüger Transverse Mercator projections that use the exact same rules and parameters as UTM with two exceptions. The Krassovsky 1940 ellipsoid is used everywhere in the former USSR, where $a = 6,378,245.0$ m and $1/f = 298.3$. The scale factor at origin is 1.0 rather than the 0.9996 scale factor used for UTM. The single unifying datum used in all the former USSR is the "Coordinate System 1942," a consistent marvel that is the largest classical datum in the world! This datum is often improperly referred to after its origin, Pulkovo Observatory, where $\Phi_o = 59° 46' 18.55''$ North, $\Lambda_o = 30° 19' 42.09''$ East of Greenwich. The defining azimuth at the point of origin to Signal A is: $\alpha_o = 317° 02' 50.62''$. The latest three-parameter shift values published by NGA for the vicinity listed as Kazakhstan are from *System 42* to *WGS84* such that $\Delta X = +15$ m, $\Delta Y = -130$ m, $\Delta Z = -84$. The stated accuracy is 25 m in each component, which is the maximum error useable for 1:50,000 mapping, and that was based on only two points.

A recent publication by Mirmakhmudov, E., Prenov Sh., Magdiev, Kh., and Fazzilova, D. on (the) *Intermediate reference frame for Uzbekistan topographic maps,* United Nations/Russian Federation Workshop on the Applications of Global Navigation Satellite Systems, Krasnoyarsk, May 18–22, 2015, states that Fazilova found in 2002 the transformation parameters **from** CS-42 **to** WGS84 are: $\Delta X = +23$ m, $\Delta Y = -125$ m, $\Delta Z = -87$ m. However, no accuracy estimate is offered.

REFERENCES: U

A Guide to Coordinate Systems in Great Britain.pdf, www.gps.gov.uk.
Al Ain Local and Global Spatial Reference Systems, Map Middle East, 2005.
Al Marzooqi, Y., Fashir, H., Babiker, T., Establishment & Testing of Dubai Virtual Reference System (DVRS) National GPS-RTK Network, GISdevelopment.net.
An Absolute/Relative Gravity Base Net in the Emirate of Dubai.
Charles N. Claire, State Plane Coordinates by Automatic Data Processing SP 62–4, 1968.

COL Alexander Ross Clarke, Principal Triangulation of 1783–1853.
Dr. Irene Fischer, South American Datum of 1969 (SAD69), U.S. Army Map Service, 1969.
Dr. Michael Cheremshynsky of the Ukraine Main Administration of Geodesy, Cartography and Cadastre of Kiev, Ukraine, Personal Communication, May 2004.
Dr. Momchil Minchev of Sofia, Bulgaria, Personal Communication, May 2004.
G.T. McCaw, Empire Survey Review, No. 52, pp. 257–262.
https://igm.gub.uy/documentos/
https://dgm.gki.com.ua/pererahunok-po-gelmertu-(po-kljuchu)-na-ploschini
https://geodesy.noaa.gov/library/.
https://www.ordnancesurvey.co.uk/documents/resources/guide-coordinate-systems-great-britain.pdf.
Hume F. Rainsford, Survey Adjustments and Least Squares, Constable, 1957.
James E. Stem, State Plane Coordinates of 1983, 1989.
Lonely Planet, 2012.
M.Sc.T.L.Eng. Barbato, F.D. of the University of the Republic, October, 2002.
Mirmakhmudov, E., Prenov, Sh., Magdiev, Kh., and Fazzilova, D. on (the) Intermediate reference frame for Uzbekistan topographic maps, United Nations/Russian Federation Workshop on the Applications of Global Navigation Satellite Systems, Krasnoyarsk, 18–22 May 2015.
Mr. Leonardo Ferrando of Montevideo, Personal Communication, October 2002.
Mr. Rubén Rodríguez of Buenos Aires, Personal Communication, October 2002.
República Oriental del Uruguay – U.S. Army Map Service (ROU-USAMS) adjustment of 1965.
State Service of Ukraine for Geodesy, Cartography and Cadastre (StateGeoCadastre) 2022.
Webster's Geographical Dictionary.
World Factbook, 2004.
World Factbook, 2012.
Y. Okia and J. Kitaka, The Uganda Triangulation Network: Establishment and Current Status, Lands and Surveys, Entebbe, Uganda.

V

Republic of Vanuatu 869
The Bolivarian Republic of Venezuela 873
The Socialist Republic of Vietnam 877

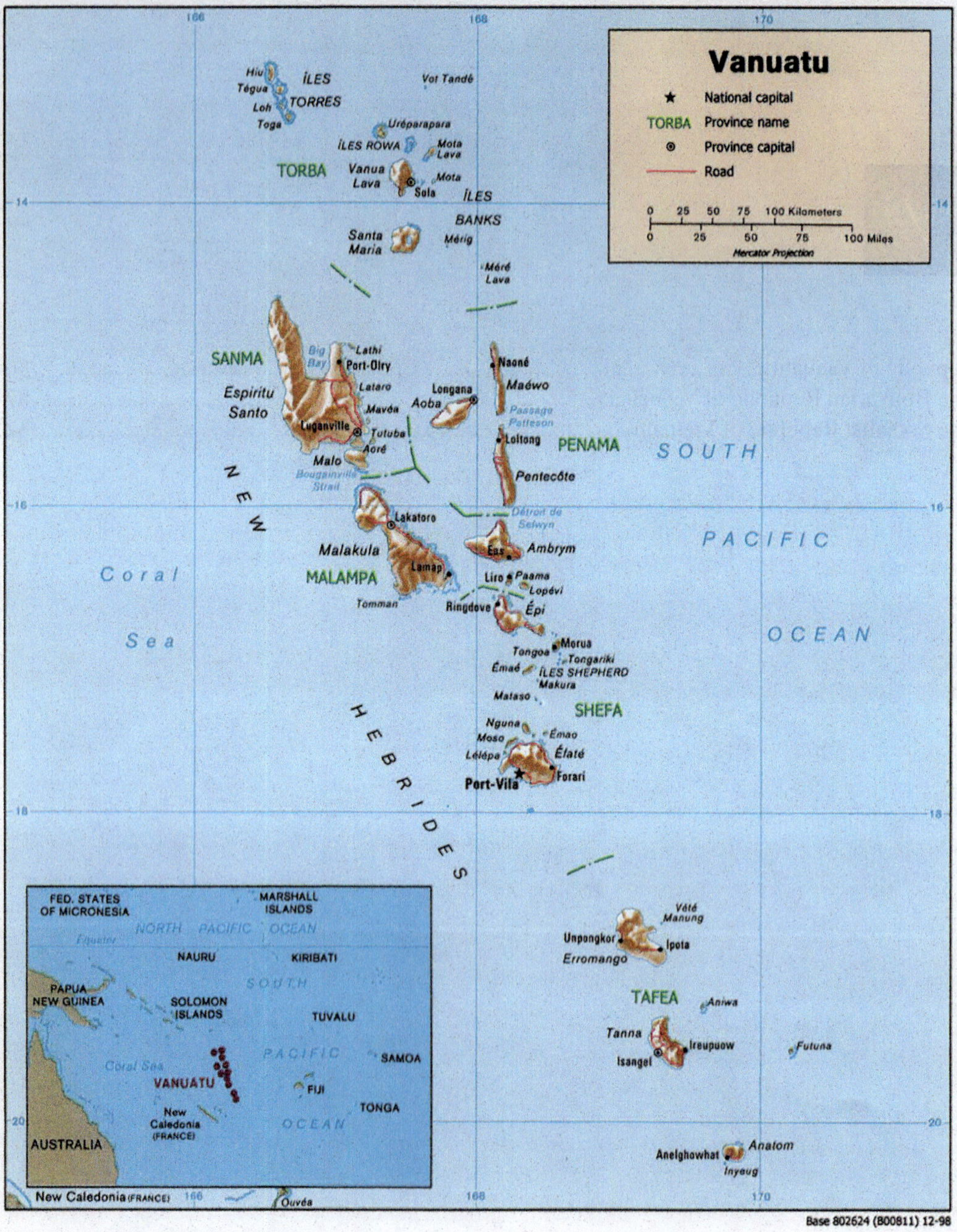
Vanuatu
National capital
TORBA Province name
Province capital
Road
0 25 50 75 100 Kilometers
0 25 50 75 100 Miles
Mercator Projection
ÎLES TORRES
Hiu
Tégua
Loh
Toga
Vot Tandé
Uréparapara
ÎLES ROWA
Mota Lava
Mota
TORBA
Vanua Lava
Sola
ÎLES BANKS
Santa Maria
Mérig
Méré Lava
SANMA
Espiritu Santo
Big Bay
Lathi
Port-Olry
Lataro
Mavéa
Luganville
Tutuba
Aoré
Malo
Bougainville Strait
Longana
Aoba
Naoné
Maéwo
Passage Patteson
Loltong
PENAMA
Pentecôte
Lakatoro
Malakula
MALAMPA
Lamap
Tomman
Détroit de Selwyn
Eas
Ambrym
Liro
Paama
Lopévi
Ringdove
Épi
Morua
Tongoa
Tongariki
Émaé
ÎLES SHEPHERD
Makura
Mataso
SHEFA
Nguna
Moso
Émao
Lélépa
Élaté
Port-Vila
Forari
Coral Sea
NEW HEBRIDES
SOUTH PACIFIC OCEAN
Vété Manung
Unpongkor
Ipota
Erromango
TAFEA
Aniwa
Tanna
Ireupuow
Isangel
Futuna
Anelghowhat
Anatom
Inyeug
New Caledonia (FRANCE)
Ouvéa
FED. STATES OF MICRONESIA
MARSHALL ISLANDS
NORTH PACIFIC OCEAN
Equator
NAURU
KIRIBATI
PAPUA NEW GUINEA
SOLOMON ISLANDS
TUVALU
Coral Sea
VANUATU
SAMOA
FIJI
New Caledonia (FRANCE)
TONGA
AUSTRALIA
166
168
170
14
16
18
20
Base 802624 (B00811) 12-98

Republic of Vanuatu

With a land area of 12,200 km², the republic is slightly larger than Connecticut. Vanuatu has a tropical climate, the terrain is comprised of mostly volcanic mountains with narrow coastal plains, the lowest point is the Pacific Ocean, and the highest point is *Tabwemasana* (1,877 m) on the island of Espiritu Santo.

Inhabited for thousands of years by Melanesians before discovery by the Portuguese navigator Pedro Fernandes de Queirós, the islands were forgotten for 160 years and were then visited by the French navigator Louis-Antoine de Bougainville in 1768. The English mariner Captain James Cook explored the islands in 1774 and named it the *New Hebrides*. "The British and French, who settled the New Hebrides in the 19th century, agreed in 1906 to an Anglo–French Condominium, which administered the islands until independence in 1980." What the *World Factbook* doesn't say is that the local people referred to it as the Pandemonium! *Vanuatu* \ vän-, -wä-tü \, is a group of more than 80 islands in the southwest Pacific Ocean northeast of *New Caledonia* and west of *Fiji*. The total coastline is 2,528 km and its maritime claims are based (naturally) on archipelagic baselines. The exclusive economic zone is 200 nautical miles (NM), the territorial sea is 12 NM, the contiguous zone is 24 NM, and the continental shelf claim is 200 NM or to the edge of the continental margin – all of these claims are customary and are recognized under the International Law of the Sea.

In the *Vanuatu Geodetic Control Network Report* by Bakeeliu, Kanas, and Kalsale in June 2001, the network that began in the 1960s was generally detailed to the present. The *Institute Géographique National* (IGN) of France started their network in the 1960s. "The IGN network was made in two blocks, one of which covers the islands of Santo, Aoba, Pentecost, Maewo, Ambrym, Malekula Epi, Éfaté in the northern part of Vanuatu while the other block covers Erromango, Tanna, Anatom and the nearby small islands in the south. The islands left out were the Banks and Torres group in the far north of Vanuatu." The report continues, "The IGN [datum – Ed.] was based on the astronomical observation made at *Bellevue* on *Éfaté*." Note that another common spelling for the island of *Éfaté* is *Île Vaté*. The *Vanuatu (IGN) 1960 Datum* origin coordinates at Bellevue are $\Phi_o = 17°\ 44'\ 17.40''$ South, $\Lambda_o = 168°\ 20'\ 33.25''$ East of Greenwich, and the ellipsoid of reference is the International 1909 (Madrid 1924), where $a = 6,378,388$ m and $1/f = 297$. The National Geospatial Intelligence Agency (NGA) lists the transformation parameters from The *Vanuatu (IGN) 1960 Datum (Bellevue)* to the WGS84 Datum as $\Delta a = -251$ m, $\Delta f = -0.14192702$, $\Delta X = -127$ m ± 20 m, $\Delta Y = 769$ m ± 20 m, and $\Delta Z = +472$ m ± 20 m. This relation is based on observations at three stations. John W. Hager, retired from what is now NGA says, "The transformation states that it is for the islands of *Éfaté* and *Erromango*, but my notes imply that the trig list applies from *Éfaté* to *Espiritu Santo*. Source for this is 'Loose Minute, MCE Ht/KHG/PL, 10 July 1970 with one page from I.G.N. Trig list.' I don't remember what all the letters mean but it was a letter from the British Military to U.S. Army Topographic Command." John Hager also commented that *"Pier Observation Spot, Vila Harbor, Éfaté Island*. I have no data but questioned for my further investigation whether it was Fila or Vila, Fila possibly being a corruption of *Éfila*." The presumed datum point is S. T. 1 (*Service Topographique*) at latitude 15° 17' 16" S, longitude 167° 58' 34" E. This was taken from "Traverse Around Perimeter of Aoba Island," 30 September 1969. He found a 1949 IGN reference to a local grid for the *Aoba Island Datum*, where False Northing = 12,000 m and False Easting = 18,000 m. According to Hager, "the only odd map projection I find is for *Nouvelles Hébrides, Fuseau Calédonie-Hébrides*, Gauss projection, [*Transverse Mercator – Ed.*], International ellipsoid, meter, latitude of origin = equator, longitude of origin = 167° E, Scale factor unknown but probably unity, false northing (y) = 2,600,000 meters, false easting (x) = 1,000,000 meters. This is from '*Catalogue de Cartes en Service Publiées par l'Institute Géographique National*,' Paris, 1 July 1949."

Referring back to the *Vanuatu Geodetic Control Network Report*, "The adjustment used by DOS [*Directorate of Overseas Surveys, UK – Ed.*] was initiated from the same points as the IGN however the astronomical observations and adjustment was done separately. The DOS adjustment covers the islands of *Santo, Aoba, Maewo, Pentecost, Ambrym, Malekula* and *Pamma* in the north and *Éfaté, Erromango, Tanna, Anatom* and *Futuna* in the south. The DOS however extended its triangulation further throughout the country covering and strengthening the network to other islands, except *Bank* and *Tores* in the far north. This adjustment was used for mapping as well as cadastral. DOS adjustment uses the same scale factor of one (1.00000) throughout the country, though each island has its own origin."

Continuing, early in the "1980s the Vanuatu Government attempted to connect the DOS north and south block using traverse methods with the introduction of Tellurometer distance measurements. However, it was found that there was some discrepancy between the two blocks. It was uncertain then that the error was in the traverse observation or the astronomical observation of the two blocks. It was also difficult to undertake alternative method of triangulation as the sights between *Epi* and *Emae islands* was very difficult. It was seen that it may be easier if a triangulation was done through the islands between the two blocks, however for some reason this was not done. The technology at that time may also be the cause of the inaccuracy of the observations. In mid 1990s the Australian Government assisted the Vanuatu Government by providing funds, technology and human resources through the Australian Defense Cooperation to run a Doppler network that covers the whole country. This has enabled the Vanuatu Government to anticipate the strengthening of the country's survey control network on the WGS72 spheroid. The network was produced to control the aerial photography of the country. For cadastral purposes the DOS geodetic adjustment is still maintained."

I asked Mr. Russell Fox, now retired from the International Geodetic Library of the Ordnance Survey International, United Kingdom, if he had anything to help me on my column on Vanuatu. To my (usual) surprise, he certainly did have something. Mr. Russell Fox worked there for three years! "The Condominium (known as the Pandemonium locally) was a strange form of government, the British and French running parallel but separate administrations in the same territory (so not analogous with *St. Maarten/St. Martin*). There were French and British police forces, hospitals, schools, *etc.* Residents had to use 'their' facilities. Citizens of countries other than Britain (& Commonwealth) or France had to opt for either honorary British or honorary French status and use the appropriate services. This split the local people also, half of whom were educated in the French milieu and half in British traditions. There was 'trouble in paradise,' as the newspapers put it, during the immediate pre-and post-independence period, as the more radical and pro-independence English-speaking ni-Vanuatu jostled for power with the French speakers and French settlers, who preferred the status quo (not least because French plantation owners would be most affected by proposed changes in land tenure)."

Mr. Russell Fox continued, "I worked in Vanuatu from 1983 to 1986. Independence had come in 1980, so I did not personally witness this, but one of the Survey Department's tasks pre-independence was to measure the heights of the flagstaffs at the British and French Residencies in *Port Vila*. There would have been a diplomatic incident if either the Union Jack or the *Tricoleur* had been flown slightly higher than the other! The Condominium was the result of Anglo-French rivalry in the Pacific during the late 19th century; I believe that the Australian colonies were particularly keen to avoid a French takeover of the *New Hebs* as well as *New Caledonia* and they lobbied the British Govt. to do something about it. The answer was condominium, if only to avoid an Anglo-French war. Another condominium was the Anglo-Egyptian Sudan. The WWII US presence in the *New Hebs* was still evident in the 1980s, with six-wheel trucks on plantations, USN dustbins [galvanized trash cans?] being used as water containers and metal plates from airfield runways being used as property fences." [*I remember seeing the same type of fences when I lived in Panamá – Ed.*]

"The main post-1978 survey activities I know of were: 1980 – A dozen Doppler stations were observed by 512 Specialist Team, Royal Engineers. 1983–1986 'Operation Algum' – major support for the Survey Department was received from the Royal Australian Survey Corps. This involved a Doppler campaign throughout the islands, new aerial photography, readjustment of the DOS and IGN trig networks on WGS84 and setting up a map production facility in the Survey Department. 1980s–1990s New Editions of the DOS 1:1,000,000 map were produced by the Survey Department, also a new 1:50,000 series. The Vanuatu Map Grid was introduced, a national TM projection to replace the assorted island grids that existed previously. The Survey Department produced a brief paper in about 1976/1977 that discussed the significant differences between DOS and IGN positions in the *New Hebs* (nearly a km in the northern islands if I recall correctly). Those discrepancies weren't solved – or circumvented – until OP Algum, but the Survey Department did develop a TM grid (called Éfaté TM 77) for the main island, Éfaté or (*Vaté*), in 1977 to improve the control situation thereby unifying disparate surveys and replacing the old Cassini grid. Both DOS and IGN used International Spheroid, but had datums in different places, and trig block boundaries in different places – the DOS North Block was islands North of *Éfaté*, and South Block was *Éfaté* and islands south. IGN had a North Block (*Éfaté* and islands North) and South Block (*Erromango* to *Aneityum*). I think the most northerly island in the *New Hebs*, the *Banks* and *Torres Islands*, were not reached by either the DOS or IGN networks and had local astro fixes only."

The National Geospatial Intelligence Agency (NGA) lists the transformation parameters from the *Santo (DOS)1965 Datum (Espiritu Santo Island)* to the WGS84 Datum as: $\Delta a = -251$ m, $\Delta f = -0.14192702$, $\Delta X = +170$ m ± 25 m, $\Delta Y = +42$ m ± 25 m, and $\Delta Z = +84$ m ± 25 m. This relation is based on observations at one station. Thanks to Mr. John W. Hager, Mr. Russell Fox, and to Mr. Tony Kanas, Surveyor, and the Vanuatu Department of Land Surveys for their generous assistance. In 2014, the U.S. Department of State published No. 137 Limits in the Seas, *Vanuatu: Archipelagic and Other Maritime Claims and Boundaries*. Coordinates are shown to four decimal places of arc seconds, and all connecting lines are defined as geodesics. No datum, ellipsoid, nor International Terrestrial Reference Frame date is stated.

Vanuatu Geodetic Control Network Report, Mike Bakeoliu, Tony Kanas, Moses Kalsale, 09 June 2001.

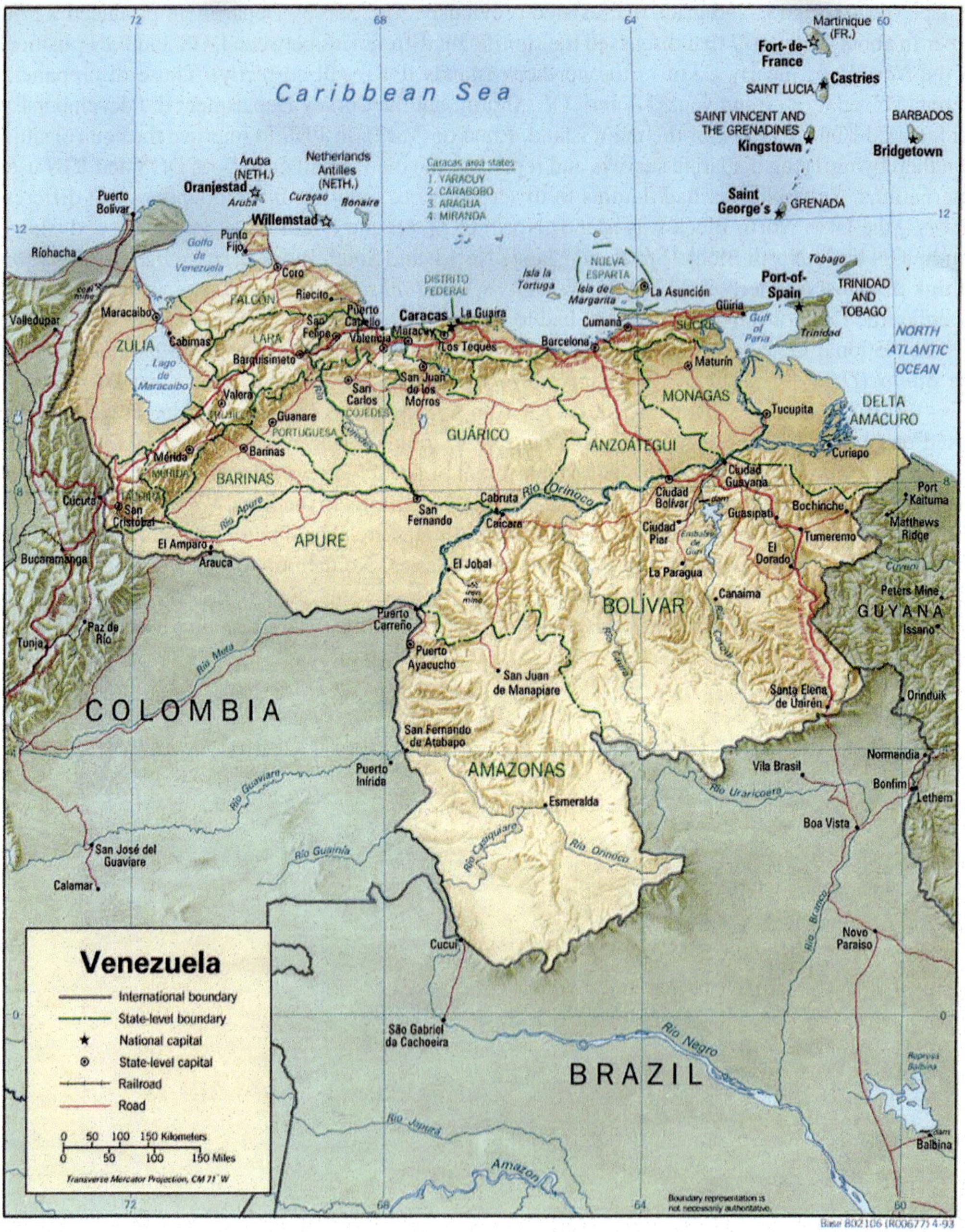

Base 802106 (R00677) 4-93

The Bolivarian Republic of Venezuela

Venezuela is more than twice the size of California; its terrain consists of the Andes Mountains and Lake Maracaibo Lowlands in the northwest; central plains (*Llanos*); and the Guiana Highlands in the southeast. The highest point is snowcapped *Pico Bolivar* (*La Columna*) at 5007 m (16,427 ft), and the *Orinoco River* is the third longest in South America.

Original inhabitants of modern Venezuela consisted of the Arawak, Caraïbe, and Chibcha peoples. Discovered by Columbus on August 1, 1498, the coast of Venezuela was traced by the Spanish navigators Ojeda and de la Cosa in 1499, and the first European settlement was established by Las Casas at Cumaná in 1520. The present capitol of *Caracas* was founded in 1567. Venezuela was included in the viceroyalty of New Granada in 1718 and was made a captaincy-general in 1731. Led by "El Libertator," Simón Bolívar, Venezuela declared its independence from Spain in 1811, but it was not assured until the battle of Campo Carabobo in 1821 near *Valencia*. Venezuela was part of Greater Colombia from 1819 to 1829 and formally separated in 1830. Bolívar's villa in the mountains is spectacular. The German town of *Colonia Tovar* was founded in 1834; *I recommend its cuisine, but the servers do not speak German.*

Longitude observations were first made in the Caribbean port city of *La Guaira* by Humboldt in 1800 and later by the Spaniard Chief of Squadron, Don José Espinoza in 1809. The *Comisión del Plano Militar* (Military Map Commission) was founded in 1904. The first geodetic surveys were initiated by the Army in 1911 in support of the railroad construction project between *Caracas* and *La Guaira*, where the international airport for Caracas is now located. The origin point for this first system was established on the plateau of the city of Caracas below the Army's *Cajigal Observatory*, probably to facilitate measurement of the initial baseline. Initial observations at *Cajigal* started in 1905, followed the next year in January through March by Dr. Luis Ugueto and Dr. Felipe Aguerrevere who determined the latitude by the Talcott Method (invented by Captain Talcott of the U.S. Army Corps of Engineers). The *Loma Quintana Datum of 1911* origin (located in the *El Mirador subdivision, 23 de Enero District of Caracas*) is: $\Phi_o = 10°\ 30'\ 24.680''$ North, $\Lambda_o = 66°\ 56'\ 02.512''$ West of Greenwich, the defining azimuth to station *Volcán* $(\alpha_o) = 316°\ 01'\ 50.33''$, $H_o = 1,077.54$ m above mean sea level at *La Guaira*. (My son André still remembers "hangover soup," the local name for bouillabaisse served in a *La Guaira* beachfront restaurant.) The ellipsoid of reference is the Hayford 1909, where the semi-major axis $a = 6,378,166.0$ m, and the reciprocal of flattening $1/f = 297$. The deflection of the vertical and geoid separation were defined to be zero at the origin. Interestingly, a later adjustment performed for the *Provisional South American Datum of 1956* only changed the azimuth from station *Loma Quintana* to station *Volcán* to $316°\ 01'\ 50.30''$, which is quite a testimonial to the superb quality of the initial work. The original *Loma Quintana Datum of 1911* was comprised of chains of quadrilaterals that eventually spanned the entire northern coast of Venezuela from the border with Colombia in the west to include the Island of Trinidad in the east. The only national projection associated with the *Loma Quintana Datum of 1911* is the Compensated Secant Conic. Originally published in 1940 by the *Dirección Cartografía Nacional* (National Cartographic Office), I became aware of the projection tables in 1980. Used for the topographic mapping, it is defined with a Central Meridian $\lambda_o = 67°\ 30'$ W, and standard parallels of $\varphi_N = 9°$ N and $\varphi_S = 4°$ N. This system baffled me for many years because I was unable to compute Lambert conformal conic developed meridional distances to match the published tables. In June 1988, the late John P. Snyder wrote to me and pointed out that he had found the same tables at the U.S. Library of Congress and the projection was actually the Equivalent Conic and not what I had thought to be the Lambert Conformal Conic! I subsequently discovered a constant +18.027-m error in all of the tabulated developed meridional distances. The nut was finally cracked with John P. Snyder's help.

Also associated with the Loma Quintana Datum of 1911 (LQD 1911) are the "Oil Grids" developed in association with petroleum exploration and development activities. This collection of Polyhedric plane projection coordinate systems has understandably baffled novices and experts alike for almost a century. There are four major families of Oil Grids referenced to the LQD 1911: *Maracaibo, Barcelona, Maturín, and Dabajuro*. The **Maracaibo Grid Family** consists of: the Maracaibo Cross Grid (1945–1981), Cruz Morillo Grid (1917–1983), Altagracia Church Grid (1920–1951), Alto de Escuque Grid (1957), Bloque B Grid (1953–1954), Boca Grita Grid, El Cubo Grid (1916–1961), Mene Grande Grid (1914–1982), El Mojón Grid (1927–1930), Monument "A" Río de Oro Grid, La Rosa Grid (1917–1982), El Mene de Marua Statue Grid (1927–1972), Iglesia de Santa Barbara Grid (1915–1986), Plaza Bólivar de San Cristóbal Grid (1939–1955), and the Barinas Grid (1953–1970). The **Barcelona Grid Family** consists of the Barcelona Grid (1913–1920), Plaza de Calabozo Grid, Chaguaramas Grid, Clarines Grid (1949–1958), Punzón Grid, Sabana de Uchire Grid, Santa Maria de Ipire Grid (1939–1974), El Sombrero Grid, Valle de la Páscua Grid (1939–1982), Areo Grid, Barbacoas Grid (1948–1980), and the Barinas West Base Grid (1946–1980). The **Maturín Grid Family** consists of the Maturín Grid (1918–1988), Irapa Grid, Tucupita Grid (1939–1966), Uracoa Grid (1966), and the Chaguamaral Grid (1955). The **Dabajuro Grid Family** consists of the Dabajuro Grid (1921–1963), La Vela de Coro Grid (1924–1981), Piritú Grid (1949), Iglesia Valera Grid (1928), and the Pedregál Grid. Every one of these grids is on the Polyhedric Projection, which is mathematically identical to the Local Space Rectangular system used in computational analytical Photogrammetry.

Other Grids associated with the LQD 1911 include the Venezuelan Gauss-Krüger Transverse Mercator, where the Central Meridian $\lambda_o = 63° \ 10' \ 48''$ W, the False Northing Latitude of Origin (φ_{FN}) = 9° 45' 02" N, the Scale Factor at Origin (m_o) = 0.9996 (same as UTM), and *both* the False Eastings and False Northings are 200 km. The Lake Maracaibo Tangent Lambert Conformal Conic Grids are defined with a Central Meridian $\lambda_o = 71° \ 36' \ 20.224''$ W, the False Northing Latitude of Origin (φ_{FN}) = 10° 38' 34.6780" N, the Scale Factor at Origin (m_o) = 1.0, where the Latitude of Origin is $\varphi_o = 10° \ 10' \ 00''$, and *both* the False Eastings and False Northings are *either* 200 or 500 km, depending on the oil company.

The *Cartografia Nacional* signed a cooperative agreement with the Inter-American Geodetic Survey (IAGS) in 1948. The following year the Director of *La Dirección de Cartografía Nacional*, Dr. Adolfo C. Romero initiated the gravimetric survey of the origin point and vicinity of the planned Provisional South American Datum of 1956 (PSAD56). The origin was in the town of *La Canoa, Anzoátegui Province*, Venezuela, where $\Phi_o = 08° \ 34' \ 17.170''$ North, $\Lambda_o = 63° \ 51' \ 34.880''$ West of Greenwich, and the defining azimuth to station *Pozo Hondo* (α_o) = 40° 22' 45.96". Subsequent relative gravity surveys were performed at many of the origin points for the Polyhedric Grids. The triangulation net of Venezuela was extended to its borders, and connections were made with Colombia to link to the *Bogotá Datum of 1948*. Subsequently the entire collections of chains were re-computed in a simultaneous adjustment that formed the basis of the PSAD56 that was conveniently referenced to the Hayford 1909 (International 1924) ellipsoid. The Compensated Secant Conic projection was dropped, and the Universal Transverse Mercator was adopted as the standard Grid for the new Datum.

In the early 1980s, an enormous hydrocarbon field was discovered off the eastern coastlines of Venezuela and Trinidad. Although a treaty existed between the two sovereign nations for the division of the mineral resources of the continental shelf, the treaty only covered the *Gulf of Paria* – not the Atlantic Ocean. The coordinate systems of the region were known to be unusually complex; Trinidad and Tobago utilized both the Old Trinidad Datum of 1903 with the Cassini-Soldner Grid and the Naparima Datum of 1972 with the UTM Grid, and Venezuela utilized both the Loma Quintana Datum of 1911 with the Compensated Secant Conic and the Provisional South American Datum of 1956 with the UTM Grid. Furthermore, PSAD56 was known to exist in parts of Trinidad also. Registering various maps of the region to each other was a cartographic nightmare of disparate Grids and Datums. The prospect of a new boundary treaty (with coordinates) being signed between the two countries without *prior* documentation of the relation of the various coordinate systems was disturbing to contemplate.

I was retained to travel to the region and acquire the necessary data to compute the relations among all existing classical coordinate systems in late 1986. Since boundary problems were (and continue to be) a sensitive subject in Venezuela, it was felt that local inquiries could be somewhat hazardous to the investigator (me). Although I was to travel with my wife and six children as a professor on a "working holiday" during the holiday season, a local prominent attorney was retained on my behalf to get my family out of the country if I was "detained" by the Secret Police. On December 13, 1986, we arrived in *Port of Spain* for a few days of stay and local travel. The morning of 17 December we arrived in Caracas, we checked into the hotel, and I telephoned the local attorney I was instructed to contact. The following morning, I went to *La Dirección de Cartografía Nacional* (DCN) and managed to get an audience with the Director. I explained my objective was to obtain the complete "Trig Lists" of the chains of quadrilaterals from both Venezuelan Datum origins to the island of Trinidad. In return I said I was willing to give a copy of all the software published by the U.S. National Geodetic Survey (NGS) and to train some of their geodetic staff in the operation of that software. I had a brand-new laptop computer with me; a new dual-floppy drive Zenith Z-181 and I demonstrated some of the NGS "MTEN" software to the Director while I spoke. The Director said that my proposal had merit and that he agreed to the "trade," but that it would take several days to amass the data that I wanted. He inquired if I would be around that long, and I replied that I was on a working holiday with my rather large family, and I expected to stay for as long as it took to acquire the data. I spent the remainder of the week downloading and installing the NGS software into the DCN computers and training some of their Geodesists. One afternoon when I returned to the hotel, I was summoned by the Desk Clerk and asked if everything was all right? I then asked why the question, and he told me that the Secret Police had been there that day and had inquired about my family. I then asked, "what happened?" Seems that they took the two members of the Secret Police outside near the pool, pointed out my family, and watched them count my children. The desk clerk said that they seemed satisfied and just left without another word. The next day I went to the DCN, nothing was said about the incident, and I did not ask. That weekend was spent sightseeing in Caracas and nearby towns (*Colonia Tovar*) in the Andes, the following Monday I met with the retired Dr. Adolfo Romero, Father of the PSAD56. The interview was fascinating, and Dr. Romero preferred to use his English rather than put up with my Spanish. The datum shift methods used by the DCN were Helmert transformations and were localized to metropolitan Caracas for cadastral changes from Loma Quintana 1911 to PSAD56. On Tuesday, December 23, 1986, I was successful in obtaining the Trig List coordinates on the last working day of the year for the DCN. I telephoned the attorney and happily informed him that I was not going to need his services and left Venezuela with my family shortly after that.

The results of the Datum relations subsequently computed were as follows: **from** Old Trinidad Datum of 1903 **to** PSAD56, $\Delta X=+84.70$ m, $\Delta Y=-180.65$ m, $\Delta Z=+96.90$ m, scale$=-17.79 \times 10^{-6}$, $R_X=-7.99''$, $R_Y=+5.66''$, $R_Z=+25.19''$, and was based on a solution of 61 co-located stations. **From** Loma Quintana Datum of 1911 **to** PSAD56, $\Delta X=-43.50$ m, $\Delta Y=+96.14$ m, $\Delta Z=-15.18$ m, scale$=+16.70 \times 10^{-6}$, $R_X=-1.43''$, $R_Y=-0.65''$, $R_Z=-0.33''$, and was based on a solution of 79 co-located stations. **From** Naparima Datum 1972 **to** PSAD56, $\Delta X=-27.69$ m, $\Delta Y=+40.01$ m, $\Delta Z=-17.56$ m, scale$=+7.85 \times 10^{-6}$, $R_X=-11.33''$, $R_Y=+4.34''$, $R_Z=+17.83''$, and was based on a solution of 61 co-located stations. The treaty between the High Contracting Parties on "the delimitation of marine and submarine areas was DONE in the City of Caracas, on the 18th day of the month of April, One Thousand Nine Hundred and Ninety in duplicate in the English and Spanish languages, both texts being equally authoritative. ... The positions of the aforementioned points have been defined by latitude and longitude of the 1956 Provisional South American Datum (International ellipsoid 1924)."

"Venezuela, as it has been established in Official gazette no. 36.653 of date 03-03-99, has a new geodetic network called **SIRGAS-ReGVen** (Venezuelan GPS Network), which is a very precise network." PATVEN-07 Transformation Parameters are **from** PSAD56 (Canoa) **to** ReGVen: $\Delta X=-270.933$ m, $\Delta Y=+115.599$ m, $\Delta Z=-380.226$ m, scale$=-5.108 \times 10^{-6}$, $R_X=-5.266''$, $R_Y=-1.238''$, $R_Z=+2.381''$.

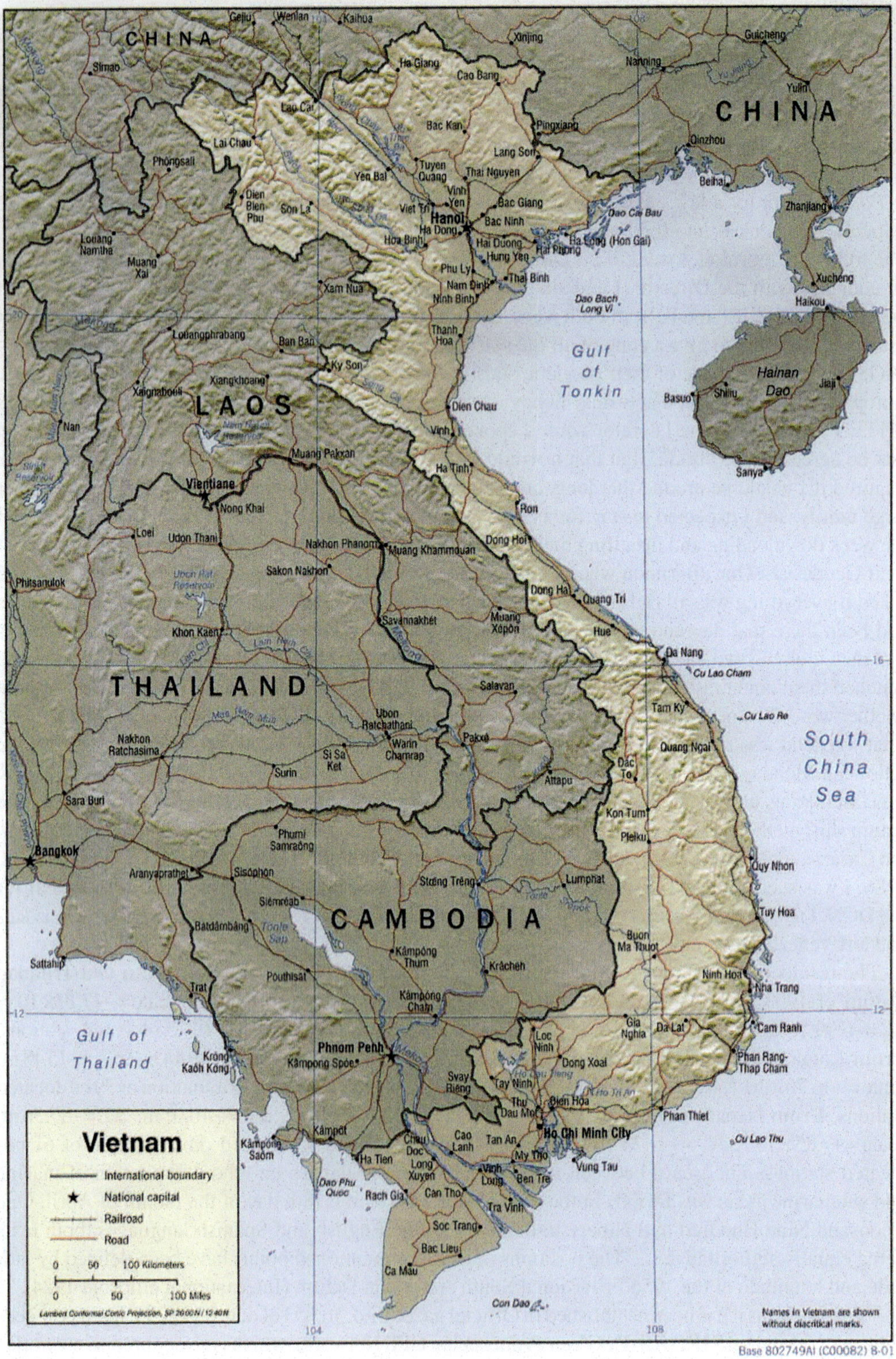

Base 802749AI (C00082) 8-01

The Socialist Republic of Vietnam

Inhabited since Paleolithic times, the beginning of Vietnamese civilization dates back to the late Neolithic or early Bronze Age. Vietnam is tropical in the south and subject to monsoons in the north. The country is low, comprised of the Mekong Delta in the south, the Red River Delta in the north, central highlands, hilly, mountainous in the far north and northwest. The lowest point in Vietnam is the South China Sea; the highest is *Ngoc Linh* at 3,143 m. The area of Vietnam is slightly larger than New Mexico; with a coastline 3,260 km long, its maritime boundary is declared according to the "Straight Baseline" principle. Vietnam is a poor, densely populated Communist state that has had to recover from the ravages of war, the loss of financial support from the old Soviet Bloc, and the rigidities of a centrally planned economy. Independence Day is 2 September (1945).

France occupied all of Vietnam by 1884 and remained for the most part until 1954. Longitude was determined in *Haïphong* by Héraud and Bouillet, Hydrographic Engineers in 1874. The longitude was deduced from the time transfer from *Saïgon*; two chronometers were used for the first traverse and five chronometers were used between the observatory at *Haïphong* and the observatory of *Saïgon*. It is interesting to note that by 1883, *Hong Kong* and *Haïphong* were connected by submarine cable. The longitude of *Hong Kong* had been determined in 1881 by telegraph, so it was deduced that the longitude of *Haïphong* was now determined to better precision than was determined by the use of chronometers. In 1886, Héraud measured the baseline in Haïphong between *Grand Mirador* and *Petit Mirador* to be 4,312.8 m. "The observatory in *Haïphong* was established to observe the longitude difference. It was essentially comprised of a masonry pillar constructed on a solid foundation, and a grass hut was constructed of bamboo." The position observed in 1874 was: $\Phi_o = 20° 51' 43.5'' N$, $\Lambda_o = 104° 20' 30''$ East of Paris or $106° 40' 43.95''$ East of Greenwich. These observations were the initial determinations that later would provide the defining Datum for the Northern part of Vietnam. *Grand Mirador* was defined as the position of the geodetic net (*Position du réseau géodesique indo-chinois*), where $\Phi_o = 21° 01' 58.415''$ and $\Lambda_o = 109° 00' 57.90''$ East of Greenwich. The azimuth from *Grand Mirador* to *Nui Deo*, $\alpha_o = 334° 29' 49.8''$. The ellipsoid of reference is the Clarke 1880, where $a = 6,378,249$ m and $1/f = 293.46$. The subsequent transferred origin of the Grid for *Signal Grand Mirador* was $\varphi_o = 20° 42' 24.337'' N$, $\lambda_o = 106° 46' 29.282''$ East of Greenwich. The False Easting and False Northing for this Hatt Azimuthal Projection was equal to zero. There were four bases measured in the 1880s by the French, they were near *Sontay*, at *Than-Hoa,* south of *Haïphong*, and near *Baria*.

Grand Mirador de Do-Son was observed in April 1887 and was a local Datum such that $\Phi_o = 20° 42' 24.9'' N$, $\Lambda_o = 106° 46' 36.15''$ East of Greenwich. The defining azimuth was $\alpha_o = 191° 33' 22''$. In 1929, the coordinates were updated to $\Phi_o = 20° 42' 24.34'' N$, $\Lambda_o = 106° 46' 29.28''$ E, and in 1932 Mr. A. Gougenheim finally changed the coordinates of *Doson* to $\Phi_o = 20° 42' 24.337'' N$, $\Lambda_o = 106° 46' 29.2282''$ E.

Another Grid was defined as *Haïphong*, the west gable of the observatory pagoda: $\Phi_o = 20° 51' 44.3'' N$, $\Lambda_o = 106° 40' 43.95''$ E. Nearby, the derivative *Hanoi* system at the "tour de la citadelle," tower of the standard, had coordinates of: $\varphi_o = 21° 01' 58'' N$, $\lambda_o = 103° 29' 52.2''$ E was used for some French surveys, but it was based on the *Haïphong System* because its coordinates are $x = 99,618.0$ m W, $y = 34,531.0$ m N. Furthermore, the derivative *Tour de la citadelle de Nam Dinh* had coordinates of: $\varphi_o = 20° 25' 30.4'' N$, $\lambda_o = 103° 49' 50.0''$ E was used for some French surveys, but was based on the *Haïphong System* because its coordinates are $x = 65,292.8$ m W, $y = 32,917.5$ m S. Note that all the azimuthal Grids used by the French in Vietnam were quadrant-based; no false origins were used.

The *Hon-Matt* Grid based on the Hatt Azimuthal projection was established in 1877 and is near *Vinh* on the *Ca River*. The coordinates of the origin are: $\varphi_o = 18°\ 47'\ 40''$ N, and $\lambda_o = 105°\ 55'\ 58.95''$ E. In 1924, the *Hon-Nieu* Grid was used, where $\varphi_o = 18°\ 48'\ 10.183''$ N and $\lambda_o = 105°\ 46'\ 33.799''$ E. It was also used in 1932 for additional French hydrographic surveys. The old *Cape Saint-Jacques* Lighthouse (*phare*) hydrographic survey was based on $\varphi_o = 10°\ 19'\ 33.220''$ N and $\lambda_o = 104°\ 44'\ 32.663''$ E.

The *Bay of Tourane* (now *Da Nang*) Grid was established in 1907 by Cot, where $\varphi_o = 16°\ 11'\ 01.44''$ N and $\lambda_o = 108°\ 06'\ 17.61''$ E; the pillar was listed at *Nui-Hoï*. The defining azimuth from *Nui-Hoï* to *Tien-Cha* is $\alpha_o = 110°\ 33'\ 32.2''$.

In 1913, a number of new hydrographic survey Grids were established, including the *Cathedral of Saïgon*, where $\varphi_o = 10°\ 46'\ 44.180''$ N and $\lambda_o = 104°\ 21'\ 25.803''$ East of Paris, as compared with the *Old Observatory of Saïgon*, where $\varphi_o = 10°\ 46'\ 42.78''$ N and $\lambda_o = 104°\ 21'\ 31.29''$ East of Paris. (Note that Paris is $2°\ 20'\ 13.95''$ East of Greenwich.) The *Bay of Natrang* was surveyed with a Hatt Grid in 1913, the origin of the coordinates was at *Signal Honheo*, where $\varphi_o = 12°\ 24'\ 13.11''$ N and $\lambda_o = 106°\ 56'\ 00.14''$ East of Paris.

By 1914, a Hatt Azimuthal Grid was observed and established at *Cana*, where $\varphi_o = 11°\ 22'\ 35.067''$ N and $\lambda_o = 108°\ 50'\ 20.137''$ E (1929 updated coordinates are listed). After World War I in Europe, the *Cana* Grid was also used for the *Îsles Catwick* in 1924. In 1923, a local survey was run from the lighthouse at *Baïkan*, where $\varphi_o = 8°\ 39'\ 59.17''$ N and $\lambda_o = 104°\ 21'\ 48.31''$ East of Paris; X_o and Y_o were equal to zero. A Datum and cadastral Grid system were established at *Long Vinh*, where $\Phi_o = 9°\ 35'\ 59.79''$ N and $\Lambda_o = 106°\ 18'\ 58.07''$ East of Greenwich. The baseline was run from *Lich Hoï Binch* to *Long Vinh*, and measured 17,805.73 m. The defining azimuth of that baseline was $\alpha_o = 230°\ 56'\ 07.426''$. That same year, the *Îles Poulo-Condore Datum of 1923* (now *Con Son*) was established, where $\Phi_o = 8°\ 41'\ 35.86''$ N, $\Lambda_o = 104°\ 14'\ 27.28''$ East of Paris, the azimuth of West Base to East Base was defined as: $\alpha_o = 191°\ 33'\ 22''$, and its length was measured to be 1,411.83 m. By 1923, the coordinates for "Mirador d' Hanoi" were published as: $\varphi_o = 21°\ 01'\ 58.415''$ N, and $\lambda_o = 103°\ 29'\ 52.120''$ East of Paris.

In the annual report of A. Gougenheim for his hydrographic mission to French Indochina of June 1930 to June 1931, he listed a projection summary. That enumerated all of the Hatt Azimuthal Equidistant projections he used that year in his mission, including the calculating machine coefficients for both the direct and the inverse for: *Origine Grand Mirador de Doson, Origine Hon Nieu, Origine Cana,* and *Origine Nui Chauvien*. Each pair of formulae for direct and inverse also included a version for units of sexagesimal seconds (degrees) as well as centisimal seconds (grads). Of course, everything was truncated at the cubic! Gougenheim later presented his own machine calculation for the ellipsoidal geodesic that was recast and presented by Paul D. Thomas of the U.S. Naval Hydrographic Office during the 1970s.

After World War II, three Hatt Azimuthal Grids were devised by the French in Vietnam. The *Signal Haïphuc* (1933) origin was at: $\varphi_o = 13°\ 26'\ 04.693''$ N and $\lambda_o = 109°\ 17'\ 44.322''$ East of Greenwich. The *Borne* (*monument*) *de Bac Lieu* (1933) origin was at: $\varphi_o = 9°\ 15'\ 42.02''$ N and $\lambda_o = 105°\ 43'\ 14.48''$ East of Greenwich. The last Grid established by the French was at Cam Ranh Bay in 1949, where the origin was published as: $\varphi_o = 11°\ 55'\ 55.85''$ N and $\lambda_o = 100°\ 49'\ 58.60''$ East of Paris rather than East of Greenwich!

John W. Hager tells me that "in 1954, the triangulation of Thailand was adjusted to Indian 1916 based on ten stations on the Burma border. In 1960, the triangulation of Cambodia and Vietnam was adjusted holding fixed two Cambodian stations connected to the Thailand adjustment of stations from the Cambodian-Vietnam adjustment. North Vietnam was also adjusted to this system but with lower standards. The details are that of the Indian Datum as defined in 1900 and labeled as Indian 1916: origin at *Kalianpur Hill Station*, $\Phi_o = 24°\ 07'\ 11.26''$ N, $\Lambda_o = 77°\ 39'\ 17.57''$ East of Greenwich, the initial azimuth to Surantal from south is: $\alpha_o = 190°\ 27'\ 05.10''$. The ellipsoid of reference is the Everest 1830 where $a = 6,377,276.345$ m, and $1/f = 300.8017$."

I was assigned to Army Map Service (later TOPOCOM) during the Vietnam War, and for a short period was a Company Commander. Some of "my" personnel were at a SECOR satellite tracking station in Thailand while establishing a precise location for a SHORAN transmitter for navigation control of airplanes. The *South Asia Datum* was used for that application and was referenced to the Modified Fisher 1960 ellipsoid, where $a = 6,378,155$ m and $1/f = 298.3$. I do not think that Datum was ever used for (at the time), unclassified applications. NGA lists two transformations for Vietnam. For Vietnam near 16° N, from Indian 1960 Datum to WGS 84 Datum: $\Delta X = +198$ m ± 25 m, $\Delta Y = +881$ m ± 25 m, and $\Delta Z = +317$ m ± 25 m, and the solution is based on two stations. For *Con Son Island* from Indian 1960 Datum to WGS 84 Datum: $\Delta X = +182$ m ± 25 m, $\Delta Y = +915$ m ± 25 m, and $\Delta Z = +344$ m ± 25 m, and the solution is based on one station.

Mal Jones of Perth, Australia, tells me that from WGS 84 Datum to *Hanoi 1972 Datum*, the ellipsoid of reference is the Krassovsky 1940, where $a = 6,378,245$ m and $1/f = 298.3$, the parameters are: $\Delta X = -21$ m, $\Delta Y = +124$ m, $\Delta Z = +68$ m, R_x axis rotation $= +0''$, R_y axis rotation $= +0''$, R_z axis rotation $= +0.814''$, and δs scale change $= +0.38$ ppm. From WGS 84 to Indian (Vietnam), $\Delta X = -199$ m, $\Delta Y = -931$ m, $\Delta Z = -321$ m, according to "Vietsovpetro," further details and accuracy are unknown. However, this is suspiciously close to VT78 parameters for the WGS 72 Datum that I received from Mr. Robert Holloway of Mt. Lawley, West Australia back in 1998. The current geodetic and mapping authority is the General Department of Land Administration in Hanoi.

Vietnam geodesy is no longer an enigma. Hoa, Dung, Thu, Huynh, and Nhung published a paper (https://doi.org/10.1051/e3sconf/20199403014) recently that detailed in exquisite detail the relationships among VN-2000, ITRS, WGS84, and PZ-90 and the various realizations of those systems. The National Coordinate Control Network consists of 71 points of "0" order, 328 points of first order, 1,177 points of second order, 160 points of second order traverse, and 12,658 points of third order. Vietnam has implemented a national network of GPS CORS sites and appears to be completely open for business development.

REFERENCES: V

Bakeeliu, Kanas, and Kalsale, Vanuatu Geodetic Control Network Report, June 2001.
Catalogue de Cartes en Service Publiées par l'Institute Géographique National, Paris, 1 July 1949.
Director, La Dirección de Cartografía Nacional (DCN), Personal Communication, December 1986.
Dr. Adolfo C. Romero, Personal Communication, December 1986.
Hoa, Dung, Thu, Huynh and Nhung, https://doi.org/10.1051/e3sconf/20199403014.
John P. Snyder, Personal Communication, June 1988.
John W. Hager, Personal Communication, 2004.
John W. Hager, Personal Communication, 2011.
Mal Jones, Perth, Australia, Personal Communication, April 2002.
Mike Bakeoliu, Tony Kanas, Moses Kalsale, Vanuatu Geodetic Control Network *Report,* 09 June 2001.
Official gazette no. 36.653 of date 03-03-99.
Paul D. Thomas, U.S. Naval Hydrographic Office, ca. 1970s.
Robert Holloway of Mt. Lawley, West Australia, Personal Communication, 1998.
Russell Fox, Personal Communication, 2004.
Tony Kanas, Surveyor and the Vanuatu Department of Land Surveys, Personal Communication.
U.S. Department of State published No. 137 Limits in the Seas, Vanuatu: Archipelagic and other Maritime Claims and Boundaries, 2014.

Y

Republic of Yemen 883

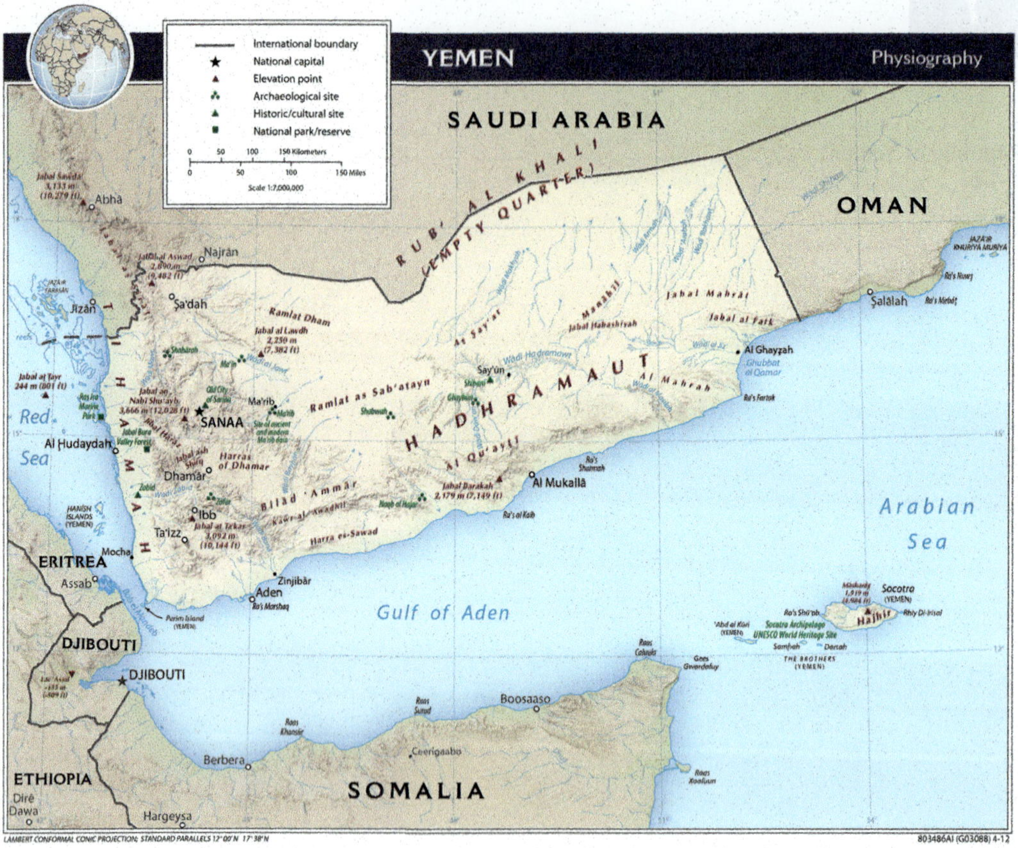
YEMEN
Physiography
International boundary
National capital
Elevation point
Archaeological site
Historic/cultural site
National park/reserve
0 50 100 150 Kilometers
0 50 100 150 Miles
Scale 1:7,000,000
SAUDI ARABIA
OMAN
RUB' AL KHALI
(EMPTY QUARTER)
Abhā
Najrān
Jīzān
Ṣa'dah
SANAA
Ma'rib
Al Ḥudaydah
Dhamār
Ibb
Ta'izz
Mocha
Aden
Zinjibār
Say'ūn
Al Mukallā
Al Ghayẓah
Ṣalālah
Ramlat Dhan
Ramlat as Sab'atayn
HADHRAMAUT
Wadi Ḥaḍramawt
Jabal Mahrāt
Jabal al Fatk
Al Mahrah
Jabal Ḥabashiyah
Jabal al Lawdh 2,250 m (7,382 ft)
Jabal an Nabi Shu'ayb 3,666 m (12,028 ft)
Jabal Barakah 2,179 m (7,149 ft)
Jabal at Tayr 244 m (801 ft)
Bilād 'Ammār
Harra es-Sawad
Harras of Dhamar
Red Sea
Gulf of Aden
Arabian Sea
Socotra (YEMEN)
Ḥadiboh
Socotra Archipelago UNESCO World Heritage Site
Samḥah
Darsah
THE BROTHERS (YEMEN)
'Abd al Kūrī (YEMEN)
Perim Island (YEMEN)
HANĪSH ISLANDS (YEMEN)
ERITREA
Assab
DJIBOUTI
DJIBOUTI
ETHIOPIA
Dirē Dawa
SOMALIA
Berbera
Hargeysa
Boosaaso
Ceerigaabo
LAMBERT CONFORMAL CONIC PROJECTION, STANDARD PARALLELS 12°00'N 17°38'N
803486AI (G03088) 4-12

Republic of Yemen

Yemen is bordered by Saudi Arabia on the north (1,458 km), Oman on the east (288 km), the Arabian Sea and Gulf of Aden to the south, and the Red Sea to the west (1,906 km). Slightly larger than twice the size of Wyoming, the country is comprised of a narrow coastal plain backed by flat-topped hills and rugged mountains with dissected upland desert plains in the center that slope into the desert interior. The lowest point is the Arabian Sea, and the highest point is *Jabal an Nabi Shu'ayb* at 3,760 m.

Around 1000 B.C. the region of present Yemen was ruled by three successive civilizations: the Minaean, the Sabaean, and the Himyarite. These three kingdoms all depended for their wealth on the spice trade, consisting mainly of frankincense and myrrh. By the 11th century *B.C.*, land routes were greatly improved throughout Arabia by using the camel as the beast of burden. Frankincense was carried from its production center at *Qana* (now *Bir'Ali*) to *Gaza*. The chief incense traders were the Minaeans, who established their capital at *Karna* (now *Sadah*), before the Sabaean era in 950 B.C., which lasted for about 14 centuries. The region was invaded by the Romans in the 1st century *A.D.*; by the 6th century *A.D.*, it was conquered first by the Ethiopians and then by the Persians. The region converted to Islam in the 7th century. North Yemen became independent of the Ottoman Empire (Turkey) in 1918. The British had set up a protectorate area around the southern port of *Aden* during the 19th century but withdrew in 1967 from what had become South Yemen. After two decades of hostilities, the two countries were formally unified as the Republic of Yemen in 1990.

Original classical triangulation of *Western Aden* was done by the British Survey of India in the early 20th century. The first large-scale map series, based on ground surveys, was published in 1917. Later updated, the Aden GSGS 3879 series at 1:126,720 scale was published as a polychrome series in 1930. Carl Rathjens and Herman von Wissmann published the *Karte des Reisegebiets in Jemen* (Map of the Region Traversed in Yemen) in 1934. There was no grid on the map at 1:100,000 scale, and it covered a limited area of Yemen between Ṣanʻāʼ and the coast. The three-part polychrome map was based on road surveys, supplemented with diaries, sketches, and maps from other travelers. Relief was represented by form lines or hachures.

In 1925, the British Survey of India established the Aden Zone Lambert Conical Orthomorphic grid where the latitude of origin $\varphi_o = 15°$ N, central meridian $\lambda_o = 45°$ E, scale factor at the latitude of origin $m_o = 0.999365678$, False Easting = 1,500 km, and False Northing = 1,000 km. This typical "British Grid" had its limits of zone listed as "North: 150,000 meter Northing grid line of the Mecca-Muscat Zone (Lambert Conical Orthomorphic projection), East: Meridian of 60°E, South: From meridian of 60°E, southwest along loxodrome defined by the points 14° 30′ N, 56° E, and 11° 30′ N, 44° E., West: Loxodrome from 44° E, 11° 30′ N toward 19° N, 39° E, to intersection of 150,000 meter Northing grid line of Mecca-Muscat Zone with the loxodrome." The ellipsoid of reference is the Clarke 1880 where $a = 6,378,249.145$ m and $1/f = 293.4663077$. Curiously, no datum was listed by the British General Staff, Geographical Section (GSGS). However, reading the Survey of India Triangulation Dossiers reveals that all British chains in the region were based on the original *Nahrwan Datum of Iraq*, where $\Phi_o = 33°$ 19′ 10.87″ N, $\Lambda_o = 44°$ 43′ 25.54″ E, and the orientation is based on the azimuth from South End Base of Nahrwan ("1M") to "2M" as $\alpha_o = 169°$ 04′ 08.2″ from south. In my opinion, the original classical datum of Yemen ("Aden Datum of 1925") is actually the British-observed Nahrwan Datum. A test point provided by the U.S. Army Corps of Engineers, Lake Survey Unit in 1943 for Aden Zone is $\varphi = 13°$ 53′ 46.728″ N, $\lambda = 37°$ 37′ 19.732″ E, $X = 703,075.269$ m, $Y = 891,245.290$ m, $\theta = -1°$ 54′ 34.3032, and $\kappa = 0.91412282$.

Numerous map series of *Aden* and Ṣanʻāʼ were produced by the British and U.S. militaries in the period between 1950 and 1961. Scales varied from 1:10,000 to 1:100,000. Other datums reported to exist include *Kamaran (Island) Datum of 1926–1927, Ras Karma (Island) Datum, Socotra (Island) Datum of 1957,* and *Socotra (Island) Datum of 1964–1965*. All of these island "astro" datums are presumably referenced to the Clarke 1880 ellipsoid.

A curiosity in large-scale map projections was developed by the U.S. Geological Survey when a mapping project was completed of *Ṣan'ā'* in the 1960s. The *Ṣan'ā'* Azimuthal Equidistant Grid System was defined by the latitude of origin $\varphi_o = 15° 37' 22"$ N, central meridian $\lambda_o = 42° 59' 32.25"$ E, scale factor at the latitude of origin $m_o = 1.0$, False Easting = 40 km, and False Northing = 20 km. The International ellipsoid was chosen for the grid, and the datum was presumably ersatz.

Yemen and Oman established their boundary in an October 1, 1992, agreement consisting of straight-line segments connecting eight turning points defined on the WGS84 Datum. The eighth point is a tri-point with Saudi Arabia. On April 18, 2001, Yemen signed a cooperation agreement based on the 1999 International Tribunal resolution of the Permanent Court of Arbitration, the Hague, the Netherlands. In 2000, Yemen and Saudi Arabia agreed to a delimitation of their common border.

Large-scale local topographic maps of Yemen are controlled by the government. However, complete topographic coverage of the country is available from commercial map sellers worldwide in the form of Russian military mapping at the scales of 1:200,000 and 1:100,000. Considering the availability of these recent up-to-date Russian maps and the phenomenal accuracy achieved nowadays with shirt-pocket-sized consumer-grade GPS receivers, such secrecy of the Yemeni government is merely restricting the economic development of their own nation.

In January 2006, a *TOTAL* (French Oil Company) document was released by the Yemen LNG Co., Ltd. about the "Geodetic Systems on Pipe Line project." The new "Official Yemen Horizontal Datum is listed as *Ṣan'ā'* – IGN reference marker, calculated on ITRF91 (epoch 92.5) YGN96 (WGS84). The document lists the *"Old South Yemen (OSY) Datum* referenced to the Krassovsky 1940 ellipsoid where $a = 6,378.245.00$ m and $1/f = 298.3$ with a transformation to *YGN96* (WGS84) Datum as: $\Delta X = -76$ m, $\Delta Y = -138$ m, $\Delta Z = +67$ m, "Datum shift provided by the French *Institute Géographique National* in October 1996. An example Transformation Test Point is offered for Geodetic marker BH-01 Datum OSY: $\phi = 13° 59' 06.145"$ N, $\lambda = 48° 10' 41.197"$ E, h = 23.89 m., and Datum YNGN96: $\phi = 13° 59' 09.5087"$ N, $\lambda = 48° 10' 40.0179"$ E, h = −0.730 m., H = 18.220 m. "Geoid model: a local geoid model computed by the French IGN in 1996 is used to convert height above WGS84 spheroid into normal heights above MSL. (*at Al Hudaydah (MSL) – Ed.*). The geoid model software is available at *TOTAL* office."

Z

Republic of Zambia 887
Republic of Zimbabwe 891

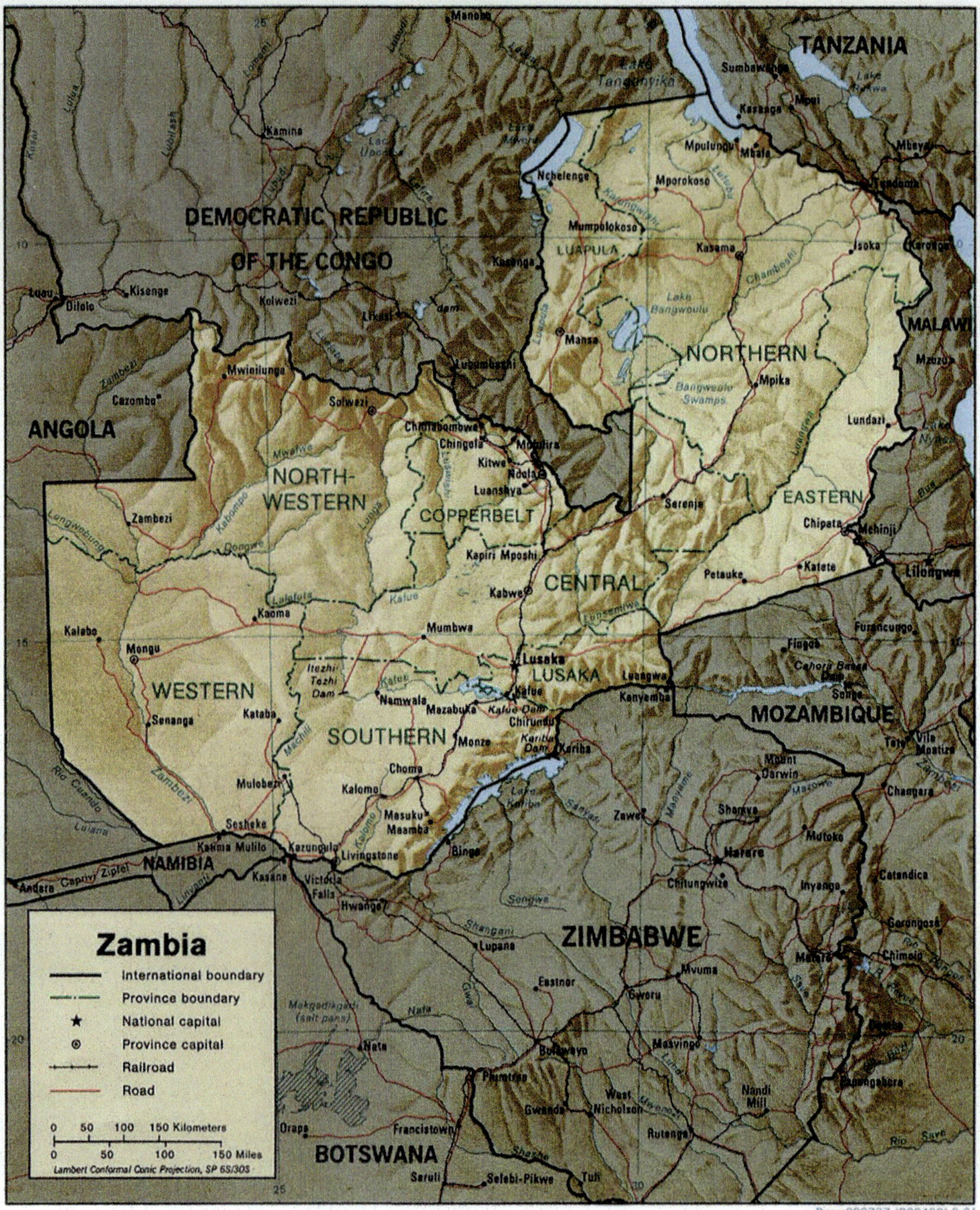

Base 802737 (B00492) 2-01

Republic of Zambia

Zambia is bordered by Angola (1,110 km), the Democratic Republic of the Congo (1,930 km), Malawi (837 km), Moçambique (419 km), Namibia (233 km), Tanzania (338 km), and Zimbabwe (813 km). The climate is tropical, modified by the altitude of the mostly high plateau with some hills and mountains. The lowest point is the *Zambezi River* (329 m), and the highest point is in the *Mafinga Hills* (2,301 m).

Evidence has been found north of *Lusaka* at *Kabwe* of human habitation that dates back 100,000 years. About 1000 *A.D.*, Swahili-Arab slave traders began intrusions into the area from the east. Between the 14th and 16th centuries, the Bantu-speaking Maravi migrated from the area presently know as *Zaïre* (*Congo* [*Kinshasa*]), and established kingdoms in eastern and southeastern Zambia. The region came under the jurisdiction of the British South Africa Company in 1889, in 1911 it became Northern Rhodesia, and in 1924 it became a British Protectorate. From 1953 to 1963, it was part of the Federation of Rhodesia and *Nyasaland* (*Malawi*) and achieved independence as a republic 40 years ago this month on October 24, 1964.

Lake Tanganyika extends into a small portion of northern Zambia, and the *Zambezi River* (used as the origin of the country's name) forms the eastern border with Malawi. The famous Arc of the 30th Meridian follows the eastern shore of Lake Tanganyika and spans the Zambezi River. The Arc of the 30th Meridian is referenced to the *Cape Datum of 1950* where the astronomic coordinates of the initial point of the Cape Datum near *Port Elizabeth* are for *Buffelsfontein* where $\Phi_o = 33°$ 59′ 32.000″ S and $\Lambda_o = 25°$ 30′ 44.622″ E. The ellipsoid of reference is the Clarke 1880, where $a = 6,378,249.145$ m and $1/f = 293.4663077$.

The northwestern border of Zambia is common with the Democratic Republic of the Congo (Zaïre), once known as the *Belgian Congo*. Zambia is adjacent to the *Katanga* province of the Congo, where a boundary commission published the results of a classical triangulation in 1954, *Comité Spécial du Katanga, Les Travaux Géodésiques du Service Géographique et Géologique*. The origin of the triangulation of Katanga (*Le Point Fondamental*) is the "A" end of the *Tshinsenda baseline* in Zambia, where $\Phi_o = 12°$ 20′ 31.568″ S and $\Lambda_o = 28°$ 01′ 02.971″ E. The altitude of the point was 1,331.31 m, as determined by trigonometric leveling from the 30th Arc triangulation performed in 1911. Subsequent double-run precise levels performed by then Major and later Brigadier Martin Hotine from *Dar es Salaam* in Tanzania necessitated a correction of +47 ft to the elevations in Zambia. Presumably, that correction was applied to the value published by the Belgians in 1954. The Tshinsenda Baseline was measured in 1912 with a length of 4,152.9912 m with the final value being adjusted with the base at *Nyanza*, both surveyed by the Katanga-Rhodesia Boundary Commission. The deflection of the vertical was constrained to zero at point "A." The projection adopted for the general map of *Katanga* was the Lambert Conical Orthomorphic with two standard parallels at $\varphi_N = 6°$ 30′ S and $\varphi_S = 11°$ 30′ S and a central meridian, $\lambda_o = 26°$ E.

Thanks to a paper in *Survey Review*, April 1997 by Dare and Mutale, a brief history of geodetic surveys in Zambia were detailed: "Between 1949 and 1964, the Directorate of Colonial Surveys, Federal Surveys, and the Directorate of Overseas Surveys, established 12 triangulation nets and three traverse loops. The main areas of primary control may be grouped as follows: Part of Arc of the 30th Meridian; *Fort Jameson (Chipata)/Malawi Network; Isoka Network; Zambia Main Network; Copperbelt Network; Solwezi/Kasama/Mumbwa Loop; Fort Rosebury (Mansa) – Congo (Zaire) link; Livingstone Memorial Area – Mansa Loop; Mwinilunga Loop Traverse; Luwingu series and Mansa loop; Mankoya loop Traverse; Kalomo Livingstone loop Traverse*. The network configuration consists of a series of (a) Triangles; (b) Braced quadrilaterals; (c) Centre point polygons; (d) Double centered polygons; (e) Traverse legs. As a rule of thumb, the orientation was controlled by azimuth observations every 10 stations and the allowable misclosure was not to exceed $2''\sqrt{n}$, where n is the number of intervening legs between astronomical stations. The side lengths in primary traverse are approximately 30 km; in other cases, the lengths of sides are approximately 60 km. For (a)-(d) a deliberate effort was made to have well-conditioned triangles by avoiding angles less than 40 degrees."

The Tshinsenda baseline is located in the *Copperbelt Province* of Zambia. Since the world market in copper has plummeted, the economy of Zambia has suffered, and great efforts are being expended to convert the economy to an agricultural base. The majority of papers on surveying and mapping topics published on Zambia are now addressed to the establishment of a national cadastre for a land registration system. Land tenure through 99-year leases is a current topic thought to be the country's economic salvation. Professor Peter Nsombo published a paper with L. Combrinck of the *Hartebeeshoek Radio Astronomy Observatory* regarding the establishment of a continuously observing reference station in *Lusaka* (ZAMB). The transformation parameters expressed in the standard American convention sign for rotations from Arc Datum 1950 to WGS84 Datum for *all* of Zambia are: $\Delta X = -152$ m ± 0.4 m, $\Delta Y = -60$ m ± 0.4 m, $\Delta Z = -297$ m ± 0.4 m, $R_x = -12'' \pm 0.4''$, $R_y = 1'' \pm 0.8''$, $R_z = 8'' \pm 1''$, $\delta s = -8.328 \pm 1.773$ ppm, and this solution was based on 11 observed points. A pilot project was undertaken in an area of *Lusaka* that developed transformation parameters that were different from the above parameters in *excess of 10 m per translation component*. As a basis of comparison, NGA lists the three-parameter transformation from Arc 1950 to WGS 84 as: $\Delta X = -147$ m ± 21 m, $\Delta Y = -74$ m ± 21 m, $\Delta Z = -283$ m ± 21 m, and this solution was based on 5 points.

Thanks to Mr. Malcolm A. B. Jones of Perth, Australia, for the Katanga data.

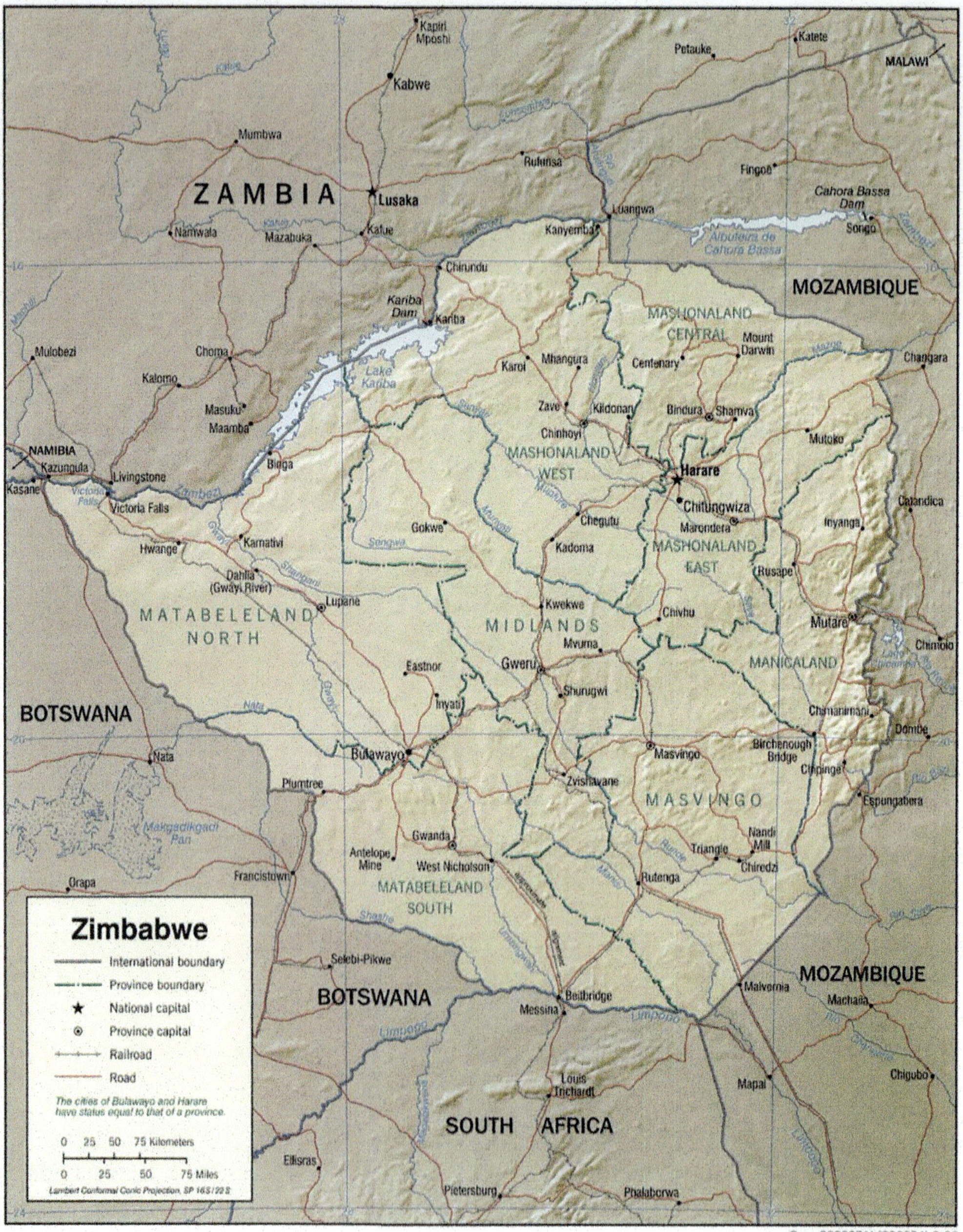

Base 802837AI (C00254) 3-02

Republic of Zimbabwe

Zimbabwe is slightly larger than Montana, it is landlocked, and it is bordered by Botswana (813 km), South Africa (225 km), Zambia (797 km), and Moçambique (1,231 km). The terrain is mostly high plateau with a high veldt with mountains in the east. The lowest point is the junction of the *Runde* and *Save Rivers* at 162 m, and the highest point is *Mount Inyangani* (2,592 m).

The republic is populated primarily by the Bantu group of peoples and is divided into two major language groups. The Shona-speaking Mashona constitute 75% of the population and the Sindebele-speaking Ndebele constitute about 20% of the population. The latter group arrived in the southwest around Bulawayo within the last 150 years and is an offshoot of the South African Zulu. They maintained control over the Mashona until the European occupation in 1890. Stone-age implements have been found in Zimbabwe, and ruins suggest an early civilization. The "Great Zimbabwe" ruins are located near *Masvingo*, and evidence suggests that it was built between the 9th and 13th centuries by Africans that established trading contacts with commercial centers on the continent's southeastern coast. In 1888, Sir Cecil Rhodes obtained a concession for mineral rights from local chiefs and later the area was proclaimed a British sphere of influence. The British South Africa Company was chartered the following year and *Salisbury* (now *Harare*, the capital) was established in 1890. In 1895, the area was formally named *Rhodesia* in honor of Sir Cecil. The UK annexed *Southern Rhodesia* from the South Africa Company in 1923. A 1961 constitution was formulated that favored Caucasians in power. In 1965, the government declared Rhodesian independence, but the UK did not recognize the act and demanded more complete voting rights for the black majority in the country. United Nations sanctions and a guerilla uprising finally led to free elections in 1979 and independence as Zimbabwe in 1980. A land redistribution policy in 2000 has caused an exodus of white farmers, it has crippled the economy, and it has caused widespread shortages of basic commodities.

In 1901, Mr. Alexander Simms completed a chain of quadrilaterals that spanned the west central part of the country. This chain started in the south near *Bulawayo* with the *Inseza Base* observed in 1898, it passed through *Gwero*, through *Salisbury* (now *Harare*) with the *Gwibi Base* observed in 1900, and it terminated in the north about 75 km east of where the *Kariba Dam* is now located on the *Zambezi River*. The geodetic coordinates of all the stations were referred to the origin point in *Salisbury*, where $\Phi_o = 17° 50' 25.440''$ S and $\Lambda_o = 31° 02' 19.000''$ E, with an azimuth to *Mt. Hampden* $\alpha_o = 273° 13' 48.456''$ and were fixed by interchange with the Royal Observatory at the Cape for longitude. After the work was published in 1905 in the *Geodetic Survey of South Africa, Volume iii*, little geodetic work was surveyed in Zimbabwe for about 25 years except by Gordon. (Prof. Charles L. Merry had some difficulty looking up those coordinates for me since his library had misplaced that volume!) Capt. H. W. Gordon, R.E., connected Simms' system to the Transvaal system of South Africa in 1906–1907. Capt. Gordon was seriously handicapped by having a budget of only £1,600 for the task, but his work was of excellent quality. The *Transvaal system* had already been completed by Col. W. G. Morris, R.E., in 1905. Gordon also later ran a short chain westward from Simms' chain at about latitude 17° 10' S to provide control for the *Copper Queen* mining area in 1928. The *Eastern Circuit* was started shortly thereafter; it ran from Harare (the capitol) eastwards to the border with Moçambique, southwards through Mutare to about latitude 20° S and then westwards, joining Simms' chain again to the east of *Bulawayo*. Another short chain was run north from Bulawayo for about 70 km to the *Lonely Mine* area. In 1936, Gordon's connecting chain with the *Transvaal* was strengthened and re-observed as well as the new *Nuanetsi Series* was observed to the east that ran north-south from the *Limpopo River* to near the Zimbabwe ruins and to *Rutenga*. The *Nuanetsi Series* was completed in December 1937.

The result of Gordon's Connection showed a difference between the *Transvaal chain* and the *Salisbury Datum* of –3.503″ in latitude, –0.602″ in longitude, –8.89″ in azimuth at point *Standhaus*, and a difference of 40 ft in the mean between the vertical heights of the two systems. The work in both Zimbabwe and Zambia suggested that Gordon's connection had introduced a swing of some 9″ into the arc, from the *northern Transvaal* upwards. In 1930 at the Stockholm Conference of the International Union of Geodesy and Geophysics, a resolution was therefore passed recommending an examination of the *Transvaal-Zimbabwe* connecting chain.

In 1932, an astronomical determination of the azimuth *Wedza – Standhaus* was made with a Wild T-3 theodolite that gave the astro-geodetic difference in azimuth between the two systems of ~4.5″. In 1936, J. E. S. Bradford observed a connection series that widened and strengthened Gordon's chain, and all angles were re-observed. Heliographs were employed throughout for the re-observations (by repetition) of this chain except for two points in the *Transvaal system*. Those exceptions were direction angles observed by F. W. J. de Roes of the South African Trigonometrical Survey. The adjustment of the revised chain was made to the Transvaal system, which was an extension of the *South African Cape Datum*. The adjusted coordinates of the tie-point *Standhaus* changed to $\varphi=20°\ 21'\ 54.654''$ S and $\lambda=29°\ 35'\ 54.631''$ E, and the azimuth to station *Filabusi* changed to $\alpha=273°\ 13'\ 48.456''$. The obsolete datum of Southern Rhodesia is now referred to as the "Old Circuit Datum" in Zimbabwe.

In 1944, D. R. Hendrikz of the South African Trigonometrical Survey wrote, "For the computation of the geographical coordinates of the stations of the Geodetic Survey, Sir David Gill adopted the numerical values of the semi-major and semi-minor axes of Clarke's 1880 figure or $a=20,926,202$ ft and $b=20,854,895$ ft. At that time this result was the most recent determination of the figure of the Earth. But, because the baselines were reduced to S.A.G. (*South African Geodetic – Ed.*) feet, the computations were really carried out on a "Modified Clarke 1880 Spheroid" defined by $a=6,378,249.145\ 326$ int. metre and $b=6,356,514.966\ 721$ int. metre. It may be remarked, in passing, that this value of the flattening for this spheroid is $1/f=293.466\ 307\ 656$ which differs slightly from the value 293.465 given by Clarke himself." Later in the document, Hendrikz went on to present relations of the "Geodetic Cape rood"=12.396 S.A.G. feet, and 1 *Cape morgen*=600 *square Cape roods*=2.116 539 816 acres. Note that the acre was originally the amount of English land that could be plowed in one day, and the morgen was roughly the amount of German land that could be plowed in a morning. Hendrikz stated that 1 *Rhynland morgen*=0.634 282 acres!

Thanks to Professor Charles L. Merry of the *University of Cape Town*, "The ellipsoid is the Clarke 1880, oriented using astronomic observations of latitude, longitude and azimuth at a point near *Port Elizabeth* in the late 1800s. The offset from the geocentre is about 350 m. An unusual feature is that it is the so-called "modified" Clarke 1880 ellipsoid, because the conversion factor "yard-to-legal metre" was used to convert Clarke's values to international metres. The legal metre is based upon a defined relationship between the toise and the metre, not a physical standard, and is about 13 ppm larger. Nevertheless, the official length standard is the international metre.

"Although local grid systems were common in the 19th and early 20th century, since the 1920s the Transverse Mercator (Gauss-Krüger) system has been exclusively used. It uses 2-degree wide panels, scale factor of unity on the central meridian and no false origin. The co-ordinate axes are directed South and West (no northings and eastings for us southerners!) and are labeled x and y respectively. It is a legal requirement for all cadastral surveys to use this grid system, and the large and medium scale national map series also uses it. The military use the UTM system, overprinted on the standard map sheets (false northing as well as false easting). ... Contrary to what DMA (*now NGA - Ed.*) believe, they do not use the Arc datum. The Arc datum is used in parts of East Africa. It is based upon the same initial point near Port Elizabeth and the same ellipsoid (modified Clarke 1880) but uses a single chain of triangulation extracted from the national networks of South Africa, ... and Zimbabwe, more or less along the 30th meridian. Close to Port Elizabeth, it is practically identical to the Cape datum, but diverges as one moves away. A GPS network is in place in Zimbabwe and the control networks are being re-adjusted. Although no final decision has been taken, it is likely that Zimbabwe will also convert to the WGS84 around the same time that South Africa does" (Personal communication, July 1997).

In April 1990, Professor Merry and J. Rens published a paper in *Survey Review* that described their solution for datum shift parameters in southern Africa that included Zimbabwe. The astronomic coordinates of the initial point of the Cape Datum near Port Elizabeth are for *Buffelsfontein*, where $\Phi_o = 33°\ 59'\ 32.000''$ S and $\Lambda_o = 25°\ 30'\ 44.622''$ E. With respect to Zimbabwe, Prof. Merry and Mr. Rens wrote, "The situation here is unique. Large rotations are evident and cannot be ignored. However, because of the relatively small size of Zimbabwe, these cannot be considered to represent a true misorientation of the Zimbabwean datum, but probably reflect regional distortions in the geodetic networks. Including the rotations in the transformation model reduces the standard deviation from near four metres to one metre – almost a fourfold reduction. Unlike the case in South Africa there is no significant scale factor. Nevertheless, we recommend a full seven-parameter transformation – neglect of the scale factor would cause significant changes in the translation components shown …, due to the high correlation between them." Those parameters recommended are $\Delta X = -121.7$ m ± 17.5 m, $\Delta Y = -121.0$ m ± 18.4 m, $\Delta Z = -258.5$ m ± 21.2 m, $R_x = +5.377'' \pm 0.527''$, $R_y = +1.857'' \pm 0.680$, $R_z = -2.989'' \pm 0.636''$, and $\delta s = +0.8 \times 10^{-6} \pm 2.3$. The *Hartebeeshoek 94 Datum* is the official coordinate system of the Republic of South Africa and presumably also may someday be of the Republic of Zimbabwe.

"Recently, Zimbabwe recognized CORS technology as an integral component for the prompt acquisition of spatial data. Consequently, the country engaged the EU and UNDP to fund the process of establishing CORS in the country. Initially, five GNSS CORS will be established primarily focusing on expediting the process of boundary mapping of farms to support security of tenure thus ensuring food security in line with the Sustainable Development Goals. Having realized the urgent need for establishing CORS in Zimbabwe, the readiness of the stakeholders and institutions which use geospatial data is yet to be evaluated. Although the primary rationale for establishing CORS in Zimbabwe was motivated by the need to regularize the land reform program to ensure tenure security, it is apparent that other stakeholders besides cadastral surveyors will use this technology." Establishment of continuously operating reference stations (CORS) in Zimbabwe, Mlambo, R., Freeman, A., *African Journal on Land Policy and Geospatial Sciences*, vol. 3, no. 3, 42, September 2020.

REFERENCES: Z

Comité Spécial du Katanga, Les Travaux Géodésiques du Service Géographique et Géologique, 1954.
D. R. Hendrikz of the South African Trigonometrical Survey, 1944.
Dare and Mutale, Survey Review, April 1997.
Geodetic Survey of South Africa, Volume iii, 1905.
Malcolm A. B. Jones of Perth, Australia, Personal Communication, September 2004.
Mlambo, R., Freeman, A., Establishment of Continuously Operating Reference Stations (CORS) in Zimbabwe, *African Journal on Land Policy and Geospatial Sciences*, vol. 3, no. 3, 42, September 2020.
Prof. Peter Nsombo, L. Combrinck, Hartebeeshoek Radio Astronomy Observatory.
Professor Charles L. Merry, University of Cape Town, Personal Communication, July 1997.

Former Yugoslavia

Appendix: Yugoslavia

Upon the collapse of Austria–Hungary the Kingdom of Serbs, Croats, and Slovenes was proclaimed in 1918. The name was changed in 1931 to Yugoslavia and divided into nine republics not based on racial lines. The principal Yugoslav mapping agency became the 1920 *Geografski Institut Jugoslovenske Narodne Armije* (GIJNA) Geographic Institute of the Yugoslav Peoples' Army, formerly the 1888 *Vojni Geografski Institut* (VGI) Military Geographic Institute. Prior to that, the land survey had been carried out by the Viennese Military Geographic Institute (1851–1908) for Serbia and by the VGI in 1871 for Croatia and Slovenia.

Between the two World Wars, the Italian *Istituto Geografico Militare* produced topographic series of the then Italian territories of *Venezia Guilia* and *Istria*. During WWII, both the German *Generalstab des Heeres* (General Staff of the Army) and the British Directorate of Military Survey, War Office (Geographical Section, General Staff) reprinted and revised many Yugoslav and Italian maps.

The coordinate systems that have been used for Yugoslavian lands have been quite diverse through history. Geographic coordinates have been based on the Prime Meridians of *Paris, France; Ferro, Canary Islands; Greenwich, England*; and also, some temporary usage of the *Vienna University Observatory* where: $\Phi_o = 48°\ 12'\ 35.50''$ N, $\Lambda_o = 16°\ 22'\ 49.98''$ E (later offset to Paris).

The names of the classical horizontal datums found in the former Yugoslavia include *Hermannskogel 1871, K.U.K. VGI Vienna University System 1892, Pulkovo 1942,* and *European 1950*. The Vienna University System 1892 used the now obsolete Zach 1812 ellipsoid and the Hermannskogel 1871 datum used the Bessel 1841 ellipsoid of revolution where the semi-major axis $(a) = 6,377,397.155$ m and the reciprocal of flattening $(1/f) = 299.1528128$. The "Parisian" system of mapping (based on the Prime Meridian of Paris, where the offset from Greenwich is accepted as $\lambda = 2°\ 20'\ 13.95''$ E was cast on the polyhedric projection from 1878 to 1959.

The polyhedric projection is aphylactic in that it is not equal-area (authalic). Also, it is not azimuthal, and it is not conformal (orthomorphic). The aphylactics used for large-scale topographic mapping (and grid systems) were adopted because of their ease of construction in the field.

For a historical perspective of how maps were produced before photogrammetry, consider that a century ago, a mapping party would leave the capital city some weeks **after** the departure of the triangulation parties. Since the lay of the land to be mapped was largely unknown, there was little opportunity to plan where control would be established, and mapping could proceed. Based on where triangulation stations were established, the topographer would then proceed to interpolate between those stations and map the topography with plane table and alidade. Since there was no prior knowledge of what and where mapping was to be accomplished, no graticules were prepared in advance. The result was that the manuscripts had to be prepared first at the base camp. Aphylactic projections were developed so that with a simple book of projection tables, the topographer could draft the graticule on a sheet of starched linen in a tent. Although we can be critical of such mathematical machinations nowadays, there was a valid and practical reason for such aphylactic projections back then.

Another aphylactic projection common in Europe is the Cassini-Soldner (rectangular spheroidal) while in the United States, the American Polyconic was used for decades by the U.S. Army Corps of Engineers for the World Polyconic Grid (predecessor of the Universal Transverse Mercator or UTM). Furthermore, the American Polyconic was also used by its originators, the U.S. Coast & Geodetic Survey as well as by the U.S. Geological Survey for the National Mapping Program.

The polyhedric projection can be characterized in a variety of ways; the easiest for the photogrammetrist to relate to is the Local Space Rectangular (LSR), a variant on the Geocentric

Coordinate System. When we consider an analytical model for our photogrammetric computations, we usually need to take the ellipsoidal curvature of the earth into account. Our defining parameters are established by our datum, be it a classical horizontal geodetic datum such as the Hermannskogel 1871 and NAD 1927, or an inertial mass-centered datum, such as NAD 1983 and WGS 1984.

The LSR equations can be found in the 3rd thru 6th editions of the *Manual of Photogrammetry*. In aerotriangulation, we use the ellipsoid height (h) and the optionally transformed LSR (Z_L) of our points as they are to compensate for ellipsoidal earth curvature. For the polyhedric projection, we force the LSR (Z_L) to be equal to zero. For a secant case, we merely assign the ellipsoid height of the point of origin (h_o) a value less than zero.

The most common classical datum (prior to the European 1950) found in the former Yugoslavia is the Hermannskogel 1871 Datum with $\Phi_o = 48°\ 16'\ 15.29"$ N, $\Lambda_o = 33°\ 57'\ 41.06"$ East of Ferro, where Ferro $= 17°\ 39'\ 46.02"$ East of Greenwich and azimuth to *Hundsheimer* is $\alpha_o = 107°\ 31'\ 41.7"$. The most common grid found on that datum is the Yugoslavia Reduced Gauss-Krüger Transverse Mercator. The scale factor at origin ($m_o = 0.9999$), the central meridians of the belts (C.M. $= \lambda_o = 15°$, 18°, 21° East of Greenwich) and the False Easting at C.M. $= 500$ km. The Ministry of Finance used the non-reduced version only between 1938 and 1940, where $m_o = 1.0$.

About 70 years ago, the Army Map Service transformed Hermannskogel 1871 Datum to the European Datum 1950. However, large data sets still survive on that old datum. The author examined the relation between the two datums recently and computed new transformations. Twenty-two points were used that are common to both datums in the former Yugoslavia and a simple three-parameter shift analysis yielded the following: $\Delta X = +770.417$ m, $\Delta Y = -108.432$ m, $\Delta Z = +600.450$ m. The accuracy of this transformation when expressed in terms of actual geodetic coordinates is Latitude change ($\Delta\phi$) $= \pm 3.74$ m, Longitude change ($\Delta\lambda$) $= \pm 4.54$ m, and Ellipsoid Height change (Δh) $= \pm 12.70$ m. On the other hand, a seven-parameter shift analysis yielded the following: $\Delta X = +758.53$ m, $\Delta Y = +259.52$ m, $\Delta Z = +542.18$ m, $\delta s = -6.0 \times 10^{-6}$, Z-rotation ($\omega$) $= +11.29$ seconds, Y-rotation (ψ) $= +2.06$ seconds, and X-rotation (ξ) $= -5.66$ seconds. The accuracy of this transformation when expressed in terms of actual geodetic coordinates is Latitude change ($\Delta\phi$) $= \pm 1.07$ m, Longitude change ($\Delta\lambda$) $= \pm 1.44$ m, and Ellipsoid Height change (Δh) $= \pm 0.64$ m. For example, station *Vel Gradiste* has the following EU50 coordinates: $\varphi = 45°\ 09'\ 17.3501"$ N, $\lambda = 18°\ 42'\ 44.9479"$ E, 0.0 m. and the following Hermannskogel 1871 coordinates: $\varphi = 45°\ 09"\ 14.4675"$ N, $\lambda = 18°\ 43'\ 00.7696"$ E, 0.0 m. The Yugoslavian Reduced Grid coordinates are Northing (X) $= 5,001,303.81$ m., Easting (Y) $= 556,359.65$ m.

REFERENCES

Carl Rathjens and Herman von Wissmann, Karte des Reisegebiets in Jemen, 1934.
Institute Géographique National, October 1996.
Manual of Photogrammetry, 3rd thru 6th editions.

Index

Aarar 407
Aarau 853
Aarbib 373
Abādan 368
Abade 680
Abaiang 422
Abashiri 402
Abassouen 219
Abayomi 459, 478
Abbasid 371, 411
Abbe 527, 726
Abdalla 844
Abdallah 405
Abdelfattah 233
Abdelkader 13
Abdelmadjid 13
Abdelrahim 752, 777
Abdoulaye 689
Abdulhamid 36
Abdullah 405, 505
Abdurraham 820
Abemama 422
Aberdeen 162, 517, 849
Abidjan 389–390
Abinader 225
Abingdon iv
Abomey 83
Abona 732
Abou 373, 775
Abraham 63
Abrantes 642
Abrisse 289
Abrolhos Lighthouse 106
Abşeron 53
Abubeker 250
Abul 411
Aburi 296
Abut 578
Abuta 402
Abuto 418, 438
Abyad 774
Abyei 749
Abyssinia 249, 251
accretion 318, 338
Aceh 361
Achaemenid 367
Achapnyak 37
Acheritan 204
Achola 418, 438
Aconcagua 31, 165
acquae 10, 560
acquis 203
Acra 296
ACSM 301
Acta 540
Activa 339, 348
acune 494
Adama 280

Adapazari 820
Addu 505
Adelaide 43
Adelino 485
Adler 382, 390
Admirantes 699
Adolf 288
Adolfo 812, 830, 874–875, 879
ADOS 566, 703
AEF 162, 187, 264, 275, 389
Aegean Sea 114–115, 819
aegis 642
Aerocarto 755
Aerodist 327–328
Aerodrome 43
Aerogeodetic 284
Aerogeopribor 637
aerotriangulation 220, 230, 514, 896
Afar 219, 722
Afdeling 215
Affini 303, 329
Afgooye 722
Afonso 641
AFREF 86, 131, 418, 437, 510, 561, 593
Afrikaansche 725
Afrikaner 725
Afrin 774
Afrique 83, 119, 158, 162, 188, 242, 275–276, 322–323, 329, 389
AGA 188
Agach 114
Agadez 587
Agadir 555, 557
Agajelu 562, 593, 603
Agalega 527
Agana 314
Aganyo 417
Agel 543
AGN 635
Agou 803
Agourai 556
Agraciada 860
Agricole 453, 479, 776, 831
AGROS 551, 695, 778
AGRPG 137
AGS 536, 635
Aguerrevere 873
Aguilar 337
Agustín 175
Ahmad 367
Ahmar 373
Ahmed 613, 752, 777
Ahmose 749
Ahunui 269
Ahurei 270
Ahvāz 368

Ahwaz 368
Aichi 402
Aigun 669
Ailinginae 518
Ailinglapalap 518
Ailuk 518
Aïn 13
Ain 64, 432, 648, 684–686, 844, 864
Ainu 669
Airy xxiv, 376–377, 850
Aitutaki 595
Aïzo 86–87, 130
Ajalooline 246
Ajman 843
Akanyaru 129, 663
Akbar 67, 356, 613
Akchanaya 825
Akiaki 269
Akija 446
Akil 452, 775
Akita 402
Akjoujt 523
Akkadians 451, 773
Aklan 635
Akmeniškiai 474
Akmolinsk 413
Akora 358
Akosso 275
Akureyri 352
Akuse-Apam chain 296
Akylbek 437
Alabama 301
Alai 437
Alain 120, 691
Alajuela 191
Alamanni 769
Alan 629
Åland 259–260
Alaouites 453, 479, 776, 831
Alawi 773
Alay 436, 787
Albany 45
Alberdi 626
Albert 543, 691, 835
Alberto 629
Alborz 367
Albrech 551, 695
Albrecht 705, 713
Albreda 279
Alcala 339
Alcide 93, 130
ALCOA 755
Aldabra 698–699
Alegre 106, 626
Alejandro 629
Aleksandar 693
Aleutian 853
Alexandria 233, 379, 751

Alexandroúpolis 114
Alexandru 539
Alfredo 625
Algeria–Niger 588
Algerie 14–15
Algiers 815–816
algorithm 234, 377, 556, 601
Algum 870–871
al-Husayn 405
Ali 179, 219, 791, 883
Alifu 505
Aliksic 695, 778
Alimi 703, 778
aliquot 853
Alistair 648
Alkmaar 573
Allah 607
Allain 220
Allegre 107
Allemand 490
Allen 536, 743
Allgemeine 792, 830
Allmänna 764
alluvium deposits (Guatemala) 318
Almagro 165, 629
Almazan 557
Almohad 523
Almoravid 523
Alofi 595
Alomalà 249
Alpen 291
Alpine 652
Alps 263, 287, 469, 769
Altagracia Church Grid 874
Altaic 113
Altay 435
"Althing" (Iceland) 351
Altiplano 93
Alto Coloane 483–485
Alto de Escuque Grid 874
Alugubuna 275
alumina (Suriname) 755
Alvarado 237, 317
Álvaro 717
Alvaro 827
Álvars 105
Alvernia 59
Alyce 317
Alzette 477
Amaa 776
Amager 216
Amami 403
Amazônia 105
Ambadu 219
Ambitus 573
Ambó 242
Amboina 363
Ambrym 869–870
Amedee 225
Amérique Méridionale 93, 130, 309
Amersfoort xix, 573
AMG 5, 44, 622
Amherst 124
AMI 684

Amin 835
AMINOIL 684
Aminoil Grid (Kuwait) 432
Amir 431, 613, 647
amiri decree 431
Amman 407
Ammiraglio 220
Amon 749
Amorites 451, 773
Ampatzidis 303, 329
Amstel 574
Amu 3, 863
Amur 669
Amuri 578–579
Anapoto 270
Anastasio 584
Anatolia 35, 819, 821
Anatom 869–870
Anawrahta 123
Anchorage 153, 183, 219, 424, 518, 595
Ancon 618
Andaman 123
Andean Plateau (Bolivia) 94, 229
Anden 217
Anderson 314
Andes 93, 165, 175, 230, 629, 860, 873, 875
Andorran 19
Andoyer 301
André 485, 652–653, 665, 691, 722, 777, 873
Andrea 90, 130, 216
Andres 175, 584, 633
Andrić 97
Andræ xxiv, 216, 351
Aneityum 871
aneroid altimeter traverses 205, 296
Angkor 135
Anglesey 850
AngloFrench convention 803
AngloGerman Boundary Commission 792
Angolan 25
Angolares 679
Anguiano 532
Anguilla 739
Anhwei 169
Anie 20, 732
Anivsky 669
Anjouan 179–180
Ankara 819–822, 830–831
Ankuanu 269
Anlier 78
ANLONG 137
Annaba 13
Annapurna 570
Annés 221, 226
Annexe 83, 119, 139, 321
Annobón 241–242
Anopheles 617
Anta 630
Antananarivo 493, 495
Antão 149

Antes 539
Anthem 531
Anthony 686, 778
Antillen 41, 55
Antilles 39, 743
Antillies vii, 1, 39, 41
Antioch 773
Antiochus 487
Antipodes 577
Antofagasta 165
Antsirana 493, 495
Antwerpen 78
Anuanu 269
Anuario 31, 36, 54, 412, 437
Anuários 106
Anuradhapura 735
Anuta 717
Anzoátegui 107, 166, 328, 630, 874
Aoba 869–870
Aomori 402
Aon 518
Aoraki 577
Aoral 135
Apam 296, 532
APB (code) 518
aphylactics xxi, 341–342, 376, 456, 600, 895
Apia 675
Apolima 675
Apollo 431
Apopata 630
apparatchiks (communist) 539
Après 268
APSG (Association of Petroleum Surveying & Geomatics) 466–467
Apsheron 53
Aqaba 379–380, 405–407
Aqmola 413
Aquileia 386
Aquitaine 513
Arabkir 37
Arad 49–50, 345, 540, 651–652
Aragats 35
Aragon 513
Arakabesan 536
Araks 53
Aral 367, 411–413, 825, 863
ARAMCO 683–684
Aranaka 422
Aras 35
Aratu 107
Araucanians 165
Araz 53
Arbat 671
ArcGIS 642
Archaemenid 53
Archipel 180, 267, 269
archipelagoes (French Polynesia) 267
ARCO 364
Arctic 145, 259, 599, 634, 657
arc-triangulation 764
Ardai 792
Arden 380

Ardennes 77, 477
Areias 107
Areo 874
Arequipa 94, 630
arête (French West Africa) 389
Argeadae 487
Argo system 188, 487
Argos Orestikon (Macedonia) 487
Argyle (UK) 849
Arholm (Sweden) 260
Arica (Chile) 94
Aripo 811
Arlit 587
Arlon 78
Armagh 255, 376
Armate 465
Armee 379, 556
Armée 13, 136, 263, 452, 774–775, 815–816, 831
Armii 658
Armije 490, 895
Arminius 287
Arno 518
Arnoldo 237
Aroangwa 560
Arorae 422
"Arq" Oil Field (Saudi Arabia) 64, 432
Arran 849
Arredores 641
arrêté (decision) 321
Arrian 431
Arsacids 35
Artavazd 35
Artemis 431
Arth 355
Artibas 860
Artigas 859
Artilharia 483–484
Aruban 39
Arula 446
Arumenii 518
Arya 355
Aryan 355, 613
Arys 412–413, 863
Asa 501, 647, 705
Asad 773
Asahi 403, 408
Asahigawa 402
Asal 219
ASCE (*Journal of Surveying Engineering*) 382, 391, 785, 832
Aseri 246
Aşgabat 825
Ashanti 804
Ashdod 381–382
Ashoka 355
Ashot 35
Ashtarak 37
Ashtech 136
Ashton 622
Asīr 683, 685
Askia 460, 509

Asmara 249
Asofa 295
Asoka 67
Asor 536
Asoteriba 752
ASPRS 200, 643, 651, 665
Assab 220
Assam 89, 123, 356
Assimi 509
Assini 389
Assoumani 179
Assyria 750
Assyrian 35, 53, 63, 367, 451, 750, 773
Astrakhan 658
astrofix 608, 698–699
astrogeodetic determinations 199–200, 209, 654
Astron 551
Astronomice 540
Astronômico 106
Astronómico 165, 242, 250
Astronomiques 241, 251
Astronomy 361, 371, 551, 854, 888, 893
Astropoint 519
Astyages (King) (Iran) 367
Asunción 625–626
Aswan 749
Ata 807, 809
Atacora 86
Atatürk 367, 819
Atkinson 327
Atlacatl 237
Atlántico 583
Atyrau 411
Auckland 577, 808, 831
Audoin 275
Aufnahme 49, 98, 345, 651
Aug 210, 250, 505, 787
Augustinian 317
Augustus 651
Aul 413, 825
Auler 849
Aumets 78
Aunap 245, 250
Aunu 675–676
Aur 518
Aurora 635
Ausgabe 447, 478
AUSLIG (Australian Land Information Group) 136–137
Aussichtsturm 291
Austen 613
Australes 267–268, 270
Austrasia 477
Austria–Hungary 97, 288
Austro–Hungarian 695
Austronesian 421, 717, 783
authalic 77, 456, 555, 895
Autograph (WILD A8) 817
Autohorika 817
Autonoma 339

Außenstelle 299, 328
Ava 125
Avabacas 630
Avan 37
Avars 713
Avenida Conselheiro Borja 483
Avestan 3
Aviação 24
Aviación 583
Avila 229
Avlabari 283
Avril 188, 210, 268, 280, 330
avulsion 318, 338
Awal 501
Awdal 721
Axum 750
Axumite 750
Aya 451, 774, 820–821
Ayabacas 630–631
Ayabelle 220–221
Ayacucho 630
Ayatollah 367
Ayina 187
Aymara 93
Ayn 774
Ayr 849
Ayribaba 825
Azabu 401, 427
Azad 614
Azali 179
Azerbaijani 53
Azeri 36
Aziz 502
Azov 839
Aztec 237, 317, 531, 583
Aztlán 531

Baa 505
Bab 775
Bab (Syria) 452
Bab al-Mandab Strait 219, 452, 775
Babar 276
Babiker 844, 864
Babur 355
Babylon 683
Babylonian 63, 451, 773
Bac 878
Bacau 800
Bačič 197, 210
Bactria 863
Badajoz 731
Badakhshan 787
Badekas 418, 478, 485
Baden 289–291
Bagan 123
Baghdad 35, 371, 411, 431, 648, 843
Bagirmi 161
Bagratuni 35
Bagre 120
Baguirmi 157
Bahamas 57, 59–60, 395
Baharu 503
Bahia 106
Bahn 291

Baie 241–242, 250–251, 270
Baigura 20, 732
Baikal 435, 657, 670
Baïkan 878
Bailey 595–596, 726
Baiocchi 387, 391
Bair 490
Baird 352
Bairiki 424
Bajao 431, 648
Baka 157
Bakar 197
Bakeeliu 869, 879
Bakem 362
Bakeoliu 871, 879
Bakhavol 5
Baki 53
Bakirkoy 820
BAKOSURTANAL 363–364, 391, 800
Bakr 371
Baku 53, 659
Balabac 634
Balad 647
Balanacan 633
Bălăneşti 539
Balboa 229, 617–618
Balchik 114
Balearic 732
BALGEOS 11, 54
Bali 362
Balicia 50
Balikesir 820–821
Ballis 81
Balochistan 613
Balodis 448, 478
Baltasar 633
Baltistan 614
Baltmane 448, 478
Baluchistan 357
Bamako 510
Bamar 123
Bamberg 114, 637
Bambou 527
Bamburg 859
Bamian 5
Ban Chiang 795
Banaba 421
Banat 694
Bancalan 633
Banda 157, 241, 275, 799, 859
Bandecas 302
Banff 849
Bang 67
Banga 364
Bangejin 428
Bangka 362
Bangkok 797
Bangla 67
Bangoma 180
Bani 843
Banja 99
Banjul 279
Banog 357

Banská 710
BAPCO 64
Baranja 195
Barbacoas 874
Barbados 57, 71, 755
Bárbara 106
barbarous slave trade 460
Barbary 815
Barbato 860, 865
Barbuda 1, 27–28
Barcelona 461, 553, 874
Bari 721
Baria 877
Barinas 874
Barnes 562, 571, 603
Barozzi 651
Barra 107, 280, 483–485
Barracks, Army (Mendoza) 31
Barre 721
Barreira 106
Barro 800
Barth 587
Barthélemy 309
Bartolomé 199
Bartolomeo 203, 559
Barú 191
Basa 635
basepoints 334, 398
Bashar 773
Bashkim 491
Basiana 717
Basotho 455
Basotholand 455
Basra 372
Bassa 459, 559
Bassac 136
Basse 309
Bassin 529
Bastion 204
Batavia 361, 363–364
Batavus 573
Batel 641
Batepá 679
Batha 373
Bathurst 279–280
bâtiments hydrographes Ammiraglio Magnaghi et Ostia 220
Batista 199
Baton Rouge xxii, 854
Batswana 101
Battalion 285, 432, 621, 670
Battambang 136
Batticaloa 735–736
Batumi 283
Baukou 803
Bauxite 327
Bavanaadittiya 505
Bavaria 289–290
Bavarian 287, 290, 713
Bavenhöe 215–216
Bavenjoj 215
Bawku 803
Bayadère 523
Baydukov 670

Bayern 290–291
Bayford 739
Baykonur 413
Bayol 85
Bazardüzü 53
Bazley 502, 705, 777, 797, 830
Beaker 847
Beaupré 242
Beautemps 242
Beaz 487
Bechuanaland 101, 562, 565, 602
Bedford 850
Beersheba 380
Beidou 172
Beik 436, 788
Beila 752
Beira 25, 54
Bekaa 451–453, 775–777
Béla 695, 777
Belarusian 73
Belavezhskaya 73
Belaya 73
Belfast 376–377
Belgique 77
Belgrade 694
Belgrano 31
Belham 553
Belitung 364
Bellevue 869
Bellingshausen 269
Belmont 527
Belorussia 73
Belorussian 73–74
Belovezhskaya 73
Bemolanga 494
Ben Nevis 847
Benahmed 16, 54
Bender 219
Benelux 477
Benes 208, 709
Bengal 67, 89, 356, 614, 705
Bengali 67
Bengasi 465
Benghachame 817, 830
Benguela 24
Beni 555
Benito 191, 241–242, 385
Benoit 114, 502
Bensalah 13
Benue 591
Bequia 743, 745–746
Ber 556
Berber 465, 523, 591
Berbera 721
Berberi 721
Berezina 73
Berg 291
Bergen 573, 599
Beri 592
Berlin-Marienkirche 290
Bermessungsamt 20, 732, 778
Berne 469, 769–770
Bernese 116, 548, 733, 770, 830, 860
Berri 819

Berthon 629
Beru 422
Berwick 849
Beset 36
Bessarabia 113, 539–540
Bessastaðaskóli 351
Betio 424
Betsimisaraka 180
Betty 13, 676
Beverloo 77
Bexton 741
BH-01 Datum OSY 884
Bharu 503
Bhat 355
Bhot 89
Bhota 89
Bhu 89
Bhutanese 89–90
BIA (Burmese Independence Army) 123
Biafra 139
Biala 638
Bialystok 638
Biao 242
Biarritz 20, 732
Bidihimba 187
Bidoon 431
Bielo 259
Bielsko 638
Bight 139
BIH 99, 625
Bihar 67, 89, 356
Bijagos 325
BIK 518
Bikar 518
Bikati 423
Bikini 517–519
Bikku 465
Bikram 569
Bilad 406, 721
Bilby 280, 690
Bilderham 574
Bill 192, 658, 793
Billeaud 822, 830
Biloxi 854
bimetallic 385
Binari 322
Binch 878
Binga 559
Bioko 241–242
Bir 220, 569, 883
Birdseye 191, 318, 337–338
Birnie 424
Biscay 263
Bishweshwar 569
Biskra 13
Bissau 149, 273, 321, 325
Bisseau 689
Bistriţa 694, 779
Bitari 321
bittereinders (Boer) 726
Bitti 465
Biukoto 257
Bizerte 815, 830

Björn 351
BKG 299, 302, 329, 710
Blagaj 97
Blake 343
Blanc 263, 523
Blanche 527
Blas 617
Bleasdale 849
Blemmyes 750
Blick 580
Blida 13
Bligh 43, 807
Bliss 657–658
Bloc 115, 431, 877
Bloemfontein 725
Bloque 874
Blumut 503
BM 154, 722
BMJC (Baltic Journal of Modern Computing) 448, 478
Bo 217, 226, 658, 703
Boa 149
Boaco 583
Boaga 386
Bobo 119, 389, 804
Bobotov 551
Bocanegra 531
Boche 490
Bochum 291
Bodin 713
Bodnârescu 540
Bodrok 709
Boer 455, 725–726, 759
Bofosso 702
Bogdan 539
Bogdania 539
Boggy 27
Bogotá xix, 175–176, 230, 874
Bogra 5
Bohemia 49–50, 207–208
Böhmen 207–208
Bohnenberger 49, 345, 637
Bohol 635
Bohuslav 208, 210–211, 709–710, 778
Bojana 10
BOK 518
Bokku 518
Bol 94
Bolama 325
Bolat 413, 825
Boletim 330
Boletin 297
Boletines 94
Bolívar 93, 229, 231, 873, 874
Bolivarian xvi, 867, 873, 875
Bolivian 93–94
Bolivien 93, 130
Bollettino Geodetica 10, 55, 303, 329
"bolsones" 338
Bombay 356
Bomford 123, 126, 131, 358, 390, 615, 797, 830
BON 78, 89, 518, 556, 815
Bonaire 39–40

Bonaparte 263, 451, 774
Bône 13
Bonin 401, 403, 523, 690–691, 778
Bonito 191
Bonne 10, 13–14, 49, 54, 77, 136, 233, 263, 363, 376–377, 381, 385, 406, 451, 469, 493–494, 527, 555–556, 573, 641, 652, 731, 769, 774, 816, 819–820, 849, 859
Bonnin 690, 777
Bonto 518
Bop 618
Bor 290, 329
Bora 267, 269
Bordone 203
Borja 483
Borkou 161
Borkowo 637
Borneo xix, 109, 362–363, 501–503, 633
Bornholm 215–216
Borno 161, 587
Bornou 157
Bornu 587
Borowa Gora 637
Boru 375
Bosiljevac 197, 210
Bosiu 455
Boskop 101
Bosman 727
Bosnian 97
Bosporus 819
Bosy 639, 643
Bothnia 259, 763
Botrange 77
bottles (buried) 114, 260, 670
Bouchilloux 310
Boudeuse 267
Bouet 157
Bougainville 220, 267, 621, 717, 869
Bougainvilleans 621
Bouguer 229, 263
Bouillet 77
Bouillon 78, 773
Bouquet de la Grye 689
Bourgoin 85, 691, 778
Bouterse 755
Boutou 588
Bowditch 27, 518
Boyer 225
Bozova 820
BP 54, 362–363
Bquna 389
Brabant 78
Brac 153–154
Bradford 892
Bragg 306
Brahman 795
Brahmand 355
Brahmaputra 123, 356
Brak 373
Bramble 553

Branco 483
Brandaris 40
Brandenberger 469–470, 478
Brandon 527, 849
Brasil 106
Brasileiro 105
Braslaw 637
Braşov 652–653
Brassaville 25, 133, 187
Brazilian 105–107, 860
Brazza 157, 183, 187
Brazzaville 23, 139, 157, 183, 187–188
Brecknock 850
Breda 755
Brega 466
Bréguet 220
Breite 658
Bremen 461, 479
Bremerle 290
Bretagne 309
Brezhnev 657
Bridgetown 755
Britannic 219
Britannica 267, 325, 328, 743
Broadfield 849
Broome 43, 45
Browne 203, 372, 390, 431, 648, 808, 831, 843
Broz 195, 693, 713
Bruce 398, 408
Brugen 113
Brunner 731, 819
Brunswick 145–146, 513
Brussels 77–78, 183–184
Bryson 103, 130
Brythonic-speaking Celtics 847
BT 109–110, 502–503
Bubi 241
BUC 540
Buchanan 759
Bucharest 487, 539–540, 652
Buchwaldt 216, 351
Buckingham 849
Bucovina 50
Bucuresti 540, 653
Budapest 49, 345–348, 541, 654, 694
Buenaventura 176
Buenos Aires 31–32, 36, 54, 860
Bugandan 835
Bugrõ 245
Bugry 474, 659
Bugufi 129–130
Buka 622, 717
Bukavu 142
Bukhara 863
Bukit 109, 362, 364, 501, 503, 633, 705
Bukovina 540, 652
Bulawayo 891
Bulgar 113, 490
Bulge 77
BULIPOS 695
Buller 578–579
Bululwe 184

Bumthang 89–90
Bungsberg 291
Buraimi 608
Burgas 113–114, 117
Burgundians 769
Burkersroda 291
Burmese 123
Burnthang 90
Burnuli 800
Bursa-Wolf 634
Burundians 129
Burundi–Rwanda 663
Buşayyir 647
Būshehr 368, 372
Bushire 705
Bushnell 676
Busoga 835
Busowa 835
Busselton 45
Butaritari 422
Bute 849
Butha 455
Buthe 455
Butler 313
Buurgplaatz 477
Buxar 67
Byarezina 73
Bydgoszcz 638
Byelaruskaya 73
Byureghavan 37
Byzantium 97

Caaguazú 626
Cabanatuan 635
Cabello 39
Cabére 20, 732
Cabinda 23–25
Cabo 150, 210, 460, 559
Caboet 19
Cabot 31
Cabral 105
Cabrion 322
Cacheu 325
Cacqueray 219
Cadastrais 559
Cadastro 485, 642–643
Cadell 123–125
Cadiz 689
Cadre 509
Cadrilaj 652
Caernarvon 850
Caesar 263, 287, 543
Cag 505
Cagayan 633
Cahora 559
Caille 527
Cairn 45, 489, 622, 849
Caitaba 800
Caithness 849
Cajigal 873
Cakchiquel 317
Calabozo 874
Cálculo 800, 830
Calcutta 67

Calder 295, 328, 562, 592, 602
Caledon 455
Caledonia 268, 869–870
Calédonie 268, 271, 869
Caliph 35, 371, 411
Calkali 490
Calvinism 288
Calypso 397
Cam 878
Camacupa 24–25
Camane 630
Cambodian 135–137, 442, 797, 878
Cambridge 670, 777
Cambridgeshire 849
Cameron 32
Cameroun 139, 141, 210
Camocio 203
Campbell 43, 577
Campeche 81
Campo 24, 32–33, 561, 800, 873
Cana 878
Çanakkale 820
Candelaria 635
Canella, Otto 543
Canigou 20, 732
Canna 849
Cannelles 741
Cannouan 745
Canonnière 220
Canouan 743
Canovas 242
Canta 630
Canton 99, 424, 769, 840
Canuel 268
Capanna 386
Capitão 483, 680, 777
Capitão-Tenente 483
Capodimonte 385
Caprivi 565
CAR 395, 635
Carabobo 873
Caracas 230, 873, 875
Caraïbe 309, 873
Carajos 527
Caranus 487
Caravelle 521
Carazo 583
Carbon 819
Cardigan 850
Carfort 219–220
Cargados 527
Caria 821
Carinola 49
Carinthia 49, 386, 713
Carl 93, 130, 208, 709, 883, 896
Carlson 517
Carmarthen 850
Carmine 684
Carnarvon 45
Carnatic 356
Carniola 386
Caroeira 498
Carolina (North & South) 283, 577, 617, 693, 701, 855

Index

Caroline Islands 535–536
Carpathian 49, 345, 539, 651, 839
Carpatho 207, 709
Carpentaria 43
Carpentier 820
Carrasco 860
Carrauntoohil 375
Carriacou 305
Carrick 849
Cartagena 176
Carthage 731, 816
Carthaginian 460, 513
Cartografia 230, 250, 485, 874
Cartografic 20, 54
Cartografico 225, 642
Cartográfico 150, 199, 532, 641
Cartographie 322, 329, 557, 815
Cartographique 135, 441, 521
cartometric 5, 303, 352, 638
Carton 13, 15
cartonné 13
Carvalho 25, 54
Casablanca 279, 556
Casamance 325, 689, 691
Casas 199, 873
Casiguran 635
Casimiro 225
Caspari 309
Castanea 385
Castello 385, 641–642
Castelo 483
Castile 513, 641
Castilla 629
Castillo 33, 242, 629
Castries 741
Castro 199
Cat Island 59
Çatalhöyük 819
Catalog 412, 438, 447, 473, 478
Catalogue 322, 328, 671, 726, 869, 879
Catalonia 731
Catalunya 20, 54
Catamarca 32
Catania 385
Catarina 107
Catastral 555
Catastro 165, 339
Catedral 859
catenary curve 743
Cathedral 27, 191, 750, 816, 820, 878
Catherine, Mount 233
Catherine, Mount Saint 305
Catherine the Great 657
Catholic 97, 288, 619, 717
Catholicism 625, 799
Catwick 878
Cavaleiro 149
Cavalhada 106
Cavalier 204
Cavalla 321, 460
Cavell 686, 778
Cay 59–60, 81, 622, 743
Cayemite 333

Cayman Islands 133, 153–154, 211, 395
CBL 169–170, 755
CC 470, 653, 770, 816
Ceará 106
Ceausescu 651
Cecil 101, 891
Celebes 362–363
Celle 291
Celsius 763
Celt 573, 641, 693
Celtic 375, 713, 847
Centraal 755
Centro 106, 618, 643
Cercetări 694, 779
Cero 630
Ceru 39
Cess 375
Cession 184, 255
Cestos 460
Ceuta 555, 731
Ceylon 501, 735–736, 778
CGPS 94, 676
CH 428, 469–470, 770, 830
Cha 791, 878
Chaafu 373
Chababiye 407
Chaca 630
Chaco 31–32, 93, 625
Chadha 123, 126, 131, 357, 391
Chadian 162
chador (black cloak) 367
Chaga 415
Chagos 505
Chaguamaral 874
Chaguaramas 874
Chai 707, 777
Chaîne 86, 333
Chalermchon 797, 831
Cham 135–136, 441
Chama 791
Chamba 187
Chambi 815
Chameau 310
Chamorro 313
Chan 429, 438
Chanakya 355
Chanda 180
Chaney 502
Chang 785, 830
Chanoîne 587
Chao 84, 140, 709, 759, 795
Chapa 485
Chapada 107
Chappal 591
Charente 309
Chari 157, 161, 187
Charlemagne 19, 287, 477, 713
Charleston 562, 602
Charlotte, Fort 743
Charrúa 859
Chateaubelair 745
Chatham 191, 577
Chato 860

Chau 483–484
Chauve 523
Chauvien 878
Chechnya 657
Cheekha 371
Cheikh 452, 775
Cheju 428
Chekiang 169
Chelm 638
Cheremshynsky 840, 865
Cherni 114–115
Chernigovskiy 659
Chernivtsi 839
Cheshire 850
Chessington 848
Chewas 497
Cheykin 436
Chharom 137, 210
Chiang 783, 795–796
Chiapas 531
Chiba 402
Chibcha 583, 873
Chiclayo 630
Chico 241
Chihli 170
Chikhachëva 671
Chilean 165–166
Chiloango 187
Chilwa 498, 560
Chimanimani 560
Chimborazo 229
Chimbote 630
chimney 305, 372, 689
Chinandega 583
Chindwin 125
Chinggis 411, 435
Chipata 887
Chiriac 541
Chiriquí 617
Chirripó 191
Chisinau 37, 54, 541
Chisumulu 498, 560
Chitrawala 615
Chiuta 498, 560
Chiyokaku 403
Cho 32–33, 429, 438
Choguel 510
Chokusu 435
Chol 81
Cholesky 652
Chonta 630
Chontal 583
Chontales 583
Chorographic 641
Chorotegano 583
Chorrillos 630
Chosŏn 427
Chott 13
Chou 169
Chowdhury 124
Chris 40, 137, 210, 643, 665
Christ 309
Christchurch 577
Christofi 205, 210

Christoph 469, 478
chronograph 125, 518–519, 676
chronometric 412
CHTRS 470, 770
Chu 411
Chua 33, 166, 625–626
Chuá 107, 860
Chuamy 863
Chulalongkorn 797, 831
Chunnel 847
Churchill 670, 672, 779
Chuuk 535
Chymyrov 437
Cidade 149
CIDG 154
Ciechanow 638
Ciências 297, 330
Científica 800, 831
Cimmeria 863
cinnamon 305
CIO 626
Ciorogârla 540, 652
circumpolar 357, 811
Cirenaica 465
Cisplatine 859
Ciudad 176, 230
Clackmannan 849
Claire 314, 329, 856, 864
Clapper, Lieutenant General 6
Clare 375
Clarines 874
Clark 399, 510
ClarkeNelson 408
Clavus, Claudius 599
Clayton, Fort (Panama) 28, 192
Clearing 124, 521, 591
Cleisham 849
Clements 796, 831
Clendinning 296, 329
Cleopatra 487
Cleuch 849
Clibrig 849
Clinton, Bill 658
Clive 67, 356
clocks 125, 599
Closer 226, 313, 418, 436, 689, 744
Coalition 9, 379, 431, 735, 755, 773
Cochabamba 93
Cocking 368, 390
Cockpits 395, 397
Coco 45, 191
cocodat 352
Coconut 421, 517, 595, 827–828
Cocotte 527
Codazzi 175
Codrington 27
Coffee 191, 305, 309, 333, 679
Coghlan 528
Cogo 242
Cohen 10, 54
Cohoha 129, 663
Coimbra 106
Colby 357, 375–376, 847
Colchis 283

Colina 483, 485
Collie 45
Collingwood 578–579
Collinson 341
collocate 672
Collocation 231, 325, 462, 784–785, 800, 832
Colméra 800
Colôane 483–485
colocated 875
Colombo 570
Colón 317, 337, 617–618
Coloneaux 322, 329
Coloniale 322, 329
Colonne 13–15, 816
Columbian 333
Columna, La 873
Comayagua 337
Combani 180
Combrinck 888, 893
Comino 513
Comissão 105, 150, 211
Comité 184, 210, 887, 893
Communism 195, 345, 441, 487, 657, 799
Como 275
Comolet 691
Comoran 180
Comore 133, 179–180
Comorin 356–357
Comoro 179–180
Compagnie 309
Compiegne 157
Comrie 836
Conakry 321–322
Concha 165, 210
Conchan 630
Concise 325, 328
Condamine 229
Conde 321, 483, 485
Conder 380
Condominium 337, 751, 869–870
Condore 878
conformality 736
conformally 494
conforme 494
conformity 14, 556, 592
Congolese 184, 663
Conly 618
Connaissance 820
Connal 527–528
Connan 727
Conoco 24
Conquistador 165, 317, 395, 625
Conquistadores 237
Conrad 287
Conselho 105
Constable 836, 865
Constanţa 113, 115, 821
Constanta 540, 652
Constantin 113, 651
Constantinople 35, 451, 774, 820
constellation 32, 733
Consul 497, 543

contras 337
Convergence 44, 314, 672, 718, 736, 777
convergency 488, 490
Conversões 643
conveyances 203
Cooke 43, 395
Coopers 396
Coordenadas 226, 643, 800, 830
Coordinating 136, 635
coordination 289, 453, 653
Coordonnées 180
Coordonnees 241, 251, 373, 390
Cooten 50, 54
Copán 317, 337
Copenhagen 215–216, 351, 599
Copperbelt 887–888
copperplate 854
copra 421, 517, 535
Coptic 749–751
Cordillera 93, 130, 191
Córdoba 31
Corfu 10
Corisco 241–242, 276
Cornejo 629
Cornelis 727
Cornell 752, 778
Cornwall 850
Cornwallis 502–503, 705–706
Coro 39, 874
CORONA 225, 413
Corozal 229
Corpe 10
Córrego 107
Correia 106
Corsair 191, 306
Corse 265
Corsica 265, 386
Corsisco 241, 251
Cortés 81, 531
Cortes 237
Corvette 220, 523, 690, 777
Cosa 199, 225, 873
Cosmodrome 413
Cot 878
Cotchesqui 229
Cote 321–322
Cotonou 83, 85–86
Cotroceni 540, 653
Courland 445–447, 473
Courteen 71
Cousteau 543
Coutinho 680, 777
COVID 375, 629
Covilhã 25, 54
CPT 358, 390
Craigowl 849
Crati 385
Crato 279
Crete 302
Crèvecœur 521
Crimea 839
Crimean 539–540, 694
Cristina 31, 150, 211

Cristo 237, 317–318, 337–338
Cristobal 175
Cristóbal 874
Crna Gora 551
Croatian 195, 197, 693
Croix, la (d'Hins Bordeaux) 221
Cromarty 850
CROPOS 695
CRTM 193
Cruach 849
Crustal 16, 94, 116–117, 718, 777, 856
Cruz 32, 532, 717–718, 874
Cryn 849
CS 54, 284, 642, 864
CTB 562, 574, 602
Cuanza 23
Cuauhtémoc 531
Cuban 199–200, 305
Cubo 874
Cucaracha 617
Cuddalore 356
Cuéllar 199
Cune 23
Cuevo 93
Cukali 10
Culebra 617
Cumaná 873
Cumberland 850
Curaçao 39–41, 55
Curi 385
Curieuse 697
Curral 149
Curta 342
Curundu 28, 192
Cush 749–750
Cushite 749–750
Cuza 539
Cuzco 629
cylindrical 346–347, 769
Cyohoha 129, 663
Cypriot 203, 205
Cyril 487, 614, 693, 726
Cyrus 53, 367
CZ 718
Czarist 3, 245, 260, 412, 445, 637, 821, 839
Czechia 207
Czechoslovak 570
Czechoslovakia 207–208, 709–710
Czechoslovakian 207, 209, 570, 710
Czechs 207–208, 709–710
Czernowitz 540
Czestochowa 638

Dabajuro 874
Dablobi 283
Dabola 323
Dachreiter 291
Dacia 539
Dacians 651
Dafour 157
d'Afrique 242, 250, 524, 691, 778
Daği 53
Dahl 601

Dahna 685
Daho 16, 54
Dahomey 83, 85, 588, 591, 803–804
DainavÚlÚ 474
Daisey 617
Daitō 401
Daito 403
Dake 402
Dakota 689, 773
Dalap 518–519
Dallaï 221
Dalmatia 50, 195, 386
Dalmatian 97–98, 195, 386
Damascus 773–774
Damāvand 367
Dana 257
Danbury 849
Dänemark 216, 226
Danes 215, 245, 375
Daniel 223, 583
Danilo 225
Danish xxiv, 215–217, 351, 549, 599, 764
Danov 113
Dans 93, 130, 325, 329
Danske 215
Dao 785
Dar 242, 371, 773–774, 791, 887
Daralsai 413, 825
Dardanelles 820
Darfur 751
Darh 373
Darién 617–619
Darius 787
Darmstadt 290–291
Darussalam 57, 109
Darvesh 5–6
Darwin 43
Darya 3, 412–413, 787, 863
Daryl 107, 131
Dashen 249
Dashnaks 36
Dashti 411
d'Assab 220
Dato 501
Daulah 67
Davao 633
Davidov 788
Davidovich 113
Davidson 848
Davin 418
Davis 415, 418, 438
Davtashen 37
d'Ayabelé 220
Dayton 196, 693
DCC 635
DCN 875, 879
Dealul 539–540, 653
Deanne 501
DeBayle 584
Debed 35
Deboyne 622
DeBrazza 157
Deby 162

DECCA 188
Deccan 355
declination 726
Decouverte 527
Dedé 114
Dedele 622
déformation 494
Degbegnon 86–87, 130
DeGraff-Hunter 615
d'Egypte 233
Dehra xix, 4, 55, 124–125, 797, 830
Dei 459
DeLambre xxiv, 77, 263, 543
Delano 855
DELF 562, 575, 603
Delgado 559
Delhi, New 90, 355, 614
Demarcación 318, 329, 562, 584, 603
Demerliac 524, 691
Demirtzoglou 303, 329
Den 215, 764, 792, 831
Denakil 249
Denali 853
Denbighshire 849
Denib 556
Denkmal 291
Dennis 306
Dentergemand 78
Denver 398
Deo 877
Deothang 90
Dépôt 13, 77, 263, 268, 451, 493, 651, 774, 816, 831
Derby 849
Derna 465
Derrington 849
Desaix 521
Desertas 642
Désirade 309
Désiré 220
d'Esnambuc 521
Det 622, 718
Detallado 318, 329, 562, 584, 603
Determinari 540
Deutches 290, 447, 638
Deutsche 208
Deutschen 792, 831
Deuxieme 78
Deve 488
Déviation 268
Devil 395
Deville 146
Devon 850
Dewa 721–722
d'Extrême 135, 441
Dezembro 800, 830
Dghamcha 523
DGN 363–364, 391
DGNSS 733, 766
Dhaalu 505
Dhabi 647, 843–844
Dhahran 685
Dhaka 67–68
Dhank 608

Dharan 570
Dharumavantha 505
Dhebel 815
DHG 208, 290, 447, 638
d'Hins 221
Dhofar 608
Dhows 180, 721
DI 153–154, 280, 303, 329, 386, 722
Diabiga 803
Diablotins 223
Diakha 689
Diamniayo 690
Dias 559
Diaz 626
Díaz 859
Dibb 211
Diego 165, 199, 493, 629
Dienst 39–41, 55, 362–363, 391, 573
Dienstgebrauch 291, 329
Dier 191
difaqane 725
differenced 362, 483
digitalization 840
Digue 699
Dili 799–800
Dilijan 36
Dilling 752
Dillon 812
Dilmun 63, 431, 607
Dimitri 302, 329
Din 63, 501, 540, 653
Dinara 195
DINDAN 250
Dingaan 725
Dingli 513
Dinh 877
Diogo 179
Dion 188
Dionysos 302
Diori 587
Dioulasso 119, 389, 804
Direcção 483, 485
Direccion 339
Directeur 83, 139, 210, 389
Direkcija 491
Diretoria 105–106
DISTAD 536–537
Distrito 105
Ditchling 849
Ditsoni 180
diurnal 296, 685
Divisão 641
d'Ivoire 389–390, 459, 509–510
diwan 67
Dixième 241, 322, 328
Djakarta 364
Djalêlo 722
Djamena 162
Djari 803
Djebiga 803
Djeneponto 362
Djerablous 775
Djibouti ix, 213, 219–221, 226, 249, 721

Djiboutian 219–220
Djinnack 280
Djogué 691
Djolof 689
Djourab 161
Djuka 755
Dlamini 759
d'lesicle 473, 479
Dlinnaya 413, 825
DLisboa 642
DLM 314
DM 381, 407, 432, 608
DMATC 502, 626, 630, 643, 797
Dmejrek 513
Dnepr 73
DNG 78
Dniester 539–540
Dnyapro 73
DOA 518
Dobara 491
d'Obock 220–221, 226
Dobodura 622
Dobrogea 540, 651, 694
Dobrudzha 113, 540
Dodo 518
Dodoma 792
Dodowa 296
Dogola 497
Doha 647
Doiran 489
Dolina 413, 825
Dolo 721–722
Dolohmwar 535
Domingo 225–226
Domingue 225, 333
Dominica 213, 223, 226
Dominican 213, 225–226, 317, 333, 799
Dominion 145, 578, 721, 750
Domning 107, 131
Domoni 180
Domturm 291
Donard 377
Donau 291
Donbass 840
Donegal 377
Donghak 428
d'Optique 220
Dorpat 245, 446–447, 473
Dorset 849
Doryā 722
Doson 877–878
Dosso 86, 588
Doton 306, 329
Douala 141–142, 210
Douglas 246, 399, 408
Doumbouya 321
Dourbes 78
Douro 641
DOV 382, 390, 407–408, 770
DP 40
DPP 783
Dracup 856
Dragoman 116

Dragomir 540
Dragon people 89
Drake 149
Drakensberg 455, 725
DRAN 116
Drangeme Chuu 89
Dravidian 67, 355
DRC 663
Driencourt 493, 495
Drina 97
droites parallèles 333
Drokpa 89
Drone 829
Drukyul 89
Državnih 491
DSLR 90, 131
DSM 103, 130, 205
DSvy 363
DTIC 306, 329
Duan 221, 684
Duane 59, 162, 200, 517
Duars 89
Duarte 225
Dubai 608, 843–844, 864
Dubayy 843
Dubica 98, 196
Dublca 98
Dublin 255, 376–377
Dubrovnik 197
Dufour 141, 769
Dukhan 63, 647
Duklja 97
Dumbarton 850
Dumeira 220
Dumfries 850
Dumpu 622
Dun xix, 4, 55, 124–125, 797, 830
Dundugbia 702
Dung 879
Dunnet 849
Dunnose 849
Dunqulah 749, 751
Dunrig 849
Duorno 446
Dupré 220
Duqm 608
Durand 613–614
Durazno 859, 861
Durban 726
Durham 562, 588, 603, 849
Durrës 10
Dušan 693
Dushanbe 787
Duty 6, 207, 299, 402, 816, 821, 854
Dvina 73
DVRS 844, 864
Dwingo 41, 55
DYNAPG 116
d'Yof 690
Dyson 314, 329, 442, 478
DZC 448
Dzerzhinskaya 73
Dzerzhinskiy 73
Dzhuchi 435

Dziani 180
Dzidefo 297, 329
Dziedzinka 446
Dzongkha 90
Dzvina 73
Dzyarzhynskaya 73

Eagle (station) 328
Eames 313
Ebon 517–518
ECC 676, 756
eccentric 676, 848
Échelon 271
Echmiadzin 35
Ecija 635
Eclipse 242, 355, 424
Ecomat 817
Ecuadorian 229–230
Ecuatorial 242
Edinburgh 84, 140
Edirne 820–821
Edney 615, 643
Edward 145–146
Eelam 735
Eesti 246
EEZ 60, 223
Éfaté 869–871
Éfila 869
Egeria 705
Eggert 788, 831
EGM 16, 418, 791
Egyetem 437
Egypte 379, 451, 774, 831
Egységes 347
Ehime 402
EHT 348
Eiao 270–271
Eidgenößische 769
Eigg 849
Eino 261
Eire 376
Eiriksson 351
Ejer 215
Ejercito 165, 630, 732
Eka 284
Ekaterine 283, 329
Ekonomiska kartan (Swedish) 763
Ekspeditsiya 670
Elam 367
Elamites 367
Elba 386
Elbag 638
Elbe 207, 287
Elblaskie 637
ELD 467
Eleanor 578
Electrotape 342
Elf 24
Elgizouli 752, 777
ELHO 117
Elhovo 117
Ellice 421, 827
Elmina 295
Elobey 241

elongation 395, 452, 775, 811
Elsaß 290, 292
Elsisi 233
elude 249
Ely 849
Emae 870
Embalo 325
Emery 43
Emiembe 759
Emir 63, 405
Emmanuel 385
Emomali 787
Enckeplatz 289
Enderbury 424
Endwerte 447, 478
Enero 873
ENG 518, 622, 718, 860, 865
Engenharia 325, 329
ENI 518
Eniibing 518
Eniirikku 518
Eniwetok xxiv, 517–519
ENLACE 339, 348
Ennedi 161
Ennylabegan 518
Enriquillo 225
Enseigne 220, 268
Ensign 220–221, 523
Entebbe 836, 865
Entrada 149
entre 318, 329, 473, 479, 562, 584, 603
entrepôts 607
Enuebing 518
Enver 9
Enyvertok 518
Eötvös 541, 654
EOV 347
Ephemeris xxiv, 216, 230, 310, 703
Epi 869–870
epiphany 35
EPN 639, 733
epochs 207, 447
Equatoguinean 241–242
Équatoriale 139, 162, 188, 242, 276
equi 357
equidistance 64, 398, 683, 764
equilateral 437
Er 412, 437, 491, 684
Erappu 518
Eratosthenes 573
Erdóhegy 347
Erdut 195
Erebuni 37
Ereğli 114
Erganzungsheft 792, 831
Ergebnisse 207
ERGNSS 732–733
Eric MASISI 101
Eridu 371
Eritrea 219, 249–250, 749, 751
Eritrean 249
Erk Kala 825
Ernesto 149, 199–200, 210
Erromango 869–871

ersatz 44, 149, 170, 217, 432, 451, 588, 631, 684, 732, 774, 884
Ertis 435
Eruwa 592
Escolar 106
Escuela 230
Escuque 874
Eshkashem 3
eshpgn Grid Pair File Prefix 676
Eskişehir 819–820
Español 317, 337
Española 242, 250
Espargos 149
Espinoza 873
Espiritu 869, 871
ESPOL 230–231
ESRI 154, 185, 602, 785
Essex 267, 847, 850
ESSO 467, 683
EST 158, 207, 247, 265, 521, 709
ESTABLECIMIENTO 339, 348
Estaciones 226
estações 800
Estadística 532
Estado 105
Estatística 105
Este 106, 175
Esteiras 241–242, 276
Estelí 583
Estonian 245–247, 250, 259, 446
Estudios 562, 584, 603
Eswatini 559
eta 654
État 263, 776, 831
Etats 453, 479
Ethiopian 249–250, 417, 721–722, 883
Etingdon 395–396
ETRF 11, 115, 302, 377, 448, 544
Etruscans 385
Eua 807–809
EUFOR 98
Eugene 290, 329
Euphrates 367, 371, 607, 775
EUPOS 285, 447–448, 639
Eurasia 16
Eurasian 448
euro 375, 477, 709, 713, 763
Eurographics 387, 390, 478, 822
Europa 216, 226
Europäische 694
Eurozone 731
Eustatius 39–40
EUVN 209
EVFR 247
EVRF 448
EVRS 247
Ewe 803
Exécutés 325, 329
Exército 105–106, 150, 641
Exercito 642
Exmouth 45
Expédition 451
Extrême 325
Extremo 25, 106–107

Eylandt 43
Eyre 43
Ezeigbo 86, 562, 593, 602
Ezulwini 759
Ezurum 820

Faafu 505
Faatasi 829–830
Faber 706
Fabián 860
Fabre 528
Fadlallah 161
Faeroes 217
Fagalii 676
Fagataufa 269
FAGS 74
Fahal 609
Fahud 608–609
Faial 642
Faidherbe 523
Failaka 431
Fairchild 5, 684
Fairey 745
Fairhead 16, 54
Fairways 381
Fais 536
Fakamaru 269–270
Fakarava 267
Fako 139
Faleolo 676
Falintil 799
Falster 215
Famagusta 203–204
Fanagua 829
Fang 170, 210, 241
Fangataufa 268
Fanjara 280
Fanning 424
Fantalli 249
Fanuatapu 675
FAO 372, 391
Faraulep 536
FARC 175
Faro 555
Farol 483
Farquhar 699
Fasching 346
fascicles 817
fascicule 372, 390, 816, 831
Fashir 844, 864
Fashoda 157
FASPRS 519
Fatu 270–271
fauna 380, 592, 618
Faure 14
Faussi 246
Faustin 158
Favé 493
Fayence 527
Fayu 536–537
Fazilova 864
Fazzilova 864–865
FBiH 99
Felda 503

Feldmeszkunst 292
Feliks 73
Felipe 629, 873
Felton 577
Fenland 847
Fennel 637
Feodosiya 839
Fer 207, 709
Ferdinand 97, 195, 267, 633, 853
Fergana 412, 787–788, 863
Ferma 659
Fernandez 31
Fernando 106, 241, 317, 337, 531, 689, 732
Fernão 241
Ferrando 860, 865
Ferrero 765
Ferro-Greenwich 292
Feuille 241, 251, 774, 831
Février 524
FG 766
FGI 260
Fidel 199
Fiducial 33, 74, 230, 398, 707
Fife 849
FIH 334
Fijian 255, 421
Fila 869
Filabusi 892
Filotécnica 680
FIN 368, 372
Financi 207
Finau 807
finial Cross 412, 446, 451, 774, 864
Finlay 153, 849
Finn 217, 226, 259, 763
Finnish 259–260, 763
FinnRef 765
Finsterwalder 93, 130, 437, 864
Firenze 249
Firestone 461, 478
first-order 257, 816
Fischer xxiv, 854, 860, 865
Fisher 32–33, 199, 879
Fisico 191
Fitzpatrick 424, 438
Fiume 386
Flamingo 59
Flamsteed 229
Fleckert 291
Fletcher 483, 807
Fleurien 43
Flinders 43
Flintshire 849
Florence 10, 54, 249
Flores 799
Florida 59, 105, 199, 860
Floridablanca, Pampanga 635
Floyd 229, 859
FMG 257
FMS 501
Foggia 385–386
Fogo 149
Foho 799

Foix 19
Folau 808–809, 830
Foley 657
FÖMI 348
FONDA 116
Fondamental 887
Fondamentale 386
Fongafale 829
Fonseca 337–338
Fonualei 809
Forceral 20, 732
Ford, Gerald 799
Formosa 32, 401–402, 783–784
Forrestierre 741
Fortaleza 106, 680, 861
Forte 106
Fortino 722
Fortran 652
Forze 465
Foula 850
Founti 555
Fourah 701
Fourcade 726
Foureau 157
Fourmi 557
four-parameter 16
Foyle 375–376
Fradique 679
franc 543
Français 23, 219, 389, 453, 479, 776, 831
Francais 225
Francés 225
Francesca 629
Francis 149
Franciscan 317, 543
Franck 257
François 270, 543, 669
Franconian 287
francophone 145
Franco-Turkish 815
Franjo 195
Frankincense 683, 883
Frankish 287, 477
Franklin 279, 670, 779, 855
Frants 216, 351
Franz 49, 97
Franziszeische 49, 345, 651
Frauenkirche 291
Frederick 4, 306, 501
Freetown 701
Frégate 242, 689, 777
Fretilin 799
Frezier 225
friars 317, 799
Frias 485
Friederich 208, 709
Friedrich 792
Friesing 49
friezes 431
Frigate 220, 267
Fritz 246
Frontera 318, 329, 562, 584, 603
Frost 456, 685

Index 909

FRY 229, 693
FSB 246, 658–659, 671–672
FSM 535
FTP 117, 425, 437–438, 733
Fua'amotu International Airport 807
Fuego 32, 577
Fuglenes 600
Fujayrah 843
Fuji 401
Fujian 172, 783
Fujianese 783
Fujimori 629
Fukushima 402
Fulani 139, 587, 591
Funafuti 827, 829
Funan 135, 795
Fur 540, 638, 691
Furada 680
Furie 385
Furnas 732
Fürstliche 469
Futuna 870
Færoe 215
Første 217, 226

Gaabla 220
Gaafu 505
gabão 275
Gabes 13
Gabés 815
Gábor 245–246, 250, 541, 551, 653–654, 665, 695, 777
Gabor 387, 391
Gaborone 101
Gadolin 259
Gaferut 536
Gagie 676
Gago 680, 777
Gaina 551
Gaisenkalns 445
Gajah 503
Galante 309–310
Galbeed 721
Galdhopiggen 599
Galicia 50
GALILEO 172, 733
Galinds 445
Galizien 291
Gallery 162
Gallia 263
Gallinhas 460
Gallipoli 487
Gama 559
Gambaga 803
Gambian 279
Gambie 267
Gambier 267, 269
GAMIT 116
Gamla 764–765
Gampo 89
Gan 184–185, 505–506
Ganālē 722
Gandajika 184–185
Gandhi 355

Gangaw 125
Ganges 67, 355, 569
Gangkar 89
Ganja 283
Gansu 172
ganwa 129
Gao 460, 509, 587
Garate 557
García 199–200, 211
Garcia 629
Garde 527
Gardner 424
Gargur 514
Garmaniski 446
Garoua 142
Garrison 63, 106, 299, 543, 663, 705, 843
Gary 424, 438, 609
Gaston 199
Gata 203, 401
Gaucho 31
Gaul 263, 287
Gauss-Krüger 234, 246, 403, 437, 505, 864
Gauss-Schreiber 139, 309, 494, 784
Gausz 292
Gautier 820
Gauß 785
Gavan 670
Gavish 382, 390, 407–408
Gavrilo 97
Gawler 578, 580
Gaza Strip 233, 379–380, 559, 883
Gbande 459
GCGD 154
GCPs 635
GDA 45
Gdansk 638
GDBD 110
GDM 109, 502–503, 529
GDP 299, 731
Geba 325
Gebrauchshöhe 769
Gedaref 752
Gedeonov 412, 437
Gedetarbo 461–462
Geisenkalns 445
Gelain 521
Gelephu 90
Gellérthegy 345–347, 694
GEMT 860
Gen 356, 438
Genedetto 203
Geneina 162
Gener 431, 648
Général 13, 77, 816, 831
Générale 93, 322
Generalna 491
Generalstab des Heeres 565, 638, 895
Generalstabe 764, 831
Generalstabens 215
Generalstabskarten 764
genesis 421
Geneva 634, 769, 774, 811

Genevieve 78
Genghis Khan 371, 411, 435, 547
Gennaro 385
Genoa 543
Genoek 361
Genoese 543
Genova 385–386
Gentil 242, 276
Genuk 362, 364
geocartography 25, 55
geoceiver 855
Geodaetica 540
Geodäsie 217, 299, 328, 387, 390, 477, 822, 830
Geodesi 363, 391
Geodesía 583
Geodesia 303, 329, 733, 777
geodesic 301, 398, 578, 641, 871, 878
GEODESICA 339, 348
Geodésica 150, 211, 297, 330, 800
Geodesie 209, 557, 710
Géodésie 816, 831
Geodèsiens 495
géodesique 877
Geodetica 10, 55, 386
Geodézia 551, 695, 777
Geodezice 540
Geodezii 658
Geodezija 25, 55
Geodezijos 474
Geodezist 436, 788, 832
Geodeziya 5, 55
Geodimeter 40, 188, 342, 368, 579, 685, 745, 797
Geodætisk 215, 351, 390
Geografia 105, 339
Geografía 532
Geografic 540
Geograficzny 638
Geografie 694, 779
Geografiske 599
Geografski 490, 895
Geographia 387, 391
geographically 23, 195, 669, 835
Geographique 690
Gèographiques 495
geoidal 559, 566, 860
GeoidUru 860
Geoinformatikai 437
Geoinformation 113, 131, 172, 210, 821, 830
Geológico 105, 225
Geologie 694, 779
Géologique 184, 210, 887, 893
geologist 93, 745
Geomatrix 399
Geometrique 816
geometriska jordeböcker 763
geometry 651
Geophysica 540
geopolitical 657
geopotential 247
GeoRepository 90
Georg 792

Georgetown 153, 328, 503
Georgian 36, 283–285
Geosentrik 363, 391
geosphere 94
geostrategic 613
GeoTrans 318, 684, 804
Gerais 105–106
Geral 106, 641
Geraldton 45
Gerhardus 199
Gerlachovsky 709
Germain 220, 263–264, 524, 815
Germanic 287, 573, 641
Germanizing 713
Gerritz 669
Getae 651
Gezira 751
Gezo 83
GG 313, 437, 718, 777
Ġgantija 513
GGIM 829–830
GGN 314
GGOS 639
GGRS 302
GGTN 314
Ghantut 844
Gharsah 815
Ghaut 553
Ghawdex 513
Ghaznī 3, 67
Ghibelline 543
GHOST 860
Ghudämis 815
Ghuzayyil 465
Gianniou 302, 329
Gibraltar 273, 299, 557, 574, 641, 731
Gifu 402
Giggiga 249
GIJNA 490, 895
GIL 59, 791–792, 830
Gilan 367
Gilbert 421–422, 827
Gildessa 219
Giles 485, 652–653, 665
Gilgamesh 63
Gilgit 614
Gillis 759, 777
GIM 101, 131
Gimie 741
Gio 461
GIPSY 800
Gisaka 130
Gitter 208, 290, 447, 638
Giuba 722
Giulia 386
Giumbo 722
Givry 523
Gjata 10, 54
GJESRM (*Global Journal of Engineering Science and Research Management*) 86, 130
Gjuhëzës 10
glacial 73

Gladstone 43
Glamorgan 850
Glasgow 497, 562, 571, 603
Glasher 849
Glavica 98
Glavnoye 658
Glendon 398, 408
Glenomara 375
GLOBK 116
GLONASS 172, 733
Glusic 10, 50, 55, 98–99, 131, 196, 210, 346, 387, 413, 437, 491, 660, 665, 671, 694, 777, 788, 830
GMT 421
Gnaviyani 505
Gnesen 291
GNIS 635
Gnomonic 32, 698
GNSSnet 348, 695
Gobabis 23, 566
Gobad 219
Gobé 242
Gobi 547
Godard 629
Godfrey 773
Godin 229
Godolphin 81
Godwin 613
Goëland 524
Goenoeng 361
Gogland 259
Goita 510
Gola 459–461
Golan 773
Goldsmid 4
Goldwater 306
Golem 9, 487
Golfe 219
Gollenberg 291
Golmberg 196, 208, 289–290, 368
Gomez 165
Gómez 225
Gon 765
Gonâve 333
Gonave 334
Gonçalves 642–643
Gonzalez 531
Goob 722
Gora Ayribaba 825
Gora El'brus 551, 657
Gorbachev 53, 436, 657, 787
Gorée 389, 689–690
Gorgas 617
Gorge 81, 791
Gori 283
Gorkha 569
Gorno 787
Gorohudzir 412
Gorzow 638
Gosling 485
Gospavic 695, 778
Gotabaya 735
Göttingen 290–291

Götzer 291
Gouga 187
Gougenheim 220, 268, 301, 690, 877–878
Goulash 345
Goulette 817
Goulter 578
Govin 452, 775
Gozo 513
GPO 90, 130
GPRS 861
GPS-campaign 352
Graciosa 642
Gradblätter 764
Gradiste 196, 896
Gradmåling 215
Gradmessung 473, 694
Graeme 580
Gran 93, 175, 625
Granada 583, 731, 873
Grand Anse Rum Distillery chimneys 305
Grandi 40
Granger 327
Grappo 609
Graundsers 599
Grauspitz 469
GRAV 856
gravimetrical 357
gravitational 726
Gray 756
Graz 49, 291
Grebo 459
Greboes 461
Green 234, 502, 705, 785, 811, 843
Greenland 215, 351, 599
Grégoire 493
Gregory 35
Greifswald 291
Grelhas 643
Grenadine 305, 667, 743–746
Grimaldi 543
Grimaldo 543
Grisslehamm 259
grist 306
Grita 874
Griži 10
Gröditzberg 291
Grodno 637
Grondingligting 456
Groote 43
Grootendorst 71, 131
Grossa 106
GROSSDEUTSCHES 291–292, 329
Grossglockner 49
Grosso 105–106
Grousson 309, 329
Großenhain 290–291
Gruda 10
Gruetzmacher 306, 329
Grundy 44–45
Grupo 106
Grye 689
GSGS 205, 255, 416, 466, 514, 883

GSI 402–403, 408
Guadagasal 622
Guadalcanal 717–718
Guahonels 318
Guaira 39, 873
Guam 273, 301, 313–314, 329, 535
Guangdong 172, 783
Guantanamo 199
Guaraní 625
Guaratinguetá 106
Guardia 385, 618
Guasopo 622
Guatemala–Honduras 317, 338
Guatemalan 317, 319
Guavitu 860
Guayamouc 333
Guayaquil 230–231
guberniyas 473
Guelph 543
Guerara 13
Guercif 556
Guerra 583, 680, 777
Guerrero 532
guerrilla 337, 629, 726, 799, 835
Guggisberg 295
GUGiK 638–639
GUGK 658
GUGKK 115
Guia 483–485
Guido 149
Guilia 895
Guillaumme 301
Guinea-Bissau 321, 325
GuineaLiberia 701
Guinean 321, 702
Guizol 556
Gulatee 357, 390, 796, 830
Gulshan 68
Gumma 402
Gunars 448, 478
Gunnar 353, 390
Gunnlaugsson 351
Gunter 853
Guntō 401, 403
Gunung 362, 364, 501, 503
Gupta 355
Gur 411
Gurkha 570
Gurley 811
Gurten 470, 770
Guru 89
Gustav 763
Gusterberg 49, 207, 291, 709
Gustin 617
GUX 718
Guy 6, 123, 126, 131, 797
Guyana x, 105, 273, 327–329, 755–756
Guyot, *M. le Capitaine de Frégate* 220, 242
Guzmán 237
Guðbrandur 351
Gwero 891
Gwibi 891

Gwynn 752
Gyamfi 297, 329
Gyumri 36

Haa 505
Ha'amonga 807
Haan 755
Ha'apai island group 807–809
Haavei 268
Habay 477–478
Habsburg 49, 97, 195, 541, 694, 713, 779
Hachirō 401
Hachyop 428
Hadendowa 752
Hadou 219
"hafa tata manoa" 313
Hafiz 773
Hagar 210, 257, 381–382, 390, 536, 812, 831
Hague 809, 884
Hahaia 180
Haidian 172, 210
Haïfa 774
Hailey 416, 438
Hainaut 78
Haïphong 877
Haïphuc 878
Haïti 225, 331, 333–334, 348
Haiti 333
Haitian 225
Haïtian 333–334
Haiwan 502
Haka 125
Hakahetau 270
Hakka 783
Hakodate 402
Halgastota 736
Hallasan 427
Halmahera 363
Haltiatunturi 259
Hamad 647
Hamadan 367
Hamani 587
Hamarin 372
Hamburg 290, 292, 532
Hamdoun 373
Hamidi 373
Hamir 647
Hammarsjon 763
Hammerfest 259, 600, 792
Hammurabi 367
Hampden 891
Hampshire 225, 245, 669, 709, 849
Hamrin 372
Hamsteen 600
Hanavave 270
Handbuch 292, 447, 788, 831
"hangover soup" (bouillabaisse) 873
Hanik 638
Hann 691
Hannah 257
Hanno 460
Hannover 290–291, 764

Hannoverska 765, 830
Hanoi 877–879
Hanover 792, 830
Hansan 428
Hanseatic 245
Hansen 477, 670
Hanta 373
Hao 268
Hapsburg 49, 207, 345, 769
Haque 68, 130
Haram 405
Harare 891
Harbel 461
Harbin 659
Härer 219
Hari 4
Harmanli 117
HARN 676, 855
Harnack 109
Harrar 219
Harris 396, 408
Harrison 527, 577
Harry 460, 479
Hart 850
Hartebeeshoek 102, 728, 760, 888, 893
Harvey 59, 398, 408, 461
Harzard 595
Hasakah 773–774
Hashemite xi, 393, 405–407, 607
Hashmie 123–124
Hassan 791
Hassen 250
Hassetché 773, 775–776
Hassler 853–854
HAT 89, 829
Hatuey 199
Hauptturm 291
Hausa 509, 587, 591
Havana 153, 199–200, 210
Hawaii 313, 421, 718, 777, 829, 853
Hawaiian 267
Hawkes 578–579
Hayden 373, 390
Haynam 428
Haytham 608
Hazel 793
headlands 465
headwaters 435
Hearst 669
Heatlie 727
Heaviside 220
Hebert 23
Hebrides 849, 869
Hébrides 869
Hebs 870–871
hectares 679
Hedgehog 600
Hedjaz 406
Hedkouk 373
Hedland 45–46
Heemskerck 807
Heerbrugg 188, 510, 517, 637, 690, 752, 849

Heeres 447, 565, 895
hegemony 288
hegy 345
heighted 114, 205
Heighting 483–484, 609, 760
Heilongjiang 669
Hein 297, 329
Heinrich 264, 587
Heinrichsthal 291
Heinz 290, 292, 329
Hejaz 406
Hejäz 685
Heketa 807
Helen 536
helio 124
heliographs 892
heliotrope 821
Heliport 800
Heliyon 16, 54
Hellenic 35, 273, 301–303, 329
Hellenistic 63, 431, 750
Hellville 493–494
Helmand 4–5
Helmertturm 196, 290, 368
Helmut 59, 162, 517
Helsinki 208, 259–260, 709
Helvetians 769
Helwân 233–234, 751
hempen rope 356
Hemus 116–117
Henan 172
Hendrik 562, 602
Hendrikz 102, 456, 479, 892–893
henge monuments 847
Henri 191
Henrique 483, 641
Hensbarrow 850
HEPOS 302–303, 329
Herat 5–6
Héraud 877
Herceg 97
Hercheretue 269
Hercule 543
Heredia 191
Herero 565
Heritage 105, 215, 405, 509, 608
Herman 883, 896
Hermannsdenkmal 291
Hermannstadt 651–652
Hermon 773
Hernán 531
Hernando 93, 531
Herres 208, 290, 638
Hertsford 850
Herz 483
Herzogthums 292
Herzogtum 290, 292
Hessen 289–290, 292
HessenDarmstadt 291
Hetau 622
Hetmanate 839
Heungmuk 429, 438
Heuvelink 562, 602
Hewanorra 741

Hey 170
hidalgos 395
Hidrográfica 24, 325
Hidrográfico 485
Hidza 99
Hierro 731
hijacked 657
Hijau 503
HIL 59
Hildebrand 216, 670, 765
Hildebrandt 114, 821
Hilki 821
Himalaya 3, 355, 357, 569, 726
Himalayan 123, 562, 569–570, 602
Himyarite 883
Hindenburg, Paul von 288
Hindu 3, 67, 109, 501, 613, 750, 795
Hinduism 569, 613
Hindukush 3
Hindustan 356
HIRAN 200, 327–328, 684–685, 756, 827
Hiroshima 402, 670
Hispania 641
Hispanic 105
Hispaniola 225, 333
Hitler 9, 207, 288–289
Hittite 819
HIV 101, 759
Hiva 267–268, 270–271
HIW 59
Hiyama 402
Hizballah 451, 773
Hjörsey 352
Hjortö 260
HK 341–342
Hkakabo Raze 123
HKPD 342–343
Hluang 797
Hmamouchi 557
HMGS 301
HMS 204, 280, 422, 697–698, 807
Hó 485
Hoa 877, 879
Hogsty 59
Hoh 547
Höhe 490
Hohen 289
Hoï 878
Hojeda 39
Hokitika 577–578, 580
Hokkaidō 401–402
Hokkaido 669
Hólar 351
Holdick 357
Holeywell 671–672, 778
Holland 43, 361, 460, 514
Hollingbourne 850
Holloway 879
Holmes 518
Holocaust 289
Holtery 306, 329
Hombori 509
Homert 291

Hominid 361, 497
Homo erectus 361, 791
homogeneous 99, 671, 808, 844
Homo sapiens 361
Homo soloensis 361
homothétie 265
Homs 773–774
Hon-Matt 878
Hondo 94, 107, 166, 230, 328, 337, 630, 874
Honheo 878
Honiara 717–718
Honoré 543
Honshū 401–403
Hoofdirektoraat 456
Hooghly 67
Hooijberg 574
Hora 839
Horatio 27
Horisha 783
Hormuz 372, 843
Horobow 854
Horrebow 473
Hortiach 487–488
Hospicio 106
Hou 485
Houël 309
Hough xxiv, 229–230, 518–519, 859
Hounslow 850
Houston 170, 671, 684
Hovas 493
Hoverla 839
Hoxha 9
Hrada 73
Hrast 98
Hristow 114–116, 638, 653, 776
HRL 59
HSGN 839–840
Hsiu 169
HTRS 197, 302–303
Huahine 269
Huan 169
Huancaramis 630
Huanghai 171
Huanshanchzuiza 659
Huascaran 629
Hudaydah 884
Huetz 859
Hugart 398
Hugh 617, 643, 759, 777
Hughes 309, 318, 337
Huirangi 578
Huka 268, 271
Hulagu 371
Hulu 503
Humala 629
Humarrasuh 220
Hümayun 819
Humboldt 873
Hume 836, 865
Hunan 169
Hundsheimer 50, 196, 208, 491, 709, 896
Hung 429, 438, 785

Hunga 809
Hungarica 540
Huns 435, 539
Huntingdonshire 849
Huntley 416
Huon 43
Hupeh 169
Huqf 608
Hurrian 35
Hurst 669
Husbergoen 599
Hussein 179, 371, 405
Huth 599
Hutu 129, 663
Huynh 879
Huyten 547
Huzi 784
Hvannadalshnúkur 351
HVRS 197
Hwang 785, 832
Hyderabad 505
Hydro 126, 827
hydrographes 220
hydrographically 215
Hydrographie 268
Hydrography 205, 220
hydropower 441, 559
Hyksos 749
Hyōgo 402
Hyperbolic 188, 257
HYTS 302, 329
Hønefoss 601

IAGBN 247
Iaşi 540
Ibadis 607
Ibaraki 402
Ibarrarte 811, 831
Iben 536
Ibenga 187
Iberia 283
Iberian 20, 641, 731–732
IBGE 105
Ibibio 591
IBM 342
Ibo 591
Ibrahim 647
IBS 804, 831
Iburi 402
ICBM 839
Ice 351, 373, 671, 683, 853
Iceland 349, 351–353, 599
Ichara 669
ICJ 338, 587
ICN 199
ICPs 635
ICSM 45–46, 54
ICT 635
ID 302, 329, 362, 364, 672
Idabelle 461
Idi Amin 835
Idlib Province 774
Idoûkâl 587
IDPs 751

Idriss 162
Idrizi 9–10, 54, 491
Ifalik 536
IFOR 97
IGAC 175–176
IGB 120
Igbo 591
IGCR 192
IGE 733
IGGU 280
Iglesia 874
IGNB 77
Igneri 811
IGP 642
IGS 74, 209, 732
iguana 741
Igurua 793, 835
IHB 680
IICT 149–150, 211, 800, 831
Ijaw 591
Ijill 523
Ijssel 574
Ikaros 431
IKBDC 432, 438
IL 265, 382
Ile 389
Ilha 483, 641–642, 680, 777, 779
Ilidža 97
Iligan 633
Illimani 93
Illyrian 9, 97, 693, 713
Illyricum 195
Ilocos 635
Iloilo 633
Ilorin 592
Iman 659
Imara 381, 407
IMF 301
Império 105
Imran 614
Inagua 59
Inca 93, 165, 629
Inchauspe 32–33
Inchon 428
Indus 3, 63, 161, 355, 607, 613, 615
INEGI 532
INETER 562, 584, 603
Informática 532
Informe 318, 329, 562, 584, 603
Ingenieros 625
Ingénieur 268
Ingus 448, 478
Inha 428
Inhambane 559
Inönü 819–820
Insaaf 614
Inseza 891
INSPIRE 717, 841
Institutul 540
Instytut 638
Inthanon 795
Inyanga 560
Inyangani 891
Ioan 539

Ionian 385
ionosphere 733
Ioüanalao 741
IPC 608, 631, 775
Ipire 874
Ipiresía 301
Ipoye 759
Iquique 165
IR 153, 474
Irapa 874
Irazú 191
IRENET 377
Irina 448, 478
iroij 517
Irtysh 435
Isaksson 601
ISCA 87, 130
ISCARJRS 130
Iselta 732
ISG 44
Ishikari 402
Ishikawa 402
Ishpushta 5
ISIS 787
Iskeul 815
Isla 165, 175, 242, 584, 717
Íslands 351
Islay 849
Îsles 878
Ismail 125
Ismet 819–820
Ismoili 787
ISN 352, 706
ISNET 352
Isoka 887
ISPRS 549, 788, 831
Issas 219, 722
Istria 895
Isufi 11, 54
Itararé 106
Iti 267, 269
Itinerária 641
Itinerario 731
ITM 377
ITRS 74, 429, 879
Itthi 797, 831
IUGG 171, 347, 654, 665, 672, 690, 770, 778
IVA 271, 358
IVB 358
Iveria 283
Ivo 97
Iwate 402
Izmayil 839
Izmir 820
Izolde 448, 478
Izundu 562, 593, 603

JA (Japanese Artillery) 402
Jaakkola 328
Jabal 63, 372, 405, 607, 683, 883
Jack 553, 669, 792, 870
Jacobus 727
Jácome 641

Jacques 229, 878
JAD 398
Jäderin 283, 362, 412–413, 489, 491, 578, 659, 788, 863
Jagladzik 488
Jahre 291
Jaipur 355
Jakarta 363, 391, 799
Jakob 259
Jaldēsa 219
Jaluit 517–518
Jamahiriya xi, 439, 465, 467
Jamaica 153, 393, 395–399, 408, 743
Jamaican 81, 398
Jamanota 39
Jamilova 446
Jammeh 279–280
Jammer 622
jamming, radio 342
Jammu 614
Jampito 630
JAN 55, 599, 601, 639, 643, 658
Janeiro 106
Janis 448, 478
Janos 345
Janszoon 577
Januario 483, 485
Jaraguá 106
jarib 356
Järjestelmä 260
Jarnefeldt 113
Jarosław 639, 643
Jarre 521
Jatvings 445
Jawaharlal 355
Jaya 361
Jebiga 803
Jebigu 803
Jeddah 685
Jedki 600
Jednotne 208
Jednotné 710
Jelenia 638
Jelgava 446
Jemen 883, 896
Jemo 518
Jenghiz 825
Jengish 435
Jensen 217, 226, 609
Jerantut 502
Jericho 379, 407
Jervis 43
Jesuit 81, 169, 625
JEU 816, 830
JGD 403
Jiangxi 172
JICA 322
Jidetaabo 461–462
Jidka 722
Jigme 90
JILS 401–402, 784
Jima 401, 403
Jin 429, 438
Jingang 171

Jinmen 785
Jinna 459
Jinotega 583
Jinubi 684
Jipe 415
Jo 591, 725
Joao 23
João 325, 329
Jobin 125
Jogorku 436
Johore 501–503
Jöns 259
Jordan-Eggert 447
Josefov 50, 208
Josephinishe 49, 345
Josip 195, 693, 713
Jouik 524
JPL 800
JTM 407
JTSK 208–209, 710
Juan 191, 199, 225, 229, 583, 859
Jubani 10
Jugoslovenske 490, 895
Juillet 290, 328
Juliana 755
Jumare 448, 478
Jumgar 435
Junnan 125
Junta 325, 329, 559, 587
Juozapine 473
Jura 769, 849
Jurien 45
Justander 259
Justino 483
JWH 519, 681
Jyväskylä 259

Ká 485
Kaabu 279, 325
Kaafu 505
Kaarappu 536
Kaart 755
kabaka 835
Kabala 703
Kabé 321
KABORE 119
Kabosi 245
Kabouba 219
Kabul 5–6, 55
Kabula 5, 55
Kabwe 887
Kaczynski 788, 831
Kadar 345
Kadhimi 371
Kadiköy 114
Kaf 406
Kagame 663
Kaganate 411
Kagawa 402
Kagera 129–130, 663–664
Kagitumba 664
Kagoshima 402
Kai 783
Kaikyō 402

Kailahun 702–703
Kainarji 114
Kainji 591
Kaitob 727
Kajaani 259
Kakheti 283
Kakitumba 664
Kala 825
Kalahari 101, 565, 727
Kalan 569
Kalemyo 124
Kalevala 259
Kalimantan 362, 364
Kalinago 223
Kaliningrad 473
Kalinka 448, 478
Kalisz 638
Kalmyks 435
Kalnas 473
Kalomo 887
Kalsale 869, 871, 879
Kaltenborn 291
Kamal 844
kamar 179
Kamaran (Island) 884
Kamerun 139
Kamikawa 402
Kamina 803
Kamiyse 373
Kammanankada 596
Kammuri 402
Kamp 671
Kampong Chhnang 136–137
Kanagawa 402
Kanaker 37
Kanas, Tony 869, 871, 879
Kancelar 207
Kanchan 569
Kanchenjunga 355
Kandahar 6
Kandavelu 736
Kandavu 255
Kandawala 736
Kandilli 820
Kandu 625
Kanem 157, 161, 587
Kangan 368
Kangari 446
K'anghsi (Emperor) 169
Kangili 835
Kanglung 90
Kangwha 428
Kanin 386
Kankan 322
Kano 427, 509, 592
Kanou 401
Kanoyama 427
Kansas 59, 73, 200, 334, 398, 532, 618
Kantara 204
Kanton 424
Kanuri 591
Kanyaru 129, 663
Kao 807
Kaolack 690

Index

Kapes 373
Kapingamarangi 536
Kapiti 579
Kar 437
Kara 436, 490
Karabakh 36, 53
Karachi 615, 643, 797, 830
Karadar 435
Karafuto chō 402, 669
Karakitai 411
Karakol 488
Karakoram 614
Karakorum 3, 547
Karamea 578–579
Karamrani 835
Karbala 372
Kardzhali 117
Karelian 259
Karen 829, 831
Kariba 891
Karisimbi 663
KARJ 117
Karlowicz, Treaty of 819
Karl-Rothe 328
Karma 884
Karmah 749
Karmek 114
Karmyz 413
Karnga 459, 478
Karpas 204
Karratha 45
Karte 540, 883, 896
Karten 93, 130, 447, 478, 764
Kartengitter 216, 226
Karthala 179
Kartli 283
Kartografi 658
Kartográfia 551, 695, 777
Kartografie 209, 710
kartografija 25, 55
Kartografijos 474
Kartografiya 5, 55
Kartvelian 283
Kartverk 601, 764
Kartverket 562, 602–603, 764, 831
Karukera 309
Kasai 402
Kasama 887
Kashmir 614
Kashta 750
Kasim 821, 831
Kasonja 98
Kassala 752
Kassel 290–292
Kastri 659, 670–671
Kasym 411
Katanga 184–185, 210, 887–888, 893
Katastra 491
Katastralni 207–208
Katastrální 710
Katerina 448, 478
Katerini 489
kath. Kirche 291
Kathmandu 569–570

Kato 116
katorga 669
Katowice 638
Katsampalos 302, 329
Kattegat 215
Katwe 835
Kauernick 291
Kaunas 473
Kaundy 411
Kavalla 115, 488
Kavieng 622
Kaynardzha 114
Kazak 411, 436
Kazalinsk 412–413, 863–864
Kazan 401, 403
Kazi 68, 130
Kazuo 403, 408
KCB 376
Keb 502
Kebir 556
Kebnekaise 763
Kedah 503, 706
Kediet 523
Keeling 45
Kei 726
Keila 622
Keita 509
Keith 396, 490
Kekes 345
Kelantan 503, 706
Kelburn 579
Kely 493
Kemal 36, 367, 373, 775, 821
Kemmel 77–78
Kemmuna 513
Kempshot 395
Kengesh 436
Kengtung 123–126
Kennedy 185, 210
KENREF 418
Kent 847, 850
Kenton 123
kepi Gjuhëzës 10
Kerch 839
Kerek 380
Kermadec 577
Kermān 368
Kern 670, 731, 769
Kérouané 322
Kertau xxiv, 502, 706–707
Kester 78
Kesztej 347
Keti 387, 391
Keuī 451, 774
Kevin 71, 131
KG 503, 817
KGB 246, 659, 671
KGD 429
Kh 864–865
Khabarovsk 659
Khalasis 124–125
Khali 684–685
Khalifa 63, 647
KHAMA 101

Khanate 411, 435–436
Khanka 658
Khao 797
Kharijite 607
Khashi 89
Khaymah 843
Khayr 411
khedive 751
Kherson 839
Khesar 90
KHG 869
Khieo 797
Khiva 863
Khmer 135, 441
Khmères 136
Khoi 565
Khoisan 23, 161, 455
Khokanda 788, 863
Khomeini, Ayatollah 367
Khor 35, 219, 684
Khorat 795
Khorazm 411
Khrebet 3
Khrushchev 657
Khuzestan 367
Kiangsi 169
Kicharere 418
Kicherere 835
Kielce 638
Kieta 622
Kiev 73, 839–840, 865
Kievan 73, 839
Kigali 663
Kigoma 129, 498
Kihoko 537
Kijabe 416
Kilambo 184
Kili 518
Kilimanjaro, Mt. 415, 791
Kimak 411
Kinabalu 501
Kinan 537
Kincardine 849
Kinki 403, 408
Kinmen 785
Kinross 849
Kintamo 183
Kipchak 411
Kirbbisi 446
Kirche 291
Kirchner 31
Kiril 116
Kiritimati 422, 424
Kirkcudbright 849
Kirkuk 372–373
Kirun 784
Kirurumu 664
Kisaka 130
KISM 417
Kismaayo 722
Kissaga 130
Kissi 459, 461, 702
Kistna 356
Kita 184, 403

Kitaka 836, 865
Kitanga 184
Kitchener 203–205, 380
Kitts 27, 223, 553, 667, 739, 779
Kivu 663
KKJ 260–261
Klafter 49, 345
Klimsch 817
KLM 755
Kloos 290, 292, 329
Kloster 671
Kluang 502
KMT 783
Knevichi 659
Knopfmitte 291
Koba 322
Kobayashi 403, 408
Kōbe 402
Kobokutei 785
Kobryn 637
Kōchi 402
Koepang 362
Kofarnihon 787
Kogo 242
Koh 3
Kohalikud 246
Koiaris 622
Koïla 321
Koirala 569
Kokalla 510
Kokand 435
Koli 825
Kolkhida 283
Köln 291
Kolomoytzeff 407
Komadugu 588
Komaritim 364
Kongo 23, 187
Kongsvinger 599
Königliche 565
Königseck 490
Konigstein 565
Konin 638
Koordinačių 474
Koordinat 5, 55
Koordinaten 289, 291, 447
Koordinatenkartei 447, 478
Koordinatsystem 765
Koordnatenkartei 447
Kopar 197
Kopciowka 637
Kopec 64, 130
Kopiec 291
Kora 13, 816
Korab 9, 487
Korabit 9, 487
Kordofan 751
Korkuteli 820
Koro 255
Korodolo 362
Korppoo 260
Korpus 245, 658
Korsakov 671
Kort 215

Kortlægning 351, 390
Kortright 703
Kosciusko 43
Koshizan 402, 783
Kosrae 535–536
Koszalin 638
Kota 503
Kotonou 275
Kotos 487
Kotromanić 97
Kotsakis 302, 329
Kotzev 115–116, 131
Kouilou 184
Koulib 407
Koumolou 86, 130
Koun 803
Koundjouare 803
Kountché 587
Kourou 119
Kousséri 161
Kovács 695, 777
Kowloon 341
Kpelle 459, 461
Kra 124, 135
Kraaij 461, 479
Krajina 195
Krakiwsky 146
Krakow 638
Kraljevine 490
Kran 488
Krapina 195
Krasnoy 658
Krasnoyarsk 864–865
Krayenhoff xxiv, 77, 573–574
Krechevo 490
Kremsmünster 49, 291
Kriegs 447, 478
Kriegskarten 20, 732, 778
kriging interpolation 642
Krikščiūnas 473, 478
Krimberg 49, 291
Krio 701
Krishna 356
Kriss 448, 478
Kristiania 599–600
Kristianstad 763
Krkonose 207
Krogh 759
Kroměříy 208
Kronshtadsky futshtok 547
Kronstadt 283, 474, 547, 652–653, 659, 840
Krosno 638
Krovak 208, 709–710, 770
Kru 459, 461
Kruger 725
Krusenstern 421, 669
Kryński 639, 643
Ksara 452, 775
KT 739
KTIMATOLOGIO 302
Kuala 502–503
Kuan 485
Kublai 123, 547

Kubu 503
Kučerina 98
Kuching 109, 503
Kuchuk 114, 367
Kucklinsberg 291
Kuczynski 629
KUDAMS 432
Kufa 371
Kuh 4
Kūh 367
Kuhtek 788
Kujustendja 113–114
Kuk 4, 551
Kulkigan 446
Kumamoto 402
Kumar 166, 230, 442, 478, 570, 736, 777
Kumba 459
Kumentaitis 474, 478
kumr 179
Kununurra 45
Kunyo 702–703
Kupang 362, 799
Kür 53
Kur 413
Kura 35, 53
Kuraymah 749
Kurdish 371, 774
Kurfurstentum 290
Kuria 422–423
Kuril 669
Kurtics 346, 652
Kusaie 535–537
Kush 3, 613
Kushiro 402
KUST 116
Kustendil 116
Kutch 614
Kutek 436
Kutushev 25, 55
Kuwamura 169, 210
Kuyi 669
KVT 245, 658
KWA 518
Kwaio 717
Kwajalein 517–518
Kwaje 703
Kwame 297, 329
Kwando 565
Kwantung 547
kwanza 23
KwaZulu-Natal 455
Kwon 429, 438
Kyeongbuk 428
Kyeonju 428
Kyichu 89
Kyivan 657
Kyōto 402
Kyrenia 204
Kyrg 437
Kyrgyz 409, 435–437
Kyu 429, 438
Kyūshū 401
Kyustendja 113, 540, 821

Kyustendyi 540, 652
Kyzyl 436, 788, 863
Københavns 351

Laamu 505
Labail 523
Laborie 741
Labourdonnais, Mahé de 697
Labuan 109, 503
LaCaille 527, 726
Lac Asal (Djibouti) 219
Lac Cohoha (*Lac Cyohoha-Sud*) 129, 663
Lac Rweru (*Lac Rugwero*) 129, 663
LAD 518, 807
Ladner 261, 562, 596, 602
Ladoga 259
Lae 518
Lafargue 241, 275–276
Lafia 592
Lafler 432, 438
Lafoole 722
Lafou 702
Lagarto 33
Laghouat 13
Lagi 10
Lago 86, 225, 583
Lagon 270
lagoon 85, 803, 827–829, 859
Lagrave 527
Laguna 531, 635, 859
Lahore 614
Laibach 49, 291
Lail 373, 390, 407–408, 796, 830
Laken 828
lam 90
Lambaréné 275
Lambasa 256
Lamberti 246
Lambton 356–357, 796
Lamlam 313
Lammerfjord 215
Lammert 675
Lamotrek 537
Lamy 157, 162
Lan 441, 676
Lanagza 489
Lanark 850
Lancashire 461, 849
Lancelin 45
Landesaufnahme 289–290, 565, 637–638
Landesdreiecknetze 290
Landestopographie 769–770
Landestriangulation 290, 292
Landesvermessung 291–292
Landesvermessungsamt 208, 291
Landgate 529
Landmælingar 351
Landt 577
Langaza 487–488
Langdon 43
Lange 361, 658
Langeland 215

Langr 659, 670–672
Langschoß 291
Lanka 67, 357, 667, 735–736, 779
Lantau 341
Lantmäteriet 90, 131, 763–766
Lapach 850
Lapid 379
Lapita 717, 807
La Porte du Lac de Bizerte 815
Laprakë 10
Larcom 376
l'Armement 268
l'Arpent 697
Lars 764, 766, 830
Larsgaard 105, 131, 532, 562, 583, 603
Larut 501, 503
Lassave 220
LAT 764, 803, 809, 811–812, 829
Lata 675
Latham 811, 830
latitudinal 85, 94, 142, 166, 264, 785
L'Atlantique 524
Latvian 445–448, 478
Laurent 689
Lausanne 819
Lautoka 256
Lawley 879
Lawrence 71, 380
Lazarev 671
LC 154
Leach 808, 830–831
Leatherdale 685, 778
Lebail 523–524, 689–690, 777
Lebedev 113–114
Lebombo 561
Lecce 9, 385
Ledersteger 99
Legaspi 633
legatee 614
Legion, French Foreign 555
Legnica 638
Leica 514, 528, 707, 844
Leicester 849
Leiden 573
Leif 351
Leigon, French Foreign 295–297
Leipoldt 727
Leipzig 116, 299, 328
Leith 850
Lelo 387, 391
Lemberg 291
LEMN 116
Lemnos 116
Lempira 337
Lemuta 675
Lenin 36, 657
Leningrad 658
Leon 583
León 641
Léon 815, 830
Leonardo 860, 865
Leopold 183–184, 187
Leopoldsberg 9, 49, 98, 345
Léopoldville 129–130, 183, 663–664

Leppan 796, 831
Lequi 800
Lerida 731
Lerrnagagat 35
Leste 106, 361, 781, 799–800
Leszno 638
Leticia 176
L'Etoile 220, 267
Lettgallen 445
Lettre 641
Leuen 518
Levato 800
Levé 242, 250
Leviathan 379
Lévrier 523
Levu 255–257
Ley 31, 93, 192, 230, 250, 255, 257, 840
Leyte 633
LGM 529
Lhaviyani 505
l'Heure 551, 625, 695
Lhomon 89
Lhotshampa 90
Liamuiga 739
Lianhuachi 172, 210
Liaotung 669
LIB 642
Liban 451, 453, 479, 776, 831
Libert 78
"Libertator, El" 873
Libia 466
Libiesi 445
Libreville 157, 241–242, 275–276
Libyan 439, 465–467
Licchavi 569
Lich 878
Licuses 20, 732
LIDAR 205
Liepaja 446
Liesganig 49, 345
Lietvos 474
Lieu 473, 574, 878
Lieve 573–574
Lifau 799
Lifou 271
Lifuka 807
Ligurian 385, 543
LIK 518
Likiep 517–518
Likoma 498, 560
l'Île 220–221
Lima 191, 629–630
Limassol 203
Limburg 78
Limes 287
Limpopo 101, 560, 891
Lindesnes 599
Lindis 578, 580
l'Indochine 135, 441
Lindsay 527–528
linen 853–854, 895
Linguetta 10
Linh 877

Linnade 246
L'institut 510, 690, 778
LINZ 579–580
Lipovica 98
LIS 626, 643, 744, 778
LIT 59
Litea 257
Littoral 49, 157, 220, 386
liurai 799
Livingstone 497, 560, 791, 887
Ljubljana 49
LKS 447–448, 474
Llangeinor 850
Llanos 175, 873
Llewellyn 424, 437
Lloyd 442, 478
LMÍ 351–352
LN 469, 770
LNEC 297, 329
LNG 884
Lo 60, 137, 170, 242, 250, 321, 658
Loanda 24
Loango 187
Lobato 150
Lobaye 187
Lobito 25
Lochmaben 43
Locodjoro 390
Lodz 638
Loema 187
Logan 145, 513–514
Logantoeng 361
Logwood 81
Loi 509
Lolland 215
Lom 114
Loma 230, 459, 701, 812, 873–875
Lombok 362
Lomé 804
Lommel 77
Lomza 637–638
Lopatin 669
Lopez 242, 275–276
López 633
LORAN 188, 601, 691
lore 14, 35, 184, 556
Loro 800
Losap 537
Losuia 622
Lothian 849
Lothringen 290, 292
Lotj 518
Lott 24, 54, 265, 284, 329, 390, 776
Louangphrabang 441
Lough 375
Lõuna 246
LOURENCO 23
Lourenço 561
Loviisa 259
Lowé 362
Löwrnburg 291
Loyada 219
LPAL 732
LPS 377

LQD 874
LSDMS 635
LSR 196, 346, 652, 895–896
LST 828
LSU 448, 654, 686, 766, 800, 856
LTM 205
LTTE 735
Luba 241
Lubinda 187
Lubirizi 664
Lublin 638
Lucayan 59, 395
Luchkartering 755
Lucia 223, 667, 741, 745, 778
Lucía 165–166
Lüderitz 565
Luglow 722
Luhwa 129, 663–664
Luis 106, 225, 532, 873
Luka 99
Lumi 415
Lumpur 503
Lunar Almanac 577
Lunda 23
Luniniec 637
Lunnay 137, 210
Lured 531
LUREF 477
Lurgan 255
Lusaka 887–888
Lusitania 641
Luso 105
Luvuvhurivier 560
Luwingu 887
Luzon 633–634
LV 448, 470, 770
Lwangasi 417
Lyons 204–205, 234
Lyoveldio 351
Lysaya 671
Lysnet 216

Maale 505
Maanmittaushallituksen 260
Maarten 39–40, 309, 574, 870
Maartin 39
Maas 573
Maasina 717
Maastricht 215
Maatshappij 809
Mabile 527
Macajna 313
Macau 169, 481, 483–485
MacCallum 705
Macedonian 487
Macenta 322
Mackenzie, Col. Colin 356
Macky 689
Maclear 726–727
MacLeod 416, 848
MACRI 31
Macuco 106
Madaha 722
Madang 622

Madeira 641–642
Madhya 357
Madīnah 683
Madol 535
Madras 4, 356, 502, 705, 796–797
Madriz 583
Madsen 217, 226
Madzanzua 561
Maewo 869–870
Mafeking 101
Mafinga 887
Magan 607
Magdeburg 291
Magdiev 864–865
Magellan 267, 633
Magelligan 375
Maggiore 386, 769
Magnaghi 220, 466
Magnavox 428, 855
Magufuli 791
Magur 537
Măgurele 113
Magyar 287, 345, 539
Magyarization 709
Mahaa 505
Mahajanga 493
Mahakam 364
Mahamane 587
Mahara 523
Mahayana 89, 135
MAHE 699
Mahé 697–699
Mahendra 569
Mahinda 418, 735
Mahira 506
Mahmud 67
Mahoré 179–180
Mähren 207–208
Mai 323, 329, 355, 502, 795–796
Maia 702–703
Maiana 423
Maiao 267, 269
Maigа 510
Maillard 309
Mailo 835
Mainassara 587
Maipo 165
Maire 807
Maison 524, 556
Maithripala 735
Maja 9, 487
Majapahit 109
Majes 630
Majlisi 787
Majorca 641
Majuba 725
Majuro 518–519
Makassar 362, 364
Makatea 267, 269
Makers 256, 565, 621, 643
Maketu 578
Makhaleng 455
Makin 423
Makkah 683

Makona 702–703
MAKPOS 695
Makri 451, 774
Maksimov 98
Makwoi 703
Mal 25, 44, 55, 137, 185, 257, 382, 390, 562, 580, 603, 879
Malabata 555
Malabo 241
Malacca 501, 795
Malagueta 460
Malaguetta 459
Malaita 717
Malaka 501
Malal 32–33
Malaria 617
Malatia 37
Malawian 497
Malayo 179
Malaysian 109, 361, 501, 503, 705
Malcolm 46, 137, 210, 438, 442, 478, 691, 778, 848, 888, 893
Maldive xii, 184, 481, 505–506
Maldonado 229, 860
Male 220, 421, 505, 625
Malebo 183
Målebordsblade 215, 217, 226
Maleh 609
Malekula 869–870
Maleolap 519
Malian 509
Malik 4
Malin 377
Malindi 418
Malinké 509
Malizia 543
Malkawi 373, 390
Mall 543
Malla 569
Malolo 808–809, 830
Malologa 829–830
Malongo 25
Malosa 498, 560
Maloti 455
Malpelo 175
Malta xii, 386, 481, 513–514
Maltépé 820
Maltese 513–514
Maluti 455
Mamadou 587
Mamady 321
Mamba 459
Mamiya 669
Mamluks 233
Managua 583
Manandhar 562, 571, 603
Manaus 107
Manche 81
Manchester 760, 831
Manchu 169, 435
Manchuria 401–402
Manchurian 169, 669
Mandab 219
Mandalay 123

Mandat 453, 479, 776, 831
Mandela, Nelson (Rolihlahla) 726
Mandingo 459
Mandinina 521
Mandjia 157
Maneapa 829
Mangapwani 793
Mangoky 494
Manica 559
Manichaeism 63
Manihi 268
Manila 313, 462, 633
Manira 505–506
Maniyur 372
Mankoya 887
Manley 398
Mannevy 241–242, 276, 524, 691
Mannheim 290–291
Mano 459, 461, 702–703
Manoca 142
Manono 675
Manrique 629
Mansa 701, 887
Mansur 371
Manto 537
Manton 537
Manu 269, 622
Manuel 150, 322, 329, 617, 629
Manuhagi 269
Manyangau 497
Manyur 368
Maori 577–578
Mapa 199, 242, 317–318, 337
Mapinduzi 791
MAPPLAN 199
Maputo 25, 54, 559, 561, 759
Mar 210, 680, 777
Maracaibo 39, 873–874
Maradu 505
Marakei 423
Maralu 505
Marash 37
Marathon 671–672, 778
Marautagaroa 269
Maravi 497, 887
Marchand, Captain 157
Marchenko 541
Marcona 630
Marcus 403
Marcussi 10, 55
Mare 539
Margai 701
Margaret 45
Margherita 835
Marguerite 183
María 27
Marian 540
Mariana 313, 535
Mariano 629
Maribor 50
Marigot 39
Marina 466, 583
Marinduque 633
Mariner 327, 565, 807, 869

Marinha 483
Marino 385
Mario 165, 325, 385–387
Maris 448, 478
Markham 796, 831
Marko 260
Markushof 291
Marlborough 578–579
Marmara 819, 821
Marmora 114
Maroc 556–557
Marosvásárhely 347, 654
Marques 559, 561
Marquesas 267
Marquises 267–268, 270
Marrakech 556–557
Marsa 816
Marseilles 769
Marshal 195, 380
Marshallese 517
Marta 176
Martello 528
Marthe 755
Marthram 90
Marti 199
Martine 396, 417
Martiniere 356
Martinskirche 291
Martinsturm 292
Märturii 540
Marua 874
Marudi 109
Marutea 269
Marxist 53, 337
Mary 191, 532, 641, 825
Marzooqi 844, 864
Masaya 583
Maseru 455
Maseyk 257
Mashona 891
Mashtots 35
MASISI 101
Maskali 221
Maskelyne 577
"maskirovka" 246, 495, 659
Massaad 452, 478, 775, 830
Massalia 543
Massawa 249
Massif 333, 493
Masvingo 891
Mat K.L.M. 39
Mât 220
Matagalpa 583
Matang 503
Mataura 268, 270
Matautu 675
Matheo 203
Mathew 401, 577
Matic 551, 695, 778
Matindas 363, 391
Matins 361
Matles 90, 130
Mato 105–106, 325, 329, 859
Matrikeldirektoratet 215

Matsu 783, 785
Matthais 622
Matti 328
Matupi 622
Maturei 269
Maturín 874
Matuwari 755
Matveyevka 659
Maui 807
Maupas 20, 732
Maupertuis, *Pierre-Louis Moreau de*
    259, 263, 763
Maupiti 267, 269
Maures 523
Maurice 43–44, 305
Mauricio 31
Mauritanian 509, 523
Maury 216, 271, 536
Maurya 355
Mauryan 67
Mauwa 702–703
Maxwell 395–396, 501
Maya 81, 237, 317, 337, 531, 570, 583
Mayan 317
Mayen 599, 601
Maynard 461–462
Mayotte 179–180
Mayreau 745
Mazen 220
Mazutian 785
Mbandzeni 759
Mbanza 23
Mbini 241
Mbua 256
Mbulembulewa 256
Mbundu 23
Mbuthaisau 256
MCAB 86
McCallum 501
McCaw xxv, 255–257, 295, 615, 703,
    722, 777, 792, 836, 865
McCleary 306
MCE 869
McGlone 643, 665
McGraw 562, 603
MCInstCES 562, 571, 603
McKean 424
McKerrow 578
McKinley 853
McMahon 4
McNally 225
McNamara 489
MDP (code) 680
Meades 59, 200, 334, 398, 532, 618
Meanguera 338
Mecanique 220
Mecca 607–608, 684, 883
Méchain 543
Mechanism 301, 453
Mecklenburg 291
Medco Energy 609
Medellín 176
Medes 53
MEDIANTE 339, 348

Medina 225
medium filium aquae 591, 653
medium scale 495
Meemu 505
Meetia 269
megalithic tombs 641
megaliths 89
Megnetismo 680, 779
Mehmed 380, 452
Mehmet 819
Mejit 519
Mekong 123, 135–136, 441, 877
Mékrou 86
Melab 13, 816
Melaka 502
Melanesians 869
Melbourne 43, 788, 796, 831
Melchior 599
Meldrum 487, 489
MELF 406
MELI 702, 732
Mélikhouré 321
Melikhoure 701
Melilla 555, 731
Melita 185, 210
Mellu 519
Melor, B. Polis 503
Melrhir 13
Melriça 641
Melzbate 373
Membership 158, 490
Memoria 318, 329, 562, 584, 603
Mémorial 816, 831
Memphis 233, 451, 774, 860
Memrhar 523
Mendaña 717, 827
Mende 459
Mendoza 31–32
Mene 874
Menezes 679
Mennbidgi 775
Mensin 447
Menton 543
Mentor 59, 192, 792
Mer 325, 329, 473, 479
Merca 722
Mercatori 246
Mercedes 861
Mercheria 13
MERCHICH 556–557
Mercury xxiv, 829
Mergui 124
Méridionale 93, 130
Merím 859
Merina 180
Merino 629
Merioneth 850
Meroe 750
Meroitic 750
Mérrické 207
Merv 825
Mescit 114
Meşedağ 821
Mesh 15, 358, 839

Meškonys 474
Mesoamerica 317, 531
Mesopotamian 371
Mesrop 35
mestizos 583, 799
Mestnaya 5, 55
Mesurado 460
Metahara 249
Metairie 793
Métameur 815
Météo 141
Météore 219
Metereológico 485
Methodius 487, 693
Metonic 115, 153, 343, 743, 817
Meuse 573–574
Mexica 531
Mexican 481, 531–532
Meyimel 555
Meza 626
MGM 498, 529
MGRS 146
MHAST 24
Miangas 634
Michaeliskirche 292
Michel 223
Microlithic 591
micrometer 124–125, 361,
    488–489, 592
microns 491
microwave 153, 314, 342
Middlesex 652, 665, 850
Midlothian 849
Midra 684
Midway 179, 256, 313, 372, 415, 848
Midžor 693
Mie 402
Miguel 24, 633, 679
Mikhail 436, 657, 787
Mil 217, 698
Milazzo 385
Mildura 43
Milenkovic 551, 695, 778
Milford 808, 831
Mili 519
Miliare 249
Militärgeographisches 50
Militari 540, 652
millet 795
Milli 787
Milligankk 751
millimeter level 33, 221, 467, 493,
    555, 680
Milne 622
Milosevic 693
Milton 701, 835
Mina 105–106, 295, 337, 609
Mināb 368
Minaean 883
Minami 401, 403
Minchev 113, 116, 131, 821, 830,
    840, 865
Mindanao 633–634
Mindaugas 473

Mindelo 149
Mindoro 634
Minerva 809
Ming 547, 669
Ministerio 583
Ministerstva 207
Minna 86, 141, 592–593
Minoan 203
Minsk 74, 131
Minster, System York 850
Minto 537
Minus 5, 49, 125, 638, 788
Miny 863
Mir 637
Mira 484
Mirador 317, 873, 877–878
Miranda 775, 831
Miri 109
Mirmakhmudov 864–865
Misamis 633
Misión 629
Missão 24, 149–150, 325, 800
Missãoes 150, 211
Mississippi 216, 271, 854
Missões 325, 329
Missouri 854
Misurata 466
Mitau 446
Mitch 238, 318, 337
Mitchell 617, 643
Mithridates 367
Mitrofanovs 448, 478
Mitspe 379
Mitt 792, 831
Mittel 694
Miyagi 402
Miyazaki 402
MK 424, 536, 622, 756
Mlambo 893
Mlanje 497
Mlapa 803
MLG 854
Mnajdra 513
Moa 577, 702
Mobray 727, 777
Moçambican 727
Moçamedes 24–25
Moco 23
Moctezuma 531
Modagor 556
Moerai 270
Mogadishu 721
Mogador Mat 557
Mogoton 583
Mohandas 355
Mohéli 179–180
Mohl 518
Mohokare 455
Mojanga 493
Mojón 874
Mokgweetsi 101
Mokil 537
Mola 732
Moldavia 539, 651

MOLDREF 541
Moleana 800
Möllehöi 216
Molles 860
Molnár 245, 250
Molo 50, 491
Mombasa 792
Mombetsu 402
Momchil 113, 116, 131, 821, 830, 840, 865
MON 89, 123, 267, 518
Mona 256
monarchs 90, 750
Monastir 490
Mondah 242, 250
Monegasque 543
Mongalla 752
Mongolian 411, 547–549
Mongpan 126
Monin 541
Monique v
monks 90, 123, 351, 543
MONMAP 548
Monmouth 850
Mono 85, 816
Monoecus 543
Monpa 89
MONREF 548
Montagne 527
Montana 165, 259, 891
Mont Bengoue 275
MONTEPOS 551, 695
Montevideo 859–861, 865
Montezuma 531
Montgomery 850
Montijo 641
Montjong 362
Montpelier 396
Montserrat 27, 481, 553
Monyul 89
Monywa 125
Moorea 267, 269
Moors 19, 523, 641, 731
MOP 268
Mopelia 269–270
Moqattam 233, 751
Moquegua 630
Mora 191
Morafenobe 494
Morag 760, 831
Morane 269
Moravia 207–208, 709
Moray 850
Morebodi 101, 103, 130–131
Moresby 621
Moreton 256
Morgan 415, 438, 617
Morillo 874
Morne 223, 697
Moro 634
Morobe 622
Moroccan 13, 459, 509, 557
Morogoro 791–792
Morondava 495

Morozova 448, 478
Morpho 203
Morris 456, 462, 592, 726, 891
Morro 153, 680, 702–703
Morton 123
mosaic 462, 523, 621
Moselle 477
Moshi 791
Moshoeshoe 455
Mosisili 456, 479
Moskvitin 669
Mossi 119
Mostar 97
Mosto 461
Mostyn 705
Mosul 371–372
Motagua 318, 338
Motu 269
Motulalo 828
Moucha 220
Mouhoun 119
Mount Apo (Philippines) 633
Mouratidis 303, 329
Mourelle 807
Moussa 219
Mowbray 727
Moyen 157, 188, 210
Mozambique 179, 679, 759, 778
MOZNET 498, 561
MozPerm 561
Mpange 498
Mpigangoma 793
MPLA 23
Mporaloko 276
Mpundweni 759
MRA 40, 684
MRICS 562, 571, 603
MrSID 602
MRT 502–503
MS 93, 131, 185, 257, 654, 800
MSA 145
MSK 548
MSL 297, 609, 635, 809, 829, 884
MSN 372
Mswazi 759
MTEN 875
MTI 10
Mtkvari 35, 283
MTM 146, 593
Mtubatuba 727
Mubarak 431
Muck 849
Muda 503
Mudry 860
Mudug 721
Mueller 570
Müffling 113
Mufumbiro 664
Müggelsberg 291
Mughal 67, 355, 613–614
Mugnier xvii, xxii, 199, 540, 775, 777, 831
Muguene 759
Muhammad 505, 607

Muharaq 63
Mühendishane 819
Mukike 129
Mukinkí 680
Mulambo 184
Mulder 755, 778
Mulenu 675
Mull 849
Mumbai 355, 614
Mumbwa 887
Munamagi 245
München 290–291
Muneendra 166, 230, 442, 478, 570, 736, 777
Mungo 587
Muni 241–242, 442
Munich 474, 478
Munkhtsetseg 549
Munkhzul 547
Muñoz 165, 210
Munsey 752, 778
Münster 291
Muntenia 540, 694
Munzinger 249, 251
Muonio 259
Muqurra 751
Murilo 537
Muroran 402
Mururoa 269–270
Musala 113
Muscat 608–609, 684, 883
Muscovite 839
Muscovy 73
Museveni 835
Musey 751, 777
Musha 219, 221
Mushah 219
Muslimié 452, 775
Musmar 752
Mussetter 241
Mussolini 9, 385
Mussoorie 357
Mustafa 36, 367, 371
Mustique 743, 745–746
Mutale 887, 893
Mutare 891
Mutene 184
Muvogero 664
Muzeal 694, 779
Mwali 179–180
Mwambutsa 129
Mwami 663
Mwanga 835
mwani 129
Mwibu 129–130, 664
Mwinilunga 887
Mwinyi 791
Myanmar xvii
Mycenaean 203
Mysia 821

Nabeba 187
Nabi 883
Nablus 380
Nabu 562, 602
NAC 497
Nachrichten 792, 830
Nacionais 643
Nansha 785
Nantwich 850
Nanumaga 828–829
Nanumea 827–829
NAP 247, 574
Naparima 812, 874–875
Napata 749–750
Naples 385
Nadap 347
NADCON 20, 548, 676, 728, 851
Naftali 379
Nagano 402
Nagarkot 570
Nagasaki 402
Nagatacho 403
Nagorno 53
Nagorno-Karabakh 53
Nagyszal 346–347
NAH 368, 372
Nahau 803
Nahl 752
Nahua 237
Náhuatl 531
Nahuatl 583
Nairn 849
Nairobi 415–418, 760, 831
Nakhchivan 36
Nakhon 795
Nam 877
Namaqualand 726
Nambicuara 105
Namgyel 90
Namhai 428
Namib 565
Namibian 565–566
Namoluk 537
Namonuito 537
Namorik 519
Namorukku 537
Namoui 596
Nampula 559
NAMRIA 634–635, 643
Namu 519, 675
Namur 78, 517
Nan 535
Nang 878
Nanking 169, 341
Nansei 401
Napoleonic 195, 215, 263, 346, 574, 769
Nar 843
Nara 402
Narach 73
Naraguta 592
Narbonensis 263
Naroch 73
Narodne 490, 895
Narver 518

NASA 44
Năsăud 694, 779
Nasekula 256
Nasenla 256
Nasional 363, 391
Nassau 290, 292
Nasser 233, 249, 752
Natal 725, 727
Natewa 256
Nath 256
Nathau 256
Natih 608
Natrang 878
Naturale 385
Nautical Almanac 4
Navarin 13
NAVD 854, 856
navigable 241, 653, 854
naviguer 131
NAVITEK 116–117
Navivvia 256
NAVO 216
Navotuvotu 256
NAVSTAR xx, 342, 855
Navua 256
Naxçıvan 36
Nayramadlin 547
Nazareth 774
Nazi 216, 385, 445, 539, 693
Nchu 725
Ncome River 725
NCOs 416
NCR 635
Ndana 257
Ndebele 891
Ndeni 718
Ndongo 23
Ndreketi 256
Neblina 105
Nederlands 573
Nederlandse 41, 55
Nee 424
Nefud 685
Negara 57, 109
Negeri 501
Negev 379
Negoescu 652
Negombo 735–736
Negri 503
Nehavand 367
Nehru 355
NEI 633–634
Neira 717
Nejd 406
Nelson 27, 399, 578–579, 726, 847
Neman 73
Nemuro 402
Nendo 718
Nepalese 562, 569, 571, 603
Nepali 90
Nepen 689
Nerchinsk 669
Nero 35

Index

Nestor 31
Netanyahu 379
Nettles 180, 358, 756, 778
Neue 49, 98, 345, 540, 651
Neusiedler 49
Neustadt 49–50
Neuve 477–478
Neuvième 241, 250
Neuwndorf bei Wilster 287
Nevado 629
Nevis 667, 739, 779, 847
NEVR 116
Nevrokopi 116
Newar 569
Newcastle 726
Newcomb 380
Newfoundland 145
Newsome 398–399, 408
Newton 16, 54, 229, 259, 263
Neyra 827
NFD 721
NFN 634
NFT 828
Ngaliema 183
Ngandong 361
Ngaoui 157
Ngatik 537
Ngau 255
Ngawang 90
NGD 442, 562, 596, 602
Ngem, Khao 797
NGI 78, 131, 428
NGII 429
NGO xxiv, 599–601
Ngoc, Linh 877
ngola (king) 23
Ngu 797
Ngulu 537
Nguni 759
Ngwane 759
Nhung 879
Niadi 184
Niangol 690–691
Niara 184
Niari 183, 187
Niassa 559
Nicaraguan 191, 337
Nicaragüense 562, 584, 603
Nicaraocali 583
Niches 531
Nicholas 578, 580, 657
Nichols 306
Nicolae 651
Nicolai 216
Nicolau 149
Nicosia 203
Nieddu 386
Nieu 878
Nieuw 577
Nieves 732, 739
Nigerian 141, 562, 587, 592–593, 602
NIGGG 116–117, 131
Nihayda 608

Niigata 402
Nikifirov 445
Nikol 36
Nikolaikirche 291
Nikolli 9–10, 54
Nikolskiy 863
Nikunau 423
Nilotic 835
Nimba 321, 389
Nippon 364
Niquirano 583
Niraj 562, 571, 603
Nistru 539
Niton 469, 769–770
NIU 595
Niuafo 809
Niuas 807
Niuatoputapu 809
NIUE xiii, 562–563, 595–596, 602
Niuean 595
Niulakita 827–829
Niutao 828–829
Nivelacni 207
Nivellement 78, 477
Nivkh 669
Niyazi 820–821
Njazidja 179
Njesuthi 725
NK 484
NKG 766
Nkhotakota 497
Nkrumah 297, 329
NOAA 86, 162, 517, 856, 865
Noah 283
Noakchott 524
Nobatae 750
Nobatia 751
Nogai 411
Noir 188
Noire 188
Nok 591
nomograms 816
Nomoi 536–537
Nomuka 808–809
Nongteng 442
non-Hindu 3
non-navigable 653
Nonouti 423
Noonu 505
NOP 593
Nóqui 23
Nordafrika 216, 226
Nordeste 106
Nordkapp 726
Nordlaguna 601
Nordlicher 540
Norges 599
Noria 165
Noriega 617
Nork 37
Normaal 247, 574
normalization 379
Norman 375, 398, 460, 513

Normandie 309
Normans 460
Noronha 24, 106
Norse 599
Norsk 601
Northumberland 849
Nottingham 849
Nouakchott 523–524
Nouamrhar 524
Nouhao 803
Nouméa 268
Nouvelle 268, 271, 527, 869
Nova 145–146, 577, 679, 701
Novgorod 657
Novo 83
Nowshak 3
Nowy 638
noxious dust stroms 411
Noya 225
Nsombo 888, 893
Nsuta 296
NT 146, 765
NTF 20, 77, 263, 265, 477, 544, 732
Nuanetsi 891
Nuba 752
Nubarashen 37
Nubia 16, 750–751
nucleus 73, 769
Nuestra 739
Nueva 583, 635
Nugaal 721
Nui 829, 877–878
Nukai 595
Nuku 267, 270–271
Nukufetau 828–829
Nukulaelae 829
Nukuoro 537
Nukutavake 269–270
Nukutipipi 269
Numes 483
Numidia 13
Numuiloa 256
Nunes 759
Nuñez 617
Nur 291, 329
Nusrah 774
Nutfield 396
Nuulua 675
Nuur 547
Nuusafee 675
Nuutele 675
NWL xxiv, 154, 310, 626, 630
Nyanjas 497
Nyanza 184, 415, 887
Nyasa 497–498, 560, 792
Nyasaland 497, 560, 887
Nyaza 415
Nyerere 791
Nyingmapa 89
Nyoman 73
Nyugat 437
NZGD 579

NZMG 579
NZTM 579
Nzwani 179

Oaxaca 532
Oberösterreich 49
Oblast 473, 787
Obock 219–220
Obote 835
Obras 483
Observatoire 77
Observatório 106, 485
Observatorio 484, 561
Observatorium 764
Observatorul 540, 653
Obuasi 296
Occidentale 83, 119, 322, 689, 777
Occidente 175
Occultation 536–537
Oceania 43, 267
Océanographique 268, 271, 521
Oder 287
Odera 418, 438
Odessa 839
Odienné 389
Oecusse 800
Oecussi 799
Oeste 175, 860
OFICIAL 317, 337, 339, 348
Oficina 31, 165, 583, 618, 625
Ofu 675–676
Ogaden 721
Ogasawara 401, 403
Oghuz 53, 411
Ogooue 157
Ogooué 183
Ogoue 241
Ōita 402
Oja 246
Ojeda 327, 873
Okarito 578, 580
Okavango 23, 565
Okayama 402
Okeana 670
Okia 836, 865
Okinawa 403
Okino 401, 403
Oldenburg 290–291
Olea 165
Olga 659
Ollanta 629
Ollikainen 260
Olosega 675–676
Olsztyn 638
Oltenia 540, 694
Olympus 203, 301
Onari 537
Onatsevich 671
ONI 381
Onotoa 423
Oostende 78
Ootomari 669
Opaki 578
Opari 752

opiate 787
Opmäling 599
Opmetings 456
Opole 638
Oppmäling 599
OPUS 265, 532
Oran 13–14
Oranjestad 39
Orbigny 93, 130
Ordinance xx, 129, 279
ore 184, 523
Orel 346
Orenburg 413
Orestikon 487
Orgil 547
Orhi 20, 732
ORIA 117
Oribi 759
oriente 229
Origine 878
Orinoco 327, 873
Oriot 435
Orissa 67
Orizaba 531
Orjen 98
Orkney 850
Ormoc 633
Ormonde 204
Orohena 267
Oroluk 537
Oronjia 493
Ortega 583
orthometric 609, 856, 860
orthophoto 176, 205, 432, 843
Ortoroids 223
ORU 540
Oryahovo 117
Orzágos 347
Ōsaka 402
Oscar 727
osculating 357
OSGB xxiv, 377, 850
Osh 4, 413, 436–437, 788, 863–864
Oslo 405, 599–600
Osmanli 819
Ostarrichi 49
Ostend 77
Ostenfeld 291
Ostia 220
Ostra 98
Ostrogoths 539
Ostroleka 638
Ostrov 670
OSY 884
Otago 578
Otaru 402
Otetou 269
Otto 49, 287–288, 543, 792
Ottomans 195, 371
Ottomar 246
Ouaddai 157
Ouagadougou 119–120
Ouargla 13
Oubangi 157

Oubangui 157, 187
Oudalfou 702
Oudemans 361
Oued 13
Oulu 259
Ouro 295
Ousmane 587
oust 341, 735
ousting 367
Outre 325, 329
Ova 256
Ovalığı 53
Owen 153–154, 493, 698, 793
Owendo 275
Oxford 849
Oxus 411
Oyapock 327
Ozieri 386

Paagoumène 268
Pablo 200, 211, 237, 629
Pacha 380, 452
Pacifique 268, 271
Pacta 195
Padang 362–363
Padmasambhava 89
Paga 621–622
Pagano 203
PAGEOS 59
Pago 675
Pagon 109
Pahang 502–503
PAIGH 229–230, 618, 643
Pailin 136
Painkhy 750
Paitia 617
Pakaraima 327
Pakhtunistan 613
Pakxan 442
Palanka 114
Palé 242
Palencia 731
Palenque 317, 531
Palermo 31
Palestinian 379, 405
Pali 10
Palk 356
Pallavas 109, 501
Palmar 106
Palmas 460–461
Paloma 861
Palyessye 73
Pama 803
Pamirsky 436
Pamma 870
Pampa 31, 33
Pampanga 635
Pamplemousses 527
Pamplona 731
Pan 94, 229, 310, 859
Panaeati 622
Panagyurishte 117
Panama 175, 191, 229, 618, 859
Panang 503

Panau 503
Panay 633, 635
Pandemonium 869–870
PANG 117
Pangai 808
Pangasinan 635
Panguna 621
Panmunjom 428
Pannonian 195
Panoramico 231
Pantelleria 466
Panthéon 544
Papagaio 680
Papatahi 578–579
Papetoai 269
Paplaka 446
Papua 361, 611, 621–623
Paraguari 626
Paramaccaner 755
Paramaribo 755
Paraná 32, 860
Parana 625
Paraoa 269
Parapara 578
Pardo 630
Pare 792
pareage 19
Parece 403
Parfond 220
Parham 755
Paria 328, 811–812, 874
Parkhurst 192, 752, 779, 792–793
Paro 89
Parry 28, 55
Parthian 35, 367
Partin 50
Pasargadae 367
Pascal 208
Pasco 630
Pascua 165
Páscua 874
pashas 36
Pashinyan 36
Pashtunistan 613
Passagem 483–484
Pastoreo 626
Patagonia 33
Patrice 477, 479
Patricia 373, 390
Patroba 418, 438
Patutahi 578
PATVEN 875
Patya 776
Paula 150, 211, 680, 778, 800, 831
Paulinen 291
Paulo 24, 105–106
Paysandú 861
Paz 93, 130, 337, 629
Pazardjik 116
Pázmány 541
PCGIAP 136, 369, 391
Peake 399, 408
Pechenegs 539
Pecny 210

Pedregál 874
Pedrosa 19
Peebleshire 849
Peenemünde 517
Pegel 477
Peggy 577
Pehuajó 32
Pei 169
Peil 247, 574
Peking 669
pelages 9
Pelée 521
Pélissier 815, 830
Pello 763
Pemba 791, 793
Pembroke 850
Penang 501–502, 705–706
Pendi 169
Pendjari 86
Peng 485
Penghu 785
Penh 135–137
Peniamina 595
Penland 93
Penn, William 395
penny-pinching 255
Pentland 93
Pepito 94
Pequeña 485
Perak 501, 503
perche 697
Perduchi 722
Pereai 503
Perez 629
Périm 220
Perkins 28, 55
Perlis 503
Permanentes 226
permutation 382
Pernambuco 106
Pernik 117
Pero 559, 625
Peronist 31
Pérouse 669
Perperidou 303, 329
Perrier 452, 775
Perry 401
Perú 611, 629–631
Perúvian 629–630
Pescadores 783, 785
Péter 541
Peterborough 849
Petit 305, 877
Petrayev 658
Petrikirche 446, 448
Petrobras 107
Pétroles 188, 242, 276
Petroškevičius 474, 478
Pevtsov 670
Pfalz 291
Pfarrturm 49, 291
Phare 188, 220, 242, 275, 310, 691, 878
Philadelphia 5, 684

Philippe 6
Philips 718, 777
Phnom 135–137
Phnum 135
Phocaeans 543
photogrammetrist 6, 93, 208, 302, 342, 398, 820, 895
photomap 422, 698–699
photomapping 421
photomosaics 510
Photostat 397
phototheodolites 600
Phou Bia 441
Phraya 795
Phuntsholing 90
physiographically 827
physiography 618
Piaui 106
Pibor 752
Picard 543
Pichus 630
Pidurutalagala 735–736
pied 697
Pierce 705
Pietermaritzburg 726
Pila 638
Pilar 106–107, 165–166, 626
Piña 860
Pinaki 269
Pinang 503
Pinder 60
Pingelap 537
Pinkwart 792, 830
Pintado 165–166
Pinto 642–643, 679
Pinzón 105, 327
Piombino 386
Piotrkow 638
Pipil 237, 583
Piquete 106
Pirat 691
Pirî 819
Piritú 874
Pisagua 94
Pisaras 537
Pisculci 653
Piscului 540, 653
Pistor 361, 731
Pital 237
Pitiahe 268
Pittier 191
Pittman 301, 672, 777
Piura 629–630
Pizarro 93, 629
Placințeanu 652
Plaine 333
Planheft 216, 226, 291–292, 329, 540, 821, 831
planimétrique, Fond Topographique ou 510
Plantamour 469, 769
Plassey 67
Plata 31
Platt 153

plebiscite 803
Pleistocene 853
Plessis xxiv, 233, 263–264, 451, 477, 774
Pleven 117
Plock 638
Ploix 309
Plovdiv 113
plumbline 357
Pndaries 562
Pobedy 435
Poděbradý 208
Podlaska 638
Pogobi 671
Põhja 246
Pohjola 259
Pohnpei 535–536
Põiksilindriline 246
Pointe 180, 188, 223, 268–269, 275
Pokaakku 519
Pokhara 570
Poku 297, 329
Polarinstitutt 601
Polatsk 73
Poliédrica 32
Polissia 840
Politecnica 230
Polnésie 271
Polonnaruwa 735
Polotsk 73
Poltava 840
Polutea 675, 778
polyeder (polyhedric) projection 346, 652
Polygonmetrischen 292
polyhedral 362, 658
Polynésie 268
polynomial 78, 261, 377
Polytechnic 10, 284
polytheistic 531
Pomare 267
Pomerania 763
Pomerantsev 412, 863
Pomeroy 395
Ponape 535, 537
Ponduine 561, 759
Pongola 759
ponies 123, 125
Pont 484, 497
Ponta 24, 106, 187, 641, 680
Póo 241
Poolbeg 377
Popomanaseu 717
Popondetta 622
Popovic 695, 778
PORALOKO 242, 276
Pord 275
Porro 816
Porta 106
portage routes 795
Portau 334
Porte 815, 817, 830
Porter 267
Porterhouse 617

Pôrto 106
Porto 83, 483, 642
Portsmouth 669, 672
Portucale 641
Portugaises 325, 329
Portugués 642–643
POSGAR 33
Possen 292
possidetis, uti 318, 337–338
Postel 301, 432
Pothenot 670
Poti 283
Poto 94, 532
Pottsdam 114, 196, 208, 290, 407
Pou 270–271
Pouce 527–528
Pougno 803
Poulo 878
Poznan 638
Pozo 94, 107, 166, 230, 328, 630, 874
Prabang 441
Praci 207
Pradesh 357
Prague 208–210, 710
Praia 149
Praniévitch 209, 710
Pranis 209, 710
Prasad 569
Praslin 699
pre-Colombian 317
Première 755, 778
Premysl 638
Prenov 864–865
Presbe 652
Presidência 107, 131
Pretoria 726, 759–760
Preuß 292
Preußische 565
Prilep 491
Princesa 807
Princeton 535
Princip 97
Principale 141, 210
Príncipe 667, 679–681, 778–779
Principe 149, 680–681
Pripyat 73
Prishtina 491
Priviglium 49
proces-verbaux 702
PROJ 200, 211, 642
Projektion 765, 830
projektionsmetoden 764, 831
Projezione 385
prolate ellipsoid xx, 229, 263
Pronto 106
Propiedad 339
protégé 162, 255, 456, 517
proto-Armenians 35, 445
Provadia 117
Providencia 584
Proyección 193
Prussia 216, 288–289, 819
Prut 539
Prußischen 289–290

Prypyats 73
Pschow 207, 291, 709
Publiées 869, 879
Publikationer 351, 390
Pucara 630
Puci 445–446
Pucusana 630
Pudweni 561
Puebla 532
Puensum 89
Puerto 39, 555
Puglia 386
Puissant 301
Pukapuka 267, 595
Pukaroa 269
Pukarua 270
Pulap 537
Pulau 634
Pulkowa 3–4
Pullen 24
Pulusuk 537
Puluwat 537
Punaauia 268–269
Puncak 361
Pundweni 561
Punic 513
Punio 803
Punjab 613–614
Punkt 40, 216
Puno 630
Punt 721
Punta 386
Punto 242, 250
Punzón 874
Puquio 630
Purana 355
Purdue 609
Puriël 41, 55
Pursat 137
Pusan 428
Pushcha 73
Putin 657
Putney 306, 329
Pweto 184
pyramide 233, 451, 774
Pyrenees 20, 263, 731–732

Qa 119, 509
Qaboos 607–608
Qadisirya 367
Qajar 367
Qala 752
Qalyûbîya 233, 751
Qana 883
Qarabağ 53
Qarakhanid 411
Qarluq 411
Qatari 647
Qattara 233
Qawasim 843
Qaywayn 843
Qim 513
Qing 669
QTRANS 648

Index

Quadrillage 264
quasigeoid 16, 54, 541
Quatorziéme 362, 390, 524
Quatorzième 221, 226, 691, 777
Quatriéme 815, 830
Queau 697
Quebec 145–146
Queensland 256, 621, 717
Queirós 869
Quemoy 783, 785
Quetzalcóatl (feathered serpent) 531
Quezon 635
Quiché 317
Quillou 184
Quilty 363, 391
Quincy 697
Quinta 165
Quintana 230, 812, 873–875
Quinti 755
Quinziéme 523
Quiros 421
Quisilla 859
Quito xix, 229–230, 263, 617
Qul 3
Qullai 787
Qundao 785
Quqon 435–436
Quraysh 607
Qurnat 451

Raa 505
Raab 49
Rabah 157
Rabat 436, 556, 788, 863
Rabaul 621
Rabih 161
RACA 800
Racal 514
Raczki 637
Radautz 49
Radcliffe 614
Radhun 505
radians 46, 601, 764
radii 14, 452, 776
radiolocation 796
Radnor 850
Radom 638
Radovec 49
Radu 540
Raevavae 270
RAF 421, 746
Rafael 191
Raghu 505
Rahmon 787
Raiatea 267
Rainier 543
Rainsford 417, 836, 865
Rairua 270
Raitea 269
Raivo 245, 250
Raja 355
Rajapaksa 735
Rajaraja 356
Rājburi 796–797

Rakovski 117
Raleigh 811
Ram 405
Rama 796
Ramaphosa 726
Ramasawmy 528
Ramleh 380
Ramón 629
Ramotar 327
Ramsden 574, 847, 849
Ramus 599
Rana 569
Rangitikei 578
Ranguana 81
Ranh 878
Ranil 735
Rann 614
Rapa 270
rapprochement 162
Raqqah 774
Raro 269
Rarotonga 595
Rasgefaanu 505
Rasht 787
Ras Karma (Island) 884
RassadIran 368–369
Rasusuva 256
Rathausturm 291
Rathjens 883, 896
Rathkrügen 291
rationale 225, 528, 559, 601, 855, 893
Rauenberg 290
Raydist 188
Raze 123
razzias 161
reambulation 114
recce 745
Rechid 556
Rechnugen 292
Rechnungen 98, 540
Recife 106
recomputation 44, 77, 488
recompute 416, 489
Rectangulaire 275, 301, 373, 390
rectangulation 751
Rectification 176, 540
rectifier 817
rectify 484, 836
rectifying 697, 817
Rede 339, 348, 800
Redonda 27
reducer 817
reestablished 53, 637, 773
REFerencenet DanmarK
  (REFDK) 217
REFERENCIA 339, 348
refraction 357, 573
RefSystems 562, 574, 602
REGAT 16
Regierung 469
Registro 193, 210
REGNA 860–861
Réguliere 556–557
ReGVen 875

REICH 216, 289–292, 329
Reichsamt 565, 638
Reichsdreiecknets 290
Reichsdreiecksnetz 208
Reichsfestpunktfeld 290
Reichsgaue 291
Reichsministerium 291, 329
Reinel 641
Reingold 322, 329
Reino 641
Reis 819
Reisegebiets 883, 896
Relatório 800, 830
religio-philosophical system 613
Rems 457, 478
Renaud 452, 775
Reng 67
Rennell 67, 356
Rentfrew 849
Repartição 483
Repère 769
Répertoire 180
Repertoire 241, 251
Reppsol 165
Repsold 502, 731
República 107, 131, 859–860, 865
Republica 193, 210, 242
Republiek 725
Republika 99
République 93
Republique 509
'REseau Géodésique de l'Atlas' 16
Réseau 544, 877
Reseñas 242, 250
Reservoir 142, 157, 705
Resolução 107, 131
Respighi 385
Resture 829
Resultados 680, 777
resurrection 35
Rete 386
Retief 725
Retriangulation 396, 513, 848, 850
RETrig (Readjustment of the
  European Triangulation
  Network) 78
Rettō 401, 403
Reunion 83–84, 139–140, 158
Réunion 180, 233, 527
Revel 245
Reynolds 356
Reza 367
Rheinpfalz 290
Rhine 287, 289, 469, 573–574
Rhodesian 792, 891
Rhone 263
Rhum 849
rhumb lines 351, 398, 428, 634
Rhynland 892
Ribeira 149
Rice 123–124, 325, 375, 701, 795
Rickett 755
RICS 399, 513
Rieng 137

Riet 469
Rift 249, 405, 415
Riga 245, 445–448, 478
rigueur, de 493
Rijeka 386
Rijksdrihoeksmeting 573
Rikets 764–765
Rikitea 269
rikstrianguleringen 765
Rikuraru 519
Rikuti 672, 778
Rimatara 267, 270
Rimmer 255–256
Rimpah 362, 364, 633
Rimpoche 89
RINEX 733
Rinner 10, 13, 54, 556, 816, 831
Río 31, 874
Riobamba 229
riparian 113, 560, 591
Rippon 850
Rissenthal 291
Ritchie 43, 55
Rittri 713
Riu 19
Rivard 306, 329
Rivas 583
Rivero 317
Rivieres 321
Rivne 839
Riyādh 685
RMSE 15, 642
Roatan 743
Robain 220
Roberos 179
Robertsfield 462
Robertson 503
Roch 119
Rocha 861
Roche 199–200, 210–211
Rodini 10
Rodney 396
Rodolfo 165, 210, 225
Rodosto 114
Rodrigues 233, 527, 561
Rodriguez 32, 55, 199–200, 210
Rodríguez 860, 865
Roes 892
Roggeveen 267, 675
Rogo 98
Rohde 442, 478
Roi 517
Roijen 573
Roja 732
Rolap 519
Rôlas 680
Rolihlahla 726
Roma 195, 386
România 649, 651–653, 839–840
Rômania 113, 115, 821
Românian 652–653
Romanian 539–540, 651, 653–654
Romanov 657
Romeijn 71

Romero 230, 812, 830, 874–875, 879
ROMPOS 695
Romuald 788, 831
Ron 382, 390
Ronde 746
Ronen 382, 391
Rongelap 519
Roosevelt 199, 279, 669–670, 672, 779, 855
Roque 680, 778
Roquebrune 543
Roraima 327
Rosa 769, 874
Roseau 223
Rosebury 887
Rosén 764–765, 831
Rosenmund 469, 769–770
Rosetta 672
Rosewood iv
Ross 376, 849–850, 865
Rosseau 727
rossiyskiy 73
Rostov 657–658
Rostovelektrosvyaz 657
Rotaru 540
Rothe 299
Rötifluh 469, 769
Rotunda 231
Rouge 20, 135, 732, 854
Rounguere 689
Rousilhe 638, 653, 755, 770, 776
Roussilhe 453
Roussin 523
Routi 615
Rovinj 197
Rovuma 560
Rowe 498
Roxburgh 850
Roy 261, 562, 596, 602, 847–849
Royale 77, 136
Royaume 562, 573, 602
Rozhen 117
RSO xxiv, 109, 462, 495, 502–503
RTK 220, 246, 398, 654, 733, 844, 864
Ruanda 129–131, 498, 791–792
Rub 684–685
Rube 519
Rubén 32, 55, 860, 865
Rubin 497–498, 792
Rudahigwa 663
Rudnitskaya 74, 131
Ruga 269
Ruggeller 469
Rugwero 129, 663
Rui 25, 54, 800, 830
Rulles 78
Rumelia 113
Rumenien 540
Rummelsberg 291
Rumoi 402
Runde 216, 891
Rune 20, 732, 765, 830
Runer 19

Ruo 498, 560
Rurutu 267, 270
Rusizi 129–130, 663–664
russkiy 73
Russo 113, 402, 427, 539
Rustam 787
Rutenga 891
Ruth 10, 54, 505
Ruthemians 207
Ruthenia 73
Rutland 849
Ruvuma 560
Ruvuvu 130
Ruwa 129–130, 663–664
Ruzizi 663–664
Rwandan 663
Rweru 129, 663
Ryder 617
Rysy 637
Ryukyu 401, 785
Rzeszow 638

Saavedra 517
Saba 39–40
Sabaean 883
Sabah xxiv, 109–110, 431, 501, 503
Sabana 874
Sabinio 663–664
Sabkhat 465
Sablino 474, 659
Saccara 233, 751
Sachsen 291
Sacz 638
Sadah 883
Saddam 371
Safra 373
Saga 402
Sagasti 629
Sahel 587
Sahelian 161
Sahrum 502
Saïdia 13
Saïgon 877–878
Saih 609
Sainte 78, 275, 309–310
Saitama 402
SAJG (South African Journal of Geomatics) 418, 438
Sakaerang 797
Sakalava 180
Sakhalinsk 669
Sakishima 403
Sal 149
Sala 165
Salaam 791, 887
Salada 531
Salalah 609
Salamanca 731–732
Salankaua 622
Salinas 630
Salisbury 891–892
Salitrera 165
Salitrero 165
Sall 689

Index

Salmoiraghi 680
Salmon, Ensign 220–221
Salonika 487–490
Salop 850
Saloum 323, 690–691, 778
Saltpeter 165
Salud 617
Salvadoran 237, 337
Salween 123–124
Salzburg 49
Sama 630–631
Samaná 225
Samarkand 863
Sami 259
Samia 791
Samnah 749
Samoan 675–676, 779
Samogitia 73
Samory 509
Sampoen 362
Samu 713
Samuel 249, 251, 267
Şan 883–884
Sanaag 721
Sanameīn 774
Sanceau 527–528
Sanchez 225
Sandanski 116
Sanday 849
Sandhope 850
Sandinistas 337, 583
Sangha 187, 428
Sanguin 460
Sanhadja 523
Sankara 119
Sansanne–Mango 803
Sansargou 804
Sanseau 528
Sanskrit 89, 135, 613
Santacilla 229
Santiago 149, 165
Santoméans 679
Santoni 249
Sao 275
Sâo 641–642
Sapitwa 497
Sapper 137
Sapporo 402
Saracens 543
Saracolé 509
Sarajevo 97–99
Saramaccaner 755
Sarandí 861
Sarawak xxiv, 109–110, 501, 503
Sarblock 436
Sarblok 788
Sardinia 385–387, 543
Sargsian 36
Sari 3
Sarmen 446
Sarmes 445–446
Sarstoon 81
Sartorio 50, 491
Sarygamysh 825

Sasanid 53
Šašeliai 474
Sassanian 367
Sassanids 35
Satawan 537
Satirapod 797, 831
Satnav 691
satraps (Persian) 53
SATREF 601, 765
Saud 406
Saudeleur 535
Sausheim 290, 292
Sauteurs 305
Sauzay 276, 691
Sauzey 322
Savada 90, 130
Savai 675
Savior 237
Savorgnan 157, 183
Savorgnon 187
Savu 799
Sawai 355
Sawba 750
Sawda 451
Sawdā 683
Sawyer 105
Saxonia 290
Saxons 287
Saxony 289
Sayan 435
Sazan 9–10
sazhen 74, 113, 637, 839, 863
Scandinavia 287, 599
Scarcies 321, 701–702
Scharnhorst 446, 863
Schasselons 859
Schaumburg 290–292
Schelde 573
Schellenberg 469
SCHÉMA 453, 776, 831
Schengen 713
Schepers 362, 391
Schellenberg 469
Schleswig 215
Schloß 291
Schloßturm 291
Schmid 59, 162, 517
Schmidt 113, 769
Schöcklberg 49, 291
Schomburgh 225
Schonckert 477, 479
Schouten 807
Schraeder 859
Schroda 291
Schubert 412, 438
Schubin 289
Schueler 297, 329
Schulgin 445
Schulte 362, 391
Schutzgebieten 792, 831
Schwartz 412
Schwarzburg 290, 292
Schwarzeck 23, 565–566
Schwerin 291
Schwyz 769

Scienze 303, 329
Scilly 269, 850
Scotia 145–146, 577, 701
Scottish 497, 577, 847, 849
Scourna 850
Scutari 9
Scythian 53, 435
Sears 109, 131, 503
Sebastia 37
Sebastian 25, 31, 55
sebkah salt flats 685
Sebkha Tah 555
Sebkhet 523
Sección 618
Secondaire 141, 210
Secours 78
Séculos 325, 329
Securitate 651
Seeberg 290, 292
Seenu 505
Sefra 13
SEGAL 25, 54
Segara 362, 364
Segovia 583
Seikan 403
Seismic 116–117, 188, 223, 276, 809
Seiziéme 524
Sek 502
Seke 792
Selangor 501, 503
Seleucid 53, 487
Selim 817, 830
Seljuk 35, 53, 411, 819
Seljuq 825
Selkirk 850
Selle 333
Selskab 215
selva (Amazon Basin) 629
Selwa 684
Semani 10
Sembilan 501–503
Semgallen 445–446
Semirech 411, 436
Semiryechenska 412
Semiryechinsk 863
Semliki 498
Semporna 503
Senawang 502
Sene 321
Sénégal 280, 322–323, 330, 523
Senegambia 691, 804
Senks 445
Sennar 752
Señora 739
Senqu 455, 725
Seoul 427–428
Septembre 188, 210, 270, 290, 328
sequitur 792, 830
Sequoia 93, 131
Seremban 502
Sergjew, O. A. 447, 479
Seria 694, 779
Serindung 362
serpent (*Quetzalcóatl*) 531

Serres 275
Service Topographique de l'Armee d'Egypte 233
Servicio 165, 229, 630, 732, 859–860
Serviçios 485
Serviço 105–106, 150, 559, 641–642
Serzh 36
Sestukalns 445
Sestya 776
Settentrionale 465
Seu 19
Sevastopol 839
seven-parameter 226, 457, 485, 714
Seville 27
Sevirne 413, 825
Sevki 380, 452
Şevki 819–821
Sèvres 124
Sevres 491
sexagesimal xxi, 84, 140–141, 878
sextant 275, 670, 680
Seychelles 667, 697–699
Seyni 587
Shaanxi 171
Shablya 114
Shafayetul 68, 130
Shah 355, 367, 569
Shalowitz 223, 226, 634, 643
Sham 406, 607, 774
Shangani 793
Shanghai 172
Shari 157–158
Sharif 405
Sharjah 843
Shashe 101
Shastra 355
Shatt 371–372, 815
Shaviyani 505
sheikhdoms 843
Sheppard 234
Shiga 402
Shiina 403, 408
Shiisi 171
Shiite 371
Shikoku 401
Shiloango 187
Shima 401–403
Shimbiris 721
Shimoda 669
Shimonoseki 783
Shinkyo 169
Shinzo 401
Shīraz 368
Shirazi 179–180
Shiribeshi 402
Shizuoka 402
Shkhara 283
Shkodër 9
Shlagim 379
SHOM 521
Shona 891
SHORAN 5, 199, 368, 462, 466, 684, 796, 809, 879

Shotō 401
Shreiber 727
Shtab 658
Shu 883
Shuaib 615, 643
Shush 367
Shushi 53
Shyngy 411
SI 565, 579, 714, 795
SIA 60
Siachen 614
Siad 721
Siah 4
Siamese 123–124
Siberia 435, 657
Sibiu 651–652
Siboney 27
Sibu 109
Sibun 81
Siburn 399
Sibutu 633
Sicily 385–387, 466, 513, 574
Siddhant 355
Sidi Ben Zekri 556
Sidi Ifni 556
Siedlce 638
Sielo 446
Siemreap 136–137
Sieradz 638
SIGD 154
Sighy 78
sigma 484, 571, 765, 770
Signaux 220, 555
Sihanoukville 137
Sikayana 759
Sikhism 613
Silabriedis 448, 478
Silberberg 291
Silesia 207, 216
Silisili 675
Silk 825, 863
Silo 413, 839
Silva 759
Silvestre 859
Simbel 249, 752
Simla 614
Simón 93, 229, 231, 873
Simplak 361
Simpson 727
Sinai 380–381, 406
Sinchulu 90
Sindebele-speaking Ndebele 891
Sindh 613–614
Sindhi 613
Sines xix–xx, 573
Singh 355
Singkawang xix, 363
Singye 90
Sinhalese 735
Sinj 50, 196
Sinkiang 3
Sino 169, 783
sinusoidal 229

Sinyum 503
Sio 485
Siraj 67
Sirdaryo 787
Siretul 539
Siri 505
Siriba 418
Sirisena 735
Sirmium 195
Sirte 465–466
Sissoco 325
Sistema 474, 643
Sixième 322
Sjöberg 764, 766, 830
Sjælland 215
Skagerrak 215, 600
Skalka 210
Skewed 741
Skierniewice 638
Skopowka 637
Skye 849
Skywaves 188
SLAR 618
Slavicized 113
Slavonia 98, 195
Slavonic 73, 487
Sleagh 849
Slesvig 215
Slieve 377
Slobodan 693
Slobozia 651
Slovakian 345
Slovenian 714
Slupsk 638
Smaalenenes 599
Smeirkaja 490
Smith 792, 831
Snell xix, 573
Snellius 77, 398, 573
Snezka 207
Sniježnica 98
Snyder 301, 495, 536, 873–874, 879
Sobhuza 759
Soborniy 671
Société 13, 188, 220, 242, 269, 276
Socotra 884
Sodano 536, 828
Söderarm 259
Soeharto 503
Soenda 362
Sofala 559
SOI 441–442, 570, 615
Sokcho 428
Söke 821
Soke 849
Sokoto 587
Sokuryobu 672, 778
Solar 242, 268, 276, 555, 671, 796, 853
Soler 532
Solís 31, 859
Solo 361
Solomón 717
Solwezi 887

Index

Somaliland 219, 249, 721
Sombrero 874
Somerset 849
Somma 386
Sommet 268
Somoni 787
Somoza 584
Sondershausen 290, 292
Sonetti 203
Songhaï 509, 523, 587
Soninke 509
Sontay 877
Sool 721
SOPAC 257
Sorachi 402
Soratte 386
Soriano 861
Sorocaba 106
Sorol 537
Sortya 776
SOSI 602
Sotho 455
Soudan 85, 509
Soudanaise 509
Soufrière 309, 741, 743
Soufriere 553, 745
Sougnatti 442, 478
Soundiata 509
Sourfiere 553
Sovietskaya 670
Soya 402, 669
Sozh 73
Spacek 192
SPAEF 242
SPAFE 276
Sparta 301
Spence 562, 570, 602
Spens 764
sperrgebiet 565
Spire 376, 386, 446, 448, 727, 849–850
Spitzbergen 600
Správa 209, 710
Srivijaya 109, 795
Srongtsen 89
Srpska 99
Staatens 601
Stadtkirche 291
Stadtkirchturms 291
Stalin 73, 657, 670, 672, 779
Stamford 705
Stamnett 601
Standaus 497
Standhaus 892
Stanford 203, 380
Stanley 183–184, 188, 791, 835
Stanovoy 669
Staritskiy 671
Staten Landt 577
Statens Lantmäteriet (Sweden) 764
Stavros 488
Stefan 539, 693
Steiermark 49
Steinberg 382, 391

stelae 750
stele 431
Stennis 216, 271
Stephen 97, 539
Stereo 15, 528, 638, 653–654, 776
stereoautograph 346, 600
stereocomparator 346
stereographique 776, 831
stéréographique 453, 776
stereophotogrammetry 630
stereoplanigraph 600, 817, 820–821
stereoplotter 249, 398, 817
Sternwarte 290–292
Sterr 456, 727
Steve 422
Steven 407
Stevenson 424, 438
Stewart 578
Stirling 850
Štít 709
Stjepan 97
Stockholm 263, 429, 438, 764–765, 831, 892
Stonehenge 847
Stork 697
Storsren 599
Stoughton 519
Strabo 651
Strachan 648
Strafford 375
Stratoú 301
Straße 299, 328
Strefa 638
Stroessner 625
Stromlo 45
Struma 488–489
Strumica 490–491
Studii și Cercetări 694, 779
Stupa 123
Stuttgart 291–292
Stuve 792
Styria 50
Suarez 493
Suba 175
Subarya 363, 391
subsidence xxii, 247
Sucre 93
Sudanese 162, 749
Sudd 792
Sudeste 860
Sudeten 207
südlicher Hauptturm 291
Sudosteuropa 540
Suel 142
Suevian 287
Suez 233
suffisant 521
Suffolk 849
Sukhothai 441
Sul 106–107, 484
Sulawesi 362, 364
Suleiman 513
Sulu 633–634
Suluhu 791

Sulutan 412, 864
Sulutem 413
Sulva 355
Sumatra 361–364, 501, 795
Sumatran 109
Sumer 371, 513
Sumeria 63
Sumerian 367, 371, 451, 773
Sunda 362–363
Sunday 223
Sunni 6, 607
Suomalaiset 259
Suomi 259
Superiore 465
Sûre 477
Suri 355
Surigao 634
Surinam 755
Surinen 327
Surrey 850
Surya 355
Sussex 849
Süsteem 246
Süsteemid 246
Susuanaisky 669
Sutherland 705, 849
Sutra 355
Suur 245
Suursaari 259
Suva 255–256
Suwalki 638
Suwon 428–429, 438
Suzanne 50, 54
Svalbard 599–601
Svanberg xxv, 259, 764
Svandberg 599, 764
Svay 137
Svenska 764, 831
Svobodnyy 658–660
Swabia 513
Swabians 287
Swahili 887
Swain 675–676
Swazi 759
Swedes 215, 319, 763
SwedeSurvey 319, 548
SWISSTOPO 470, 770, 830
Sydney 43–44, 197, 210, 424, 717
Syed 125
Sylvester 345
Syr 412–413, 787, 863
Syrdariya 411
Syrie 451, 453, 479, 774, 776, 831
Systém 710
Systema 5, 55
Systémé 275, 301, 493
systèmes 158
System South Berule 850
Syurkum 670–671
Szczecin 638
Széchényihegy 346
Székesfehérvár 437
Szólóhegy 347
Søkortarkivet 215

Tabeller 765, 830
Tabiteuea 424
Tabola 322
Taboriya 322
Tabou 390
Tabu 406
Tabuaerau 424
Taburao 422
Tabwemasana 869
Tacheograph 859
tachéomètre 85
tacheometric 204
tacheometry 346
Tacloban 633
Tacuarembó 859–860
Tadjoura 219
Tadjura 219–220
Taf 40
Tafaraoui 14
Tagalog 633
Tagaytay 635
Taghès 587
Taguig 635
Tagus 641
Tah 555
Tahaa 269
Taharqa 750
Tahat 13
Tāherī 368
Tahiti 43, 263, 267–269, 421, 577
Tahitians 267
Tahrir 774
Tahuata 267, 271
Tai 341
Taieri 578–580
Taif 683
Taimur 608
Taino 199, 333
Taiohae 270
Taipa 483–485
Taipei 783–784
Taiping 501, 503
Taiwanese 783–784
Taiwan-Fukien 785
Tajik 787
Tajikistani 787
Tajín 531
Tajourra 219
Tajumulco 317
Takakoto 269–270
Takowa 519
Takowaka 519
Talamanca 191
Talas 411
Talcott 473, 854, 873
Taliban 3, 5, 614
Talkessels 93, 130
Tallinn 245–246, 562, 575, 602
Tamana 424
Tamanrasset 16
Tamar 343, 379
Tamara 321
Tamaulipas 532
Tambores 860

Tambovskoye 659
Tambralinga 795
Tamerlane 863
Tamil 735
Tamim 647
Tanague 321
Tananarive 495
Tandja 587
Tanekas 83
Tanga 791–792
Tanganyikan 791
Tánger 555
Tangier 555
Tangiri 411
Tangsil 361
Tanjore 356
Tann 684
Tanna 869–870
Tano 268
tantras 89
Tanvenui 257
Tany 493
Tanzanian 791, 793
Taongi 519
Tapuarava 269
Tapuarva 270
Taputapuatea 267
Taraboš 10
Tarai 569
Taranaki 578–579
Tararidas 860
Tarawa 421–422, 424
Tarbu 245, 446
Tardi 84, 140
Tariq 608
Tårn 216
Tarnobrzeg 638
Tarnow 50, 638
Tarnyba 474
taro swamps 829
Taroa 519
Taroudant 556
Tarqui 229, 630
Tartary 669
Tartto 260
Taruga 591
Tarut 684
Tash 487
Tasman 255, 577, 807
Tasmania 43–44
Tasso 629
Tata 162, 313
Tataa 268
Tatamailau 799
Tatar 73, 539, 670
Tatary 670
Tau 341, 436, 788
Taungyyi 125
Tavakoli 369, 391
Tavastians 259
Tavava 269–270
Taveita 415
Tavenui 255
Taveuni 255–256

Tavistock 395, 752
Tbilisi 283–285, 329
Tebboune 13
Tecamachalco 532
Technica 387, 391
Téchnica 483
Técnica 318, 329, 562, 584, 603
Ted 285
Teddy 199, 669
Tegucigalpa 337, 339, 348
Teheran 705
Tehreek 614
Teide 731
Teil 540
Teixeira 642–643
Teke 183, 187
Tel Aviv 381
Tele 455
Teleapa 676
Telegrafo 385–386
telegraphically 4
Télégraphique 24
Telekitokelau 809
Teleti 283
Tellancourt 78
Telurometer 205, 608–609
Temanufaara 269–270
Tematagi 269
Tembikundu 702
Temoe 269
Tena 119
Tendera 752
Tenerife 731
Teng 502
Tengea 702–703
Tengri 411
Tenkrugu 803
Tenner 245, 445–446, 473
Tennessee 113, 337, 459
Tenochtitlán 531
Teotihuacán 531
Terara 249
Terengganu 503
Teriaroa 267
Terkidağ 114
Ternopil 839
Terra 328, 591
Terraco 484
Terrae 573
Terre 309–310
"terreno" 484
Terrestres 135, 441
terriers 375
terrigenous 3
Territoire 219
Territoriales 562, 584, 603
Terrones 629
Tesla 551, 695, 778
Tessala 14
Tete 498, 559, 561
Tetiaroa 269
Tétouan 555
Teutoburg 287
Teutons 215

Index

Tewodros 249
Thaa 505
Thaba 455, 725
Thaddeus 35
Thai 123, 135, 795–797
Thakaundrove 256
Thames 578
Thammarat 795
Thani 647
Thar 614
Thebes 750
Theodor 551, 695
Theodosius 693
Theresa 49
Theresia 345
Thessaloniki 115
Thiès 691
Thimhu 90, 130
Thimphu 90
Thompson 146, 191, 578
Thomson 578
Thouin 268
Thrace 114, 819
Thuirangi 578
Thüringen 290, 292
Tiahuanaco 93
Tian 435
Tiao 690–691
tibe 179
Tiberias 773
Tibesti 161, 588
Tibet 3, 89, 435, 569
Tien 435, 878
Tierra 32, 577
Tiflis 368
Tiglamor 347
Tigran 35
Tigre 338
Tigris 367, 371, 607
Tihama 683
Tikal 317
Tikituru 256
Tikopia 717
Timah 705
Timár 245–246, 250, 540–541, 551, 653–654, 665, 694–695, 777, 779
Timaru 578, 580
Timbalai 109–110, 503, 706
Timbali 109
Timbekundu 701–702
Timbuctou 389
Timbuktu 509–510
Timiris 523–524
Timok 693
Timorese 799
Timur 3
Tingourkou 803
Tinto 318, 338
Tiranë 10
Tiree 849
Tiridates 35
Tirman 691
Tissot 263

Tisza 345
Tithe Commutation Act 847
Titicaca 93
Tito 195, 487, 693, 713
Tiv 591
Tlang 67
Tlaxcala 532
Tmor 98
Toafa 808
Toamasina 493
Tochigi 402
Tocopilla 165
Todar 355
Tofua 807
Togcha 313
Togdheer 721
Togo 83, 85, 119, 295, 803–804, 830
Togoland 803–804
Togolese 781, 803
Togoville 803
"toise" of Peru 229, 697, 892
Tokacha 863
Toku 809
Tokushima 402
Tokyo xix, 169, 401–403, 427–429
Tōkyō 402
Toledo 629, 731
Toliara 493
Toloa 807
Tolui 825
Tomanivi 255
Tomás 532
Tomasz 498
Tombouctou 510, 804
Tommy 618
Tondano 362
Tondo 509
Tong 588
Tongan 421, 807–808
Tongatapu 807–809
Tongo 142
Tonino 669
Tonle 136
Topasses 799
Topazovaya 671
Topcon M505AX 654
Topnav 24
Topocartografico 465
TOPOCOM 162, 816, 879
Topográfica 31
Topograficheskoye 658
Topografico 385
Topografisch 573
Topografische 363
Topografiska 764, 831
Topografiske 215
Topografov 245, 547, 658
topographically 81, 125, 346, 694
Topographie 477, 479, 815
Topographische 362, 391, 769
Topographiske 599
Topona 149
Tor 850
TORAN 188

Tordesillas 105
Tores 870
Torgau 291
Tori 401, 403
Torne 763
Torneå 763
Tornedalen 764
Tornio 259
Torno 259
Tōrō 670
Toro 835
Torokina 622
Toronto 68, 325, 329
Torrens 501
Torres 869, 871
Tortue 333
Tortuga 334
Toru 800, 832
Torun 638
Toshiyuki 403, 408
Totolom 535
Tottore 402
Touadera 158
Toubkal 555
Toucouleur 689
Tourane 878
Touré 509
Tovar 873, 875
Townsend 618–619
Townsville 43
Toyama 402
Toyohara 669–670
tracts 73, 317
Trafalgar 847
Traité 495
Tranchée 241, 275
Transcaspia 412, 788, 863
Transcaucasia 284
Transcaucasian 54, 283
Transcontinental 855
Transjordan 405–407
translocation 566
transmitter 879
Transnistria 539
Transnistrian 540
Transoxania 411
Transversal 541
Transylvanian 49, 540, 653
Trasconi 386
Trashichhodzong 90, 130
Traversed 125, 296, 595, 685, 726, 795, 883
traversers 124
Travis 255
Trazzonara 386
Trdat (King) 35
Treizième 220, 226
Trelawny 395
Treng 137
Trengganu 503
Trentino 386
Tretya 776
Trevis 722
Trgoviški 693

Triangulaçã 680, 779
Triangulacni 207
triangular 357, 395, 488–489
triangulate 229
triangulated 32, 125, 205, 373, 608, 736, 859
triangulators 412–413
Triangulierungen 207
Tricoleur 870
Tridentina 386
Trieste 50, 115, 386, 491
Triglav 713
Trignometrical 356
Trigonometrica 209
trigonometrically 4, 114, 204
Trigonometricke 208
Trigonometrické 710
Trigonometrieskais 445
Trigonometrischen 292
trilithon 807
Trimble NetR5 GNSS systems 86, 154, 548
trimetrogon 119
Trinil 361
tripod xx, 124, 484
Tripoli 451, 453, 608, 815
Tripolitania 466
Trisirisatayawong 797, 831
Tristao 321
triumvirate 36
Triune 565
Troll 93, 130
Trom 503
Tromeur 690, 778
Trondheim 599–600
troposphere 733
Trotter 4
Troughton 395, 592
Trovoada 679
TROY 117
Trujillo 230, 337
Truk 535–537
truncating 33, 176
truncation 175, 593, 776
Tsai 784, 831
Tsakhkadzor 37
Tsaoueni 180
Tsaratanana 493
Tsarist 36, 74, 283, 637
tsars 657
Tsendenjong 89
Tseng 785, 830, 832
tsetse 161, 791
Tshinsenda 184, 887–888
Tshinzinga 664
Tsinger 670
Tsodilo 101
Tsugaru 402
tsunami 185, 401, 621
Tua 256
Tuamotu 267, 269–270
Tuareg 587
Tubingen 290
Tübingen 291

Tubuai 267–270
Tucupita 874
Tudjman 195
Tugela 455
Tuhirangi 578–579
Tulagi 717
Tuléar 493
Tumbes 630
Tumen 428
Tumukuro 269
Tumukuru 270
Tungi 703
Tungsha 785
Tunis 386, 815, 817
Tunisian 13, 466, 781, 815–817
Tupai 269
Tupí 105
Tupou 807
TurboRogue 428
Turco 694
Tureia 268–270
Turfan 169
Turin 543
Turkestan 412–413, 437, 787–788, 830, 863
Turkic 3, 53, 411, 825
Turkmen 436, 825
Türkmenbaşy 825
Turkmenia 863
Turkmenistan 3, 5, 367, 411, 781, 825
Turko 412, 787
Turkoman 3
Turku 259–260
Turm 49, 77, 98, 291–292, 345
Turmberg 291
Turnu 113
Turpan 169
Turquino 199
Turuk 436, 788
Tutsi 129, 663
Tutuila 675–676
Tuvalu 421, 781, 827–831
Tuvaluans 827
Tuwaiq 685
Tuwayyir 647
Tyk 670
Tylos 63
Tyre 368, 451, 553
Tyrol 49–50, 386
Tyrrhenian 385
Tyuratam 413
Tzarist 284
Tzu 402, 784–785
Tzur 382, 391
Tørshavn 217

Ua 268, 270–271, 841, 865
UAE 63, 647, 683, 843
Ubaid 431, 647
Ubajay 860
Ubangi River 187–188
Ubangi-Shari 157
Uberaba 107, 626, 860
Uberaga 33

Übersichten 216, 226
Überwasserkirche 291
ubugabire 129
Ucciali 249
Uccle 77–78
Uchire 874
Učka 386
Udgave, Anden 217, 226
Udine 385–386
Ufficio 465
Ugandans 835
Ugarte 165
Ugric 259, 445
Uguene 561
Ugueto 873
Uhuru 791
Uikkanen 261
uitlanders 725
Ujelang 519
Ujiji 791
Ujjain 355
Ujungpadang 362
Uka 270
UKLAD 638
Ukrainian 73, 284, 436, 539–540, 839–841
Ulaanbaatar 547–549
Ulete 792
Ulithi 535–537
Uljin 428
Ulloa 229
Ultramar 325, 329, 559
Ulul 537
Ulysses 325
Umaro 325
Umba 415
Umm Er Rus 684, 843
undulating 83, 119, 389, 705, 721
undulation 45, 347, 635
Unij 291
Universelle 83, 139, 210, 389
Universidad 225, 339
Universidade 25, 54, 860
Universitario 225
unmapped 157, 658
Unterwalden 769
Upolu 675
Upravleniye 547, 658
Uracoa 874
Urad 207, 209
Urady 207
Urakawa 402
Ural 113, 411, 657
Urartu 35
Urdaneta 635
Urdu 613
Urgell 19
Uri 769
Urim 382
Urkaste 446
Urrutia 317
Ursa 395
Uruguaiana 860
Urundi 129–130, 498, 791–792

Index

USAF 518–519
Usambara 792
USAMS 859–860, 865
USGS 518, 536, 676, 854
USHER 812
USHO 537
USMC 285
USN 401, 421, 518, 870
Usrlurim 382
USS 424, 536, 827
Ussuri 669
Ustasha 195
Ústav 207
Ustrední 209, 710
Usu 402
Usuel 275, 301, 493
Usutu 561, 759
Utah 441
Utilizando 643
Utirik 519
Utrecht 299, 361, 739
Utupua 718
Uygur 435
Uyuni 94
Uzbek 411, 435–436, 787

Vaalserverg 573
Vaavu 505
Vache 333
Vaduz 469
Vahaga 269
Vahitahi 269–270
Vai 459
Vaik 345
Vaipae 270
Vairaatea 269–270
Vaisseau 219–220, 268, 270, 690, 755, 778
Vaitua 270
Vaitupu 829
Vakhan 3
Vakhanskiy 3
Vakhsh 787
Vakhushti Batonishvili 283
Val de Serra 1947 107
Valandovo 117
Valdemar 245
Valdivia 165
Valencia 873
Valentin 115–116, 131
Valera 874
Vasco 559, 617
Vasconcelos 149
Vasile 540
Vaté 869, 871
Vatia 256
Vatu 256
Vatuluma 717
Vauclin 521
Vava 807–809
Vavao 269
Vaz 325
Vecchia 385
Vedic 355

Veis 302
"veiviseren" (Wizard) 602
Vel 196, 896
Vela 403, 874
Velazco 200, 211
Velázquez 199
veldt 455, 891
Velha 149
Venables 395
Venetian 195
Venezia 386, 895
Valerio 387, 391
Valery 658
Vallaux 268, 270
Valle 874
Valsts 445
Valstybiné 474
Valtion 260
Vanavana 269
Vancouver 43, 256
Vanga 415
Vanha 260
Vanikolo 718
Vantroys 390
Vanua 255–257
Vanuatu 718, 777, 867, 869–871, 879
Vardar 487–488, 490, 819
Vardges 37, 54
Vardupe 446
Varesmäe 246
Varna 114–115, 117
Varsovic 637
Varus 287
Vasa 763
Venezuelan 27, 175, 230, 328, 812, 874–875
Venice 385, 535, 713
Venkataraman 505
Vera 532, 573
Veracruz 531–532
verbaux 321, 702
verboten 672
Verdal grasslands 599
verdant 587
Verdeans 149
Vereeniging 726
Verejnych 207
Vermessungs 792, 830
Vermessungsamt 447, 478
Vermessungskunde 292, 447, 788, 831
Vermessungswesen 13, 55, 816, 831
Vermont 245, 487
Verpli 659
Versailles 279, 769
Verst 113, 284, 658
Verticale 220, 268
Verzeichnis 447
Vespucci 327
Vespúcio, Américo 106
vestibule 446, 448
Vethleti 347
Veverka 208, 210–211, 709–710, 778
Vey 732

Vice xvii, 84, 129, 140, 358, 629, 791
Vicente 105, 149, 229, 327
Victor 309, 385, 487
Victorian 847
Vidal 153
Videnskabernes 215
Vidin 113, 117
Viegas 680, 777
Viennese 651, 895
Vientiane 441–442
Vietnamese 877
Vietsovpetro 879
Vieux 741
Vigaga 664
Vigan 633
Vila 679, 869–870
Vilayet 406, 815
Villares 200, 211
Villavicencio 176
Ville 141, 210, 493
Vilnius 474
Vincente 229
Vincentians 744
Vinh 878
Vinland 351
Virap 35
Virgil 540
Viscounde 483, 485
Višegrad 97
Viski 446
Visokoye 863
Vistula 73, 287
Viteaz 653–654, 665
Viti 255–257
Vitogo 256
Vitolnieki 446
Vitória 106
Vitosha 114
Vivas 199
Viviate 629
Vizakna 651
Vizcarra 629
Vlaanderen 78
Vladimir 114, 116, 657
Vladivostok 659
VLBI 136
VOIROL 13–15, 816
Vojensky 207
Vojni 490, 895
Vojvodina 694
Volage 203
Volcán 317, 617, 873
Volcan 663–664
Volcanoes 93, 237, 361, 583
Volkoff 114
Volontat 452, 775
Volta 85–86, 119, 588, 803–804, 830
Volumen 412, 437
Voortrekker 455, 725
Vorarlberg 49
Vorder 469
Vorderasien 216, 226
Vornetvure 540
Vostochnoga 670

Voulet 587
Voyenno 658
Voyennykh 245, 658
Vpadina 411, 825
Vries 669
Vrouwe 573–574
Vrŭkh 114–115
Vryburg 101
Vsetin 208, 709
Vucic 693
Vukčić 97
Vunta 792

Wabern 770
Wabutima 622
WACs 532
Wadai 161
Waddi 591
Wade 689
Wafer 191
Wahab 4
Wairarapa 578–579
Waitukubuli 223
Wakayama 402
Wakefield 751–752, 778
Wakhan 3–4, 613
Walachia 539
Walbrzych 638
Waldo 59, 200, 334, 532, 618
Wales 43–44, 46, 398, 847, 849–850
Walker 94, 191, 356, 626, 797, 816, 831
Wallace 81
Wallachia 651, 694
Wallis 267
Walshe 579
Walter 811
Walvis 565
Wanga 415
Wanganui 578–579
Wanschaff 361, 765
Waqf 405
Wareika 396
Waremme 78
Warren 796–797
Warszawa 638
Warwick 849
Watanabe 800, 832
Waterloo 77
Watson 396–397, 408
Wau 622
Wauhope 4
Wawat 749
Weber 524
Wedza 497, 892
Wegener 361
Wehrmach 821, 831
Weightman 685, 779
Weimar 288
Weiner 50
Weipa Mission Astro 43
Wekker 755, 778
Wellesley 501

Wellington 578–579
Wels 49
Wemperhardt 477
Werhmach 820
Westminster 145
Westmorland 395, 850
Weti 793
Wewak 622
Weyn 722
Whitehouse 752, 792
Whittingdale 416
Wickremesinghe 735
Wien 291
Wiener 49
Wiesbaden 292
Wight 849
Wigtown 849
Wilfred 609
Wilhelm 259–260, 288, 292, 621, 792, 807
Wilingili 505
Wilkes 255
Willaumez 157
Willebrord xix, 573
Willem 727
Willemstad 39
Willerzie 78
Williamson 796, 831
Willis 5
Willoughby 755
Wilno 473
Wilson 36, 124–125, 380
Wilster 287
Wiltshire 849
Windberg 291
Winston 670, 779
Winterbotham 204, 727, 751, 812
Wisconsin 6, 67, 146, 176, 787
Wissmann 883, 896
Witwatersrand 725
Wloclawek 638
Wojskowy 638
Wojtek 638
Woleai 537
Wolkau 690, 777
Wolkeau 216, 289, 329, 574
Woodland 73
Woodlark 622
Woodrow 36
Woodside 524
Woore 501
Woqooyi 721
Worcester 849
Worden 685
World Aeronautical Chart (W.A.C.) 141, 532, 681
Wotho 519
Wotje 519
WPA 855
Wroclaw 638
WSKTRANS 601
Wulafu 702
Württemberg 290, 291
Wusail 647

Wuteve 459
Wyoming 795, 883

Xagħram 513
Xang 441
Xayamaca 395
Xilungo 759
Xinjiang 411

Yaakou 381
Yacaré 859–860
Yahya 279, 501
Yair 379
Yalta 670, 672, 779, 839
Yama 401–402
Yamaguchi 402
Yambol 114
Yamoussoukro 389
Yáñez 327
Yang 363, 391, 785, 832
Yannis 302
Yao 497
Yaoundé 139
Yaroslav 73
Yasaqlığı 53
Yaşar 821, 831
Yayer 523–524
Yaylası 53
Yeltsin, Boris 657–658
Yemel 'yanov-Nikifirov survey 445
Yemeni 523, 884
Yenisey 435
Yeografikí 301
Yerevan 36–37
Yesim 413, 825
Yobe 588
Yoff 691
Yokohama 669
Yongle 171
Yorkshire 850
Yoruba 591
Yorupikku 537
Yoweri 835
Yüan 547
Yubari 402
Yūbari 402
Yucatán 81
Yugo 99
Yugoslavism 97
Yundola 116
Yunis 381
Yunnan 123, 125
Yuryev 445–446, 448
Yuryukh 863
Yuwan 402
Yuzhno 669

Zabahonski 74, 131
Zaby 843
Zaferi 113
Zagreb 197
Zagros 367
Zahra 233, 751
Zahrān 685

Zaïre 887
Zakhodnyaya 73
Zakiewicz 498
Zakladni 209, 710
Zalesky 4, 413
Zaliessky 3
Zaliv 671
Zamberti 203
Zambesi 560
Zambezia 559
Zambian 24
Zamboanga 633
Zamosc 638
Zande 157
Zanderij 328, 756
Zanzibar 415, 791–793
Zapadnaya 73
Zapiski 412, 437–438, 540, 788, 821, 832
Zaria 592
Zariņš, Ansis 448
Zaroa 592
Zealanders 257
Zeehaen 807
Zeeland 577
Zeila 219
Zeiss 346, 600, 817, 820–821
Zeitschrift 13, 55, 816, 831
Zekri 556
Zememericky 209
Zemepisny 207
Zenith 45, 135, 295, 355, 448, 654, 726, 875

Zervas 302, 329
Zeta (province) 551
Zetland 850
Zeyla 219
Zeytun 37
Zhabdrung 90
Zhdanko 670
Zhdanov 113
Zhejiang 172
Zhemchuznikov 863
zhuz 411
Zhylinskiy 473
Ziada 219
Zielona 638
Ziggah 297, 330
Zigida 462
Zikhron 381
Zilhão 483
Zimbabwean 893
Zimmerwald 470, 770
Zimnicea 540
zincography 848
Zinder 389, 804
Zipfel 565
Ziyād 299
Zoll 49, 345
Zoroaster 367
Zoroastrian 35
Zoroastrianism 63
Zouîtât 523
Zugspitze 287
Zuid 725
Zuiderzee 574

Zuidplaspolder 573
Zulu 455, 725, 759, 891
Zum 540
Zuurberg 23, 559, 664, 727, 760

Þorbergsson, Gunnar 353, 390
Þorláksson 351

Ḥa Ġar 513

ΑΖΙΜΟΥΘΙΑΚΗ 302, 329

ΗΑΤΤ 302, 329

ΙΣΑΠΕΧΟΥΣΑ 302, 329

ΠΡΟΒΟΛΗ 302, 329

ΤΟΥ 302, 329

безопасности 671

Гора 657

ДФНИ 117

служба 671

Федеральная 671

Эльбрус 657